International Conference on Smart Grids and Power Systems (ICSGPS 2025)

Journal of Physics: Conference Series Volume 3012

Wuhan, China
17-20 January 2025

ISBN: 979-8-3313-2255-7
ISSN: 1742-6588

Printed from e-media with permission by:

Curran Associates, Inc.
57 Morehouse Lane
Red Hook, NY 12571

proceedings.com

Some format issues inherent in the e-media version may also appear in this print version.

Institute of Physics
Dirac House, Temple Back
Bristol BS1 6BE UK

Phone: 44 1 17 929 7481
Fax: 44 1 17 920 0979

techtracking@iop.org

Additional copies of this publication are available from:

Curran Associates, Inc.
57 Morehouse Lane
Red Hook, NY 12571 USA
Phone: 845-758-0400
Fax: 845-758-2633
Email: curran@proceedings.com
Web: www.proceedings.com

TABLE OF CONTENTS

Journal of Physics: Conference Series **3012** (2025) 011001 doi:10.1088/1742-6596/3012/1/011001

Preface

As the world grapples with the challenges of climate change and energy sustainability, the development of smart grids and advanced power systems has become a critical focus for both academia and industry. Carrying the latest scientific research results, the International Conference on Smart Grids and Power Systems (ICSGPS 2025) was held in Wuhan, China from January 17th to 20th , 2025. The conference brought together scholars and engineers in these fields to share cutting-edge research, discuss emerging trends and explore opportunities for collaboration in this rapidly evolving field.

The conference would not have been possible without the generous support of our esteemed supporter: Shanghai Jiao Tong University, Wuhan University of Technology, Hunan University, Central South University, Guangxi University, Wuhan University. The conference was mainly conducted by keynote speeches, invited speeches, scholar forum , workshop and oral presentation. This gathering provided a platform for in-depth discussions and exchanges on the latest advancements and innovations in the field. We would like to extend our sincere gratitude to all the esteemed speakers who contributed to the success of the conference with their insightful presentations. Their participation not only enriched the academic content but also inspired meaningful interactions among attendees.

Under a rigorous peer-review process, the conference proceedings include more than 100 high-quality papers, which encapsulates a diverse range of innovative studies and insights into the latest advancements in smart grid technologies, power system optimization, renewable energy integration, and energy management strategies. We also express our deepest appreciation to our guest editors, Haoran Ji and Chengxi Liu, who have ensured the quality and relevance of the papers presented in this proceedings. Their dedication and expertise have been instrumental in maintaining the high standards of academic excellence that this conference strives to uphold.

All in all, the success of ICSGPS2025 has benefited from the collaboration of all parties. We extend our gratitude to all contributors, participants and committee members for their dedication to advancing the fields. We look forward to continuing our collaboration and exploring new frontiers in the field of smart grids and power systems.

 doi:10.1088/1742-6596/3012/1/011001

Committee

General Chair
Yang Xiao, University of Alabama, USA

Co-Chairs
Qian Ai, Shanghai Jiao Tong University, China
Tao Hai, Guangxi University, China

Technical Chairs
Tao Yu, South China University of Technology, China
Wenxuan Yao, Hunan University, China
Pei Zhang, Tianjin University, China
Heng Li, Central South University, China

Publication Chairs
Mohan Lal Kolhe, University of Agde, Norway
Fujun Ma, Hunan University, China
Tongdan Jin, Ingram School of Engineering, Texas State University, USA
Anan Zhang, Southwest Petroleum University, China

Program Chairs
Haoming Liu, Hohai University, China
Zifa Liu, North China Electric Power University, China
Tao Wang, Xihua University, China

Organization Chairs
Tao Jiang, Northeast Electric Power University, China
Yuyu Zhu, Southwest University of Science and Technology, China

Scholars Forum Chair
Chengxi Liu, Wuhan University, China

Technical Program Committee
Abraham Lomi, National Institute of Technology, Indonesia
Zhonghua Gou, Wuhan University, China
Hua Li, Inner Mongolia University of Technology, China
Hao Lu, Xinjiang University, China
Dazhi Yang, Harbin Institute of Technology, China
Quanrui Hao, Shandong University, China
Yang Han, University of Electronic Science and Technology of China, China
Dayi Li, Huazhong University of Science and Technology, China
Chenyang Xia, China University of Mining and Technology, China

Peer Review Statement

All papers published in this volume have been reviewed through processes administered by the Editors. Reviews were conducted by expert referees to the professional and scientific standards expected of a proceedings journal published by IOP Publishing.

- **Type of peer review:** Single Anonymous
- **Conference submission management system:** Morressier
- **Number of submissions received:** 151
- **Number of submissions sent for review:** 151
- **Number of submissions accepted:** 100
- **Acceptance Rate (Submissions Accepted / Submissions Received × 100):** 66.2
- **Average number of reviews per paper:** 1
- **Total number of reviewers involved:** 37
- **Contact person for queries:**
  Name: Chengxi Liu
  Email: icsgps@163.com
  Affiliation: Wuhan University

Design and Finite Element Simulation of Structured Parameters of Vibration-ultrasonic Fusion Sensor

Shengnan Zhao[0009-0007-9495-4444], **Quan Dong**[0009-0001-6986-2552], **Qingjun Zhang***[0009-0002-8171-9289]

NO.70, Tengda Rd, Economic Development Zone, Henan Pilot Free Trade Zone PR

*Corresponding author's e-mail: 1320086227@qq.com

Abstract. The potential insulation and mechanical defects in gas-insulated switchgear (GIS) are the main causes of failures. Synchronous detection of partial discharge and mechanical vibration signals within GIS equipment is crucial for the assessment of power equipment status. This paper innovatively fills the damping material of the sound-absorbing backing into a high-quality stainless steel conical shell, based on the structural similarity between accelerometers and ultrasonic sensors. It designs a new internal structure for the vibration-ultrasonic fusion sensor and establishes an equivalent model to conduct a simulation analysis of the sensor's structural parameters.

1. Introduction

GIS offers advantages such as compact structure, environmental friendliness, and high reliability, making it widely used in power systems[1]-[3]. However, as its usage increases, failures occur frequently, primarily comprising mechanical failures, insulation failures, secondary circuit failures, body leakage failures, and other issues. Among these, mechanical and insulation failures account for a significant proportion[4]-[5].

To address these challenges, researchers have developed various detection methods. Norwegian scholar Asle Schei developed a portable ultrasonic partial discharge detection system. The device includes continuous sampling and pulse-triggered modes, enabling accurate identification of discharge defects in GIS[6]. German scholar Sacha M. Markalous designed an acoustic-electric combined detection device based on ultrasonic and ultra-high frequency (UHF) methods. The system incorporates a wavelet packet algorithm for noise reduction and uses iterative algorithms based on average oil temperature and sound velocity to locate partial discharge sources[7]. Chinese scholar Chen Hongfu proposed an overall scheme for an ultrasonic-based GIS partial discharge detection system. They designed a preamplifier and data acquisition device and developed data processing and analysis software on the Visual Studio platform. The system identifies defect patterns through features such as signal amplitude, periodicity, and waveform dispersion. Experiments have demonstrated its excellent detection performance for various discharge defects[8].

Despite these advances, current research is largely limited to single-physical-quantity detection methods. Most existing systems focus solely on measuring either the electrical or mechanical state of GIS equipment. This limitation fails to account for the mechanical-electrical interaction characteristics of GIS, making it difficult to achieve comprehensive detection and evaluation of its operational state. Furthermore, the variability of failure types and differences in degradation states impose additional

Journal of Physics: Conference Series 3012 (2025) 012001 doi:10.1088/1742-6596/3012/1/012001

constraints on the accuracy and scope of these methods. A single detection approach is inadequate for meeting the demands of complex defect conditions[9]-[10].

Given these limitations, there is an urgent need to develop fusion detection technology that combines mechanical vibration and partial discharge measurements. This approach enables simultaneous monitoring of internal insulation and mechanical conditions, allowing for the timely identification of potential internal defects. Such technology is of significant practical importance for ensuring the safe and reliable operation of power networks. To this end, this paper proposes a novel vibration-ultrasonic fusion sensor that addresses the shortcomings of existing methods and provides a comprehensive solution for GIS operational state detection.

2. Structure Design of Vibration-Ultrasonic Fusion Sensor

This paper primarily focuses on the measurement of vibration acceleration and ultrasonic partial discharge in GIS. Due to the complex propagation chain of abnormal vibration signals caused by mechanical faults within the numerous mechanical structures of GIS, the transmission of these signals to the outer shell exhibits an exponential decay pattern[11]-[12]. Consequently, the accelerometer must possess high output sensitivity; thus, the structure of the fusion sensor should draw inspiration from compression-type accelerometers. For the acquisition of ultrasonic partial discharge signals, contact ultrasonic sensors are recommended to ensure minimal signal attenuation by closely adhering to the equipment's shell.

Analyzing the internal structures of the aforementioned compression-type accelerometers and contact ultrasonic sensors reveals many similarities between the two. The primary structure of the accelerometer consists of a base, piezoelectric element, and mass block from bottom to top, while the ultrasonic sensor is structured with a matching base, piezoelectric element, and backing layer from bottom to top. Based on the characteristics of both the accelerometer and the ultrasonic sensor, this paper designs a novel vibration-ultrasonic fusion sensor, as illustrated in Figure 1. The internal structure mainly comprises a matching base, piezoelectric element, backing mass block, upper preloading layer, spring, shell, and leads.

This design innovatively incorporates the damping material of the ultrasonic sensor as the load mass block into the structure of the compression-type accelerometer. The mass backing block not only provides inertial force during the measurement of acceleration signals but also acts as a sound-absorbing backing when receiving ultrasonic signals, absorbing the ultrasound that penetrates the piezoelectric ceramic and preventing reflections that could cause secondary excitation, which would negatively affect the sensor's receiving performance and signal resolution. This integration of the mass block of the accelerometer and the backing of the ultrasonic sensor achieves structural synergy. During the integration process, the mutual influence of both components is considered, ensuring that the matching base satisfies the stiffness requirements of the basic structural unit of the accelerometer and the acoustic impedance matching requirements of the basic structural unit of the ultrasonic sensor simultaneously.

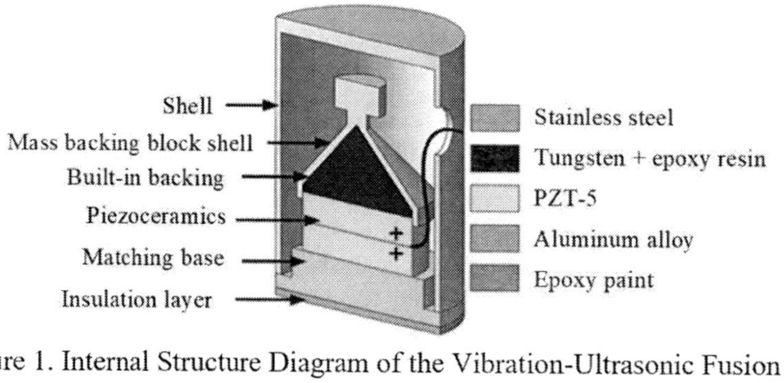

Figure 1. Internal Structure Diagram of the Vibration-Ultrasonic Fusion Sensor

3. Finite Element Simulation of Structural Parameters of Vibration-Ultrasonic Fusion Sensor

The design of the internal structure of the fusion sensor described above meets the basic structural unit requirements of both the accelerometer and the ultrasonic sensor, enabling simultaneous and co-located perception of fusion signals. However, in the practical application of the new sensor, it is essential to thoroughly consider the impact of the structural parameters and materials on the individual output performance of the fusion sensor. Additionally, the cross-coupling effects between the two output parameters during the fusion process must be adequately addressed. Therefore, finite element simulation is necessary to investigate how structural parameters affect the primary performance, with theoretical results providing effective guidance for the design and fabrication of the sensor.

3.1. Finite Element Simulation Setup

The simulation model is constructed based on the designed fusion sensor model, as shown in Figure 2. The model omits the preloading spring and sensor shell for clarity. Due to the axially symmetrical structure of the sensor itself, the model is set up in a two-dimensional axisymmetric format to streamline the computational process and accelerate convergence. In Figure 2.4, piezoelectric element 1 is polarized in the "-Z direction," while piezoelectric element 2 is polarized in the "+Z direction," ensuring that both piezoelectric elements output positive voltage at the coupling interface. During the finite element calculation, the built-in piezoelectric effect module of the software is used to couple solid mechanics and electrostatic multiphysics, simulating the influence of the structural parameters of the vibration-ultrasonic fusion sensor on its main performance and the cross-coupling of output parameters.

According to the calibration input, calibration output, and dynamic range specifications outlined in "GBT 20485-2007 Vibration and Shock Sensor Calibration Method" and "GBT 19800-2005 Nondestructive Testing - Acoustic Emission Detection Transducer," the vibration input is set as a unit acceleration applied at the bottom of the matching base, and the ultrasonic input is configured as a specified dynamic displacement. The output voltage at the coupling surface of the piezoelectric element is measured, and the frequency response of the ultrasonic output is calculated.

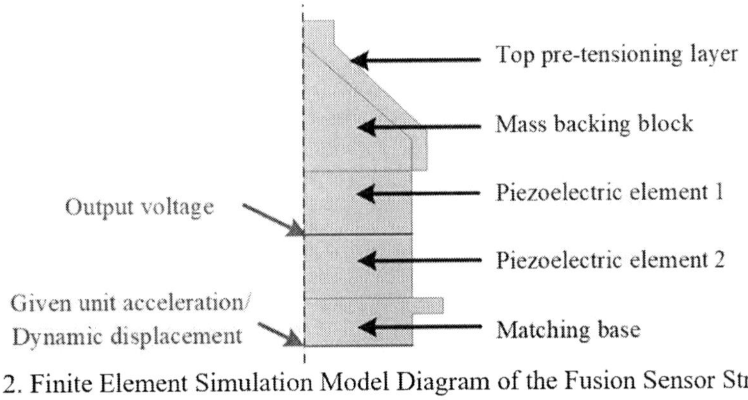

Figure 2. Finite Element Simulation Model Diagram of the Fusion Sensor Structure

The structural parameter variables for the sensor are set as follows: matching base thickness, thickness of piezoelectric elements, backing thickness, structural radius, matching base material, and backing material. The performance parameters of the sensor include vibration output sensitivity, vibration output natural frequency, 5% effective bandwidth for vibration output, ultrasonic resonance frequency, and ultrasonic frequency response. The vibration output sensitivity characterizes the nominal output of the sensor under a standard acceleration input, reflecting the response level of the vibration sensor. The inherent frequency of vibration output, commonly referred to as the natural resonance frequency, indicates that vibration sensors generally exhibit optimal acquisition sensitivity and linearity when operating within one-fifth of their natural frequency. Thus, it serves to define the operational frequency range of the vibration sensor. The 5% effective bandwidth of the vibration output characterizes the linearity of the output signal of the vibration sensor. The ultrasonic resonance frequency represents the center frequency of the maximum output response of the ultrasonic sensor, necessitating that the resonance frequency aligns with the ultrasonic frequency band of the detection equipment to ensure effective measurement of ultrasonic signals. The ultrasonic frequency response reflects the output response level of the ultrasonic sensor. The specific parameter settings for the structural variables in the simulation are detailed in Table 1, wherein the base materials include common magnesium alloy, aluminum alloy, titanium alloy, stainless steel, and alumina, while the mass backing materials are chosen from aluminum alloy, titanium alloy, gray cast iron, copper alloy, and tungsten.

Table 1. Structural Variables Parameter Table

Structural name		Parameter value
Matching base thickness		1/2/3/4/5/6 mm
Piezoelectric element thickness		1/2/3/4/5/6 mm
Backing thickness		1/2/3/4/5/6 mm
Piezoelectric element radius		1/2/3/4/5/6 mm
Matching base material	Young's modulus	45/71/96/193/390 GPa
	Density	1800/2770/ 3500/4620/7750 kg/m3
Mass backing block material	Young's modulus	71/96/110/120/344 GPa
	Density	2770/4620/7200/8300/19250 kg/m3

3.2. Simulation Result Analysis

Using finite element simulation software, a parametric scan of the aforementioned structural variables was conducted to obtain the output response of the integrated sensor under given unit acceleration or dynamic displacement. Characteristic extraction of performance parameters such as inherent frequency, 5% effective bandwidth of vibration, vibration sensitivity, ultrasonic resonance frequency, and ultrasonic frequency response was performed on the output response curves.

This analysis summarized the effects of structural parameters on the main performance and output parameter cross-coupling of the vibration-ultrasonic integrated sensor. The output performance variation curves obtained under different structural parameters are shown in Figure 3. and Figure 4. Since the simulated ultrasonic frequency response amplitudes were all greater than 100 dB with minimal differences, frequency response characteristics were not included in the analysis.

As seen in Figure 3(a), increasing the matching base thickness slightly decreases the inherent frequency, significantly reduces the 5% effective bandwidth, and drastically lowers the ultrasonic resonance frequency, while the vibration sensitivity remains unchanged. A thinner matching base allows for higher inherent frequency and a broader vibration bandwidth, but excessive thickness causes high-frequency signal attenuation.

Figure 3(b) shows that increasing the piezoelectric element thickness reduces both the inherent frequency and 5% bandwidth while boosting vibration sensitivity and lowering ultrasonic resonance frequency. Thicker piezoelectric elements produce more charge under the same acceleration, enhancing sensitivity but lowering the ultrasonic resonance frequency due to their inverse relationship.

In Figure 3(c), as the mass backing block thickness increases, the 5% bandwidth decreases, vibration sensitivity rises, and ultrasonic resonance frequency drops. A thicker backing increases mass, improving vibration sensitivity but increasing damping on high-frequency ultrasonic signals.

Figure 3(d) shows that increasing the overall structure radius lowers the inherent frequency while leaving the 5% bandwidth, vibration sensitivity, and ultrasonic resonance frequency largely unchanged.

The larger size reduces the inherent frequency but doesn't affect the force on the piezoelectric element or ultrasonic polarization, keeping sensitivity and resonance frequency stable.

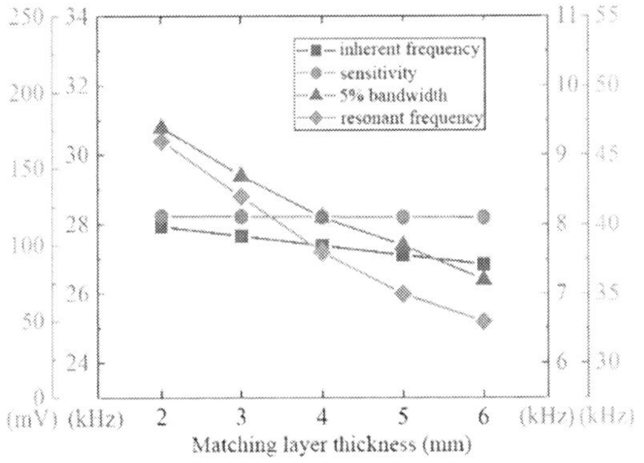

(a)Matching base thickness

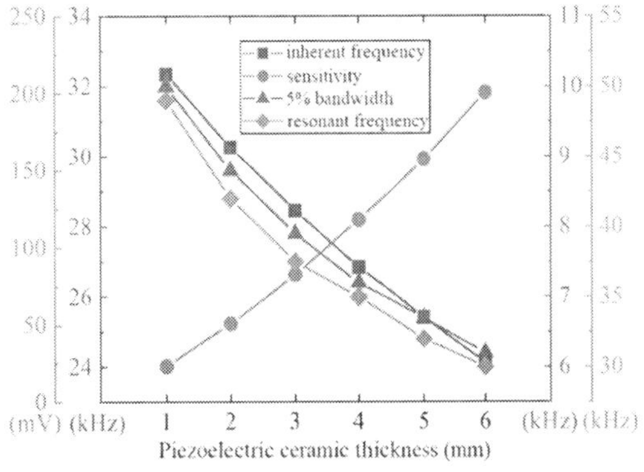

(b)Piezoelectric ceramic thickness

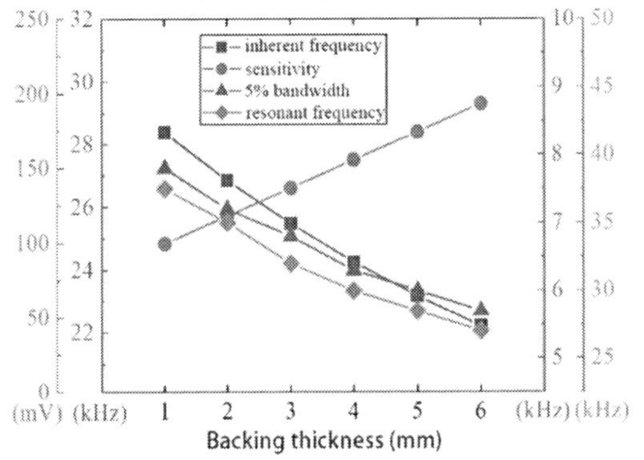

(c)Backing thickness

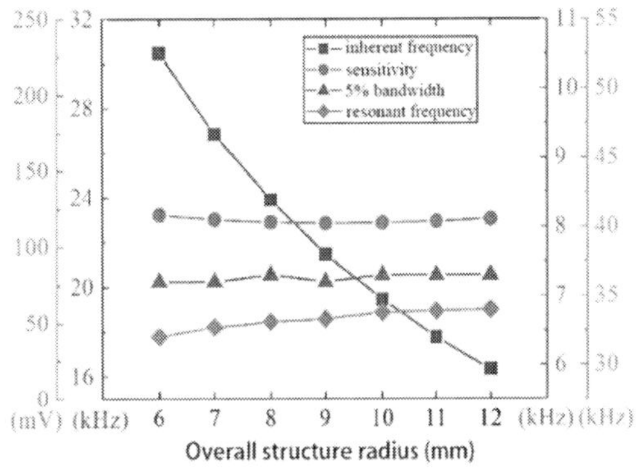

(d)Overall structure radius

Figure 3. The Effect of Different Structural Parameters on the Output Performance of the Vibration-Ultrasonic Integrated Sensor

From Figures 4(a) and (b), it can be observed that the primary material characteristic of the matching base is Young's modulus. A larger Young's modulus allows the integrated sensor to achieve a broader vibration operating bandwidth and a higher ultrasonic resonance frequency. From Figures 2-6(c) and (d), it can be seen that the primary material characteristic of the backing is density. A higher density leads to greater inertial force, enhancing the vibration output sensitivity of the integrated sensor. However, this also narrows the effective vibration operating bandwidth and reduces both the ultrasonic resonance frequency and frequency response. Therefore, it is necessary to select an appropriate backing material based on a comprehensive consideration of various output responses.

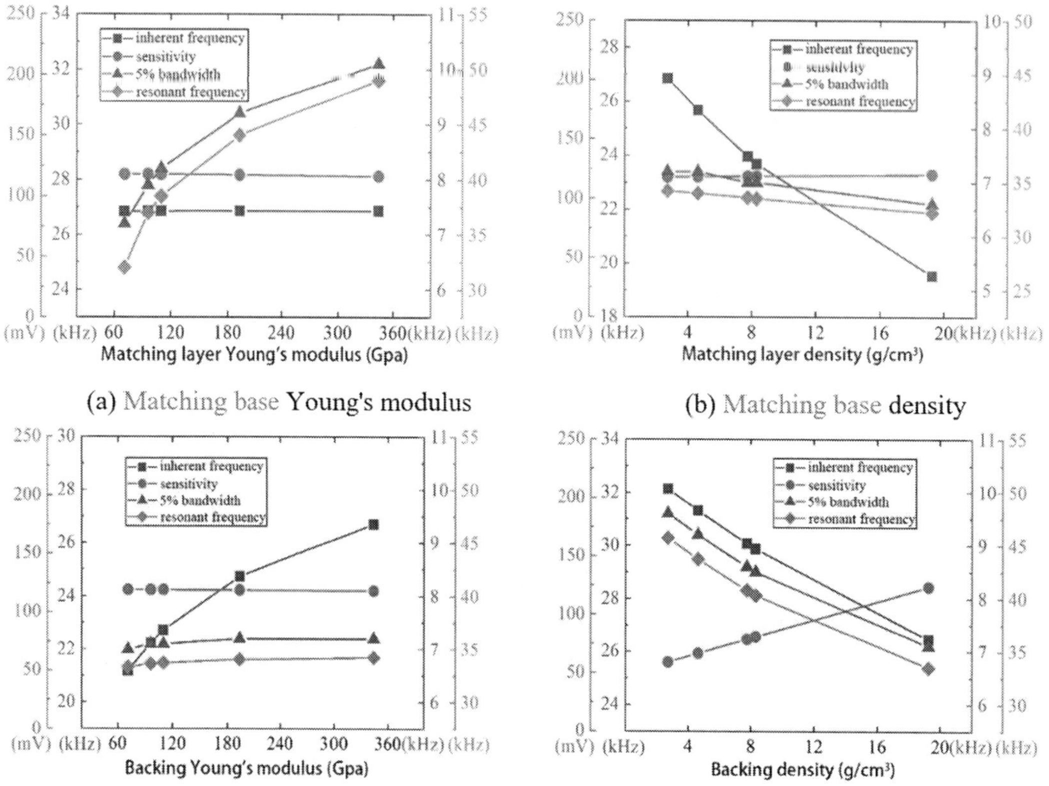

(a) Matching base Young's modulus

(b) Matching base density

(c) Backing Young's modulus

(d)Backing density

Figure 4. The Effect of Material Parameters of the Matching Base and Mass Backing on the Output Performance of the Integrated Sensor

4. Conclusion

In summary, the simulation results indicate that a lightweight and compact structure of the sensor results in a broader vibration operating bandwidth and a higher ultrasonic resonance frequency. However, reducing the mass of the backing block leads to a decrease in vibration output sensitivity, while decreasing the thickness of the piezoelectric ceramics results in lower vibration and ultrasonic output responses. By employing a multi-objective genetic optimization algorithm, the combination of structural parameters that yields the highest comprehensive score is selected as the final solution, representing the optimal configuration for the integrated sensor structure, as shown in Table 2. The final physical appearance and installation of the sensor is shown in Figure 5.

Table 2 Optimal Integrated Sensor Structural Parameter Combinations

Parameter	Piezoelectric Element Radius	Piezoelectric Element Thickness	Matching Base Radius	Matching Base Thickness
value	20.2mm	4.3mm	22.7mm	6.1mm
Parameter	Base Young's Modulus	Mass Backing Radius	Mass Backing Thickness	Mass Backing Density
value	393.4GPa	22.2mm	14.8mm	19.5g/cm^3

(a) Vibration-Ultrasonic Fusion Sensor Physical Diagram

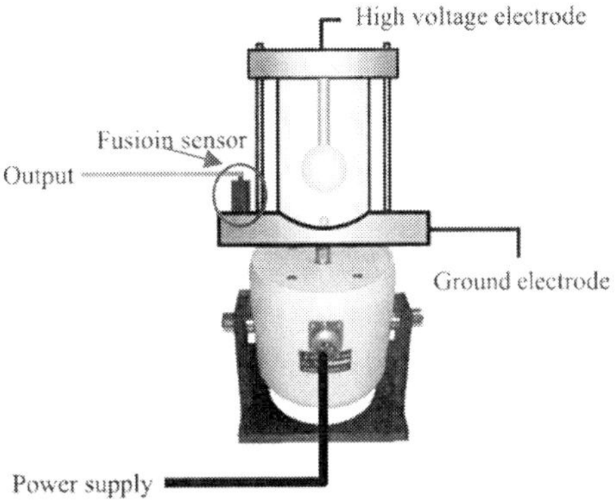

(b) Fusion Sensor Installation in the Simulation Experiment

Figure 5. Schematic Diagram of the Fusion Sensor and Its Installation

References

[1] Chen Fei. Research on the Development and Application of GIS Equipment [D]. Zhejiang University, 2007.

[2] Wang Feng, Qiu Yuchang. Recent Development Trends of Gas-Insulated Switchgear (GIS) [J]. Electric Power Technology, 2003(02): 54–57.

[3] Han Xiancai, Sun Xin, Chen Haibo, et al. Overview of the Development of Ultra-High Voltage AC Transmission Technology in China [J]. Proceedings of the CSEE, 2020, 40(14): 4371-4386+4719.

[4] Wang Linyuan, Bai Di. Study on the Characteristics of Mechanical Vibration Signals of GIS Abnormal Noise [J]. Shandong Industrial Technology, 2017(21): 171.

[5] Cui Yangliu, Ma Hongzhong, Wang Taoyun, et al. Fault Analysis of GIS Based on Fault Tree Theory [J]. High Voltage Apparatus, 2015, 51(07): 125-129+135.

[6] A. Schei, S. Kyrkjeeide, V. Larsen. Acoustic insulation analyzer for periodic condition assessment of Gas Insulated Substations[A]. IEEE/PES Transmission and Distribution Conference and Exhibition[C]. Yokohama, Japan: IEEE, 2002, 2: 919-924.

[7] Sacha Markalous, Stefan Tenbohlen, Kurt Feser. Detection and location of partial discharges in power transformers using acoustic and electromagnetic signals[J]. IEEE Transactions on Dielectrics and Electrical Insulation, 2008, 15(6): 1576-1583.

[8] Chen Hongfu. Research on Partial Discharge Detection in GIS Using Ultrasonic Methods [D]. Shanghai Jiao Tong University, 2008.

[9] Han Xutao, Liu Zehui, Li Junhao, et al. Research on GIS Partial Discharge Detection Method Based on Photoelectric Composite Sensors [J]. Proceedings of the CSEE, 2018, 38(22): 6760–6769.

[10] Zhang Zhaoyu, Hu Yidan, Song Yanfeng, et al. Development and Application of a Mechanical Vibration-Ultrasonic Fusion Detection Sensor for Power Equipment [J]. Proceedings of the CSEE, : 1–12.

[11] Du Zhiye, Hao Zhaoyang, Hao Qian, et al. Analysis of Vibration Propagation Characteristics of the Complex Shell Structure of 220kV GIL [J]. Journal of Hunan University (Natural Sciences), 2022, 49(02): 125–134.

[12] Shen Li, Huang Yulong, Qian Jiali. Phase Frequency Characteristics of Circuit Breaker Vibration Signals and Their Application in Monitoring [J]. Journal of Electrical Engineering, 1997(03): 42–45.

[13] Li Chunfeng, Song Pengfei. Application Analysis of Fiber Optic EFPI Ultrasonic Sensors in Partial Discharge Monitoring[J]. Light Source & Lighting, 2024, (12): 89-91.

Autonomous Navigation Method for UAV Based on the Fusion of Laser SLAM and AprilTag

Jie Zhou[1][0009-0000-9416-3721], Kai Mao[1*][0009-0008-0465-5780], Rui Wang[1][0009-0000-3122-6959], Tao He[1][0009-0006-6121-8822], Liang Huang[1][0009-0003-1636-5733]

[1] State Grid Chongqing Electric Power CO. Yongchuan Power Supply Branch Company, Yongchuan, Chongqing Municipality, 402160, P.R. China
maokai_023@163.com

Abstract: To improve the accuracy and efficiency of autonomous navigation for indoor UAV's (Unmanned Aerial Vehicle) in complex substation environments, this paper proposes a multi-sensor fusion-based autonomous navigation method. An indoor UAV platform integrating multi-dimensional LiDAR, depth cameras, and edge computing devices was designed and built. The method uses AprilTag markers as visual references and constructs maps based on laser SLAM(Simultaneous Localization And Mapping) technology. A relocation scheme that incorporates AprilTag information is proposed, and Dijkstra and DWA (Dynamic Window Approach) algorithms are used to plan the global and local navigation paths of the UAV, respectively. In a GPS-denied environment, the system achieves precise UAV localization and environmental point cloud mapping, ensuring accurate positioning and path planning. Multi-scenario experiments in a ROS (Robot Operating System) environment demonstrate that this method provides high positioning accuracy and strong robustness.
Keywords: Indoor positioning, UAV, Laser SLAM, AprilTag, Relocation

1. Introduction

With the advancement of the national smart grid and the promotion of unmanned inspection in substations, UAV inspection has become an inevitable trend. Autonomous UAV inspections not only enhance the automation level of substations but also significantly reduce manual labor, improve inspection efficiency, minimize safety risks, and bring considerable economic and social benefits to power maintenance departments [1-2]. Current research mainly revolves around laser SLAM technology, combined with multi-sensor fusion, to achieve autonomous UAV positioning and navigation in facilities [3-4]. Reference [5] proposed an autonomous navigation system integrating GPS, visual odometry, and a barometer, ensuring reliable UAV positioning even when certain sensors fail. Reference [6] successfully implemented UAV flight in GPS-denied environments through visual camera positioning. Reference [7] designed a system combining RGB-D sensors and inertial sensors, utilizing closed-form covariance to limit visual uncertainty and achieve more accurate positioning. Reference [8] proposed a UAV autonomous flight system that adapts to dynamic environments, integrating fast local environment sensing algorithms and sliding window path replanning. Reference [9] introduced a collision avoidance scheme based on the fusion of millimeter-wave radar and monocular cameras, using extended Kalman filters to build 3D obstacle coordinates and employing efficient path planning algorithms to avoid obstacles, with experiments validating its feasibility. However, laser radar obtains limited environmental information, especially in complex environments where positioning performance is poor, and sensor fusion can result in localization loss.

To address these issues, this paper proposes a relocation scheme based on laser SLAM and AprilTag information. The positions of AprilTag markers are incorporated into the environmental map, creating a more precise 2D grid map enriched with visual target information. Meanwhile, global and local planners are combined to achieve UAV path planning and control, aiming for efficient autonomous UAV inspections in indoor environments such as substations.

2. UAV Components and System Framework

2.1. UAV Hardware System

The UAV serves as a fundamental platform for monitoring substation environments and equipment. Its autonomous navigation functionality requires the capability to sense spatial information, perform control calculations, and execute autonomous flight. To this end, we developed an inspection UAV platform (as shown in Figure 1). The hardware system mainly consists of three components: the UAV body, the multi-sensor system, and the main controller. The UAV used is a DJI M3T, and the multi-sensor system is responsible for perceiving spatial environmental information, primarily configured with LiDAR and a depth camera. The main controller is the Nvidia Jetson AGX Xavier edge computing device, which provides high performance and powerful computing capabilities, meeting the requirements for running complex navigation algorithms.

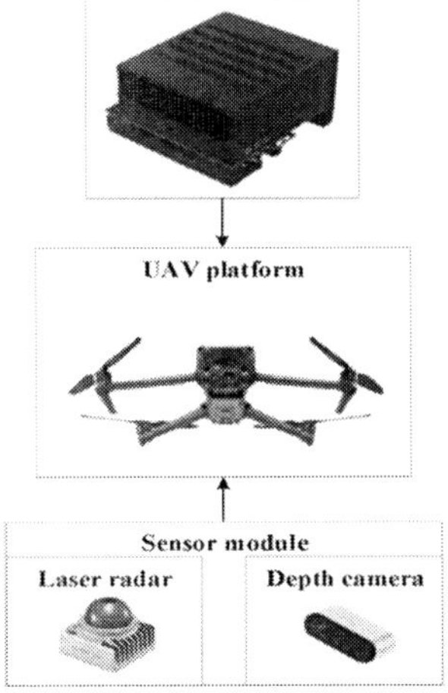

Figure 1. Hardware system of the indoor UAV platform

2.2. *UAV Software System*

The mobile UAV software system is divided into the upper and lower computers. The upper computer runs in the ROS Melodic environment, handling sensor data processing and navigation functions. It collects and fuses data, providing corrected pose, 2D LiDAR, and odometry information. The navigation package manages tasks such as mapping, localization, and path planning, with data transmitted in real-time using a "publish/subscribe" model. The lower computer runs on an embedded controller, receiving speed commands from the upper computer and outputting control signals to drive the UAV. Communication between the upper and lower computers is done via the CAN/2.0B protocol.

3. AprilTag-Based SLAM Localization Optimization

3.1. *LiDAR SLAM Localization Technology*

Currently, the mainstream SLAM optimization algorithms mainly include the following three types: SLAM based on the Extended Kalman Filter (EKF), SLAM based on the Rao-Blackwellized Particle Filter (RB-PF), and SLAM based on scan matching, such as Hector_SLAM. Among these, Hector_SLAM is widely used due to its higher robustness and lower computational resource requirements, making it the algorithm of choice for the system design in this paper [10], as shown in Figure 2.

The map model is set as a grid, and localization is performed using scan matching. The least squares method is used to solve for the optimal UAV pose S_i, as shown in equation (1):

$$\varsigma_i^* = \arg_{S_i} \min \sum_{i=1}^{n} [1 - M(S_i(\varsigma_i))]^2 \tag{1}$$

In equation (1), ς_i^* represents the optimal position of the drone, and M is the occupancy value function of the grid map at position $S_i(\varsigma_i)$, which indicates the likelihood of that grid being occupied.

In equation $S_i(\varsigma_i)$, as shown in (2), the laser scan endpoint is transformed from the LiDAR frame to the global system:

$$S_i(\varsigma_i) = \begin{pmatrix} \cos\theta & -\sin\theta \\ \sin\theta & \cos\theta \end{pmatrix} \begin{pmatrix} S_{i,x} \\ S_{i,y} \end{pmatrix} + \begin{pmatrix} C_x \\ C_y \end{pmatrix} \tag{2}$$

In equation (2), c_x c_y θ is the positional parameter.

ς_i For the grid map value $M(S_i(\varsigma_i + \Delta\varsigma_i))$ at coordinates $S_i(\varsigma_i + \Delta\varsigma_i)$, a Taylor expansion is performed with respect to $\Delta\varsigma_i$, then bilinear interpolation is used to estimate the occupancy value of grid P_m, as shown in equation (3):

$$M(P_m) \approx \frac{y - y_0}{y_1 - y_0}\left(\frac{x - x_0}{x_1 - x_0} M(P_{11}) + \frac{x_1 - x}{x_1 - x_0} M(P_{01})\right) + \frac{y_1 - y}{y_1 - y_0}\left(\frac{x - x_0}{x_1 - x_0} M(P_{10}) + \frac{x_1 - x}{x_1 - x_0} M(P_{00})\right) \tag{3}$$

In equation (3), P_{00}、P_{01}、P_{10}、P_{11} is the four adjacent points around point $P_m(x,y)$ in the raster map. In the actual calculation, replacing P_m with $S_i(\varsigma_i)$ yields the optimal estimate of S_i.

Inspection UAVs require coordinate transformations to achieve overall coordination, which necessitates a unified coordinate system to determine the relative positions between different components, such as the UAV body and the SLAM LiDAR. The map (map coordinate system) is used as the parent node to provide precise global coordinates, continuously updated by sensor data sources. Based on this, the displacement of the odom (odometry coordinate system) relative to the map is calculated. The orientation of the odometry coordinate system does not change with UAV movement, but its position accumulates drift due to the UAV's motion, providing short-term local precise positioning.

3.2. AprilTag-Assisted Localization Technology

During the UAV inspection, LiDAR SLAM may cause positioning errors in environments with sparse features. When such errors occur, sensor data needs to be reacquired to accurately determine the UAV' s current position. The commonly used method for this is scan matching (Scan-Match). The Scan-Match process includes: 1) obtaining a local submap m'_1 ofthe current position through LiDAR scanning, 2) comparing this submap with all other submaps m_i,i=1,....n to determine their similarity, and 3) determining the geometric relationships to find the best matching position T' in the environment. However, due to the limitations in the accuracy of Scan-Match, this paper introduces AprilTag markers to optimize the relocation process.

In the process of recognizing AprilTag, it is first necessary to extract the AprilTag marker from the image and then decode it. The extraction process starts with using Canny edge detection to identify edges in the image. Based on the detected edges, a quadrilateral detection algorithm is applied to filter out quadrilateral shapes. Once a quadrilateral is detected, the internal information is extracted and decoded according to AprilTag's encoding rules to determine if the quadrilateral is an AprilTag.

When an AprilTag is detected, its pixel coordinates in the image can be obtained from the camera, along with the pre-calibrated vertex coordinates of the AprilTag in the world coordinate system. Using the known pixel coordinates and world coordinates of the AprilTag, a homography matrix can be calculated through a linear transformation algorithm. By using this homography matrix, the detected pixel coordinates of the AprilTag in the image are mapped to the camera coordinate system. Combined with the intrinsic parameter matrix of the camera, the pose of the UAV can be determined.

3.3. SLAM Relocation Scheme Combined with AprilTag

This paper adopts an AprilTag-Scan-Match relocation scheme, which combines AprilTag markers and scan matching for relocation purposes. Different ID AprilTag markers are placed directly below indoor equipment to assist in localization.

The process flow is shown in Figure 3. A ROS message type named "AprilTag Mail" is defined, encapsulating the pose (x,y,yaw) of the AprilTag li identified by the camera, along with tag ID information (1,2, … , n), into "AprilTag_Pose". During mapping, this information is saved along with the environmental map, constructing a 2D grid map containing AprilTag markers. A distance threshold and an angle threshold θ are set; when the UAV's position relative to tag li reaches a threshold, relocation is performed. The camera identifies the AprilTag and determines its pose in the camera coordinate system lib(xib,yib,θib), comparing the tag li's information with the ID database to obtain its specific pose in the world coordinate system liw(xiw,yiw,θiw). Using this tag's pose in both the camera and world coordinate systems, the UAV's world coordinate pose can be calculated. This calculated UAV pose is then used to create a new local submap around the UAV, which is compared with previously established submaps to determine the optimal pose, thus accurately localizing the UAV in the grid map. Each time the camera detects AprilTag information, the UAV pose is updated based on previous calculations and the observed AprilTag data, thereby reducing cumulative error.

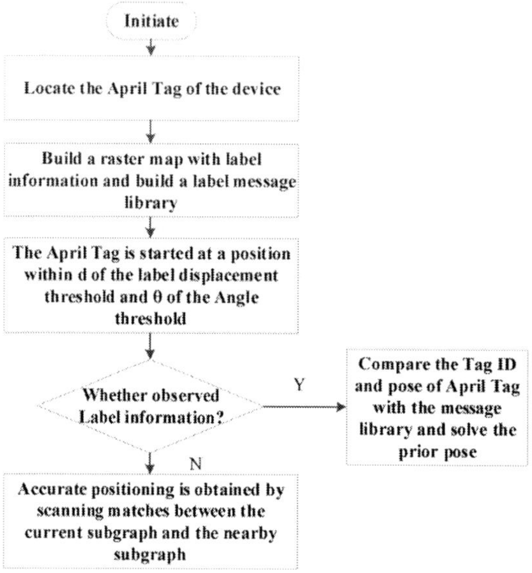

Figure 2. AprilTag-scan-match relocation flowchart

Traditional SLAM algorithms consist of two parts: the front-end and the back-end. The front-end performs data association through sensor fusion and completes pose estimation via laser scan matching. The back-end optimizes the pose through loop closure detection, reducing errors. However, in environments with sparse features, traditional SLAM struggles with insufficient localization accuracy, and positioning drift increases over time. To address this issue, this paper introduces multi-sensor fusion technology to improve the accuracy of pose estimation. During the relocation phase, AprilTag markers are incorporated to construct a grid map that includes AprilTag information, enabling more precise UAV localization. A schematic of the SLAM localization optimization based on AprilTag is shown in Figure 3.

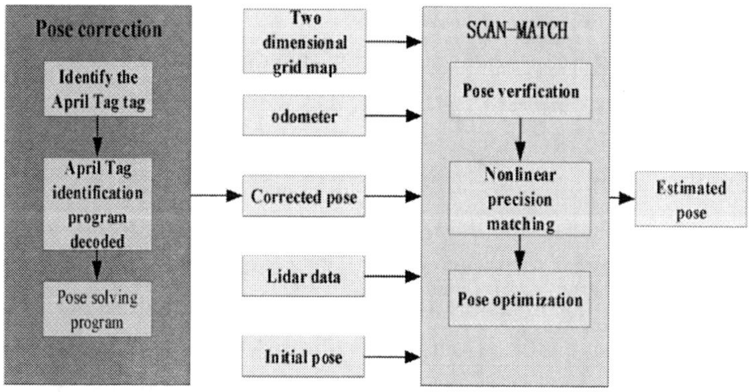

Figure 3: Schematic diagram of SLAM localization optimization based on apriltag

4. UAV Path Planning

Path planning uses a combination of global and local planners to enhance UAV obstacle avoidance. Both planners rely on cost maps, including static, obstacle, and inflation layers. The static map is a 2D grid generated by SLAM, while the inflation layer prevents collisions. The move_base package in ROS handles planning and control.

As shown in Figure 4 (a), the global planner uses Dijkstra's algorithm to calculate the shortest path, generating waypoints that the local planner processes to control the UAV. The local planner employs the DWA algorithm Figure 4 (b) to evaluate and select optimal trajectories, ensuring real-time obstacle avoidance and navigation.

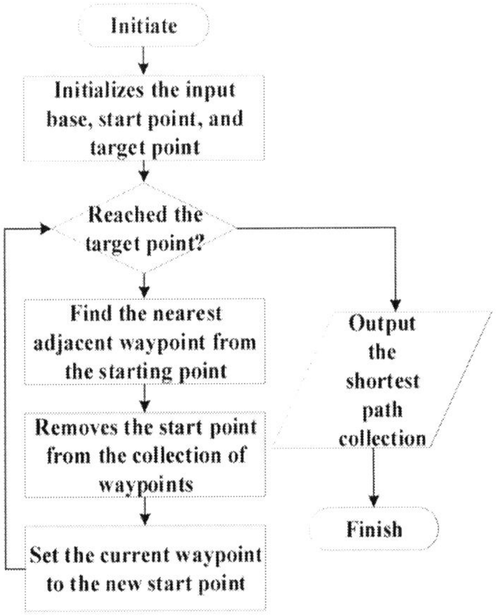

(a) Global path planning

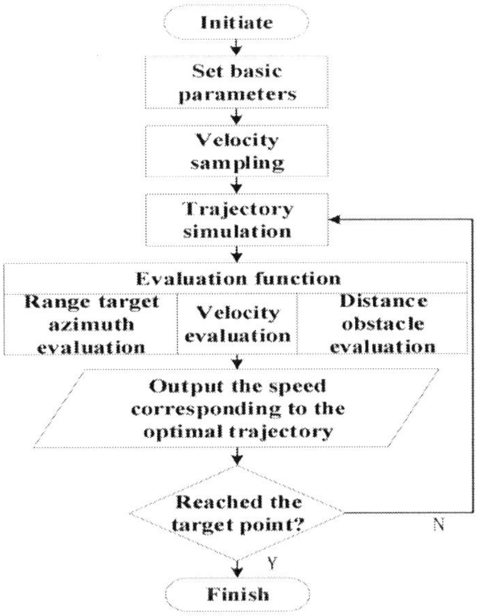

（b）Local path planning

Figure 4: Path planning algorithm flowchart

5. Experimental Validation

In this paper, the Scan-Match method and AprilTag-Scan-Match method were tested and compared in the Gazebo simulation environment. A grid map with AprilTags was constructed in the Gazebo environment, as shown in Figure 5.

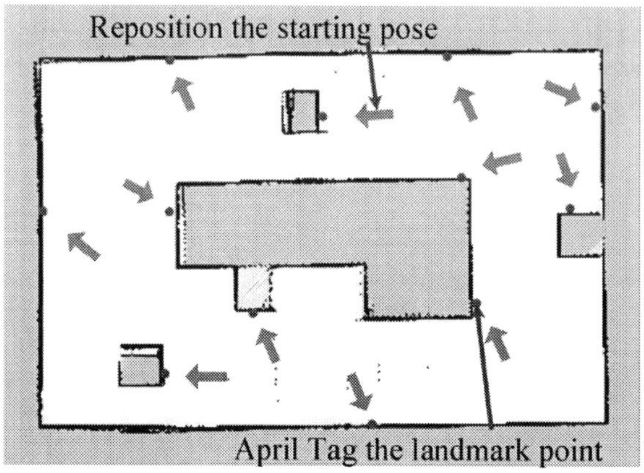

Figure 5. Grid map with apriltags

Using the aforementioned SLAM methods, a map of the simulated environment was created, with the poses of AprilTags observed by the UAV visualized on the map. Once the map was completed, the UAV was placed at 12 different AprilTag points, and two methods were used for relocation. In the experiment, the UAV underwent relocation tests at these 12 positions, and the performance was evaluated in three aspects: the number of successful initializations, the time required for relocation, and the need for manual intervention. The experimental results are shown in Table 1.

Table 1. Comparison of Relocation Initialization Results

Relocation Error/m	Scan-Match	April Tag-Scan-Match
Position 1	0.095	0.065
Position 2	0.156	0.079
Position 3	0.179	0.096
Position 4	0.204	0.106
Position 5	Relocation Failed	0.114
Position 6	0.269	0.126
Position 7	0.296	0.137
Position 8	0.320	0.152
Position 9	0.361	0.179
Position 10	Relocation Failed	0.195
Position 11	0.416	0.225
Position 12	0.436	0.259
Average Relocation Time	6.3s	1.5s
Need for Human Intervention	Yes	Yes

Based on the data in Table 1, the following conclusions can be drawn: 1. The Scan-Match method can achieve automated initialization without needing to move the UAV; however, it has longer localization times and occasional failures. The proposed algorithm, which incorporates AprilTag information, significantly reduces the time required for Scan-Match relocation and offers greater reliability in similar environments. The localization error is also notably reduced, with a final error of 0.259 m compared to the 160 m loop path in the simulation environment, resulting in a localization error ratio of 0.16%, which is 40% lower than that of Scan-Match.

To validate the proposed UAV navigation method, we conducted multi-scenario experiments using multi-sensor fusion in the ROS environment to create a point cloud map. Figure 6 shows the UAV's path in a complex indoor environment. The green line represents the flight path, blue points are waypoints for precise control, and the red polygon indicates the photography area. This setup ensures full coverage of required views during inspection, enhancing monitoring in complex environments.

Figure 6. UAV Navigation Path and Photography Area

6. Conclusion

This study focuses on a substation in Chongqing, using a DJI Mavic 3T UAV with the Payload SDK V3 development kit, combined with a Mid360 LiDAR. Through the SLAM and visual fusion algorithm, the UAV achieved local path planning and autonomous inspection without GPS. A visual-based indoor SLAM scheme was designed, aligning odometry data with LiDAR scans within the Scan-Match localization framework, and incorporating the AprilTag visual reference system for pose correction. This ensures long-term stability of autonomous navigation inside the substation.

Acknowledgments

Supported by the Key Science and Technology Project to 2024 YuDian Technology No. 58#, State Grid Chongqing Electric Power Company.

Reference

[1] Liu Z, Du Y, Chen Y, et al.: Simulation and experiment on the safety distance of typical±500 kV DC transmission lines and towers for UAV inspection. High Voltage Engineering, 2019, 45(02): 426-432.
[2] Tang M, Dai L, Lin C, et al.: Application of unmanned aerial vehicle in inspecting transmission lines. Electric Power, 2013, 46(3): 35-38.

[3] Li J, Duan Y, Wang C, et al.: Application of the Unmanned Aerial Vehicle in the Transmission Line Inspection. Power System and Clean Energy,2017,33(08):62-65+70.

[4] Li Y, Zhu S, Yu Y.: Improved Visual SLAM Algorithm in Factory Environment.ROBOT, 2019,41(01):95-103.

[5] Chambers A, Scherer S, Yoder L, et al.: Robust multi-sensor fusion for micro aerial vehicle navigation in GPS-degraded/denied environments,(2014).

[6] Scaramuzza D, Achtelik M C, Doitsidis L, et al.: Vision-controlled micro flying robots: from system design to autonomous navigation and mapping in GPS-denied environments. IEEE Robotics & Automation Magazine, 2014, 21(3): 26-40.

[7] Li D, Li Q, Cheng N, et al.: Combined RGBD-inertial based state estimation for MAV in GPS-denied indoor environments, (2013).

[8] Yu H, Zhang F, Huang P, et al.: Autonomous obstacle avoidance for uav based on fusion of radar and monocular camera,(2020).

[9] Gao F, Wang L, Zhou B, et al.: Teach-repeat-replan: A complete and robust system for aggressive flight in complex environments. IEEE Transactions on Robotics, 2020, 36(5): 1526-1545.

[10] Kohlbrecher S, Von Stryk O, Meyer J, et al.: A flexible and scalable SLAM system with full 3D motion estimation, (2011).

Journal of Physics: Conference Series 3012 (2025) 012003 doi:10.1088/1742-6596/3012/1/012003

Configurable Study on the Characteristics of Combined Defects of Free Metal Particles and Suspended Discharge

Lin Yan[1][0009-0008-1377-2567], **Congwei Yao**[2][0000-0001-9364-2798], **Hongwei Ding**[1][0009-0004-2851-2515], **Xutao Han**[1*][0000-0002-5980-4500]

[1] Xi'an Jiaotong University, No.28, Xianning West Road, Xi'an, Shaanxi, 710049, P.R. China

[2] Guangdong Key Laboratory of Electric Power Equipment Reliability, Electric Power Research Institute of Guangdong Power Grid Co., Ltd., Guangzhou 510080, P.R. China

*Corresponding author's e-mail: 1320086227@qq.com

Abstract. Since the practical application of Gas Insulated Switchgear (GIS) in the 1960s, it has been widely adopted worldwide. However, during long-term operation, the complex internal structure and high enclosure of GIS devices inevitably lead to various forms of defects. These defects may result in partial discharge, posing a threat to the operational safety and reliability of the equipment. To better understand the discharge mechanisms under complex defects, this study focuses on the discharge characteristics and interrelationships of combined defects involving free metal particles and suspended discharge. By designing a configurable experimental setup, different vibration and voltage conditions were applied in an environment filled with SF6 gas to systematically analyse the partial discharge behaviour under multi-defect conditions. The experimental results indicate that both vibration and voltage significantly influence the discharge characteristics, and a certain regularity is observed in the discharge behaviour between free metal particles and suspended discharge.

1. Introduction

GIS is widely used in high-voltage power systems due to its compact structure, high reliability, and low maintenance requirements, making it one of the key components of modern power transmission and distribution systems [1],[2],[3]. However, the complex operating environment within GIS and the various defects that may arise during long-term operation directly affect the insulation performance and operational safety of the equipment [4],[5] . Partial Discharge (PD), as one of the main representations of insulation defects in GIS, plays a crucial role in monitoring and analyzing its characteristics and development process for evaluating the operational status of the equipment and preventing major accidents [6],[7].

Common defect types encountered in actual operation of GIS include free metal particles, suspended potential bodies, and air gaps. These defects can not only independently trigger partial discharge but may also interact under specific conditions, leading to increased discharge intensity and complex discharge types, which can accelerate the aging of insulation materials and even induce breakdown[8]. Particularly, free metal particles, as typical pollution defects, are prone to movement under the influence of electric fields and vibrations, resulting in arcing and discharge phenomena[9].

Journal of Physics: Conference Series **3012** (2025) 012003 doi:10.1088/1742-6596/3012/1/012003

Additionally, suspended potential bodies, due to their suspended characteristics, have a complex electric field distribution that can easily lead to partial discharge[10]. Therefore, studying the combined effects and discharge characteristics of these two types of defects is of significant theoretical and practical value for a deeper understanding of the evolution mechanisms of complex defects within GIS.

This paper presents a configurable model for the combined defects of free metal particles and suspended potential bodies. Through experimental methods, the influence of vibration and voltage changes on discharge characteristics and particle motion properties was systematically investigated. By applying different vibration and voltage excitations, key parameters such as particle jumping acceleration (PVIA) and partial discharge inception voltage (PDIV) were obtained. Additionally, the analysis of mechanical vibration signals and ultrasonic partial discharge signals revealed the discharge laws of the combined defects and their correlations and complementarities under various excitation conditions. This study not only enriches the understanding of the discharge mechanisms of composite defects in GIS but also provides experimental evidence for the diagnosis and prevention of multiple defects in practical engineering applications.

2. Device Introduction

To investigate the correlation of discharges between defects under multi-defect conditions, a configurable setup for combined defects of free metal particles and suspended discharge was designed, as shown in Figure 1. To better simulate the actual operating conditions of GIS, the setup includes a suspended electrode configuration. The upper electrode is made of copper, while the lower electrode is an aluminum concave bowl. A cylindrical hollow protective cover (PTFE) is used to fix the electrodes, allowing the metal particles to freely rest on the grounded side within the protective cover. This electrode model is placed inside an airtight organic glass container filled with SF6 gas at a pressure of 0.55 MPa. The outer structure of the container is bolted together using four nylon rods to connect the upper and lower electrode plates. Experimental results indicate that the device maintains airtightness at a pressure of 0.55 MPa.

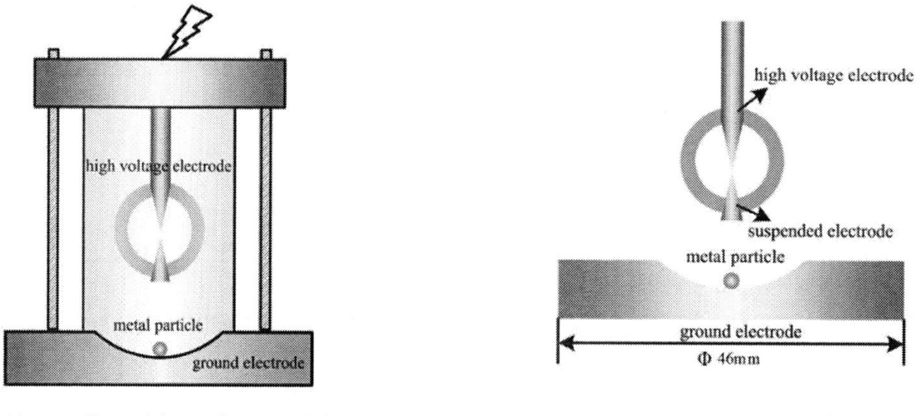

(a) Configurable Defect Model (b) Electrode Model

Figure 1. Configurable Defect Combining Free Metal Particles and Suspended Discharge

3. Analysis of Basic Characteristics

The experimental apparatus is shown in Figure 2. The oscilloscope has a bandwidth of 300-1500 MHz, the power amplifier provides input to the vibration table, and the partial discharge meter measures the PD signal and generates the phase-resolved partial discharge (PRPD) spectrum. An independent vibration signal was first applied to observe and record the jump behavior of the particles, resulting in the determination of PVIA. An independent voltage signal was then applied to observe and record PDIV. Finally, different vibration excitations were applied to study the variations in the initial discharge voltage and motion characteristics of the particles. In this section, the particles were maintained in a spherical shape, made of aluminum, with a diameter of 3 mm and a vibration frequency of 100 Hz. The amplitude of the vibration excitation was varied, and the measured mechanical vibration signal and ultrasonic signal are shown in Figure 3.

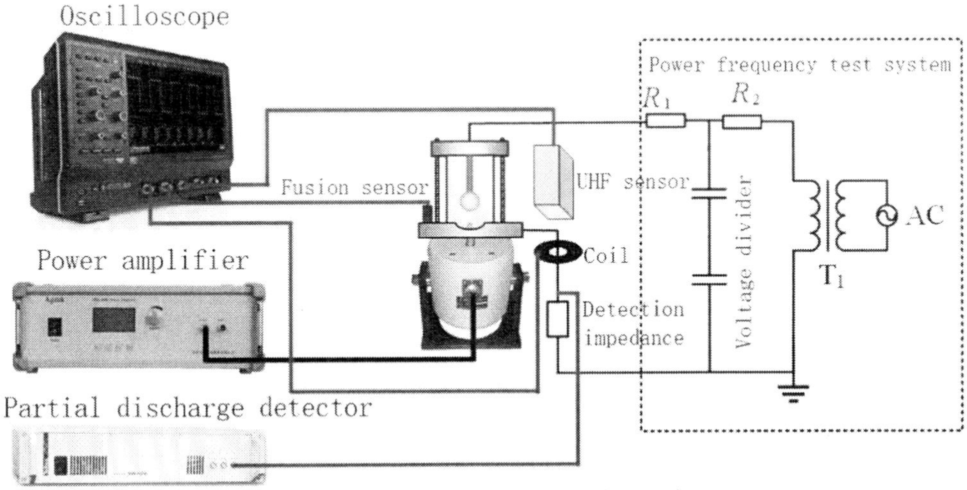

Figure 2. Schematic diagram of the experimental arrangement

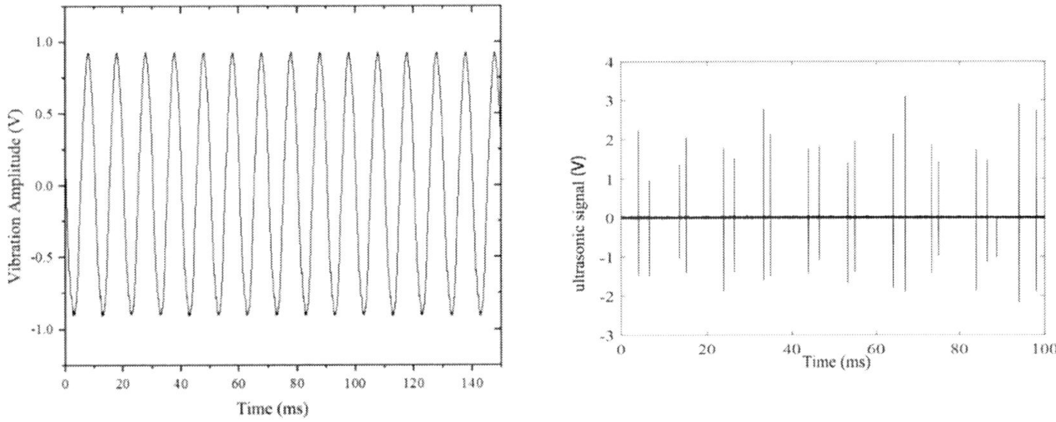

(a) Mechanical Vibration Signal (b) Ultrasonic Partial Discharge Signal

Figure 3. Time Domain Waveforms of Mechanical Vibration and Partial Discharge Ultrasound Under the Combination Defect of Free Metal Particles and Suspended Discharge

From the figure, it can be seen that the combination defect of free metal particles and suspended discharge exhibits significant mechanical vibration signals and ultrasonic partial discharge signals under the joint action of mechanical and electrical effects. Moreover, the vibration amplitude is positively correlated with the output of the vibration table, with a frequency typical of the normal bus vibration at 100 Hz. Therefore, the research on the characteristics of discharge under combined defects influenced by electromechanical effects mainly focuses on the partial discharge signals under different vibration states and voltage conditions, and characteristic maps are plotted for regularity analysis. The motion of free metal particles under the combined influence of electric fields and mechanical vibrations can be described by the following dynamic model:

$$F_{\text{total}} = F_{\text{Coulomb}} + F_{\text{vibration}} = qE + m \cdot a_{\text{vib}} \cdot \sin(2\pi f t) \quad (1)$$

Journal of Physics: Conference Series **3012** (2025) 012003 doi:10.1088/1742-6596/3012/1/012003

where:

F_{total} is the total force acting on the particle;

$F_{Coulomb}$ is the Coulomb force, with q being the particle charge and E the electric field strength;

$F_{vibration}$ is the inertial force induced by vibration;

m is the particle mass;

a_{vib} is the vibration acceleration;

f is the vibration frequency;

t is time.

This model explains why particle jumping is suppressed under suspended defects, as the Coulomb force near the particles is insufficient to overcome the gravitational and inertial forces. Specifically, when the electric field strength E near the particles is low (due to the dominant field distribution around the suspended electrode), the Coulomb force $F_{Coulomb}$ becomes negligible compared to the inertial force $F_{vibration}$. As a result, the particles remain stationary, and no jumping phenomenon is observed."

4. Correlation and Complementarity Analysis

4.1. Influence of Vibration Amplitude on the Initial Discharge Characteristics and Motion Characteristics of Configurable Defects

The experiment began with the application of an independent vibration signal to observe and record the jump behavior of the particles, resulting in the determination of PVIA. An independent voltage signal was then applied to observe and record PDIV. Finally, under different vibration excitations, variations in the initial discharge voltage and motion characteristics of the particles were studied. In this section, the particles were maintained in a spherical shape, made of aluminum, with a diameter of 3 mm and a vibration frequency of 100 Hz. The amplitude of the vibration excitation was varied, and statistics on PDIV, discharge quantity, discharge repetition rate, etc., were collected.

During the experiment, it was found that within twice the suspended discharge PDIV, the particles did not exhibit any jumping phenomenon. This can be explained by the electric field distribution around the suspended electrode, which can be modeled as:

$$E(r) = \frac{V}{r ln\,(\frac{R_{outer}}{R_{inner}})} \tag{2}$$

Where:

$E(r)$ is the electric field strength at a radial distance;

r from the electrode center;

V is the applied voltage;

R_{outer} and R_{inner} are the radii of the outer and inner electrodes, respectively.

The electric field is highly concentrated near the suspended electrode, resulting in a weak Coulomb force on the free metal particles. This force is insufficient to overcome the gravitational and inertial forces, as described by the threshold condition for particle jumping:

$$a_{vib} > \frac{qE}{m} \tag{3}$$

Where:

a_{vib} is the vibration acceleration;

q is the particle charge;

E is the electric field strength, and m is the particle mass.

In experiments, the vibration acceleration did not meet this threshold, which explains the absence of particle jumping behavior.

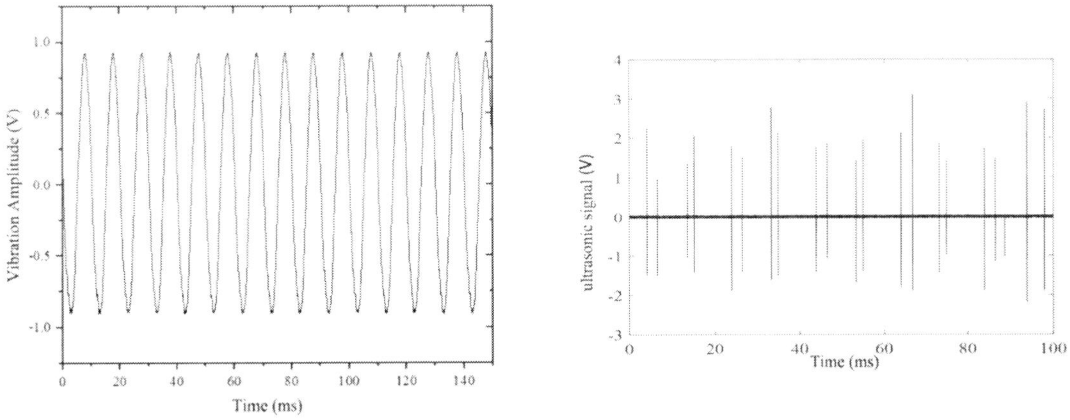

(a) Mechanical Vibration Signal (b) Ultrasonic Partial Discharge Signal

Figure 4. Vibration and Ultrasonic Signals During Discharge of Free Metal Particles and Suspended Discharge Combination Defect

4.2. Influence of Voltage Amplitude on Discharge Characteristics and Motion Characteristics of Configurable Defects

Under the same vibration excitation and condition, different voltage excitations were applied to study the variations in the discharge characteristics of configurable defects and to observe the correlation and complementarity between suspended discharge and particle discharge. In this section, the particles were maintained in a spherical shape, made of aluminum, with a vibration frequency of 100 Hz, while keeping the relative distance of the suspended electrode constant. The voltage excitation amplitude was varied, and statistics on discharge quantity were collected, as shown in Figure 5.

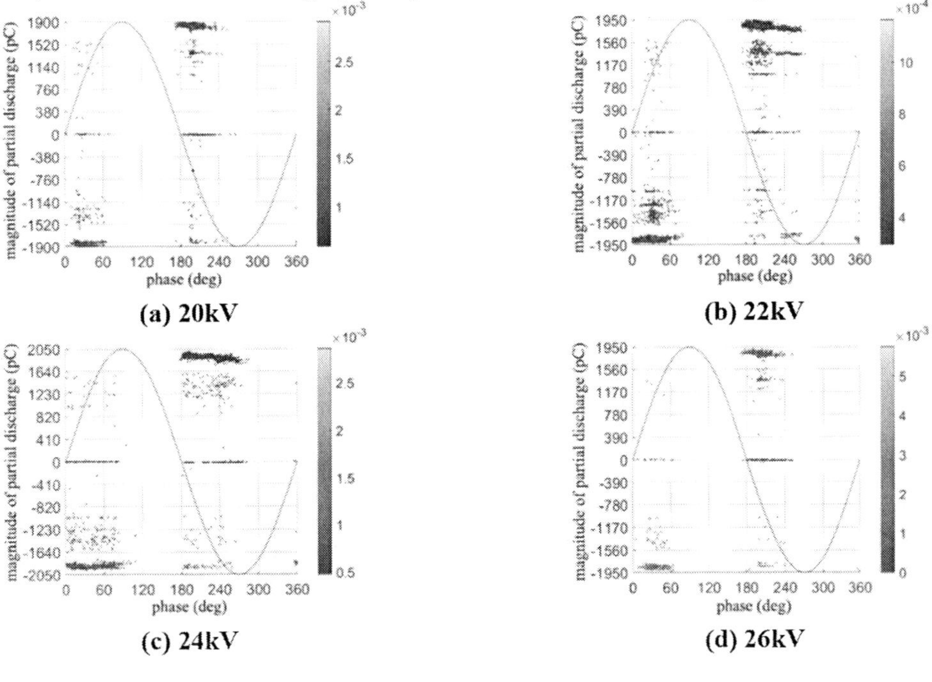

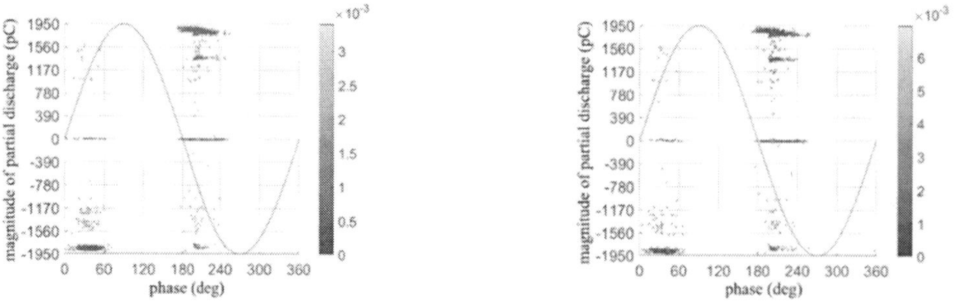

(e) 28kV (f) 30kV

Figure 5. PRPD Spectra of Particle Discharge Under Different Voltage Conditions

During the experiment, the particles did not exhibit any jumping behavior, and no ultrasonic pulses from the particles were detected. As shown in the figure, with the continuous increase of voltage excitation, the maximum discharge quantity in the PRPD spectra has exceeded the range of the partial discharge instrument, reaching its maximum value. Therefore, the PRPD spectra of the partial discharge instrument cannot represent the discharge characteristics of the suspended defect as voltage changes. To better explore the variations in discharge characteristics of suspended defects under different voltage amplitudes, this project collected the maximum amplitude of the ultrasonic partial discharge signal as a characteristic quantity to represent the discharge situation, as shown in Figure 6.

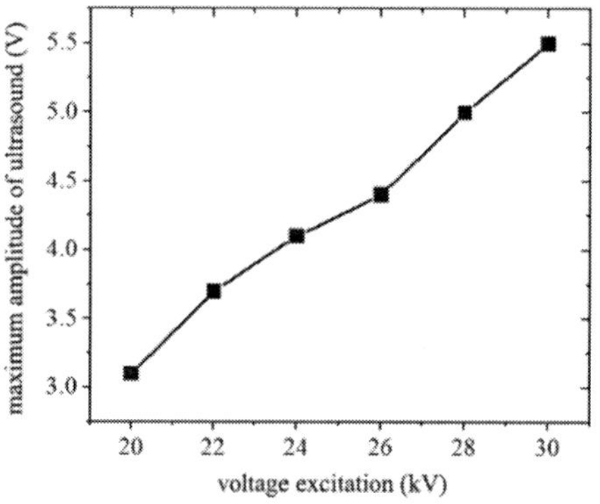

Figure 6. Vibration and Ultrasonic Signals During Suspended Discharge

From the above figure, it can be observed that for suspended defects under different voltage excitations, as the voltage increases, the maximum amplitude of the ultrasonic signal shows a linear increasing trend, indicating a positive correlation between the intensity of discharge and voltage. This suggests that voltage plays a dominant role in the development of discharge, with the intensity of discharge development continuously enhancing with increasing voltage. The PRPD spectra at multiple voltage levels maintain a similar spectral pattern, indicating that within this voltage range, it still represents a typical suspended discharge defect type. The discharge characteristics are stable in amplitude, and the intervals between adjacent discharges are consistent.

5. Conclusion

This study systematically investigates the discharge characteristics of combined defects involving free metal particles and suspended discharge in SF6-insulated GIS. Key findings include:

The electric field distribution dominated by suspended defects significantly suppresses the Coulomb force on free metal particles ($F_{Coulomb}$<0.1 mN at 30 kV), preventing particle jumping (PVIA > 2g).

The ultrasonic signal amplitude of suspended discharge exhibits a strong linear correlation with voltage (R^2=0.98), confirming voltage as the primary driver of discharge intensity.

The PRPD spectra remain consistent with typical suspended discharge patterns across all tested voltage levels (20–30 kV), indicating minimal interaction between the two defect types.

These results suggest that in GIS maintenance, priority should be given to detecting and mitigating suspended defects. Future work will explore higher voltage regimes and hybrid gas environments to further validate these findings

Acknowledgments

This work was supported by the Guangdong Provincial Key Laboratory of Electric Power Equipment Reliability, Electric Power Research Institute of Guangdong Power Grid Co., Ltd., Guangzhou, Guangdong, 510080, China, and by the 2023 Open Fund of Guangdong Provincial Key Laboratory of Electric Power Equipment Reliability (Grant No. GDDKY2023KF02).

References

[1] LIU Junhua,XU Minhua,HUANG Chengjun,et al. Investigation on the attenuation characteristics of electromagnetic waves in GIS[J]. Transactions of China Electrotechnical Society,2010,25(8):52-58.

[2] WANG Yanbiao,CHEN Zhenyong,GUO Wenping,et al.PD pattern recognition for GIS based on CNN architecture optimized by the double attention mechanism[J]. Journal of Electric Power Science and Technology,2022,37(2):22-29.

[3] RIECHERT U, HOLAUS W. Ultra high-voltage gas-insulated switchgear - A technology milestone[J]. European Transactions on Electrical Power, 2011, 22(2011): 60-82.

[4] LI Junhao,HAN Xutao,LIU Zehui,et al. Review on partial discharge measurement technology of electrical equipment[J]. High Voltage Engineering,2015,41(8):2583-2601.

[5] LIU Hangbin,LIN Houfei,CHU Jing,et al.Recognition of partial discharge patterns of GIS based on CWGAN-div and Mi-CNN[J]. Zhejiang Electric Power,2023,42(8):75-83.

[6] Song, H., Dai, J.J., Li, Z., et al. Assessment Method for Severity of Partial Discharge in GIS Under Operating Conditions. Transactions of China Electrotechnical Society, 2019, 39(04): 1231-1241.

[7] Dong, B.B., Gao, C.S., Xu, Y., et al. A Partial Discharge Recognition Algorithm for GIS Equipment Based on Modified Time-Frequency Clustering Boundary. High Voltage Engineering, 2023, 49(11): 4736-4746.

[8] WANG Caixiong,TANG Zhiguo,CHANG Wenzhi,et al.Experimental study on development characteristics of point discharge in GIS[J]. Suspended discharge Power System Technology,2011,35(11):157-162.

[9] Ma Feiyue, Niu Bo, Zhang Tao, Tian Yi, Cao Wen. Study on the Movement Process of Foreign Metal Particles in GIS Under High Voltage. High Voltage Apparatus, 2021, 57(11): 35-41.

[10] Wang Fei. Research on Partial Discharge Monitoring and Discharge Type Identification in Switchgear [D]. Nanjing University of Information Science & Technology, 2024:9-10.

Optimal Low-Carbon Power System Scheduling with Environmental Sustainability

Wenxin Guo[1], Yiquan Li[1], Ruifeng Zhao[1], Qingyu Tu[1*], Yueting Lin[1], Tianlong Wu[1], Zifei Wang[2]

[1] Guangdong Power Dispatch and Control Center, Guangzhou Guangdong 510000, China

[2] Southern Power Grid Digital Grid Research Institute Co., Ltd, Guangzhou Guangdong 510663, China

[*]Corresponding author's e-mail: 2994084591@qq.com

Abstract. This paper presents a low-carbon economic dispatch methodology for power systems, aiming to minimize both carbon emissions and operational losses while ensuring system reliability. The proposed approach leverages the Beluga Whale Optimization (BWO) algorithm, a bio-inspired optimization technique, to achieve an efficient and sustainable power dispatch solution. The BWO algorithm combines global exploration and local exploitation to effectively navigate the complex, multi-objective optimization space, addressing the trade-offs between economic and environmental goals. The effectiveness of the model is validated through simulations on the IEEE 57-node power system, demonstrating that the BWO algorithm successfully reduces carbon emissions and line losses, while offering a significant advantage in terms of computation time. The results highlight the potential of BWO for real-time power system optimization, providing a promising solution for the integration of renewable energy sources and the transition towards low-carbon grids.

1. Introduction

Carbon emissions, particularly in the form of carbon dioxide (CO_2), have reached unprecedented levels due to human activities over the past century [1]. These emissions primarily arise from the combustion of diesel fuels in power sources, agricultural processes, transport, deforestation and agriculture. As of today, atmospheric CO_2 levels are higher than at any time in recorded history, surpassing 415 parts per million (ppm). The dramatic increase in greenhouse gases is the main drive for climate variability globally, causing increases in local air temperatures, rising ocean floors, dramatic climate events, and devastation of existing fragile ecosystems and biodiversity [2].

Power system dispatch refers to the process of determining how much electricity each generator in the grid should produce to meet the current demand for electricity, while optimizing various factors such as cost, reliability, and environmental impact. In the context of decarbonization, power system dispatch is of critical importance because it directly affects the ability to integrate renewable energy sources, reduce carbon emissions, and transition to a low-carbon grid [3].

Dispatch decisions directly affect the carbon intensity of electricity generation. Carbon-intensive power plants, such as coal and gas plants, are typically dispatched as "last resort" units when renewable and lower-emission sources (e.g., nuclear, hydro, or biomass) cannot meet demand. Power plants with high carbon emissions are dispatched when renewable and low-emission energy sources (e.g., nuclear, hydropower, or biomass) are unable to meet customer demand for the load. Minimizing the use of these fossil fuel power plants through this optimal dispatch process can effectively reduce energy consumption [4]:

- Prioritize renewables: When wind or solar power is available, dispatch decisions should favor these sources over fossil fuel generation.
- Minimize the use of peaking plants: Gas-fired peaking plants are often used to meet demand during periods of high consumption, but they are highly carbon-intensive. Dispatch strategies should aim to minimize their use by leveraging renewable energy and storage technologies.
- Implement carbon pricing: Incorporating carbon prices into the dispatch decision-making process can make fossil fuel generation more expensive and less competitive compared to renewables, further incentivizing the use of clean energy sources.

In particular, the essay explores the usefulness of renewal power sources in modern grid dispatching: How can we ensure that renewable energy sources are prioritized and effectively integrated into the dispatch process while accounting for their intermittent nature? Through the development and application of this new scheduling decision framework, the study aims to provide actionable insights and strategies for system operators, policymakers, and energy planners seeking to achieve a cleaner and more resilient energy future [5].

2. Literature Review

The optimization of power system dispatch to minimize carbon emissions while ensuring system reliability and efficiency has been a focal point in the transition towards sustainable energy systems. Several approaches have been explored in the literature to address this challenge, including conventional optimization techniques and bio-inspired algorithms.

Traditional economic dispatch (ED) [6] models aim to minimize the operational cost of power systems, focusing primarily on the efficient allocation of generation resources. These models have evolved to incorporate environmental considerations, such as reducing emissions and optimizing fuel consumption. However, traditional methods often fall short in addressing the complexities of modern electric energy networks, including the consolidation of regenerative power generation and dynamic load conditions. Moreover, these approaches struggle with a multiobjective optimum where both targets for calculation of losses and emissions need to be minimised at the same time [7]. Over recent years, the inclusion of environmental objectives, particularly carbon emission reduction, in economic dispatch has gained attention. Researchers have proposed modifications to conventional ED models to incorporate carbon emission factors, such as the carbon intensity of different generation sources. Several algorithms, including mixed-integer linear programming (MILP) [8] and genetic algorithms (GA) [9], have been used to solve these complex optimization problems. These methods, while effective, often require significant computational resources, especially for large-scale systems.

In response to the limitations of conventional methods, bio-inspired algorithms have been increasingly applied to power system optimization problems. These algorithms, such as genetic algorithms (GA), particle swarm optimization (PSO) [10], and cuckoo search, offer greater flexibility in handling complex, nonlinear, and multi-objective problems. They have shown promise in improving the efficiency of economic dispatch, particularly in terms of minimizing both operational costs and environmental impacts. The Beluga Whale Optimization (BWO) [11] algorithm, a recent bio-inspired optimization technique, has demonstrated significant potential in solving complex optimization problems. Inspired by the hunting and social behaviors of beluga whales, BWO balances exploration and exploitation to navigate vast search spaces and avoid local optima. This algorithm has shown superior performance in a variety of applications, including engineering design and machine learning. Its unique combination of global search capabilities and local refinement makes it particularly suitable for power system optimization, where both the exploration of diverse solutions and the fine-tuning of the best options are essential. While the use of BWO in power system dispatch has not been extensively explored in the literature, its potential for addressing low-carbon economic dispatch problems is promising. BWO's ability to minimize carbon emissions while optimizing grid losses and operational costs in power systems offers a significant advantage, especially when integrated with renewable energy sources [12]. Studies have shown that BWO can effectively handle the trade-offs between environmental goals and operational efficiency, ensuring both sustainable and economically viable power system operations [13].

3. Materials and Methods

The core of the low-carbon economic dispatch is the inclusion of carbon emission limits as part of the dispatch optimization process. This model aims to minimize not only the total generation cost but also the carbon emissions from fossil-fuel plants. To achieve this, carbon emissions from each power plant are quantified based on the type of fuel used and the operational conditions of the plant. Renewable energy sources and low-emission resources are prioritized in the dispatch process due to their minimal or zero emissions. Meanwhile, higher-emission fossil-fuel plants are dispatched only, when necessary, typically during periods of high demand or when renewable energy generation is insufficient.

3.1. Objective Functions

In the context of a low-carbon economic dispatch framework, the objective functions for minimizing carbon emissions and grid line loss are crucial components. These objective functions help the system operator optimize the dispatch of generation units to achieve both environmental goals and efficient grid operation. Below is a detailed outline of these objective functions. The objective of minimizing carbon emissions aims to reduce the total carbon output from power generation. This function can be defined as follows:

$$\text{Minimize } C_{\text{emissions}} = \sum_{i \in G} E_i \cdot P_i \tag{1}$$

where:

- $C_{\text{emissions}}$: Total carbon emissions from all generating units.
- G: Set of all generating units in the system.
- E_i : Emission factor of unit i (typically measured in tons of CO_2 per MWh), which indicates the amount of CO_2 emitted per unit of electricity generated.
- P_i: Power output of generating unit i (in MW).

Minimizing grid line loss is essential for increasing the transmission effectiveness of electricity and minimizing energy wastage. Line loss in the transmission network occurs due to the resistance in transmission lines, which increases with the distance and amount of current flow. The objective function for minimizing line loss can be represented as:

$$\text{Minimize } L_{\text{loss}} = \sum_{(i,j) \in L} R_{ij} \cdot I_{ij}^2 \tag{2}$$

where: L_{loss} means the total line loss across the transmission network; L means the set of all transmission lines (i,j) between buses i and j in the grid; R_{ij} is the resistance of transmission line (i,j), which depends on the material and length of the line; represents the current flowing through transmission line (i,j), which is a function of current and voltage levels of power flow.

Since the power flow P_{ij} and voltage V_{ij} are related to the current I_{ij} by $I_{ij} = \frac{P_{ij}}{V_{ij}}$, the line loss objective can also be expressed in terms of power flow:

$$L_{\text{loss}} = \sum_{(i,j) \in L} \frac{R_{ij} \cdot P_{ij}^2}{V_{ij}^2} \tag{3}$$

where: P_{ij} is the power flow from bus i to bus j (in MW); V_{ij} is the voltage at the buses i and j (in kV).

This formulation minimizes the resistive losses by optimizing the power flow across lines, ensuring that power is transmitted in a manner that reduces losses. By minimizing line losses, the framework indirectly contributes to emission reduction, not only by improving grid efficiency, as less generation is required to meet the same demand.

For a low-carbon dispatch strategy, the two objectives—minimizing carbon emissions and minimizing line loss—can be combined into a multi-objective optimization problem:

$$\text{Minimize } Z = \omega_1 \cdot C_{\text{emissions}} + \omega_2 \cdot L_{\text{loss}} \tag{4}$$

where: Z means overall objective function representing the combined minimization of carbon emissions and line losses; ω_1 and ω_2 are the weighting factors that reflect the relative importance of carbon emissions reduction and line loss minimization in the dispatch decision.

The weights ω_1 and ω_2 can be adjusted based on policy priorities, operational costs, or environmental targets, enabling a flexible approach to dispatch that balances emission reductions and grid efficiency.

These objective functions are subject to typical power system constraints, including:

- Power Balance Constraint: Ensuring that total generation equals total demand plus losses.

$$\sum_{i=1}^{N_g} P_{g,i} - \sum_{j=1}^{N_b} P_{d,j} - P_l = 0 \tag{5}$$

where $P_{g,i}$ is the power generated by generator i; $P_{d,j}$ is the power demand at bus j; P_l is the total system loss; n is the total number of buses in the system; N_g is the total number of generators; N_b is the total number of buses.

- Generator Limits: Each generator i operates within its minimum and maximum power output limits.

$$P_{\min,i} \le P_{g,i} \le P_{\max,i}, \forall i = 1,2,\ldots,N_g \tag{6}$$

where $P_{\min,i}$ is the minimum power output of generator i; $P_{\max,i}$ is the maximum power output of generator i.

- Line Capacity Constraints: Ensuring power flows do not exceed the capacity of transmission lines.

$$\left|P_{i,j}\right| \le S_{\max,ij}, \forall (i,j) \in \Gamma \tag{7}$$

where $P_{i,j}$ is the power flow from bus i to bus j; $S_{\max,ij}$ is the maximum power flow capacity of the transmission line between i and bus j; Γ represents the set of all transmission lines in the system.

- Voltage Limits: Maintaining bus voltage levels within acceptable bounds to ensure stability and security.

$$V_{\min,i} \le V_i \le V_{\max,i}, \forall i = 1,2,\ldots,N_b \tag{8}$$

where V_i is the voltage magnitude at bus i; $V_{i,\min}$ and $V_{i,\max}$ are the minimum and maximum voltage limits at bus i.

These objective functions and the corresponding constraints form the core of the optimization model for low-carbon economic dispatch. By minimizing both carbon emissions and grid line losses, the model provides an efficient way to balance the goals of decarbonizing the power system and ensuring grid reliability, while optimizing for economic efficiency. The combined objective function Z could by solved by an optimisation technology like online programming, affine programming or mixed integer planning, depending on the complexity of the model and the system configuration.

3.2. Optimization Techniques

The Beluga Whale Optimization (BWO) algorithm offers several unique advantages that make it well-suited for low-carbon economic dispatch in power systems. By mimicking the natural behaviors of beluga whales, such as exploration and exploitation, BWO can effectively balance the objectives of minimizing carbon emissions and grid line losses in a complex power system environment. Below are the main principles of BWO:

3.2.1. Position Update Formula (Exploration Phase). During the discovery stage, the BWO updates each whale's position in response to stochastic movements, simulating the beluga's behavior in a wide search for food.

The position of each whale X_i is updated as:

$$X_i^{t+1} = X_i^t + \alpha \cdot r \cdot (X_{\text{best}}^t - X_i^t) \tag{9}$$

where X_i^{t+1} means the position of the i-th whale at the next iteration t+1; X_i^t means the current position of the i-th whale at iteration t; X_{best}^t represents the position of the best whale (the best solution found so far); α is a scaling factor that controls the exploration intensity; r is a random vector within [0,1] to introduce stochasticity and ensure diverse search directions. This formula allows the whales to explore the solution space by moving toward the best-known solution while maintaining random variability.

3.2.2. Encircling Behavior (Exploitation Phase). In the exploitation phase, the whales focus on a promising area in the search space to refine their position. This phase is based on the idea of "encircling" the best solution, a common behavior in natural hunting where animals converge on prey. The position update in this phase is given by:

$$X_i^{t+1} = X_{\text{best}}^t - \beta \cdot \left| \mathrm{D} \cdot X_i^t - X_{\text{best}}^t \right| \tag{10}$$

where β is a contraction factor (typically a random value between 0 and 1), controlling the degree of attraction toward the best solution; D is a vector that represents the distance between the current whale position and the best solution; X_{best}^t is the position of the best whale (the best solution found so far). This formula encourages the whales to move closer to the best solution, enhancing local search in promising areas.

3.2.3. Whale Falling (Fine-Tuning Phase. The whale falling or "diving" behavior represents a fine-tuning mechanism where whales adjust their positions in a final push toward the optimal solution, adding further precision to the search. The position update can be expressed as:

$$X_i^{t+1} = X_i^t + \gamma \cdot (X_i^t - X_{\text{best}}^t) + \delta \cdot \gamma \tag{11}$$

where γ is an adaptive scaling factor that gradually decreases over iterations, ehancing the convergence rate; δ is another factor controlling the influence of random movements, typically reduced in later iterations to focus on exploitation; γ denotes a random vector for stochasticity. This formula helps the algorithm fine-tune the solution in later stages, where convergence to the global optimum is more critical.

3.2.4. Fitness Evaluation. After each position update, the algorithm evaluates the fitness $f(X_i)$ of each whale (solution) based on the objective function. The best solution is updated according to:

$$X_{\text{best}}^{t+1} = \arg\min_i f(X_i^{t+1}) \tag{12}$$

where X_{best}^{t+1} is the updated best solution at iteration t+1 and means the fitness of the i-th solution (lower values indicate better fitness for minimization problems). This ensures that an optimal resolution discovered to date is retained and provides guidance for subsequent iterations.

3.2.5. Termination Condition. The algorithm continues updating positions and evaluating fitness until until stopping criteria are satisfied, for example, the largest iteration count $T_{\max}$ or a satisfactory solution is obtained $f(X_{\text{best}}) \le \varepsilon$, where ε is a small tolerance value.

4. Performance Evaluation and Discussions

Using the IEEE 57-bus test system provides a robust framework to validate the low-carbon economic dispatch model optimized by the BWO algorithm. This test system, commonly employed in power systems research, offers a realistic simulation environment with multiple buses, generators, and loads. It allows for comprehensive assessment of both the algorithm's performance and the model's effectiveness in reducing emissions and line losses under varying operational conditions. The specific parameters set are shown in Table 1.

Table 1. IEEE 57 Node System Simulation Parameter Settings

Parameters	D	t	β_{min}	β_{max}	K	X_{min}	X_{max}
Value	30	`50	0.2	1.2	1000	-100	100

4.1. Variations in load analysis for the IEEE 57 nodes throughout a 24-hour period.

For the IEEE 57-node system, develop a configuration file that captures the load variations throughout the day, including base load, peak load, off-peak load, and intermediate load conditions.

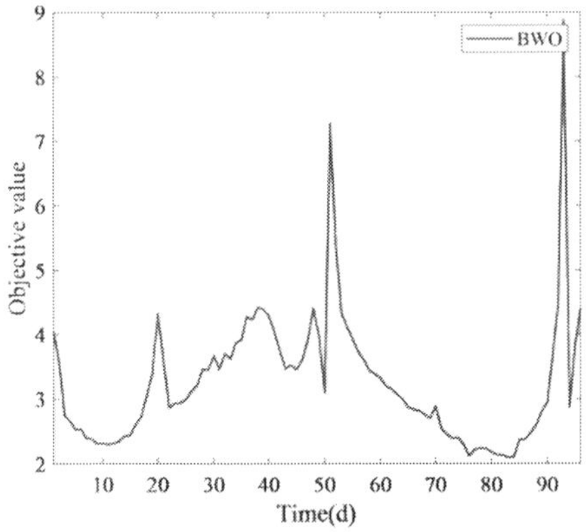

Figure 1. Objective function values of BWO applied to IEEE 57 nodes systems at different moments in a 24-hour period.

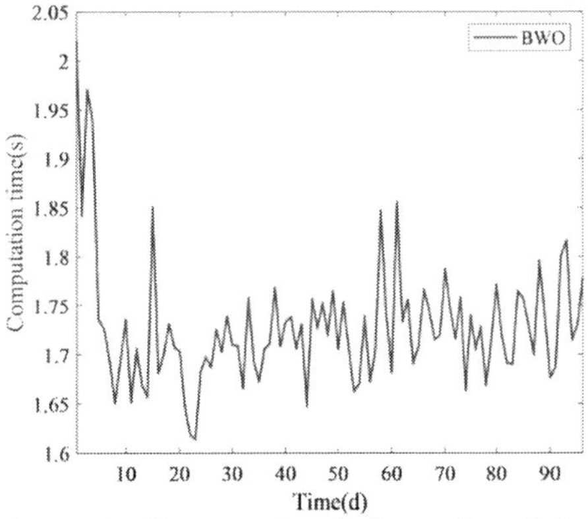

Figure 2. Computational time of applying BWO to IEEE 57 nodes systems at different moments in a 24-hour period.

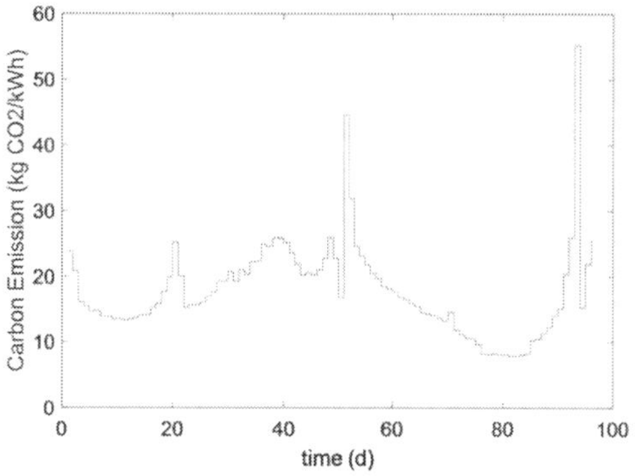

Figure 3. Carbon emissions data from the application of BWO to the IEEE 57 node system at different points in time over a 24-hour period.

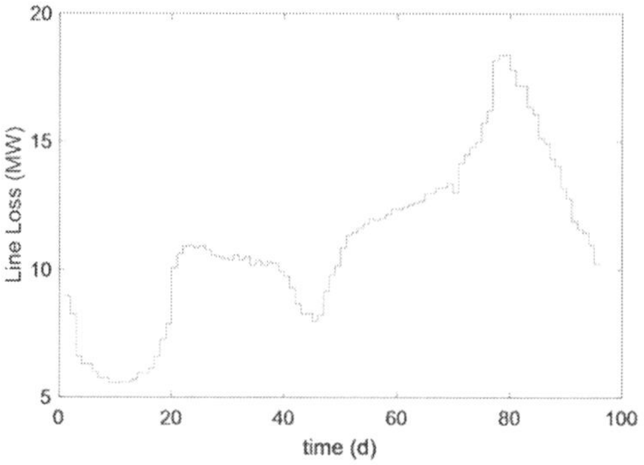

Figure 4. Line Losses of IEEE 57 node systems with BWO at different moments in a 24-hour period.

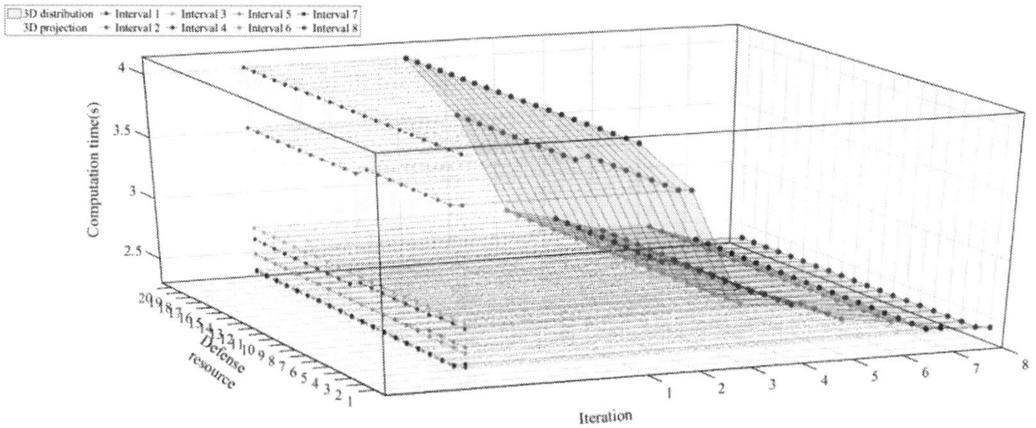

Figure 5. Changes in the goal of applying BWO to the IEEE 57 node system with 20 iterations at different moments over a 24 hour period.

From the above figures can be seen that the application of the BWO algorithm in this study offers a powerful approach for achieving low-carbon economic dispatch by leveraging its bio-inspired exploration and exploitation balance, adaptability to complex functions, and efficient convergence. These features allow BWO to manage the dual objectives of minimizing emissions and grid losses effectively, while ensuring scalability, robustness, and adaptability. By simulating the natural behaviors of beluga whales, BWO brings a novel and effective optimization strategy to this challenging area, providing valuable solutions to enhance efficiencies in electrical systems as well as sustainability.

5. Conclusion

This study proposes a low-carbon economic dispatch methodology for power systems that focuses on improving reliability and environmental performance by minimising carbon emissions and operational losses. The proposed model utilises the Beluga Whale Optimisation (BWO) algorithm to achieve an optimal dispatch solution that effectively balances economic and environmental objectives. The effectiveness of this approach is confirmed through simulation validation on an IEEE 57-node power system, showing that the BWO algorithm not only significantly reduces carbon emissions and line losses, but also has an extremely short computation time. This computational efficiency is a critical advantage for real-time or near-real-time applications, making the BWO algorithm highly suitable for practical power system dispatch.

Overall, this methodology demonstrates significant potential in supporting the transition toward low-carbon power systems, providing a calculable and adaptable solution to meet the challenges of sustainable energy management in modern grids.

Acknowledgments
The authors gratefully acknowledge the support of the technology project of China Southern Power Grid Company Limited [030000KC23070019(GDKJXM20230813)]

References

[1] Ripple W J, Wolf C, Gregg J W, et al. The 2023 state of the climate report: Entering uncharted territory[J]. BioScience, 2023, 73(12): 841-850.
[2] Goldstein B, Gounaridis D, Newell J P. The carbon footprint of household energy use in the United States[J]. Proceedings of the National Academy of Sciences, 2020, 117(32): 19122-19130.
[3] Zhang X, Cui X, Li B, et al. Immediate actions on coal phaseout enable a just low-carbon transition in China's power sector[J]. Applied Energy, 2022, 308: 118401.
[4] Rahman M M, Sultana N, Velayutham E. Renewable energy, energy intensity and carbon reduction: Experience of large emerging economies[J]. Renewable Energy, 2022, 184: 252-265.
[5] Tian X, An C, Chen Z. The role of clean energy in achieving decarbonization of electricity generation, transportation, and heating sectors by 2050: A meta-analysis review[J]. Renewable and Sustainable Energy Reviews, 2023, 182: 113404.
[6] Abttan, Rana Ali, Adnan Hasan Tawafan, and Samar Jaafar Ismael. "Economic dispatch by optimization techniques." International Journal of Electrical and Computer Engineering 12.3 (2022): 2228-2241.
[7] Cerda-Flores, Sandra C., Arturo A. Rojas-Punzo, and Fabricio Nápoles-Rivera. "Applications of multi-objective optimization to industrial processes: a literature review." Processes 10.1 (2022): 133.
[8] Kumar, Jasvinder, et al. "Solver-based mixed integer linear programming (MILP) based novel approach for hydroelectric power generation optimization." IEEE Access 8 (2020): 174880-174892.
[9] Katoch, Sourabh, Sumit Singh Chauhan, and Vijay Kumar. "A review on genetic algorithm: past, present, and future." Multimedia tools and applications 80 (2021): 8091-8126.
[10] Shami, Tareq M., et al. "Particle swarm optimization: A comprehensive survey." Ieee Access 10 (2022): 10031-10061.
[11] Houssein E H, Sayed A. Dynamic candidate solution boosted beluga whale optimization algorithm for biomedical classification[J]. Mathematics, 2023, 11(3): 707.
[12] Taghizad-Tavana K, Alizadeh A, Ghanbari-Ghalehjoughi M, et al. A comprehensive review of electric vehicles in energy systems: Integration with renewable energy sources, charging levels, different types, and standards[J]. Energies, 2023, 16(2): 630.
[13] Farghali M, Osman A I, Chen Z, et al. Social, environmental, and economic consequences of integrating renewable energies in the electricity sector: a review[J]. Environmental Chemistry Letters, 2023, 21(3): 1381-1418.

Journal of Physics: Conference Series **3012** (2025) 012005 doi:10.1088/1742-6596/3012/1/012005

Research on Multi-source Data Fusion Methods in Distribution Networks

Xiting Guo*, Youqin Wang, Tiliang Wang, Siyao Zhang, Jie Li

State Grid Fujian Electric Power Company Ningde Power Supply Company, 352100, P.R. China

*Corresponding author's e-mail: 3829414024@qq.com

Abstract. Multi-source data represents a complex data type. This article proposes a method for the comprehensive utilization of multi-source data suitable for distribution network line loss calculation. Firstly, to facilitate data fusion, multi-source measurement data is converted. Secondly, to ensure consistency in measurement time snapshots, a specific measurement time from a Distribution Phasor Measurement Unit (D-PMU) is selected as the benchmark. SCADA data undergoes time registration and data interpolation, while smart meter data is time-aligned using a combination of "measured values + predicted values." Subsequently, the time-series data is filtered to obtain more accurate distribution network data, and then multi-source data fusion is achieved based on the Dempster-Shafer (D-S) evidence theory. By conducting power flow state analysis based on the high-quality fused data, distribution network data can be utilized in a more refined manner, addressing the shortcomings of traditional methods. Finally, a 10kV distribution network is selected as a case study to obtain the power flow state using the multi-source data fusion method, upon which the distribution network line loss is calculated. The results demonstrate that the proposed method can effectively accomplish the comprehensive utilization of multi-source data and improve the accuracy of distribution network data applications.

1. Introduction

The Distributed Synchronous Phasor Measurement Units (D-PMUs), SCADA systems, and AMI (Advanced Metering Infrastructure) deployed in distribution networks hold significant value for their respective departments[1-4].

The original time scales and measured quantities collected by various measurement systems are different, making it impossible to compare and match them with each other.

In response to the aforementioned issues, numerous scholars have conducted research on the integrated application of multi-source data in distribution networks, focusing on the design of fusion mechanisms and the optimization of fusion algorithms. Literature[5] utilizes the linear characteristics of PMU measurement equations to decompose SCADA measurement equations into linearized equations, thereby simplifying calculations.

In summary, most scholars typically consider data fusion for two types of measurement without exploring the expansion of its application scenarios. In the future, D-PMU, SCADA systems, and AMI systems will coexist in distribution networks for a long time.

Considering the issues of traditional line loss calculation methods, this paper proposes a new data preprocessing mechanism that comprehensively utilizes information from the D-PMU, SCADA, and AMI measurement systems in distribution networks. Based on this, the D-S evidence theory method is selected for multi-source data fusion in distribution networks. Driven by both massive data and theoretical models, line loss calculation and analysis in distribution networks are conducted based on multi-source data fusion. This approach makes use of data information from multiple perspectives and in-depth, improving calculation accuracy and aligning more closely with actual operations.

2. Conversion of Multi-Source Data in Distribution Networks

2.1. Comparison of Measurement Systems

Currently, the measurement terminals equipped in distribution networks mainly include SCADA systems, D-PMUs, and AMI measurement systems. These three measurement systems have certain differences in terms of their access system locations, measured state quantities, transmission delays, measurement errors, sampling frequencies, and so on. A comparison of the three measurement systems is shown in Table 1[6].

Table.1 Comparison of measurement systems

Measurement Systems	D-PMU	SCADA	AMI
Measurement Category	Node Voltage Phasor Branch Current Phasor	Node Voltage Magnitude Branch Current Magnitude Node Injection Power Branch Power	User Electricity Consumption Information Node Voltage Magnitude and Branch Current Magnitude Node Injection Power and Branch Power
Time Stamp	Exist	Non-exist	Exist
Time Delay	Small	Larger	Large
Measurement Error	Mag < 0.2% Pha < 0.5°	1%	1%
Sampling Period	5-20ms	2-4s	15min

2.2. D-PMU Measurement Conversion

The node voltage phasor and the associated branch current phasor can be directly measured through D-PMU, he associated branch active power P_{ij}, reactive power Q_{ij} as shown below respectively:

$$\begin{aligned} P_{ij} &= U_i I_{ij}\left(\cos\theta_i \cos\theta_{ij} + \sin\theta_i \sin\theta_{ij}\right) \\ Q_{ij} &= U_i I_{ij}\left(\sin\theta_i \cos\theta_{ij} - \cos\theta_i \sin\theta_{ij}\right) \end{aligned} \tag{1}$$

In this formula, U_i, θ_i is the voltage magnitude and phase measured for Node i, respectively; I_{ij} is the current magnitude and phase measured for branch ij,respectively.

When converted to the rectangular coordinate system, the real and imaginary parts of the node voltage phasor are transformed as follows:

$$\begin{aligned} U_{ir} &= U_i \cos\theta_i \\ U_{ii} &= U_i \sin\theta_i \end{aligned} \tag{2}$$

Similarly, the real and imaginary parts of the measured current phasor can be transformed as follows:

$$\begin{aligned} I_{ir} &= I_i \cos\theta_i \\ I_{ii} &= I_i \sin\theta_i \end{aligned} \tag{3}$$

2.3. Measurement Conversion of SCADA and AMI

According to Equations (2) and (3), the active and reactive powers measured by SCADA or AMI for a branch can be transformed into the real and imaginary parts of the branch's current:

$$\begin{aligned} I_{ijr} &= \frac{P_{ij}U_{ir} + Q_{ij}U_{ii}}{U_{ir}^2 + U_{ii}^2} \\ I_{iji} &= \frac{P_{ij}U_{ii} - Q_{ij}U_{ir}}{U_{ir}^2 + U_{ii}^2} \end{aligned} \tag{4}$$

In this formula, P_{ij} and Q_{ij} represent the active power and reactive power, respectively, measured by SCADA or AMI for the branch ij.

Similarly, transform P_{ij} and Q_{ij}, which are measured at a node by SCADA or AMI, into the real and imaginary parts of the node's current.

$$\begin{aligned} I_{ir} &= \frac{P_iU_{ir} + Q_iU_{ii}}{U_{ir}^2 + U_{ii}^2} \\ I_{ii} &= \frac{P_iU_{ii} - Q_iU_{ir}}{U_{ir}^2 + U_{ii}^2} \end{aligned} \tag{5}$$

3. Data fusion strategy

3.1. Data pre-processing mechanism

This paper focuses on D-PMU measurements, with SCADA and AMI measurements serving as auxiliary measurements.

This paper aims to integrate and utilize data from these three sources and proposes the following strategy.

A D-PMU measurement moment is taken as the benchmark for multi-source measurement data fusion, assuming a PMU sampling period of 10ms. This paper assumes that there will be no abrupt changes in the state of the distribution network.

This paper proposes a novel approach for time synchronization of SCADA data. Not every sampling interval contains SCADA data. To address this problem, this paper selects the Lagrange low-order linear interpolation algorithm for data imputation[7]. This interpolation method mainly obtains the variation pattern through piecewise linear fitting of the data., as shown in Figure 1.

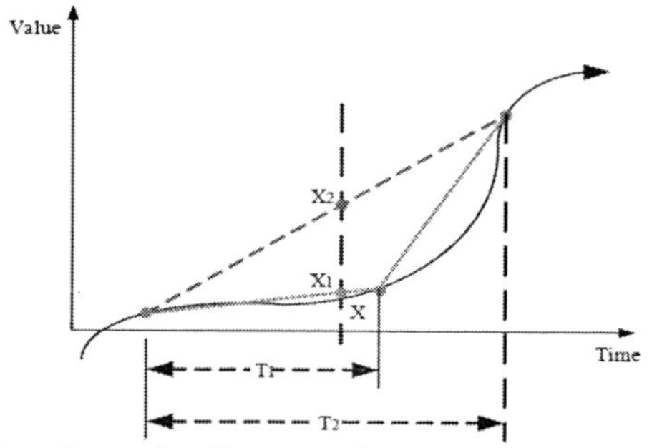

Figure.1 The schematic diagram of Lagrange linear interpolation

As shown in Table 1, the sampling period of the SCADA system is 2-4 seconds, and in this paper, the interpolation interval for SCADA data is set to 2 seconds. Accordingly, the interval interpolation formula is:

$$M_{\text{SCADA}}^{\varphi,T_{\text{i}}} = M_{\text{SCADA}}^{\varphi,T_{\text{ib}}} \frac{T_{\text{i}} - T_{ie}}{T_{\text{ib}} - T_{ie}} + M_{\text{SCADA}}^{\varphi,T_{ie}} \frac{T_{\text{i}} - T_{\text{ib}}}{T_{ie} - T_{ib}} \tag{6}$$

In this formula, φ is the phase label, T_{i} is the interpolation time point for the target, T_{ib} is the time point at the beginning of the interpolation interval, T_{ie} is the time point at the end of the interpolation interval, and $T_{\text{ib}} < T_{\text{i}} < T_{\text{ie}}$; M is the value under the aforementioned relevant subscript.

Below is a detailed description of the data processing steps for the AMI system in this paper.

Smart meters are an important component of the AMI system. In this paper, relying on historical load data, we adopt the method of "measured value + predicted value" for smart meters to achieve timestamp alignment.

For the AMI measurement data required at time t, this paper utilizes historical data and employs a Long Short-Term Memory (LSTM) neural network to predict user power, taking comprehensive consideration of the credibility of both the predicted data and the AMI sampling data[8-9]used measurement value z_t obtained after data fusion is:

$$z_t = (1-q)z_c + q \cdot z_v^f \tag{7}$$

In this formula, $q = e^{-\left|z_c^f - z_c\right|}$, is an exponential function with base "e", used to measure the credibility of the predicted value. The larger the prediction error, the lower the credibility, and thus the less reliance on the predicted value. z_c^f is the predicted value of moment "c", z_c is the measured value of moment "c", v represents the first acquisition time point of AMI after time t, z_v^f is the predicted value of moment "v". This method can greatly exploit the value and potential of historical load data.

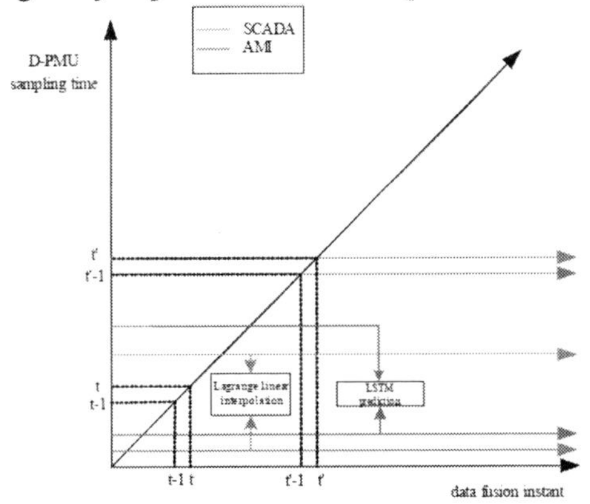

Figure 2. The schematic diagram of multi-source data fusion strategy

As shown in Figure 2, based on the D-PMU sampling time points, when performing data fusion within the interval [t-1, t], since there are no SCADA system and AMI sampling data within this interval, it is necessary to perform Lagrange linear interpolation on the SCADA measurement data according to Equation (6) and combine the AMI measurement data according to Equation (7) to obtain the required data. When performing data fusion within the interval [t'-1, t'], both SCADA and AMI data exist within this interval, and only time stamping is required for the SCADA data, while the AMI data is processed in the same way as above.

Noise reduction is performed on the time series data, which can be expressed using the formula:

$$\begin{cases} Mea_i = \alpha y_i + (1-\alpha)(Mea_{i-1} + trend_{i-1}) \\ trend_i = \beta(Mea_i - Mea_{i-1}) + (1-\alpha_2)t_{i-1y} \\ y_{i+h} = Mea_i + h \times trend_i \end{cases} \tag{8}$$

In this formula, Mea_i represents the true value of the time series data; y_i represents the first smoothed value of the data; $trend_{i-1}$ represents the second smoothed value of the data; α and β are the smoothing parameters for the trend of the data, respectively h represents the prediction step length of the data.

The data obtained after noise filtering through the aforementioned data preprocessing mechanism can provide a high-quality dataset that is urgently needed for multi-source data fusion in distribution networks.

3.2. Data fusion strategy

Based on the multi-source data of the distribution network that have undergone operations such as time scale alignment and data completion in Section 2.1, this paper selects the D-S evidence theory method to fuse and utilize the multi-source data of the distribution network.

The D-S evidence theory method has significant advantages such as a solid and complete theoretical foundation, simple structure, and ease of implementation. It can synthesize the information represented by multi-source data and reasonably represent "uncertainty," making it widely used in the field of data fusion[10]. The data fusion process using the D-S evidence theory method is as follows:

3.2.1. Frame of discernment

Assuming the existence of a hypothetical empty set $\mathbf{\Theta}$ as the frame of discernment for the evidence theory, it contains $\mathbf{N}$ independent and mutually exclusive hypotheses.

$$\mathbf{\Theta} = \{\mathbf{H}_1, \mathbf{H}_2, \cdots, \mathbf{H}_N\} \tag{9}$$

The power set of $\mathbf{\Theta}$, denoted as $2^{\mathbf{\Theta}}$, consists of all events of the frame of discernment $\mathbf{\Theta}$.

3.2.2. Basic Probability Assignment Function

Basic Probability Assignment (BPA) is a function under mapping $m: 2^{\mathbf{\Theta}} \to [0,1]$, satisfying the following constraints:

$$\begin{cases} m(\varnothing)=0 \\ \sum_{\mathbf{A}\subseteq\mathbf{\Theta}} m(\mathbf{A})=1 \end{cases} \tag{10}$$

In the formula, event $\mathbf{A}$ is any subset of $\mathbf{\Theta}$, $m(\mathbf{A})$ is used to quantify the degree to which evidence $\mathbf{E}$ supports event $\mathbf{A}$. Any event $\mathbf{A}$ that satisfies $m(\mathbf{A}) > 0$ is referred to as a focal element.

3.2.3. Belief Function

Belief Function (Bel), represents the lower bound of the interval for the degree of belief in the truth of event $\mathbf{A}$, defined as:

$$Bel(\mathbf{A}) = \sum_{\mathbf{B}\subseteq\mathbf{A}} m(\mathbf{B}) \qquad \mathbf{A},\mathbf{B} \subseteq \mathbf{\Theta} \tag{11}$$

In the formula, both events $\mathbf{A}$ and $\mathbf{B}$ are any subsets of $\mathbf{\Theta}$, and m is the basic probability assignment function.

Journal of Physics: Conference Series **3012** (2025) 012005 doi:10.1088/1742-6596/3012/1/012005

3.2.4. Plausibility Function

Plausibility Function (Pl), represents the upper bound of the interval for the degree of belief in the truth of event $\mathbf{A}$, defined as:

$$Pl(\mathbf{A}) = \sum_{\mathbf{B} \cap \mathbf{A} \neq \varnothing} m(\mathbf{B}) \qquad \mathbf{A}, \mathbf{B} \subseteq \mathbf{\Theta} \tag{12}$$

3.2.5. Combination Rule

$m_i (i = 1, 2, \cdots, n)$ is the basic probability assignment function for a certain evidence $\mathbf{E}$ under the framework of discernment. Specifically for multi-source information in power distribution networks, m_i represents the probabilities of the four events "D-PMU measurement is the true value", "SCADA system measurement is the true value", "smart meter measurement is the true value", and "uncertainty" occurring under the support of evidence $\mathbf{E}$. The fusion result is:

$$\mathrm{m}(\mathbf{A}) = \begin{cases} \frac{1}{1-K} \sum_{\mathbf{A}_1 \cap \mathbf{A}_2 \cap \mathbf{A}_3 \cap \mathbf{A}_4 = \mathbf{A}} m_1(\mathbf{A}_1) \cdot m_2(\mathbf{A}_2) \cdot m_3(\mathbf{A}_3) \cdot m_4(\mathbf{A}_4) \\ 0, \mathbf{A} = \varnothing \end{cases} \tag{13}$$

In this formula, $\mathbf{K}$ is the evidence conflict factor,

$$\mathbf{K} = \sum_{\mathbf{A}_1 \cap \mathbf{A}_2 \cap \mathbf{A}_3 \cap \mathbf{A}_4 = \varnothing} m_1(\mathbf{A}_1) \cdot m_2(\mathbf{A}_2) \cdot m_3(\mathbf{A}_3) \cdot m_4(\mathbf{A}_4) \tag{14}$$

The larger the value of $\mathbf{K}$, the deeper the contradiction between the evidences. When $\mathbf{K} = 1$, there is complete conflict between the evidences, and the combination is meaningless.

In summary, the multi-source data fusion process for power distribution networks proposed in this paper is shown in Figure 3.

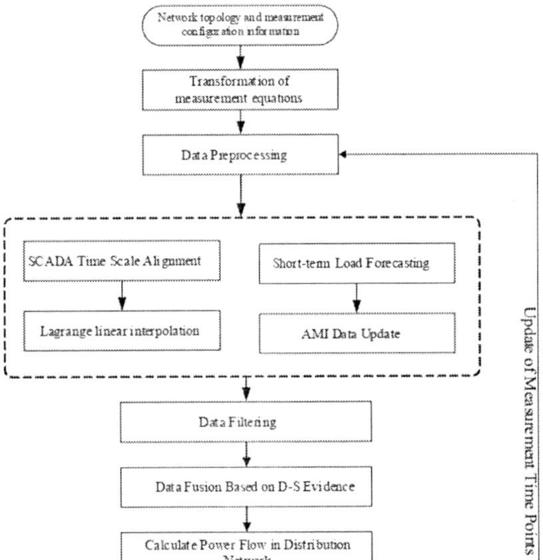

Figure 3. The flow chart of distribution network multi-source data fusion

3.3. Verification of Effectiveness

To verify the effectiveness and reliability of the multi-source data processing and fusion technology for power distribution networks proposed in this paper, we take measured voltage data from power systems as an example and select actual voltage operation data and measurement terminal data from a power distribution network in a certain region for data fusion and analysis of multi-source data processing results.

To ensure that the experimental results are more illustrative, we select the more common weighted average method, Kalman filtering method, and D-S evidence theory method for performance comparison. The following three indicators are chosen to measure the effectiveness of the three data fusion methods: data processing accuracy, which evaluates the ability of the preprocessed data to retain the original information; fusion time consumption, which measures the time taken to complete multi-source data fusion; and data fusion accuracy, which assesses the discrepancy between the data processed by multi-source fusion techniques and the actual values. The comparison results of the three methods are shown in Table 2.

Table.2 Comparison of effectiveness of the three data fusion methods

	Data Processing Accuracy	Time Consumption	Data Fusion Accuracy
Weighted Average	83%	2.6s	55%
Kalman Filtering	93%	3.3s	83%
D-S Evidence	97%	3s	90%

The fusion error was defined as the difference between the fused data and the true value, and the fusion accuracy was represented by the magnitude of its variance. The results obtained are shown in Figure 4.

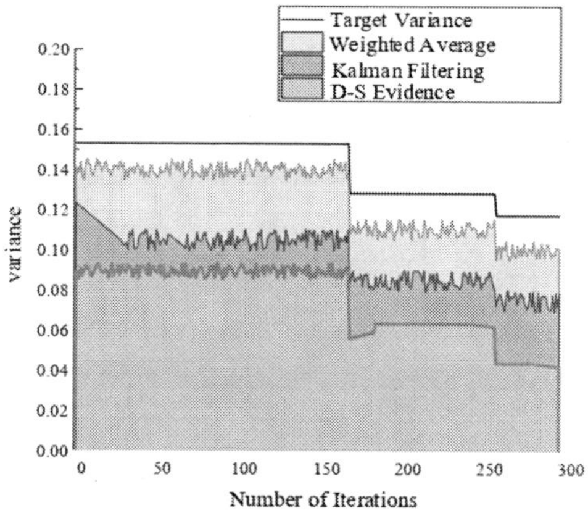

Figure 4. Comparison of the three data fusion methods variance

As shown in Figure 4, all three data fusion methods can achieve multi-source data fusion in power distribution networks and meet the variance requirements. However, the weighted average method, which is greatly influenced by the setting of weight values, experiences frequent fluctuations in its variance around a certain value, resulting in unstable data with high dispersion and large errors. The Kalman filtering method has low requirements for data accuracy and loses some valid information, leading to a slight reduction in variance but with equally obvious waveform oscillations. Its fusion effect is slightly better than the weighted average method. In contrast, the D-S evidence theory method selected in this paper has the smallest variance, with a relatively stable curve and stable and accurate fusion results. Experiments show that the D-S evidence theory method, from different perspectives, effectively fuses and utilizes multi-source data in power distribution networks, fully tapping the potential of the data and enhancing its application value.

4. Conclusion

Based on D-PMU, operations such as time synchronization, data completion, and noise filtering are performed on data obtained from the SCADA system and AMI measurement system to construct a high-quality multi-source data set for power distribution networks. Subsequently, fusion of multi-source heterogeneous data is achieved based on the D-S evidence theory method. The research shows that:

The preprocessing mechanism designed in this paper for multi-source data in power distribution networks can provide a large amount of high-quality data for data fusion, increasing the accuracy of the fusion.

The D-S evidence theory method selected in this paper can better retain the original information of multi-source data in power distribution networks, with high data fusion accuracy. By conducting power distribution network line loss analysis based on multi-source fused data, the value of the data is fully exploited, the calculation accuracy is improved, and it aligns more closely with actual operational conditions.

Acknowledgments

The research in this thesis was supported by the "Science and Technology Project of State Grid Fujian Electric Power Co., Ltd.", with the project name being "Research and Development of Key Technologies and Equipment for Active Voltage Support by Renewable Energy Stations in Distribution Networks," and the project code being 52139024000B.

References

[1] Wei X C, Qin L Y, Fang C, et al ,2023, Review of the synchronous measurement technology and its application in distribution networks[J]. Electric Power Construction, 44(05): 13-22.

[2] Zhang G, Deng X Z, Ma X W, et al ,2023, Online prediction of extreme deviation of physical-data fusion frequency based on wide-area measurements[J]. Electric Power Construction, 44(08): 1-12.

[3] Deng X Y, Wang X B, CAO H Z, et al ,2023, PMU abnormal data identification algorithm based on stream clustering[J]. Electric Power Engineering Technology, 42(04): 167-174.

[4] MA L, Lou J H, Dong P Y, et al ,2023, Location of voltage sag source in radial distribution network based on μPMU and binary search method[J]. Electric Power Engineering Technology, 42(05): 177-184.

[5] Li C, Wei Z N, Ni M, et al ,2014, A PMU measurement based stepwise linear state estimation model[J]. Power System Technology, 38(06): 1700-1704.

[6] Meng Y T,2020, Research on distribution network state estimation in hybrid measurement environment[D]. Shanghai: Shanghai Jiao Tong University.

[7] Shi C H, Dai W Y, Wan J Y, et al ,2022, Station area line loss prediction method based on long short-term memory network autoencoder[J]. Distribution & Utilization, 39(01): 52-57.

[8] Zhu R Z, Wang D J ,2023, Power prediction of photovoltaic system based on LSTM neural network[J]. Electric Power Technology and Environmental Protection, 39(03): 201-206.

[9] Wang H B, Wu S T, Zhou W H, et al ,2023, Research on distribution network state estimation based on different measurement data fusion[J]. Mechanical and Electrical Information (04): 12-16.

[10] Li Q, Zhang T Y ,2022, Research on Multi-source data fusion technology of energy internet[J]. Electric Drive, 52(15): 68-73.

Journal of Physics: Conference Series **3012** (2025) 012006 doi:10.1088/1742-6596/3012/1/012006

Research on Harmonic State Estimation in Distribution Networks Based on Optimal Configuration of D-PMU

Chuanliang Zheng*, **Jiawei Ye, Xinhui Fu, Lanlan Wei, Tao Zheng**

State Grid Fujian Electric Power Company Ningde Power Supply Company, 352100, PR China

*Corresponding author's e-mail: 3957331805@qq.com

Abstract. Along with the increase of a massive number of power electronic equipment, the problem of harmonics in the distribution network is becoming increasingly serious. The accuracy of harmonic state estimation directly affects the subsequent harmonic control effect. The distribution phasor measurement unit (D-PMU) can measure the node voltage and branch current in real time, and can be used to estimate the harmonic state. However, at present, the price of D-PMU is truly high. How to carry out reasonable optimization placement to ensure that the harmonic state of the whole distribution network is considerable, while reduce errors of harmonic state estimation, is an urgent problem to be solved. Firstly, a D-PMU optimal configuration model with the highest accuracy of D-PMU economic configuration and harmonic state estimation accuracy is constructed, and an improved binary particle swarm-genetic mixing algorithm is proposed for solving. Then, the IEEE14 node model is built in the real-time simulator. The mean interpolation method and Vondrak filter method are used for data processing and the influence of optimized D-PMU configuration scenarios on harmonic state estimation is analyzed. The results show that the proposed algorithm can give a reasonable D-PMU configuration scheme from the perspective of reducing investment cost and reducing harmonic state estimation error, which is helpful to support engineering decision-making.

1. Introduction

In recent years, the rapid development of power electronics technology has brought about a large amount of harmonic pollution, which affects the normal and stable operation of the distribution network[1]. Harmonic state estimation is the foundation and key to implementing harmonic control[2]. As an advanced measuring device in the current distribution network, D-PMU can measure node voltage and branch current in real time, effectively supporting harmonic state estimation.

The article constructs a D-PMU optimization configuration model with the observable harmonic state of the entire network as the constraint. Improved the binary particle swarm optimization algorithm and built an IEEE 14 node model for calculation and analysis in a real-time simulator.

2. Harmonic state estimation model

The mathematical model for harmonic state estimation is shown in the following equation[1]:

$$\mathbf{Z}(h) = \mathbf{H}(h)\mathbf{X}(h) + \boldsymbol{\eta}(h) \quad \text{* MERGEFORMAT (1)}$$

In the formula, $\mathbf{Z}(h)$ is the measurement matrix for subharmonic quantities; $\mathbf{H}(h)$ is the measurement matrix; $\mathbf{X}(h)$ is the estimated state quantity; $\boldsymbol{\eta}(h)$ is the measurement errors.

In the distribution network, the measurement equations for each variable are as follows[1]:

1) Node harmonic voltage measurement equation:

$$\mathbf{U}_{\mathrm{m},i} = \mathbf{E}\mathbf{U}_{\mathrm{S},i} + \boldsymbol{\eta}_i \quad \text{* MERGEFORMAT (2)}$$

In the formula, $\mathbf{U}_{\mathrm{m},i}$ is the measurement of the harmonic voltage at the node, $\mathbf{U}_{\mathrm{S},i}$ is the state variable of the harmonic voltage at the node, $\boldsymbol{\eta}_i$ is the measurement error.

2)Measurement equation for injecting harmonic current into nodes:

$$\mathbf{I}_{\mathrm{m},ii} = y_{ii}\mathbf{U}_{\mathrm{S},i} + \sum_{j=1}^{\mathrm{n_c}} y_{ij}\mathbf{U}_{\mathrm{S},j} + \boldsymbol{\eta}_{ii} \quad \text{* MERGEFORMAT (3)}$$

In the equation, $\mathbf{I}_{\mathrm{m},ii}$ represents the measurement of harmonic current injected into node i, y_{ii} represents the self admittance of node i, y_{ij} represents the mutual admittance between nodes i and j, and n_c represents the number of nodes connected to node i.

3) Measurement equation for branch harmonic current

$$\mathbf{I}_{\mathrm{m},ij} = y_{ij}\mathbf{U}_{\mathrm{S},i} - y_{ij}\mathbf{U}_{\mathrm{S},j} + \boldsymbol{\eta}_{ij} \quad \text{* MERGEFORMAT (4)}$$

In the formula, $\mathbf{I}_{m,ij}$ represents the harmonic current side of the branch from node i to node j, and the rest is the same as above.

When solving the measurement equation, the article uses weighted least squares method for calculation. The objective function can be written as:

$$\mathbf{J}(\mathbf{X}) = (\mathbf{Z} - \mathbf{H}\mathbf{X})^T \mathbf{R}^{-1}(\mathbf{Z} - \mathbf{H}\mathbf{X}) \quad \text{* MERGEFORMAT (5)}$$

In the equation, R is the variance matrix of measurement error, as shown in equation (6)

$$\mathbf{R} = \begin{bmatrix} \sigma_1^2 & & & \\ & \sigma_2^2 & & \\ & & \ddots & \\ & & & \sigma_m^2 \end{bmatrix} \quad \text{* MERGEFORMAT (6)}$$

To minimize measurement error, then:

$$\frac{\partial \mathbf{J}(\mathbf{X})}{\partial \mathbf{X}} = 0 \quad \text{* MERGEFORMAT (7)}$$

By solving equation (7):

$$\left(\mathbf{H}^T\mathbf{R}^{-1}\mathbf{H}\right)\hat{\mathbf{X}} = \mathbf{H}^T\mathbf{R}^{-1}\mathbf{Z} \quad \text{* MERGEFORMAT (8)}$$

$$\hat{\mathbf{X}} = \left(\mathbf{H}^T\mathbf{R}^{-1}\mathbf{H}\right)^{-1}\mathbf{H}^T\mathbf{R}^{-1}\mathbf{Z} \quad \text{* MERGEFORMAT (9)}$$

In the formula, $\hat{\mathbf{X}}$ is the estimated value.

Journal of Physics: Conference Series **3012** (2025) 012006 doi:10.1088/1742-6596/3012/1/012006

3. D-PMU Optimization Configuration Based on Improved Binary Particle Swarm Optimization Algorithm

3.1. D-PMU Optimization Configuration Theory

3.1.1. Working principle and measurement characteristics of D-PMU. D-PMU is a synchronous phasor measurement device designed based on GPS, which has measurement characteristics such as precise data synchronization, direct measurement of corresponding node data, high sampling frequency, and fast data upload and download speed[3]. The working principle of D-PMU is shown in Figure 1.

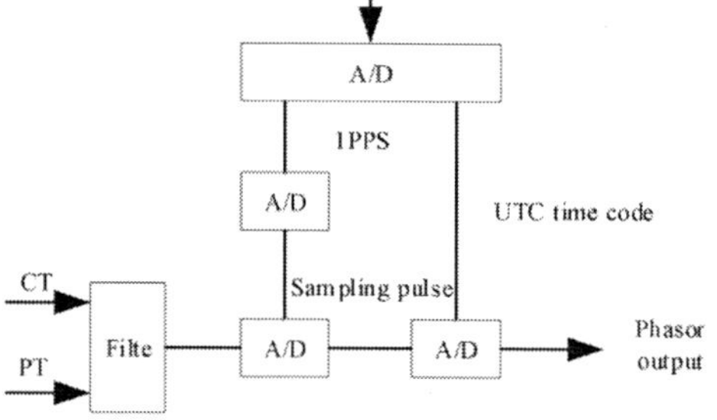

Figure 1. The working principle of D-PMU

*3.1.2. Theoretical basis for D-PMU optimization configuration.*Ohm's law states that the voltage drop between two nodes is equal to the product of the current between the nodes and the line impedance. As shown in equation (10):

$$V_A - V_B = I_{AB} \times Z_{AB} \qquad \text{* MERGEFORMAT (10)}$$

If a D-PMU is installed at node A , it can measure the voltage V_A and branch current I_{AB} of node A , and the impedance Z_{AB} is a constant value. If these three variables are known, the node B voltage V_B can be indirectly calculated.

Due to the unique measurement characteristics of D-PMU, the theoretical basis for optimizing D-PMU configuration can be obtained as follows[4]: The voltage of nodes equipped with D-PMU and the current of their adjacent branches can be directly measured by D-PMU; If one end node of a branch is equipped with a D-PMU, the voltage of the other node can be indirectly obtained through equation (1); If the voltage at both ends of a branch is known, the current can be indirectly obtained through equation (1); If there is only one unknown current flowing in and out of a certain node, the magnitude and direction of the current can be indirectly calculated by Kirchhoff's current law.

3.2. Observable rules for harmonics

The nodes connected to the load are designated as "normal nodes", while nodes not connected to the load or generator are designated as "zero injection nodes". Based on the D-PMU measurement rules, it is reasonable to infer the rules for harmonic observability analysis[5]:

1) The harmonic state of ordinary nodes is observable: directly known; One or more busbars connected to the busbar are directly known.

2) Zero injection node harmonic state is observable: directly known; All nodes connected to the zero injection node are directly or indirectly known.

3) The harmonic state of nodes connected to zero injection nodes is observable; Zero injection nodes and other busbars connected to the zero injection bus are directly or indirectly known.

3.3. D-PMU Optimization Configuration Model

$$\min f(x) = \min\left(a\sum_{i=1}^{n} \varepsilon_i + b\sum_{i=1}^{n} \left|\mathbf{X}_i - \hat{\mathbf{X}}_i\right| \right) \qquad \text{* MERGEFORMAT (11)}$$

$$\begin{cases} \text{s.t. } \mathbf{Y}*\varepsilon \geq \mathbf{I} \\ \mathbf{Z}(h) = \mathbf{H}(h)\mathbf{X}(h) + \boldsymbol{\eta}(h) \end{cases} \qquad \text{* MERGEFORMAT (12)}$$

$$\mathbf{Y}(i,j) = \begin{cases} 1, i = j \\ 1, i \textit{ connects with } j \\ 0, \textit{others} \end{cases} \qquad \text{* MERGEFORMAT (13)}$$

In the equation, variable ε_i represents whether node i is equipped with a D-PMU. If node i is configured with a D-PMU, then $\varepsilon_i = 1$, otherwise $\varepsilon_i = 0$; $\tilde{\mathbf{X}}_i$ and $\mathbf{X}_i$ represent actual measured values and estimated values obtained through harmonic states, respectively; A and b are two weight factors, set different weights for the number of D-PMU installed and the harmonic state estimation error; $\mathbf{I}$ is the identity matrix.

3.4. Improved binary particle swarm genetic hybrid algorithm

This article considers combining binary particle swarm optimization algorithm with genetic algorithm to solve D-PMU optimization model. Kennedy et al. proposed the Binary Particle Swarm Optimization (BPSO) algorithm, and the particle velocity update formula is as follows:

$$V_{id}^{t+1} = \omega \times V_{id}^{t} + c_1 \times r_1 \times \left(\mathbf{X}_{id-\text{pbest}}^{t} - \mathbf{X}_{id}^{t}\right) + c_2 \times r_2 \times \left(\mathbf{X}_{id-\text{gbest}}^{t} - \mathbf{X}_{id}^{t}\right) \qquad \text{* MERGEFORMAT (14)}$$

In the BPSO algorithm, as the name suggests, particle position values can only be 0 or 1. The formula for updating particle position is:

$$\mathbf{X}_{id}^{t+1} = \begin{cases} 1, r < S\left(V_{id}^{t}\right) \\ 0, r \geqslant S\left(V_{id}^{t}\right) \end{cases} \qquad \text{* MERGEFORMAT (15)}$$

In the formula, r is a random number in the interval of $[0,1]$; $S(x)$ is the Sigmoid function:

$$S(x) = \frac{1}{1+e^{-x}} \qquad \text{* MERGEFORMAT (16)}$$

The BPSO algorithm can easily lead to local optima in search and solution. To overcome this limitation, the binary particle swarm algorithm process is improved. Firstly, change the expression of the inertia coefficient to a function that varies with the number of iterations, as shown in equation (17):

$$\omega(t) = \omega_{\max} - (\omega_{\max} - \omega_{\min}) \times e^{\left(\frac{t}{n_{\max}} - 1\right)} \qquad \text{* MERGEFORMAT (17)}$$

In the formula, $\omega_{\max}$ and $\omega_{\min}$ represent the upper and lower limits of the inertia coefficient, which are set to 0.95 and 0.45 respectively in this paper; t, $n_{\max}$ represents the current number of steps and the total number of steps in the iterative program.

The improved inertia coefficient can avoid falling into local optima. Specifically in the process of searching for solutions, in the early stages of iteration, particles can quickly search for possible solutions and find the neighborhood where the global optimal solution is located. The BPSO algorithm, which introduces time-varying weight coefficients, greatly expands the search space of the algorithm and improves convergence speed.

Meanwhile, in the binary particle swarm optimization algorithm, the idea of genetic algorithm is referenced, and crossover and mutation operations are introduced.

The flowchart of the hybrid algorithm is shown in Figure 2, and the calculation steps are as follows:

①Input the topology information of the distribution network;

②Set particle swarm size N, crossover mutation probability, and randomly initialize D-PMU;

③Calculate the fitness value, position, and velocity of particles;

④Compare the results from the previous step to obtain the optimal individual particle in the group;

⑤ Determine if it meets the requirements for observability across the entire network. If so, continue;

⑥Determine if the maximum number of iterations has been reached, if so, continue;

⑦If the maximum number of iterations has not been reached, the binary encoding string of each particle is cross operated with the encoding binary strings of the individual optimal particle and the group optimal particle based on probability;

⑧The particles at the end of the crossover will mutate according to probability h;

⑨Decoding the latest generation of particle binary strings that meet the requirements;

⑩Output the optimal fitness value (number of D-PMU configurations) and D-PMU configuration node position (the position of node 1 in the binary string).

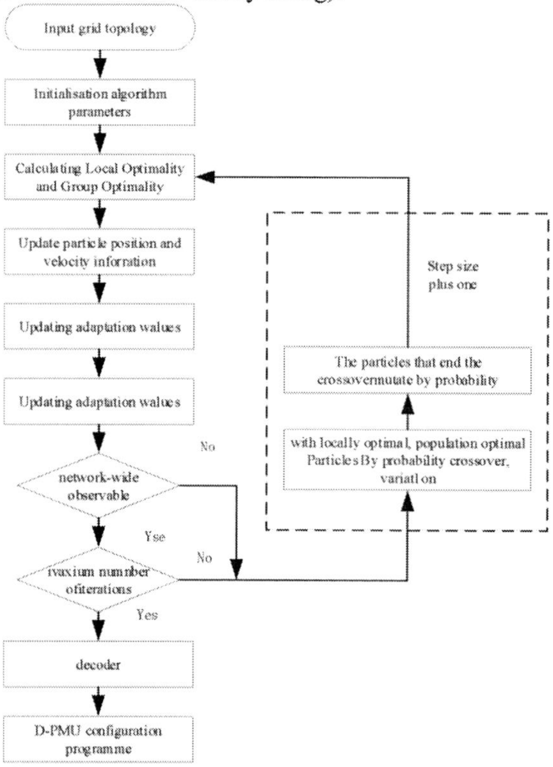

Figure 2. The flow chart of binary particle swarm and genetic hybrid algorithm

4. Simulation analysis

4.1. Basic parameters

This chapter conducts a detailed study on the IEEE 14 node system by measuring node harmonic voltage, branch harmonic current, and node injected harmonic current. Build a circuit topology model in the PXI real-time simulator, and simulate to obtain the required dataset. The simulation environment is shown in Figure 3. The system topology structure is shown in Figure 4.

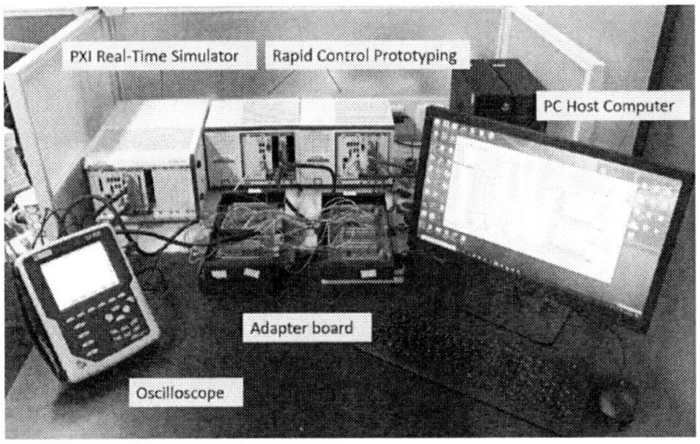

Figure 3. Real-time simulation platform of IEEE14 nodes

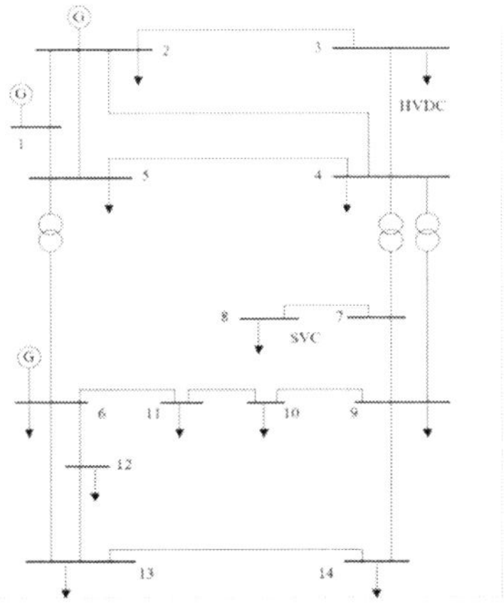

Figure 4. IEEE14 nodes system diagram

This system contains two harmonic sources, one located at the high-voltage direct current (HVDC) terminal of node 3; Static Var Compensator (SVC) located at node 8. Distribution lines, transformers, load parameters can be found in reference[6]. Preprocess data to eliminate the influence of bad data.

(1) Handling missing values: Use the mean imputation method to solve the problem of missing data and complete the dataset. The principle is to replace each missing value with the mean of the non missing part of the dataset, and its model is[7]:

$$\bar{y} = \frac{\sum_{i=1}^{n} a_i y_i}{n_1} \qquad \text{* MERGEFORMAT (18)}$$

In the formula, $\bar{y}$ is the mean of the sample dataset, a_i is the data discrimination factor, $a_i = 0$ represents missing data, y_i is the i value in the sample dataset, and n_1 is the number of non missing parts in the dataset.

(2) Vondrak filtering: The advantage of Vondrak filtering is that it does not require pre-defined fitting functions to obtain filtered data. This applies to both equidistant and non equidistant time series data, and can be used to separate signals of different frequencies. The basic criterion is[8]:

$$Q = F + \lambda^2 S = \frac{1}{n}\sum_{i=1}^{n} p_i \left(y_i - y_i'\right)^2 + \frac{\lambda^2}{n-3}\sum_{i=1}^{n-3}\left(\Delta^3 y_i'\right)^2 = \min \qquad \text{* MERGEFORMAT (19)}$$

In the formula, $y_i^{'}$ is the filtering value corresponding to the initial time series y_i, p_i is the corresponding weight of y_i, and $\Delta^3 y_i^{'}$ is the third-order difference calculated based on a third-order Lagrangian polynomial for filtering.

(3) Measurement redundancy calculation: The definition of measurement redundancy is:

$$R_n = \sum_{i=1}^{n} \frac{l_i + 1}{n} \times \varepsilon_i \quad \text{* MERGEFORMAT (20)}$$

In the formula, n represents the total number of nodes in the distribution network; l_i is the number of nodes connected to node i.

4.2. *Simulation analysis*

The article sets the particle swarm population $N = 1000$, maximum iteration times $n_{max} = 200$, $a = 0.8$, $b = 0.2$, crossover probability of 0.8, and mutation probability of 0.05 in the method described earlier. The optimization model is solved, and multiple configuration schemes are obtained as shown in Table 1, all of which can achieve observation of the harmonic state of the entire network.

Table 1 D-PMU configuration scenarios of IEEE 14 nodes

	Number of D-PMU	Configure nodes	Measurement redundancy R
Scenario 1	3	2,6,9	1.071
Scenario 2	4	2,6,7,9	1.357
Scenario 3	4	2,6,8,9	1.214

To evaluate the accuracy difference of harmonic state estimation in optimizing configuration schemes, two indicators are used for evaluation. The smaller these two indicators are, the better the harmonic state estimation results are. The specific details are as follows:

$$\alpha = \frac{1}{n} \sum_{i=1}^{n} \left| \frac{\tilde{\mathbf{X}}_i - \mathbf{X}_i}{\mathbf{X}_i} \right| \times 100 \quad \text{* MERGEFORMAT (21)}$$

$$\beta = \max \left| \tilde{\mathbf{X}}_i - \mathbf{X}_i \right| \quad \text{* MERGEFORMAT (22)}$$

Simulation shows that harmonic sources HVDC and SVC mainly inject 5th, 7th, and 11th harmonics into the system, with the 5th harmonic having the highest current[9]. Taking scenario 3 configuration scheme as an example, the weighted least squares method is used to estimate the harmonic state of the 6k ± 1st harmonic. The voltage amplitude and phase angle estimation errors of each harmonic at each node of the system are shown in Figures 5 and 6.

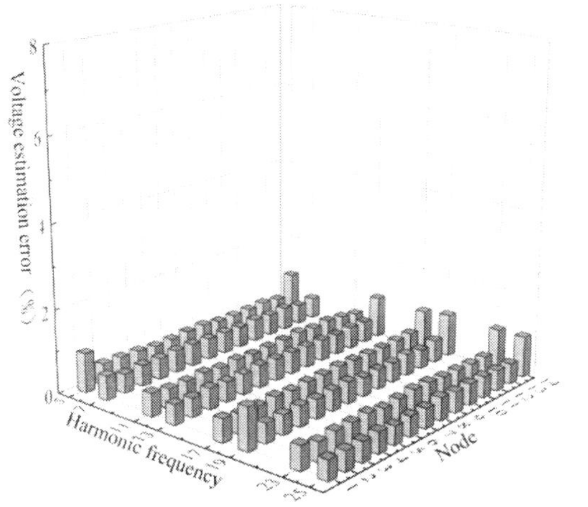

Figure 5. Harmonic volatge amplitude estimation precision

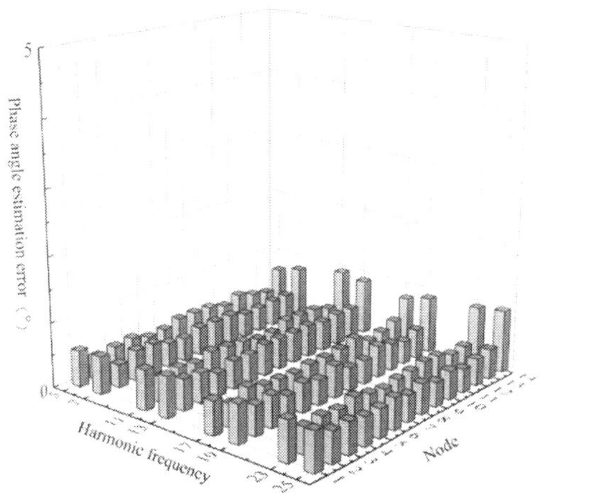

Figure 6. Harmonic volatge phase angle estimation precision

The comparison results of harmonic state estimation in three scenas are shown in Figure7 and Figure8.

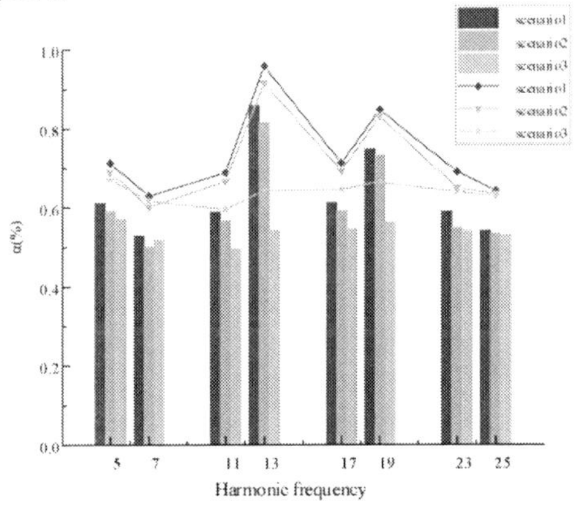

Figure 7. the comparision of scenario 1, 2, 3 about α(%)

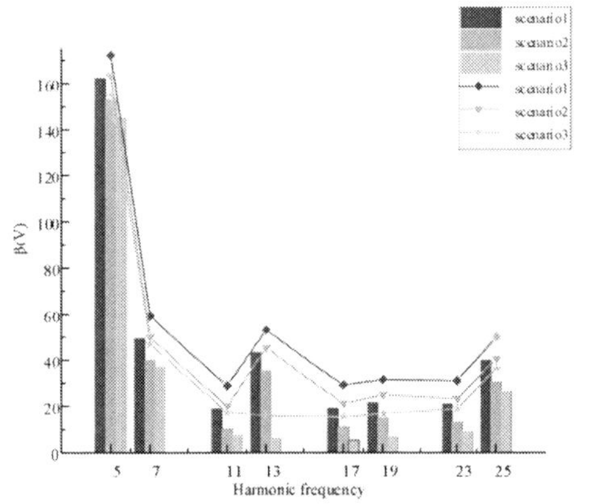

Figure 8. the comparision of scenario 1, 2, 3 about β(V)

The improvement in harmonic state estimation accuracy in scenarios 2 and 3 indicates that adding D-PMU can effectively enhance the system's state estimation accuracy. By comparison, it can be seen that the accuracy of harmonic state estimation fluctuates greatly in scenario 2. This is due to the large estimation error of the harmonic state at node 8. In the IEEE 14 node topology, node 8 is only associated with node 7, making it difficult to effectively estimate through the states of other nodes. Scenario 2 lacks measurement configuration for node 8 and is susceptible to measurement errors. Scenario 3 has configured D-PMU at node 8, effectively reducing the impact of errors and improving the accuracy of harmonic state estimation in the system. After comprehensive consideration, it can be concluded that option 3 is a better configuration result.

5. Conclusion

The article constructs an optimization model that comprehensively optimizes the economic efficiency and harmonic state estimation accuracy of D-PMU configuration, and conducts simulation calculations using the IEEE 14 node standard test system as an example.

(1) The improved binary particle swarm genetic algorithm proposed in the article is used to solve

the D-PMU optimization configuration model, which can ensure the accuracy of harmonic state estimation under observable global harmonic states in the distribution network, and the solution is relatively fast.

(2) The mean interpolation method and Vondrak filtering method used in the article can effectively process PMU measurement data and harmonic estimation data.

(3) To enhance the accuracy of harmonic state estimation and reduce errors generated during algorithm iteration, higher weights should be assigned to the measurement information of harmonic sources and their nearby nodes in the weighted least squares method.

This article only discusses static harmonic state estimation. The next step will continue to study dynamic harmonic state estimation.

Acknowledgments

The research in this thesis was supported by the "Science and Technology Project of State Grid Fujian Electric Power Co., Ltd.", with the project name being "Research and Development of Key Technologies and Equipment for Active Voltage Support by Renewable Energy Stations in Distribution Networks," and the project code being 52139024000B.

References

[1] Qu ZQ, 2020, Research on the method of harmonic state observability analysis and optimal measuring devices placement in distribution networks[D]. Qingdao: China University of Petroleum (East China).

[2] Wang YS, Han MY, Tan ZY, 2010, Study on the effect of PMU allocation on the quality of harmonic state estimation[J]. 38(21), Power System Protection and Control ,233-236+240.

[3] Zhang SW, 2013, Research on power system harmonic state estimation based on optimal PMU placement and measurement information[D]. Baoding: North China Electric Power University, .

[4] Liu YP, Li X, Zeng SQ, et al, 2021, PMU optimal configuration of distribution system for interval state estimation[J], 15(07), Southern Power System Technology, 67-75.

[5] Mandich M, Xia T, Sun K, 2019, Optimal PMU placement using stochastic methods[C]//2019 IEEE Power & Energy Society General Meeting (PESGM). Atlanta, GA, USA: IEEE, 1-5.

[6] Abu-hashim R, Burch R, Chang G, et al, 1999, Test systems for harmonics modeling and simulation[J], 14(2), IEEE Transactions on Power Delivery. 579-587.

[7] WangM, 2016, Imputation methods of missing values for compositional data[D]. Taiyuan: Shanxi University.

[8] Yu WX, 2023, Application of Vondrak filter in GNSS time series analysis[J], 39(01), Standardization of Surveying and Mapping, 73-78.

[9] Zhang YQ, 2012, Research on power system harmonic state estimation[D]. Beijing: North China Electric Power University.

Application of Multi-type Reactive Power Regulation Resource Co-optimization Method Based on Linearized Voltage Sensitivity in the Environment of New Energy Access

Zhenhua You[1], Shihan Yan[2*], Liu Bin[1], Lv Hui[3], Liwei Duan[2]

[1] Yuxi Power Supply Bureau of Yunnan Power Grid Co., Ltd., Yuxi 653199, China

[2] Wuhan Longde Control Technology Co., Ltd., Wuhan 430073, China

[3] Beijing Yongshang Technology Co., Ltd., Beijing 100085, China

*Corresponding author'se-mail: 939334771@qq.com

Abstract. The article proposes a cooperative optimization method based on linearized voltage sensitivity for multiple types of reactive voltage regulation resources, which calculates the node voltage sensitivity by constructing a Zbus linearized tidal model without moving point iteration, and realizes the resolution of the relationship between voltage and injected power at each node. Then the voltage regulation range and power regulation capability of the substation are evaluated based on the forecast data of PV and load. The simulation results show that the method can effectively improve the voltage stability of the grid, which provides a theoretical basis and technical support for the voltage control of the grid under the new energy environment.

1. Background

Currently, along with the construction of new power systems and the continuous promotion of energy transition, the new energy industry is booming. Especially in the distribution network, the rapid growth of distributed new energy installed capacity significantly increases the complexity of grid operation and control[1].Due to the volatility and randomness of new energy, the active power balance of the system is often affected, and the wide distribution of distributed new energy often leads to voltage overrun at the end of the feeder.

To address these challenges, studies have suggested broadening the grid's regulation tools and improving the source-load interaction capability, with research focusing on how to manage the aggregation of a large number of distributed resources. However, regardless of whether demand-side response is used, or through the establishment of load aggregators or virtual power plants, the control authority of these methods is not on the grid side, and faces the problems of regulation compensation and upfront investment. Therefore, there is an urgent need to explore reactive power regulation resources on the grid side that can be directly controlled. Substations often utilize OLTCs as primary grid-side devices for direct control, achieving voltage regulation through tap adjustments.

Since the resistance of the distribution network is greater than the reactance, there is a strong coupling relationship between voltage and active power, so the control of active power can be realized by regulating the voltage. In order to ensure that the substation in the distribution network can stably participate in the regulation of active power, it is also necessary to describes the overall ability of the substation to regulate voltage and give full consideration to the security of the network voltage.

However, there is a lack of systematic research on this issue at present. Therefore, how to determine the taps of on-load voltage regulating transformers in distribution networks and the adjustable range of their corresponding voltage levels has become a key issue to be solved.

2. Research Method

The aim of this paper is to provide a cooperative voltage regulation method for multiple types of reactive power regulation resources based on linearized voltage sensitivity. By constructing a Zbus linearized tidal model with no moving point iterations, the model can be utilized to compute the skewness of each node voltage with respect to the injected power, and thus obtain an expression for the voltage sensitivity. This approach enable streamline the computational process to reduce complexity. Subsequently, the short-term source-load prediction data are comprehensively considered, and the time scales of the PV and load prediction data are aligned by combining the cubic spline interpolation method, thus assessing the voltage regulation range and power adjustment range of the substation, aiming to improve the voltage control capability of the grid under the new energy access environment [2].

2.1. Zbus linearized tidal current model

The Zbus trend model, as an effective trend calculation method, is able to determine the steady state operating state of the system by analyzing the parameters. In this study, the Zbus model is linearized by an iterative method without moving points to establish the linear relationship between voltage and injected power at each PQ node[3].

The objective is to solve for the voltage and phase angle at each node in the power system so that the various components of the system (generators, transformers, lines, loads, etc.) satisfy the constraints such as power balance and tidal balance. Here, a Zbus current model based on a single immobile point iteration is proposed to boost the solution efficiency.

In standard distribution networks, the balancing node, often referred to as the root node, is assumed to maintain a constant voltage, while other nodes are treated as PQ nodes.

The voltage Ui at node i can be considered as the sum of two voltage components: one is the voltage Ui1 caused by the reference node at node i, and the other is the voltage Ui2 caused by the other PQ nodes at node i. In a system containing 1 reference node and N PQ nodes, the form of the power flow equation is:

$$\bar{\boldsymbol{S}} = \bar{\boldsymbol{U}}\boldsymbol{I} \tag{1}$$

$$\boldsymbol{I} = \boldsymbol{YU} \tag{2}$$

By partitioning each matrix based on balanced nodes and PQ nodes, we can derive formulas:

$$\begin{bmatrix} \bar{S}_0 \\ \bar{S} \end{bmatrix} = \begin{bmatrix} \bar{V}_0 & \\ & \mathrm{diag}(\bar{V}) \end{bmatrix} \begin{bmatrix} I_0 \\ I \end{bmatrix} \tag{3}$$

$$\begin{bmatrix} I_0 \\ I \end{bmatrix} = Y \begin{bmatrix} V_0 \\ V \end{bmatrix} = \begin{bmatrix} Y_{00} & Y_{0L} \\ Y_{L0} & Y_{LL} \end{bmatrix} \begin{bmatrix} V_0 \\ V \end{bmatrix} \tag{4}$$

Where matrix chunking is based on whether it is a balanced node or not, I_0, S_0 and V_0 are the balanced node injected current, node injected power and node voltage, respectively; I, S and V are the injected current vector, node injected power vector and node voltage phasor of the PQ node, respectively, and Y is the node conductance matrix of the system, Y_{00} is the balanced node self-conductance, Y_{0L}, Y_{L0} are the mutual conductance matrices between the balanced node and the PQ node, and Y_{LL} is the conductance matrix of all PQ nodes. conductance matrix of the balance node and the PQ node, and Y_{LL} is the conductance matrix of all PQ nodes.

According to the above equation, the implicit expression of PQ node voltage can be obtained:

$$V = Y_{LL}^{-1}\mathrm{diag}\left(\bar{V}\right)^{-1}\bar{S} - Y_{LL}^{-1}Y_{L0}V_0 \tag{5}$$

The voltage generated at the root node is approximated as a constant set to W. So the Zbus current model is:

$$V = Y_{LL}^{-1}\mathrm{diag}\left(\bar{V}\right)^{-1}\bar{S} + W \tag{6}$$

The physical concept of the Zbus tidal model is clear, and the use of the Y matrix with sparse properties and the equivalent current source injection form can both reduce the computational memory and increase the computational efficiency.

For for any nonlinear function y = f (x), in the solution of its zeros have f (x_0) = 0, at this time it can be written in another equivalent form = Ψ (x_0), known as the function Ψ (x_0) of immovable point,through the calculation of immovable point of the process of calculating the zeros of a function known as immovable point of the iterative process, iterative formula shown in the formula, the iterative process As shown in Figure:

$$X(k) = \varphi\left(x(k-1)\right) \tag{7}$$

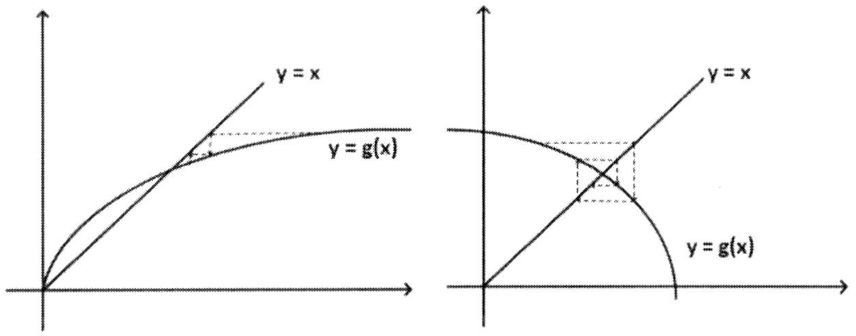

Figure 1. Iterative convergence process for immovable points

Zbus current equation has obvious iterative function characteristics. Therefore, it can be solved by means of immovable point iteration, but immovable point iteration needs to go through many calculations, and the algorithmic complexity is high, so we use a linear approximation approach involving a single fixed-point iteration to develop the Zbus nonlinear model, based on the linearized tidal current model at the reference tidal current point[4]. At this point, the latest operating point $U^{0'}$ of the system is selected as the reference tidal point, and the node power is updated when the system tidal current changes, and the node voltage V is obtained according to the following equation.

$$V = A\bar{S} + W \tag{8}$$

Where $A = Y_{LL}^{-1}\mathrm{diag}\left(\bar{V}\right)^{-1}$, $W = -Y_{LL}^{-1}Y_{L0}V_0$,the equation above effectively simplifies the solving process and improves computational efficiency.The single fixed-point iteration method retains high accuracy over larger intervals, unlike traditional methods[5].

2.2. *Voltage sensitivity calculation*

Voltage sensitivity is the response of the node voltage to the power change, which reflects the key characteristic of voltage stability. The traditional method obtains the voltage sensitivity by inverting the Jacobi matrix, but in this study, the analytical acquisition of the sensitivity expression is realized by the bias derivation calculation of the Zbus model, which reduces the computational complexity[6].

Voltage sensitivity is the degree to which the nodal voltage in a power system responds to changes in power. Specifically, it measures the degree to which changes in power in a power system affect the nodal voltage. Power-voltage sensitivity is derived by computing the derivatives of the nodal voltage with respect to various factors, including load power changes, generator output variations, and line parameter adjustments. Unlike traditional methods, which calculate voltage sensitivity by inverting the Jacobi matrix—an approach that is computationally demanding and not suitable for real-time control—the proposed method offers a more efficient solution, and at the same time it is not possible to obtain the voltage sensitivity about the balanced nodes.

$$\frac{\partial V}{\partial \boldsymbol{P}} = \boldsymbol{Y}_{LL}{}^{-1} diag(\bar{\boldsymbol{V}}^{*})^{-1} \tag{9}$$

$$\frac{\partial V}{\partial \boldsymbol{Q}} = -j\boldsymbol{Y}_{LL}{}^{-1} diag(\bar{\boldsymbol{V}}^{*})^{-1} \tag{10}$$

In addition to the injected power of each PQ node will lead to the change of each node voltage, the voltage change of the root node will also have an impact on each node voltage, in the linearized trend model based on ZBus, the node voltage of each PQ node and the voltage vector of the balance node are also in a linear relationship, so it is also possible to derive the voltage V0 of the balance node based on the linearized trend equation.The sensitivity of the PQ node to the balance node voltage is quickly calculated:

$$\frac{\partial V}{\partial V_0} = -\boldsymbol{Y}_{LL}{}^{-1}\boldsymbol{Y}_{L0} \tag{11}$$

This relationship enables a quantitative evaluation of how variations in the balance node voltage influence the voltage changes at individual PQ nodes.

In distribution networks, two primary factors influence the voltage at a PQ node. One of them is the adjustment of the OLTC in the substation, which modifies the balancing node voltage and consequently affects the voltage at each PQ node.The second is that the change of load power also leads to the change of node voltage.

Through the analysis based on the Zbus trend model, the voltage sensitivity can be obtained, and then the linear relationship between these variables can be established, which provides theoretical support for the boundary calculation of substation voltage regulation.

2.3. Load and PV data processing at different time scales

To ensure the accuracy of the system's voltage regulation, the cubic spline interpolation method is used to unify the time scales of the PV and load data. The processing period of the predicted data is set to 15 minutes to ensure accurate assessment of the voltage regulation range and power adjustment range.

2.3.1. Calculation of power at the point of electrical coupling between the distribution network and the main network. The output power of distributed PV is characterized by randomness and volatility due to the influence of the external environment. If there is a lack of measures to smooth PV fluctuations within the distribution network, the fluctuation of PV power will be directly reflected in the coupling point between the distribution network and the main network, i.e., the downgrid point. Therefore, effectively smoothing the dramatic fluctuation of power at the down-grid point is a key means to improve the local consumption of PV, ensure the utilization of equipment and reduce the impact of the distribution network on the main network. The power calculation formula for the downgrid point is as follows:

$$P_t^{pcc} = \sum_{f=1}^{N_F} P_{f,t}^{load} + \sum_{m=1}^{N_B} P_{m,t}^{pv} + P_t^{loss} \tag{12}$$

Where P_t^{pcc} is the active power at the lower grid point at time t; $P_{f,t}^{load}$ is the active power of the fth feeder load; P_t^{loss} is the internal network loss of the distribution network; N_F is the total number of feeders in the distribution network; N_B is the total number of nodes in the distribution network; $P_{m,t}^{pv}$ is the distributed PV active in the mth node in the distribution network, in order to unify the standard with the direction of the load power when the PV outputs power outward $P_{i,t}^{pv} \leq 0$, which is equivalent to the negative power loads.

2.3.2. Processing of photovoltaic data according to cubic spline interpolation. The pre-dispatch phase depends on the short-term forecast of load and PV output, under the premise of guaranteeing the accuracy, the short-term forecast period of load and PV is 15 min at this stage, so this paper selects the pre-dispatch period as 15 min. real-time control is controlled by the actual data taken by the grid measurement equipment for the feedback control, so the real-time control step size Tc is related to the sampling frequency of the measurement device in the system.

The power of the lower network is obtained by summing the load power, PV power and distribution network loss, the distribution network loss is relatively fixed and accounts for a small ratio of the power of the lower network, which can be regarded as a fixed value under a certain accuracy, the load power has a strong regularity, and the accuracy of the load prediction at the present stage can reach a time scale of 1min, but due to the strong fluctuation and stochasticity of the PV power, the mature prediction level at the present stage can only reach a time scale of 15min. 15min time scale. Therefore, in order to ensure the accuracy of pre-scheduling, the pre-scheduling period Tp is selected to be 15 min. However, due to the non-uniformity of the prediction steps of PV and load, in order to obtain the power fluctuation rate of the downstream outlets in each minute of the pre-scheduling period, it is necessary to align the time scales of the prediction data of PV and load.

Cubic spline difference is the main processing means of PV data, firstly, the cubic spline interpolation is applied to the PV prediction data of 15min, so that it is transformed into the data of 1min time scale, and then subsequently, the power prediction value of the lower grid point is calculated, and the calculation process is as follows:

$$\begin{cases} Sp\left(P_{15\min}^{pv,for}\right) = P_{1\min}^{pv,for} \\ P_{1\min}^{load,for} + P_{1\min}^{pv,for} + P_{1\min}^{loss} = P_{1\min}^{pcc,for} \end{cases} \tag{13}$$

Sp is the cubic spline interpolation function; $P_{15\min}^{pv,for}$ and $P_{1\min}^{pv,for}$ are the 15-min and 1-min predicted values of PV output, respectively; $P_{1\min}^{load,for}$, $P_{1\min}^{pcc,for}$ and $P_{1\min}^{loss}$ are the 1-min predicted value of load active power, the 1-min predicted value of active power of the downstream network point, and the network loss of the distribution network, respectively.

3. Methodological applications

In this paper, the system voltage is regulated by the OLTC of the substation, and the upper and lower limits of the substation voltage regulation are calculated. The coupling relationship between load and photovoltaic (PV) generation affects the voltage regulation boundary, and the voltage sensitivity analysis based on the Zbus model can realize the quantitative assessment of the substation voltage regulation capability.

3.1. Calculation of substation voltage regulation range

The on-load regulator transformer is a typical traditional discrete voltage regulator that needs to satisfy the constraints of its tap regulation range. Through the operating constraints of the on-load regulator transformer, its voltage regulation can be calculated, and the corresponding constraints are as follows:

$$\Delta V^{\mathrm{OLTC}} = \Delta k \Delta V_{tap} \tag{14}$$

$$k_{\min} \le k_2 \le k_{0,t} \le k_1 \le k_{\max} \tag{15}$$

Where k represents the tap position after OLTC regulation in time period t, which is taken as k_1 and k_2 in the two scenarios; ΔV^{OLTC} is the amount of voltage regulation of the on-load regulator transformer; $k_{\min}$ and $k_{\max}$ are the minimum and maximum stalls of the on-load regulator taps, respectively.

The coupled relationship between load active power and PV generation significantly influences the voltage regulation process in the grid. PV systems are usually equipped with inverters that can provide active power support. In power systems, the voltage regulation capability of the substation is realized by OLTCs, which can adjust the output voltage by adjusting its taps to control the voltage level of the distribution network[7] .Since the resistance in the distribution network is much larger than the reactance, the coupling between voltage and active power is more obvious, so the substation can indirectly realize the regulation of active power by adjusting the voltage. Therefore, through the adjustment of OLTC can not only adjust the bus voltage, but also indirectly realize the control of active power. In order to ensure the safe and stable operation of the power grid, it is necessary to calculate the upper and lower limits of substation voltage regulation, i.e., the voltage regulation range. In practice, to assess the voltage regulation range of the substation, it is usually necessary to calculate the voltage sensitivity of each node to assess, the node voltage is affected by the load power and PV power changes. The dynamics of load active power and PV output puts new requirements on maximizing the voltage regulation range. To quantitatively assess the voltage regulation capability of a substation, the following equation can be used:

$$\Delta V_{range} = \Delta V^{\mathrm{OLTC}} + \frac{\partial V}{\partial P} \cdot (\Delta P_{PV} + \Delta P_{load}) \tag{16}$$

In the above equation, ΔV^{OLTC} represents the voltage regulation of the on-load regulator transformer, $\frac{\partial V}{\partial P}$ represents the voltage to active power sensitivity, and ΔP_{PV} and ΔP_{load} are the active power change of photovoltaic power generation and the load active power change, respectively.

3.2. *Calculation of substation capacity regulation range*

The CVR coefficient is a key parameter used to measure the impact of voltage variation on load power demand in the voltage regulation control of a substation[8].The CVR coefficient is typically represented as:

$$CVR = \frac{\Delta P_{load}}{\frac{\Delta V}{V}} \tag{17}$$

The CVR coefficients allow quantification of the impact of voltage regulation on load power. When the voltage regulation range of a substation is known, the power adjustment range of the substation within that regulation range can be calculated by combining the CVR coefficients. When the voltage regulation range ΔV_{range} of the substation is known, then the power adjustment range ΔP_{range} of the substation can be obtained based on the CVR coefficient by using the following formula:

$$\Delta P_{range} = CVR \cdot \Delta V_{range} \cdot P_{base} \tag{18}$$

4. Example analysis

The relevant model is constructed based on the above theory, and the changes in voltage and capacity at the regulation point before and after PV participation in co-optimization are compared according to the simulation results. The system solution environment is Intel Core i5-8300H CPU and 16GB memory configuration, which is realized by MATLAB 2022a.

By applying the K-means clustering algorithm, the PV output data are finely classified and grouped to reveal their intrinsic distribution characteristics and regularity. This method can effectively extract the characteristic patterns of PV output data, and also provides more targeted and structured data support for subsequent analysis and processing. After completing the clustering of the PV output data, it is combined with the load forecast data for further processing to form a complete analytical framework, which can more accurately simulate the operational status and characteristics in the actual power system.

On this basis, the changes in key performance indicators of the PV system before and after its participation in grid co-optimization are analyzed in depth, focusing on comparing the dynamic response characteristics of voltage level and capacity at the regulation point. Simulation experiments are carried out by establishing arithmetic cases to compare and verify the operational performance of the system in different scenarios, especially the significant differences and potential advantages of the PV system before and after the introduction of the co-optimization method.

As shown in Figure 2 and Figure 3, the voltage and capacity at the regulation point before the PV participates in the co-optimization.As shown in Figure 4 and Figure 5, the voltage and capacity at the regulation point after the PV participates in the co-optimization.Table 1 and Table 2 show the values of voltage and power at the regulation point before and after PV participation in co-optimizatio.

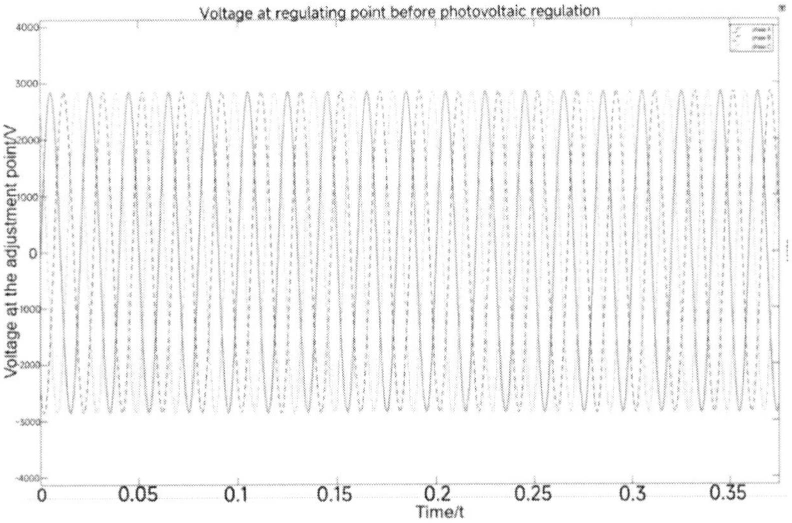

Figure 2. Voltage at the regulation point prior to PV participation in co-optimization

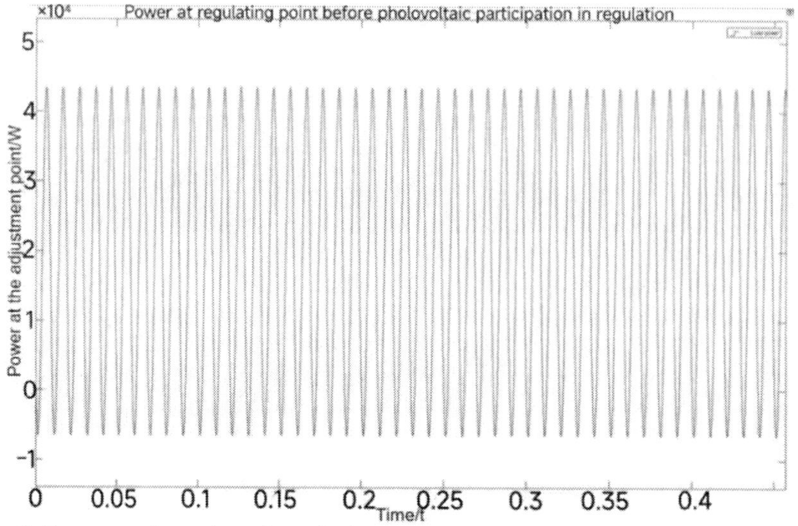

Figure 3. Power at the point of regulation prior to PV participation in co-optimization

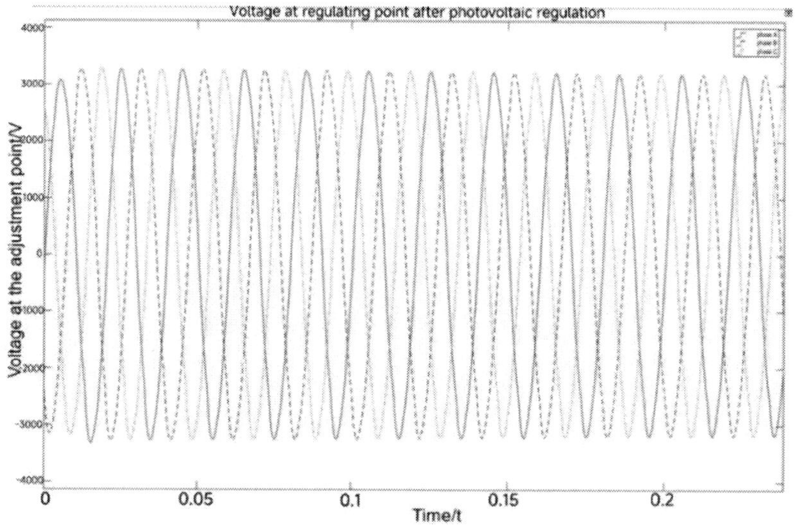

Figure 4. Voltage at the regulation point after PV participation in co-optimization

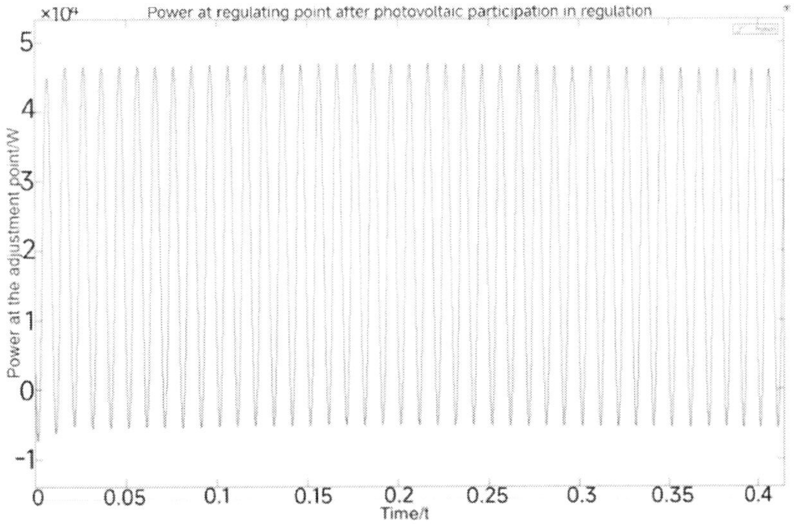

Figure 5. Power at the regulation point after PV participation in co-optimization

Table 1. Values of voltage and power at the regulation point prior to PV participation in co-optimization

	Maximum value	Minimum value	RMS
Voltage value	2.852/kV	-2.852/kV	2.019/kV
Power value	43.47/kW	-65.01/kW	25.57/kW

Table 2. Voltage and power values at the regulation point after PV participation in co-optimization

	Maximum value	Minimum value	RMS
Voltage value	3.273/kV	-3.316/kV	2.256/kV
Power value	46.99/kW	-72.95/kW	27.84/kW

5. Conclusion

The simulation study show that the cooperative optimization method of multiple types of reactive power regulation resources based on linearized voltage sensitivity analysis can effectively improve the operation efficiency and stability of new energy access power systems. This paper discusses the potential application of this optimization method in the context of high proportion of new energy access, and proves its practicality and advancement in improving the voltage stability of the grid, enhancing the flexibility of system scheduling, and reducing the operational risk.

Acknowledgments

This work was supported by China Southern Power Grid Corporation Science and Technology Project YNKJXM20230248.

References

[1] Liu Hongbo,Liu Shencheng,Gai Xueyang,Liu Yongfa & Yan Yutong.(2024).Overview of Active Distribution Network Planning With High Proportion of New Energy Access[J]. Power Generation Technology, 45(1): 151-161.

[2] WANG Ling, SHENG Chao, XU Boyu. Operating Characteristics of Photovoltaic Power

Generation System and Its Equivalent Description[J]. Guangdong Electric Power,2012,25(3): 57-64.

[3] Chen Yifeng, Xiong Junhua & Zheng Bingxiao. (2023). Optimal Siting and Sizing of Distributed Generators in Distribution Networks Using Gradient Method Based on Zbus Matrix [J]. Information & Computer (Theoretical Edition), 35(12): 16-19.

[4] Hu Shiyang, Yang Meng & Wang Fangzong. (2017). Fast Decoupled Power Flow Calculation Method for Distribution Systems Based on Coordinate Rotation Transformation [J]. Power System Protection and Control, 45(4): 40-44.

[5] Wu Chengguo & Xiao Shiwu. (2022). Quasi-Newton Iterative Calculation Method for Short-Circuit Current in Power Grids with High Proportion of Distributed Generators [J]. Power System Technology, 46(12): 4581-4591. DOI: 10.13335/j.1000-3673.pst.2022.1357.

[6] Ma Chenglian, Li Chuang, Xue Bing, et al. (2024). Partitioned Voltage Coordination Control Strategy for Distribution Networks with High Proportion of Photovoltaics [J]. Journal of Northeast Electric Power University, 44(4): 77-85. DOI: 10.19718/j.issn.1005-2992.2024-04-0077-09.

[7] Sun Wenhui, Pan Ming, Xu Mengxiao, et al. (2024). Study on Factors Affecting the Critical Point of OLTC "Negative Voltage Regulation Effect" under Photovoltaic Integration Conditions [J]. Transformer, 61(1): 40-46. DOI: 10.19487/j.cnki.1001-8425.2024.01.015.

[8] Xing Haijun, Xie Baojiang, Qin Jian, et al. (2021). Optimal Operation Method for Active Distribution Networks Considering CVR [J]. Electric Power, 54(11): 76-81.

Collaborative Optimization Method of Active Distribution Network Based on Conservation Voltage Reduction

Zhenhua You[1], Liwei Duan[2*], Bin Liu[1],Hui Lv[3],Shihan Yan[2]

[1] Yuxi Power Supply Bureau of Yunnan Power Grid Co., Ltd, Yuxi 653199, China

[2] WuhanLongde Control Technology Co., Ltd, Wuhan 430073, China

[3] Beijing Yongshang Technology Co., Ltd, Beijing 100085, China

*Corresponding author's e-mail: 1551538609@qq.com

Abstract. The increasing use of distributed resources and flexible loads with fast active power regulation in the distribution network provides a new approach to power grid operation control. Addressing the challenge of real-time control of user-side loads, conservation voltage reduction (CVR) is introduced to directly manage the feeder load power. Firstly, the K-means clustering algorithm combined with KL divergence is used to analyze the distributed photovoltaic characteristics. Secondly, the feeder load power regulation characteristics are identified based on the load modeling method. Then, considering the active management measures of various regulating equipment for power and voltage, a multi-objective optimization model for the collaborative operation of the distribution network is established, using comprehensive operation cost and node voltage deviation rate as indicators. Finally, using the IEEE 33-node distribution network as an example, the results show that the proposed method demonstrates good energy-saving and voltage regulation performance.

1. Introduction

With the continuous growth of global energy demand and the increasingly serious environmental problems, distributed photovoltaics are used in power systems as an important clean and renewable energy source. The active distribution network is a key hub for coordinating the power generation end and the user end [1]. However, due to the uncertainty and fluctuation of the power supply and demand mode of distributed energy, and for the distribution network, the high permeability distributed energy changes the power flow from radial waterfall to random two-way power flow, which is easy to lead to safety problems such as node voltage exceeding the limit [2]. How to balance the supply and demand relationship of each energy unit in the electric power system by optimizing the scheduling and coordination strategy, reducing energy consumption and improving the security of the system has important research significance.

Typical power controllable resources such as energy storage, diesel engine and micro gas turbine in distribution network [3], but the economy of energy storage is often poor. Although diesel engine and micro gas turbine can participate in peak shaving optimization, they need to run at a low power level for a long time to ensure sufficient regulation capacity, which seriously reduces equipment efficiency. There are also studies that show that the potential of load side regulation can be tapped [4], but the control right of this method does not belong to the grid side, and there are limitations such as uncontrollable response and uncertain response speed, which makes it difficult to effectively participate in the real-time regulation of the distribution network [5].

Under the above background, this paper proposes a distributed resource collaborative optimization method based on conservation voltage reduction. Aiming at the uncertain influence of photovoltaic, the characteristic scene is analyzed in the form of clustering [6]. Aiming at the defect that the user side load is difficult to control in real time, the conservation voltage reduction is introduced [7], and the feeder load power is directly controlled by common direct control equipment such as on-load tap changer (OLTC). The operation is simple and easy to implement, which has a positive impact on energy saving and security and stability, and effectively cooperates with the collaborative interaction of active distribution network.

2. Photovoltaic uncertainty processing

Using the historical light intensity and temperature data, the K-means algorithm is used to cluster the generated large amount of distributed photovoltaic output data to obtain representative feature scenarios and their probability distributions. Since there may be a certain error between the probability distribution of the feature scene after scene clustering and the real probability distribution, the probability distribution uncertainty set of the feature scene is constructed by combining the KL divergence to constrain the error between the real scene probability distribution and the initial scene probability distribution obtained by clustering Ω:

$$\Omega = \begin{cases} \sum_{k=1}^{K} p_k \ln \frac{p_k}{p_k^0} \le \zeta \\ \sum_{k=1}^{K} p_k = 1, \ p_k \in [0,1] \end{cases} \tag{1}$$

Where, K is the number of feature scenes after K-means clustering; p_k is the real probability value under scene K ; p_k^0 is the initial probability under the scene K after K-means clustering; ζ is the KL divergence value between the initial probability distribution and the real scene probability distribution.

According to the daily solar irradiance, as shown in Fig.1, the historical operation data of the photovoltaic power station are divided into four typical daily scenarios under different light intensities in spring, summer, autumn and winter. It can be found that the photovoltaic output power is significantly different due to the influence of weather.

3. Feeder load regulation characteristics and regulation cost based on ZIP load model

The steady-state characteristics are represented by the constant impedance (Z)-constant current (I) - constant power (P) load model, as follows :

$$\begin{aligned} & P_i^L = P_{i,0}^L \left[a_Z \left(\frac{U_i}{U_{i,0}} \right)^2 + a_I \left(\frac{U_i}{U_{i,0}} \right) + a_P \right] \\ & a_Z + a_I + a_P = 1 \end{aligned} \tag{2}$$

Where, $P_{i,0}^L$ and $U_{i,0}$ is the initial active reactive power and voltage of the node; a_Z、a_I、a_P is the active component ratio coefficient of three static loads;Therefore, the sensitivity of feeder load power response to voltage change is defined as voltage-active power coupling coefficient, namely CVR coefficient:

$$f_i^{CVR} = \frac{\Delta P_i^{feeder}}{\Delta U_i^{feeder}} \tag{3}$$

The voltage deviation rate is used as an index to judge the power quality level, and the cost model of feeder load power regulation based on power quality is constructed [8]:

$$\min \quad C_{i,t}^{feeder} = c_{i,t}^{L} \Delta P_{i,t}^{feeder} \tag{4}$$

Where, $c_{i,t}^{L}$ is the compensation coefficient under different power quality at different times; $\Delta P_{i,t}^{feeder}$ is the active power variation of node i feeder load at time t ,which can be quickly obtained based on voltage sensitivity and CVR coefficient.

4. Multi-objective optimization model for collaborative interaction of distribution network

4.1. Comprehensive operation cost

The objective function with the minimum network loss cost, light abandonment cost, total operating cost and feeder load power regulation cost is as follows :

$$\min \quad F_1 = C_{loss} \sum_{t=1}^{T} \sum_{ij \in \Omega_L} l_{ij,t} r_{ij} + C_{pva} \sum_{t=1}^{T} \sum_{i \in \Omega_p} (P_{i,t,0}^{pv} - P_{i,t}^{pv}) + \sum_{t=1}^{T} \sum_{s=1}^{N_G} b_s P_{G,s,t} + C_{\sup,t} \sum_{t=1}^{T} P_{\sup,t} + \sum_{t=1}^{T} \sum_{i=1}^{N_{bus}} C_{i,t}^{feeder} \tag{5}$$

Where, $l_{ij,t}$ is the square of the current value of the line ij at time t , Ω_L is the system line set, and r is the resistance on the line ij ; C_{pva} is the cost coefficient of abandoning light, $P_{i,t,0}^{pv}$ is the active power generated by photovoltaic on the node i at time t ,and $P_{i,t}^{pv}$ is the absorbed photovoltaic power; N_G is the number of micro-turbines, $P_{G,s,t}$ is the output value of micro-turbine s at time t , and b_s is its cost consumption parameter. $C_{\sup,t}$, $P_{\sup,t}$ are the power purchase cost coefficient and power purchase power of the active distribution network to the upper power grid under the time-of-use price.

4.2. The node voltage deviation rate is the smallest

$$\min \quad F_2 = \frac{1}{T} \sum_{t=1}^{T} \left(\frac{1}{N_{bus}} \sum_{i=1}^{N_{bus}} \left| \frac{U_0 - U_{i,t}}{U_0} \right| \right) \times 100\% \tag{6}$$

In the formula, N_{bus} is the total number of nodes in the system; U_0 is rated voltage; $U_{i,t}$ is the actual voltage of node i at time t .

4.3. Operation constraints of each power regulation equipment

$$0 \le P_{i,t}^{pv} \le P_{i,t}^{pv,\max} \tag{7}$$

$$(P_{i,t}^{pv})^2 + (Q_{i,t}^{pv})^2 \le (S_i^{pv})^2 \tag{8}$$

Equations (7)-(8) are photovoltaic constraints. S_i^{pv} is the installed capacity of photovoltaic at node.

$$U_{o,t} = U_{i,t} - 2(r_{ij} P_{ij,t} + x_{ij} Q_{ij,t}) + [(r_{ij})^2 + (x_{ij})^2] l_{ij,t} \tag{9}$$

$$\sum_{n=0}^{N_{ij,t}} 2^n \lambda_{ij,n,t} \le K_{ij} \tag{10}$$

$$\begin{cases} m_{ij,t} = k_{ij}^{\min} U_{j,t} + \Delta k_{ij,t} \sum_{n=0}^{K_{ij}} 2^n x_{ij,n,t} \\ 0 \le U_{j,t} - x_{ij,n,t} \le (1-\lambda_{ij,n,t})M \\ 0 \le x_{ij,n,t} \le \lambda_{ij,n,t} M \\ U_{o,t} = k_{ij}^{\min} m_{ij,t} + \Delta k_{ij,t} \sum_{n=0}^{K_{ij}} 2^n y_{ij,n,t} \\ 0 \le m_{ij,t} - y_{ij,n,t} \le (1-\lambda_{ij,n,t})M \\ 0 \le y_{ij,n,t} \le \lambda_{ij,n,t} M \\ x_{ij,n,t} = \lambda_{ij,n,t} U_{j,t} \\ y_{ij,n,t} = \lambda_{ij,n,t} m_{ij,t} \end{cases} \tag{11}$$

Equations (9)-(11) are the constraints of the OLTC.OLTC is incorporated into the feeder in the form of injected power. K_{ij} is the maximum tap of OLTC of ij on the line.Let $\lambda_{ij,n,t}$ be a binary variable. $m_{ij,t}, x_{ij,n,t}, y_{ij,n,t}$ are intermediate variables. M is the constant of Big-M method.

$$\begin{cases} P_{G,s,\min} \le P_{G,s,t} \le P_{G,s,\max} \quad P_{G,s,t} - P_{G,s,t-1} \le P_{G,s}^{up} \\ Q_{G,s,\min} \le Q_{G,s,t} \le Q_{G,s,\max} \quad P_{G,s,t-1} - P_{G,s,t} \le P_{G,s}^{down} \end{cases} \tag{12}$$

Equation(12) is the constraint of micro gas turbine, including output and climbing.

$$Q_j^{svc,\min} \le Q_j^{svc} \le Q_j^{svc,\max} \tag{13}$$

$$Q_{j,t}^{CB} = N_{j,t}^{CB} Q_{j,t}^{CB,step} \tag{14}$$

$$\begin{cases} \sum_{t=1}^{T} (N_{j,t}^{CB,+} + N_{j,t}^{CB,-}) \le B^{CB,\max} \\ N_{j,t}^{CB} - N_{j,t-1}^{CB} = N_{j,t}^{CB,+} - N_{j,t}^{CB,-} \\ 0 \le N_{j,t}^{CB} \le N^{CB,\max} \end{cases} \tag{15}$$

Equation(13)-(15) are constraints for reactive power compensation devices, including Static Var Compensater (SVC)and capacitorbank(CB).

$$\begin{cases} P_{j,t} = \sum_{k:j\to k} P_{jk,t} - \sum_{i:i\to j} (P_{ij,t} - r_{ij} l_{ij,t}) \\ Q_{j,t} = \sum_{k:j\to k} Q_{jk,t} - \sum_{i:i\to j} (Q_{ij,t} - x_{ij} l_{ij,t}) \\ U_{j,t} = U_{i,t} - 2(r_{ij} P_{ij,t} + x_{ij} Q_{ij,t}) + ((r_{ij})^2 + (x_{ij})^2) l_{ij} \end{cases} \tag{16}$$

$$\left\| \begin{matrix} 2P_{ij,t} \\ 2Q_{ij,t} \\ l_{ij,t} - U_{i,t} \end{matrix} \right\|_2 \le l_{ij,t} + U_{i,t} \tag{17}$$

$$\begin{cases} U^{\min} \le U_{j,t,0} + \Delta U_{j,t}^{feeder} \le U^{\max} \\ l_{ij,t} \le l_{ij}^{\max} \end{cases} \tag{18}$$

Equations (16)-(18) adopt the optimal power flow model based on Distflow, and establish a power flow constraint model that is converted into a mixed integer second-order cone programming(SOCP) by second-order cone convex relaxation.

5. Example analysis

Based on the improved IEEE33 node standard test system, this paper uses mixed integer second-order cone programming and Cplex commercial optimization package to solve the collaborative optimization problem of distribution network. The system data comes from Reference [9], and its solution environment is Intel Core i5-8300H CPU and 16GB RAM configuration, which is realized by MATLAB 2018b.

The improved IEEE33 node topology diagram is shown in Figure 1.It includes OLTC, reactive power compensation device,distributed photovoltaic,distributed micro-combustion unit, and 32 feeders.

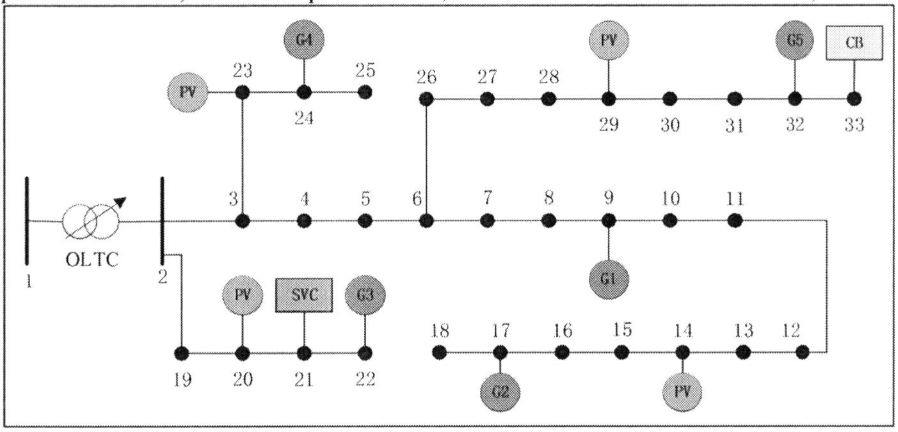

Figure 1. Improved IEEE33 node system

The photovoltaic output obtained by K-means clustering combined with KL divergence, according to the daily solar irradiance, as shown in Figure 2, the historical operation data of the photovoltaic power station are divided into four typical daily scenarios under different light intensities in spring, summer, autumn and winter.

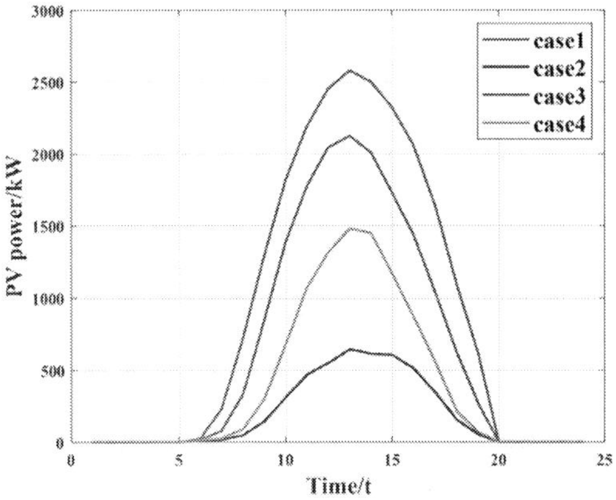

Figure 2. Photovoltaic typical scene output

Journal of Physics: Conference Series 3012 (2025) 012008 doi:10.1088/1742-6596/3012/1/012008

Figure 3 shows the 24-hour load curve and the time-of-use electricity price curve in the example.After obtaining the photovoltaic output data, the photovoltaic typical case3 is used as the photovoltaic data. According to the optimization model and constraints given in the fourth section, the power of the generator and the superior power purchase is shown in Figure 4(a). Figure 4(b) is the comparison of the average node voltage deviation rate before and after adding CVR.

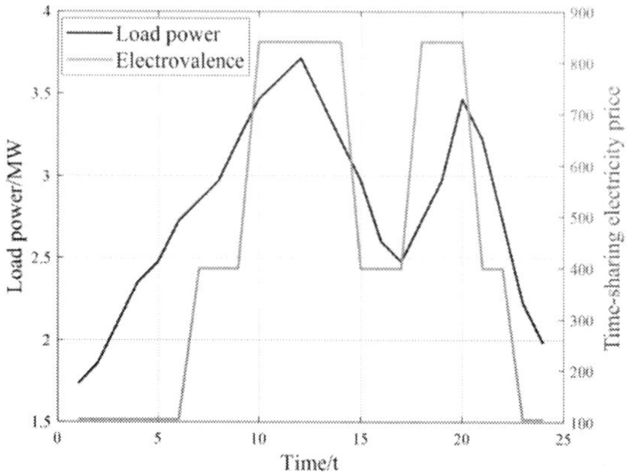

Figure 3. Load and electricity price curve

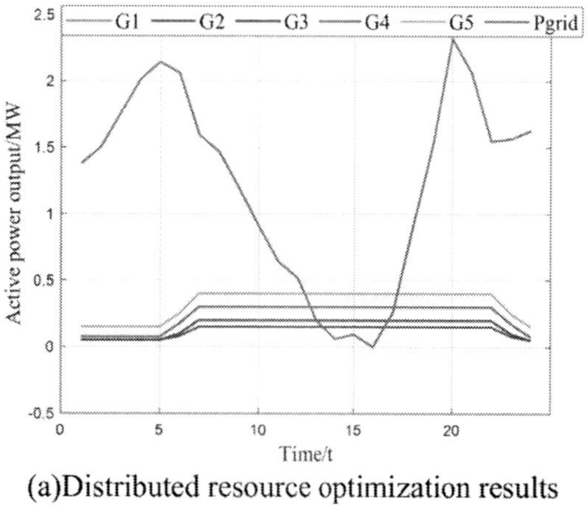

(a)Distributed resource optimization results

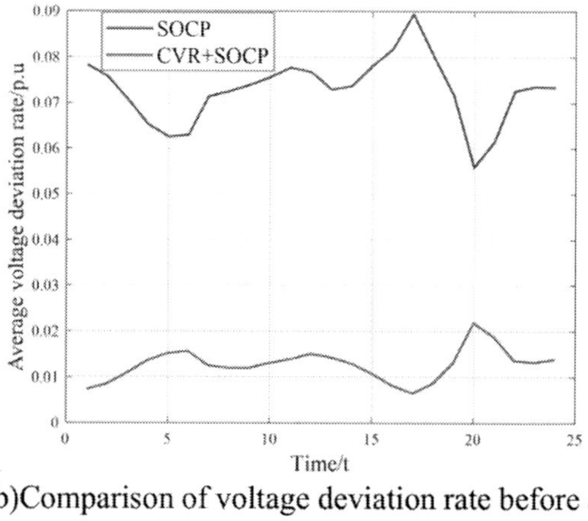

(b)Comparison of voltage deviation rate before and after adding CVR

Figure 4. Collaborative optimization results

Table 1. Comparison of energy saving effect before and after adding CVR

	SOCP	SOCP+CVR
Cost	1.6374e+04/yuan	1.6222e+04/yuan
power consumption	68117.7/kW·h	67137.7/kW·h

Figure 4 (b) shows that after adding the CVR technology combined with SOCP, the average node voltage deviation rate is improved compared with the traditional SOCP solution method.Table 1 also gives the comparison of the total cost and daily power consumption after the two methods are solved. It can be found that the method proposed in this paper can better save costs and power consumption.

6. Conclusion

In this paper, firstly, the K-means clustering algorithm combined with KL divergence is used to construct the photovoltaic output scenario, which outputs the influencing factors of photovoltaic random uncertainty. Then, the load modeling method of feeder load power is adjusted and identified, and the direct control of feeder load power is introduced by introducing conservation voltage reduction. The results effectively show that the method plays a better role in regulating voltage in the collaborative optimization of active distribution network. In addition, the simulation in the example also has a better energy-saving effect.

Acknowledgments
This article is supported by the China Southern Power Grid Technology Project ' Research and Application of Multi-level Collaborative Control Technology for Directly Controlled Flexible Loads for Power Tight Balance Supply Scenarios ' (YNKJXM20230248). Thank you.

References

[1] Liu Hongbo,Liu Shencheng,Gai Xueyang,Liu Yongfa & Yan Yutong.(2024).Overview of Active Distribution Network Planning With High Proportion of New Energy Access[J]. Power Generation Technology,, 45(1): 151-161.
[2] Liu Yingru. (2024). Voltage control and optimal scheduling of distribution network cluster with high permeability distributed photovoltaic (master 's degree thesis, Northeast Electric Power University).
[3] Yang Shibo, Sun Liang, Chen Lidong & Liu Jiayu. (2021). Coordinated optimal dispatch of distribution network system with CCHP microgrid considering time-of-use electricity price. Electric Power Automation Equipment (04), 15-23.
[4] Tian Shiming, Wang Beibei, & Zhang Jing. (2014). Key technologies of demand response under smart grid conditions. Chinese Journal of Electrical Engineering, 34 (22), 3576-3589.
[5] Li Sisi, Jin Li, Yu Long, Du Lishi, Yue Liang, & Zhang Xirun. (2024). Cloud-side collaborative load resource allocation strategy for multiple controllable load regulation. Telecom Science, 40 (8), 52-62.
[6] Chen Kai, Lei Qi, & Li Doumeng.(2024). Typical scenario generation and output characteristics analysis of distributed photovoltaic power generation in rail transit infrastructure based on improved K-means clustering. Journal of Electrical Engineering, 19 (2), 364-372.
[7] Alaei, M. H., & Latify, M. A. (2024). Dynamic investment planning of CVR implementation considering PEVs'reactive power compensation capability. IET Generation, Transmission & Distribution, 18(4), 793-808.
[8] Xu Jian, Lei Xichun, Liao Siyang, Ke Deping & Sun Yuanzhang.(2024).Automatic power control of direct-controlled virtual power plant with multi-feeder load coordination in distribution network.1-16.
[9] Zhang Shixu, Miao Shihong, Yang Weichen, Liu Zhiwei, Wang Peng & Cao Kan. (2021). Distributed robust optimal scheduling strategy for distribution network based on adaptive step size ADMM. High Voltage Technology (01), 81-93.

Journal of Physics: Conference Series 3012 (2025) 012009 doi:10.1088/1742-6596/3012/1/012009

The Distributed Resource Clustering Method Based On Indicator-weighted K-means++

Qiang Ye[1], Wuxiao Chen[1], Shian Zhan[1], Yuqing Cai[1], Zhijun Jiang[1], Han Lin[1], Hongfeng Xie[2*], Siyang Liao[2]

[1] Marketing Service Center State Grid Fujian Electric Power Co., Ltd, Fuzhou, China

[2] School of Electrical and Automation Wuhan University Wuhan, China

*Corresponding author's e-mail: 2875590664@qq.com

Abstract. The large number of distributed resources in the distribution network provides new resources for grid interactive regulation. First, the regulation characteristics of distributed resources are analyzed, and a method for calculating adjustable capacity is established. Then, the k-means++ algorithm with indicator weights is used to cluster the distributed resources. Finally, a simulation is conducted on a photovoltaic cluster with 60 units, and the results are compared with traditional methods.

1. Introduction

With the continuous growth of electricity demand in China, the traditional power supply model can no longer fully meet this demand. In response to China's carbon neutrality and green development goals, the share of renewable energy in the power grid is steadily increasing, promoting the transformation and upgrading of the power system. However, due to the volatility and uncertainty of new energy, traditional grid regulation methods are no longer sufficient. In this context, the distribution network is evolving from a traditional system that only receives and distributes electricity to users, into a new type of distribution network that actively interacts with the higher-level grid, with an integrated approach of source-grid-load-storage (SGLS). Therefore, there is an urgent need to deepen the exploration and utilization of adjustable resources to participate in grid interaction.

Currently, research on adjustable resources in the distribution network mainly focuses on distributed energy resources (DER), which include distributed photovoltaic (PV), energy storage system (ESS), electric vehicles (EV), and flexible loads with energy storage characteristics. However, the regulation characteristics of different types of DER vary, requiring in-depth research. The large-scale integration of PV power into the grid causes the distribution network to exhibit fluctuations in both the magnitude and direction of power flow, which are randomly influenced by photovoltaic output. Therefore, coordinated control with energy storage is often considered[1]. Reference [2] considers the priority index of various indicators for sequential planning and optimization of energy storage, promoting the absorption of new energy in the grid. However, introducing energy storage into the distribution network is often economically inefficient and faces issues related to capacity configuration and site selection[3]. As for flexible loads such as electric vehicles located on the demand side, they have energy storage-like characteristics and are often treated as generalized loads to participate in grid regulation[4-5]. By establishing mathematical models for them, their regulation characteristics and capabilities under different operating conditions can be evaluated to respond orderly to grid control instructions[6-7].

Distributed resources in the distribution network typically have the characteristics of being numerous, having small individual capacities, and being widely distributed in terms of location[8]. Direct centralized management of a large number of control resources often leads to difficulties in solving and a sharp increase in variable dimensions. Therefore, these resources need to be aggregated appropriately[9]. This not only reduces the complexity of solving the problem but also allows for the performance differences of control resources to be assessed through cluster analysis, providing decision support for formulating distributed resource regulation strategies in the distribution network. Currently, resource aggregation management is commonly implemented through virtual power plants (VPP), which participate in grid interaction[10-11].

In summary, this paper will conduct an in-depth study of the regulation characteristics and regulation capabilities of various distributed resources. Based on this research, it proposes a k-means++ distributed resource aggregation method with indicator weighting, which reasonably aggregates distributed resources to participate in grid demand response in a clustered manner.

2. Calculation of Adjustable Capacity of Distributed Resources

This paper considers DER such as PV, energy storage, and electric vehicles, and studies their response characteristics and adjustable capacity calculation methods when responding to grid control instructions.

2.1. Indicators of distributed resources

2.1.1. Adjustable capacity. Adjustable capacity represents the control ability of each distributed resource. When faced with demand for upward or downward adjustment, calculating the adjustable capacity of each distributed resource. For PV, in order to ensure PV absorption, it should operate using maximum power point tracking (MPPT), considering only the downward adjustment capability, with no upward adjustment capability. For distributed energy storage and electric vehicles, due to their two states of charging and discharging, both upward and downward adjustment capabilities are considered. The upward and downward adjustment capacities of distributed resources can be calculated using equations (1)-(2):

$$\alpha_{up} = P_{max} - P_t \#(1)$$

$$\alpha_{down} = P_t - P_{min} \#(2)$$

In the equations, P_{max} and P_{min} represent the maximum and minimum operating boundaries of the adjustable resources, respectively, while P_t represents the real-time operating power of the adjustable resources.

2.1.2. Regulation time. Regulation time represents the ability of each distributed resource to respond to control instructions. A smaller regulation time indicates a stronger response capability of the distributed resource.

2.1.3. Unit regulation cost. Unit regulation cost, as an important economic indicator, it is affected by various factors, including the construction investment and regulation pricing of different stakeholders, the amount of power regulation, and the duration of participation in regulation.

2.2. Calculation of Adjustable Capacity for Distributed Photovoltaic

Since the output power of solar panels is influenced by factors such as light intensity, temperature, and other environmental conditions, the operating point (the combination of voltage and current) of photovoltaic cells changes with these variations.

Therefore, the photovoltaic system must constantly adjust its operating point. In practice, distributed photovoltaic systems often operate using MPPT, and it is generally believed that they only have active power downward adjustment capability. According to national standards, under normal operating conditions, the rate of change in active power for a PV station should not exceed 10% of the installed capacity per minute. Therefore, the downward adjustable capacity $\alpha_{r,\text{down}}^{PV}$ is considered to be 10% of the current power $P_{r,t}^{\text{PV}}$.

$$\alpha_{r,\text{ down}}^{PV} = 0.1P_{r,t}^{\text{PV}} \#(3)$$

2.3. Calculation of Adjustable Capacity for Energy Storage

This paper calculates the adjustable capacity based on the APC command time granularity. For electrochemical energy storage, the regulation response time is first calculated using parameters such as charging and discharging power, state of charge (SOC), and charging/discharging efficiency, as shown in equations (4) and (5). Then, the adjustable capacity of the ESS is determined, as shown in equations (6) and (7).

$$\tau_{s,\text{down}}^{\text{ES}} = \frac{(S_{s,\max}^{\text{ES}} - S_{s,t}^{\text{ES}})E_{s,\text{N}}^{\text{ES}}}{P_{s,\text{cN}}^{\text{ES}}\eta_{s,\text{c}}^{\text{ES}}} \#(4)$$

$$\tau_{s,\text{up}}^{\text{ES}} = \frac{(S_{s,t}^{\text{ES}} - S_{s,\min}^{\text{ES}})E_{s,\text{N}}^{\text{ES}}\eta_{s,\text{d}}^{\text{ES}}}{P_{s,\text{dN}}^{\text{ES}}} \#(5)$$

$$\alpha_{s,\text{down}}^{\text{ES}} = \begin{cases} P_{s,\text{cN}}^{\text{ES}} - P_{s,t}^{\text{ES}} & \tau_{s,\text{down}}^{\text{ES}} \geq \Delta t \\ 0 & \tau_{s,\text{down}}^{\text{ES}} < \Delta t \end{cases} \#(6)$$

$$\alpha_{s,\text{up}}^{\text{ES}} = \begin{cases} P_{s,\text{t}}^{\text{ES}} + P_{s,dN}^{\text{ES}} & \tau_{s,\text{down}}^{\text{ES}} \geq \Delta t \\ 0 & \tau_{s,\text{down}}^{\text{ES}} < \Delta t \end{cases} \#(7)$$

In the equations, $S_{s,\max}^{\text{ES}}$ and $S_{s,\min}^{\text{ES}}$ represent the maximum and minimum SOC of the ESS, respectively. $S_{s,t}^{\text{ES}}$ denotes the SOC of ESS s at time t, $E_{s,\text{N}}^{\text{ES}}$ is the rated capacity of ESS s, $P_{s,\text{cN}}^{\text{ES}}$ and $P_{s,dN}^{\text{ES}}$ are the rated charging and discharging power of energy storage s, $\eta_{s,\text{c}}^{\text{ES}}$ and $\eta_{s,\text{d}}^{\text{ES}}$ are the rated charging and discharging efficiencies of energy storage s, and Δt is the APC cycle.

2.4. Calculation of Adjustable Capacity for Electric Vehicles

By satisfying the user's demand, the SOC S_n^{EV} of the electric vehicle at the expected departure time τ_n^{EV} is calculated. Based on this, the SOC $S_{n,t+\Delta t}^{\text{EV}}$ that the electric vehicle needs to reach in the next period is determined.

$$S_{n,t+\Delta t}^{\text{EV}} = S_n^{\text{EV}} - \frac{(\tau_n^{EV} - t - \Delta t)P_{n,\text{cN}}^{\text{EV}}\eta_{n,\text{c}}^{\text{EV}}}{E_{n,\text{N}}^{\text{EV}}} \#(8)$$

In the equations, $P_{n,\text{cN}}^{\text{EV}}$ and $\eta_{n,\text{c}}^{\text{EV}}$ represent the rated charging power and efficiency of EV, and $E_{n,\text{N}}^{\text{EV}}$ represents the rated capacity of EV.

When the EV load power needs to be increased, it is sufficient to ensure that the SOC does not exceed the maximum SOC $S_{n,max}^{\text{EV}}$. In this case, the power upward adjustment boundary is:

$$P_{n,+}^{\text{EV}} = \min\left(P_{n,cN}^{\text{EV}}, \frac{(S_{n,max}^{\text{EV}} - S_{n,t}^{\text{EV}})E_{s,\text{N}}^{\text{ES}}}{\eta_{n,\text{c}}^{\text{EV}}\Delta t}\right) \#(9)$$

In the equation, $S_{n,t}^{\text{EV}}$ represents the SOC of EV n at time t.

When the electric vehicle load power needs to be decreased, if the SOC at time t is greater than $S_{n,t+\Delta t}^{\text{EV}}$, and the user demand is satisfied, the charging power can be reduced to zero. Otherwise, charging power is provided at a certain rate to meet the user's demand. In this case, power downward adjustment boundary is:

$$P_{n,-}^{\text{EV}} = \begin{cases} \min\left(P_{n,cN}^{\text{EV}}, \frac{E_{s,\text{N}}^{\text{ES}}\Delta S_{n,t}^{\text{EV}}}{\eta_{n,\text{c}}^{\text{EV}}\Delta t}\right) & \Delta S_{n,t}^{\text{EV}} \geq 0 \\ 0 & \Delta S_{n,t}^{\text{EV}} < 0 \end{cases} \#(10)$$

In the equation, $\Delta S_{n,t}^{\mathrm{EV}} = S_{n,max}^{\mathrm{EV}} - S_{n,t}^{\mathrm{EV}}$

Based on the above, the upward adjustable capacity $\alpha_{n,\mathrm{up}}^{EV}$ and downward adjustable capacity $\alpha_{n,\mathrm{down}}^{EV}$ of the electric vehicle can be obtained.

$$\alpha_{n,\mathrm{up}}^{EV} = P_{n,+}^{\mathrm{EV}} - P_{n,t}^{\mathrm{EV}} \#(11)$$

$$\alpha_{n,\mathrm{down}}^{EV} = P_{n,t}^{\mathrm{EV}} - P_{n,-}^{\mathrm{EV}} \#(12)$$

3. Weighting Configuration Method

This paper uses the Analytic Hierarchy Process (AHP) to calculate the indicator weights.

3.1. Construct a hierarchical evaluation model

Hierarchical evaluation model is constructed with the control performance of the control resources as the goal layer, and the technical criteria (adjustable capacity, regulation time) and economic criteria (unit regulation cost) are set as the criterion layer. The three aggregation feature indicators are set as the indicator layer.

3.2. Construct the judgment matrix

$$\boldsymbol{E} = (a_{ij})_{n\times n} = \begin{bmatrix} a_{11} & a_{12} & \cdots & a_{1n} \\ a_{21} & a_{22} & \cdots & a_{2n} \\ \vdots & \vdots & & \vdots \\ a_{n1} & a_{n2} & \cdots & a_{nn} \end{bmatrix} \quad (13)$$

Where a_{ij} represents the comparison result of the importance between two indicators, with larger values indicating greater importance. As shown in Table 1.

Table 1. Judgment matrix assignment reference table

Importance Level	Description
1	Equal importance
3	Moderately more important
5	Strongly more important
7	Very strongly more important
9	Absolutely more important
2,4,6,8	Intermediate values

3.3. Calculate the indicator weights

Calculate the nth power of the product of each row to obtain an n-dimensional vector. Then, normalize the results to obtain the indicator weights w_i and the eigenvector $\boldsymbol{F}$.

$$\bar{w_i} = \sqrt[n]{\prod_{j=1}^{n} a_{ij}} \#(14)$$

$$w_i = \frac{\bar{w_i}}{\sum_{i=1}^{n} \frac{\bar{w_i}}{w_i}} \#(15)$$

3.4. Consistency test

Calculate the maximum eigenvalue $\lambda_{\max}$:

$$\lambda_{max} = \frac{1}{I} \sum_{i=1}^{I} \frac{(\boldsymbol{EF})_i}{w_i} \#(16)$$

Calculate the consistency index H_E:

$$H_E = \frac{\lambda_{max} - I}{I - 1} \#(17)$$

Combine the order of the judgment matrix n and the random consistency index G_F to calculate the consistency ratio H_F. When $H_F \leq 0.1$, the judgment matrix is considered reasonable:

$$H_F = \frac{H_E}{G_F} \#(18)$$

4. Consider the k-means++ aggregation method with weighted indicators

K-means++ algorithm has excellent aggregation capabilities and shows significant results when handling large-scale distributed resource data. This paper adopts the k-means++ aggregation method with weighted indicators to achieve better aggregation results.

4.1. Normalize the indicators

The clustering indicators have different dimensions and magnitudes, with significant differences between them, so normalization is required. The selected indicators will be classified into positive and inverse indicators, as shown in equations (19) - (20).

$$X_{ij}^{+} = \frac{X_{ij} - X_{ij,min}}{X_{ij,max} - X_{ij,min}} \#(19)$$

$$X_{ij}^{-} = \frac{X_{ij,max} - X_{ij}}{X_{ij,max} - X_{ij,min}} \#(20)$$

In the equation, X_{ij}^{+} represents the positive indicator, and X_{ij}^{-} represents the inverse indicator.

4.2. The elbow method

This paper uses the elbow method to determine the number of clusters K, with the core metric being the sum of squared errors:

$$n_{\mathrm{SSE}} = \sum_{i=1}^{K} \sum_{p \in C_i} |p - m_i|^2 \#(21)$$

4.3. Selection of initial cluster centers

Step 1: Randomly select one sample from the input dataset as the first cluster center.

Step 2: For each sample in the dataset, calculate the shortest distance d_i to the already selected cluster centers.

Step 3: Select a new sample as the next cluster center. The probability of selection is positively correlated with d_i, meaning the farther the sample is from the existing centers, the higher the probability of being selected.

Step 4: Repeat steps 2 and 3 until K cluster centers have been selected.

4.4. Weighted Euclidean distance calculation

After selecting the K cluster centers, considering the impact of indicator weights, calculating the Euclidean distance between the sample and each cluster center.

$$p(U_n, U_{o,k}) = \sqrt{\sum_{j=1}^{3} w_j (x_j - x_{o,kj})^2} \#(22)$$

In the equation, k = 1, 2, ···, K; U_n represents the sample, $U_{o,k}$ represents the k-th cluster center, x_j represents the indicators of the sample, and $x_{o,kj}$ represents the indicators of the k-th cluster center.

4.5. Sample clustering
Based on the Euclidean distance calculation results, assign each sample feature vector to the cluster corresponding to its nearest cluster center, and then update the cluster centers:

$$U_{o,k} = \frac{1}{N_k}\sum_{C_k} U_{k,n} \#(23)$$

In the equation, C_k represents the cluster, N_k represents the total number of samples in the cluster, and $U_{k,n}$ represents the n-th sample in the k-th cluster.

4.6. Determine the final distribution of cluster centers
Repeat steps 4 and 5 above until the cluster centers no longer update, then determine the final cluster centers and complete the clustering process.

4.7. Clustering validity test
This paper uses the Davies-Bouldin Index (DBI) to test clustering validity. The DBI, also known as the classification accuracy index, reflects the maximum similarity between clusters based on the average distance between samples. It is specifically defined as:

$$I_{DBI} = \frac{1}{K}\sum_{k=1}^{K} \max_{k\neq j}\frac{d(X_k) + d(X_j)}{\|c_k - c_j\|} \#(24)$$

In the equation, $d(X_k)$ represents the average distance between points within cluster X. The smaller the I_{DBI}, the lower the dispersion, and the better the clustering performance.

5. Simulation example

This article presents a simulation case study of a grid-connected photovoltaic cluster with a scale of 60. The selected indicators include adjustable capacity, regulation response time, and unit adjustment cost. The relevant indicators of this photovoltaic cluster are shown in Table 2.

Table 2. Resource clustering indicators

	C(CNY)	α (kW)	T (s)
PV	0.3-0.6	0-15	1.8-6.5

5.1. Distributed photovoltaic clustering analysis

5.1.1. Normalization of indicators. The clustering indicators exhibit different dimensions and magnitudes, with considerable variations among them, necessitating normalization. Selected indicators are classified into positive and negative categories for calculation. The positive indicator includes adjustable capacity, while the negative indicators include regulation response time and unit adjustment cost. Indicators are processed separately, as described in equations (19) and (20).

5.1.2. Determination of K. Using the method demonstrated in Section 3 (the elbow method), the optimal nu mber of K for the distributed photovoltaic cluster is calculated, as shown in Figure 1. When k is smaller than the true number of clusters, the increase in k will significantly improve the aggregation within each cluster, leading to a large decrease in n_{SSE}. However, when k reaches the true number of clusters, further increases in k will result in diminishing returns in terms of aggregation, causing the rate of decrease in n_{SSE} to sharply drop. As k continues to increase, the decrease in n_{SSE}

will gradually level off. In other words, the relationship between n_{SSE} and k forms an "elbow" shape, and the k value at the "elbow" corresponds to the true number of clusters in the data.

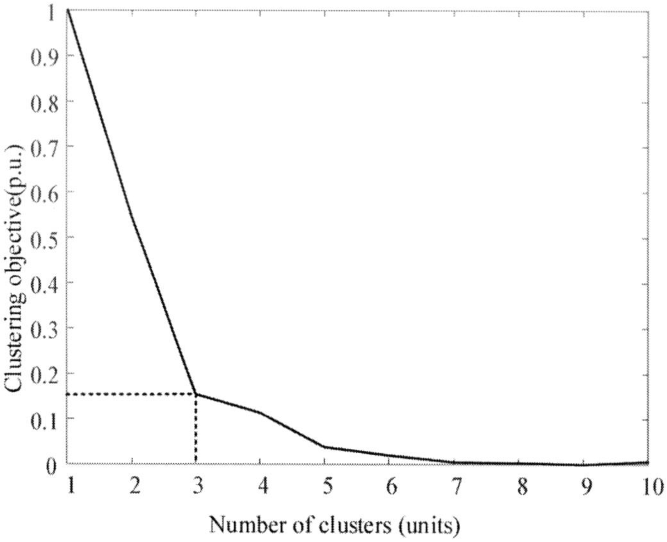

Figure 1. Relationship between clustering objective and number of clusters

It can be inferred from the diagram that the optimal number of clusters for the distributed photovoltaic group is 3.

5.1.3. Indicator weighting. Using the AHP for indicator weighting configuration, the judgment matrices for the criterion layer and the indicator layer are shown in Tables 3 and 4, respectively. Since the economic criterion only considers unit adjustment cost, it is not necessary to construct a judgment matrix. The results of the indicator weighting configuration are shown in Table 5.

Table 3. Judgment matrix of the criterion layer

Criteria	Technical criterion	Economic criterion
Technical criterion	1	3
Economic criterion	1/3	1

Table 4. Judgment matrix of the indicator layer

Technical indicator	Adjustable capacity	Regulation time
Adjustable capacity	1	7/3
Regulation time	3/7	1

Table 5. Indicator weighting configuration results

indicator	Adjustable capacity	Regulation time	Unit adjustment cost
Adjustable capacity	0.56	0.24	0.2

5.1.4. Clustering analysis. Clustering analysis is performed based on the method demonstrated in Section 3, and the specific clustering results are shown in Table 6.

Table 6. Clustering analysis results

	Cluster	C(CNY)	α (kW)	T (s)
Cluster 1	29	0.31	147.1	2.1
Cluster 2	20	0.39	106.6	3.8
Cluster 3	11	0.46	97.9	4.6

The clustering results of K-means, K-means++, and the weighted K-means++ algorithm used in this paper are compared and verified separately, with analysis and comparison conducted using the DBI index. The results are shown in Table 7.

Table 7. Comparison results of different clustering algorithms

	Cluster 1	Cluster 2	Cluster 3	DBI
K-means	21	26	13	1.7431
K-means++	23	25	12	1.7327
This paper	29	20	11	1.7217

It can be seen that the clustering method used in this paper performs better in terms of the DBI index, indicating an improvement in clustering performance.

6. Summary

The K-means++ algorithm considering indicator weights, along with the clustering analysis of distributed resources under different subjective indicator weights, reduces the dimensionality of regulation resources with small capacity and large quantity into resource clusters with large capacity and small quantity. This improves the feasibility and economy of distributed resources participating in grid regulation. Compared to traditional K-means and K-means++ algorithms, it offers greater advantages in meeting grid regulation requirements.

References

[1] Ou, Y, Jiang, T, Zhang, Y, et al. (2024). Voltage control method for high-proportion distributed photovoltaic distribution networks with multi-time-scale control considering inverter reactive power regulation [J/OL]. *Electric Power Automation Equipment*, 1-16. [Online] Available: 2024-11-11.

[2] Shi, Q, Tang, B, Yu, G, et al. (2024). Distribution network distributed energy storage sequencing planning based on priority index [J]. Automation of Electric Power Systems, 48(13): 89-99.

[3] Liu, H, Xu, L, Hao, S, et al. (2020). Distributed hybrid energy storage optimization method based on distribution network partitioning [J]. Electric Power Automation Equipment, 40(5): 137-143.

[4] Zhai, D, Yao, L, Cheng, F, et al. (2024). "Vehicle-hydrogen" flexible controllable load coordinated grid control method to enhance system frequency support capacity [J/OL]. Grid Technology, 1-16. [Online] Available: 2024-11-11.

[5] Zhang, H, Wu, J, Cai, J, et al. (2022). Robust optimization configuration of energy storage in integrated energy systems considering air conditioning and flexible thermal load responses [J]. Grid Technology, 46(07): 2733-2744.

[6] Wang, X, Xu, J, Liao, S, et al. (2021). Flexible load control margin assessment method considering virtual energy storage characteristics [J]. Huadian Technology, 43(09): 37-45.

[7] Long, H, Zhao, S, Wu, Z, et al. (2023). Study on air conditioning load adjustable potential assessment and control strategy considering rebound effect [J]. Grid Technology, 47(09): 3638-3648.

[8] Xu, J, Lei, X, Liao, S, et al. (2024). Automatic power control of direct-controlled virtual power plants for multi-feeder load coordination in distribution networks [J/OL]. Automation of Electric Power Systems, 1-16. [Online] Available: 2024-11-11.

[9] Wu, Y, Yao, L, Liao, S, et al. (2022). Peak shaving method for aggregating distributed photovoltaics and energy storage based on the improved K-means++ algorithm [J]. *Grid Technology*, 46(10): 3923-3931.

[10] Ge, H, Xia, Y, Zou, W, Hao, S, & Ma, G. (2022). Multi VPP bi-level optimization scheduling method based on complementary water-light-storage system. *2022 IEEE 3rd China International Youth Conference on Electrical Engineering (CIYCEE)*, Wuhan, China, 2022, pp. 1-6.

[11] Xu, T, Chen, T, Zhang, X, et al. (2024). Wide-area aggregation and control method for virtual power plant distributed resources based on the Chino polytope [J]. *Automation of Electric Power Systems*, 48(18): 139-148.

doi:10.1088/1742-6596/3012/1/012010

A Method For Modelling the Power Regulation Characteristics of Electrolytic Lead Loads

Qiang Ye[1], Wuxiao Chen[1], Peng Zheng[1], Lingling Zhu[1], Yuqing Cai[1], Tianfang Si[2*], Siyang Liao[2]

[1] Marketing Service Center State Grid Fujian Electric Power Co., Ltd, Fuzhou, China

[2] School of Electrical and Automation Wuhan University Wuhan, China

*Corresponding author's e-mail: 2939107003@qq.com

Abstract. As the proportion of new energy power generation in the power system gradually increases, the requirement for the regulation flexibility of the power system is also improved, relying only on the traditional generation side of the regulation resources can no longer meet the demand for regulation flexibility of the grid, and tapping the regulation potential of the load-side regulation resources has become a major research hotspot. As a typical high energy-consuming load, before guiding the lead load to participate in grid power regulation, whether its power can be regulated and how to regulate it have become an urgent problem to be solved. Based on this, this paper proposes a power regulation characteristic modelling method of electrolytic lead load, by analysing the production links of the process of electrolytic lead load and the coupling relationship between the links, clarifying the key production equipment needed in each link, analysing the adjustable parameters of the key power equipment and the overall impact of parameter adjustment on production, establishing a power regulation characteristic model of high energy-consuming electrolytic lead load, and thereby helping to solve the problem of unknown regulation characteristics of electrolytic lead load, and lay the foundation for the participation of electrolytic lead load in grid power regulation.

1. Introduction

In order to achieve the dual-carbon goal, i.e., carbon peaking and carbon neutrality, the penetration of new energy sources in the power system is increasing at an unprecedented rate [1].However, new energy generation, such as solar and wind, is highly dependent on natural conditions, and its output presents extreme uncertainty and volatility [2-3].This characteristic brings unprecedented challenges to the frequency stability and safe operation of power systems [4].In the traditional mode, the power system relies on the output of FM units to maintain the stability of system frequency [5].However, with the gradual increase of the proportion of new energy sources, it has been difficult to ensure the stable operation of the power system by relying only on the generation-side FM resources [6].Therefore, we need to tap more load-adjustable resources to participate in the power regulation of the grid, so as to enhance the resilience of the power system [7].In this process, high energy-consuming loads have gradually become a high-potential and high-quality regulation resource due to their advantages of high regulation flexibility and large adjustable capacity [8].By reasonably dispatching high energy-consuming loads, we can not only effectively balance the supply and demand relationship of the power system, but also further promote the coordinated development of source, network, load and storage

[9].In recent years, the research on the power regulation characteristics of high energy-consuming loads has become a hot issue of concern in academia and industry. Through in-depth study of the regulation mechanism and potential of high energy-consuming loads, we can make better use of this resource to provide strong support for the stable operation of the power system and the consumption of new energy. This will make an important contribution to the realisation of the dual-carbon target and the promotion of energy transition and sustainable development.

Currently, academia and industry have conducted relatively in-depth studies on typical high energy-consuming loads that account for a relatively large share of energy consumption, such as aluminium electrolysis and polysilicon loads [10]. However, there are relatively few studies on those high energy-consuming loads that account for a relatively small share of the overall loads nationwide, but a relatively high share of the loads in specific provinces or regions. Taking the electrolytic lead load as an example, although it does not account for a prominent proportion of high energy-consuming loads in the whole country, it occupies a considerable proportion in some coastal cities rich in lead and zinc ore resources [11].

If these electrolytic lead loads can be effectively guided to participate in the power regulation of the local grid, they can not only alleviate the pressure on the frequency regulation of the local grid, but also promote the consumption of new energy and the stability of the grid frequency in the context of the increasing proportion of new energy. This is not only of great significance for guaranteeing the safe operation of the power system, but also provides new ideas and ways for future energy transition and sustainable development. Therefore, the research and utilisation of such high energy-consuming loads deserve more attention and investment.

However, there are still a lot of uncertainties about the power regulation capability of lead electrolytic loads, their regulation methods, and the specific impact of power regulation on the lead electrolytic production process. In order to fully exploit the potential of the electrolytic lead load as a high-quality regulation resource, we need to carry out an in-depth study of its power regulation characteristics and establish a corresponding model of power regulation characteristics. This will not only help us better understand the regulation mechanism of electrolytic lead load, but also provide theoretical support for its participation in grid power regulation, so as to jointly maintain the frequency stability of the power system and ensure the safe and reliable operation of the power grid.

2. Modelling the power regulation characteristics of electrolytic lead loads

Electrolytic lead load is a typical high energy-consuming load, and its production process is very complex and delicate. In order to model its power regulation characteristics, it is necessary to analyse its various production links, analyse the production cooperation relationship between the links, derive the key production equipment in each link, and establish a model of its power regulation characteristics.

The production of electrolytic lead load includes four basic links: raw material preparation and crushing, melting and refining, electrolysis, and product separation and collection. The first step of raw material preparation and crushing is mainly to separate the lead-containing ores from other impurities by beneficiation, and then the selected lead-containing ores are crushed for subsequent smelting and refining. The second step of smelting and refining is to enter the crushed lead ore into the smelting furnace for smelting. During the smelting process, lead is extracted and crude lead is produced. Crude lead is then subjected to preliminary refining by fire, with the purpose of removing some of the metal impurities in the crude lead, such as copper, tin, etc., and adjusting the content of impurities such as arsenic and antimony in the crude lead, in order to cast the anode plate adapted to electrolytic refining. The third step of electrolysis is carried out in the electrolysis tank, where the electrolytically refined anode plate is put into the electrolysis tank, the appropriate amount of electrolyte is added, and a current of dozens or even hundreds of amperes is passed in. The fourth step of the product separation and collection session means that after the electrolysis is completed, the anode sludge (containing mainly precious metals and unreacted lead) and cathode lead (high purity electrolytic lead) need to be separated. The cathode lead is washed and dried to obtain a high purity electrolytic lead product. The anode sludge requires further treatment to recover the precious metals. There are a series of coupling relationships such as material interaction in the above four key production stages, and the power regulation characteristic model of the electrolytic lead load needs to be built on the four basic production stages.

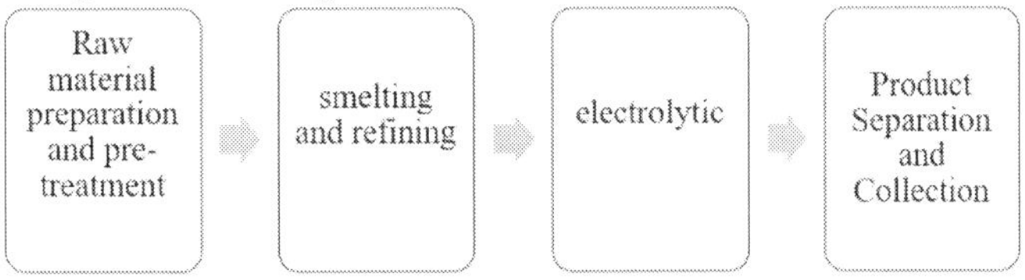

Figure 1. Electrolytic lead load production flow chart

The raw material preparation link mainly involves the screening, washing and pre-treatment of lead ores or recycled lead raw materials, which usually do not involve the continuous operation of high-power equipment, so the need for power regulation is not obvious. In this link, more attention is paid to the quality and uniformity of raw materials, as this link has to provide raw materials for the subsequent smelting and refining process, the quality and speed of raw material provision will also indirectly affect the power consumption of the entire electrolytic lead load. Among them, the crushing link for lead ore crushing treatment, involving the key equipment stone crusher, modern crusher usually adopts frequency conversion speed control technology, by changing the motor power supply frequency to change the speed of the motor. This method can achieve smooth power regulation and avoid impact and wear on the equipment, and its power characteristic can be expressed as:

$$P_{set1} \propto f_{set1} \tag{1}$$

It should be noted that when adjusting the power of the crusher, in order to ensure the normal subsequent production and the stable operation of the entire production line, the frequency should not be adjusted beyond its safe production range:

$$f_{set1\min} \leq f_{set1} \leq f_{set1\max} \tag{2}$$

Where, f_{set1} is the frequency of the crusher inverter, $f_{set1\min}$ is the minimum value of the frequency when the crusher power is adjusted within the safety range, and $f_{set1\max}$ is the maximum value of the frequency when the crusher power is adjusted within the safety range.

In the melting and refining process involves two key production equipment, melting furnace and refining furnace, melting furnace using high temperature will be pre-treated lead ore melting, refining crude lead, melting furnace generally need to maintain high temperature, but can also be adjusted according to the actual production situation in a small range of melting furnace power, and the power of the furnace is usually through the adjustment of the current of the heating elements, such as resistor wires, electrodes and so on to achieve, and its the current can be realised by slowly adjusting the current regulator on the power supply control panel, then the power characteristic of the melting furnace during the production of electrolytic lead load can be expressed as:

$$P_{set2} = I_r^2 R_r \tag{3}$$

Where, I_r is the current through the furnace resistance wire, R_r is the resistance of the furnace resistance wire.

When the melting furnace current decreases, the temperature will be reduced, when the temperature is reduced beyond a certain range, the melting reaction process will not continue, affecting the production process; and when the melting furnace current increases, the temperature increases to more than a certain range, it will affect the service life of the equipment, and at the same time reduce the efficiency of the reaction, therefore, the melting furnace current regulation should meet the following constraints:

$$I_{r\min} \leq I_r \leq I_{r\max} \tag{4}$$

Where, $I_{r\min}$ is the minimum value of current through the furnace under safe production of electrolytic lead load; $I_{r\max}$ is the maximum value of current through the furnace under safe production of electrolytic lead load.

Refining furnace will be smelting the crude lead for further purification, to remove the impurities, refining furnace power adjustment process is similar to the smelting furnace, can also be adjusted by adjusting the size of the current through the refining furnace resistance wire to adjust its power, so its power characteristics can be expressed as:

$$P_{set3} = I_s^2 R_s \tag{5}$$

Where, I_s is the current through the furnace resistance wire, R_s is the resistance of the furnace resistance wire.

Since the temperature and pressure in the refining furnace need to be strictly controlled within a certain range to ensure production efficiency and product quality, the following constraints should be met when regulating the production current of the refining furnace:

$$I_{s\min} \leq I_s \leq I_{s\max} \tag{6}$$

Where, $I_{s\min}$ is the minimum value of current through the refining furnace under safe production of electrolytic lead load; $I_{s\max}$ is the maximum value of current through the refining furnace under safe production of electrolytic lead load.

After completing the refining of the anode plate needs to be put into the electrolysis tank for electrolysis, the electrolysis tank contains a higher temperature electrolyte to provide the reaction environment required for electrolysis, due to the electrolysis tank has a certain degree of heat storage capacity, a short period of time the power change will not have a greater impact on the production of the electrolysis link is also a power-adjustable link. The electrolysis tank is equivalent to a series connection of resistance R and reverse electromotive force E, the power size of the electrolysis tank and the size of the DC current through the electrolysis tank, its power characteristics can be expressed as:

$$P_{set4} = I_d^2 R + I_d E \tag{7}$$

where P_{set4} is the power of the electrolyser and I_d is the DC current through the electrolyser.

Electrolytic lead plants can generally change the voltage of the electrolyser by changing the ratio of the transformer taps, which in turn changes the DC current through the electrolyser:

$$I_d \propto k \tag{8}$$

Where k is the tap ratio of the on-load regulator transformer connected to the electrolyser.

Since the electrolysis reaction needs to be carried out at high temperatures, too low a temperature may cause the reaction to stop and too high a temperature may damage the electrolyser, so the power regulation of the electrolyser also needs to be maintained within the safe production range, so the regulation of the on-load regulator transformer should satisfy the following constraints:

$$k_{\min} \leq k \leq k_{\max} \tag{9}$$

Where, $k_{\min}$ is the minimum value of the ratio of the on-load regulator transformer that enables safe production operation of the electrolytic cell; $k_{\max}$ is the maximum value of the ratio of the on-load regulator transformer that enables safe production operation of the electrolytic cell.

Finally, it is the product separation and collection link of electrolytic lead load, the filter is used to separate the lead and other impurities after electrolysis, and the adjustment of the pressure in the filter can change its power consumption, and the power characteristic of the filter can be expressed as follows:

$$P_{set5} \propto F_{set5} \tag{10}$$

Where, P_{set5} is the power consumption of the filter, F_{set5} is the value of the pressure in the filter.

The pressure values in the filter should be controlled within safe production limits, satisfying the following constraints:

$$F_{set5\min} \le F_{set5} \le F_{set5\max} \tag{11}$$

Where, $F_{set5\min}$ is the minimum value of pressure in the filter under safe production, and $F_{set5\max}$ is the maximum value of pressure in the filter under safe production.

In summary, this paper combines the production process of electrolytic lead load to derive the four major production links of its production process, including raw material preparation and pretreatment, melting and refining, electrolysis, and product separation and collection, and analyses the relationship between the production of each link, and then conducts an in-depth analysis of the links to derive the key production of each link of power equipment, analyses the adjustable parameters, and establishes the high-power power equipment of the power regulation characteristic model is used to complete the modelling of the overall power regulation characteristics of the electrolytic lead load, and this power regulation model is of certain reference significance for the participation of energy-consuming loads in the power regulation of the power grid.

3. Numerical experiments

In order to verify the effectiveness of a power regulation characteristic modelling method of electrolytic lead load proposed in this paper, this paper takes an electrolytic lead load in Fujian Province as an example to carry out an arithmetic analysis. According to the method proposed in the text, the power supply frequency of the crusher, the current of the melting furnace, the current of the refining furnace, the current of the electrolysis tank and the pressure value of the filter of the electrolytic lead load are all adjustable parameters, and the initial value of each adjustable parameter in the article is recorded through the on-site research, and the results are shown in Table 1.

Table 1 Initial values for each adjustable parameter

f_{set10}/Hz	I_{r0}/A	I_{S0}/A	I_{d0}/A	F_{set50}/Pa
50	200	300	800	0.3

The power consumed by the crusher, melting furnace, refining furnace, electrolysis tank, and filter for the electrolytic lead load before parameter adjustment is performed is shown in Table 2.

Table 2 Initial power values for each adjustable device

P_{set10}/kW	P_{set20}/kW	P_{set30}/kW	P_{set40}/kW	P_{set50}/kW
60	20	40	500	10

The adjustable parameters of each adjustable device were appropriately adjusted according to the methods mentioned in the text, field experiments were conducted, the values of the adjusted parameters were recorded, and the power values of each adjustable device after parameter adjustment were measured using a power meter.

Table 3 Adjusted values of each adjustable parameter

f_{set1}/Hz	I_r/A	I_S/A	I_d/A	F_{set5}/Pa
54	205	310	805	0.35

Table 4 Adjusted power values for each adjustable device

P_{set1}/kW	P_{set2}/kW	P_{set3}/kW	P_{set4}/kW	P_{set5}/kW

67	24	45	510	12

Analysing the data in Table 1-Table 4, it can be seen that the power of the electrolytic lead load can be changed by adjusting the power supply frequency of the crusher of the electrolytic lead load, the current of the melting furnace, the current of the refining furnace, the current of the electrolysis tank, and the value of the pressure of the filter, which verifies the power adjustable characteristic of the electrolytic lead load. Moreover, observing the adjustable parameters and their corresponding equipment power in the table, it can be seen that the power consumed by the production equipment increases with the increase of adjustable parameters, which corresponds to the relationship between the power change and the parameter change proposed in the paper, thus verifying the validity of the modelling method of the power regulation characteristics of the electrolytic lead load proposed in this paper.

References

[1] Qiu Yan. Research on the sustainable development strategy of new energy power generation industry under the goal of "double carbon"[J]. China Equipment Engineering, 2024,(20):261-263.

[2] Zhang Xu. Characteristics and application of new energy photovoltaic power generation system[J].Petrochemical Technology, 2024,31(06):384-385.

[3] LU Caiyun, ZHU Wen, ZHANG Haitian, et al. Optimisation research and application of calculation strategy for new energy power generation prediction accuracy[J]. Electrical Application, 2024,43(06):54-59.

[4] Qiu Jun.Construction of new power system and its optimisation strategy under the goal of "double carbon"[J] .Network Security and Informatisation, 2024,(10):6-8.

[5] Paulus M, Borggrefe F. The potential of demand-side management in energy-intensive industries for electricity markets in Germany[J]. Applied Energy, 2011, 88(2): 432-441.

[6] Liao S, Xu J, Sun Y, et al. Control of Energy-Intensive Load for Power Smoothing in Wind Power Plants[J]. IEEE Transactions on Power Systems, 2018, PP(6): 1-1.

[7] Jiang H, Lin J, Song Y, et al. MPC-Based Frequency Control With Demand-Side Participation: A Case Study in an Isolated Wind-Aluminum Power System[J]. IEEE Transactions on Power Systems, 2015, 30(6): 3327-3337.

[8] Xu J, Chen Y, Liao S, et al. Demand side industrial load control for local utilization of wind power in isolated grids[J]. Applied Energy, 2019, 243(JUN.1): 47-56.

[9] Zhao, Z. H., et al. "A new energy rational utilisation assessment method taking into account source-grid-load-storage flexible regulation resources." Grid Technology 1-12. doi:10.13335/j.1000-3673.pst.2024.1360.

[10] Xueyi Jiang, Arrow Xu, Siyang Liao, et al. A multi-type load synergistic control method to smooth wind power fluctuation in industrial parks considering the different response willingness of industrial users[J]. Journal of Wuhan University (Engineering Edition), 2022, 55(09): 876-885.

[11] Li A-Xin, Wang Zai-Hong, Yan Na, et al.A review of fire-processed refined lead and electrolytic lead for lead-acid batteries[J]. Storage Battery,2023,60(05):236-239+250.DOI:10.16679/j.cnki.21-1121.2023.05.002.

Journal of Physics: Conference Series 3012 (2025) 012011 doi:10.1088/1742-6596/3012/1/012011

On-Load Transformer Regulation-Based Power Fluctuation Smoothing Strategy for Electrolytic Lead Loads on Park Contact Lines

Wuxiao Chen[1], **Yuqing Cai**[1], **Zhijun Jiang**[1], **Xuan Deng**[1], **Zhexin Lin**[1], **Yihao Zou**[1], **Tianfang Si**[2*], **Siyang Liao**[2]

[1] Marketing Service Center State Grid Fujian Electric Power Co., Ltd, Fuzhou, China

[2] School of Electrical and Automation Wuhan University Wuhan, China

[*]Corresponding author's e-mail: 2939107003@qq.com

Abstract. Green and low-carbon operation of industrial parks has become the main direction of future development, and the randomness and volatility of new energy output cause power fluctuations in the contact line of industrial parks, which brings great challenges to the frequency stability of the power grid. High energy-consuming loads themselves have certain power regulation capabilities, if enough to fully tap the regulation potential of high energy-consuming loads, will be able to help suppress the contact line power fluctuations. Based on this, this paper takes the high energy-consuming electrolytic lead load in industrial park as an example, and puts forward a strategy of smoothing power fluctuation of contact line based on on-load transformer regulation of electrolytic lead load, firstly, establishes the power smoothing target of the contact line, and analyses the relationship between on-load voltage regulator transformer ratio and the amount of power change of electrolytic lead load, and then, with the goal of minimizing the cost of power regulation for electrolytic lead load, and with the goal of minimising power control costs for all the energy-consuming electrolytic lead loads, we propose that all the on-load voltage regulator transformer ratios of all the energy-consuming electrolytic lead loads are the same. With the objective of minimising the power regulation cost of electrolytic lead load, the on-load regulator transformer ratios of all energy-consuming electrolytic lead loads are taken as decision variables, and the rated on-load regulator transformer ratio is selected as the initial value of the iteration. Solving the optimisation model using Matlab will result in the new energy fluctuation smoothing support power that should be provided by each electrolytic lead load and the on-load regulator transformer ratio of each electrolytic lead load when the regulation cost of the electrolytic lead load is minimised.

1. Introduction

The development of industrial parks towards a green and low-carbon mode of operation has become a dominant trend for future progress [1].China's industrial electricity demand is rapidly migrating to areas rich in new energy resources, such as Xinjiang and Yunnan, where industrial parks integrating wind and solar energy are regarded as a major mainstream direction for the development of industrial enterprises in the future [2].However, it is worth noting that the power generation capacity of new energy sources is not stable, but has intermittent, fluctuating, and unpredictable characteristics [3].This characteristic leads to frequent power fluctuations in the contact line between industrial parks and power grids, which brings unprecedented challenges and great pressure on the safe and stable operation of power grids..

With the aggravation of the global greenhouse effect and the frequent occurrence of extreme weather, the Chinese government solemnly promises the world to strive to achieve "carbon neutrality" by 2030 and "carbon peaking" by 2060 as the "double carbon target" [4]. The "dual-carbon" goal invariably puts heavy pressure on industrial parks, an important carrier of low-carbon energy transition, to accelerate carbon emission reduction [5]. Under the "dual-carbon" strategy, the green and low-carbon operation of industrial parks has become the main direction of future development. Literature [6] points out that industrial parks will enter the energy Internet era, the need to effectively explore the application and value of the industrial park system framework, the construction of industrial park energy system model; literature [7-8] pointed out the need to build a distributed wind power photovoltaic industrial park. China's industrial load accelerates to Xinjiang, Yunnan and other new energy high incidence areas, wind and light integrated industrial parks become the main direction of the future development of industrial enterprises.

Within industrial parks, electrolytic lead loads, as a typical high energy-consuming load, offer the possibility of solving this problem by their power adjustable characteristics. This type of load not only has a huge regulation capacity, but also is flexible and convenient to regulate [9].Through reasonable scheduling and management, high energy-carrying loads can be adaptively adjusted according to the fluctuation of wind power within the park. In this way, the high energy-carrying loads can balance the fluctuation of wind power in the park to a certain extent, which can effectively alleviate the huge pressure on the power grid in the process of promoting the green and low-carbon transformation of industrial parks [10].The implementation of this strategy not only helps to improve the energy use efficiency of the industrial park itself, but also contributes to the safe and stable operation of the power grid.

In this paper, we will study the contact line power smoothing strategy for electrolytic lead loads based on on-load regulating transformer regulation, firstly, establish the contact line power smoothing objective, analyse the relationship between the on-load regulating transformer ratio and the amount of power change of electrolytic lead loads, and establish the optimization model based on the on-load transformer by taking the rated on-load transformer ratio as the initial value of iteration with the objective of minimizing the cost of power regulation of electrolytic lead loads, and taking the rated on-load regulating transformer ratio of all the high-energy-consuming electrolytic lead loads as the decision variable. The day-ahead optimisation model is established based on on-load transformer regulation, and the rated on-load regulator transformer ratio is selected as the initial value for iteration. Finally, the amount of active power regulation and the corresponding on-load transformer ratio of each electrolytic lead load that can suppress the power fluctuation of the contact line in the day-ahead time scale are solved. In order to accelerate the high proportion of new energy access to industrial parks, and promote the green and low-carbon transformation of industrial parks.

2. Day-ahead optimal scheduling model based on on-load transformer regulation

In traditional power systems, distribution networks often play the role of "passive receivers", i.e. they exist as "passive" networks. The core responsibility of the distribution network control centre focuses on the flexible switching of the network topology, the optimal allocation of reactive power, and the maintenance of voltage stability, with the core objective of ensuring that the distribution network can fully meet the load demand, while safeguarding the power quality, reducing the network loss, and continuously improving the reliability of power supply. However, the traditional distribution network scheduling strategy mostly relies on experience and established rules, which is not satisfactory in the face of the increasingly complex power market environment and the emergence of new energy technologies. With the rapid development of renewable energy and the profound changes in the power market structure, the traditional distribution network scheduling model is undergoing a profound transformation, gradually moving towards the direction of intelligence, digitalisation and automation. In this context, the concept of "active consumption of new energy by distribution networks" has gradually become a hot topic in the energy industry. Specifically, power grid enterprises need to continuously improve their flexibility and adaptability, through active scheduling and control, actively accept and effectively consume new energy to ensure the security and stability of the entire power system. Industrial parks, as an important part of the distribution network, are constantly connected to renewable energy sources such as wind power, making the power fluctuations between the park and the power grid. In order to smooth out such fluctuations, improving the flexibility and adaptability of industrial parks has become an urgent problem. This will not only help the industrial park itself to realise green and low-carbon transformation, but also has great significance for the safe and stable operation of the power grid. Through the use of advanced control technology and intelligent management means, the industrial park can more accurately regulate the internal new energy generation and load demand, thus effectively reducing the fluctuation of the power of the contact line and improving the stability and reliability of the entire power system.

The power regulation method of the electrolytic lead load in industrial parks mainly includes regulating the voltage of the high-voltage side of the lead plant bus, the voltage drop of the transformer taps or the saturation reactor. The regulation of the high-voltage side of the bus generally needs to be adjusted on the power generation side, which affects all the node voltages, and it is not applicable to industrial parks with multiple loads; the transformer taps are discrete regulation, and the response time is in the order of seconds; and the saturation reactor is generally used to suppress the anodic effect in the lead load in the electrolyzer. The saturation reactor is generally used to suppress the anodic effect in the electrolysis tank, and adjusting the voltage of the saturation reactor will affect the suppression effect of the anodic effect and thus affect the production quality of the product. Therefore, this paper does not use the method of adjusting the voltage drop of the saturation reactor to regulate the power of the electrolytic lead load, and instead, it adopts the method of adjusting the ratio of the transformer to achieve the power control of electrolytic lead load, and it designs a power fluctuation of electrolytic lead load levelling contact line based on the transformer regulation. Strategy

Firstly, based on historical power consumption data, weather forecast and other information to predict the active output of renewable energy, active demand of electrolytic lead load and active demand of conventional load for the coming day, according to the power smoothing target of the liaison line between the industrial park and the large power grid, consider the control cost of electrolytic lead load participating in demand side response to minimise the impact on production, and formulate the day-ahead operation plan of the transformer for electrolytic lead load on-load regulator for 24h.The proposed control strategy in this paper is to control the loads by means of the demand side response.

The control strategy proposed in this paper is the day-ahead hourly optimal scheduling through on-load transformer regulation, taking the electrolytic lead load on-load regulator as the scheduling object, and formulating a 24h day-ahead scheduling plan for electrolytic lead on-load regulator transformers with a time resolution of 1h. Using the day-ahead predicted power value of the wind power clusters, and taking into account the power constraints on the regulation of the liaison line and the constraints on the lead load in a holistic way, and with the least impact on the production, we optimize. The hourly on-load regulator operation plan is developed with the objective of minimising the impact on production.

2.1. Contact line power smoothing target

The power of the liaison line between the power supply of an industrial park and the main grid is determined by a combination of factors, including the output of renewable energy sources, the demand of industrial loads, and the demand of users' electrical and thermal loads. However, due to the significant uncertainty in the output of renewable energy sources such as wind power and photovoltaic, as well as the complex influence of load characteristics, the power of the liaison line between the park and the power grid tends to fluctuate considerably. Such fluctuations not only bring challenges to the scheduling and operation of the grid, but also may affect the safe and stable operation of the entire grid. Therefore, how to effectively cope with the uncertainty of renewable energy sources and optimise the contact line power between industrial parks and the large power grid has become an urgent problem to be solved. Figure 1 shows the topology schematic of an industrial park in a region, the new energy fluctuations in the park only consider the wind power fluctuations, the total power is recorded as *Pwind*, the power of electrolytic lead loads is recorded as $P_{Pb,i}$, the total power of the rest of the loads is recorded as P_L, the power of the liaison line between the industrial park and the large grid is recorded as Pg, and the number of electrolytic lead loads is N_{Pb}, which is recorded as $i = 1, 2, \ldots, N_{Pb}$.

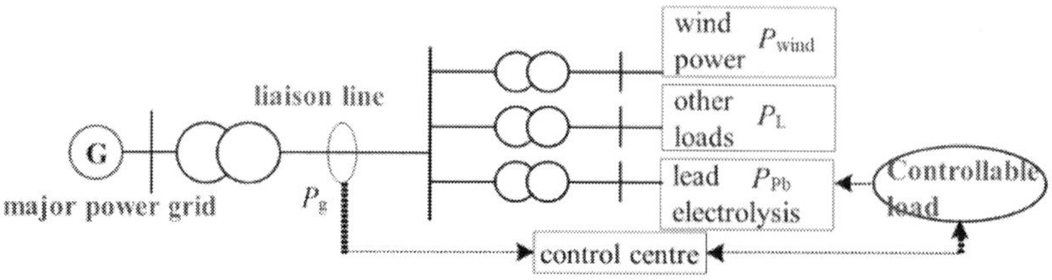

Figure 1 Schematic diagram of a typical industrial park topology

Neglecting line losses, the power balance equation for a typical industrial park is:

$$P_g = \sum_{i \in N_{Pb}} P_{Pbi} + P_L - P_{\text{wind}} \tag{1}$$

If random load fluctuations are not taken into account, the park contact line power fluctuation smoothing is equivalent to wind power fluctuation smoothing. With the increase of wind power penetration in the park, the contact line power fluctuation increases, which is not conducive to the safe and stable operation of the main grid. In this paper, the LPF principle is applied to smooth the contact line power, according to which the recursive form of the contact line power smoothing objective is:

$$P_{\text{gLPF}}(t) = \alpha P_{\text{gLPF}}(t-1) + (1-\alpha) P_{g0}(t) \tag{2}$$

Where, $P_{\text{gLPF}}(t)$ is the contact line active power smoothing target at time t; $P_{g0}(t)$ is the active power of the contact line at time t when no load control is applied; $\alpha = \tau/(\tau+\Delta t)$ is the smoothing parameter, τ is the filtering time constant, Δt is the control period, the smaller α is, the smoother the target contact line power curve is, and the greater the smoothing difficulty is.

Therefore, the typical industrial park unbalanced power ΔP_w is:

$$\Delta P_w(t) = P_{\text{gLPF}}(t) - P_{g0}(t) \tag{3}$$

2.2. *On-load Regulator Transformer Control Strategy*

Day-ahead optimal scheduling of power grids is a key part of the power system, which can effectively optimise the utilisation efficiency of power resources and improve the reliability and economy of power supply. Day-ahead optimal scheduling in industrial parks is to develop the operation plan of on-load regulator transformers based on the power smoothing objective of the contact line to minimise the impact of power regulation on production. Through the previous analysis, it is known that when the sampling time interval of Δt=1h, the amount of active regulation is different, and the impact on production is also different, considering that there is a significant difference in the control cost of the electrolytic lead load participating in the demand-side response, and at the same time this time, saturated reactor does not participate in the regulation, and it can effectively inhibit the anode effect. Therefore, the control cost of day-ahead optimal scheduling for hourly scheduling is mainly to consider the impact of regulating the power of the electrolytic lead load on the production. Then the minimum impact of electrolytic lead electrothermal energy conversion on production is the optimisation objective, and the objective function is:

$$\min \sum_{i=1}^{N_{AL}} (C_{Ti}(\Delta P_{\text{Pb-T}i})) \tag{4}$$

Where, C_{Ti} ($\Delta P_{Pb\text{-}i}$) is the ith electrolytic lead load participation in demand side response control cost; $\Delta P_{Pb\text{-}Ti}$ is the ith electrolytic lead load active regulation.

Constraints:

(1) Industrial parks

Active Balance Constraints:

$$P_{gLPF}(t)=\sum_{i=1}^{N_{Pb}}[P_{Pbi}(t)+\Delta P_{Pbi}(t)]+P_{L}(t)\text{-}P_{wind}(t) \tag{5}$$

In the formula, $P_{gLPFi}(t)$ is the power target value of the contact line at time t; $P_{Pb\text{-}Ti}$ is the active power prediction value of load i at time t when there is no load control; $\Delta P_{Pb\text{-}Ti}$ is the active power adjustment amount of load i at time t; $P_L(t)$ is the active power prediction value of other loads at time t; and $P_{wind}(t)$ is the predicted value of wind power at time.

Contact line transmission power constraints:

$$P_{g,min\text{-}}\le P_{gLPF}(t)\le P_{g,max+} \tag{6}$$

Where $P_{g,min\text{-}}$ is the maximum transmission power in the negative direction of the contact line, i.e. the maximum transmission power from the distribution system to the industrial park; $P_{g,max+\text{-}}$ is the maximum transmission power in the positive direction of the contact line, i.e. the maximum transmission power from the industrial park to the distribution system.

(2) Electrolytic lead load

Through the previous analysis, it can be seen that the regulation of the on-load regulator transformer is discrete regulation, the regulation range is ±3*2.5%, the response time is 7s, then the on-load regulator transformer ratio can be expressed as:

$$k_{Pbi}(t)=(1+0.025n)k_{PbN} \quad n=-3,-2,-1,0,1,2,3 \tag{7}$$

Where, k_{Pbi}(t) is the on-load regulator transformer ratio for moment load i; k_{PbN} is the on-load regulator transformer ratio rating.

The coupling relationship between the on-load regulator transformer ratio and the active power of the electrolytic lead load is:

$$P_{Pb\text{-}Ti}=g(k_{Pbi})=(\frac{[1.35(U_{Pb}-k_{Pbi}U_{SRi})]^2-1.35k_{Pbi}E_{Pb}(U_{Pb}/k_{Pbi}-U_{SRi})}{k_{Pbi}R_{Pbi}}) \tag{8}$$

Then the coupling relationship between the amount of active regulation of the electrolytic lead load and the ratio of the on-load regulator transformer is:

$$\Delta P_{Pbi}(t)=g(k_{Pbi})-g((1+0.025n)k_{PbN}) \quad n=-3,-2,-1,0,1,2,3 \tag{9}$$

It can be seen through the previous analysis, the sampling time interval of Δt=1h, the amount of active regulation will seriously affect the electrolytic lead load electrolyzer temperature, the coupling relationship between the temperature indicators of the production equipment for the electrolytic lead load and the active power is shown below:

$$\Delta P_{Pb-Ti}=cm\Delta T \tag{10}$$

where c is the specific heat capacity of the electrolyte in the electrolyzer, m is the mass of the electrolyte in the electrolyzer, and ΔT is the amount of temperature change in the electrolyzer due to power regulation of the electrolytic lead load.

When the amount of industrial load regulation is $\Delta P_{\mathrm{Pb\text{-}T}i}$, the temperature of the production equipment shall not exceed its upper and lower temperature limits, temperature constraints, during the sampling period $\Delta t = 1\mathrm{h}$:

$$-0.5 \le \mathrm{SOE}_{\mathrm{LPF}i}(t+1) \le 0.5 \tag{11}$$

Where $\mathrm{SOE}_{\mathrm{LPF}i}(t+1)$ is the temperature index of the production equipment during the Δt sampling time interval when the control objective is $\Delta P_{\mathrm{Pb\text{-}T}i}$.

So the day-ahead optimal scheduling model is a discrete nonlinear planning problem, and the co-optimisation model of electrolytic lead loads can be solved for nonlinear integer optimisation using the Yalmip toolbox of MATAB, in which the on-load regulator transformer ratios of all the high-load-energy industrial loads are the decision variables, and the rated on-load regulator transformer ratios can be selected as the initial values of the iterations. Solving this optimization model will yield the new energy fluctuation smoothing support power $\Delta P_{\mathrm{Ab}i}$ that should be provided by each electrolytic lead load when the cost of regulating the electrolytic lead load is minimized and the on-load regulator transformer ratio $k_{\mathrm{AL}i}$ for each electrolytic lead load.

3. Numerical experiments

In order to verify the effectiveness of the on-load transformer regulation based on the electrolytic lead load to suppress the power fluctuation of the park contact line, this paper is based on the RTDS real-time simulation platform to build an industrial park grid in Fujian Province, and the verification is carried out through the real-time interaction of MATLAB online calculation and RTDS simulation. Figure 2 shows the schematic diagram of an industrial park network in Fujian Province, the high energy-carrying adjustable loads in the industrial park considered in this paper are electrolytic lead (Pb1, Pb2, Pb3, Pb4, Pb5, Pb6), and their rated capacities in the industrial park are 450+j280MVA and 200+j100MVA, respectively; the grey shaded solid box represents the new energy stations, which are mainlywind power, with a total installed capacity of 300 MW; conventional loads have a rated capacity of 20+j10MVA.

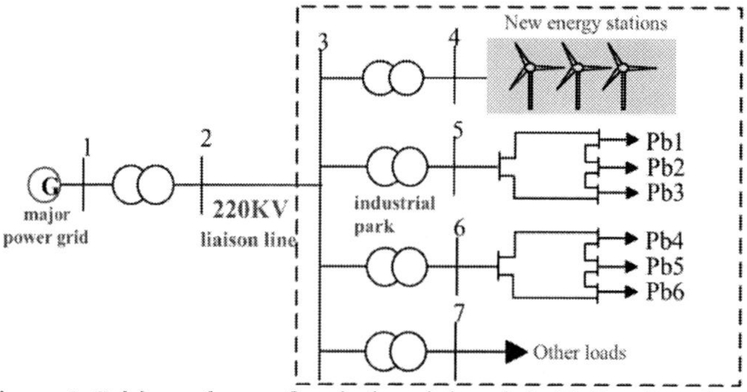

Figure 2 Grid topology of an industrial park in Fujian Province

The predicted power of each industrial load on the simulation day is 410, 400, 390, 160, 150, 140 (MW); the conventional load on the simulation day is 20 MW. The simulation step size t=30s. The predicted data of wind power on the simulation day is shown in Fig. 3.

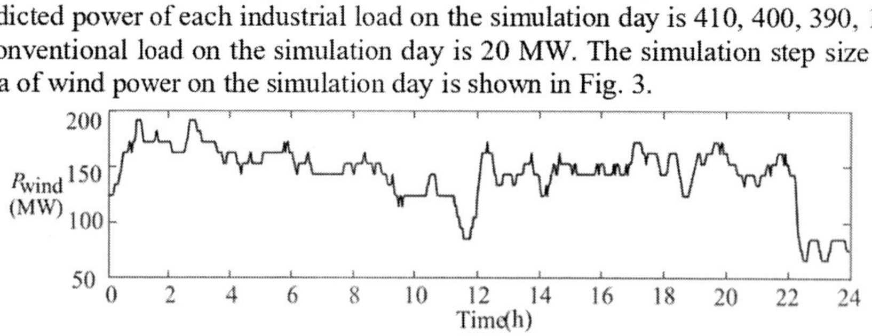

Figure 3 Wind power forecast

According to the strategy proposed in this paper, the on-load regulator ratios for each electrolytic lead load are calculated using matlab and the results are shown in Table 1.

Table 1 Work programme for on-load regulator transformers

	0:00-1:00	1:00-2:00	2:00-3:00	3:00-4:00	4:00-5:00	5:00-6:00	6:00-7:00
Pb1	-1	0	0	-1	-1	0	0
Pb 2	0	0	0	-1	-1	1	1
Pb 3	1	0	0	-2	-1	0	2
Pb 4	-1	0	0	-1	-2	0	0
Pb 5	0	0	0	-1	-1	1	1
Pb 6	1	0	0	-2	-1	0	2
	7:00-8:00	8:00-9:00	9:00-10:00	10:00-11:00	11:00-12:00	13:00-14:00	15:00-16:00
Pb 1	-2	-1	0	0	-3	-2	0
Pb 2	-2	-1	1	0	-1	-1	0
Pb 3	-1	-1	0	0	-2	-2	0
Pb 4	-2	-1	2	0	-3	-2	0
Pb 5	-2	-1	1	0	-1	-1	0
Pb 6	-1	-1	0	0	-2	-2	0

The contact line power smoothing effect is shown in Fig. 4.In the figure, P_{gLPF} is the contact line smoothing target; P_{g0} is the contact line free power when the industrial load is uncontrolled; P_g is the contact line free power under the strategy of this paper. It can be seen that under the control strategy proposed in this paper, the wind power is effectively smoothed, thus proving the effectiveness of the strategy proposed in this paper.

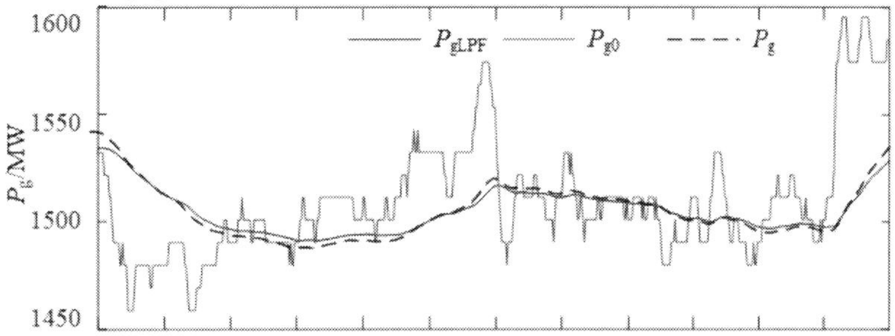

Figure 4 Power smoothing effect of liaison line

References

[1] DING Xin, XU Arrow, SUN Yuanzhang, et al.Control strategy for source-load coordination smoothing wind power fluctuation in networked energy-consuming electrolytic aluminium industrial power grid[J]. Power Automation Equipment, 2022,42(11):47-55.DOI:10.16081/j.epae.202207015.

[2] CHEN Yu-Wei, WU Xi-Chun, ZHOU Lin. Exploration on the participation of energy-consuming industrial loads in grid peaking auxiliary service[J]. Power Safety Technology, 2023,25(11):1-4.

[3] LI Yixin, WU Weijie, ZHANG Yining, et al. Inertia deficit mechanism and inertia level assessment of power system with large-scale new energy access[J]. Southern Energy Construction, 2024,11(05):132-139.DOI:10.16516/j.ceec.2024.5.14.

[4] Huang Yuhan, Ding Tao, Li Yuting, et al. A review of energy decarbonisation technologies in the context of carbon neutrality and implications for the development of new power systems[J]. Chinese Journal of Electrical Engineering, 2021, 41(S1): 28-51.

[5] Chen Lujun. Industrial Parks Must Contribute to Peak Carbon Neutrality[J]. Resource

Recycling, 2021, (02): 15-20.
[6] WANG Qiang, CUI Yong, ZHANG Xiaowei. Design of energy internet system architecture for industrial parks under "dual carbon" strategy[J]. Electricity Technology, 2022(14): 180-183.
[7] Liu Y, Li B and Deng Y. Real-Time Control Strategy of Hybrid Energy Storage to Smooth Out Fluctuations in PV Industrial Parks[C]//2021 IEEE 5th Conference on Energy Internet and Energy System Integration (EI2), Taiyuan, China: IEEE, 2021: 1668-1673.
[8] Jiang C, Lin H, Wen E, et al. A Demand Side Response Strategy for Large Industrial Customers Considering the Uncertainty of Renewable Energy Generation[C]//2020 International Conference on Smart Grids and Energy Systems (SGES), Perth, Australia: IEEE, 2020: 285-290.
[9] Ma Guiwei. Optimisation method of wind power consumption in source-load system taking into account high energy-consuming loads [D].Shenyang University of Technology, 2020.DOI:10.27322/d.cnki.gsgyu.2020.000566.
[10] Bao Y. Research on the participation of high energy-consuming electrolytic aluminium loads in power system frequency regulation and auxiliary service strategy[D]. Wuhan University, 2019.

A Fault Diagnosis Method for UHV DC Control and Protection Based on Dynamic Time Window

Ningmin Zheng[1], Wujie Chao[2], Lei Gan[3], Qiannan Zhang[3], Chunsheng Wu[4*]

[1] Minyue Interconnection Power Operation Co., Ltd,Zhangzhou, China

[2] State grid Fujian Electric Power Research Institute,Fuzhou, china

[3] State Grid Fujian Ultra-high Voltage Company,Fuzhou, china

[4] HVDC Transmission Branch of XJ Group Co., Ltd,Xuchang, China

*Corresponding author's e-mail:wuchunsheng_xj@163.com

Abstract. In this paper, a fault diagnosis method based on a dynamic time window is proposed to enhance the accuracy and efficiency of fault diagnosis in a DC control system protection mechanism. In the DC transmission system, the normal operation of the control and protection device is very important to ensure the stability and security of the power grid. However, traditional fault diagnosis methods frequently encounter difficulties in promptly and precisely pinpointing faults, especially when dealing with complex and variable fault information. The core of the proposed method is to use the time window technology to efficiently process the fault information in the DC control and protection system.Specifically, an initial time window is established to capture and filter the event list pertaining to the fault. Subsequently, a multi-dimensional and multi-level analysis of this event list is conducted, enabling the precise identification of the fault's type, location, and cause, which provides strong support for subsequent fault handling and system recovery. The experimental results show that the fault diagnosis method based on dynamic time window has high accuracy and practicability, significantly enhancing both the efficiency and effectiveness of fault diagnosis.

1. Introduction

As an important part of the UHV power grid, the UHV DC transmission system has the advantages of long transmission distance, large transmission power, fast start-up and adjustment speed, strong controllability, flexible active power regulation capability, and the ability to quickly improve the operating characteristics of AC systems. It efficiently transmits clean energy over wide areas with low loss, and represents an important technical solution for long-distance, large-capacity power transmission. As more and more UHV DC transmission projects are completed and put into operation, the workload of debugging , operation and maintenance of the UHV DC control and protection system is increasing. How to improve the debugging, operation and maintenance capabilities and levels of the UHV DC control and protection system, while ensuring the operational safety of UHV DC transmission projects, has become a key issue that needs to be resolved urgently in the operation and maintenance process of UHV DC transmission projects.

Manual DC fault diagnosis has been the standard practice for many years. However, with the advancement of technology and the increasing complexity of DC systems, this method has begun to show its limitations. The following are some of the main drawbacks of manual DC fault diagnosis:

Inefficiency: Manual detection tends to consume a considerable amount of time, especially in large and complex DC systems.

Accuracy issues: Human judgment may be influenced by factors such as subjectivity, fatigue, and insufficient experience, which may lead to misdiagnosis of faults.

Handling Big Data Challenge: As modern sensor technology advances, vast amounts of data are swiftly acquired, rendering manual analysis both difficult and time-consuming.

Slow response time: In power systems, rapid response to faults is critical, yet manual diagnosis may not be able to make decisions quickly at crucial moments.

Difficulty in detecting complex faults: For some complex or intermittent faults, in-depth technical knowledge and analysis may be required, which may exceed the capabilities of most operators.

This paper proposes a dynamic time-window-based fault diagnosis method for DC control and protection systems. When a fault occurs in the system (such as ground faults, short circuit faults, etc.), it generates significant disturbances and may result in power loss. At this time, both primary and auxiliary equipment will produce a large number of messages, wave recording data, and SER information. The fault diagnosis method proposed in this paper, based on the mechanism of information filtering and subscription, can quickly obtain all alarm events that occur during the fault. By dividing the alarm events at the fault moment according to the time window and based on the set fault diagnosis strategy library, it can automatically identify and save the operating conditions before and after the fault, collect SER data and wave recording files at the fault moment, and initiate automatic analysis. Traditional manual fault diagnosis methods require operational and maintenance personnel to search and filter key information from a large amount of complex messages and collect related wave recordings. Due to the experience of personnel, there may be incomplete or inaccurate information collection, leading to untimely reporting to relevant departments. Moreover, the analysis of events and wave recordings heavily relies on the experience of the personnel, resulting in low efficiency in accident handling. The fault diagnosis method proposed in this paper for DC control and protection systems achieves the replacement of manual labor and automated fault analysis. It aims to improve the efficiency and accuracy of fault diagnosis through automated and intelligent means.

2. DC control and protection fault diagnosis method based on dynamic time window

Addressing the aforementioned issues, this paper introduces a dynamic time window-based fault diagnosis method for DC control and protection systems, aiming to enhance the efficiency and accuracy of fault diagnosis through automation and intelligent approaches.

1) Real-time Data Acquisition and Dynamic Time Window Construction: Real-time acquisition of protection events and recording their occurrence times. Using these occurrence times as the starting point, a dynamic time window is established. The length of this time window can be adjusted based on actual conditions, ensuring that all protection events pertaining to the current fault are captured. The protection events collected within the time window are stored in a list for subsequent analysis.

2) FaultId Query: Based on the corresponding relationship between protection events and faultId (this relationship is established through historical experience analysis), the system queries the fault ID corresponding to each protection event in the list. This correspondence may include various scenarios such as one protection event corresponding to a single fault, one protection event corresponding to multiple faults, or multiple protection events corresponding to single fault.

3) Fault Type Determination and Diagnosis.

If the queried faultId belongs to a parent fault type, meaning it encompasses multiple sub-faults, further queries are made from the protection event list to determine if there are corresponding sub-fault protection events.If such events exist, the description corresponding to the identified sub-fault is designated as the current fault. If they do not exist, the current fault is determined to be at least one of the sub-faults within the parent fault, in order to specify the exact scope of the fault.

If the queried faultId corresponds to a sub-fault type, then the description associated with that sub-fault is directly identified as the description of the current fault.

Dynamic Time Window Update: The dynamic time window algorithm is employed to update the protection event list in real-time, ensuring the timeliness and accuracy of the data. This algorithm boasts low computational complexity, enabling efficient updates to the protection event list and enhancing the real-time and accurate nature of fault diagnosis.

Exhaustive Query and Comprehensive Analysis: When querying sub-fault protection events, an exhaustive traversal of the list is performed to ensure that every protection event is thoroughly considered. Additionally, a comprehensive analysis of the logical relationship between protection events and faultId is conducted to arrive at the most accurate fault diagnosis results.

3. Fault diagnosis of DC control and protection based on dynamic time window Method example application

In this paper, during the execution of real-time querying for protection events, the precise time of occurrence for each event is recorded. This time of occurrence serves as the initial point for defining a temporal window for the respective protection event, with a preestablished duration specifying the window's length. Protection events gathered within this temporal window are then compiled into a list. Subsequently, the corresponding faultId for the protection events within this list are queried. If the faultId is of a parent fault type, which encompasses multiple sub-faults, the list of protection events is queried to determine if there are any corresponding protection events generated by the sub-faults. If such events are identified, the description corresponding to the sub-fault that triggered the protection event is designated as the current fault. Conversely, if no corresponding protection events from sub-faults are found, the current fault is deemed to be at least one of the sub-faults within the parent fault category. If the faultId is of a sub-fault type, the description corresponding to that sub-fault is directly determined as the current fault. The following will illustrate this process with specific examples.

3.1. Real-time acquisition of protection events is performed, with all events occurring within a specified time frame being systematically stored in a dedicated list.

The diagnosis of the faulty component requires a comprehensive judgment based on multiple events that occur at the same time. These events usually complete their actions within a certain time frame, but there is a possibility of delayed event actions or simultaneous failures in new component. The use of a "dynamic time window" algorithm can encompass these events and seamlessly integrate them into the fault diagnosis framework.

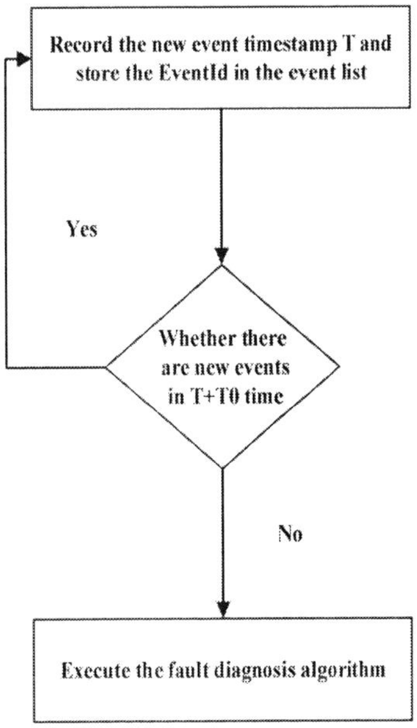

Figure 1. Decision event list.

The flowchart illustrating the "Dynamic Time Window" algorithm is presented in Figure 1. When a new protection event occurs, it is promptly captured in real-time, and the timestamp of this event is recorded as the initial point for the corresponding time window. A predefined duration, denoted as T, serves as the width of the time window, within which all protection events are captured and stored in a dedicated list. It is then judged whether any new protection events occur within the set duration T (time interval T0). If so, the information of the new protection event is recorded into the list, and the process repeats, adopting the timestamp of the latest event as the new starting point for the time window, and the predetermined duration T serving as the length of the new time window for subsequent judgments.

3.2. Query the faultId corresponding to the protection events in the list.

When a component fails, a series of protective action events usually occur within a certain time frame. Based on these events, control and protection experts can analyze the possible faulty components through experience. The correspondence between protection events and faults includes: one protection event corresponding to one fault, one protection event corresponding to multiple faults, and multiple protection events corresponding to one fault. The relational table outlining the correspondence between protection events and fault components is presented in Table 1.

Table 1. Relationship table between events and fault parts

EventId	Description	Fault Position
101	Valve group overcurrent protection switching system	Short circuit fault in valve area
102	Valve group over-current protection Y lock	
103	Single bridge commutation failure protection switching system	Valve area short-circuit fault, pulse loss, AC system fault, etc
104	Single bridge commutation failure protection S latches	

Table 1 is split into Table 2 and Table 3, and the faultId corresponding to the protection event is queried through the event and faultId relationship table.

Table 2. Event and fault id relationship table

EventId	Description	FaultId
101	Valve group overcurrent protection switching system	1
102	Valve group over-current protection Y lock	1
103	Single bridge commutation failure protection switching system	2
104	Single bridge commutation failure protection S latches	2

Table 3. Event and fault id relationship table

FaultId	Description	ParentId
1	Short circuit fault in valve area	2
2	Short circuit fault in valve area, pulse loss, AC system fault, etc	4

3.3. Fault diagnosis

The flowchart of the fault diagnosis algorithm is presented in Figure 2. If the faultId corresponding to the protection event is found to be the parent fault type, which encompasses multiple sub-faults, then the corresponding sub-fault is traversed and inquired whether the corresponding event is generated. If not, the fault component is determined to be the fault description corresponding to at least one sub-fault in the parent fault. On the other hand, If the faultId corresponding to the protection event is found to be a sub-fault type, the fault component is judged to be the description corresponding to the sub-fault.

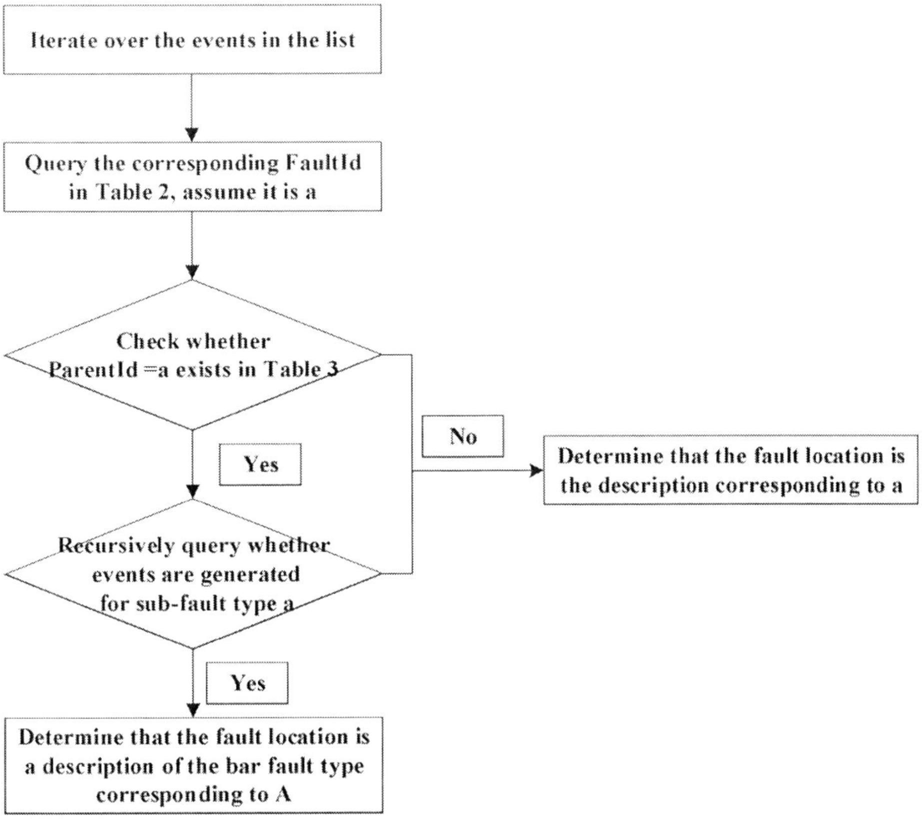

Figure 2. DC control protection fault diagnosis.

If the faultId corresponds to the parent fault type and an event is generated in one of the corresponding sub-fault as a result of the query, the fault component is initially determined to be the description corresponding to the parent fault. However, it should be noted that other sub-fault within the parent fault may also have faults occurring simultaneously, albeit with a lower probability of being the actual fault. Therefore, these other sub-fault are not immediately considered as the source of the fault, but should be investigated further if necessary.

As shown in Table I, events 101 and 102 exhibit a many-to-one relationship with "valve area short circuit fault". event 103 and 104 demonstrates a one-to-many relationship, as it can correspond to multiple fault types such as "valve area short circuit fault, pulse loss, AC system fault," among others.

If only events 101 and 102 are present in the event list, querying Table 2 reveals that the corresponding faultId for both events 101 and 102 are 1. Further querying Table 3 shows that faultId equal to 1 is a sub-fault id, indicating that the fault can be diagnosed as "valve area short-circuit fault".

However, if the event list includes 103 and 104, querying Table 2 indicates that the corresponding faultId for both events 103 and 104 are 2. Querying Table 3 then reveals that faultId equal to 2 is a parent fault id. In this case, it is necessary to check whether any events have occurred within the child faults of the parent faultId equal to 2. If the event list also includes 101 or 102, the fault can be primarily diagnosed as "valve area short-circuit fault", with "loss of pulses" or "AC system fault" considered less likely, therefore not taken into account. Thus, the fault is definitively diagnosed as "valve area short-circuit fault".

On the other hand, if the event list only contains 103 and 104, the fault can be diagnosed as one of the faults within the set "valve area short-circuit fault, loss of pulses, AC system fault, etc.", effectively narrowing down the scope of possible faults.

4. Experimental result

Based on the data from the station control and protection system of Taizhou, we have summarized and organized the fault data from the DC filter area, valve area, pole area, bipolar area, DC line area, and converter transformer area. A total of 83 types of faults were identified, as detailed in the Table 4.

Table 4. Fault type list

Serial number	Fault type
1	AC/DC collision line
2	BPS switch fault
3	Single bridge commutation failure protection switching system
4	F101 ~ F107: Internal ground fault of DC filter
5	F108 ~ F117: high voltage capacitor C11 ground fault
6	F108 ~ F117: high voltage capacitor C12 ground faul
7	F118 ~ F127: high voltage capacitor C21 ground fault
8	F301, F302, F303: Ground fault in the pole bus area
9	F304: valve group connection line ground fault
10	F305 or F306: Neutral bus zone ground fault
......	......

From the various fault types mentioned above, we conducted the following simulation experiments. We randomly simulated 100 fault scenarios in the control and protection system, and meticulously recorded the accuracy of the fault diagnosis method proposed in this paper for each simulation, thereby obtaining the accuracy of the diagnostic results for this set of experiments. To ensure the reliability and generality of the results, we did not stop there but further expanded the scale of the experiments by conducting an additional 10 sets of similar experiments. In each set of experiments, we detailed the results, and the specific outcomes are shown in the figure 3.

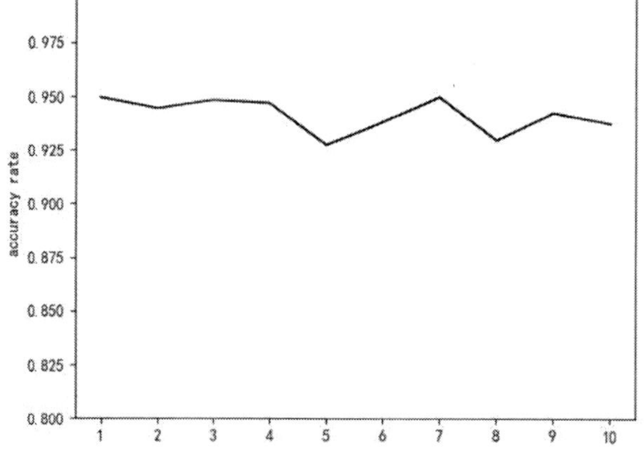

Figure 3. Accuracy of algorithm diagnosis resul.

To validate the effectiveness of the fault diagnosis method proposed in this paper, simulations of typical faults in the DC filter, converter valve, pole area, bipolar area, and DC line fault areas were conducted based on data from the Taizhou Ultra High Voltage (UHV) converter station. The experimental data are shown in Table 5

Table 5. Experimental data

Fault area	Number of Fault types	Sample data
Dc filter	9	1242
Converter valve	10	2022
Polar region	7	1248
Bipolar region	5	166
Dc line	3	12960

Ten simulation experiments were conducted using the fault diagnosis method proposed in this paper. Ten personnel were selected from the field operation and maintenance staff to manually judge the faults. The average time and average accuracy of fault determination by both methods are statistically summarized in the Table 6 and Table 7

Table 6. Result of traditional manual decision method

Fault area	Accuracy	Time(Hour)
Dc filter	93.25%	5.08
Converter valve	96.64%	8.26
Polar region	92.37%	6.26
Bipolar region	98.29%	2.51
Dc line	90.32%	15.86

Table 7. The result of fault determination method is presented in this paper

Fault area	Accuracy	Time(Seconds)
Dc filter	90.34%	327.94
Converter valve	92.33%	301.76
Polar region	89.62%	324.38
Bipolar region	94.78%	269.77
Dc line	86.68%	330.98

As clearly shown in the figure 3, the innovative fault diagnosis method proposed in this paper achieves an accuracy rate of over 90% in fault type identification consistently. This data fully demonstrates the superior performance of the method in terms of accuracy and reliability. Compared to the traditionally labor-intensive and time-consuming fault diagnosis process that relies solely on manual effort, the fault diagnosis method proposed in this paper exhibits significant advantages. It not only rapidly and accurately identifies faults but also guarantees a high level of accuracy, significantly reducing the time required for fault recognition and localization, and improving overall work efficiency.

5. Conclusion

Based on the above experimental results, the fault diagnosis method for control and protection systems proposed in this paper demonstrates strong practical feasibility and promotional value. Based on this fault diagnosis method, we have further developed a DC control and protection fault diagnosis system, which has been successfully deployed and implemented at multiple important converter stations such as Taizhou and Huaian. The application of this system has greatly reduced the workload of field operation and maintenance personnel, who no longer need to spend a lot of time and effort manually troubleshooting and diagnosing faults. Instead, they can rely on the system to quickly and accurately locate and solve problems.

Additionally, the application of this system has significantly improved the efficiency of fault diagnosis and maintenance for control and protection systems among field operation and maintenance personnel. They can respond to faults more quickly, reducing the outage time and economic losses caused by faults, thereby ensuring the stable operation and power supply quality of the power system. In summary, the fault diagnosis method for control and protection systems proposed in this paper and its application in practical scenarios not only have important theoretical significance but also play a huge role in actual production, providing powerful technical support for the safe and stable operation of the power system.

With the increasing importance of UHVDC technology in the global power system, the research on its fault diagnosis capability is particularly critical. Current research has made some progress, but there are still many challenges and untapped potential. Looking ahead, we can foresee technological innovation and development trends in several fields, which will provide new directions and opportunities for further research and application of UHVDC fault diagnosis. With the improvement of algorithm efficiency, real-time fault diagnosis will become possible. This can not only respond to system failures in a timely manner, but also reduce the reliance on manual intervention and improve the reliability and autonomy of the system through automated processing. In this paper, the dynamic time window algorithm and automatic analysis process are introduced to realize the fast and accurate fault diagnosis of DC control and protection system. This method can not only significantly improve the efficiency and accuracy of fault diagnosis, but also reduce the work burden of technical experts, and provide a strong guarantee for the safe and stable operation of power system.

References

[1] ZY, Wu. JG, Zheng. DY, Li. FS, Han, and XX, Zheng. (2023) Design of intelligent checking software for HVDC control and protection program based on Xuji route. Mechanical and Electrical Information, 06:12-16.

[2] XS, Wang. (2022) Fault analysis of DC telecontrol system in a converter station. Qinghai Electric Power,41(01): 45-49.

[3] HT, Liu. QW, Ji. and LL, Jia. (2021) Construction of dynamic test platform for UHV DC system control and protection device. China Measurement & Test, 47(S2): 144-148.

[4] YX, Long. XG, Wang. ZW, Zhou. RS, Wang. and HF, Yin. (2024) Research on text time window division based on LDA model. Science Focus, 19(02): 34-45.

[5] X, Shi. ZL, Feng. HP, Zhao. and XS, Qin. (2023) Fast calculating method of satellite time window to Earth based on polygon distance. Computer Measurement & Control, 31(09): 290-296.

[6] XG, Qi. HR, Zhang. JH, Li. and MM, Shan. (2023) NFV network fault detection based on dynamic time distortion. Journal of Nanjing University(Natural Science), 59(02): 343-350.

[7] MD, Zhang. Q, Xu. ZH, Liu. CY, Ma. and P, Gao. (2022) Water quality time series prediction based on dynamic sliding window BP neural network. Journal of Environmental Engineering, 12(03): 809-815.

[8] QQ, Liao. (2021) Research on alarm correlation based on dynamic time window and alarm level weight. Digital Technology & Application, 39(01): 37-40.

[9] Lin, Sheng. (2023) Analysis of grounding fault location in substation intelligent DC system. Electronic Technology, 52(11): 288-289.

Journal of Physics: Conference Series **3012** (2025) 012013 doi:10.1088/1742-6596/3012/1/012013

Day-Ahead Optimal Scheduling of Wind-Solar-Hydro-Storage Complementary System Based on Deep Reinforcement Learning

Xinzhi Ding[1,2*], Jiaxu Shen[1], Juncong Sai[1], Qi Jin[1], Yongsheng Ding[1], Zhengcheng Dong[3]

[1] Yunnan electric power test and Research Institute (Group) Co. Kunming, 650217

[2] State Key Laboratory of Advanced Electromagnetic Engineering and Technology, School of Electrical and Electronic Engineering, Huazhong University of Science and Technology, Wuhan, 430074

[3] School of Automation, Wuhan University of Technology, Wuhan, China

[*]Corresponding author's e-mail: D202080565@hust.edu.cn

Abstract. The access to wind and solar with volatility increases the difficulty of power system regulation. Recently, reinforcement learning has been widely used in power system scheduling, and hydropower has been increasingly emphasized as an effective way to meet the demand for flexible resource. In this paper, a day ahead scheduling method based on Soft Actor-Critic algorithm for wind-solar-hydro-storage complementary system is proposed to deal with the uncertainty and enhance the economic benefits.

1. Introduction

During the development of modern industry, the excessive use of traditional fossil fuels has given rise to a series of environmental problems, and the shortage of fossil fuels also leads to energy security concerns and tensions in the supply of needs [1]. Recently, to cope with these tough challenges, some countries have vigorously promoted the development of the green energy sources such as wind power and photovoltaics to transform the energy structure. Among them, in order to realize the target of zero carbon emissions, China has gradually increased the proportion of renewable energy and accelerated the construction of efficient and clean new energy system [2].

However, due to the influence of weather factors, wind power and photovoltaics exhibit obvious volatility and intermittency, which lead to the intensification of the peak and frequency regulation pressure during the direct grid integration, and also limit the capacity of renewable energy consumption, thus generating lots of wind and solar abandonment. Therefore, there is a need for adjustable generation resources with greater flexibility to operate in a complementary manner with them. Since hydropower has the advantages of high storability and fast regulation, its role in the power system has gradually changed from power generation to energy regulation. The joint transmission of wind power and photovoltaics can effectively balance the fluctuation of renewable energy generation. In addition, due to the time-shift and regulation characteristics of energy storage technology, it has also become an effective means of controlling wind and solar power generation. Therefore, the research on optimal scheduling of wind-solar-hydro-storage complementary system is an important prerequisite and guarantee to play the advantages of multi-energy system as well as to improve the penetration rate of new energy generation in the power grid.

Essentially, the combined operation of wind-solar-hydro-storage is a stochastic sequential decision-making problem centered on hydropower station and storage device. Currently, researchers have conducted various studies about the optimal scheduling of multi-energy complementary systems. These studies can be mainly classified into mathematical planning methods, heuristic optimization methods and deep reinforcement learning methods at the algorithmic level. Due to the access of multiple distributed energy sources, the optimization objectives are increasingly rich and the operation constraints become more complex. The use of mathematical planning methods to solve faces the problems of dimensional catastrophe and difficult derivation process, while the use of heuristic algorithms is susceptible to the local optimum, and the optimization productivity decreases significantly with the increase of the problem scale [3]. In addition, these mathematical planning methods and heuristic algorithms hardly consider the uncertainties of renewable energy sources and loads comprehensively. In recent years, deep reinforcement learning has demonstrated its powerful decision optimization capability in the power sector [4]. As a model-free machine learning method, deep reinforcement learning can effectively utilize the feedback information obtained from interaction with the environment by setting appropriate actions and rewards to continuously learn the complex laws and empirical knowledge of the system. The policy network trained with relevant historical data can effectively adapt to the stochastic changes in the multi-energy complementary system, and has stronger generalization ability and solution speed.

2. Related Works

There have been some literature studies regarding complementary system of multiple energy sources including wind power and photovoltaics. Reference [5] proposed a day-ahead optimal scheduling model for wind-solar-storage system considering carbon emission, which transforms the consumption of new energy and load shedding into economic indexes and adopts particle swarm algorithm to solve. The experimental results show that the model can sufficiently take into account the actual operating characteristics and realize the optimal operating economy by regulating the charging and discharging power of the storage battery and the exchange power with the upper power grid. Reference [6] proposed a wind-solar control strategy incorporating energy storage system based on double deep Q network (DDQN), which realizes the cooperative operation of wind and solar power generation through end-to-end control, effectively accommodating the uncertainties and improving the benefits of the system. These methods utilize the regulation capability of the energy storage system to deal with the uncertainties and improve the level of new energy consumption as well as the system stability.

In addition, some studies have combined the hydro with wind power and photovoltaics to meet their demand for flexible resources. Reference [7] developed a wind-solar-hydro dispatch model with multi-objective optimization by considering peak curtailment and downstream navigation demand. This model takes the standard deviation of the residual load and the downstream water level as indicators to evaluate the multi energy complementary system, and then solves the pareto optimal problem by using the ε-constraint method. Reference [8] established a short-term optimal dispatch model for wind solar–hydro hybrid pumped storage system solved by mixed integer linear programming (MILP). The model combines cascaded hydropower, wind power and solar power for complementary generation, which effectively increases the total power generation of the complementary system and reduces the net load fluctuation as well as the abandoned wind and solar energy. Reference [9] proposed a wind-solar-hydro system based on conditional generative adversarial network and inner and outer nested particle swarm algorithm-dynamic programming (PSODP). The conditional generative adversarial network can effectively consider the uncertainties of wind and solar power, and the PSO-DP algorithm can more reasonably realize the water allocation of the cascade hydropower station and improve the power generation capacity. Considering that most of the existing regulation methods for wind-solar-hydro hybrid systems are traditional optimization methods, which are difficult to accurately model the existing large amounts of uncertainties, Reference [10] proposed a short-term optimal scheduling model for hydro-wind-solar multi-energy power systems based on the deep Q-network method with the goal of maximizing the power generation benefits. The model fully considers the prediction uncertainties of wind and solar outputs, and has higher power benefits and computational efficiency than the traditional methods. However, the simple DQN network will discretize such continuous problems as scheduling of multi energy complementary systems, resulting in increased error and computation. From the above, it can be seen that there are few studies that consider the optimal scheduling of wind-solar complements combining energy storage and hydropower at the same time, and also rarely apply deep reinforcement learning techniques to it. Therefore, in this paper, we take the wind-solar hydro-storage complementary system as the target to conduct the research on the day-ahead optimal scheduling based on deep reinforcement learning.

3. Wind-Solar-Hydro-Storage Complementary Optimal Scheduling Model

3.1. *Wind-solar-hydro-storage complementary system structure*

The multi-energy complementary system studied in this paper is mainly composed of wind station, photovoltaic station and energy storage station, and the structure is shown in Fig. 1. The wind power, photovoltaics and hydropower are the main sources of power supply in the multi-energy complementary system, in which the hydropower undertakes base load and peak regulation, and regulates the output power together with the energy storage.

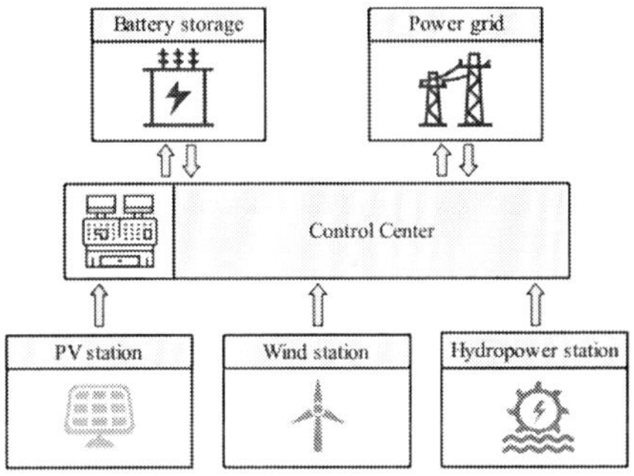

Fig. 1. The complementary system structure.

3.2. *Objective function*

During the scheduling, the objective function takes economy as the optimization target. Specifically, the overall revenue from the purchase and sale of electricity between the system and the upper-level grid is taken as the target, and the costs of abandoned power and load shedding are taken into account in order to promote the utilization of new energy sources and improve the stability of the system's power supply, which can be expressed as follows:

$$\max F_1 = (R_b - R_e - F_c) \tag{1}$$

$$R_b = \begin{cases} \sum_{t=1}^{T} C_t^{\text{buy}} \cdot P_t^{\text{buy}}, P_t^{\text{buy}} \le 0 \\ \sum_{t=1}^{T} C_t^{\text{sale}} \cdot P_t^{\text{grid}}, P_t^{\text{grid}} > 0 \end{cases} \tag{2}$$

$$R_e = \sum_{t=1}^{T} C_t^{e} \cdot P_t^{e} \tag{3}$$

$$R_c = \sum_{t=1}^{T} C_t^{c} \cdot P_t^{c} \tag{4}$$

Where, T denotes the scheduling time period, R_b denotes the benefits from the purchase and sale of electricity, C_t^{buy} and C_t^{sale} are the purchase price and sale price of electricity, respectively. R_e denotes the penalty of power abandonment, and R_c indicates the penalty of load shedding. C_t^e and C_t^c are the abandonment cost and shedding cost, respectively. P_t^e and P_t^c are the amount of abandoned power and load shedding, respectively.

3.3. Constraints

In order to guarantee the safety of wind-solar-hydro-storage complementary system's operation, it is necessary to establish the constraints according to the system model, which mainly includes the wind-solar output constraints, hydropower constraints, battery storage station constraints, power balance constraints, and the exchange power constraints between the system and the higher grid, etc. The specific equations are as follows.

1) Wind and solar power constraints

$$0 \le P_w^t \le P_{w,max} \tag{5}$$

$$0 \le P_{pv}^t \le P_{pv,max} \tag{6}$$

The output of the wind and photovoltaic stations in the system must be within the agreed boundaries. $P_{w,max}$ and $P_{pv,max}$ indicate the maximum values of wind and PV generation, respectively.

2) Battery storage constraints

$$SOC_t = SOC_{t-1} + (P_t^{B,ch} \cdot \beta_{ch} - P_t^{B,dis} / \beta_{dis}) \cdot \Delta t / W_E \tag{7}$$

$$SOC^{min} \le SOC_t \le SOC^{max} \tag{8}$$

$$SOC_1 = SOC_T \tag{9}$$

Where, SOC_t represents the amount of charge state at moment t, $P_t^{B,ch}$ and $P_t^{B,dis}$ represents the charge power and discharge power. β_{ch} and β_{dis} represent the charge and discharge efficiency respectively. W_E represents the rated capacity of the battery. SOC^{min} and SOC^{max} are the minimum and maximum charge state amounts, respectively. The initial and final charge state amounts in a scheduling period should remain the same.

3) Hydropower constraints

Hydropower constraints include water balance constraint, hydropower output constraint and reservoir capacity constraint. The formulas are as follows:

$$V_t^H = V_{t-1}^H + (I_t^H - Q_t^H) \cdot \Delta t \tag{10}$$

$$P_t^{H,min} \le P_t^H \le P_t^{H,max} \tag{11}$$

$$V_t^{H,min} \le V_t^H \le V_t^{H,max} \tag{12}$$

Where I_t^H represents the inflow of the hydropower station at moment t, Q_t^H represents the power generating flow at moment t, $V_t^{H,min}$ and $V_t^{H,max}$ are the minimum and maximum values of reservoir capacity, respectively. $P_t^{H,min}$ and $P_t^{H,max}$ represent the minimum and maximum values of hydropower output, respectively.

4) Exchange power constraint

$$P_t^{grid,min} \le P_t^{grid} \le P_t^{grid,max} \tag{13}$$

Where, $P_{grid,min}^t$ and $P_{grid,max}^t$ denote the minimum and maximum values of the allowed power exchange between the complementary system and the higher grid at moment t, respectively.

5) The power balance constraint

$$P_t^{grid} + P_t^w + P_t^{pv} + P_t^H + P_t^{grid} + P_t^{B,dis} = P_t^l + P_t^{B,ch} \tag{14}$$

Where, P_l^t denotes the load demand at moment t. When the left side of Eq. (14) is greater than the right side, abandonment operation is required. When the right side is greater than the left side, load shedding operation is needed.

4. Deep reinforcement learning framework

4.1. Soft actor-critic algorithm

The proposed framework based on the soft actor-critic (SAC) algorithm is shown in Fig. 2. The SAC algorithm uses the experience replay mechanism to store the generated historical data. When the amount of data in the reply buffer reaches a certain level, the model parameters are trained by randomly sampling the required data in small batches. After training, the new policy network in turn generates new data by interacting with the environment again and deposits it in the reply buffer. Through the repeated alternation of interaction and training, the reward value of the agent can finally be realized to converge and the optimal policy network is obtained.

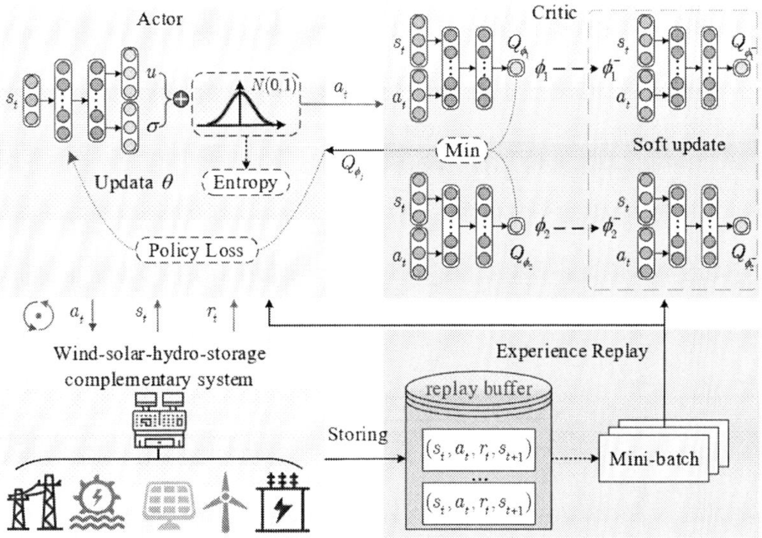

Fig. 2. Scheduling framework based on SAC algorithm.

SAC is an offline learning reinforcement learning algorithm in the actor-critic framework that uses a maximum entropy strategy [11]. The optimal policy π^* learned by the SAC algorithm can be expressed as:

$$\begin{cases} \pi^* = \arg\max_{\pi} E_{(s_t,a_t)\sim\rho_\pi}[\sum_t r(s_t,a_t) + \alpha H(\pi(\cdot \mid s_t))] \\ H(\pi(\cdot \mid s_t)) = E_{a\sim\pi(\cdot\mid s_t)}[-\log \pi(a_t \mid s_t)] \end{cases} \tag{15}$$

Where, $r(s_t,a_t)$ denotes the reward function, α is the entropy regular term, and $H(\pi(\cdot \mid s_t))$ is entropy at state s_t. By maximizing the entropy, SAC has stronger environment exploration and generalization capabilities, which can avoid converging to local optimal solution too early.

The SAC model consists of a policy network with parameter θ, two value networks with parameters ϕ_1 and ϕ_2, and two target value networks with parameters ϕ_1^- and ϕ_2^-. The value network concatenates the state vector s_t and action sets a_t as inputs, and outputs the corresponding state-action values $Q(s_t,a_t)$ through multiple hidden layers. Based on the idea of TD3, SAC uses two value networks and two corresponding target value networks to alleviate the problem of value overestimation and improve the stability of the algorithm. The loss function of the value network is expressed as:

$$J_Q(\omega) = E_{(s_t,a_t)\sim D}[\frac{1}{2}(Q_\phi(s_t,a_t) - \hat{Q}(s_t,a_t))]^2 \quad (16)$$

$$\hat{Q}(s_t,a_t) = r(s_t,a_t) + \gamma E_{s_{t+1}\sim p(s_{t+1}|s_t,a_t)} V_{\phi^-}(s_{t+1}) \quad (17)$$

$$V_{\phi^-}(s_{t+1}) = E_{a_{t+1}\sim \pi_\theta(\cdot|s_{t+1})}[\min_{j=1,2} Q_{\phi_j^-}(s_{t+1}, a_{t+1}) - \alpha \log \pi(a_{t+1} \mid s_{t+1})] \quad (18)$$

Then, SAC adopts soft update method to update the parameter weights of the target value network.

$$\begin{cases} \phi_1^- \leftarrow \tau\phi_1 + (1-\tau)\phi_1^- \\ \phi_2^- \leftarrow \tau\phi_2 + (1-\tau)\phi_2^- \end{cases} \quad (19)$$

Where, ε_t denotes the noise sampled from normal distribution N. Leveraging the loss function $J_\pi(\theta)$, we can train the policy network to learn how to choose the optimal successive actions in a given state.

In addition, the entropy regularity term α needs to be automatically adjusted according to different states, and SAC defines it to be updated as follows:

$$V(s_t) = E_{a_t\sim\pi, s_t\sim R}[-\alpha \log \pi(a_t \mid s_t) - \alpha H_0] \quad (20)$$

4.2. Algorithm design

The variables in the state space of the wind-solar-hydro-storage complementary system based on SAC algorithm include timestep, electricity price, storage charge state, current output of the hydro, load, solar generation, wind generation and the inflow of the hydropower station. The expression of the state space is as follows:

$$s_t = (t, SOC^t, V_H^t, P_w^t, P_{pv}^t, P_l^t, C_{buy}^t, C_{sale}^t) \quad (21)$$

The action space includes the power generation flow of the hydropower station as well as the charge power and discharge power of the energy storage, which can be expressed as:

$$a_t = (Q_H^t, P_{B,dis}^t, P_{B,ch}^t) \quad (22)$$

The reward function consists of the operation benefits from the purchase and sale of electricity and the rule penalties of energy storage, expressed as:

$$r_1^t = w_1 F_1 + w_2 F_2 \quad (23)$$

Where w_1 and w_2 represent the weighting coefficients. The overall training process of the SAC algorithm is shown in Table 1.

5. Case Study

In order to validate the effectiveness of the proposed reinforcement learning approach, a case of wind-solar-hydro-storage complementary system is built for simulation. The case contains an energy storage station with 500kW installed capacity, a photovoltaic station with 800kW installed capacity and a wind power station with 1000kW installed capacity. The normal water level of the configured hydropower station is 1179.6 m, and the dead water level is 1175.8 m. Only a single 800 kW unit operated during the study period, with a maximum generation flow of 3.56 m/s. During the moment of normal-water period, the inflow to the reservoir is almost uniform on each day. In addition, the dispatch cycle of system is 24 hours, and this paper considers the effect of time-of-use electric price in one day to obtain higher power generation benefits by rationally controlling the action of energy storage and hydro generation. Each day is divided into peak, valley and flat periods with different purchase and sale prices.

Table 1. SAC training process for scheduling problem

Input: load, wind power output, solar power output, electricity price, etc.
Output: Output of energy storage and hydropower station
1. Initialize neural network parameters θ, ϕ_1, ϕ_2, ϕ_1^-, ϕ_2^- and reply buffer D

2. **For** episode $n=1$ to max episode N **do**
3. Initialize state s_1
4. **For** each environment step $t=1:T$ **do**
5. Select the action $a_t = \pi_\theta(s_t)$ according to the current state s_t and policy π
6. Perform the action a_t, observe the reward r_t and new state s_{t+1}
7. Store the quadruples (s_t, a_t, r_t, s_{t+1}) in D
8. **End for**
9. If updating network parameters **do**
10. Sample a mini-batch B from D
11. Update the value parameters θ with gradient descent via Eq. (16)- Eq. (18)
12. Soft update the target value parameters ϕ_1^-, ϕ_2^- via Eq. (19)
13. Update the policy parameters θ with gradient descent via Eq. (20)
14. Update the entropy regularity term α via Eq. (21)
15. **End for**

The SAC algorithm is executed in pytorch environment to solve the intelligent scheduling problem and trained on a single NVIDIA GeForce RTX 3070Ti GPU. The batchsize is set to 128, the capacity of replay buffer is 2000, the hidden dimension of neural network is 512, the learning rate of the actor network and critical network both are 0.001. The discount factor γ and Soft-updated parameter τ is set to 0.99 and 0.002, respectively. This paper splits the data of one month in a certain region into training set and test set in a ratio of 2:1.

Given that the load and wind-solar output of different dispatch cycles may cause the fluctuation of the rewards in the iterative process, this paper uses the average reward value of different scheduling cycles in the training set to carry out the corresponding smoothing process, and the convergence curve of the total reward is shown in Fig.3. It can be noticed that the rewards value obtained by the agent through the continuous learning from the environment is increasing. In the later stage of training, the reward gradually converges, and the policy network can effectively realize the decision-making of the complementary system.

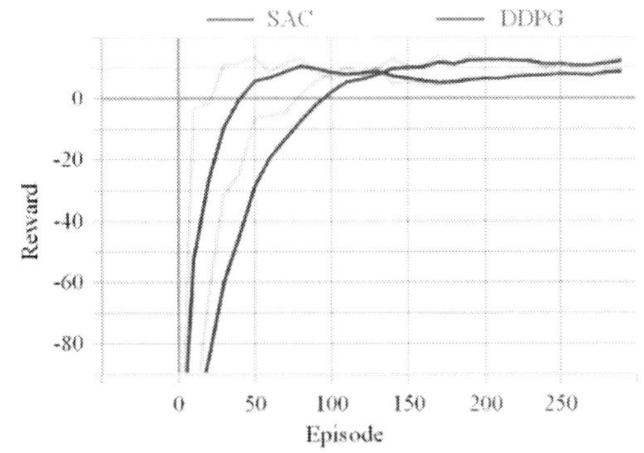

Fig. 3. The reward value of the training process.

In addition, in order to verify the superiority of SAC algorithm in the task of optimal scheduling of wind-solar-hydro storage, we choose another reinforcement learning algorithm DDPG for comparison, which is an actor-critic framework based on deterministic strategy [12]. We can see from the results in Fig. 3 that although SAC does not converge as fast as DDPG, it can eventually obtain a higher reward value due to its stronger exploration ability granted by the stochastic strategy. This strongly suggests the advantages of the SAC algorithm during the iteration procedure.

We further analyzes the performance on the proposed algorithm in day ahead scheduling by randomly selecting a day from the test set. Table 2 shows the results of SAC and DDPG. It can be found that SAC has better generalization performance and robustness compared to DDPG. The total benefits obtained by SAC increased by 13.5% compared to DDPG. The policy network of SAC obtained through training on the same historical data is able to significantly reduce the costs of wind and solar abandonment compared to DDPG, and achieve a more rational regulation of energy storage and hydropower.

Table 2. Experimental results of SAC and DDPG

Methods	**SAC**	**DDPG**
Total benefits	448.0	394.4
Sale benefits	629.7	673.8
Purchase costs	108.5	77.8
Abandonment costs	73.2	201.6
Shedding costs	0	0

In addition, from Fig. 4, it can be seen that when the generation capacity of wind and solar is low, the hydropower can effectively take the role of baseload, reducing the costs of purchasing power from the upper grid. When the generation capacity is large, hydropower can acquire impressive profits from

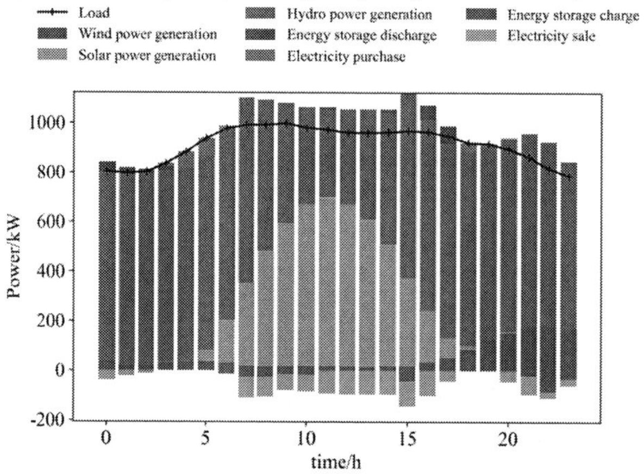

Fig. 4. Scheduling results based on SAC algorithm.

electricity selling. In addition, energy storage can further reduce the amount of wind and solar abandonment as well as load shedding penalty, which improves scheduling flexibility and stability in cooperation with hydropower. For example, at 17:00 when the sale price of electricity is high and the load demand is large, the energy storage can both meet the load demand by discharging and further profit by selling electricity. The above findings demonstrate that the approach based on SAC algorithm has superior scheduling performance.

6. Conclusion

In this paper, a complementary optimal scheduling model based on SAC algorithm is proposed, which fully takes into account the uncertainty of wind-solar output and utilizes the regulating effects of hydropower and storage to maximize the economic benefits.

The case simulation results show that the reinforcement learning method is able to train a policy network that can effectively deal with the uncertainty of wind and solar power through historical data and also indicate that the SAC algorithm based on maximum entropy has unique advantages in scheduling of multi-energy complementary system, which can explore more action space.

Acknowledgments

This research was Supported by Yunnan Technology Innovation Center of New Power Systems, Yunnan Engineering Research Center for Efficient Construction and Utilization and Industrial Support of Green Energy and the Joint Funds of the Natural Science Foundation of Hubei Province, China (No. 2024AFD364).

References

[1] Fan W, Hao Y. An empirical research on the relationship amongst renewable energy consumption, economic growth and foreign direct investment in China[J]. Renewable energy, 2020, 146: 598-609.

[2] Aktas A, Erhan K, Özdemir S, et al. Dynamic energy management for photovoltaic power system including hybrid energy storage in smart grid applications[J]. Energy, 2018, 162: 72-82.

[3] Ding Z, Wen X, Tan Q, et al. A forecast-driven decision-making model for long-term operation of a hydro-wind-photovoltaic hybrid system[J]. Applied Energy, 2021, 291: 116820.

[4] Xu W, Zhang X, Peng A, et al. Deep reinforcement learning for cascaded hydropower reservoirs considering inflow forecasts[J]. Water Resources Management, 2020, 34: 3003-3018.

[5] Zhu Y, Lan Z, Wei Z, et al. Research on optimal operation of wind-PV-ES complementary system considering carbon emission cost[J]. Power system protection and control, 2019, 47(10): 127-133.

[6] Huang S, Li P, Yang M, et al. A control strategy based on deep reinforcement learning under the combined wind-solar storage system[J]. IEEE Transactions on Industry Applications, 2021, 57(6): 6547-6558.

[7] Liu B, Peng Z, Liao S, et al. A multi-objective optimization model for the coordinated operation of hydropower and renewable energy[J]. Frontiers in Energy Research, 2023, 11: 1193415.

[8] Liu Y, Zhang H, Guo P, et al. Optimal Scheduling of a Cascade Hydropower Energy Storage System for Solar and Wind Energy Accommodation[J]. 2024.

[9] Li F, Chen S, Ju C, et al. Research on short-term joint optimization scheduling strategy for hydro-wind-solar hybrid systems considering uncertainty in renewable energy generation[J]. Energy Strategy Reviews, 2023, 50: 101242.

[10] Jiang W, Liu Y , Fang G ,et al. Research on short-term optimal scheduling of hydro-wind-solar multi-energy power system based on deep reinforcement learning[J].Journal of Cleaner Production, 2023, 385:135704-.

[11] Haarnoja T, Zhou A, Abbeel P, et al. Soft actor-critic: Off-policy maximum entropy deep reinforcement learning with a stochastic actor[C]//International conference on machine learning. PMLR, 2018: 1861-1870.

[12] Lillicrap T P, Hunt J J, Pritzel A, et al. Continuous control with deep reinforcement learning[J]. arXiv preprint arXiv:1509.02971, 2015.

Sampling Time Error Correction Method in Synchronized Phasor Measurement Using Akima Piecewise Cubic Hermite Interpolation

Xiaodong Yin[1],Changxi Yue[1],Siyuan Liang[1,*],Li Yao[2],Wei Liu[2],Jicheng Yu[1]

[1]China Electric Power Research Institute, Wuhan 430070, China
[2]State Grid Zhejiang Marketing Service Center, Hangzhou 310007, China

E-mail: *Corresponding author: liangsiyuan@epri.sgcc.com.cn

Abstract. This paper introduces a novel method for synchronized phasor measurement, employing soft synchronized sampling with temporal pulse signals like Pulse Per Second (PPS) from GPS as the time reference. The method begins by detecting and characterizing sampling time errors in the local oscillator relative to the GPS pulse signals. These errors are then corrected using Akima piecewise cubic Hermite interpolation, effectively aligning the sampling instances to an ideal timing grid. The corrected data are processed using a Recursive DFT algorithm to extract synchronized phasors with high precision. Compliance with the C37.118.1 standard is demonstrated by evaluating critical metrics such as frequency, phase angle, and Total Vector Error (TVE). The results highlight the accuracy and efficiency of this software-centric approach, offering a cost-effective and scalable solution compared to conventional hardware-intensive methods for power grid phasor measurement.

1. Introduction

With the acceleration of the construction of new power systems, the synchronized phase measurement accuracy in the grid wide-area monitoring system (WAMS) has become a key technical bottleneck restricting the development of smart grid. The traditional synchronous sampling technique is limited by fixed parameter configurations, which makes it difficult to guarantee the accurate perception of dynamic power quality under complex working conditions, and the cumulative effect of its steady state error seriously restricts the reliability of power grid state estimation and stability control. Aiming at this core problem, a dynamic adaptive sampling parameter adjustment mechanism is proposed to provide theoretical support for improving the grid panoramic sensing accuracy and realizing the precise control of smart grid. This research direction not only has important engineering application value, but also provides a new methodological framework for the improvement of the theoretical system of dynamic synchronous measurement of power system.

Advancements in Synchrophasor Measurement Techniques; recent works have explored innovative methods to address challenges in synchrophasor measurement, particularly for fault detection and transient event localization. Izadi and Mohsenian-Rad introduced a method leveraging synchronous waveform measurements to locate transient events and incipient faults in distribution networks, highlighting the importance of high temporal resolution in phasor systems

[1]. AbdelRaheem et al. proposed a lightweight sampling time error correction technique tailored for micro phasor measurement units, demonstrating significant improvements in computational efficiency[2].

Error Correction and Signal Processing Innovations; the accuracy of phasor measurements often depends on effective sampling time error correction and signal interpolation. Chen et al. proposed using resampling algorithms for precise amplitude and phase determination, which has been applied to calibrate sampled value instruments in power systems [3]. Lin et al. further enhanced wide-frequency measurement devices with resampling methods, ensuring accuracy over a broad range of frequencies[4]. Advanced interpolation techniques, such as those using Diophantine approximation-based filters, have also been employed to refine synchronized power grid sampling[5].

Soft synchronized sampling has emerged as a viable alternative. It eliminates the need for continuous ADC adjustments by leveraging interpolation to correct timing errors within consecutive PPS intervals. While earlier interpolation techniques, such as linear interpolation and polynomial interpolation , have been investigated, they often fall short in terms of accuracy and adaptability. This study addresses these shortcomings by employing Akima piecewise cubic Hermite interpolation, which offers superior error correction, smoothness, and computational efficiency[?]. These features make it well-suited for real-time applications requiring high-fidelity synchronized sampling.

The main contributions of this paper can be outlined as follows:

(i) A novel sampling error correction approach based on Akima piecewise cubic Hermite interpolation is proposed, offering an effective yet simplified method for addressing STEs without frequent ADC adjustments.

(ii) An in-depth comparative analysis highlights the advantages of Akima interpolation over conventional methods, emphasizing its ability to deliver precise synchronized sampling for power grid monitoring.

The remainder of this paper is structured as follows: Section II details the Methodology. Section III describes the experimental setup and methodology for validation. Finally, Section IV concludes the paper and outlines potential directions for future research.

2. Methodology

2.1. Crystal oscillator frequency error monitoring

Accurate monitoring of the crystal oscillator's frequency is essential for mitigating Sampling Time Errors (STEs) in synchronized phasor measurement systems. This process utilizes the Pulse Per Second (PPS) signal from GPS as a stable time reference and an MCU's internal counter for frequency measurement.

The MCU counter, driven by the crystal oscillator, increments continuously at the oscillator's frequency f_{osc}. Upon receiving a PPS signal, the counter value C_n is recorded, where n represents the index of the PPS pulse. The elapsed time between two consecutive PPS signals is ideally 1 second, enabling the calculation of the crystal oscillator frequency as:

$$f_{osc} = \frac{C_{n+1} - C_n}{T_{PPS}}, \tag{1}$$

where $C_{n+1} - C_n$ represents the number of counter increments within the interval T_{PPS}, which is nominally 1 second.

In an ideal system, the counter increment over this interval should match the expected value $C_{ideal} = f_{nom} \cdot T_{PPS}$, where f_{nom} is the nominal frequency of the oscillator. However, due to factors such as temperature variations and component aging, the actual counter increment deviates from C_{ideal}. This deviation reflects a relative frequency error Δf, which can be expressed as:

$$\Delta f = \frac{f_{osc} - f_{nom}}{f_{nom}} = \frac{C_{n+1} - C_n - C_{ideal}}{C_{ideal}}. \tag{2}$$

This error provides a direct measure of the crystal oscillator's deviation from its nominal frequency. Over time, the cumulative effect of these deviations contributes to STEs. For an observation period spanning N PPS intervals, the total time deviation ΔT caused by the frequency error is given by:

$$\Delta T = \sum_{n=1}^{N} \frac{\Delta f_n}{f_{nom}} \cdot T_{PPS}. \tag{3}$$

This deviation in time leads to non-uniform sampling intervals, significantly affecting the accuracy of synchronized phasor measurement. By continuously monitoring C_n values and comparing them with the expected ideal increments, the method provides a dynamic characterization of the oscillator's frequency behavior relative to the GPS reference. This information is critical for subsequent correction steps, ensuring that the synchronized sampling process adheres to the desired accuracy standards.

2.2. Sampling Data Correction Based on Akima Piecewise Cubic Hermite Interpolation

The cumulative time deviation ΔT derived in the previous section introduces non-uniformity in the sampling intervals, resulting in observed sampling values $\{x_i^{obs}\}$ at times $\{t_i^{obs}\}$ that deviate from their ideal counterparts. To correct this and align the data with the ideal uniform time grid, Akima piecewise cubic Hermite interpolation is employed. This method ensures smooth and accurate estimation of the correct sampling values at the desired time points.

In a synchronized sampling system, the ideal sampling times are given by:

$$t_i^{ideal} = t_0 + i \cdot T_s, \tag{4}$$

where t_0 is the initial time, T_s is the uniform sampling interval, and i is the sample index. However, due to frequency drift and other inaccuracies, the observed sampling times t_i^{obs} are shifted by the time deviation $\Delta T(t_i^{ideal})$:

$$t_i^{obs} = t_i^{ideal} + \Delta T(t_i^{ideal}). \tag{5}$$

The observed sampling values $\{x_i^{obs}\}$ are recorded at these shifted times $\{t_i^{obs}\}$. To estimate the correct values $\{x_i^{ideal}\}$ at the ideal times $\{t_i^{ideal}\}$, an interpolation function $x(t)$ is constructed from the observed data pairs $\{(t_i^{obs}, x_i^{obs})\}$ using Akima piecewise cubic Hermite interpolation.

The interpolation function $x(t)$ is defined piecewise over the interval $[t_k, t_{k+1}]$ as:

$$x(t) = a_k + b_k(t - t_k) + c_k(t - t_k)^2 + d_k(t - t_k)^3, \quad t \in [t_k, t_{k+1}], \tag{6}$$

where a_k, b_k, c_k, d_k are the coefficients for the k-th segment. These coefficients are determined based on the slopes of the neighboring intervals, ensuring smooth transitions and avoiding oscillations. Akima interpolation provides superior flexibility and accuracy compared to traditional interpolation methods, particularly for non-uniformly spaced data.

Once the interpolation function is constructed, the correct sampling values at the ideal times are calculated as:

$$x_i^{ideal} = x(t_i^{ideal}). \tag{7}$$

By combining the time deviation correction derived from the cumulative frequency error and the Akima interpolation, this method effectively maps the observed sampling data onto the ideal time grid. The resulting corrected data align with the synchronized sampling requirements, eliminating distortions caused by non-uniform sampling intervals. This approach ensures high accuracy and smoothness, making it a robust solution for synchronized phasor measurement in power grid monitoring systems.

2.3. Phasor Estimation Utilizing Corrected Sampling Data

Building on the corrected sampled data obtained through Akima piecewise cubic Hermite interpolation in Section 3.2, this section presents the process of phasor measurement using the Recursive Discrete Fourier Transform (RDFT) algorithm. The corrected data, aligned with uniform sampling intervals, serve as the input for accurate phasor estimation, eliminating the distortions introduced by sampling time errors.

Phasor measurement involves determining the magnitude and phase of a sinusoidal signal, typically represented as:

$$x(t) = X\cos(2\pi f t + \phi), \tag{8}$$

where X is the signal amplitude, f is the frequency, and ϕ is the phase angle. Using the corrected sampled data $\{x_i^{corr}\}$, the RDFT algorithm estimates the phasor X_k corresponding to the fundamental frequency f_0 within a sliding window of N samples. The phasor is expressed as a complex number:

$$X_k = \frac{2}{N}\sum_{n=0}^{N-1} x_n e^{-j2\pi\frac{n}{N}}, \tag{9}$$

where x_n represents the n-th corrected sample in the window, and N is the number of samples covering one cycle of the fundamental frequency f_0. The phasor's magnitude and phase are subsequently derived as:

$$|X_k| = \sqrt{\mathrm{Re}(X_k)^2 + \mathrm{Im}(X_k)^2}, \tag{10}$$

$$\phi_k = \arctan 2\left(\mathrm{Im}(X_k), \mathrm{Re}(X_k)\right), \tag{11}$$

where $\mathrm{Re}(X_k)$ and $\mathrm{Im}(X_k)$ are the real and imaginary components of X_k, respectively. The RDFT algorithm efficiently updates the phasor estimation by leveraging previously computed results, making it particularly suitable for real-time applications.

The accuracy of the phasor measurement is assessed using the Total Vector Error (TVE), a standard metric defined by IEEE C37.118.1. The TVE quantifies the deviation between the measured phasor $\hat{X}$ and the reference phasor X_{ref} as:

$$\mathrm{TVE}(\%) = \frac{\sqrt{\left(\mathrm{Re}(\hat{X}) - \mathrm{Re}(X_{ref})\right)^2 + \left(\mathrm{Im}(\hat{X}) - \mathrm{Im}(X_{ref})\right)^2}}{\sqrt{\mathrm{Re}(X_{ref})^2 + \mathrm{Im}(X_{ref})^2}} \times 100. \tag{12}$$

By integrating the corrected sampled data into the RDFT-based phasor estimation process, this method ensures that the measured phasors meet the precision requirements, thereby demonstrating the effectiveness of the proposed sampling error correction approach in synchronized phasor measurement systems.

3. Performance evaluation

To evaluate the effectiveness of the proposed Akima piecewise cubic Hermite interpolation combined with Recursive DFT for synchronized phasor estimation, simulations were carried out under realistic power grid conditions. MATLAB/Simulink was employed to model a 50 Hz sinusoidal waveform representing voltage and current signals, adhering to the IEEE C37.118.1 standard. Sampling time errors were introduced by varying the local oscillator frequency within a range of ±100 ppm, mimicking the impacts of frequency drift in practical scenarios. The performance of the proposed method was benchmarked against two-point interpolation and unadjusted data, with an emphasis on metrics such as phasor accuracy, robustness to frequency variations, and compliance with standard requirements. The results demonstrated that the proposed method provides improved precision and stability compared to the alternatives.

3.1. Steady-State Conditions Experiment

In frequency offset experiment, the power system frequency was varied from 48 Hz to 52 Hz in 0.1 Hz steps. Figure 1 shows that the proposed method achieves the lowest Total Vector Error (TVE), consistently staying well below the 1% threshold, while two-point interpolation and the uncorrected case exhibit higher errors. For Frequency Error (FE), the proposed method outperforms the others with an average error of 2.5×10^{-5} Hz, as opposed to higher values observed for the alternatives. In terms of Rate of Frequency Error (RFE), all algorithms meet the standard, but the proposed method demonstrates greater stability across the tested range, confirming its superior accuracy and robustness.

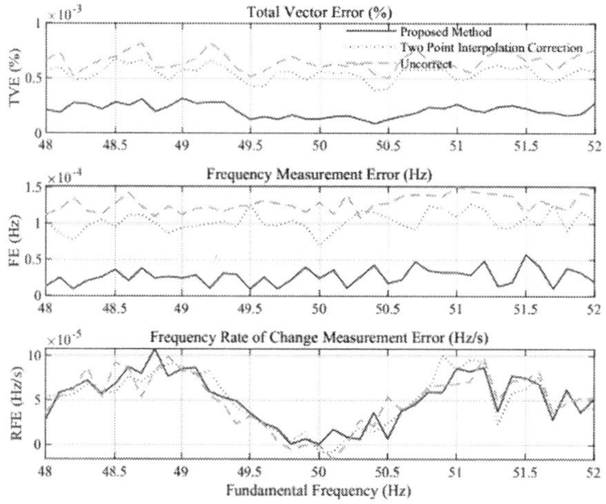

Figure 1. Fundamental Frequency Offset Testing

An additional steady-state experiment was conducted to assess the algorithm's effectiveness under harmonic distortion, incorporating harmonics from the 2nd to the 50th order into the fundamental signal. As illustrated in Figure 2, all three methods comply with the standard limits of 1% TVE, 25 mHz FE, and 0.1 Hz/s RFE. Among them, the proposed method demonstrates the highest accuracy, achieving the smallest errors in both TVE and FE across all harmonic orders. While all methods remain within acceptable RFE limits, the proposed approach exhibits slightly greater consistency.

3.2. Dynamic Conditions Testing

To assess the performance of the proposed method in handling oscillations within power systems, experiments were conducted under both amplitude and phase modulation scenarios. Figure 3

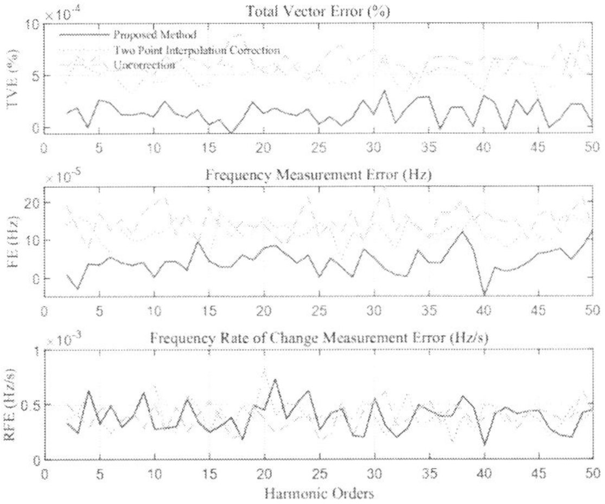

Figure 2. Harmonic Distortion Testing

and Figure 4 present the results, showing that the proposed algorithm consistently outperforms the alternatives across varying modulation frequencies.

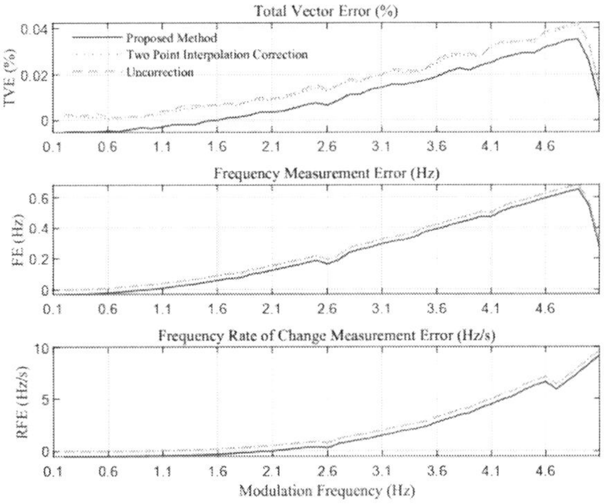

Figure 3. Amplitude Modulation Testing

In amplitude modulation tests, as modulation frequency increases, errors in Total Vector Error (TVE) and Frequency Error (FE) grow for all algorithms. However, the proposed method maintains lower TVE and FE values compared to the two-point interpolation and uncorrected cases, demonstrating its effectiveness in minimizing errors even under high-frequency modulation. The stability of the proposed method across the tested range highlights its robustness in challenging conditions.

Similarly, during phase modulation tests, the proposed method achieves superior TVE performance, maintaining a consistent edge over the other algorithms. While Frequency Error (FE) and Rate of Frequency Error (RFE) values remain within acceptable limits for all methods, the proposed algorithm displays better stability and precision.

To evaluate measurement performance under frequency fluctuations, tests were conducted with varying rates of frequency change around a nominal 50 Hz system frequency. As shown

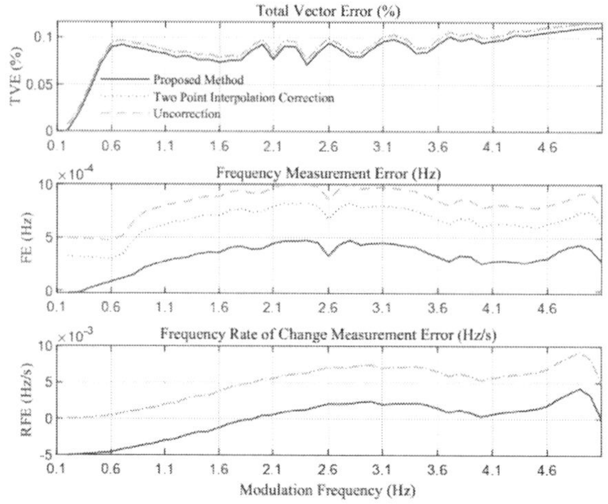

Figure 4. Phase Angle Modulation Testing

in Figure 5, the proposed method consistently achieves the lowest Total Vector Error (TVE) and Frequency Error (FE), demonstrating superior robustness compared to the two-point interpolation and uncorrected methods. While all algorithms meet the IEEE Standard limits for Rate of Frequency Error (RFE), the proposed method exhibits greater stability and accuracy across the tested range.

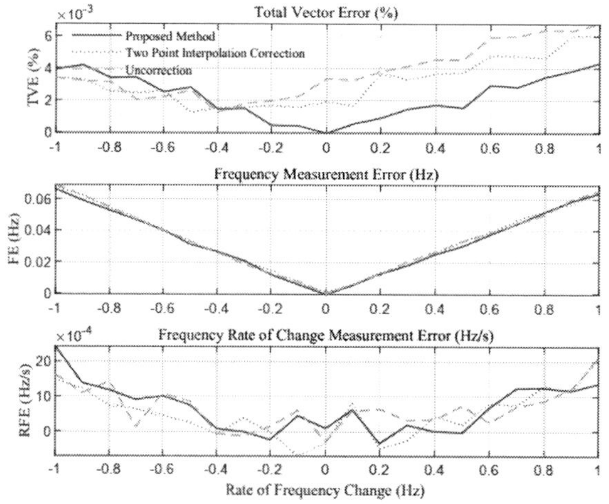

Figure 5. Frequency Ramp Testing

4. Conclusion

This study introduces a novel phasor measurement approach for power grid monitoring, leveraging soft synchronization. By utilizing time pulse signals from GNSS or mobile stations, the method accurately identifies frequency drift and sampling time errors in the local oscillator. These discrepancies are corrected using Akima piecewise cubic Hermite interpolation, ensuring the sampled data aligns with an ideal uniform sampling rate. The adjusted data is then processed with a Recursive DFT algorithm to achieve precise synchrophasor estimation.

This research not only advances the state-of-the-art in phasor measurement techniques but also demonstrates significant practical relevance. By validating the proposed method against the IEEE C37.118.1 standard and comparing it with alternative approaches, including two-point interpolation and the uncorrected baseline, the study illustrates its superior accuracy and robustness. Specifically, the Akima interpolation method achieves more precise correction of sampling time errors compared to two-point interpolation, particularly in scenarios involving non-linear deviations caused by frequency drift. The uncorrected baseline, by contrast, exhibits substantial degradation in phasor measurement accuracy, underscoring the necessity of effective error correction.

The proposed method simplifies the phasor measurement process by eliminating the need for frequent adjustments to ADCs or reliance on complex hardware-based synchronization solutions. As the requirements for real-time and high-accuracy monitoring in power systems continue to grow, this method offers a scalable solution for enhancing the reliability and efficiency of synchronized phasor measurement in practical applications. The findings of this study highlight its potential to support the evolving demands of modern power grids, ensuring robust and accurate monitoring under challenging conditions.

5. Acknowledgment

This work was supported by the State Grid Cooperation of China Science and Technology project(5700-202319273A-1-1-ZN).

References

[1] M. Izadi and H. Mohsenian-Rad, "Synchronous waveform measurements to locate transient events and incipient faults in power distribution networks," IEEE Trans. Smart Grid, vol. 12, no. 5, pp. 4295–4307, Sep. 2021.

[2] M. AbdelRaheem, M. Hassan, and H. Selim, "A lightweight sampling time error correction technique for micro phasor measurement units," IEEE Trans. Instrum. Meas., vol. 71, 2022, Art. no. 9004408.

[3] Chen Y, Mohns E, Seckelmann M, et al. Precise amplitude and phase determination using resampling algorithms for calibrating sampled value instruments[J]. Sensors, 2020, 20(24): 7345.

[4] I. Niazazari, H. Livani, A. Ghasemkhani, Y. Liu, and L. Yang, "Event cause analysis in distribution networks using synchro waveform measurements," in Proc. North Amer. Power Symp., Tempe, AZ, USA, 2021, pp. 1–5.

[5] M. Izadi and H. Mohsenian-Rad, "A synchronized Lissajous-based approach to achieve situational awareness using synchronized waveform measurements," in Proc. IEEE PES Gen. Meeting, Washington, DC, USA, 2021, pp. 1–5.

[6] Phadke, A. G., T. Bi, and V. Tech. "Phasor Measurement Units WAMS and Their Applications in Protection and Control of Power Systems." Journal of Modern Power Systems and Clean Energy 6, no. 4 (April 2018): 619-629. Also, Phadke, A. G., T. Bi, and V. Tech. Power System Grid Operation Using Synchrophasor Technology. Springer, 2018.

[7] Umunnakwe, A., A. Alimi, K. R. Davis, and K. L. Butler-Purry. "Improving Situational Awareness in Power Grids of Developing Countries: A Case Study of the Nigerian Grid." In 2023 IEEE Power & Energy Society Innovative Smart Grid Technologies Conference (ISGT), 1-5. Washington, DC, USA, 2023. doi: 10.1109/ISGT51731.2023.10066359.

[8] Meegahapola L G, Bu S, Wadduwage D P, et al. Review on oscillatory stability in power grids with renewable energy sources: Monitoring, analysis, and control using synchrophasor technology[J]. IEEE Transactions on Industrial Electronics, 2020, 68(1): 519-531.

[9] Sahoo A, Mahmud K, Ravishankar J. An enhanced frequency-adaptive single-phase grid synchronization technique[J]. IEEE Transactions on Instrumentation and Measurement, 2021, 70: 1-11.

[10] Xu W, Huang Z, Xie X, et al. Synchronized waveforms–a frontier of data-based power system and apparatus monitoring, protection, and control[J]. IEEE Transactions on Power Delivery, 2021, 37(1): 3-17.

[11] Ahmed H, Biricik S, Benbouzid M. Linear Kalman filter-based grid synchronization technique: An alternative implementation[J]. IEEE Transactions on Industrial Informatics, 2020, 17(6): 3847-3856.

The Stable Control of Gird Isolated Island based on Wind Power

Qitao Wang[1], Guangkai Huang[1], Zhengyang Wu[2], Cheng Cao[3], Wanqing Song[1*]

[1] School of Electronic and Electrical Engineering, Minnan University of Science and Technology, Quanzhou 362700, China.

[2] Research Department, Minnan University of Science and Technology, Quanzhou 362700, China.

[3] HCL Technologies, Santa Clara, CA 95054, USA

[*]Corresponding author: swqls@126.com

Abstract. When the relation between the interconnected grid and the main system is weak, once a serious failure occurs in the tie line both the local small grid and the large grid, the local weakly connected grid will disconnect from the main grid and into island operation. Due to the small number of island operating units and the overall capacity and inertia are very small, the system frequency and voltage are easy to fluctuate in a large range, and operational stability is very prominent. This paper studied the effects the stability of isolated island grid for gusty wind and gradual wind. At the same time, the island system stable operation for different ratio of wind generator is also studied. Finally, by analyzing the regulating effect of reactive energy compensation device, voltage stability of wind electricity is improved.

1. Introduction

Once this wind generation is connected to the electricity grid, the current flow of this grid will change, and each node is equivalent to the load where active power output is positive, and reactive power output is negative. Consequently, the uncertainty of wind speed results in fluctuations in wind farm power output and demand [1]. Reference [2] implements intelligent model optimization govern of doubly-fed induction generators based on the principle of maximum wind energy capture. Literature [3] realizes intelligent model optimization govern of doubly-fed induction generators.

When the wind electricity proportion integrated into the electricity network is small, the injected wind generator has minimal impact on the electricity network, but the transient stability of this network will deteriorate as the wind power output increases[4]. The effect of wind generator on the electricity grid becomes more significant when wind farms are connected in concentrated clusters, and with the continuous expansion of wind farm scale, it is necessary to research the effect of the wind farm integration proportion on the stable isolated island grid. In addition, the connected position of the wind generator will also affect the transient stability of this electricity grid, the closer wind generator connected to grid devices, the more significant the impact will be. Under the grid same fault, the wind power equipment is more possible to lose stability than the grid equipment[5]. When this wind generator is connected to the electricity network at different locations, the grid's transient stability may be improved or decreased [6].

The grid's transient stability for isolated island systems is also affected by the grid's structure. Generally, the stronger the grid structure, the better the grid's transient stability[7]. The degree of connection both the wind generator and the electricity network also is a positive correlation for the transient stability. The closer the connection both the wind generator and the electricity network, the better the grid's transient stability.

From the above, it can be concluded that wind speed disturbances, the integration capacity and connected location of wind farms, and different the grid network structure has different transient stability for the isolated island grid. In practice, these factors often have combined effects, making it impossible to draw a general conclusion about whether wind power integration enhances or weakens transient stability. Instead, it requires a comprehensive consideration of the characteristics of the wind electricity, the original network system, and the specific fault conditions.

Furthermore, this paper proposes the use of a SVC to enhance voltage stability in wind farms, offering a practical solution to improve the operational stability of isolated island grids with high wind power penetration.

2. Wind Power Model

Fig.1 shows the curve both the wind speed and the active energy output for variable pitch wind turbine. In the figure, v_r is the rated value for the wind speed, P_r is the rated value for the active power output, v_{cut-in} and $v_{cut-out}$ are the cutin and cutout wind speed, respectively.

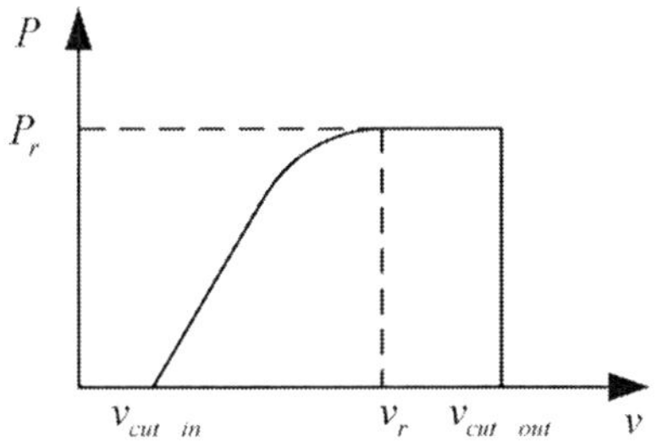

Fig.1. The curve both wind speed and output power

In general, the manufacturer will provide the wind power characteristics curve which also can be measured by experiment, the calculating formula express in equation[1].

$$P_M=\begin{cases} 0, \ v<v_{cut-in} \\ \dfrac{v-v_{cut-in}}{v_r-v_{cut-in}}P_r, v_{cut-in}<v<v_r \\ P_r, v_r<v<v_{cut-out} \\ 0, v>v_{cut-out} \end{cases} \tag{1}$$

Ignoring the dynamic characteristics of the blade, the torque is shown the equation (2):

$$M_w=\frac{1}{2}\rho\pi C_p R^3\frac{V_w^2}{\lambda}\cdot\frac{\Omega_N}{BMVA}\times10^{-6} \tag{2}$$

where ρ represents the air density, λ represents the blade tip rate ratio, C_p represents the blade wind energy conversion efficiency coefficient, V_w represents the wind speed for hub height, R represents the blade sweep radius, Ω and Ω_N represent the mechanical angular velocity and rated value of the blade, and BMVA represents the system reference capacity.

Substituting $\lambda=\frac{\Omega R}{V_w}$ into equation (2), the equation (3) is derived:

$$M_w = \frac{1}{2}\rho\pi C_p R^3 \frac{V_w^3}{\Omega R} \cdot \frac{\Omega_N}{BMVA} \times 10^{-6} = \frac{1}{2}\pi R^2 \rho V_w^3 C_p \frac{1}{\omega} \cdot \frac{1}{BMVA} \times 10^{-6} \tag{3}$$

where the electromagnetic power is the equation (4)

$$P_M = \frac{1}{2}\pi R^2 \rho V_w^3 C_p \frac{1}{BMVA} \times 10^{-6} \tag{4}$$

C_p is usually denoted by the function $C_p = f(\lambda, \beta)$ with λ and pitch angle β. Fig. 2 shows the C_p characteristics of a typical wind turbine.

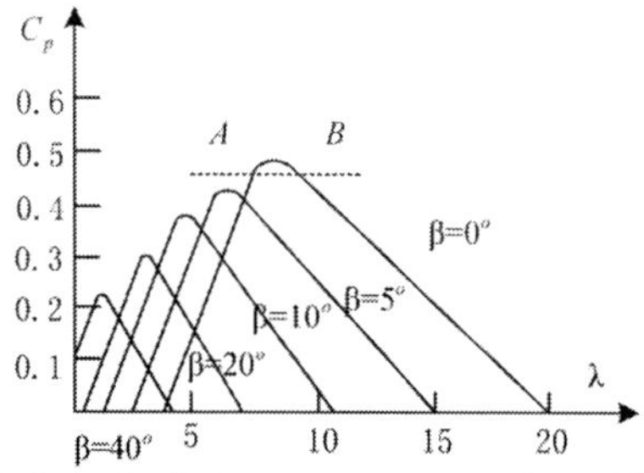

Fig. 2. Characteristics of wind energy conversion efficiency coefficient

Obviously in Fig. 2, the same C_p value, the wind turbine may have two operating points A and B, which correspond to the two states for the higher wind speed and the lower wind speed, respectively. Therefore, according to the actual change of wind speed, by adjusting the large (small) turbine operating, the blade can still have the same C_p. The C_p characteristics of fixed-pitch wind turbines can be expressed as the equation (5):

$$C_p = \frac{16}{27} \frac{\lambda}{\lambda + \frac{1.32 + \left[(\lambda - 8)/20\right]^2}{B}} - 0.57 \frac{\lambda^2}{\frac{L}{D}\left(\lambda + \frac{1}{2B}\right)} \tag{5}$$

where parameter, B represents the number of blades and L/D represents the lift ratio.
In addition, the C_p characteristics of variable pitch wind turbines can be expressed as the equation (6):

$$C_p = \frac{1}{2}\left(\frac{RC_f}{\lambda} - 0.022\beta - 2\right) e^{-0.255\frac{RC_f}{\lambda}} \tag{6}$$

where parameter C_f stands for blade design constant.

3. Wind velocity Modelling

Because the wind's speed will affect the output power of the wind generator, properly to analyze and study the input and output of the wind's generator, an appropriate wind's model should be modeled. The wind speed can be decomposed by basic, random, gradual and gusty wind velocity, respectively. When this wind turbine is in normal operating, it has a basic wind velocity. Considering the stochastic disturbance of the wind velocity, the acting wind velocity is below equation (7):

$$V_w = \bar{v}_w + v_{wn} \tag{7}$$

where $\bar{v}_w$ is the basic wind velocity and v_{wn} represents this stochastic small disturbance for the wind speed. Combining gust wind and gradient wind become equation (8):

Journal of Physics: Conference Series **3012** (2025) 012015 doi:10.1088/1742-6596/3012/1/012015

$$v_w = \bar{v}_w + v_{wn} + v_{wg} + v_{wr} \tag{8}$$

where v_{wg} and v_{wr} represent gust and gradient wind, respectively.

Random wind speed is unpredictable, and the impact on the entire wind farm is very small and negligible, so the other three wind speed models are as follows.

3.1. Basic Wind velocity

According to the calculation of wind field wind related data, basic wind basically meets Weibull distribution:

$$\bar{v}_w = C \cdot \Gamma\left(1 + \frac{1}{k}\right) \tag{9}$$

where $\Gamma(\cdot)$ represents the gamma function, constant coefficient k and C are the shape parameters and the distribution scales, respectively. In general, when the second is taken as a reference unit, the basic wind speed can be treated as a constant, and its simulation can be shown Fig.3(a), where the wind speed is always maintained at 14m/s within 0s to 7s.

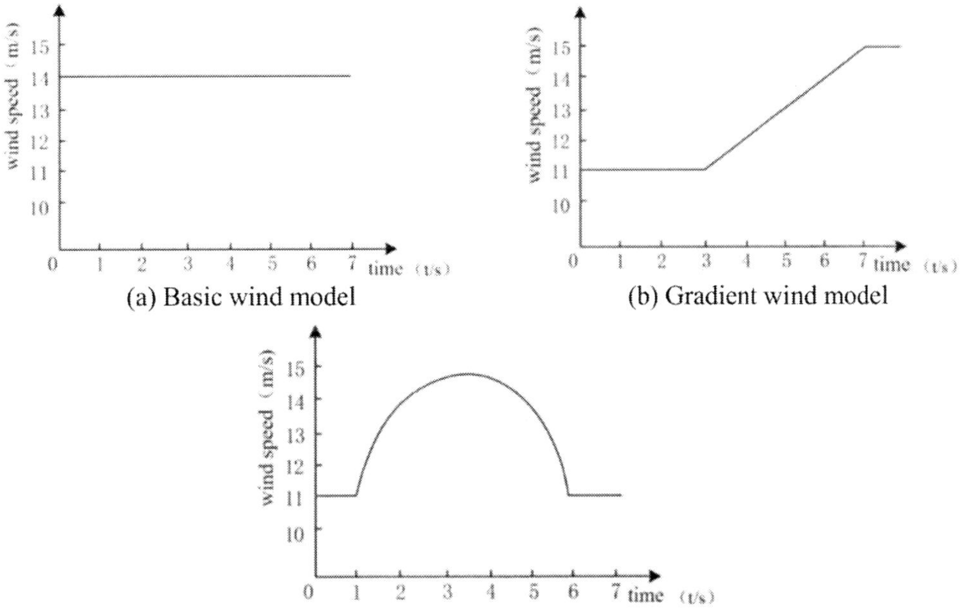

(a) Basic wind model (b) Gradient wind model

(c) Gust model

Fig. 3. Simulation of different wind speed models

3.2. Gradual Wind velocity

This kind of gradual wind velocity is calculated by the equation (10), and its simulation diagram is shown Fig.3(b). The wind speed is 11m/s at the initial moment, and remains unchanged until 3s, starts to rise from 3s to 7s, and remains this constant after the wind speed reaches 15m/s.

$$v_{wr} = \begin{cases} 0, t < T_{1R} \\ v_{ramp}, T_{1R} < t < T_{2R} \\ 0, t > T_{2R} \end{cases} \tag{10}$$

$$v_{ramp} = MaxR\left[1 - (t - T_{2R}) / (T_{1R} - T_{2R})\right] \tag{11}$$

where v_{wr} and $MaxR$ represent gradual wind speed and maximum, respectively, T_{1R}, T_{2R} and T_R gradual wind start, end and hold times, respectively.

3.3. *Gust Speed*

The gust can be calculated by the equation(12), and its simulation can be shown in Fig.3(c). The wind speed is 11m/s in the period from 0s to 1s, then rises and decreases in the period from 1s to 6s, and is 11m/s in the period from 6s to 7s.

$$v_{wg} = \begin{cases} 0, t < T_{1G} \\ v_{\cos}, T_{1G} < t < T_{1G} + T_G \\ 0, t > T_{1G} + T_G \end{cases} \tag{12}$$

$$v_{\cos} = (MaxG/2)\{1 - \cos 2\pi[(t/T_G) - (T_{1G}/T_G)]\} \tag{13}$$

where v_{wg} and $MaxR$ represent gust speed and maximum, respectively. T_G and T_{1G} represent gust wind and start time, respectively.

4. Stability Analysis with Different Wind Power Ratio

Beside wind velocity fluctuations, the proportion for the wind generator integration also affect wind electricity output.

When the wind power integration capacity arrived 60% of the rated capacity of the grid, a three-phase permanent fault occurs under an N-1 contingency. In case to wind velocity fluctuations, the variation in wind turbine output is also influenced by the proportion of wind turbine integration. Therefore, this section examines the different wind generator integration proportion both the safe and stable operation of an isolated island grid, with comparative analysis conducted. Before the operation of the island grid, set the active power boundary of the tie line is 31MW, then the power grid can maintain stable operation within this boundary.

Compared with the active power for the tie line both 20MW and 40MW, respectively, both the frequency deviation and the positive sequence voltage curve for the bus are shown in Fig.4. At the same time, when the current flow of the tie line is the same before the operation of the island grid, the comparison analysis is made with the wind power capacity for 30% and 80% of the rated capacity. Fig.5 shown the comparison with the bus frequency deviation and positive sequence voltage curve.

When the system frequency decreases, the generator voltage decreases, leading to a drop in voltage. Once a fault happen, to change in the wind turbine integration capacity ratio will affect the island grid stability. The voltage wave is relatively small; however, when the inter tie current flow is fixed, an excessively high wind turbine integration capacity will lead to frequency instability after a fault. Moreover, the higher the wind turbine integration ratio, the more unstable the frequency becomes.

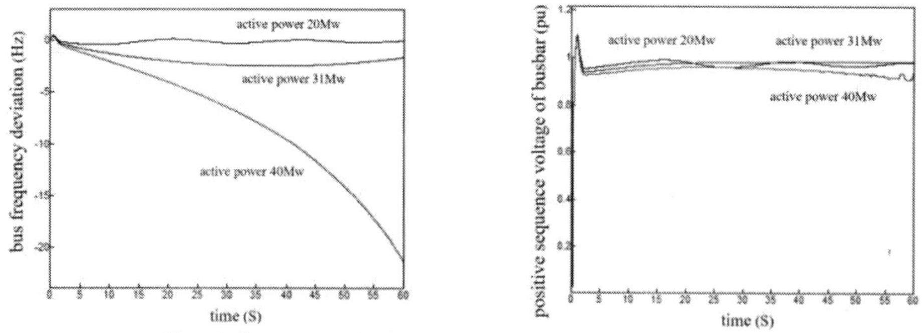

Fig. 4. frequency deviation and positive sequence voltage

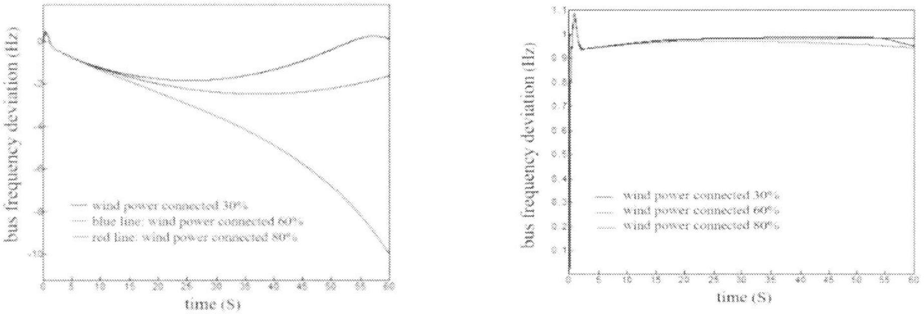

Fig. 5. Bus frequency deviation and positive sequence voltage with different wind ratio

5. Application of Reactive Power Compensation Device

Because most of the traditional reactive power compensation devices are manually controlled for capacitor switching, can not be adjusted continuously, the efficiency and quality can not meet the needs of the grid. Since SVC is a new type of reactive power compensation device, and the SVC has been widely used because of its fast response and the advantages of providing dynamic reactive power in time during disturbance, the SVC is often added to the electricity network to keep that reactive energy requirements improve the network safety and stability.

In general, the SVC is installed at the lower voltage side of the main transformer. when N-1 three-phase permanent fault occurs, the wind generator is compared and analyzed whether SVC or not, as shown in Fig.6.

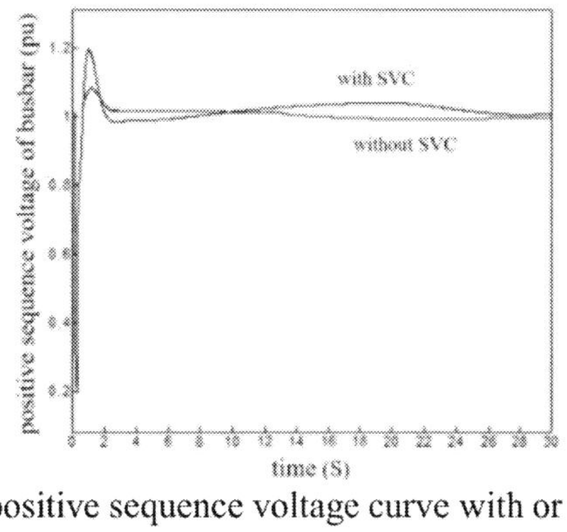

Fig. 6. Bus positive sequence voltage curve with or without SVC

The Fig.6 shown under the same ratio of the wind turbine integration, the bus voltage is improved when an SVC is installed at the wind farm, with the maximum value not exceeding 1.2 pu, which is beneficial for the stable isolated island network.

6. Conclusion

The different wind velocity affect the grid frequency's stability, with longer duration of disturbance having a greater effect for the stable isolated island network. Additionally, from the perspective of the wind generator output, the impact of the wind generator integration ratio on the safe and stable operation of the isolated island grid is also considered. The results indicate that a higher wind turbine integration ratio has a greater affected stability for the isolated island grid. Finally, by comparing this scenarios with and without SVC at the wind farm, it is concluded that the installation of an SVC is beneficial for improving isolated island grid's stability.

References

[1] Dahiya P, Mukhija P, Saxena A R. Design of sampled data and event-triggered load frequency

controller for isolated hybrid power system, International Journal of Electrical Power & Energy Systems, 2018, 100:331-349.
[2] Mele F M, Milano F, Cashman D, et al. Impact of Wind Farm Frequency Control on the Dynamic Response of the All-Island Irish System, IEEE Power & energy Society, 2018, 1(8):56717423
[3] H Khouidmi, A Benzouaoui, B Bessedik, Nonlinear Predictive Control Method for Maximizing Wind Energy Extraction of Variable Speed Wind Turbines under Turbulence, Journal Europeen des Systemes Automatises, 2021, 54(5):661-670
[4] RS Camargo, AEA Amorim, EJ Bueno, Novel multilevel STATCOM for power system stability enhancement on DFIG-based wind farms, Electric Power Systems Research, 2021, 197(8):107316
[5] Amel Brik, Nour El, Yakine Kouba and Ahmed Amine Ladjici, Power System Transient Stability Analysis Considering Short-Circuit Faults and Renewable Energy Sources, engineering proceeding, 2024, 67, 42
[6] Shi L, Dai S, Yao L, et al. Impact of Wind Farms of DFIG Type on Power System Transient Stability, Journal of Electromagnetic Analysis & Applications, 2010, 02(8):475-481.
[7] A Mitra, D Chatterjee, Active Power Control of DFIG-Based Wind Farm for Improvement of Transient Stability of Power Systems, IEEE Transactions on Power Systems, 2016,

Quantitative Evaluation of Flexible Mutual Aid Matching Metrics for Low voltage transformer area

Dan Zhu[1], Min Li[1], Hengchao Zhang[1], Zhiying Zhong[2], Mingzhi Huang[2]*

[1] Suzhou Power Supply Branch of State Grid Jiangsu Electric Power Co., Suzhou, Jiangsu Province, 215004, P.R. China

[2] Jiangsu Fantian PowerTechnology Co., Ltd, Nanjing,Jiangsu Province, 210096, P.R. China

*Corresponding author's e-mail: 1608371008@qq.com

Abstract. In order to solve the problems of load imbalance or seasonal load fluctuation between stations, distribution substation overload caused by electric vehicle load, power supply failure that cannot be load transferred in time, reverse overload and power quality caused by a high proportion of distributed PV accessed to the station area, and so on. Taking Suzhou as an example, based on the OLP model system, we sort out the problem labels of Suzhou station areas and establish a corresponding judging system, calculate the residual shortage of station areas by comparing the differences in the labels of the sample station areas, and screen out the stations suitable for carrying out flexible interconnection, connecting adjacent stations flexibly, realizing the power inter-calculation between the station areas and supporting the coordinated and optimized operation of low-voltage stations, so as to provide basis for the precise selection of the point of the coordinated control of the flexible and direct low-voltage stations.

1. Introduction

Distributed photovoltaic-based distributed power supply has entered a new period of rapid development, with the distributed power supply, electric vehicle charging piles, coal to electricity and other zero-carbon, low-carbon elements access ratio is increasing, the distribution network, especially the low-voltage distribution network, the role of the "hub platform" more and more prominent, the distribution network to consume and grid-connected service pressure is increasing, and is facing the double challenge of ensuring a stable supply of electricity and accelerating the clean and low-carbon transformation. It is facing the dual challenges of ensuring sustainable and stable power supply and accelerating clean and low-carbon transformation [1]. By the end of 2023, Suzhou Company has a total of 833 28 stations, of which 392 heavy-loaded stations, 69 overloaded stations, 1.72% of photovoltaic backward transmission stations, and 11.85% of outage stations have occurred. Therefore, it is particularly important to propose an innovative method to accurately and quickly screen out the suitable flexible interconnected stations through modeling algorithms.

Flexible DC transmission technology is introduced into the flexible interconnection between stations to realize flexible inter-computing between low-voltage stations, optimize the allocation of power resources, improve the stability and reliability of the distribution system, reduce load loss, reduce power costs, and guarantee the safety of power consumption in the distribution network.

2. Research Programs and Innovations

Based on the real energy consumption data of the station area in Suzhou area in 2023, an innovative algorithm is constructed to focus on analyzing the impact of differentiated load characteristics on the stability of the station area, and the specific research contents are as follows:

1)Analyze the technical characteristics and operation mechanism of DC flexible interconnection equipment, and obtain the data of relevant indexes of energy security in the station area through the use of mining system;

2)Analyze the acquired historical energy-use data indicators of the station area through the cluster analysis method, and divide the multiple indicators of the station area into four indicator sets of geographic location, power supply reliability, power supply quality, and power-use characteristics in accordance with their business functions, and screen out the problematic information of the station area, such as frequent power outages, heavy overloads, backward transmission, and three-phase imbalance, and establish the problematic labeling system;

Based on the problem of specificity of energy use in the station area, a special red line label is established, which contains indicators that may affect the operational stability between stations, such as whether it is the same line or not, and the number of three-phase users;

Through the hierarchical analysis method, the station area problem labeling system is divided into three tiers, the comparison matrix between the first and second tier indicators is constructed, the relative weights are calculated, and based on the calculation results, the second tier labeling grades A, B, and C are divided into three categories, with the grades decreasing in order, and the severity of each type of indicator is divided.

3)According to the problem labels labeled on the stations, select the target stations that need flexible interconnection, and determine the sample range of screening stations by latitude and longitude.

4)Construct an innovative label comparison method, using the labeling system and big data analysis technology to compare the feature points with the load characteristics of the stations, construct a feature library, and establish a station relationship comparison model. The model is applied to actual DC flexible interconnection station scenarios, and the effect of the model is evaluated through simulation and actual operation data to further improve the comprehensive research and judgment model, optimize the parameters, provide the list of flexible interconnections and the optimal recommended stations, and accurately locate and match the interconnected stations of flexible DC equipment.

The overall architecture of the above program is shown in Figure 1:

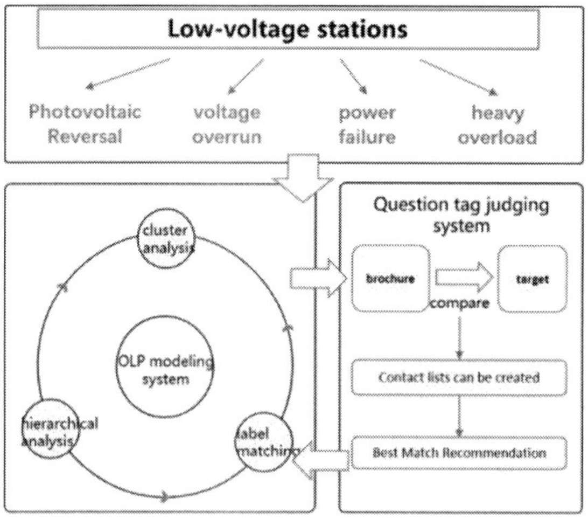

Figure 1. OLP Model Architecture Diagram

3. Analysis of energy use data in the station area

3.1. Power Data Research
The data indicators related to energy use in the station area are divided into classes according to the degree of similarity using cluster analysis, so that the similarity between indicators in the same class is stronger than the similarity of elements in other classes, and the homogeneity of indicators between classes is maximized and the heterogeneity of indicators between classes is maximized [2]. Based on the above methodology, the energy use-related indicators of a station area are divided into four major indicator sets: geographic location, quality of energy use, stability of power supply, and characteristics of power use.

3.2. Data preprocessing

3.2.1. Anomaly data cleansing. By using the mining system to obtain the outage information of all stations in Suju District in 2023 for analysis, including the number of outages, outage duration and the number of outage users, and the outage information of the whole year is summarized and analyzed. The analysis found that some of the outage duration records in the system are negative, such as the 3# box variable of Nanyuan Xincun had an outage record of -84,270,778.0 minutes in the month of November 23, which is not in line with the outliers in common sense. An outlier is an extreme value or deviation in a set of data that is significantly inconsistent with the majority of the data, and the current outage duration is a value that does not meet the definition of common sense.

3.2.2. Analysis of energy use discrepancies. Based on the statistical analysis of the maximum load ratio and average load ratio of station areas in Suzhou 2023, as shown in Fig. 2 below, there is a big difference in the energy load between station areas, and some of them are in the state of serious light load or even empty load while some of them are in the state of over-heavy load, which makes the allocation of power resources between station areas very irrational, resulting in a great waste. Existing 65.59% of the average annual load ratio of the station area is less than 10%, this part of the station area and the overloaded station area also need to pay attention to, in the establishment of flexible interconnection pairing will increase the matching priority, for the subsequent station labeling comparison to provide a multi-faceted entry point.

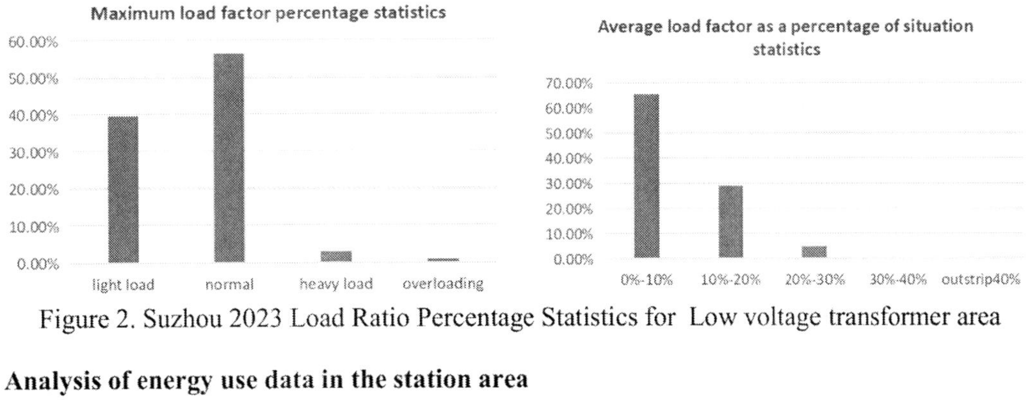

Figure 2. Suzhou 2023 Load Ratio Percentage Statistics for Low voltage transformer area

4. Analysis of energy use data in the station area

4.1. Power Data Research
Using a semi-supervised to supervised machine learning method, machine learning training and optimization are carried out to find out the correspondence between the characteristic values of each energy consumption index of the station area (input data) and the matching degree of flexible interconnection of the station area (output data) and generate a function (solidified model algorithm) by taking the energy consumption characteristic values of the station area to be measured in 2023 and

the station area latitude and longitude coordinate data data and Inputting into this model, the problem label of the corresponding station, the list of stations where flexible interconnection can be established and the name of the best matching station are obtained, as shown in Fig. 3 below.

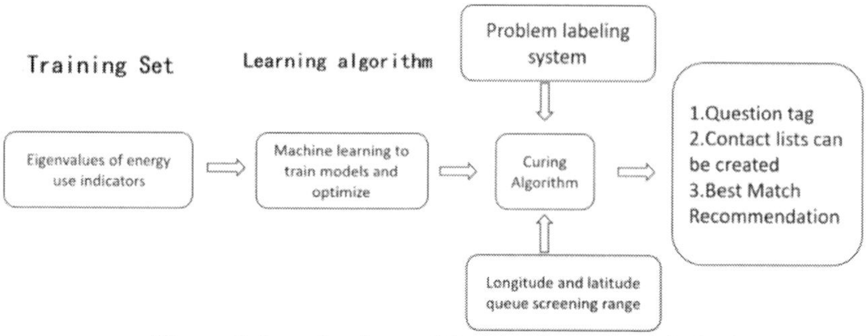

Figure 3. Introduction to Machine Learning Methods

In the initial learning of the model algorithm, there is no clear corresponding target value generated between the training set formed by the eigenvalues of the various energy-use indicators of the stations and the output of the algorithm, and at this time, a semi-supervised learning method is used: the model is without a clear target, and the algorithm automatically discovers the structure and patterns in the data without knowing the correct output in advance.

After the algorithm performs the initial evaluation of all station samples in the state of semi-supervised learning, the machine learning method is changed to supervised learning, where the model is trained by learning the relationship between the inputs and the corresponding labels; the algorithm optimizes the model by minimizing the gap between the predicted values and the actual labels.

4.2. Power Data Research

*4.2.1. Establishment of a problem labeling system.*Hierarchical analysis method (AHP) is used to establish the inter-scalar evaluation system of the problem, by analyzing the factors contained in the complex problem and their interconnections, the problem is decomposed into different elements and these elements are grouped into different levels [4], so as to form a multilevel structure, and at each level a judgment matrix can be established for the pair-by-pair comparisons of elements of that level according to a certain prescribed criterion, and the weights of elements of that level for that criterion can be derived by calculating the maximal eigenvalue of the judgment matrix and the corresponding orthogonalized eigenvectors to derive the weights of the elements of that layer for that criterion [5]. On this basis, the combined weights of the elements of each layer for the overall objective are calculated, so as to derive the weights of different scenarios [6], which provides the basis for selecting the optimal program.

By analyzing the acquired historical energy consumption data indicators of the station area, the abnormal problem elements of each indicator of the station area are enumerated, and the problem labeling system is established from multiple indicators in three aspects of geographic location, power supply reliability, power supply quality, and power consumption characteristics, respectively, and the special comparison red line labels between the station areas. As shown in Figure 4 below:

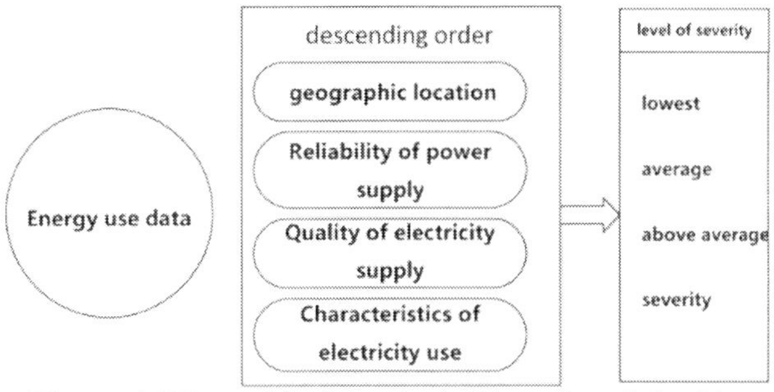

Figure 4. Hierarchical Analysis Three Level Architectured

*4.2.2. Determine the labeling hierarchy.*1)Construct judgment matrix by hierarchical pairwise comparison.To evaluate the quality of energy use in the station area, B1 is the reliability of power supply, B2 is the quality of power supply, and B3 is the characteristics of power use. The judgment matrix B made by two-by-two comparison of B1, B2 and B3 is shown in Table 1:

Table 1. Judgment matrix B

	B1	B2	B3
B1	1	1/2	1/4
B2	2	1	1/2
B3	4	2	1

The quality of the judgment matrix is measured by the satisfactory consistency of the judgments in the matrix, and in order to examine the reasonableness of the results produced by the AHP decision analysis method, the consistency test of the judgment matrix is required [7].

2)Find the largest eigenroot λmax for the judgment matrix, and then find its corresponding eigenvector W [8], i.e.

$$BW=\lambda maxW_i \tag{1}$$

Where the components of W (W1, W2, W3) are the relative importance corresponding to the three indicator elements, i.e., the weight coefficients [9].

3)The square root method is used to find the principal eigenvector W of the matrix B, which is the weight vector consisting of the weights of the sibling elements, and the product of each row of the judgment matrix is calculated [10].

$$W_i=\prod_{j=1}^{n} b_{ij},\ j=1,2,3 \tag{2}$$

Calculate the nth root.

$$\overline{W_i}=\sqrt[n]{M_i}\quad (i=1,2,3) \tag{3}$$

Normalize the vector =$[\overline{W_1},\ \overline{W_2},\ \overline{W_3}]T$:

$$\overline{W}=\frac{\overline{W_i}}{\sum_{i=1}^{n}\overline{W_i}}\quad (i=1,2,3) \tag{4}$$

Then $W=[W_1,W_2,W_3]T$ is the desired eigenvector.

4)Calculate the maximum characteristic root using the square.

$$\lambda max=\sum_{i=1}^{n}\frac{(AW)_i}{nW_i} \tag{5}$$

Where denotes the ith element of AW. Judgment matrix consistency indicator CI, average random consistency indicator RI and consistency ratio CR are introduced to judge consistency [11].

Finally, the relative importance weights of the second and third levels of the hierarchical analysis method are solved according to the above method, and the labels of A, B, and C are divided according to the degree of influence of the indicators on the operation of the stations

4.2.3. Determining label severity. Analyzing the historical energy consumption data of Suzhou station area in 2023, the severity range of each problem indicator of the station area is counted to generate the label intensity division. For example, statistics on the percentage of load abnormalities, three-phase unbalance abnormalities, and voltage abnormalities in the station area are shown in Figures 6 and 7 below:

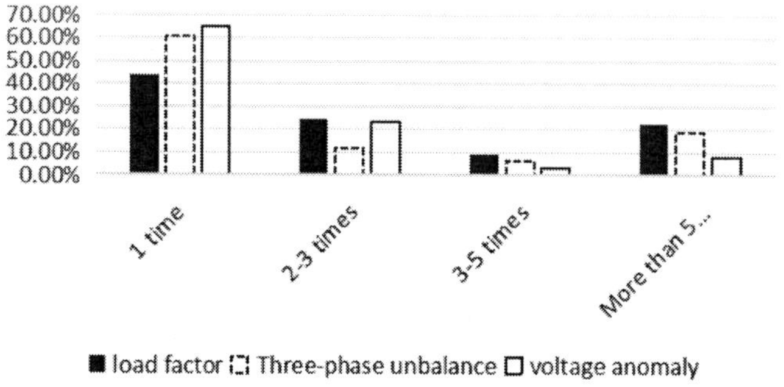

Figure 6. Statistics of various anomalies in Suzhou's 2023 station area

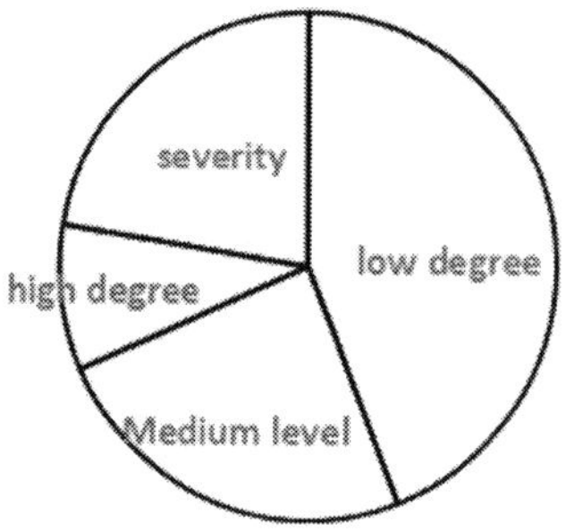

Figure 7. Statistics on the degree of each anomaly in the station area of Suzhou 2023

The range of the number of times each anomaly occurs is filtered to ensure a more balanced number of anomalous stations within each interval. According to the bar chart, these three types of anomaly indicator labels can be categorized into low, medium, high, and severe degrees according to the above range division.

According to the above statistical method, each type of indicator can be divided to generate labeling levels of different degrees.

4.3. Establishment of Judging System for Problem Labeling in Terrace Areas

4.3.1. Determine the sample station screening area. Based on the established station labeling system, select the station that needs to build a flexible direct-connected intermeter, and obtain the latitude and longitude values of the current target station. Take the latitude and longitude of the current station area as the center, and determine the range of transformers that can be screened for the associated station area according to the set latitude and longitude value que. For example, if the latitude and longitude of the target station area A is (j,ω) and the set value of latitude and longitude i, then the four latitude and longitude coordinates of the screening range of (j+i,ω+i), (j+i,ω-i), (j-i,ω+i), (j-i,ω-i) are connected to form a rectangle of area.

4.3.2. Calculate the exact distance of two points. The formula for calculating the physical distance between two distribution transformers based on the latitude and longitude coordinates of the two distribution transformers in the GIS, for the latitude and longitude coordinates of the transformers within the screening range determined by matching the latitude and longitude data of the transformers in the GIS, is as follows:

$$d_{AB}=\mathrm{R}\times cos^{-1}[\cos\omega_{\mathrm{A}}\times\cos\omega_{\mathrm{B}}\times\cos(j_A-j_B)+\sin\omega_{\mathrm{A}}\times\sin\omega_{\mathrm{B}}] \quad (6)$$

Where the latitude and longitude coordinates of transformer A and transformer B are $(j_A, \omega_{\mathrm{A}})$ and $(j_B, \omega_{\mathrm{B}})$ respectively, and the radius of the earth is R, with an average of 6371km; the actual distance between all the transformers of the stations within the current screening range and the transformers of the target station is calculated by using Eq.

4.3.3. Comparison of problematic labels in the station area. Based on the problem labels formed on the station area energy consumption index data, a set of problem label judgment system is established, which mainly includes the following steps:

1) Eliminate sample stations containing special red line labels: red line labels are defined as indicators that affect the operational stability between stations, such as belonging to the same line, the number of the same single-phase users accounting for more than 50% of the three items, and unbalance occurring on the same phase of the three-phase, etc., Eliminate the sample stations that have the same red line labels with the target station, the purpose of which is to exclude the same weaknesses of the interconnection of stations based on the specificity of the station's operation, and the risk that may arise from the same weakness. possible risks.

2) Screening out serious sample stations containing Class A labels: Class A labels are defined as indicators that can directly affect the operational stability of a station, defined as the degree of severity, basically major problem stations that experience frequent power outages, excessive number of various anomalies, and so on. It is determined in the comparison mechanism that this type of station has more serious problems and is not suitable for flexible interconnection with the target station that already has problems.

3) Sample station's own problem label scoring: After excluding special red line labels and Class A serious labels, the problem label degree scoring is performed for stations that do not contain the above labels, and the different degrees of labels are assigned values by comparing the different degrees of importance between the labels and labeling grades in the previous stage.

The main way of scoring the overall capability of the station is that the default base score of the station is the percentage value, and the corresponding type and degree of problem labels will be deducted from the corresponding value of the labels, and the remaining score at the end will be the scoring of the station's own capability.

Straight-line distance scoring between sample stations: Calculate the straight-line distance between the remaining stations and the latitude and longitude of the target station, and evaluate according to the distance.

4.4. Model training

Suzhou 8w stations are selected as the sample objects, based on the established algorithmic model, a supervised machine learning method is applied to analyze these data in depth, and the model algorithm is verified and optimized.

According to the results of the matching degree trial calculation, Suzhou Power Supply Company's professionals conduct field visits to the station area to verify the accuracy of the matching degree results, and feed the results back to the algorithm model, which continuously optimizes the weight of each label according to the feedback.

5. Research Programs and Innovations

Based on the above algorithm, a trial calculation is performed for the Suzhou station area:

1) Determine the target station area:The samples of Suzhou station area are labeled by the station area problem labeling system to visualize the station area problem. The target station area is selected for the station area problem to form a list of station areas, as shown in Table 2 below:

Table 2. List of target stations (partial)

	serial number	labels
Binjiang East Zhangshi Township 65# Xibaqiao Distribution Substation	0590000460000	Severe overloading of the annual maximum load factor High annual average load factor
Haiyu Xiandang 29# diagonal road north distribution substation	06220205010000	Annual maximum load ratio overload severe High annual average load ratio
Binjiang Dongzhangshi Township 65#Xibaqiao	10590000500000	Annual maximum load ratio overload severe Low annual outage number Low annual outage duration Low annual outage number of users

2) Screening sample station list:From the above target stations, the station: 29# diagonal road north distribution substation in Xiandang, Haiyu is randomly selected as a trial case. The latitude and longitude coordinates of this station area are (120 .809 97, 31.7 600 585), because 1 degree of latitude and longitude represents 111 .199km, and through the Suzhou station grouping case, the initial setting of the que value i=0.00 5 degrees will divide a square range with a side length of about 1.11 2km, and screen the sample station area in this range, and the results are shown in Table 3:

Table3. List of target stations (partial)

Name	serial number
Haiyu Shouchuang 9#Xiaoyao distribution substation	0590000460000
Haiyu Shouchuang 2#Yaoma Bay substation	06220205010000
Haiyu Shouchuang 10#Yaoma Bay North substation	0590000500000

3) Problem label comparison:The stations in the sample list are compared with the target stations by label matching method.

4) Optimal station recommendation:Based on the relative weight of the labels analyzed in the previous stage, the value is assigned to each type and degree label, and the scores of the problem labels are deducted through the deduction method to calculate the evaluation of the sample stations and recommend the stations with the highest scores.

Based on the above four steps of sample label comparison, the output results are shown in Table 4 below.

Table 4. List of interconnectable stations

	Serial number	Labels	Compatibility
Haiyu Shouchuang 9#Xiaoyao distribution substation	0590000470000	Average annual load factor light load medium	94
Haiyu Shouchuang 2#Yaoma Bay distribution substation	3090102130000	Annual average load factor light load severe Low annual number of outages Low annual outage duration Low annual number of outages	88
Haiyu Shouchuang 10#Yaoma Bay North Distribution Substatio	0590000460000	Average annual load factor light load severe Low annual number of outages Medium annual outage length Annual number of customers without power	81
Haiyu Beitang 4#Chenjiatang distribution substation	3090102130000	Low number of three unbalances Low number of outages per year Number of outages per year of the annual number of customers without power	76

6. Conclude

Based on the OLP modeling system, an evaluation system for the labeling of station area problems is established through hierarchical analysis and cluster analysis. The system is used to evaluate the actual energy use data of more than 80,000 station areas in Suzhou, and the problems of station areas are labeled. An innovative label evaluation system is established to score the energy-use stability of stations, and finally the actual straight-line distance between stations is introduced as a correlation factor to analyze the correlation between the target and sample stations, and it is found that the sample stations that are within 500 meters of the target station, and are more stable in operation and have no problematic labels, will have a higher degree of match with the target station.

References

[1] Liu Haitao,Dong Xinzhou,,Xiong Xiong,et al.Energy interconnection and energy microcirculation system based on low-voltage flexible straight end-grid[J]. Power System Automation, 2023,47(06) p 40-47.
[2] Jin Hao. Application of cluster analysis in real estate market positioning[J]. Logistics Engineering and Management, 2014,36(07) p 259-260.
[3] Cao Qiuqin, Zhang Ruilong. Research on the evaluation of enterprise creditworthiness based on Ali microfinance model[J]. China Township Enterprise Accounting, 2016,(06) p 49-51.
[4] WANG Yi, LIU Xiaoyun, HE Jun. Research on the Application of E-commerce Platform by Agricultural Enterprises in Anhui Province[J]. Modern Agricultural Science and Technology, 2012,(21) p 290-292.
[5] Zhou Y. Research on TOPSIS method in multi-attribute decision making [D]. Wuhan: Wuhan University of Technology, 2009 p 8-9.
[6] CHENG Qiuhua, CHEN Xieci, LI Suxia. Evaluation of the status quo of tourist agricultural parks based on hierarchical analysis method (AHP)--Taking Guangxi Yulin "Colorful Fields" as an example[J]. Gansu Agriculture, 2019,(06) p 103-108.

Journal of Physics: Conference Series **3012** (2025) 012016 doi:10.1088/1742-6596/3012/1/012016

[7] LIU Pan, HE Qilin, PENG Wei. Research on the application of EOC method in coal mine safety production accidents[J]. China Production Safety Science and Technology, 2012,8(10) p 135-139.
[8] Wang Hui,He Xun, Jiang Qun. Research on the provincial landing of China Mobile's excellence index and the evaluation system of provincial company's comprehensive ability. China New Communications, 2009,11(17) p 13-18.
[9] Yu Xiaopeng, Zeng Zhong, Wan Kun, et al. Internal Impact Analysis of Power Grid Technical Rehabilitation Project Based on Balanced Scorecard[J]. Science and Technology, 2013,(14) p 145-147.
[10] HU Hong, LU Hui, LIU Meizhao, et al. Evaluation and analysis of comprehensive effectiveness of low-voltage distribution station area[J]. China Equipment Engineering, 2019,(05) P 225-227.
[11] HAN Zhongmin. Knowing latitude and longitude to calculate the exact distance between two points[J]. Science and Technology Communication, 2011(11) p 196.

Journal of Physics: Conference Series 3012 (2025) 012017 doi:10.1088/1742-6596/3012/1/012017

A Method To Promote New Energy Consumption Based On Optimizing Multi-generation Capacity Settlement

Peiguang Chen[1], Zihao Tian[1], Zehao Zhao[2], Ziheng Wang[2], Jizhe Dong[2*]

[1] State Grid Jilin Electric Power Co. Ltd., Changchun City, Jilin Province, 130000, China

[2] Changchun University of Technology, Changchun City, Jilin Province, 130000, China

[*]Corresponding author's e-mail: jizhedong@yeah.net

Abstract. This paper proposes a method to promote new energy consumption based on the optimization of multi-generation capacity settlement for the current new energy consumption problems. By analyzing the trading mechanism of a provincial power market, it explores its impact on the rate of new energy consumption and proposes a model to optimize the transaction settlement method. The model incentivizes new energy power plants to improve the accuracy of power generation forecasts by optimizing the allocation of base power and the price settlement of multi-generation capacity and introducing a compensation mechanism. Through the analysis of examples, it is verified that the optimized scheme can effectively reduce the phenomenon of wind and light abandonment, improve the rate of new energy consumption, alleviate the pressure of thermal power units' peaking, and then promote the market-oriented development of new energy power.

1. Introduction

1.1. Current status of research at home and abroad

New energy represented by wind power is developing explosively, and the promotion of clean and low-carbon energy transformation has become the trend of China's energy development [1-3]. And, with the accelerated development of new energy, the wind and light abandonment phenomenon has appeared in local areas of China, and the contradiction of new energy consumption has gradually appeared [4]. In China, the existing power market trading mechanism makes it difficult to give full play to the advantages of new energy, there is a certain amount of abandoned wind, and abandoned light phenomenon. Literature [5] through the construction of a power generation alternative mechanism to solve the new energy incremental power abandonment due to market bidding.

In foreign countries, literature [6] pointed out that Spain through the establishment of a special scheduling department and trading center with a clear division of labor, the promulgation of tariff policy, improved the wind power forecasting mechanism and increased the cross-border power trading four measures to promote new energy consumption.

1.2. The research content of this paper

This paper mainly studies the new energy consumption method based on the power transaction deviation settlement compensation mechanism proposed in the literature [7] on the proposed multi-generation capacity settlement optimization. Firstly, the impact of a provincial power market trading mechanism on new energy consumption is analyzed. Then, a multi-generation capacity settlement optimization model is proposed, aiming to reduce the wind and light abandonment phenomenon by adjusting the settlement methods of base power and contract power. Finally, the effect of the optimization model in practical application is verified through example analysis, which assesses its potential to enhance the new energy consumption rate, alleviate the pressure of thermal power units' peak adjustment, and optimize the revenue of the power market. This paper makes certain improvements to the base power allocation problem in the current trading mechanism, which can make up for the uneven distribution of base power in the current market, and at the same time, the proposal e^{λ} can further optimize the distribution of benefits in the new energy market based on the deviation settlement compensation mechanism to improve the competitiveness of the market and stimulate the vitality of the market.

2. Analysis of the correlation between trading mechanism and new energy consumption

2.1. Overview of trading mechanism

The types of power market transactions in a province are mainly divided into electric energy transactions and power auxiliary transactions, of which, electric energy transactions can be divided into medium- and long-term power transactions and power spot transactions according to the transaction cycle, and power auxiliary service transactions refer to frequency regulation, standby and other paid power auxiliary services.

In the medium- and long-term trading of electricity, trading methods include centralized trading and bilateral negotiation trading. New energy power plants also participate in power trading in the power market mainly through medium- and long-term trading. At the same time, new energy power plants will also participate in some power dispatch institutions organized by the power auxiliary transactions, to complete some of the peak, standby, and other services, to maximize the power plant's power generation capacity, as far as possible to improve the plant's income.

2.2. Impact of trading mechanism on new energy consumption

In the power market, the new energy consumption rate is often subject to power generation plans, scheduling arrangements market price fluctuations, and other factors, among which, the flexibility of the trading mechanism is a key factor affecting the new energy consumption, can be used to complete the new energy power plant's affluent power grid peaking services, reduce the pressure of the thermal power generation unit's peaking to increase the new energy consumption rate, but also play a role in incentivizing the new energy power plants to improve their power generation capacity, to meet the role of the market. It can also serve as an incentive for new energy power plants to improve their power generation capacity and cater to the market. In addition, the mutual constraints of contract power and base power, compensation mechanism, and other factors also have an important impact on new energy consumption. Through the optimization of the trading mechanism, such as the establishment of the multi-generation capacity settlement optimization model, the relationship between the power generation of new energy power plants and the market demand can be better coordinated, and while ensuring the safe and stable operation of the electric power market, it can, to a certain extent, tap the regulatory potential of the system through the market-based mechanism and promote the consumption of new energy [8]. In the optimization of multi-generation capacity settlement, the compensation coefficient is introduced to compensate for the lesser power plants, to incentivize the new energy power plants to improve the accuracy of predicted power generation, and to reduce the power deviation. At the same time, the compensation mechanism can optimize the settlement price in the case of multi-generation capacity, encourage new energy power plants to actively participate in market transactions, and improve the market competitiveness of new energy power plants.

3. Optimization of multi-generation settlement to promote new energy consumption

3.1. Model description

To reduce the abandonment rate of new energy power generation in a province, encourage new energy power generation enterprises to actively participate in power market transactions, increase their share in the market, and ease the constraints of contract power and base power, while increasing the income from more power generation and ensuring that the settlement is on time and accurate, in the settlement of the transaction, the first settlement of the contract power, and then according to the new energy power plant power generation and the difference between the contract power, according to the installed capacity of the allocation of the base power. The base power is allocated according to the installed capacity. The base power is settled at a fixed tariff, and the contracted power is settled at the contracted tariff. The part of the unfinished power generation plan is replaced by other power plants, and the settlement price of the replacement power is determined by the original model and the compensation coefficient.

3.2. Settlement Process

Firstly, the allocation of base power quantity, then the determination of more and less power generation and the settlement of the price of more power generation, and finally the calculation of the compensation payment to the less power plant.

The intra-month generation capacity L_a of the new energy plant is:

$$L_a = \sum_{x=1}^{m} L_{ax} \tag{1}$$

Where: m is the number of new energy power plants, and L_{ax} is the intra-month power generation of each new energy power plant.

Intra-month contracted power from new energy plants L_h is:

$$L_h = \sum_{x=1}^{m} L_{hx} \tag{2}$$

Where: L_{hx} is the amount of electricity contracted within a month for each new energy plant.

The intra-month baseload L_j of the new energy plant is:

$$L_j = L_a - L_h \tag{3}$$

The intra-month base power L_{jx} of a single new energy plant can be expressed as:

$$L_{jx} = L_j \frac{R_x}{\sum_{x=1}^{m} R_x} \tag{4}$$

Where: R_x is the installed capacity of a single new energy plant.

After determining the base power of each new energy plant, the new energy plant's power generation, contracted power, and base power are used to determine the new energy plant's over-generation and under-generation, and to determine whether the new energy plant is an over-generation or under-generation plant. Among them, the over-generation L_{Fx} can be expressed as:

$$L_{Fx} = L_{ax} - L_{jx} - L_{hx} \tag{5}$$

The degeneration L_{Sx} can be expressed as:

$$L_{Sx} = L_{jx} + L_{hx} - L_{ax} \tag{6}$$

For the less generation of fewer power plants, for the sake of simple calculation, assuming that there is only one type of trading variety for new energy power plants, the comprehensive settlement price D_{Sx} of this part of electricity can be directly expressed as:

$$D_{Sx} = D_{hx} \tag{7}$$

Where: D_{hx} is the contracted tariff for the plant.

Based on the consolidated clearing price for less generation, the consolidated clearing price D_F for more generation from more power plants can be expressed as:

$$D_F = \frac{\sum_{x=1}^{m} L_{Sx} D_{Sx}}{\sum_{x=1}^{m} L_{Fx}} \cdot e^{\lambda} \tag{8}$$

Where: e^{λ} refers to the index factor of the actual power generation of the new energy plant, which can be specified as:

$$e^{\lambda} = e^{\frac{L_{ax}}{\sum_{x=1}^{m} L_{ax}}} \tag{9}$$

The introduction of the index factor can optimize the settlement price of multi-power plants for multi-generation capacity, and improve the results of new energy power plants' revenue on multi-generation capacity.

Finally, the compensation price B_x of the multi-power plant to less power plant can be expressed as:

$$B_x = \omega L_{Sx} D_{hx} \tag{10}$$

Where: ω is the compensation coefficient, the general range of values is 0~1. D_{hx} Refers to the contractual tariff corresponding to the multi-power plant.

As shown in Fig. 1, the computational flowchart of this deviation power settlement optimization method is shown.

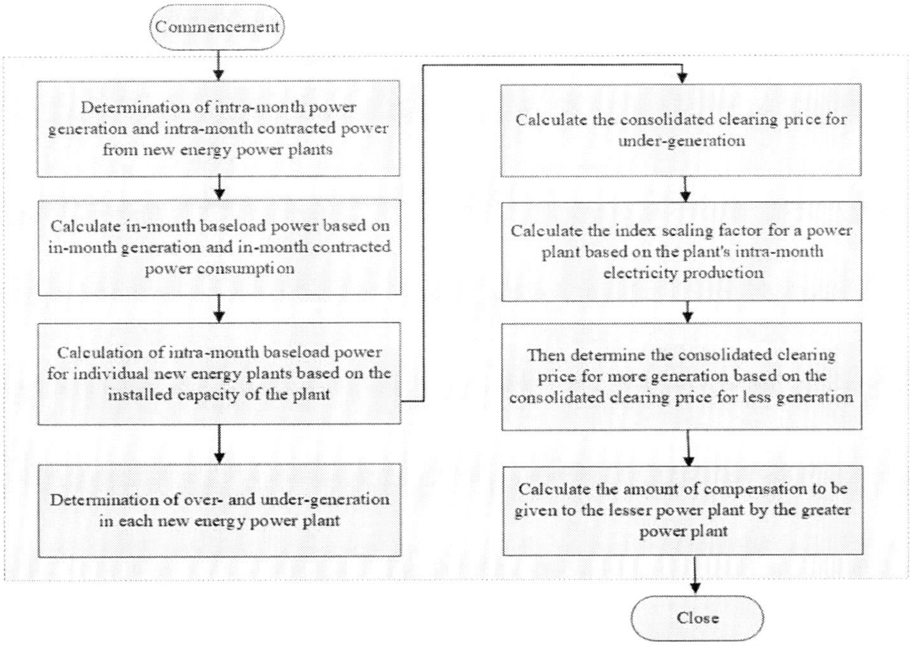

Figure 1. Deviation power settlement optimization method calculation process

4. Example Analysis

Assuming that there are four power plants, A, B, C, and D, with installed capacities of 8.6GW, 8.4GW, 8GW, and 5GW respectively, the total power generation capacity of the four power plants is 50GW-h, and the total contracted power is 10GW-h. Meanwhile, assuming that the feed-in tariffs of the four power plants are 0.48 Yuan/(kW-h) that the contracted power transaction price is executed by the provisions of the contract, and that the value of the compensation coefficient is set as 8% for the time being. 8 per cent. The calculation of the total revenue is retained in one decimal place.

Table 1. Optimization methods for deviated power settlement

Electric power plant	A	B	C	D
L_{ax} / $(GW\cdot h)$	13.5	11.5	10	5
L_{hx} / $(GW\cdot h)$	4	2.5	2	1.5
D_{hx} / $CNY/(kW\cdot h)$	0.39	0.38	0.36	1.35
L_{jx} / $(GW\cdot h)$	8.6	8.4	8	5
L_{Fx} / $(GW\cdot h)$	0.9	0.6		
D_F / $CNY/(kW\cdot h)$	0.49.5	0.4666		
L_{Sx} / $(GW\cdot h)$				1.5

B_x / CNY	46800	45600		
Z_{hx} / CNY	1560000	950000	720000	525000
Z_{jx} / CNY	4128000	4032000	3840000	2400000
Z_{Fx} / CNY	441450	279960		
Z_x / CNY	6083000	5216000	4560000	3017000

Comparing the over-generation gain and total generation gain of the four power plants, it can be seen that the deviation power settlement method can reasonably optimize the generation gain according to the power generation of each power plant, which improves the competitiveness of the new energy power generation in the power market to a certain extent, encourages and guides the new energy power plants to actively renovate, and at the same time, it also exerts a certain market-based regulating mechanism, utilizing the abundant power generation of the power plants to complete the power market's Peaking demand. Utilizing the abundant power generation capacity of each power plant to complete the peak demand of the power market and improve the rate of new energy consumption. Compared with the current rolling treatment of contracted power deviation in the power market, this method can maximize the power generation capacity of new energy power plants by reasonably allocating the base power and actively mobilizing the participation of each power plant in the power market.

5. Conclusion

Based on the current trading mechanism and new energy consumption situation in a province, this paper makes further development suggestions on the treatment of deviation power generated by new energy power plants. Through the introduction of the new energy power plant's actual power generation index factor e^{λ} , the price settlement of each new energy power plant's excess power generation is further optimized, and at the same time, a certain amount of price compensation is provided to the lesser power plants, which motivates the lesser power plants to improve the accuracy of forecasting their power generation capacity. At the same time, the deviation power processing method can also solve the mutual constraints between contracted power and power generation capacity to a certain extent, alleviate the peak pressure of thermal power units, and make use of the affluent power generation capacity of each new energy power plant for peak shifting to improve the rate of new energy consumption. In addition, this paper does not take the value range of ω and its specific value for detailed demonstration, in the future also need to be more specific research, improve the compensation price, and further improve the settlement optimization method.

Acknowledgments
This work was supported by the Research Program of State Grid Jilin Electric Power Co. Ltd. under Grant SGJLJY00ZLJS2400021.

References

[1] Aatola P, Ollikainen M, Toppinen A . Impact of the carbon price on the integrating European electricity market[J].Biochemistry, 2013, 46(43):12152 -63.
[2] Ahamada I, Kirat D . Non-linear Pass-Through of the CO2 Emission-Allowance Price onto Wholesale Electricity Prices[J]. Assessment, 2018, 23(5):1-14.
[3] Acworth W, de Oca M M, Boute A, et al. Emissions trading in regulated electricity markets[J].

Climate Policy, 2020, 20(1): 60-70. [4] Zhang ZL.
[4] Zhang ZL. 'China's '13th Five-Year Plan' New Energy Grid Integration and Consumption Situation, Countermeasures Research and Multi-Scenario Operation Simulation Analysis.' China Power 51.1 (2018): 8.
[5] Chen Yihua, Zhang Wei, Zhang Chenggang, Zhang Yao, and Wu Pengtao. Research on power spot market trading mechanism to promote new energy consumption in new power system[J]. Intelligent Power,2022,50(2):97-104
[6] Xu Shasha. 'Research on market mechanism practice and strategy for promoting clean energy consumption.' Engineering Construction and Design 2(2020):2.
[7] Liu Dunnan, Tang Honghai, Yang Mo, et al. Power trading deviation settlement compensation mechanism for promoting new energy consumption[J]. Power system automation,2017,41(24):105-111.
[8] XU Gaoyuan,WANG Xiaojing. A study on the price of new energy consumption market by counting and dynamic carbon emission[J]. Journal of Solar Energy,2024,45(02):360-367.

Journal of Physics: Conference Series **3012** (2025) 012018 doi:10.1088/1742-6596/3012/1/012018

Discharge Severity Identification of Suspended Insulators Based on Improved SSD Algorithm

Hai Huang*, Xun Zhang, Jun Wu, Yi Shi, Baisong Li, Xiaohua Liu

State Grid Hubei Electric Power Co., Ltd. Ultra High Voltage Company

*Corresponding author's e-mail:1048768378@qq.com

Abstract. In order to realize the intelligent evaluation of UV solar-blind imaging of pollution discharge of suspended insulators based on edge detection, this paper establishes a test platform for pollution discharge of suspended insulators, defines UV parameters such as relative spot area, proposes a criterion for discharge severity, and establishes a database of UV images of different discharges. The improved SSD model is used to train the model, which realizes the reduction of model parameters and the improvement of training convergence speed. The results show that the model file size of the improved SSD is 4.5 Mb, which has good embedded detection performance. In the process of model training, a self-updating learning rate mechanism is designed, which improves the convergence speed by about 3.5 times and reduces the convergence value by 1/2. The intelligent assessment of insulator discharge severity can be divided into four levels.

1. Introduction

Suspended insulator is one of the significant elements in transmission lines. Its insulation performance will gradually deteriorate under the action of long-term electrical, thermal, mechanical and environmental factors, which may lead to the increase of local electric field and the generation of surface discharge and ultraviolet radiation[1]. Ultraviolet imaging method has the advantages of long-distance detection, non-contact and accurate positioning, and is widely used in discharge detection of electrical equipment[2].

With the development of artificial intelligence, deep learning algorithms are gradually applied to the intelligent evaluation of insulator operation status[3-5]. In the literature, a variety of insulator fault detection methods based on deep learning are proposed, including two-stage detection of collaborative deep learning, single-stage image recognition framework, YOLOV3 detection, etc[6-8]. The UV imager mounted on the UAV can improve the detection efficiency and reduce the labor intensity[9-10]. However, after the inspection, the image and video are imported into the back-end for analysis, resulting in poor real-time diagnosis. Therefore, the deployment of edge computing platform on UAV can reduce data transmission and improve real-time diagnosis[11].

The SSD (Single Shot Multibox Detector) deep learning algorithm has higher target detection speed and accuracy, but its model is larger. In order to adapt it to edge computing devices, this paper presents an enhanced SSD convolutional neural network framework. It replaces the high-dimensional feature extraction backbone network with a low-dimensional core convolution kernel cluster. And it uses a support function sub-network to do target prediction. A porcelain suspension insulator test platform was built to obtain leakage current and ultraviolet image video at different discharge stages.

A self-learning rate update mechanism was designed to improve the training performance of the model. Finally, the trained model is downloaded to the Jetson nano platform to realize the intelligent evaluation of discharge severity.

2. Test wiring and research methods

2.1. Test principle

In order to simulate the different discharge severity of insulator surface, experimental research in artificial fog chamber is carried out in this paper. The test transformer model is TDTW-1200/300, the fog chamber size is 1.5×1.5×2.7m, the humidity is controlled by SJC-3000 ultrasonic humidifier, and the maximum relative humidity can reach 95%~97%, a 1.2 × 0.6m ultraviolet transparent glass is installed on the side, which has a transmittance of 95%~97% and can effectively transmit ultraviolet light in the wavelength range of 240nm~280nm.

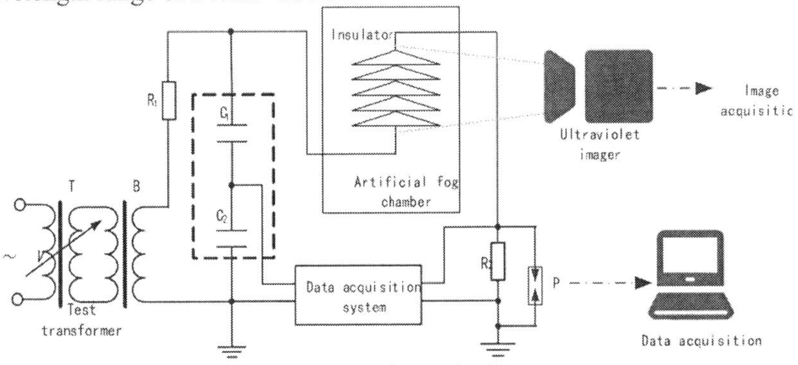

Figure 1. Test schematic diagram.

The test object is XWP2-70 suspension porcelain insulator. The discharge image of insulator string with 7 insulators is collected by CoroCAM504 ultraviolet imager. The video is stored in the acquisition card. The leakage current is sampled by a 10 Ω non-inductive precision resistor, and the signal is collected by the HS5 data acquisition card.

2.2. Test method

The artificial contamination method is used to simulate the surface contamination. The insulator is washed with pure water and dried for 12 hours, and then the surface is contaminated with a mixture of sodium chloride and kaolin. Control ESDD is 0.4 mg/cm^2, NSDD is 2mg/cm^2, the applied voltage is 70 kV, the gain of the ultraviolet imager is set to 60 %, the observation distance is 14 meters.

2.3. Relative spot area

With the increase of the observation distance, the area of the discharge spot will decrease, so it is difficult to establish a unified criterion directly according to the size of the spot area. In order to overcome the shortcomings of traditional spot area parameters, the relative spot area parameter is proposed in this paper, and the calculation is shown in equation (1):

$$S_{re} = \frac{S}{S_{ins}} \tag{1}$$

Where, S_{ins} is the area of the rectangular area corresponding to the contour of the insulator in the captured image, that is, the pixels contained in the area, and the unit is pixel, as shown in Figure 2. S represents the spot area of discharge, and its calculation method is as follows (2):

$$S = \sum_{i=1}^{M}\sum_{j=1}^{N} B(x,y) \tag{2}$$

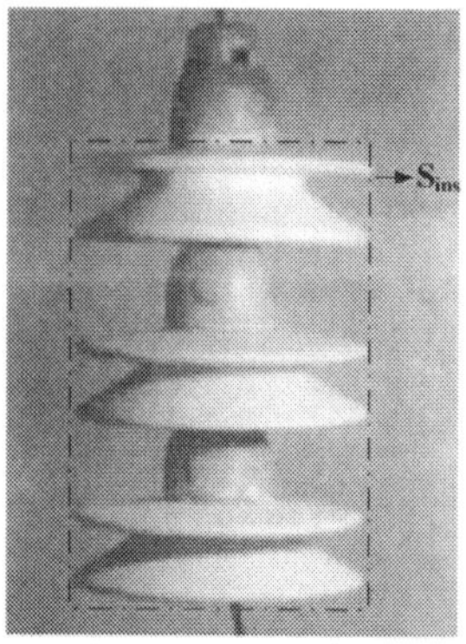

Figure 2. Insulator area S_{ins}

In Equation (2), M and N denote the row and column values of the two-dimensional pixel matrix, respectively. $B(x, y)$ is a binary image processed by threshold segmentation, morphological filtering and small area algorithm [2]. Similarly, the area of the discharge spot is the number of pixels contained in the discharge image, and its unit is also defined as a pixel in this paper.

The ultraviolet imager adopts the principle of dual-channel superposition imaging. The size of the discharge image of the ultraviolet channel and the image of the insulator body will change with the distance according to the same law. Therefore, at different detection distances, the ratio of formula (1) will be approximately unchanged.

2.4. Discharge severity classification

Based on the above experiments, the typical UV images of discharge under different humidity can be obtained, as shown in Figure.3.

(a)$0 \leq S_{re} < 0.50$ (b)$0.5 \leq S_{re} < 1.0$ (c)$1.0 \leq S_{re} < 2.0$ (d)$S_{re} \geq 2.0$

Figure 3. Discharge UV images under different humidity.

According to the characteristics of corona, spark and arc in the discharge process, combined with the relative spot area under different humidity, the operation state of porcelain suspension insulator is analyzed. There are four types of operating states :

(a) When the humidity reaches about 70 %, when $0 \leq S_{re} < 0.50$, the small air gap around the insulator cap first appears slight corona discharge, and scattered light spots appear on the ultraviolet image, which is marked as " slight " ;

(b) When the humidity reaches about 75 % and $0.5 \leq S_{re} < 1.0$, there is a blue-purple brush-like small spark discharge accompanied by obvious "discharge sound ". The spot area shows an irregular state, large and small, and the state is marked as " moderate ".

(c) When the humidity reaches about 80 % and $1.0 < 2.0$, a small arc discharge occurs with a length of no more than 2cm. The discharge area completely covers the body of the porcelain suspension insulator, and the state is marked as " heavy ".

(d) When the humidity reaches about 85 %, $S_{re} \geq 2$, the porcelain suspension insulator body appears bright arc, enters the state of large arc discharge, flashover may occur at any time, and the state is marked as " worse ".

3. Assessment of discharge severity of porcelain insulator

3.1. Edge platform construction

It can be seen from the above tests that the ultraviolet image can effectively reflect the discharge condition of the insulator surface, which can be identified and evaluated by image and video analysis. In this paper, an edge computing and recognition platform based on NVIDIA Jetson nano is established. The operating system of the computing platform is Ubuntu18.04, and the parameters are shown in Table 1.

Table 1. Parameters of platform.

GPU	NVIDIA MaxwellTM 128 CUDA
CPU	ARM Cortex-A57
memory	4 GB
computing power	472 GFLOPS
power loss	5 W

3.2. UV image data set and corresponding image annotation

Based on the experimental research of the artificial climate chamber mentioned above, this article obtained 10000 ultraviolet images from the discharge ultraviolet video stream through video analysis. For the convenience of subsequent classification, according to the size of the ultraviolet image spot, the intensity of the discharge, and the phenomenon of the discharge, this article divides it into four states: "mild discharge", "moderate discharge", "severe discharge", and "poor". The images of each discharge state account for 25% of the total discharge images. Using Labellmg image annotation software to create PASCAL datasets and generate corresponding XML files, XML is an important way of data transmission that can span various languages and platforms. Divide the above UV image dataset into a 4:1 ratio, with one part being the training set and the other part being the testing set, consisting of 8000 images for network training and 2000 images for model performance evaluation.

3.3. Improvement of SSD model

To make the model smaller, the improved detection network model in this paper is shown in Figure 4. Its parts include two things: "highly optimized core convolution subnet" and "SSD-based auxiliary feature extraction subnet". After the core convolution kernel does the first extraction, many feature maps are made. After that, certain specific feature maps are selected and transferred to the auxiliary feature extraction sub-network for prediction.In this manner, the model can be minimized while the precision of target detection is guaranteed.

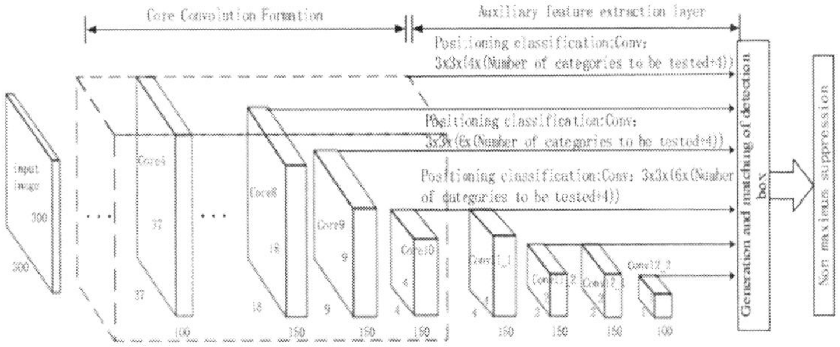

Figure 4. Improved detection network model.

3.3.1. Core Convolution Group Subnetwork. Traditional object detection algorithms mostly use models such as VGG16 for low-level feature extraction. These models often use 3 × 3 convolution kernels as filters for feature maps, which have two drawbacks: One problem is the lack of convolution depth. Also, too many convolution kernels lead to a large number of channels in the output feature map, creating a huge amount of parameters and demanding high hardware computing power and memory. In order to facilitate the deployment of the neural network model to embedded edge computing devices, this paper designs a smaller core convolution group, and uses this network architecture to achieve more efficient feature extraction.The design concept of Core convolution cluster is presented as follows: minimize the number of 3x3 convolution kernels as much as possible; Significantly reduce the number of channels in the input feature map of a 3x3 convolutional kernel[12].

The Core convolution group in this article has 10 Core convolution kernels, from Core1 to Core10. Each Core convolution kernel has two convolutional layers:a squeezed convolutional layer S and a dilated convolutional layer E.We minimize the quantity of convolution kernels in the S and E layers while maintaining the model's accuracy. Also, a Pool downsampling layer was added after several Core convolutions to cut down the number of channels in the feature map.

The Core convolution cluster consists of low dimensional convolution kernels. This can condense the computational parameters of the neural network to a great extent.When the input channel number of the convolution kernel is M and its output channel number is N, the parameter number generated by using standard convolution for feature extraction is:

$$M \times K \times N \times K = K^2 MN \tag{3}$$

When the optimized convolution put forward in this paper is adopted, the parameter number is:

$$M \times s_{1\times1} + M \times e_{1\times1} + 9 \times M \times e_{3\times3} = \left(\frac{s_{1\times1} + 8e_{3\times3}}{N} + 1 \right) MN \tag{4}$$

Compared with the traditional convolution formula, the convolution of our designed calculation parameters after appropriate optimization is greatly reduced in number, and it can be seen from the analysis results that the larger *N*, the better the effect can be achieved.

3.3.2. Sub network of Auxiliary feature extraction. The SSD algorithm is a widely used target detection network in recent years. It detects objects of different sizes in the scene through a set of non-uniform convolutional feature layers with gradually reduced dimensions, and then uses these convolutional feature layers to obtain the position and category of the target to be detected. In this paper, Core4, Core8, Core9, Core10, Conv11-2 and Conv12-2 convolution layers are selected from the core convolution kernel sub-network to form an auxiliary feature extraction sub-network. The overall architecture and the parameters of each layer are shown in Table 2. After the initial extraction of the corresponding convolution layer features, a set of convolutions is used to predict the position of each prediction box on the feature layer (loc) and the category of the target in the prediction box (conf).

Table 2. Architecture of feature extraction network.

Network layer	Convolutional kernel size	Feature drawing size
Core4-loc	3×3×16	37×37
Core4-conf	3×3×84	37×37
Core8-loc	3×3×24	18×18
Core8-conf	3×3×126	18×18
Core9-loc	3×3×24	9×9
Core9-conf	3×3×126	9×9
Core10-loc	3×3×24	4×4
Core10-conf	3×3×126	4×4
Conv11-2-loc	3×3×24	2×2
Conv11-2-conf	3×3×126	2×2
Conv12-2-loc	3×3×16	1×1
Conv12-2-conf	3×3×84	1×1

4. Training Model Optimization

The NVIDIA deep learning server is used to train the model. The server is equipped with Zhiqiang 4200 series central processor and NVIDIA P6000 series graphics card, which can complete the complex neural network training process.

4.1. Ultraviolet discharge image video stream detection

This article uses OpenCV software to read and form a video stream from the video signal collected by the ultraviolet imager. The image frame rate is 28.2 FPS, and the resolution of the image is 1280 × 720 pixels. Firstly, the above model is trained on the server side, and after training, the model is downloaded to the edge side hardware device to achieve real-time recognition of discharge images in the captured video. This recognition model can obtain six feature map parameters of the image to be detected, and then use a deep network to identify the type of discharge video image. The principle is shown in Figure 5.

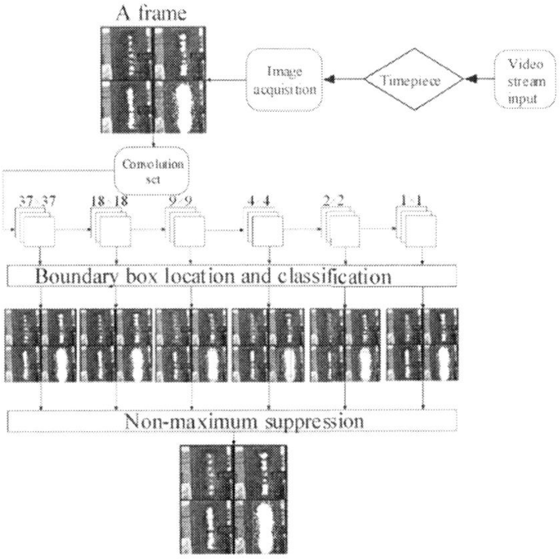

Figure 5. Video image detection process .

After two convolution calculations, the SSD deep learning network will continuously change the relative position of the predicted box in the ultraviolet image when the "predicted box position" is set. In addition, the estimated category of each prediction box is constantly changing, and six feature maps will provide six detection results. Finally, this article processed the six results using non maximum suppression to find the position and type of the bounding box with the highest confidence, which is also the final predicted value of the input UV image. Write the predicted results into a new video image using OpenCV software. Repeat the above process to complete the recognition of ultraviolet discharge images of ceramic suspension insulators.

5. Model performance test and analysis

The verification images are loaded into the improved SSD model, the original SSD model and the model in Reference [13], and the parameter calculation model proposed in Reference [14] is used to analyze the calculation parameters. The calculation method is shown in formula (5).

$$N = \sum_{i=1}^{n} \alpha_i c_i \tag{5}$$

Among them, the number of parameters for each convolutional layer, the total number of parameters generated for model training, train 1000 times and analyze the relationship between the number of intermediate parameters generated during the training process, as shown in Figure 6.

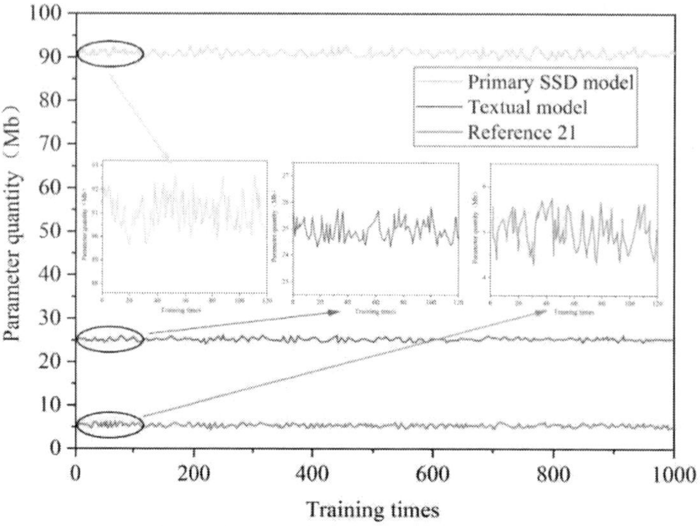

Figure 6. Model parameter comparison.

The parameters generated by the original SSD model are about 91 Mb. Reference [13] used MobileNet instead of the underlying feature extraction network, and the number of parameters after simplifying the model is about 25Mb ; the model in this paper uses multiple low-dimensional convolutions to generate only 1 / 20 of the intermediate parameters of the original SSD network, which greatly reduces the model. The prediction accuracy and size of the model are shown in Table 3, and the model file is only 4.5 Mb. In this paper, we further simplify the related network structure. Comparing the simplified network structure with the lightweight model MobileNet + SSD, the following conclusions can be drawn : the simplified network structure truly achieves the goal of model miniaturization.

Table 3. Model comparison.

Model	Model size	Average prediction accuracy
SSD	91.6Mb	78.46%
reference [14]	26.3Mb	76.61%
this article model	4.5Mb	83.53%

6. Model performance test and analysis

In this paper, the discharge video taken in the experiment is loaded into the model observation recognition effect, and the accuracy rate is 83.53 %, as shown in Figure 7.

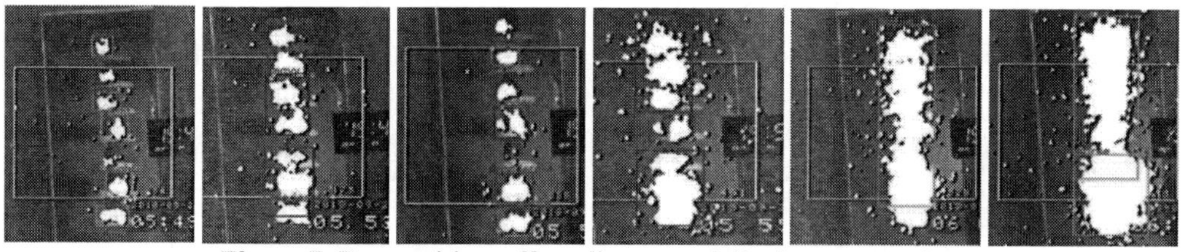
Figure 7. Recognition effect diagram of discharge severity.

It can be seen from Figure.3 that the proposed algorithm can accurately identify most of the discharges. When $0 \leq S\text{re} < 0.50$, the phenomenon of missing identification will occasionally occur ; when $0.50 \leq S\text{re} < 1.0$ and $1.0 \leq S\text{re} < 2.0$, it is prone to repeated alternating recognition, and it is necessary to combine other on-site information to improve the accuracy of evaluation. When $S\text{re} \geq 2.0$, the porcelain suspension insulator is close to flashover, the arc is more continuous, and the discharge spot area is larger. The algorithm can accurately identify it.

7. Conclusion

(1) The test platform of polluted insulator is built, and the relative spot area parameter is proposed to classify the severity of insulator discharge, so as to facilitate the training of subsequent neural network.

(2) The underlying structure of SSD model is improved. On the precondition of guaranteeing the model's accuracy, the model file size is only 4.5 MB, which has good embedded detection performance.

(3) The training strategy of the model is optimized, and the self-updating learning rate mechanism is designed. The convergence speed of the model is increased by about 3.5 times, and the convergence value is reduced by 1 / 2. The training curve is smooth and the convergence value is better.

(4) The discharge video collected in the experiment is loaded into the model, and the recognition accuracy is 83.53 %. Most of the discharges can be accurately identified.

References

[1] Zhang, Y, 2018, Research on infrared detection blind area and diagnosis strategy of high voltage porcelain insulator. Master's thesis, Hunan University, Hunan.

[2] Wang, S.; Jiang, T.; Li, W.; Niu, L.; Lv, F, 2021, Infrared and UV Imaging Characteristics of Insulator Strings Containing a Zero-Value Insulator Based on Experiment and Multi-Physics Simulation. IEEJ T. Electr. Electr. 2021, 16, 374-382.

[3] Wang, Z.; Wang, Y.; Wang, Q.; Kang, S.; Mikulovich, V, 2021, Two Stage Insulator Fault Detection Method Basedon Collaborative Deep Learning. Trans. China Electrotechnical Soc. 2021, 36, 3594-3604.

[4] Tang, R.; Zhang, M.Y.; Xu, H.; Li, W.; Li, W.B, 2021, Insulator recognition of transmission line inspection image based on deep learning. Adv. Power Syst. Hydroelectric Eng. 2021, 37, 41-46.

[5] Zhang, X.Y.; Wang, Y.C.; Wang, S.L, 2021, Research on insulator fault inspection algorithm based on deep learning. J. Zhongyuan Univ. Tech. 2021, 32, 45-49.

[6] Zhang, M, 2021, Insulator detection and defect recognition of transmission line inspection image based on deep learning. Master's thesis, Xi'an University of technology, Xi'an, 2021.

[7] Lv, F.C.; Niu, L.L.; Wang, S.H.; Chu, Y.P, 2021, Evaluation of discharge severity of porcelain suspension insulator based on UV imaging and improved yolov3. High Volt. Eng. 2021, 47, 377-386.

[8] Dong, X.; Wu, J.; Wang, H, 2017, Airborne ultraviolet inspection and fault diagnosis technology for transmission lines. J. Elec. Power Sci. Tech. 2017, 32, 117-122.

[9] Hu, Z.K.; Zhao, C.Y.; Wang, Z.D.; Lin, C, 2020, Intelligent identification of hidden dangers in key parts of transmission tower based on edge calculation and UAV inspection image. Zhejiang Electr. Power 2020, 39, 21-27.

[10] Huang, D.M.; Wang, Y.Q.; Hu, A.D.; Sun, J.Z.; Shi, S.; Sun, Y.; Fang, L.F, 2022, Insulator state edge recognition method based on multi-dimensional feature fusion. China Power 2022, 55, 133-141.

[11] Redmon, J.; Farhadi, A, 2017, YOLO9000: better, faster, stronger. In IEEE Conference on Computer Vision & Pattern Recognition, Honolulu, HI, USA, 21-26 July 2017.

[12] Simonyan, K.; Zisserman, A, 2015, Very deep convolutional networks for large-scale image recognition. IEEE Trans. Pattern Anal. Mach. Intell. 2015, 20, 45-56.

[13] Lu, Y.Q.; Sun, C.Y.; Cao, H.W.; Yan, H.W, 2020, Foreign body detection method for transmission equipment based on edge computing and deep learning. Electr. Power 2020, 53, 27-33.

[14] Nagasaka, H.; Maruyama, N.; Nukada, A.; Endo, T.; Matsuoka, S, 2018, Statistical power modeling of GPU kernels using performance cou. IEEE Trans. Pattern Anal. Mach. Intell. 2018, 38, 235-245.

Journal of Physics: Conference Series **3012** (2025) 012019 doi:10.1088/1742-6596/3012/1/012019

Communicationless Flexible Interconnection Technology For Wide Access Of Charging Piles

Xiaofeng Dong[1], Junting Li[1], Qiong Zhu[1], Jianqiao Zhou[2*], Yajun Jia[2], Bin Liu[2]

[1] State Grid Suzhou Power Supply Company, Suzhou, Jiangsu Province, 215004, PR China

[2] Shanghai JunShi Electric Technology Co.,Ltd, Shanghai, 200240, P.R. China

*Corresponding author's e-mail: jianqiaozhou@sjtu.edu.cn

Abstract. With the wide access of electric vehicles (EVs), the problem of low voltage at the end of LVDN will be caused. Soft open point (SOP) is a new type of flexible interconnection device (FID) that can be used for voltage regulation at the end of feeder in distribution network. However, the existing SOP voltage regulation scheme needs to know the voltage value of all the feeder terminal nodes, which requires good communication conditions between the feeders of the distribution network. In this paper, a voltage control strategy for large-scale charging piles connected to FID network is proposed. Based on VSC active-reactive power coordinated voltage regulation, the flexible interconnection system at the end of the feeder knows the voltage and power information of the PCC node through the change of the DC bus voltage. Through the coordinated control of the AC and DC line voltage of the flexible interconnection system, it can realize the voltage limit control of the multi-terminal flexible interconnection distribution system without relying on the system communication conditions, and effectively improve the system 's ability to accept EVs.

1. Introduction

With the rapid increase in the number of charging stations, EVs will become a new type of important load in the future of the power grid [1]. Considering the randomness of the charging behavior of a large number of EVs, the impact of EVs access to the power grid charging on the power grid is gradually highlighted, such as load balance, power capacity, power quality, environment and other aspects will have an impact. Among them, the extensive use of large-scale charging piles may cause voltage fluctuations at the end of the distribution network, especially in the areas where charging piles are concentrated, and voltage fluctuations may affect the stability of the power grid and the quality of power supply[2,3].

In order to solve the problem of low voltage or even voltage over-limit at the end of the feeder of the distribution network caused by the access of large-scale charging piles, the existing solutions mainly include optimizing the distribution network architecture[4], installing reactive power compensation devices[5], accessing distributed power sources[6], and applying ESS[7]. However, these traditional methods have limited adjustment range and effect, and poor economy. Soft open point (SOP) is a new type of power electronic device, which can replace the mechanical contact switch at the end of the line to realize the flexible interconnection of different feeders, and can quickly adjust the active and reactive power of the connected feeder[8]. Reference [9] proposed that SOP uses the

Journal of Physics: Conference Series **3012** (2025) 012019 doi:10.1088/1742-6596/3012/1/012019

information between feeders to optimize the distribution of power flow in distribution network by using optimization algorithm, so as to realize the functions of feeder capacity balance, power mutual aid and network loss optimization, and then improve the overall performance of low-voltage distribution network. In reference [10], a new voltage regulation strategy is proposed by using the coordinated control of SOP and DG inverter in distribution system, which can quickly and efficiently solve the problem of minimizing voltage deviation. In [11] and [12], a control strategy combining SOP centralized and local voltage is proposed. According to the information of the whole system, the active power transmission and reactive power compensation of SOP are adjusted at the same time, so as to realize the regulation of global power by SOP. In the above research, SOP centralized control strategy is widely used, which requires good communication conditions between feeders of distribution network. However, the scale of low-voltage distribution network is large, the communication network is not perfect, the communication reliability is low, and this situation can not be significantly improved in a short time in the future. With the communication conditions of the existing low-voltage distribution network, the distribution network cannot achieve real-time and safe regulation of the local voltage in the face of the randomness, rapidity and real-time characteristics of the terminal voltage.

In this paper, a voltage control strategy for large-scale charging piles connected to flexible interconnected distribution networks is proposed. It can quickly realize the flexible adjustment of active and reactive power of the connected feeders when there is no communication in the distribution network, and then realize the problem of low voltage or even voltage over-limit at the end of the distribution network feeder caused by large-scale charging piles.

2. Analysis of the influence of EV_S wide access on the terminal voltage of LVDN

Figure 1 is the equivalent circuit diagram of the low-voltage distribution network with EV_S.

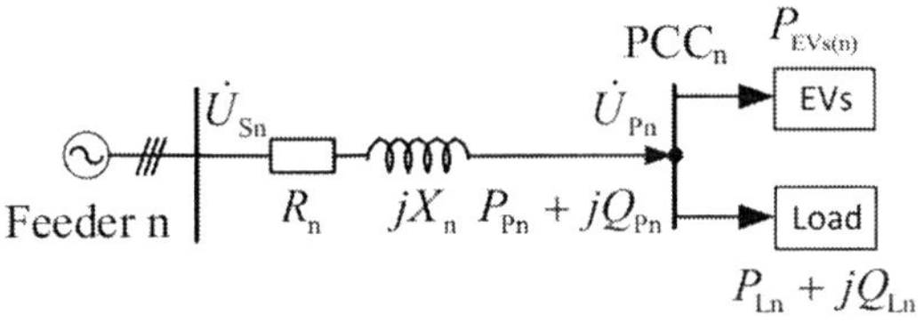

Figure 1. Equivalent circuit diagram of LVDN with EV_S

Among them, the first terminal voltage of the feeder n is $\dot{U}_{Sn}$,the terminal PCC voltage is $U_{Pn}\angle 0$, the power transmitted from the first terminal to the end of the line is P_{Pn}+jQ_{Pn} , R_n+jX_n is the line impedance of the feeder n, and the power flow equation of the power system can be obtained. The expression of the first terminal voltage $\dot{U}_{Sn}$ of the feeder is (j represents the imaginary part) :

$$\dot{U}_{Sn}=U_{Pn}+\frac{P_{Pn}R_n+Q_{Pn}X_n}{U_{Pn}}+j\frac{P_{Pn}R_n-Q_{Pn}X_n}{U_{Pn}} \tag{1}$$

For the power grid of 110 kV and below voltage level, ignoring the transverse component of voltage drop, ignoring the imaginary part of (1).The charging power of the electric vehicle is $P_{EV_s(n)}$, the feeder load is P_{Ln}+jQ_{Ln}, and the PCC node voltage U_{Pn} expression is further :

$$U_{Pn}=\frac{U_{Sn}}{2}+\sqrt{\left(\frac{U_{Sn}}{2}\right)^2+P_{EVs(n)}R_n-P_{Ln}R_n-Q_{Ln}X_n} \tag{2}$$

Among them, when the load at the PCC node does not change, with the increase of the charging power of the electric vehicle, the PCC voltage at the end of the feeder of the distribution network is low or even the voltage exceeds the limit, which can be adjusted by the active power and reactive power provided by the FID for feeder n. Figure 2 shows the equivalent circuit diagram of VSC_n connected to the feeder end of distribution network.

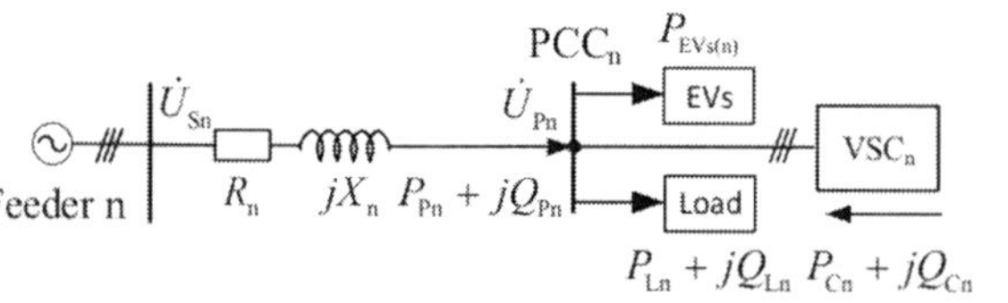

Figure 2. Equivalent circuit diagram of VSC_n connected to feeder end of distribution network

The active power and reactive power regulated at the end of the feeder are P_{Cn} and Q_{Cn}, respectively. The PCC node voltage U_{Pn} expression is :

$$U_{Pn} = \frac{U_{Sn}}{2} + \sqrt{\left(\frac{U_{Sn}}{2}\right)^2 + (P_{EVs(n)} - P_{Ln})R_n - Q_{Ln}X_n - (P_{Cn}R_n + Q_{Cn}X_n} \tag{3}$$

Among them, when the n-terminal of the feeder is connected to the FID, the active interaction between different feeders can be realized on the basis of local reactive power regulation, and the voltage level of the feeder PCC node can be effectively adjusted.

3. Voltage control strategy of large-scale charging piles connected to flexible interconnected distribution network

The above discussion discusses how to improve the VSC 's ability to adjust the node voltage and the feeder 's ability to absorb distributed energy. However, the power regulation of VSC will affect the power flow distribution of other feeders, resulting in changes in the voltage of grid-connected nodes at the end of other feeders. This paper will combine the active-reactive power coordinated voltage regulation scheme, through the coordinated control of VSC at the end of different feeders, to solve the problem of voltage violation at the end of the feeder. This paper takes the flexible interconnected low-voltage distribution system at both ends as an example to illustrate the coordinated control strategy of the system in the voltage violation of different feeder nodes. Figure 3 is the equivalent circuit of the flexible interconnected distribution network at both ends.

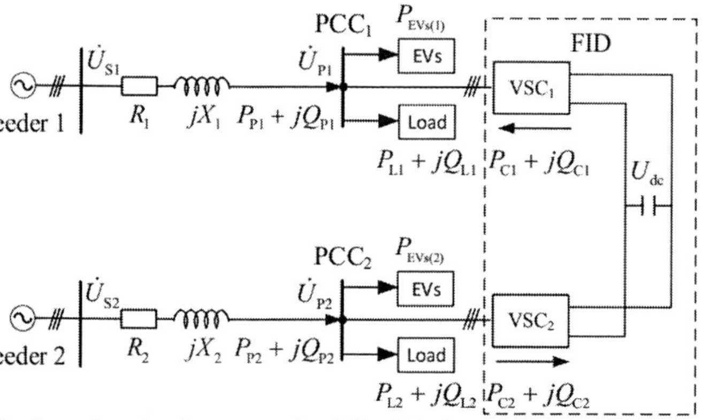

Figure 3. Equivalent circuit of two-terminal flexible interconnected distribution network

3.1. *Voltage regulation strategy when single feeder node voltage exceeds the limit*

When the charging power $P_{Evs(1)}$ of the electric vehicle with feeder 1 is large, the lower limit of U_{P1} generation voltage is exceeded. Figure 4 is the VSC_1 control block diagram, and Figure 5 is the VSC_2 control block diagram. VSC_1 adopts the ' active-reactive ' coordinated voltage regulation control strategy ($i_{d1,ref} = \alpha$), and the constant U_{P1} is $U_{P1,ref}$, ref. VSC_2 is a ' U_{dc}-Q ' control strategy ($i_{d2,ref} = \gamma$), which can transfer the excess power of Feeder 2 to Feeder 1.

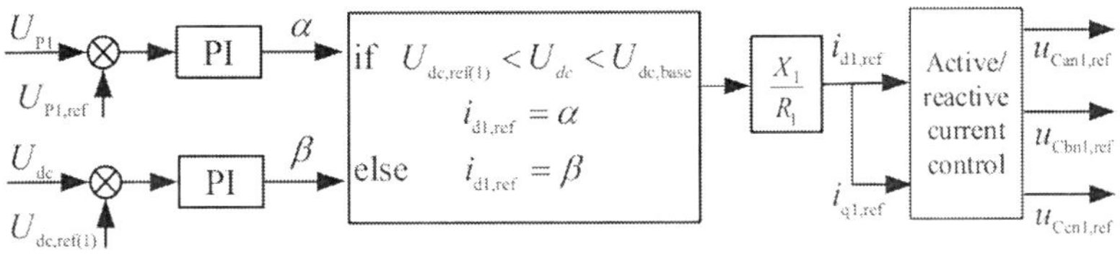

Figure 4. VSC$_1$ control block diagram

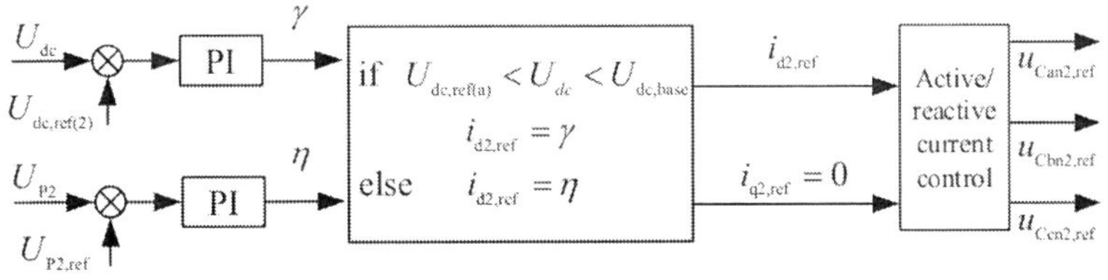

Figure 5. VSC$_2$ control block diagram

Figure 6 is the reference value determination method of DC bus voltage in VSC$_2$ control strategy. VSC$_2$ can map its own node voltage to DC bus. At this time, the global system can know the remaining capacity of feeder 2 by using the information of DC bus voltage. The ' U_{dc}-U_{P2} ' mapping relation is :

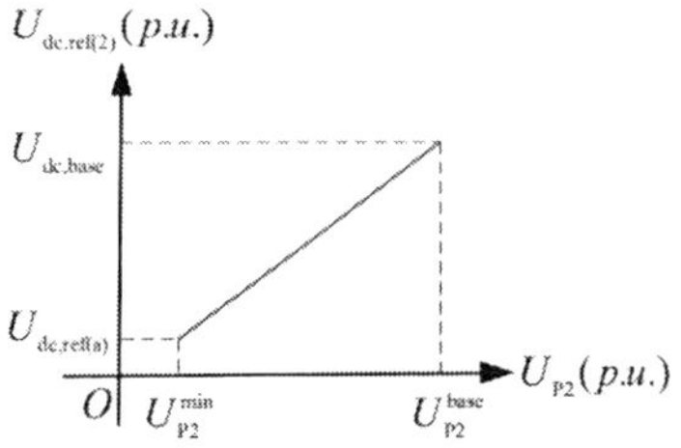

Figure 6. Determination method of DC bus voltage reference value

$$U_{dc,ref(2)} = kU_{P2} + b \tag{4}$$

It can be seen from Eq. (4) that as the PCC node voltage U_{P2} of feeder 2 changes, the DC bus voltage U_{dc} controlled by VSC$_2$ will also change with the reference value $U_{dc,ref(2)}$. When the PCC node voltage of feeder 2 reaches the lower limit U_{P2}^{min} , the DC bus voltage reaches the threshold $U_{dc,ref(a)}$. Therefore, when the DC bus voltage is in the range of $U_{dc,base}$ to $U_{dc,ref(a)}$, it means that feeder 2 has enough capacity to bear the shortage power of feeder 1.

Due to the randomness of the charging power of the electric vehicle, when $P_{Evs(1)}$ is greatly reduced, the active power transmitted from feeder 2 to feeder 1 will be reduced. According to the ' U_{dc}-U_{P2} ' mapping relationship, the DC bus voltage U_{dc} controlled by VSC$_2$ increases with U_{P2}. The active power P_{C1} andP_{C2} regulated by VSC$_1$ and VSC$_2$ gradually decreases. When P_{C1} and P_{C2} transmit in reverse, it shows that feeder 1 can fully bear the charging power of electric vehicles. In order to prevent VSC from switching control mode frequently, when P_{C1} reaches the threshold, the system exits VSC$_1$ and VSC$_2$ control strategies.

3.2. Voltage regulation strategy for two feeder nodes when the voltage exceeds the limit

When the charging power $P_{Evs(1)}$ of the electric vehicle of feeder 1 is large, a large amount of energy of feeder 2 flows into feeder 1, so that U_{P2} exceeds the lower limit of overvoltage U_{P2}^{min} , and the corresponding DC bus voltage will be less than $U_{dc,ref(a)}$, indicating that the capacity of feeder 2 is not enough to bear the short power of feeder 1. Therefore, in a non-communication environment, when the DC bus voltage is set to $U_{dc,ref(a)}$, VSC$_2$ will switch from the original ' U_{dc}-Q ' control strategy to the U_{P2} voltage control strategy, and its reference value is $U_{P2,ref}$= U_{P2}^{min} . That is : VSC$_1$ adopts ' active-reactive ' coordinated voltage regulation control strategy ($i_{d1,ref}$= α), VSC$_2$ adopts U_{P2} voltage control strategy ($i_{dm,ref}$=η), to ensure that feeder 1 and feeder 2 are not over-limit. However, as $P_{Evs(1)}$ increases, the lack of power in feeder 1 will cause the DC bus U_{dc} to exceed the threshold $U_{dc,ref(1)}$. At this time, VSC1 switches to the ' U_{dc}-Q ' control strategy ($i_{d1,ref}$=β), and the reference value of the DC bus voltage is $U_{dc,ref(1)}$. After the above control switching, VSC_1 and VSC_2 control U_{dc} and U_{P2} as fixed values respectively. If the insufficient power of Feeder 1 leads to the decrease of U_{P1} to the lower limit U_{P1}^{min}, the electric vehicle of Feeder 1 is converted from disorderly charging to orderly charging control to avoid U_{P1} exceeding the lower limit.

At a certain moment, as the charging power $P_{Evs(1)}$ of the electric vehicle decreases, the node power U_{P1} of the feeder 1 will rise to a certain value, and the VSC_1 will be switched from the ' U_{dc}-Q ' control strategy ($i_{d1,ref}$=β) to the ' active-reactive ' coordinated voltage regulation control strategy ($i_{d1,ref}$= α). With the further decrease of $P_{Evs(1)}$, when the DC bus voltage U_{dc} rises to $U_{dc,ref(a)}$, VSC_2 will switch from the U_{P2} voltage control strategy ($i_{d2,ref}$=η) control strategy to the ' U_{dc}-Q ' control strategy ($i_{d1,ref}$= γ), and the active power P_{C1} and P_{C2} regulated by VSC_1 and VSC_2 will gradually decrease. When P_{C1} and P_{C2} are transmitted back, it shows that feeder 1 can fully bear the charging power of electric vehicles. In order to prevent VSC from frequently switching control mode, when P_{C1} reaches the threshold. The system exits VSC_1 and VSC_2 control strategies.

4. Simulation result

In order to verify the feasibility of the control strategy proposed in this paper, the flexible interconnected distribution network shown in Figure 3 is built in MATLAB / Simulink simulation. It is assumed that the voltage at the head of the two feeders is 1.00 p.u.(380V), the impedance and capacity of the feeders are (0.33 + j0.345)Ω and 1.00p.u.(200 kW), and the DC bus voltage is 1.0 p.u.(750V), where the lower limits of the node voltage U_{P1} and U_{P2} of the two feeders are set to 0.9 p.u. When VSC_1 and VSC_2 control the AC side voltage, the reference values of the two feeder voltages are set to 0.95 p.u., $U_{P1,ref}$ is set to 0.9 p.u., and the DC side voltage reference value $U_{dc,ref(1)}$= 0.96p.u., $U_{dc,ref(2)}$ =0.192U_{P2}+0.807, when U_{P2} = 0.9 p.u., $U_{dc,ref(a)}$ = 0.98 p.u.

The simulation results of single feeder over-limit are shown in Figure 7. Before 1s, the $P_{EVs(1)}$ of Feeder 1 is 0.1p.u.(20kW), and the U_{P1} and U_{P2} do not exceed the limit. After 1s, due to the increase of $P_{EVs(1)}$, the U_{P1} at the end of feeder 1 reaches the lower limit of 0.9p.u. At 2s, VSC_1 adopts 'active-reactive' coordinated control strategy to make U_{P1} operate at 0.95p.u., VSC_2 adopts 'U_{dc}-Q' control strategy to provide active power to feeder 1. At 5s, the $P_{EVs(1)}$ of feeder 1 gradually decreases from 0.28 p.u.(56kW), resulting in a decrease in the power regulated by VSC_1 and VSC_2, which in turn increases U_{P2} and U_{dc}. 7 s, the active power P_{C1} controlled by VSC_1 reversely reaches the threshold of 0.02p.u. (1kW), indicating that feeder 1 can fully bear the charging power of electric vehicles, and the system exits voltage regulation.

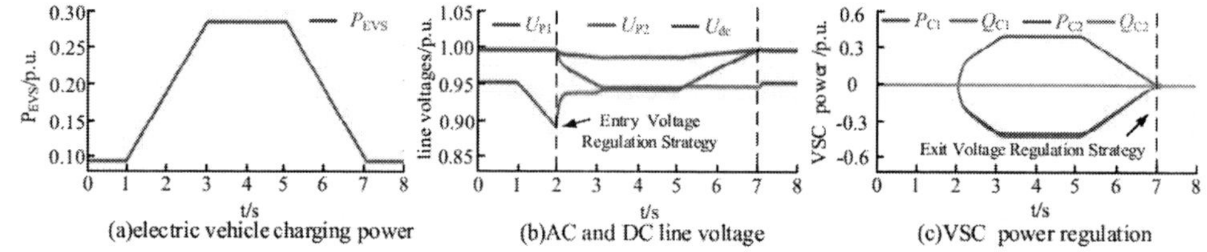

Figure 7. The simulation waveform when the node voltage of a single feeder exceeds the lower limit

The simulation results of the two feeders entering the voltage regulation are shown in Figure 8 (a) ~ (c). VSC_2 uses 'U_{dc}-Q' control strategy to provide active power to feeder 1 before 2s. After 2s, due to the increase of $P_{EVs(1)}$, the U_{P2} reaches the lower limit of 0.9p.u., and the U_{dc} is also reduced to 0.98p.u.from the '$U_{dc,ref(2)}$-U_{P2}' mapping relationship, indicating that Feeder 2 has reached its adjustment capacity limit. At 5 s, VSC_2 is switched from 'U_{dc}-Q' control mode to U_{P2} voltage control mode. Due to the continuous increase of $P_{EVs(1)}$, the power shortage of feeder 1 is provided by the DC bus. At 5.2 s, when U_{dc} decreases to 0.96 p.u., VSC_1 is switched from ' active-reactive ' coordinated control strategy to ' U_{dc}-Q ' control strategy, and U_{dc} is controlled to 0.96 p.u. At 6 s, the lack of power in Feeder 1 leads to the decrease of U_{P1} to 0.9 p.u., and the electric vehicle of Feeder 1 is converted from disorderly charging to orderly charging control to avoid U_{P1} exceeding the lower limit.

The simulation results of the two feeders out of the voltage regulation are shown in Figure 8 (d) ~ (f). At 2s, the $P_{EVs(1)}$ of Feeder 1 begins to decrease from 0.5p.u.(100kW). At 2.7s, the U_{P1} of Feeder 1 will rise to 0.95p.u. At this time, the VSC_1 control U_{P1} is 0.95 p.u. At 2.9s, as $P_{EVs(1)}$ further decreases, U_{dc} will also rise to 0.98p.u. When VSC_2 senses that U_{dc} reaches the threshold of 0.98 p.u., VSC_2 will switch from U_{P2} voltage control strategy control strategy to 'U_{dc}-Q' control strategy. When 6s, P_{C1} and P_{C2} reversely reach the threshold of 0.02p.u.(1kW), indicating that feeder 1 can fully bear $P_{EVs(1)}$, and the system exits voltage regulation.

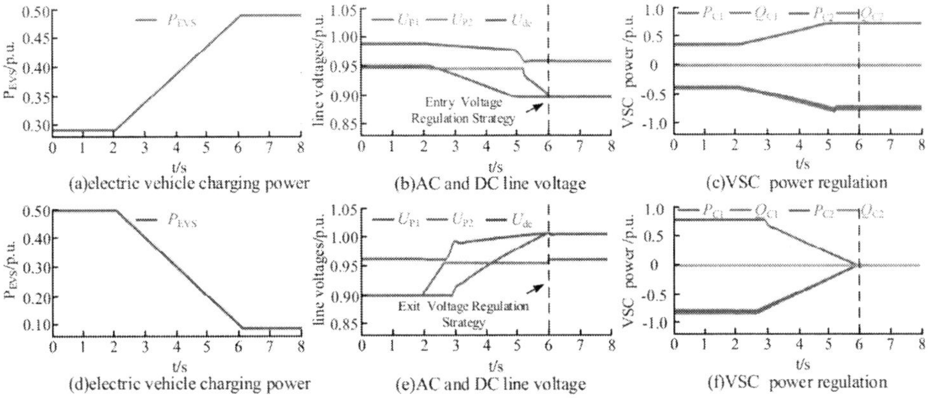

Figure 8. The simulation waveform of two feeder nodes when the voltage is lower than the lower limit

5. Conclusion

This paper proposes a voltage control strategy for large-scale charging piles connected to flexible interconnected distribution networks. Firstly, the influence of electric vehicle access on the terminal voltage of low-voltage distribution network is analyzed, and a flexible interconnection scheme is proposed. A power corridor is established between different feeders, so that the interconnected feeders can jointly bear the charging power of electric vehicles. Secondly, under the condition of weak communication in low-voltage distribution network, SOP uses the change of DC bus voltage to know the voltage and power information of PCC node, which overcomes the risk of voltage over-limit at the end of distribution feeder caused by electric vehicle access and effectively improves the system 's ability to accept electric vehicles

Acknowledgments

Project Supported by the State Grid Company Limited Science and Technology Program (31020247FZC), Shanghai Committee of Science and Technology (23DZ1201202) and National Natural Science Foundation of China (Grant Number: 52107200).

References

[1] Karmaker A K,Roy S and Ahmed M R 2019 Analysis of the Impact of Electric Vehicle Charging Station on Power Quality Issues International Conference on Electrical, Computer and Communication Engineering (ECCE), pp 1-6

[2] Tong B, Zhang L, Li G, Zhang B, Xie F and Tang W 2024 An Overvoltage-Averse Model for Renewable-Rich AC/DC Distribution Networks Considering the Sensitivity of Voltage Violation Probability , IEEE Transactions on Sustainable Energy, vol. 16, no. 1, pp. 613-626

[3] Guan X, Shao D, Jiang W, Shao B and Ma Y 2024 A SVC Voltage Regulation Control Strategy of Active Distribution Network Based on Improved Particle Swarm Algorithm, *IEEE 7th Student Conference on Electric Machines and Systems (SCEMS)*, pp 1-6

[4] Yu H Lei X Niu S Shao Z and Jian L 2023 Enhancing electric vehicle penetration and grid operation performance in old residential communities through hybrid AC/DC microgrid reconstruction, Applied Energy, pp 121-459
[5] Cai A Yu Y Xu L Niu Y and J. Yan 2019 Review on reactive power compensation of electric vehicle charging piles, International Conference on Electrical Machines and Systems (ICEMS), pp 1-4
[6] Wang Y Hu Z Dai P Zhu C Ye C and Hu J 2022 An Electric Vehicle Charger Control Strategy for Reactive Power Compensation," Asia Conference on Power and Electrical Engineering (ACPEE) pp 638-642
[7] JAMROEN C DECHANUPAPRITTA S 2019 Coordinated control of battery energy storage system and plug-in electric vehicles for frequency regulation in smart grid, IEEE PES GTD Grand International Conference and Exposition Asia （GTD Asia） Bangkok Thailand 286-291
[8] Yang M Pei X and Li Y 2020 Multi-port Coordinated Control Strategy of SOP in Distribution Network, IEEE Applied Power Electronics Conference and Exposition (APEC) pp 2333-2337
[9] Zhu Q 2017 Synergetic control strategy for interconnected distribution network via AC/DC/AC converter improving renewable energy consumption, IEEE Conference on Energy Internet and Energy System Integration (EI2), pp 1-6
[10] Zheng Y Song Y and Hill D J 2020 A general coordinated voltage regulation method in distribution networks with soft open points, International Journal of Electrical Power & Energy Systems, vol. 116, pp 105571
[11] Li P et al 2017 Coordinated Control Method of Voltage and Reactive Power for Active Distribution Networks Based on Soft Open Point, IEEE Transactions on Sustainable Energy, vol. 8 no 4, pp 1430-1442
[12] Li P et al 2019 Combined decentralized and local voltage control strategy of soft open points in active distribution networks, Applied Energy, vol. 241, pp 613-624

Journal of Physics: Conference Series 3012 (2025) 012020 doi:10.1088/1742-6596/3012/1/012020

Active Support Control for Hybrid Renewable Energy Stations with Grid-Following and Grid-Forming Capabilities Considering Grid Strength

Xin Shan*, Kai Hou, Liang Guo, Chao Niu, Guoyang Cai

NARI Technology Co.,Ltd. State Grid Electric Power Research Institute, Nanjing, China

*Corresponding author's e-mail: shanxin1@sgepri.sgcc.com.cn

Abstract. This paper proposes an active support control method for hybrid renewable energy stations with grid-following (GFL) and grid-forming (GFM) capabilities, addressing the issue of poor matching between GFL and GFM coordination strategies when grid strength fluctuates. Firstly, the short-circuit ratio (SCR) of the station is evaluated based on the operating status data, and the control mode of the units is switched between grid-following and GFM. Secondly, the GFM units autonomously respond to inertia, and the station control system issues virtual inertia frequency regulation instructions to the GFL units. When the frequency exceeds the deadband, the target value of the primary frequency regulation power for the entire station is calculated and allocated synergistically among the units based on their active power margins. When the voltage exceeds the deadband, the GFM units autonomously respond to reactive power, and the GFL units are coordinated for active voltage regulation in conjunction with the reactive power support effect of the GFM units. Finally, simulation examples verify that the proposed active support control method for hybrid renewable energy stations with GFL and GFM capabilities.

1. Introduction

With the large-scale development of wind power, the synchronous power supply and active support capabilities in future power systems will continue to decline [1-2]. The utilization of grid-forming (GFM) technology to retrofit wind turbines, thereby enhancing the stability of wind power integration systems, has become a mainstream research direction both domestically and internationally [3-4]. Actually, grid-following (GFL) and GFM grid-connection technologies are both popular grid-connection control technologies in the renewable energy industry. GFL technology controls the converter output current using phase-locked loop (PLL) synchronization, featuring higher maturity, diverse product types, and low costs. It is currently the mainstream grid-connection control technology in the renewable energy industry [5]. However, it exhibits high dynamic performance and strong stability in strong grids but lacks active frequency and voltage support capabilities and struggles to operate stably in weak grids [6]. In contrast, GFM technology does not use PLL for grid synchronization, adopting power self-synchronization control. In weak grids, GFM technology has strong support capabilities for grid frequency and voltage, large control stability margins, and can operate independently of the large grid [7], which has developed rapidly in recent years. However, GFM n control has issues including slow power response in weak grids, difficulty in accurately achieving maximum power point tracking and poor economics, risk of low-frequency oscillation at high grid strength. Therefore, hybrid GFL and GFM grid connection is an important grid formation method in the future, and hybrid renewable energy stations are an important form of renewable energy stations in the future [8].

The hybrid renewable energy stations with both GFL and GFM capabilities still face issues such as switching logic during significant fluctuations in grid strength, coordination of active support regulation amounts, and synchronization issues due to the autonomy of GFM units, seriously affecting the stable operation of the grid. Despite the ongoing research efforts by scholars both domestically and internationally, a clear technical system has yet to be established [9-10]. Therefore, this paper proposes an active support control method for hybrid renewable energy stations with GFL and GFM capabilities, addressing the issue of poor matching between GFL and GFM coordination strategies when grid strength fluctuates. The proposed method can successfully addressing the coordination challenges between GFL and GFM units under fluctuating grid strength, and validating its efficacy through simulation examples.

2. Assessment of SCR and control mode switching of renewable energy units

2.1. Assessment of SCR of Renewable Energy Stations

The assessment of the short-circuit ratio (SCR) of renewable energy stations is used to measure the relative strength of the AC system, serving as the premise for determining whether the station adopts GFL or GFM control mode. For this purpose, based on the measured voltage, the SCR of the station is assessed [11]. The formula is expressed as

$$OSCR = \left|U_N \dot{U}_{oc}\right| / \left|\Delta \dot{U} \dot{U}\right| \tag{1}$$

where $OSCR$ is the SCR measured online; U_N is the rated voltage of the PCC; $\dot{U}_{oc}$ is the open circuit voltage; $\Delta \dot{U}$ is the voltage change; $\dot{U}$ is the voltage of PCC points of renewable energy.

2.2. Control Mode Switching of Renewable Energy Units Between GFL and GFM

The switching of renewable energy units between GFL and GFM control modes should be based on the comparison between the SCR and the critical SCR for system stability. When the SCR is less than the critical SCR, the operating mode of the units is switched to GFM mode to enhance the active support capability of the system under weak grids. Conversely, the operating mode of the units can be switched to GFL mode. The switching formula between GFM and GFL modes is given as

$$\begin{gathered} GF_{L-M} = \begin{cases} 1, OSCR < CSCR \\ 0, OSCR > SSCR \end{cases} \\ GF_{M-L} = \begin{cases} 0, OSCR < CSCR \\ 1, OSCR > SSCR \end{cases} \end{gathered} \tag{2}$$

where GF_{L-M} is flag bit for switching from GFL mode to GFM mode in renewable energy stations; GF_{M-L} is flag bit for switching from GFM mode to GFL mode; CSCR is the critical short-circuit ratio; SSCR is the stable operation short-circuit ratio.

By comparing the SCR of the renewable energy station with the critical SCR for system stability and autonomously switching the operating mode of the renewable energy units, the system stability and dynamic response are optimized under different grid strength conditions, effectively enhancing the overall operation quality and reliability of the grid. This dynamic switching mechanism not only enhances the adaptability of the system but also maximizes the utilization of renewable energy resources, reducing unstable factors caused by changes in grid conditions.

3. Inertia and primary frequency regulation control methods for hybrid GFL and GFM renewable energy stations

3.1. Analysis of active power regulation margin of wind units

When hybrid renewable energy stations perform frequency regulation, it is necessary to first clarify the regulation margin range of the units. The up-regulation active power margin of wind turbine units, under both unlimited and limited power conditions, can be expressed as

$$P_{capup}^{WT}=\begin{cases}K_s\times E=K_s\times\frac{1}{2}J\omega^2, & \text{without Limitation}\\ K_d^{WT}\times P_{opt}^{WT}, & \text{with Limitation}\end{cases} \tag{3}$$

where P_{capup}^{WT} up-regulation active power margin of wind turbine units; K_s is the utilization coefficient of rotor kinetic energy; E represents the kinetic energy stored in the rotor, while J represents the moment of inertia. ω is the rotor speed; K_d^{WT} is the power curtailment, typically ranging from 10% to 20%; P_{opt}^{WT} output power of a wind turbine during maximum power point tracking (MPPT).

The down-regulation active power margin of wind turbine units is

$$P_{capdown}^{WT}=P_t^{WT}-P_{\min}^{WT} \tag{4}$$

where $P_{capdown}^{WT}$ is down-regulation active power margin of wind turbine units; P_t^{WT} is the output power and $P_{\min}^{WT}$ is the minimum output power.

3.2. Active frequency support control method for hybrid GFL and GFM renewable energy stations

3.2.1. Coordinated inertia frequency regulation control for hybrid renewable energy stations

GFM units adopt a virtual synchronous machine-like approach for autonomous frequency perception and execution in inertia response, with the control equation being

$$2H\frac{d\omega}{dt}=P_{mec}-P_e-K_d\left(\omega-\omega_{grid}\right) \tag{5}$$

where H represents the inertia time constant of the virtual synchronous machine, ω denotes the synchronous angular frequency, ω_{grid} is the grid angular frequency, P_{mec} and P_e stand for mechanical power and electromagnetic power, respectively, and K_d signifies the damping coefficient.

When frequency disturbance occurs, the control system issues virtual inertia frequency regulation instructions to the GFL units. The calculation of the target value of the active power required for inertia response of the GFL units is

$$\Delta P_t=-\frac{T_J}{f_N}\times\frac{\mathrm{d}f}{\mathrm{d}t}\times\sum_{j=1}^{N}P_{tj} \tag{6}$$

where ΔP_t is the target value of active power required for inertia-based frequency regulation by GFL units in renewable energy stations; T_J denotes the equivalent inertia time constant; $\frac{\mathrm{d}f}{\mathrm{d}t}$ represents the rate of change of frequency and f_N is the rated frequency; P_{tj} is the active power of the j-th (j=1…N) GFL unit.

When the active power margin of the GFM units meets the inertia frequency regulation requirements of the station, the GFM units undertake the task of system inertia response regulation. The formula is expressed as

$$\Delta P_i^{\lim}=\Delta P_t\times\left(P_{capi}/P_{cap}^{GFM}\right) \tag{7}$$

where $\Delta P_i^{\lim}$ represents the active power limit for inertia-based frequency regulation that the i-th GFM unit needs to undertake; ΔP_t denotes the active power adjustment for inertia-based frequency regulation at the current moment; P_{capi} stands for the active power regulation capability of the i-th GFM unit, and P_{cap}^{GFM} represents the total active power regulation capability of all GFM units.

When the active power margin of the GFM units does not meet the inertia frequency regulation requirements of the station, both GFL and GFM units need to collaboratively participate in the inertia response of the system frequency. The formula is expressed as

$$\begin{cases} \Delta P_i^{\lim} = P_{capi} \\ \Delta P_j = \Delta P_{left} \times \left(P_{capj} / P_{cap}^{GFL} \right) \end{cases} \tag{8}$$

where $\Delta P_i^{\lim}$ denotes the active power limit for inertia-based frequency regulation that the i-th GFM unit needs to undertake, which is determined by (5). P_{capi} stands for the active power regulation capability of the i-th GFM unit. ΔP_j represents the inertia-based frequency regulation command value that the j-th GFL unit needs to undertake. ΔP_{left} denotes the active power adjustment required by the GFL units to assist in, excluding the inertia-based frequency regulation active power undertaken by the GFM units, which is determined by (6). P_{cap}^{GFL} represents the total active power regulation capability of all GFL units.

3.2.2. *Coordinated primary frequency control*

When the frequency exceeds the deadband, the target value of primary frequency regulation power is calculated and allocated synergistically among the units based on their active power margins. The calculation of the target value of active power required for primary frequency regulation is

$$\Delta P_t = -K_f \times \frac{\Delta f}{f_N} \times P_t \tag{9}$$

where ΔP_t denotes the target value of active power required for primary frequency regulation; K_f is the active power regulation coefficient for primary frequency regulation; Δf represents the frequency deviation of the power system; f_N is the rated frequency of the power system, and P_t is the real-time active power.

Based on the active power margins of the units and the target value of active power for primary frequency regulation, frequency regulation power is allocated equally among the units. The formula is

$$\Delta P_i = \Delta P_t \times \left(P_{capi-inert} / \left(P_{cap-inert}^{GFM} + P_{cap-inert}^{GFL} \right) \right) \tag{10}$$

where ΔP_i denotes the active power command for primary frequency regulation that the i-th unit needs to undertake; ΔP_t is the active power adjustment for primary frequency regulation at the current moment; $P_{capi-inert}$ stands for the remaining active power regulation capability of the i-th unit after executing inertia-based frequency regulation; $P_{cap-inert}^{GFM}$ represents the remaining active power regulation capability of all GFM units after executing inertia-based frequency regulation, and $P_{cap-inert}^{GFL}$ denotes the remaining active power regulation capability of all GFL units after executing inertia-based frequency regulation. This scheme promptly responds to changes in PCC frequency, calculates the target value of active power required for primary frequency regulation, and allocates it reasonably according to the real-time active power margins of each unit. This ensures that when the frequency exceeds the set deadband, the renewable energy station can quickly and effectively participate in frequency regulation, maintaining grid frequency stability. It not only improves the speed and accuracy of frequency regulation response but also achieves load balancing among the units through the equal margin allocation mechanism, thereby enhancing the stability and efficiency of the entire renewable energy station.

4. Voltage control methods for hybrid GFL and GFM renewable energy stations

4.1. Analysis of reactive power regulation margin

This section analyzes the reactive power regulation margins of wind units. For DFIG units, it can be obtained as

$$\begin{cases} Q_{g\min} = -\dfrac{3U_S^2}{2X_S} - \sqrt{\left(\dfrac{3X_m}{2X_S}U_S I_{r\max}\right)^2 - P_S^2} - \sqrt{S_c^2 - \left(\dfrac{s}{1-s}P_s\right)^2} \\ Q_{g\max} = -\dfrac{3U_S^2}{2X_S} + \sqrt{\left(\dfrac{3X_m}{2X_S}U_S I_{r\max}\right)^2 - P_S^2} + \sqrt{S_c^2 - \left(\dfrac{s}{1-s}P_s\right)^2} \end{cases} \tag{11}$$

where U_S denotes the stator voltage, X_S represents the stator reactance; X_m is the excitation impedance; $I_{r\max}$ is the maximum rotor current, P_S stands for the stator active power; S_C is the apparent power of the wind power converter, s represents the slip; $Q_{g\max}$ denotes the maximum reactive power capability, and $Q_{g\min}$ denotes the minimum reactive power capability.

For PMSG units, their reactive power regulation capability range can be expressed as

$$\begin{cases} Q_{s\max} = \sqrt{\left(\dfrac{U_g U_c}{X}\right)^2 - P^2} - \dfrac{U_g^2}{X} \\ Q_{s\min} = -\sqrt{\left(\dfrac{U_g U_c}{X}\right)^2 - P^2} - \dfrac{U_g^2}{X} \end{cases} \tag{12}$$

where U_g represents the grid voltage; U_c is the voltage at the grid side of the grid-side converter; P stands for the output active power of the wind power converter; $Q_{S\max}$ denotes the maximum reactive power capability, and $Q_{S\min}$ denotes the minimum reactive power capability.

4.2. Voltage active support control method for hybrid GFM renewable energy stations

4.2.1. Reactive power regulation target value for renewable energy stations

When the difference between the voltage at the high-voltage side of the PCC and the present voltage reference value exceeds the voltage dead-band range, the reactive power target value during system voltage disturbance is calculated based on the SCR, the voltage value at the high-voltage side of the PCC, and the reactive power value. The calculation equation is

$$\Delta Q = Q^{t+1} \text{-} Q^{t} = \text{OSCR} * V_1^{t+1}\left(V_1^{t+1} \text{-} V_1^{t}\right) + Q^{t}\left(\frac{V_1^{t+1}}{V_1^{t}} - 1\right) \tag{13}$$

where ΔQ denotes the reactive power target value; OSCR represents the online-measured SCR of the station; Q^{t+1} and V_1^{t+1} are the compensated reactive power and voltage at the high-voltage side of the PCC, respectively, while Q^{t} and V_1^{t} are the uncompensated reactive power and voltage.

4.2.2. Analysis of reactive power margin and voltage regulation target matching for GFM units

When the GFM units meets the voltage regulation requirements of the station, the GFM units undertake the task of system voltage regulation, which has

$$\Delta Q_i^{\lim} = \Delta Q_t \times \left(Q_{capi} / Q_{cap}^{GFM}\right) + \Delta Q_{lossi} \tag{14}$$

where $\Delta Q_i^{\lim}$ denotes the voltage regulation limit that the i-th GFM unit needs to undertake; ΔQ_t is the fast voltage regulation adjustment at the current moment; Q_{capi} is the regulation capability of the i-th

GFM unit; Q_{cap}^{GFM} is the total regulation capability of all GFM units, and ΔQ_{lossi} is the reactive power loss of i-th GFM unit.

When the reactive power margin of GFM units does not meet the voltage regulation requirements of the station, GFL units need to collaboratively participate in system voltage regulation, which has

$$\begin{cases} \Delta Q_i^{\lim} = Q_{capi} \\ \Delta Q_j = \Delta Q_{left} \times \left(Q_{capj} / Q_{cap}^{GFL} \right) + \Delta Q_{lossj} \end{cases} \tag{15}$$

where $\Delta Q_i^{\lim}$ denotes the fast voltage regulation limit that the i-th GFM unit needs to undertake; Q_{capi} is the regulation capability of the i-th GFM unit; ΔQ_j is the fast voltage regulation command value that the j-th GFL unit needs to undertake; ΔQ_{left} is the reactive power adjustment required by GFL units to assist in, excluding the fast voltage regulation reactive power undertaken by GFM units; Q_{cap}^{GFL} is the total regulation capability of all GFL units, and ΔQ_{lossj} is the reactive power loss of the j-th GFL unit.

By comparing the reactive power margin of GFM units with the target value for voltage regulation in the station, the above method achieves coordinated voltage regulation. When the GFM units is sufficient to meet the voltage regulation requirements, the units can independently execute voltage regulation actions. This ensures voltage stability while improving the operational flexibility and reliability of the entire renewable energy station, effectively balancing the autonomy of GFM units and the coordination of GFL units, and enhancing the grid's response capability and overall stability in the face of voltage fluctuations.

5. Simulation results

The hybrid renewable energy stations with GFL and GFM capabilities is built using the PSCAD simulation platform in Fig. 1, with the simulation parameters in TABLE I. In the case of these parameters, a disturbance of power change on the PCC point is set in the model at 3s and 5s respectively. The simulation results of the proposed control methods are shown in Fig.2.

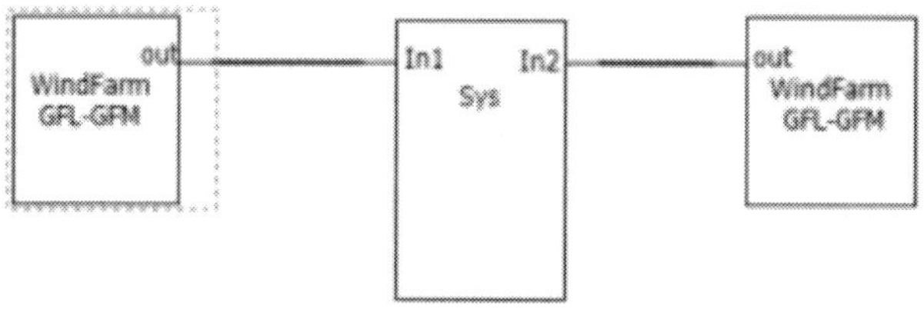

Figure.1 Simulation model at PSCAD/EMTDC platform

TABLE I. The main parameters

Item	value
Rated ac voltage	380V
Rated dc voltage	700V
System frequency	60Hz
Rated active power of DG	40kW/60kW

The simulation results demonstrate that when a disturbance occurs, the system, under the control scheme proposed in this paper, transitions both its frequency and voltage to a new equilibrium point, exhibiting excellent dynamic response performance. Hence, it is evident that this proposed strategy exhibits swift adaptability to fluctuations in power system. It calculates the requisite active power target for primary frequency regulation and distributes this target judiciously, taking into account the instantaneous active power reserves of individual units. This ensures that upon exceeding the predefined frequency deadband, the renewable energy station promptly and effectively engages in primary frequency control of the grid, thereby safeguarding grid frequency stability. This strategy not only elevates the speed and precision of the frequency regulation response but also fosters load balancing across units through an equitable reserve allocation mechanism, ultimately enhancing the overall stability and efficiency of the renewable energy station.

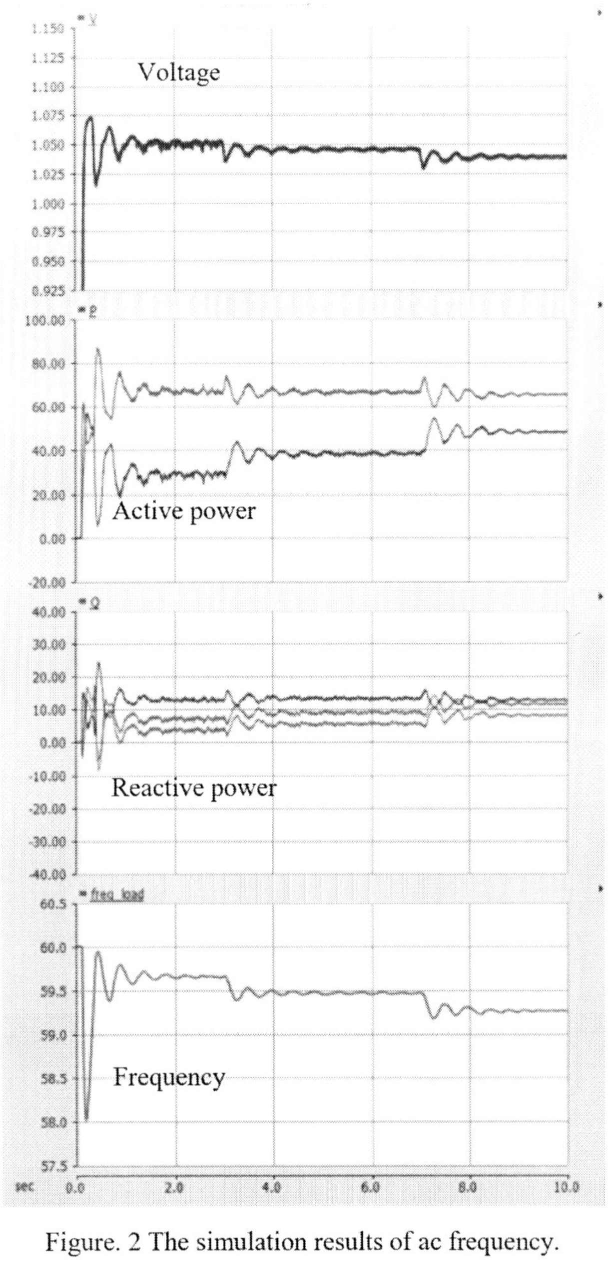

Figure. 2 The simulation results of ac frequency.

6. Conclusion

This paper proposes an active support control method for hybrid renewable energy stations with both GFL and GFM capabilities. Firstly, the SCR of the station is assessed based on the operational status data of the high-voltage side at the PCC, enabling the switching of control modes between GFL and GFM for the station's units. Secondly, the GFM units exhibit autonomous inertia response, while the station control system issues virtual inertia frequency regulation instructions to the GFL units. When the frequency exceeds the deadband, the target value for primary frequency regulation power across the entire station is calculated and allocated synergistically among the units, based on their respective active power margins. In cases where the voltage surpasses the deadband, the GFM units autonomously respond to reactive power, while the GFL units coordinate for active voltage regulation in conjunction with the reactive power support provided by the GFM units.

Acknowledgment

This work is supported by Science and Technology Project of State Grid Corporation of China. (5108-202218280A-2-306-XG).

References

[1] International Renewable Energy Agency. Global energy transformation: a roadmap to 2050[EB/OL]. https://www.irena.org/publications/2019/Apr/Globalenergy-transformation-A-roadmap-to-2050-2019Edition.

[2] LI C, WANG S, COLAS F, et al. Dominant instability mechanism of VSI connecting to a very weak grid[J]. IEEE Transactions on Power Systems, 2022, 37(1): 828 – 831

[3] ZHONG Q, NGUYEN P, MA Z, et al. Self-synchronized synchronverters: inverters without a dedicated synchronization unit[J]. IEEE Transactions on Power Electronics, 2014, 29(2): 617 - 630.

[4] XIE Z, CHEN Y, WU W, et al. Modeling and control parameters design for grid-connected inverter system considering the effect of PLL and grid impedance[J]. IEEE Access, 2020(8): 40474 - 40484.

[5] LIU Z, MIAO S H, FAN Z H, et al. Improved power flow control strategy of the hybrid AC/DC microgrid based on VSM[J]. IET Generation, Transmission and Distribution, 2019, 13(1): 81 - 91.

[6] WU H, WANG X F. Control of grid-forming VSCs: a perspective of adaptive fast/slow internal voltage source[J]. IEEE Transactions on Power Electronics, 2023, 38(8): 10151-10169.

[7] CHEN J R, O' DONNELL T. Parameter constraints for virtual synchronous generator considering stability[J]. IEEE Transactions on Power Systems, 2019, 34(3): 2479-2481.

[8] YU G, HU Y, LIU Chenxi, et al. Research on the Coordinated Optimization Configuration Method for Hybrid Multi-infeed Systems with Grid-Following or Grid-Forming [J]. Proceedings of the CSEE, 2024, 1-15.

[9] WANG Y, XIANG W, ZHANG H, et al. Analysis of Optimal Proportion Configuration of Grid Forming Direct Drive Wind Turbine[J]. Power System Technology, 2024, 1-13.

[10] LIU P, LIU H, WU L,et al. Grid-following and Grid-forming Combined Control and Optimization Configuration to Satisfy Oscillatory Constraints Across Whole Operating Region[J]. Automation of Electric Power Systems,2024,48(12):139-146.

[11] YU L, SUN Huadong, ZHAO B, et al. Short Circuit Ratio Index Analysis and Critical Short Circuit Ratio Calculation of Renewable Energy Grid-connected System [J]. Proceedings of the CSEE, 2022, 42(03): 919-929.

Journal of Physics: Conference Series 3012 (2025) 012021 doi:10.1088/1742-6596/3012/1/012021

Assessment of Lithium-ion Battery Controllable Capacity Based on CNN-BiLSTM-Attention

Qingliang Wang[1], Shuling Zhang[2], Qiaoyi Yang[1], Chaohui Zhou [1], Liuhui You[3]

[1] State Grid Fujian Electric Co., Ltd., Fuzhou, China

[2] Electric Power Science Research Institute, State Grid Fujian Electric Co., Ltd. Fuzhou, China

[3] College of Electrical, Energy and Power Engineering. Yangzhou University.

Yangzhou, China

*Corresponding author's e-mail: zsl0591@qq.com

Abstract. To overcome the limitations of low accuracy in lithium-ion battery controllable capacity assessment using a single data-driven method, this paper introduces a novel CNN-BiLSTM-Attention model. Firstly, health features pertinent to battery cycle aging are extracted from the charging/discharging process and capacity incremental analysis. Secondly, a convolutional neural network (CNN) is employed to capture battery time-series data features, while a bidirectional long short-term memory (BiLSTM) network handles forward and backward data dependencies, enhancing temporal correlations and capturing long-term dependencies. An attention mechanism further strengthens input data correlations. The proposed CNN-BiLSTM-Attention hybrid model achieves precise battery state-of-health (SOH) estimation. Comparative simulations across multiple batteries demonstrate the model's high estimation accuracy and stability, outperforming single-model approaches.

1. Introduction

Lithium-ion batteries are notable for their extended lifespan, minimal self-discharge, and high safety and reliability. These characteristics render them a dependable source of sustained energy security for a range of critical applications, particularly in the field of new energy vehicles and energy storage systems [1]. The controllable capacity of lithium-ion batteries is primarily defined by two key parameters: State of Charge (SOC) and State of Health (SOH). Given that SOC is readily measurable within the battery management system, precise SOH estimation emerges as a pivotal factor in assessing the battery's controllable capacity [2].

Accurate assessment of SOH is essential for predicting the SOC and determining the operational range of the battery. It provides vital insights into the energy management of battery packs and plays a pivotal role in strategies designed to extend battery lifespan. Moreover, a comprehensive understanding of the aging mechanisms of batteries, along with precise forecasting of their aging trajectories and accurate SOH evaluations, is crucial for facilitating timely maintenance and replacement. This proactive approach not only significantly enhances the safety of lithium-ion battery systems but also supports the advancement of electric vehicle technology [3]. Currently, two primary methodologies are employed to define battery SOH: impedance measurement and capacity assessment. However, the direct measurement of internal resistance poses challenges, leading researchers to predominantly favor the capacity-based approach. This method defines SOH as the ratio of a battery's actual available capacity to its nominal capacity.

The assessment techniques for evaluating the health status of lithium-ion batteries can be classified into three primary categories based on their foundational principles: experimental-based, model-based, and data-driven methods. Experimental-based methods utilize specialized measurement equipment to directly assess battery parameters, providing a clear and intuitive evaluation. However, these techniques are characterized by high costs and low efficiency, rendering them impractical for large-scale applications due to their reliance on offline analysis. In contrast, model-based methods utilize the physical characteristics of the battery to indirectly deduce its SOH by employing an appropriate battery model. Reference [4] introduces an enhanced Thevenin model for electric vehicles lithium-ion batteries, proven superior in dynamic performance and SOC accuracy through Hybrid Pulse Power Characterization, Dynamic Stress Test, and Federal Urban Driving Schedules. Reference [5] proposes a cooperative estimation approach utilizing a fractional-order battery model. This method concurrently evaluates variations in internal resistance and capacity by employing an iterative model observer, enabling adaptive cooperative estimation of both battery resistance and capacity. Reference [6] employs a convolutional neural network to assess battery health status using the current-voltage curve as input and health status as output, enabling SOH estimation. Reference [7] utilizes a BiLSTM neural network model, achieving enhanced estimation accuracy. These data-driven methods excel in real-time tracking of the battery's complex dynamics and in extracting health characteristics closely associated with battery aging for health status assessment.

This paper concentrates on the estimation of the SOH as a critical component in assessing battery controllability. It investigates the correlation between the aging state of lithium-ion batteries and quantifiable cycle data. Recognizing the limitations of relying solely on a single data-driven model for accurate SOH evaluation, and considering the persistent reliance on timing during the battery's cycle aging process, this study introduces an integrated model. The CNN-BiLSTM-Attention model is designed to estimate the actual SOH of the battery, facilitating a precise assessment of its controllability.

2. Methodology

2.1. Battery Health Feature Extraction

Essential health characteristics for SOH assessment of lithium-ion batteries are derived from battery data, simplifying model input complexity and enabling accurate aging status evaluation. As the battery undergoes more charge-discharge cycles, critical parameters including current, voltage, and temperature exhibit clear, consistent trends. Existing literature has shown that the Incremental Capacity (IC) curve [8] effectively describes the battery's SOH. To elucidate the relationship between these parameters and battery aging, this analysis will examine charging and discharging processes, temperature profiles, and the IC curve. This comprehensive examination aims to uncover health characteristics that are closely associated with the battery's cycle aging behavior, ultimately pinpointing key parameters that precisely reflect the battery's aging state.

During the constant current discharge phase of a lithium-ion battery, discharge occurs at a stable 2A rate until completion, followed by a gradual current decrease to nearly 0A. With increasing charge-discharge cycles, the discharge duration decreases. Consequently, the equal voltage drop interval C1 and discharge voltage peak-to-valley interval C2, from 4.2V to 2.8V, are selected as health characteristics.

During charging and discharging, the internal temperature of a lithium-ion battery peaks and then gradually declines, primarily due to internal resistance and heat from electrochemical reactions. As charge-discharge cycles increase, battery aging shortens the time to peak temperature. Therefore, the time to peak charging temperature (D1) and average discharge temperature (D2) are selected as indicators of battery health status.

To ease analysis, the capacity increment method converts the voltage-capacity curve during charging into a more detailed IC curve. By examining changes in charge amount during charging and discharging, a deeper understanding of the battery's current working state and internal information can be gained. In the process of battery charging, the calculation formula of capacity and voltage is as follows:

$$Q = It \tag{1}$$

$$V = f(Q) \tag{2}$$

where, Q is the battery capacity, I is charging current, t is charging time, and V is the battery voltage.

The partial derivative of the capacity to the voltage is obtained, and the calculation formula of the IC curve is:

$$\frac{\mathrm{d}Q}{\mathrm{d}V} \approx \frac{\delta Q}{\delta V} = \frac{\Delta t \cdot I}{\Delta V} = \frac{\Delta t \cdot I}{V_t - V_{t-1}} \tag{3}$$

The IC curve of the NASA battery dataset's B5 battery in the 30th cycle is depicted in Figure 1. Various features of the IC curve are intimately linked to the lithium-ion battery's working and internal states. Notably, peak height and position are highly relevant, reflecting the battery's internal phase changes and accurately depicting SOH variations. Consequently, peak height (E1) and peak position (E2) of the IC curve are chosen as health indicators.

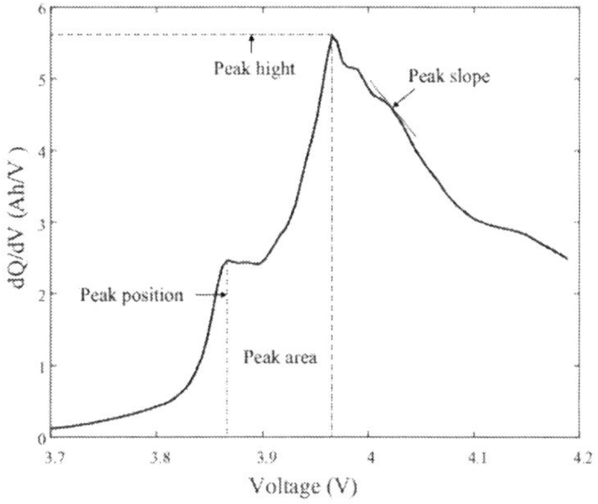

Figure 1. IC curve of battery

2.2. *Correlation Analysis*

The following four health characteristics are extracted: constant current charging time (A1), constant voltage charging time (A2), discharge voltage equal pressure drop interval (C1), discharge voltage peak-valley interval (C2), charging temperature peak time (D1), average discharge temperature (D2), peak height (E1) of the IC curve, and peak position (E2) of the IC curve. To evaluate their effectiveness in accurately reflecting battery health status changes during SOH estimation, Pearson and Spearman correlation analysis methods were used to assess the correlation between these features and battery available capacity.

Pearson and Spearman correlation coefficients assess the correlation between two datasets, ranging from -1 (perfect negative correlation) to 1 (perfect positive correlation). A coefficient of 0 indicates no correlation. The formulas for calculating these coefficients are provided below:

$$\text{Pearson} = \frac{E(\alpha\beta) - E(\alpha)E(\beta)}{\sqrt{E(\alpha)^2 - (E(\alpha))^2}\sqrt{E(\beta)^2 - (E(\beta))^2}} \tag{4}$$

$$\text{Spearman} = \frac{\sum_i (\alpha_i - \bar{\alpha})(\beta_i - \bar{\beta})}{\sqrt{\sum_i (\alpha_i - \bar{\alpha})^2 \sum_i (\beta_i - \bar{\beta})^2}} \tag{5}$$

where, α and β are two variables of correlation analysis, $E(\alpha)$ and $E(\beta)$ are the mean values of the two variables, $\bar{\alpha}$ and $\bar{\beta}$ are the mean values of the variable α_i and β_i.

Health characteristics from B5, B6, and B7 batteries in NASA's dataset are extracted into four sets, and their Pearson correlation coefficients with battery available capacity are computed (Table I). Additionally, Spearman correlation coefficients are calculated and presented in Table II.

Table I. Pearson correlation coefficient between health characteristics and battery capacity

Num	A1	A2	C1	C2	D1	D2	E1	E2
B5	0.892	0.819	0.818	0.816	0.910	0.873	0.955	0.959
B6	0.864	0.795	0.812	0.788	0.906	0.828	0.942	0.961
B7	0.856	0.834	0.837	0.829	0.923	0.869	0.951	0.968

Table II. Spearman correlation coefficient between health characteristics and battery capacity

Num	A1	A2	C1	C2	D1	D2	E1	E2
B5	0.883	0.876	0.873	0.877	0.906	0.859	0.937	0.964
B6	0.871	0.838	0.868	0.848	0.893	0.843	0.958	0.971
B7	0.869	0.860	0.862	0.881	0.901	0.864	0.931	0.956

Tables I and II demonstrate that the extracted health features exhibit strong correlations with battery capacity, with correlation coefficients spanning from 0.7 to 1.0. This finding underscores the effectiveness of the selected health features in characterizing the battery's cycle aging characteristics. To mitigate the risk of overfitting in the estimation model, it is crucial to maintain a manageable number of health features. Therefore, only the most critical features that are highly correlated with battery capacity degradation should be included as inputs. Conversely, overly simplistic models may neglect complex data relationships, compromising estimation accuracy. Hence, an optimization process is crucial to determine the optimal health feature dimension via simulation for peak performance.

2.3. Framework for Estimating the SOH of Lithium-ion Batteries Using CNN-BiLSTM-Attention Model

Upon estimating battery SOH, the CNN-BiLSTM-Attention model undergoes a structured process encompassing data preprocessing, feature extraction, training, estimation, and validation. The architecture, illustrated in Figure 2, comprises an input layer, a CNN layer for local feature extraction and mapping a BiLSTM layer to mine temporal information, and an attention layer for computing the weights of the BiLSTM outputs.

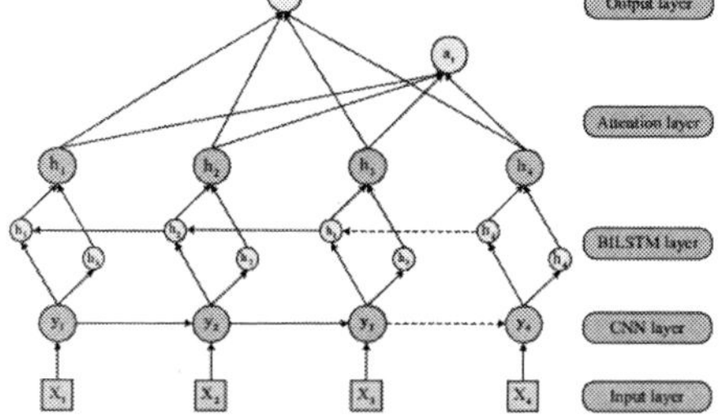

Figure 2. CNN-BiLSTM-Attention model structure

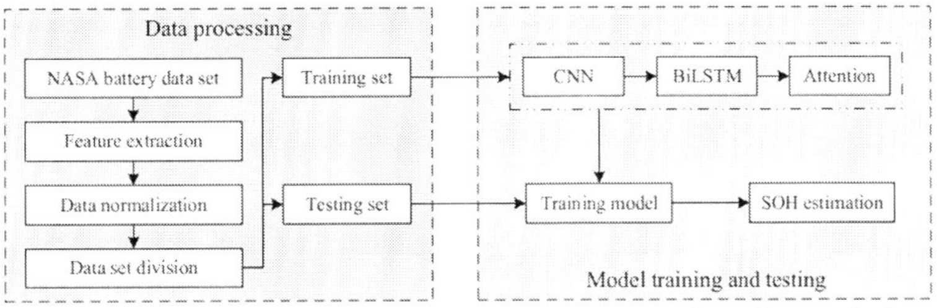

Figure 3. SOH estimation process of lithium-ion battery based on CNN-BiLSTM-Attention

As illustrated in Figure 3, the process of estimating the SOH of a lithium-ion battery using the CNN-BiLSTM-Attention model is primarily divided into three stages.

2.4.1 Data preprocessing

Through simulation analysis of aging battery data, health features notably associated with battery aging, such as charge/discharge time, temperature, and IC curve peak, are extracted. After simulation verification, when the input health feature is four-dimensional, the accuracy of the battery health state estimation result is the highest. Therefore, the four health features with the largest correlation coefficient are extracted: constant current charging time A1, charging temperature peak time D1, and IC curve principal components E1 and E2 are denoised and normalized as model input.

2.4.2 Model training

The four health features are input into a one-dimensional CNN layer, where convolution operations extract different local features. Pooling operations then perform dimensionality reduction and feature selection to simplify the model. The BiLSTM layer uses two LSTM networks to process forward and reverse sequences, capturing context from both directions. The attention layer computes weights based on the similarity between the BiLSTM outputs and the features, aggregating key information for the final output.

2.4.3 Model testing phase

The trained model is employed to assess the SOH of lithium-ion batteries, and its superiority over a single data-driven model is demonstrated through comparison.

3. Example Analysis

3.1. Dataset Description

In this study, we selected the cycle aging data of B5, B6, B7, and B18 batteries from the battery dataset published by NASA's Prognostics Center of Excellence (PCoE) [12]. There are two main factors: 1) different discharge cut-off voltages are set in the experiment to simulate the uncertainty of battery aging in actual use, which makes the data set closer to reality; 2) the lithium-ion battery data under different experimental conditions are selected to effectively test the wide applicability of the algorithm, and the public dataset is utilized to verify the effectiveness of the method. The details of NASA's battery data cited in this article are shown in Table III.

Table III. Pearson correlation coefficient between health characteristics and battery capacity

Num	temperature /°C	charging current /A	discharge current /A	cut-off voltage /V
B5	24	1.5	2	2.7
B6	24	1.5	2	2.5
B7	24	1.5	2	2.2
B18	24	1.5	2	2.5

B5, B6, and B7 are 18650 NCA battery cells rated at 3.7V and 2A·h. These batteries were charged at 24°C using a protocol that involved constant current charging at 1.5A until the voltage reached 4.2V, followed by constant voltage charging until the current dropped below 20mA. During discharge, a constant current of 2A was applied until the voltage of B5, B6, and B7 dropped to 2.7V, 2.5V, and 2.2V, respectively. Figure 4 depicts the capacity degradation curves of B5, B6, and B7 during cyclic aging.

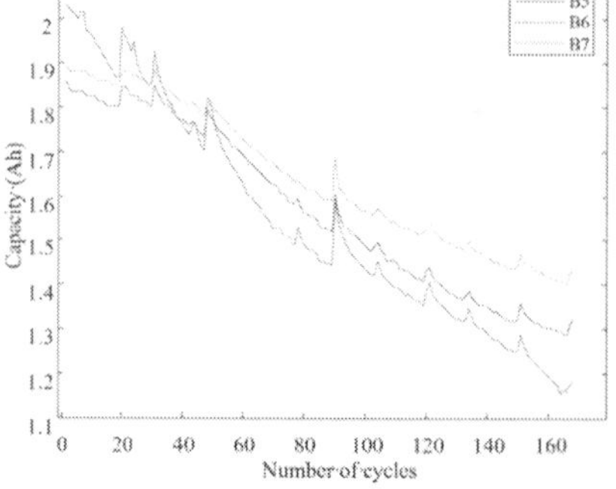

Figure 4. Battery capacity degradation curve

3.2. Battery SOH Estimation Results

Table III offers detailed information on the B5, B6, B7, and B18 batteries from the NASA battery dataset, which plays a crucial role in the simulation analysis conducted in this paper. To assess the estimation performance of the CNN-BiLSTM-Attention model for lithium-ion batteries introduced here, the B7 battery serves as a case study to estimate its SOH using the proposed estimation method. Additionally, the health state estimations for the B5 to B18 batteries are compared and analyzed to demonstrate the advantages of the combined model over individual data-driven approaches discussed in this chapter.

Taking B7 as an example, Figs. 5 and 6 illustrate the estimation results and errors of the CNN-BiLSTM-Attention model, respectively. The model demonstrates accurate SOH estimation with an error within 0.5%, indicating a satisfactory performance.

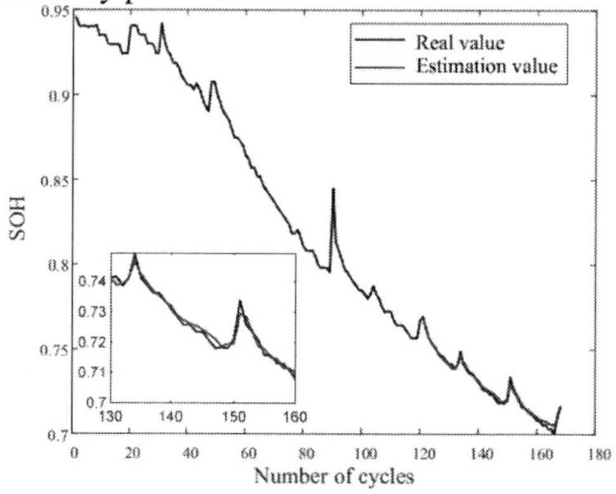

Figure 5. SOH estimation results of B7 battery

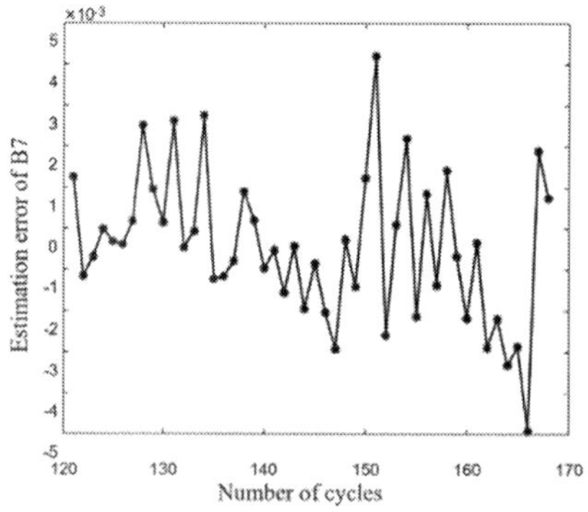

Figure 6. SOH estimation error of B7 battery

To demonstrate the superior estimation performance of the combined model proposed in this chapter, comparative simulations are conducted on B5, B6, B7, and B18 batteries, with results presented in Figure 7. In this figure, M1, M2, and M3 represent the estimation results of the BiLSTM model, CNN-BiLSTM model, and the proposed CNN-BiLSTM-Attention model, respectively.

The estimation results in Figure 7 show that the proposed CNN-BiLSTM-Attention model accurately estimates the SOH of the four batteries, closely matching the real values and timely responding to battery capacity regeneration. Across most cycle periods, precise estimation is achieved. Compared to the BiLSTM and CNN-BiLSTM models, the proposed combined model offers higher estimation accuracy and broader applicability. Additionally, it outperforms single data-driven models in tracking battery capacity regeneration.

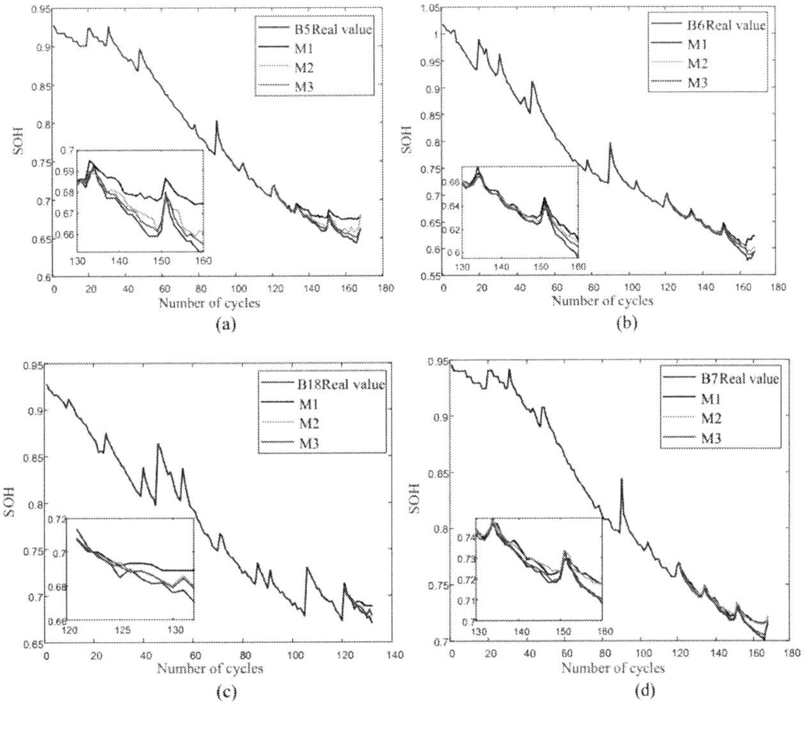

Figure 7. SOH estimation results of battery with different estimation models

As demonstrated by the SOH estimation error statistics in Table IV, the CNN-BiLSTM-Attention model surpasses both the BiLSTM and CNN-BiLSTM models across all evaluation metrics. With maximum MAE and RMSE values below 0.5%, the model exhibits high stability. This verifies that the CNN-BiLSTM-Attention model proposed in this paper provides superior SOH estimation for lithium-ion batteries under various operating conditions.

Table IV. Summary of statistical results of SOH estimation error of each battery

	Method	RMSE/Ah	MAPE（%）	MAE/Ah
	M1	0.0048	0.0057	0.0038
B5	M2	0.0032	0.0045	0.0030
	M3	0.0029	0.0035	0.0024
	M1	0.0124	0.0141	0.0086
B6	M2	0.0071	0.0078	0.0056
	M3	0.0044	0.0053	0.0032
B7	M1	0.0051	0.0059	0.0042

	M2	0.0046	0.0051	0.0036
	M3	0.0025	0.0026	0.0018
	M1	0.0913	0.0115	0.0087
B18	M2	0.0079	0.0094	0.0065
	M3	0.0044	0.0053	0.0039

4. Conclusion

In conclusion, this study proposes an innovative SOH estimation approach for lithium-ion batteries utilizing the CNN-BiLSTM-Attention combination model. By harnessing one-dimensional CNN for feature extraction, BiLSTM for temporal analysis, and an attention mechanism to prioritize intermediate features, we overcome limitations of traditional time series methods. Initially, battery health characteristics are extracted using charging/discharging processes and capacity incremental analysis. Correlation analysis validates the link between these features and battery health. Comparative SOH estimation results confirm that the CNN-BiLSTM-Attention model exhibits high accuracy and broad applicability.

Acknowledgment

This work is supported by the Program of Science and Technology Project of State Grid Fujian Electric Power Co., Ltd. (Research on Operation Control Strategy and Online Evaluation Technology of Grid-related Performance for New-type Electrochemical Energy Storage Power Station in Provinces with High Proportion of Clean Energy.) Grant No. 521304240009.

References

[1] K. Zhang, L. Y. Lou, H. X. Feng, et al., "Study on energy-saving techniques of the lithium-ion batteries cooling system using a backup battery," Case Studies in Thermal Engineering, vol. 61, pp. 104931-104931, 2024.

[2] Y. Gao, K. Liu, C. Zhu, et al., "Co-estimation of state-of-charge and state-of-health for Lithium-ion batteries using an enhanced electrochemical model," IEEE Transactions on Industrial Electronics, vol. 69, no. 3, pp. 2684-2696, 2022.

[3] D. Li and Y. Lin, "Remaining useful life prediction of lithium battery based on sequential CNN-LSTM method," Journal of Electrochemical Energy Conversion and Storage, vol. 1, pp. 1-26, 2021.

[4] H. He, R. Xiong, and J. Fan, "Evaluation of Lithium-Ion Battery Equivalent Circuit Models for State of Charge Estimation by an Experimental Approach," Energies, vol. 4, no. 4, pp. 582-598, 2011.

[5] S. Hosseininasab, C. Lin, S. Pischinger, et al., "State-of-health estimation of lithium-ion batteries for electrified vehicles using a reduced-order electrochemical model," Journal of Energy Storage, vol. 52, pp. 104684, 2022.

[6] G. Lee, D. Kim, and C. Lee, "A convolutional neural network model for SOH estimation of Li-ion batteries with physical interpretability," Mechanical Systems and Signal Processing, vol. 188, 2023.

[7] Y. Yu, C. H. Hu, X. Si, et al., "Averaged BiLSTM networks for RUL prognostics with non-life-cycle labeled dataset," Neurocomputing, vol. 402, pp. 134-147, 2020.

[8] Y. Wu, Q. Xue, J. W. Shen, et al., "State of health estimation for lithium-ion batteries based on healthy features and long short-term memory," IEEE Access, vol. 8, pp. 28533-28547, 2020.

[9] B. Saha and K. Goebel, "Battery Data Set," [Online]. Available: https://ti.arc.nasa.gov/tech/dash/groups/pcoe/prognostic-data-repository/. [Accessed: Aug. 2024].

Journal of Physics: Conference Series 3012 (2025) 012022 doi:10.1088/1742-6596/3012/1/012022

An Adaptive Noise Reduction Approach for Direct Current Electricity Meter Signal Processing

Lihua Zhong, Yuyao Yang*, Yilin Ji, Lei Feng, Feng Pan

Metrology Center, Guangdong Power Grid Co., Ltd., Guangzhou, Guangdong Province, 510080, PR China

*Corresponding author's e-mail: 313529747@qq.com

Abstract. DC loads typically comprise DC/DC converters and other power electronic components, which exhibit notable nonlinear characteristics. This nonlinearity introduces substantial ripple signals into DC output voltage and current, complicating the accuracy of DC energy measurements. Therefore, it is essential to enhance measurement precision through effective filtering and noise reduction techniques. To address the limitations of traditional filter-based noise reduction—specifically the suboptimal parameter settings and limited denoising effectiveness—this study proposes an adaptive noise reduction method for DC meter signals, employing adaptive variational mode decomposition (VMD) combined with wavelet threshold denoising.This method first establishes a parameter optimization model for the VMD of DC voltage and current signals. Through mutual information analysis, the model adaptively distinguishes between effective mode components and noise mode components within the original signals. Building on this, an optimization model for wavelet threshold denoising parameters is developed, utilizing a composite evaluation metric that incorporates signal-to-noise ratio, root mean square error, smoothness, and correlation coefficient to achieve optimal noise filtering of the identified noise modal components. The proposed method's effectiveness is validated through the calculation and analysis of measured data, demonstrating its potential to significantly improve DC energy measurement accuracy.

1. Introduction

Driven by the "dual carbon" policy goals and the development of a modernized power system, large-scale integration of distributed renewable energy sources into distribution networks and the acceleration of end-user energy substitution are becoming prevalent[1]. Among these sources, distributed photovoltaics and electric vehicles key elements of distributed renewable energy have experienced significant growth in recent years. Compared to traditional AC distribution networks, DC distribution networks offer advantages such as higher power supply capacity, lower equipment losses, and enhanced control flexibility, positioning them as a promising framework for the safe integration of large-scale distributed renewable energy sources[2-3]. In this context, energy measurement is shifting from traditional AC systems to DC systems.

Presently, DC energy measurement methods primarily include the average value method, the root mean square (RMS) value method, and the time-domain integration method[4]. However, most distributed renewable energy devices connect to the grid through power electronic converters like DC/DC converters, which introduce significant nonlinearities and result in considerable ripple within the voltage and current signals at the DC output. These ripples contain high-frequency AC harmonic components that bias measurements with the average value and RMS methods; as ripple content increases, so does the measurement error, compromising accuracy and the fairness of energy billing. While the time-domain integration method can mitigate ripple effects by increasing the sampling frequency, limitations in the processing capacity of measurement chips restrict this frequency enhancement. Consequently, advancing high-quality noise reduction techniques for DC energy measurement data is critical to improving measurement accuracy and ensuring reliable billing practices[5-6].

Filter settings are a common approach for reducing noise in direct current (DC) energy measurement data. By incorporating filter circuits, ripple content in the output signal can be minimized at the hardware level. However, the size of inductors and capacitors in these filter circuits directly influences filtering performance: insufficient inductor and capacitor sizing results in poor filtering, leaving high ripple content in the output signal and impacting measurement accuracy; conversely, oversized components improve filtering but risk distorting the output signal, also compromising accuracy[7]. To address these limitations, digital filtering techniques have emerged as a solution. Conventional digital filtering techniques, such as median filtering, limit amplitude filtering, and sliding average filtering, are effective in standard conditions but often fall short in dynamic environments with rapid fluctuations and multiple interferences. In contrast, modern filtering techniques—such as wavelet denoising, wavelet threshold denoising (WTD), and empirical mode decomposition (EMD)—offer enhanced performance[8]. For instance, Reference [9] proposes a DC charging energy measurement algorithm for electric vehicles that leverages discrete wavelet denoising and trapezoidal integration, delivering a simpler structure and higher measurement accuracy compared to traditional methods. Reference [10] introduces a novel wavelet threshold denoising method based on local mean decomposition, achieving superior denoising by enhancing signal-to-noise ratio (SNR) more effectively than basic WTD, EMD-based denoising, and local mean decomposition techniques. Reference [11] describes a denoising method for DC charging pile energy data using an improved wavelet threshold function, which effectively filters ripple from the DC charging pile's output signal and increases measurement accuracy.

Wavelet denoising and WTD methods exhibit strong denoising performance but involve multiple parameters, such as wavelet basis selection, decomposition levels, threshold function, and threshold selection criteria. EMD, on the other hand, approaches the original signal directly, decomposing it into intrinsic mode functions (IMFs). By denoising and reconstructing these IMFs, the original signal can be effectively purified. EMD bypasses wavelet basis selection issues inherent in wavelet-based methods but can suffer from modal aliasing, and the number of decomposition levels is often hard to confirm precisely. Variational mode decomposition (VMD) improves on EMD by addressing the modal aliasing problem; however, it requires the prior determination of two key parameters: the quadratic penalty factor and the number of decomposition modes. Adaptive variational mode decomposition (AVMD), an evolution of VMD, has gained attention due to its adaptability. Reference [12] describes a transient disturbance waveform denoising algorithm combining VMD and sliding mean filtering, optimizing the number of VMD decomposition layers by analyzing the Pearson correlation coefficient of each modal component. In Reference [13], Fourier transform-based spectral analysis of the detected signal aids in adaptively determining VMD modes. While AVMD has seen extensive application in fields like mechanical vibration analysis, fault diagnosis, and harmonic detection, its use in DC energy measurement data denoising remains underexplored.

This paper presents a denoising method for DC energy metering signals by combining Adaptive Variational Mode Decomposition (AVMD) with Wavelet Threshold Denoising (AWTD). Initially, an AVMD parameter optimization model is developed, aimed at minimizing the local signal envelope entropy of the DC output voltage and current signals. Solving this model enables optimal decomposition of each Intrinsic Mode Function (IMF). Using the maximum mutual information coefficient between each IMF and the original sampled data sequence, effective components are reliably differentiated from noise components. For identified noise components, Wavelet Threshold Denoising (WTD) is applied to achieve effective filtering[14].

To further ensure the denoising performance of WTD, an AWTD parameter optimization model is formulated, evaluated with a composite index encompassing signal-to-noise ratio, root mean square error, smoothness and correlation coefficient. This model considers key parameters, including the wavelet basis function, decomposition level, threshold function, and threshold selection criteria. By solving this model, the optimal parameter combination is determined, enabling enhanced filtering and denoising of noise components. Finally, case simulation analysis is conducted to verify the effectiveness of the proposed method.

2. AVMD Parameter Optimization Model

Variational Mode Decomposition (VMD) is a novel signal processing method proposed by K. Dragomiretskiy et al., with the core idea being the variational problem. This method differs from the Empirical Mode Decomposition (EMD) algorithm's iterative sifting process when obtaining Intrinsic Mode Functions (IMFs). Instead, it transfers the signal decomposition process into a variational framework, determining the frequency center and bandwidth of each component by iteratively searching for the optimal solution of the variational model. This enables adaptive signal frequency domain partitioning and effective separation of each component, highlighting the local characteristics of the data, showing better noise robustness, and possessing good sampling effects[15].

2.1. Objective Function of AVMD Parameter Optimization Model

The objective function of the AVMD parameter optimization model is to minimize the local signal envelope entropy of all IMFs after VMD of the output DC voltage and current original signals. The specific formula is as follows:

$$\min f = \min\left(\min_{i=1,2,\cdots,K} E_i\right) \tag{1}$$

$$E_i = -\sum_{t=1}^{T} p_{i,t} \lg p_{i,t} \tag{2}$$

$$p_{i,t} = \frac{y_i(t)}{\sum_{t=1}^{T} y_i(t)} \tag{3}$$

$$y_i(t) = \frac{1}{\pi}\int_{-\infty}^{+\infty} \frac{x_i(\tau)}{t-\tau} d\tau \tag{4}$$

In the formulas, $x_i(\tau)$ is the sequence of the *i-th* mode component; E_i is the signal envelope value of the *i-th* mode component after VMD of the output DC voltage/current original sampling data sequence; $p_{i,t}$ is the proportionality coefficient of the *i-th* mode component after Hilbert transform; T is the new sequence of the *i-th* mode component after Hilbert transform; T is the total number of points in the new sequence formed by the Hilbert transform of each mode component; K is the number of mode decompositions.

2.2. Objective Function of AVMD Parameter Optimization Model

The quadratic penalty factor and the number of modal decompositions in the model are optimization variables that need to satisfy their respective upper and lower limit constraints:

$$K_{\min} \le K \le K_{\max},\quad K \in \boldsymbol{N}_+ \tag{5}$$

$$0 \le \alpha < +\infty,\quad \alpha \in \boldsymbol{R}_+ \tag{6}$$

where α and K are the quadratic penalty factor and the number of modal decompositions, respectively; Kmin and Kmax are the lower and upper limit values of the number of modal decompositions; N+ is the set of positive integers; R+ is the set of positive real numbers.

2.3. Discrimination of Effective and Noise Components Based on Mutual Information Analysis

This paper distinguishes effective components from noise components by calculating the maximum mutual information coefficient between each mode component and the original signal. The formula for calculating the maximum mutual information coefficient is as follows:

$$\gamma_{\text{MIC}}\left(X_i;O\right)=\max_{a\times b<N^{0.6}}\left\{\frac{\sum_{x_i\in X_i}\sum_{o\in O}p\left(x_i,o\right)\log\frac{p\left(x_i,o\right)}{p\left(x_i\right)p\left(o\right)}}{\log\left[\min\left(a,b\right)\right]}\right\} \tag{7}$$

where N is the number of sample points of each decomposed mode sequence and the original signal; $p(x_i,o)$ is the joint probability density of the i-th decomposed mode sequence and the original signal; $p(x_i)$ and $p(o)$ are the marginal probability densities of the i-th decomposed mode sequence and the original signal, respectively; this paper uses the grid method to calculate the joint probability density and the marginal probability density between each decomposed mode sequence and the original signal; a and b are the number of rows and columns of the grid, respectively.

Based on the preset maximum mutual information coefficient threshold, the components after AVMD of the original signal are distinguished into effective components and noise components, as follows:

$$\begin{cases}\gamma_{\text{MIC}}\left(X_i;O\right)\geq\gamma_0, X_i \text{ represents the effective component}\\ \gamma_{\text{MIC}}\left(X_i;O\right)<\gamma_0, X_i \text{ represents the noise component}\end{cases} \tag{8}$$

Where γ_0 is the threshold for the maximum mutual information coefficient.

3. AVMD Parameter Optimization Model

3.1. Objective Function of AWTD Parameter Optimization Model

This paper aims to optimize the composite evaluation index of wavelet threshold denoising performance as the objective function of the AWTD parameter optimization model. The specific formula is as follows:

$$\max f=\max\lambda \tag{9}$$

In the formula, λ represents the composite evaluation index for wavelet threshold denoising that takes into account the signal-to-noise ratio (SNR), root mean square error (RMSE), smoothness, and correlation coefficient. The specific calculations are as follows:

The formula for calculating the signal-to-noise ratio (SNR) is as follows:

$$\eta_{\text{SNR}}=10\lg\left[\frac{\frac{1}{N}\sum_{i=1}^{N}O^2(i)}{\frac{1}{N}\sum_{i=1}^{N}\left[O(i)-G(i)\right]^2}\right] \tag{10}$$

In the formula, $o(i)$ represents the original signal data; $G(i)$ represents the reconstructed signal after wavelet threshold denoising; N is the total number of data points in the sequence.

The Root Mean Square Error (RMSE) calculation formula is as follows:

$$\eta_{\text{RMSE}}=\sqrt{\frac{1}{N}\sum_{i=1}^{N}\left[O(i)-G(i)\right]^2} \tag{11}$$

The smoothness calculation formula is as follows:

$$\eta_r = \frac{\sum_{i=1}^{N}\left[G(i)-G(i-1)\right]^2}{\sum_{i=1}^{N}\left[O(i)-O(i-1)\right]^2} \tag{12}$$

The correlation coefficient calculation formula is as follows:

$$\eta_R = \frac{\sum_{i=1}^{N}\left(G(i)-\frac{1}{N}\sum_{i=1}^{N}G(i)\right)\left(O(i)-\frac{1}{N}\sum_{i=1}^{N}O(i)\right)}{\sqrt{\sum_{i=1}^{N}\left(G(i)-\frac{1}{N}\sum_{i=1}^{N}G(i)\right)^2\sum_{i=1}^{N}\left(O(i)-\frac{1}{N}\sum_{i=1}^{N}O(i)\right)^2}} \tag{13}$$

$$\eta = w_1'\eta_{SNR} + w_2'\eta_{RMSE} + w_3'\eta_r + w_4'\eta_R \tag{14}$$

In the formula, w_1', w_2', w_3', and w_4' represent the weight coefficients for the signal-to-noise ratio (SNR), root mean square error (RMSE), smoothness, and correlation coefficient, respectively. This paper uses the Analytic Hierarchy Process (AHP) to solve for these weight coefficients.

3.2. The constraints of the AWTD parameter optimization model

The decision variables of the model are the wavelet basis function, decomposition level, threshold function, and threshold criterion, with corresponding constraints set as follows:

$$\psi(t) \in \Psi,\ d_{\min} \le d \le d_{\max},\ T \in \Omega,\ H \in \Gamma \tag{15}$$

In the formula, $\Psi(t)$、d、T and H represent the wavelet basis function, decomposition level, threshold function and threshold criterion respectively; Ψ、Ω、Γ represent the set of wavelet basis functions, the set of threshold functions, and the set of threshold criteria respectively; $d_{\max}$ and $d_{\min}$ represent the upper and lower limits of the decomposition level.

4. Case Simulation Analysis

4.1. Simulation Results Analysis

This paper verifies the method using simulated data of DC voltage and current at the output end of a DC charging pile. Typically, DC charging piles operate in either constant voltage or constant current charging modes. In addition to the DC fundamental component, the output signals contain higher harmonics, specifically the 11th, 13th and 15th harmonics with harmonic contents of 5%, 3% and 1% respectively. Therefore, the simulated data of the original DC voltage and current signals at the output end are calculated based on equations (16) and (17).

In the formulas, U_0 and I_0 represent the fundamental components of the output DC voltage and DC current, respectively, with values of 500V and 60A; r_{11}, r_{13}, and r_{15} represent the percentage content of the 11th, 13th, and 15th harmonics, respectively, with values of 5%, 3%, and 1%; the harmonic orders are k_1=11, k_2=13, k_3=15; the power frequency f=50Hz; the sampling frequency t=1ms. Based on the aforementioned parameters, the simulated sequences of the output DC voltage and DC current original signals are shown in Figure 1.

$$U(t) = U_0 + \sqrt{2}r_{11}U_0\cos(2\pi k_1 ft) + \sqrt{2}r_{13}U_0\cos(2\pi k_2 ft + 4\pi/5) + \sqrt{2}r_{15}U_0\cos(2\pi k_3 ft + \pi/3) \tag{16}$$

$$I(t) = I_0 + \sqrt{2}r_{11}I_0\cos(2\pi k_1 ft + \pi/2) + \sqrt{2}r_{13}I_0\cos(2\pi k_2 ft + \pi/3) + \sqrt{2}r_{15}I_0\cos(2\pi k_3 ft + 3\pi/4) \tag{17}$$

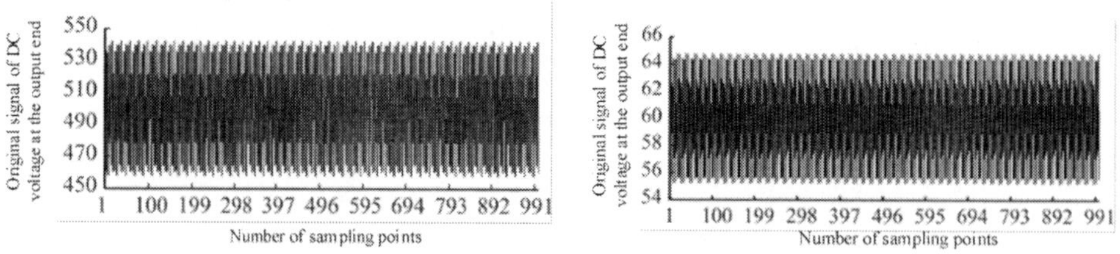

Figure 1. Output DC voltage/curren t original signal

The Improved Harmony Search algorithm is used to solve the AVMD parameter optimization model, with the following algorithm parameters set: HM=100,

$HMCR$=0.95, PAR=0.9, PAR=0.5, bw_{max}=1, bw_{min}=0.01. The range of modal numbers is 1 to 20, and the range of the quadratic penalty factor is 0 to 1000, to obtain the optimal quadratic penalty factor and the number of mode decompositions. Among them, the optimized parameters for the AVMD of the original DC voltage signal are: the optimal number of modes is 6, and the quadratic penalty factor is 416. Similarly, for the original DC current signal, the AVMD is performed, and the calculated optimized parameters are: the optimal number of modes is 10, and the quadratic penalty factor is 3. Based on the above parameter settings, the AVMD of the original DC voltage and current signals is realized, as shown in Figures 2 and 3.

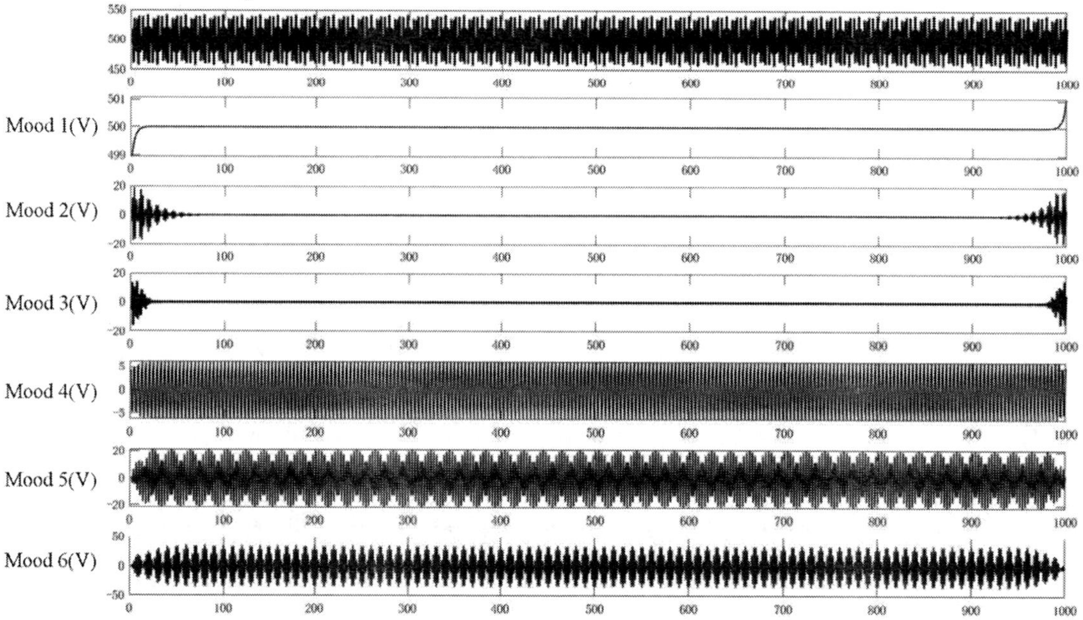

Figure 2. IMF of the DC voltage original signal by AVMD

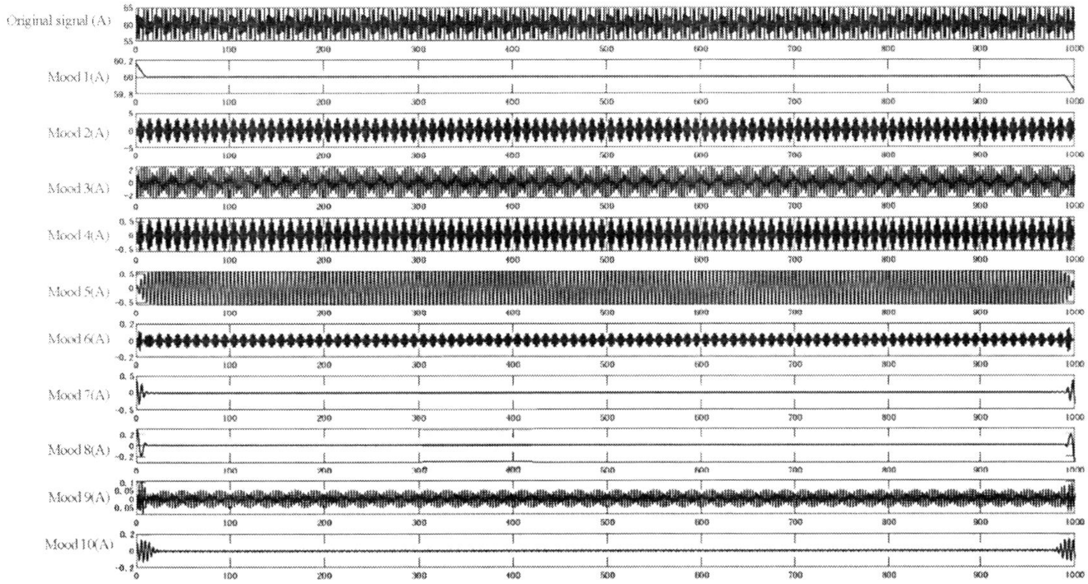

Figure 3. IMF of the DC current original signal by AVMD

Based on the results of Figures 2 and 3, the maximum mutual information coefficients between each Intrinsic Mode Function (IMF) and the original signal are calculated. With the maximum mutual information coefficient threshold set to 0.8, the effective components and noise components are screened out. Among them, the effective components after AVMD of the original DC voltage signal are modes 1 to 4, and the noise components are modes 5 and 6; for the original DC current signal, the effective components after AVMD are modes 1, 3, 5, 7 to 10, and the noise components are modes 2、4 and 6.

Further, the Improved Harmony Search algorithm is used to solve the WTD parameters and optimize the model to achieve the optimal filtering and noise reduction for the noise components. The wavelet basis functions include haar, db, bior, coif, and sym; the upper and lower limits for the decomposition level are set to 20 and 2, respectively; the threshold functions include soft thresholding and hard thresholding; the threshold criteria include the rigrsure criterion, sqt-wolog criterion, heursure criterion, and minimaxi criterion. Based on the optimized parameter combination, AWTD is performed on the noise components, and the filtered and noise-reduced components are superimposed on the effective components to obtain the denoised DC voltage and current data, as shown in Figure 4.

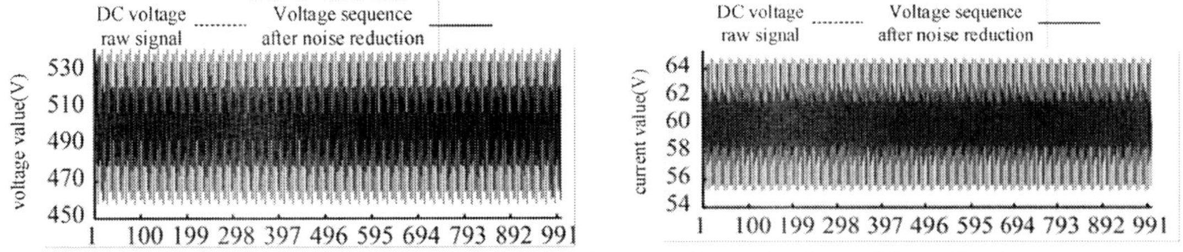

Figure 4. Data comparison before and after noise reduction of the DC voltage

4.2. *Method Effectiveness Verification*

To verify the effectiveness of the method proposed in this paper, this section sets up a comparison of four methods. Method 1 is the classic WTD, Method 2 is the combination of classic VMD and AWTD, Method 3 is AVMD with classic WTD, and Method 4 is AVMD-AWTD (the method of this paper). In Methods 1 and 3, when using WTD for filtering and noise reduction, the db4 wavelet basis function is chosen, with a decomposition level of 4, and a soft threshold function with the heursure criterion is applied. In Method 2, when using classic VMD for modal decomposition of the original DC voltage and current signals, the quadratic penalty factor and the number of mode decompositions are both set to 1000 and 5, respectively. Based on the above parameter settings, the signal-to-noise ratio (SNR) and root mean square error (RMSE) indicators of the noise-reduced signals for each method are calculated, as shown in Tables 1 and 2.

Table 1. Noise reduction results comparison for DC voltage original signal by method 1 to method 4

Method	SNR/dB	RMSE
Method 1	17.215	21.242
Method 2	20.682	19.672
Method 3	25.344	18.215
Method 4	28.526	16.983

Table 2. Noise reduction results comparison for DC current original signal by method 1 to method 4

Method	SNR/dB	RMSE
Method 1	21.532	5.958
Method 2	23.667	4.482
Method 3	26.728	3.017
Method 4	30.945	1.842

Comparing the results of Method 2, Method 3 with Method 1, it can be seen that applying VMD can improve the filtering and noise reduction effects of WTD. Comparing the results of Method 4 with Method 2, Method 3, it is observed that after introducing the AVMD mechanism, the average improvement in signal-to-noise ratio (SNR) and root mean square error (RMSE) is 34.34% and 36.29%, respectively; after introducing the AWTD mechanism, the average improvement in SNR and RMSE is 14.17% and 22.86%, respectively. It is evident that by introducing a parameter optimization mechanism to improve the classic VMD and WTD, the filtering and noise reduction effects on DC measurement signals can be significantly enhanced.

5. Conclusion

In response to the issues of suboptimal parameter settings and the need for improved filtering and noise reduction effects in existing DC load voltage and current signal filtering methods, this paper proposes a noise reduction method for DC energy metering signals based on AVMD-WTD. By optimizing the computational parameters of VMD and WTD, including the quadratic penalty factor, the number of mode decompositions, wavelet basis functions, decomposition levels, threshold functions, and threshold criteria, the proposed method further enhances the filtering and noise reduction effects on DC energy metering signals compared to traditional VMD and WTD methods, as demonstrated through the analysis of actual measured data. This improvement, in turn, ensures the accuracy of DC energy metering.

Acknowledgments

This work is supported by the National Key Research and Development Program of China (No. 2023YFF0614804).

References

[1] Zhang L, Chen H. 2021, A comparative analysis of DC energy measurement methods for renewable energy systems. Journal of Electrical Measurement and Instrumentation (vol 35), pp. 89-102.

[2] Smith T, Brown R. 2020, Effects of ripple in DC circuits on energy billing accuracy: Analysis of RMS and average value methods. International Journal of Energy Systems Engineering (vol 12), pp. 356-365.

[3] Yu X, Zhang Y. 2022, Advancements in DC distribution networks: Integration challenges and solutions for distributed renewable energy. IEEE Transactions on Smart Grid (vol 11), pp.

1785-1793.
[4] Li S, Gao M. 2022, Ripple reduction in DC energy measurement through time-domain integration and sampling frequency enhancement. Measurement Science and Technology (vol 33), pp. 543-552.
[5] Kim J, Lee T. 2018, Digital filtering methods in DC energy measurement: A comparative study of wavelet and empirical mode decomposition. Energy and Power Engineering Journal (vol 14), pp. 720-728.
[6] Wang H, Zhao Q. 2023, Discrete wavelet denoising and trapezoidal integration in electric vehicle DC charging energy measurement. IEEE Transactions on Power Electronics (vol 38), pp. 5503-5512.
[7] Cheng F, Li X. 2021, Enhanced wavelet threshold denoising for DC energy signals: Application in power measurement. Journal of Applied Signal Processing (vol 44), pp. 2015-2024.
[8] Zhou Y, Sun L. 2020, Denoising methods for DC charging pile data using wavelet threshold functions. Energy Systems Analysis (vol 29), pp. 133-142.
[9] Zhao L, Xu W. 2022, High-frequency ripple filtering in DC energy measurement: A case study using local mean decomposition and wavelet thresholding. IEEE Transactions on Industrial Electronics (vol 39), pp. 923-931.
[10] Fang J, Wang P. 2019, A novel method for adaptive mode decomposition: Application to denoising DC signals in energy meters. Measurement Science and Technology (vol 30), pp. 1156-1163.
[11] Lin H, Yang D. 2021, Empirical mode decomposition and variational mode decomposition in DC signal denoising: Methods and applications. IEEE Access (vol 10), pp. 5030-5038.
[12] He Q, Zhang M. 2020, Spectral analysis for adaptive VMD mode selection in DC transient waveform denoising. International Journal of Electrical Power & Energy Systems (vol 118), pp. 105791.
[13] Sun T, Li J. 2021, Application of AVMD and sliding mean filtering in transient disturbance denoising for DC energy data. IEEE Transactions on Power Systems (vol 36), pp. 1873-1881.
[14] Chen R, Huang Z. 2023, Comparative study of adaptive denoising techniques for mechanical vibration and DC energy signal processing. Mechanical Systems and Signal Processing (vol 178), pp. 109184.
[15] Liu Y, Zhu Q. 2022, Multi-parameter optimization in adaptive wavelet denoising for DC metering data. Signal Processing and Renewable Energy Applications (vol 14), pp. 345-356.

Grid Load Forecasting Based On Spatio -temporal Graphical Information

Wenshan Feng[1], Zhuo Long[1*], Gongping Wu[1], Liwei Xie[1], Jinping Xie[2], Fengqin Huang[1]

[1] School of Electrical and Information Engineering,Changsha University of Science and Technology,Changsha 410114, China

[2] College of Mechanical and Intelligent Manufacturing,Central South University of Forestry and Technology,Changsha 410114, China

[*]Corresponding author's e-mail: longzhuo601@csust.edu.cn

Abstract. The forecast of power grid load can predict future power demand. This helps power companies plan their power generation effectively. Additionally, it aids in allocating power resources wisely. Thus power companies ensure that the supply capacity meets the demand. This prevents power shortages or overloads caused by supply and demand imbalances. This paper proposes a spatio-temporal graph information-based load forecasting method that extracts and analyzes the spatio-temporal correlation properties of the graph information by converting time-series data into Markov field graph information and using a multivariate spatio-temporal graph neural network. Electric load data from Fuzeyuan District in Yuxi City, Yunnan Province, is used to confirm the forecasting method's viability and accuracy. The findings demonstrate that the approach can improve load forecasting outcomes.

1. Introduction

China's economy and industry are developing at a rapid pace, which is driving up demand for electricity [1] and raising power system requirements. As the foundation of China's military, economy and livelihood and other important areas, the power industry, its power supply and power scheduling is the core link of the power system. Against the backdrop of continued growth in social electricity consumption, power supply and scheduling face many challenges. Despite the efforts of power plants to meet the strong social demand for electricity, the uneven distribution of power generation capacity and the irrationality and lagging of grid scheduling still exist. Therefore, scientific and accurate power load forecasting is crucial for the stable operation of the power system. Through load forecasting, the load trend in the coming period can be roughly determined, so as to more finely formulate the starting and stopping plan of the units and the grid operation programme, ensure the balance between power generation and power consumption, and ensure the electricity system operates steadily.

The advantage of Time series method, regression forecasting method[2] and other traditional forecasting methods that the model is simple, the model building speed is faster. but the disadvantage is that the power load influencing factors are complex, the power load data exists randomness and uncertainty. So it will produce a large error in the prediction process. As artificial intelligence evolves at a fast pace, deep learning algorithms provide more algorithms for power load prediction[3]. For example: BP neural network [4] has strong nonlinear mapping ability and generalisation ability, and has been widely used in practical power load prediction. Researchers applied RNN

(RecurrentNeuralNetwork) to power load prediction, and found that its performance is significantly superior to the traditional back propagation neural network. Some RNN versions have been presented and have outperformed RNN in time series prediction [5]. In load prediction tasks, LSTM (Long Short-Time Memory Networks) and GRU (Gated Recurrent Unit Networks) are typically more effective than conventional RNNs [6]. Although existing deep learning algorithms have some applications in grid load forecasting, most of the methods learn directly from time-series data, which results in limited state information of the obtained signals and lacking explicit mechanisms for spatial dependency modeling.

To solve this issue, this paper suggests a prediction model based on MTGNN (Multivariate Spatial-Temporal Series Graph Neural Network). It incorporates Graph Neural Networks (GNN) to model spatial dependencies among multivariate time-series data. Each node represents a variable, and edges encode relationships, enabling it to capture complex spatial interdependencies. Compared to traditional method, this approch handles noisy and irregular multivariate data effectively through its end-to-end learning framework and uses temporal convolutions, which are faster than recurrent structures. This method can extract rich structured knowledge features by transforming the time-domain signals into graphical information, and show the detailed characteristics of the changing states of the grid load state more intuitively and comprehensively than the traditional prediction methods[7]. The study of the interactions between system status and the changing time trends of the system can benefit greatly from the use of graphical information to differentiate between various dynamic mechanisms in sequences of time. Another promising method is the construction of multivariate time series data prediction models using spatio-temporal graphical neural networks[8][9][10].The approach preserves the time series' temporal progression while fully utilizing the dependency between time series.

2. Module building

2.1. Information on Markov Field Graphs

The main method for expressing the information in a time series using information of Markov field graphs [11] is to quantize the sequences of time and determine its Markov transition probabilities.The following steps are primarily included, as shown in Figure. 1.

2.1.1. Quantification of time series. Using the Markov field method, the state at any given time can be precisely described. A given time series $X_t=[x_{1t},x_{2t},\cdots x_{nt}]$ is divided into B regions equally according to its peak-to-peak amplitude and described as quantile bins of B. By (1), x_{it} gets mapped to its matching quantile value Q_i based on the quantile bin at each moment,where $x_{\min}$ and $x_{\max}$ represent the time series's minmum and maximum values, $j=[1,2,\cdots b,\cdots B]$.

$$\begin{cases} Q_i=\frac{j}{B}, if\ \frac{j\text{-}1}{B}<\frac{x_{it}-x_{\min}}{x_{\max}-x_{\min}}\leq B \\ Q_i=\frac{1}{B}, if\ X_{it}=X_{min} \end{cases} \tag{1}$$

2.1.2. Markov transfer matrix caculation. The mathematical characteristics of the entire time series can be summed up in terms of the possibility of the B quantile bin through counting all of the number N_b of time points that correspond to the b_{th} quantile bin value Q_b. This yields in the possibility P_b that the b_{th} quantile bin value will occur throughout the time sequence, which is determined by (2).

$$P_b=\frac{N_b}{n} \tag{2}$$

Based on formula (3),the transfer probability w_{ij} from the quantile bin Q_i to the first quantile bin Q_j is determined by N_{ij} and N_i. N_{ij} is the total number of the value Q_i of the ith quantile bin transferred to the value Q_j of the jth quantile bin. And N_i is the whole number that correspond to the value Q_i of the ith quantile bin. Then according to formula (4), the complete time series' Markov transfer matrix W can be found.

$$w_{ij}=\frac{N_{ij}}{N_i} \tag{3}$$

$$W = \begin{pmatrix} w_{11}|p(x_t \in Q_2|x_{t-1} \in Q_1) & \dots & w_{1B}|p(x_t \in Q_1|x_{t-1} \in Q_B) \\ w_{21}|p(x_t \in Q_2|x_{t-1} \in Q_B) & \cdots & w_{2B}|p(x_t \in Q_2|x_{t-1} \in Q_B) \\ \vdots & \ddots & \vdots \\ w_{B1}|p(x_t \in Q_2|x_{t-1} \in Q_1) & \cdots & w_{BB}|p(x_t \in Q_B|x_{t-1} \in Q_B) \end{pmatrix} \tag{4}$$

2.1.3. Markov Field Graph Information. The continuing time series is translated to graphical information $G_t = [V_t, E_t]$ in conjunction with the preceding steps. The B quantile bins' probabilistic - based observation obtained from (2) are represented by the nodes $V_t = [P_{t1}, P_{t2}, \cdots P_{tB}]$, which are limited sets of vertices. By converting the Markov transfer matrix produced by (3) and (4), the edges E_t represent the vertex transfer probabilities of B-quantile bins. The procedures listed above can be obtained the Markov field map data for every period in the time series.

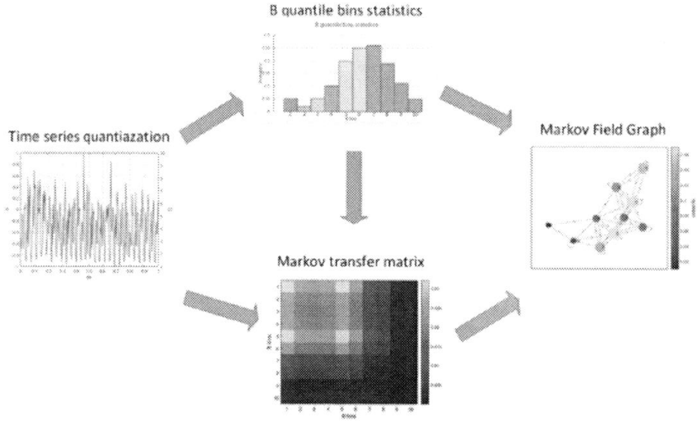

Figure 1. The Markov Field Graph Information method's framework (B=10)

2.2. Multivariate Spatial-Temporal Series Graph Neural Net-work

MTGNN is a technique that offers notable benefits for mining the intrinsic dependencies between multiple time series. In order to mine the hidden linkages between linking relations among the map information sequences, the graph structure extraction section primarily computes the graph adjacent matrix, which is then fed into the graph convolution section. The temporal and graph convolution sections are linked together to capture temporal and geographical connections, respectively.Each temporal convolution module is followed by skip connections, and the output module projects the hidden features to the required output dimension to produce the final output. The following are the method's main elements.

2.2.1. Graph Structure Extraction Module. The graph structure extraction module is used to statistically discover the connectivity relations of graph information sequences. The neighbor adjacency matrix $A = [A_1 \cdots A_t \cdots A_T]$ of the input graph dataset $G = [g_1, \cdots g_t, \cdots g_T]$ is statistically obtained to obtain its average statistical connectivity matrix A_V, in accordance with (5)(6), after the edge relation E_t of the input graph information sequence is converted to A_t. Where α is the threshold coefficient.

$$A_{t(i,j)} = \begin{Bmatrix} 1, if\ E_{t(i,j)} \neq 0 \\ 0, otherwise \end{Bmatrix} \tag{5}$$

$$A_v = soft\max\left(RELU\left(\sum_{t=1}^{T} A_t / T - \alpha \right) \right) \tag{6}$$

A connection relation is regarded as valid when the value of the relation between nodes is greater than the α. If not, it is regarded as invalid.

Journal of Physics: Conference Series **3012** (2025) 012023 doi:10.1088/1742-6596/3012/1/012023

2.2.2. Temporal Convolution Module. This module focuses on extracting time-dependent features of graphical information using a set of expanded one-dimensional convolutional filter time modules. It consists of two main expanded starting layers. One is a sigmoid formula that is employed as a selective section to regulate the information flow, and the other is a tangent hyperbolic activation formula for filtering. Various filter sizes are used in the extended beginning layer to guarantee that the module can be deployed in various temporal modes. In addition to accurately characterizing the short-term signal modes, the regularities and traits concealed within the long term signal modes can be discovered by integrating the various modes of the filter sizes. (7) can be used to indicate the expansion's initial layer. The four filters' outputs will cascade throughout the channel dimensions after being truncated in accordance with the maximum filter. (8) represents the dilation convolution, where the dilation parameter is denoted by d.

$$z = concat\left(conv\left(z, f_{1\times2}\right), conv\left(z, f_{1\times3}\right), conv\left(z, f_{1\times6}\right), conv\left(z, f_{1\times7}\right)\right) \quad (7)$$

$$conv\left(z, f_{1\times k}\left(t\right)\right) = \sum_{s=0}^{k-1} f_{1\times k}\left(t\right) z\left(t - d \times s\right) \quad (8)$$

2.2.3. Graph Convolution Module. In order to address spatial correlations in the network, the graph convolution section attempts to fuse a node's information with those of its nearby nodes. Each node's intake and outflow data are processed by two hybrid hop propagation layers that make up the graph convolution module. The outputs of the two hybrid hop propagation layers are added to determine the net influx information. The hybrid hop propagation layers and graph convolution module architecture are depicted in Figure 2. A hybrid hop propagation layer is used to process the information flow on geographically dependent nodes when a graph adjacency matrix is provided. Information propagation and selection are the two primary phases that make up the hybrid hop propagation layer. Information is first propagated horizontally, and then it is selected vertically. The following is a definition of the steps involved in the spread of information.

$$H^{(k)} = \beta H_{in} + \left(1-\beta\right) A H^{(k-1)} \quad (9)$$

Where the percentage of maintaining the root node's initial statu is controlled by the hyperparameter β. The following is the definition of the information selection step:

$$H_{out} = \sum_{i=0}^{K} H^{(k)} W^{(k)} \quad (10)$$

Where H_{in} is the input concealed status of the output of the current layer, H_{out} is the output concealed status of the present layer, and K is the propagation level.And $H^{(0)} = H_{in}$, $A = \tilde{D}^{-1}\left(A+I\right)$, $\tilde{D}_{ii} = 1 + \sum_j A_{ij}$.

2.2.4. Skip connection layers and output modules. The $1 \times L_i$ the norm convolutional layer is actually the size of the skip-connect layer, where L_i is the ith skip-connect layer's input length. Normalizing the information skipped to the output modules in order for their output sequence lengths to be equal is its primary purpose. Two 1x1 standard convolutional modules make up the output module, which transforms the output into the required output dimension sequence.

Combining the previous basic theory, Figure.2 shows the framework of the prediction model based on Markov field map and MTGNN. The inputs are first projected using a 1 × 1 norm convolution into an potential space. Interleaving temporal and graph convolution modules allows for the capturing of geographical and temporal connections, respectively. An exponential rate of q is used to enhance the dilation factor d, a hyper-parameter that regulates the magnitude of the receptive field of a temporal convolution module. The concealed graph adjacency matrix, which is utilized by graph convolution sections, is learned by layer of the graph learning. To prevent the issue of gradient disappearing, residual connections and skip connections are incorporated into the model. To obtain the final results, the output module maps concealed qualities to the appropriate dimension.

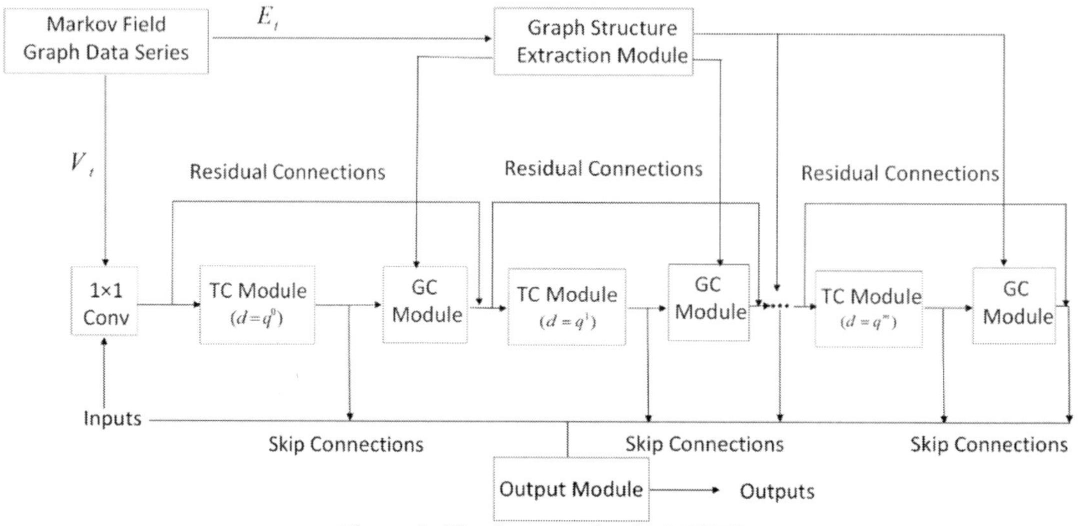

Figure 2. The framework of MTGNN

3. Modeling practice and results analysis

In order to verify the validity of the prediction model proposed in this paper, the active power values in the terminal power data from June 2022 to September 2023 for a total of 17 months in Fuzeyuan Community,Yuxi City, Yunnan Province,were selected as the load sample data and collected at 15-minute intervals.

3.1. Experiment.

To facilitate the simulation calculations, the data are first normalized. Time series signals are analyzed quantitatively and described as probability. The Markov transfer probability matrix is used to produce graph adjacency matrix information for time series signals. The data is then transformed into a sequence of Markov field map information. Figure.3 shows the Markov transfer matrices corresponding to the load data for different periods. And it can be seen that there is remarkable difference between the three periods.

The Markov field graph information was grouped at 7 intervals. The prediction experiment uses the initial 7 intervals of data as input to forecast the status of the lagging 7 intervals. A training set, a valid set, and a test set were separated from the Markov field map dataset in an 8:1:1 ratio. A multivariate spatio-emporal graph neural network will be utilized to get the training set's hidden spatiotemporal association data.The spatial convolution component captures and picks up the corrsponding spatial information of every each signal.The temporal convolution module learns and infers the dynamic deterioration law for continuous time sequence signals in the time dimension. The valid set is then used to optimize and modify the model parameters. Lastly, using the test set, specific prediction results are acquired and the model's prediction accuracy is confirmed.

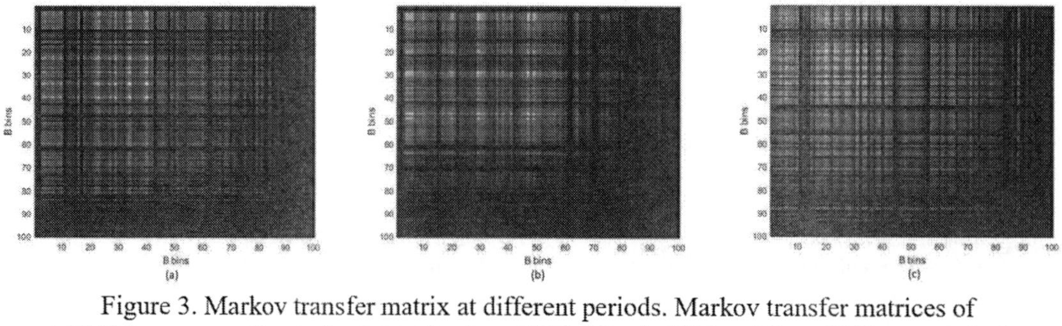

Figure 3. Markov transfer matrix at different periods. Markov transfer matrices of (a)(b)(c)corresponding to load data for June 2022, March 2023 and June 2023, respectively.

3.2. Predicted evaluation indicators.

The following three prediction evaluation metrics wereselected to assess the accuracy of the load prediction model:

MAE (Mean Absolute Error)

$$MAE = \frac{\sum_{i=1}^{M}|y_i - \hat{y}_i|}{M} \quad (11)$$

MAPE (Mean Absolute Percentage Error)

$$MAPE = \frac{1}{M}\sum_{i=1}^{M}\left|\frac{(y_i - \hat{y}_i)}{\hat{y}_i}\right| \quad (12)$$

RMSE (Root Mean Square Error)

$$RMSE = \sqrt{\frac{1}{M}\sum_{i=1}^{M}(y_i - \hat{y}_i)^2} \quad (13)$$

Where M is the quantity of samples, y_i is the reality value, and $\hat{y}_i$ is the value that was predicted.

3.3. Predicted Results and Analyses.

Based on the prediction results presented in Figure.4, it is evident that the approach is highly effective at predicting the power load state's trend, and the prediction evaluation indexes MAE, MAPE and RMSE are chosen to assess the accuracy of the prediction model. The table shows that the indexes of horizon2, horizon4 and horizon6 are lower than1, which further indicates that MTGNN performs in the power load forecasting task very well.

Table 1. Indicators for evaluating the impact of prediction

Horizon	MAE	RMSE	MAPE
2	0.3328	0.4494	0.0412
4	0.4117	0.5363	0.0655
6	0.3712	0.5039	0.0646

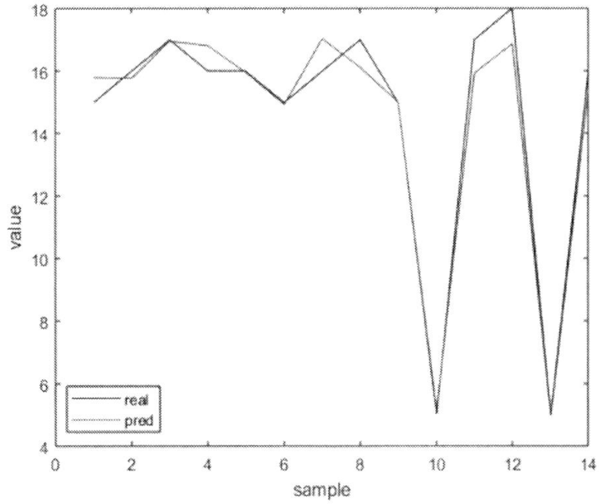

Figure 4. The Result of MTGNN

4. Conclusion

This research proposes a method for estimating grid loads using Markov field graphs and MSTGNN. The method uses Markov field theory to extract graphical information from time series data acquired throughout each monitoring cycle, which represents a rather steady state signal. The MSTGNN uses graphical information sequences to identify and learn spatial-temporal correlations in monitoring data, resulting in reliable predictions of power load. In contrast to traditional prediction methods, this method transforms the time series into Markov field graphs, and the Markov field graph information has strong expressive ability for the state of the time series signals, which not only retains the rich spatial information of the operating state of the timing signals, but also contains the characteristic information of the dynamic evolution of time. Finally, the forecasting method is validated using the power load data of Fuzeyuan District in Yuxi City, Yunnan Province, and the method's effectiveness and viability is proved by calculating the parameters such as MAE, RMSE and MAPE. The method has a good application prospect in electric load forecasting.

Acknowledgments
This work was supported in part by the Natural Science Foundation of China under Grant 52407037 and Grant 62403077, in part by Changsha Natural Science Foundation Youth Project under Grant kq2402022, in part by the Natural Science Excellent Youth Fund of Hunan Provincial under Grant 2024JJ4001, in part by the General Projects of the Hunan Provincial Education Department under Grant 22C0138,and in part by the Excellent Youth Projects of the Hunan Provincial Education Department under Grant 22B0306.

References

[1] C. Xu, Y. Zhang, X. Yang, X. Liu, C. Wang and S. Chen, "Grid Power Load Forecasting Mechanism Based on Neural Network," 2023 IEEE 5th Eurasia Conference on IOT, Communication and Engineering (ECICE), Yunlin, Taiwan, 2023, pp. 548-551

[2] L. Jiahao, G. Yuan and L. Dunnan, "Research on power medium and long-term load forecasting based on forecasting empirical partial least squares regression," 2021 International Conference on ArtificialIntelligence and Electromechanical Automation (AIEA), Guangzhou, China, 2021, pp. 395-398.

[3] H. Shahinzadeh, H. Sadrarhami, M. M. Hayati, H. Majidi-Gharehnaz, M. Abapour and G. B. Gharehpetian, "Review and Comparative Analysis of Deep Learning Techniques for Smart Grid Load Forecasting," 2024 20th CSI International Symposium on Artificial Intelligence and Signal Processing (AISP), Babol, Iran, Islamic Republic of, 2024, pp. 1-9.

[4] H. Zhang, Y. Yu, D. Lv, Q. Li and Z. Zhang, "Research on Short-term Load Forecasting Model Based on iForest-GA-BPNN," 2024 IEEE 7th Advanced Information Technology, Electronic and Automation Control Conference (IAEAC), Chongqing, China, 2024, pp. 1115-1118.

[5] M. Nickpasand, S. W. Jensen and H. M. Gaspar, "LSTM RNN in support of Manufacturing Simulation-based Digital Twin; A Case from Offshore Wind Turbine Production," 2024 19th Annual System of Systems Engineering Conference (SoSE), Tacoma, WA, USA, 2024, pp. 281-287. 684-689.

[6] R. Raman, V. Kumar, B. G. Pillai, D. Rabadiya, R. Divekar and H. Vachharajani, "Forecasting Bitcoin Value with Hybrid LSTM-GRU Neural Networks," 2024 Second International Conference on Data Science and Information System (ICDSIS), Hassan, India, 2024, pp. 1-5.

[7] Q. Li, P. Xu, X. Yang, Y. Wu, H. He and D. He, "Spatial-Temporal Traffic Prediction With an Interactive Spatial-Enhanced Graph Convolutional Network Model," in IEEE Transactions on Intelligent Transportation Systems, vol. 25, no. 12, pp. 20767-20778.

[8] G. Costantino, S. Giffard-Roisin, M. Dalla Mura and A. Socquet, "Denoising of Geodetic Time Series Using Spatiotemporal Graph Neural Networks: Application to Slow Slip Event Extraction," in IEEE Journal of Selected Topics in Applied Earth Observations and Remote Sensing, vol. 17, pp. 17567-17579, 2024,

[9] A. Zhang and Z. l. Gu, "Traffic Speed Prediction Based on Multi Spatial-Temporal Graph Fusion Convolution Network," 2023 IEEE International Conference on High Performance Computing & Communications, Data Science & Systems, Smart City & Dependability in Sensor, Cloud & Big Data Systems & Application (HPCC/DSS/SmartCity/DependSys),

Melbourne, Australia, 2023, pp. 846-851

[10] S. Liu, H. Xu, F. Meng and Q. Ren, "MDSTGCN : Multi-Scale Dynamic Spatial-Temporal Graph Convolution Network With Edge Feature Embedding for Traffic Forecasting," 2024 IEEE 24th International Symposium on Cluster, Cloud and Internet Computing (CCGrid), Philadelphia, PA, USA, 2024, pp. 284-290.

[11] S. Colonnese, G. Pagliari, M. Biagi, R. Cusani and G. Scarano,"Compound Markov Random Field Model of Signals on Graph: An Application to Graph Learning," 2018 7th European Workshop on Visual Information Processing (EUVIP), Tampere, Finland, 2018, pp. 1-5.

[12] X. Zhang et al., "Fault Prediction for Electromechanical Equipment Based on Spatial-Temporal Graph Information," in IEEE Transactions on Industrial Informatics, vol. 19, no. 2, pp. 1413-1424.

[13] Zonghan Wu, Shirui Pan, Guodong Long, Jing Jiang, Xiaojun Chang, and Chengqi Zhang. 2020. Connecting the Dots: Multivariate Time Series Forecasting with Graph Neural Networks. In Proceedings of the 26th ACM SIGKDD International Conference on Knowledge Discovery & Data Mining (KDD '20). Association for Computing Machinery, New York, NY, USA, 753–763

Journal of Physics: Conference Series **3012** (2025) 012024 doi:10.1088/1742-6596/3012/1/012024

Mobile Energy Storage Spatio-temporal Dispatch in Low-carbon Distribution Networks

Jinsong Liu[1], Bo Huang[2*], Shu Liu[1], Chen Fang[1], Xing Lei[1]

[1] State Grid Shanghai Electric Power Research Institute, NO. 171 Handan Rd Shanghai, 200433, P.R.China

[2] University of Shanghai for Science and Technology, NO. 516 Jungong Rd, Shanghai, 200093, P.R.China

*Corresponding author's e-mail: 18258075206@163.com

Abstract. As the proportion of new energy generation increases, optimizing the scheduling of energy storage devices to achieve decarbonized grid operation becomes a key issue. This paper proposes a low-carbon joint dispatch optimization model based on mobile energy storage. By constructing a spatio-temporal network model of the energy storage device, the charging and discharging behaviors, mobility characteristics, distribution network current constraints, and carbon emission factors of the energy storage device are considered comprehensively to optimize its operation. With the objective of minimizing the operating cost and carbon emission, the model adopts the mixed integer programming method to optimize the variables such as the movement path, charging and discharging power, and the state of the energy storage device. The model introduces the carbon emission factor of gas turbine and grid power purchase, combines the scheduling of energy storage devices with the low-carbon objective, and ensures the feasibility of the scheme through the tidal current constraints, SOC limits and path continuity constraints. The example results show that the proposed model effectively reduces the grid operation cost and carbon emission, while enhancing the capacity of renewable energy consumption, providing a new idea for the application of mobile energy storage in low-carbon grid scheduling.

1. Introduction

With the global carbon neutrality goal [1], the proportion of new energy power generation is gradually increased in various countries, and the proportion of renewable energy sources, such as wind power and photovoltaic (PV), in the power system has grown significantly. However, the volatility and intermittency of new energy generation pose challenges to the stability and economy of power grids [2]. Energy storage technology [3], as an important means to regulate the balance between power supply and demand, has gradually become the key to solve this problem. In particular, mobile energy storage devices are considered as an important tool to promote the decarbonized operation of distribution grids because of their flexible scheduling and multi-Case adaptation [4].

Mobile energy storage devices can optimize the charging and discharging behaviors through flexible mobile strategies, which can enhance the capacity of renewable energy consumption, reduce the phenomenon of wind and light abandonment, and reduce the operating costs of the power grid [5]. Compared with stationary energy storage [6], mobile energy storage has higher flexibility and can select suitable stations for charging and discharging according to demand. In addition, its scheduling optimization can effectively alleviate regional grid current constraints and improve voltage stability and line losses [7].

In recent years, research on mobile energy storage in grid optimization and scheduling has made some progress. Existing research mainly focuses on charging station siting and capacity optimization distribution grid economic scheduling optimal scheduling under the new energy consumption and low carbon goals. Reference [8] improves the grid's ability to cope with the challenges posed by renewable energy integration by optimizing the configuration of modular mobile battery storage in urban distribution grids. Reference [9] proposed an optimization strategy for energy sharing in building clusters considering the mobile energy storage characteristics of electric vehicles, which improves the energy sharing efficiency and security of smart building clusters. Reference [10] optimizes the distributed energy resource scheduling in the distribution network by considering the transportation time, power loss and operation mode of MESS to effectively reduce the PV power curtailment due to overvoltage. Reference [11] proposed to improve the resilience of multi-energy microgrids under natural disasters through mobile energy providers and explored the impacts of different scheduling strategies on resilience and cost.

Although current research on mobile energy storage has made some progress, challenges remain in a number of key areas. First, many studies still regard energy storage devices as static distributions and fail to fully utilize their spatio-temporal flexibility. Second, most of the existing studies focus on the construction of theoretical models and the application of optimization algorithms, and the scheduling problems in the actual power grid need further improvement. In addition, existing models are still deficient in the refinement modeling of carbon emission factors and computational efficiency improvement.

In view of this, this paper proposes a low-carbon joint dispatch optimization model for power grids based on mobile energy storage with the following main contributions:

1) Constructing a spatio-temporal network model for mobile energy storage: an optimization framework applicable to low-carbon objectives is proposed by considering the movement paths and charging/discharging strategies of energy storage devices.
2) Decarbonization optimization objective: the carbon emission factors of gas turbine and grid power purchase are considered comprehensively and incorporated into the objective function, which improves the model's ability to control carbon emission in a fine-grained manner.
3) The Cat Swarm Optimization (CSO) algorithm [12] is used to solve the problem, which can effectively optimize the scheduling strategy of the energy storage device by simulating the foraging behavior of cat swarms.

The structure of this paper is as follows: the second part introduces the specific construction of the model, including the objective function, constraints and variable definitions; the third part describes the solution method and its efficiency improvement strategy; the fourth part verifies the validity of the model and the application prospect through case study; the fifth part summarizes the research of this paper and discusses the direction of future work.

Through the research of this paper, we hope to provide theoretical support and technical reference for the application of mobile energy storage devices in low-carbon distribution network scheduling, and further promote the intelligence and efficiency of low-carbon grid scheduling.

2. Distribution Network Low-carbon Operation

In this article, a low-carbon joint dispatch optimization model based on mobile energy storage is proposed, aiming to achieve the dual minimization of system operation cost and carbon emission by optimizing the spatio-temporal scheduling of mobile energy storage devices as well as the distribution grid operation strategy. The model fully considers the mobile characteristics of energy storage devices, power status, distribution grid current constraints and carbon emission factors of gas turbine power generation and grid power purchase, and the construction of the model is described in detail from the optimization objectives, decision variables and constraints in the following.

2.1. Optimization objectives

The optimization objective function of the model integrates the power purchase cost and carbon benefit of the distribution network and achieves the low-carbon objective through carbon emission optimization. The specific objective function is:

$$\min C_{total} = \sum_{t} (w_1 \cdot P_{1,\ t} \cdot \vartheta + w_2 \cdot C_{eminsion,t}) \tag{1}$$

where: ϑ is the unit price of power purchased from the grid; $P_{1,\ t}$ is the active tidal current of power purchased at the first node; $C_{eminsion,t}$ is the total carbon emission in time period t, which is determined by the power purchased from the grid and the gas turbine power generation; and w_1, w_2 is a weighting factor reflecting the importance of the cost of power purchased and the carbon emission in the objective function.

Overall carbon emissions from the grid system $C_{eminsion}$ is the sum of carbon emissions from gas turbines and grid purchased electricity, expressed as follows:

$$C_{eminsion} = \sum_{t} \left(C_t^{gt} + C_t^{grid} \right) \tag{2}$$

$$C_t^{gt} = P_t^{GT} \cdot EF_{gt}, \forall t \tag{3}$$

$$C_t^{grid} = P_t^{grid} \cdot EF_{grid,t}, \forall t \tag{4}$$

where: C_t^{gt} is the carbon emissions from gas turbines at each time step; EF_{gt} is the carbon emission factor for gas turbines; C_t^{grid} is the carbon emissions from grid purchased electricity at each time step; $EF_{grid,t}$ is the carbon emission factor for grid purchased electricity at each time step.

2.2. Mobile Energy Storage Model

To ensure the practicality of the proposed model, certain assumptions have been made to simplify the complexity of real-world applications. The simplification of assuming feasible journeys is based on the role of mobile energy storage as a grid asset. The dispatch center centrally plans the movement paths, ensuring that transit constraints such as time and energy consumption are within acceptable limits.

During the charging time, the mobile storage energy starts from the terminal station, travels to the appropriate charging station for charging or discharging and returns to the starting station before the start of the next duty cycle. The path planning and movement of mobile energy storage devices are centrally managed by the grid dispatch center. To simplify the model, we assume feasible journeys and focus on optimizing spatial and temporal positions for charging and discharging.

In this study, the optimization focuses on minimizing the overall operating cost of the power grid, with an emphasis on the cost of purchasing electricity from the upper grid. This assumption simplifies the model while being consistent with actual grid management strategies, allowing the study to highlight the role of mobile energy storage in reducing total operating costs and enhancing grid flexibility.

Consider a simple transportation network where the charging stations are named S_1, S_2, S_3. Assume that the mobile storage energy travels between S_1 -S_2 , S_1 -S_3 , S_2 -S_3 at timeslots t_1 , t_2 , and t_3, respectively, and the set of charging stations is N_b, and the mobile storage energy can move between the charging stations in the set N_b. The 0, 1 integer variable $\zeta_{ij,t}^{(b)}$ denotes the movement trajectory of a class of mobile storage energy b at the time slot t. If $\zeta_{ij,t}^{(b)} = 1$, then b is traveling from node i to node j at the time slot t , i.e., b is at the node i at the time slot t , at the node j at the time slot t+1, and $i, j \in N_b$.

A transportation network with travel time can be represented by the spatio-temporal network model shown in Figure 1, where the dashed lines indicate the movement paths between stations and the solid lines indicate the stops of the mobile storage energy at a particular station.

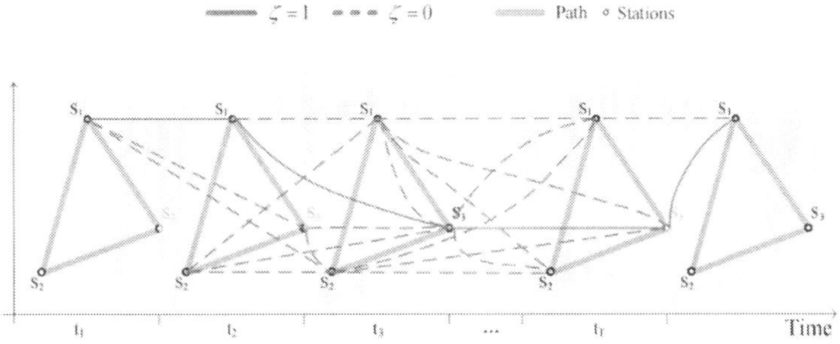

Figure 1. Time-space network model

The movement cost of mobile energy storage is represented by the energy consumption during transit between stations. Specifically, the energy loss incurred by the device while moving from one charging station to another is incorporated into the power constraints. This ensures that the energy used during transit is deducted from the available energy, reflecting the practical cost of movement in the optimization process. By modeling movement cost through energy consumption, the proposed framework balances computational simplicity with real-world operational considerations.

The above definition of the movement path already covers all the possible operating states of the mobile energy storage during the rechargeable time, i.e., if the mobile energy storage b is necessarily in a certain movement state during the time period, or moves between two stations, or stays at a certain charging station point, then there is:

$$\sum_{(i,j)\in R} \zeta_{ij,l}^{(b)} = 1, \forall b \in B, t \in T_b^c \tag{5}$$

where: B is the set of mobile energy storage types.

Mobile storage traveling has continuity, i.e., the end point of a time slot must coincide with the start point of the next time slot, and the end point of a time slot should be the start point of the $t+1$ time slot. R_k^- denotes the set of all moving paths with end point k, i.e., ; $R_k^- = (\cdot, k)$ R_k^+ denotes the set of all moving paths with start point k, i.e., $R_k^+ = (k, \cdot)$. The mobile storage movement continuity constraint can be expressed as:

$$\sum_{(i_1,j_1)\in R_k^-} \zeta_{i_1,j_1,t}^{(b)} = \sum_{(i_2,j_2)\in R_k^+} \zeta_{i_2,j_2,t+1}^{(b)}, \forall b \in B, k \in N_b, t \in T_b^c \tag{6}$$

The start and end positions of mobile energy storage in the rechargeable time period should be constrained to be at the starting station, and the node where the starting station is located is denoted as k , then there is:

$$\sum_{(i,j)\in R_{k_E}^+} \zeta_{ij,t_b^{c,S}}^{(b)} = 1, \forall b \in B \tag{7}$$

$$\sum_{(i,j)\in R_{k_E}^-} \zeta_{ij,t_b^{c,E}}^{(b)} = 1, \forall b \in B \tag{8}$$

where: $t_b^{c,S}$ and $t_b^{c,E}$ are the first and last rechargeable periods, respectively.

During the operating hours of the mobile energy storage, its position can be constrained in the model to the k_E node, i.e.

$$\xi_{k_E k_E,t}^{(b)} = 1, \quad \forall b \in B, t \in T_b^{nc} \tag{9}$$

The power $E_t^{(b)}$ constraint of mobile energy storage b at period t is:

$$E_{\min}^{(b)} \le E_t^{(b)} \le E_{\max}^{(b)}, \forall b \in B, t \in T \tag{10}$$

where: $E_{\max}^{(b)}$ is the battery capacity; $E_{\min}^{(b)}$ is set to 0.1 $E_{\max}^{(b)}$.

Mobile energy storage exchanges power with the distribution grid via charging posts. During this process, the mobile energy storage must stay at the charging station point. When the mobile energy storage is in operation time, its charging and discharging power is 0, denoted as:

$$0 \le P_{i,t}^{(b),c} \le \zeta_{ii,t}^{(b)} \xi_{b,t}^{c} P_{\max}^{(b)}, \forall b \in B, i \in N_b, t \in T_b^c \tag{11}$$

$$0 \le P_{i,t}^{(b),dc} \le \zeta_{i,i}^{(b)} \xi_{b,t}^{dc} P_{max}^{(b)}, \forall b \in B, i \in N_b, t \in T_b^c \tag{12}$$

where: $P_{\max}^{(b)}$ is the maximum charging/discharging power of the mobile energy storage charging pile; $P_{i,t}^{(b),c}$, $P_{i,t}^{(b),dc}$ are the charging and discharging power of the mobile energy storage b at the node i in the time period t respectively; $\xi_{b,t}^{c}$, $\xi_{b,t}^{dc}$ are Boolean variables, which are the charging and discharging states of the mobile energy storage b in the time period t respectively, then there are:

$$\xi_{b,t}^{c} + \xi_{b,t}^{dc} = 1, \forall b \in B, t \in T_b^c \tag{13}$$

Mobile energy storage needs to consume power to move between charging stations, the power consumed between every two stations i, j can be derived empirically, set to a fixed value $M_{ij}^{(b)}$, then the power consumption during the movement is:

$$E_{M,t}^{(b)} = \xi_{ij,t}^{(b)} M_{ij}^{(b)}, \forall b \in B, (i,j) \in R, t \in T_b^c \tag{14}$$

The variation of power during the rechargeable hours of mobile energy storage is expressed as:

$$E_{t+1}^{(b)} = E_t^{(b)} + \sum_{i \in N_b} (P_{i,t}^{(b),c} \eta - P_{i,t}^{(b),dc} / \eta) \Delta t / \eta - E_{M,t}^{(b)} \tag{15}$$

At the beginning of the rechargeable period, the mobile energy storage capacity is the power at the beginning of the operating period minus the power consumed during the operating period; at the end of the rechargeable period, the mobile energy storage capacity shall meet its operating power requirements:

$$E_{t_b^{c,S}}^{(b)} = E_{t_b^{c,E}}^{(b)} - E_{cons}^{(b)}, \forall b \in B \tag{16}$$

$$E_{t_b^{c,E}}^{(b)} \ge E_{need}^{(b)}, \forall b \in B \tag{17}$$

where: $E_{cons}^{(b)}$ is the amount of power consumed during the operation of the mobile energy storage b; $E_{need}^{(b)}$ is the amount of power that the mobile energy storage b needs to reach before the start of the next section of the operation.

2.3. *Distribution network DistFlow Model*

In this article, a linear tidal current model is used to construct a complete expression for the tidal current constraints by combining the spatio-temporal characteristics of mobile energy storage, as detailed below:

$$\sum_{(i,j)\in L} P_{ij,t} = -P_{i,t}^{D} + P_{i,t}^{L} + \sum_{ESS}(P_{i,t}^{(b),dc} - P_{i,t}^{(b),c}) + P_{i,t}^{PV} + P_{i,t}^{GT} \tag{18}$$

$$\sum_{(i,j)\in L} Q_{ij,t} = -Q_{i,t}^{D} + Q_{i,t}^{L} \tag{19}$$

where: $P_{ij,t}$, $Q_{ij,t}$ are the active and reactive currents of the line i - j, $P_{i,t}^{D}$, $Q_{i,t}^{D}$ are the active and reactive loads of node i; $P_{i,t}^{L}$, $Q_{i,t}^{L}$ are the external grid power injections at node i; $P_{i,t}^{PV}$ are the distributed PV power injections; $P_{i,t}^{GT}$ denotes the output power of the gas turbine; $P_{i,t}^{(b),c}$, $P_{i,t}^{(b),dc}$ are the charging and discharging power of the mobile energy storage equipment; and η are the energy storage charging and discharging efficiencies.

The node voltage needs to be kept within the allowed range. The simplified linear model is as follows:

$$V_j = V_i - (r_{ij}P_{ij} + x_{ij}Q_{ij}) / U_s \tag{20}$$

$$0.95 \le V_i \le 1.05 \tag{21}$$

where: V_i, V_j are the voltages at nodes i and j; r_{ij}, x_{ij} are the resistances and reactances of the lines i - j; P_{ij}, Q_{ij} are the active and reactive currents of the lines i - j; U_s is the reference value of the rated voltage; V_i is the voltage per unit value of node i.

The power output of the gas turbine at different time steps is limited by the start-stop condition and the physical properties of the equipment, with the following operational constraints:

$$P_{gt}^{\min} \cdot y_{gt\_status,t} \le P_{i,t}^{GT} \le P_{gt}^{\max} \cdot y_{gt\_status,t}, \forall t \tag{22}$$

where: $y_{gt\_status,t}$ denotes the start/stop state of the gas turbine, which is a Boolean variable; $P_{gt}^{\min}$, $P_{gt}^{\max}$ denote the minimum and maximum output power of the gas turbine.

3. Solution Methods

In this article, we apply to the CSO to optimize a low-carbon dispatch model based on mobile energy storage. The Cat Swarm Algorithm is a natural heuristic optimization algorithm that simulates the hunting behavior of a cat colony. The algorithm operates in two stages:

1. Exploration Phase: Cats perform global searches across the solution space, identifying potential solutions. In this phase, dynamic adjustment of search radius ensures a balance between exploration and computational efficiency.
2. Exploitation Phase: Cats focus on refining potential solutions through local searches. A gradient-based refinement is used to enhance solution accuracy.

The two phases alternate with a predefined switching probability to dynamically balance global exploration and local exploitation. The algorithm stops when either the fitness function stabilizes, or the maximum number of iterations is reached.

The fitness function in this study is designed to evaluate candidate solutions by comprehensively considering system operating costs and constraint penalties. The system operating costs include electricity purchase costs, carbon emission costs, and energy losses incurred during the movement of mobile energy storage devices. Meanwhile, constraint penalties are applied to solutions that violate operational constraints such as energy storage continuity, power flow limits, and SOC boundaries, ensuring the feasibility of candidate solutions.

This design ensures that the algorithm converges toward optimal solutions that minimize total costs while satisfying operational constraints, effectively balancing economic and environmental objectives.

The flow chart of CSO algorithm operation is shown in Figure 2.

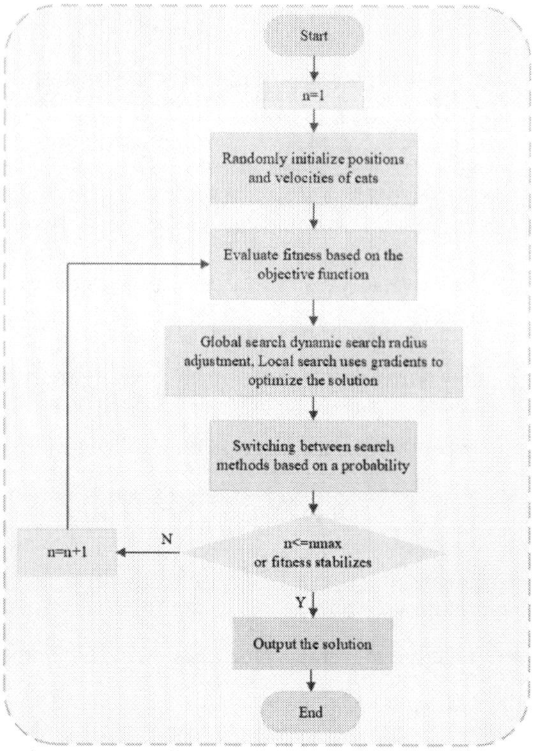

Figure 2. The flow chart of CSO solving model

4. Case Study

This case study is based on the proposed low-carbon joint dispatch optimization model, which utilizes the dispatch flexibility of mobile energy storage devices in the distribution network to minimize the grid operation cost and carbon emission cost through spatio-temporal network optimization. The case background, data setup, simulation cases, solution results and analysis are described in detail below.

4.1. Case Parameters

The distribution grid uses a modified IEEE 33-node system as shown in Figure 3, containing 33 nodes and 32 lines with a 24-hour dispatch time. The unit price of power purchased from the grid is 0.6 ¥/kW. The mobile energy storage start and end stations are located at node 16, and the remaining charging stations are located at nodes 3, 25, and 33. Nodes 17, 18, 22, 25, and 33 are connected to photovoltaic power generation. The weighting factor w_1, w_2 is 0.5, and the parameters of mobile energy storage equipment and gas turbine are shown in Table 1 and Table 2. The load profile is shown in Figure 4, and the PV output profile and the carbon emission factor of grid power purchase are shown in Figures. 5 ,6.

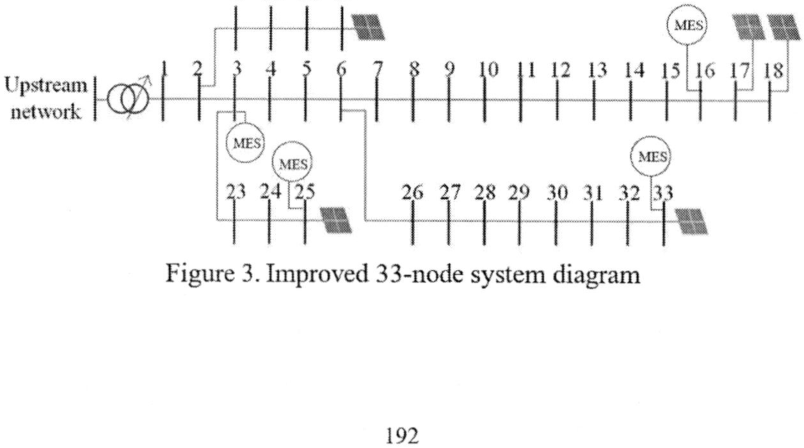

Figure 3. Improved 33-node system diagram

doi:10.1088/1742-6596/3012/1/012024

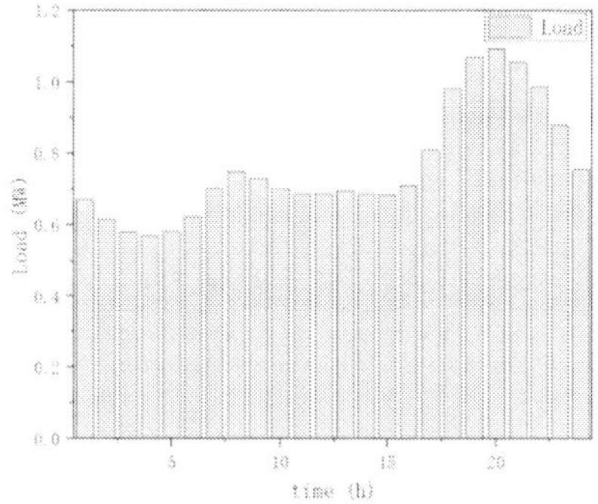

Figure 4. Load curve

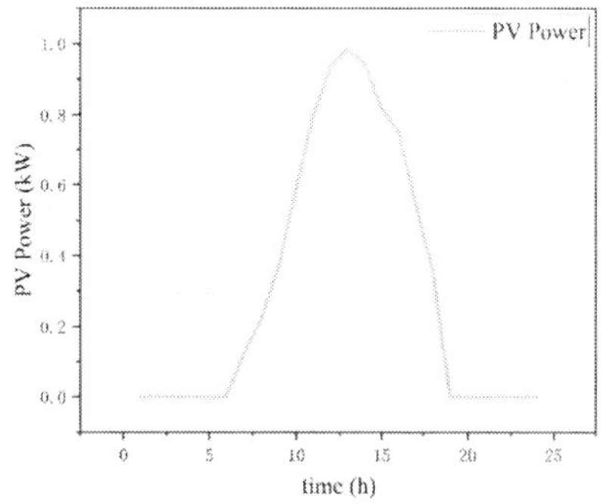

Figure 5. PV Output

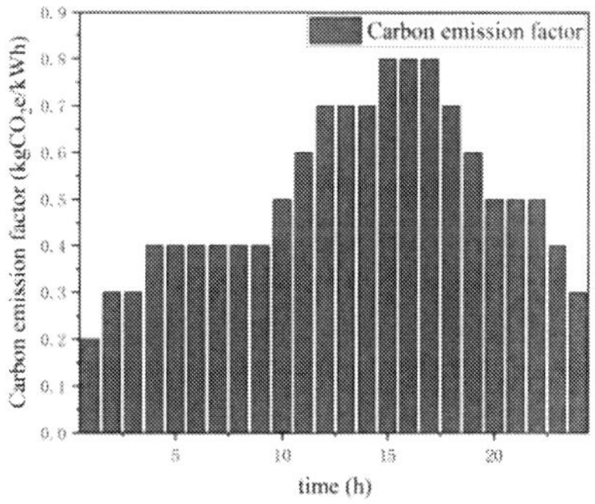

Figure 6. Carbon emission factors

Table 1. Parameters of mobile energy storage devices

Parameters	Numerical value
Number of energy storage devices	3
Quantitative	2.5MWh
SOC Range	0.1-0.9
Charge and discharge efficiency	0.95
Maximum charge/discharge power	0.5MW
Initial position	Node 16
Initial SOC	0.5

Table 2. Gas turbine parameters

Parameters	Numerical value
Number of gas turbines	2
Maximum output power	50 kW (node 15) 80 kW (node 30)
Minimum output power	2 kW

Journal of Physics: Conference Series **3012** (2025) 012024 doi:10.1088/1742-6596/3012/1/012024

In this article, the following three Cases are designed for comparison:

Case 1: No mobile energy storage is used, relying only on gas turbines and fixed energy storage devices. The optimization objective is to minimize grid operating costs.

Case 2: Mobile energy storage devices are introduced to reduce operating costs by optimizing charging, discharging and paths. The optimization goal is to minimize the grid operation cost.

Case 3: Based on Case 2, the optimization objective is to minimize the total cost of grid operation and carbon emission costs.

4.2. Simulation and Analysis

In this case study, the parameters of the CSO Algorithm are set as follows: the cat swarm size is 30, the maximum number of iterations is 1000, the exploration factor is 0.8, and the exploitation factor is 1.2, and these parameters are tuned to ensure that the algorithm performs an effective global search and local exploitation within the solution space, and is ultimately able to find a better solution and apply it to the low-carbon dispatch problem of mobile energy storage.

The results of the three Case study cases are shown in Table 3. The charging and discharging power of mobile storage under Cases 2 and 3 is shown in Figure 7, and the mobile storage motion trajectory is shown in Figure 8.

Table 3. Case results

Case	Total operating costs (¥)	Carbon emissions (kg)
Case 1	15289	725
Case 2	12867	643
Case 3	13452	582

In terms of grid operation cost, Case 1 has the highest total operation cost; Case 2 significantly reduces the power purchase cost through the participation of mobile energy storage devices; and Case 3 further reduces the power purchase demand in the high carbon emission time period under the influence of the weight of carbon emission optimization. In terms of total carbon emissions, Case 3 is optimal in terms of total carbon emissions, with a reduction of 19.4% compared to Case 1, indicating that the joint optimization of mobile energy storage devices and carbon emission factors effectively improves the system's low-carbon operation capability. The energy storage device presents obvious path optimization results in Case 3, and its moving strategy is highly correlated with the peak load area, indicating that the model can effectively utilize the location flexibility of the energy storage device.

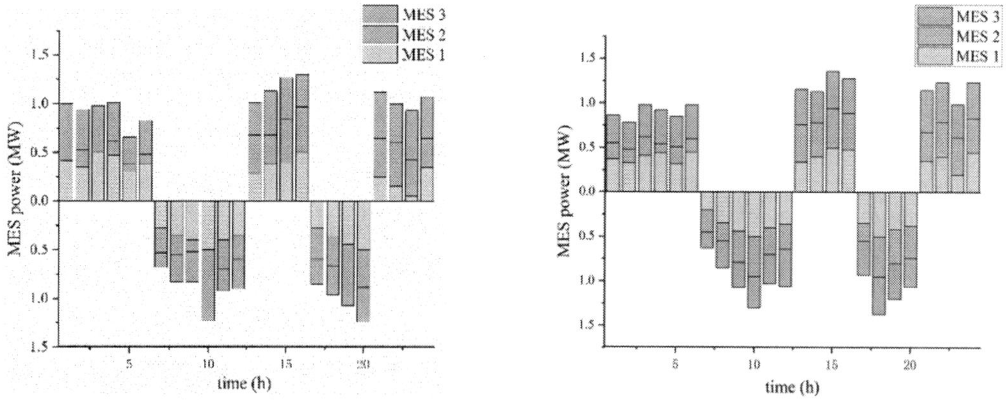

(a) case 2 (b) case3

Figure 7. Charging and discharging power of mobile energy storage in different cases

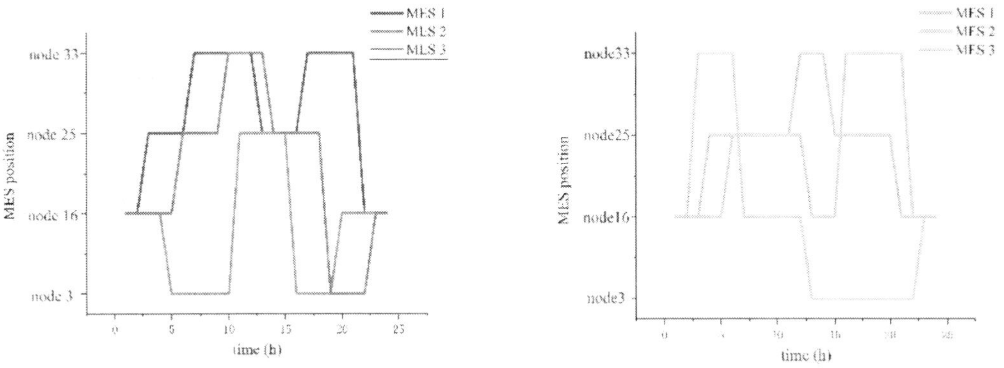

(a) case 2 (b) case3

Figure 8. Mobile energy storage moving trajectory in different cases

In Case 2, the mobile energy storage device significantly reduces the cost of purchased power, and the path optimization of the device reflects the effect of supporting high load nodes.

Case 3 reduces carbon emissions by dynamically adjusting power purchases during high-emission hours while maintaining low operating costs. The involvement of mobile energy storage improves the operational reliability of the distribution network.

5. Conclusion

In this study, a low-carbon dispatch optimization model based on mobile energy storage is proposed, aiming to achieve the dual minimization of grid operation cost and carbon emission by optimizing the charging and discharging strategy, and movement path selection of energy storage devices. By incorporating carbon emission factors into the objective function, the model not only improves the energy efficiency of the grid, but also effectively reduces the environmental impact.

The CSO algorithm avoids optimal local solutions through global search and performs well in large-scale nonlinear problems, but its solutions are approximate. The results show that mobile energy storage devices can effectively regulate the grid load and reduce the dependence on traditional energy sources, thus realizing the dual goals of grid economic operation and low-carbon dispatch. Future research can further optimize the computational efficiency of the model, explore multi-objective optimization and large-scale example applications, and provide more theoretical and practical support for low-carbon grid dispatch.

Acknowledgments

This study is supported by the Science and Technology Projects of State Grid Corporation of China (No. 52094023002U).

References

[1] Dong B, Zhang Z, Zhou C. Towards a just Chinese energy transition: socioeconomic considerations in China's carbon neutrality policies[J]. Energy Research & Social Science, 2025, 119: 103855.

[2] Guanpeng L, Yi Y, Zhuangzhuang L, et al. Optimal energy portfolio allocation method for regulable hydropower plants considering the impact of new energy generation[J]. Frontiers in Energy Research, 2023, 11: 1114949.

[3] Kunatsa T, Myburgh C H, Freitas D A. A Review on State-of-Charge Estimation Methods, Energy Storage Technologies and State-of-the-Art Simulators. Recent Developments and Challenges[J]. World Electric Vehicle Journal, 2024, 15(9): 381.

[4] Lu Z, Xu X, Yan Z, et al. Mobile Energy-Storage Technology in Power Grid: a Review of Models and Applications[J]. Sustainability ,2024, 16(16): 6857.

[5] Chenyang Z, Ying Y, Xuan L, et al. Mobile energy storage technologies for boosting carbon

neutrality[J]. The Innovation, 2023, 4(6): 100518.

[6] Zhang X, Liu Y, Zhan Y, et al. Multi-scenario optimization and performance evaluation of integrated energy system considering co-scheduling of EVs and stationary energy storage[J]. Renewable Energy, 2024, 237: 121927.

[7] Yan J, Liu S, Yan Y, et al. How to choose mobile energy storage or fixed energy storage in high proportion renewable energy scenarios: Evidence in China[J]. Applied Energy, 2024,376:124274.

[8] Fu D, Li B, Yin L, et al. Research on optimal configuration of mobile energy storage in distribution networks considering various energy utilization efficiencies[J]. Frontiers in Energy Research, 2024, 12: 1388681.

[9] Lin S, Li T, Shen Y, et al. Energy sharing optimization strategy of smart building cluster considering mobile energy storage characteristics of electric vehicles[J]. Electric Power Systems Research, 2025, 23: 8111067.

[10] Kim S, Lee J. Optimal Scheduling Strategy for Distribution Network with Mobile Energy Storage System and Offline Control PVs to Minimize the Solar Energy Curtailment[J]. Energies, 2024, 17(9): 2234.

[11] Milad G Z, Mehrdad H T, Saeid Z G. Preventive scheduling of a multi-energy microgrid with mobile energy storage to enhance the resiliency of the system[J]. Energy, 2023, 263: 125597.

[12] Suresh P, Keerthika P, Devi M R, et al. Optimized task scheduling approach with fault tolerant load balancing using multi-objective cat swarm optimization for multi-cloud environment[J]. Applied Soft Computing, 2024, 165112129.

Journal of Physics: Conference Series **3012** (2025) 012025 doi:10.1088/1742-6596/3012/1/012025

Transient Synchronization Stability Analysis of the Muti-VSCs System Considering Voltage-Dependent Current Injection During Low Voltage Ride-Through

Jingpeng Yue[1*], Xiaobing Liang[1], Jie Zhou[1], Rongcai Pan[1], Zhi Wang[2]

[1] Electric Power Research Institute of Guangdong Power Grid Co., LTD, Guangzhou, China

[2] Tianjin University, Tianjin China

*Corresponding author's e-mail: 659613777@qq.com

Abstract. The integration of renewable energy has made voltage source converters (VSCs) essential in power systems. However, during grid faults, the point of common coupling voltage, used for synchronization by VSCs, is affected by interactions between the injected current and grid impedance, posing significant challenges to the transient stability of multi-VSC (MVSC) systems. This paper addresses the gap in research on PLL synchronization transient stability in MVSC systems, particularly considering voltage-dependent current injection (VDCI) required by low-voltage ride-through (LVRT) standards. First, this paper develops the mathematical model of the MVSC systems that integrates both PLL dynamics and VDCI. Then, a transient stability assessment method combining the equal-area criterion (EAC) and numerical integration is applied to evaluate the transient stability of the MVSC systems under grid fault conditions. Finally, the simulation verifies the validity of the theoretical analysis.

1. Introduction

As renewable energy continues to gain prominence, voltage source converters (VSCs) have become indispensable components in power systems. Typically, VSCs employ phase-locked loops (PLLs) for synchronization by aligning with the terminal voltage of the converter. However, during grid faults, the point of common coupling (PCC) voltage used for synchronization is susceptible to fluctuations caused by the interaction between the VSC's injected current and the voltage drop across the grid impedance, which poses a risk of PLL synchronization instability in VSCs. Hence, it is essential to analyse the stability analysis of the grid-tied multi-VSC (MVSCs) system under fault.

Existing studies on the synchronization stability of grid-tied multi-VSC systems can be broadly divided into two main areas: determining the existence of equilibrium points (EPs) during grid faults and evaluating the system's ability to transition from pre-fault EPs to fault-on EPs. While substantial research has been conducted on identifying the conditions for EP existence [1]-[2], the aspect of transient stability, which examines the system's capability to reach fault-on EPs, has received comparatively less attention. This paper aims to address this gap by focusing on the transient stability of the MVSCs systems, specifically investigating their ability to achieve stable operation during grid faults.

In the context of transient stability evaluation, various analytical and numerical approaches have been explored. Equal-area criterion (EAC)-based methods have been applied to analyse PLL synchronization transient stability of the MVSCs system in [3]-[5], often neglecting dynamic interactions between PLLs or oversimplifying these interactions, which leads to overly conservative results. Meanwhile, energy function-based techniques have been developed, but limitations such as ignoring the proportional coefficients of PLLs or key interaction terms reduce their accuracy [6]. A recent method integrating EAC and numerical integration has shown promise for more precise stability analysis [7], but it assumes constant current injection and does not account for voltage-dependent current injection (VDCI), which is a critical requirement for low-voltage ride-through (LVRT) compliance.

Some recent studies have begun to explore the dynamic coupling of PLLs and VDCI, focusing on single-VSC systems connected to the grid [8]-[9]. These works analyse transient stability using phase-trajectory or equal-area approaches, but the findings are limited to single-converter systems and cannot be directly applied to the more complex MVSCs systems. The challenges of modelling and analysing multi-converter interactions remain unresolved, necessitating further research.

In summary, almost all the existing research works on multi-VSC grid-connected systems have not considered the effect of voltage dependent current injection. The research works that consider the effect of voltage dependent current injection have focused on single VSC grid-connected systems. There is a research gap in analysing the transient stability of MVSCs systems, considering both PLL dynamics and voltage-dependent current injection. Addressing this gap is the primary motivation of this paper. The main contributions of the paper lie in the two aspects:

1) Considering both PLL dynamics and voltage-dependent current injection, this paper develops the mathematical model of the MVSCs system for the first time.

2) Using the transient stability assessment method by combining EAC and numerical integration, this paper analyses the transient synchronisation stability of the MVSCs systems for the first time considering the effects of voltage-dependent current injection.

The structure of this paper is organized as follows: Section 2 introduces the MVSCs system. Section 3 formulates a nonlinear model of the MVSCs system, incorporating the VDCI dynamics. Section 4 proposes a novel PLL-STS assessment method, integrating the EAC with a numerical integration approach, followed by simulation results that validate the theoretical analysis. Finally, Section 5 concludes the paper.

2. Description of the MVSCs system considering the VDCI dynamic

The MVSC system studied in this paper is illustrated in Figure 1. The system consists of N converters, each connected to a common point of common coupling (PCC) via an LC filter and a line impedance $\boldsymbol{Z}_{Lk}$. These converters are then interconnected with the infinite bus. Notably, the converters synchronize at distinct points of synchronization (POS). In Figure 1, $\boldsymbol{V}_g$ and $\boldsymbol{V}_{tk}$ represent the voltages of the infinite bus and POS$_k$, respectively. The impedances $\boldsymbol{Z}_{Lg1}$ and $\boldsymbol{Z}_{Lg2}$ denote the equivalent impedances between the POS and the fault point, and between the fault point and the infinite bus, respectively, while $\boldsymbol{Z}_F$ indicates the ground impedance.

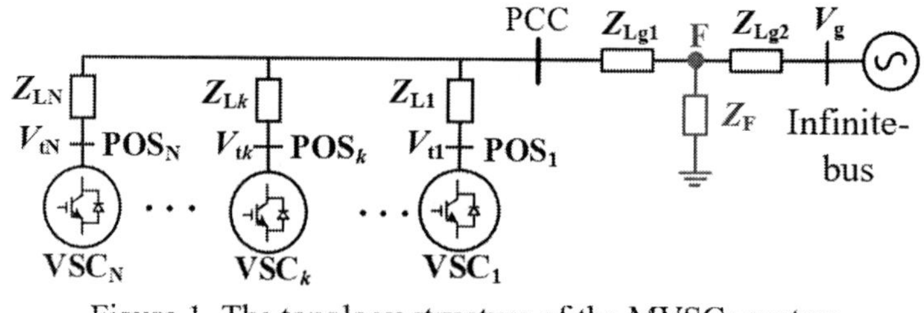

Figure 1. The topology structure of the MVSCs system

The fault control strategy for each converter VSC$_k$ during faults consists of three components: includes three primary components: phase-locked loop (PLL), voltage-dependent current injection (VDCI), and a current control loop, as depicted in Figure 2. For current control, a proportional-integral

(PI) controller is employed to rapidly regulate the injected current of VSC$_k$ following the current references I_{tdrefk} and I_{tqrefk}.

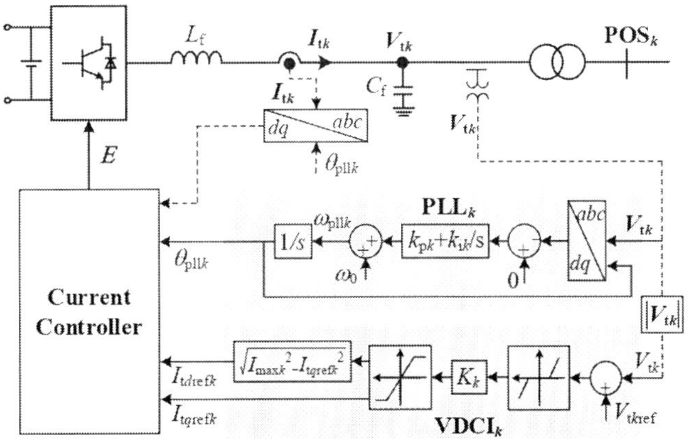

Figure 2. The control strategy for VSCk during the grid fault

The PLL$_k$ tracks the phase of the POS$_k$ voltage $\boldsymbol{V}_{tk}$, ensuring synchronization by providing the angular frequency ω_{pllk} and phase angle θ_{pllk}. According to the control structure of PLL$_k$, the dynamic behavior follows the equations outlined in (1).

$$\theta_{pllk} = -\int\left(k_{pk}V_{tqk} + k_{ik}\int V_{tqk}dt + \omega_0\right)dt \tag{1}$$

Using VSC$_k$ as an example, this paper explains the implementation of VDCI based on grid code requirements. When a grid fault occurs, the associated voltage drop at the POS of VSC$_k$ triggers a voltage dip. According to grid codes, when this voltage dip exceeds 10%, the converter must inject reactive current to support the affected voltage [10]. The amount of reactive current injected is directly proportional to the voltage drop, as described in Figure. 3.

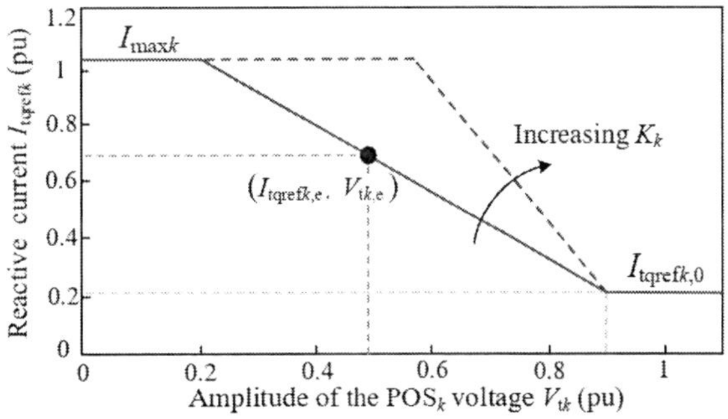

Figure 3. Voltage-dependent reactive current injection (VDCI) defined by grid code

In this paper, $I_{tqrefk,0}$ represents the pre-fault reactive current of VSC$_k$. The reactive current injection proportional coefficient K_k (referred to as the K-factor) is in the range of 0-10 typically falls within a range of 0 to 10 [8]-[9]. A steady-state operating point during low-voltage ride-through (LVRT) is indicated by the solid dot in the figure. For small K_k values, the reactive current I_{tqrefk} may remain below the maximum allowable current I_{maxk} allowed by VSC$_k$ during the low-voltage ride-through. This study specifically addresses the PLL synchronization transient stability (PLL-STS) of the MVSC system under such conditions.

Regarding the active current injection I_{tdrefk}, no strict grid code requirements exist. Typically, any remaining current capacity of the converter is used for this purpose. Consequently, under LVRT conditions, the dynamic expressions for I_{tdrefk} and I_{tqrefk} during LVRT can be expressed as follows.

$$\begin{cases} I_{\mathrm{tqref}k} = I_{\mathrm{tqref}k,0} + K_k(0.9 - V_{\mathrm{t}k}) \\ I_{\mathrm{tdref}k} = \sqrt{I_{\max k}{}^2 - I_{\mathrm{tqref}k}{}^2} \end{cases} \tag{2}$$

where V_{tk} is the magnitude of the POS_k voltage.

3. Nonlinear model of the MVSCs system considering the VDCI dynamic

This paper focuses on exploring the PLL synchronization transient stability of the MVSCs system within PLL synchronization time scales. To simplify the modeling, the dynamics of the current control are disregarded. This simplification is justified by the fact that the bandwidth of PLL is much lower than that of the current control loop in this study.

With the above assumptions, the circuit of the MVSCs system is obtained, as illustrated in Figure.4. In Figure 4, $\boldsymbol{V}_g$ and $\boldsymbol{V}_{tk}$ denote the infinite bus and POS_k voltage, respectively. $\boldsymbol{Z}_{Lk}$, $\boldsymbol{Z}_{Lg1}$, and $\boldsymbol{Z}_{Lg2}$ represent the equivalent impedance between the POS to PCC, PCC to fault point, and fault point to the infinite bus, respectively. $\boldsymbol{Z}_F$ is ground impedance. The specific expressions for $\boldsymbol{I}_{tk}$ in Figure.4 are provided in (3).

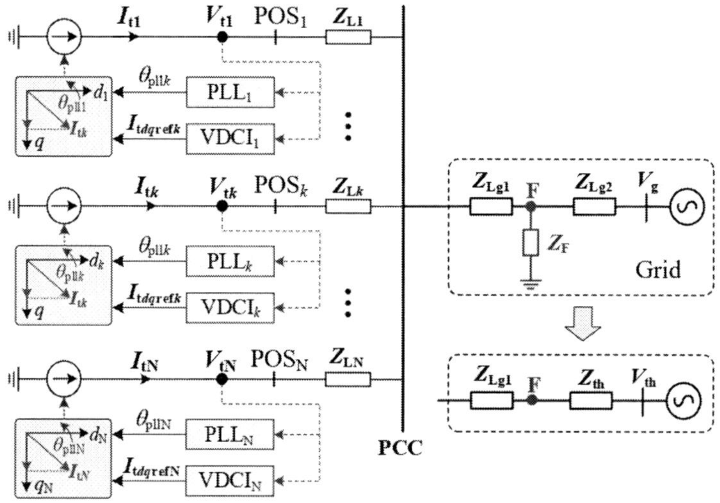

Figure 4. The equivalent circuit of the MVSCs system considering the VDCI dynamic

$$\boldsymbol{I}_{\mathrm{t}k} = (I_{\mathrm{tdref}k} - jI_{\mathrm{tqref}k})e^{j\theta_{\mathrm{pll}k}} = I_{\mathrm{tref}k}e^{j(\theta_{\mathrm{pll}k} - \theta_{Ik})} \tag{3}$$

where $I_{\mathrm{tref}k}=\{(I_{\mathrm{tdref}k})^2+(I_{\mathrm{tqref}k})^2\}^{1/2}= I_{\max k}$. The expression for θ_{Ik} is given in (4).

$$\theta_{Ik} = \arctan\left(\frac{I_{\mathrm{tqref}k}}{I_{\mathrm{tdref}k}}\right) = \arctan\left(\frac{I_{\mathrm{tqref}k,0} + K_k(0.9 - V_{\mathrm{t}k})}{\sqrt{I_{\max k}{}^2 - \left(I_{\mathrm{tqref}k,0} + K_k(0.9 - V_{\mathrm{t}k})\right)^2}}\right) \tag{4}$$

From Figure 4, the POS voltage of VSC_k can be expressed as follows:

$$\boldsymbol{V}_{\mathrm{t}k} = K_g V_g e^{j(\theta_g + \varphi_g)} + Z_k I_{\max k} e^{j(\theta_{\mathrm{pll}k} - \theta_{Ik} + \varphi_k)} + Z_e \sum_{j=1, j\neq k}^{\mathrm{N}} I_{\max j} e^{j(\theta_{\mathrm{pll}j} - \theta_{Ij} + \varphi_e)} \tag{5}$$

where V_g and θ_g are the amplitude and phase of $\boldsymbol{V}_g$, respectively. The expressions for K_g, Z_k, Z_e, φ_g, φ_k, and φ_e are given in (6).

$$\begin{cases} K_{\mathrm{g}} = \left| \boldsymbol{Z}_{\mathrm{F}} / \left(\boldsymbol{Z}_{\mathrm{F}} + \boldsymbol{Z}_{\mathrm{Lg2}} \right) \right| \\ Z_k = \left| \boldsymbol{Z}_{\mathrm{L}k} + \boldsymbol{Z}_{\mathrm{Lg1}} + \boldsymbol{Z}_{\mathrm{th}} \right| \\ Z_{\mathrm{e}} = \left| \boldsymbol{Z}_{\mathrm{Lg1}} + \boldsymbol{Z}_{\mathrm{th}} \right| \\ \varphi_{\mathrm{g}} = \angle \left(\boldsymbol{Z}_{\mathrm{F}} / (\boldsymbol{Z}_{\mathrm{F}} + \boldsymbol{Z}_{\mathrm{Lg2}}) \right) \\ \varphi_k = \angle \left(\boldsymbol{Z}_{\mathrm{L}k} + \boldsymbol{Z}_{\mathrm{Lg1}} + \boldsymbol{Z}_{\mathrm{th}} \right) \\ \varphi_{\mathrm{e}} = \angle \left(\boldsymbol{Z}_{\mathrm{Lg1}} + \boldsymbol{Z}_{\mathrm{th}} \right) \end{cases} \tag{6}$$

Transforming V_{tk} to the dq_k coordinate system, the d-axis and q-axis components of POS$_k$ voltage, denoted as V_{tdk} and V_{tqk}, can be obtained as:

$$\begin{cases} V_{tdk} = K_{\mathrm{g}} V_{\mathrm{g}} \cos\left(\theta_{\mathrm{pll}k} - \theta_{\mathrm{g}} - \varphi_{\mathrm{g}}\right) + Z_k I_{\max k} \cos\left(\theta_{Ik} - \varphi_k\right) + Z_{\mathrm{e}} \sum\limits_{j=1, j \neq k}^{\mathrm{N}} I_{\max j} \cos\left(\delta_{kj} + \theta_{Ij} - \varphi_{\mathrm{e}}\right) \\ V_{tqk} = K_{\mathrm{g}} V_{\mathrm{g}} \sin\left(\theta_{\mathrm{pll}k} - \theta_{\mathrm{g}} - \varphi_{\mathrm{g}}\right) + Z_k I_{\max k} \sin\left(\theta_{Ik} - \varphi_k\right) + Z_{\mathrm{e}} \sum\limits_{j=1, j \neq k}^{\mathrm{N}} I_{\max j} \sin\left(\delta_{kj} + \theta_{Ij} - \varphi_{\mathrm{e}}\right) \end{cases} \tag{7}$$

The magnitude of the POS$_k$ voltage V_{tk} can be calculated as:

$$V_{tk} = \sqrt{V_{tdk}^{\ 2} + V_{tqk}^{\ 2}} \tag{8}$$

By combining (1)-(8), we can obtain the nonlinear model for the MVSCs system considering the VDCI dynamic, as formulated in (9).

$$\begin{cases} \dot{\delta}_k = \omega_{\mathrm{pll}k} - \omega_0 \triangleq \Delta\omega_k \\ J_k \Delta\dot{\omega}_k = \underbrace{(P_{\mathrm{m}k} + P_{\mathrm{I}k})}_{P_{\mathrm{m}k\_\mathrm{I}}} - P_{\mathrm{e}k} - D_k \Delta\omega_k + \underbrace{(P_{\mathrm{c}k} + P_{\mathrm{cI}k})}_{P_{\mathrm{c}k\_\mathrm{I}}}, \ k = 1, \ldots, \mathrm{N} \\ \Delta\omega_k(\mathrm{t}_0^{\ +}) = -k_{\mathrm{p}k} V_{tqk}(\mathrm{t}_0^{\ +}) \end{cases} \tag{9}$$

where $\delta_k = \theta_{\mathrm{pll}k} - \theta_{\mathrm{g}}$, denotes the phase angle difference between PLL$_k$ and infinite bus voltage; J_k and D_k are the equivalent inertia and equivalent damping of VSC$_k$, respectively; P_{mk} is the equivalent mechanical power of VSC$_k$; P_{ek} is the equivalent electromagnetic power between VSC$_k$ and grid; P_{ck} is the equivalent mechanical interactive power between VSC$_k$ and other VSCs, reflecting the impacts of the transient interaction between VSC$_s$ on the VSC$_k$. $\Delta\omega_k(\mathrm{t}_0^+)$ represents the angular frequency mutation, i.e., at the instant of faults (t=t$_0^+$), $\Delta\omega_k$ abruptly jumps from 0 to $-k_{pk}V_{tqk}(\mathrm{t}_0^+)$. The expressions of J_k, D_k, P_{mk}, P_{ek}, P_{ck}, and $V_{tqk}(\mathrm{t}_0^+)$ are shown in (10).

$$\begin{cases} J_k = 1/k_{\mathrm{i}k}; \quad D_k = k_{\mathrm{p}k} K_{\mathrm{g}} V_{\mathrm{g}} \cos(\delta_k - \varphi_{\mathrm{g}})/k_{\mathrm{i}k} \\ P_{\mathrm{m}k} = Z_k I_{\mathrm{tref}k} \sin(\varphi_k - \theta_{Ik}); \quad P_{\mathrm{e}k} = K_{\mathrm{g}} V_{\mathrm{g}} \sin(\delta_k - \varphi_{\mathrm{g}}) \\ P_{\mathrm{c}k} = Z_{\mathrm{e}} \sum\limits_{j=1, j\neq k}^{\mathrm{N}} \left(-I_{\mathrm{tref}j} \sin\left(\delta_{kj} + \theta_{Ij} - \varphi_{\mathrm{e}}\right) - \dfrac{k_{\mathrm{p}k}}{k_{\mathrm{i}k}} I_{\mathrm{tref}j} \cos\left(\delta_{kj} + \theta_{Ij} - \varphi_{\mathrm{e}}\right) \Delta\omega_{kj} \right) \\ V_{tqk}(\mathrm{t}_0^{\ +}) = K_{\mathrm{g}} V_{\mathrm{g}} \sin\left(\delta_{ke,0} - \varphi_{\mathrm{g}}\right) + Z_k I_{\mathrm{tref}k} \sin\left(\theta_{Ik}(\mathrm{t}_0^{\ +}) - \varphi_k\right) + Z_{\mathrm{e}} \sum\limits_{j=1, j\neq k}^{\mathrm{N}} I_{\mathrm{tref}j} \sin\left(\delta_{kje,0} + \theta_{Ij}(\mathrm{t}_0^{\ +}) - \varphi_{\mathrm{e}}\right) \end{cases} \tag{10}$$

Compared with not considering the VDCI dynamic, the system model considering the VDCI dynamic has additional P_{Ik} and P_{cIk}. P_{Ik} reflects the effect of the interaction between the VDCI dynamic of VSC$_k$ and the grid on the transient stability of VSC$_k$; P_{cIk} reflects the effect of VDCI dynamics of other VSCs on the transient stability of VSC$_k$. The expressions for P_{Ik} and P_{cIk} are given in (11).

$$\begin{cases} P_{\mathrm{I}k} = -k_{\mathrm{p}k} Z_k I_{\mathrm{tref}k} \cos\left(\theta_{Ik} - \varphi_k\right)\theta_{Ij}' / k_{\mathrm{i}k} \\ P_{\mathrm{cI}k} = -Z_{\mathrm{e}} \sum\limits_{j=1, j\neq k}^{\mathrm{N}} k_{\mathrm{p}k} I_{\mathrm{tref}j} \cos\left(\delta_{kj} + \theta_{Ij} - \varphi_{\mathrm{e}}\right)\theta_{Ij}' / k_{\mathrm{i}k} \end{cases} \tag{11}$$

4. Transient stability analysis of the MVSCs system considering the VDCI dynamic

4.1. Transient Stability Assessment Method by Combining Equal Area Criterion and Numerical Integration

According to (9), the mathematical model of the MVSCs system considering A the VDCI dynamic has the same form as the rotor motion equations. Thus, the transient stability of the system can be analyzed using the equal area criterion. However, considering VDCI dynamic and transient interactions between VSCs, the unbalanced equivalent power term becomes very complex and difficult to solve analytically. To solve this problem, this paper introduces a transient stability analysis method by combining equal area criterion and numerical integration method. By numerically solving the acceleration and deceleration areas, the transient stability of the system can be accurately evaluated. This method is described in detail as follows.

In pre-fault conditions, the VSC$_k$ operates at point a, the initial phase angle is denoted as δ_{ka}, as depicted in Figure 5. This subsection introduces the acceleration and deceleration area of VSC$_k$, exemplified by the case $\delta_{ke}>\delta_{ka}$, where δ_{ke} is the PLL$_k$ phase angle component of the SEP during grid fault.

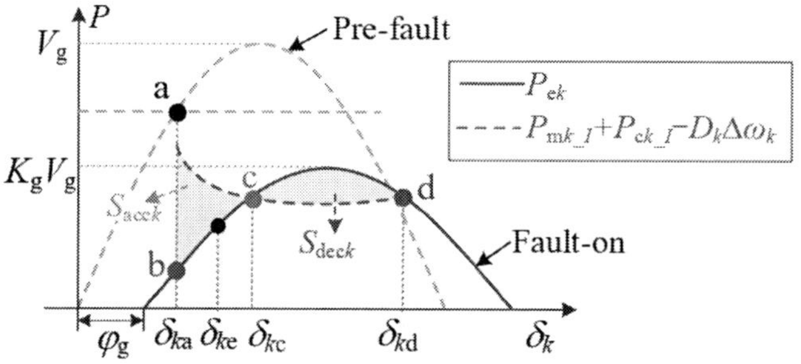

Figure 5. Acceleration and deceleration area schematic

After a ground fault occurs, the electromagnetic torque curve of VSC$_k$ drops to P_{ek}. At the same time, the angular frequency $\Delta\omega_k$ undergoes an instantaneous jump from 0 to $\Delta\omega_{\mathrm{init}k}$. Notably, the phase angle δ_k does not change abruptly, so the operating point shifts to point b, leads VSC$_k$ to accelerate from point b to point c. At point c, δ_k reaches the value δ_{kc}, and $\Delta\omega_k$ reaches its maximum value $\Delta\omega_{kc}$. The acceleration area $S_{\mathrm{acc}k}$ for VSC$_k$ during this transient process is quantified as follows:

$$S_{\mathrm{acc}k} = \int_{\delta_{ka}}^{\delta_{kc}} \left(P_{\mathrm{m}k\_\mathrm{I}} - P_{ek} - D_k\Delta\omega_k + P_{ck\_\mathrm{I}}\right) d\delta_k = \int_{\delta_{ka}}^{\delta_{kc}} J_k \frac{d\Delta\omega_k}{dt} d\delta_k = \frac{1}{2} J_k \Delta\omega_{kc}^{\ 2} - \frac{1}{2} J_k \Delta\omega_{\mathrm{init}k}^{\ 2} \tag{12}$$

After crossing the point c, the VSC$_k$ undergoes a deceleration as it moves from point c to point d. The relevant deceleration area $S_{\mathrm{dec}k}$ is given by the following:

$$S_{\mathrm{dec}k} = \int_{\delta_{kc}}^{\delta_{kd}} -\left(P_{\mathrm{m}k\_\mathrm{I}} - P_{ek} - D_k\Delta\omega_k + P_{ck\_\mathrm{I}}\right) d\delta_k = \frac{1}{2} J_k \Delta\omega_{kc}^{\ 2} - \frac{1}{2} J_k \Delta\omega_{kd}^{\ 2} \tag{13}$$

Given the nonlinear damping and nonlinear transient interaction, the acceleration and deceleration area in (12) and (13) cannot be analytically obtained. Thus, this paper introduces a numerical integration method to precisely calculate these areas. For more detailed information, please see [7] and [11].

By combining (12) and (13), the transient stability criterion for VSC$_k$ can be derived, as given in (14):

$$S_{\mathrm{acc}k} + E_{\mathrm{k0}k} \le S_{\mathrm{dec}k} \tag{14}$$

where E_{k0k}=0.5$J_k\Delta\omega_{kinit}^2$ represents the initial kinetic energy of VSC$_k$ due to the angular frequency jump. The jump notably leads to an increase of the acceleration area, consequently degrading the transient stability of VSC$_k$. Hence, this factor needs to be considered.

As can be seen from (9), the dynamics of the converters are coupled with each other. Consequently, if any converter, say the k-th converter, does not satisfy criterion (14) during faults, its δ_k diverges. This divergence in turn causes instability in other converters. Thus, the MVSCs system is stable only when all converters satisfy the criterion (14).

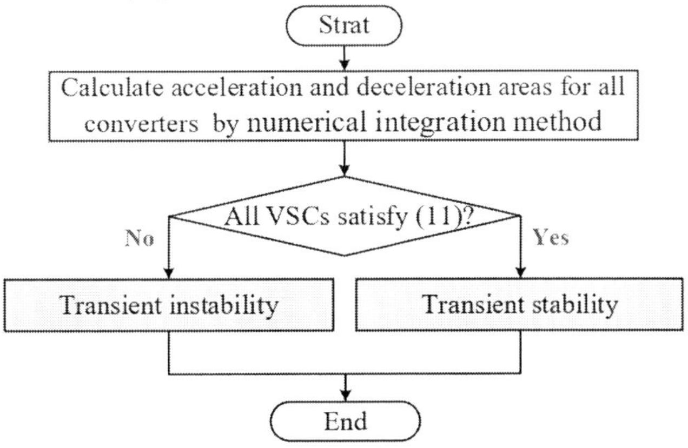

Figure 6. Transient stability assessment method for the MVSCs systems

4.2. Simulation verification

To verify the theoretical analysis above, a three grid-tied VSCs system is built in the PSCAD/EMTDC simulation platform, with detailed parameters in Table 1. Two cases were considered, referred to as Case 1 and Case 2. For Case 1 and Case 2, the $\boldsymbol{Z}_F$ is set at (0.024+0.08j) pu and (0.021+0.074j) pu, respectively.

Table 1. System parameters

System	Parameter name	Value
Base values	Base value of Power	5MW
	Base value of AC voltage	10kV
	Base value of frequency	50Hz
	Capacity of VSC$_1$/VSC$_2$/VSC$_3$	1.5/1.5/2MW
Hardware parameter	Switching frequency	10kHz
	LC filter of VSC$_1$, VSC$_2$, and VSC$_3$ are same	0.1/0.04 pu
	X_{Lg1}/ R_{Lg1}	0.2/0.06 pu
	X_{Lg2}/ R_{Lg2}	0.2/0.06 pu
	X_{L1}/ R_{L1}	0.2/0.06 pu
	X_{L2}/ R_{L2}	0.3/0.09 pu
	X_{L3}/ R_{L3}	0.5/0.15pu
	$I_{tdref1,0}$/$I_{tqref1,0}$	0.3/0.06pu
	$I_{tdref2,0}$/$I_{tqref2,0}$	0.3/0.06pu
	$I_{tdref3,0}$/$I_{tqref3,0}$	0.4/0.08pu
	I_{max1}/ I_{max2}/ I_{max3}	0.375/0.375/0.5pu
	$V_{g,0}$	1pu
Control parameter	PLL parameters of VSC$_1$ k_{p1}/k_{i1}	20 / 250
	PLL parameters of VSC$_2$ k_{p2}/k_{i2}	40/ 1000
	PLL parameters of VSC$_3$ k_{p3}/k_{i3}	30/500

	VDCI coefficient of VSC_1 K_1	0.39
	VDCI coefficient of VSC_2 K_2	0.39
	VDCI coefficient of VSC_3 K_3	0.52
	Current control of VSC_1, VSC_2, and VSC_3 are same	2 / 200

For Case 1, taking VSC_3 as an example, Figure. 7 illustrates the P_{e3}-δ_3 and $(P_{m3\_I}+P_{c3\_I}-D_3\Delta\omega_3)$ -δ_3 curves. Using the algorithm proposed in Figure. 6, the specific values of acceleration and deceleration areas for VSC_1, VSC_2, and VSC_3 can be calculated and are listed in Table 2. From Table 2, the system meets the PLL synchronization transient stability criterion, i.e., all three VSCs meet the criterion (14). Thus, the three grid-tied VSCs system must be transient stable for Case 1.

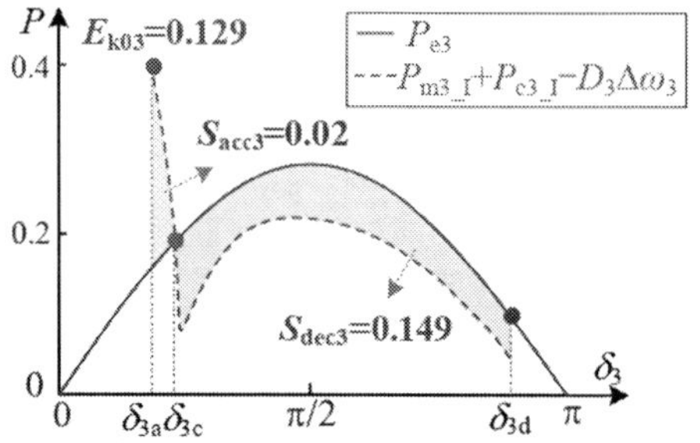

Figure 7. Acceleration and deceleration area of VSC_3 for Case 1

Table 2. Acceleration and deceleration area of Case 1

VSC_k	S_{acck}	E_{k0k}	$S_{acck}+E_{k0k}$	S_{deck}	Meet (14)?
VSC_1	0.0107	0.06	0.0706	0.0706	Yes
VSC_2	0.0184	0.0556	0.074	0.074	Yes
VSC_3	0.02	0.129	0.149	0.149	Yes

The simulation results for Case 1 are presented in Figure 8. These results clearly show that the system retains transient stability throughout the simulation. The observed stability behavior aligns closely with the predictions from the theoretical analysis, which demonstrates the accuracy and effectiveness of the proposed transient stability assessment method. This consistency between simulation and theory confirms that the method provides reliable results for evaluating transient stability in grid-connected systems.

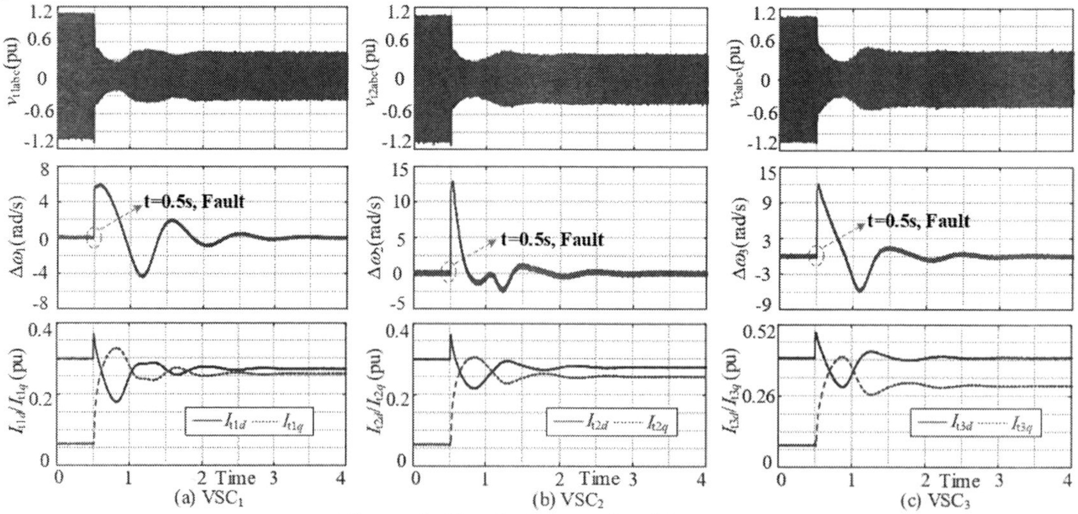

Figure 8. Simulation result of Case 1

For Case 2, the P_{e3}-δ_3 and $(P_{m3\_I}+P_{e3\_I} - D_3\Delta\omega_3)$-$\delta_3$ curves for VSC_3 are presented in Figure. 9. Specific values of acceleration and deceleration areas for three VSCs under Care A 2 are detailed in Table 3. As shown in Table 3, VSC_3 did not satisfy the stability criterion (14), as indicated by $S_{acc3}+E_{k03}>S_{dec3}$. Thus, the system failed to meet the transient stability criteria, resulting in instability in Case 2.

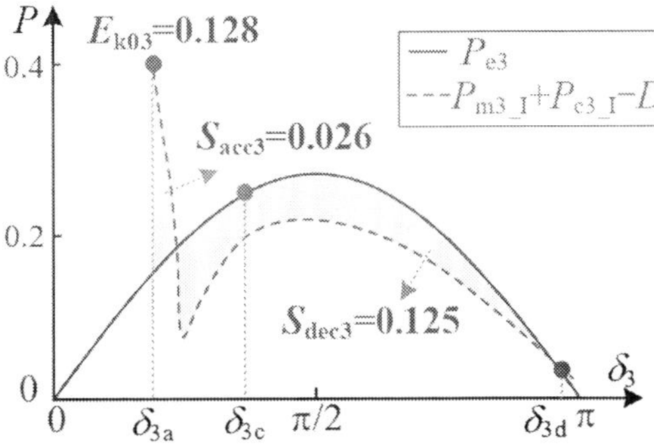

Figure 9. Acceleration and deceleration area of VSC_3 for Case 2

Table 3. Acceleration and deceleration area of Case 2

VSC_k	S_{acck}	E_{k0k}	$S_{acck}+E_{k0k}$	S_{deck}	Meet (14)?
VSC_1	0.015	0.0617	0.0767	0.0767	Yes
VSC_2	0.02	0.057	0.077	0.077	Yes
VSC_3	**0.026**	**0.128**	**0.154**	**0.125**	**No**

The simulation results for Case 2 are shown in Figure 10. As depicted, the system exhibits instability in Case 2, which is in agreement with the theoretical analysis. This further validates that, even when considering the VDCI dynamics, the transient stability of the MVSCs system can be accurately and effectively assessed using the method proposed in this paper. The consistency between the simulation results and theoretical predictions underscores the reliability of the proposed approach for analyzing transient stability in such systems.

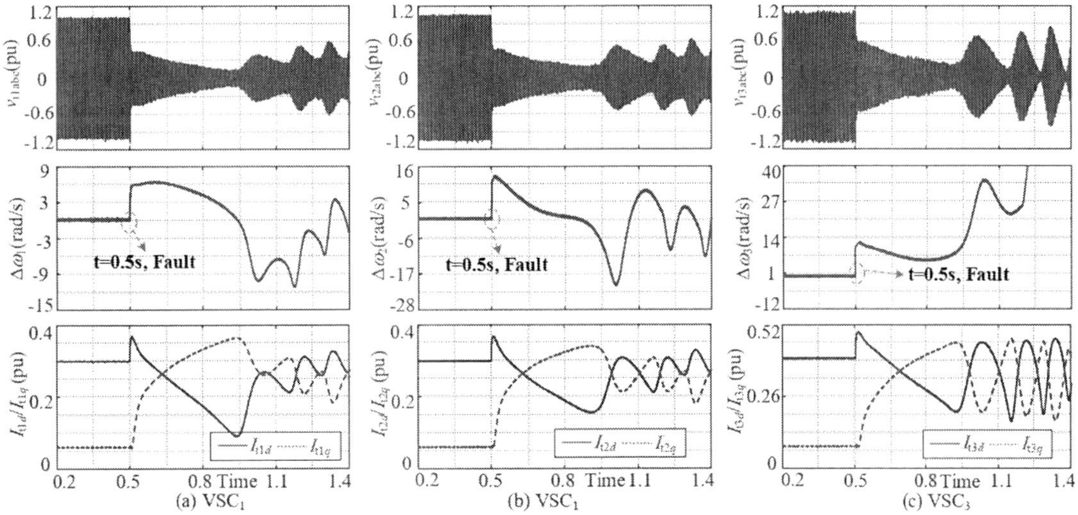

Figure 10. Simulation result of Case 2

5. Conclusion

This study investigates the transient stability of multi-VSC (MVSC) systems considering voltage-dependent current injection during low voltage ride-through (LVRT). By incorporating voltage-dependent current injection (VDCI) required for (LVRT) compliance and phase-locked loop (PLL) dynamics, the paper develops a comprehensive mathematical model for MVSC systems. Then, using a method that combines the equal-area criterion (EAC) and numerical integration, the transient synchronization stability of MVSC systems is analyzed. The simulation validate the theoretical analysis and demonstrate the effectiveness of the proposed method for evaluating the stability of MVSC systems during faults. This work provides a foundation for further research and practical improvements in the stability of grid-connected converter systems.

Acknowledgments
This work was supported by China Southern Power Grid Co., Ltd.' science and technology program (GDKJXM20231008).

References

[1] S. Chen, J. Yao, Y. Liu, et al, "Coupling Mechanism Analysis and Transient Stability Assessment for Multi-paralleled Wind Farms During LVRT," *IEEE Trans. Sustainable Energy*, vol. 12, no. 4, pp. 2132-2145, 2021.
[2] X. He, H. Geng, "PLL Synchronization Stability of Grid-Connected Multi-converter Systems," *IEEE Trans. Industry Applications*, vol. 58, no. 1, pp. 830-842, Jan.-Feb. 2022.
[3] X. Yi, Y. Peng, Q.Zhou, et al, "Transient Synchronization Stability Analysis and Enhancement of Paralleled Converters Considering Different Current Injection Strategies", *IEEE Trans. Sustainable Energy*, vol.13, no.4, pp.1957-1968, 2022.
[4] S. Huang, J. Yao, Y. Lin, et al, "Interaction Mechanism Analysis and Synchronization Stability Control for Paralleled PLL-VSCs System during LVRT", *2023 IEEE International Conference on Power Science and Technology* (ICPST), pp.426-431, 2023.
[5] X. Fu, M. Huang, S. Pan, et al, "Cascading Synchronization Instability in Multi-VSC Grid-Connected System", *IEEE Trans. Power Electronics*, vol.37, no.7, pp.7572-7576, 2022.
[6] X. Fu, M. Huang, C. K. Tse, et al, "Synchronization Stability of Grid-Following VSC Considering Interactions of Inner Current Loop and Parallel-Connected Converters," IEEE *Trans.* Smart Grid, Early Access.
[7] Z. Wang, L. Guo, X. Li, et al, "Transient Synchronization Stability of Grid-Tied Multi-VSCs System Considering Nonlinear Damping and Transient Interactions," *IEEE Trans. Power Electronics*, vol. 39, no. 12, pp. 16775-16791, Dec. 2024.
[8] X. Wang, H. Wu, X. Wang, et al, "Transient Stability Analysis of Grid-Following VSCs Considering Voltage-Dependent Current Injection During Fault Ride-Through," *IEEE Trans. Energy Conversion*, vol. 37, no. 4, pp. 2749-2760, Dec. 2022.
[9] Q. Hu, F. Ji, F. Ma, et al, "Matching Analysis of LVRT Grid Code and Injection Current Dependent Voltage Response of WTC Connected to High Impedance AC Grid," *IEEE Trans. Energy Conversion*, vol. 37, no. 3, pp. 2236-2239, Sept. 2022.
[10] Requirements for Offshore Grid Connections in the Grid of TenneT TSO GmbH, 21st December 2012, Tennet.
[11] X. Zhou, L. Guo, X. Li, et al, "Transient Stability Analysis of Renewable Power Generations via VSC-HVDC," *IEEE Trans. Industrial Electronics*, Early Access.

Research on the Identification of Incorrect Wiring of Energy Meter Based on YOLO Modeling

Aihua Chen[1*], Qinghua Li[1], Junyi Liu[1], Ziyuan Pan[1], Jinrong Huang[1], Yu Lin[1], Jiawei Ou[1], Xiaoqi Xia[2], Yingqi Liang[3]

[1] Zhongshan Power Supply Bureau of Guangdong Power Grid Co., Ltd, Zhongshan, Guangdong Province, 528400, P.R. China

[2] Guangdong Power Grid Co., Ltd, Guangzhou, Guangdong Province, 510000, P.R. China

[3] Electric Power Dispatching and Control Center of Guangdong Power Grid Co., Ltd, Guangzhou, Guangdong Province, 510000, P.R. China

[*]Corresponding author's e-mail: chenaihua188@163.com

Abstract. To address the growing demand for smart grid development and enhance the efficiency of energy meter wiring detection in grid metering operations, as well as to overcome the inefficiencies and human errors inherent in traditional manual wiring inspections, this paper proposes an intelligent recognition approach for energy meter wiring based on the YOLOv5 model. This paper collects on-site photos of electric energy meters and constructs and trains a dedicated electric energy meter wiring recognition model to achieve automation of wiring status detection. Experimental results demonstrate the model's strong performance in energy meter terminal recognition tasks, achieving an accuracy rate of 96%, a recall rate of 99%, and scores of 96% and 49% under the mAP50 and mAP50-95 metrics, respectively. The model is capable of efficiently and accurately identifying both the energy meter terminal number and wire sequence, significantly reducing errors that are common in manual inspections. Additionally, this paper examines the occurrence of misdetections and omissions during the detection process and proposes strategies to enhance detection accuracy by optimizing the network model's learning capabilities. The study highlights the significant potential of the YOLOv5-based energy meter wiring identification method in improving detection efficiency, reducing labor costs, and ensuring the security of power systems. This approach, particularly in the context of smart grid construction and implementation, is expected to play an increasingly vital role in the future.

1. Introduction

With the rapid development and widespread adoption of smart grids in China, the energy meter, as an essential metering device in the power system, plays a critical role in ensuring the accuracy and reliability of power consumption measurements[1]. These measurements are vital for fair energy trading, effective energy management, and the stable operation of the power system. Furthermore, the accuracy of energy consumption data impacts the reputation and credibility of power supply companies. Traditional methods of energy meter wiring inspection primarily rely on manual labor, which is time-consuming, labor-intensive, and prone to human error[2-4]. These factors contribute to difficulties in promptly detecting wiring issues, which can lead to inaccurate power measurements and potential electrical safety hazards.

The current energy meter wiring inspection typically utilizes a multimeter. While multimeters are simple and cost-effective for basic inspections, they face significant limitations when dealing with complex wiring systems, large-scale installations, or situations that demand high-precision detection. Multimeters are unable to automatically identify common wiring errors such as reverse connections or incorrect wiring sequences, and the manual operation process is cumbersome and inefficient. This is particularly problematic in multi-loop or multi-phase meters, where human error can lead to missed or misdiagnosed faults. Moreover, the dynamic detection capabilities of multimeters are insufficient for capturing potential issues in wiring configurations, and improper operation, especially in high-voltage or high-power circuits, could pose significant safety risks, including electrocution accidents. These challenges highlight the urgent need for a safer and more efficient method of identifying wiring errors in energy meters.

Recent advancements in artificial intelligence (AI), machine learning, and computer vision technologies have begun to gain traction in the power industry, offering new solutions to the complex problem of energy meter wiring identification. For instance, Wu et al.[5] utilizedmachine learningto identify the state of wiring in energy meters, thereby advancing technological support for energy metering fieldwork. This paper aims to develop a large-scale model for energy meter wiring identification, facilitating automated and intelligent detection. The goal is to significantly enhance detection efficiency, improve accuracy, reduce labor costs, and ensure the safety and reliability of the power system.

YOLO (You Only Look Once) is a deep learning-based object detection model renowned for its efficiency and accuracy in real-time detection tasks. The fundamental approach of YOLO involves transforming object detection into a regression problem, where all target objects and their respective locations are predicted in a single forward pass. Over successive versions, the YOLO model has undergone continuous refinement, enhancing its detection accuracy, processing speed, and adaptability to diverse applications. Its key attributes—high efficiency, rapid processing, optimized accuracy, flexibility, and ease of use—make YOLO well-suited for real-time image recognition tasks across various domains.

The YOLO model has been successfully applied in several industries. Raza et al. [6] employed YOLOv5 to build a collision avoidance system for vehicles in foggy conditions, effectively preventing traffic accidents; Dewi et al. [7] developed a traffic sign recognition system with a 99% accuracy rate, designed for autonomous driving applications; Adibhatla et al. [8] utilized YOLOv5 for defect detection in printed circuit boards, significantly improving inspection efficiency; Liu et al. [9] integrated YOLO with a power grid insulator detection system, enhancing power grid patrol efficiency; Gao et al. [10] incorporated YOLOv5s into a robotic tomato picking system, achieving a 92% detection accuracy for tomato localization; and Lu et al. [11] combined YOLO with Generative Adversarial Networks (GANs) to enhance image resolution, thereby improving obstacle avoidance capabilities in unmanned aerial vehicles (UAVs). Ren et al. [12] proposed the MRS-YOLO model to accelerate garbage sorting, achieving a mAP50 of 74.5%. These successful applications demonstrate the versatility and effectiveness of YOLO in various fields, making it an ideal choice for the intelligent recognition of energy meter wiring in this paper. At present, the wiring detection of electric energy meters mainly relies on manual experience or the use of multimeters for manual line-by-line inspection. Although these traditional methods can complete the judgment of wiring status to a certain extent, they are less convenient and the detection process takes a long time. Since the operation may involve live work, there is a high safety risk. In addition, manual inspection is easily affected by experience level and subjective judgment, and it is difficult to ensure the consistency and accuracy of inspection results. This method is particularly limited when facing complex or large-scale power equipment wiring scenarios. In contrast, the intelligent identification method based on YOLOv5 proposed in this article uses deep learning technology to realize automatic detection of the wiring status of electric energy meters. It not only greatly improves the detection efficiency, but also significantly improves safety and accuracy, providing a basis for smart grids. The construction provides a more efficient and reliable technical solution.

2. Materials and methods

2.1. Data processing

The field image collection of energy meters was conducted in metering cabinets in Zhongshan City, Guangdong Province, resulting in 1,000 images of three-phase three-wire and three-phase four-wire energy meters. After manual screening, 500 images with complete features were selected. To enhance dataset diversity and model generalization, various data augmentation techniques were applied, including geometric transformations and pixel-level processing. Geometric transformations involved random flipping (horizontal and vertical), rotation (±15°), scaling (0.8 to 1.2), and cropping (up to 20% of the image area) to simulate variations in orientation, distance, and localized features of the energy meters. Pixel-level processing techniques included Gaussian noise injection (mean 0, variance 0.01) to simulate noise interference, random erasure (occluding up to 10% of the image area) to mimic partial information loss, and brightness adjustments (factor range of 0.8–1.2) to address challenges caused by inconsistent lighting conditions. These augmentation strategies expanded the dataset to include 980 three-phase three-wire images and 620 three-phase four-wire images, totaling 1,500 samples. The dataset was split into a 9:1 ratio, with 1,200 images used for training and 300 images for validation and testing.

Annotations were conducted using the labelimg tool to label the wires and their corresponding connection points on the energy meter terminals, focusing on terminal numbers. The wires were labeled sequentially from left to right as AIi, AU, AIo, BIi, BU, BIo, CIi, CU, CIo, and UZ, and the annotations were saved in .txt format for seamless integration into the training pipeline. The annotated images were then categorized into three-phase three-wire and three-phase four-wire groups to structure the dataset. Subsequently, the YOLOv5 model was employed to train the network using the prepared dataset, enabling automated recognition of energy meter wiring states. This methodology effectively addresses the challenges in traditional manual inspection and provides a robust foundation for efficient and accurate wiring state detection in energy meters.

2.2. Evaluation indicators

In order to evaluate the energy meter miswiring detection model, this paper chooses accuracy (B), recall (R) and mean accuracy (mAP) as evaluation indexes, where the mean accuracy is evaluated using mAP50 and mAP50-95.

2.2.1. Precision. Precision (B) is an assessment of the model's accuracy in the detection task and refers to the proportion of all boxes predicted as positive samples that are actually positive samples. That is, how many of the boxes predicted by the model as target objects are correct, with the following formula:

$$\mathrm{Precision} = \frac{TP}{TP + FP} \tag{1}$$

Where TP is the True Positives and FP is the False Positives. The higher the precision, the less false positives the model will make in its prediction.

2.2.2. Recall. Recall (R) is an assessment of the model's recall ability, which refers to the proportion of all boxes that are true for positive samples that are correctly predicted by the model:

$$\mathrm{Re}call = \frac{TP}{TP + FN} \tag{2}$$

where FN is the False Negatives. The higher the recall, the fewer target objects are missed by the model.

2.2.3. mAP50. The mAP is a commonly used evaluation metric in target detection that indicates the overall performance of the model over multiple categories. mAP50 is the mean Average Precision of the model under different categories with an IoU threshold of 0.5.

$$mAP50 = \frac{1}{N}\sum_{i=1}^{N} Ap_i \tag{3}$$

where Ap_i is the average precision of category i and N is the number of categories.

2.2.4. mAP50-95. The mAP50-95 is the mean accuracy average calculated over multiple IoU thresholds (from 0.5 to 0.95, in 0.05 increments). This assessment is more rigorous than mAP50 because the formula takes into account prediction frames with different degrees of overlap to comprehensively assess the accuracy of the model.

$$mAP50-95 = \frac{1}{10}\sum_{i=1}^{10} Ap_{IoU} \tag{4}$$

Where Ap_{IoU} is the average accuracy calculated at a given IoU threshold, i from 1 to 10, representing a change in IoU threshold from 0.5 to 0.95.

2.3. Algorithmic framework

The intelligent recognition method for energy meter wiring proposed in this paper combines YOLOv5 and IJ (Intelligent Judgement) modules to form an efficient and intelligent wiring detection system. The whole framework is divided into two main parts: the YOLOv5 module and the IJ module, each of which plays a different but crucial role in energy meter wiring identification, and both of them work together to realize the automated identification and intelligent judgment of energy meter wiring.

First, the YOLOv5 module, as the core part of the deep learning target detection technology, is responsible for feature extraction and analysis of the input wiring image of the energy meter.The YOLOv5 model can accurately identify the wiring location of the energy meter through the detailed processing of the wiring image by convolutional neural network (CNN). The module not only automatically recognizes the location information of the terminal blocks, but also assigns a unique label to each wiring. These labels include the number of the terminal and cable classification information, which can not only clearly mark the location of the terminal, but also clearly identify the terminal connected to the specific line type. This process greatly improves the automation level of image analysis and avoids the problem of misjudgment or omission caused by human negligence in traditional manual inspection. Through the YOLOv5 module, the system can quickly complete the initial extraction from the image input to the wiring location and cable identification, providing accurate data support for subsequent intelligent judgment.

After the YOLOv5 module completes the initial identification, the IJ module further plays its role of intelligent judgment. the core task of the IJ module is to process and analyze the results of the wiring image annotation output from the YOLOv5 module, to ensure that the wiring conforms to the standardized three-phase four-wire wiring sequence. Specifically, the IJ module analyzes each bounding box coordinate output by YOLOv5 and sorts and matches the wiring identifiers according to the preset wiring rules (e.g., AIi, AU, AIo, BIi, BU, BIo, CIi, CU, CIo, UZ and other wiring sequences). Through this process, the IJ module is able to accurately verify that the wiring is in accordance with the specified order and specifications, so that wiring errors or omissions can be detected in a timely manner. The IJ module not only recognizes errors in wiring, but also categorizes them according to different types of errors. In addition, the system will also trigger an audio alarm to remind the operator to correct the operation in time. This intelligent feedback mechanism not only improves the efficiency of the operator, but also significantly reduces the wiring errors caused by human negligence, thus effectively guaranteeing the security and reliability of the power system.

Through this efficient workflow, the combination of YOLOv5 and the IJ module not only improves the accuracy of energy meter wiring identification, but also effectively shortens the time for wiring inspection, thus dramatically improving overall work efficiency. In addition, the system is able to detect and feedback wiring errors in real time, which enables the operator to correct potential problems in time during the work process and reduces the risk of system failures and safety hazards.

In summary, the intelligent identification method of energy meter wiring based on YOLOv5 and IJ module proposed in this paper has significant advantages, which not only improves the accuracy and efficiency of wiring inspection, but also provides reliable technical support for the construction and application of smart grid. With the development and application of intelligent technology, the system will be further expanded to more fields in the future, helping the power industry to innovate and develop in digitalization, automation and intelligence.

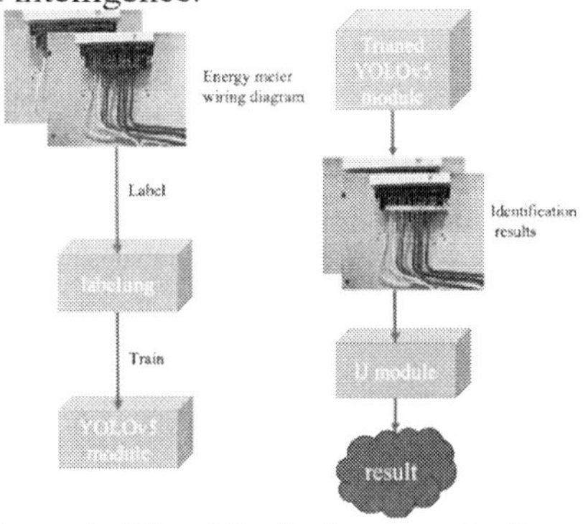

Figure 1. Algorithmic framework diagram

2.4. Model training

The computer configuration used in this experiment is as follows: the processor is Intel(R) Xeon(R) CPU E5-2698 v3 with 2.30GHz; the graphics card is NVIDIA GeForce RTX 2080 Ti; the operating system is Linux; the deep learning framework is PyTorch 1.13.1, and the Python version is 3.8.

Before training, set the paths to the training set and test set folders of the yaml file, set the batchsize to 16, and train for a total of 500 epochs. After training, the model converges at 252 epochs with box loss, cls loss, and dfl loss, as shown in Figure 2.

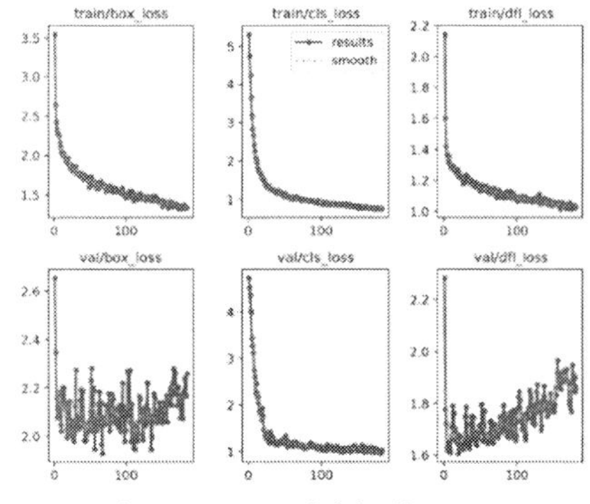

Figure 2. Model indicators

After 252 epochs of training, each loss tends to converge, at the same time, according to the Precision-Confidence Curve in Figure 3, it can be seen that the curve decreases slowly, so the accuracy of the model can also achieve a desirable effect.

Journal of Physics: Conference Series **3012** (2025) 012026 doi:10.1088/1742-6596/3012/1/012026

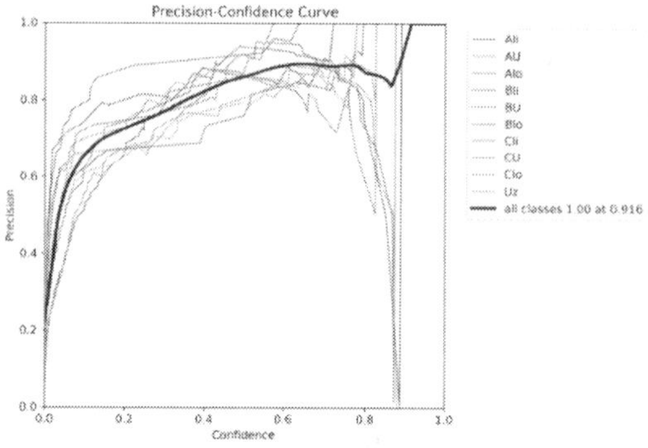

Figure 3. Precision-Confidence Curve

3. Experimental results and analysis

The three-phase four-wire system is a commonly employed configuration for providing a neutral line, predominantly used in residential, commercial, and small to medium-sized industrial applications. This system enables the simultaneous supply of three-phase power and a neutral line, making it highly suitable for most standard power distribution systems. Consequently, a significant portion of grid metering operations focuses on the three-phase four-wire energy meter, which is the primary subject of this paper's investigation into intelligent recognition of wiring errors in such meters.

The three-phase four-wire system consists of three conductors—Phase A, Phase B, and Phase C—and a neutral conductor (N). The operational principle of this system involves measuring the phase voltage (the voltage relative to the neutral wire) and the corresponding phase current for each of the three phases. The electrical energy for each phase is then calculated, and the total electrical energy is obtained by summing the energy of all three phases. The three-phase four-wire energy meter uses three individual power elements to independently measure the power of each phase. This enables the accurate calculation of total power, even in the event of an unbalanced load across the phases. The key advantages of this system include its ability to support both single-phase and three-phase loads, its adaptability to unbalanced load conditions, and its high measurement accuracy. As a result, it is widely utilized in residential and commercial power systems, as well as in small and medium-sized industrial applications. Figure 4 illustrates the physical wiring diagram of a three-phase four-wire energy meter.

Figure 4. Three-phase four-wire model energy meter wiring physical drawing

3.1. Model training results

3.1.1. Subjective evaluation. After the model is fully trained, the validation set is input to the trained model for testing, and its accuracy can reach 96%, and in the 100 test sets, 95 pictures of energy meter wiring can be accurately recognized, including all ten wires and the wire sequence of the ten wires. As shown in Figure 5:

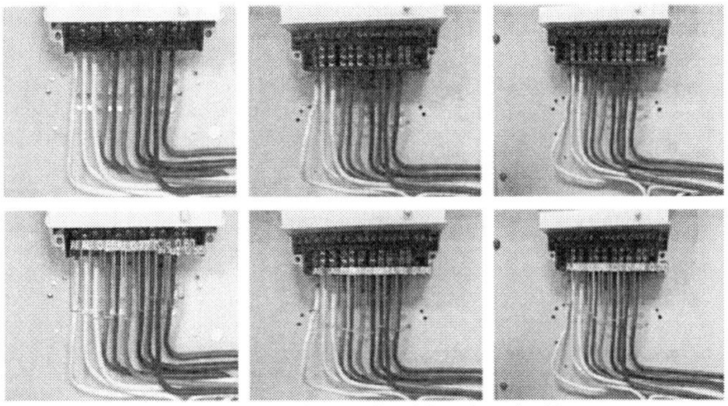
Figure 5. Evaluation of subjective indicators

As shown in Figure 5, the first row is the original picture, and the second row is the picture after recognition by yolov5, it can be seen that the algorithm is able to accurately identify the location of each line, and while accurately and without error framed, it is able to match the markings on the line as well as the numbering on the terminal block, thus accurately identifying the line sequence represented by the 10 line locks.

3.1.2. Evaluation of objective indicators. In order to evaluate the model performance more objectively, this paper takes precision, recall, mAP50, and mAP50-95 to rate the model, and the results are shown in Table 1.

Table 1. Evaluation table of objective indicators

	Precision/%	Recall/%	mAP50/%	mAP50-95/%
Yolov5	96	99	96	49

From the table, it is easy to see that the YOLOv5 algorithm performs well in the task of visual algorithm for energy meter terminal recognition, in which the accuracy (Precision) can reach 96%, which indicates that 96% of the boxes predicted by the model are correct, i.e., the model's misdiagnosis rate is low, and the prediction result is of high quality, so the YOLOv5 algorithm is able to identify very accurately the terminal number and identify ten wires respectively. The recall rate (Recall) can reach 99%, indicating that among all the real terminal block targets, 99% of them are correctly recognized by the model, and the model has a very low miss detection rate. In the mAP50 metric, the algorithm is able to reach 96%, indicating that the model is more effective in detecting when the IoU threshold is 0.5, while in the more stringent mAP50-95 metric, the algorithm is able to reach 49%, indicating that with the increase of the IoU threshold, the model is still capable of strong detection, but with a slight decrease in precision.

3.2. Misdetection and leakage analysis

In order to provide guidance for the practical application of subsequent models and optimize the performance of the algorithm, this paper conducts an in-depth analysis of the false detection and missed detection phenomena of the YOLOv5 model during the detection process. As shown in Figure 6, the current false detection of the algorithm is mainly manifested in the repeated identification of wires as multiple targets. The main reasons for this situation include overlapping wires, bending wires, and tilted shooting angles caused by the high position of the meter, which makes it difficult for the model to accurately distinguish wire targets. In addition, poor lighting conditions and complex background interference further aggravate the false detection and missed detection problems of the model. In order to solve the above problems, future work will introduce more advanced image enhancement technology to improve the impact of shooting angles and insufficient lighting on detection results, and at the same time optimize the feature extraction ability of the network model so that it can more efficiently learn and identify the effective features of the electricity meter wiring in various complex environments, thereby improving the overall recognition accuracy and robustness.

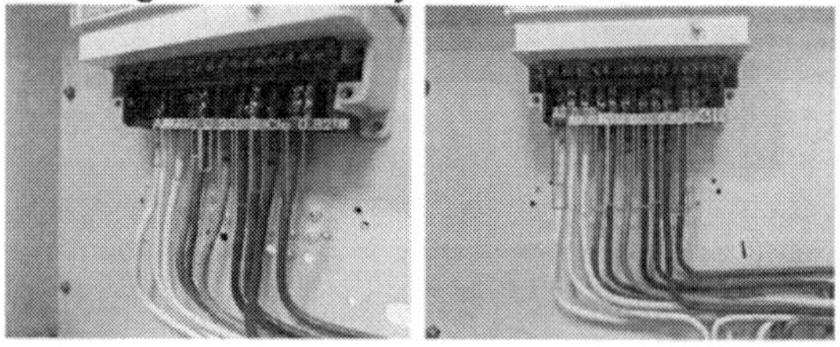

Figure 6. Diagram of the analysis of the situation of wrong and missed inspections

4. Conclusion

The rapid development of China's smart grid highlights the crucial role of energy meters in modern power systems. As core equipment for energy metering, their accuracy and reliability are essential for fair trade in the energy market, effective energy management, and grid stability. However, traditional wiring inspections rely on manual labor, which is inefficient and prone to human error. This paper proposes an intelligent identification method using YOLOv5-based deep learning models to enhance automation and smart grid development.

This approach constructs and trains a model tailored for energy metering, enabling automated detection of wiring conditions through advanced image recognition. The YOLOv5 model excels in real-time detection tasks due to its efficient object detection capabilities, achieved by training on extensive datasets. It accurately identifies terminal block positions, wiring count, sequence, and evaluates wiring configuration's correctness.

Experiments demonstrate the model's high performance: Precision of 96%, Recall of 99%, mAP scores of 96% for mAP50 and 49% for mAP50-95. These metrics showcase its exceptional capability in energy meter wiring recognition, potential applications, and robustness.

To address the limitations of conventional methods, the YOLOv5-based approach reduces manual inspection tasks, minimizes human error, and ensures wiring follows standards. The model optimizes wire sequence detection by automating terminal numbering identification, improving process efficiency and providing reliable data for management.

While the model performs well in theory, issues like misdetection remain. Analysis reveals errors from multiple identifications of wires in complex environments, such as those with overlapping terminals or varied backgrounds. Future research could focus on enhancing network architectures and training robustness, potentially using deeper models or multi-scale approaches to improve performance in diverse scenarios.

Overall, the YOLOv5-based method offers significant advantages for energy meter automation and smart grid development, reducing labor costs and ensuring safety and reliability. Its potential expansion into other power equipment tasks aligns with broader grid automation efforts, poised to play a key role in future smart grid evolution.

References

[1] Adibhatla V A, Chih H C, Hsu C C, Cheng J, Abbod M F and Shieh J S 2021 Applying deep learning to defect detection in printed circuit boards via a newest model of you-only-look-once Mathematical Biosciences and Engineering vol 18 p 4411-4428.

[2] Dewi C, Chen R C and Yu H 2020 A Weight analysis for various prohibitory sign detection and recognition using deep learning Multimedia Tools and Applications vol 79 p 32897-915.

[3] Gao G H, Shuai C Y, Wang S Y and Ding T 2024 Using improved YOLO V5s to recognize tomatoes in a continuous working environment Signal Image and Video Processing vol 18 p 4019-4028.

[4] Zhang J, Zheng K, Zhou F, Du J, Xiao J, Hu J and Tan S 2021 Intelligent investigating technology for abnormal operation of wiring terminals of electric energy metering device based on deep

learning 2021 IEEE 2nd International Conference on Information Technology, Big Data and Artificial Intelligence (ICIBA) p 859-864.

[5] Wu D, Jiang J, Li J and Chen L 2024 A Least Squares Dual Support Vector Machine Based Algorithm for Identifying Abnormal Behavior in Energy Meter Installation and Removal Field Operations 2024 International Conference on Artificial Intelligence and Power Systems (AIPS) p 434-437.

[6] Liu C Y, Wu Y Q, Liu J J, Sun Z and Xu H J 2021 Insulator Faults Detection in Aerial Images from High-Voltage Transmission Lines Based on Deep Learning Model Applied Sciences-Basel vol 11 p 20.

[7] Lu Q, Guo L and Sun X 2024 Application of improved deep learning algorithm in obstacle avoidance of UAV Journal of Physics: Conference Series p 7-19.

[8] Raza N, Habib M A, Ahmad M, Abbas Q, Aldajani M B and Latif M A 2024 Efficient and Cost-Effective Vehicle Detection in Foggy Weather for Edge/Fog-Enabled Traffic Surveillance and Collision Avoidance Systems Cmc-Computers Materials & Continua vol 81 p 911-931.

[9] Ren Y M, Li Y Z and Gao X Y 2024 An MRS-YOLO Model for High-Precision Waste Detection and Classification Sensors vol 24 p 22.

[10] Shen Y, Huang R, Zhu W and Feng J 2023 Quantitative evaluation method for software quality of smart electricity meter based on analytic hierarchy process Proceedings of SPIE p 12.

[11] Xing Y, Sun Y, Dong X, Zhai X and Sun K 2022 Research on a non-stop replacement technology of electric energy meter connected with standard meter 2022 IEEE Conference on Telecommunications, Optics and Computer Science (TOCS) p 1391-1394.

[12] Zhang J M, Xiao J, Liu N T, Tan S S, Li W Q and Ieee 2024 Research on the Error Analysis of Electric Energy Metering Devices in AC Power Transmission Systems In 6th Asia Energy and Electrical Engineering Symposium (AEEES) p 88-92.

Journal of Physics: Conference Series 3012 (2025) 012027 doi:10.1088/1742-6596/3012/1/012027

Low-cost Flexible Interconnection Technology for a new Type of Magnetic Control Switch based on Controllable Magnetic Flux

Wei Luo[1], ZheYu Zhang[2*], MingLei Xie[3], JinFeng Wang[4],WanLing Du[5], YuanZhe Zhu[6],Yan Le[7]

[1] Meizhou Power Supply Bureau of Guangdong Power Grid., No. 48, Binfang Avenue,Kuangzhou 514000,P.R,China.

[2] Hubei University of Technology., No. 28, Nanli Rd,Wuhan 430068,P.R,China.

[3] China Southern Power Grid Research Institute Co., Ltd., No. 11, Kexiang Rd,Guangzhou 510700,P.R,China.

[4] Electric Power Research Institute of Guangdong Power Grid.,Guangzhou,Guangdong,Province, 510000,P.R.China.

[5] Electric Power Research Institute of Guangdong Power Grid., Guangzhou,Guangdong,Province, 510000,P.R.China.

[6] Meizhou Power Supply Bureau of Guangdong Power Grid., No. 48, Binfang Avenue,Kuangzhou 514000,P.R,China.

[7] Huazhong University of Science and Technology., No. 1037, Luoyu Road,Wuhan 430074,P.R,China.

*Corresponding author's e-mail: 17764266905@163.com

Abstract. For the use of new energy power distribution systems, a novel kind of flexible interconnection technology for magnetic control switches based on controllable magnetic flux is suggested. This technology is inexpensive. A linear transformer with several secondary windings is used in the technology. When the secondary winding is connected to the inverter, its primary winding is connected in series with the harmonic-producing load. through the use of an inverter with a fixed compensation coefficient to track the primary fundamental wave current after it has been detected. The transformer exhibits a low impedance characteristic with respect to the fundamental wave current and a high impedance characteristic with regard to the harmonic current after the fundamental frequency component current compensation condition is met. The passive filtering branch will receive the power system's harmonic current in this manner. The working principle and control scheme of a novel kind of high-performance, maintenance-free, and cost-effective control strategy for magnetic control switches are also thoroughly explained. Lastly, experimental results and simulated waveforms are presented to show how effective this technology is.

1. Background

In recent years , new energy consumption becomes a big challenge in many countries, this is particularly prominent in China[1]. Meanwhile, with the continuous progress of new energy technologies and the gradual reduction of costs, the proportion of new energy power generation in the power system is getting higher and higher. Microgrid has good environmental and economical benefits and has attracted more and more attentions of power researchers[2]-[4]. New energy forms such as solar photovoltaic-based power generation and wind-driven power generation. are being connected to the distribution grid in large quantities, and the scale of connection continues to expand. On the one hand, this provides a new way to meet the ever-increasing energy demands. On the other hand, it also creates new challenges for the planning, operation, and management of the distribution grid. Among them, the differences in the characteristics of new energy compared to traditional power grids may cause problems such as harmonics and voltage fluctuations to the power grid. The widespread use of power electronic devices and nonlinear loads has deteriorated power quality in recent years because of its harmonic contamination and low power factor phenomenon[5]-[7]. Flexible interconnection devices can provide a good grid-connected interface to achieve efficient connection. However, in the past, the need for flexible interconnection in the distribution grid was not extremely urgent. Most flexible interconnection devices adopted the topology and control strategies of back-to-back power electronic inverters. Traditional switching equipment is expensive, which increases the construction and operation costs of the power system, and its operating speed is relatively slow, making it challenging to fulfill of rapid switching. In the event of a power system failure, it may lead to the expansion of the fault range and affect the reliability of the power system. Meanwhile, the connection method is relatively fixed, making it difficult to achieve flexible adjustments to the topology of the power system.

In today's power field, with the ongoing expansion of electricity demand and developing requirements for power supply reliability, and meanwhile, in order to solve above-mentioned issue, it is need to have some new technologies to combine and integrate the available systems to optimize energy systems[8].This paper proposes a new type of low-cost flexible interconnection technology of magnetic control switches based on controllable magnetic flux. The main parts of this technology are the hybrid active power filter of the injection circuit and the series active power filter with fundamental magnetic flux compensation (FMFC). In the active power filter with FMFC, only the fundamental current is detected and tracked, and the bandwidth requirement of the control system is low and the control scheme is easy[5]-[7]. Therefore, this technology can achieve rapid switching actions with a short response time, thus meeting the requirements of rapid switching in the power system. It can reduce the energy loss during the switching process and improve the efficiency of the device. Meanwhile, by controlling the magnitude and direction of the magnetic flux, precise control over the switching state can be achieved, enhancing the flexibility and reliability of the device. Traditional voltage - source converters (VSC) or hard - switching topologies rely on high - frequency and high - voltage - withstand power devices (such as SiC MOSFET, IGBT modules), and the cost accounts for as high as 30% - 50%. By achieving soft - switching (ZVS/ZCS) through controllable magnetic flux, the switching loss is reduced, allowing the use of low - frequency and low - voltage - withstand silicon - based IGBT or MOSFET, and the device cost is reduced by more than 40%. Traditional high - frequency transformers and inductors need to use high - cost ferrite or nanocrystalline materials, and they are large in size and high in loss. The controllable magnetic flux technology reduces the risk of core saturation by dynamically adjusting the magnetic circuit, allowing the use of low - cost silicon steel sheets or amorphous alloys, and the material cost is reduced by 30% - 50%. Traditional multi - level converters (such as MMC) require a large number of sub - modules (including capacitors and switching devices), which have high costs and complex control. The new technology replaces some power electronic switches through magnetic flux modulation, reduces hardware complexity, and the soft - switching technology reduces losses, thus reducing the heat dissipation requirements. Therefore, it can achieve low cost.

In recent years, with the rapid development of new energy and the progress of smart grid construction, the demand for flexible and efficient power control technologies in the power system has been continuously increasing. As an innovative power control technology, the flexible interconnection technology of the new type of magnetic control switch with controllable magnetic flux is gradually becoming a research hotspot in the power field.

Drawing on the outcomes of previous pertinent research, this paper focuses on the adjustable reactor with controllable magnetic flux and puts forward a scheme of a flexible interconnection device applied between distribution networks. Meanwhile, the principle of this scheme is also introduced. Eventually, through simulation means, the effectiveness and rationality of this scheme are verified.

2. Theory

This chapter examines the principle of the transformer when the main magnetic flux is controllable in a multifunctional power quality controller. Specifically, the harmonic suppression principle and voltage deviation theory of a reactor based on a controllable magnetic field were analyzed. A topology diagram of the multifunctional power quality controller was elucidated, and two special cases of this adjustable reactor were investigated. A series hybrid active power filter based on basic magnetic flux compensation can be produced by applying these two special cases of the adjustable reactor to the active power filter.

This chapter presents a comprehensive description of the novel active power filter. Upon implementation of this active power filter, only the fundamental wave is detected and tracked, and fundamental magnetic flux compensation is executed for the series transformer. Consequently, the series transformer exhibits primary-side leakage impedance to the fundamental wave and excitation impedance to harmonics. This active power filter accomplishes the harmonic isolation function by directing the harmonic currents into the passive filter branch by simply raising the system impedance to the harmonics.

2.1. Principle of Adjustable Reactor with Controllable Magnetic Flux

A transformer with an air gap serves as the foundation for the adjustable reactor, as seen in Figure 2.1. The transformer's primary winding (AX) has N_1 turns, whereas the secondary winding (ax) has N_2 turns. Consequently, this transformer's transformation ratio can be written as $K = N_1/N_2$.

Suppose this transformer is serially connected between power supply and load which is a harmonic source. At this time, the main current of the whole system will contain the fundamental current and the nth harmonic current, that is, $i_1 = i_1^{(1)} + \sum i_1^n$. Detection is carried out on the fundamental current i_1 on the primary side AX of the transformer. Then a voltage source inverter is used to generate a current i_2 by tracking this current, and i_2 is fed into the secondary side ax of the transformer after being reversed. This transformer exhibits two different situations for the fundamental wave and harmonics.

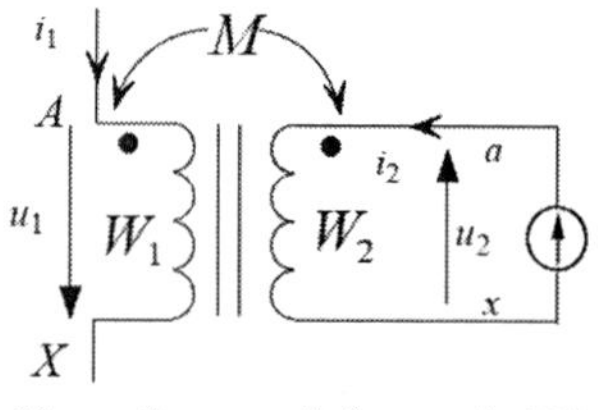

Figure 1. Transformer Schematic Diagram

2.1.1. For the fundamental wave current. The voltage equation at the transformer's two ends can be written as follows using the transformer T-type equivalent circuit.

$$U_1 = (r_1 + j\omega L_{11})I_1 + j\omega M I_2 \tag{1}$$

$$-U_2 = (r_2 + j\omega L_{22})I_2 + j\omega MI_1 \quad (2)$$

The fundamental angular frequency is denoted by ω in this case. The self-inductance of the transformer's primary side coil is denoted by L_{11}, and the self-inductance of the secondary side coil by L_{22}. M stands for the mutual inductance magnitude between the transformer's primary and secondary side coils. Thus, the formula above can also be expressed in an equivalent way as:

$$U_1 = [r_1 + j\omega(L_{11} - kM)]I_1 + j\omega kM(I_1 + I_2/k) \quad (3)$$

$$-U_2 = [r_1 + j\omega(L_{22} - M/k)]I_2 + j\omega M(I_1 + I_2/k) \quad (4)$$

In the formula, $L_{11} - kM = L_{S1}$ represents the leakage inductance of the transformer's primary - side winding., $L_{22} - M/k = L_{S2}$ represents the leakage inductance of the transformer's secondary - side winding., $x_1 = \omega L_{S1}$ represents the leakage reactance of the primary side winding of the transformer, $x_2 = \omega L_{S2}$ represents the leakage reactance of the secondary side winding of the transformer, and $Z_m = j\omega kM$ represents the excitation reactance of the transformer. When the current injected into the secondary side of the transformer satisfies $I_1 + I_2/k = 0$, we can see from the T-type equivalent circuit of the transformer that the current flowing through the excitation impedance branch of the transformer is zero, and there is no voltage difference across the excitation impedance branch of the T-type equivalent circuit, which is equivalent to being short-circuited. The basic side's equiv impedance of this transformer, as seen from the primary side, is currently Z_1, and Z_1 is the leakage impedance of the transformer's primary side winding. Similarly, when we look from the residual side, the equivalent impedance of the residual side of this transformer is Z_2, and Z_2 is leakage impedance of the winding on the residual side of the transformer. At this time, the equivalent circuit of the series transformer is shown in Figure 2. Below.

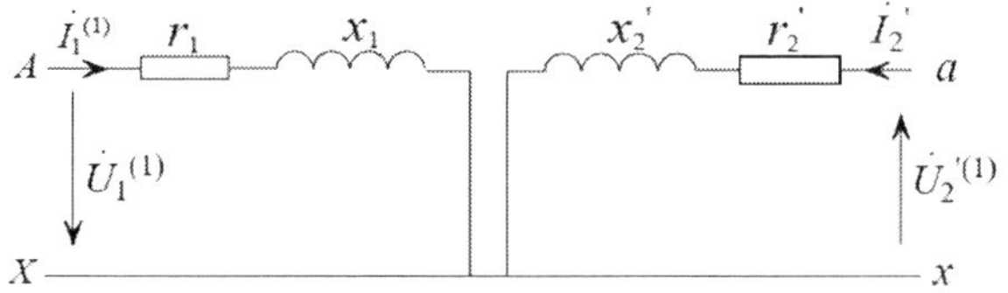

Figure 2. Fundamental Wave Equivalent

2.1.2. For the nth harmonic current. Because the inverter only produces the fundamental current that is connected to the transformer's secondary side winding, there is no harmonic current present in the secondary side current. At this moment, the transformer's secondary winding side is equal to an unloaded condition for the nth harmonic, and the secondary side's harmonic current is zero. At this point, the expression of U_1 can be derived using the transformer's T-type equivalent circuit. The transformer's equivalent impedance, as viewed from the primary side AX, is $Z_n = Z_1 + Z_m$. Figure 3. currently displays the transformer's equivalent circuit.

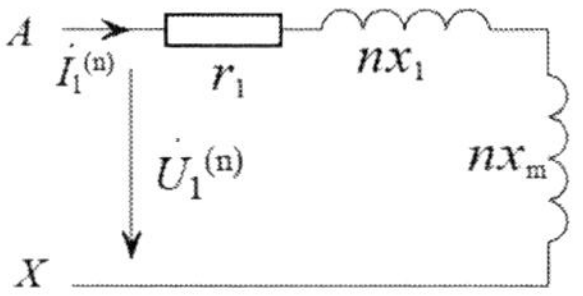

Figure 3. Harmonic equivalent circuit

According to the analysis above, the impedance displayed by the transformer connected in series in the circuit for the fundamental component will be very low (that is, the leakage impedance Z_1 on the primary side) when the fundamental current generated by the inverter and injected into the secondary side of the transformer and the fundamental current on the primary side of the transformer satisfy a certain relationship. However, the impedance displayed by the transformer for the n-th harmonic component will become very high (that is, the excitation impedance $Z_m$along with the leakage impedanceZ_1). We all know that, in general, a transformer contains two types of impedance: the magnetizing impedance and the short-circuit impedance. The sum of $Z_1$and $Z_2$is the short-circuit impedance. Generally speaking, the short-circuit impedance is substantially lower than the magnetizing impedance, and the leakage impedance is nearly as low. During the transformer design process, the magnetizing impedance is often greater than 100 times the leakage impedance. Therefore, when facing the n - th harmonic component, the harmonic impedance is more than 100 times larger than the leakage impedance. Therefore, we can appropriately change the control parameters to increase the harmonic impedance of the transformer instead of the fundamental impedance, thereby achieving harmonic isolation. Due to the different impedances exhibited for the harmonic and the fundamental respectively, the passive LC filter branch will have to accept the harmonic., thereby achieving better harmonic isolation.

3. System and Control Design

During the process of power flow control, the complex power output by the power sources on both sides of the feeder contains two components. One is the part determined by the combined action of branch loads and line impedance, and this part is defined as the inherent power of the line. The other portion, known as the circulating power, is established by the line impedance and the voltage differential between the power sources on either side. Examine the dual-power closed-loop distribution network's power flow control law in more detail using the dual-power distribution ring network's power flow distribution law as a guide.

Taking the feeder on side A as an example, the complex power output by the power source on side A can be expressed as

$$S_A = S_{stA} + S_{loopA} \tag{5}$$

Suppose the complex power of the load on side A isS_{Ad}, then the complex power transmitted between the feeder on side A and the feeder on side B is

$$S_{tr} = S_A - S_{Ad} \tag{6}$$

Changing the output magnitude of the power sources of the feeders on both sides, that is altering the magnitude of S_{tr} through regulation is the goal of power flow control in the dual-power distribution ring network. S_{stA} or S_{loopA} must be altered when S_{Ad} stays constant. As previously discussed, branch loads or line impedance can be used to changeS_{stA}, whereas line impedance or the voltage differential between the power sources on either side can be used to changeS_{loopA}.

The above analysis shows that by controlling the line impedance of the dual-power distribution ring network, the inherent power and the circulating power of the line can be changed simultaneously, thus realizing the regulation of the output power of the power sources at both ends and achieving the power flow control goal under the steady state of the system. Therefore, configuring variable impedance devices within the ring network can effectively realize the power flow control function of the ring network.

Based on variable reactance, flexible interconnection devices can be separated into two categories, namely series type and parallel type, according to the deployment position of variable reactance devices. Specifically, if the variable reactance is directly installed between the feeders, this connection mode is called parallel connection; on the contrary, if the variable reactance is directly arranged at the head end of a certain feeder and there is direct interconnection between the feeders, then this connection mode is called series connection. Figure 4(a)-(b). shows their respective connection modes.

Journal of Physics: Conference Series **3012** (2025) 012027 doi:10.1088/1742-6596/3012/1/012027

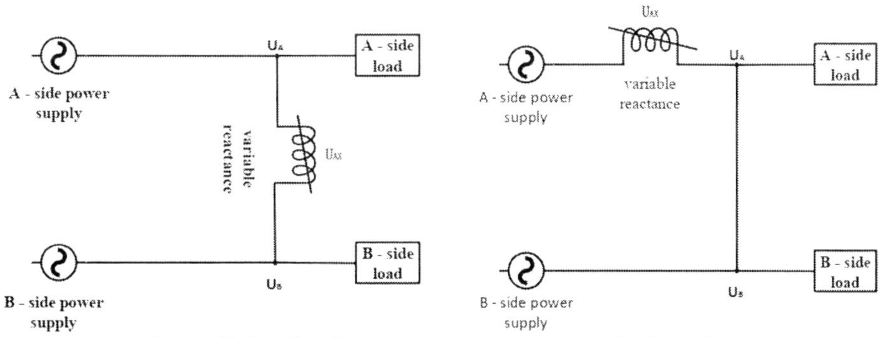

(a) parallel – circuit (b) in series

Figure 4. Parallel and series circuit diagrams

4. Simulation

To help prove the efficacy of the new flexible interconnected device with flux-controllable magnetic switch, based on the above theory, simulation verification was carried out using MATLAB. The flux-controllable reactor was connected in series at the feeder line's head end. Through the flux-controllable reactor, the current and power of the feeder line were adjusted after the load changed. Through fluctuation situation of the load, the feeder line was adjusted indirectly and a rapid and precise response was made.

Figure 5(a)-(e) shows the voltage that the power supply output, steady - state load voltage, steady - state load voltage, the current after the load change, the voltage after the load changes. The output current of the power supply on the feeder side and its magnitude will also change accordingly when the load fluctuates. Figure 5(f) shows the current waveforms on both sides of AX and ax. When the load fluctuates and the flux-controllable reactor is in different states, the current on the ax side can accurately and rapidly follow the current on the AX side according to the corresponding control instructions.

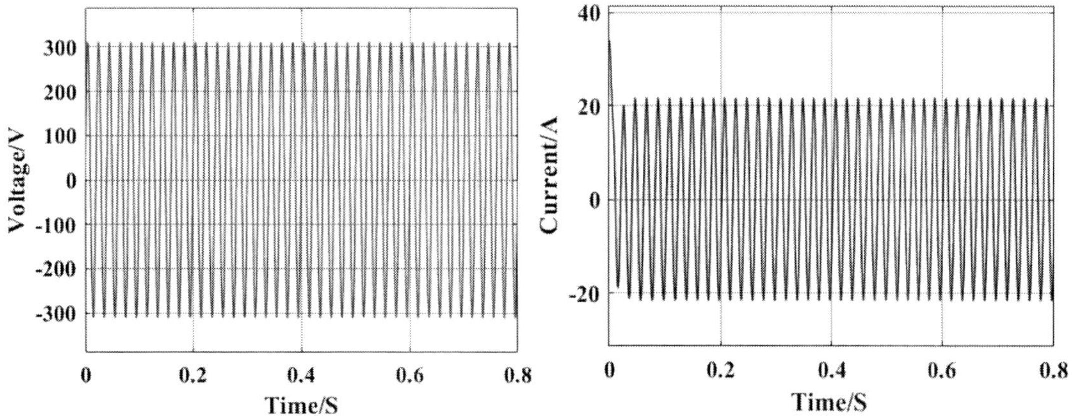

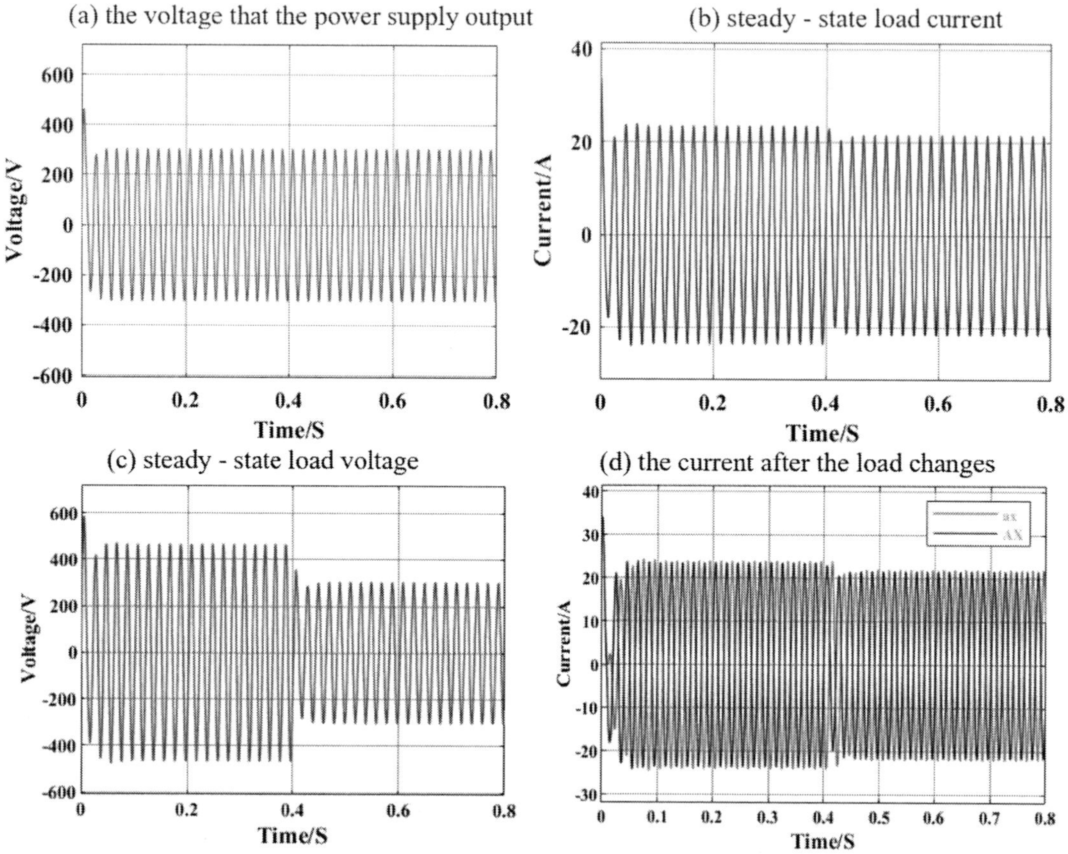

(e) the voltage after the load changes (f) current on both sides of AX and ax

Figure 5. Waveform of steady-state power flow regulation

5. Conclusion

This paper proposes a new type of flexible interconnection device with magnetic control switch based on controllable magnetic flux. In view of the characteristics of transformers, a flexible interconnection with controllable magnetic flux is achieved. By connecting this reactor in series or in parallel between feeders, under this method, we can dynamically and continuously change the active power flow between feeders by changing the equivalent impedance value of the working winding of the adjustable reactor in real time.

Therefore, we can regard the variable reactance value as a parameter characterizing the connectivity between feeders in the distribution ring network, and realize the flexible interconnection between feeders through continuous and real-time control of this parameter.

Project fund

Supported by the Southern Power Grid Science and Technology Project in China (Project No. 031400KC23120011).

References

[1] J. Zhu, P. Xie, P. Xuan, J. Zou and P. Yu, "Renewable energy consumption technology under energy internet environment," *2017 IEEE Conference on Energy Internet and Energy System Integration (EI2)*, Beijing, China, 2017, pp. 1-5.

[2] C. Marnay and O. Bailey, "The CERTS microgrid and the future of the macrogrid," Lawrence Berkeley National Laboratory, Berkeley, CA, USA, Rep. LBNL-55281, Aug. 2004.

[3] R. H. Lasseter et al., Integration of Distributed Energy Resources: The CERTS Microgrid Concept. Berkeley, CA, USA: Consortium for Electric Reliability Technology Solutions, 2002.

[4] C. Marnay, F. J. Robio, and A. S. Siddiqui, "Shape of the microgrid," inProc. IEEE Power Eng. Soc. Winter Meet., 2001, vol. 1, pp. 150–153.

[5] H. Akagi, E. H. Watanabe, and M. Aredes, Instantaneous Power Theory and Applications to Power Conditioning. Piscataway, NJ, USA: IEEE Press, 2007, pp. 1–17.

[6] A. Baggini, Handbook of Power Quality. New York, NY, USA: Wiley, 2008, pp. 187–257.

[7] R. C. Dugan, Electrical Power Systems Quality. New York, NY, USA: McGraw-Hill, 2003, pp. 167–223.

[8] C. Xing, W. Renbo, T. Jiapeng, H. Chenhui, Z. Yun and L. Zhen, "Optimal Dispatch Method of Regional Energy Internet with New Energy," *2022 5th International Conference on Energy, Electrical and Power Engineering (CEEPE)*, Chongqing, China, 2022, pp. 1293-1298.

Journal of Physics: Conference Series 3012 (2025) 012028 doi:10.1088/1742-6596/3012/1/012028

Research on the Extraction Method of Quantitative Parameters for Discharge Ultraviolet Imaging Detection of Polluted Insulators based on Digital Image Processing

Shenli li Wang*, Yong Du, Jun Wu, Baisong Li, Xiangkui He, Quan Fang

State Grid Hubei Electric Power Co., Ltd. Ultra High Voltage Company

*Corresponding author's e-mail: 17762199790@163.com

Abstract. Solar-blind ultraviolet(UV) imaging is an effective discharge detection method. To quantitatively describe discharge characteristics, a new method using digital image processing algorithms to extract image parameters for characterizing discharge characteristics is proposed. The discharge areas were extracted by threshold segmentation, morphological filtering, and small area elimination algorithms on UV images. The boundary point coordinates of each discharge area are obtained using a multi area boundary tracking algorithm, and the spot area parameter is defined to quantify the discharge intensity, and the total running time for the above image processing is 0.59 seconds. The parallel line projection algorithm in the Radon algorithm was used to obtain the spatial distribution curve of discharge intensity along the surface of the insulator.

1. Introduction

Pollution flashover accident of insulators seriously threatens the safe and stable operation of power systems. The surface of contaminated insulators is often accompanied by discharge phenomenon and radiated ultraviolet signal[1]. Solar blind ultraviolet imaging is a visual discharge detection method[2], which can detect the ultraviolet signal of 240-280nm band of discharge radiation and output ultraviolet video or image[3-4]. Compared with traditional discharge detection methods, solar-blind UV imaging can display the discharge contour directly, and has the advantages of long detection distance, high sensitivity and accurate discharge location. In recent years, ultraviolet imaging method has been widely used in the discharge detection and equipment maintenance of power system [5-6].

The detection of discharge using UV imaging method requires the extraction of relevant quantitative parameters. The UV video and image carry relevant information reflecting the discharge characteristics, such as the position, the approximate shape, the number of discharge points, and the size of the discharge area. From the working principle of UV imaging, it can be known that the discharge image in UV images is actually the discharge ionization luminescence area. Therefore, it is feasible to extract parameters from the perspective of image processing to characterize the discharge characteristics. In recent years, some scholars have conducted relevant research. For example, reference [7] divides discharge into four different stages by measuring the actual diameter of the discharge area in UV images. However, due to the fact that the shape of the discharge area is not necessarily circular, there are some limitations to measuring the diameter. Reference [8] uses the sum of grayscale values of the discharge area to characterize the discharge intensity, but does not obtain information on the position of the discharge on the insulator string.

Based on the above analysis, in order to quantify the discharge intensity, the digital grayscale transformation, binarization, morphological filtering, and small area elimination algorithms are used in this paper to segment the discharge image, and the multi area boundary tracking algorithm is sued to extract the boundary points of each area. On this basis, the parameter of spot area is defined, and the area values of each contour region are obtained. To reflect the spatial distribution characteristics of discharge intensity, the parallel line projection Radon algorithm is used to obtain the intensity distribution curve of discharge intensity along the surface of the insulator.

2. Extraction of discharge intensity quantification parameter

2.1. UV image preprocessing

The typical UV image of pollution discharge on the surface of suspended porcelain insulator taken on site is shown in Figure 1, the air relative humidity is 90%, and the model of ultraviolet imager is Israel's superB.

Figure 1. UV image of insulator surface pollution discharge

By analyzing the above images, it can be seen that there are four main discharge points on the insulator string. The brightness of the discharge area image is higher, while the brightness of the background image is lower, and there are also some discrete white interference points around the discharge area.

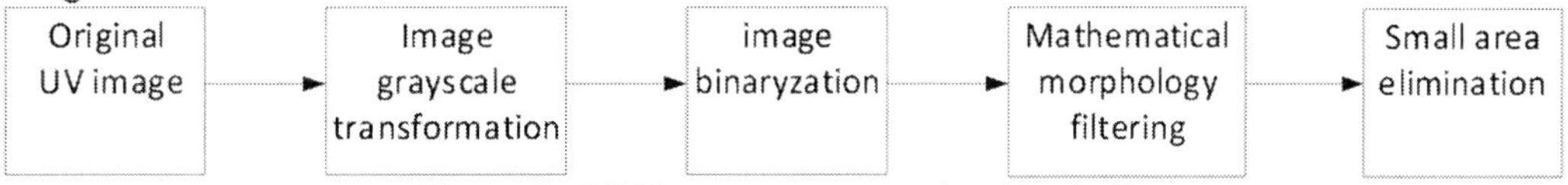

Figure 2 . UV image preprocessing block diagram

In order to extract the regional parameters of the discharge image, it is necessary to preprocess the UV image and segment discharge regions from the original image. The image preprocessing algorithm used in this paper is shown in Figure 2.

2.1.1. UV image graycale transformation. The output image of the UV imager is an RGB color image, which contains three color components. In order to facilitate subsequent processing, a grayscale transformation algorithm is used to convert it into a grayscale image. The algorithm is shown in the formula (1):

$$Y = 0.299R + 0.587G + 0.114B \tag{1}$$

Where Y is the grayscale value, which ranges from 0 to 255; R, G and B are the three color component values of red, green, and blue in the original color image,and the image processing result is shown in Figure 3 (a).

Now the main algorithms for image segmentation include edge operator method, threshold method, and region growing method. For the characteristics of UV images, a threshold segmentation algorithm is used in this paper[9]. The basic principle is to select an appropriate threshold and scan each pixel in the image matrix sequentially. If the grayscale value of a pixel is higher than the threshold, the grayscale value is set to 1, otherwise, it is set to 0. The algorithm is shown in formula (2).

$$g(x,y)=\begin{cases}1 & f(x,y)\geq T\\0 & f(x,y)<T\end{cases} \tag{2}$$

Where T is the threshold; $g(x, y)$ is the binary grayscale value, $f(x, y)$ is the grayscale value before binarization. The grayscale values of each pixel in the image segmented by the threshold method are only "1" and "0", hence it is also known as a binary image.

The histogram analysis shows that the gray value of the discharge area image is concentrated near 0.9, while the gray value of the background image is concentrated below 0.8, so the threshold value of 0.85 is selected in this paper, and the image processing effect is shown in figure 3 (b).

2.1.2. Mathematical morphology filtering. As shown in Figure 3(b), after threshold segmentation, the interference point images with grayscale values close to the discharge area are also retained in the binary image. The above images can be regarded as noise images and need to be filtered out. Analysis shows that the size of noisy images is generally much smaller than the discharge spot area, this paper adopts the binary mathematical morphology opening and closing algorithm to filter the noise images[10].

The basic operations of mathematical morphology are erosion and dilation. Simply performing erosion and dilation operations on an image can filter out noisy images, but it can cause the discharge spot area to shrink or expand, which has a significant impact on subsequent parameter calculations. In this paper, the opening and closing operations of mathematical morphology are used to filter the image, and they are defined in formula (4) and (5).

$$A\circ B=(A\Theta B)\oplus B \tag{3}$$

$$A\bullet B=(A\oplus B)\Theta B \tag{4}$$

Where A is the image to be processed, B is the structural element, symbols Θ and $\oplus$ represent erosion and expansion operations; symbols $\circ$ and $\bullet$ represent opening and closing operation respectively. The opening operation can eliminate the scattered points with small pixel size in the image without affecting the target area, and can smooth the outer boundary of the image. The closing operation can eliminate some holes in the image area, and also smooth the inner boundary of the image.

The shape and size of the structural element have a significant effect on the filtering effect. Since the shape of the discharge area image is mostly close to a circle, this paper selects the disc-shaped structural element, and the radius of the structural element is 2. After the opening and closing operations of Figure 3(b), the results are shown in figure 3(c).

2.1.3. Small area elimination From Figure 3(c), it can be seen that there are still some interference points with large area remaining in the image, and the small area elimination algorithm is used to remove the remaining noise points. The basic steps of the algorithm are as follows [11]:

(1) The region labeling algorithm is used to label the connected image regions for the image of Figure 3(c), and the pixels in each connected region are given the same label value;

(2) Count the number of pixels contained in each connected region ;

(3) Set a threshold to compare the number of pixels contained in each region with the size of the set area threshold. The pixel values in the region greater than the threshold remain unchanged, and the pixel values in the region less than the threshold are set to 0 (set to black colour).

When the threshold value is set to 150, the result image processed by the small region elimination algorithm is shown in Figure 3(d).

(a)GrayScale image

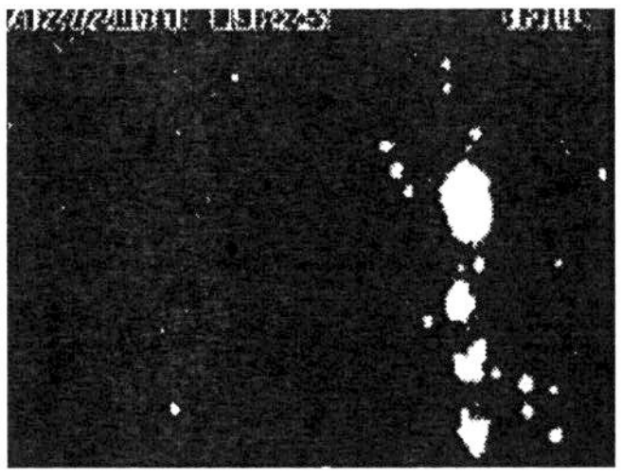

(b)Binary image

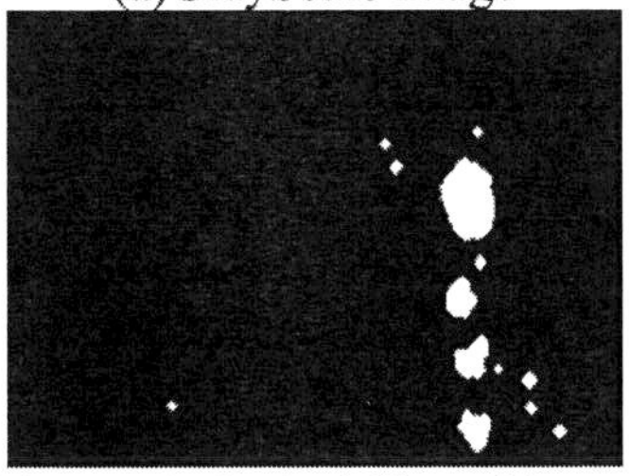

(c) Morphological filtered image

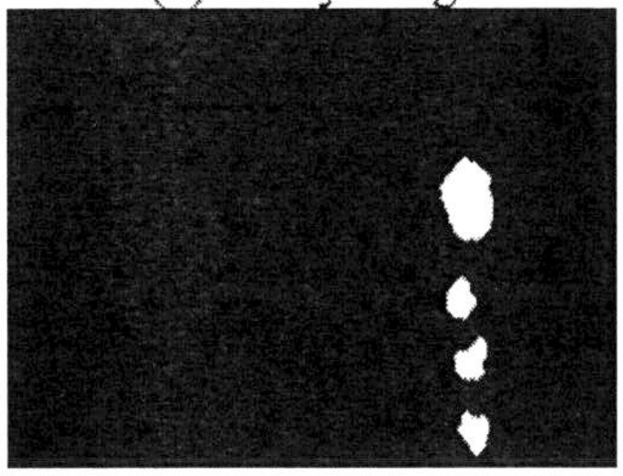

(d) Image processed by small area elimination

Figure 3. UV image preprocessing results

2.2. *Calculation of image spot area parameters*

2.2.1. Multi-region boundary tracking. In order to obtain the quantitative parameters of the discharge region image, it is necessary to extract the boundary points of each discharge region. Since there are often more than one discharge point in the UV image, a multi-region boundary tracking algorithm is used in this paper, and the basic principle and steps of the algorithm are as follows:

(1) Firstly, scan the pixel values of each pixel in the binary image matrix in order from left to right and from bottom to top. The first white pixel scanned is the first boundary point of the first area, as marked as "1" in Figure 4, and record the coordinates of this point;

(2) Starting from this point, the next boundary point is searched step by step, and the initial search direction is defined as the upper left. If the point on the upper left is a white point, it is a boundary point, otherwise the search direction rotates 45° clockwise, if the pixel value of the upper point is still not a white point, the search direction is rotated 45° clockwise until the white pixel point is found. This point is the second boundary point, as indicated by the pixel labeled "2" in Figure 4;

(3) Taking the second boundary point as the base point, the search direction is rotated counterclockwise 90° on the basis of the current search direction, which is used as the current initial search direction, and the next white pixel point is continued to be searched by the above method until the initial boundary point is returned;

(4) Repeating the above three steps can search the boundary of all discharge areas in turn and obtain the coordinate values of the boundary points.

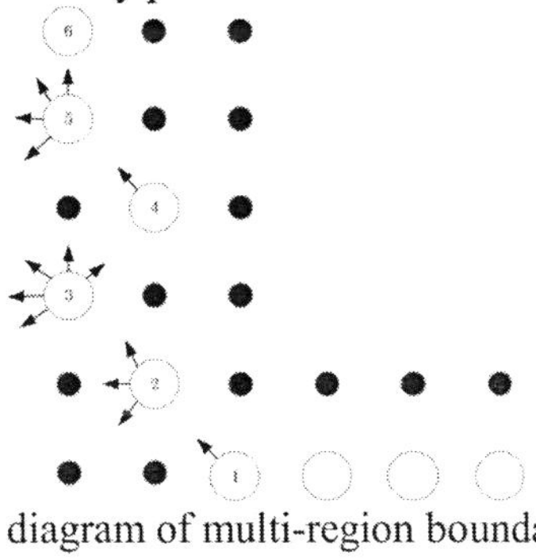

Figure 4. Schematic diagram of multi-region boundary tracking algorithm

Using the above algorithm, the coordinate values of the boundary points of the four discharge regions can be obtained. After connecting the coordinate values to form closed red curves and superimposing them on the original image, the boundary extraction result is shown in Figure 5.

Figure 5. Discharge region boundary

From the figure 5, it can be seen that the shape and size of the extracted boundary are in good agreement with the actual contour of the discharge region image. It shows that the algorithms proposed in this paper can effectively filter out noise while maintaining the size and shape of the original discharge region image as much as possible.

2.2.2. Quantitative parameter extraction. In order to characterize the size of each discharge spot area, relevant parameters need to be extracted. According to the characteristics of the UV image, the spot area quantitative parameter s is extracted in this paper.

Spot area s is to count the number of pixels contained in each discharge spot area, that is, the number of pixels "1" contained in each spot area, and the s is calculated withe the equation (5) :

$$s = \sum_{(x,y)\in Q} 1 \tag{5}$$

Where Q is the connected region, x and y is are the coordinates of each pixel, based on the definition, the four discharge regions in Figure 5 are shown in Table 1.

Table 1 The calculation results of each spot area

Spot area	The number of the spot area			
	1	2	3	4
spot area/pixel.	638	176	197	174

It can be seen from Table 1 that the discharge strength of each discharge point can be well quantified by using the spot area parameter.The total running time for the above image processing is 0.59 seconds.

3. Extraction of spatial distribution curve of discharge intensity

The spatial distribution of discharge along the surface of the insulator is also an important parameter reflecting the discharge characteristics of pollution discharge. It can be seen as a curve obtained by projecting the discharge image perpendicular to the surface of the insulator. In practical engineering, some high-voltage equipment is not arranged vertically. Here, this paper adopts the parallel line projection algorithm in Radon transform[12].

The basic principle of parallel projection transformation is to rotate the coordinate axis by different angles, and then project the image in different rotation directions to obtain its projection distribution characteristics, the basic transformation principle is shown in Figure 6 .

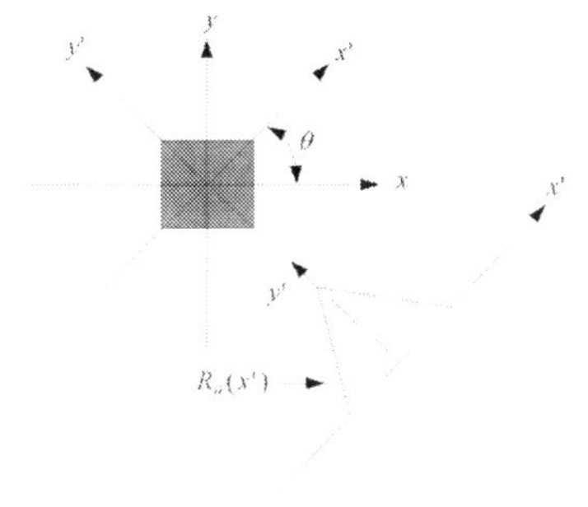

Figure 6. Schematic of Radon transforms projection calculation

Where $f(x, y)$ is the grayscale values of each pixel in the image matrix, (x, y) is the coordinate value of the pixel, then the Radon transform of $f(x, y)$ is the integration along a line parallel to the y' axis.

$$R_\theta(x') = \int_{-\infty}^{\infty} f(x'\cos\theta - y'\sin\theta, x'\sin\theta + y'\cos\theta)dy' \quad (6)$$

$$\begin{bmatrix} x' \\ y' \end{bmatrix} = \begin{bmatrix} \cos\theta & \sin\theta \\ -\sin\theta & \cos\theta \end{bmatrix} \begin{bmatrix} x \\ y \end{bmatrix} \quad (7)$$

Where θ is the angle between the rotated x' axis and the original horizontal x axis, x' and y' are the coordinate values in the rotated coordinate axis.

In order to facilitate processing, a software is develop this paper with Matlab GUI. Taking Figure 1 as an example, to obtain the spatial distribution of discharge intensity along the direction of the insulator string, the UV image needs to be projected along the direction perpendicular to the insulator string, the algorithm steps are as follows:

(1) Image preprocessing. still using the above image preprocessing algorithm to segment the discharge area;

(2) Calculate the rotation angle θ. Draw a straight line segment along the direction of the insulator string in Figure 1 using the mouse. The two ends of the straight line segment are the starting and ending points to be analyzed. Record the coordinates of the two points and calculate the angle θ between the straight line segment and the horizontal line.

For example, if analyzing the distribution of discharge space between the high-voltage side of the insulator string, the straight line segment is shown in Figure 7(a), and the θ is 91.9198 degrees, and the spatial distribution curve of discharge intensity is shown in Figure 7(b).

(a)schematic diagram of Projection analysis path

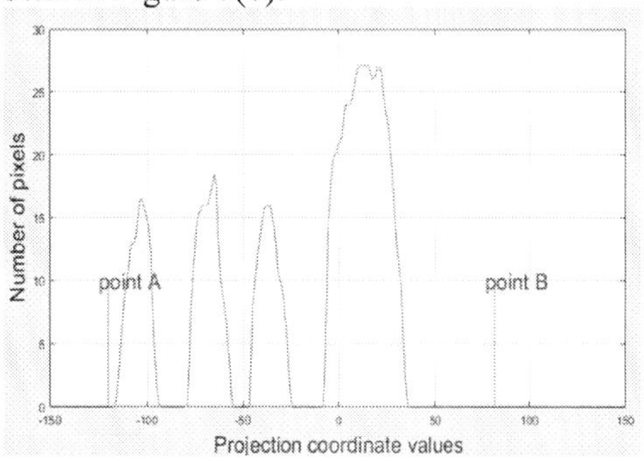

(b) Spatial distribution curve of discharge intensity

Figure 7. Extraction of discharge intensity curve along insulator surface

doi:10.1088/1742-6596/3012/1/012028

4. Conclusion

(1) According to the characteristics of UV image, the discharge area image can be effectively segmented with threshold segmentation, morphological opening and closing algorithm and small area image elimination algorithm.

(2) The multi-region boundary tracking algorithm is used to extract the coordinates of each discharge region. The spot area parameter is defined, which can be used to quantitatively analyze the discharge intensity,the total running time for the above image processing is 0.59 seconds.

(3) The parallel line projection algorithm in radon transform is used to obtain the discharge intensity distribution curve along the surface of the equipment, which can be used to characterize the spatial distribution characteristics of discharge.

References

[1] Zhang Haifeng, Pang Qichang, Li Hong, et al. High-voltage corona discharge detection based on UV spectroscopy technology [J]. Journal of Photons, 2006,35 (8): 1162-1166.

[2] WANG S H, LI W, JIANG T Y, et al. Study on the influence factors of rod plate gap corona discharge environment based on UV imaging[J]. IEEJ Transactions on Electrical and Electronic Engineering, 2021,16(1): 35-43.

[3] Wenjun Zhou, Han Li, Xiao Yi, et al. A criterion for UV detection of AC corona inception in a rod-plane air gap [J]. IEEE Trans. on Dielectrics and Electrical Insulation, 2011,18(1):232-237.

[4] LIU Yunpeng, LI Yonglin, PEI Shaoton, et al. Evaluation of the Insulation State of Contaminated Porcelain Insulators Based on Characteristics of Ultraviolet Irradiance[J]. High Voltage Engineering, 2023,49 (4): 1622-1631.

[5] Dai Rijun. Research on discharge detection method of high voltage electrical equipment in power plant based on ultraviolet signal [D]. Baoding: North China Electric Power University. 2012.

[6] Ye Huisheng, Lei Hongcai, Zang Chunyan. UV imaging detection characteristics of corona discharge in an extremely inhomogeneous field [J]. High-voltage electrical appliances, 2010,46 (1): 35-39.

[7] Xiao Meng, Wen Zeng. A new live detection method for insulators-UV imaging method [J]. High voltage technique, 2006,32 (6): 42- 44.

[8] Ma Bin, Zhou Wenjun, Wang Tao, et al. Ultraviolet digital image processing of AC corona discharge in rod-plate model and its application criteria [J]. Power system automation, 2008, 32 (24) : 74-79.

[9] Lin Kaiyan, Wu Junhui, Xu Lihong. Summary of color image segmentation methods[J]. Chinese Journal of Image and Graphics, 2005,10 (1): 1-10.

[10] YU jing, XIAO Chuangbai, LIAO QINmin. Digital Image Processing[M]. Beijing: Tsinghua University Press, 2023.

[11] Li Shiying, Yang Gaobo. Feature extraction and image processing [M]. Beijing: Electronic Industry Press, 2010.

[12] Zhang Defeng. MATLAB Digital Image Processing [M]. Beijing: Mechanical Industry Press, 2012.

Design and Application of Power Wireless Private Network Communication Terminals for Electricity Information Collection

Pengguo He[1], He Zhang[1], Leitao Li[2], Xingtao Wang[3*]

[1] Beijing Fibrlink Communications Co., Ltd., Beijing 102211, China, P.R. China

[2] Beijing Autohome Technologies Co., Ltd., Beijing 100080, P.R. China

[3] State Grid Information & Telecommunication Group Co., Ltd., Beijing 102211, P.R. China

*Corresponding author's e-mail: wangzi10060@163.com

Abstract. State Grid Corporation has carried out the construction of power wireless private network for the needs of power grid business. The power wireless private network is an important wireless network for terminal communication access networks. The remote communication terminal is the primary method for power grid equipment and facilities to access the communication system. Based on the access requirements of the electricity information collection service terminal, this paper developed a remote communication terminal for the 230 MHz power wireless private network. The hardware design, software design, timing synchronization algorithm optimization, and performance testing were carried out. The pilot verification was carried out under the actual grid operation conditions. The results showed that the performance index of the terminal meet the requirements of the test specification, and can carry the electricity information collection business.

1. Introduction

The terminal communication access network is an important part of the network layer for the State Grid Corporation to build a strong and smart grid. It is the main bridge for the power business terminal and equipment to connect to the power grid backbone network and business main station. In addition to the public network of telecommunications operators and power carrier communication, the power wireless private network is one of the important networks of the terminal communication access network, which can break through the "last mile" of power grid communication and effectively cover remote areas [1-3]. At present, the power wireless private network can use two frequency bands of 230 MHz and 1800 MHz. After the 230 MHz frequency band resource was authorized for the State Grid Corporation, the 1800 MHz frequency resource is no longer approved in principle. Compared with the 1800 MHz frequency band, the 230 MHz power wireless private network is closely integrated with the power business needs, has the advantage of wide coverage, can reduce the cost of network operation and maintenance, and has broadband transmission capabilities. The State Grid Corporation has formulated enterprise standards for 230 MHz power wireless communication systems, and has carried out power business load testing and verification of the 230 MHz power wireless private network in many provinces. The results verified that the 230 MHz power wireless private network can be an effective communication method for power business, including the distribution automation, measurement automation, precision load control, and so on [4-6].

Li Yongliang, et al. and Wang Yundi, et al. carried out the development and verification of the 1800MHz power wireless private network remote communication terminal[7, 8]. Lu Bing, et al. researched the safety of the LTE-G 230 MHz wireless private network terminal, and carried out the terminal development and application[9]. Zheng Xiaoqing, et al. proposed an LTE-G 230 power equipment transparent access technology and communication terminal development[10]. At present, the development and application of remote communication terminal for 230 MHz power wireless private network is in the test and verification stage. The 230 MHz frequency band is divided into 480 frequency points, each with a bandwidth of 25kHz. Due to the special management of this frequency band, the configurable bandwidth of the system is relatively special, resulting in that the physical layer frame structure of 230 MHz is significant different with that of the LTE technology. A special wireless frame structure is designed for the 230 MHz technology based on the discrete spectrum. Through the discrete spectrum aggregation technology, the carrier aggregation at any frequency point of 230 MHz full bandwidth power spectrum is achieved to meet the diverse requirements of the power business. According to the needs of the distribution and utilization business, this paper designed and implemented the 230 MHz power wireless private network remote communication terminal, which was compatible with different power business processes and communication interfaces. For the electricity information collection business in power grid, the communication access and functional adaptation of the business equipment was completed with the 230 MHz remote communication terminal. The reliability and business load-bearing capacity of the terminal are verified by the performance testing and demonstration application.

2. Design of The Remote Communication Terminal

2.1. Hardware Design

According to the functional division, the remote communication terminal can be composed of three parts: data processing unit, communication module, and power supply module, as shown in Fig. 1. The data processing unit and the communication module interact through the Transistor-Transistor Logic (TTL) serial port, and the power supply module supplies power to each part after level conversion.

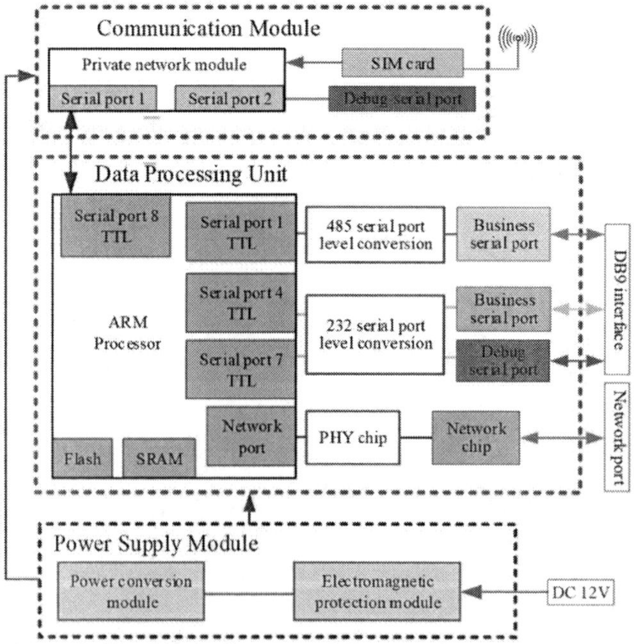

Figure 1. The block diagram of the overall hardware design

(1) Data Processing Unit: the data processing unit connects and communicates with the power business equipment and the communication module, providing the RS232 serial ports, the RS485 serial ports, the network port, and other external interfaces. The business serial port and debug serial port share the same one DB9 interface by using different pins for external communication, which is convenient for on-site installation and maintenance. According to the 230 MHz enterprise standard specification, the type and quantity of the external interfaces can be adjusted for different forms of the embedded terminal and the external terminal. The ARM processor, embedded real-time multi-tasking µC/OS III operating system, and the LWIP protocol stack are used in the terminal to achieve the multi-task processing mechanism, compatible with different working modes of various power business equipment.

(2) Communication Unit: the communication unit uses a commercial 230 MHz module with the wireless data communication function of the discrete narrowband aggregation. The communication module interacts with the data processing unit through the TTL serial port, and provides the TTL debug serial port, the SIM card seat, and the 230 MHz wireless private network antennas, and other external interfaces.

(3) Power Supply Unit: the power supply unit uses DC 12V as input power. The anti-reverse function of positive and negative polarity was designed. The electromagnetic compatibility (EMC) protection module was developed to achieve EMC-4 level protection. The DC 12V can be converted into DC 5V, and other required voltages through the power conversion module.

The hardware circuits of the external terminal and embedded terminal are shown in Fig. 2.

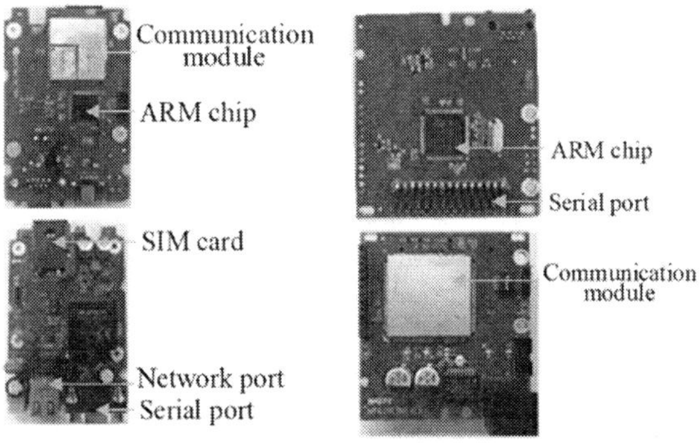

Figure 2 The hardware circuits of the external terminal (left) and embedded terminal (right).

2.2. *Optimization of The Timing Synchronization Algorithm*

In the general master synchronization sequence generation method of the power wireless private network, the frequency domain characteristics of the sequence are needed. Due to the sub-band resource limitations, each sub-band only contains 11 sub-carriers. It is necessary to segment the Zadoff-Chu (ZC) sequence in the frequency domain and perform Fourier transform for each segment to obtain time-domain data. This method is computationally complex and causes the characteristics of the ZC sequence to deteriorate. In this paper, the ZC sequence were mapped directly as the time-domain signal of the master synchronization signal, and the received time-domain data was used for correlation with the local sequence. The specific method is as follows:

(1) The primary synchronization signal (PSS) is generated by the ZC sequence, as formula (1).

$$d_u(n)=\begin{cases} e^{-j\frac{\pi un(n+1)}{449}} & n=0,1,\dots 223 \\ e^{-j\frac{\pi u(n+1)(n+2)}{449}} & n=224,225,\dots 448 \end{cases} \tag{1}$$

(2) Take $N_{ID}^{(2)}=0/1/2$, the first synchronization signal sequence PSS of 448-point length is generated by using table 1. The relationship between the sector ID and the ZC sequence u parameter is shown in table 1.

Table 1. The Relationship Between Sector ID and ZC Sequence u Parameter

$N_{ID}^{(2)}$	Root index u
0	1
1	224
2	448

(3) The ZC sequence generated in the previous step is used as the baseband time-domain signal of the master synchronization signal. The last 15 data points of the sequence are added to the head of the sequence as the cyclic prefix.

2.3. *Design of The Application Software*

The terminal runs the μC/OS III real-time multi-tasking operating system. The multi-task modular design was adopted in the application software according to different power business needs for rapid software development. For the electricity information collection business, the concentrator connects to the remote communication terminal, communicate to the business main station through the base station and the core network. The overall design of the terminal software is shown in Fig. 3.

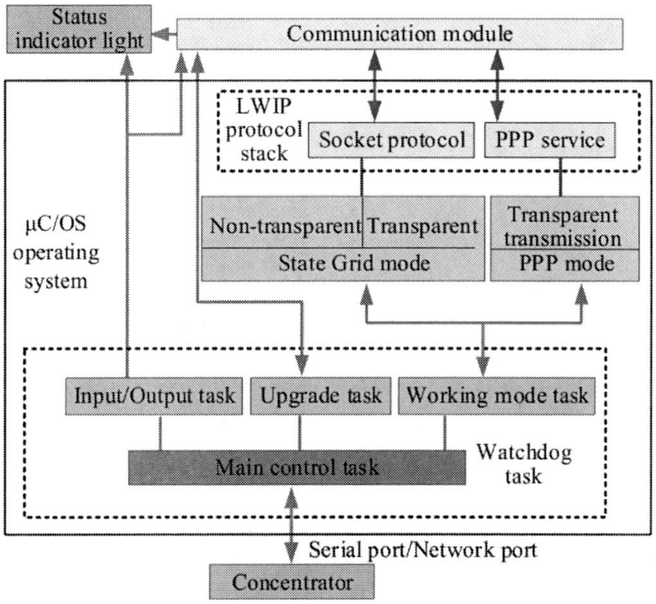

Figure 3. The Design of The Application Software

The "main control task" interacts with the concentrator, identifies the working mode, and manages other tasks. The "working mode task" identifies the working mode of the concentrator, which includes the "State Grid mode" and the "PPP (Point to Point Protocol) mode". The "State Grid mode", which can be divided into transparent mode and non-transparent mode according to the Q/GDW 1376.3-2013 standard specification, establishes a Socket connection with the communication module. The "PPP mode" establishes a PPPoE (Point-to-Point Protocol over Ethernet) server and establishes a PPP dial-up connection with the communication module, implementing data transparent transmission mode. The terminal transplants the LWIP protocol stack to provide Socket and PPP functions. The "upgrade task" interacts with the communication module to achieve remote terminal hardware upgrades through the air interface. the "I/O task" can control the start and restart of the module, as well as the on and off operations of the terminal status indicator light. The module can directly control the status indicator light. The "watchdog task" monitors the running status of each task in real-time. If there is an exception of any task, the "watchdog task" will restart the terminal and report the error log to the main station.

3. Testing of The Communication Performance

The communication performance testing and pilot verification was carried out at the Hunan Power Grid. There were three 230 MHz base stations in the pilot area. The power information acquisition terminals were connected to the laboratory. The test network topology diagram is shown in Fig. 4. According to the actual construction situation of the on-site wireless private network. The test bandwidth was configured to 5MHz. The electricity information collection business was configured with 160×25KHz frequency points. The RSRP at the test position was -69, and the SINR is 13. The wireless environment of the test area was well, and the theoretical business rate of the 8-carrier terminal in this environment could reach 55Kbps.

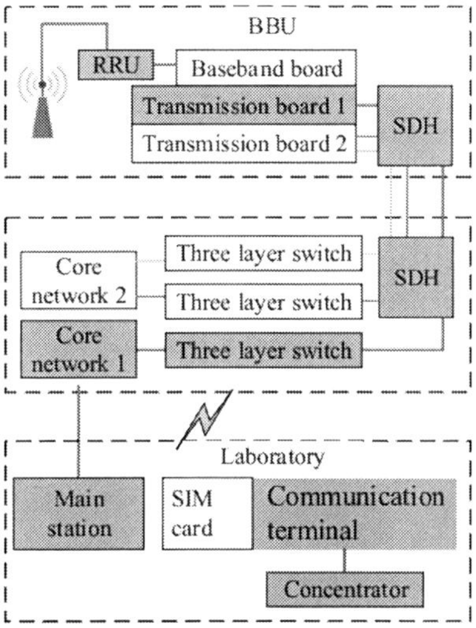

Figure 4. The Network Topology Diagram of The Test Environment

According to the tests, the one-way delay of the downlink was 49.219ms. The one-way delay of the uplink was 79.439ms. The business didn't fall off the line continuously. The uplink rate was >86% of the theoretical value. The downlink rate was >88% of the theoretical value. The packet loss rate of system was $<10^{-5}$. The results meet the requirements of the 230 MHz technical specification.

4. Pilot Verification of The Terminal

Based on the network environment of the communication performance test, the regulation function and adaptability test of the terminal for power consumption information collection business were carried out, including the basic functions test, the one-time collection success rate test, the remote control accuracy test, and the data collecting response time tests. The pilot used the WT-F100 power consumption information collection system of the State Grid of Hunan Province. Six single-phase cost control smart meters and two three-phase four-wire cost control smart meters were hanged with the concentrator. The power business equipment and the remote communication terminal were connected through the Ethernet port.

In the pilot verification, the parameters of the communication terminal and the 230 MHz wireless private network were correctly configured. The process of the network bearing and activation was completed, and the main station of the power consumption information collection business was successfully accessed. During the pilot process, the remote communication terminal communication was stable and reliable, with no disconnection or interruption. The data collection and remote control of the business main station were normal. The online operation effect was well. The remote communication terminal can effectively adapt to the on-site environment. The results are shown in table 2, which can meet the requirements of the 230 MHz technical specification.

Table 2. The Results of the Pilot Verification

Item	Method	Result
Basic function	① The main station sends timed or random control commands (tripping/allowing closing/emergency power limitation, etc.), allows/prohibits active reporting, observe whether the command is issued normally, and whether the operation is effective. ②The main station sends data call-measurement commands, observes whether the command is issued normally, and whether the data reporting is normal.	The function is normal
Remote control accuracy	Repeat the basic function test 20 times and record the accuracy rate.	The remote control accuracy is ≥99.99%
One-time collection success rate	The main station sets the collection cycle and collects the real-time data of the electricity meter, and records the one-time collection success rate.	The one-time collection success rate is ≥95%
Data call-measurement response time	①Set the SINR value to ±0.1dB, and the RSRP value to -110±5dB. ②The main station sets the collection cycle and collects the real-time data of the electricity meter, records the data call-measurement time. ③Repeat the above steps 20 times and calculate the average delay.	The conventional data query and response time is <5s

5. Conclusions

Based on the 230 MHz power wireless private network, this paper carried out the design and implementation of the 230 MHz power wireless private network remote communication terminal. The research and application status of the remote communication terminal were overviewed. The design and functional implementation of the hardware, software, and timing synchronization algorithms were carried out. For the power consumption information collection business, the communication performance of the terminal was tested, and the pilot verification was carried out in the actual power grid operation scenario. The results of the performance testing and pilot verification showed that the remote communication terminal designed in this paper could meet the communication requirements of the 230 MHz power wireless private network and can carry the power consumption information collection business. The other performance indicators will be tested in the follow-up work. The business protocol and functional adaptability tests of the distribution automation business will be carried out in future to verify the carrying capacity of the remote communication terminal.

Acknowledgments

This work is supported by the Science and Technology Project of State Grid Information & Telecommunication Group CO, LTD (No. 546826240006).

References

[1] Zheng Menhua, Rao Hua, Wang Kai, et al. Research and Application of Self-built 5G Core Private Network Based on Shared Wireless Access Network Mode[J]. Electric Power Information and Communication Technology, 2023, 21(7): 67-74.

[2] Zhao Hongda, Chen Chen, Wang Zhe, et al. Evolution Strategy of 4G Wireless Private Network to 5G in Power CPS Environment [J]. Electric Power, 2021, 54(3): 68-79.

[3] Shang Li, Li Yingmin, Liu Wenzhao, et al. Application and Prospect Analysis of 5G Key Technologies in Power Wireless Private Network[J]. Process Automation Instrumentation, 2023, 44(8): 106-110.

[4] Wu Qing, Wang Rining, Wang Shengxin, et al. Evolution strategy of 230 MHz power wireless

private network construction for dual carbon target[J]. Distribution & Utilization, 2023, 40(11): 45-53.

[5] Yao Yanyan, Sun Shijie, Zhao Qunhui. Design of Communication Terminal Equipment for LTE 230 Power Wireless Private Network[J]. Microcontrollers & Embedded Systems, 2022, (3): 83-87.

[6] Qiu Shuqi. Research on hybrid network of power wireless private network[J]. Communication & Information Technology, 2024, (4): 10-13.

[7] Li Yongliang, Feng Jiebo, Zheng Xiaoqing, et al. Load control terminal access technology based on 1.8GHz TD-LTE wireless private network[J]. Electrical Measurement & Instrumentation, 2019, 56(24): 128-132.

[8] Wang Yundi, Wang Xingtao, Wu Qing, et al. The design of telecommunication terminal for 1.8GHz power wireless private network[J]. Application of Electronic Technique, 2020, 46(5): 88-92.

[9] Lu Bing, Yue Huiqing. Research on terminal security of LTE230 system based on national secret algorithm[J]. Distribution & Utilization, 2018, 35(12): 14-20,28.

[10] Zheng Xiaoqing, Ying Zhanhuang, WANG Qiang, et al. Research on Wireless Communication Access Technology for Ubiquitous Power Internet of Things[J]. Electric Power Construction, 2019, 40(11): 16-23.

Journal of Physics: Conference Series **3012** (2025) 012030 doi:10.1088/1742-6596/3012/1/012030

Optimal Operation of PV-Integrated Energy Storage and Charging Stations in LV Distribution Networks with DC Intertied Systems

Xiugao Pei*, **Junpeng Liu, Xinhua Zhao, Tingli Yan, Huiyong Jia, Hao Wang**

Laiwu Power Supply Company of State Crid Shandong Electric Power Corporation, Laiwu, P.R.China

*Corresponding author's e-mail: 503808571@qq.com

Abstract. This paper presents an optimization framework for integrating photovoltaic (PV) systems with energy storage and electric vehicle (EV) charging stations in low-voltage (LV) distribution networks, with a focus on reducing urban traffic carbon emissions and enhancing energy utilization efficiency. The research tackles the issues arising from the variability and unpredictability of solar power production. due to environmental factors and the growing demand for EV charging infrastructure. Three key contributions are made: First, an operational model for photovoltaic energy storage integrated charging stations suitable for low-voltage distribution networks is proposed., based on an analysis of their structural and operational characteristics. Second, an optimization model is developed for these systems within low-voltage DC intertied networks, ensuring both the stability of the LV network and the operational requirements of the DC system. Third, an optimization algorithm is introduced to minimize daily operating costs and reduce power exchange fluctuations with the LV network. The algorithm employs a stochastic model for charging pile loads and PV generation, utilizing a Genetic Algorithm (GA) to solve hierarchical optimization objectives. The proposed methods are validated through simulations on a rural LV distribution network in Shandong Province. Results demonstrate a reduction in the average daily operating cost from 5287 yuan to 4989 yuan and a decrease in voltage fluctuation variance by approximately 0.2, confirming the effectiveness and practicality of the proposed approach.

1. Introduction

Transportation methods in urban areas, such as traditional vehicles and trucks, are considered the core source of carbon emissions, which occupy 10% of total emissions in China. Thus, the transition of traditional vehicles to electric vehicles (EVs) can lead to huge progress in decarbonization. In the past decade, The electric vehicle (EV) sector in China has witnessed a significant surge in development. As of December 2020, the number of new energy vehicles (NEVs) sold has reached a substantial figure., predominantly electric vehicles, had surged to 1.37 million units [1]. This high level of market penetration is deeply rooted in China's robust manufacturing capabilities and the continuous optimization of critical supporting infrastructure, such as charging stations. Notably, the development of charging infrastructure has emerged as a fundamental pillar driving market expansion.

In metropolitan areas, conventional automobiles and freight trucks play a major role in greenhouse gas emissions, representing around 10% of China's overall carbon footprint. [1]. The transition from traditional vehicles to electric vehicles (EVs) presents a substantial opportunity for decarbonization.

Over the past decade, China's EV market has experienced rapid growth, with new energy vehicle (NEV) sales reaching 1.37 million units by the end of 2020 [1]. This growth is supported by robust manufacturing capabilities and the continuous development of critical infrastructure, such as charging stations, which are essential for market expansion.

Photovoltaic (PV) energy is a promising renewable energy source that can power EV charging stations, reducing reliance on the public grid and increasing the use of distributed energy sources [2]-[4]. Nevertheless, the performance of PV systems can be significantly influenced by external conditions, including sunlight intensity and surrounding air temperature, leading to fluctuations in output power [5]. Without integrating auxiliary systems like energy storage, these fluctuations can destabilize charging station operations.

The combination of photovoltaic (PV) systems with energy storage units and charging infrastructure has facilitated the utilization of energy storage technologies in managing power generation processes [6]. Through the regulation of energy storage system operations, including both charging and discharging cycles, surplus energy can be stored to smooth power fluctuations, and stored energy can be released when PV generation is insufficient to meet EV charging demand [7]-[10]. This approach reduces dependency on the public grid, alleviates peak load pressures, enhances the economic efficiency of charging stations, and improves grid stability.

Current studies focusing on the combination of photovoltaic systems with energy storage and charging infrastructure have thoroughly investigated various aspects of system architecture., energy management, and control strategies, significantly enhancing the efficiency of distributed PV utilization and the stability of system operations in distribution grids. For instance, [11] proposed a scheme to integrate electric vehicles (EVs) as flexible loads into microgrids with photovoltaic (PV) power generation. By combining PV generation systems with EV charging infrastructure, this approach enhances the local utilization of renewable energy. To address the mismatch between EV charging load peaks and peak shaving requirements, [12] introduced a comprehensive construction plan for solar-powered energy storage and charging stations, detailing the key components of the station and its energy management strategies on the supply side. Additionally, [13] provided technical guidance for the system design of such stations, offering methodologies for line design, equipment selection, and layout optimization. The study also proposed a facility design scheme aimed at maximizing the benefits of PV generation, energy storage charging and discharging, and EV charging. Furthermore, [14] developed a microgrid system for solar-powered energy storage and charging stations, focusing on the operation and control mechanisms of the distribution power grid to facilitate seamless transitions between islanded and grid-connected modes.

1) An operational model for PV-integrated energy storage and charging(PESC) stations in low-voltage (LV) distribution networks is proposed. This model is developed by summarizing the structural and operational characteristics of such systems, providing a comprehensive framework for their integration into LV networks.

2) An optimization model for solar-powered energy storage and charging stations is introduced, specifically designed for low-voltage DC intertied systems. The proposed optimization strategy ensures both the operational stability of the LV distribution network and the functional requirements of the DC intertied system.

3) An optimization algorithm is proposed to minimize daily operating costs and reduce fluctuations in power exchange with the LV distribution network. The planning model incorporates a stochastic model of the DC load from charging piles and PV generation. A Genetic Algorithm (GA) is employed to solve the hierarchical optimization objectives, addressing both upper and lower-layer optimization challenges.

The remainder of the paper is organized as follows: Section II describes the construction of the operational model for solar-powered energy storage and charging stations in LV distribution networks. Section III presents the optimization model. Section IV validates the proposed methods through simulations on a rural LV distribution network , Shandong Province. Section V concludes the paper.

2. Construction of PESC Station Operation Model in LV Distribution Network

In solar-powered energy storage and charging stations, EV charging load is influenced by a variety of factors, including vehicle performance parameters and user travel patterns. These factors exhibit considerable randomness in both temporal and spatial dimensions, leading to fluctuating and unpredictable charging demands. To address this variability, it is essential to incorporate real-world scenarios and derive typical operating patterns for photovoltaic-storage-charging stations. This section offers a detailed overview of the architecture and functional attributes of solar-powered energy storage and charging facilities integrated into low-voltage (LV) distribution grids. Building upon this examination, an operational model for photovoltaic storage charging stations in LV distribution networks is proposed, offering a framework to optimize their performance and integration into the grid.

2.1. PESC Stations System Architecture

The internal architecture of solar-powered energy storage and charging stations primarily consists of a photovoltaic energy production system, an ESS, fast charging piles, EVs, and other electrical loads within the station. Externally, the station is connected to the DC bus of the distribution network through DC-DC converters, enabling the efficient delivery of electrical energy to the system. Internally, the photovoltaic energy production system and the EV charging station are interconnected via the DC bus, allowing PV-generated power to be directly transmitted to the charging piles. The ESS plays a critical role in storing surplus PV energy, thereby enhancing the system's economic efficiency and energy transmission performance. Furthermore, the charging station is linked to the DC bus of the distribution network through DC-DC converters, serving as an auxiliary power supply to ensure stable and reliable energy provision [15]. This integrated design not only optimizes energy utilization but also supports the seamless operation of the charging station within the broader distribution network.

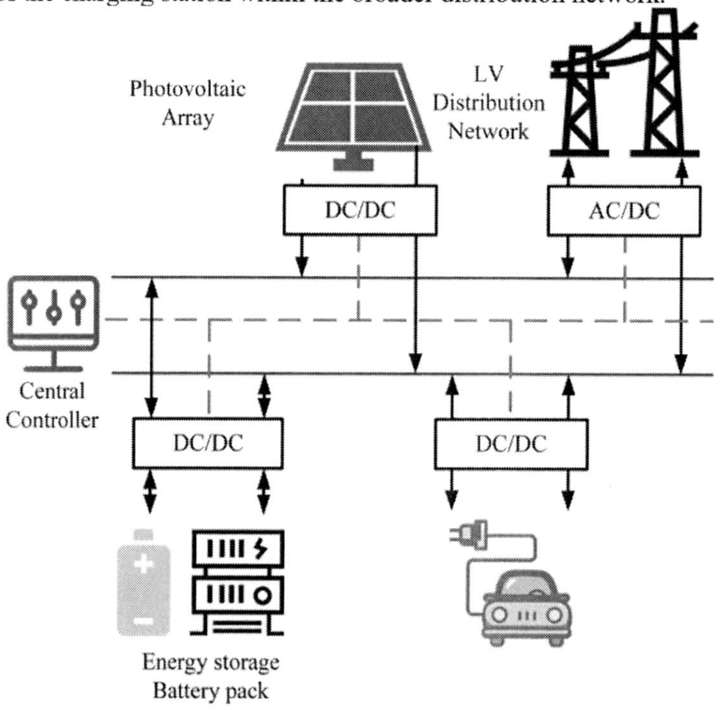

Figure 1. PESC stations system architecture

2.2. Characteristic Analysis of PESC Station System

The working performance of a PV module is affected by many factors, mainly including ambient temperature, solar irradiation intensity, and module surface temperature, and its output power is shown in Eq 1

$$P_{PV} = P_s \frac{G_r}{G_s}\left[1 + k\left(T_r - T_s\right)\right] \tag{1}$$

where PPV represents the generated power of the solar panel; *Gs* and *Ts* correspond to the solar irradiance and surface temperature under actual operating conditions, where Gs=1000W/m2 and Ts =25°C; *Gr* and *Tr* represent the solar irradiance and operating temperature of the solar panel under real-world conditions; k represents the power temperature coefficient, which is determined by the material and manufacturing process of the photovoltaic module[17]. This relationship highlights the influence of environmental factors on PV module performance and underscores the importance of considering real-world operating conditions in system design and optimization.

The battery storage unit within a combined solar-storage-charging facility serves a vital function in regulating power flow and maintaining energy balance across the system. It absorbs and stores excess electricity produced by the station's grid during periods of PV curtail. Conversely, when PV output is insufficient to meet demand, the ESS discharges stored energy to compensate for the shortfall, ensuring a stable and continuous power supply. This dual functionality enables the ESS to act as a critical regulatory mechanism, balancing energy supply and demand in real time. By smoothing power fluctuations and enhancing grid stability, the ESS significantly contributes to the efficient and reliable operation of the charging station, making it an indispensable component of the integrated system.

Throughout the battery's operational cycle, SOC serves as an indicator of its available energy reserves. SOC indicates the ratio of the available charge in the battery at a given moment to its rated capacity, and it can be expressed as:

$$S_t = \frac{E_t}{E_0} \times 100\% \tag{2}$$

where *St* represents the SOC of the battery at time *t*; *Et* denotes the remaining charge of the battery at time *t*; *E0* refers to the rated capacity of the battery. Thus, the battery's charge and discharge cycles can be characterized by the SOC in the following:

$$S_t = \left(1 - \lambda\right) S_{t-1} + \left(\frac{P'_{Ein}\eta_{Ein}}{E_0} - \frac{P'_{Eout}}{E_0 \eta_{out}} \right) \tag{3}$$

St and *St−1* indicate the SOC of the battery at a specific time *t* and *t−1*, respectively; Pt_{in} and Pt_{out} denote the charging power and discharging power of the battery at time *t*, where only one state charging or discharging can exist at the same time. η_{in} and η_{out} denote the battery's charging and discharging efficiencies, respectively; λ indicates the battery's self-discharge rate when idle. Furthermore, the operational power range of the battery plays a vital role in determining its capacity loss over time. Keeping the State of Charge (SOC) between 20% and 80% during charge and discharge cycles can greatly enhance the battery's lifespan [17]-[18]. This strategy ensures the energy storage system achieves both peak performance and extended durability.

2.3. *Operating Model for PESC Stations*

PV, energy storage, and charging stations are interconnected through electrical links to establish a microgrid. This microgrid operates by harnessing renewable energy sources, such as PV and wind power, to facilitate energy storage. Additionally, during off-peak hours at night, surplus electricity from the grid can be stored for later use.

2.3.1. *High daytime PV generation model*

During periods of PV generation at a high level, the operational strategy of an LV distribution network integrated with a photovoltaic storage charging station prioritizes the direct utilization of solar energy. Specifically, the generated PV power is first allocated to EV charging or other immediate loads. This method decreases reliance on the public electricity network while improving the overall efficiency of solar energy utilization. When the PV system generates more power than the immediate load requires, the excess energy is used to charge the storage system. The energy storage unit serves a vital function in smoothing power variability, ensuring a stable equilibrium between energy supply and demand. Additionally, it compensates for the natural variability of solar power generation, guaranteeing a consistent and reliable energy supply [20]. To further maximize the utilization of clean energy, the photovoltaic-storage charging station employs intelligent scheduling systems to optimize charging loads. These systems encourage EVs to charge during periods of peak PV output, dynamically adjusting the charging power to align with energy availability while prioritizing user charging requirements. If the energy storage system reaches full capacity or is unable to absorb additional surplus energy, the remaining power can be exported back to the public grid. This not only generates additional revenue through electricity sales but also enhances the integration and consumption of photovoltaic power within the grid.

2.3.2. Absence of PV generation during the nigh model

At night, when PV generation ceases, the operation of a photovoltaic-storage charging station shifts to rely on energy storage and grid dispatch. During this period, the ESS supplies power to the charging stations and other loads, ensuring uninterrupted operation of the facility. When the ESS has sufficient stored energy, it prioritizes meeting the EV charging demand. However, if the energy stored in the system is inadequate or the charging needs surpass the capacity of the ESS, the system compensates by sourcing additional power from the grid to ensure all charging demands are fully satisfied [20]. To optimize cost-effectiveness, the charging facility can implement smart scheduling approaches that respond to dynamic grid electricity pricing., optimizing electricity costs and improving overall operational efficiency. This approach ensures reliable service while minimizing operational expenses.

3. Operation Model in LV Distribution Network Optimization Model for PESC Station

This section introduces the operational model of a low-voltage distribution network incorporating solar-powered energy storage and charging stations. An optimization strategy is proposed, including the calculation of the daily operating costs of the solar-powered energy storage and charging stations and the quantification of fluctuations in energy exchange with the low-voltage distribution network. The main optimization problem is formulated using a genetic algorithm (GA), which solves the optimization objectives in upper and lower layers.

This section presents the operational model of a low-voltage (LV) distribution network integrated with PV-based energy storage and charging stations. An optimization strategy is proposed, focusing on two key objectives: (1) minimizing the daily operating costs of the solar-powered energy storage and charging stations, and (2) quantifying and reducing fluctuations in energy exchange with the LV distribution network. The optimization problem is formulated using a Genetic Algorithm (GA), which addresses the hierarchical optimization objectives across both upper and lower layers.

3.1. Daily Operating Cost

Assume that the daily operational cost of the photovoltaic-storage integrated charging station is based solely on the electricity exchange with the low-voltage distribution network. This cost is calculated by subtracting the revenue generated from electricity sales from the total electricity purchase cost. The resulting cost can be expressed as follows:

The daily operational expenses of the solar-storage combined charging facility are determined by its power interactions with the low-voltage (LV) distribution grid. This expense is computed by deducting the income earned from supplying electricity to the grid from the overall expenditure on purchasing electricity from the grid. The daily operational cost can be formulated as follows:

$$F_{\mathrm{Cost}} = \sum K_{\mathrm{E}}^{t} \cdot (P_{\mathrm{Gin}}^{t} - P_{\mathrm{Gout}}^{t}) \cdot \Delta t \tag{4}$$

where F_{Cost} is the daily operating cost function of the charging station; K_E^t denotes the electricity price at time t; Δt refers to the sampling time step; P^t_{Gin} and P^t_{Gout} are the power drawn from and supplied to the public grid at time t, respectively.

3.2. The Fluctuation of Power Exchange with LV Distribution Network.

The exchange of electrical energy between a photovoltaic-storage integrated charging station and the public grid includes both electricity purchasing and selling activities. These interactions inevitably impose certain impacts on the public grid . To quantify the fluctuation caused by this energy exchange, this study evaluates the sum of the standard deviations of the daily electricity bought from and supplied to the public grid. The resulting metric is defined as the standard deviation of exchanged power and can be mathematically expressed as follows:

$$\begin{cases} F_\sigma = \mu F_{cost}(P_{Gin,\sigma} + P_{Gout,\sigma}) \\ P_{Gin,\sigma} = \sqrt{\frac{1}{T}\sum_{t=1}^{T}\left(P^t_{Gin} - \bar{P}_{Gin}\right)^2} \\ P_{Gout,\sigma} = \sqrt{\frac{1}{T}\sum_{t=1}^{T}\left(P^t_{Gout} - \bar{P}_{Gout}\right)^2} \end{cases} \tag{5}$$

F_σ represents the function for the standard deviation of exchanged power. $P_{Gin,\sigma}$ and $P_{Gout,\sigma}$ denote the standard deviations of the daily electricity procurement and supply, respectively. $\bar{P}_{Gin}$ and $\bar{P}_{Gout}$ represent the average values of the daily electricity procurement and supply, respectively. T refers to the total number of sampling periods.

Based on the above, the objective function for the active power scheduling optimization model of the photovoltaic-storage integrated charging station can be expressed as:

$$\min F = \lambda_1 F_{Cost} + \lambda_2 F_\sigma \tag{6}$$

in the equation, λ_1 and λ_2 represent the weights, indicating the preferences for the optimization objectives. This study prioritizes reducing the standard deviation of the exchanged power fluctuations. Based on the above scenario assumptions, setting λ_1=0.25 and λ_2=0.75.

3.3. Optimization Constraints.

1) Input-output balance constraints

The PIEC stations in the LV distribution network need to maintain input and output when exchanging power with the coupling point Power balance, which satisfies the equality constraints, can be expressed as:

$$P_C + (P_{Ein} - P_{Eout}) = (P_{Gin} - P_{Gout}) + P_{PV} \tag{7}$$

2) Operational constraints of photovoltaic power generation systems

The PV output power of the charging station will not exceed the upper and lower power limits during normal operation, which can be expressed as:

$$P_{PV,min} \le P_{PV} \le P_{PV,max} \tag{8}$$

where $P_{PV,max}$, $P_{PV,min}$ represent the maximum and minimum power values of the PV during normal operation, respectively.

3) Operational constraints of energy storage systems

The ESS battery pack operates under the constraint that it can only be in one state—either charging or discharging—at any given time. Additionally, the charging and discharging power must remain within the allowable operational limits. To ensure the longevity of the battery and optimize its cycle life, it is essential to control the SOC within a specified range.

This principle can be mathematically expressed as follows:

$$\begin{cases} 0 \le P_{\mathrm{ESin}} \le P_{\mathrm{ESin,max}} \\ 0 \le P_{\mathrm{ESout}} \le P_{\mathrm{ESout,max}} \\ S_{\mathrm{ES,min}} \le S_{\mathrm{ES}} \le S_{\mathrm{ES,max}} \end{cases} \tag{9}$$

4) Charging station grid-connected power distribution system operation constraints

The charging station's power grid is interconnected with the public power grid via a grid-connected distribution system. During power exchange operations, both the power purchased from and the power sold to the public grid must adhere to the rated power limits of the grid-connected distribution system. This constraint can be expressed as follows:

$$\begin{cases} P_{\mathrm{Gin}} \le P_{\mathrm{Sin}} \\ P_{\mathrm{Gout}} \le P_{\mathrm{Sout}} \end{cases} \tag{10}$$

where P_{Sin} and P_{Sout} denote the rated input and output power capacities of the grid-connected distribution system, respectively. For the purposes of this study, it is assumed that the equipment capacity of the grid-connected distribution system is sufficient to support the subsequent research and operational requirements.

3.4. Planning Model Procedures

The primary objective of the planning model is to minimize both the daily operating costs and the fluctuations in power exchange with the public grid. The model incorporates a stochastic representation of the DC load from charging piles and PV power generation within the system. The optimization problem is formulated using GA, which addresses the hierarchical objectives across upper and lower layers. The flowchart of the proposed planning model is illustrated in Fig. 2. The procedural steps are outlined as follows:

Initialization: A Genetic Algorithm (GA) population is created, where each chromosome corresponds to a possible schedule setup for the solar-powered energy storage and charging facilities. Steps 2 through 6 are performed for every chromosome in the current GA iteration.

Scenario Selection: A specific scenario is chosen, which includes randomized values for uncertain variables such as load requirements and outputs from renewable distributed generation (DG) sources.

Sub-Optimization: The sub-optimization task is resolved for the chosen scenario.

Stopping Criterion Evaluation: The termination condition for the photovoltaic-charging stochastic process is assessed. The constrained outcomes for various random scenarios are denoted by the random variable CPCR. The stochastic process can conclude either after a predetermined number of iterations or when the stopping criterion outlined in [21] is achieved. Steps 2 to 4 are repeated until the stopping condition is fulfilled.

This iterative approach guarantees that the optimization model converges to a solution that effectively balances operational expenses and power exchange stability.

$$\frac{\sigma(C_{\mathrm{PCR}})}{E(C_{\mathrm{PCR}})} \le \varepsilon \tag{11}$$

$\delta(C_{\mathrm{PCR}})$ is the standard deviation of the variable, while $E(C_{PCR})$ is the expected value of the stochastic variable. ε is a selected small tolerance.

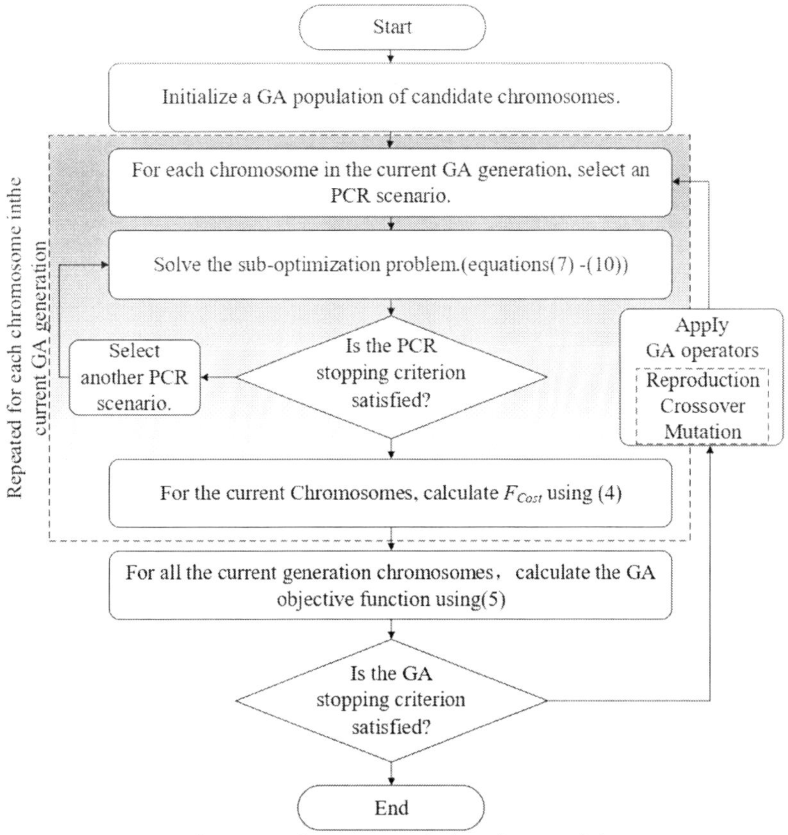

Figure 2. The proposed planning model

4. Simulation Verification

In this paper, the effectiveness and practicality of the proposed optimization strategy for the optical storage charging station are verified using a real rural LV distribution network in Laiwu, Shandong Province, as a test case. The model was modeled simulated and tested using PSCAD/EMTDC.

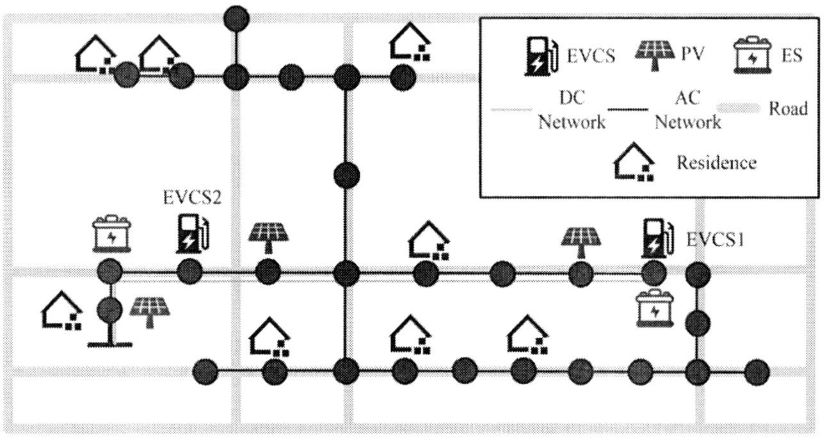

Figure 3. Topology of a country in Laiwu

Figure 3 shows the topology of a country in Laiwu, base voltage is 10kV; as can be seen in the figure, it is a cluster of countries, and there are also power facilities such as factories on the outskirts of the country and DC lines with optical storage charging stations are planned.

4.1. Performance Verification of Operation Model in LV Distribution Network Optimization Model for PESC Station

Through multiple simulation results, the average one-day operating cost has been reduced from 5287 yuan to 4989 yuan, while the voltage fluctuation variance has been reduced by about 0.2.

4.1.1. Simulation of high daytime PV generation

During this time, household activities such as lighting, cooking, and operating some agricultural machinery will consume electricity. In the presence of distributed PV systems, power generation is relatively low in the early morning, so power demand during this period may rely on stored energy from the battery systems or supplementary power from the grid.

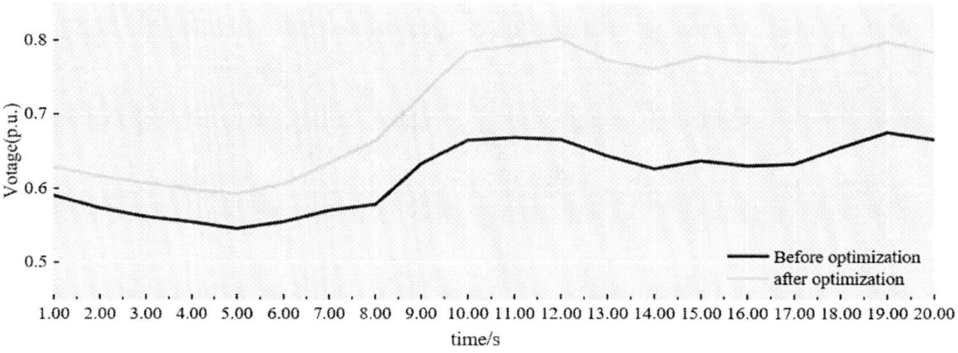

Figure 4. AC/DC grid connection point voltage (daytime)

4.1.2. Simulation of the absence of PV generation during the night

Overall, electricity demand stabilizes in the evening, but the demand for charging for electric vehicles may persist. Users requiring extended charging times might take advantage of nighttime off-peak electricity rates. Battery systems also release energy to meet electricity needs during this period.

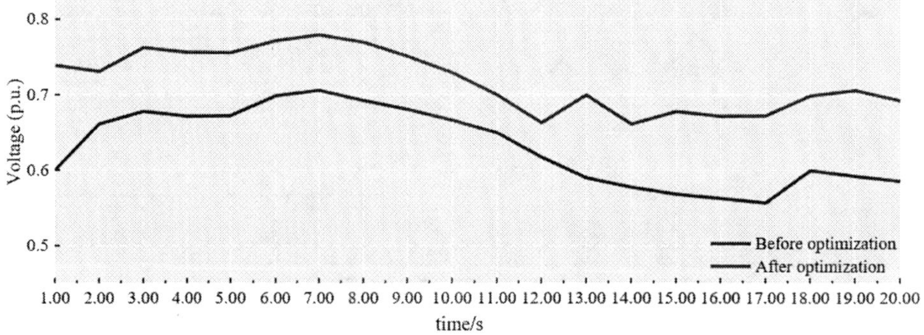

Figure 5. AC/DC grid connection point voltage (night)

According to Figure 4 and 5, the introduction of optimization can significantly increase the access point node voltage and reduce voltage fluctuation during critical system operations, which is beneficial to charging stations and stable operation of the power grid system.

5. Conclusion

A novel optimization method is proposed in this paper for the cooperation of LV distribution networks and opical storage charging stations considering a DC intertied system. Firstly, an operational model for photovoltaic (PV)-integrated energy storage and electric vehicle (EV) charging stations is proposed, specifically designed for low-voltage (LV) distribution networks. This model is developed through a detailed analysis of the structural and operational characteristics of such systems. Then, an optimization model is formulated for systems within low-voltage DC intertied networks. The model ensures the stability of the LV network while also satisfying the operational requirements of DC systems. An optimization algorithm is introduced with the objective of minimizing daily operational costs and reducing power exchange fluctuations between the system and the LV network. This approach utilizes a stochastic model to account for variability in charging pile loads and PV generation. To solve the hierarchical optimization problem, a Genetic Algorithm (GA) is employed. The proposed methods are validated through simulations performed on a rural low-voltage distribution network in Laiwu, Shandong Province, demonstrating their practical applicability and effectiveness. The simulation results indicate that, after optimization, the average daily operating cost has decreased from 5287 yuan to 4989 yuan, and the voltage fluctuation variance has been reduced by approximately 0.2, thereby validating the effectiveness and practicality of the proposed method.

References

[1] Electric vehicle market in China." [Online]. Available: https://www.statista.com/topics/5623/electric-vehicle-market-in china/#topicOverview. [Accessed: Dec. 21, 2024]

[2] Chao-bo Zhou, Shao-zhou Qi, Ji-hong Zhang, Si-yan Tang,Potential Co-benefit effect analysis of orderly charging and discharging of electric vehicles in China,Energy,Volume 226,2021,120352,ISSN 0360-5442.

[3] G. Yesilyurt, C. Andic, E. Aydin and B. Turkay, "Optimum Allocation and Sizing of Storages with Multi-Objective Grey Wolf Optimizer," 2024 8th International Artificial Intelligence and Data Processing Symposium (IDAP), Malatya, Turkiye, 2024, pp. 1-7.

[4] A. A. Khan, R. S. Niloy and N. Mohammad, "Role of Power System Operator in Initiating Peer-to-Peer Energy Trading in the Distribution Networks," 2024 International Conference on Advances in Computing, Communication, Electrical, and Smart Systems (iCACCESS), Dhaka, Bangladesh, 2024, pp. 01-05.

[5] C. Puganesa,G. S. Ashwinraj,M. Kugendran,K. R. Nair. "Effect of environmental factors on the photovoltaic module performance: A short review."

[6] Tan K M, Babu T S, Ramachandaramurthy V K, et al. Empowering smart grid: A comprehensive review of energy storage technology and application with renewable energy integration[J]. Journal of Energy Storage, 2021, 39: 102591.

[7] Liu Z, Chen Y, Zhuo R, et al. Energy storage capacity optimization for autonomy microgrid considering CHP and EV scheduling[J]. Applied Energy, 2018, 210: 1113-1125.

[8] N.R.Tummuru,M.K.Mishra,andS.Srinivas,"Dynamicenergymanage ment of renewable grid integrated hybrid energy storage system," IEEE Trans. Ind. Electron., vol. 62, no. 12, pp. 7728–7737, Dec. 2015.

[9] Chaudhari K, Ukil A, Kollimalla S K, et al. Energy storage management for EV charging stations: Comparison between uncoordinated and statistical charging loads[C]//IECON 2016-42nd Annual Conference of the IEEE Industrial Electronics Society. IEEE, 2016: 4109-4114.

[10] H.S. Das, M.M. Rahman, S. Li, C.W. Tan,

[11] Electric vehicles standards, charging infrastructure, and impact on grid integration: A technological review,Renewable and Sustainable Energy Reviews,Volume 120,2020,109618,ISSN 1364-0321.

[12] Kayal P, Chanda C K. Placement of wind and solar based DGs in distribution system for power loss minimization and voltage stability improvement[J]. International Journal of Electrical

Journal of Physics: Conference Series **3012** (2025) 012030
doi:10.1088/1742-6596/3012/1/012030

Power & Energy Systems, 2013, 53: 795-809.

[13] Yang M, Zhang L, Zhao Z, et al. Comprehensive benefits analysis of electric vehicle charging station integrated photovoltaic and energy storage[J]. Journal of Cleaner Production, 2021, 302: 126967.

[14] Yao M, Da D, Lu X, et al. A review of capacity allocation and control strategies for electric vehicle charging stations with integrated photovoltaic and energy storage systems[J]. World Electric Vehicle Journal, 2024, 15(3): 101.

[15] Maghami M R, Pasupuleti J, Ekanayake J. Energy storage and demand response as hybrid mitigation technique for photovoltaic grid connection: Challenges and future trends[J]. Journal of Energy Storage, 2024, 88: 111680.

[16] Richardson P, Flynn D, Keane A. Optimal charging of electric vehicles in low-voltage distribution systems[J]. IEEE Transactions on Power Systems, 2011, 27(1): 268-279.

[17] Inoa E, Wang J. PHEV charging strategies for maximized energy saving[J]. IEEE Transactions on Vehicular Technology, 2011, 60(7): 2978-2986.

[18] D. I. Stroe, V. Knap, M. Swierczynski, A. I. Stroe, and R. Teodorescu, "Operation of a grid-connected lithium-ion battery energy storage system for primary frequency regulation: A battery lifetime perspective," IEEE Trans. Ind. Appl., vol. 53, no. 1, pp. 430–438, Jan./Feb. 2017.

[19] Tummuru N R, Mishra M K, Srinivas S. Dynamic energy management of renewable grid integrated hybrid energy storage system[J]. IEEE Transactions on Industrial Electronics, 2015, 62(12): 7728-7737.

[20] K. Kumar, B. Sivaneasan, P. Cheah, P. So, and D. Wang, "V2G capacity estimation usingdynamicEVscheduling,"IEEETrans.SmartGrid,vol.5, no. 2, pp. 1051–1060, Sep. 2014.

[21] He Y, Venkatesh B, Guan L. Optimal scheduling for charging and discharging of electric vehicles[J]. IEEE transactions on smart grid, 2012, 3(3): 1095-1105.

[22] H. Doagou-Mojarrad, H. Rastegar, and G. B. Gharehpetian, "Probabilistic multi-objective HVDC/AC transmission expansion planning considering distant wind/solar farms," IET Sci. Meas. Technol., vol. 10, no. 2, pp. 140–149, Mar. 2016.

Journal of Physics: Conference Series 3012 (2025) 012031 doi:10.1088/1742-6596/3012/1/012031

A Multi-winding Grading Reactor for Distributed Capacitance Compensation of Urbanized Cables

Wei Luo[1], JingYu Huang[2]*, Yi Liu[2][0000-0003-1806-3045], Minglei Xie[3], Jinfeng Wang[4], Yuanzhe Zhu[5], Dayi Li[2][0000-0001-6717-1072], Wanlin Du[4], Fang Chen[5]

[1] Guangdong Power Grid Co., Ltd. Meizhou Power Supply Bureau, Meizhou 514000, Guangdong, China

[2] School of Electrical and Electronic Engineering, Huazhong University of Science and Technology, Wuhan 430074, Hubei, China

[3] China Southern Power Grid Electric Power Research Institute Co. LTD, Guangzhou 510000, Guangdong, China

[4] Electric Power Research Institute of Guangdong Power Grid Co. LTD, Guangzhou 510000, Guangdong, China

[5] China Southern Power Grid Energy Development Research Institute Co. LTD, Guangzhou 510000, Guangdong, China

*Corresponding author's e-mail: hjingyu@hust.edu.cn

Abstract. With the widespread use of underground cables, a large amount of distributed capacitance is added to the urban power grid circuit, which brings great challenges to the transmission capacity and stability of the system. The mainstream suppression control method of distributed capacitance of power grid cable mostly depends on the characteristics of power grid cable material itself. In practical engineering applications, the characteristics of cable material itself are easily affected by environmental factors, which affects the suppression and compensation effect. In view of the above problems, this paper proposes a multi-winding transformer-type graded continuously adjustable reactor for distributed capacitance compensation of urbanized cables. The distributed capacitance of urbanized cables is compensated through practical application requirements. The thyristor and voltage-type full-bridge PWM inverter are used as the load on the secondary side of the transformer, and the switching control of multiple thyristors realizes the coarse adjustment of the adjustable reactance. The output current of the inverter is controlled to achieve fine adjustment of the reactance value. The device can improve the compensation accuracy and improve the power quality of the power grid. The validity and correctness of the proposed method are preliminarily verified by MatLab software.

1.Background

With the continuous development of power grid and power electronics technology, adjustable reactance has developed from mechanical type, electromagnetic type to today 's electronic type, and has become a product with different structures, various control schemes and different uses. [1] - [2] Among them, the adjustable-turn and air-gap reactors are the earliest and most widely used adjustable reactors. They change the reactance value of the reactor based on the most basic adjustment coil turns and by adjusting the air gap length of the core to change the magnetic path permeance of the reactor, but this type of adjustable reactance cannot be continuously adjusted. [3] The magnetically saturated controllable reactor uses DC control to adjust the core permeability to change the AC effective reactance value. However, due to the operation in the magnetic flux saturation zone, the core magnetic flux density is large and the noise problem is serious. [4] The reactance values of thyristor phase-controlled reactor, high leakage impedance transformer-type adjustable reactor and adjustable capacitor-type adjustable reactor are also continuously adjustable, but when the conduction angle is small, this adjustable reactor will produce serious harmonics.

In order to suppress the harmonics generated by the adjustable reactance itself, some scholars have proposed a multi-winding high-leakage reactance transformer reactor, but its device structure is too complex. In order to improve its structure, References [5] - [6] proposed a hybrid transformer-type adjustable reactance based on inverter compensation and thyristor switching control. However, the design of the transformer body of this structure is more complicated. It is necessary to establish a multilateral equivalent model of a multi-winding transformer, and the cost is high when the capacity is large, so it is not suitable for use.

In view of the defects of the above adjustable reactance device, combined with the demand for real-time compensation of the distributed capacitance of the urban cable, this paper proposes a new type of multi-stage continuously adjustable multi-winding reactor [7] - [8]. Through the switching control of the thyristor and the control of the output current of the inverter, the arbitrary adjustment of the equivalent impedance of the primary side of the transformer is realized, which reduces the output capacity of the inverter and reduces the cost. The complexity of the mathematical model is reduced by only one bidirectional thyristor switch and a full-bridge inverter module in each operating condition. The validity and correctness of the proposed method are preliminarily proved by MatLab software.

2.principle analysis

The multi-winding transformer is shown in Fig.1, including primary winding and N secondary windings, $N \geq 3$; the two ends of the primary winding are connected in series in the power grid cable line ; the M secondary windings and the corresponding M bidirectional thyristors are connected in series or / and in parallel to form the passive series / parallel shunt reactor shunt, $M < N$; the shunt of the active inverter fine-tuning reactor includes a single-phase full-bridge conversion circuit, whose AC side is connected to any of the remaining N-M secondary windings in the multi-winding transformer.

The controller is used to control the number k of inductive reactance loops composed of secondary windings and corresponding bidirectional thyristors put into operation in the shunt of passive series / parallel coarse-tuning reactor by controlling the switching state of M bidirectional thyristors, and then the inductive impedance value output by the primary winding is adjusted step by step, so that the inductive impedance value is roughly matched with the distributed capacitance impedance of the cable, $0 \leq K \leq M$; the controller is also used to generate a control signal to control the on-off of the switching tube of the full-bridge converter circuit, so as to control the output voltage and current of the AC side of the full-bridge converter circuit, and then continuously adjust the inductive impedance value after coarse matching, so as to realize the ' fine adjustment ' of the reactance value, so that the inductive impedance value after continuous adjustment is completely matched with the distributed capacitance impedance of the cable, and the distributed capacitance compensation of the cable is realized.

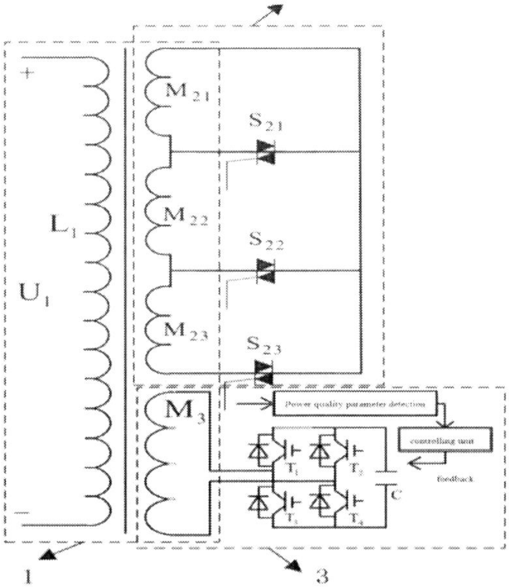

Figure 1. Multi-winding grading reactor circuit structure diagram

2.1 transformer equivalent circuit

The number of turns and physical parameters of each winding in Figure 1 are converted to the primary side, and M thyristors can divide the adjustable reactance into M + 1 working conditions. K = 0 is the first condition, and the voltage and current equations of the multi-winding transformer are as follows.

$$U_1 = \mathrm{jw}(I_1 L_1 + I_3 M_3) \tag{1}$$

At this time, all thyristors are open, and only the full-bridge inverter is put into operation.

When 0< K≤M, the thyristor is only S_{2K} closed, and the thyristor switching windings M_{21}, M_{22} and M_{2K} are put into operation. The voltage and current equations satisfied by the multi-winding transformer are

$$U_1 = \mathrm{jw}(I_1 L_1 + (I_{21} M_{21} + I_{22} M_{22} + \cdots + I_{2K} M_{2K}) + I_3 M_3) \tag{2}$$

According to the different K values, a total of M working conditions are included.

The equivalent circuit of the adjustable reactance is equivalent to the T-type equivalent circuit of the two-winding transformer under the first working condition, and equivalent to the three-winding transformer under the second to M + 1 working conditions. The equivalent circuit diagram of the three-winding transformer can be obtained as shown in Figure 2.

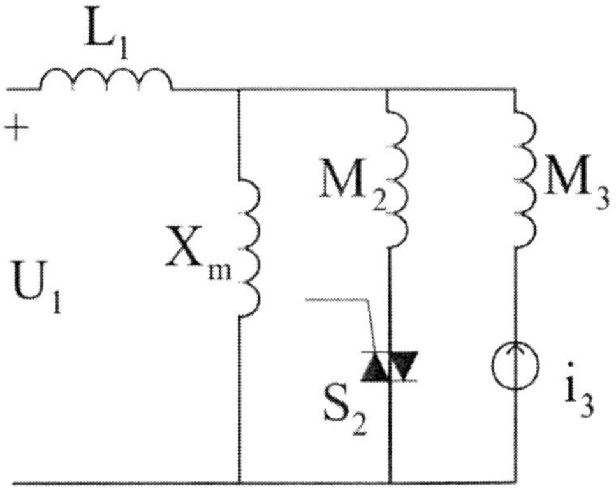

Figure 2. Equivalent circuit diagram of three-winding transformer

The capacity of the adjustable reactance mainly comes from the thyristor switch winding. The inverter control winding only plays a connecting role between various working conditions. When only the thyristor switch winding $M_{21}$is put into operation, the equivalent reactance is the largest and the capacity is the smallest. When all the switching windings are put into operation, the equivalent reactance is the smallest and the rated capacity is output.

[9] In order to simplify the analysis, the equivalent circuit of the three-winding transformer can be simplified into a two-port network model by only caring about the parameters of the primary side and the full-bridge conversion circuit, as shown in Figure 3.

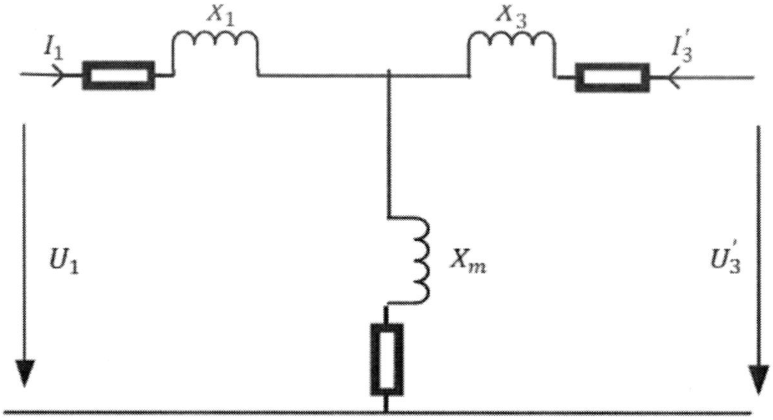

Figure 3. Two-port network

The corresponding potential equation can be obtained

$$U_1 = Z_{11}I_1 + Z_{12}I_3' \tag{3}$$

Among them, X_1 is the equivalent impedance of one side winding, $X_3{}'$is the equivalent impedance of two side winding, andX_m is the equivalent excitation impedance.

If the output current of the secondary side inverter tracks the primary side current, that is, $I_3 = \beta K I_1$, $I_3 = \beta K I_1$, $K$is the ratio between the primary winding and the fine-tuning reactor winding, then the primary side port impedance of the transformer can be obtained as

$$Z_1 = \frac{U_1}{I_1} = X_1 + (1 + \beta)X_m \tag{4}$$

From the formula (4), it can be seen that the multi-winding grading reactor can realize the continuous linear adjustment of the reactance under various working conditions. The two-port network model can simplify the system.

2.2. *The continuously adjustable grading reactor*

The capacity adjustment range of the transformer is S_{min}~S_{max}, and the change range of the equivalent impedance of the primary side is

$$\frac{U_N^2}{S_{max}} \leq Z_1 \leq \frac{U_N^2}{S_{min}} \tag{5}$$

In the M operating conditions when the thyristor switch winding is put into operation, the rated capacity adjustment range is divided into M intervals on average, and the capacity range of each interval is

$$\left[S_{min} + (\mathrm{m} - 1) * \frac{S_{max} - S_{min}}{M}, S_{min} + \mathrm{m} * \frac{S_{max} - S_{min}}{M}\right] \tag{6}$$

Where m=1、2...M。

If the adjustable reactance needs to be continuously and steplessly adjustable, the full-bridge inverter module works in two conditions of inductive capacitance, that is, β can be positive or negative, and the inverter outputs the rated capacity at this time.

$$S_{3N} = \frac{S_{max} - S_{min}}{2M} \tag{7}$$

The capacity is provided by the coarse adjustment reactor shunt

$$S_2 = S_{min} + \left(m - \frac{1}{2}\right) * \frac{S_{max} - S_{min}}{M} \tag{8}$$

It can be seen from formula (4) (7) (8). When $\beta = 0$, that is, the output capacity of the inverter is 0, the primary side impedance of the adjustable reactance is provided by the coarse-tuning reactor shunt. At this time, it can be obtained that

$$X_1 + X_m = \frac{U_N^2}{S_2} = \frac{U_N^2}{S_{min} + \left(m - \frac{1}{2}\right) * \frac{S_{max} - S_{min}}{M}} \tag{9}$$

BecauseX_1 is far less thanX_m, it can be considered that $X_m \approx \frac{U_N^2}{S_2}$. Therefore, by making X_1 value as far as possible less than X_m, the multi-winding transformer can be continuously adjustable by controlling the output current of the full-bridge inverter circuit to control the size of $(1+\beta)X_m$.

3 Simulation prove

3.1. *Reactor simulation parameter design*

In the matlab simulation, a reactor with a rated capacity of 1000KVA and a rated voltage of 10KV is designed. The rated current is 100A and the excitation impedance is 2350Ω. Under the rated voltage, the reactor capacity is continuously adjustable in the range of 200 ~ 1000KVA.

The M value is equal to 4, at this time, the adjustable reactance is divided into 5 working conditions, and the capacity difference between each working condition is 200KVA.At this time, the output rated capacity of the full-bridge inverter module is 100KVA, and the output rated voltage of the inverter is 500V. The maximum output current of the inverter is 100KVA / 500V is 200A.Considering the need to retain the margin, the IGBT full-bridge used for fine-tuning the reactor can adopt the specification of 1200 / 500A.Considering the switching loss and the effect of the inverter, the switching frequency is set to 10KHZ.

As shown in Table 1, the capacity range and short-circuit impedance of the five working conditions, as well as the working conditions of the thyristor switching windings, are known. At this time, the reactor is graded and adjustable between 100 and 2500 Ω.

Table 1. Graded reactor short-circuit impedance at all levels

working condition	range of capacity	Short-circuit impedance	Working thyristor winding	X_m	X_1
1	0~200KVA	2500Ω	None	2350Ω	145Ω
2	200~400KVA	333Ω	M21	297.5Ω	32Ω
3	400~600KVA	200Ω	M21~M22	178.5Ω	19.1Ω
4	600~800KVA	142.9Ω	M21~M23	126Ω	14.4Ω
5	800~1000KVA	111.1Ω	M21~M24	98.2Ω	11.3Ω

According to the design principle of thyristor switch winding, the rated current of the short circuit of the switch winding should be consistent under each working condition, and it is selected as 500A.Then the ratio of the total number of turns of the primary side winding to the thyristor switch winding should be 500 * 111.1 / 10000 = 5.55, and the maximum voltage of the thyristor is about 10000 / 5.55 = 1800V.Considering the need to retain the margin, the thyristor specification can be selected as 6KV / 800A.

According to the above conditions, the winding structure of the grading reactor can be designed. The number of turns of the primary side winding is designed to be 1000 turns ; according to the difference of primary side short circuit current in each working condition, the number of turns of thyristor winding M_{21} is designed to be 60, and the number of turns of M_{22}、M_{23}、$M_{24}$is designed to be 40. The fine-tuning reactor winding is designed to be 50 turns, and in order to stabilize the DC-side voltage value of the full-bridge conversion circuit, the DC-side capacitance value is designed to be 5000 μF.

3.2. simulation result

In order to prove the effectiveness of the grading reactor to realize the continuous grading adjustment of the reactance, based on the above theory, the output current of the full-bridge inverter circuit is used to track the primary side current and maintain the proportional relationship under five working conditions. The equivalent impedance that changes linearly with β can be obtained. Figures 4 ~ 8 are the waveforms of the primary side voltage U_1, the primary side current I_1, and the output current I_3 of the full-bridge inverter circuit under working conditions 1 ~ 5, respectively. It is based on the different compensation requirements of the distributed capacitance of the urban cable Z_{nd}, and then the corresponding β value is calculated by the X_m and X_1 under the working condition, and simulated under the corresponding reactor parameters.

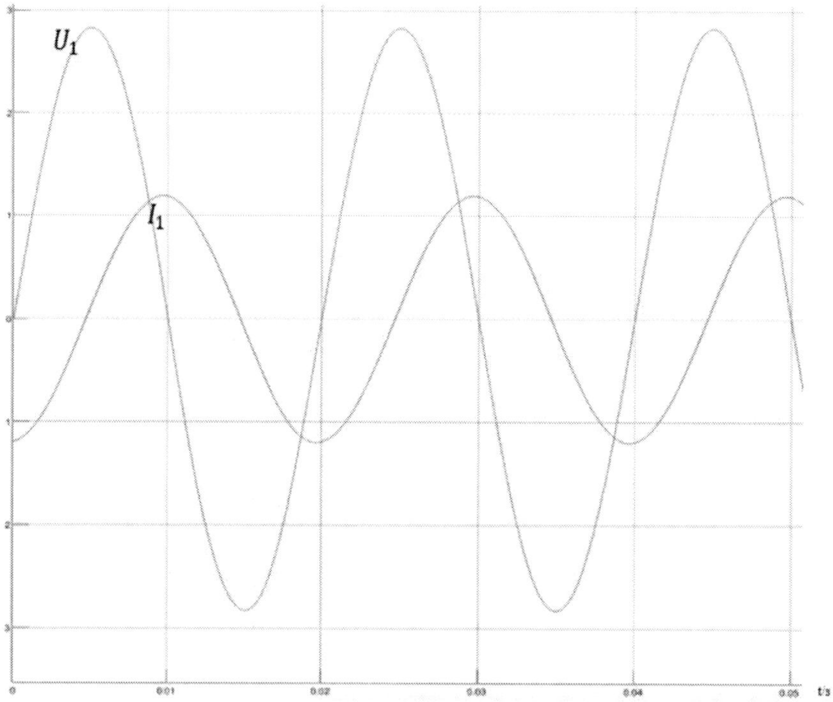

Figure 4. U_1: 5KV/div; I_1:5A/div;
Working condition 1（Z_{nd}=2450Ω，β=-0.31）

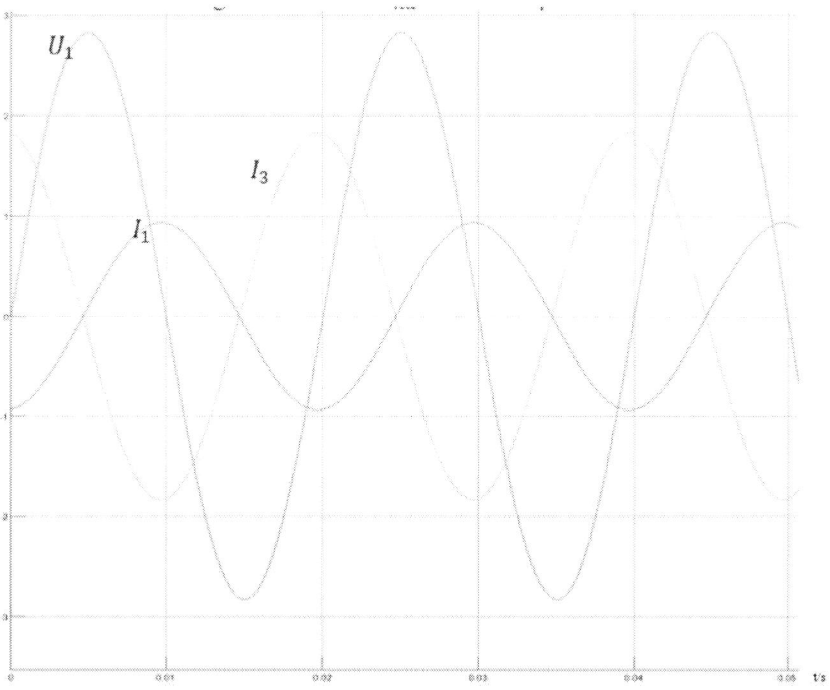

Figure 5. U_1： 5KV/div; I_1:50A/div; I_3： 150A/div
Working condition 2（Z_{nd}=320Ω， β=-0.30）

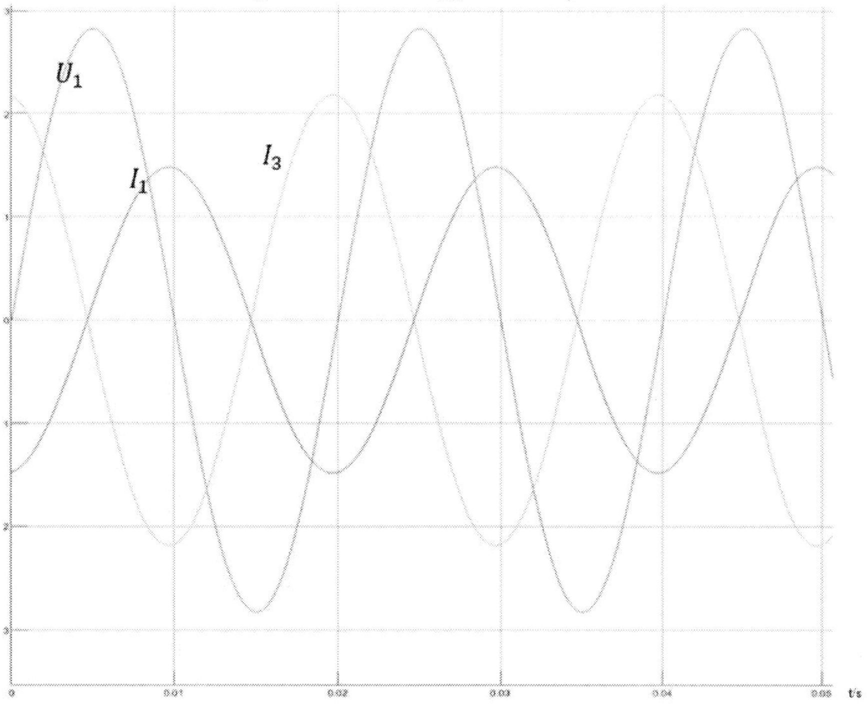

Figure 6. U_1： 5KV/div; I_1:50A/div; I_3： 150A/div
Working condition 3（Z_{nd}=193Ω， β=-0.24）

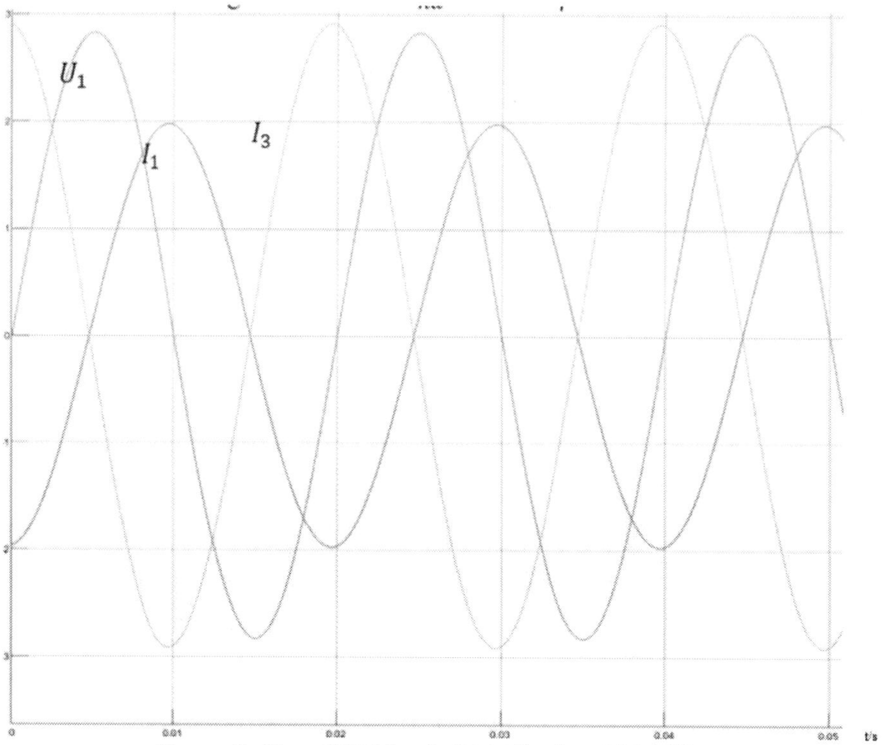

Figure 7. U_1: 5KV/div; I_1:50A/div; I_3: 200A/div
Working condition 4（Z_{nd}=136Ω，β=-0.31）

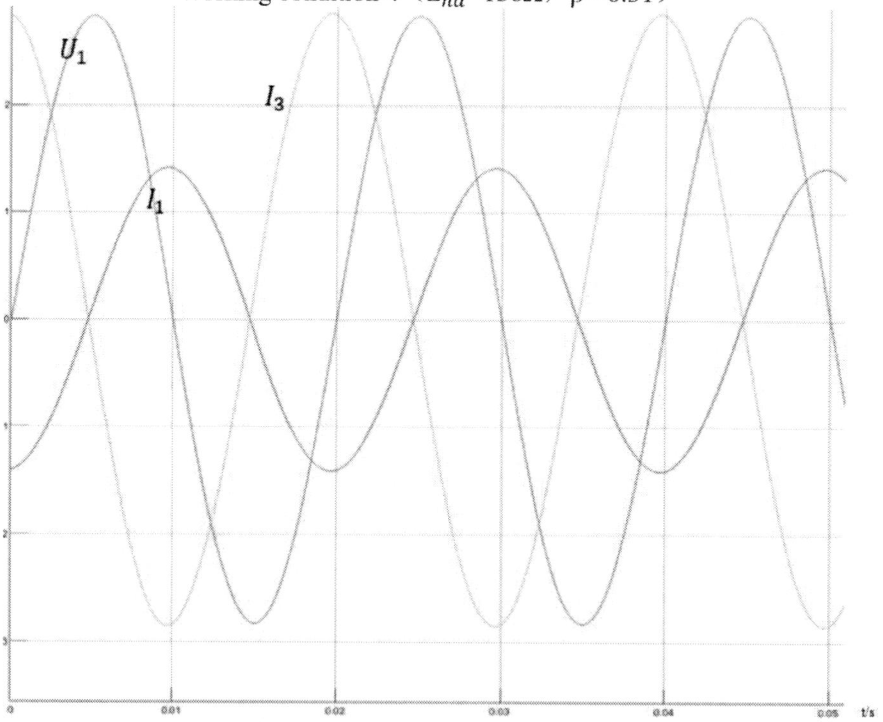

Figure 8 U_1: 5KV/div; I_1:100A/div; I_3: 300A/div
Working condition 5（Z_{nd}=106Ω，β=-0.31）

The harmonic content of voltage and current waveforms in figures 5 to 9 is small, and the current I_1 of the primary side of the grading reactor lags behind U_1close to the 90 degree angle, and the equivalent impedance is approximately pure inductance, which meets the needs of urban cable capacitance compensation. Comparing the actual equivalent impedance U_1/I_1 the simulation legend with the compensation impedance Z_{nd} of the distributed capacitance of the cable, the error is less than 5 % under five working conditions, which meets the requirements of high compensation accuracy of the distributed capacitance of the urbanized cable.

4.Conclusion

In this paper, a multi-winding grading reactor for distributed capacitance compensation of urbanized cables is proposed. The device reduces the cost by reducing the output capacity of the full-bridge conversion module. The coarse adjustment function of the thyristor switch winding and the fine adjustment function of the full-bridge inverter module realize the adjustable grading on-demand compensation. In the principle analysis, the design idea of the model and parameters is expounded, and the feasibility and excellent performance of the reactor scheme are verified by simulation.

。

Acknowledgments

This work was supported by the Key Science and Technology Project of China Southern Power Grid Co., Ltd. under Grant 031400KC23120011

References

[1] Y. Zhang. 1997, Automatic tuning of arc suppression coil. East China Electric Power (vol 8), pp.48-49.
[2] X. Cai. 2002, New bias magnetic arc suppression coil and its control. Power system automation (vol 10), pp.32-35.
[3] Ma Y H, Wang X L, Li Q F. 2007, Principle and characteristic analysis of magnetically saturated controllable reactor. Journal of Shenyang University of Technology (vol 5), pp. 47-49 + 61.
[4] Chang S Z. 2004, Application research of thyristor switched capacitor (TSC) type arc suppression coil control device. Tsinghua University (vol 3).
[5] Ding H F, Zhu Q C. 2005, Hybrid transformer controllable reactor and its simulation. Relay (vol 10), pp. 25-30.
[6] Liu H T, Zhang W X, Y. Liu, et al. 2006, Analysis of reactive power capacity of control winding of controllable reactor with transformer type. Journal of Power System and Automation (vol 6), pp. 70-72 + 112.
[7] Li D Y, Chen Q F, Jia Z C. 2003, New principle of adjustable reactor based on controllable magnetic flux. Chinese Journal of Electrical Engineering (vol 2), pp.120-124.
[8] Y. Zhang, Chen Q F, J Tian, etc. 2009, Controllable reactor based on transformer port regulation. Chinese Journal of Electrical Engineering (vol 18), pp.113-118.
[9] Xu Y C, Fan S R, Xie S S, etc.2022, A power quality conditioner suitable for high permeability active distribution networks. Journal of Motor and Control(vol 26), pp. 46-56 + 67

Journal of Physics: Conference Series **3012** (2025) 012032 doi:10.1088/1742-6596/3012/1/012032

Research on the Design of Enclosed Protective Discharge Gap for Transformer Neutral Point Based on Energy Aggregation Characteristics

Xiaohui Li[1*], Jun Kang[2], Jianwu Li[2], Shanghu Zhou[3], Fujie Wang[1], Guoxiang Ma[2], Zhiwu Qin[1], Huifang Dong[1], Yuanyue Zhang[1], Junjun Yang[1], Chengxiao Tong[1], Shifeng Liu[1]

[1] Haibei Power Supply Company of State Grid Qinghai Electric Power Company, Haibei, Qinghai Province, 812200, P.R. China

[2] State Grid Qinghai Electric Power Company, Xining, Qinghai Province, 630100, P.R. China

[3] State Grid Qinghai Electric Power Co.Ltd. Research Institute, Xining, Qinghai Province, 630100, P.R. China

*Corresponding author's e-mail: 972286316@qq.com

Abstract. Power transformers are very important components of the power system, playing a crucial role in controlling the voltage of electrical energy and driving the transmission of electrical energy. However, in transformers above 110kV, the breakdown voltage of the protection gap is highly dispersed. Due to the influence of altitude and other factors, the protection effect of the neutral point, especially in Northwest China, is poor, prone to protection mal-operation, which seriously affects the lives of residents and social production. Based on this the development of neutral point protection equipment is of great significance. This article focuses on the energy concentration characteristics of discharge gaps in enclosed spaces. By calculating the arc energy value and the number and duration of energy super-positions within the protective equipment's enclosed casing, the enclosed shell structure and design pressure boundary are determined. An enclosed shell that meets the usage requirements is designed, and the accuracy and rationality of the design are verified through various performance tests.

1. Introduction

Power transformers are electrical equipment that are extremely numerous and important in the power system. They serve the pivotal role of increasing or decreasing voltage of electrical energy in the power system, facilitating the rational transmission, distribution, and use of electrical energy. Therefore, the operation state of the transformer in the power system, especially the high-voltage transformer, directly determines the safe and stable operation of the power grid system[1-2].

At present, transformers above 110kV generally adopt graded insulation. When the neutral point is not grounded, in order to protect the neutral point insulation, protection gap is usually set at the neutral point of the transformer to avoid continuous overvoltage damage to the transformer at the neutral point of the main transformer. However, in recent years, it often occurs that the protection effect of the neutral point is poor due to the large dispersion of the breakdown voltage of the neutral point protection gap of the main transformer, resulting in protection mal-operation events and even total station shutdown accidents, especially in Northwest China[3-4]. Due to the different altitudes of the substations in the region, high-voltage electrical test professionals are required to continuously correct the neutral point discharge gap distance through on-site tests. Due to the influence of atmospheric humidity, temperature, salt spray, high and low temperature, atmospheric pressure and other external environment, the corrected discharge gap distance has a large error, resulting in many local and municipal power supply companies and ultra-high voltage companies are afraid to put the discharge gap into operation or set the discharge gap. There are great hidden dangers in the safe operation of the grid voltage device. Once the neutral point steady-state overvoltage occurs, the transformer burning accident is likely to occur, which will cause huge impact and threat to people's production and life. Therefore, it is of great theoretical and practical significance to study how the neutral point discharge gap of transformer can operate correctly under steady-state overvoltage, and to put forward the development of intelligent closed neutral point protection equipment.

A lot of research has been done on transformer neutral point protection device at home and abroad. At present, the mainstream neutral point protection devices in engineering applications mainly have three structures: surge arrester, discharge gap and parallel connection of arrester and discharge gap. Due to the inherent thermal stability problem of arrester, when the steady-state over-voltage condition of continuous operation occurs at the neutral point of transformer, the device is prone to loss of thermal stability and damage or explosion. Discharge gaps are severely influenced by external factors, leading to significant discharge variability and unstable gap withstand voltage values. In practical engineering applications, this can result in "failure to operate" or "misoperation" scenarios. When discharge gaps are paralleled with arresters, the arresters primarily protect against lightning overvoltages, while the discharge gaps mainly protect against power frequency overvoltages and other internal overvoltages. Since the lower limit of discharge voltage under lightning condition of discharge gap is close to the residual voltage of zinc oxide arrester, the gap is very easy to misoperate and cause the arrester to lose its function[5-8].

When designing neutral point protection devices for transformers, the enclosed shell is a component that requires special attention. Due to the harsh and harsh working environment of the protective device, the shell will be subjected to various mechanical, thermal, electrical and other hazards and challenges during normal operation, which can easily lead to malfunctions or even explosions, seriously affecting the efficiency of the protective device.

This paper studies the energy accumulation characteristics of the enclosed space of the discharge gap, aiming at determining the air pressure boundary conditions for the design of the protective device cavity. Through the calculation of the electric energy value of the arc in the enclosed shell and the number and time of energy superposition, the enclosed shell that meets the use requirements is designed, and through the performance tests, the closed shell structure and the design air pressure boundary are finally determined.

2. Enclosed shell design

In practical applications, the enclosed shell of protective devices needs to meet various complex conditions to achieve stable long-term operation. Starting from the energy accumulation characteristics of enclosed spaces, this chapter analyzes the requirements for shell pressure, shell mechanical load, and shell sealing, and designs the enclosed shell of protective devices according to actual work needs.

The design object of this article is the closed discharge gap structure shown in Figure 1, according to the existing design experience of protection device, its design parameters are as follows: The neutral point protection device has a rated voltage of 110kV. At a transformer capacity of 120 M/W, the neutral point short-circuit current does not exceed 2.01kV for the vast majority of cases, this paper takes the maximum value of 2.01kA(considering the 1.1 times margin, the short-circuit current is 2.211kA). The cavity is a circular cavity with a cylinder length of 1330mm, a diameter of 320mm and a capacity of $0.107m^3$. The cavity is filled with 0.1MPa SF_6 insulating gas, the shell is made of aluminum alloy.

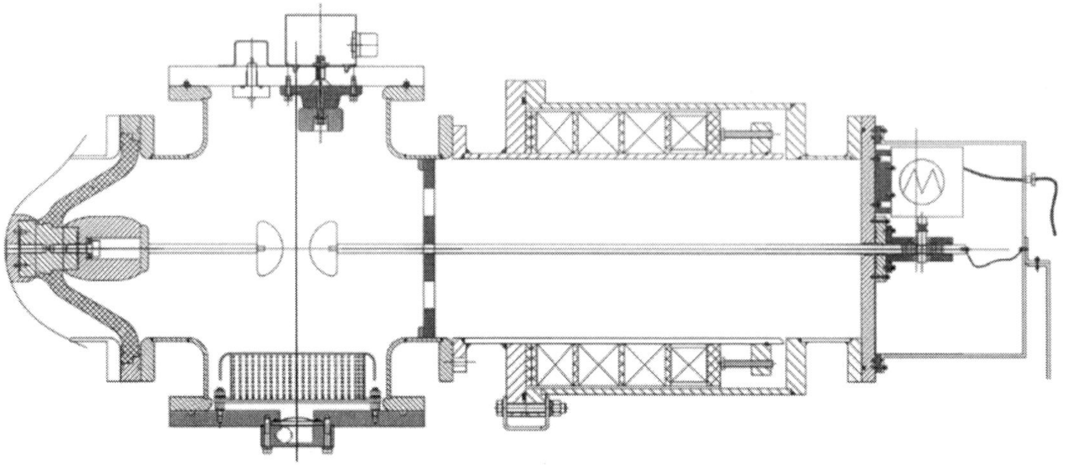

Figure 1. Schematic diagram of protective device structure

2.1. Shell wall thickness and protection time

The wall thickness of the shell should be designed based on the internal pressure of the shell and the shortest endurance time for the shell to not burn through, which needs to meet the following conditions:

(1)It should be able to withstand 2.211kA (root mean square value) fault current arc under rated pressure, without burning through within 900ms, SF6 gas leakage, and no external effects;

(2)It can burn through within 1000ms, SF6 gas can leak, but there must be no fragments.

To avoid shell burning through, the duration of the fault current should be coordinated with the shell design and compartment size. It is analyzed according to the power system standard documents. According to the standard documents, "The primary current setting of the neutral point discharge gap zero sequence current protection for 110kV transformers can generally be set to 40-100A, with a 0.3-0.5s delay tripping of the circuit breakers on each side of the transformer after the protection action. For zero sequence voltage protection of a semi insulated 110kV transformer with a neutral point grounded through a discharge gap, the 3U0 setting is generally set to 150-180V (rated at 300V), and after the protection action, it is equipped with 0.3-0.5s delay tripping of the circuit breakers on both sides of the transformer".

Therefore, the continuous discharge time of the neutral point protection device is $\geq$0.5s, and according to relevant literature, some faults may last for 1 second due to issues such as inability to extinguish the arc and inadequate coordination of circuit breaker protection.

Based on the above requirements, the wall thickness of the enclosed shell is designed to be 6mm, and the action protection time, that is, the duration of the fault current, is set to 1 second.

2.2. Internal structure

The following factors shall also be considered in the design of the shell:

(1) The enclosure is evacuated as part of the normal inflation process;

(2) The total pressure difference may be applied to the shell wall;

(3) The pressure generated by accidental leakage between adjacent compartments under different operating pressures;

Based on the above requirements, the main technical parameters of the device shell are calculated as shown in Table 1[9-11].

Table 1. Main technical parameters table of shell

Name	Mark	Data	Name	Mark	Data
Shell internal diameter	D_i	320mm	Bolts allowable stress deign	$[\sigma]_b^t$	105MPa
Flange outer diameter	D_0	475mm	Flange allowable stress	$[\sigma]_f$	68MPa
Sealing ring diameter	D_{G1}	373mm	Flange allowable stress design	$[\sigma]_f^t$	68MPa
Bolt center circle diameter	D_b	407mm	Allowable stress of cylinder at design temperature	$[\sigma]t$	47.5MPa
Bolt hole diameter	d_b	12mm	Welded joint coefficient	φ	0.9
Calculation pressure	P_c	0.2MPa	Shell corrosion allowance	C2	0mm
gasket factor	m	0.5	Negative deviation of shell wall thickness	C1	0.6mm
Gasket pressure ratio	y	0MPa	Bolts allowable stress	$[\sigma]_b$	117MPa

After obtaining the main technical parameters, use these parameters to calculate the required number of bolts and flange thickness for the shell.

2.2.1. Bolt quantity. It is known that the stress sectional area of bolts in the shell M20 is 245mm^2. Just calculate the minimum bolt cross-sectional area A_p under the operating state, and then divide A_p by M20 to obtain the number of bolts.

The total axial force caused by internal pressure F is:

$$F = 0.785 D_{G_1}^2 \cdot P_c = 21843N \tag{1}$$

Minimum gasket clamping force required in operating state F_P is:

$$F_p = 6.28 D_{G_1} bmP_c = 1756.8N \tag{2}$$

Due to gasket pressure ratio being set to 0, the minimum clamping force of the gasket in the pre tightened state is:

$$F_a = 3.14 D_G by = 0 \tag{3}$$

Through the above calculated parameters, the minimum bolt load under operation W_p can be calculated by adding F and F_P. Minimum bolt cross-sectional area in operating state can be calculated as:

$$A_p = W_p / [\sigma]_b^t = 225mm^2 \tag{4}$$

The required number of bolts is calculated as 1 based on the minimum bolt cross-sectional area and bolt stress cross-sectional area. Considering the harsh on-site installation and operation conditions, it is recommended to have 12 bolts.

2.2.2. Flange thickness. Total axial force caused by internal pressure is:

$$F = \pi D_{G_1}^2 P_c / 4 = 24843N \tag{5}$$

Axial force caused by internal pressure acting on the inner diameter section of the flange is:

$$F_D = \pi D_i^2 P_c / 4 = 16076N \tag{6}$$

Flange torque is:

$$M_0 = F_D L_D + (F - F_D) L_T + F_P L_P = 1624724mm \tag{7}$$

Flange thickness is:

$$\delta f = \sqrt{6M_0 / [\sigma]_f^t (3.14 D_b - n d_b)} \tag{8}$$

By calculation, the flange thickness is 11.2mm. Considering the harsh installation and operation conditions on site, it is recommended that the flange thickness be 25mm.

2.3. Shell sealing

Due to the leakage of the shell, the insulation performance of the protective device will decrease, partial discharge will increase, and even damage the equipment. Therefore, when designing the device shell, its sealing performance needs to be considered, mainly focusing on the design of its flange docking sealing structure and basin sealing structure.

2.3.1. Seal structure design. The protective device designed in this paper uses a flange connection with an O-ring seal to reduce the leakage rate of SF_6 gas. This design effectively delays the aging process of rubber materials affected by the environment during long-term operation, greatly enhancing the sealing reliability. The O-ring has a shore hardness of 70, which is relatively lower, resulting in less contraction of the cross-section under the same stretching conditions, leading to better sealing performance and a longer lifespan. With a cross-sectional diameter of Φ7mm, the sealing band width is larger under the same compression amount, resulting in less permanent compression deformation. Compared to rectangular sealing grooves, using a "T"-shaped groove can provide greater contact pressure, thereby improving sealing performance.

The sealing of the shell's pot-type insulator also uses an O-ring to enhance the sealing reliability. The design scheme of flange butt sealing structure and basin sealing structure are shown in figures 2 and 3.

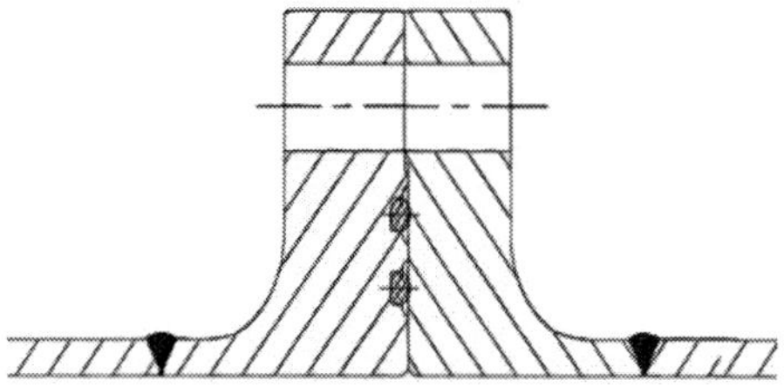

Figure 2. Flange sealing structure diagram

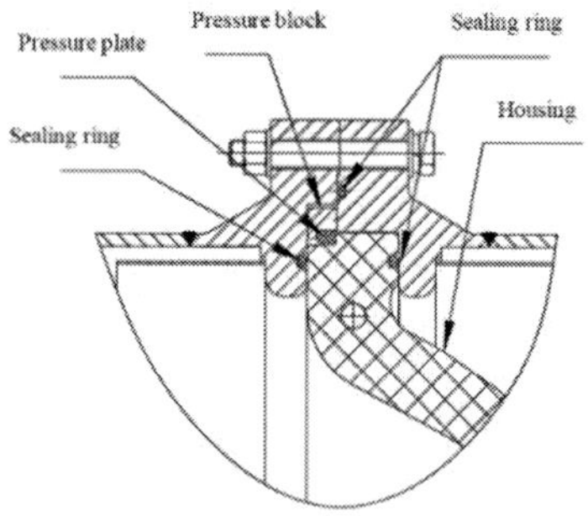

Figure 3. Basin sealing structure diagram

2.3.2. Technical parameters. After completing the design of the shell structure, the main technical parameters of the sealing structure were calculated as shown in Table 2.

Table 2. Main technical parameters of shell sealing

Parameter	Data
Sealing structure	Double seal
Groove position of sealing groove	Φ 373mm
Seal ring wire diameter	Φ 6mm
Seal ring material	EPDM rubber
Compression ratio of sealing ring	22%
Seal ring hardness (shore a type)	70 ± 5

2.4. Parameter summary

According to the above calculation and design, the key parameters and verification setting conditions related to the shell are summarized in Table 3 for clearer and easier reference.

Table 3. Experimental parameters and set conditions

Parameter	Data	Parameter	Data
Flange thickness	25mm	Shore hardness	70
Bolts number	12	Transformer capacity	120 M/W
Shell wall thickness	6mm	Gas filling volume	0.1MPa
Cylinder capacity	0.107m^3	Fault current duration	1s

Rated voltage	110kV	Cylinder diameter	Φ 320mm
Short-circuit current	2.211kA	Bursting pressure	2.5MPa

3. Soundness verification

After completing the design of the key parameters of the shell, verify the rationality and accuracy of the design parameters according to the requirements of the standard documents.

3.1. Shell damage and burning through

Under the action of internal fault arcs, the change process of gases in the gas chamber can be divided into two stages: before and after the casing is punctured. Due to the difficulty in determining the shape of the casing puncture in the arc root area and the drastic changes with time and arc current, it is challenging to simulate the state parameters of the gas chamber after the casing puncture. However, since the temperature, pressure, and specific mass in the gas chamber at a certain moment after the casing puncture are lower than the state parameters at the same moment without puncture, simplifications are made in the design according to the most severe conditions, that is, without considering the casing puncture.

The SF_6 gas in the gas chamber is considered an ideal gas, and it is assumed that all the energy dQ_e of the arc during the dt time is used to increase the internal energy of the gas. Additionally, chemical reactions that release energy will occur between the SF_6 gas and the casing after the arc is generated, and this chemical energy dQ_c will also increase the thermal energy in the chamber, increasing with temperature T. It can be set that $dQ_c = hTdQ_e$, where h is a coefficient that can be taken as $h = 5\times10^{-4}$ (K^{-1}). Therefore, the total thermal energy produced by the arc in the gas chamber at time t during the dt period is:

$$dQ = (1 + hT)UI\sqrt{2}\left|\sin(\varpi t)\right|dt \tag{9}$$

where U is arc voltage, I is effective value of arc current, $\omega=100\pi$.

The arc voltage is calculated according to the calculation of domestic and foreign manufacturers and relevant data according to 20~25V/cm. Combined with the design conditions in this paper, the maximum arc voltage is 400V, which is taken as 400V in the subsequent calculation.

Since all the energy of the arc is used to increase the internal energy of the gas, the relationship between the temperature rise of the shell and the increase in internal energy is:

$$d(mC_vT) = mC_vdT = dQ \Rightarrow dT = dQ/mC_v \tag{10}$$

where C_v is specific heat of constant volume SF_6 gas obtained based on thermodynamic principles, m is gas mass.

Thus, according to the ideal gas law, we can deduce gas pressure P is:

$$\frac{dT}{T} = \frac{dP}{P} \tag{11}$$

For the convenience of simulation, discretize the above equation:

$$\begin{cases} \Delta Q = (1 + hT)UI\sqrt{2}\left|\sin(\varpi t)\right|\Delta t \\ T_1 = T_0 + \Delta Q/mC_v \\ P_1 = P_0 + P_0(T_1 - T_0)/T_0 \end{cases} \tag{12}$$

There are many physical factors that cause the shell to burn through in the arc root area of the arc. Currently, it is difficult to theoretically obtain the arc burning time before the shell burns through. The burning time of aluminum shells is related to the fault current and shell wall thickness as follows:

$$t = 87.4\frac{d^{1.77}}{I^{0.67}} \tag{13}$$

where d is shell wall thickness, I is effective value of fault current.

Based on the above calculations, under a fault current of 2.21kA, the air pressure value is much lower than the shell failure pressure value 2.5MPa. When the wall thickness is 6mm, the shell burning through time is 1224ms, which is significantly higher than the specified 900ms. The design in this article meets the requirements.

3.2. *Mechanical load*

When designing the wall thickness of the shell, it should be considered that it will not produce permanent deformation and affect the product performance when it is subjected to external forces during transportation and installation. Using software to calculate shell stiffness parameters, simulation setting conditions and simulation results are shown in figures 4, 5 and 6.

According to the national standard documents, the bursting pressure of the aluminum shell is 2.5MPa. it was found that a pressure of 2.5MPa was applied to the inner surface of the shell, resulting in a maximum deformation of 1.89 mm and a maximum equivalent stress of 260.54MPa.The national standard requires that the minimum tensile strength of the pipe is 270MPa, so the design in this paper meets the application requirements.

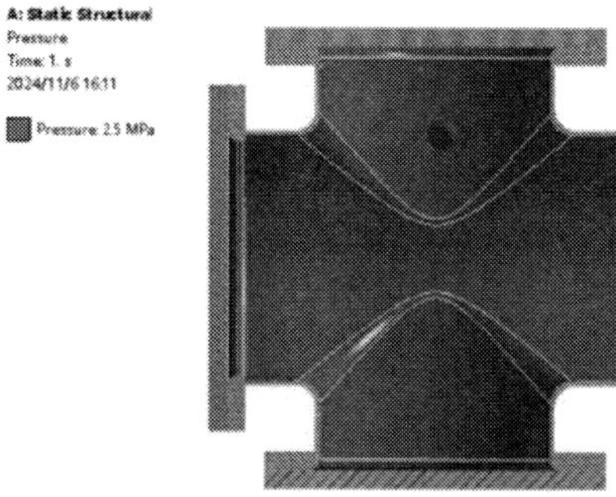

Figure 4. Internal loading diagram of the shell

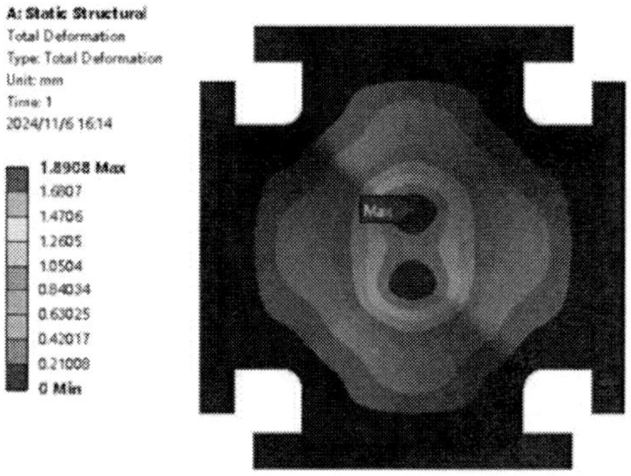

Figure 5. Shell deformation diagram

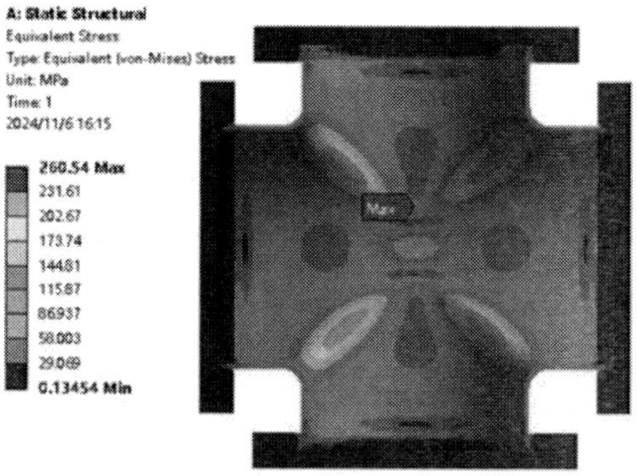

Figure 6. Shell stress distribution diagram

3.3. *Shell leakage*

When checking the leakage of the product with the cover, the leakage rate f is calculated according to the concentration of the leakage gas in the cover, the volume of the cover, the volume of the test object and the absolute pressure of the test site is:

$$F = \frac{\Delta C(V_m - V_1)P}{\Delta t} \tag{14}$$

where ΔC is increment of the leakage gas concentration, V_m is volume of the enclosed chamber, V_l is volume of the object being detected, P is environmental atmospheric pressure during the detection time.

Relative annual air leakage rate is:

$$F_y = \frac{FT}{V(P_r + P)} \times 100\% \tag{15}$$

where V is volume of gas sealing system of test object, P_r is Inflation pressure.

According to equations 14 and 15, the increment of leaked gas concentration from the beginning to the end of the experiment is:

$$\Delta C = \frac{F_y V(P_r + P)\Delta t}{\Delta V \cdot PT} \tag{16}$$

The inner diameter of the design shell is 320mm, the outer diameter is 332mm, the diameter of the conductive rod is 70mm, and the total length is 1.4m, thus the calculated volume V of the test sample is 1.73mm^2. Referring to the conventional situation, the inflation pressure is 0.1MPa, the ambient atmospheric pressure P is 0.1MPa, and the gap volume between the leak detection tool and the product is 0.07.

Based on the test object of the designed detection cover, we carried out the sealing performance verification test of double "O" ring. During the test, no SF6 leakage was detected inside the hood, and the concentration increase was 0 ppm. Based on an annual leakage rate of 0.01%, it can be calculated that the SF6 leakage detected within the leak detection hood is 37.2ppm. This design clearly meets the requirements.

4. Conclusion

Starting from practical engineering requirements, this article aims to study the energy accumulation characteristics of discharge gaps in enclosed spaces, calculate the energy values and energy superposition times and times of arc electricity inside the enclosed shell, and design a closed shell that meets the requirements of use to solve practical engineering application problems. Conduct various performance tests according to the requirements of the national standard documents, and ultimately determine the closed shell structure and design pressure boundary. The inspection results indicate that the design in this article meets the requirements of the national standard document.

References

[1] Yan Z and Zhu DH 2015 High Voltage Insulation Technology (Beijing: China Electric Power Press)

[2] Yang ZY 2022 Research on Analysis of Overvoltage at the Neutral Point of Power Transformers and Insulation Protection Technology (Shandong: Shandong University of Technology) pp 1-15

[3] Xu XJ 2019 Analysis of Tripping Fault Caused by Breakdown of Neutral Point Gap of Main Transformer (Electric Power System Equipment vol 24) pp 142-143

[4] Wang J , Ma LT and Wang XT 2020 Research on the Parameter Matching Characteristics of Gapped Lightning Arresters for the Neutral Point of 110kV Transformers (High Voltage Apparatus vol 56) pp 0228-0232

[5] Wang PF 2023 Study of Ball Gap Triggered Discharge Characteristics of Transformer Neutral Point Protection (Harbin University of Science and Technology) pp 1-10

[6] Winands G J J, Liu Z, Pemen A J M 2005 Long Lifetime, Triggered, Speek-gap Switch for Repetitive Pulsed Power Applications(Review of Scientific Instruments) pp 085107/1-085107/6

[7] Xu MC 2016 Study on Controlled Gap of 110kV Transformer Neutral Point (BeiJing: North China Electric Power University)

[8] Yang X, Dai HL, Shi HP 2016 Transformer Neutral Protection Method of Composite Horn Gap Parallel to Surge Arrester (HeBei: Power System Technology) pp 23-37

[9] Huang JH, Wang WG, Shou BN 2008 Comparison on Each Country's Method of Confirming Allowable Stress of Pressure Vessels Materials (Pressure Vessel Technology vol 4) pp 38-44
[10] Wang ZY, Wu J, Xue MD 2018 A Design Method Based on Analysis of Cylindrical Pressure Vessels with Intermediate Spherically Dished Head (Pressure Vessel Technology vol 1) pp 20-28
[11] Mollmann H 1981 Introduction to the Theory of Thin Shells (Chichester: John Wiley & Sons)

Journal of Physics: Conference Series 3012 (2025) 012033 doi:10.1088/1742-6596/3012/1/012033

Comparison of Breakdown Characteristics of SF6 in Extremely Non-Uniform Fields under Oscillation and Aperiodic Lightning Impulses

Jun Kang[1*], Xiaohui Li[3], Jianwu Li[1], Shanghu Zhou[2], Fujie Wang[3], Guoxiang Ma[1], Qingmin Di[3], Yang Jin[3], Jiale Li[3], Ke Wang[3], Jingzhi Yang[3]

[1] State Grid Qinghai Electric Power Company, Xining, Qinghai Province, 630100, P.R. China

[2] State Grid Qinghai Electric Power Co.Ltd.Research Institute, Xining Qinghai Province, 630100, P.R. China

[3] Haibei Power Supply Company of State Grid Qinghai Electric Power Company, Haibei, Qinghai Province, 812200, P.R. China

[*]Corresponding author's e-mail: 782204970@qq.com

Abstract. SF6 gas has good insulation properties and is widely used in power equipment. The impulse withstand voltage test is an important measure to ensure the safe and reliable operation of power equipment. Among them, oscillating lightning impulse (OLI) and aperiodic lightning impulse (ALI) are both suitable for the field, but the understanding of the clarity of the similarities and differences between the two types of waveforms is still not evident.. Therefore, it is essential to initially examine the discharge characteristics of insulation defects under OLI and ALI. The article develops a high-voltage conductor protrusion defect, typical of insulation defects in power equipment, and analyses its breakdown characteristics under OLI and ALI First, a generator for impulse voltage has been developed to create two types of OLI and two types of ALI; then a high-voltage conductor protrusion defect is constructed in a 363kV GIS, with a protrusion length of 12mm and an end curvature radius of 0.4mm, and then its 50% breakdown voltage and voltage-time characteristics are studied through experiments. The study demonstrates that the 50% breakdown voltage difference under OLI and ALI is quite insignificant. and from the perspective of detecting protrusion defects, they are equivalent; with the increase of wavefront time, the breakdown points under OLI are concentrated near each wave crest, and the voltage-time characteristics are dispersed, while under ALI they are continuously distributed near the wave crest. The study's outcomes increase our grasp of the extreme non-uniform field discharge phenomenon in SF6 under oscillating lightning impulses, and also offer guidance for the scientific selection of waveform parameters for on-site impulse voltage withstand tests of power equipment....

1. Introduction

SF_6 gas has excellent physical and chemical properties. Due to its excellent electrical insulation performance, chemical stability and arc extinguishing ability, it is widely used in power equipment as an insulating medium[1][2]. However, when there are sharp points or other gaps or defects in the power equipment, it will cause the distortion of the electric field and the increase of the unevenness of the electric field, which will eventually lead to the reduction of the insulation performance of SF_6, and in severe cases, it will lead to the loss of insulation and cause serious consequences[3]. At present, SF_6 gas is widely used in various power equipment, but due to improper operation or unintentional entry of metal particles or debris during the production and transportation of power equipment. In order to ensure the life cycle and safety and reliability of power equipment, the on-site acceptance test of power equipment is particularly important.

Combining the AC withstand voltage test with partial discharge detection is seen as a reliable approach to detect the majority of insulation defects and is commonly applied in practical tests. However, as the voltage level of power equipment rises, the sensitivity of partial discharge detection diminishes, and the power frequency withstand voltage test is no longer sufficient to expose enough defects[4]. For insulation defects such as installation errors, high-voltage protrusions, and insulator contamination, the lightning impulse withstand voltage test can play a more effective detection effect[5]. Oscillating lightning impulse (OLI) is different from the traditional lighting impulse with different excellent properties. The standard IEC 60060-3 defines and requires on-site tests for high-voltage electrical equipment, indicating that OLI and aperiodic lightning impulse (ALI) waves are suitable for on-site high-voltage tests[6]. Currently, factory tests for equipment use standard lightning tests. The characteristics of OLI waves in impulse tests, as well as the differences and connections with aperiodic lightning impulse waves of equal front time, are issues worth studying and analyzing.

Domestic and foreign scholars have examined the discharge behavior of SF6 gas when subjected to impulse voltage, usually focusing on standard lightning and very fast transient over-voltage (VFTO). In the 20th century, Okabe S. and others conducted extensive research on the breakdown characteristics of SF_6 in slightly non-uniform and highly non-uniform field gaps under steep-fronted VFTO[7]. Domestic researchers, including Yanming Li, initially investigated the discharge behavior of SF6 under OLI, with a main emphasis on the partial discharge characteristics of different defects[8]. Chengqian Jia et al. studied the insulation properties of SF_6 and the mixed gas of SF_6 and N_2, mainly focusing on power frequency and standard lightning impulse breakdown voltage[9]. Xuanrui Zhang et al. studied the use of oscillating lightning waveforms in transformer partial discharge measurements and found that simultaneous partial discharge measurements during oscillating lightning impulse tests can achieve early diagnosis of transformer faults and a comprehensive assessment of insulation status[10]. From the above, it can be seen that there is still a lack of domestic and international research on the breakdown characteristics of gaps in SF_6 gas under OLI waves.

This paper concentrates on investigating the breakdown characteristics of high-voltage conductor protrusion defects in SF6 gas under OLI waves, and contrasts these characteristics with those observed under ALI waves. First, an impulse test platform is constructed, capable of generating OLI and ALI waveforms with a front time of 1.2μs. Then, simulating extremely non-uniform electric fields with sharp protrusion defects. Protrusion defects are simulated on high-voltage conductors within an actual 363kV GIS, and the breakdown characteristics under both types of Lightning Impulse waveforms are studied, including the 50% breakdown voltage and the voltage-time characteristics. The research can deepen the understanding of discharge at Protrusion Defects in SF_6 under OLI waves and provide a basis and support for waveform selection in field withstand tests.

2. Platform construction

As illustrated in Figure 1, the experimental platform includes an impulse voltage generation system, a GIS test section, and a measurement system. The impulse voltage is generated by a MARX circuit, and OLI waves are produced by incorporating an inductor. The voltage is fed into the GIS test section through a high-voltage bushing, and a capacitive voltage divider is employed to measure the voltage waveform.

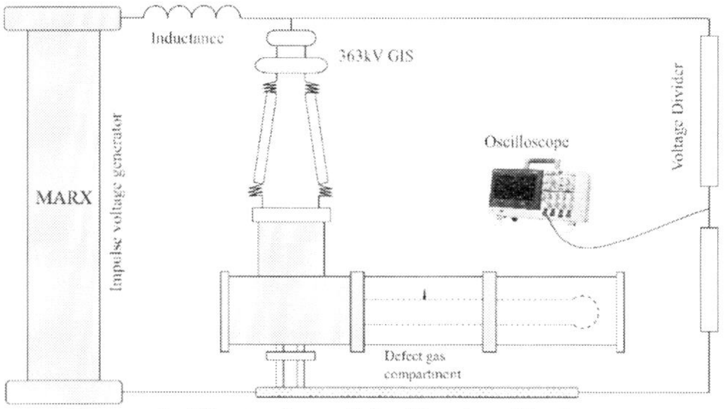

Figure 1. Illustration of the Testing Platform

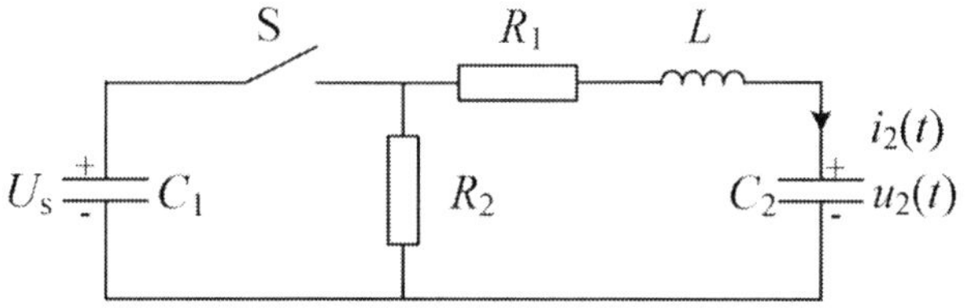

Figure 2. Discharge equivalent circuit of impulse voltage generator

In the circuit configuration, C_1 is the capacitor associated with the impulse voltage generator, C_2 is the total load capacitance, R_1 is the front resistance, R_2 is the tail resistance, and L is the wave-tuning inductor. The voltage across the load capacitance is denoted as $u_2(t)$.

$$u_2 = Ae^{-\alpha_1 t} + Be^{-\alpha_2 t}\sin(\omega t + \theta) \tag{1}$$

Where α_1, α_2, ω, A, B, θ are all constants, and α_1, α_2 and ω are determined by circuit parameters.

$$\begin{aligned} &\alpha_1^3 - a\alpha_1^2 + b\alpha_1 - c = 0 \\ &8\alpha_2^3 - 8a\alpha_2^2 + 2\left(a^2 + b\right)\alpha_2 + c - ab = 0 \end{aligned} \tag{2}$$

$$\omega = \sqrt{3\alpha_2^2 - 2a\alpha_2 + b} \tag{3}$$

Among them, a, b, and c are composed of various parameters in the circuit, as follows.

$$\begin{aligned} a &= \frac{C_1 R_1 R_2 + L}{C_1 R_2 L} \\ b &= \frac{C_1 R_2 + C_2 R_2 + C_2 R_1}{C_1 C_2 R_2 L} \\ c &= \frac{1}{C_1 C_2 R_2 L} \end{aligned} \tag{4}$$

The oscillation frequency can be derived by the following formula.

$$f = \frac{\omega}{2\pi} = \frac{\sqrt{3\alpha_2^2 - 2a\alpha_2 + b}}{2\pi} \tag{5}$$

By analyzing the circuit and properly adjusting the parameters of the circuit components, it is possible to generate a specific OLI waveform. The comparison between the measured and calculated OLI waveforms are depicted in the Figure 3, and the two curves match very well, proving the correctness of the principle calculation.

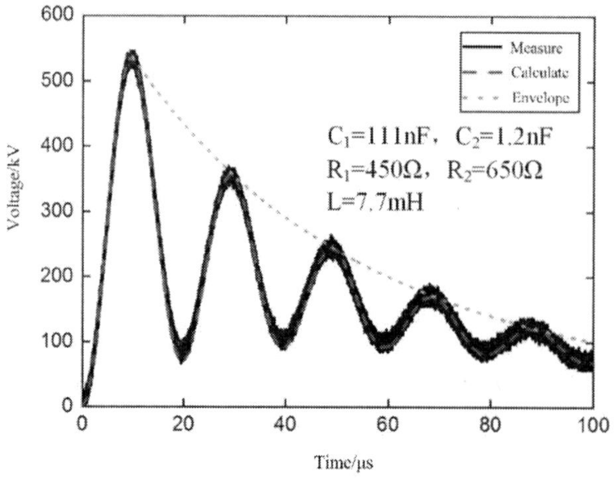

Figure 3. Comparison of measured and calculated OLI waves

The creation of oscillating impulse voltage is primarily governed by the LC resonance between the loop series capacitor and the modulating inductor., which also determines the oscillation frequency of the waveform, while the loop series resistor mainly affects its damping characteristics, and the wave tail resistor affects the envelope characteristics. Theoretical calculations allow for the accurate acquisition of impulse voltage waveform parameters. Therefore, when generating oscillating lightning impulse voltage, the circuit component parameters are first quickly determined based on the calculation results, and then fine-tuned based on the measured results.

The lightning impulse voltage waveform parameters refer to the requirements of standard IEC 60060-3 and national standard GB/T 16927.3. The parameter requirements for ALI and OLI waveforms are depicted in the Table 1.

Table 1. Waveform parameter requirements for on-site impulse withstand voltage test of power equipment

Voltage Type	T_f/μs	T_t/μs	f/kHz
OLI	0.8-20	40-100	15-400
ALI	0.8-20	40-100	——

The specific voltage waveform parameters are selected according to the lightning impulse waveform parameter requirements, and the specific values are depicted in the Table 2.

Table 2. Parameters of Lightning Impulse Voltage Waveforms

Voltage Type	T_f/μs	T_t/μs	f/kHz
OLI	1.2	50	296
	3	50	118
ALI	1.2	50	——
	3	50	——

The OLI waveforms have front times of 1.2μs and 3μs, resulting in two oscillation frequencies, which are also used for comparison. The ALI waveforms similarly have front times of 1.2μs and 3μs, with all waveforms having a half-peak time of 50μs. The OLI and ALI voltage waveforms are depicted in the Figure 4.

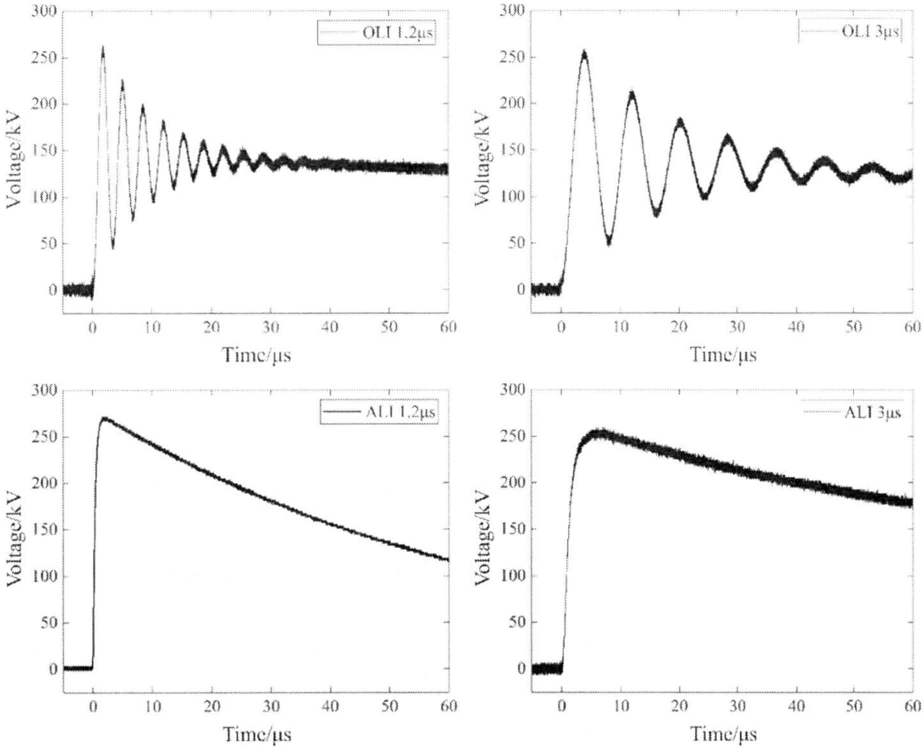

Figure 4. Waveforms of OLI and ALI Voltage

The 363kV GIS test section is shown in the Figure 5, which mainly consists of high-voltage bushings, disconnectors, circuit breakers, and multiple gas compartments. One of the gas compartments is a defect compartment, used to arrange high-voltage conductor protrusion defects. Simulating extremely non-uniform electric fields with sharp protrusion defects. In the test, the impulse high voltage is transmitted to the high-voltage conductor through the high-voltage bushing, exciting discharge within the gas compartment's defect.

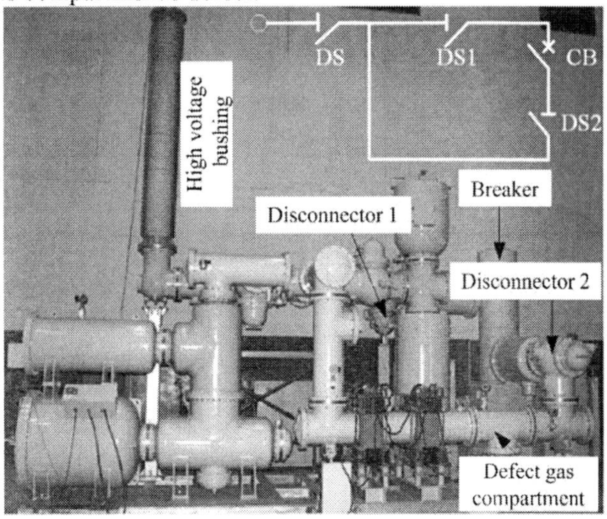

Figure 5. 330kV GIS test section physical picture

The test defects selected are typical insulation defects in GIS, specifically high-voltage conductor protrusion defects. To enhance the arc-resistant performance, they are made from tungsten-copper rods with a diameter of 3mm. The overall shape is conical, with the end polished into a hemispherical shape. The defect is threaded at the lower part and fixed onto the high-voltage conductor. The arrangement of the defect is shown in the Figure 6,and the protrusion defect physical picture is depicted in the Figure 7.

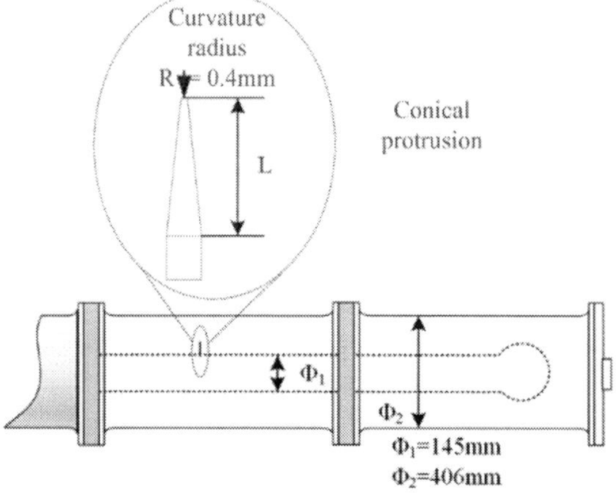

Figure 6. Illustration of High-Voltage Conductor Protrusion Defects

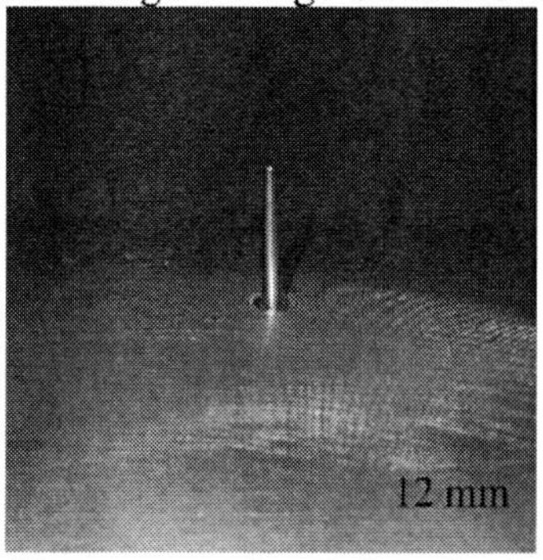

Figure 7. 12mm protrusion defect physical picture

3. Results and analysis

3.1 50% Breakdown voltage

Under ALI and OLI voltages, the 50% breakdown voltage of a 12mm tip is depicted in Table 3.

Table 3. 50% breakdown voltage of protrusion defect under ALI and OLI voltage

Voltage Type	Front Time (T_f) /μs	50% Breakdown Voltage /kV	
		Positive	Negative
ALI	1.2	256.6	559.6
	3	252.6	553
OLI	1.2	257.2	560.6
	3	253.1	552.6

From Table 3, the breakdown voltage under positive polarity is observed to be between 252.6kV and 257.2kV, while under negative polarity, it spans from 552.6kV to 560.6kV, which is 2.2 to 2.3 times that of the positive polarity, exhibiting a significant polarity effect. At the same time, it is apparent that there is no notable difference in the 50% breakdown voltage between OLI and ALI voltages.

Journal of Physics: Conference Series **3012** (2025) 012033 doi:10.1088/1742-6596/3012/1/012033

The experimental study investigated the 50% breakdown voltage and V-t characteristics under different waveforms. To ensure the effectiveness of the experiment, a 10-minute interval was maintained between two consecutive tests to allow the gap insulation to recover to its initial level.

3.2 Voltage-time characteristics

Under OLI voltage, the V-t (voltage-time) characteristics of the 12mm protrusion are depicted in Figure 8. Breakdown typically occurs in the vicinity of the peak of the applied voltage wave. With different oscillation frequencies, the V-t characteristics exhibit different features. For the higher oscillation frequency OLI 1.2μs, under negative polarity, breakdown first occurs at the third voltage peak. Under positive polarity, only one breakdown is measured, occurring at the second voltage peak, hence the maximum discharge delay is 5.91μs. For OLI 3μs, under negative polarity, breakdown first occurs at the second voltage peak, and under positive polarity, breakdown only occurs near the first voltage peak, with a distribution similar to the V-t characteristics under ALI.

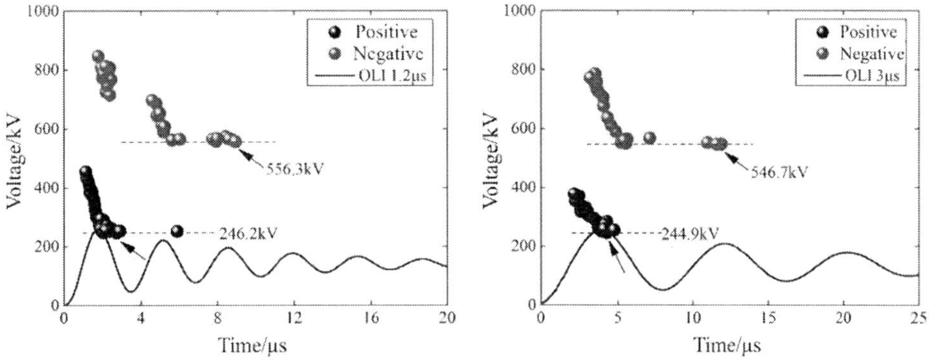

(a) OLI 1.2μs (b) OLI 3μs

Figure 8. 12mm protrusion OLI voltage-time characteristics

Under ALI, the V-t characteristics of the 12mm protrusion are depicted in Figure 9. Under ALI, the discharge is concentrated around the peak value of the applied voltage, and the negative polarity breakdown point is noticeably positioned higher than the positive polarity., and the maximum discharge delay is greater. With low applied voltage, the breakdown point distribution is even, and as the voltage rises, the discharge delay shortens, leading to an upward trend in the voltage-time characteristic.

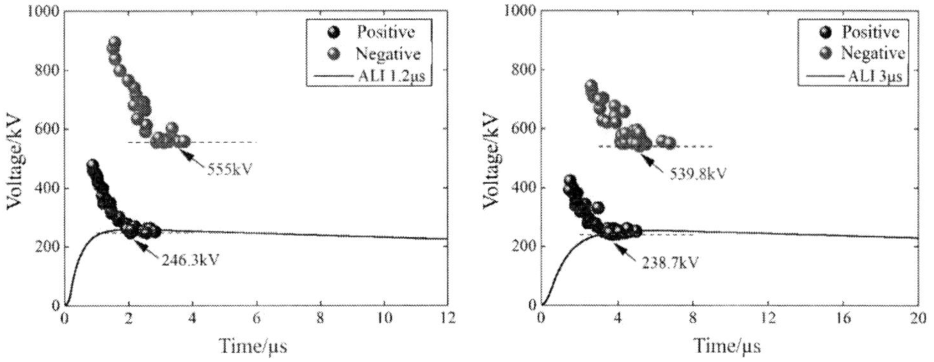

(a) OLI 1.2μs (b) OLI 3μs

Figure 9. 12mm protrusion ALI voltage-time characteristics

The maximum discharge time delay and the lowest breakdown point in the volt-time characteristics are depicted in Table 4.

Table 4. Characteristic values of V-t characteristics of protrusion defects under OLI and ALI

Voltage Type	Maximum Breakdown Delay Time T_{max}/μs		Minimum Breakdown Voltage V_{min}/kV	
	Positive	Negative	Positive	Negative
OLI 1.2μs	5.91	8.94	242.6	556.3
OLI 3μs	4.75	11.6	244.9	546.7
ALI 1.2μs	2.57	3.76	243.5	555.0
ALI 3μs	5.20	6.80	238.7	539.8

It can be seen from Table 4 that under OLI 1.2μs and OLI 3μs, the maximum breakdown delay under negative polarity is significantly higher than under positive polarity, but the difference decreases as the wavefront time increases. As the wavefront time increases, the positive polarity V_{min} changes very little, while the negative polarity obviously increases.

Under ALI 1.2μs and ALI 3μs, the maximum breakdown delay of negative polarity is greater than that of positive polarity, but it is not obvious under ALI, and the difference is smaller when the wavefront time is longer. As the wavefront time increases, V_{min} does not change much under positive and negative polarity, and shows a decreasing trend.

Under the same wavefront time and voltage polarity, the maximum breakdown delay of OLI is larger than that of ALI, and the difference is larger under negative polarity. However, when comparing the minimum breakdown voltage of OLI and ALI, the difference is not much.

4. Conclusion

By comparing the breakdown characteristics of different wavefront times and impulse voltage waveforms, observations show that with an increase in wavefront time, breakdown under OLI voltage increasingly focuses from the second and third wave peaks towards the first wave peak, especially under negative polarity voltage. Additionally, the maximum breakdown delay is greater under OLI compared to ALI, while the minimum breakdown voltage remains relatively similar. This results in more dispersed breakdown points and a larger discharge delay under oscillating lightning, which is particularly evident under negative polarity.

According to the rules obtained from the experiment, suggestions are made for actual defect detection: It is preferred to select a shorter wavefront time (such as 1.2μs) because this can more effectively stimulate the breakdown at the defect, especially under negative polarity, the distribution of the breakdown point is more dispersed, which helps to detect more potential defects. For high-sensitivity defect detection, it is recommended to select a higher oscillation frequency (such as 296kHz), especially under negative polarity, because this can increase the dispersion of the breakdown point, making it easier to detect tiny defects.

Acknowledgments

Science and Technology Project Funded by State Grid Qinghai Electric Power Company （522805240001）

References

[1] XW Li, H Zhao, Murphy AB. 2018 SF6-alternative Gases for Application in Gas-insulated Switchgear (Journal of Physics D: Applied Physics) vol 51.
[2] YS Ji, M Zhang, CY Wang. 2020 Detection Methods and Experimental Study on Decomposition Products of SF6/N2 Mixed Gas (High Voltage Apparatus) vol 56 p 97-102.
[3] FZ Kong, JT Ma, et al. 2023 Effect of SF6 Pressure on the Sensitivity of Partial Discharge Detection of Typical Defects in GIS （Electric Power Engineering Technology） vol 42 p 233-241.
[4] SL Zhang. 2020 Intelligent Diagnosis Method for Discharge Fault Mode of High Voltage SF6

Gas Insulated Switch-gear（High Voltage Engineering）vol 46 p 432-440.
[5] HC Sun, XA Li, J L, et al. 2013 Research on the Effectiveness of Detecting Metal Spike Defects in GIS Under Lightning Impulse Superimposed Power Frequency Voltage（Power System Technology）vol 46 p 795-802.
[6] IEC 60060−3 2006 (High-voltage test techniques. part 3:Definitions and requirements for on-site testing)
[7] S Okabe, S Yuasa, S Kaneko. 2008 Evaluation of Breakdown Characteristics of Gas Insulated Switchgears for Non-standard Lightning Impulse Waveforms-breakdown Characteristics for Non-standard Lightning Impulse Waveforms Associated with Disconnector Switching Surges (IEEE Transactions on Dielectrics and Electrical Insulation) vol 15 p 721-729.
[8] XF Zhao, X Yao, ZF Guo, et al. 2011 Characteristics and Development Mechanisms of Partial Discharge in SF6 Gas under Impulse Voltages (IEEE Transactions on Plasma Science) vol 39 p 668-674.
[9] CQ Jia, YL Xia, Y Ni, et al. 2022 Research on Insulation Characteristics of SF6/N2 Gas Mixture (Sichuan Electric Power Technology) vol 45 p 33-37.
[10] XR Zhang, HT Wang, YD Hu, et al. 2023 Oscillating Lightning Impulse Voltage Withstand Test and Partial Discharge Detection Technology for Power Transformers (High Voltage Engineering) vol 49 p 5032-5041.

Parameter Identification for Reconfigurable Battery Systems Based on Differential Model and Adam Optimizer

Changyou Geng, Dezhi Ren, Xinyi Zheng, Enkai Mao, Hao Yv, Weiji Han*

China-UK Low Carbon College, Shanghai Jiaotong University, 3 Yinlian Road, Pudong New Area, Shanghai, 201306, China

E-mail: w.han@sjtu.edu.cn

Abstract. Reconfigurable battery systems (RBSs) have emerged as an innovative solution to advanced battery management by dynamically reconfiguring battery cell connections to optimize system performance. For the modeling and simulation of RBSs, online battery parameter identification remains a critical challenge, particularly under time-varying operational conditions. Thus, based on a differentiable system-level RBS model we recently developed, this work proposes to employ the Adam optimizer to solve a least squares problem for the RBS parameter identification. The efficient parameter identification can be conducted using real-time RBS operational data without restricting dynamic reconfiguration. The test results show that, as compared with the experimental measurement data, low mean absolute errors can be achieved when simulating the RBS's terminal voltage, cell currents, and cell terminal voltages using the parameter values identified by the proposed method.

1. Introduction

Reconfigurable battery systems (RBSs) [1] allow dynamically adjusting the connections between battery cells during operation. Then, it becomes possible to optimize the RBS performance on the fly based on various battery states. Advanced control strategies empower RBSs to enhance fault tolerance, extend energy delivery, balance cell states, optimize thermal management, and improve power efficiency [2, 3, 4, 5, 6]. RBSs also support a wide range of battery chemistries, types, and ages, allowing for diagnostics, repairs, and cell replacements to enhance system reliability and adaptability. [7].

Parameter identification is a critical prerequisite for model-based simulation and control. The dynamic reconfiguration in RBSs can be deployed to enhance the accuracy of parameter identification. Schneider et al. [8] demonstrated that dynamic battery reconfiguration can help to improve the state and parameter estimation accuracy, especially under low dynamic current excitation. Michael et al. [9] investigated the state and parameter estimation in reconfigurable battery modules, and employed Kalman filtering to enhance prediction accuracy across varying conditions. This research highlighted the importance of input signal analysis and switching operations in boosting system excitation for improved parameter identification. Pinter et al. [10] addressed challenges in parameter identification for lithium-iron phosphate cells caused by aging and noise, and proposed a single-cell reference model to improve the reliability of online

Journal of Physics: Conference Series **3012** (2025) 012034 doi:10.1088/1742-6596/3012/1/012034

SOC estimation, while noting that further refinements are required for precise applications. Tashakor et al. [11] applied feedforward neural networks to modular battery systems, utilizing machine learning to adaptively enhance parameter identification precision based on operational data.

Rather than specifying system configurations to collect data for parameter identification, this work aims to leverage real-time operational data of the RBS. Experimental results showed that even milliohm-level line and switch conduction resistances have a notable impact on RBS performance, especially in configurations with parallel branches. As a result, the combined resistance of the switch and its corresponding branch must be considered an additional RBS parameter to be identified, along with the battery parameters. Then, based on a differentiable system-level RBS model accounting for all possible configurations, the widely used Adam optimization algorithm is utilized to solve the least squares problem for parameter identification.

2. Method

2.1. Differentiable model

This differentiable system-level RBS model allows for efficient and flexible modeling of RBSs by maintaining a consistent system configuration despite dynamic reconfigurations. By preserving all mesh structures and incorporating switch resistance variables that reflect the on/off states of switches, frequent reconstruction of KCL and KVL equations can be avoided. This approach enhances the modeling efficiency and supports the analysis of various system configurations. To account for the switching behavior of switches, distinct resistances are set for the on and off states of switch. Consequently, a resistance variable $R_n^{S_m}$ is introduced to represent the m-th switch resistance of n-th cell. The value of this resistance is determined based on the binary switch state s_n^m, and is represented as:

$$R_n^{S_m} = R_{w,n}^m + (1 - s_n^m) R_{\mathrm{ds(off)},n}^m + s_n^m R_{\mathrm{ds(on)},n}^m. \tag{1}$$

Here, $R_{w,n}^m$ represents the wire resistance associated with the corresponding switch branch. Together with the battery cell parameters, all $R_{w,n}^m$ form the set of parameters to be identified, denoted as θ.

The on/off states of all switches form a vector, denoted as SSV. The cells are modeled as voltage sources, resistances, and conductances based on their equivalent circuit models (ECMs). By formulating KCL and KVL equations for the circuit that includes all switches and cells, state space equations describing the relationship between the switches and the system's electrical behavior are established as follows.

$$X = [V^T, I_{R_1}^T, I_{R_2}^T]^T. \tag{2}$$

$$\dot{X} = A(\theta; SSV)X + B(\theta; SSV)i_t. \tag{3}$$

$$v_t = C(\theta; SSV)X + D(\theta; SSV)i_t. \tag{4}$$

The vector $V = [v_1, v_2, \dots, v_N]^T$ represents the open-circuit voltages (OCVs) of the battery cells. I_{R_1} specifies the vector of currents through the R_1 components in the battery cell equivalent circuit model (ECM), and I_{R_2} specifies the vector of currents through the R_2 components in ECM. The v_t and i_t denote the terminal voltage and input current of system, respectively. All battery parameters and switch branch resistances to be identified, θ, and switch states, S, are included in the coefficient matrices A, B, C, and D.

An RBS can have a large number of potential switch state combinations, but only a small fraction of them lead to acceptable system configurations. In this study, we leverage a predefined set of valid switch combinations. More details about the differentiable model and valid switch state combinations can be found in our recent work [12].

2.2. Adam Optimizer

Thanks to the differentiable model, the least squares problem for parameter identification can be solved using gradient-based methods. Common gradient-based methods, such as the Gauss-Newton method and quasi-Newton methods, typically require the computation of the Jacobian matrix of the output with respect to the input. However, when the data points used for identification reach a certain volume, calculating the Jacobian matrix becomes highly time-consuming, especially in the context of this complex differentiable RBS model.

Therefore, in this study, we employ the Adam optimizer [13] to solve the least squares problem. The loss function is defined as follows.

$$\mathcal{L}(\theta) = \frac{1}{2}\sum_{i=1}^{N}\left(v_t^i - f(i_t^i;\theta)\right)^2, \tag{5}$$

where f denotes the RBS model, and N denotes the number of data points. Since the loss to be minimized is a scalar, the Adam optimizer only needs to compute the partial derivatives of this scalar with respect to all parameters during each update. For automatic differentiation frameworks, such as PyTorch, this calculation is much faster than calculating the Jacobian matrix for each output with respect to the inputs individually.

Furthermore, Adam only requires the computation of first-order gradients for each parameter, eliminating the need for second-order derivatives (such as the Hessian matrix). This makes it more computationally efficient when handling high-dimensional parameters and large datasets. Because it incorporates the Momentum method, Adam is not limited to local infinitesimal changes and is better suited for complex non-convex problems. Widely used in the training of deep neural network parameters, Adam has demonstrated excellent performance in numerous practical applications, offering strong adaptability and fast convergence.

3. Result

The tested RBS type and the testbed used in the experiment are shown in Fig.1 and Fig.2, respectively.

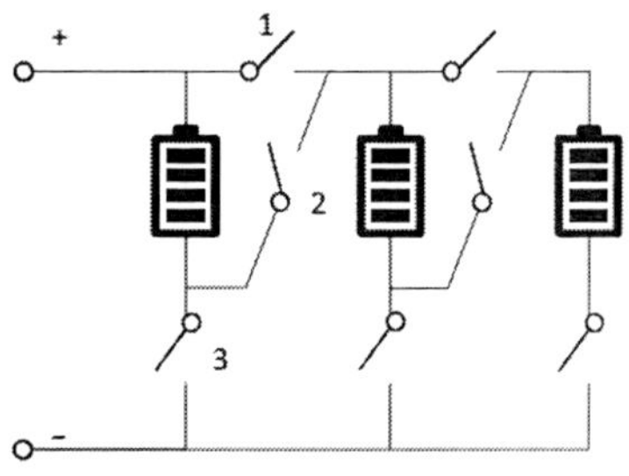

Figure 1. Tested RBS type.

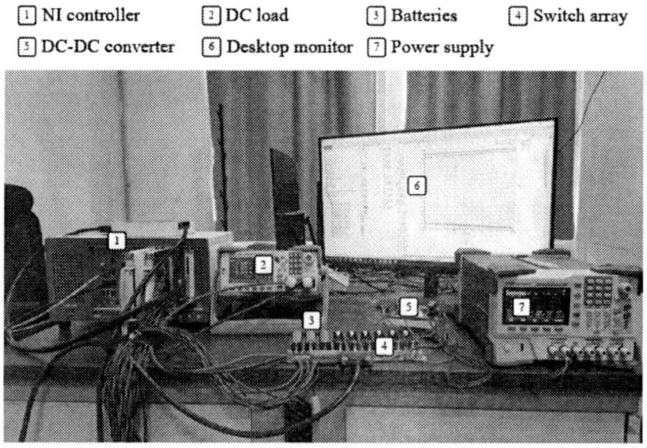

Figure 2. RBS testbed.

Three ICR 18650 battery cells and seven MOSFET switches are employed in this RBS. A constant power load is employed, and the system configuration switches among all 12 valid configurations to generate dynamic operational data. The parameters θ are first fitted using data under an 8 W constant power load and tested on data from a 4 W load.

The fitting and testing mean absolute errors (MAEs) of the system terminal voltage, cell current, and cell terminal voltage are listed in Table 1.

Table 1. MAE of fitting and testing.

Metric	Fitting MAE	Testing MAE
System terminal voltage (V)	0.0158	0.0141
Cell current (A)	0.0027	0.0018
Cell terminal voltage (V)	0.0069	0.0339

The comparison of the system terminal voltage between simulation and experiment under a tested 4 W load is shown in Fig. 3. The results demonstrate that these parameters are fitted accurately, showing good agreement between simulation and experimental data.

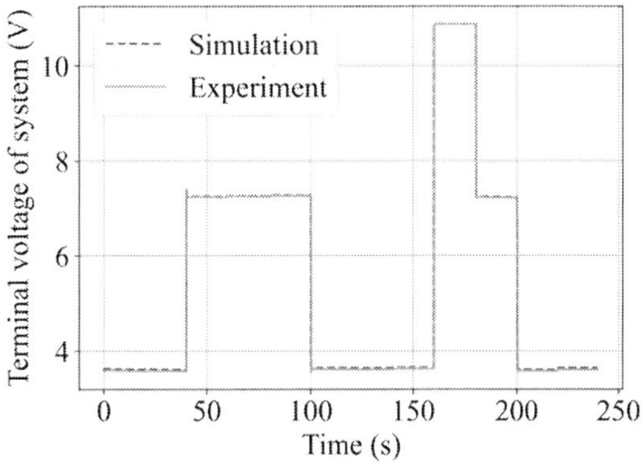

Figure 3. Comparison of system terminal voltage between simulation and experiment under a tested 4 W load.

To validate the significant impact of wire resistance on system outputs, a sensitivity analysis is conducted using the Morris method [14] on three key outputs: system terminal voltage, cell current, and cell terminal voltage. In the sensitivity analysis, the output is the average of the errors between the simulated values of these system outputs and the experimental data, and the input is θ, the parameters to be identified. Figure 4 illustrates the Morris method's results, where the bar heights represent the μ^* values and the error bars indicate σ. Notably, the μ^* of $R_{w,n}^{m}$ for the system terminal voltage remains consistently high—comparable to that of R_0—and exceeds the other battery parameters, underscoring its pronounced effect on the system terminal voltage. Apart from the second switch in each cell, $R_{w,n}^{m}$ also demonstrates a high μ^* for cell current, minimal μ^* for cell terminal voltage, and large σ in relation to cell current, pointing to strong interactions with other parameters. In summary, $R_{w,n}^{m}$ exerts a strong impact on the system's outputs and should therefore be carefully considered in parameter identification.

4. Conclusion

This study presents a system-level parameter identification framework for RBSs using the Adam optimizer based on a differentiable RBS model enabling efficient analysis without frequent circuit

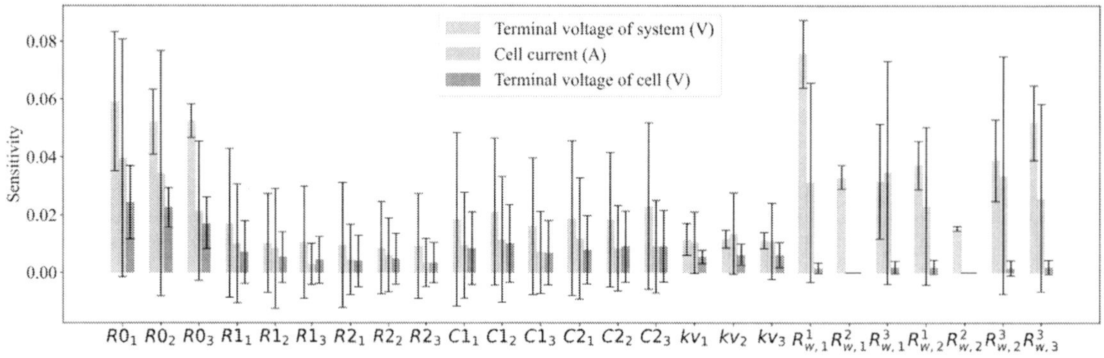

Figure 4. Sensitivity result.

equation reconstruction. Experimental validation demonstrates high accuracy in parameter identification and highlights the important impact of milliohm-level connection and switch conduction resistances on the system performance. By leveraging normal operational data, the proposed parameter identification method has the potential for online applications, enabling the online parameter update and accurate simulation for developing advanced battery management solutions.

References

[1] Ci S, Lin N and Wu D 2016 *IEEE Access* **4** 1175–1189 ISSN 2169-3536
[2] Alahmad M, Hess H, Mojarradi M, West W and Whitacre J 2008 *Journal of Power Sources* **177** 566–578
[3] Zhang Z, Cai Y Y, Zhang Y, Gu D J and Liu Y F 2015 *IEEE Transactions on Power Electronics* **31** 304–317
[4] Han W, Zou C, Zhang L, Ouyang Q and Wik T 2019 *IEEE Transactions on Smart Grid* **10** 6954–6964
[5] Altaf F, Egardt B and Johannesson Mårdh L 2017 *IEEE Transactions on Control Systems Technology* **25** 47–62 ISSN 1558-0865
[6] Taesic Kim, Wei Qiao and Liyan Qu 2012 *IEEE Transactions on Power Electronics* **27** 4723–4733 ISSN 0885-8993, 1941-0107
[7] Badam A, Chandra R, Dutra J, Ferrese A, Hodges S, Hu P, Meinershagen J, Moscibroda T, Priyantha B and Skiani E 2015 *Proceedings of the 25th Symposium on Operating Systems Principles* (Monterey California: ACM) pp 215–229 ISBN 978-1-4503-3834-9
[8] Schneider D, Liebhart B, Endisch C and Kennel R 2022 *2022 12th International Conference on Power, Energy and Electrical Engineering (CPEEE)* (IEEE) pp 71–77
[9] Theiler M, Schneider D and Endisch C 2023 *Batteries* **9** 145 ISSN 2313-0105
[10] Pinter Z M, Engelhardt J, Rohde G, Traeholt C and Marinelli M 2022 *2022 International Conference on Renewable Energies and Smart Technologies (REST)* (Tirana, Albania: IEEE) pp 1–5 ISBN 978-1-66540-971-1
[11] Tashakor N, Amirrezai M, Furqan M S, Shah S R, Pourhadi P and Goetz S 2023 *2023 8th IEEE Workshop on the Electronic Grid (eGRID)* pp 1–6 ISSN 2831-3658
[12] Geng C, Ren D, Zheng X and Han W 2025 *arXiv preprint arXiv:2501.03028*
[13] Kingma D P 2014 *arXiv preprint arXiv:1412.6980*
[14] Morris M D 1991 *Technometrics* **33** 161–174 ISSN 0040-1706, 1537-2723

Data-driven Adaptive Robust Optimization of Self-Scheduling Problem of Multi-Energy Virtual Power Plant Under Uncertainties

Jian Wang[1*], Zakiud Din[2], Run Cao[3], Xingyuan Song[1], Zibo Wu[3]

[1] College of Smart Energy, Shanghai Jiao Tong University, Shanghai, PR China

[2] College of Electrical and Mechanical Engineering, National University of Sciences and Technology, Islamabad, Pakistan

[3] Department of Electrical Engineering, Shanghai Jiao Tong University, Shanghai, PR China

*Corresponding author's e-mail: wangjianee@sjtu.edu.cn

Abstract: To improve the stability and economy of power system operation under the deep penetration of renewable energy, a trilevel adaptive robust optimization model is proposed for the self-scheduling of the multi-energy virtual power plant, which is solved by the column-and-constraint generation algorithm. The uncertainties of the available photovoltaic capacities are addressed by the data-driven scenario-based uncertainty sets, which are developed based on the historical operational scenarios of the power system with the help of data-driven technology. Using the actual operational data from a chosen campus, the results show that the proposed method can achieve a synergistic improvement of operational stability and economy of the multi-energy virtual power plant. It can also improve the accuracy of uncertainty modeling while reducing the conservatism of robust optimization, saving by about 7.3% of the operating costs.

Keywords: data-driven scenario-based uncertainty set, trilevel adaptive robust optimization, self-scheduling problem, multi-energy virtual power plant.

1. Introduction

With the increasing proportion of renewable energy in the power system, how to achieve efficient operation of a large number of distributed resources under uncertainties is currently a significant research problem. Multi-energy virtual power plant (MEVPP) can aggregate renewable energy sources (RESs) with various flexible multi-energy distributed energy resources (DERs) to mitigate the uncertainty and participate in various types of electricity markets.

Paper [1] considers the uncertainty of renewable energy generation outputs, market prices, electricity demands, and reserve demands, and proposes a day ahead self-scheduling model in the day-ahead market (DAM) and ancillary service market (ASM). The two-stage optimization model can be applied to the self-scheduling of MEVPP. Paper [2] makes bidding decisions considering the uncertainty of wind power generation, and energy and reserve services are jointly cleared in the first stage optimization. In the second stage, the transaction in the real-time balancing market is optimized for the deviation caused by uncertainty factors.

Stochastic programming (SP) and robust optimization (RO) are two of the most commonly used methods for uncertain factors in optimal scheduling problem [3]. In Paper [2], SP models uncertain factors by generating a large number of operational scenarios. SP relies on the precise probability distribution of uncertain factors, which is difficult to obtain in practical applications [4]. Paper [5] utilizes RO to optimize the worst-case scenario of uncertainty, including energy price uncertainty, wind power output uncertainty, etc. Paper [6] uses adaptive robust optimization (ARO) to alleviate the excessive conservatism of traditional RO.

This paper proposes a data-driven ARO scheduling method for a MEVPP considering market transactions. The main contributions include: (1) A trilevel optimization model based on ARO is proposed to achieve coordinated scheduling of MEVPP in market environments, which is iteratively solved through column-and-constraint generation (C&CG) algorithm; (2) A data-driven scenario-based uncertainty set (DSBS) is proposed, which improves the construction method of the uncertainty set by utilizing a large amount of historical operational data through data-driven technology.

2. Trilevel Self-Scheduling Model of MEVPP

Data-driven ARO approach is used to deal with the optimal scheduling problem with uncertainties, which formulates a trilevel optimization problem. In the first level, the energy schedule is decided with the target of minimizing the overall cost of the MEVPP. After the decision, in the second level, the worst realization of the uncertain available photovoltaic (PV) capacity occurs in the real-time stage. After the scheduling problem is solved under the worst realization of the uncertainties, in the third level, corrective actions can be made to minimize the cost of MEVPP in the real-time operation. Sets Ω and Ξ are the uncertainty and the feasibility sets, respectively, with the constraints of the corresponding variables in the second-level and third-level, which are going to be analyzed in the following subsections.

Variables in sets $\Phi^1=\left\{I_{chp,t}^{\mathrm{DA,su}},I_{chp,t}^{\mathrm{DA,sd}},I_{chp,t}^{\mathrm{DA}},p_{vpp,t}^{\mathrm{DA}},g_{vpp,t}^{\mathrm{DA}}\right\},\forall t$; $\Phi^2=\left\{\tilde{P}_{pv,t}^{\mathrm{ca}}\right\},\forall t$; and $\Phi^3=\begin{Bmatrix}p_{vpp,t}^{\mathrm{IB+}},p_{vpp,t}^{\mathrm{IB-}},g_{vpp,t}^{\mathrm{IB+}},g_{vpp,t}^{\mathrm{IB-}},p_{chp,t},g_{chp,t},\\ h_{chp,t},h_{gb,t},g_{gb,t},c_{cc,t},p_{cc,t},c_{ac,t},h_{ac,t},\\ E_{k,t},s_{k,t}^{\mathrm{ch}},s_{k,t}^{\mathrm{dch}},d_{de,t}^{\mathrm{p}},d_{de,t}^{\mathrm{m}},d_{de,t},p_{pv,t}\end{Bmatrix},\forall t$, are for the first-level, second-level, and third-level optimization, respectively.

$$\min_{\Phi^1}\sum_{t\in T}\left(\lambda_t^{\mathrm{E}}\cdot p_{vpp,t}^{\mathrm{DA}}+\lambda_t^{\mathrm{G}}\cdot g_{vpp,t}^{\mathrm{DA}}+CO_t^{\mathrm{sud}}\right)+\max_{\Phi^2\in\Omega}\min_{\Phi^3\in\Xi}\sum_{t\in T}\left(\lambda_t^{\mathrm{EIB+}}\cdot p_{vpp,t}^{\mathrm{IB+}}+\lambda_t^{\mathrm{EIB-}}\cdot p_{vpp,t}^{\mathrm{IB-}}+\lambda_t^{\mathrm{GIB+}}\cdot g_{vpp,t}^{\mathrm{IB+}}+\lambda_t^{\mathrm{GIB-}}\cdot g_{vpp,t}^{\mathrm{IB-}}+CO_t^{\mathrm{op}}\right) \tag{1}$$

s.t.

$$-\overline{P_{vpp,t}^{\mathrm{DA,wi}}}\le p_{vpp,t}^{\mathrm{DA}}\le\overline{P_{vpp,t}^{\mathrm{DA,in}}} \tag{2}$$

$$0\le g_{vpp,t}^{\mathrm{DA}}\le\overline{G_{vpp,t}^{\mathrm{DA}}} \tag{3}$$

$$I_{chp,t}^{\mathrm{DA,su}}-I_{chp,t}^{\mathrm{DA,sd}}=I_{chp,t}^{\mathrm{DA}}-I_{chp,t-1}^{\mathrm{DA}} \tag{4}$$

The first and second terms of Eq. (1) are the cost of purchasing electricity and natural gas from DAM and natural gas market (GM), respectively. Variables $p_{vpp,t}^{\mathrm{DA}}$ and $g_{vpp,t}^{\mathrm{DA}}$ represent the day-ahead electric energy and natural gas schedules, respectively. The third term is the turn-up and shut-down cost. The fourth and fifth terms are the electric energy imbalance costs. Variables $p_{vpp,t}^{\mathrm{IB+}}$ and $p_{vpp,t}^{\mathrm{IB-}}$ are the electric energy imbalance of injection and withdrawal directions of MEVPP, respectively. MEVPP in this thesis is modeled as a price taker. Parameters $\lambda_t^{\mathrm{EIB+}}$ and $\lambda_t^{\mathrm{EIB-}}$ are the imbalance penalties of the injection and withdrawal directions, respectively. The sixth and seventh terms are the costs of natural gas imbalance. Variables $g_{vpp,t}^{\mathrm{IB+}}$ and $g_{vpp,t}^{\mathrm{IB-}}$ are the natural gas imbalance of injection and withdrawal directions, respectively. Parameters $\lambda_t^{\mathrm{GIB+}}$ and $\lambda_t^{\mathrm{GIB-}}$ are the imbalance penalties of two directions, respectively. The last term is the operational cost of MEVPP.

The limits of the electricity trading in DAM are given in Eq. (2), which affect the day-ahead electric energy schedule. Parameters $P_{vpp,t}^{\mathrm{DA,in}}$ and $P_{vpp,t}^{\mathrm{DA,wi}}$ are the maximum amount of power injection and withdrawal in the day-ahead electric energy schedule. The limits of the natural gas trading are given in Eq. (3), which affect the day-ahead natural gas schedule. The start-up and shut-down states of Combined heat and power unit (CHP) are calculated from its on/off state in Eq. (4).

2.1 Definition of Uncertainty Sets

There exist uncertainties in the available PV capacity. Because it is determined by the solar radiation which is very difficult to be forecasted in advance accurately.

(1) Conventional budget-constrained uncertainty set (BCUS)

The uncertainties of the available PV capacity can be modeled by the corresponding fluctuation intervals in BCUS. The polyhedral uncertainty set Ω in Eq. (1) is modeled as:

$\Omega = \{$

$$\tilde{P}_{pv,t}^{\mathrm{ca}} = P_{pv,t}^{\mathrm{ca}} + \gamma_{pv,t}^{+} \cdot \hat{P}_{pv,t}^{\mathrm{ca+}} - \gamma_{pv,t}^{-} \cdot \hat{P}_{pv,t}^{\mathrm{ca-}} \tag{5}$$

$$0 \le \gamma_{pv,t}^{+} \le x_{pv,t} \tag{6}$$

$$0 \le \gamma_{pv,t}^{-} \le \left(1 - x_{pv,t}\right) \tag{7}$$

$$\sum_{pv,t} \left(\gamma_{pv,t}^{+} + \gamma_{pv,t}^{-}\right) \le \Gamma^{\mathrm{pv}} \tag{8}$$

$\}$

In Eq. (5), uncertain parameter $\tilde{P}_{pv,t}^{\mathrm{ca}}$ represents the uncertain available PV capacity parameter $P_{pv,t}^{\mathrm{ca}}$ represents the best forecast of the available PV capacity; parameter $\hat{P}_{pv,t}^{\mathrm{ca+/-}}$ is the upward/downward fluctuation interval, which represents the maximum upward/downward deviation of the uncertainty; variables $\gamma_{pv,t}^{+}$ and $\gamma_{pv,t}^{-}$ are the control factors of two directions. Eq. (6)-(7) limit that the control factors can vary between 0 and 1 and impose the complementary of two directions through a binary variable $x_{pv,t}$. Parameter Γ^{pv} in Eq. (8) is defined as the budget of uncertainties, which is used to model the conservativeness level. A risk averse MEVPP would choose higher value of Γ^{pv}, while a risk neutral MEVPP would choose lower value of Γ^{pv}. Parameter Γ^{pv} varies within $\left[0, |PV| \times |T|\right]$, where $|PV|$ represents the number of PVs and $|T|$ represents the number of time intervals.

(2) Data-driven scenario-based uncertainty set (DSBS)

The shape of BCUS is defined by the upper and lower bounds of the uncertainties, which cannot accurately capture their real range of fluctuation, and the corresponding decision-making outcomes are also too conservative. To construct an effective uncertainty set that is flexible, practical, and easy to specify, the DSBS is developed. DSBS is constructed as the convex hull of a collection of multivariate points that represents different scenarios. The historical power profiles of the energy system can be directly utilized as scenarios.

$\Omega = \{$

$$\mathbf{P}^{\mathrm{ca}} = \sum_{k^{\mathrm{pv}}=1}^{K^{\mathrm{pv}}} \alpha_{k^{\mathrm{pv}}}^{\mathrm{pv}} \cdot \mathbf{P}_{k^{\mathrm{pv}}}^{\mathrm{ca}} \tag{9}$$

$$\sum_{k^{\mathrm{pv}}=1}^{K^{\mathrm{pv}}} \alpha_{k^{\mathrm{pv}}}^{\mathrm{pv}} = 1 \tag{10}$$

$$\alpha_{k^{\mathrm{pv}}}^{\mathrm{pv}} \geq 0 \tag{11}$$

$\}$

where $\mathbf{P}^{\mathrm{ca}} \in \mathbb{R}^{|\mathrm{PV}|\times|\mathrm{T}|}$ in Eq. (9) represents the uncertain available PV capacity. $\left\{\mathbf{P}_{k^{\mathrm{pv}}}^{\mathrm{ca}}\right\}_{k^{\mathrm{pv}}=1,\dots,K^{\mathrm{pv}}}$ is the set of daily available PV capacity profiles (scenarios) corresponding to K^{pv} days, collected from the historical data of power system operation. The conservativeness level is modeled by the number of scenarios K^{pv}. Variable $\alpha_{k^{\mathrm{pv}}}^{\mathrm{pv}}$ represents the convex combination weight of scenario $P_{k^{\mathrm{pv}}}^{\mathrm{ca}}$. A risk averse MEVPP would choose a higher value of K^{pv}, while a risk neutral MEVPP would choose a lower value of K^{pv}.

2.2 Definition of Feasibility Sets

Considering known the day-ahead electric energy schedule and natural gas schedule, i.e., $p_{vpp,t}^{\mathrm{DA}*}$ and $g_{vpp,t}^{\mathrm{DA}*}$; the start-up and shut-down states and on/off state of CHP, i.e., $I_{chp,t}^{\mathrm{DA,su}*}$, $I_{chp,t}^{\mathrm{DA,sd}*}$ and $I_{chp,t}^{\mathrm{DA}*}$; and the realizations of the uncertain available PV capacities, i.e., $\tilde{P}_{pv,t}^{\mathrm{ca}}$, set $\Xi\left(p_{vpp,t}^{\mathrm{DA}*}, g_{vpp,t}^{\mathrm{DA}*}, I_{chp,t}^{\mathrm{DA,su}*}, I_{chp,t}^{\mathrm{DA,sd}*}, I_{chp,t}^{\mathrm{DA}*}, \tilde{P}_{pv,t}^{\mathrm{ca}}\right)$ in Eq. (1) is considered to define the feasibility of the remaining variables, i.e., variables in set Φ^3:

$\Xi\left(p_{vpp,t}^{\mathrm{DA}*}, g_{vpp,t}^{\mathrm{DA}*}, I_{chp,t}^{\mathrm{DA,su}*}, I_{chp,t}^{\mathrm{DA,sd}*}, I_{chp,t}^{\mathrm{DA}*}, \tilde{P}_{pv,t}^{\mathrm{ca}}\right) = \{\Phi^3 :$

$$p_{vpp,t}^{\mathrm{DA}*} + p_{vpp,t}^{\mathrm{IB}+} - p_{vpp,t}^{\mathrm{IB}-} = \sum_{\substack{pv,chp,se,cc,\\ de\in DV\cup DS\cup DM}} \left(-\tilde{P}_{pv,t}^{\mathrm{ca}} - p_{chp,t} - \left(s_{se,t}^{\mathrm{dch}} - s_{se,t}^{\mathrm{ch}}\right) + p_{cc,t} + d_{de,t}\right) \tag{12}$$

$$g_{vpp,t}^{\mathrm{DA}*} + g_{vpp,t}^{\mathrm{IB}+} - g_{vpp,t}^{\mathrm{IB}-} = \sum_{chp,gb\in DV} \left(g_{chp,t} + g_{gb,t}\right) \tag{13}$$

$$0 = \sum_{chp,gb,st,ac,dh\in DV\cup DS\cup DM} \left(-h_{chp,t} - h_{gb,t} - \left(s_{st,t}^{\mathrm{dch}} - s_{st,t}^{\mathrm{ch}}\right) + h_{ac,t} + D_{dh,t}\right) \tag{14}$$

$$0 = \sum_{cc,ac,dc\in DV\cup DM} \left(-c_{cc,t} - c_{ac,t} + D_{dc,t}\right) \tag{15}$$

$$0 \leq p_{vpp,t}^{\mathrm{IB}+} \leq \overline{P_{vpp,t}^{\mathrm{IB}+}} \tag{16}$$

$$0 \leq p_{vpp,t}^{\mathrm{IB}-} \leq \overline{P_{vpp,t}^{\mathrm{IB}-}} \tag{17}$$

$$0 \leq g_{vpp,t}^{\mathrm{IB}+} \leq \overline{G_{vpp,t}^{\mathrm{IB}+}} \tag{18}$$

$$0 \leq g_{vpp,t}^{\mathrm{IB}-} \leq \overline{G_{vpp,t}^{\mathrm{IB}-}} \tag{19}$$

$$\begin{cases} p_{chp,t} = \eta_{chp}^{e} \cdot g_{chp,t}, \ \sigma_{chp,t}^{e} \\ h_{chp,t} = \left[\left(1-\eta_{chp}^{e}-\eta_{chp}^{loss}\right)\cdot\eta_{chp}^{hrs}\right] / \eta_{chp}^{e} \cdot p_{chp,t} \\ h_{gb,t} = \eta_{gb} \cdot g_{gb,t}, \ \sigma_{gb,t} \\ c_{cc,t} = COP_{cc} \cdot p_{cc,t}, \ \sigma_{cc,t} \\ c_{ac,t} = COP_{ac} \cdot h_{ac,t}, \ \sigma_{ac,t} \end{cases} \tag{20}$$

$$\begin{cases} 0 \le h_{gb,t} \le \overline{H_{gb}} \\ 0 \le c_{cc,t} \le \overline{C_{cc}} \\ 0 \le c_{ac,t} \le \overline{C_{ac}} \end{cases} \tag{21}$$

$$I_{chp,t}^{DA*} \cdot \underline{P_{chp}} \le p_{chp,t} \le I_{chp,t}^{DA*} \cdot \overline{P_{chp}} \tag{22}$$

$$p_{chp,t} - p_{chp,t-1} \le \left(1 - I_{chp,t}^{DA,su*}\right)\cdot R_{chp}^{u} + I_{chp,t}^{DA,su*} \cdot \underline{P_{chp}} \tag{23}$$

$$p_{chp,t} - p_{chp,t-1} \ge -\left(1 - I_{chp,t}^{DA,sd*}\right)\cdot R_{chp}^{d} - I_{chp,t}^{DA,sd*} \cdot \underline{P_{chp}} \tag{24}$$

$$0 \le s_{k,t}^{ch} \le I_{se,t} \cdot \overline{S_{k}^{ch}}, \forall k \in DS \tag{25}$$

$$0 \le s_{k,t}^{dch} \le \left(1 - I_{se,t}\right)\cdot \overline{S_{k}^{dch}}, \forall k \in DS \tag{26}$$

$$E_{k,t} = E_{k,t-1} + \left(s_{k,t}^{ch}\cdot\eta_{k}^{ch} - s_{k,t}^{dch} / \eta_{k}^{dch}\right)\cdot\Delta t, \forall k \in DS \tag{27}$$

$$\underline{SOC_{k}} \le E_{k,t} / CAP_{k} \le \overline{SOC_{k}}, \forall k \in DS \tag{28}$$

$$E_{k,\mathrm{Nt}} = E_{k,\mathrm{INI}}, \forall k \in DS \tag{29}$$

$$d_{de,t}^{p} - d_{de,t}^{m} = d_{de,t} - D_{de,t} \tag{30}$$

$$0 \le d_{de,t}^{m} \le I_{de,t} \cdot \overline{D_{de,t}^{m}} \tag{31}$$

$$0 \le d_{de,t}^{p} \le \left(1 - I_{de,t}\right)\cdot \overline{D_{de,t}^{p}} \tag{32}$$

}

In the real-time operational stage, the deviations from the day-ahead energy schedule are represented by imbalances. The power balance and natural gas balance of the MEVPP in the real-time operation are given by Eqs. (12)-(13), respectively. The heat balance and cooling balance are given by Eqs. (14)-(15), respectively. Eqs. (16)-(17) limit the amount of power imbalance and Eqs. (18)-(19) limit the amount of natural gas imbalance. Parameters $\overline{P_{vpp,t}^{IB+}} / \overline{G_{vpp,t}^{IB+}}$ and $\overline{P_{vpp,t}^{IB-}} / \overline{G_{vpp,t}^{IB-}}$ are the maximum electricity/natural gas imbalance of the injection and withdrawal directions, respectively.

For the backpressure CHP, the heat-to-electricity ratio is a constant parameter given by $\left[\left(1-\eta_{chp}^{e}-\eta_{chp}^{loss}\right)\cdot\eta_{chp}^{hrs}\right] / \eta_{chp}^{e}$, where parameters η_{chp}^{e}, η_{chp}^{loss} and η_{chp}^{hrs} represent the electricity production efficiency, heat loss coefficient and heat recovery efficiency of CHP, respectively. Therefore, the efficiencies of electricity generation and heat generation of CHP can be described by Eq. (20), respectively. Variable $g_{chp,t}$ is the natural gas consumed by CHP. Parameter η_{gb} represents the efficiency of heat generation of gas boiler (GB). Variables $h_{gb,t}$ and $g_{gb,t}$ are the heat produced and natural gas consumed by GB, respectively. Parameter COP_{ac} / COP_{cc} represents the efficiency of the cooling generation of absorption cooler (AC)/ Compression cooler (CC). In Eqs. (23)-(24), Parameters R_{chp}^{u} and R_{chp}^{d} are the ramp-up and ramp-down limits of CHP, respectively.

Eqs. (25)-(26) limit the charging and discharging power of SE. Binary variable $I_{se,t}$ characterizes the complementarity between the charging and discharging states. Eq. (27) presents the energy stored in Electric energy storage (SE) at the end of each time interval. Variables $s_{se,t}^{\mathrm{ch}}$ and $s_{se,t}^{\mathrm{dch}}$ are the charging and discharging powers, respectively. Eq. (28) represents the state of charge (SOC) limit. Eq. (29) forces that the final energy storage should go back to the initial energy storage of the day, so that a new cycle can be started on the following day. Reduceable electric demand (DE) is modeled as Eqs. (30)-(32). Eq. (30) calculates the demand increase and curtailment.

2.3 Solution of Data-driven ARO

To address the trilevel optimization problem, the C&CG method is employed, decomposing the problem into a master problem and a subproblem. By iteratively solving these two problems, convergence toward the optimal solution is achieved.

(1) Master Problem

The master problem is provided by Eqs. (33) - (35):

$$\min_{\Phi^1,\Phi^{1M}_{t,v'},\vartheta} \sum_{t\in T}\left(\lambda_t^{\mathrm{E}} \cdot p_{vpp,t}^{\mathrm{DA}} + \lambda_t^{\mathrm{G}} \cdot g_{vpp,t}^{\mathrm{DA}} + CO_t^{\mathrm{sud}}\right) + \vartheta \tag{33}$$

s.t.

$$\text{Eqs. (2) - (4)} \tag{34}$$

$$\vartheta \geq \sum_{t\in T}\left(\lambda_t^{\mathrm{EIB+}} \cdot p_{vpp,t,v'}^{\mathrm{IB+}} + \lambda_t^{\mathrm{EIB-}} \cdot p_{vpp,t,v'}^{\mathrm{IB-}} + \lambda_t^{\mathrm{GIB+}} \cdot g_{vpp,t,v'}^{\mathrm{IB+}} + \lambda_t^{\mathrm{GIB-}} \cdot g_{vpp,t,v'}^{\mathrm{IB-}} + CO_{t,v'}^{\mathrm{op}}\right) \tag{35}$$

where variables in set Φ^1 and $\Phi^{1M}_{t,v'} = \left\{\begin{matrix} p_{vpp,t,v'}^{\mathrm{IB+}}, p_{vpp,t,v'}^{\mathrm{IB-}}, g_{vpp,t,v'}^{\mathrm{IB+}}, g_{vpp,t,v'}^{\mathrm{IB-}}, p_{chp,t,v'}, \\ g_{chp,t,v'}, h_{chp,t,v'}, h_{gb,t,v'}, g_{gb,t,v'}, c_{cc,t,v'}, p_{cc,t,v'}, c_{ac,t,v'}, \\ h_{ac,t,v'}, E_{k,t,v'}, s_{k,t,v'}^{\mathrm{ch}}, s_{k,t,v'}^{\mathrm{dch}}, d_{de,t,v'}^{\mathrm{p}}, d_{de,t,v'}^{\mathrm{m}}, d_{de,t,v'} \end{matrix}\right\}$, $\forall v' \leq v$ are for the optimization and v is the iteration counter. $\tilde{P}_{pv,t,v'}^{ca*}$ with a star in the superscript is considered as given input data, which are obtained from the subproblem described in the following subsection. Master problem is a relaxed version of the original problem (1) - (4), which is progressively reconstructed with the help of the auxiliary variable represented by ϑ at each iteration. By solving the master problem, the day-ahead energy schedule is decided with the target of minimizing the overall cost of the MEVPP.

(2) Subproblem

The bilevel subproblem is provided below:

$$\max_{\Phi^2\in\Omega} \min_{\Phi^3\in\Xi} \sum_{t\in T}\left(\lambda_t^{\mathrm{EIB+}} \cdot p_{vpp,t}^{\mathrm{IB+}} + \lambda_t^{\mathrm{EIB-}} \cdot p_{vpp,t}^{\mathrm{IB-}} + \lambda_t^{\mathrm{GIB+}} \cdot g_{vpp,t}^{\mathrm{IB+}} + \lambda_t^{\mathrm{GIB-}} \cdot g_{vpp,t}^{\mathrm{IB-}} + CO_t^{\mathrm{op}}\right) \tag{36}$$

where sets Ω and Ξ represent the uncertainty set and the feasibility set, respectively, with the constraints of the corresponding variables. The upper-level optimization determines the worst-case scenario for the uncertain available PV capacity during the real-time stage. In the lower-level optimization, corrective actions are taken to minimize the cost of MEVPP considering the worst-case realization of the uncertainties in real-time operation. To solve the bilevel subproblem, the lower-level problem can be replaced by its KKT conditions.

Journal of Physics: Conference Series **3012** (2025) 012035 doi:10.1088/1742-6596/3012/1/012035

3. Case Study

There exists a MEVPP in one university campus named Smart Multi-Energy District [7]. The uncertainty set of available PV capacity described by BCUS and DSBS are given by Fig.1 and Fig.2, respectively.

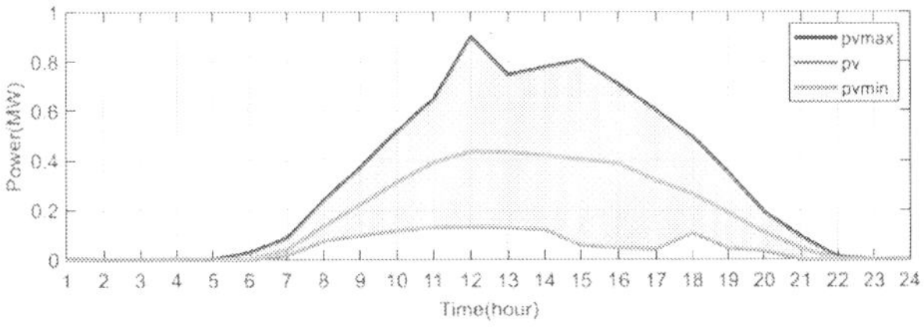

Figure 1. Uncertain available PV capacity described by BCUS

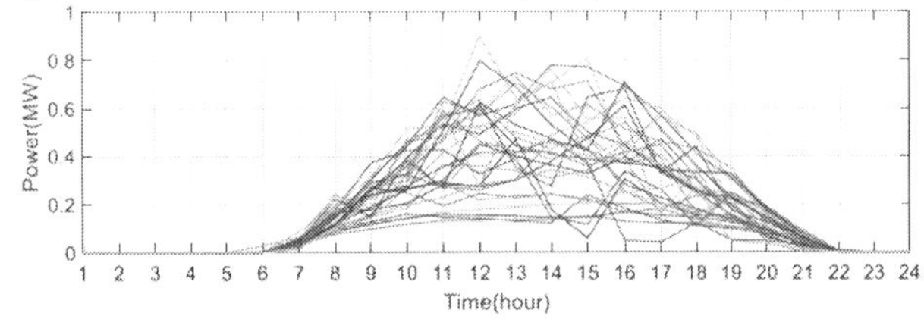

Figure 2. 31 scenarios of uncertain available PV capacity described by DSBS

The risk preference of MEVPP exerts a great impact on the economic results. When MEVPP is more risk-averse, in other words, when the budget of uncertainty is large in BCUS or the number of scenarios is large in DSBS, the cost of MEVPP is higher, as manifested in Fig.3. Because MEVPP needs to tackle larger power deviations caused by PV uncertainties, which prevents it from achieving the optimal strategy. But a higher budget of uncertainty parameter/number of scenarios comes with a conservative cost with less commercial risk. Moreover, comparing Fig.3 (b) to Fig.3 (a), one can find that when both the numbers of scenarios and the budget of uncertainty parameters reach their maximum values, using DSBS can reduce the overall cost from 5215.25 € when using BCUS to 4834.58 €, significantly saving the overall cost by about 7.3%. This indicates that using DSBS is more cost-effective than using BCUS.

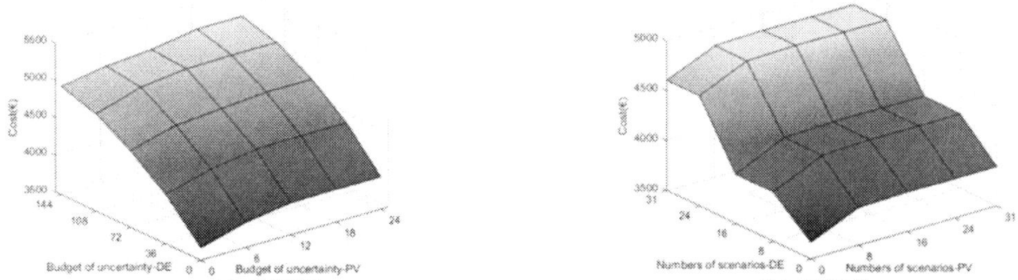

(a) BCUS (b) DSBS

Figure 3. Cost of MEVPP with different budget of uncertainty parameters/numbers of scenarios

When using DSBS, the electric energy schedules of MEVPP under different numbers of scenarios are given in Fig. 4. The numbers of scenarios can vary within a certain range, $K^{pv} \in [0,31]$, $K^{pv} \in Z$. $K^{pv} = 0$ is solely applied to indicate the situation that the available PV capacity is with the corresponding values of the best forecast. MEVPP buys more electricity from the public network when the numbers of scenarios are higher. Because it results in a lower PV output, which needs strong power support from the public network. The differences in the electric energy schedules caused by different numbers of scenarios are larger during 9:00 a.m.-19:00 p.m., because the PV works with higher production, hence uncertainty, during this period. During other periods, the PV output is 0 MW and is, therefore, free from the impact of uncertainty.

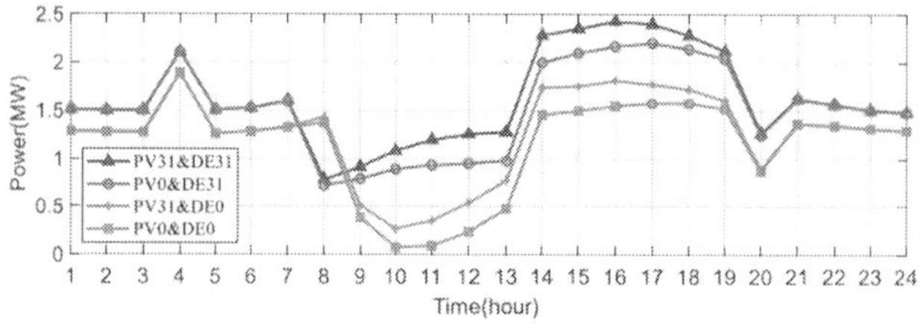

Figure 4. Day-ahead electric energy schedule of MEVPP using DSBS

4. Conclusion

This paper proposes an optimal self-scheduling model of MEVPP under uncertainties. ARO is applied to handle the uncertainties of available PV capacities, which formulates a trilevel optimization model solved by the C&CG method. The fluctuations of uncertainty factors are developed by two types of uncertainty sets, BCUS and DSBS. Case studies are set according to the working Smart Multi-Energy District in the chosen campus. It is confirmed that using DSBS is less conservative and more cost-effective than using BCUS. The total cost can be decreased by about 7.3%. Utilizing the coordinated interaction among multiple distributed resources in MEVPP can decrease the fluctuation of renewable energy and achieve a synergistic improvement of operational stability and economy.

Acknowledgments

This work is supported by the National Natural Science Foundation of China (No. 52407124).

Reference

[1] Wang J, Xie N, Huang C, et al. Two-stage stochastic-robust model for the self-scheduling problem of an aggregator participating in energy and reserve markets[J]. Protection and Control of Modern Power Systems, 2023, 8(3): 1-20.

[2] Dimitriadis C N, Tsimopoulos E G, Georgiadis M C. Strategic bidding of an energy storage agent in a joint energy and reserve market under stochastic generation [J]. Energy, 2022, 242: 123026.

[3] Yan C, Tang Y, Dai J, et al. Uncertainty modeling of wind power frequency regulation potential considering distributed characteristics of forecast errors [J]. Protection and Control of Modern Power Systems, 2021, 6(1): 1-13.

[4] Khojasteh M, Faria P, Vale Z. A robust model for aggregated bidding of energy storages and wind resources in the joint energy and reserve markets [J]. Energy, 2022, 238: 121735.

[5] Jiang S, Wu C, Gao S, et al. Robust frequency risk-constrained unit commitment model for AC-DC system considering wind uncertainty [J]. Renewable Energy, 2022.

[6] Shen F, Zhao L, Wang M, et al. Data-driven adaptive robust optimization for energy systems in ethylene plant under demand uncertainty [J]. Applied Energy, 2022, 307: 118148.

[7] Bovera F. United we stand: How aggregates of distributed energy resources can shape the future energy system[D]. Mialn, Italy: Politecnico di Milano, 2021.

Journal of Physics: Conference Series 3012 (2025) 012036 doi:10.1088/1742-6596/3012/1/012036

Research Progress on Typical Defects and its Online Monitoring Technology for UHVDC Bushing

Xiaochen Yang[0000-0001-5635-0681], **Baoquan Wan***, **Xiang Gao**[0009-0000-3498-5950]

State Key Laboratory of Power Grid Environmental Protection, China Electric Power Research Institute, Wuhan 430074, China

*Corresponding author's e-mail: wanbaoquan@epri.sgcc.com.cn

Abstract. Ultra-high voltage direct current (UHVDC) bushing is the important equipment in UHV transmission system. At present, the resin impregnated paper (RIP) bushing is widely used because of the prominent performance. However, the UHVDC RIP bushing faces many challenges during long-term operation. On the one hand, there may be primary defects like bubbles and cracks due to the poor workmanship in the production process. For the other hand, the interaction of electric, thermal, mechanical stress may cause various serious failures in the bushing, which could occur in the core, the end shield and connection system, and the conductor and connecting parts. Thus, in order to realize defect perception, UHF method, infrared thermal imaging, online monitoring technology based on the end shield, and SF_6 gas analysis method are applied to the live detection and online monitoring of RIP bushing.

1. Introduction

Ultra-high voltage direct current (UHVDC) transmission technology is an important physical basis for achieving large-scale clean energy access and cross-regional power transmission [1]. By the end of 2023, there have been 20 UHVDC transmission projects operated in China. The UHVDC bushing is one of the key equipment to realize circuit connection, electrical insulation, and mechanical support in the system. There are two main types of UHVDC bushings, converter transformer valve side bushing and wall bushing. Compared with traditional oil impregnated paper (OIP) bushing, resin impregnated paper (RIP) bushing has prominent advantages in electrical performance, mechanical properties, environmental friendliness, and no explosion risk [2,3]. Therefore, RIP bushing has become the mainstream structure of DC bushing [4,5].

The UHVDC bushing also faces multiple security challenges and exposes some potential safety hazards in long-term operation. Firstly, some undetected primary defects may become hidden danger due to the large size and complex manufacturing process [6-8]. Secondly, the insulation system is subjected to electrical stress of DC superimposed high order harmonics, complex thermal stress, and high load mechanical stress [9]. Moreover, the interface charge accumulation of SF_6-RIP composite insulation under DC electric field will also be intertwined with metal particle defects [10]. All these factors will lead to the excessive value of the partial discharge (PD), interface discharge, the end shield fault, and other problems occur from time to time.

In order to ensure safety and stability, the defect detection technology of the bushing has become the focus [11,12]. Electrical parameters, thermodynamic properties, mechanical characteristics, and gas chemical components are the data foundations for defect identification and fault location.

Generally, electrical parameters are represented by PD and dielectric properties, thermodynamic properties are mainly symbolized by temperature distribution, and mechanical characteristics usually refer to vibration signals. By designing multi-dimensional technical means such as offline detection, live detection, and online monitoring, the quality can be strictly controlled by optimized steps including handover, daily operation, and status evaluation.

In this paper, the typical defects of UHVDC bushings and their evolution mechanism, defect detection technology, fault diagnosis and analysis methods are reviewed. The key problems that need to be urgently solved for UHVDC bushing defect detection technology are pointed out.

2. Typical defects of UHVDC bushings

2.1. Defects caused by poor production process

The manufacturing process of the capacitive shielding structure in the RIP bushing is complicated, and the main process is shown in Fig. 1. Firstly, the dried insulating crepe paper and the surface-treated aluminum foil are alternately wound at intervals set on the central tube into a cylinder, thus forming a layered capacitor shield, as the Fig.2 shows. Then, it is impregnated and casted with epoxy resin under vacuum environment. After the steps under stepwise-temperature curing, drying, degassing, and others. Finally, it is machined, sprayed, turned into shape, and assembled with outdoor insulation.

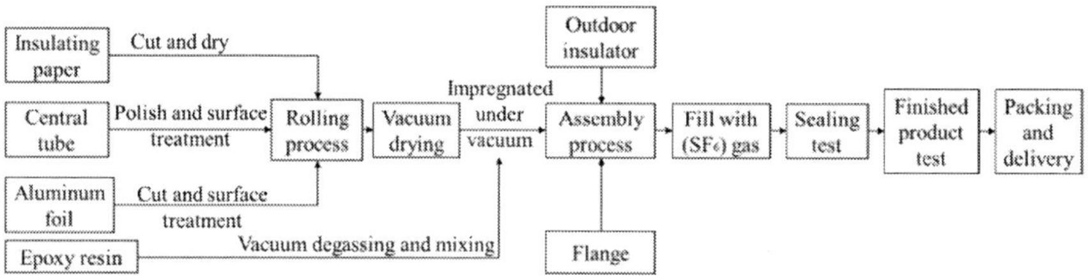

Figure 1. The manufacturing process flow of RIP bushing

As the reason of the large size and high weigh, the manufacturing of UHVDC bushings faces two major outstanding difficulties. Firstly, during the rolling process, the insulating paper and aluminum foil are easy to wrinkle because the conductive metal electrodes (plates) are prone to uneven stress, which become weak areas of insulation. Therefore, the constant tension rolling process is crucial for the large-sized RIP bushings quality. Secondly, during the vacuum impregnation process, the important prerequisite for good insulation performance is that there is no trapped air and residual gap in the insulation. At the meanwhile, release internal stress uniformly and discharge the generated gas inside the bushing is also necessary because there are complex thermodynamic processes when the epoxy core is cured. Thus, the heat dissipation during the curing process of the epoxy core must to be optimized and controlled to manufacture a flawless core.

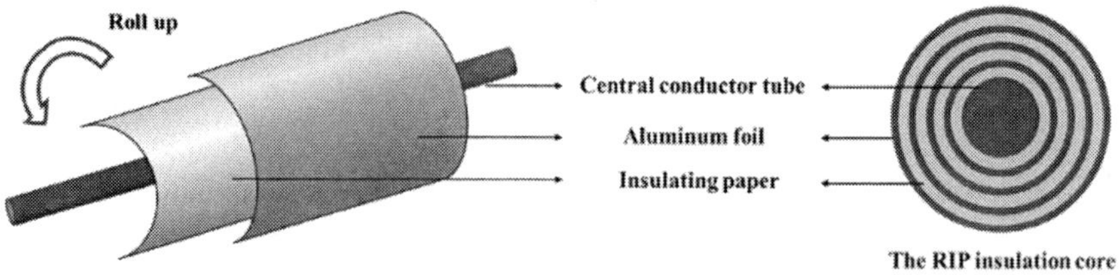

Figure 2. The rolling process of RIP bushing

The most common defects in the production process are bubbles and cracks. These subtle defects are generally located in hidden places inside the epoxy core, and it is very difficult to identify even if the bushing is disassembled for inspection. Fig. 3 (a) shows the bubbles inside the core due to the poor vacuum impregnation process, while Fig. 3 (b) shows the local cracks caused by the poor curing process because of the differences in thermodynamic properties of epoxy resin, paper, and foil.

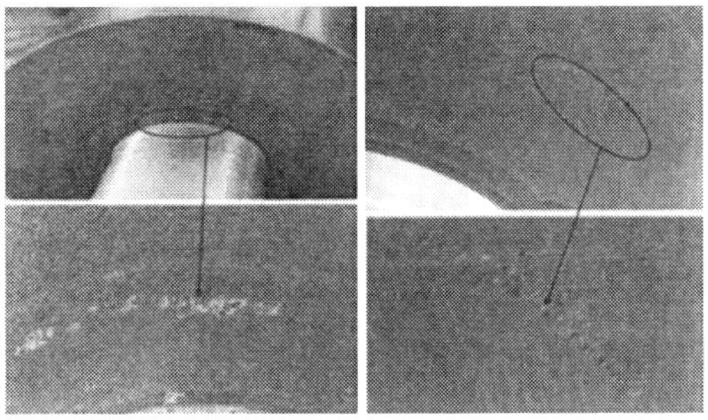

Figure. 3. Typical bubbles and cracks inside the UHVDC bushing. (a) Bubbles in the core of RIP bushing. (b) Cross section of the core with cracks.

2.2. Defects arising from long-term service conditions

2.2.1. Composite stress accelerates the occurrence and development of faults. The UHVDC bushing is subjected to multi-field coupling stress of mechanical, electrical, and thermal during operation, as shown in Fig. 4. It puts forward higher requirements for RIP bushing insulation performance.

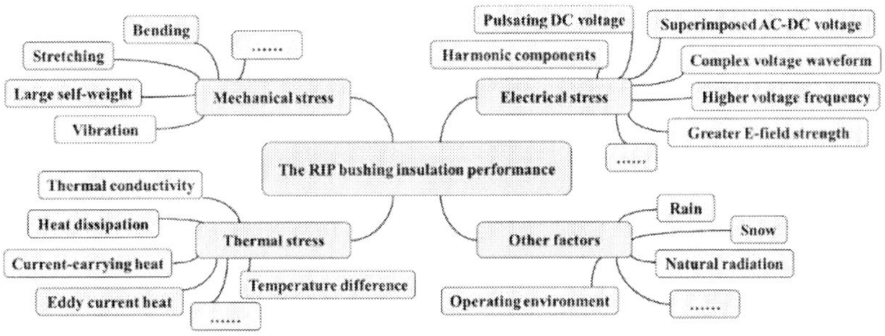

Figure. 4. The UHVDC bushing is subjected to multi-field coupling stress during operation

Since it is directly connected to the converter valve components, the bushing is affected by the rapid and reciprocating change between the on/off operation of converter devices. For the wall bushing, its main insulation is subjected to pulsating DC voltage. As for the valve side bushing, it bears more complex AC and DC superimposed voltage. The current it carries contains a large number of harmonic components in addition to the 50Hz power frequency current. Additionally, the RIP insulation faces more complex voltage waveform, higher voltage frequency, and greater E-field strength compared with traditional AC bushing.

Compared with OIP insulation, the internal insulation of the RIP bushing has lower thermal conductivity and heat dissipation. The current-carrying heat in the central conductive tube and the eddy current heating caused by high-order harmonics continuously accumulate heat in the RIP material, making the temperature inside the bushing constantly rise. Due to the large thickness of the of the RIP insulation, taking the ±800 kV converter transformer valve side bushing as an example, the maximum temperature of the internal fault point inside the core may exceed 120℃, and the radial temperature gradient can reach 90°C. Besides, there may be large differences in ambient temperature between the inside and outside of the valve hall. Accordingly, the thermodynamic optimization of heat dissipation in dry bushing is the bottleneck problems restricting its development.

2.2.2. Typical defects in service condition. The structure of UHVDC bushing is complex, as shown in Fig. 5. Through combing and analyzing domestic and international cases, we concluded that the typical faults of converter transformer valve side bushing and DC wall bushing can be mainly divided into 5 categories, as summarized in Table. 1.

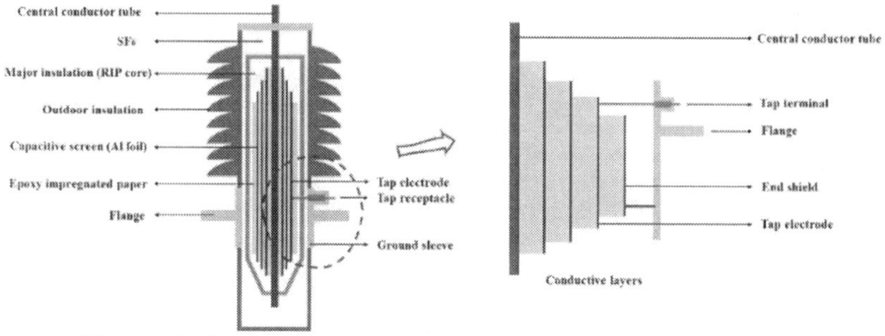

Figure. 5. Structure diagram of capacitance dry-type bushing

Failures of the RIP core. The main defects of the core are mainly bubbles, cracks, and moisture. These defects are often the primary defects from the production process, or secondary defects generated and developed during service process. The RIP insulation material will deteriorate under the electric-thermal-mechanical stress for a long time. After a series of physical and chemical reactions, the smooth and flawless core first appears a tiny crack, then it expands into a larger size crack, and finally develop to PD and even breakdown. The surface charge will accumulate on the RIP-SF6 interface under the DC electric field, which will distort the electric field on the insulating material surface and reduce the flashover voltage. At the same time, the slow dissipation of charges will bring them participate in surface discharge. It is worth noting that the metal particle or protrusion caused by incomplete grinding on the core surface will aggravate the surface charge accumulation. In particular, when it works in a harsh condition, such as polarity reversal voltage, the surface insulation strength will be greatly decreased.

Failure of the end shield and connection system. The end shield of the valve side bushing is a crucial structure for grounding and dielectric property measurement. The grounding of the end shield is frequently untied to measure the capacitance, dielectric loss, and PD of the bushing. Thus, the end shield often appears poorly grounded and even grounding failure, which is one of the weak points of the bushing quality. The elasticity reduction or deformation of the spring plate may cause the inadequate contact between the spring plate and the lead post. That is the common cause of grounding fault for the built-in end shield device. While for the end shield with welded lead, the loosening of the connection is usually caused by body vibration during transportation and operation, and the electrodynamic impact of the short circuit current. Furthermore, moisture, rust, and other reasons will also contribute to the fracture, failure, and discharge at the end shield.

Failure of the current-carrying conductor and the connecting parts. The temperature inside the bushing is always in a dynamic change. Especially, because of the skin effect, overheating is highly likely to occur inside the bushing when there is a defect in the conductor as well as its connecting parts. In practical application, even if the SF_6 gas-insulated valve side bushing that passed the temperature rise test, it may still discharge caused by overheating under the long-term current load. Consequently, for the valve side bushing, it is indispensable to pay attention to the local overheating due to the uneven shunt of the elastic connecting parts.

Failure of the gas seal and measurement system. Defects in the sealing structure will bring about SF_6 gas leakage and pressure drop inside the bushing, such as the sealing material aging, the plastic deformation caused by improper coordination. Besides, aging, corrosion, and the insufficient material strength of the external structure like the elevated seat may also lead to gas leakage. Failures of the SF_6 density measurement relay will result in pressure alarms, misreporting or missing the blocking signal. While the defects of the meter itself and the hidden dangers of the secondary measurement loop will lead the pressure alarm function of the density relay to fail.

Failure of the outdoor insulation. Dirt and dust are easily absorbed on the bushing surface under DC electric field. The distortion of the local field strength on the insulation shed may bring about the unpredictable weak discharge, further accelerating the insulation aging. Insulation aging, surface fouling, and charge accumulation of the silicone rubber shed will lead to surface flashover and outdoor insulation failure.

Table 1. Typical faults of converter transformer valve side bushing and DC wall bushing

Number	Type		Typical faults
1	RIP core	core	Bubbles, cracks, and moisture
		SF6-RIP interface	Metal particle or protrusion on the surface
2	End shield and connection system		Unreliability of the end shield wiring, loose connection caused by vibration, moisture, rust and other reasons
3	Current-carrying conductor and connecting parts		Local overheating of bushing caused by uneven shunt of elastic electrical contact parts
4	Gas seal and measurement system		Defects of the bushing sealing structure and failure of the SF_6 measurement
5	Outdoor insulation		Insulation aging, fouling of silicone rubber shed

3. The live defect detection and online monitoring technology of the UHVDC bushing

The traditional inspection methods of the UHVDC bushing rely on the regular maintenance. But as the number of devices increase, the security risk of various defects and failures also rises. Higher requirements have been put forward for live detection and online monitoring technology, in order to achieve faults advanced perception, warning, and governance. However, advantages and disadvantages of these technologies and means are both prominent, and further breakthroughs and innovations are needed to meet the future development needs of UHVDC bushing.

3.1. Ultra-high frequency detection technology

Ultra-high frequency (UHF) method is a common electromagnetic detection method for PD. Antenna sensors are used to receive high frequency electromagnetic signals from several kHz to GHz of PD radiation. The UHF method can filter the frequency of electromagnetic interference on the valve side, and has the advantages of strong anti-interference ability and convenient positioning. Literature [12] had measured PD with contactless UHF method, further, they had used the minimum energy method to extract time-difference-of-arrival (TDOA), and adopted Chan algorithm to locate the UHF points. As shown in Fig. 6 (Fig. 9-10 in [12]). However, there is no clear correspondence between the signal amplitude and the apparent discharge, making it impossible to calibrate the discharge amount. The waveform and amplitude of the detection signal are also affected by many factors, such as the detection location, and the frequency band of the discharge signal. Accordingly, this method is suitable for online monitoring, but how to achieve precise identification of defect types and accurate positioning is the focus of attention in the future.

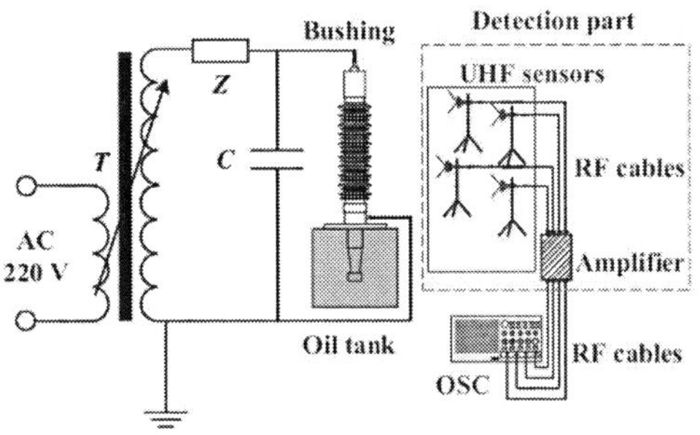

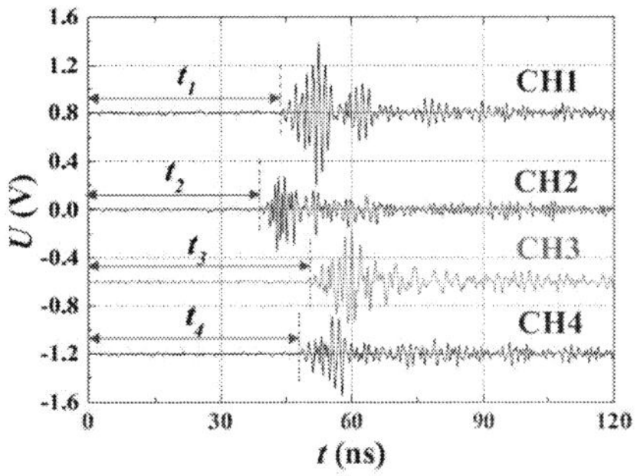

Figure 6. (a) Experimental setup (b) The PD-induced electromagnetic signals. (Fig. 9-10 in [12])

3.2. *Infrared thermal imaging technology*

The temperature distribution of bushing in normal operation is uniform. However, once the insulation goes wrong, temperature of the faulty area will rise caused by the current leakage or local overheating [14]. The Infrared spectrum of the converter transformer valve-side bushing with abnormal temperature is given in Fig. 7. The conduction relationship model is established by analyzing the temperature conduction mechanism of discharge heating and defect overheating inside the DC bushing. In order to determine the threshold of external temperature change during infrared monitoring, factors such as ambient temperature and material surface emissivity need to be considered. It could find defect hotspots and fault sources, and realize defect diagnosis and fault early warning. Based on bushing temperature monitoring, this method has the advantage of being free from electromagnetic interference, and is suitable for live detection and online monitoring. Similarly, building accurate and reliable fault mapping model is the focus of future development.

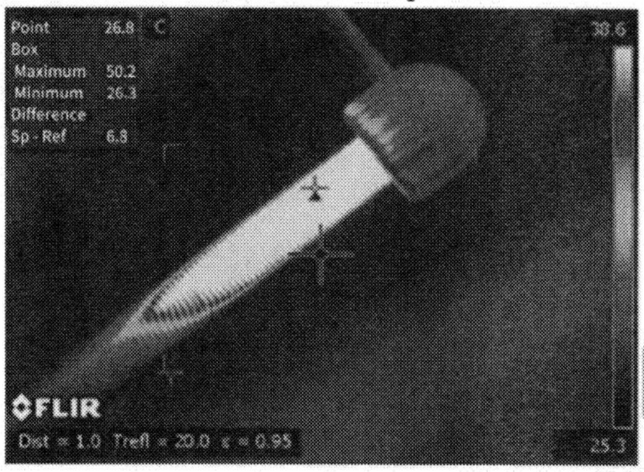

Figure 7. Infrared spectrum of the converter transformer valve-side bushing with abnormal temperature. (Fig. 9 in [14])

3.3. *Online dielectric characteristics monitoring technology based on bushing end shield*

When the bushing is interruption maintenance, the insulation condition of the bushing could be evaluated by measuring the dielectric characteristics through the end shield, including dielectric loss, capacitance capacity, the end shield leakage current, and the end shield voltage [12]. The typical measurement based on bushing end shield is given in Fig.8. Domestic scholars have proposed a high efficiency magnetic shielding method for the microcurrent sensor on the end shield of bushing based on the zero-flux space technology. On the premise of not changing the original grounding of the end shield, they have achieved an integrated design of the end shield leakage current sensor, high-frequency pulse current sensor, and effective grounding. They have established a high-precision analysis algorithm based on fusion of time-frequency domain features. This method can efficaciously suppress the influence of frequency leakage, fence effect, and complex harmonics on accuracy, and realize the accurate online monitoring of multiple parameters under complex operating conditions.

Figure 8. The typical measurement based on bushing end shield

3.4. SF_6 gas composition analysis technology

The basic gas generation law of SF_6 discharge decomposition products has been widely studied, which provides an important basis for fault identification. As mentioned above, metal particle or protrusion on conductor and suspension potential could cause PD. The researcher analyzed the relationship between PD and SF_6 discharge decomposition components, and found that, the main decomposition products include SO_2F_2, SOF_2, SO_2, CF_4, etc. Among them, the amount of SO_2F_2 and other products increased linearly, and some products showed saturation characteristics. The output of characteristic gas rises with the enhance of discharge intensity, while the heighten of pressure will inhibit the discharge intensity and lead to the decrease of gas production. Carbonaceous gas especially CO_2, CS_2, have been widely regarded as the characteristic gases produced by the cracking of epoxy resin at the gas-solid interface. Moreover, according to relevant studies, the influence of micro-water and micro-oxygen on SF_6 discharge decomposition cannot be ignored. Some scholar clarified the mechanism of H_2O and O_2 participating in the reaction and further pointed out that the different content of H_2O and O_2 would also lead to obvious differences in the content, gas production rate, ratio and other aspects of SF_6 decomposition components. Literature [15] gave the panoramic reaction course of SF6-epoxy pyrolysis process, as shown in Fig. 9.

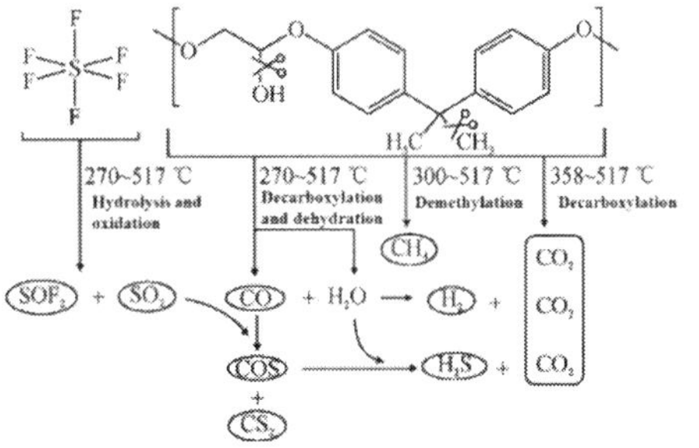

Figure 9. Panoramic reaction course of SF6-epoxy pyrolysis process. (Fig. 5 in Reference [15])

Combined with relevant historical data, SF_6 gas composition analysis technology could effectively support the systematic analysis of defects and fault detection. This technology could analyze and identify the defects types by electrically sampling gas, but cannot locate them.

4. Conclusion

In this paper, the typical types of defects that may occur in the production and operation of UHVDC RIP bushing are reviewed. The key factors like electrical, thermal, mechanical, which have significant influence on the occurrence and development of defects during operation are summarized. These multiple factors interwoven with each other make RIP bushing face many challenges. Defects or failures in important parts will lead to serious consequences, particularly on the core, the end shield, the conductor and connecting parts, the gas seal and measuring system, as well as the outdoor insulation. Further, the live defect detection and online monitoring technologies used in UHVDC bushing are summarized, and both the advantages and shortcomings of various methods are pointed out.

In the future, the fault generation and development mechanism of UHVDC RIP bushing should be deeply researched. It is necessary to optimize the live defect detection and online monitoring schemes and develop the technology with both quantitative analysis and precise positioning functions. Improving the accuracy and validity of measurement data is the future development directions and technological trends. Through efficient identification and precise positioning of defects, differentiated operation and maintenance for the bushing can be supported.

Acknowledgments

This work was supported by the Science and Technology Project of State Grid Corporation of China (5108-202218280 A-2-3 64-XG).

References

[1] S. Zhang, Z. Peng, P. Liu, and N. Li: Design and dielectric characteristics of the ±1100 kV UHVDC wall bushing in China. IEEE Transactions on Dielectrics & Electrical Insulation, 22(1), 409-419. (2015).
[2] W. Cao, W. Shen, B. He and K. Wu: Structural design for dc bushing core based on the material life, IEEE Transactions on Dielectrics and Electrical Insulation, vol. 20, no. 1, pp. 281-288. (2013).
[3] P. Yin, Z. Xu, X. Luo, W. Hu, X. Xie and B. Wan: Study on Insulation Characteristics of Oil Impregnated Paper Bushings under Long-term Electrothermal Combined Test, 2020 IEEE International Conference on High Voltage Engineering and Application (ICHVE), Beijing, China, pp. 1-4. (2020).
[4] H. Zhang, P. Liu, H. Feng, S. Zhang, D. Li and Z. Peng: Space charge dynamics in epoxy resin impregnated crepe paper multilayer under voltage polarity reversal, IEEE Transactions on Dielectrics and Electrical Insulation, vol. 26, no. 1, pp. 253-260. (2019).
[5] X. Xie, B. Wan, W. Hu, Z. Xu, Y. Wang, Y. Deng and Y. Zeng: Influence of Thermal Effect on Insulation Performance of Resin Impregnated Paper Bushing, 2022 4th International Conference on Smart Power & Internet Energy Systems (SPIES), Beijing, China, pp. 149-154. (2022).
[6] J. Zhang, X. Li, X. Wei, Z. Dang, Z. Xie, J. Feng, J. Wei and G. Yue: Mechanical Damage Properties of Dry Bushing Core Composites, Insulators and Surge Arresters, 6, 316, pp. 226-233. (2023).
[7] X. Li, H. Tang, F. Yang, H. Cheng and S. Zhang: Internal Discharge Characteristics of HVDC Wall Bushing Model with Resin Impregnated Paper Capacitor Core under Long-term DC Voltage of Different Polarity, 2022 IEEE International Conference on High Voltage Engineering and Applications (ICHVE), Chongqing, China, pp. 1-4. (2022).
[8] Y. Zhou, H. Zhang, Z. Bai, X. Xie and T. Liu: Study on Surface Charge and Mechanical Properties of Epoxy Impregnated Paper Composites, 2024 IEEE 7th International Electrical and Energy Conference (CIEEC), Harbin, China, pp. 2567-2571. (2024).
[9] Q. Wang, G. Xie, H. Tian, Z. Peng and X. Yang: Impact of Temperature on the Transient DC Field Distribution of ± 1100 kV UHVDC Wall Bushing, IEEE Transactions on Power Delivery, vol. 36, no. 3, pp. 1449-1457. (2020).
[10] S. Zhang, J. Li, H. Tang, X. Li, Y. Zhang, B. Qi and J. Yan: Long-time Flashover Discharge Characteristics of Gas-solid Interface of UHVDC Bushing and its Influencing Factors, Proceedings of the CSEE, 44, 14, pp. 5776-5786.
[11] B. Qi, Q. Dai, C. Li, Z. Zeng, M. Fu and R. Zhuo: The Mechanism and Diagnosis of Insulation Deterioration Caused by Moisture Ingress into Oil-Impregnated Paper Bushing, Energies, 11, 6, 1496. (2018)
[12] W. Zong, H. Deng and X. Liu: Diagnosis and Analysis of Abnormal Secondary Voltage at End Shield of Valve Side Bushing of Converter Transformer, Transformer, 61, 11, pp. 53-61. (2024).
[13] J. Jiang, J. Chen, J. Li, X. Yang, Albarracín - Sánchez Ricardo, & P. Ranjan, et al: Propagation and localisation of partial discharge in transformer bushing based on ultra - high frequency technique. High Voltage. (2021).
[14] A. Zhou, L. Gao, X. Ji, M. Zhang and H. Tang: Analysis and Field-Repair of Current Overheating Faults on Dry Type SF_6 Gas-Insulated Valve Side Bushing in Converter Transformer, Power System Technology, 42, 5, pp. 1401-1409. (2018).
[15] R. Yang, D. Lin, H. Yang, J. Xue, X. Zhang, W. Zhang: Research on Thermal Decomposition of SF_6-epoxy Resin, High Voltage Apparatus, 60.4, pp. 139-145. (2024).

Power supplying unit Delineation Method Using an Equal-Angle Centerline Clustering Algorithm

Zhi Guan, Hejun Yang*, Jingjing Yu, Jindong Huang, Anxin He

Anhui Province Key Laboratory of Renewable Energy Utilization and Energy Saving, Hefei University of Technology, Hefei 230009, China

*Corresponding author's e-mail: cquyhj@126.com

Abstract. Aiming at the existing medium-voltage (MV) distribution grid power supplying unit delineation without considering load complementarity, and the delineation scheme cannot reflect the optimization. This paper proposes a power supplying unit delineation method using an equal-angle centerline clustering algorithm. Firstly, this paper proposes a power supplying unit delineation method based on a centerline clustering algorithm. Secondly, the centerline clustering algorithm is improved based on Particle Swarm Optimization (PSO) to seek the optimal result of block delineation. Then, the feeder blocks are combined into power supplying units by the maximum weight matching method. Finally, a power supplying area with the integration of three substations is used to verify the rationality and effectiveness of the proposed method.

1. Introduction

With the acceleration of urbanization, the power supply is not only related to the quality of urban development, but also affects people's daily life. The inclusion of flexible resources such as electric vehicles and energy storage exacerbates the difficulty of grid planning [1], [2]. Only scientific and reasonable planning of urban power grid can meet the growing demand for electricity in society. At present, many regions use a grid delineation method to divide the planning area, and the results of the delineation are continuously adjusted with the restriction conditions of roads, rivers, total load requirements, and voltage requirements, to improve the reliability of the distribution grid [3]. The significance of grid delineation is to classify the loads of the grid and to enable the line in each grid of the distribution grid to independently undertake the power supplying load in their own grid, which is the key to realize refined planning [4].

The power supplying units are subdivided from the power supplying grid. The delineation of power supplying units is planned according to the power supplying grid delineation of MV distribution grid and the grid-connected capacity configuration for energy storage [5]. Current planning methods tend to forecast the maximum loads of electricity consumers and predict the simultaneous rate empirically. That's an inaccurate statement, and when the simultaneous rate is too high, it can easily to cause the overload phenomenon [6]. Therefore, it is important to consider the load characteristics of in the grid firstly, and then the power supplying units are divided according to the load characteristics. Ref. [7] proposes a distribution grid planning method based on density peak clustering for urban power supplying area segmentation. Ref. [8] proposes a centerline swing distance weighted clustering algorithm for power supplying unit segmentation and line deployment. Ref. [9] uses a clustering algorithm to process the spatial distribution information of massive power loads in distribution grids, which provides a theoretical basis for the delineation of power supplying units. Ref. [10] proposes a cluster performance index that keeps the voltage in the operating range, and set the partition model of distribution grid. Ref. [11] clustered the characteristics of load electricity consumption, and portrayed the electricity consumption behavior of users on this basis.

However, although the above literatures have proposed some methods on the delineation of power supplying units or power supplying grids, the process of finding the optimal partition result is complex. Therefore, this paper proposes a power supplying unit delineation method using an equal-angle centerline clustering algorithm, which is easier to find the optimal location of the initial centerline. Firstly, this paper proposes a power supplying unit delineation method based on a centerline clustering algorithm. Secondly, the centerline clustering algorithm is improved based on PSO to seek the optimal result of block delineation. Then, the feeder blocks are matched to form power supplying units. Finally, the proposed method is validated with examples.

2. The Centerline Clustering Algorithm

In urban, suburban and city administrative areas, the planning of power supplying areas is particularly important due to the need for urban construction and grid development. Multiple substations are interconnected to form one based on consideration of multiple functional subdivisions and administrative levels. The interconnection between substations requires the feeder drawn from the substation as a medium, which forms an inter-station power supplying unit, and the delineation of feeder blocks provides the basis for the formation of power supplying units. The traditional distribution grid planning area delineation does not take into account the impact of load changes on the delineation of blocks. Therefore, it is necessary to explain the basic definition of the delineation of power supplying units in more detail.

2.1. Basic definition of power supplying unit.

The power supplying unit is the smallest unit of a distribution grid planning, which is further divided on the basis of power supply grid. Under normal mode, each power supplying line in the power supplying unit should only supply power for the load of their own unit [12]. When dividing the power supplying unit, the complementary characteristics of various loads and the development needs in the power supplying unit should be taken into account, so as to improve the utilization rate of equipment. The power supplying unit is divided into intra-station power supplying unit and inter-station power supplying unit, and its structure is shown in Figure 1 and Figure 2.

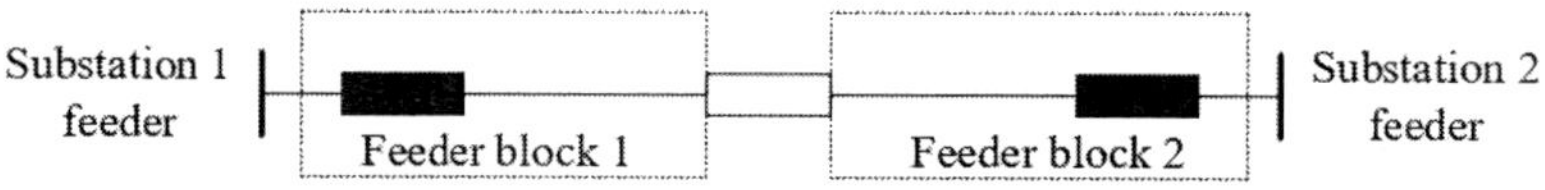

Figure1. Inter-station power supplying units

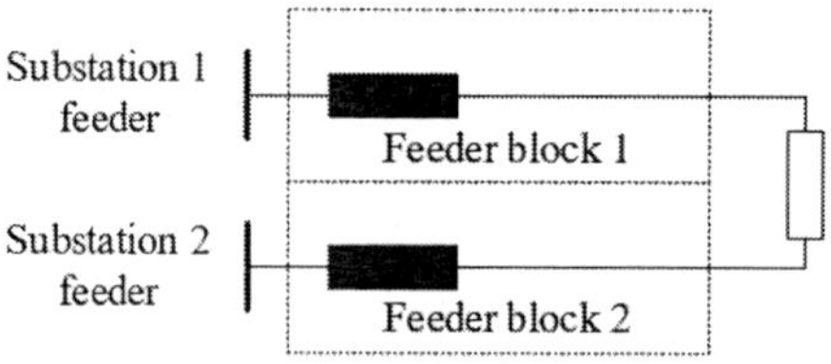

Figure 2. Intra-station power supplying units

2.2. Block segmentation based on equal-angle centerline clustering algorithm

The core of the centerline clustering algorithm is to set the initial equal-angle centerline and obtain new delineation schemes through continuous iteration [13], and finally select the scheme with the optimal comprehensive indicators. Its calculation flow is shown in Figure 3.

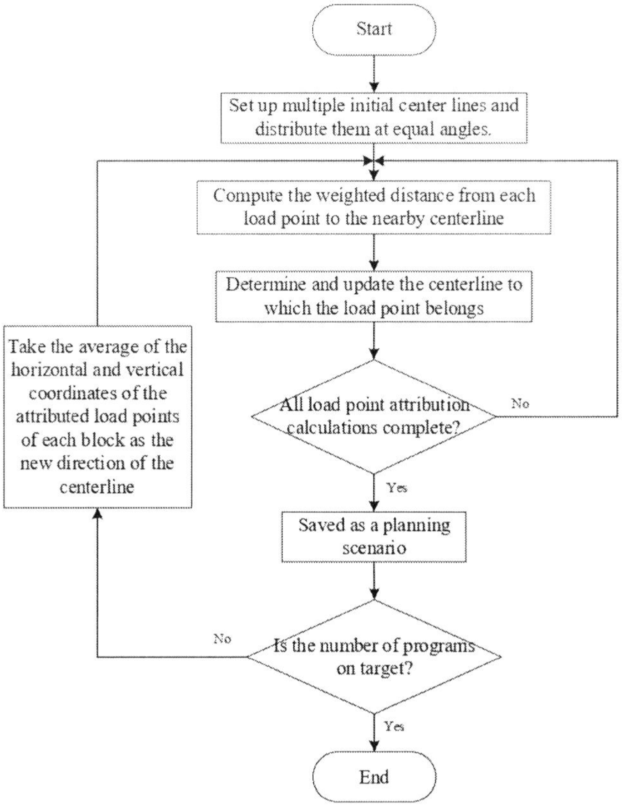

Figure 3. Flowchart of equal-angle centerline clustering algorithm

3. Block Segmentation Optimization Method Based on PSO Algorithm

The centerline clustering algorithm can iteratively derive the centerline belonging to each load point, and divide the feeder blocks accordingly. However, the number of partition schemes is too large, and it is difficult to find the optimal solution from them. It can only obtain as many results as possible, and then select a more reasonable solution according to the comprehensive index.

As a global search algorithm, PSO can find the approximate solution of the global optimal solution, even though it may not be able to find the global optimal solution. Its essence is to update the velocity and position of the next particle through the individual and group historical optimal solution of the particle, so as to obtain the optimal solution.

3.1. PSO-based feeder block delineation approach

In the delineation process of the centerline clustering algorithm, the initial centerline position needs to be selected firstly. If the selected position is too random, it will lead to the complexity of the calculation, and the delineation result is biased towards the local optimum. Therefore, PSO can be used to optimize the initial position of the centerline, the specific steps are as follows:

1) Determine the number of initial centerlines N, and the interval of each line is $360°/N$.

2) Set the number of particles in the PSO algorithm, the number of iterations, the weight coefficient, the learning factor and the speed range. The range of the first centerline is set to $0°\sim360°/N$, and this range is the range of particle movement. The rest of the centerlines are sequentially spaced $360°/N$.

3) Calculate the load between the two initial centerlines and the distance between these two

centerlines in turn, then cluster the load to the centerline with a closer distance, thus forming the feeder block.

4) Updating the centerline positions, take the average load location within each block as the new direction of the centerline, and calculating the load peaks and load peak-to-valley differences for each block.

5) Correcting the position of the load in each block with a weighted distance, calculated by the following formula:

$$\varepsilon_{1,ij} = \left(\frac{\sum_i P_{ij}}{\sum_i P_{ij} - P_{ij}} \right)^k \tag{1}$$

$$\varepsilon_{2,ij} = \left(\frac{\sum_i \Delta P_{pv,ij}}{\sum_i \Delta P_{pv,ij} - \Delta P_{pv,ij}} \right)^k \tag{2}$$

$$l_{ij} = d_{ij} \varepsilon_{1,ij} \varepsilon_{2,ij} \tag{3}$$

where P_{ij} and $\Delta P_{pv,ij}$ denote the load peak and load peak-valley difference of block i after the load point j is added to block i; l_{ij} denotes the weighted distance from node j to centerline i; d_{ij} denotes the vertical distance from node j to centerline i; k denotes the amplification factor of the weighting factor.

6) The loads are clustered to the weighted closer centerlines, and the load balance and the load peak-to-valley difference are calculated for each block:

$$\lambda_{1,i} = \left(\frac{P_{i,\max} - P_{ij,\max}}{P_{i,\max}} \right)^q \tag{4}$$

$$\lambda_{2,i} = \left(\frac{L_{i,\max} - L_{i,\min}}{L_{i,\max}} \right)^q \tag{5}$$

$$\lambda_i = \omega_1 \lambda_{1,i} + \omega_2 \lambda_{2,i} \tag{6}$$

where $\lambda_{1,i}$ denotes the load balance index of the ith feeder block; $P_{i,max}$ denotes the number with the largest value in the maximum load of the ith feeder block; $P_{ij,max}$ denotes the maximum load of the jth load point in the ith feeder block. $\lambda_{2,i}$ denotes the load peak-valley difference indicator of the ith feeder block; $L_{i,max}$ and $L_{i,min}$ denote the maximum and minimum load of the ith feeder block, respectively; q denotes the amplification factor of the weighting factor. ω_1, ω_2 are the weights of the indicators, which sum to 1; and λ_i is the overall assessment indicators.

7) Calculate the sum of the overall evaluation metrics of the block, take the minimum value of the sum as the optimization objective. The objective function, constraints and update process are shown as follows:

$$\min \sum_{i=1}^{N} \lambda_i \tag{7}$$

$$\begin{aligned} v_i^{d+1} = {} & \omega v_i^d + c_1 r_1 \left(\theta_{i,p} - \theta_i^d \right) \\ & + c_2 r_2 \left(\theta_{i,g} - \theta_i^d \right) \end{aligned} \tag{8}$$

$$\theta_i^{d+1} = \theta_i^d + v_i^{d+1} \tag{9}$$

$$s.t. \begin{cases} v_{\min} \le v_i^d \le v_{\max} \\ 0 \le \theta_i^d \le \dfrac{360°}{N} \end{cases} \tag{10}$$

where ω denotes the inertia factor of global search; c_1, c_2 denote the acceleration constant; r_1, r_2 are random numbers between 0 and 1; v_i^d, θ_i^d denote the current speed and the current position of particle optimization, respectively; $\theta_{i,p}$ and $\theta_{i,g}$ denote the individual and population optima, respectively; v_{max}, v_{min} denote the maximum and minimum values of the velocity of particle motion, respectively.

8) The optimal feeder block segmentation strategy is determined, and ultimately, the feeder blocks are merged to create the power supplying unit based on the maximum weight matching method.

3.2. Optimal matching of power supplying units

As can be seen from section 1.1, a power supplying unit is formed by combining two feeder blocks through a connection switch, in order to make the inter-block connection closer. This paper uses the maximum weight matching method to combine the blocks into the intra-station and inter-station power supplying units. The specific steps are as follows:

1) For all feeder blocks, take the geometric center of its load, calculate the average value of all load coordinates in the block.

2) Calculate the distance between the geometric centers of the blocks to form a distance matrix. The matrix is a symmetric matrix, and the main diagonal elements are all zero.

3) List all possible combinations, calculate the sum of the geometric center distances of each combination of feeder blocks, take the combination with the smallest sum of distances as the optimal match, and divide the power supplying units accordingly.

4. Analysis of Examples

4.1. Overview of the algorithms

This calculation example selects three substations with the same capacity and their load areas within the power supply range to be analyzed, the capacity of all three substations is 2×50MVA, and the number of load points within the power supply range are 101, 86 and 80, respectively, with an average load of 0.64MW, the main transformer load ratio is 62%, and the simultaneous rate of all loads in the region is 72%. The load types include residential loads, commercial loads, industrial loads and educational loads, and their distribution is shown in Figure 4.

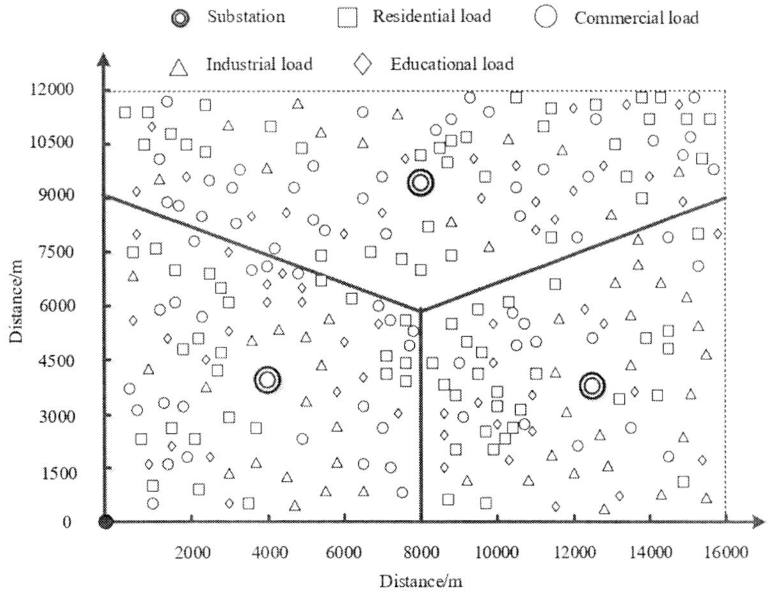

Figure 4. Distribution of load points

The power supplying area of the substation is bounded by the red solid line. The coordinates of each load point are derived from the distribution of load points in the figure, and some of the load point coordinates are shown in Table 1.

Table 1 Partial load point data

Serial number	**Substation**	**Load type**	**Horizontal coordinate/m**	**Vertical coordinate/m**	**Load/MW**
0	1	Residential	300	11400	0.63
1	1	Residential	900	11400	0.62
2	1	Commercial	1400	11700	0.71
3	1	Residential	2400	11600	0.57
4	1	Education	1000	11000	0.46
5	1	Residential	800	10500	0.61
6	1	Commercial	1200	10100	0.71
7	1	Industries	1200	9500	0.78
……	……	……	……	……	……
266	3	Industries	15500	600	0.77

4.2. *Example results*

4.2.1. Feeder block delineation results. According to the load size in the power supply range of the substation, it can be predicted that 10 feeders come out of each substation. In order to make the feeder distribution more uniform, the initial centerlines can all be spaced 36° apart, then only need to set the direction of the first centerline. The angle between the first centerline and the x-axis direction is regarded as a particle, 50 particles are taken, and the range of particle motion is 0°~36°. The maximum number of iterations is set to 100, and the results of feeder block delineation derived from the optimization of PSO algorithm are shown in Figure 5.

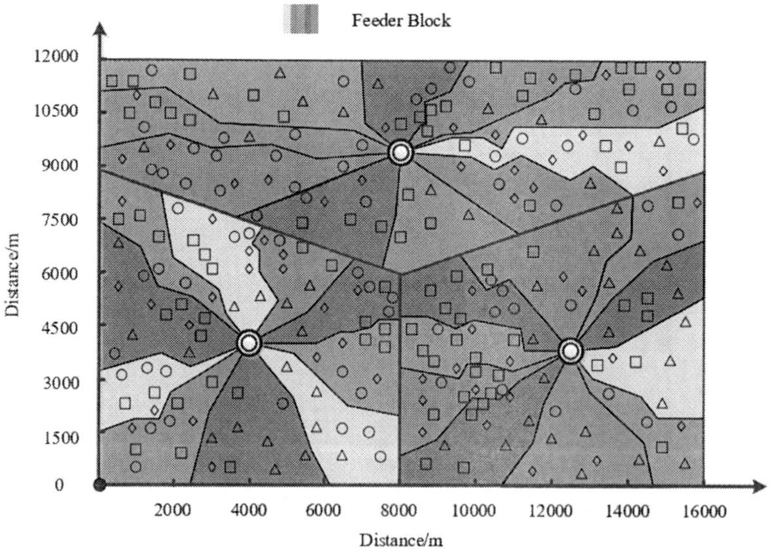

Figure 5. Feeder block delineation results

The location of the initial centerline is crucial to the final optimization result, the PSO algorithm can record the individual and population optimal values of the particles in each iteration. And it is easier to find the optimal position of the initial centerline when the particles move in the direction of the individual and population optimal in the next iteration.

4.2.2. Power supplying unit delineation results. In order to enhance the load transfer capacity of power supplying area, the priority should be given to inter-station power supplying units during the delineation of power supplying units, and the optimal delineation of power supplying units derive from the maximum weight matching method is shown in Figure 6.

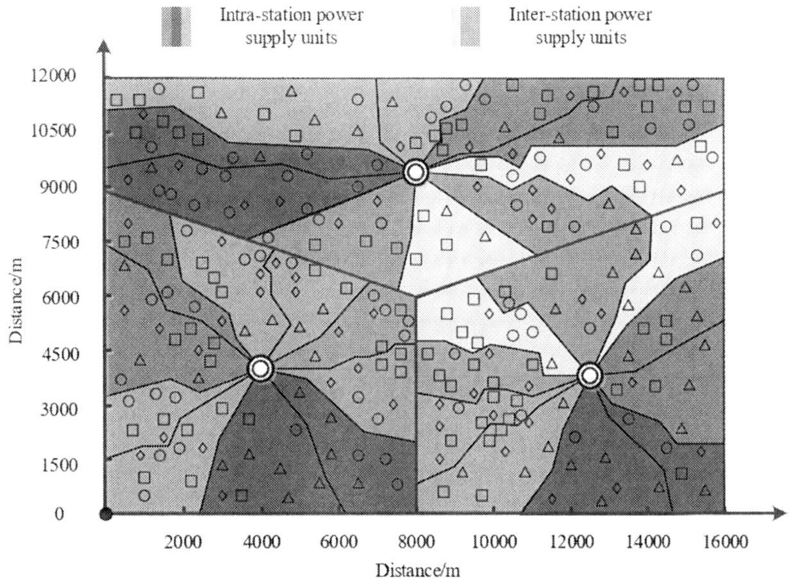

Figure 6. Power supplying unit delineation results

4.2.3. The comparative analysis. In this paper, the load side effectiveness of MV distribution grid is portrayed in terms of load balance and load peak-to-valley difference. The weighting ratio of the indicator weights ω_1, ω_2 in this calculation is 1:1, and the amplification factor of the weighting factor is taken as 1. The proposed method is compared with the delineation scheme proposed in the Ref. [13], the results are shown in Table 2. It can be seen that the method proposed in this paper is superior to the method proposed in Ref. [13] in terms of comprehensive assessment of load equalization and load peak-to-valley difference.

Table 2 Comparison of integrated assessment indicators

Optimization algorithm	Substation1	Substation2	Substation3
Proposed method	6.8625	6.8458	6.3624
Compared method	7.1179	7.3439	6.8729

5. Conclusion

This paper uses PSO to improve the equal-angle centerline clustering algorithm to obtain the result of optimal block delineation, and then the feeder blocks are combined to form the power supplying units according to the maximum weight matching method. The conclusions are obtained as follows:

1) The load balancing degree and the peak-to-valley load difference should be fully considered when

dividing feeder blocks and power supplying units.

2) The equal-angle centerline clustering algorithm combined with PSO optimization can find out the position of the initial centerline more accurately.

3) Considering the distance between blocks in the process of combining feeder blocks to form a power supplying unit can improve the tightness of the power supplying unit.

Although this paper proposes a method for dividing power supplying areas, it does not consider the impact of flexible resources such as energy storage and electric vehicles on the delineation results, so, the future work will focus on these aspects.

References

[1] Yang H J, Yang Z C, Gong M C and Tang K 2025 "Commercial Operation Mode of Shared Energy Storage System Considering Power Transaction Satisfaction of Renewable Energy Power Plants", Journal of Energy Storage, vol. 105, pp. 114738.

[2] Yang H J, Chu Y X, Ma Y H and Zhang D B 2024 Operation Strategy and Optimization Configuration of Hybrid Energy Storage System for Enhancing Cycle Life [J], Journal of Energy Storage, vol. 95, pp. 112560.

[3] Shu Z, Yang W F and Liao W 2023 Reliability Evaluation Method of Urban Distribution grid Power Supply Considering Automatic Partitioning of Distribution grid, Southern Energy Construction, vol. 10, pp. 98-104.

[4] Zhang J X, Wang Z Y, Tang J X and Chong G 2019 Mesh Method for Distribution grid Planning Model Based on Bi-level Planning, Proceedings of PURPLE MOUNTAIN FORUM 2019-International Forum on Smart Grid Protection and Control, vol. 585, pp. 205-216.

[5] Zhang Z and Yang J 2024 Power supplying unit Planning of Distribution grid Including Energy Storage based on N-1 Criterion, Academic Journal of Science and Technology, vol. 11.

[6] Liu W S, Liu H, Wang F and Wang C Y 2020 "Practical Automatic Planning for MV Distribution grid Considering Complementation of Load Characteristic and Power supplying unit Partitioning", in IEEE Access, vol. 8, pp. 91807-91817.

[7] Li Y, Du M Q, Xie W and Yang B Z 2018 "Method for delineation of urban load power supply district based on cluster analysis", IET Generation, Transmission & Distribution, vol. 12, pp. 4577-4581.

[8] Liu X O 2021 "Automatic routing of medium voltage distribution grid based on load complementary characteristics and power supplying unit delineation" International Journal of Electrical Power & Energy Systems, vol. 133, p. 106467.

[9] Xiao B, Guo B and Jiang Z 2018 Distribution grid Planning Method Based on Clustering Blocks of Load Nodes[J]. Electric Power Construction, vol. 39, pp 85-95.

[10] Chai Y Y, Guo L, Wang C S and Zhao Z Z 2018 "Network Partition and Voltage Coordination Control for Distribution grids With High Penetration of Distributed PV Units", IEEE Transactions on Power Systems, vol. 33, pp. 3396-3407.

[11] Li H, Hu B and Liu Y B 2021 Classification of Electricity Consumption Behavior Based on Improved K-Means and LSTM[J], Applied Sciences, vol. 11, pp. 7625-7625.

[12] Wei Z C, Zhao F Z and Wang J H 2016 "Gridding evaluation index system and method of MV and LV intelligent distribution grid", Power System Technology, vol. 40, pp. 249-255.

[13] Ge S Y, Cai Q Y and Liu H 2020 Practical automatic planning for MV distribution grid considering complementation of load characteristic and power supplying unit partitioning[J], Proceedings of the CSEE, vol. 40, pp. 790-802.

Journal of Physics: Conference Series 3012 (2025) 012038 doi:10.1088/1742-6596/3012/1/012038

Design of Fast Frequency Voltage Active Support Method and Testing System for Flywheel Array

Lingfeng Wei[0009-0004-6151-431X], **Guanjun Li***[0009-0001-4951-1149], **Haoyuan Li**[0000-0001-7539-2422], **Guowei Cheng**[0009-0009-5917-392X], **Mou Huang**[0009-0007-0775-2642], **Bo Sun**[0009-0009-3782-6205]

China Electric Power Research Institute Co., Ltd.,Nanjing 210003, China

*Corresponding author's e-mail: 649624218@qq.com

Abstract. The use of flywheel array energy management system and frequency/voltage active support control device is an effective way to enhance the active support capability of flywheel array to the power grid. Firstly, a flywheel array frequency/voltage active support control system was constructed based on an energy management system and simulation server. Then, research was conducted on control strategies such as peak shaving and frequency regulation through the control system. Then, based on the dynamic response characteristics of the output power of the flywheel array port, optimization methods for parameters such as inertia and damping of the flywheel array control system and energy management optimization strategies are proposed. Finally, a regional power grid model and a flywheel array grid connected model were constructed, and the active support effect of the control strategy on the power grid was verified on the joint simulation platform during frequency/voltage fluctuations in the regional power grid.

1. Introduction

In recent years, intermittent energy has developed rapidly and its penetration rate in the power grid has gradually increased, posing huge challenges to power grid peak regulation, safe and stable operation, and power supply quality. In addition, the increasingly severe peak valley difference problem not only seriously affects the power supply quality on the user side, but also brings great losses to the national economy. [1]In response to the above issues, energy storage technology can penetrate into various aspects of the power system, including generation, transmission, transformation, distribution, and utilization, and play different roles.

With the breakthrough of core technologies, the development of flywheel energy storage technology will show a trend of large-scale development, playing an important role in improving the stability of the power system. The direct inertia flywheel will present the development direction of ultra-high inertia, large capacity, and high overload technology. The development of high-speed advanced flywheel technology will have the technological direction of intrinsic safety, low rotor loss, and single machine high power/large energy storage.

At present, China has achieved some results in the participation of flywheel arrays in primary frequency regulation of the power grid, but there is still relatively little research on using the inertia support of large-scale flywheel arrays to improve the safety and support capacity of the power system. It is urgent to carry out breakthroughs in key technologies such as new inertia flywheel technology, flywheel intrinsic safety, inertia support, and frequency/voltage support for large-scale applications in

Journal of Physics: Conference Series **3012** (2025) 012038 doi:10.1088/1742-6596/3012/1/012038

the power grid, laying a theoretical and technical foundation for improving the safe operation of high proportion new energy power systems.[2]

This article proposes the flywheel array frequency/voltage active support control technology to address the risks of insufficient inertia and frequency/voltage transient instability in the new power system. The flywheel array energy management and collaborative optimization strategy are developed, and the flywheel array energy management system and frequency/voltage active support control device are designed. A testing system based on the energy management system and simulation server is built, and the control strategy is verified.

2. Flywheel Array Energy Management System and Frequency/Voltage Active Support Control Device

The flywheel array energy management system and frequency/voltage active support control device consist of a flywheel energy storage array simulation system and an energy management system. The system topology is shown in Figure 1.

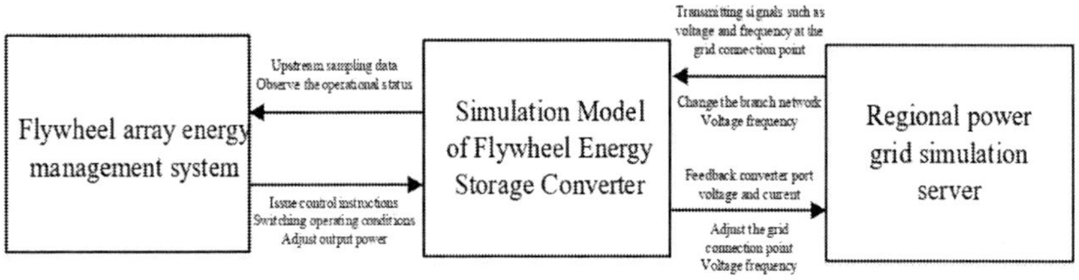

Figure 1. Topology of control device

2.1. *Flywheel energy storage inverter control chassis*

The control chassis of the flywheel energy storage converter consists of a power module, a sampling module, an input/output module, a PWM wave generation module, a photoelectric conversion module, an operation module, and an upper computer controller. This chassis is used to simulate various operating conditions of flywheel energy storage inverters. The control chassis of the flywheel energy storage inverter is shown in Figure 2.

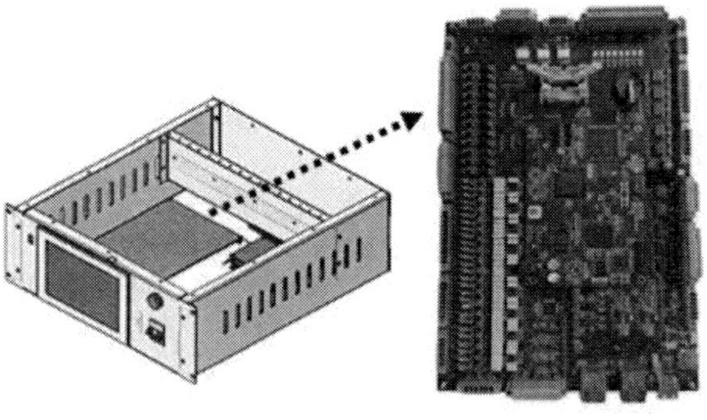

Figure 2. Flywheel energy storage inverter control chassis

Real time monitoring and control of the flywheel energy storage inverter control chassis are carried out using an upper computer. [3]The upper computer issues control instructions to the flywheel energy storage converter control chassis to switch operating conditions. The flywheel energy storage converter transmits the current operating status, sampling data, etc. to the upper computer. The upper computer control interface is shown in Figure 3.

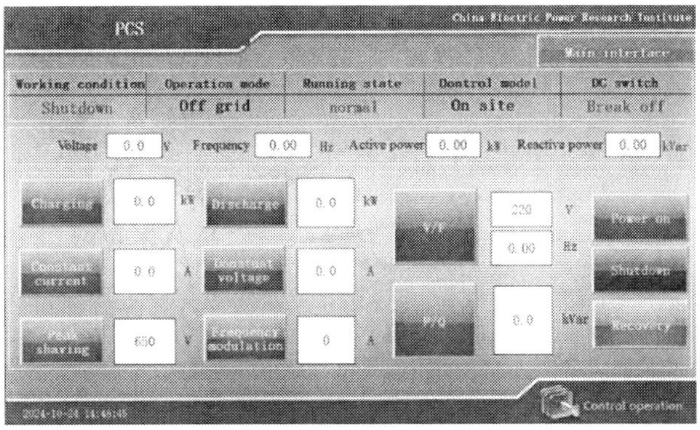

Figure 3. Control interface of the upper computer of the wheel energy storage inverter

2.2. Semi physical simulation system for flywheel energy storage converter

The flywheel energy storage converter simulation system is used to simulate the grid connection of flywheel arrays through low-voltage parallel connection. [4]The system consists of a flywheel energy storage converter control chassis and a simulation server. The topology of the single machine model of the flywheel energy converter is shown in Figure 4.

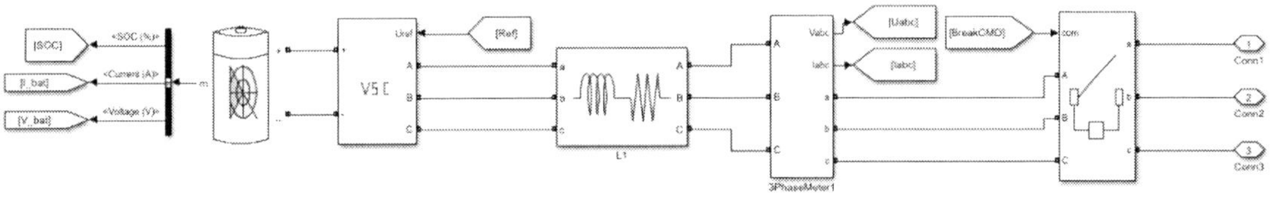

Figure 4. Single machine model topology of flywheel energy converter

Firstly, a mathematical model of 50 flywheel energy storage converters in low-voltage parallel was built based on a simulation server to simulate the application scenario of flywheel array grid connection. Then, control commands are issued to the flywheel energy storage inverter control chassis, and the sampling module of the control chassis receives voltage and current data of the electrical topology during the simulation process and generates PWM wave signals based on the above control commands. Finally, the simulation server receives PWM signals, and the model calculates and generates new sampling signals to input into the flywheel energy storage converter control chassis, forming a hardware in the loop.

2.3. Energy Management System

Build a flywheel energy storage array energy management system as shown in Figure 5. This system is used to display key data of the low-voltage parallel model of the flywheel energy storage converter in real-time during simulation operation, and issue control commands such as peak shaving and frequency regulation. The energy management system and flywheel energy storage converter simulation system are implemented through data exchange by the communication management machine.

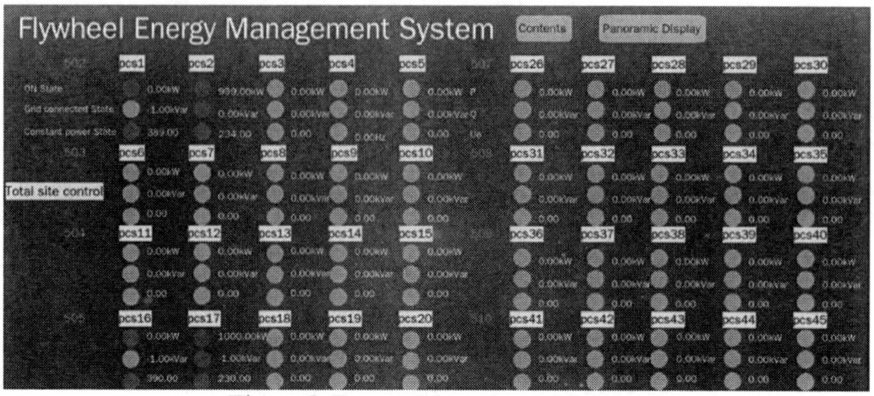

Figure 5. Energy Management System

3. Control Strategy Verification Platform

3.1. *Joint simulation platform*

Build a joint simulation testing platform based on flywheel energy storage controller and regional power grid simulation server, for verifying the frequency and voltage active support control strategy of flywheel energy storage converter on the power grid. [5]The system topology is shown in Figure 6. This system connects the AC port of the flywheel energy storage converter in the semi physical simulation system to the grid connection point in the regional power grid model through transmission line decoupling. Joint simulation of the application scenario of energy storage power stations connected to the actual regional power grid.[6]By adjusting the load, creating grounding faults, and regulating the power generation of new energy, the voltage and frequency fluctuations of the grid connection point can be generated. Then, the semi physical simulation system of the flywheel energy storage converter actively supports the voltage frequency fluctuation.

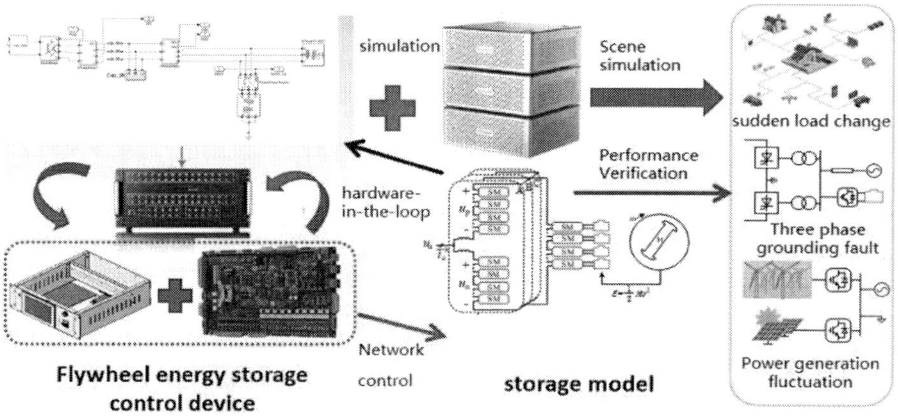

Figure 6. Topology of the Joint Simulation Platform System

3.2. *Regional Power Grid Model*

Build a regional power grid model based on the collection of funds. The model mainly consists of synchronous generators, AC busbars, transformers, transmission lines, circuit breakers, loads, new energy stations, measurement modules, etc. By switching loads and adjusting the output of new energy station models, it is possible to simulate power grid frequency fluctuations. Simulating voltage fluctuations in the power grid can be achieved by manufacturing grounding short circuit faults. The regional power grid model constructed is shown in Figure 7.

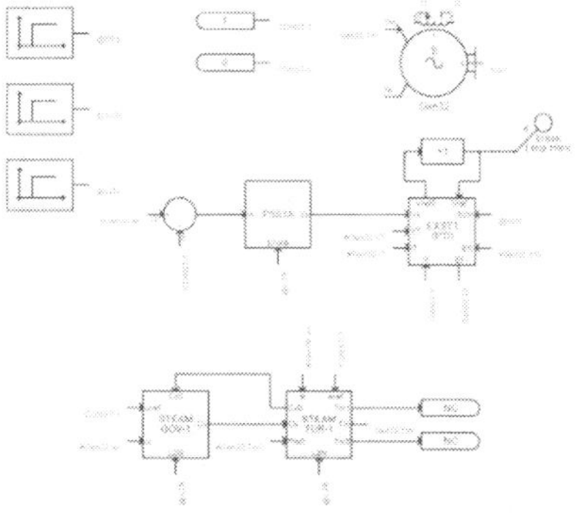

Figure 7. Regional power grid model

4. Energy management and collaborative optimization strategy for flywheel array

This article studies the frequency and voltage regulation control strategies of energy management systems and conducts verification work on the above-mentioned testing platform. Taking the grid connection of 50 flywheel arrays as an example, the 50 flywheel energy storage converters are divided into two groups of 25 flywheel energy storage converters for low-voltage parallel systems. Two systems are respectively integrated into different grid connection points of the regional power grid model to simulate the grid connected operation scenario of two flywheel energy storage stations. One group conducts voltage regulation control strategy validation, while the other group conducts frequency regulation strategy validation.

4.1. Frequency support

The primary frequency modulation active power control of the flywheel energy storage converter satisfies the following formula.

$$\Delta P = -\frac{1}{\delta} \times \frac{\Delta f}{f_n} \times P_n \tag{1}$$

In the formula, ΔP represents the change in active power of the flywheel energy storage converter, and δ represents the primary frequency modulation and differential rate; Δf is the deviation between the AC port frequency of the energy storage converter and the rated frequency of the system,f_n is the rated frequency of the grid type energy storage converter, f is the grid connection point frequency of the flywheel energy storage converter, and P_n is the rated active power of the flywheel energy storage converter.[7]

The flywheel energy storage inverter has inertia response capability during the charging and discharging process, providing inertia response under the premise of satisfying formula (2).

$$\Delta P = -\frac{T_J}{f_N} \cdot \frac{df}{dt} P_N \tag{2}$$

In the formula, ΔP represents the change in active power of the flywheel energy storage converter, T_J represents the equivalent inertia time constant of the flywheel energy storage converter, f_N represents the rated frequency of the grid type energy storage converter, f represents the grid connection frequency of the flywheel energy storage converter, t represents time, and P_N represents the rated active power of the flywheel energy storage converter.[8]

4.2. Voltage support

The dynamic reactive power support of the flywheel energy storage converter satisfies formula (3) during grid voltage faults.

Journal of Physics: Conference Series **3012** (2025) 012038 doi:10.1088/1742-6596/3012/1/012038

$$\Delta I_t = K_1 \times (0.9 - U_t) \times I_N \ (0 \leq U_t \leq 0.9) \quad (3)$$

In the formula: ΔIt is the increment of dynamic reactive current output by the energy storage inverter; K1 is the ratio coefficient of dynamic reactive current output by the energy storage inverter; Ut is the ratio of the actual voltage at the AC port of the energy storage inverter to the rated voltage; IN is the rated output current value of the AC port of the energy storage inverter. During the voltage drop period, the output reactive current of the energy storage inverter is the sum of the output reactive current I0 during normal operation before the voltage drop and the dynamic reactive current increment ΔIt. The maximum output capacity of the reactive current is not less than 1.05 times the rated current of the energy storage inverter, and the control deviation of the dynamic reactive current is not more than ± 10%.[9]

5. Experimental results

Running a joint simulation platform, the flywheel energy storage converter controller operates in grid connected constant power mode, adjusting the power of new energy stations in the grid model to change the voltage frequency of the grid connection point. Verify the supporting effect of flywheel energy storage on grid frequency. The voltage and current waveforms of the inverter port are shown in Figures 8 and 9. The support of flywheel energy storage for grid connection frequency is shown in Figure 10. The curve indicates that during the transient process of power grid frequency fluctuations, the output of flywheel energy storage has a supporting effect on the power grid frequency, and the power grid frequency is ultimately regulated and tends to stabilize. Through the analysis of experimental results, the joint simulation platform can effectively verify the operational effectiveness of the control algorithm.

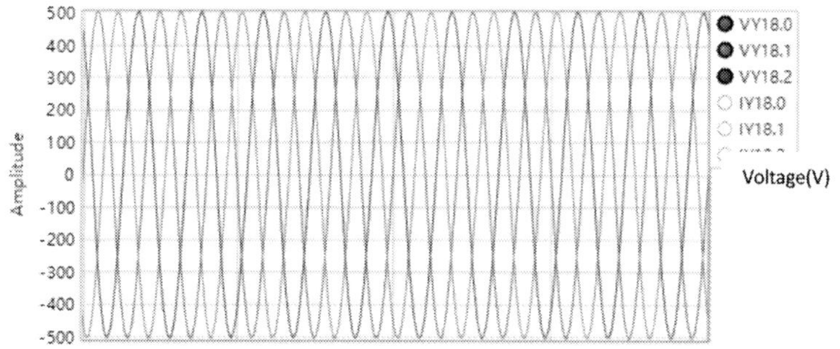

Figure 8. Port voltage waveform of flywheel energy storage converter

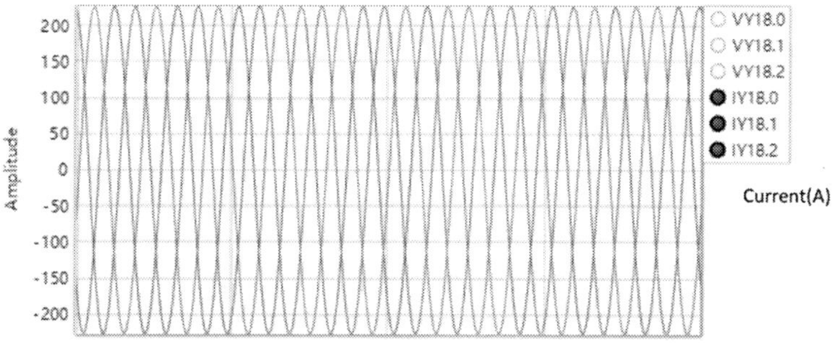

Figure 9.Port current waveform of flywheel energy storage converter

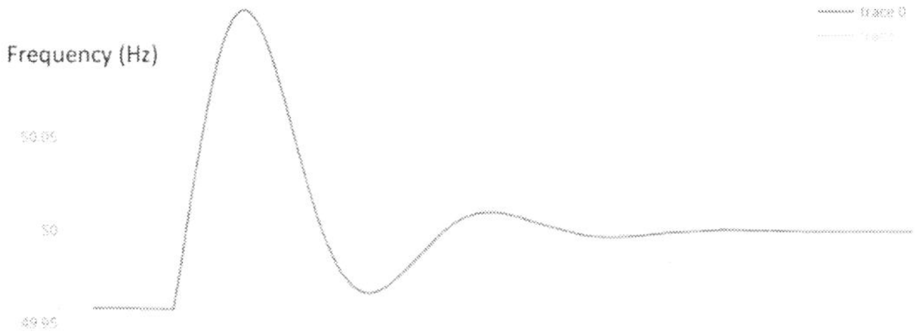

Figure 10.Frequency waveform of grid connection point

6. Conclusion

The use of flywheel array energy management system and frequency/voltage active support control device can effectively enhance the active support capability of flywheel array to the power grid. From an economic perspective, algorithm validation and functional testing through simulating the working conditions of flywheel array grid connection can greatly save testing costs; From the perspective of safety, this testing method can significantly shorten the development cycle of the voltage frequency active support control algorithm for the flywheel device and reduce the risks of damage to the switch tube and hardware circuit during the development and debugging process, thereby improving the safety of the experiment; In terms of the impact on the power grid, the simulation system is isolated from the actual power grid through transformers, and the harmonics generated by the flywheel response and the impact on the power grid will be reduced; From an operational perspective, this article simplifies the complex operation process. The energy management system can issue control instructions in real time and facilitate the observation of data and waveform curves during system operation.

Acknowledgement.

This work was supported by National Key Research and Development Program of China, i.e., Key Technologies for 10MW Magnetic Suspension Flywheel Energy Storage (2023YFB2406400).

References

[1] Chen Lei, Zhuang Jun, Wang Zhiyuan et al. DC voltage auxiliary control strategy for variable speed pumped storage units in flexible DC transmission systems [J]. Power System Protection and Control, 2022, 50 (04): 154-162. DOI: 10.19783/j.cnki. pspc.210486

[2] ILIEV I, TRIVEDI C, DAHLHAUG O G. Variable-speed operation of Francis 72 turbines: A review of the perspectives and challenges [J]. Renewable and Sustainable Energy Reviews, 2019, 103:109-130.

[3] Zhang Huiqiang. Dynamic characteristics analysis of doubly fed wind turbines based on MATLAB [J]. China Equipment Engineering, 2023, (20): 109-112

[4] Zhang Jingjing, Wei Shiyu, Yang Feifan, et al. Simulation and analysis of wind power generation systems based on wind turbines and doubly fed induction generators [J]. Journal of Hebei Institute of Architecture and Engineering, 2023,41 (03): 182-188

[5] Jing Haoran, Li Jia, Zhao Hongsheng, et al. Modeling and simulation of the full cycle conversion process of doubly fed variable speed pumped storage [J]. Electric Power Construction, 2023,44 (10): 41-50

[6] M. Chen, D. Zhou and F. Blaabjerg, "Multivariable grid-forming converters with direct states control," IEEE Transactions on Industry Applications, vol. 59, no. 4, pp. 4334-4341, July-August 2023(06): 16-18

[7] N. A. Ashtiani, S. A. Khajehoddin and M. Karimi-Ghartemani, "Modeling and stability analysis of single-phase microgrids controlled in stationary frame," IEEE Transactions on Power Electronics, vol. 37, no. 7, pp. 7759- 7774, July 2022(08): 26-28

[8] K. Jiang, D. Liu, K. Cao, P. Xiong, X. Ji, "Small-signal modeling and configuration analysis of

grid-forming converter under 100% renewable electricity systems," Electronics, vol.12, no. 19, pp. 4078, September 2023(15): 66-58
[9] A. Bidram, A. Davoudi, F. L. Lewis and J. M. Guerrero, "Distributed cooperative secondary control of microgrids using feedback linearization," IEEE Transactions on Power Systems, vol. 28, no. 3, pp. 3462-3470, August 2013.(21): 102-106

Journal of Physics: Conference Series **3012** (2025) 012039 doi:10.1088/1742-6596/3012/1/012039

An Adaptive Interactive Multi-model Electrical Load Modeling Approach with Center-of-mass Fixation

Cheng Li, Zhenshu Wang*

School of Electrical Engineering, Shandong University, Jinan, Shandong Province, 250000, PR China

*Corresponding author's e-mail: zhenshuwang@sdu.edu.cn

Abstract. The time-varying load problem of power systems poses significant challenges to load modeling, which in turn affects the accuracy of digital simulation of electrical systems. The interactive multi-model algorithm (IMM) can effectively deal with the time-varying problem of electrical loads. In this paper, for the application of the IMM algorithm in the field of load modeling, we propose an adaptive interactive multi-model load modeling method for a center-of-mass-fixed model set in a high-dimensional space (CF-AIMM), and through the division of the high-dimensional space, we design a model set that covers the real model space, which is aimed at characterizing the dynamic characteristics of electrical loads with more accuracy. load dynamics, thus improving the adaptability and accuracy of the algorithm in coping with load time variability.

1. Introduction

Load modeling plays a key role in the accuracy of power system calculations. With the changes of load characteristics and environmental factors, accurate load modeling can significantly improve the accuracy of system simulation and stability analysis, and avoid potential risks in system operation. Load modeling provides accurate load models for power system simulation, fault analysis, and trend calculation, which plays an important role in ensuring the dependability and robustness of power systems [1-3].

With the development of electric power technology, such as electric vehicles, air conditioning heat pumps, distributed energy storage, and other user-side flexible and adjustable resources aggregation, the role of electrical loads has become more and more important[4], At the same time, this also brings the time-varying and stochastic problems of electrical loads, which bring challenges to improve the accuracy of electrical load modeling[5]. To address the time-varying nature of the load components, literature [6] introduces a reliable method for identifying time-varying parameters in induction motor and ZIP load modeling, effectively managing the outliers present in the measurements. Literature [7] presents an LSTM-driven generative framework to identify load model parameters that change over time, using a probabilistic approach.. Literature [8] introduced a strong tracking filter utilizing the orthogonality principle to convert the load model parameter estimation task into a tracking problem. Literature [9] adopted two two-stage filtering methods to determine the changing parameters of the load over time, and investigated the stochastic variation of the load. Literature [10] proposed to decompose the load time-variation into the time-variation of load magnitude and the time-variation of load constituent components, and improved the accuracy of the load model by an improved integrated load model structure and a multi-curve fitting parameter identification method. Literature [11] improves the traditional load model structure by introducing dynamic compensation to effectively enhance the accuracy of load modeling.

In the field of target tracking, the interactive multi-model algorithm is the mainstream choice, which characterizes the possible existence of the target's motion state through multiple models, and then adopts the model that is closest to the target according to the model selection and input interaction, so as to be able to effectively deal with the target's maneuvering problem, and its excellent performance has been fully demonstrated in the field of missile control and aerospace[12]. With the deep development of new power systems, the components of power loads are becoming more and more complex, showing great stochasticity and time-varying nature, which poses a great challenge to the accurate modeling of loads[13]. Although the IMM algorithm shows strong performance and promise in dealing with the problem of time-varying loads, since the multiple model sets of the IMM algorithm are artificially designed to automatically select the most matching model set during the target maneuver, however, when the design of the model sets does not adequately cover the real model space, the accuracy of the load model obtained in the end will be reduced; and if we increase the behaviors of the loads in order to simulate more If the number of model sets is increased in order to simulate the behavior of more loads, it will lead to an increase in computational burden and competition among model sets, which will affect the accuracy of load modeling. Therefore, the model set of the interactive multi-model algorithm can be adequately designed to cover the real motion space of the dynamic system as much as possible, so as to further effectively deal with the time-varying problem of the load parameters.

In order to obtain the most realistic load model as possible, this paper proposes the center-of-mass-fixed adaptive interactive multi-model load modeling method (CF-AIMM), i.e., the center-of-mass-fixed model set design method is adopted to reduce the number of model sets as much as possible under the condition of guaranteeing to cover the possible motion space of the load model, and the appropriate distribution of the model sets will significantly enhance the precision of load modeling.

2. Adaptive Interactive Multiple Model Algorithm

The IMM algorithm has been widely used to solve the problem of estimating complex systems and time-varying parameters. The core idea of the IMM algorithm is to map all the dynamic behaviors of a system into a set of models, each of which corresponds to a different motion model[14]. Through a soft-switching mechanism, the algorithm can switch between different motion models, a process that follows a Markov process. For state estimation, the algorithm configures a filter for each model and assigns different weights to each model set by comparing the estimation errors to indicate how well it matches the actual system. The pre-set model sets are selected in the input interaction session and the optimal model set parameter operations are extracted, thus constituting the AIMM algorithm.

2.1. Basic Steps of Adaptive Interacting Multiple Model Algorithm

2.1.1. Load Model Selection. All the model sets are estimated after the corresponding filter, thus obtaining the error between the actual measurement data and the simulated data, and selecting a number of models that best match the actual system load characteristics according to the error to form the optimal load model set.

2.1.2. Interaction of Optimal Model Set. The transfer probabilities between the optimal model sets obey a Markov process, and the mixing probabilities at the current moment are calculated based on the weights of the optimal model sets at the previous moment and the predefined Markov transfer probabilities. Using the mixing probability, the states and covariances of the optimal model sets are mixed and input to the corresponding filters for estimation, and the result after interaction is obtained.

2.1.3. Information Fusion. After the results of input interactions are obtained from the previous process, the weights between model sets are adjusted by the calculation of the likelihood function to increase the probability of the model set that matches the behavior of the actual system, and finally the updated weights are used to calculate the state and covariance of the fused set of optimal models.

2.2. Improvement of Adaptive Interactive Multi-Model Load Modeling Algorithm
Literature [15] proposes an interactive multi-model algorithm based on the center-of-mass fixed structure, which significantly improves the tracking performance of the algorithm through the design of the interacting model sets. The paper emphasizes that the set of models involved in the interaction should be located as far as possible inside the real model space. Taking the two-dimensional plane as an example, the expected distance between the center of mass of a region and any point within the region is minimized, and a similar principle applies to high-dimensional spaces. Therefore, when the model set is selected as the center of mass of a region in the high-dimensional space, the accuracy of the interactive multi-model load modeling algorithm can be improved to a great extent.

3. Comprehensive Load Modeling Based on CF-AIMM Algorithm

3.1. Model Set Design in High-Dimensional Space Based on Centroid-Fixed Structure

*3.1.1. Normalization of load model parameters.*Since the variation range of each parameter in the integrated load model is not the same, in order to facilitate the application of the subsequent algorithm, these parameters need to be normalized to ensure that the variation range of each parameter is consistent. Assuming that the range of variation of the first parameter is $[p_{i\min}, p_{i\max}]$, the result after parameter normalization is:

$$p_i^{'} = \frac{p_i - p_{i\min}}{p_{i\max} - p_{i\min}} \tag{1}$$

After the above normalization, the range of variation of each parameter is scaled to [0,1].

3.1.2. Model set design in high dimensional space. After parameter normalization, the range of parameter variations in each dimension is compressed to [0,1], resulting in a 22-dimensional hypercube space. In order to ensure that the optional model set in the interaction algorithm can fully cover the real model space, this 22-dimensional hypercube space needs to be appropriately cut. The specific cutting method is as follows: take the uniform cutting method, each dimension of the 22-dimensional hypercube is uniformly cut into k segments, and the number of cutting blocks thus obtained is k^{22}. Let the coordinates of the load parameter in the hypercube at a certain moment be $(x_1, x_2, x_3, \cdots, x_{22})$, assuming that there is a random point $(y_1, y_2, y_3, \cdots, y_{22})$ in the block, the total of the squared distances between the true parameter of the load and the random points at that moment is:

$$d^2 = (x_1 - y_1)^2 + (x_2 - y_2)^2 + \ldots + (x_{22} - y_{22})^2 \tag{2}$$

The resulting expectation formula for the total squared distance between two points is given below:

$$E(d^2) = \frac{\int_{x_{1\min}}^{x_{1\max}} \int_{x_{2\min}}^{x_{2\max}} \int_{x_{3\min}}^{x_{3\max}} \ldots \int_{x_{22\min}}^{x_{22\max}} d^2 dx_1 dx_2 dx_3 \ldots dx_{22}}{V^2} \tag{3}$$

In Eq. (1), V represents the volume of the block obtained after the hypercube cut; $x_{1\min}, x_{2\min}, \ldots, x_{22\min}$ represents the minimum value in each dimension of each cut block, respectively; and $x_{1\max}, x_{2\max}, \ldots, x_{22\max}$ represents the maximum value in each dimension of each cut block, respectively. In order to find the minimum value of the distance sum of squares expectation, the partial derivatives of the parameters of each dimension, respectively, are taken and set to 0 to find the extreme point:

$$\frac{\partial E(d^2)}{\partial y_i}=0 \tag{4}$$

Solving the partial derivative equation in (2) yields the solution::

$$y_i=\frac{x_{i\min}+x_{i\max}}{2} \tag{5}$$

The above equation represents the minimal value solution in the ith dimension and it can be shown that the solution is unique. According to equation (3), the following conclusion can be deduced: when the true coordinates of a moment in the load parameter space are known, if a random point is chosen as the center of mass of the region, it is able to minimize the expected value of the distance squared. This conclusion is consistent with the results in the literature [13]. Therefore, the center-of-mass model set design method proposed in the literature [13] can be adopted by using the centers of mass in the high-dimensional space as the elements of the model set. This approach will significantly improve the accuracy of the interaction algorithm and reduce the deviation from the real modal pattern.

3.2. *Comprehensive load modeling*

The comprehensive load model adopts the third-order induction motor model under electromechanical transients and the static load model, and the mathematical expressions of the specific load model are as follows.

3.2.1. Third-Order Induction Motor Model. The three-phase induction motor is modeled by Eqs. (6)-(8), and the corresponding active and reactive powers are Eqs. (9)-(10).

$$\frac{dE'_d}{dt}=\frac{-E'_d-(X-X')I_q}{T'_{d0}}+s\omega_s E'_q \tag{6}$$

$$\frac{dE'_q}{dt}=\frac{-E'_d+(X-X')I_q}{T'_{d0}}-s\omega_s E'_d \tag{7}$$

$$\frac{ds}{dt}=\frac{T_M-T_E}{T_j} \tag{8}$$

Where $E'd$ and $E'q$ represent the d-axis and q-axis transient voltages of the induction motor, respectively; s denotes the slip; X represents the synchronous reactance; X' indicates the transient reactance; T_J is the inertia time constant; T_M denotes the mechanical torque; T_E represents the electromagnetic torque; T_{d0}' is the circuit time constant; and I_d , I_q represent the d-axis and q-axis current components, respectively.

$$P=\frac{R_s}{R_s^2+X'^2}U\left(U-E'_d\right)-\frac{X'}{R_s^2+X'^2}UE'_q \tag{9}$$

$$Q=\frac{X'}{R_s^2+X'^2}U\left(U-E'_d\right)+\frac{R_s}{R_s^2+X'^2}UE'_q \tag{10}$$

Where the input variables are $u = [u_d, u_q]^T$, the output variables are $Y = [P, Q]^T$, the state variables are $x = [E'_d, E'_q, s]^T$, and the parameter variables are $\theta = [R_s, X_s, R_r, X_r, X_m, T'_{d0}, T_j, K_L, a, MP]^T$.

3.2.2. Static Load Model. The active and reactive power of the static load are described by the models in Equations (11) and (12).

$$P_{ZIP} = P_0[a_p(\frac{V_i}{V_0})^2 + b_p(\frac{V_i}{V_0}) + c_p] \tag{11}$$

$$Q_{ZIP} = Q_0[a_q(\frac{V_i}{V_0})^2 + b_q(\frac{V_i}{V_0}) + c_q] \tag{12}$$

Where P_{ZIP} and Q_{ZIP} represent the active and reactive power of the static load, respectively. V_0 denotes the rated voltage, and V_i is the actual bus voltage. The parameter variables need to satisfy: $a_p + b_p + c_p = 1$, $a_q + b_q + c_q = 1$ and all parameters are between 0 and 1.

4. Case Simulation

The above load model is simulated in MATLAB 2022a platform to obtain the dynamic response curve of the integrated load model by inputting the voltage curve with perturbation. The total simulation time is set to 10 s. The input bus voltages are as follows: short circuit faults are set at t = 2 s and 7 s. The fault durations are 0.05 s for both.

In the model selection stage, three optimal model sets are selected to participate in the subsequent interaction process by comparing the normalized sum of new interest squares of each model set. During operation, the optimal model sets are adjusted accordingly as the load components change, as shown in Fig. As the soft-switching mechanism of Markov process is followed among the model sets, in the interaction phase, these three optimal model sets are assigned different weights according to their matching degree with the actual state, so as to realize the dynamic soft-switching. The weights of the optimal model sets change at each moment, as shown in Fig.

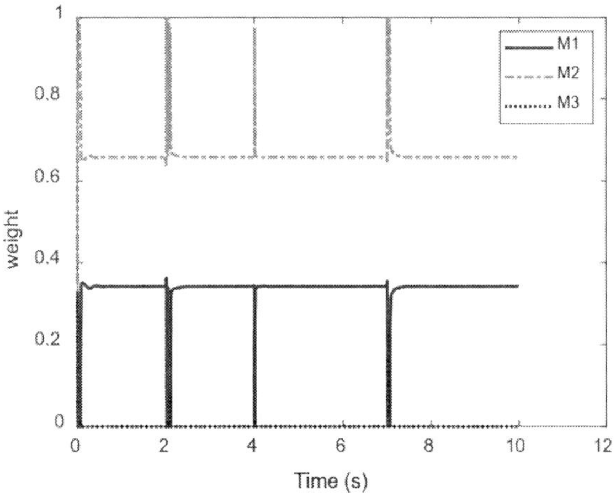

Figure 1. Weights of the optimal set of models at each moment

Journal of Physics: Conference Series 3012 (2025) 012039 doi:10.1088/1742-6596/3012/1/012039

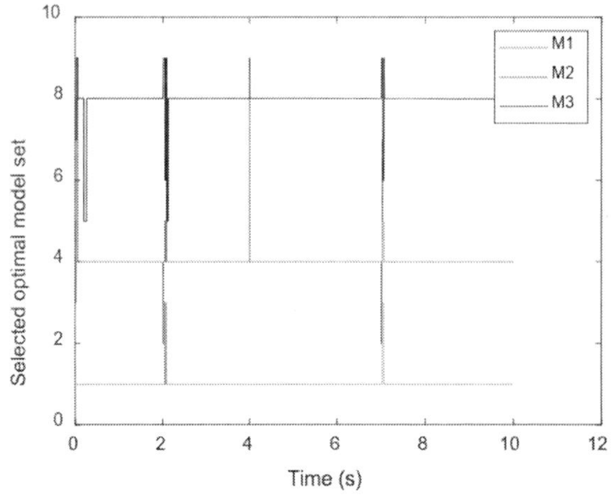

Figure 2. The optimal set of models for each moment in time

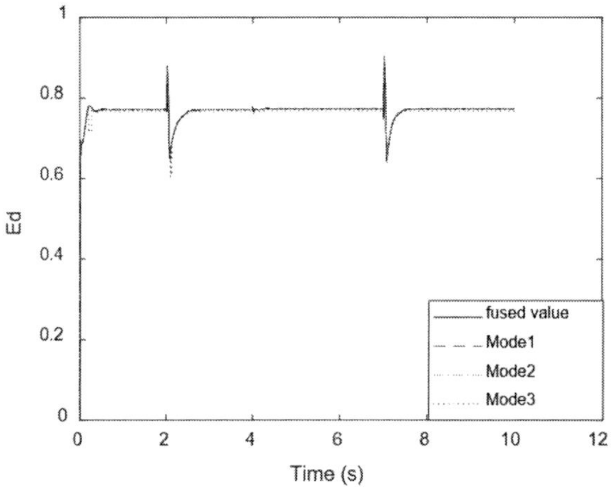

Figure 3. The evolution of the state variable Ed' within the optimal model set.

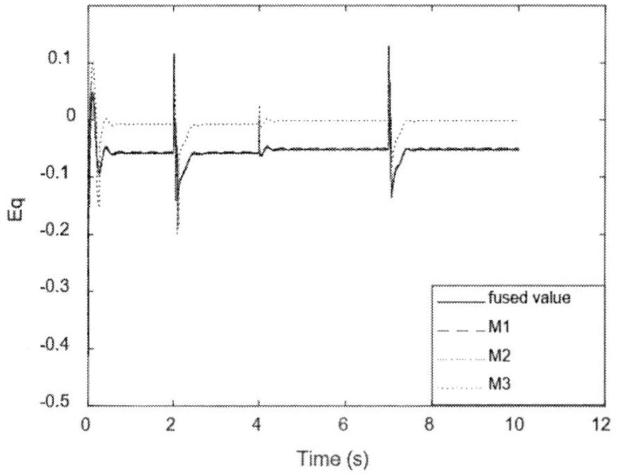

Figure 4. The evolution of the state variable Eq' within the optimal model set.

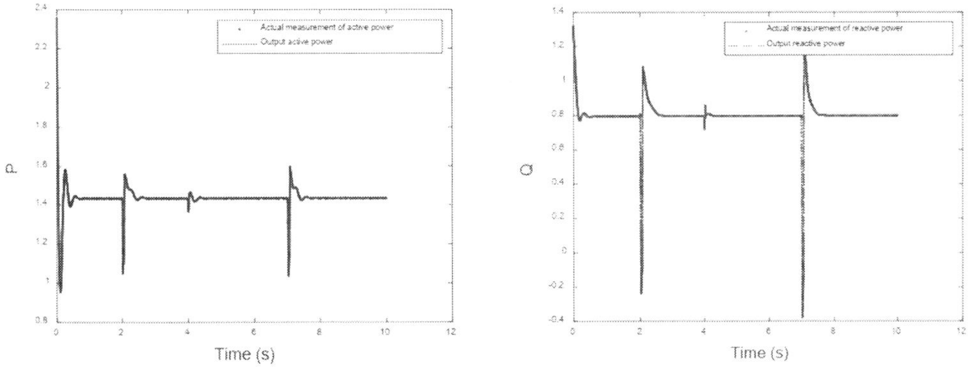

Figure 5. The comparison of active power output with actual measured values with respect to (a) real power and (b) reactive power.

By comparing Fig. 1 and Fig. 2, It can be noted that as the input voltage fluctuates, the most suitable model set and their associated model weights are adjusted. In contrast, during the stage when the voltage is more stable, the most suitable model set and its weights remain largely unchanged. Furthermore, under the effect of the interactive multi-model algorithm, the model that matches the actual system will receive a higher weight value. After the information fusion process is completed, the calculated active and reactive power values can be obtained using the fused state variables. The variation of the state variables of the most suitable model set is shown in Fig. 3 and Fig. 4, and the final power comparison curve is presented in Fig. 5.

Table 1. Error between output result and actual measurement

	CF-AIMM			IMM		
	Maximum Error	Mean Error	Relative Error	Maximum Error	Mean Error	Relative Error
P	9.0132×10^{-6}	9.0132×10^{-6}	6.4098×10^{-6}	2.9976×10^{-5}	4.3702×10^{-6}	1.0132×10^{-5}
Q	4.8223×10^{-6}	4.8223×10^{-6}	1.0500×10^{-6}	2.0132×10^{-5}	5.0132×10^{-6}	1.4132×10^{-5}

From the error comparison in Table 1, it can be seen that the interactive multi-model algorithm with a center-of-mass fixed design model set achieves high accuracy in the application of electrical load modeling. The proposed model set algorithm can sufficiently cover the real space of the model set, avoiding the issue of reduced estimation accuracy caused by inappropriate model set selection. Additionally, the proposed method effectively addresses the problem of time-varying loads and contributes to the development of load modeling.

5. Conclusion

Accurate modeling of electrical loads will improve the practicality of electrical digital simulation, and the interactive multi-model algorithm can effectively cope with the time-varying nature of electrical loads, and this paper applies the design method of center-of-mass model set design to electrical load modeling. The main research results are as follows: in order to ensure that the model set of the interactive multi-model algorithm can fully cover the real state space, inspired by the model set design method of the center-of-mass fixed structure, it is applied to the high-dimensional parameter identification of the load model, and the designed model set can effectively cover the high-dimensional parameter of the real space, which can provide a strong support for the solution of the problem of load time-variation.

The later work will enrich the load model based on the traditional three-phase induction motor and static load to solve the load time-variation problem of modern new power systems. And the problem of excessive computation under high dimensional space will be further studied.

References

[1] Y. Zhu and J. V. Milanović, "Automatic Identification of Power System Load Models Based on Field Measurements," in IEEE Transactions on Power Systems, vol. 33, no. 3, pp. 3162-3171, May 2018, doi: 10.1109/TPWRS.2017.2763752.

[2] C. Wang, Z. Wang, J. Wang and D. Zhao, "SVM-Based Parameter Identification for Composite ZIP and Electronic Load Modeling," in IEEE Transactions on Power Systems, vol. 34, no. 1, pp. 182-193, Jan. 2019, doi: 10.1109/TPWRS.2018.2865966.

[3] Wang Z S 2019 Theory and Methods of Power System Load Modeling (Beijing: China Electric Power Press)

[4] Dong K, Zhao J F, Sun R C, et al. 2024 Research Status and Prospects of Power Load Modeling with New Load Components [J/OL]. Automation of Electric Power Systems

[5] Melanie Bennett, Yilu Liu,The impact of large-scale dynamic load modeling on frequency response in the U.S. Eastern Interconnection[J],International Journal of Electrical Power & Energy Systems,2020,120:105983

[6] C. Wang, Z. Wang, J. Wang and D. Zhao, "Robust Time-Varying Parameter Identification for Composite Load Modeling," in IEEE Transactions on Smart Grid, vol. 10, no. 1, pp. 967-979

[7] M. Khodayar and J. Wang, "Probabilistic Time-Varying Parameter Identification for Load Modeling: A Deep Generative Approach," in IEEE Transactions on Industrial Informatics, vol. 17, no. 3, pp. 1625-1636.

[8] Z. Wang, Z. Li and S. Bian, "Adaptive Fuzzy Logic Strong Tracking Based Load Modeling," in IEEE Transactions on Power Delivery, vol. 37, no. 3, pp. 1530-1538, June 2022.

[9] M. S. Hossan and B. Chowdhury, "Comparison of Time-Varying Load Models for Estimating CVR Factor and VSF Using Dual-Stage Adaptive Filter," in IEEE Transactions on Power Delivery, vol. 34, no. 3, pp. 1001-1010, June 2019.

[10] Shi J H, He R M. 2004 Research on Load Time-Variability in Dynamic Load Modeling [J]. Proceedings of the CSEE.

[11] Yang Q, Hao W B, Hong X L, et al. 2011 Power System Load Modeling Considering Load Time-Variability and Dynamic Compensation [J]. Power System Protection and Control

[12] J. A. Besada, J. Garcia, G. De Miguel, A. Berlanga, J. M. Molina and J. R. Casar, "Design of IMM filter for radar tracking using evolution strategies," in IEEE Transactions on Aerospace and Electronic Systems, vol. 41, no. 3, pp. 1109-1122, July 2005, doi: 10.1109/TAES.2005.1541457.

[13] Zhang Z G, Kang C Q. 2022 Challenges and Prospects of Building a New Power System Under the Carbon Neutrality Goal [J]. Proceedings of the CSEE.

[14] H. A. P. Blom and Y. Bar-Shalom, "The interacting multiple model algorithm for systems with Markovian switching coefficients," in IEEE Transactions on Automatic Control, vol. 33, no. 8, pp. 780-783, Aug. 1988, doi: 10.1109/9.1299.

[15] G. Li, Q. Wang, S. Zhang, W. Sheng and T. Kirubarajan, "Interacting Multiple-Mode Estimation Using Centroid Fixed Structure for High Precision," in IEEE Transactions on Aerospace and Electronic Systems, vol. 60, no. 1, pp. 802-818, Feb. 2024.

Analysis of Fiber Orientation in the Buffer Layer of High-Voltage Cables Based on Microscopic Images

Yang You[1], Yong Xu[2], Zhaoguo Wu[3], Xiong Liu[3], Xiao Ren[3], Weisheng He[1*], Wenke He[1]

[1] Chongqing University of Posts and Telecommunications, Chongqing, 400065, P.R China

[2] Urban Power Supply Company of State Grid Chongqing Electric Power Company, Chongqing 400010, P.R. China

[3] State Grid Chongqing Electric Power Research Institute Chongqing, Chongqing 404100, P.R. China

[*]Corresponding author's e-mail: hwsxjtu@126.com

Abstract. With the accelerated progress of urbanization, the application of high-voltage(HV) cables in HV transmission networks has become increasingly widespread. Analyzing, detecting, and evaluating the fiber characteristics of the buffer layer has become a hot topic in current research. In this study, a series of original microscopic images were obtained for multiple samples using image analysis methods. Basic image analysis techniques such as binarization, morphological closing, filtering, and edge extraction were employed to obtain continuous and complete images of the buffer layer fibers. The Hough transform was then utilized to determine the orientation distribution characteristics of polyester fibers. This study provides an effective approach for obtaining the fiber distribution characteristics of the buffer layer based on microscopic images, offering valuable references for future performance testing and quality supervision of buffer layers.

Keywords—HV cables; buffer layer; fiber characteristics; Hough transform; orientation distribution.

1. INTRODUCTION

With the rapid development of the electric power industry, high-voltage(HV) cables have become essential carriers for power transmission, drawing significant attention to their safety and reliability. In recent years, the issue of ablation defects in the buffer layer of HV cables has become increasingly prominent, emerging as a critical factor affecting the performance and lifespan of the cables. Ablation defects not only lead to localized temperature increases in the cables, accelerating aging but also potentially cause cable failures, posing severe threats to the stable operation of the power system [1].

Research has shown that ablation defects are closely related to the conductive properties of the buffer layer. As a critical component of HV cables, the buffer layer's conductive performance directly influences the electrical performance and thermal stability of the cable. The conductive properties of the buffer layer are affected by the characteristics of the internal conductive fillers and the polyester fiber substrate. Therefore, a deep understanding of the morphological characteristics of fibers in the buffer layer is of great significance for studying the dynamic conductive properties of the entire HV cable buffer layer and detecting ablation defects [2,3].

Currently, studies on the morphological characteristics of fiber materials usually use methods such as fluorescence and X-ray diffraction to obtain key parameters such as fiber diameter, length, angle and twist. These methods and parameters not only reflect the morphological characteristics of fibers comprehensively, but also provide strong support for the analysis and optimization of the properties of fiber materials. However, the above methods are costly and require a lot of time, and cannot obtain effective buffer layer fiber distribution in a timely manner. Therefore, in this paper, the morphological characteristics of polyester fibers in the buffer layer of HV cables are investigated using microscopic image processing methods, which are fast and low-cost. The results of this study will provide basic data and theoretical support for subsequent studies on the dynamic conductive properties and ablation defect detection in the buffer layer of HV cables. These contributions will improve the safety and reliability of HV cables and promote the sustainable development of the power industry [4,5].

2. STRUCTURAL CHARACTERISTICS OF HV CABLES AND MATERIAL PROPERTIES OF THE BUFFER LAYER

2.1. Structural Characteristics of HV Cables

HV cables are composed of several components, including the conductor, insulation layer, buffer layer, and sheath. Among these, the buffer layer plays a crucial role. Its primary functions are electrical conduction and transition, mechanical buffering, thermal expansion compensation, and water resistance.

Firstly, the buffer layer acts as a medium for transitioning from high potential to low potential within the cable, ensuring electrical continuity between the insulation layer and the metallic sheath, which effectively prevents leakage and breakdown of the cable. Secondly, the buffer layer possesses excellent elasticity and cushioning properties, enabling it to mitigate the impact of external mechanical stress on the insulation layer, thereby protecting the cable from damage [6]. Thirdly, when moisture enters the cable, the buffer layer quickly absorbs it and expands to an appropriate size in a short period, thus preventing further axial penetration of moisture along the cable. Finally, during cable operation, the current generates heat, causing thermal expansion of the cable. The buffer layer can absorb the stress generated by this thermal expansion, preventing separation between the insulation layer and the metallic sheath due to excessive thermal stress, thereby ensuring the safe and stable operation of the cable.

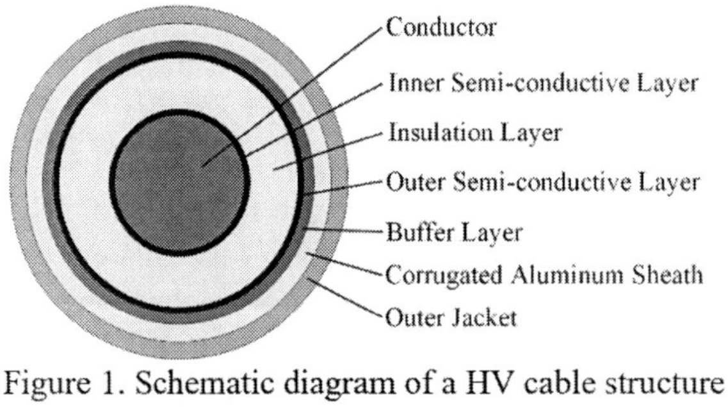

Figure 1. Schematic diagram of a HV cable structure

2.2. Material Properties of the Buffer Layer in HV Cables

In the design of cable buffer layers, a common three-layer structure typically consists of, from the inside out, fluffy cotton, water-blocking powder, and non-woven fabric. This structure combines the unique properties of different materials to provide multiple layers of protection and buffering for the cable. Additionally, the incorporation of polyester fibers and carbon black fillers further enhances the overall performance of the buffer layer [7].

The inner layer of fluffy cotton, made from natural cotton fibers or synthetic fibers, offers softness that effectively reduces friction and damage during cable installation and use. The water-blocking powder, composed of high molecular materials like sodium polyacrylate, acts as an efficient water-absorbing material. It quickly absorbs moisture and expands, forming a waterproof barrier with the external aluminum sheath, effectively preventing moisture from penetrating along the cable's length. The outermost non-woven fabric, created by bonding fibers through physical or chemical methods without weaving, is characterized by its lightweight, softness, and durability. It effectively supports the buffer layer structure, preventing deformation of the cable under external pressure, while also allowing air circulation to maintain a dry internal environment [8].

Moreover, the addition of polyester fibers and carbon black plays a crucial role in the cable buffer layer. Polyester fibers, known for their high strength and excellent mechanical properties, enhance the structural integrity and abrasion resistance of the buffer layer. They tightly bond with the substrate, improving the overall strength and stability of the buffer layer. The elasticity of polyester fibers also ensures that the cable can quickly return to its original shape after being subjected to impact or pressure.

Carbon black, as an excellent filler, increases the strength and wear resistance of the buffer layer and enhances its antiaging properties. Additionally, the conductivity of carbon black helps prevent static electricity accumulation. Finally, carbon black improves the thermal conductivity of the buffer layer, aiding in heat dissipation and preventing performance degradation caused by overheating.

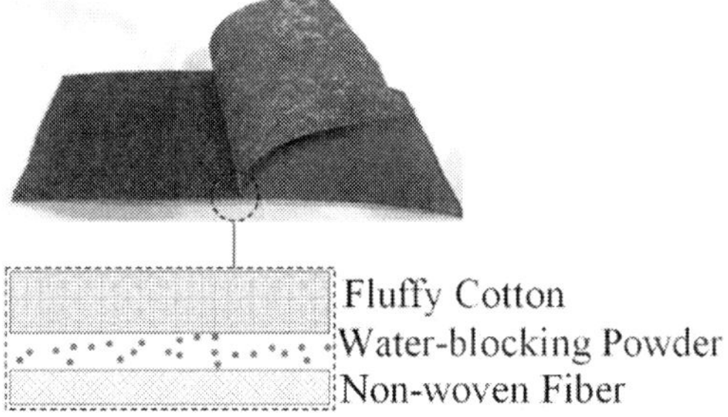

Figure 2. Three-Layer structure of the buffer layer

3. MICROSCOPIC IMAGE ACQUISITION OF THE BUFFER LAYER

High-performance curing materials such as epoxy resin or crystal drop adhesive are used to accurately cure small sections of the buffer layer, facilitating subsequent slicing operations.

3.1. Sample Preparation and Slicing

A new buffer layer is cut to an appropriate size and placed flat into a mold, then epoxy resin is poured over it. The sample is left to cure at an ambient temperature of 30°C for 12-24 hours, resulting in a cured buffer layer sample, as shown in Figure 3. A total of 10 samples were prepared [9].

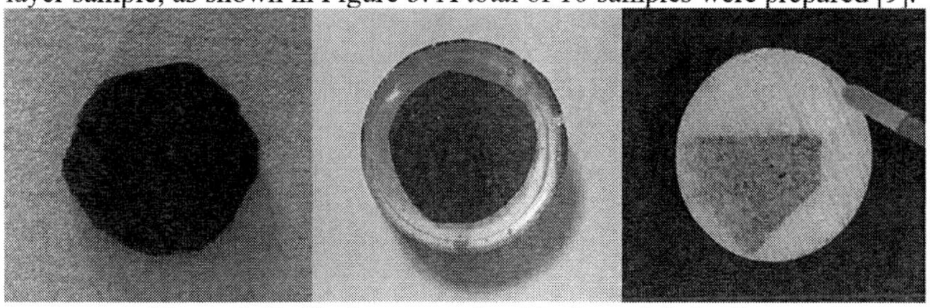

Figure 3. Buffer layers, curing blocks and their slices

The epoxy-cured block was securely fixed using a cable insulation layer slicer, with the slicing thickness set between 0.2 mm and 0.5 mm to ensure uniform slicing from top to bottom. Starting from the successful slicing of the buffer layer section with a complete pattern, approximately 10 quality-approved slices were obtained from each sample. A total of 118 valid slices were obtained from 10 samples.

3.2. Image Acquisition
After softening the slices with glycerin, they are placed on a slide and observed using an optical microscope. An electronic eyepiece is used on the microscope to record images of the buffer layer. By applying appropriate light intensity beneath the microscope, clear fiber images can be obtained.

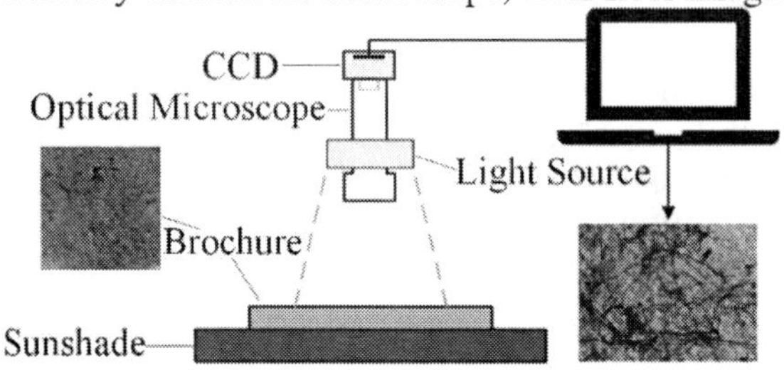

Figure 4. Buffer layer micrograph acquisition process

A total of 590 fiber images were obtained under the microscope from the observation of 118 valid slices.

4. MICROSCOPIC IMAGE PROCESSING OF THE BUFFER LAYER
The images are first converted to grayscale, followed by preprocessing, which includes filtering, enhancement, binarization, closing operations, and skeleton extraction [10]. Finally, the orientation information in the images is analyzed using the Hough transform.

4.1. Image Preprocessing
Minimum Image preprocessing is a crucial step in image data analysis. It involves a series of operations on the raw images to improve their quality, enhance useful information, and eliminate or reduce noise and irrelevant information. Preprocessing methods typically include filtering, image enhancement, binarization, and closing operations.

Filtering techniques are used to eliminate noise and interference in images. By applying various filters, such as median filters, images can be smoothed and edges sharpened, thereby improving the clarity and accuracy of the images. Image enhancement techniques aim to improve the contrast, brightness, and clarity of images, making the details more prominent and facilitating subsequent analysis and processing. Binarization is the process of converting a grayscale image into a binary image. By setting a threshold, the pixels in the image are divided into black and white, allowing the extraction of the target regions in the image. Closing operations are typically implemented through a series of dilation followed by erosion operations, which help eliminate small holes or black spots within objects while maintaining the overall shape and size of the objects. Skeleton extraction enables the extraction of the main contours of the target objects from the image, which are essentially the centerlines or medial axes of the objects. These preprocessing techniques have a significant impact on subsequent tasks such as image analysis, object detection, and feature extraction [11].

In Fig. 5, a and b, the non-shaded areas (black) are the main structure of the fibers as well as agglomerates of carbon black particles, and the shaded areas (grey) are portions of the sliced epoxy powder produced during the slicing process due to the high degree of drying of the epoxy curing block, which are formed by combining with a small portion of carbon black.

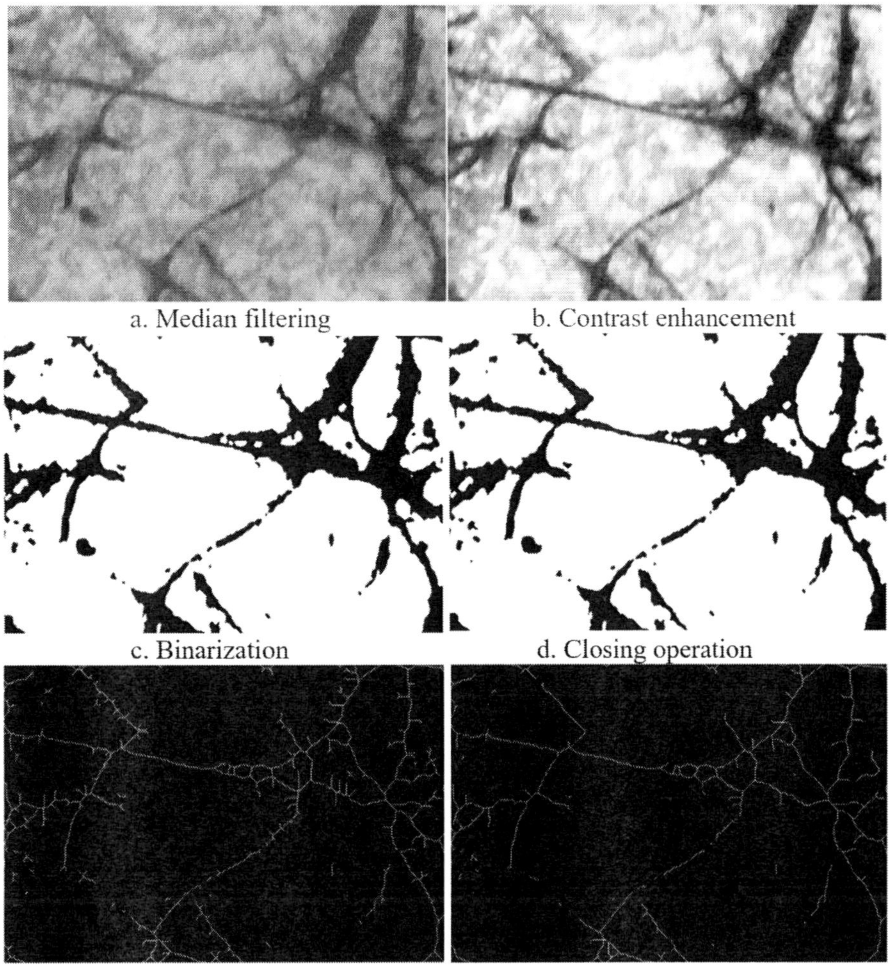

a. Median filtering b. Contrast enhancement

c. Binarization d. Closing operation

e. Skeleton extraction f. Skeleton refinement

Figure 5. Image preprocessing process

4.2. Hough Transform

Hough transform is a common image analysis method for detecting simple geometric shapes in images. The parameters of these shapes can be accurately extracted in complex image environments by means of the Hough transform. The basic principle is to utilize the duality of points and lines to map the shape transformations in the original image space into the parameter space, thus transforming the problem of straight lines detection into the problem of finding peaks in the parameter space [12-14].

Considering the problem of detecting straight lines, Hough chose the parametric equation for the slant section of a straight line:

$$f(x,y) = y - mx - c = 0 \tag{1}$$

A point in image space is then mapped to a sinusoidal curve in parameter space, clearly solving the unboundedness problem:

$$f(x,y) = \rho(\theta) = x\cos(\theta) + y\sin(\theta) \tag{2}$$

Specifically, for straight line detection, the Hough transform maps each point in the image to a curve in the parameter space (usually ρ-θ space in a polar coordinate system), and the intersection of these curves in the parameter space corresponds to a straight line in the image.

As a result, the Hough transform can be regarded as a continuous form of the continuous form of the Radon transform. It can be written in the form ρ-θ:

Journal of Physics: Conference Series **3012** (2025) 012040 doi:10.1088/1742-6596/3012/1/012040

$$Rf = g(\rho,\theta) = \int_{-\infty}^{+\infty}\int_{-\infty}^{+\infty} f(x,y)\delta(\rho - x\cos\theta - y\sin\theta)\mathrm{d}x\mathrm{d}y \quad (3)$$

In addition, the Hough transform is robust and can tolerate unfavorable factors such as noise, deformation or occlusion in the image to a certain extent, which improves the accuracy and reliability of shape detection [15].

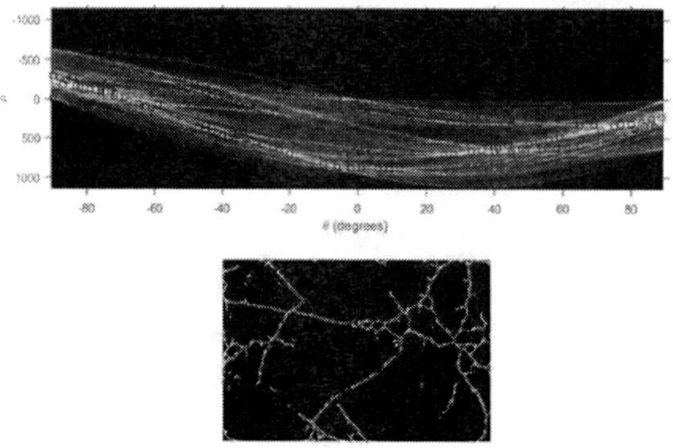

Figure 6. Line extraction using the Hough transform

4.3. *Statistical Results*

For the 590 fiber images obtained under the microscope, the statistics after applying the Hough transform are shown in Figure 7.

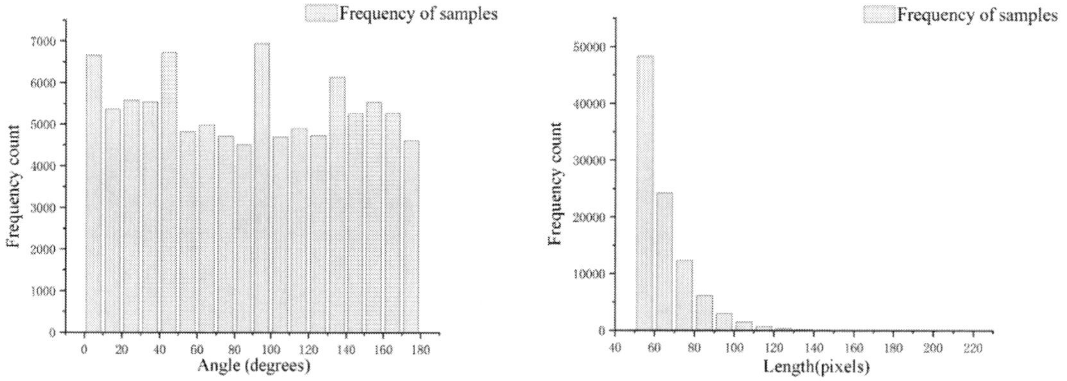

Figure 7. Fiber orientation degree and length statistics

"0-180°" is the angle formed by rotating counterclockwise to the negative direction of the y-axis in the ordinary right-angle coordinate system, using the positive direction of the x-axis of the right-angle coordinate system as a reference.

From the theoretical level, the fiber distribution in the nonwoven fabric is more uniform, but the thickness of the nonwoven fabric is smaller, and the number of slices is smaller. The fiber distribution in fluffy cotton is chaotic, but the thickness is larger, and the number of slices is large. As can be seen from the figure, the fiber orientation distribution is roughly in line with the law. The length distribution of fibers is more intuitive, with short fibers dominating and the number of long fibers distributed less.

5. CONCLUSION

In this study, 590 images were acquired by curing and slicing the fibers, with 118 valid slices obtained from 10 cured blocks. Nearly 100,000 fiber angles and lengths were analyzed. The results show that the orientation distribution of fibers in two dimensions is in accordance with the theoretical reality, with the horizontal and vertical directions of the nonwoven fabrics and their corresponding equidistant angles being more prominent, but the overall number of fibers is chaotically distributed with the dominance of the fibers of fluffy cotton. The length values are dominated by short fibers, with a few long fibers present.

Frequency counts of fiber orientation and length were performed using the Hough transform. During the skeleton extraction process, some of the fibers were entangled together in the low microscopic state and could not be effectively distinguished, resulting in many small bifurcations during the skeleton extraction process. The extra small bifurcations have a great impact on the subsequent Hough transform extraction of linear directionality. In future studies, increasing the microscopic resolution and magnification should be considered to obtain more accurate and reliable orientation distribution information.

National Natural Science Foundation of China (52307162)
Natural Science Foundation of Chongqing (CSTB2023NSCQ-MSX0667)

References

[1] Jiang L, Xin Y, Yan W, et al 2019 Study on ablation between metal sheath and buffer layer of high voltage XLPE insulated power cable (2019 2nd International Conference on Electrical Materials and Power Equipment (ICEMPE). IEEE) pp 372-375.
[2] DU B X, DU X Y, KONG X X, et al 2021 An investigation on discharge fault in buffer layer of 220 kV XLPE AC cable (IET Science Measurement & Technology) vol 15 pp 508-516.
[3] Tian Y, Zhu G X, Miao Q Q, et al 2024 Study on the ablation process and failure mechanism of the buffer layer in high-voltage XLPE cable (International Journal of Electrochemical Science) vol 19(8) pp 100662.
[4] Zhang H, Li J 2024 Intrinsic Current Concentration of Buffer Layer Material for Cable Ablation Failure: Role of Random Fiber Networks (arXiv preprint) arXiv: 2405. 04430.
[5] Liu J J, Ma M G, Liu X, et al 2024 High-Voltage Cable Buffer Layer Ablation Fault Identification Based on Artificial Intelligence and Frequency Domain Impedance Spectroscopy (Sensors) vol 24(10) pp 3067.
[6] Chen Y, Hui B J, Cheng Y T, et al 2021 Effects of connection conditions between insulation screen and Al sheath on the buffer layer failures of high-voltage XLPE cables (Engineering Failure Analysis) vol 122 pp 105263.
[7] Huang Z Q, Lu F, Tong Y F, et al 2022 Technology for the Detection of Ablation Defects in Buffer Layers of High-Voltage Cables (Access, IEEE) vol 10 pp 92843-92853.
[8] Zhang H R, Gao J, Wu K N, et al 2023 Effects of the inhomogeneous microstructures on buffer layer ablation failure in high voltage cables (Engineering Failure Analysis) vol 152 pp 107491.
[9] Chen Y, Zhou K, Kong J, et al 2023 A Review of Failure Analysis and Detection Methods for Buffer Layers in High-Voltage Cables (IEEE Electrical Insulation Magazine) vol 39(6) pp 7-15.
[10] Battseren B, Tudevdagva U, Hardt W 2018 Image Processing Algorithms for High Voltage Power Line Detection (ISSN) pp 2618-7558.
[11] Meng S, Pan R, Gao W D, et al 2022 Automatic recognition of woven fabric structural parameters: a review (Artificial Intelligence Review) vol 55(8) pp 6345-6387.
[12] Mukhopadhyay P, Chaudhuri B B 2015 A survey of Hough Transform (Pattern Recognition) vol 48(3) pp 993-1010.
[13] Aggarwal N, Karl W.C 2006 Line detection in images through regularized hough transform (IEEE transactions on image processing) vol 15(3) pp 582-591.
[14] Breuer K, Stommel M, Korte W 2019 Analysis and evaluation of fiber orientation reconstruction methods (Journal of Composites Science) vol 3(3) pp 67.
[15] Chandrasekar L, Durga G 2014 Implementation of Hough Transform for image processing applications (2014 International Conference on Communication and Signal Processing. IEEE) pp 843-847.

Journal of Physics: Conference Series 3012 (2025) 012041 doi:10.1088/1742-6596/3012/1/012041

Experiment on Early Fire Detection of Medium-voltage Switchgear Based on Pyrolytic Particles

Zhaoguo Wu[1], Qian Wang[1], Ke Wu[2], Weisheng He[4], Huixian Huang[3], Yong Li[1], Yingkai Long[1], Zixiong Cai[4*]

[1] State Grid Chongqing Electric Power Research Institute Chongqing, China

[2] State Grid Chongqing Electric Power Company Ultra-High Voltage Branch

[3] State Grid Chongqing Electric Power Company, Chongqing, China

[4] Chongqing University of Posts and Telecommunications, Chongqing, China

[*]Corresponding author's e-mail: cquptczx@163.com

Abstract. Fire accidents frequently occur in medium-voltage switchgear, causing huge economic losses, so it is particularly important to ensure its safe and stable operation. This paper proposes a early fire detection method based on pyrolytic particle sensor. First, this paper outlines the characteristics of MV switchgear electrical fires and the limitations of existing detection methods. Based on the pyrolysis characteristics of common materials in switchgear, an experimental platform for early fire detection has been constructed, and experimental studies have been carried out for polyethylene (PE) and polyvinyl chloride (PVC) materials. Experiments were conducted to compare the response rates of methane, ethylene, ethane and acetylene sensors with those of pyrolysis particles. The results of the study showed that during pyrolysis of the PVC material, as the temperature increased from 33°C to 136°C, the methane, ethylene, ethane, and acetylene gas sensors did not show a response, while the pyrolysis particle sensor began to respond at 90°C; During the pyrolysis of PE materials, as the temperature increased from 28°C to 360°C, the gas sensor also showed no response, while the pyrolysis particle sensor began to respond at 90°C.

1. Introduction

Medium-voltage switchgear and ring main units are widely used in substations, distribution stations and power distribution networks in industrial and commercial facilities, mainly responsible for the control and protection of medium voltage power lines. The MV switchgear ensures the reliable transmission of electric energy and the safe operation of the system by integrating circuit breakers, disconnectors and protection relays, while the ring main unit uses the ring network structure to support multi-point power supply. It can quickly switch the power path when a local fault occurs, improving the reliability and flexibility of the distribution network. MV switchgear ensures reliable transmission of power and safe operation of the system through integrated circuit breakers, disconnect switches and protection relays, while ring main units support multi-point power supply by utilizing the ring network structure, which is capable of rapidly switching power supply paths in the event of localized faults, enhancing the reliability and flexibility of the power distribution network. However, in the 6kV-35kV MV power supply system, switchgear cabinets are exposed to the risk of arc faults, which mainly originate from the compactness of the cabinet structure, damage to insulation materials or product defects, leading to partial discharges, insulation breakdowns and even explosions, which have become the main cause of fires and large-scale power outages. Statistics show that cable joint failures account for more than 65% of switchgear failures, which has become a key factor in the frequent occurrence of power outages. The hidden, cumulative and sudden nature of electrical fires has increased the risk of fires, which are mainly caused by factors such as aging and wear of equipment and external environmental influences. These problems often accumulate over a long period of time and suddenly break out under high load or extreme environmental conditions, leading to fire accidents[1-3].

Currently, the commonly used fire detection methods mainly include smoke detectors, temperature detectors, flame detectors and gas detectors[4]. Smoke detectors are based on the optical or ionization properties of smoke, which are sensitive to smoke but susceptible to interference from dust, steam, etc. Temperature detectors determine fires by detecting the ambient temperature or the rate of temperature rise, which are low-cost but insensitive to early-stage fires. Flame detectors respond quickly to open fires by detecting ultraviolet or infrared rays from flames but are susceptible to interference from other light sources. Gas detectors detect fires by detecting the resulting Hazardous gas concentration change to alarm, suitable for enclosed space, but the response speed is slower.

S. FOUZIYA SULTHANA et al [5] proposed sensing smoke concentration through optical or electromagnetic settling elements to warn of fires, but it is not effective in detecting tiny particles and dust and aerosols can trigger false alarms; TONG Rui-tao et al [6] studied the resistive leakage detection technique, but its judgment based on residual current is susceptible to electromagnetic interference and current shock, and the false alarm rate is high; Yilmaz et al [7] pointed out that conventional point temperature measurements require a large number of sensors and that assumptions about load and environment limit their warning reliability; Fawad Khan et al. [8] believed that at different stages of a fire, the gases produced by combustion can be used as a reliable basis for detection, especially CO sensors, which not only have the advantage of low power consumption but can also effectively detect smoldering fires.

This paper studies the pyrolysis particle sensor for the early detection of electrical fires in switch cabinets. The pyrolysis products of cable materials in simulated switch cabinets are collected by pyrolysis particle sensors and characteristic gas sensors, and the response characteristics of pyrolysis particle sensors are systematically analyzed, and the response speed is compared with that of characteristic gas sensors.

2. Electrical fire characteristics of switchgear

The interior of the switchgear consists of several functional areas: trolley compartment, busbar compartment, cable compartment and instrument compartment. The insulating materials (PVC and PE) wrapped around the surface of busbars and cables are susceptible to high temperatures that can cause them to melt and burn; The epoxy resin in the insulator where the busbar is connected to other equipment will also burn first under high temperature or arc action[9-10].

PVC first undergoes a pyrolysis reaction at 120°C to 140°C to produce gases such as ethylene, ethane, propylene, propane, methyl chloride, carbon monoxide and carbon dioxide. As the temperature increases, the rate of gas production accelerates and the concentration increases with pyrolysis time. PE pyrolyzes at around 350°C, producing gases mainly methane, ethylene, ethane, propane and butane. These gases start to be released before the material produces open flame and smoke, and accumulate with increasing temperature and time [11-14].

3. Detection Principle of Pyrolysis Particle Sensor

The pyrolytic particle sensor measures mass concentration based on the Mie scattering theory of particles. When light strikes suspended particles in the air, the light is scattered, with the size of the particles being approximately equal to the wavelength of the light. For some particle properties, the intensity of light scattered by a particle is proportional to its mass concentration. The mass concentration of a particle can be obtained by measuring the intensity of the scattered light and applying a conversion factor [15]. The principle of operation is shown in Figure 1.

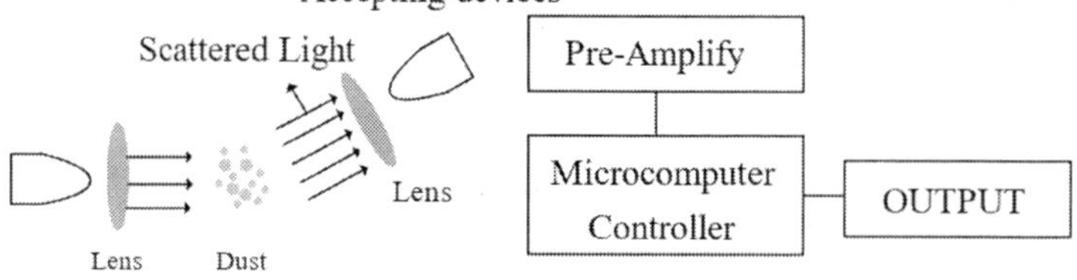

Figure 1. Schematic diagram of pyrolytic particle sensor

4. Comparative test of pyrolysis particle detection of typical materials

4.1. *Construction of test platform*

4.1.1. Sensor Performance. Based on the pyrolysis gas characteristics of typical materials such as PVC and PE, this paper selects methane, ethane, ethylene and acetylene as the characteristic gases, and uses the corresponding characteristic gas sensors and pyrolysis particle sensors to conduct comparative tests. The performance parameters of each sensor are shown in Table 1.

Table 1 Sensor parameters

Sensor type	Principle	Resolution	Range
PT100	resistance thermometry	0.1℃	0-600℃
CH4	catalytic combustion	1%LEL	0-100%LEL
C4H4	catalytic combustion	1%LEL	3-100%LEL
C2H2	catalytic combustion	1ppm	0-1000ppm
Pyrolysis particles	semiconductor	1ug/m3	0-2000ug/m3

4.1.2. Simulation test platform. This experimental platform uses a 60cm × 60cm × 60cm acrylic box to simulate the working environment inside a switchgear cabinet. As shown in Figure 2. A gas sensor and a pyrolysis particle sensor are installed on the top of the box to collect the pyrolysis particle concentration and characteristic gas concentration in the environment. A heating pad is placed at the bottom of the box, with pyrolysis material on its upper surface and a PT100 temperature sensor on its lower surface for real-time monitoring of the pyrolysis temperature.

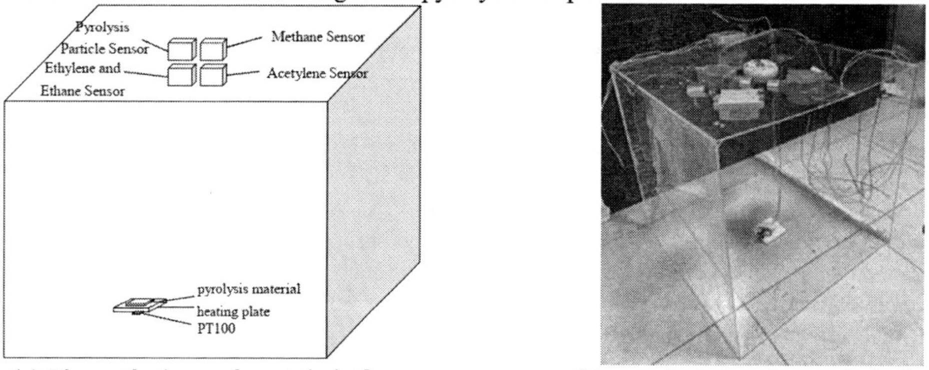

(a) Theoretical experimental platform (b) practical experimental platform

Figure 2. Theoretical experimental platform and practical experimental platform

4.1.3. Experimental content. First, fix the PVC material board and PT100 temperature sensor on the upper and lower surfaces of the heating sheet respectively, install pyrolysis particle sensor, methane sensor, ethylene-ethane sensor and acetylene sensor on the top, and check whether the test environment is airtight. Second, after the sensor preheating operation for 5 minutes, the heating pad starts to energize and work, and stops heating when the pyrolysis temperature stabilizes at 120℃ to 140℃.Third, record the pyrolysis temperature, pyrolysis particle concentration, methane, ethylene-ethane and acetylene concentrations and their response times during the test. Finally, replace the pyrolysis material with a PE material plate, repeat the above test steps, and stop heating when the pyrolysis temperature rises to 360°C.

4.2. Results. When the pyrolysis material is PVC material plate, the curves of pyrolysis particle concentration and characteristic gas concentration with time are shown in figure. 4. As can be seen from the figure 3, in the preheating stage, the pyrolysis particle and characteristic gas concentrations are relatively stable; after the heating pad is energized, the pyrolysis particle concentration gradually increases with the increase of pyrolysis temperature, and finally reaches a stable state at 136°C. And the rest of the characteristic gas sensors did not show obvious response during the whole test.

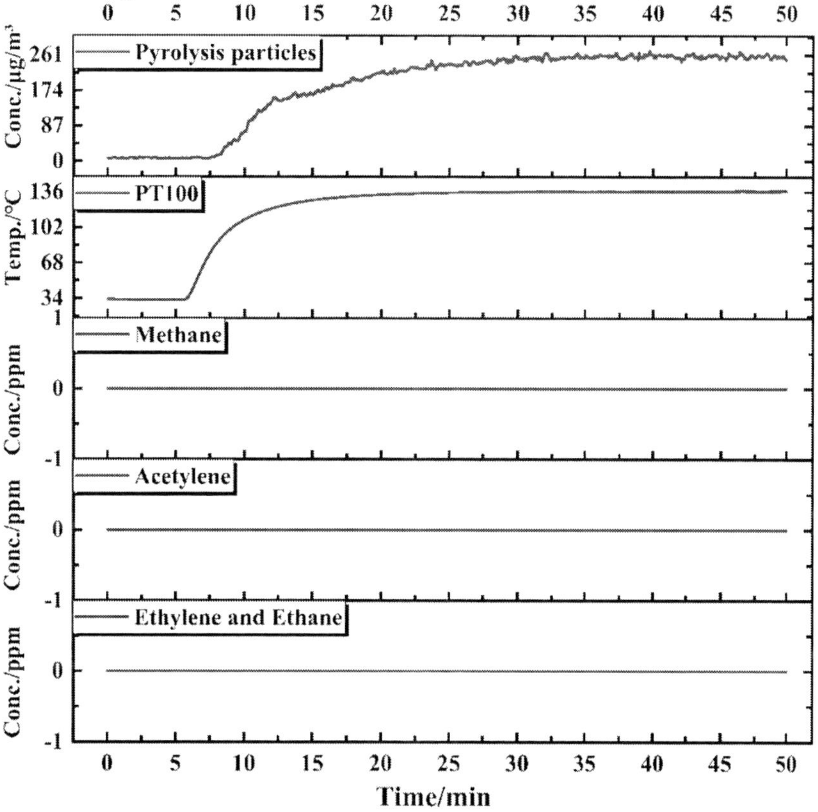

Figure 3. Response of each sensor during pyrolysis of PVC material

The trends of pyrolysis particle concentration and characteristic gas concentration with time during the pyrolysis process after replacing the plate with PE material are shown in Figure 4. The figure shows that in the preheating stage, the concentrations of pyrolysis particles and characteristic gases are kept relatively stable. When the heating pad was energized, the pyrolysis temperature increased rapidly, and the pyrolysis particle concentration showed a trend of a smooth rise and then a slow decline, followed by a sharp rise to the maximum value of the sensor's range, when the pyrolysis temperature had already reached 360°C. However, none of the other characteristic gas sensors showed a significant response throughout the experiment.

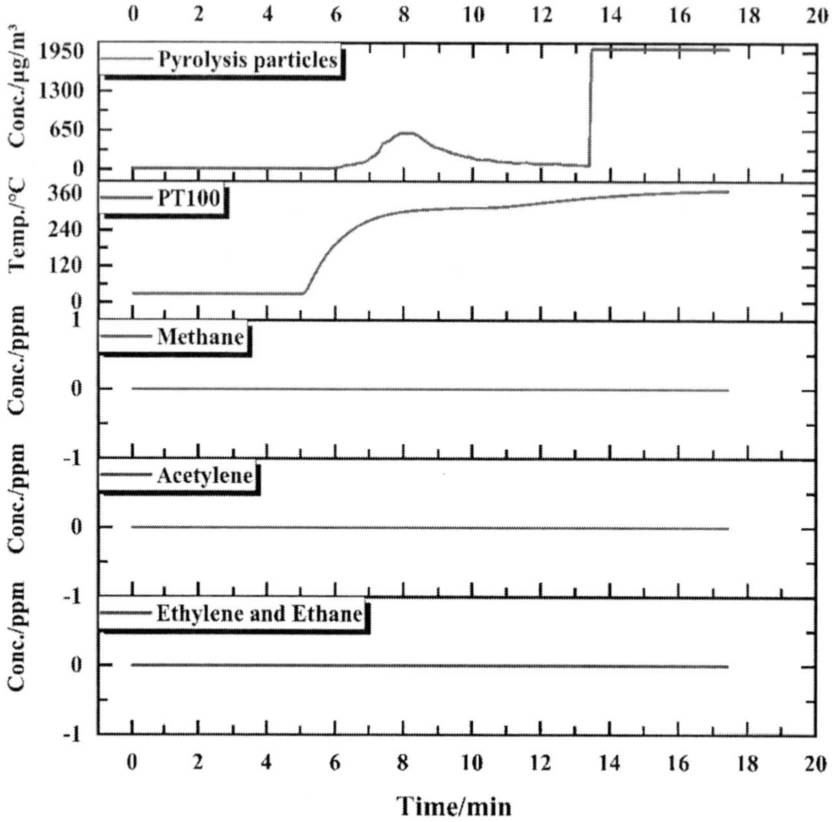

Figure 4. Response of each sensor during pyrolysis of PE material

Through the analysis of the above test results, it can be seen that in the pyrolysis test in the simulated switch cabinet environment, the pyrolysis particle sensor shows higher sensitivity and its response time is better than that of the characteristic gas sensor.

5. Conclusion

In this paper, a study was conducted for the early fire detection of medium voltage switchgear, carrying out test experiments on the pyrolytic properties of typical materials and obtaining the following conclusions: Although methane, ethylene, ethane and acetylene gases should be produced at specific temperature stages based on the pyrolysis properties of PVC and PE materials, the concentrations of these gases were not detected under the gas sensor resolution conditions used in this experimental platform (1% LEL and 1 ppm). This suggests that the concentrations of these gases were not sufficient to trigger a sensor response under the conditions of this experiment. The Pyrolysis Particle Sensor not only effectively detects pyrolysis reactions in PVC and PE materials, but also responds quickly.

Acknowledgments

2024 Science Project of State Grid Chongqing Power Company (Yudian Science Project, No.11)

References

[1] M. R. Vallerotonda, S. Ansaldi, A. Pirone, et al, "Accident Triggered by Electrical Failures in Seveso Sites," DOAJ (DOAJ: Directory of Open Access Journals), vol. 91, Jun. 2022.
[2] M. Ivanova, R. Dimitrova, and A. Filipov, "Analysis of Power Outages and Human errors in the Operation of Equipment in Power Grids," in 2020 12th Electrical Engineering Faculty Conference (BulEF), Sep. 2020.

[3] K. M. Alkar, Moad. Meto, M. Jamjum, et al, "Performance Evaluation of The Blackout and Power Outages in Libyan Power Grid - Al-Zawia Combined Cycle Power Plant Case Study," in 2019 1st International Conference on Sustainable Renewable Energy Systems and Applications (ICSRESA), Dec. 2019.
[4] R. Bogue, "Sensors for fire detection," Sensor Review, vol. 33, no. 2, pp. 99–103, Mar. 2013.
[5] S. Fouziya Sulthana, A. Wise, Ravikumar Chinthaginjala, et al, "Review Study on Recent Developments in Fire Sensing Methods," IEEE Access, vol. 11, pp. 90269–90282, Jan. 2023.
[6] R. Tong, L. Guo, and Z. Cao, "The Analysis in Several Application Issues for Leakage Current Electrical Fire Monitoring System," Procedia Engineering, vol. 52, pp. 342–348, 2013.
[7] G. Yilmaz and S. E. Karlik, "A distributed optical fiber sensor for temperature detection in power cables," Sensors and Actuators A: Physical, vol. 125, no. 2, pp. 148–155, Jan. 2006.
[8] F. Khan, Z. Xu, J. Sun, et al, "Recent Advances in Sensors for Fire Detection," Sensors, vol. 22, no. 9, p. 3310, Apr. 2022.
[9] Y. Zhao, L. Chen, and R. Qiao, "Temperature Field Simulation and Structure Improvement of 12kV Switchgear," Journal of Physics Conference Series, vol. 2179, no. 1, pp. 012010–012010, Jan. 2022.
[10] W. Ruan, Y. Yao, X. Ouyang, et al, "Study on the reliability of 10kV uninterrupted maintenance outgoing switchgear under the condition of main device switching to the back-up device," in 2020 IEEE International Conference on High Voltage Engineering and Application (ICHVE), Sep. 2020.
[11] Y. Wang, N. Kang, J. Lin, et al, "On the pyrolysis characteristic parameters of four flame-retardant classes of PVC sheathless cable insulation materials," Journal of Analytical and Applied Pyrolysis, vol. 170, pp. 105901–105901, Feb. 2023.
[12] I. Ahmad et al., "Pyrolysis Study of Polypropylene and Polyethylene Into Premium Oil Products," International Journal of Green Energy, vol. 12, no. 7, pp. 663–671, Mar. 2014.
[13] S.-Y. Mun and C.-H. Hwang, "Experimental and Numerical Studies on Major Pyrolysis Properties of Flame Retardant PVC Cables Composed of Multiple Materials," Materials, vol. 13, no. 7, p. 1712, Apr. 2020.
[14] Z. Wang, R. Wei, X. Wang, et al, "Pyrolysis and Combustion of Polyvinyl Chloride (PVC) Sheath for New and Aged Cables via Thermogravimetric Analysis-Fourier Transform Infrared (TG-FTIR) and Calorimeter," Materials, vol. 11, no. 10, p. 1997, Oct. 2018.
[15] J. Han, X. Liu, M. Jiang, et al, "An improved on-line measurement method of particulate matter concentration using tri-wavelength laser light scattering," Fuel, vol. 302, p. 121197, Oct. 2021.

Journal of Physics: Conference Series 3012 (2025) 012042 doi:10.1088/1742-6596/3012/1/012042

RSFCL-based Fault Current Limiting and Protection Method for Shipboard DC Microgrids

Hao Ding[1], Yiran An[2], Yayu Yang[1,3*], Moduo Yu[4], Binjie Jin[1], Yuteng Zhuo[1]

[1] Department of Electrical Engineering Shanghai Maritime University Shanghai, 200120, P.R. China

[2] Pudong Electric Power Supply Company, Shanghai, 200120, P.R. China

[3] Department of Electrical Engineering and Computer Science,The University of Tennessee,Knoxville, TN, USA

[4] Key Laboratory of Control of Power Transmission and Conversion Ministry of Education Shanghai Jiao Tong University, Shanghai, 200240, P.R. China

*Corresponding author's e-mail: yyyangsjtu@163.com

Abstract. In recent years, DC power grids have been increasingly adopted in ship power systems. The complexity and variability of ship working conditions bring challenges to the fault protection of the system. Shipboard DC microgrids in the short-circuit fault between the poles has a rapid rise in fault current, amplitude and other hazards, may cause damage to power electronics, resulting in the converter station blocking affect the reliability of power supply. Therefore, studying protection strategies for short-circuit faults in shipboard DC microgrids is essential. This paper proposes a protection scheme using resistive superconducting fault current limiter (RSFCL), utilizing its terminal voltage characteristics for fault identification. The proposed method demonstrates high accuracy and speed, and simulations in PSCAD/EMTDC are conducted to validate its effectiveness.

1. Introduction

DC microgrids, particularly medium-voltage DC (MVDC) systems with modular multilevel converters (MMC), offer significant advantages: simplified power management without frequency synchronization, efficient integration of diverse power sources and energy storage systems, reduced footprint, and enhanced reliability[1][2]. These benefits make DC power architecture particularly suitable for shipboard DC microgrids[3].

In shipboard DC microgrids, fault currents can quickly escalate to dangerous levels within milliseconds, which can cause severe damage to power electronics and system infrastructure, so current limiters need to be added to limit the escalation of fault currents. In[4], the feasibility of designing fault detection algorithms using RSFCL is established through examining the current characteristics in half-bridge MMC MTdc grids and utilizing an equivalent RSFCL quenching model. In[5], a DC Vacuum Circuit Breaker (DC VCB) is combined with an RSFCL for superconducting aircraft applications, where the fault current is limited by the RSFCL, and the VCB interrupts the limiting current by creating an artificial current zero. It can be seen that RSFCL is mostly used in land-based grids and aerospace industry, and that while traditional protection methods such as DC circuit breakers and mechanical switches have been proposed for shipboard DC microgrids, they tend to be limited in terms of responsiveness and cost-effectiveness. In this paper, the application of RSFCL to shipboard DC microgrids is able to quickly limit fault currents, improve system stability and equipment protection, while maintaining low energy consumption and high reliability.

In [6], a transient current-based protection scheme is proposed, utilizing transient current characteristics after a fault occurs to determine the fault area and distinguish between external and internal faults. In [7], the propagation characteristics of line-mode inverse travelling-waves (TW) are analysed to differentiate between external and internal faults, but the dependence on travelling wave characteristics and complex threshold settings increases engineering complexity. In [8], faults are distinguished by comparing the transient energy on the line side as well as the valve side of the reactor. For internal faults, the line-side energy is significantly higher than the valve-side energy; the opposite is true for external faults. Accordingly, the fault protection method can be designed starting from the voltage characteristics of RSFCL.

Based on the unique advantages of RSFCL, this paper proposes it use in shipboard DC microgrids for effective fault current limitation. Additionally, a novel protection scheme is developed using the RSFCL's voltage characteristics. This boundary-based principle leverages the voltage difference across the RSFCL to enable fast and reliable fault detection, enhancing the overall protection performance of shipboard DC microgrids.

The rest of the paper is organized as follows: In section II, a brief overview of the topology is given. In section III, the inter-pole short-circuit fault current is modelled analytically. In section IV, simulation verification is carried out after adding RSFCL to verify the effectiveness of the scheme. In section V, the fault protection method based on RSFCL voltage characteristics is validated. In section VI, the conclusions are presented.

2. Simplified topology analysis of MMC-based shipboard DC microgrids

Figure 1 shows the MMC topology, where SM denotes submodule, L_0 is the bridge arm reactor and R_g is the clamp resistance. A converter consists of six bridge arms, each of which contains N submodules as well as a bridge arm reactor. The upper and lower bridge arms are merged into a single-phase unit, with the three-phase units connected in parallel. Without considering the presence of submodules redundancy, a phase unit in the MMC contains 2N series-connected sub-modules, with N submodules for each of the upper and lower bridge arms. A single bridge arm can put in at least zero and up to N sub-modules, which enables (N+1) levels to be realised.

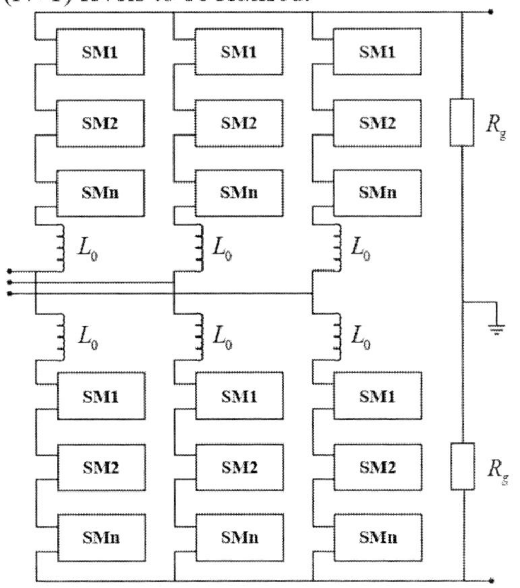

Figure 1. The topology of MMC

By simplifying the topology and retaining only the core power equipment and protection components, it is possible to focus more effectively on the current limiting characteristics of RSFCL. Figure 2 shows a simplified topology of shipboard DC microgrids. The deployment of 4 RSFCLs in this topology is designed to enhance system reliability and fault protection capability. RSFCLs can effectively manage faults occurring in different locations and directions. Each RSFCL serves to limit the magnitude of fault currents, thereby minimizing the impact on other parts of the system. Additionally, the voltage across each RSFCL can be utilized for fault identification, allowing for precise differentiation between external and internal faults. Based on the topology of Figure 2, a shipboard DC microgrid simulation model is built in PSCAD/EMTDC[9]. The relevant parameters of converter stations are shown in Table 1.

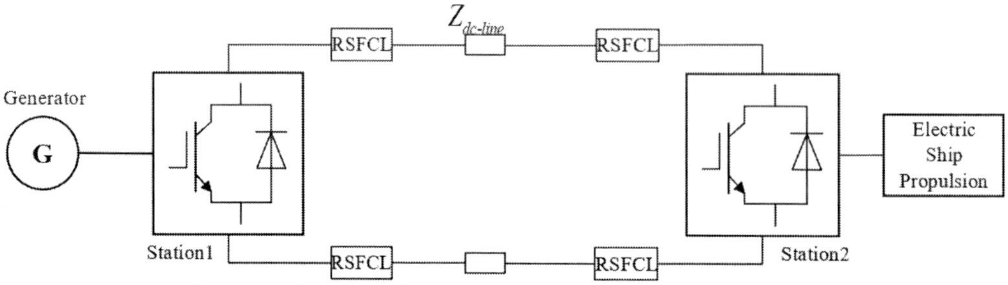

Figure 2. The simplified topology of shipboard DC microgrids

Table 1. Parameters of converter stations.

	Parameters
Submodule on-State resistance/mohm	0.001361
Submodule capacitance/mF	50
Bridge arm inductors/mH	1
Rated capacity/MVA	10
DC voltage/kV	±3.3
Electron level(N+1)	21

3. Fault current analysis and calculation

3.1. *Inter-pole short circuit fault analysis*

When an inter-pole short circuit occurs in the MMC-based shipboard DC microgrids, the DC side distributed capacitor of the submodule generates a transient charge and discharge current, which is quickly fed into the fault point.

When the converter station is in normal working condition, the resistance, inductance and capacitance of each bridge arm connected to the line remain unchanged, so the MMC can be transformed into the circuit on the left side of Figure 3. Further parallel connection can be equivalent to an RLC circuit. The equivalent parameters of the converter station are depicted in formula (1). The equivalent circuit is shown in the right side of Figure 3.

$$\begin{cases} R_{con} = 2R_{arm} / 3 \\ L_{con} = 2L_{arm} / 3 \\ C_{con} = 3C_{eq} \end{cases} \tag{1}$$

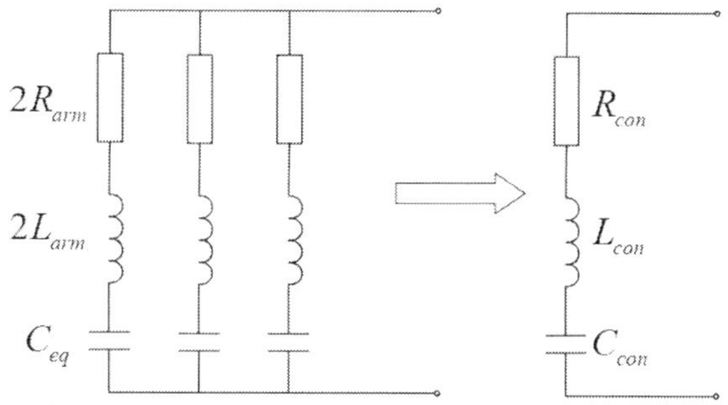

Figure 3. The equivalent model of converter station

The bridge arm equivalent resistance R_{arm} is the on-resistance of the N switching devices on the bridge arm plus the resistance of the bridge arm reactance, L_{arm} is the bridge arm reactor value, and C_{eq} is the equivalent capacitance of all submodules in each phase, which is calculated by formula (2) according to the principle of equal energy.

$$\frac{1}{2}C_{eq}(NU_{sm}^2) = 2N\times\frac{1}{2}C_{sm}U_{sm}^2 \Rightarrow C_{eq} = \frac{2C_{sm}}{N} \quad (2)$$

Where: C_{eq} is the equivalent capacitance value of a phase unit; U_{sm} is the submodule capacitance voltage; C_{sm} is the DC side capacitance value of the submodule.

3.2. *Fault characteristic analysis*

The transient process before locking can be expressed by Kirchhoff's voltage law as in formula (3).

$$L_{con}C_{con}\frac{d^2u_c}{dt^2} + R_{con}C_{con}\frac{du_c}{dt} + u_c = 0 \quad (3)$$

Where: u_c is the voltage across the equivalent capacitor.

Solve the characteristic equation of formula (3), and the two conjugate complex numbers are solved to be the characteristic roots of the equation. The change process of the submodule capacitor voltage value u_c is a second-order underdamped oscillation process. The characteristic roots of the equation are shown in formula (4).

$$\lambda_{1,2} = -\frac{R_{con}}{2L_{con}} \pm j\sqrt{\frac{1}{L_{con}C_{con}} - (\frac{R_{con}}{2L_{con}})^2} = -a + jb \quad (4)$$

Let the DC voltage and DC current at the moment of fault be U_0 and I_0 respectively, and the transient solution of capacitor voltage u_c can be solved:

$$u_c = Ae^{-at}\sin(bt+\beta) \quad (5)$$

Where: $A = \sqrt{U_0^2 + (\frac{aU_0C_{con} - I_0}{bC_{con}})^2}$, $\beta = \arcsin(\frac{U_0}{A})$.

The transient solution of the fault current i_{dc} can be obtained from the transient solution of the capacitor voltage u_c:

$$i_{dc} = A\sqrt{\frac{C_{con}}{L_{con}}}e^{-at}\sin(bt+\theta) \quad (6)$$

Where: $\theta = \arcsin(\frac{I_0}{A}\times\sqrt{\frac{L_{con}}{C_{con}}})$.

From the analysis of formula (5) and (6), it can be observed that during the pre-blocking stage of the converter station, the submodule distributed capacitance voltage (u_c) is rapidly decreased while the fault current is significantly elevated. The theoretical fault current value is determined by three key parameters: the submodule distributed capacitance, the submodule quantity, and the bridge arm reactor impedance. Specifically, the fault current magnitude is found to be directly proportional to the submodule capacitance value, while being inversely proportional to both the submodule quantity and the bridge arm reactor magnitude.

3.3. Simulation verification of steady state and fault conditions

The DC voltage is positive and negative 3.3kV, the frequency is 60Hz, and it is a double-terminal MMC pseudo-bipolar shipboard DC microgrid.

Using Nearest Level Modulation (NLM), described as a voltage approximation modulation strategy, the AC-side voltage is matched to the modulation voltage by controlling the number of submodules inserted or bypassed in the MMC phase units. The steady state operating characteristics of the simulated system are illustrated in Figure 4.

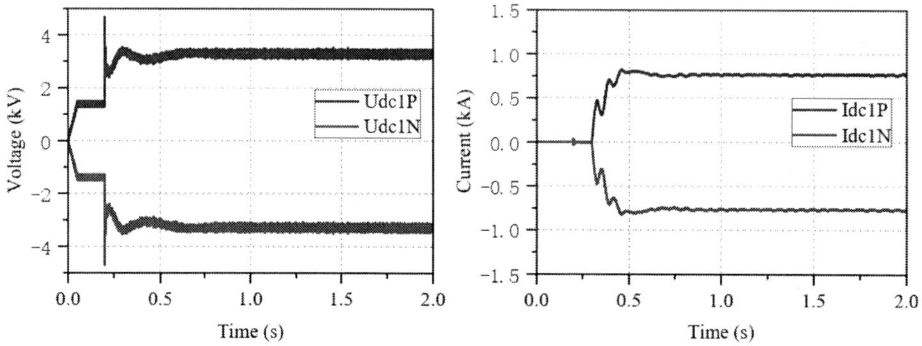

Figure 4. Steady state operation characteristics of the system

At 2 seconds, an inter-pole short circuit fault is configured between the converter stations, and the DC fault current Idc and the theoretical fault current of converter station A are shown in Figure 5. It is verified that the above fault analysis is met.

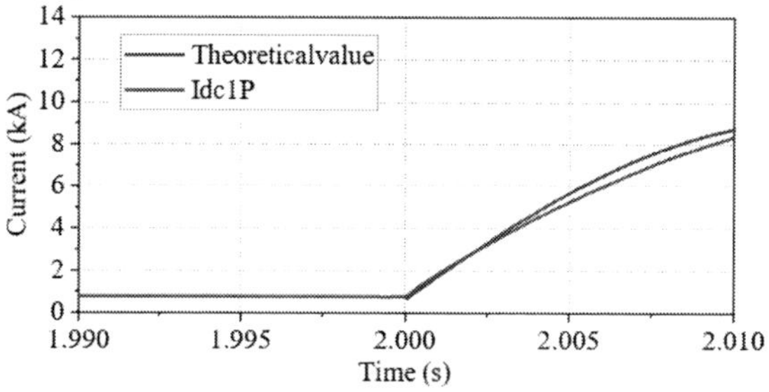

Figure 5. Inter-pole short circuit fault simulation verification of the system

4. RSFCL-based fault current limiting

4.1. Mathematical model of RSFCL application

Superconductors transition from a zero-resistance state to a resistive state when any of their critical parameters (current, temperature, or magnetic field) exceeds its threshold.

The RSFCL exploits this characteristic in its operation. During normal system operation, all parameters remain within their critical boundaries, maintaining the superconductor in a zero-resistance state. In this condition, the RSFCL generates no losses and does not affect the system's steady-state operation. However, when a fault occurs and the current flowing through the limiter exceeds the critical threshold, the superconductor transitions to its quenched state, manifesting high resistance characteristics. RSFCL minimises the impact on system efficiency compared to conventional current limiters, which incur additional power losses due to inherent impedance.

In addition, the superconducting current limiter has obvious advantages in terms of power density and relatively small size, which makes it suitable for application in space-constrained shipboard DC microgrids.

A distinctive advantage of the RSFCL is that its activation and deactivation require no external control. The fault current exceeding the critical threshold serves as signal itself. Once the current surpasses the critical value, the limiter automatically functions as a system-integrated resistor, enhancing the damping of the DC system post-fault and thereby restricting the fault current's rise.

The mathematical model of the RSFCL is given as follows[10].

$$R_{SFCL}(t)=\begin{cases}0 & (t<t_0)\\ R_m(1-e^{-\frac{t-t_0}{T}})^{\frac{1}{2}} & (t_0\le t)\end{cases} \tag{7}$$

4.2. Simulation verification after adding RSFCL

As illustrated in Figure 6, an inter-pole short-circuit fault is set at 2 seconds and is set as a permanent fault. It can be observed that prior to 2 seconds, the integration of RSFCL has no impact on the system's steady-state current, which remains constant at 0.8kA during normal operation. After the fault occurs, while the current without RSFCL surges to approximately 11kA, the RSFCL transitions to its quenched state within 1ms and effectively limits the fault current to approximately 1.3kA, demonstrating excellent fault current limitation performance and rapid response characteristics.

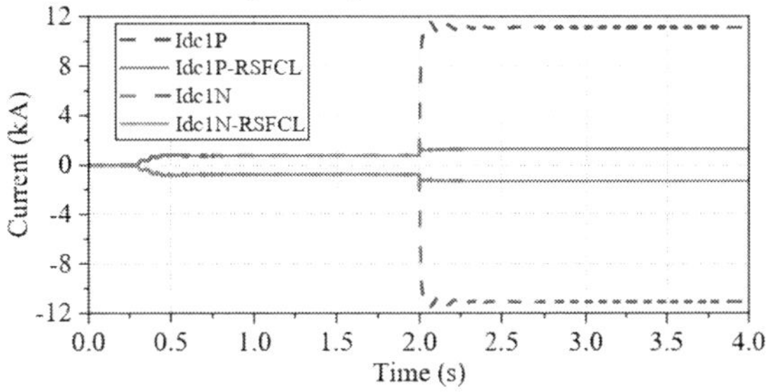

Figure 6. Comparison of fault current before and after adding RSFCL

5. Fault protection method based on RSFCL voltage

5.1. RSFCL transient voltage characterisation

The use of DC circuit breaker to isolate the faulted line is the most widely used fault isolation method at present, and this paper also uses DC circuit breaker to deal with the fault. The theoretical analysis of the equation is similar to formula(3) and will not be described again, and from formula(5), u_c is always positive, so whether the sender or receiver RSFCL, the voltage under the forward fault is always a positive value, while the voltage under the reverse fault is always a negative value.

The method used here references fault current limiters(FCL) commonly used for fast protection in modern power grids[11]. Comparing the two cases of internal and external faults: in the internal faults, RSFCLs at both the transmitter and receiver ends are positive. On the contrary, in the case of external faults, the two RSFCLs are not both positive. Therefore, the fault can be identified by the positive and negative RSFCL voltages at the sending and receiving ends. If both ends are identified as positive faults, the system is determined to be an internal fault and the DC circuit breaker operates. In summary, the RSFCL voltage characteristics under different fault position can be derived as follows Table 2.

Table 2.The RSFCL voltage characteristics under different fault position

Fault position	Rectifier RSFCL voltage	Inverter RSFCL voltage
Internal fault	>0	>0
Sending end external fault	<0	>0
Receiving end external fault	>0	<0

5.2. DC line protection methods

Protection criteria are established based on RSFCL terminal voltage characteristics during different faults positions. For external faults, at least one RSFCL exhibits non-positive voltage. The fault discrimination is implemented using fomula(8), with a protection threshold of 1kV•ms and a detection window of 0.5ms to ensure both reliability and speed. In summary, the protection scheme used in this paper simultaneously analyses the two RSFCLs at both ends and uses their voltage polarity to determine if the fault is positive; if both ends are identified as positive, the system is determined to be an internal fault and the DC circuit breaker operates.

$$\int_0^{t_{set}} u_R dt \geq I_{set} \tag{8}$$

The f1 fault is set between converter station 1 and the RSFCL on the side of converter station 1, and the f2 fault is set at 2km of the cable between the two RSFCLs, the results are shown in Figure 7. Figure 7(a) is an external fault, and Figure 7(b) is an internal fault. The protection criterion is shown in Table 3, 4.

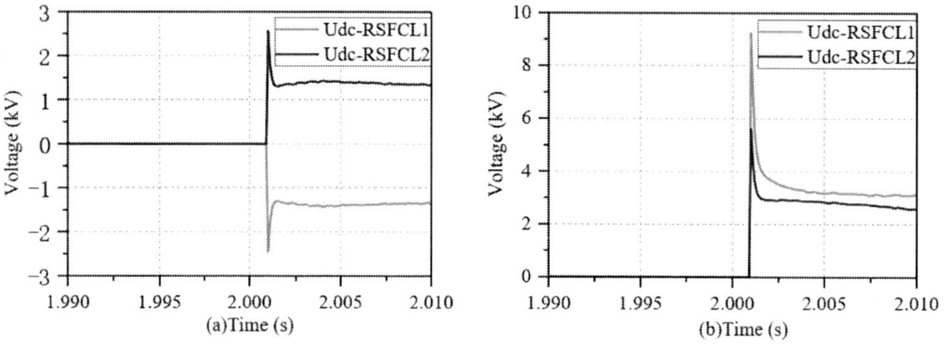

Figure 7. RSFCL voltage in an external fault and an internal fault

Table 3. Rectifier side protection criterion

Position	$\int_0^{t_{set}} u_R dt$	$\int_0^{t_{set}} u_R dt - I_{set}$	Direction of fault
f1	-1.25	<0	Reverse
f2	4.164	>0	Forward

Table 4. Inverter side protection criterion

Position	$\int_0^{t_{set}} u_R dt$	$\int_0^{t_{set}} u_R dt - I_{set}$	Direction of fault
f1	1.28	>0	Forward
f2	2.8	>0	Forward

The fault identification results, presented in Table 5, align with the predetermined fault types, confirming the viability of the protection method.

Table 5. Faults identification results

Position	Sending end	Receiving end	Type of fault
f1	Reverse	Forward	External
f2	Forward	Forward	Internal

6. Conclusion

By analyzing the characteristics and expressions of fault current, this paper finds that during a fault, the current rises sharply and may damage equipment. Leveraging the critical properties of superconductors, a fault protection scheme for shipboard DC microgrids using RSFCL is proposed, which locates the fault by analyzing the voltage at its terminals. Simulations are conducted in PSCAD/EMTDC to validate the proposed method. The results indicate that adding RSFCL effectively responds to inter-pole short-circuit faults on the converter station side, quickly identifying the fault and inhibiting the fault current rise, thereby providing effective system protection.

Acknowledgments

1. This work was supported by National Natural Science Foundation of China under Grant 52407123.
2. This work was supported by Shanghai 2023 Science and Technology Innovation Action Plan-Star Cultivation (Sailing Program) under Grant 23YF1416000.

References

[1] Y. Guan, Y. Yang, X. Xu, J. Ma and Y. Liu, "Research on Shore DC Grids Considering Increasing Penetration of Renewables," 2023 IEEE International Conference on Advanced Power System Automation and Protection (APAP), Xuchang, China, 2023, pp. 1-4, doi: 10.1109/APAP59666.2023.10348492.

[2] Y. Yang, X. Jiang, M. Yu, S. Ye and C. Cai, "A review of key technologies of shore power systems," 12th International Conference on Renewable Power Generation (RPG 2023), Shanghai, China, 2023, pp. 1162-1167, doi: 10.1049/icp.2023.2451.

[3] L. Xu et al., "A Review of DC Shipboard Microgrids—Part I: Power Architectures, Energy Storage, and Power Converters," in IEEE Transactions on Power Electronics, vol. 37, no. 5, pp. 5155-5172, May 2022, doi: 10.1109/TPEL.2021.3128417.

[4] G. Zhou, M. Han, S. Filizadeh, Z. Geng and X. Zhang, "A Fault Detection Scheme in MTDC Systems Using a Superconducting Fault Current Limiter," in IEEE Systems Journal, vol. 16, no. 3, pp. 3867-3877, Sept. 2022, doi: 10.1109/JSYST.2021.3122890.

[5] M. Junaid, X. Yu, S. Cao, W. Yu, D. Zuo and J. Wang, "Simulation Analysis of DC Fault Interruption Characteristics of Superconducting Electric Aircraft Propulsion," in CSEE Journal of Power and Energy Systems, vol. 10, no. 4, pp. 1834-1842, July 2024, doi: 10.17775/CSEEJPES.2023.03650.

[6] Y. Yang, N. Tai, C. Fan, X. Zheng, J. Liu and B. Lv, "Novel transient current-based protection scheme for DC line faults in UHVDC system," 2017 IEEE Power & Energy Society General Meeting, Chicago, IL, USA, 2017, pp. 1-6, doi: 10.1109/PESGM.2017.8274208.

[7] S. Yang, W. Xiang and J. Wen, "An Improved DC Fault Protection Scheme Independent of Boundary Components for MMC Based HVDC Grids," in IEEE Transactions on Power Delivery, vol. 36, no. 4, pp. 2520-2531, Aug. 2021, doi: 10.1109/TPWRD.2020.3029308.

[8] Z. Dai, N. Liu, C. Zhang, X. Pan and J. Wang, "A Pilot Protection for HVDC Transmission Lines Based on Transient Energy Ratio of DC Filter Link," in IEEE Transactions on Power Delivery, vol. 35, no. 4, pp. 1695-1706, Aug. 2020, doi: 10.1109/TPWRD.2019.2950350.

[9] T. Wang, Z. Yu, F. Xie, Z. Hao, A. Monti and F. Ponci, "Protection of Line Faults in HVDC Grids Through Convexity Detection in Backward Traveling Wave Voltages," in IEEE Transactions on Power Delivery, vol. 38, no. 5, pp. 3660-3676, Oct. 2023, doi: 10.1109/TPWRD.2023.3286863.

[10] G. -W. Kim and H. -S. Choi, "Applicability Analysis of a Multi-Filar Meander R-SFCL and an RSC-DCCB in an MMC-MTDC Grid," in IEEE Transactions on Applied Superconductivity, vol. 33, no. 5, pp. 1-7, Aug. 2023, Art no. 5600507, doi: 10.1109/TASC.2023.3249639.

[11] S. Ouali and A. Cherkaoui, "Integration of Fault Current Limiters in Electric Power Systems: A Review," in IEEE Systems Journal, vol. 15, no. 3, pp. 4470-4479, Sept. 2021, doi: 10.1109/JSYST.2020.3012345.

Journal of Physics: Conference Series **3012** (2025) 012043 doi:10.1088/1742-6596/3012/1/012043

Optimal Dispatch Method of Controllable Loads with Consideration of Distribution Network Constraints

Yuan Ren[1], Guomin Luo[1], Pei Zhang[1,2*], Xuegang Lu[3], Zhenyi Wang[3]

[1] Beijing Jiaotong University, 100044, China

[2] Tianjin University, 300072, China

[3]Yunnan Power Grid Co., Ltd., Yunnan Kunming, 650000, China

[*]Corresponding author's e-mail: 23121468@bjtu.edu.cn

Abstract. Controllable loads provide flexibility to handle fluctuations and randomness of renewable generation. However, controlled loads are distributed at low voltage levels in the distribution system. How to dispatch controllable loads to meet transmission system operation without violating distribution network operating limits is a challenge. This paper proposes an optimal dispatch method with consideration of distribution network constraints. First, sensitivity factors between controllable loads and line flows, and nodal voltages are computed. Second, the optimal dispatch of controllable loads is formulated as an optimization problem. The proposed method is validated using T30 - DF6 and T57 - DUK systems. Study results show that the proposed method can quickly dispatch the controllable loads to meet transmission system operation requirements.

1. Introduction

The usual way to handle congestion in the power system is to change generator outputs or cut down on loads. However, this often leads to high congestion costs [1]. As new power system infrastructure keeps improving, there are now more and more different types of controllable load resources in the distribution network [2]. These loads can be encouraged to help manage congestion, offering a green solution to lower these costs [3]. Thus, it is crucial to consider how to incorporate controllable loads in distribution network dispatch to ease network congestion.

Researchers have studied how to use controllable loads to reduce network congestion. From a demand response view, there are mainly two ways of managing congestion: price or incentives [4-5]. In reference [6], the author created a situation with a lot of renewable energy and changed how controllable loads are used based on prices to reduce congestion. In reference [7], the authors introduced a bi-level optimal demand response scheme that uses thermally controllable loads to lower peak loads in the distribution network. In reference [8], dynamic tariffs and daily power based on network tariffs were used to manage congestion by coordinating controllable electric loads like electric vehicles (EVs) and heat pumps (HPs). These studies fail to fully solve the influence of node voltage and power flow changes on dispatching results. In addition, they did not consider the role of the load side in the coordination process [9-10].

To solve the above problems, this paper develops an optimal dispatch model for controllable loads in order to reduce operational costs. Using the Newton-Raphson method and sensitivity analysis method, we establish sensitivity relationships among controllable load outputs, transmission power flows, and node voltages within the operating system. Constraints are placed on both transmission lines and nodes to preserve system stability. Two scheduling strategies have been formulated based on different methods of incorporating these network constraints. The effectiveness of the proposed approach is validated by the T30-DF6 and T57-DUK systems.

2. Optimal Scheduling Method of Controllable Loads Considering Network Constraints

2.1. Controllable Loads Optimal Scheduling Model Considering Network Constraints

The optimal scheduling model for controllable loads can handle network constraints effectively. If the internal adjustments cause an imbalance in the transmission network, a scheduling command will be sent through the interface to the distribution network. As for this command, the distribution network acts upon it and adjusts the controllable loads within its scope accurately. Figure 1 presents the corresponding adjustment model.

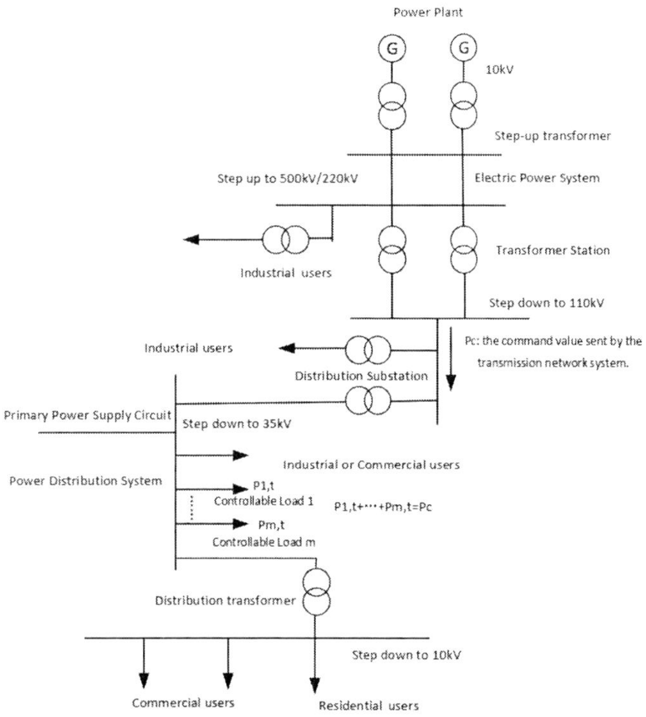

Figure 1. Optimal Adjustment Model

2.1.1. Objective function

The objective function aims to minimize total cost. To achieve this, the primary objective is to reduce the adjustment cost of different controllable loads in each dispatch interval.

$$\min\left(\sum_{i=1}^{N_{DR}} P_{i,t} * Ct_{DRi,t}\right) \qquad (1)$$

N_{DR} represents the number of controllable loads. $P_{i,t}$ represents the active adjustment of the controllable load subject i at the moment t. $Ct_{DRi,t}$ represents the adjustment cost of controllable load subject i in t period. t represents the total number of periods considered.

2.1.2. Constraints

(1) Load Balance Constraints

For optimal controllable load scheduling, the aggregate scheduling values of the controllable loads must match the instruction values from the transmission network.

$$\sum_{i=1}^{N_{DR}} P_{i,t} = P_{c,t} \tag{2}$$

$P_{c,t}$ is the dispatching instruction value.

(2) Load Power Constraint

The overall load value is calculated by adding the adjustment amounts of controllable loads to the basic load value, and this value must be within the specified maximum and minimum constraint ranges.

$$P_{\min,i,t} \le P_{i,t} + P_{base,i,t} \le P_{\max,i,t} \tag{3}$$

where $P_{base,i,t}$ is the basic capacity of controllable load at the moment t, $P_{\min,i,t}$ is the minimum capacity value of the controllable load i at the moment t, and $P_{\max,i,t}$ is the maximum capacity value of the controllable load i at time t.

(3) Load Adjustment Upper and Lower Limit constraints

In each adjustment period, it is required that the magnitude of controllable load adjustments cannot exceed the maximum limit set for that specific adjustment. A positive command value sent to the system increases the output of the controllable loads within the distribution network system, denoting a positive adjustment amount.

$$0 \le P_{i,t} \le UL_{i,t} \tag{4}$$

$UL_{i,t}$ is the upper limit of upward adjustment of controllable load subject i in the time period t.

Negative command value leads to a reduction in the output of the controllable loads within the distribution network system by a negative adjustment amount.

$$DL_{i,t} \le P_{i,t} \le 0 \tag{5}$$

$DL_{i,t}$ is the lower limit of downward adjustment of controllable load subject i at time t.

(4) Voltage Constraint

The node voltages in both the transmission and distribution systems must remain within their respective permissible ranges:

$$V_{\min} \le V_{node,0} + \Delta V_{node} \le V_{\max} \tag{6}$$

where $V_{\min}$ is the minimum allowable operating voltage of the network node, $V_{\max}$ is the maximum allowable operating voltage of the network node, $V_{node,0}$ is the network node voltage, and ΔV_{node} is the change of system node voltage after controllable loads participate in the regulation in the time period t.

(5) Line Constraint

$$-S_{\max} \le S_{line,0} + \Delta S_{line,0} \le S_{\max} \tag{7}$$

In the formula, $S_{line,0}$ is the apparent power transmission on the line at the current moment. $\Delta S_{line,0}$ represents the change of apparent power of the line after the controllable loads participate in the adjustment at time t. $S_{\max}$ is the maximum allowable transmission apparent power of the line.

The network constraints include voltage constraints and line constraints.

2.2. Network Constraint Solving Method Based on Sensitivity Analysis Method

Before utilizing the sensitivity analysis method, it is important to establish the initial point of power flow. This entails calculating the AC power flow of the entire system using the Newton-Raphson method.

The following correction formula (8) is obtained according to the power flow calculation steps.

$$\begin{bmatrix} \Delta P \\ \Delta Q \end{bmatrix} = \begin{bmatrix} H & N \\ M & L \end{bmatrix} \begin{bmatrix} \Delta\theta \\ \Delta U \end{bmatrix} \tag{8}$$

ΔP and ΔQ represent the change of active power and reactive power of the node respectively. H, N, M, and L are the block matrix corresponding to the Jacoby matrix.

The inverse of formula (8) can be obtained:

$$\begin{bmatrix} \Delta\theta \\ \Delta U \end{bmatrix} = \begin{bmatrix} A & B \\ C & D \end{bmatrix} \begin{bmatrix} \Delta P \\ \Delta Q \end{bmatrix} \tag{9}$$

where A, B, C, and D are the block matrix of the inverse matrix of the Jacoby matrix.

According to equation (9), by ignoring the reactive power of the loads, equation (10) can be derived:

$$\Delta U = C\Delta P \tag{10}$$

In the model, the change of ΔP is caused by the change of controllable load outputs, so ΔP is approximately equal to $\Delta P_{i,t}$.

$$\Delta V_{node,i} = C_{i,1}\Delta P_{1,t} + C_{i,2}\Delta P_{2,t} + \ldots + C_{i,N-1}\Delta P_{N-1,t} \tag{11}$$

$\Delta V_{node,i}$ is the change of system node voltage after the controllable loads participate in regulation.

$$\begin{bmatrix} \Delta V_{node,1} \\ \Delta V_{node,2} \\ \ldots \\ \Delta V_{node,PQ} \end{bmatrix} = \begin{bmatrix} C_{1,1} & \ldots & C_{1,N-1} \\ \ldots & \ldots & \ldots \\ C_{1,PQ} & \ldots & C_{PQ,N-1} \end{bmatrix} \begin{bmatrix} \Delta P_{1,t} \\ \Delta P_{2,t} \\ \ldots \\ \Delta P_{N-1,t} \end{bmatrix} \tag{12}$$

In the formula, $\Delta P_{i,t}$ represents the active adjustment of the controllable load on node i during time t period. According to formula (11) and formula (12), the voltage constraint in formula (6) becomes:

$$V_{\min} - V_{node,0} \le \Delta V_{node} \le V_{\max} - V_{node,0} \tag{13}$$

The following formula (14) can be obtained by extending formula (13).

$$\begin{cases} V_{\min,1} - V_{node,0,1} \le C_{11}\Delta P_{1,t} + C_{12}\Delta P_{2,t} + \ldots + \\ C_{1,N-1}\Delta P_{N-1,t} \le V_{\max,1} - V_{node,0,1} \\ \ldots \\ V_{\min,PQ} - V_{node,0,PQ} \le C_{PQ1}\Delta P_{1,t} + C_{PQ2}\Delta P_{2,t} + \ldots + \\ C_{PQ,N-1}\Delta P_{N-1,t} \le V_{\max,PQ} - V_{node,0,PQ} \end{cases} \tag{14}$$

PQ is the total number of PQ nodes in the system.

In power flow calculation, the relationship between system line transmission power and node power is as follows:

$$\begin{cases} P_{ij} = U_i U_j (G_{ij} \cos\theta_{ij} + B_{ij} \sin\theta_{ij}) - t_{ij} G_{ij} U_i^2 \\ Q_{ij} = U_i U_j (G_{ij} \sin\theta_{ij} - B_{ij} \cos\theta_{ij}) + (t_{ij} B_{ij} - b_{ij0}) U_i^2 \end{cases} \tag{15}$$

P_{ij} and Q_{ij} are the active and reactive power transmitted from node i to node j, respectively. t_{ij} is the transformer ratio on branch i to j. If the branch is only a wire, t_{ij} is 0; If there is a transformer in this branch, t_{ij} is the ratio of the reference voltage at the beginning and end of the bus; b_{ij} is the half susceptance of branch i to j. G_{ij} is the real component of the admittance matrix. B_{ij} is the imaginary component of the admittance matrix.

Formula (16) is derived through the Taylor expansion of formula (15).

$$\begin{bmatrix} \Delta P_l \\ \Delta Q_l \end{bmatrix} = \begin{bmatrix} E & F \\ G & K \end{bmatrix} \begin{bmatrix} \Delta\theta \\ \Delta U \end{bmatrix} = \begin{bmatrix} E & F \\ G & K \end{bmatrix} \begin{bmatrix} A & B \\ C & D \end{bmatrix} \begin{bmatrix} \Delta P \\ \Delta Q \end{bmatrix} = \begin{bmatrix} R & S \\ L & T \end{bmatrix} \begin{bmatrix} \Delta P \\ \Delta Q \end{bmatrix} \tag{16}$$

ΔP_l and ΔQ_l are the changes of active and reactive power transmitted by each branch respectively. E ,F ,G ,and K are the sub-block matrix of the sensitivity matrix representing the sensitivity relationship between the active and reactive power transmitted by each branch and the system voltage and phase angle. R, S, L and T are the block matrix of the sensitivity matrix that represents the sensitivity relationship between the active and reactive power of each branch and the active and reactive power of each node.

Equation (17) is derived from equation (16) by neglecting the reactive power variations of controllable loads in system regulation.

$$\begin{aligned} \Delta P_l &= R \Delta P_{i,t} \\ \Delta Q_l &= L \Delta P_{i,t} \end{aligned} \tag{17}$$

According to the formula of apparent power, the line constraint can be transformed into equation (18).

$$(R \Delta P_{i,t})^2 + (L \Delta P_{i,t})^2 \le (S_{\max} - S_{line,0})^2 \tag{18}$$

By expanding the above formula, the following formula can be obtained.

$$\begin{cases} (R_{1,1} \Delta P_{1,t} + \ldots + R_{1,N-1} \Delta P_{N-1,t})^2 + (L_{1,1} \Delta P_{1,t} + \ldots + L_{1,N-1} \Delta P_{N-1,t})^2 \\ \le (S_{\max,1} - S_{line,1,0})^2 \\ \qquad \ldots \\ (R_{t,1} \Delta P_{1,t} + \ldots + R_{t,N-1} \Delta P_{N-1,t})^2 + (L_{t,1} \Delta P_{1,t} + \ldots + L_{t,N-1} \Delta P_{N-1,t})^2 \\ \le (S_{\max,t} - S_{line,t,0})^2 \end{cases} \tag{19}$$

Equations (12)-(19) perform the sensitivity calculations concerning the initial power flow solution as the baseline system. The sensitivity of each load regulation for line power and node voltage is determined. Constraints on voltage and lines are converted and placed on the appropriate nodes and lines of the system.

2.3. Model Solving Method Considering Network Constraints

According to the different steps of the solution method, this paper puts forward two strategic models. In strategy one, the solution process of this model is as follows:

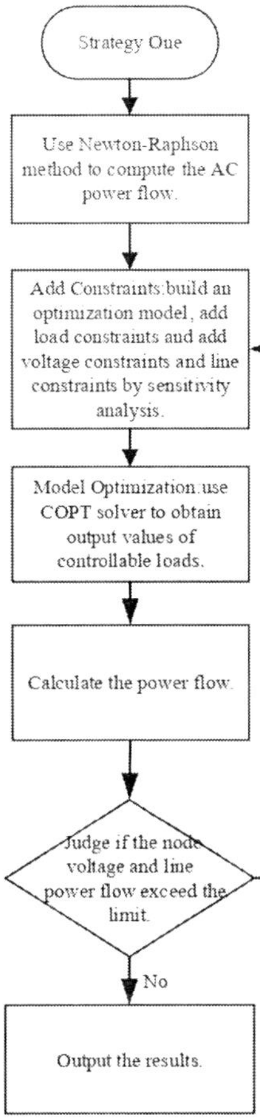

Figure 2. Strategy One

The above is the proposed strategy one. The second strategy is to separate the process of optimal solution from the process of adding network constraints. In strategy two, the solution process of this model is as follows:

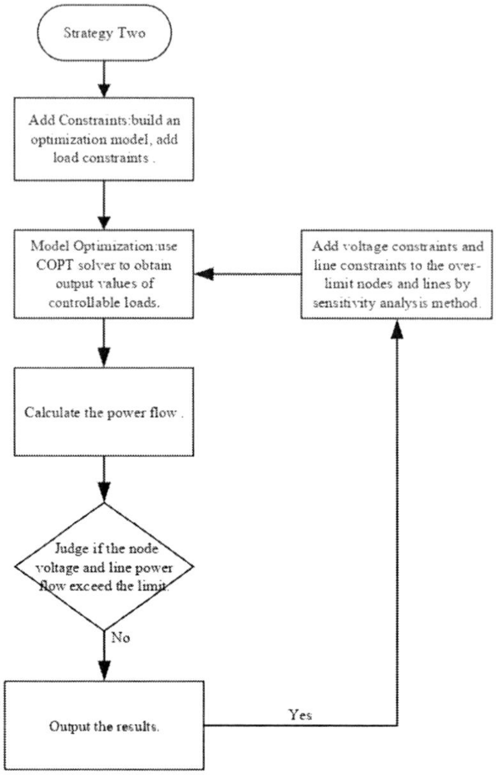

Figure 3. Strategy Two

3. Case study

3.1. Model Parameter

The T30-DF6 is a power transmission and distribution system. Different distribution networks are connected to nodes in the transmission network through transformers. Tables 1 and 2 show the structures of the transmission and distribution networks, respectively.

Table 1. T30 Transmission Network System Data Sheet

Transmission network	Number of nodes	Number of lines	Number of generators	Voltage class (kV)
T30	30	41	6	135

Table 2. District Distribution Network System Data Sheet

Sub-district distribution network	Nodes with adjustable loads	Line number
DF1	40, 41, 43, 44, 46, 47	42, 43, 44, 45, 46, 47, 48
DF2	48, 49, 51	53, 54, 55, 56
DF3	53, 54, 56, 58	57, 58, 59, 60, 61, 62, 63
DF4	60, 61, 62, 63, 64, 74	64, 65, 66, 67, 68, 69, 70
DF5	66, 67, 68, 69, 70, 71, 72	71, 72, 73, 74, 75, 76, 77, 78
DF6	73	76, 77, 78, 79

The T57-DUK system has a transmission network with 57 nodes. Table 3 shows its configuration.

Table 3. T57 Transmission Network System Data Sheet

Transmission network	Number of nodes	Number of lines	Number of generators
T57	52	70	7

3.2. *Case Analysis*

The T30-DF6 system has 27 adjustable loads. Figures 4(a) and 4(b) show the changes in node load amounts in the DF3 distribution network area for two different methods. Figure 4(a) shows the changes when the generator output at the 13th node in the transmission network is increased by 25 MW. Figure 4(b) shows the changes when the generator output is decreased by 10 MW.

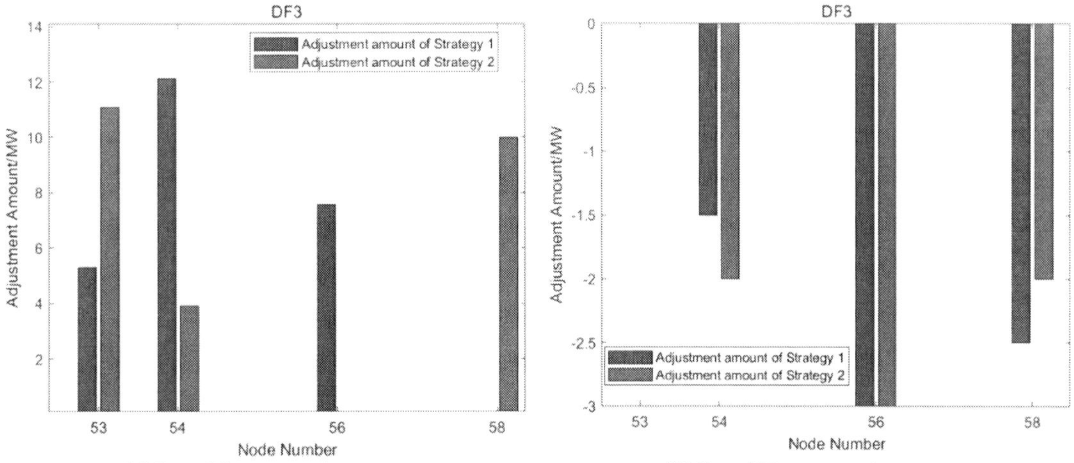

(a) Load Increase Amount	(b) Load Decrease Amount

Figure 4. DF3 Load Regulation Results

When the generator output at the 27th node in the transmission network goes up by 10 MW, Figure 5(a) shows the node load changes in the DF5 distribution network area. Figure 5(b) shows the node load changes in DF5 when the generator output goes down by 10 MW.

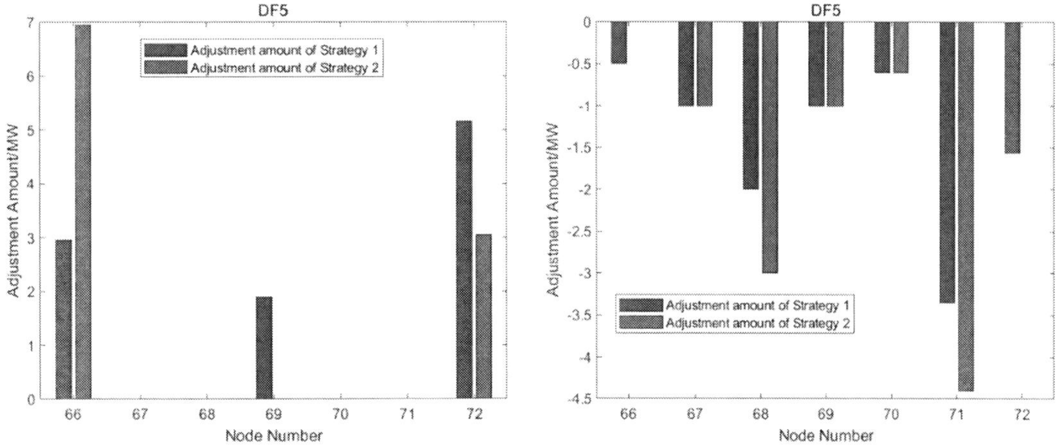

(a) Load Increase Amount	(b) Load Decrease Amount

Figure 5. DF5 Load Regulation Results

When the generator output at the third node of the T57 system is increased by 30 MW, certain node loads need to be adjusted. Table 4 below lists the load adjustments, considering the network limits, for the first period.

Table 4. Controllable loads regulation

Load node serial number	18	19	20	27	28	29	52	53	54
Strategy 1 Adjustment Amount /MW	6.95	0	0	0	3.89	6.28	6.41	6.47	0
Strategy 2 Adjustment Amount /MW	3.95	0	0	0	1.69	8.65	8.92	6.79	0

When the generator output at the third node of the T57 system is lowered by 30 MW, the load adjustments considering network constraints in the second period are shown in Table 5 below.

Table 5. Controllable loads regulation

Load node serial number	18	19	20	27	28	29	52	53	54
Strategy 1 Adjustment Amount /MW	0	-3.3	-2.3	-6.16	-4.6	-6.28	-4.9	-2.46	0
Strategy 2 Adjustment Amount /MW	-14.4	-3.3	-2.3	0	0	-10	0	0	0

The numbers in Table 5 show how the controllable loads changed after the T57-DUK distribution network received the dispatch orders. These changes help reduce network congestion in the system.

The solution time under different strategies is shown in Table 6.

Table 6. Solution time

System	Adjustable load quantity	Adopt strategy	Solution time/s
T30-DF6	4	Strategy 1/2	2.488/2.39`
	7	Strategy 1/2	3.149/2.286
T57-DUK	9	Strategy 1/2	83.512/73.129

A comparison of the results shows that strategy two is better at fixing problems quickly. Conversely, strategy one consistently utilizes the controllable loads to their maximum capacity during each scheduling period, significantly reducing network congestion within the system.

4. Conclusion

This paper aims to explore the impact of generator output changes in the transmission network on controllable load outputs in the distribution network. To enhance the optimization framework, a controllable load optimization model that is coupled with network constraints with the help of sensitivity analysis method is used. The effectiveness of this method is tested on the T30-DF6 and T57-DUK systems. However, as a result of the growing participation of renewable energy resources in the transmission and distribution grids, the interactive power flow of these systems is characterized by high uncertainty. Hence, the development of an optimal scheduling model that considers the uncertainty of new energy output will be an important topic for future study.

Acknowledgments

The Science and Technology Project of China Southern Power Grid Corporation supported the key technology research and application of coordinated regulation of "source, network and storage" based on physical information system (YNKJXM20222463).

References

[1] L. Zhu, H. Tang and D. Fang, "Modeling and Comprehensive Assessment of Flexible Resource for Power Dispatch," *2022 9th International Forum on Electrical Engineering and Automation (IFEEA),* Zhuhai, China, 2022, pp. 449-454, doi: 10.1109/IFEEA57288.2022.10038058.

[2] Z. Hui, H. Yan, B. Li, J. Zhou and X. Wu, "A Flexible Resource Planning Model Considering Balance between Power Supply and Renewable Energy Consumption," *2024 9th Asia Conference on Power and Electrical Engineering (ACPEE)*, Shanghai, China, 2024, pp. 65-71.

[3] T. Jin, Z. Gao, S. Hu, H. Sun, X. Chen and Z. Wang, "Transmission and Distribution Cooperative Congestion Scheduling Strategy Based on Dynamic Integration of Flexible Resources in Distribution Network," *2023 IEEE 7th Conference on Energy Internet and Energy System Integration (EI2)*, Hangzhou, China, 2023, pp. 1864-1869

[4] Ning Zhang, Juan Yan, Cungang Hu, Qiuye Sun, Lingxiao Yang, David Wenzhong Gao, Josep M. Guerrero, Yushuai Li, "Price-Matching-Based Regional Energy Market With Hierarchical Reinforcement Learning Algorithm", *IEEE Transactions on Industrial Informatics*, vol.20, no.9, pp.11103-11114, 2024

[5] J. Hu, X. Liu, M. Shahidehpour and S. Xia, "Optimal Operation of Energy Hubs With Large-Scale Distributed Energy Resources for Distribution Network Congestion Management," in *IEEE Transactions on Sustainable Energy*, vol. 12, no. 3, pp. 1755-1765, July 2021

[6] Yingjun Wu, Zhiwei Lin, Yijun Xu, Gianfranco Chicco, Tao Huang, Junjie Shao, Zhaorui Chen, "Demand Response Scheduling Considering Pollutant Diffusion Uncertainty of Industrial Customers", *IEEE Transactions on Sustainable Energy*, vol.15, no.1, pp.187-200, 2024.

[7] Hua Feng, Shiqi Sun, Hanbing Zhang, Jichao Ye, Ning Ding, Yonghai Xu, "A Cost-benefit Analysis Method for Multimarket Resources Considering Peer-to-peer Transactions", *2023 IEEE Sustainable Power and Energy Conference (iSPEC)*, pp.1-7, 2023

[8] Yue Fan, Wei Cui, Wei Wang, Ying Wang, "Collaborative Real-Time Operation for Long-Term and Short-Term Energy Storage: A Scheduling-Corrective Approach", *2023 IEEE 7th Conference on Energy Internet and Energy System Integration (EI2)*, pp.129-134, 2023.

[9] Ye Chang, Miao Shihong, Li Chao, et al. Collaborative optimization of transmission anddistribution two-stage power grid based on improved parallel subspace algorithm [J]. Journal of Electrotechnical Technology, 2018,33 (23): 5509-5522.

[10] Reddy S S. Multi-Objective Based Congestion Management Using Generation Rescheduling and Load Shedding [J]. IEEE Transactions on Power Systems, 2017, 32(2): 852-863

Journal of Physics: Conference Series **3012** (2025) 012044 doi:10.1088/1742-6596/3012/1/012044

A Method for Insulator Classification and Fault Detection Based on Improved YOLOv10

Jiayi Wan[1],Qinkang Pang[1],Meng Tian[1*] and Zhengcheng Dong[1]

[1] School of Automation, Wuhan University of Technology, Wuhan, China

E-mail: tm@whut.edu.cn

Abstract. With the continuous development of the power industry, the increase in electrical equipment makes power inspection increasingly important. Insulators, as key components of the power system, directly affect the safe operation of the power system. Traditional manual inspection methods suffer from low efficiency and insufficient accuracy. Therefore, this study proposes an improved YOLOv10 object detection model based on deep learning, specifically for the category recognition and fault detection of insulators. The model introduces the MultiSEAMHead module to enhance the detection capability for occluded targets and employs the normalized Gaussian Wasserstein distance (NWD) to improve the recognition accuracy of small targets. Experimental results on the insulator dataset show that the improved model achieved an mAP50 of 84.5%, reduced FLOPs to 7.8, and achieved a FPS of 93.5 frames per second, demonstrating significant advantages in detection accuracy and real-time performance. Additionally, the improved model's capability in handling small and occluded targets has been significantly enhanced. Therefore, the model has significant practical application value and broad prospects in actual power inspection.

1. Introduction

With the continuous development of the power industry, the number of electrical devices is also increasing. To ensure the normal operation of electrical equipment, power inspection schemes are particularly important. Insulators, which are key components in the power system used to support and insulate conductors, directly affect the safe operation of the power system. Insulators are exposed to the outdoors for long periods and are affected by various harsh environments. Failures such as inclined insulator string, self-explosion, and sheds damage can lead to equipment damage and personal casualties with serious consequences. Therefore, the identification of different insulators and their various faults provides important protection for subsequent processing and accident prevention.

In the field of insulator classification and fault detection, object detection technology has become a key technique due to its exceptional performance. This technology relies on deep learning algorithms, particularly Convolutional Neural Networks (CNN) [1] and Region Proposal Networks (RPN) [2]. CNN enables automatic feature learning across multiple levels which progress from basic visual elements like edges to complex semantic representations, thereby improving image recognition and detection accuracy.RPN, on the other hand, generates candidate regions directly on feature maps and performs bounding box and classification, achieving rapid and efficient generation of object candidate regions. This significantly reduces the computational burden associated with candidate region generation in traditional methods, thereby increasing the speed and accuracy of object detection.Compared to traditional techniques, this methodology shows enhanced effectiveness in classifying insulators and detecting operational faults.

During the development process of object detection models, several outstanding architectures have emerged. YOLO (You Only Look Once) [3, 4, 5] transforms the object detection task into a single regression problem, thereby significantly enhancing the inference speed and accuracy of object detection. SSD (Single Shot MultiBox Detector) [6, 7, 8]has advanced a deconvolutional approach for single-shot object detection. It improves the recognition of smaller objects by estimating bounding boxes and class probabilities within a unified network architecture. FCOS (Fully Convolutional One-Stage Object Detection) introduces a fully convolutional approach, eliminating the need for anchor boxes, which simplifies the detection process. Among these options, YOLO stands out as a widely popular one-stage model known for its rapid inference capabilities and superior accuracy. The recently introduced YOLOv10 [9] provides robust technical support for insulator detection.

However, in the task of insulator category and fault detection, issues such as small target sizes, inconspicuous features, severe occlusions and overlaps among targets still persist, affecting detection metrics. Therefore, proposing an algorithm specifically tailored for the detection of insulator targets holds significant importance. This paper proposes a series of targeted enhancements to the YOLOv10 model, which include the following aspects:

(1) Incorporating the MultiSEAMHead module, inspired by the occlusion-aware attention transformation from YOLO-Face V2 [10], to effectively address occlusion scenarios.

(2) Replacing the original loss function of YOLOv10 with NWD (Normalized Weighted Distance) [11], which is based on normalized Gaussian Wasserstein distance to improve recognition performance for small target objects.

Experimental results show that the detection accuracy of the improved module can reach 84.5%, the number of floating-point operations (FLOPs) in the neural network is reduced to 7.8, and the frames per second (FPS) reach 93.5f/s, which can effectively be applied to the tasks of insulator type identification and fault inspection.

2. Improved model

2.1. Architecture of the improved model

The YOLOv10 model incorporates a consistent dual-assignment strategy for training without the need for NMS, which markedly decreases inference latency. It also features a Compact Inverted Residual Block (CIB) structure that bolsters feature extraction and preserves more information, offering strong technical support for insulator detection. In light of the high-speed demands of power patrol tasks, the most lightweight version, YOLOv10n, has been chosen for enhancement. On this basis, to address the issues of mutual occlusion and the difficulty in accurately detecting small objects in real-time shooting, this paper proposes an improved algorithm. While retaining part of the YOLOv10n structure, this paper designs a MultiSEAMHead structure to improve the Head part of the One-to-one Head, and simultaneously uses the NWD based on the normalized Gaussian Wasserstein distance to improve the loss function. This allows the detection algorithm to achieve a balance between detection accuracy and computational complexity, effectively solving the tasks of insulator type and fault identification.

2.2. MultiSEAMHead Module

The MultiSEAMHead module utilizes the MultiSEAM (Separated and Enhancement Attention Module) mentioned in YOLO-FaceV2. This mechanism tries to boost the response from unoccluded faces to make up for the reduced response from occluded faces. Applying the MultiSEAM module to the insulator dataset can effectively address the issue of mutual occlusion

Journal of Physics: Conference Series **3012** (2025) 012044 doi:10.1088/1742-6596/3012/1/012044

between insulators and between insulators and the environment, which is challenging for identification.

On the left side of Figure 1 showcases the MultiSEAM architecture, integrating three CSMM modules that differ in size. The module outputs are processed via average pooling, then subjected to a channel expansion phase, concluding with a multiplication step to refine the feature representation. On the right side is the detailed structure of the CSMM module, which not only addresses the issue of mutual occlusion between targets but also effectively handles targets of varying sizes. This is designed to tackle the challenges in insulator inspection tasks, where there are numerous targets of different sizes and distances.

This module design aims to enhance the network's attention and capture capabilities for occluded features by meticulously processing spatial dimensions and channels.

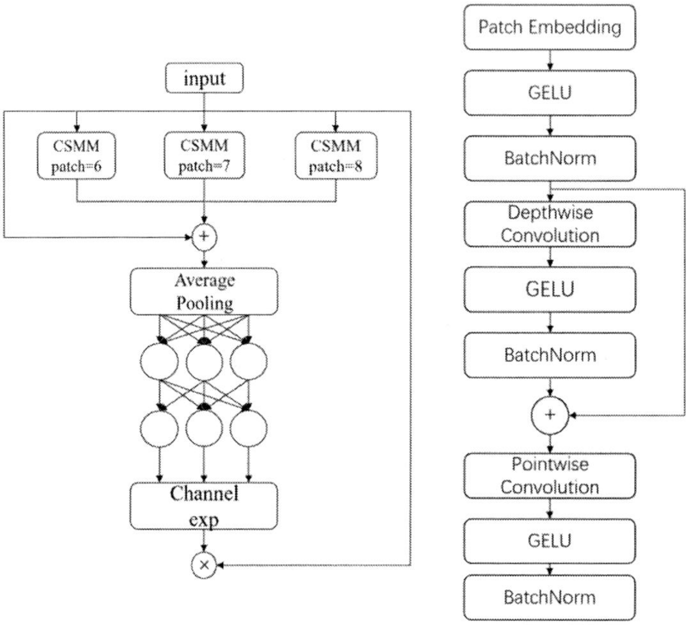

Figure 1. Architecture of MultiSEAM.

2.3. Normalized Wasserstein Distance

The standard Intersection over Union (IoU) metric is highly sensitive to objects of varying scales, with minor positional deviations leading to a significant decrease in IoU. This impact is even more pronounced for small objects. Consequently, NWD is employed, measuring the distribution distance through the Wasserstein distance derived from optimal transport theory. The NWD measure is designed as a loss function:

$$\mathcal{L}_{NWD} = 1 - NWD(\mathcal{N}_p, \mathcal{N}_g) \tag{1}$$

Here, $\mathcal{N}_p$ represents the Gaussian distribution model corresponding to the predicted box P, while $\mathcal{N}_g$ denotes the Gaussian distribution model associated with the ground truth box G.

NWD offers a significant advantage over the traditional IoU metric in small object detection by modeling bounding boxes as 2D Gaussian distributions. This approach overcomes the scale sensitivity problem inherent in IoU, providing a robust measure of similarity that is not only scale-invariant but also smooth to positional deviations. NWD effectively addresses issues that

IoU-Loss cannot, such as providing optimization gradients for the network when there is no overlap between predicted and true bounding boxes, or when the predicted box completely contains the true box. Compared to IoU, NWD demonstrates excellent characteristics in detecting small objects, including handling scale uncertainty, offering smoothness to positional deviations, and measuring the similarity between non-overlapping or inclusive bounding boxes. These attributes make NWD more suitable for small object detection, where traditional IoU metrics fall short due to their sensitivity to scale and position.

3. Experients

3.1. Experimental environment configuration

The experiments were performed on an Ubuntu 20.04 machine, utilizing the PyTorch 2.2.1 deep learning library and coded in Python 3.9. The hardware setup consisted of an RTX3090 GPU along with 30GB RAM. Training and testing were performed in the PyCharm development environment. The training parameters are shown in Table 1.

Table 1. Experimental model parameter settings.

Parameter	Value	Parameter	Value
imagsz	640×640	lr0	0.01
epochs	450	lrf	0.01
batch	32	patience	100

3.2. Dataset

The dataset is derived from manual inspections and on-site drone photography, encompassing various angles, lighting, and background conditions, which align with the diverse scenarios that may be encountered during actual power inspections. Figure 2 illustrates a dataset comprising six distinct varieties of objects for detection, which are categorized into three types of insulator detection objects: composite insulators, glass insulators, and porcelain insulators, as well as three types of insulator fault detection: inclined insulator strings, self explosion, and shed damage. These diverse data types and scenarios provide relatively rich references and training resources for power inspections.

The collection of 11,915 images is split into training and validation sets at a 9:1 ratio, with the detailed breakdown presented in Table 2.

3.3. Experimental results analysis

3.3.1. Comparison experiment of improved loss function In object detection tasks, the 'iou_ratio' parameter is used to balance the weights of the NWD loss and the traditional Intersection over Union (IoU) loss. The improved loss function L is expressed as:

$$L = \alpha \cdot \mathcal{L}_{IoU} + (1 - \alpha) \cdot \mathcal{L}_{NWD} \tag{2}$$

$$\mathcal{L}_{IoU} = 1 - IoU(A, B) \tag{3}$$

Here, $\mathcal{L}_{IoU}$ is the loss function used to measure the similarity between non-intersecting boxes. $\mathcal{L}_{NWD}$ is the loss function used to measure the similarity between non-intersecting boxes. α is the iou_ratio, with a value range of [0, 1], which balances the weights of $\mathcal{L}_{IoU}$ and $\mathcal{L}_{NWD}$.

Adjusting the 'iou_ratio' parameter can affect the model's detection capability for small targets. When there are more small targets in the dataset, reducing the 'iou_ratio' can increase

Figure 2. Dataset categories.

Table 2. Dataset distribution.

Classes	Number of training images	Number of testing images
Composite insulator	2087	232
Glass insulator	1960	218
Porcelain insulator	1786	198
Inclined insulator	1328	148
Self explosion	2398	266
Sheds damage	1165	129

Table 3. Ablation study of improved model.

iou_ratio	P/%	R/%	mAP50
0.2	86.5	77.0	83.6
0.3	87.1	76.3	83.8
0.4	87.5	77.0	84.5
0.5	85.7	77.5	83.8
0.6	85.7	77.0	82.9

the NWD measurement, which in turn boosts the detection accuracy for smaller objects. Considering that the used insulator dataset has more small target objects, the experiment was designed to increase the 'iou_ratio' from 0.2 to 0.6 to explore the impact of the coefficient 'iou_ratio' in NWD on model performance. The experimental results are shown in Table 3.

It can be observed that when the 'iou_ratio' is too small or too large, all indicators decrease. When the 'iou_ratio' coefficient is 0.4, both P and mAP50 indicators are significantly better than in other cases. Overall, it is concluded that setting 'iou_ratio' to 0.4 should be applied in subsequent experiments.

3.3.2. Ablation study of improved model To verify the effectiveness of the improved model in this paper, it was tested under the same environment and training parameters as the original YOLOv10n model. The corresponding indicators were compared as shown in Table 4. When the MultiSEAMHead module was used alone, the recall rate increased by 2.6%, and the FLOPs decreased by 0.6, showing some improvement over the original model. When the NWD loss function was introduced alone, the model's overall indicators improved, but the changes were not significant. By combining all improvements, the corresponding indicators showed further comprehensive improvement over the original model and the model with individual improvements, with mAP50 increasing by 1.4% compared to the original indicators, and FLOPs decreasing by 7%.

Table 4. Ablation study of improved model.

MultiSEAMHead	NWD	P/%	R/%	mAP50/%	FLOPs	FPS/$f \cdot s^{-1}$
×	×	86.9	75.7	83.1	8.4	135.5
✓	×	87.0	78.3	83.9	7.8	91.9
×	✓	86.1	76.9	83.5	8.4	179.6
✓	✓	87.5	77.0	84.5	7.8	93.6

3.3.3. Mainstream algorithm performance comparison experiment To explore the effect of model improvements, the same dataset was used to train YOLOv5, YOLOv6, YOLOv8, and YOLOv10 for 450 rounds, with results as shown in Table 5. Analysis of Table 5 data indicates that the improved model's mAP50 increased by 3.1%, 8.3%, 2.9%, and 1.5% compared to YOLOv5n, YOLOv6n, YOLOv8n, and YOLOv10n, respectively, making the model more accurate in detection. FLOPs decreased by 34%, 4%, and 7% compared to YOLOv6n, YOLOv8n, and YOLOv10n, respectively, indicating that our model reduces computational resource consumption while maintaining high precision, which has significant practical application value.

Table 5. Performance comparison of mainstream algorithms.

Models	P/%	R/%	mAP50/%	FLOPs	$FPS/f \cdot s^{-1}$
YOLOv5n	85.2	75.6	81.5	7.2	214.7
YOLOv6n	80.2	72.8	76.3	11.9	246.3
YOLOv8n	86.9	75.8	81.7	8.2	233.1
YOLOv10n	86.9	75.7	83.1	8.4	135.5
Ours	87.5	77.0	84.5	7.8	93.6

By comparing the specific performance of the improved model and the original model for each recognition object, the mAP50 values were statistically analyzed, with the results shown in Figure 3. The left panel of Figure 3 illustrates the detection results of the initial model, while the right panel shows the results of the enhanced model. The improved model showed an increase in mAP50 for the composite insulator, glass insulator, porcelain insulator, and sheds damage categories.In the context of the sheds damage class, which is characterized by a prevalence of small targets, mAP50 has seen an enhancement of 4.4% relative to the baseline model.

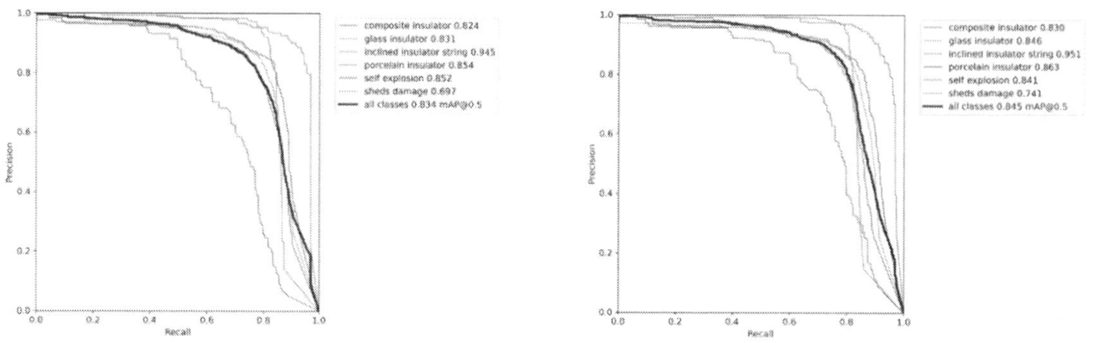

(a)The P-R curve of the original model (b)The P-R curve of the improved model

Figure 3. Comparison of mAP50 between YOLOv10n and the improved model.

3.3.4. Validation phase Accurately identifying the types and faults of insulators is essential for taking appropriate maintenance actions. Figure 4 illustrates the superior performance of our enhanced network model in specific identification tasks. The model can not only successfully detecting three different types of insulators—composite, glass, and porcelain—but also accurately recognizing three common insulator faults, including inclined insulator strings, self explosion, and shed damage. Moreover, it can simultaneously label both the fault type and the corresponding insulator category on the same image. This intuitive dual labeling greatly assists maintenance personnel in quickly understanding the fault conditions and devising more rapid and effective solutions based on this information.

As shown in the first image in Figure 4, when comparing the improved model with other mainstream algorithms, the confidence in detecting composite insulators is significantly enhanced in conditions with dim lighting, reaching 0.91 for the improved model compared to 0.67 for the YOLOv6n-based model. In the second image, the improved model shows a 10%-30% increase in detection accuracy for small targets compared to YOLOv6n. In the fourth image, while YOLOv6n missed the detection of a fault, the improved model accurately identified the location of the fault while maintaining a certain level of precision. The improved model has great detection accuracy under various conditions compared to other models. So it shows great potential for practical application.

4. Conclusion

This study addresses the challenges in the detection of insulators within the field of power inspection by proposing an improved YOLOv10 object detection model. The model integrates the MultiSEAMHead module and the NWD loss function, significantly improving the detection performance for small and occluded targets. The experimental results indicate that the improved model achieved an mAP50 of 84. 5% and reduced the FLOPs to 7.8 GFLOPs. The improved model shows a clear advantage in the case of occlusions between insulators and their environment, significantly enhancing the detection ability for occluded targets. In addition, the improved model can accurately identify different types of insulators and their faults, and label the fault type and the corresponding insulator category on the same image. It can provide valuable fault information for maintenance personnel, helping them quickly formulate repair plans and reduce the risk of accidents. Therefore, the improved YOLOv10 model offers considerable practical significance and demonstrates extensive application potential in the field of power system inspection.

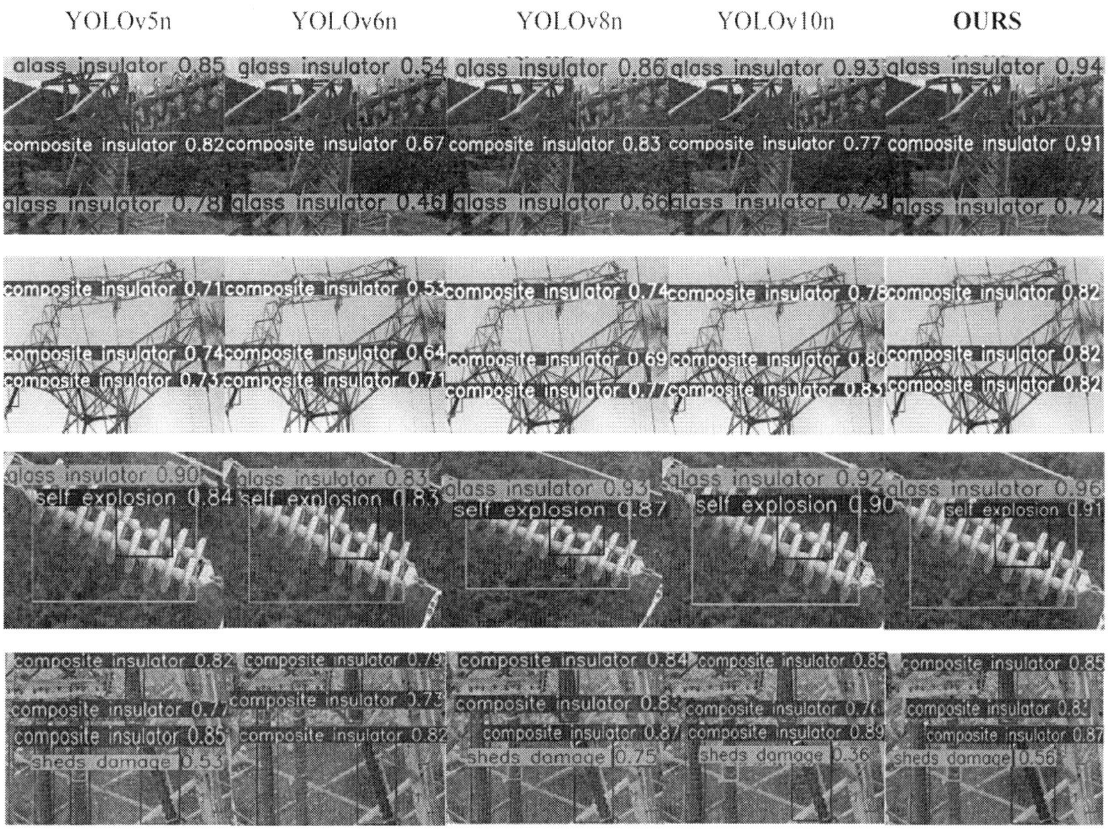

Figure 4. Validation results of different algorithms.

References

[1] Li Z, Yang W, Liu F and Peng S 2021 A survey of convolutional neural networks: analysis, applications, and prospects *IEEE transactions on neural networks and learning systems* **33** 6999-7019.

[2] Ren S, He K, Girshick R and Sun J 2016 Faster R-CNN: Towards real-time object detection with region proposal networks *IEEE transactions on pattern analysis and machine intelligence* **39** 1137-1149.

[3] Redmon J, Divvala S, Girshick R and Farhadi A 2015 You only look once: unified, real-time object detection *2015 IEEE Conf. on Computer Vision and Pattern Recognition (CVPR)* (Las Vegas) 779-788.

[4] Bochkovskiy A, Wang C Y and Liao H Y M 2020 Yolov4:optimal speed and accuracy of object detection *arXiv preprint arXiv:2004.10934*.

[5] Kaur R and Singh S 2023 A comprehensive review on object detection with deep learning *Digital Signal Processing* **132**.

[6] Liu W, Anguelov D, Erhan D, Szegedy C, Reed S, Fu C Y and Berg A C 2015 Ssd: single shot multibox detector *Computer Vision-ECCV 2015* (Amsterdam: Springer).

[7] FU C Y, Liu W, Ranga A, Tyagi A and Berg A C 2017 Dssd: Deconvolutional single shot detector *arXiv preprint arXiv:1701.06659*.

[8] Li Z, Yang L and Zhou F 2024 Fssd: feature fusion single shot multibox detector *arXiv preprint arXiv:1712.00960*.

[9] Wang A, Chen H, Liu L, Chen K, Lin Z, Han J and Ding G 2024 Yolov10: real-time end-to-end object detection *arXiv preprint arXiv:2405.14458*.

[10] Yu Z, Huang H, Chen W, Su Y, Liu Y and Wang X 2022 Yolo-facev2: a scale and occlusion aware face detector *arXiv preprint arXiv:2208.02019* .

[11] Wang J, Xu C, Yang W and Yu L 2021 A normalized gaussian wasserstein distance for tiny objection detection *arXiv preprint arXiv:2110.13389*.

Journal of Physics: Conference Series **3012** (2025) 012045 doi:10.1088/1742-6596/3012/1/012045

Effect of metal particle attachment on electric field distortion of high voltage AC GIL post insulators

Chenxi Fu*, Jiyu Bi, Haoran Ding, Daochun Huang

School of Electrical Engineering and Automation, Wuhan University, Wuhan, China

*Corresponding author's e-mail: 1986674656@qq.com

Abstract: The tri-post insulator is an essential insulating component in an insulated Transmission Line (GIL). The tri-post insulator is susceptible to defects and failures due to various factors. These factors include the adhesion of metal particles and even cracking and breakdown. To analyze the effect of metal particles adhering to the insulator surface on electric field distortion, this essay will establish a GIL 1:1 electrostatic field model simulation. The model will be based on finite element simulation and adopt the control variable method . We control the irrelevant variables to be the same and optimal. This helps analyze the distortion ability of metal particles' shapes, placing positions, and suspension heights on the electric field. The study aims to explore the failure mechanism of their flashover along the surface. It also draws conclusions based on the analysis. The results can help optimize the GIL tri-post insulator. They provide theoretical guidance for on-site maintenance and accident analysis. Additionally, the study offers a reference for optimizing GIL's structural and electrical performance. It also supports intelligent operation and maintenance.

1. Introduction

In the 1970s, GIL entered the product practicality stage as an independent technology, such as a GIS lead wire for power plants and substations. Long-distance GIL transmission lines have been built in North America (United States, Canada), Europe (Germany) and Asia (Japan). AZZ, ABB, Siemens and Alstom have mastered the manufacturing technology of GIL and successively developed GIL products of 550 kV and below. Japan in Asia developed and operated in 2000 between the new Nagoya thermal power plant and the East China Sea substation, the world's most extended double-circuit total length of 3250 meters of 275 kV GIL. 2004, the first 550kV GIL was applied to the Pingyao Ultra-high Voltage Substation of Hangzhou Electric Power Bureau. In 2013, seven 550kV GILs were applied at Xiluodu Hydropower Station in Yunnan Province. Among them, 475 meters and 480 meters of vertical high drop structure GILs were installed on the left and right banks, and the highest voltage level, 800kV GIL, was applied for the first time in Laxiwa Hydropower Station on the Yellow River[1].

China has built the world's largest and highest voltage level AC-DC hybrid power grid, the main transmission network for traditional overhead lines[2]. In today's international community, clean energy and "double carbon emission reduction" goal, especially the rapid development of large-scale photovoltaic power plants and offshore wind farms, the grid transmission voltage level of the new requirements. The GIL high-voltage transmission lines can reduce the heat loss caused by current transmission and reduce the material cost of power transmission; with large transmission capacity, ease of installation, high reliability and other advantages, it has been gradually in the field of large-capacity, long-distance power transmission has more applications [3].

As a key part in GIL insulation, the tri-post insulator plays the role of electrical insulation and mechanical support ,and is affected by a variety of factors ,such as electric field, heat, particles, etc. [4].They are prone to produce insulation defects, reduce insulation performance, and are prone to sudden flashover along the surface, which can lead to safety accidents. According to the failure statistics of GIL of State Grid in the past 5 years, the failure caused by insulation failure accounts for 74% of all the shortcomings of GIL. Insulation failure caused by metal particles leads to 38% of insulation failures[5].On September 29, 2016, a station's 500 kV third central group AC filter Waveform 8M bus protection tripped. The site was disassembled and inspected, and it found that the point of failure of the tri-post insulator left the lower pillar broken and scattered to the side of the tee tank; the inability of the tri-post where the shell particle trap deformation from the site disassembled situation, the failure of the insulator outer surface ablation is relative.From the on-site disassembly situation, the outer surface of the faulty insulator is lightly rubbed, and the cracked surface is more seriously abraded.Internal breakdown of the tri-post insulator, tri-post insulator aluminium metal surface and the insulator between the existence of minor gap defects in the production, transportation, installation, operation and other processes of the gradual expansion of small gaps, deterioration, and ultimately lead to the breakdown of the discharge.

Therefore, it is of great significance to study the influence of insulator insulation performance, such as the shape, position and suspension height of metal particles, to improve the insulation performance of GIL and ensure the safety of power transmission.

2. Model description

2.1. Model parameters

GIL is composed of standard straight line segment, connecting head and so on. The internal conductor of GIL is generally designed with a hollow tube type, and its support mainly uses one or more insulators. During the normal operation of GIL equipment, SF6 is often used as insulating gas, and the electric field distribution of the tri-post insulator is extremely uneven. The normal field strength is concentrated in the abdomen area, and the tangential field strength is concentrated in the leg. The interface between the metal insert and the epoxy resin is large. This uneven distribution will cause the local charge accumulation of the tri-post insulator to be more serious, which is easily affected by the attached metal particles and causes insulation flashover. The GIL model is shown in Figure 1.

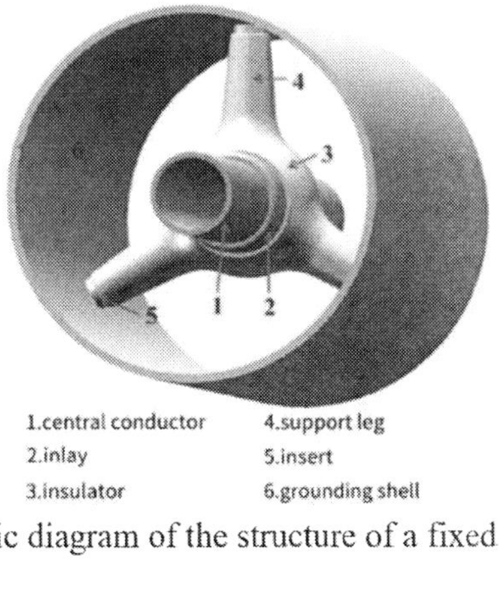

Figure 1 Schematic diagram of the structure of a fixed tri-post insulator

The end structure components of the typical fixed tri-post insulator are particle isolator, aluminum ring, screw hole, fixed welding plate, insulation pad, and fixed nut. The particle isolator and the fixed welding plate are fastened to the screw holes on the integrally cast insulation parts by the fixed nut, which is mainly used to support and fix the standard unit conductor, and the equipotential connection with the GIL shell is realized by the welding of adjacent components [6]. Based on the reading of tri-post insulator insulator modeling papers in recent years, the key parameters of tri-post insulator insulator modeling are summarized, including the parameters of the ventral zone, the back zone, and the transition zone. Therefore, in the modeling of 500kVGIL1: 1, the low-voltage bending plate, shielding cover, conductor, and other parts with less influence on the calculation results are simplified, and the particle trap is ignored.

2.2. *Material setting*

To ensure the accuracy of the simulation results, the critical contour curves of the tri-post insulator legs, bellies, and transition zones are restored during the modeling process. Table 1 shows the key parameters based on which the modeling is performed.The insulator is a commonly used epoxy-alumina composite material, and its dielectric constant is 1, so the relative dielectric constant of the insulator material is defined as 6.5. The dielectric constant of sulfur SF6 and metal is 1, and the simulation is defined as the relative dielectric constant of the material properties of the inner airbag,and shell. The conductor structure is 1 [7].

Table 1 Key parameters for modeling tri-post insulator

Parameters	(be) worth	Parameters	(be) worth
Maximum radius of insulator center web	131mm	Outrigger radius	60mm
Insulator abdomen width	220mm	Distance from the center of the leg circle to the center of the insulator	361.84mm
Conductor Radius	116mm	Rounded corners of legs and abdominal transition area	80mm

2.3. *Particle parameters*

Common metal particles in GIL are spherical, bar-shaped and sheet-shaped. According to the actual situation of the standard size of the metal particles present in GIL, the experimental setup of sheet-shaped metal particles with a length of 5mm, a width of 1mm, and a height of 0.5mm, the diameter of the upper and lower two hemispheres of the bar-shaped metal particles is 1mm, and the height of the cylinder that connects the two hemispheres is 10mm; the size of common spherical metal particles is in the range of 0.5-2.0mm in diameter. The particle model is shown in Figure 2.The shape, position, and suspension height of the metal particles will have different degrees of influence on the electric field distribution on the surface of the post[8].

According to the structural characteristics of the tri-post insulator, four typical positions are extracted to place the particles, which are at the legs of the tri-post insulator, the abdomen, the frontal transition position, and the transition position of the abdomen of the right leg of the insulator. As shown in Figure2, the surface shape of the four typical positions is very different, so the metal particles in four different positions may have different distortion ability to the electric field distribution on the insulator surface.

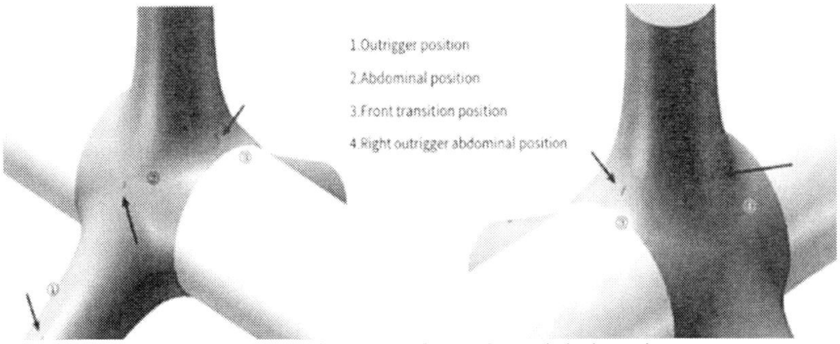

Figure 2. Distribution of metal particle locations

3. Calculation results and analysis

3.1. The effect of particles' position

Four spherical particles with a diameter of 0.5 mm are placed at four typical locations. Their suspension height is set to 0.25 mm, where the spherical particles are set on the left side of the abdomen of the tri-post insulator.The voltage is loaded and the electric field simulation is carried out and the simulation results are shown in Figure 3.From the figure, it can be seen that particles with a diameter of 0.5 mm have the strongest ability to distort the surface electric field of the tri-post insulator at the left abdomen position.The surface electric field strengths of four spherical particles are shown in Figure 4.

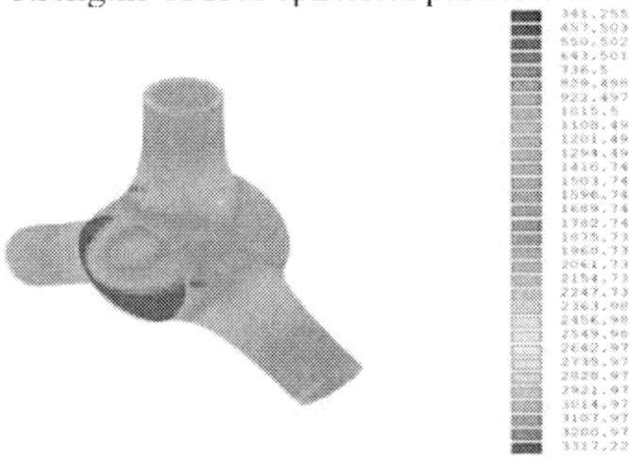
Figure 3 Diameter 0.5mm spherical particles suspended at 0.25mm height

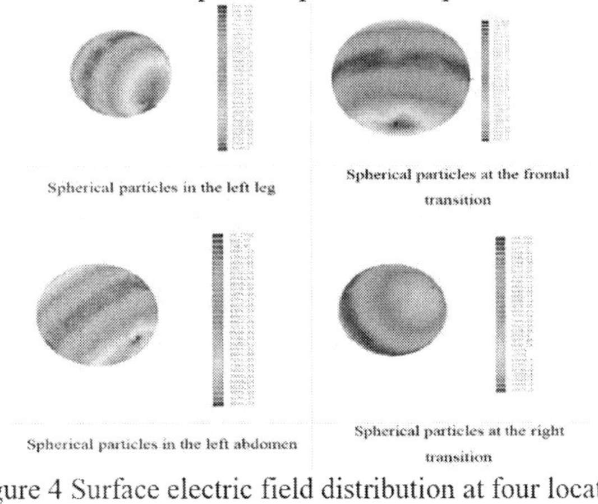

Figure 4 Surface electric field distribution at four locations

By comparing the four spherical particles' surface electric field strength maximum value, it can be seen that the insulator's left abdomen at the electric field strength is the largest, but the left abdomen at the placement of the spherical particles of their own surface electric field strength maximum value is not the largest of the four particles, the largest but the right transition at the particles, so the initial speculation that the particles of their surface electric field strength of the maximum size of the corresponding three posts of the insulator surface position of the electric field strength .The value of the electric field strength on the surface of the particle itself and its corresponding electric field strength at the surface position of the tri-post insulator do not have a single linear relationship, which will be confirmed later in the simulation[9].Then the sheet particles are placed in four locations to control the suspension height of 0.25 mm. The simulation results are shown in Figure 5.As can be seen from the figure, the maximum electric field strength of the tri-post insulator occurs at the frontal transition.

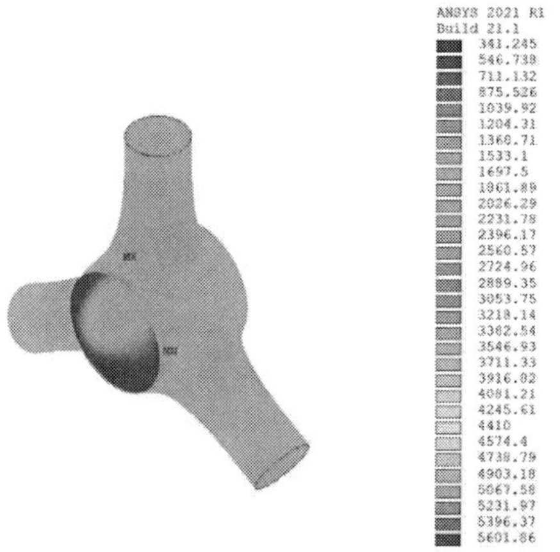

Figure 5 Flake metal particles suspended at a height of 0.25mm

For particles of different shapes, the maximum value of the electric field strength on the surface of the tri-post insulator appears at different positions. The possible reason for this is that the surface shape is different at different positions, so for the same shape of particles, their ability to distort the electric field strength on the surface of the insulator at different positions is different, and at the same position, different shapes of particles due to the differences in the shape of the particle surface will also show different ability to distort the electric field.[10]

Therefore, for the same shape of the particles, its ability to distort the electric field at different locations should have the same law.The larger the particle size, the greater the maximum value of the surface. To verify this, Set up the following simulation experimental group: In each of the four typical locations of 1 mm diameter spherical particles, control the suspension height of 0.25 mm in the 500 kV electric field under the simulation. It can be seen that the spherical particles with a diameter of 1.0 mm, when placed at four different locations, exhibit the same pattern of aberration ability to the surface electric field at different locations.The above simulation results are organized as shown in Table 2.

From the data in the table, it can also be seen that the spherical particles, strip particles, and sheet particles have very different laws for the distortion ability of different positions of the electric field, while the strip particles and sheet particles have similar laws, which may be related to the shape of the strip particles and sheet particles being similar.

Table 2 Effect of particulate position on electric field distribution

Experimental group	*Electric field strength of the insulator surface(kv/mm)*			
	DIA0.5mm Spherical particles	*Bar particles*	*Tablet particles*	*DIA1.0mm Spherical particles*
Left leg	2.36	8.61	4.41	3.03
Left Abdomen	3.32	5.12	4.08	5.25
Positive transition	2.64	7.51	5.60	3.35
Location of maximum field strength	*left abdomen*	*left leg*	*positive transition*	*left abdomen*

The following conclusions can be obtained from the above simulation results:

1) The particle position has great influence on the particle's ability to distort the electric field on the insulator surface.

2) For spherical particles, their ability to distort the electric field is the strongest when they are in the abdomen; for bar-shaped metal particles, their ability to distort the electric field is the strongest when they are in the left leg; for lamellar metal particles, their ability to distort the electric field is the strongest when they are in the frontal transition between the leg and the abdomen.

3) The structure of the sheet-shaped particles and the strip-shaped particles have certain similarities, and their ability to distort the electric field at different locations has certain similarities.

4) There is no single linear relationship between the size of the maximum value of the electric field strength on the surface of the particle itself and the size of the maximum value of the electric field strength at its corresponding position on the surface of the tri-post insulator.

3.2. Effect of particle shape on the electric field distribution on the surface of tri-post insulator

Organizing the above simulation results can be obtained in Table 3, the experimental group only changed the shape of the particles and controlled their suspension height constant, both 0.25mm.The following conclusions can be drawn from the data in the table and the graphical results:

1) Bar-shaped particles have the strongest ability to distort the electric field on the surface of the tri-post insulator, and spherical particles are the weakest

2) The maximum value of the electric field strength on the surface of different particles is also very different. Compared with spherical metal particles and sheet particles, bar-shaped particles have greater influence on the electric field strength on the insulator surface, and the electric field strength on their surfaces is also larger, which may cause gap discharge, leading to the insulator surface local field strength being too large and insulation failure.

Table 3 Effect of particulate shape on electric field distribution

Experimental group	***Maximum field strength (kV/mm)***			
	Left leg	***Left abdomen***	***Frontal transition***	***Right transition***
Spherical particles	2.36	3.32	2.64	3.11
Bar particles	8.61	5.12	7.51	7.25
Flaky particles	4.41	4.08	5.60	4.90
Shape of the particle when	*Bar particles*	*Bar particles*	*Bar particles*	*Bar particles*

the maximum occurs				

The electric field distribution on the surface of different particles themselves is shown in Figure 6

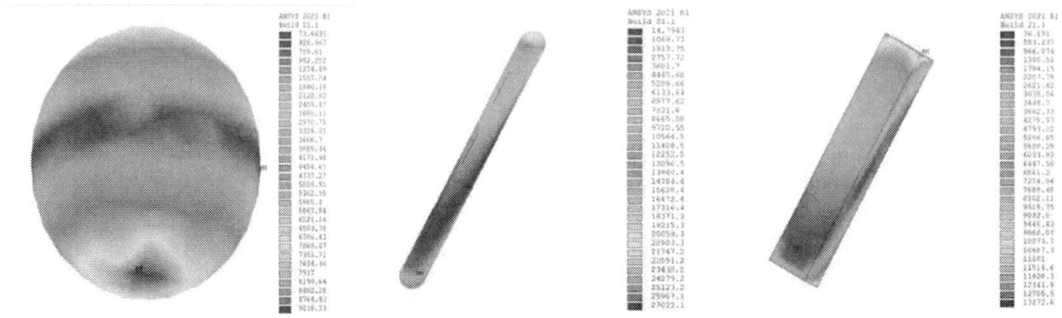

Figure 6 Distribution of electric field on the surface of strip particles

3.3. The effect of particle suspension height on the surface electric field distribution of tri-post insulator

Explore the influence of particle suspension height on the surface electric field distribution of the three posts and the selection of the tri-post insulator surface electric field distribution of the strongest ability to influence the bar insulator, control its position for the left leg, and change the particle suspension height. Different suspension heights of the strip particles on the tri-post insulator surface electric field distribution of the influence of the tri-post insulator are shown in Figure 7, and different suspension heights of the strip particles on the surface electric field distribution are shown in Figure 8.

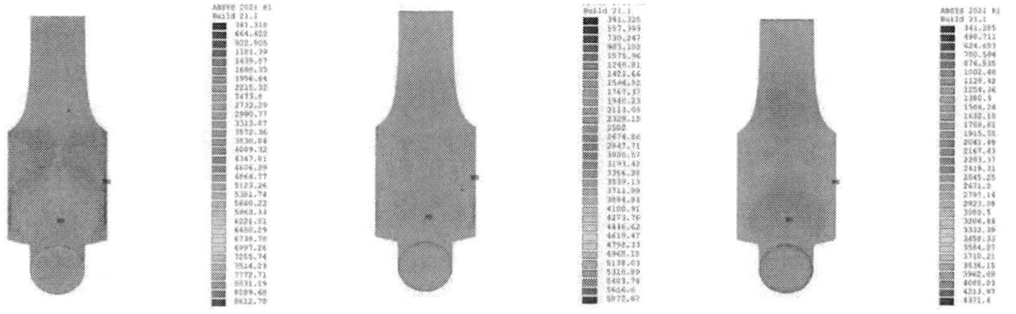

0.25mm 0.50mm 0.75mm

Figure 7 Electric field distribution of tri-post insulator

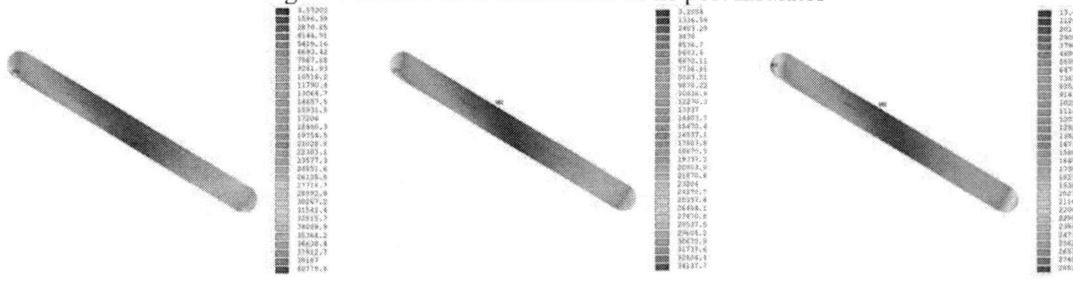

0.25mm 0.50mm 0.75mm

Figure 8 Electric field distribution on the surface of the bar microparticles

The following conclusions can be obtained from the above simulation results:

1) The effects of bar particles with different suspension heights on the electric field distribution on the surface of tri-post insulator are very different, and the larger the suspension height is, the smaller its effect on the electric field distribution on the insulator surface is.

2) The larger the suspension height, the smaller the strength of the electric field on the surface of the bar particles themselves.

4. Conclusion

This article explores the distribution of electric field on the attached surface of metal particles of tri-post insulator, according to the different forms of defects that exist, that is, according to the shape, size, suspension height, and location of particles. To obtain the distribution of defects that exist on the surface of the particles of the field strength, through the simulation results, the analysis can be made clear that the shape of the particles, the location, the size of the particles, and the suspension height of the impact on it are different. The larger the particle size, the greater the maximum value of the surface electric field intensity of the particle itself.

1)The larger the particle suspension height, the smaller the influence on the electric field distribution on the insulator surface. When the suspension height is greater than a certain value, it can be considered that the metal particles have little effect on the electric field distribution on the surface of the insulator and do not cause surface discharge.

2)The larger the suspension height, the smaller the surface electric field intensity of the particles themselves.

3)The strip particles have the most robust distortion ability to the surface electric field of the tri-post insulator, and their surface electric field strength is also significant, which may cause gap discharge, resulting in excessive local field strength on the surface of the insulator and insulation failure.

Attention should be paid to the prevention and control of strip particles in the prevention and control of particles.

References

[1] Fei Ye. Research on temperature rise characteristics and ultrasonic fault location technology of UHV GIL [D].North China Electric Power University (Beijing).

[2] Zhang Yunke. Protection for HVDC Transmission Line and Filters in AC-DC Hybrid grids[D]. Tianjin University, 2021.

[3] Chen Jing, Zang Chunyan, GONG Yulu, et al. Research Review on Tri-post Insulator for EHV/UHV GIL[J]. High Voltage Apparatus,2024,60(02):143-155.

[4] Jia Yunfei, GAO Lu, KI Shengchang, et al. Electrical Performance Optimization for Tri-Post Insulator on 1100KV Gas-Insulated Metal-Enclosed Transmission Line[J]. Journal of Xi'an Jiaotong University, 2020, 54(7): 168-179.

[5] Liu Peng, Wu Zehua, Zhu Sijia, et al. Simulation on Electric Field Distribution of 1 100kV AC Tri-Post Insulator Influenced by Defects [J]. Journal of Electrotechnology, 2022, 37(02) : 469-478.

[6] Shao Qianqiu, SUN Paitao, SIMA Wenxia, et al. Effect of a Copper Particle on DC Surface Flashover Characteristics of Epoxy Resin[J]. High Voltage Technology,2018,44(05):1651-1659.

[7] Yao Yongqi, WANG Zhijun, CHEN Sheng, et al. Optimization Design on Insulating Spacer and Its Assembly Structure of 550 kV GIL[J]. High Voltage Electrical Apparatus, 2021, 57(05):50-57..

[8] Li Peng, Yan Xianglian, Wang Hao, et al. Research and Application of UHVAC Gas-Insulated Transmission Line[J]. Grid Technology,2017,41(10):3161-3167.

[9] Han Shuai, GAO Fei, LIAO Sizhuo, et al. Review on Surface Defects and Diagnosis Methods of

Basin-type Insulator in GIS[J]. Insulation Materials,2022,55(02):12-22.
[10] Chen Jing, Wang Dianlang, Pi Tianman.Study on multifield coupled finite element modeling and solution of 500 kV GIL[J]. Science and Innovation, 2024,(16):6-9.

Energy Reconfiguration-based Multi-Objective Predictive Power Control Strategy for Reconfigurable Series Power System

Zhiheng Ai, Xu Yang [*], Hangjie Cen, Yi'ao Liu

School of Mechanical Engineering, Beijing Institute of Technology, Beijing 100081, China

[*]Corresponding author's e-mail: waiiicha@outlook.com

Abstract. The reconfigurable heavy-duty vehicle (RHV) is an automated heavy-duty flat transporter designed with reconfigurable features to boost cargo capacity. The reconfigurable series power system (RSPS) serves as the power source for the RHV. The reconfiguration process of the RHV involves not only the mechanical coupling but also the integration of the RSPS units, requiring consistency in the state of charge (SOC) of the battery packs. This paper establishes a mathematical model of the RSPS, considering relatively constant auxiliary power loads during the reconfiguration process and a predefined longitudinal velocity sequence. An adaptive multi-objective predictive power control strategy is proposed to balance and optimize multiple power control objectives during the reconfiguration process, including the fixed variation of SOC, velocity tracking, and fuel consumption, with fluctuations in the power requirements of vehicle auxiliaries taken into account. Simulation results indicate that the proposed strategy can reach the specific SOC and track the longitudinal velocity while enhancing fuel economy under certain reconfiguration conditions, with a deviation in the fixed SOC variation of 8% and a reduction in engine power fluctuations by 48.2%.

1. Introduction

The reconfigurable heavy-duty vehicle (RHV) is an automated heavy-duty flat transporter, belonging to the category of self-reconfiguring ground vehicles [1] which can be connected in large numbers through reconfigurable docking devices to multiple cargo capacity. The energy systems of multiple RHVs undergo "energy reconfiguration" via the electrical connectors of the reconfiguration docking device as the mechanical systems are reconfigured to form a larger power source capable of driving higher-tonnage vehicles, as illustrated in Figure 1(a). The reconfigurable series power system (RSPS) of RHV adopts a typical series hybrid vehicle powertrain widely used in heavy-duty vehicles, including the engine generator set (EGS) and battery pack [2]. The energy reconfiguration entails balancing the output voltages of various energy storage units to establish interconnection, presenting a challenge to its power control strategy (PCS) for the fixed variation of the battery pack's state of charge (SOC), which maps the output voltage, to attain a specific value.

Implementing PCS enables the efficient coordination of dual-power sources, ensuring stable power output and achieving fuel savings [3]. Existing PCS can be classified into Rule-based PCS (RB-PCS) and optimization-based algorithms [4]. RB-PCS is simple and easy to implement but often results in poor system performance and significant oscillations [5]. Optimization-based PCS can further be classified into online PCS and offline PCS. Dynamic programming and the Pontryagin maximum principle are typical offline optimization algorithms that can achieve optimal control results. However, they require high computational resources and necessitate certainty in driving cycles [6]. Online optimization methods, such as the equivalent consumption minimization strategy (ECMS) and model predictive control (MPC), are more commonly used. Both ECMS and MPC can integrate longitudinal speed tracking of the vehicle and allocate power to reduce goals like instantaneous equivalent fuel consumption or energy consumption [7], [8]. In addition, FMPC was developed to control the reconfiguration of multiple power sources, proving that the MPC method can effectively achieve the target of energy reconfiguration [1]. However, the power consumption model of the auxiliary systems should be integrated into the system model for more accurate prediction in MPC-based PCS [9]. Prevalent PCS methods mainly focus on optimizing fuel economy and maintaining the battery's SOC level, with limited research on the fixed variation of SOC.

Therefore, it is crucial to develop a PCS for RHV that enhances fuel economy and meets the fixed variation of SOC and reconfiguration process-related auxiliary power demand of energy reconfiguration. The key contributions of this paper are outlined below: 1) Development of a mathematical model of the RSPS, incorporating the power consumption of vehicle auxiliaries and longitudinal velocity tracking. 2) Formulation of an optimization problem for energy reconfiguration, considering fixed SOC variation, velocity tracking, and fuel consumption. 3) Introduction of an adaptive multi-objective model predictive control (AMOMPC) methodology to effectively optimize the power flow during the energy reconfiguration process.

The structure of this paper is as follows: In Section 2, the mathematical model of the RSPS is established. Section 3 proposes an AMOMPC strategy for energy reconfiguration. The results of the simulation are discussed in Section 4. Section 5 summarizes the conclusions of the study.

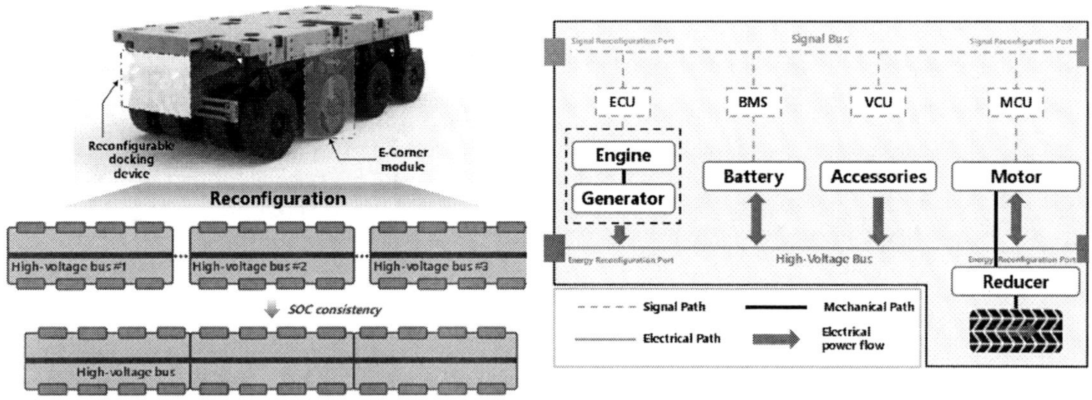

(a) Concept of RHV and reconfiguration (b) Architecture of RSPS

Figure 1. Diagram of the studied system.

2. Modeling of RSPS

RSPS constitutes the power system of the RHV, which is a typical series hybrid energy system as shown in Figure 1(b). This section presents the development of a control-oriented model for the RSPS, with auxiliary power considered as a parameter to describe the power flow between the load and power sources and the longitudinal vehicle dynamics.

2.1. *The Power flow of RSPS*

In the series hybrid configuration, the RSPS model is designed to encompass all electric power demands of the RHV actions during the reconfiguration docking process. This includes the power demands of the electric motor for driving and braking, the static auxiliary of the vehicle, the hydraulic suspension system for leveling, and the hydraulic steering system for all-wheel steering.

On the load side of the RSPS model, the total power demand consists of the cumulative auxiliary power demand P_{aux} and the electric motor power demand P_{mot}. The cumulative auxiliary power demand primarily consists of the static auxiliary power demand of the vehicle $P_{\text{aux,veh}}$, the power demand of the hydraulic suspension system $P_{\text{aux,sus}}$, and the power demand of the hydraulic steering system $P_{\text{aux,str}}$. Here, $P_{\text{aux,veh}}$ is considered constant, while $P_{\text{aux,sus}}$ and $P_{\text{aux,str}}$ are time-dependent functions reflecting the primary operating conditions of the RSPS during reconfiguration. P_{mot} corresponds to the electric power consumed by the electric motors, which can be derived indirectly from the motor speed and torque.

On the source side, the power is supplied by the battery pack P_{bat} and the EGS P_{EGS}. The EGS outputs the required power based on control inputs, while the battery pack functions as a passive charging/discharging unit, with its power determined by P_{EGS}, P_{mot}, and P_{aux}.

The power flow of RSPS can be expressed as:

$$P_{\text{mot}} + P_{\text{aux}} = P_{\text{bat}} + P_{\text{EGS}} \tag{1}$$

$$P_{\text{aux}} = P_{\text{aux,veh}} + P_{\text{aux,sus}}(t) + P_{\text{aux,str}}(t) \tag{2}$$

2.2. RHV Longitudinal Dynamics

During the reconfiguration docking process, the RHV utilizes its electric drive and regenerative braking systems to perform acceleration and deceleration maneuvers. The electric motor generates torque, which is transmitted to the wheels through the gearbox, serving as the primary source of torque for the vehicle. Given the RHV's omni-directional steering capabilities, the velocity tracking for the reconfiguration docking process can be effectively represented using a longitudinal dynamics model. This method allows for an in-depth analysis of the vehicle's longitudinal dynamics and energy allocation during the reconfiguration phase. Accordingly, a simplified longitudinal vehicle model is formulated as follows:

$$\delta m \dot{v} = \frac{1}{R}\sum_{i=1}^{8} i_g T_{\text{mot},i} - \left(f_r mg\cos\theta + mg\sin\theta + \frac{1}{2}C_D A\rho v^2\right) \tag{3}$$

Where $T_{mot,i}$ is the output torque of the i-th motor, m, g, f_r, θ, C_D, A, ρ, i_g, R, δ, and v represent the RHV mass, the gravitational acceleration, the rolling resistance coefficient, the road incline angle, the air resistance coefficient, the frontal area, the air density, the reducer ratio, the driving wheel radius, the rotational inertia coefficient, and the vehicle velocity respectively.

2.3. Electric Motor Model

The RSPS achieves velocity tracking through the motor's electric drive torque control, with the motor serving as the primary load of the RSPS. The drive motor model is as follows:

$$\omega_{\text{mot},i} = \frac{v}{i_g R} \tag{4}$$

$$P_{\text{mot}} = \sum_{i=1}^{8} \eta_{\text{mot},i}^{-\text{sgn}(T_{\text{mot},i})} T_{\text{mot},i}\omega_{\text{mot},i} = \eta_{mot}^{-\text{sgn}(T_{\text{mot}})} T_{\text{mot}}\omega_{\text{mot}} \tag{5}$$

Where $\omega_{\text{mot},i}$ is the rotational velocity of the i-th motor, and η_{mot} is the motor efficiency. Assuming the vehicle's total torque is distributed evenly, only the total torque T_{mot} of the eight motors needs to be considered.

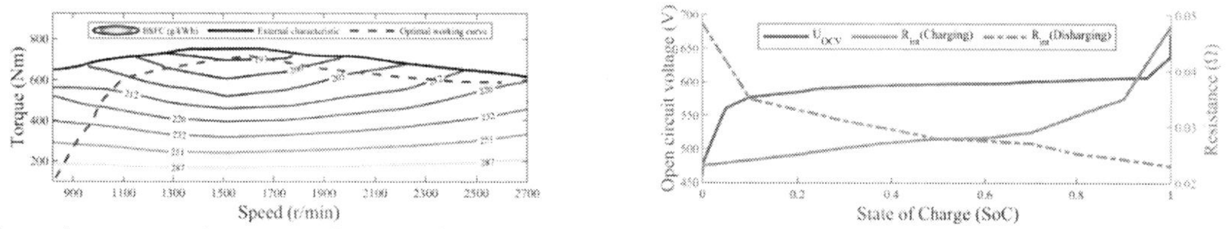

(a) Fuel consumption map of the engine. (b) OCV and internal resistance of battery pack.

Figure 2. Parameters of the RSPS model.

2.4. EGS Model

The EGS integrates the engine and the generator, which are rigidly connected. Power commands are directly transmitted to the controller of the EGS as it controls the cooperative operation of the engine and generator to output the corresponding power. The brake-specific fuel consumption (BSFC) is obtained from the optimal working curve calibrated in advance as is shown in Figure 2(a). During each sampling period, the instantaneous fuel consumption rate $\dot{Q}$ can be determined as follows:

$$\dot{Q} = -\frac{P_{\text{EGS}} f_{\text{BSFC}}(P_{\text{EGS}})}{367.1\rho_{\text{f}} g \eta_{\text{gen}}} \tag{6}$$

Where f_{BSFC} represents the mapping functions for BSFC, η_{gen} is the generator efficiency, P_{EGS} is the output power of the EGS, and ρ_{f} is the fuel density.

2.5. Battery Pack Model

This paper uses the internal resistance model utilized for battery electrothermal modeling [10] to describe the battery pack, as the battery pack of RSPS is composed of 190 battery cells connected in series. The balancing between individual battery cells is not the focus of this paper. In this model, the open-circuit voltage (OCV) U_{ocv} and internal resistance R_{int} of the battery are represented as functions of SOC, obtained through offline experimental tests, as depicted in Figure 2(b).

$$U_{\text{ocv}} = f_{\text{ocv}}(SOC) \tag{7}$$

$$R_{\text{int}} = f_{\text{int}}(SOC) \tag{8}$$

Where f_{ocv} and f_{int} represent the mapping functions for OCV and internal resistance, respectively.

Based on the internal resistance model, the battery model is established as follows:

$$I = \frac{U_{\text{ocv}} - \sqrt{U_{\text{ocv}}^2 - 4P_{\text{bat}}R_{\text{int}}}}{2R_{\text{int}}} \tag{9}$$

$$SOC(t) = SOC_0 - \frac{\int I\,\text{d}t}{C_{\text{n}}} \tag{10}$$

Where I is the battery current, C_{n} is the battery capacity, and SOC_0 is the initial SOC.

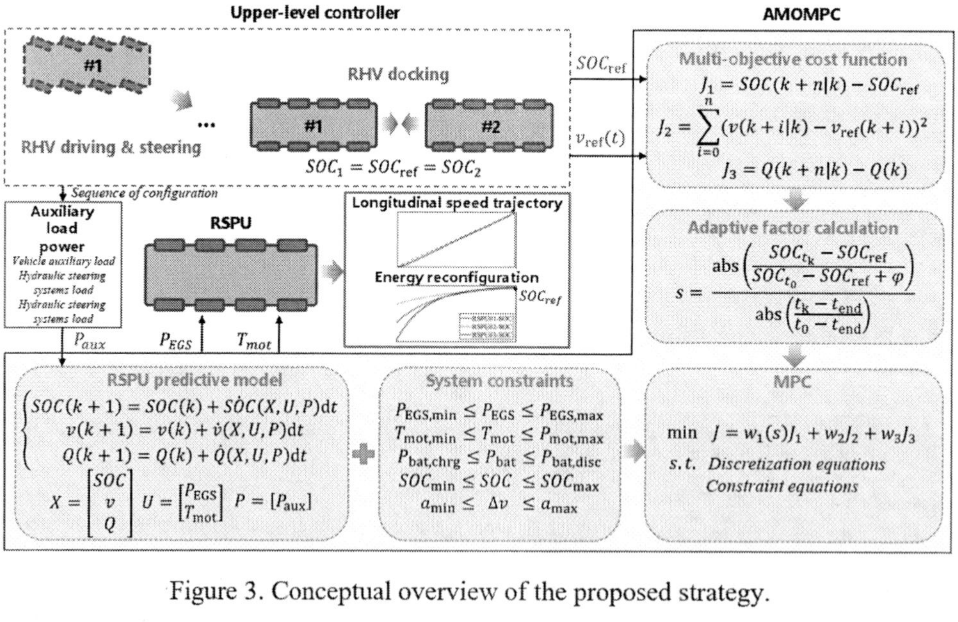

Figure 3. Conceptual overview of the proposed strategy.

3. Proposal of Power Control Strategy

An AMOMPC strategy is introduced to address the power control of the RHV throughout the reconfiguration process. The PCS developed is part of the vehicle control unit. The reference of velocity, SOC, and auxiliary power demand can be seen as given by the upper-level controller. The conceptual overview of the developed strategy is shown in Figure 3. First, the MPC problem is structured to comprehensively describe the PCS goals of the RSPS during the reconfiguration process. Then an AMOMPC strategy is proposed to solve the studied problem.

3.1. Predictive Model Formation

The motion planning for the RHV automated reconfiguration docking process is handled by the upper-level controller, which orchestrates a series of actions, including driving, remaining, steering, and leveling, while also determining the SOC target for energy reconfiguration. These coordinated actions are translated into corresponding longitudinal velocity $v_{\text{ref}}(t)$, power demand for the hydraulic steering systems $P_{\text{aux,str}}(t)$ and hydraulic suspension systems $P_{\text{aux,sus}}(t)$, and the target SOC value SOC_{ref}. Based on these inputs, an appropriate control sequence can be formulated. P_{EGS} of the control sequence is influenced by P_{aux} as shown in equation (1). T_{mot} of the control sequence is related to $v_{\text{ref}}(t)$. The fixed variation of SOC is set by SOC_{ref}.

$P_{\text{aux}}(\text{t})$ is incorporated as a parameter in the power flow of the RSPS model. Combining equations (1), (3), (6), and (10), the nonlinear RSPS predictive model is expressed as:

$$\begin{cases} S\dot{O}C = -\dfrac{U_{\text{ocv}} - \sqrt{U_{\text{ocv}}^2 - 4R_{\text{int}}(T_{\text{mot}}\dfrac{v}{R \cdot i_{\text{g}}}\eta_{\text{mot}}^{-sgn(T_{\text{mot}})} + P_{\text{aux}} - P_{\text{EGS}})}}{2C_{\text{n}}R_{\text{int}}} \\ \dot{v} = \dfrac{1}{\delta m}(\dfrac{T_{\text{mot}}i_{\text{g}}}{R} - f_{\text{r}}mg\cos\theta - mg\sin\theta - \dfrac{1}{2}C_{\text{D}}A\rho v^2) \\ \dot{Q} = -\dfrac{P_{\text{EGS}}b}{367.1\rho_{\text{f}}\eta_{\text{gen}}g} \end{cases} \tag{11}$$

The model can be discretized as:

$$X(k+1) = X(k) + F(X(k), U(k), P(k))\Delta t \tag{12}$$

Where

$$X = \begin{bmatrix} SOC \\ v_{\text{x}} \\ Q \end{bmatrix}, U = \begin{bmatrix} P_{\text{EGS}} \\ T_{\text{mot}} \end{bmatrix}, P = [P_{\text{aux}}] \tag{13}$$

$v_{ref}(t)$ and SOC_{ref} are incorporated as reference variables within the cost function. Longitudinal speed tracking, fixed SOC variation, and fuel economy are simultaneously concerned in the energy reconfiguration-based PCS, resulting in a multi-objective optimization problem as the influence between these objectives is both intricate and tightly linked. The cost functions of the studied multi-objective optimization problem are outlined below:

$$J_1 = (SOC(k+n|k) - SOC_{\text{ref}})^2 \tag{14}$$

$$J_2 = \sum_{i=1}^{n}(v(k+i|k) - v_{\text{ref}}(k+i|k))^2 \tag{15}$$

$$J_3 = (Q(k+n|k) - Q(k))^2 \tag{16}$$

Where J_1 represents the error between the predicted final SOC and the target SOC upon the prediction horizon's completion, J_2 represents the tracking performance of v at each time step within the prediction horizon, and J_3 represents the fuel consumption within the prediction horizon.

Finally, the MPC problem to describe the power control of RSPS for the energy reconfiguration process can be expressed as:

$$\begin{aligned} \min. \quad & J = \{J_1, J_2, J_3\} \\ \text{s.t.} \quad & \text{Eq. 12} \\ & P_{\mathrm{EGS,min}} \le P_{\mathrm{EGS}} \le P_{\mathrm{EGS,max}} \\ & T_{\mathrm{mot,min}} \le T_{\mathrm{mot}} \le P_{\mathrm{mot,max}} \\ & P_{\mathrm{bat,chrg}} \le P_{\mathrm{bat}} \le P_{\mathrm{bat,disc}} \\ & SOC_{\mathrm{min}} \le SOC \le SOC_{\mathrm{max}} \\ & a_{\mathrm{min}} \le \Delta v \le a_{\mathrm{max}} \end{aligned} \tag{17}$$

3.2. Adaptive Multi-objective MPC

A weighted method is applied to formulate an MPC-based PCS, creating a multi-objective MPC (MOMPC) to provide a baseline. The cost function of MOMPC combines the three cost functions into a single objective. The weighted cost function can be characterized by:

$$J = w_1 J_1 + w_2 J_2 + w_3 J_3 \tag{18}$$

where w_1, w_2, and w_3 are the weights corresponding to J_1, J_2, and J_3 respectively. These weights ensure that each cost function has a feasible solution.

By using a weighting weighted method, potential conflicts between multiple cost functions can be addressed. For instance, the fixed variation of SOC may be disrupted by a sudden increase in auxiliary power demand or a surge in driving power during acceleration, leading to battery discharge. Conversely, increasing the EGS output power to compensate may adversely impact fuel economy. However, MOMPC uses the same weight coefficients across different prediction horizons when addressing these issues, which reduces its robustness in handling variations in auxiliary power demands and limits its ability to utilize global information effectively. Abrupt changes in accessory power can lead to frequent shifts in the EGS operating point. Meanwhile, the goal of reaching the target SOC may limit MOMPC's feasible solution range in the prediction horizon. To address this challenge, AMOMPC is employed to manage the multi-objective power control issue within the energy reconfiguration process.

Considering that J_1 and J_3 may interfere with each other when the variance between the present SOC and the desired SOC is large or small, an adaptive factor s is introduced:

$$s = \mathrm{abs}\left(\frac{\dfrac{SOC_{t=t_k} - SOC_{\mathrm{ref}}}{SOC_{t=t_0} - SOC_{\mathrm{ref}} + \varphi}}{\dfrac{t_k - t_{\mathrm{end}}}{t_{\mathrm{start}} - t_{\mathrm{end}}}}\right) \tag{19}$$

Where $SOC_{t=t_k}$ and SOC_{ref} represent the SOC of prediction start time and the target SOC respectively, t_k, t_{start}, and t_{end} denote the prediction start time, the global start time, and the global end time, φ serves as a constant coefficient introduced to preclude division-by-zero conditions.

Since MOMPC solves the optimization problem within the prediction horizon, it will try to keep the final SOC of the prediction horizon as close as possible to the target value. This may cause a significant waste of the battery's output capacity and frequent fluctuations in the EGS output power. The adaptive factor s is quantified as the ratio of the remaining SOC balancing process to the remaining control process. A higher ratio indicates a greater priority for fixed variation of SoC, while a lower ratio indicates reduced priority. With the introduction of s, the process of obtaining the fixed SOC variation is informed to every prediction horizon as global information, mitigating the short-sightedness of MOMPC in handling it in each prediction horizon. The coefficient of J_1 is formulated as a function of s and is designed to be positively correlated with it. When s is large, the optimization task within the prediction horizon favors achieving the fixed SOC variation. Conversely, when s is small, the solution prioritizes other optimization goals. Notably, when the SOC approaches SOC_{ref}, the influence of s may lead to continuous battery discharge in response to sudden power demand changes. In summary, the adaptive-weighted cost function of AMOMPC can be expressed as:

$$J = (ks + c)J_1 + w_2J_2 + w_3J_3 \tag{20}$$

Where k and c denote the linear adaptive adjustments in the AMOMPC.

4. Results of Simulation

The simulation in this paper is executed with the help of Matlab and CasADi [11]. The typical working condition of energy reconfiguration is depicted in Figure 4(a). Parameters related are shown in Table 1. Several simulations were conducted to demonstrate AMOMPC's solution capability and adaptive optimization effect.

Table 1. Parameters

Component	Variable	Symbolic	Value	Units
RHV	Mass	m	40000	kg
	Rolling resistance coefficient	f_r	0.015	-
	Frontal area	A	2.2	m2
	Air resistance coefficient	C_D	0.3	-
	Reducer ratio	i_g	13	-
	Driving wheel radius	R	0.65	m
	Max&Min acceleration	a_{max}, a_{min}	5, -5	m/s
	Rotational inertia coefficient	δ	1.1	-
RSPS	EGS maximum power	$P_{EGS,max}$	200	kW
	Generator coefficient	η_{gen}	0.92	--
	Motor coefficient	η_{mot}	0.91	--
	Motor maximum power	$P_{mot,i,max}$	140	kW
	Motor maximum torque	$T_{mot,i,max}$	1200	Nm
	Battery Capacity	C_n	40	Ah
	Battery maximum discharge power	$P_{bat,disc}$	200	kW
	Battery maximum charge power	$P_{bat,chr}$	60	kW
AMOMPC/MOMPC	Reconfiguration time	--	300	s
	Sample time	Δt	1	s
	Prediction horizon	--	10	s
	Control horizon	--	1	s
	Weight of cost function	w_1, k, c w_2 w_3 φ	5×10^5, 5×10^5, 1 1×10^5 1 1×10^{-3}	--

The MPC-based PCS effectively manages SOC regulation and velocity tracking during the RHV reconfiguration process, as demonstrated below. The velocity tracking results during the RHV docking process are illustrated in the following Figure 4(b). Since both methods achieve satisfactory velocity tracking performance, their optimization effects on fuel economy are comparable and thus not analyzed.

Journal of Physics: Conference Series **3012** (2025) 012046 doi:10.1088/1742-6596/3012/1/012046

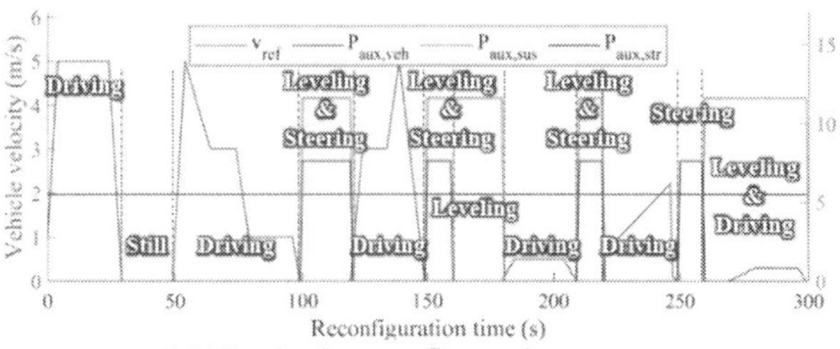

(a) Typical reconfiguration process.

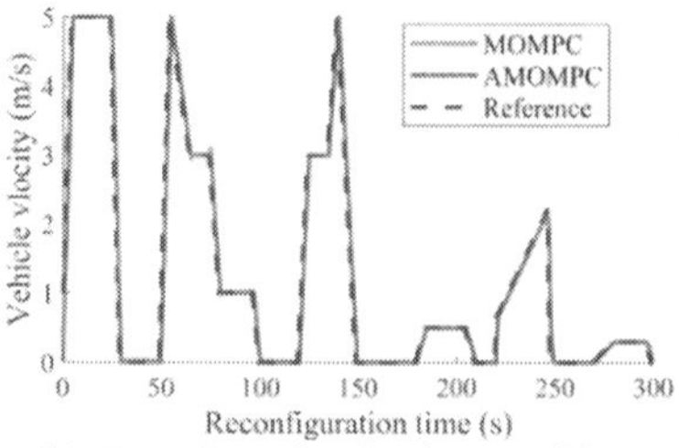

(b) Results of velocity tracking.

Figure 4. Simulation conditions and partial outcomes.

A comparison of the energy regulation performance between MOMPC and AMOMPC was conducted under scenarios where SOC decreases, maintains, and increases. A comprehensive analysis of the results across different scenarios shows that both AMOMPC and MOMPC can track the target SOC at the final time, converging within an acceptable range of the target value. The average error in controlling the fixed SOC variation using AMOMPC is 8%, which represents a 46.7% improvement compared to MOMPC. When comparing the EGS power control commands, AMOMPC reduces the number of operating point transitions by 48.2% and significantly mitigates abrupt changes in the operating points compared to MOMPC. In Figure 5(c) and Figure 5(d), The AMOMPC method illustrates scenarios where the battery continues discharging during the energy regulation process, a result not observed in MOMPC. This is because, in response to large power fluctuations, AMOMPC adjusts the range of battery SOC variation within the prediction horizon based on the adaptive factor level. When sufficient time is available for SOC adjustment, AMOMPC allows the EGS to operate within a more stable range, thus improving fuel economy to some extent and facilitating better control of the EGS.

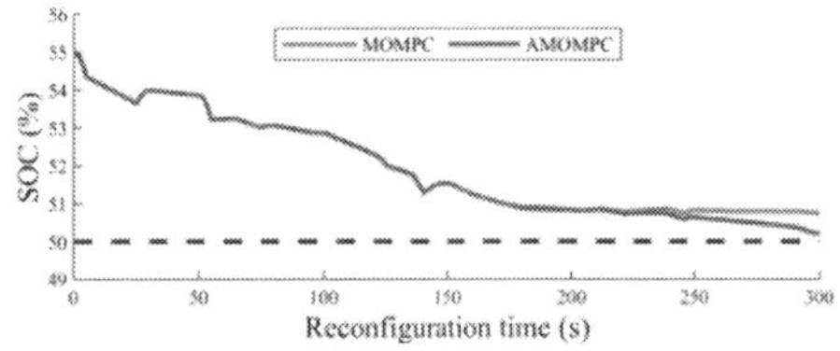

(a) The SOC curve in the decreasing condition.

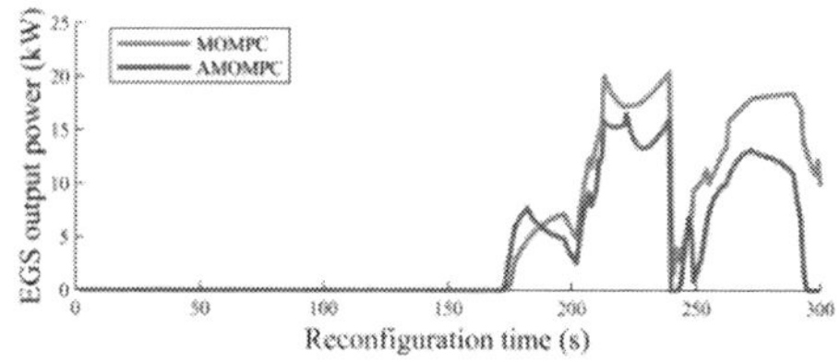

(b) The P_{EGS} curve in the decreasing condition.

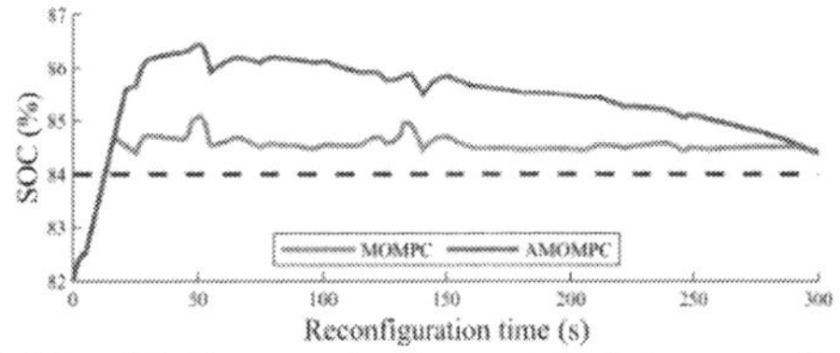

(c) The SOC curve in the maintaining condition.

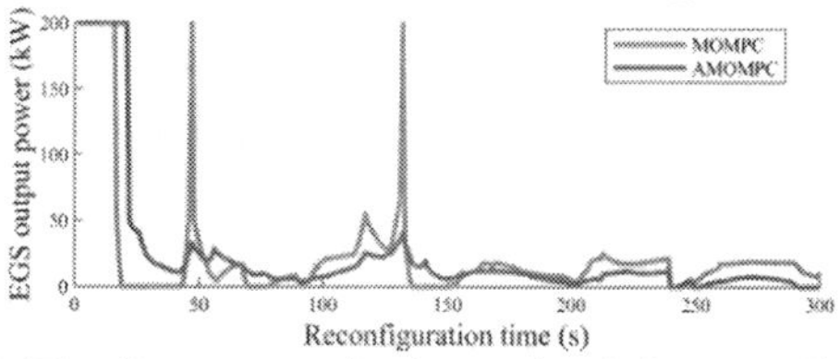

(d) The P_{EGS} curve in the maintaining condition.

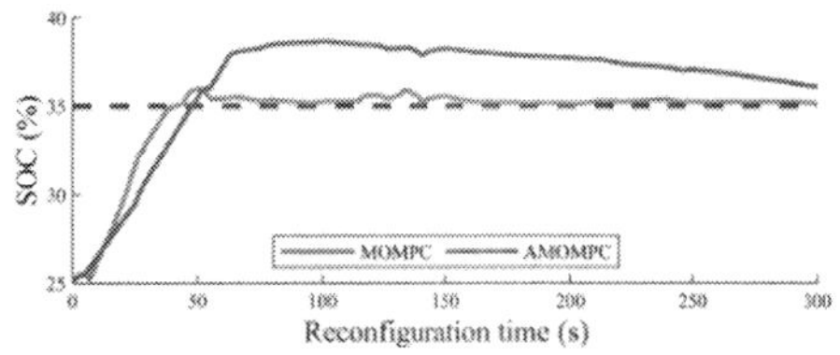

(e) The SOC curve in the increasing condition.

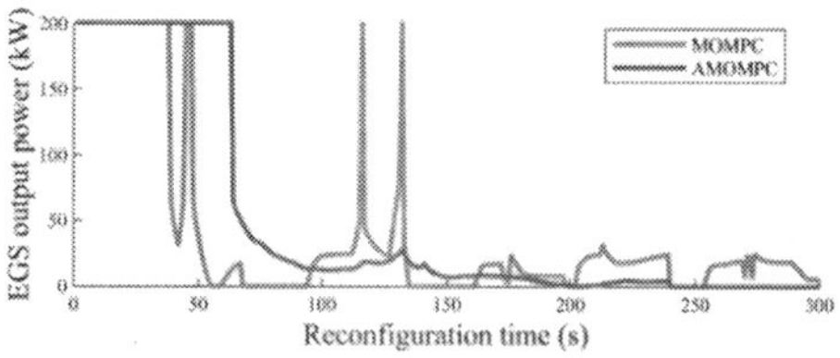

(f) The P_{EGS} curve in the increasing condition.

Figure 5. Simulation results of the AMOMPC and MOMPC strategies.

Within the framework of energy reconfiguration using AMOMPC, three RSPS units are designed to converge to the same SOC level by the conclusion of the control horizon. Figure 6(a) and Figure 6(b) show scenarios where the SOC of the three RSPS units increases, remains stable, or decreases, respectively. AMOMPC successfully achieves consistent SOC regulation while maintaining smooth transitions in the EGS operating points. Figure 6(c) and Figure 6(d) present a scenario where the SOC of all three RSPS units increases simultaneously, demonstrating the effect of the linear adjustments of the adaptive factor.

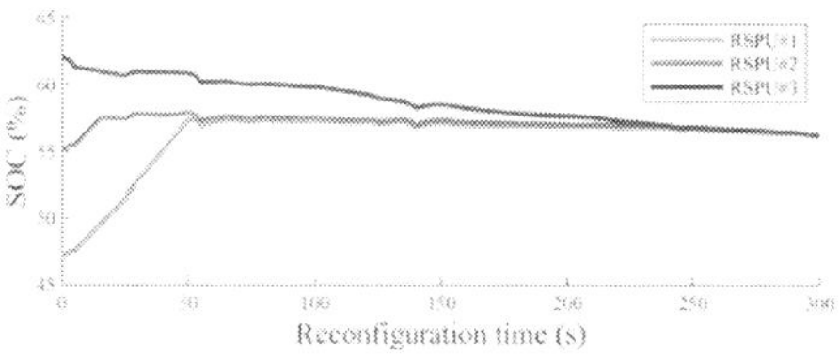

(a) The SOC curve in the balanced condition.

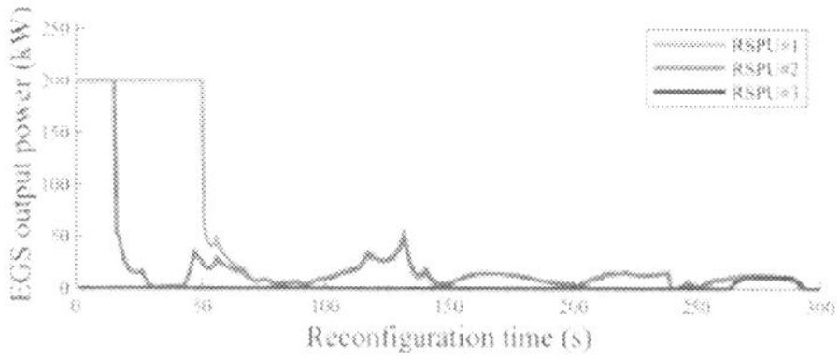

(b) The P_{EGS} curve in the balanced condition.

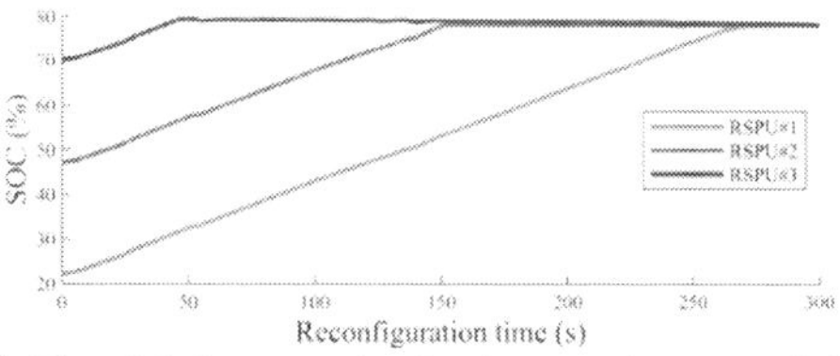

(c) The SOC curve in the increasing condition.

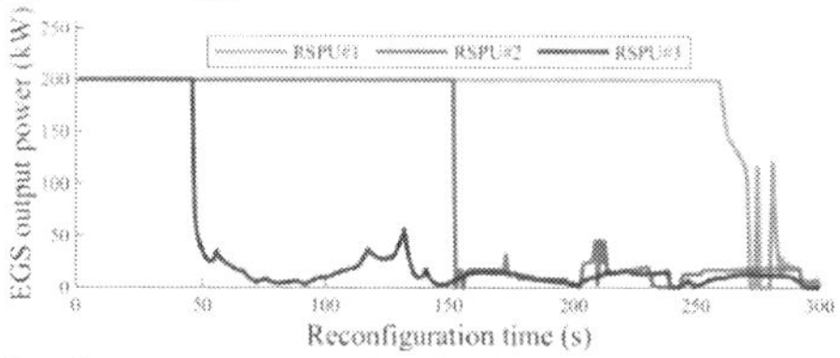

(d) The P_{EGS} curve in the increasing condition.

Figure 6. Simulation results of multiple AMOMPC strategies under different initial conditions.

5. Conclusion

This paper develops an energy system model specifically designed for energy reconfiguration and proposes a corresponding implementation method. The developed RSPS model integrates longitudinal velocity tracking and accounts for auxiliary power fluctuations. The energy reconfiguration process is formulated as a multi-objective optimization problem, encompassing fixed SOC variation, velocity tracking, and fuel consumption objectives. To address this multi-objective MPC problem efficiently, an AMOMPC approach is proposed to improve solution performance. Simulation results show that AMOMPC successfully achieves the energy configuration-based fixed variation of SOC while ensuring precise vehicle velocity tracking. The adaptive factor optimizes the selection of EGS operating points and minimizes their fluctuations. Future research could explore the integration of real-time data analytics into the AMOMPC framework to enhance energy reconfiguration adaptability.

References

[1] Yang, X., & Ni, J. (2024). The energy management strategy of the multi-source parallel power system for the self-reconfigurable ground vehicle. Proceedings of the Institution of Mechanical Engineers, Part D: Journal of Automobile Engineering, 09544070241245466.

[2] Yang, C., Wang, M., Wang, W., Chen, R., Ma, Y., Xiang, C., & Zeng, G. (2024). A Power Preconditioning-Based Power Flow Predictive Control Strategy for Hybrid Electric Vehicle Using Fast Iteration Optimization Algorithm. IEEE/ASME Transactions on Mechatronics, 29(2), 1465–1476.

[3] Wang, M., Yang, C., Wang, W., Lu, Z., Yang, L., & Chen, R. (2024). An Efficient Power Control Scheme for Heavy-Duty Hybrid Electric Vehicle With Online Optimized Variable Universe Fuzzy System. IEEE Transactions on Fuzzy Systems, 32(5), 2725–2737.

[4] Wang, F., Xia, J., Zhu, X., Xu, X., & Ni, Y.-Q. (2024). An Online Predictive Energy Management Strategy for Multi-Mode Plug-in Hybrid Electric Vehicle With Mode Transition Schedule Optimization. IEEE/ASME Transactions on Mechatronics, 29(3), 1972–1984.

[5] Moghimi, M., Teymoori, M. M., & Chitsaz, I. (2024). New vehicle control strategy for fuel consumption and battery state of health improvement in a series hybrid vehicle utilizing fuzzy logic. Journal of the Brazilian Society of Mechanical Sciences and Engineering, 46(5), 270.

[6] Hu, D., Huang, C., Yin, G., Li, Y., Huang, Y., Huang, H., Wu, J., Li, W., & Xie, H. (2024). A transfer-based reinforcement learning collaborative energy management strategy for

extended-range electric buses with cabin temperature comfort consideration. Energy, 290, 130097.

[7] Radrizzani, S., Panzani, G., & Savaresi, S. M. (2024). Simultaneous Energy Management and Speed Control in a Hybrid Tractor With Experimental Validation. IEEE Transactions on Control Systems Technology, 32(4), 1285–1297.

[8] Guo, L., Nie, S., Liu, H., Han, L., Ruan, S., Liu, R., Yang, N., & Xiang, C. (2024). Bi-Level Optimization of Speed Trajectory and Power Management for Autonomous HEVs in Off-Road Scenarios. IEEE Transactions on Intelligent Vehicles, 9(2), 4191–4205.

[9] Huang, Y., Wang, H., Khajepour, A., He, H., & Ji, J. (2017). Model predictive control power management strategies for HEVs: A review. Journal of Power Sources, 341, 91–106.

[10] Upadhyaya, A., & Mahanta, C. (2024). An Overview of Battery Based Electric Vehicle Technologies With Emphasis on Energy Sources, Their Configuration Topologies and Management Strategies. IEEE Transactions on Intelligent Transportation Systems, 25(2), 1087–1111.

[11] Andersson, J. A., Gillis, J., Horn, G., Rawlings, J. B., & Diehl, M. (2019). CasADi: a software framework for nonlinear optimization and optimal control. Mathematical Programming Computation, 11, 1-36.

Optimal Power Flow Control of Synchronous Generators for Short Circuit Ratio Enhancement in Power Systems with High-Proportion Wind Power Level

Xiaoqiang Wei[1][0009-0007-7187-5903], **Jiayi Wang**[2*][0009-0009-9910-1322], **Xinyan Cui**[2][0009-0000-9176-8020], **Shilong Li**[2][0009-0007-7535-7011], **Hao Yang**[2][0009-0009-1041-1317]

[1] State Grid Heilongjiang Electric Power Company Limited, Haerbin 150000, Heilongjiang Province, China

[2] Northeast Electric Power University, Jilin 132012, Jilin Province, China

*Corresponding author's e-mail: 2202300099@neepu.edu.cn

Abstract. With the integration of large-scale wind turbines, the short-circuit ratio (SCR) of the grid decreases, the voltage regulation ability weakens, and the fault easily leads to the transient overvoltage of renewable energy units and the off-grid accident. For the power grid with high-proportion of wind power, the optimal configuration method of synchro taking into account short-circuit ratio increase is studied. Firstly, the multiple renewable energy stations short circuit ratio (MRSCR), which can account for the interaction between sites, is analysed. Secondly, based on the topology of the power grid, an optimal power flow control model of synchronous generators is developed with system security constraints. The Beetle Swarm Optimization Algorithm (BSO) is then applied to calculate the optimal MRSCR, yielding the best voltage and power settings for the synchronous generators. Finally, based on PSD-BPA, an example analysis in a power grid with high-proportion wind power level is carried out to verify that the proposed optimization method can significantly improve the MRSCR of wind turbines.

1. Introduction

In recent years, in response to the slogan of "peak carbon emissions and carbon neutrality", efforts have accelerated to develop a new power grid dominated by renewable energy sources [1]. With the increasing presence of renewable energy in the grid, the voltage support strength of the system weakens, posing significant challenges to grid voltage stability. The extensive integration of wind energy units reduces the SCR level of renewable energy stations in the grid, further weakening voltage support capability. Currently, the national compulsory standard GB38755-2019 "Guidelines for Power System Reliability and Stability" clearly mandates that the SCR value for renewable energy sites must meet a reasonable threshold to ensure voltage stability and security within the power system. [2-3].

The traditional definition of SCR is based on physical principles, where the SCR at a specific point in the power system is the proportion of the short-circuit capacity that reflects the voltage strength of the system relative to the rated power of the electrical equipment. However, this calculation overlooks the interactions between different renewable energy DC transmission stations, rendering it unsuitable for grids with a high share of renewable energy. Therefore, for power grids with substantial renewable integration, a new concept of SCR has been introduced to assess voltage support strength. References

[4, 5] consider the interactions between multiple branches post-renewable integration by treating the integrated renewable energy as a whole, proposing weighted SCR and composite SCR metrics. These methods use weighting or summation to roughly estimate interactions among feed-in branches. Reference [6] develops two SCR indicators for renewable energy-integrated systems: the capacity-based SCR (SCR-S) and the voltage-based SCR (SCR-U). Reference [7] considers short circuit current under transient conditions and proposes the multi-infeed transient SCR, deriving the relationship between AC side short circuit capacity and voltage based on system structure and known magnitude-phase characteristics of renewable energy. This metric reflects the transient changes in voltage support capability across system nodes. Reference [8] presents the concept of the SCR, considering factors such as the distribution of electrical quantities and phase differences between various nodes MRSCR is suitable for assessing voltage strength in power systems with multi-station integration across various scenarios. MRSCR can also be simplified based on impedance ratios, offering a more straightforward calculation method, and is the indicator used in this paper to evaluate voltage support strength.

Currently, static voltage var controllers (SVC), static synchronous controllers (STATCOM), and synchronous compensating machines are widely used in engineering to enhance the voltage support capability of renewable energy stations. SVC and STATCOM have limited reactive power response and may introduce delays in reactive response, sometimes exacerbating transient overvoltage issues. Synchronous condensers, a type of synchronous generator, can effectively increase system short circuit capacity, strengthen grid support, and improve voltage support during disturbances. Reference [9] uses simulations to analyze how the integration of synchronous condensers enhances the SCR for DC networks, thereby improving voltage stability. Reference [10] assesses the impact of synchronous condensers on MRSCR, based on MRSCR's role in quantifying grid voltage support. It proposes an optimization strategy for locating and sizing centralized and distributed condensers in renewable-heavy grids to enhance voltage support strength. Reference [11] suggests a strategy of installing small synchronous condensers on the low-voltage side of renewable energy collection stations to mitigate transient overvoltage issues, while simultaneously improving the cost-effectiveness of the system. Reference [12] introduces a placement method for synchronous condensers that considers DC multi-infeed interaction factors and effective multi-infeed SCR, aimed at mitigating commutation failure in multi-infeed systems. Reference [13] examines the voltage support provided by SVC, STATCOM, and synchronous condensers for transient voltage issues at the sending end and proposes reactive power compensation strategies to raise grid voltage levels. The output of synchronous generators significantly impacts the SCR of renewable energy stations. An optimization model for synchronous generators power flow control is proposed to meet the SCR requirements for adequate voltage support remains a major challenge.

Based on the aforementioned studies and addressing gaps in current research, this paper proposes a method for optimizing synchronous generators power flow control model with consideration for enhancing the SCR. First, the MRSCR, which accounts for interactions between renewable energy stations, is analyzed. Next, an optimization configuration model for synchronous generators power flow control is developed based on grid topology, incorporating system operational safety constraints. Using the BSO algorithm, the model optimizes the active power output and node voltage values of synchronous generators, with the objective of maximizing MRSCR. This approach aims to achieve synchronous generator power flow control that ensure the voltage reliability and stability of the power system. Finally, simulations are performed on the Nordic system to validate the efficacy of the proposed approach.

The remaining paper is organized as follows: Chapter 1 provides an overview of the development and selection of SCR-related indicators. Chapter 2 develops a synchronous generator power flow control optimization model aimed at maximizing SCR, which solved by the BSO algorithm. Chapter 3 conducts simulation verification based on the Nordic test system.

2. Definition of SCR indicator

To help achieve the "dual carbon" goal, the proportion of wind energy in the transmitting grid has been rising steadily each year, which can, however, lead to a reduction in voltage support capability. Due to its clear physical significance and computational simplicity, SCR stands out as a key metric for assessing voltage support capability and grid strength. SCR is used to compare the relative magnitude between new energy facilities and the AC system, making it a standard for evaluating grid strength in systems with multiple new energy stations, thus effectively ensuring the stability of new energy grid integration. To ensure that large-scale power electronic devices can still operate safely and stably after connecting to the AC system, the Guidelines for Power System Reliability and Stability explicitly require that the SCR indices for both sending and receiving systems must meet reasonable levels [14]. The following section will introduce the methodology for constructing the SCR metric.

2.1. New energy equivalent grid connection model

Calculation of Equivalent Short-Circuit Capacity of AC System:

$$S_{ac,i} = \frac{U_N E_i}{Z_{t,i}} \tag{1}$$

$Z_{t,i}$ is the Thevenin equivalent reactance of the system solved according to the Thevenin equivalent potential at time t, U_N is the nominal system voltage, E_i is the Thevenin equivalent potential of the system obtained at time t.

When accounting for the mutual influence among multiple new energy sources, the grid-connected capacity of a new energy source includes not only the capacity of its own directly connected units but also requires factoring in the influence of other branches. At this point, the equivalent capacity of the AC system connected to the grid can be determined using the following calculation.

$$S_{eq,i} = S_i + \sum \frac{Z^*_{ij}}{Z^*_t} \frac{U^*_i}{U_j} S_j \tag{2}$$

Where S_i and S_j represent the capacities of the new energy sources connected to nodes i and j; Ui and U_j are the voltages at the node i and j; $Z_{t,i}$ is the equivalent reactance at t, derived using equivalent potential of Thevenin; Z_{ij} is the impedance between the connection points i and j.

2.2. Selection of SCR

In a single-feed renewable energy system, the conventional SCR can be utilized to evaluate strength of the grid. It is defined by the short-circuit capacity and the capacity of the renewable energy source connected to the grid.

$$SCR = \frac{S_{ac}}{P_N} \tag{3}$$

Where the S_{ac} represents the short-circuit capacity at the grid connection point of the station, and P_N denotes the nominal active power output of the wind farm connected at that grid point. However, this SCR only considers the grid strength in a single-feed system and does not account for the interactions between different systems, leading to significant errors in practical applications.

For new energy multi-feed systems, the MRSCR effectively accounts for the interactions between different new energy stations. It considers the amplitude and phase variations of various electrical parameters between the grid connection points of different stations, while also incorporating the effect of reactive power generated by each renewable energy station. This makes it effective for evaluating grid strength in renewable energy-dominant power systems across various scenarios.

Figure 1 illustrates the simplified model of a multi-new energy station grid. The impedance matrix of the sending-end grid system at node N is represented by the matrix Z, where S_{NE}, P_{NE}, Q_{NE}, and U_{NE} represent the apparent power, active power, reactive power, and grid-connected voltage of the new energy generation station i respectively. Z_{ij} represents the equivalent impedance between the grid connection points i and j, whereas Z_i is the impedance on the system side, measured between the equivalent power source i and its relevant grid connection point.

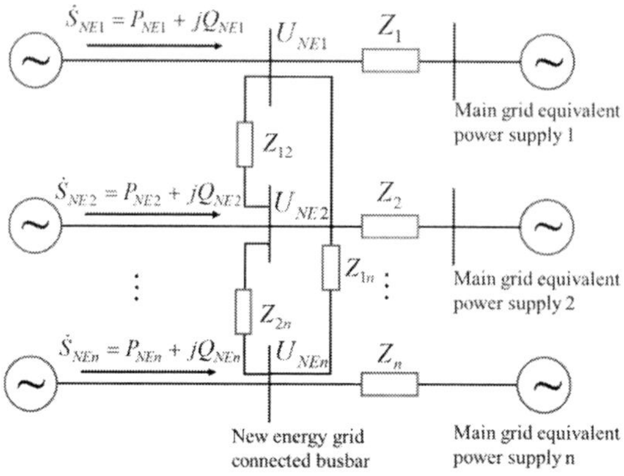

Figure 1. The simplified model of a multi-new energy station grid.

Using the node voltage equation, the at the grid connection point can be expressed as:

$$\begin{bmatrix} \dot{U}_{NE,1} \\ \dot{U}_{NE,2} \\ \vdots \\ \dot{U}_{NE,m} \\ \vdots \\ \dot{U}_{NE,n} \end{bmatrix} = \begin{bmatrix} \dot{Z}_{11} & \cdots & \dot{Z}_{1m} & \cdots & \dot{Z}_{1n} \\ \dot{Z}_{21} & \cdots & \dot{Z}_{2m} & \cdots & \dot{Z}_{2n} \\ \vdots & & \vdots & & \vdots \\ \dot{Z}_{m1} & \cdots & \dot{Z}_{mm} & \cdots & \dot{Z}_{mn} \\ \vdots & & \vdots & & \vdots \\ \dot{Z}_{n1} & \cdots & \dot{Z}_{n,m} & \cdots & \dot{Z}_{nn} \end{bmatrix} \begin{bmatrix} \dot{I}_{NE,1} \\ \dot{I}_{NE,2} \\ \vdots \\ \dot{I}_{NE,m} \\ \vdots \\ \dot{I}_{NE,n} \end{bmatrix} \tag{4}$$

Where $\dot{I}_{NE,i}$ represents the current supplied by the renewable energy bus i to the AC system; $\dot{U}_{NE,i}$ refers to the voltage at the connection point i of the grid.

The SCR can also reflect the comparison between the rated voltage of the system and the voltage produced by the renewable energy equipment., expressed as:

$$R_{MRSCRi} = \frac{|\dot{U}_{Ni}|}{|\dot{U}_{NEi}|} = \frac{|\dot{U}_{Ni}|}{\left|\dot{Z}_{ii}\dot{I}_{NE,i} + \sum_{j=1,j\neq i}^{n} \dot{Z}_{ij}\dot{I}_{NE,j}\right|} \tag{5}$$

U_{Ni} represents the nominal voltage of the bus i and U_{NEi} is the voltage produced by the renewable energy generation unit.

Make the actual voltage at grid connection point i be $\dot{U}_i$, by simplifying the above equation, we get:

$$R_{MRSCRi} = \frac{|\dot{U}_i\dot{U}_N / \dot{Z}_{ii}^*|}{\left|\dot{U}_i\dot{I}_{NE,i}^* + \sum_{j=1,j\neq i}^{n} \frac{\dot{Z}_{ij}}{\dot{Z}_{ii}}\dot{U}_i\dot{I}_{NE,j}^*\right|} = \left|\frac{\dot{U}_i\dot{U}_N}{\dot{Z}_{ii}^*}\right| / \left|\dot{S}_{NE,i} + \sum_{j=1,j\neq i}^{n} \prod ij\dot{S}_{NE,j}\right| \tag{6}$$

Where $S_{NE,i}$ denotes the actual apparent power supplied by the renewable energy to the grid-connected bus node i; $\prod ij = \dot{Z}_{ij}^*\dot{U}_i/\dot{Z}_{ii}^*\dot{U}_j$ is a power conversion factor which reflect the phase and amplitude differences of electrical values at the grid connection points i and j.

The MRSCR effectively characterizes the grid strength of a power system with multiple renewable energy stations. Since MRSCR comprehensively considers electrical variables such as grid connection point voltage magnitudes and phase angles, as well as node active power and reactive power, it provides an intuitive and accurate representation of the voltage stability level. This allows MRSCR to reflect voltage support capability of the grid and overall stability.

3. Optimal power flow control of synchronous generators

3.1. Model construction for synchronous generator power flow control

To maximize the total SCR of renewable energy facilities, a synchronous generator power flow control optimization model is established. This model comprehensively considers constraints on wind farm and overall SCR balance, synchronous generator voltage, active power output, and power flow. The optimization variables include the voltage magnitude and active power of the synchronous generators. The model is expressed as follows.

The objective is represented by maximizing the MRSCR of renewable energy stations, which is:

$$J = \max \sum_{i=1}^{n} R_{MRSCRi} \tag{7}$$

Where i represents the renewable energy facility i; n is the overall number of renewable energy facilities; R_{MRSCRi} represents the MRSCR of the renewable energy facilty i. This objective function aims to enhance the overall MRSCR level of system by solving the optimal power flow control model of synchronous generators, strengthening the voltage regulation capability and stability of the grid.

The following are the constraints.

1) Constraint on the SCR of Wind Farms in the Power Grid:

$$R_{MRSCRi} > R_{MRSCRi,\min} \tag{8}$$

$R_{MRSCRi,\min}$ is the minimum threshold of the SCR for wind farm i.

2) Constraint on the Balance of SCR among Wind Farms in the Power Grid:

To ensure a balanced distribution of MRSCR values across different wind farms in the power grid and prevent extreme differences in SCR among stations, a constraint on SCR balance is defined as shown in equation (9).

$$\sum_{i=1}^{n} \left(\frac{R_{MRSCRi} - R_{MRSCRav}}{R_{MRSCRav}} \right)^2 \le R_{MRSCRD} \tag{9}$$

$R_{MRSCRav}$ is the average MRSCR of the wind farms in the power grid; R_{MRSCRD} is the upper limit of the SCR balance degree.

3) Voltage magnitude constraint of synchronous generators:

$$V_{\min} \le V \le V_{\max} \tag{10}$$

V is the voltage magnitude of each synchronous generator; $V_{\min}$ and $V_{\max}$ represent the lower and upper voltage limits of the synchronous generator nodes respectively. In this paper, $V_{\min}$ and $V_{\max}$ are set to 0.9 and 1.1 respectively.

4) Active power output constraint of synchronous generators:

$$P_{g,i,\min} \le P_{g,i} \le P_{g,i,\max} \tag{11}$$

$P_{g,i}$ is the active power of the synchronous generator i; $P_{g,i,\min}$ and $P_{g,i,\max}$ represent the lower and upper active power of synchronous generator i respectively.

5) Power flow constraint:

$$\begin{cases} P_i = U_i \sum_{j=1}^{n} U_j \left(G_{ij} \cos\theta_{ij} + B_{ij} \sin\theta_{ij} \right) \\ Q_i = U_i \sum_{j=1}^{n} U_j \left(G_{ij} \sin\theta_{ij} - B_{ij} \cos\theta_{ij} \right) \end{cases} \quad i = 1, 2, \text{K} \;, n \tag{12}$$

Where P_i and Q_i represent the active power and reactive power at node i respectively; G_{ij} and B_{ij} are the real and imaginary components of the admittance matrix at position (i, j); θ_{ij} is the phase difference between nodes i and j.

3.2. Solution to the optimal power flow control model of synchronous generator

For the optimal power flow control model of synchronous generator, considering a swarm intelligence method for solving the model, this paper adopts the BSO method, which combines PSO with Beetle

Antennae Search (BAS). This method is known for its robustness and high optimization efficiency, making it suitable for solving complex nonlinear optimization problems.

For the selection of active power and voltage magnitude of synchronous generators in this paper, the BSO algorithm is used for power flow optimization and tuning. The initial positions of the beetles are obtained through random sampling, and an adaptive inertia weight strategy is applied during the iteration process. In a beetle swarm with N_p beetles, the position of the beetle i in the element j is denoted as $X_{i,j}^{\kappa+1}$, and the speed of beetle $v_{i,j}^{\kappa+1}$ and position $X_{i,j}^{\kappa+1}$ in the iteration (κ+1) can be expressed as:

$$\begin{aligned} \upsilon_{i,j}^{\kappa+1} = \omega^{\kappa}\upsilon_{i,j}^{\kappa} + c_1 r_1\left(\rho_{i,j}^{\kappa} - X_{i,j}^{\kappa}\right) \\ + c_2 r_2\left(g_j^{\kappa} - X_{i,j}^{\kappa}\right) \end{aligned} \tag{13}$$

$$X_{i,j}^{\kappa+1} = X_{i,j}^{\kappa} + \lambda\upsilon_{i,j}^{\kappa+1} + (1-\lambda)\xi_{i,j}^{\kappa} \tag{14}$$

$$\omega^{\kappa+1} = \omega_{\max} - \frac{\kappa+1}{\kappa_{\max}}\left(\omega_{\max} - \omega_{\min}\right) \tag{15}$$

$\rho_{i,j}^{\kappa}$ represents the element j of the individual best position of beetle i during the iteration κ; g_j^{κ} represents the element j of the globally optimal position during the iteration κ; r_1 and r_2 are random numbers in range [0, 1]; ω^{κ} denotes the inertia weight value, while $\omega_{\max}$ and $\omega_{\min}$ are the upper and lower inertia weights; c_1 and c_2 are the individual and global learning factors respectively. $\kappa_{\max}$ represents the maximum allowable total iterations; λ is the position-speed factor, and $\xi_{i,j}^{\kappa}$ is the position increment factor of beetle i in the element j during the iteration κ.

In the iteration (κ+1), the position increment factor is updated as follows:

$$\delta^{\kappa} = \eta_{\delta}\delta^{\kappa-1} \tag{16}$$

$$X_{r,i,j}^{\kappa} = X_{i,j}^{\kappa-1} + \frac{d}{2}\upsilon_{i,j}^{\kappa-1} \tag{17}$$

$$X_{l,i,j}^{\kappa} = X_{i,j}^{\kappa-1} - \frac{d}{2}\upsilon_{i,j}^{\kappa-1} \tag{18}$$

$$\xi_{i,j}^{\kappa+1} = \delta^{\kappa}\upsilon_{i,j}^{\kappa}\operatorname{sign}\left(f\left(X_{r,i,j}^{\kappa}\right) - f\left(X_{l,i,j}^{\kappa}\right)\right) \tag{19}$$

$X_{l,i,j}^{\kappa}$ and $X_{r,i,j}^{\kappa}$ represent the position of the left and right sensory antennae of the beetle i at element j in the iteration κ respectively. $f(X_{l,i,j}^{\kappa})$ and $f(X_{r,i,j}^{\kappa})$ represent the fitness value of the left and right sensory antennae respectively. δ^{κ} denotes the iteration search step size; while sign( ) represents the signum function; n_{δ} is the step size factor, and d is the gap between the antennae of beetle.

4. Example Implementation

4.1. Nordic power system

This paper uses the modified Nordic equivalent external region and the northern region to form a sending-end power grid for the case analysis, as shown in Figure 2 [15]. The system consists of 7 double-fed wind turbine generators and 5 synchronous generators, one of which is a balancing generator, and power is delivered through two external transmission channels. According to the SCR requirements for renewable energy stations in reference [8], the SCR of the wind turbine generators must be greater than 1.5. Based on the above SCR standard, the optimal power flow control model of synchronous generator is carried out. The model is developed and integrated on both the PSD-BPA simulation platform and the Matlab, and the efficiency of the proposed optimization approach for synchronous generator power flow is validated.

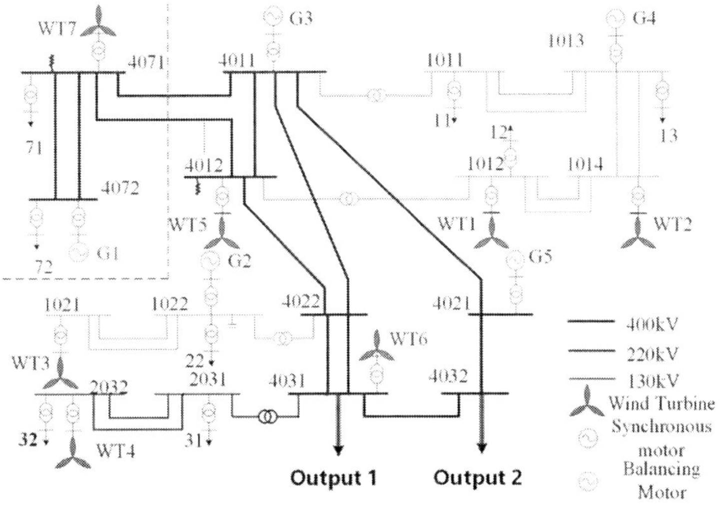

Figure 2. Nordic case structure

4.2. *Implementation results*

Before optimizing the synchronous generator power flow model, the MRSCR values for each wind turbine generator are shown in Table 1. Among them, generators WT2, WT3, WT5, WT6, and WT7 do not meet the requirement of an MRSCR greater than 1.5, with WT7 having the lowest value at 1.117, indicating weaker voltage support capability in the grid. Using joint simulation in Matlab and BPA software, Matlab is used to model and optimize the active power and voltage of each synchronous generator. The optimal power flow model is then applied in BPA software for SCR calculation.

Table 1. MRSCR of wind turbines before optimal power flow control of synchronous generators

Wind turbine	MRSCR	Wind turbine	MRSCR
WT1	1.658	WT5	1.453
WT2	1.447	WT6	1.164
WT3	1.369	WT7	1.117
WT4	2.831		

The BSO algorithm is adopted to solve the optimal power flow control model of the synchronous generators. The optimized variables include the voltage V1 of synchronous generator G1, the voltages V2, V3, V4 and V5 of synchronous generators G2-G5, and the active power outputs P2, P3, P4 and P5. The variables of the synchronous generator before and after optimization are shown in Table 2. After applying the optimal power flow model to the test system, the SCR results for each wind turbine power generator are presented in Table 3.

Table 2. Variables of Synchronous Generator Before and After Optimal Control

Variables	Before Optimal Control	After Optimal Control
V1	1.05 pu	1.1 pu
V2	1.05 pu	1.07 pu
V3	1.05 pu	1.08 pu
V4	1.05 pu	1.1 pu
V5	1.05 pu	1.09 pu
P2	200 MW	257 MW
P3	668 MW	705 MW
P4	300 MW	359 MW
P5	250 MW	288 MW

As shown in Table 3, after optimizing the power flow control model of synchronous generator, the SCR for each wind turbine generator is greater than 1.5, meeting the SCR requirement for renewable energy stations. The minimum MRSCR is 1.721, with an overall MASCR of 17.498, indicating an improvement in the SCR level of the wind power stations in the grid. Comparing the SCR results after optimal power flow control of synchronous with those before optimal power flow control, we can find that the voltage regulation capability of wind farms has increased after optimal power flow control.

Table 3. MRSCR of wind turbines after optimal power flow control of synchronous generators

Wind turbine	MRSCR	Wind turbine	MRSCR
WT1	2.564	WT5	2.517
WT2	2.322	WT6	1.721
WT3	2.109	WT7	1.897
WT4	4.368		

4.3. Comparative example

To confirm the advantages of the proposed method, the PSO algorithm is employed for solving the optimal power flow control model of the synchronous generators. The iteration count, objective function, and constraints are set identically to those used in the BSO algorithm.

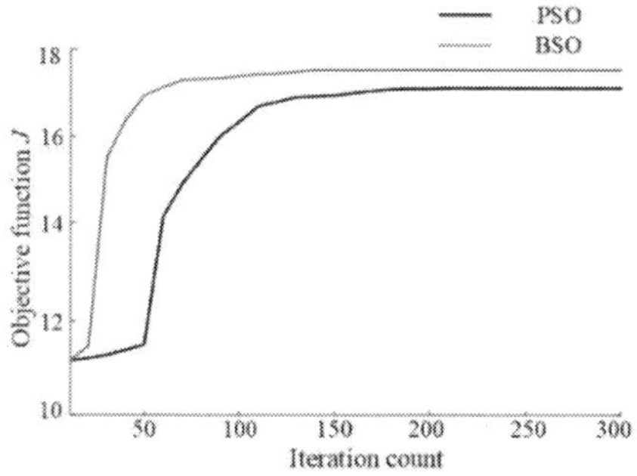

Figure 3. Iteration curves of different algorithms

To illustrate the convergence process of the BSO algorithm, Figure 3 shows the changes in the objective function value *J*. Compared to the traditional PSO algorithm, the BSO method adopted in this paper demonstrates superior convergence performance and optimization capability.

Table 4. MRSCR of wind turbines with PSO algorithm

Wind turbine	MRSCR	Wind turbine	MRSCR
WT1	2.413	WT5	2.432
WT2	2.251	WT6	1.521
WT3	1.967	WT7	1.695
WT4	4.158		

The result of the optimal power flow control is served to the synchronous generators in the power grid. The SCR results for each wind turbine power generator after optimal power flow control of synchronous generators are presented in Table 4.

According to the Table 4 and Table 3, it can be observed that the MRSCR values obtained through the BSO algorithm are larger, demonstrating better convergence performance and optimization capability. This effectively proves the advantage of the proposed method in increasing the SCR and boosting the voltage stabilization ability of wind farms.

5. Conclusion

The MRSCR takes into account the interactions between different stations to measure the voltage strength in the power grid. In this paper, a method is proposed for solving the model of optimal power flow control of synchronous generators in grids with significant wind power penetration, aiming to improve the SCR by solving the optimal power flow control model of the synchronous generators, thus boosting the voltage stabilization capacity of wind farms. This Beetle Swarm-based computational approach offers good convergence accuracy and strong global search ability, effectively avoiding local optima. It provides robust technical support for improving the operational efficiency and stability of the power grid. The effectiveness of proposed method in enhancing the voltage management ability of wind farms is validated using the BPA software on an improved Nordic system.

Acknowledgments
The work was supported in part by The Science and Technology Project of State Grid Heilongjiang Electric Power Co.,Ltd. (52240123000D).

References

[1] Shi W.H., Qu J.X., Luo K., et, al. 2022, Grid-Integration and Operation of High Proportioned New Energy. Strategic Study of CAE, pp.1-12.
[2] 2019, GB 38755-2019 Code on Security and Stability for Power System. Beijing: Standardization Administration of the People's Republic of China.
[3] Grid Operation and Control Standardization Technical Committee. Interpretation and Guidance of Code on Security and Stability for Power System and Guide in Technology for Power System. 2020, Beijing: China Electric Power Press.
[4] Zhang Y., Huang S.H., J. Schmall, J. Conto, J. Billo and E. Rehman. 2014, Evaluating System Strength for Large-Scale Wind Plant Integration. In 2014 IEEE PES General Meeting| Conference & Exposition, pp. 1−5.
[5] 2014, Minnesota Renewable Energy Integration and Transmission Study. Minnesota Department of Commerce.
[6] YU L, SUN H, ZHAO B, et al. 2022, Short Circuit Ratio Index Analysis and Critical Short Circuit Ratio Calculation of Renewable Energy Grid-Connected System. Proceedings of the CSEE (vol 3), pp. 919-929.
[7] HUANG Q, CAO S, WU Y, et al. 2024, A New Multi-Infeed Short Circuit Ratio Index Taking into account the Amplitude and Phase Characteristics of New Energy. Proceedings of the CSEE (vol 15): pp. 5960-5973.
[8] SUN H, XU S, XU T, et al. 2021, Definition and Index of Short Circuit Ratio for Multiple Renewable Energy Stations. Proceedings of the CSEE (vol 2), pp. 497-506.
[9] WANG Y, ZHANG Y, ZHOU Q, et al. 2017, Study on Application of New Generation Large Capacity Synchronous Condenser in Power Grid. Power System Technology (vol 1), pp. 22-28.
[10] ZHOU Y, SUN H, XU S, et al. 2022, Synchronous Condenser Optimized Configuration Scheme for Power Grid Voltage Strength Improvement. Power System Technology (vol 10), pp. 3848-3856.
[11] SUO Z, LIU J, JIANG W, et al. 2019, Research on Synchronous Condenser Configuration of Large-Scale Renewable Energy DC Transmission System. Electric Power Automation Equipment (vol 9), pp. 124-129.
[12] WANG Q, LI T, TANG X, et al. 2019, Method of Site Selection for Synchronous Condenser Responding to Commutation Failures of Multi-Infeed DC System. Automation of Electric Power Systems (vol 1), pp. 222-227.
[13] LIN A, HUANG Y, LIN W, et al. 2020, Comparison of Suppression Effect of Different Dynamic Reactive Power Compensation Devices for Transient Overvoltage Caused by HVDC System Faults. Power Capacitor & Reactive Power Compensation (vol 4), pp. 116-122.
[14] KANG Y, LIN X, ZHENG Y, et al. 2020, The Static Stable-Limit and Static Stable-Working Zone for Single-Machine Infinite-Bus System of Renewable-Energy Grid-Connected Converter. Proceedings of the CSEE (vol 14), pp. 4506-4515+4730.
[15] T. V. Cutsem, 2014 “The Nordic Test System for Voltage Stability Assessment” [Online].

Research on Low-Energy Consumption Strategies for Substation Buildings

Weiyu Zou[1], Xi Luo[1], Zhuoyan Huang[1], Mengya Yin[2, 3*]

[1] Guangzhou Power Supply Bureau, Guangdong Power Grid Co., Ltd, Guangzhou, 510062, China

[2] Guangdong Provincial Academy of Building Research Group Co., Ltd, Guangzhou, 510500, China

[3] Guangdong Jianke Innovation Technology Research Institute Co., Ltd, Zhongshan, 528403, China

*Corresponding author's e-mail: ymy445225248@163.com

Abstract. The significance of energy conservation in the design of substation buildings is escalating. This study examines energy-saving design strategies for substation buildings in summer hot winter warm climates. This research utilizes a 500kV substation located in Guangzhou, China, as a case study to dissect its energy consumption patterns and inform strategies for energy-efficient operations. The assessment of cold-load demand, energy-saving potential, and energy use characteristics incorporates enclosure structure, fresh air load, solar photovoltaic (PV) generation, and lighting electricity consumption. The methodology combines on-site testing with simulation analysis. The study reveals significant energy reduction potentials: (1) Optimizing enclosure thermal performance can decrease summer cold loads by 25-30%; (2) Fresh air load reduction of approximately 28% is achievable through zone-specific volume control; (3) Lighting electricity consumption can be slashed by 72% by enhancing natural lighting utilization; (4) Solar PV electricity generation can elevate the building's integrated energy savings to 43.63 %, demonstrating substantial energy conservation. The research of this project can provide energy-saving design reference for the low-energy transformation of substations in summer hot winter warm areas.
Keywords. summer hot winter warm area; substations; low energy consumption; renewable energy; natural lighting

1. Introduction

The power sector, essential for national energy and low-carbon development, exemplifies a new energy-based system through green transmission and transformation projects. Substations are critical conduits bridging electricity production to end-use consumers [1]. The advancement of the high-quality, green and low-carbon construction of substations is the core action of promoting the concept of sustainable development, helping to achieve the national "double-carbon" goal. Substation buildings are gradually developing in the direction of low energy consumption and high efficiency [2].

Amidst societal endeavors for energy conservation and carbon reduction, the imperative for substation buildings to enhance energy efficiency and effectuate energy savings assumes greater significance. Many scholars have conducted exploratory research on energy conservation in substation buildings. M. Lu, J. Wang, J. Chen, and X. Chen [1] performed a computational analysis focusing on the thermal dynamics of substation main control rooms. The study elucidated the correlation between the total heat consumption index and the external wall's heat transfer coefficient, considering diverse indoor heat dissipation densities. X. Chen, J. Wang, M. Lu, J. Chen, and C. Jing [3] conducted a simulation study on the dynamic annual variation of cooling and heating loads for typical buildings in three high-altitude cold region substations. The analysis aimed to dissect the heat load demands across varying regional climates. Y. Liu, Y. Zhang, Z. Cai, Jun Zhang, and Q. Liu [4] conducted an analysis of energy consumption patterns within an indoor substation distribution unit building in Guangzhou, China. They observed that the building's air-conditioning energy consumption is inversely proportional to the heat transfer coefficient of the enclosure structure. Conversely, it increases with the rise of the shading coefficient of external windows. T. Xiong, G. Li, Y. Liu, and S. Li [5] identified that the ventilation and air-conditioning system are a major part of the electricity consumption in the substation. Their field research and simulation analysis of a cold region substation revealed that effective indoor heat exclusion is pivotal for achieving near-zero energy consumption goals in substations. C. Jia [6] confirmed the substantial energy and carbon reduction potential of photovoltaic (PV) power generation systems. This was demonstrated through a case study of a 110kV substation PV energy saving and emission reduction system in Xi'an, China. C. Liu, X. Mei, and Y. Yang [7] designed solar PV panels for substation buildings in regions with hot summers and cold winters. They similarly discovered that solar PV panels significantly reduce carbon emissions.

The above studies show that energy conservation has become an important concern for substation buildings, and a variety of energy-saving technologies are widely used. However, a number of challenges remain in the field of substation building energy conservation research: (1) The focus is disproportionately skewed towards thermal insulation and energy-saving technologies, with scant attention given to their integration and optimization; (2) The majority of investigations have been conducted in isolation, neglecting the development of a comprehensive, multifaceted strategy for energy efficiency. In summary, this paper takes a 500kV substation in Guangzhou, China, a hot summer warm winter area, as a research object, and analyzes the energy-saving effect of the substation building under the combination of several low-energy strategies with thermal insulation and energy saving as the main line, so as to provide reference and strategies for the low-energy design and transformation of the substation in regions with similar climatic conditions.

2. Characterization of climate and substation building energy use in Guangzhou area

2.1. Climatic characteristics

Guangzhou, Bordering the South China Sea, has significant marine climate characteristics and belongs to the summer hot winter warm area of building thermal climate design zones [8]. It is hot and rainy in summer, lasting from May to September, with average temperatures above 28°C. The maximum temperature in July and August often exceeds 35°C, when air-conditioning equipment becomes a necessity, leading to a significant increase in energy consumption. However, it has a short winter lasting only one or two months, during which high humidity and frequent rain often cause discomfort.

Guangzhou's climate, marked by high temperatures, humidity and rainfall, poses challenges for building design. The high temperatures in summer require air-conditioning systems to operate for long periods of time, while heating is largely unnecessary in winter. Therefore, the low-energy strategy for substations in the Guangzhou area of China focuses on reducing the energy consumption associated with air conditioning during the summer months and optimizing the lighting system.

2.2. Characteristics of energy use in substations

Substation buildings have special characteristics that differ from conventional industrial buildings [4]. Although substation buildings belong to the category of industrial buildings, their energy use characteristics are closer to civil buildings, especially office buildings [9]. This is mainly due to the fact that the substation enclosure construction is similar to public buildings in that the energy efficient design is based on thermal insulation [10]. There is fewer productive equipment inside the building, and the power consumption and heat production of the equipment themselves are small. The main control communication building in the substation serves as the primary workspace and rest area for station operators. It must maintain an appropriate indoor thermal environment and provide adequate office lighting. Consequently, its energy consumption profile closely resembles that of typical office buildings. The main control and communication building in the substation differs from traditional offices as its equipment generates significantly more heat and operates continuously throughout the year. To address the heat dissipation challenges from continuous operation and power transmission, the air-conditioning system must operate 24 hours a day.

In summer hot winter warm areas, the energy use of substation buildings compared to ordinary office buildings has the following significant features:

(1) High share of air-conditioning energy consumption: as a result of the high temperature and high humidity environment in summer hot winter warm areas, maintaining optimal temperature and humidity within office space and control rooms has become a major challenge. For this reason, air-conditioning equipment needs to run for long periods of time, lead to a significant proportion of air-conditioning load in total energy consumption. The primary determinant of air-conditioning energy consumption is the heat transfer between indoors and outdoors through the enclosure structure, so the thermal insulation design of the enclosure structure plays a pivotal role in reducing the energy use. (2) High lighting demand: influenced by the nature of substation work, equipment rooms and control rooms require continuous lighting, which results in persistent energy consumption. (3) Ventilation and heat dissipation system: the ventilation of equipment rooms in substation is primarily dependent on fans and air-conditioning fresh air systems which are used for smoke exhaustion, heat dissipation and dehumidification. To meet the thermal energy generation and dissipation needs of the equipment, a large amount of energy must be consumed for cooling [11]. It was found that approximately 75% of the total energy consumption in substation buildings is attributed to air conditioning, 7% to ventilation, and 5% to lighting [5]. Therefore, the improvement of insulation, ventilation and lighting performance of the enclosure structure has become the main way to save energy in substation buildings in this region.

2.3. Project overview

A 500kV substation in Guangzhou area of China is located in the northeastern part of Guangzhou city, with excellent geographical location and convenient transportation. The substation has a reasonable layout and is divided into several functional areas, including the main control building, the inspection building and the power distribution building. The main control building has a floor area of 2125.96m^2, 2 floors above ground, and a building height of 8.20m. Each functional area is equipped with corresponding air-conditioning and lighting systems according to the actual needs.

To cope with the high temperature and high humidity environment, the building envelope adopts highly efficient thermal insulating materials, which reduces the cooling load of the building and, consequently, the energy consumption of the air-conditioning system. In addition, about 3000m^2 of PV panels were installed on the roof of the substation to meet the electricity demand of the station's office and living areas. At the same time, photoconductive non-electric lighting technology is introduced in the station, which optimizes the use of natural light sources and results in an estimated annual electricity savings of approximately 3000kWh. Through these energy-saving measures, the substation realizes high energy-saving benefits. The specific measures are analyzed as follows.

3. Analysis of low energy consumption strategies for substations

This study presents a multifaceted comprehensive energy-saving effect enhancement strategy by analyzing the characteristics of substation energy use in summer hot winter warm area and combining them with the specific conditions of the substation buildings under study. The strategy involves several key measures: reducing air-conditioning energy consumption by enhancing the thermal performance of the enclosure structure; reducing ventilation energy consumption by optimizing the spatio-temporal organization of air supply; reducing lighting energy consumption by improving the lighting strategy and introducing renewable solar PV power generation to replace fossil power generation.

3.1. Energy efficiency analysis of the enclosure structure

The substation adopts an all-steel structure and prefabricated construction, which is not only more convenient in the construction process, but also has good thermal insulation performance, thereby reducing the air-conditioning load. The substation's main control building enhances thermal insulation by implementing measures that reduce heat transfer coefficients for roof, walls and external windows, and by optimizing the solar heat gain coefficients for external windows, as detailed in Table 1.

To demonstrate the inherent energy-saving benefits of the research object's enclosure structure, its parameters were compared against the Unified Standard for Industrial Building Energy Efficiency Design and the Design Standards for Public Building Energy Efficiency. The monthly air-conditioning energy consumption for three types of enclosure structures was then simulated using PKPM-GBP (version 20210930) under identical building space conditions, as illustrated in Figure 1. The enclosure structures' optimized design markedly cuts summer air conditioning energy use, especially from June to September when the cooling load is considerably lower than the comparison building under both standards. Notably, in August, the peak cooling load is reduced by approximately 2000kWh compared to the industrial building. In comparison to industrial buildings, the main control building has a cumulative cold load demand reduction of approximately 30% following an upgrade to the thermal performance of the envelope. Furthermore, when compared to civil public buildings, there's a noticeable 25% reduction in the same metric. This indicates that the energy efficiency of the building envelope of the case substation is optimal.

Table 1. Comparison of energy saving parameters of the enclosure structure

	Case substation energy efficiency design programmed	Unified standard for energy efficiency design of industrial buildings GB51245-2017	Design Standards for Energy Efficiency of Public Buildings GB50189-2015
Roof heat transfer coefficient $W/(m^2 \cdot K)$	0.39	0.9	0.8
Heat transfer coefficient of external walls $W/(m^2 \cdot K)$	0.7	1.5	1.5
Heat transfer coefficient of external windows $W/(m^2 \cdot K)$	2.94	4.0	5.2
Solar heat gain coefficient (SHGC)	0.14	0.52/—	0.52/—

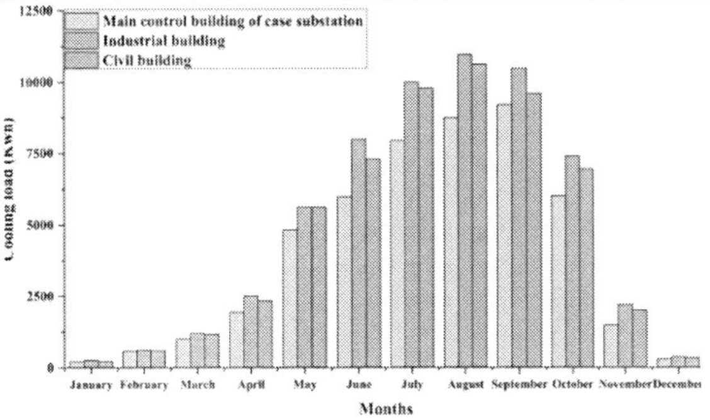

Figure 1. Comparison diagram of the monthly cooling load of the building envelope

The purpose of using air conditioning in buildings is to create a comfortable indoor thermal environment. To understand the indoor thermal conditions under the current energy consumption level of air conditioning, the temperature and humidity were measured on-site from July 19 to July 21, 2024. The data are shown in Figure 2-3. It is evident that the outdoor temperature and humidity fluctuate considerably, exhibiting a discernible cyclical pattern between day and night. In the event of the highest outdoor temperature approaching 40°C, the temperature within the office remains stable at 25-26°C. Meanwhile, the temperature between the equipment is consistently maintained at 23-24°C, and the relative humidity is also observed to be stable, oscillating between 57-65%. The office temperature is marginally higher than that of the equipment rooms. The average relative humidity levels in both spaces are nearly identical, but the fluctuations in relative humidity in the office are greater than in the equipment room, which is because the respiration and perspiration of people in the office will influence the relative humidity. Overall, the indoor temperature and humidity under air-conditioning conditions are in line with the needs of the human body and the use of equipment.

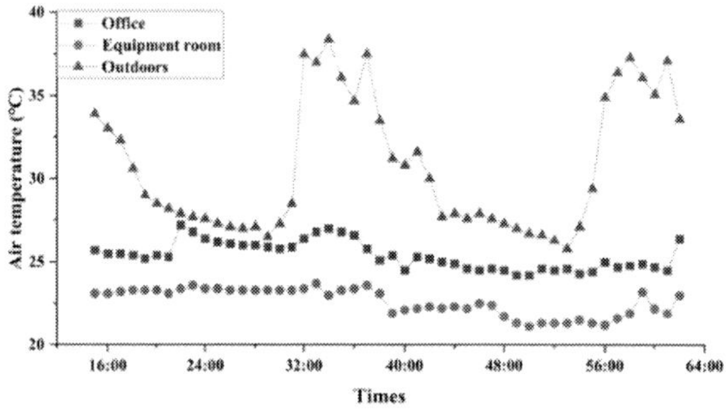

Figure 2. Comparison of indoor and outdoor air humidity in substations

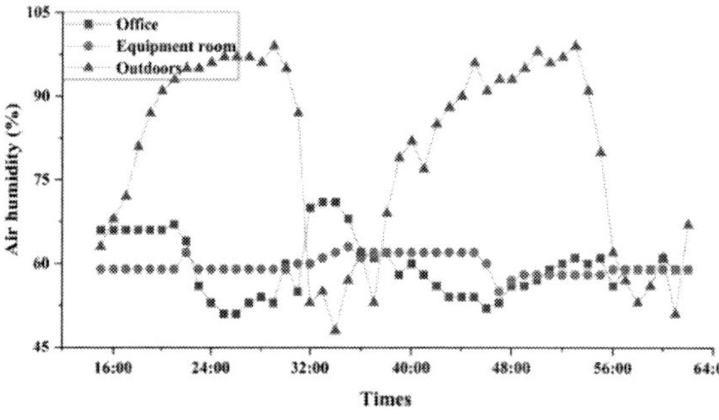

Figure 3. Comparison of indoor and outdoor air humidity in substations

3.2. Analysis of fresh air energy consumption

The fresh air equipment in the air-conditioning system of the substation provides fresh air to ensure the healthy quality of indoor air [12]. which is also a major energy-using component of the air-conditioning system. Generally, the larger the fresh air volume, the higher the energy consumption. The indoor carbon dioxide (CO_2) concentration reflects the strength of fresh air. In the substation, the indoor CO_2 concentration is required to be less than 900ppm. Additionally, the small inertia of indoor CO_2 concentration results in faster fresh air conditioning and lower energy consumption per unit hour of the fresh air unit [13]. Figure 4 displays the indoor CO_2 concentrations in the case building, which meet design standards for both the office and equipment room. The office shows significantly higher CO_2 levels due to higher personnel density and accumulated respiratory emissions, coupled with warmer air temperatures. In contrast, the equipment room, with fewer people, exhibits lower and more stable CO_2 concentrations and cooler air temperatures.

Journal of Physics: Conference Series 3012 (2025) 012048 doi:10.1088/1742-6596/3012/1/012048

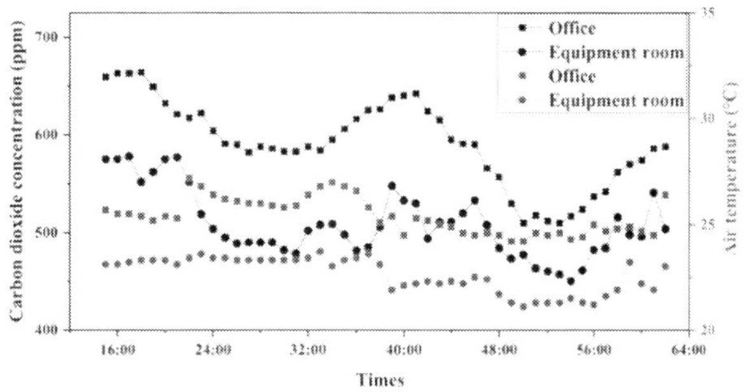

Figure 4. Carbon dioxide concentration between offices and equipment in the main control building

A comparison of CO_2 concentrations in offices and equipment rooms, shown in Figure 4, indicates that unoccupied spaces exhibit significantly lower CO_2 concentrations, thereby necessitating lower fresh air volumes. The measured CO_2 concentration in the occupied office is significantly less than the required 900ppm, indicating that lower fresh air volumes can also be considered. Reducing the fresh air load by controlling the amount of fresh air through CO_2 concentration is essentially achieved by reducing the amount of fresh air delivered to dilute the CO_2 concentration. The ventilation required to dilute CO_2 is calculated according to the steady state dilution equation [14], which is given below:

$$v = q/(c - co) \tag{1}$$

Where, q——CO_2 generation rate per person, L/s; v——required fresh air volume per person for diluting CO_2, $m^3/(s{\cdot}p)$; c——allowable indoor CO_2 concentration, L/m^3; co——outdoor air CO_2 concentration, L/m^3, which is generally taken as $0.3L/m^3$.

The permissible concentration of indoor CO_2 is 0.07~0.15%, and if the midpoint of this range, 0.1%, is used, the minimum fresh air volume required for each person is calculated to be $24m^3/(h{\cdot}p)$. At the same time, if the office fresh air supply is changed from a continuous air supply to an intermittent air supply, such as changing the air supply period from 8:00-18:00 to 9:00-11:00, 11:30-13:00, 13:30-15:00 and 15:30-17:30, the case substation can be reasonably partitioned according to the needs of different areas and dynamically adjusts fresh air supply, which effectively reduces the amount of unnecessary fresh air supply. To demonstrate the energy-saving effect of zoned fresh air control, a benchmark building with continuous supply and a fresh air volume of $30m^3/(h{\cdot}p)$ is set up as a comparison. The simulation shows that the cumulative fresh air load of the air-conditioning system in the case substation is reduced by about 28% compared with the benchmark building, as shown in Figure 5.

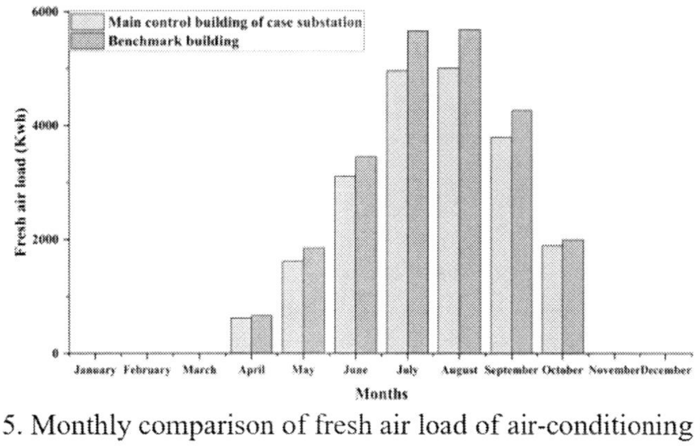

Figure 5. Monthly comparison of fresh air load of air-conditioning system

3.3. Natural light

Dynamic energy simulations reveal that lighting is a significant component of the case substation's total energy use, underscoring the need for lighting energy conservation. By strategically placing main functional rooms such as offices, meeting rooms, and control rooms in areas with the best natural light, the design minimizes lighting needs and maximizes natural light use. Guangzhou, being in the IVth light climate zone, has no top daylighting requirements for the substation. According to "Standard for daylighting Design of building" GB/T 50033, the daylighting requirements of each main functional room are shown in Table 2 [14]. Lighting simulation was performed using PKPM-GBP. The indoor wall and ceiling reflectance of the substation building are both 0.75, the floor reflectance is 0.5, and the visible light transmittance of the exterior windows is 0.57. Other simulation parameters are shown in Table 3.

Table 2. Standard value of daylighting in buildings

Daylighting level	Name of the premises	Side daylighting	
		Standard value of daylight factor (%)	Standard value of natural light intensity(lx)
II	Main control room, Cable room	4.0%	600
III	Offices, Conference room	3.0%	450
V	Corridors, Bathroom, staircase	1.0%	150

Table 3. Substation lighting simulation conditions

CIE Sky Model	Spatial grid spacing/m	Number of grids/pcs	Ground reflection coefficient	Number of light reflections/times	Computation time
whole cloudy sky	0.50	8314	0.30	5	January 1 to December 31

For the main functional rooms on the first floor, the proportion of the area with an average illumination value of not less than 450lx for at least 4h/d in the spaces of offices and conference rooms is 100%. Additionally, the average number of hours of illumination meeting the standard is more than 6h/d. Therefore, so the indoor lighting requirements can be met through natural lighting. For the auxiliary rooms, such as the equipment room 1 and cable room 2, which are well oriented, the proportion of the area with an average number of hours of illumination meeting both the standards of at least 3h/d and at least 4h/d is 91.6%. This level of illumination can adequately meet the demand for indoor lighting. The statistical results of the number of hours of illuminance compliance for the main functional space on the first floor are shown in Table 4. Daylighting simulation analysis of the case substation was conducted using the software PKPM-Daylight. The simulation results of the first floor are shown in Figure 6.

Table 4. Statistical results of the number of hours of compliance with the illuminance standard for the main functional spaces on the first floor

Room Name	Daylighting level	Illuminance standard value (lx)	Average number of hours of illumination compliance (h/d)	Proportion of area meeting the illuminance standard with an average number of hours of not less than 4h/d (%)	Whether it meets the requirements
Office	III	450	6	100	yes
Conference room 1	III	450	7	100	yes
Conference room 2	III	450	6	100	yes
equipment room 1	II	600	3	91.6	yes
equipment room 2	II	600	1	8.2	no
Cable room 1	II	600	0	6.6	no
Cable room 2	II	600	5	85.5	yes

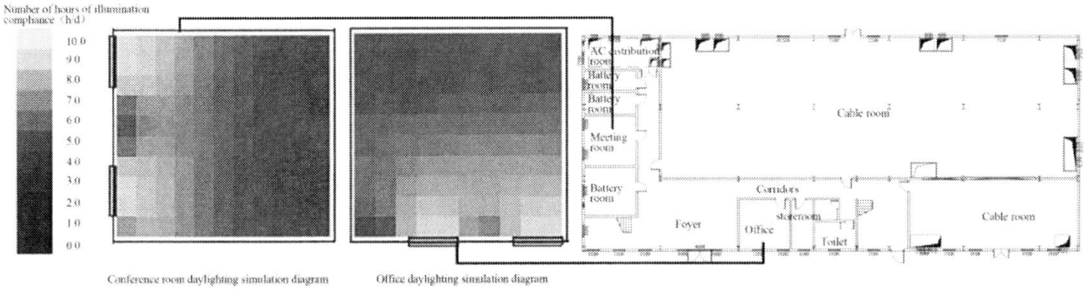

Figure 6. Distribution of the number of hours of illuminance compliance on the first floor of the main control room of the case substation

For the main functional rooms on the second floor, the average number of hours of illumination in the office space that meets the standard is more than 4h/d. Additionally, the proportion of the area that meets the standard with an average number of hours of illumination of at least 4h/d is 86.88%. Therefore, the indoor daylighting needs can be met through natural lighting.

The case substation enhances lighting efficiency by choosing energy-saving products that meet illumination standards, optimizing lighting power density, and employing high-efficiency intelligent lighting systems. It also implements zoning control for public areas and uses energy-saving measures such as timing and sensing to reduce unnecessary lighting, thereby effectively lowering energy consumption. By simulating and estimating the 24-hour lighting duration for each functional and corridor space, the following lighting power densities and durations were determined: the main control room has a lighting power density of 6.37 W/m^2 with a duration of 7 hours; the relay room has a lighting power density of 1.5 W/m^2 with a duration of 4 hours; and the aisle lighting has a power density of 1.6 W/m^2 with a duration of 8.5 hours. Compared to the benchmark building power consumption, the lighting power density of the case substation building is 70%, which is lower than the standard limit. Consequently, the lighting power consumption has been reduced by 72%, as shown in Figure 7.

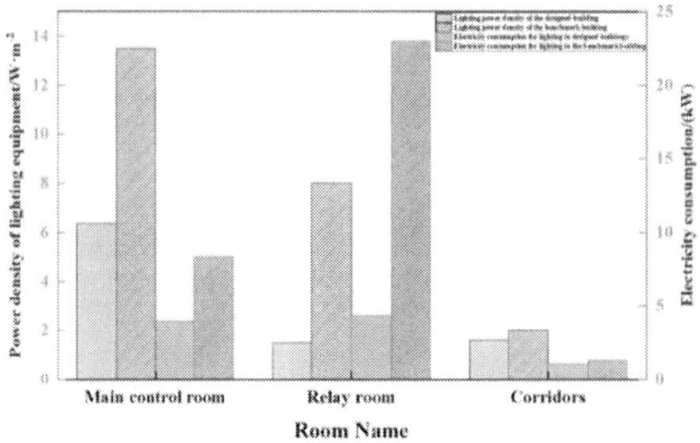

Figure 7. Comparison of substation lighting energy savings

3.4. *Solar PV applications*

The substation utilizes the roof of the main control building and the hillside on the northeast side of the site to install a solar PV power generation system, directing the generated power primarily to the air conditioning and lighting systems within the main control building. A steel structure was constructed as the main frame on the roof of the main control building. With this structure, a total of 140 standard 655Wp monocrystalline silicon modules (module size 2384*1303*35mm) were installed in the form of 20° inclination, covering an area of about 1,050m^2, with an installed capacity of 91.7kWp. Meanwhile, 280 monocrystalline silicon modules were installed in the northeastern slope in accordance with the direction of the mountain, covering an area of about 2,800m^2, with an installed capacity of 183.40kWp. Of this capacity, 15% of the PV power supply is used for the main control building. The arrangement of solar PV modules is shown in Figure 8.

Figure 8. Solar photovoltaic distribution map

In this paper, the hour-by-hour horizontal solar radiation test data from July 19 to July 21 in Guangzhou is used to calculate the PV system's power generation capacity on building. According to "Code for design of photovoltaic power station" GB 50797-2012, the PV power generation can be calculated according to the following formula [15]:

$$E_P = H_A * \frac{P_{AZ}}{E_s} * K \tag{2}$$

Where, E_P——electricity generation, kW; H_A——total solar irradiance on the horizontal plane, kWh/m^2; P_{AZ}——module installation capacity, kWp; E_S——irradiance under standard conditions, constant 1kWh/m^2; K —— integrated efficiency coefficient. PV panel integrated efficiency coefficient based on the standard "Specification of photovoltaic power generation efficiency" GB/T39857-2021, monocrystalline silicon PV module efficiency should not be less than 17.8%.

Hourly electricity consumption simulations for the main control building's air-conditioning and lighting systems over a 24-hour period were compared with solar radiation intensity and PV power generation curves, as depicted in Figure 9, excluding the PV panels' energy storage capacity. PV generation peaks between 10:00-13:00, sufficiently covering the cooling and lighting demands for three midday hours.

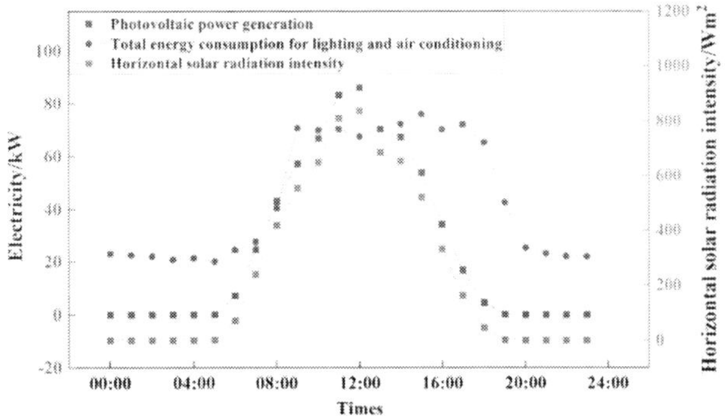

Figure 9. Solar radiation intensity, photovoltaic power generation and building energy consumption

Building energy consumption excludes renewable energy generation. The case substation has no domestic hot water system energy, elevator system energy and heating system energy. Therefore, only the sum of air-conditioning energy, lighting energy and ventilation energy is considered. The annual PV power generation of the building can be estimated through the formula (2). Considering the probability of cloudy and rainy weather in Guangzhou area, the average annual solar radiation rate is taken as 843.5kWh/m^2·a. The renewable energy generation is 62,642.1kWh, and the energy consumption supplied by the PV system is 29.47kWh/m^2. The main control building of the case substation has a floor area of 2125.96m^2, and there is no elevator equipment in the building. Through the simulation calculation of the annual energy consumption of the building body, including air conditioning, lighting and ventilation energy consumption, the annual power consumption of each system is as follows: lighting accounts for 9.56kWh/m^2, with a total annual consumption of 20324.18kWh; air conditioning consumes 30.28kWh/m^2, amounting to an annual total of 64374.07kWh; ventilation uses 2.95kWh/m^2, with an annual power consumption of 6271.58kWh. Based on the simulation results, the energy consumption of the building body is estimated at 43.33kWh/m^2, with a total annual electricity consumption of 92117.85kWh for the building body, and an integrated electricity consumption of 29475.75kWh. Therefore, the integrated value of the building energy consumption without renewable energy power generation is 112.66kWh/m^2·a, compared to 36.04kWh/m^2 for the case substation.

The comprehensive energy-saving rate of a building is the ratio of the difference in the comprehensive energy consumption values between the design building and the benchmark building to the comprehensive energy consumption value of the benchmark building. The energy consumption of the benchmark building is 16.08kWh/m^2 for lighting, 43.54kWh/m^2 for air conditioning, and 4.32kWh/m^2 for ventilation. Additionally, the combined value of the building's energy consumption is 63.94kWh/m^2, which results in a combined energy saving rate of 43.63% for the case substation building.

4. Conclusion

This study investigates low-energy strategies for a substation building in Guangzhou, utilizing field research and PKPM simulations to understand its energy, load and climate profiles. The analysis focuses on major energy consumers, including air-conditioning and lighting. We analyzed the building ontology energy-saving strategy of the main control building and the favorable impact of renewable energy utilization on building energy consumption in the case substation, completing a feasibility study on the technical application of low-energy substations. The study yielded the following conclusions:

(1) By optimizing the thermal insulation of the enclosure structure, the substation can significantly reduce its cooling load by about 25-30% during typical summer month compared with similar industrial

buildings and similar office buildings, thus helping to realize the low energy consumption of the substation building. (2) With regard to the design of fresh air systems, reasonable design ensures good indoor natural ventilation, while dynamic adjustment of the fresh air volume minimizes excess supply. These measures effectively reduce the fresh air load of the air-conditioning system, enabling the main control building to save about 30% on its fresh air load. (3) With regard to lighting energy consumption, the case substation employs natural lighting to the fullest extent possible, reduces the lighting power density, and minimizes lighting duration through a combination of rational design and active energy efficiency. The lighting power density of the building is 70% below the standard limit, while the lighting power consumption has been reduced by 72%. (4) With regard to the utilization of renewable energy, the deployment of solar PV technology enables the effective fulfilment of the building's cooling and lighting power requirements for a period of three hours at noon. This results in a comprehensive energy saving rate of 43.63%.

References

[1] Lu, M., Wang, J., Chen, J., Chen, X., 2017. Influence of Heat Release Density in Converting Station Monitor Room on Energy Saving Performance of Building Envelope. Build. Energy Effic. 45, 128–131.

[2] Liu, M., 2020. RESEARCH AND APPLICATION OF ENERGY SAVING TECHNOLOGY FOR POWER SUBSTATIONS (Master). Shanghai Jiao Tong University.

[3] Chen, X., Wang, J., Lu, M., Chen, J., Jing, C., 2017. Cooling and Heating Load Characteristics of Typical Building for Converting Station in High Altitude Cold Region. Build. Energy Effic. 45, 126–131.

[4] Liu, Y., Zhang, Y., Cai, Z., Zhang, Jun, Liu, Q., 2020. The Influence of Thermal Performance of Building Exterior Envelope on Building Energy Consumption of Substation. Build. Energy Effic. 48, 29–33.

[5] Xiong, T., Li, G., Liu, Y., Li, S., 2021. Research on Method of Near—zero Energy Consumption Substation Building. Build. Energy Effic. 49, 39–44.

[6] Jia, C., 2024. Photovoltaic Energy Saving and Emission Reduction System of 110 kV Substation. Energy Energy Conserv. 86–88.

[7] Liu, C., Mei, X., Yang, Y., 2024. Exploring the carbon reduction effect of solar photovoltaic systems on substations in hot summer and cold winter regions. Theor. Res. Urban Constr. 10–12.

[8] Zhao, Y., 2022. Analysis of Typical Urban Climate Characteristics Based on Outdoor Thermal Comfort Control (Master). Xi'an University of Architecture and Technology.

[9] Zhang, Z., Nie, J., Sarenengaowa, Zhang, Q., Wang, A., Song, R., 2021. Building Energy Saving Analysis of 110 kV Substations Based on Climate Regional Characteristics in Inner Mongolia. Inn. Mong. Electr. Power 39, 33–39.

[10] Liu, Y., 2021. Study on energy efficient design of substation buildings in the hot-humid area of China (Master). South China University of Technology.

[11] Li, Z., 2019. Introduction to substation energy saving and consumption reduction measures. Resour. Econ. Environ. Prot. 44+46.

[12] Zhang, D., Zhang, L., Zhang, Y., 2024. Effects of Living Walls in the Removal of Indoor CO2 and Energy Conservation in Office Buildings. Build. Sci. 40, 100–110.

[13] Luo, Z., Zhang, J., Zhang, T., Liu, X., 2024. Potential Evaluation and Influencing Factors of Fresh Air System Demand Response. Power Syst. Technol. 48, 2775–2783.

[14] GB/T50033—2013, Standard for daylighting Design of building, 2013.

[15] GB 50797—2012, Code for design of photovoltaic power station, 2012.

Journal of Physics: Conference Series **3012** (2025) 012049 doi:10.1088/1742-6596/3012/1/012049

Study on the Impact of Particle Trap Structural Parameters on Electric Field Distribution in 500kV GIL

Xiyue Yu*, Daochun Huang, Yiming Wang

School of Electrical Engineering and Automation, Wuhan University, Wuhan, China

*Corresponding author's e-mail: 2484732642@qq.com

Abstract: The motion of metal particles plays a significant role in the failure of insulation in gas insulated transmission lines (GIL). So, as an important structure to capture particles and limit their movement, the design of particle traps is one of the key technologies of GIL. To study the influence of the structural parameters of the particle trap on the nearby electric field and optimize the electric field distribution of the particle trap, in this paper, a FEM simulation model was constructed first. Two paths near the upper and lower surfaces of the particle trap were selected in the model to calculate the electric field distribution on them. Further, parametric scanning is added to the research module to analyze the influence of trap height, thickness, slot width, and spacing on the electric field distribution. The results show that increasing the thickness of the trap can effectively enhance the shielding effect of the internal electric field. Increasing the trap height can significantly reduce the field strength below the trap, and increasing the slot width can effectively reduce the field strength above the trap. When the trap thickness increases from 2 mm to 6 mm, the electric field intensity on the lower surface of the trap decreases from 0.2 kV/mm to 0.05 kV/mm, and the optimization degree is as high as 75%. When the trap height increases from 6 mm to 10 m, the electric field intensity on the lower surface of the trap decreases from 0.5 kV/mm to 0.18 kV/mm, and the optimization degree can reach 64%. This study can provide a reference for optimizing particle trap structure parameters.

1. Introduction

During the manufacturing, transportation, installation, and operation of the GIL, due to poor internal cleaning, equipment vibration, mechanical friction of the contacts, or other factors, metal particles may be generated and accumulated, the presence of which can lead to a localized electric field strengthening and a significant reduction in the withstand voltage of the equipment[1]. At present, the measures for inhibiting the movement of metal particles mainly include particle traps[2], surface coating coverage[3], and optimization of insulation support structures[4]. Among them, particle traps are a structure that inhibits the movement of particles by constructing a local low field strength area, which is a crucial measure of particle suppression[5].

To effectively capture metal particles, researchers around the world have been working since the 1970s to design different structures for particle traps, leading to successively proposing various particle trap design schemes[6]. Westinghouse Electric of the United States developed the grid-type particle trap. It proposed the inner buckle slot and wing baffle to prevent the particles entering the trap area from escaping[7]. Wang Jian of North China Electric Power University proposed a wedge-shaped particle trap, which can shield the electric field at the bottom of the trap more efficiently than the traditional trap, and can block the escape of spherical particles[8]. It can be seen that most of the studies are based on the motion characteristics of metal particles, and are dedicated to optimizing the shape of particle traps.

Journal of Physics: Conference Series **3012** (2025) 012049 doi:10.1088/1742-6596/3012/1/012049

There are few discussions on the influence of other structural parameters (such as trap thickness and height) of particle traps and the study of the electric field distribution around particle traps. Since the particle trap structure will change the distribution of the surrounding electric field and reducing its electric field can effectively improve the shielding effect of the trap on the internal field strength, it will affect the capture and suppression of metal particles. Therefore, it is necessary to study the influence of the change in the structural parameter of the particle trap on the electric field distribution.

In this paper, the conjoined particle trap in a typical 500 kV GIL is taken as the research object, and the influence of other parameters of the slot on the electric field near the upper and lower surfaces of the particle trap is studied. The influence law and degree of parameter change on the electric field distribution are summarized, which provides a reference for optimizing the structural parameters of the particle trap in high-voltage GIL.

2. The model and structural parameters of the particle trap

Typically, particle trap structures can be described by the following structural parameters: slot shape, slot width(length), slot thickness, slot spacing, and number of slots. The length and width of the groove represent the width and width and length of the opening groove (usually grid-shaped), which are two shape parameters of the particle trap opening. The groove thickness is consistent with the thickness of the metal plate of the particle trap, and its value is mostly given by actual engineering experience. Slot spacing refers to the spacing between two adjacent slots, and the slot array represents the distribution of slots.

The initial structure parameters of the particle trap are shown in Table 1.

Table 1 Initial structural parameters of particle traps

The name of the parameter	Parameter value
Slot width (W)	4 mm
Slot length	24 mm
Particle trap thickness (D)	2 mm
Particle trap height (H)	10 mm
Slot spacing (L)	5 mm
Number of slots	84 pcs

The initial structure of the particle trap is shown in Figure 1.

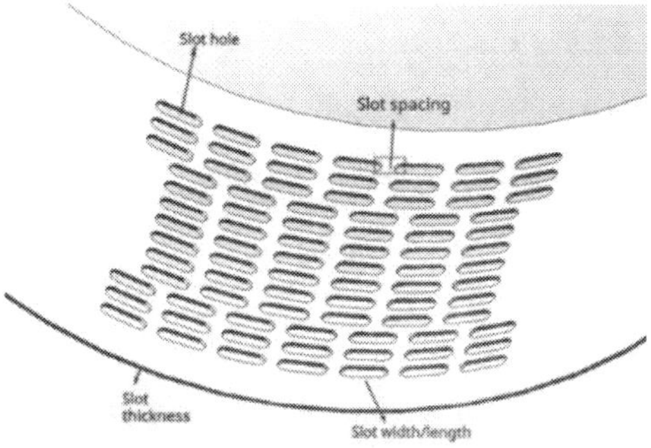

Figure 1 The model of the particle trap and the description of the structural parameters

To study the influence of structural parameters on the electric field distribution on the upper side and lower side of the particle trap, path 1 and path 2 were taken out at 4 mm near the upper and lower surfaces of the particle trap (Figure 2), and the electric intensity on them is calculated.

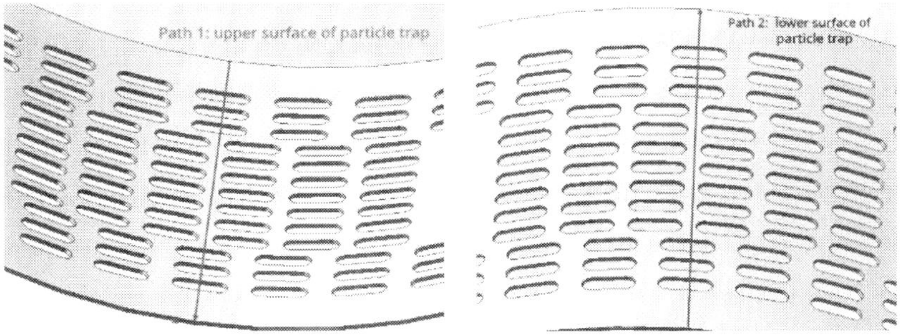

Figure 2 The path of the electric field value on the upper and lower surfaces of the particle trap

The radial section of the particle trap is represented by the blue box at the bottom, as seen in Figure 3. The particle trap without openings is represented by the overall rectangle, like the one on the far left; the presence of a blank space between the two rectangles indicates that this is a particle trap slot. Paths 1 and 2 are 4mm from the upper and lower surfaces of the particle trap, respectively. Along paths 1 and 2, the influence of the change of particle trap structure on the electric field distribution in four typical regions: the trap unslotted area, the trap slotted area, the particle trap area, the particle trap edge, and the outside of the particle trap can be observed, which can reflect the degree of influence of each region on the change of structural parameters.

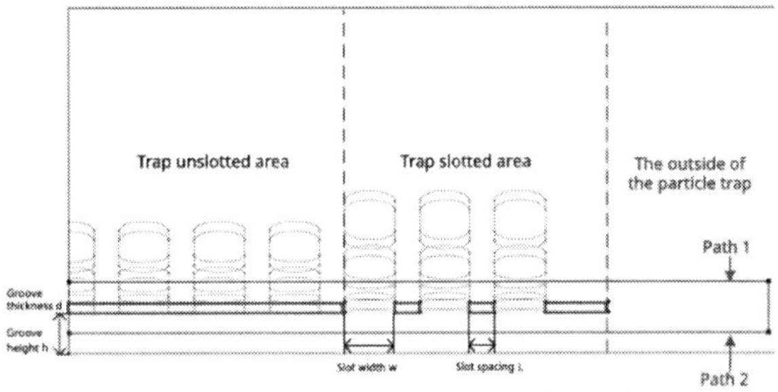

Figure 3 Typical region on the path

3. The influence of particle trap structural parameters on electric field distribution

3.1. Effect of trap height and thickness on electric field distribution

The arrangement of the particle trap is equivalent to changing the position height of the inner wall of the GIL, so increasing the height of the particle trap and increasing the thickness of the particle trap need to be discussed together. For the conjoined particle trap structure in this paper, an excessive increase in the height an d thickness of the particle trap will cause a certain volume crossover between the tri-posted insulator and the particle trap, so it is necessary to make certain restrictions on the height and thickness.

After testing, when the height and thickness of the trap are less than or equal to 10 mm, they will not cross with the tri-posted insulator. So, the following example is designed to study the influence of the height and thickness of the particle trap on the electric field distribution, as shown in Table 2.

Table 2 Particle trap height and thickness settings

The name of the parameter	Initial value(mm)	Step size(mm)	Stop value(mm)
Particle trap height *(h)*	6	1	10
Particle trap thickness *(d)*	2	1	6

3.1.1. Trap thickness. As shown in Figure 4, the height of the trap(h=10mm) remains unchanged, and the thickness of the trap is gradually increased from 2mm to 6mm in steps of 1 mm.

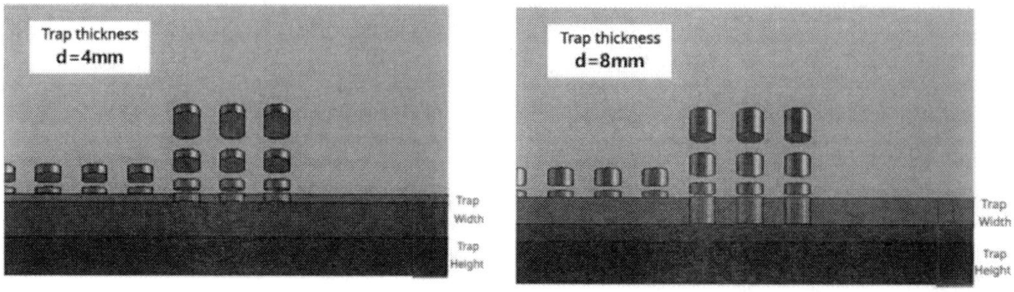

Figure 4. Schematic diagram of the increase in the thickness

The height of the trap remains the same, and the increase in thickness does not change the distance between the upper surface of the trap and the tri-posted insulator and the high-voltage conductor, but only the distance between the lower surface of the particle trap and the grounded shell. So, the change of the thickness has almost no effect on the electric field on path 1, only the influence of the electric field on path 2 with the thickness of the trap is analyzed. As shown in Figure 5, as the thickness of the trap increases, the volume of the low field strength region below the trap decreases, and the electric field strength inside the trap decreases.

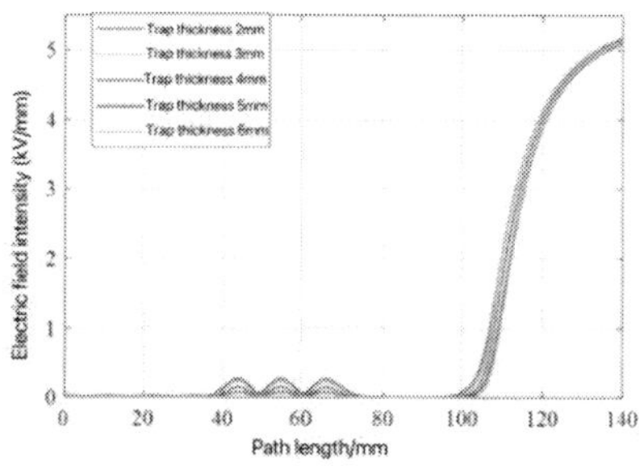

Figure 5. Distribution of the electric field on path 2 (lower surface)

3.1.2. Trap height. The thickness of the trap(d=2mm) remains unchanged, and the trap height is gradually increased from 6mm to 10 mm in steps of 1 mm.

With the increase of the trap height, the volume of the low field strength area below the particle trap will also increase, which makes the trapped charged metal particles subjected to low electric field force for a longer time (Figure 6). The charged metal particles receive almost only gravity and gas resistance in the low field strength region, and their activity will be rapidly lowered. So, increasing the trap height is beneficial to enhance the trap's ability to capture particles.

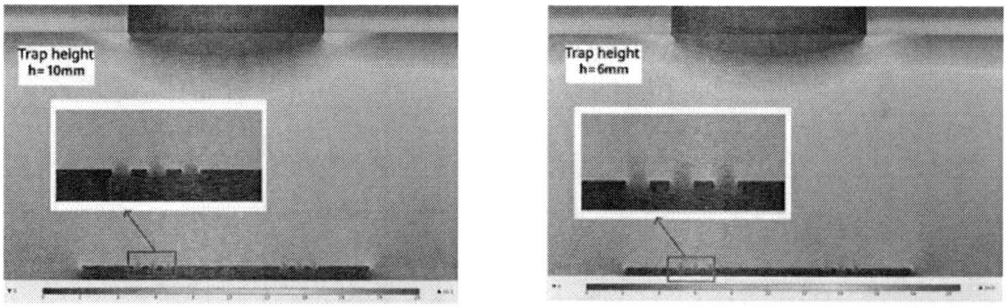

Figure 6. Distribution of electric field in axial section when the trap height is 10 mm and 6 mm

Further, the electric field strengths on paths 1 and 2 defined above near the upper and lower surfaces of the particle trap were extracted, as shown in Figure 7.

As indicated in Figure 7(a), the electric field on path 1 increases with the height of the trap when the thickness of the trap is constant. The appearance of the slot changes the distribution of the original electric field line. The electric field line passes through the trap slot and stops on the grounded metal shell. As a result, the electric field shows a wavy change in the trap slot area. However, in other areas without trap slots (e.g. x=0~40mm), the electric field distribution is more uniform. Others, at the edge of the particle trap (x=110mm), the electric field increases rapidly due to the distortion caused by the corner of the edge, and then rapidly decreases, transitioning to the low field strength region (x=120~140mm) near the ground electrode.

The opposite of the electric field variation law on path 1, the electric field on path 2 decreases with the increase of the trap height, as shown in Figure 7(b). Path 2 is located below the particle trap. The electrostatic shielding effect of the particle trap makes the area below the particle trap in a low field strength. However, due to the appearance of the trap slot, the electrostatic shielding effect is destroyed to a certain extent, a part of the electric field line passes through the slot and terminates on the grounding shell. The electric field in the trap slot area (x=40~70mm) is significantly greater than the field strength in other areas in the particle trap. Else, at the edge of the particle trap (x=110mm), because of the disappearance of the electrostatic shielding effect, the electric field will increase rapidly, and transition to the high field strength area outside the trap (x=110~140mm).

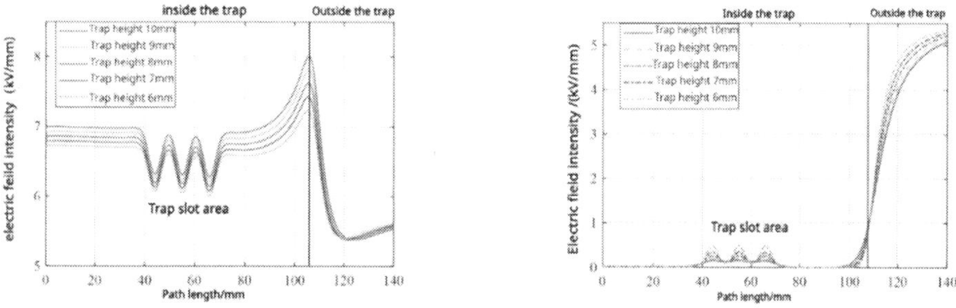

(a) Path 1 (b) Path 2

Figure 7 Distribution of electric fields on path 1 and 2

3.2. *Effect of trap slot shape on electric field distribution*

Controlling the width and spacing of the particle trap hole will change the distribution of the hole on the whole metal plate. The larger the width of the trap hole, or the wider the spacing between the trap holes, the gap between the outermost trap hole and the edge of the particle trap will be smaller. To ensure that the particle trap slots don't extend past the particle trap's edge, specific limitations on the slots' width and spacing must be put in place. Accordingly, the following example is designed to study the effects of particle trap slot width and spacing on the electric field distribution, as shown in Table 3.

Table 3 Particulate trap slot width and spacing settings

The name of the parameter	Initial value(mm)	Step size(mm)	Stop value(mm)
Slot width (W) (L=5mm)	3	0.5	5.5
Slot spacing (L) (W=4mm)	5	1	10

3.2.1. Trap slot width. The trap slot spacing (L=5 mm) remains unchanged, and the width of the particle trap slot gradually increases from 3 mm to 5.5 mm in 0.5 mm steps.

The electric field strengths on paths 1 and 2 defined above near the upper and lower surfaces of the particulate trap are extracted, shown in Figure 8.

As can be seen from Figure 8(a), the electric field intensity in the trap slot area (x=40~100mm) on path 1 decreases with the increase of trap width. In addition, since the width of the trap will reduce the distance between the outermost slot and the edge, in x=0~40mm, the trap width of 5.5mm will also have a unique wavy electric field distribution near the slot, while the trap width of 3mm will not appear wavy.

The variation of the electric field intensity with the width of the trap in path 2 is just the opposite of the law in path 1(Figure 8(b)). The larger the width of the trap, the greater the electric field intensity on the lower surface. This increases the electric field force experienced by particles entering the trap and increases the amount of charge gained by colliding with the grounded shell. According to the figure, the maximum field strength under the trap slot is less than 1 kV/mm, which is much lower than the lifting field strength at this voltage, so it can be considered that the particulate activity can be effectively suppressed in the wide trap.

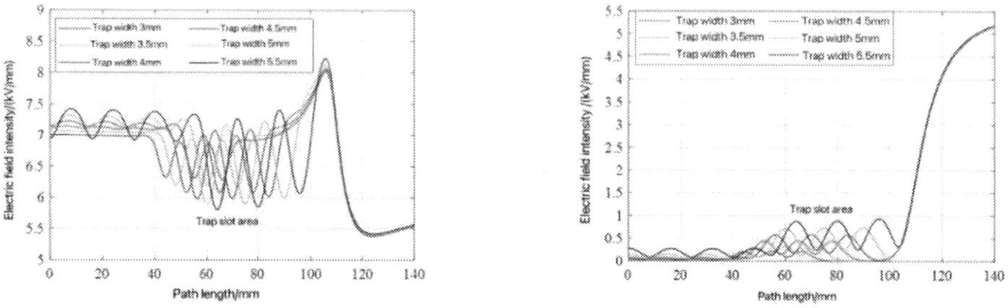

(a) Path 1 (b) Path 2

Figure 8 Distribution of the electric field on paths 1 and 2

3.2.2. Trap slot spacing. The trap slot width (W=4mm) remains unchanged, and the slot width is gradually increased from 5mm to 10mm in steps of 1mm.

The electric field strengths on paths 1 and 2 are extracted as follows.

From Figure 9(a), the electric field strength on path 1 decreases slightly as the spacing increases. From Figure 9(b), the electric field at path 2 (bottom of the trap) hardly changes with the trap spacing. Therefore, it can be concluded that the spacing of the trap slots has so little effect on the electric field intensity near the particle trap that can be ignored.

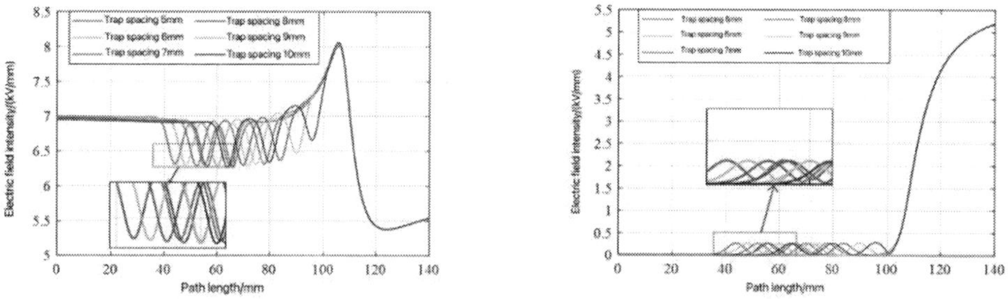

(a) Path 1 (b) Path 2

Figure 9 Distribution of the electric field on paths 1 and 2

4. Conclusion

In this paper, the electric field near the upper and lower surfaces of the particle trap is selected as the research object. The influence of the change in the structural parameters of the particle trap (thickness, height, slot width, and spacing) on the electric field distribution was studied. Conclusions are summarized below:

(1) The field strength above the trap will not be changed by increasing the thickness of the particle trap, but the shielding ability of the shielding area below the trap can be enhanced. When the trap thickness increases by 4mm, the electric field intensity above the trap remains unchanged at about 3 ~

4kV, while the degree of electric field optimization below the trap reaches 75%. So, if the construction conditions permit, the thickness of the trap can be appropriately increased in exchange for better electric field shielding performance.

(2) The change in trap height will affect the electric field on the upper and lower surfaces of the trap. Raising the trap height from 6mm to 10mm results in a 47.8% increase in the upper surface's maximum field strength and a 64% decrease in the lower surface's field strength. As the height of the trap increases, the closer the upper surface of the trap is to the high-voltage conductor, the degree of distortion of the electric field on the upper surface increases. At the same time, the volume of the electric field shielding area under the trap is expanded, and the electric field strength under the trap is reduced, which facilitates particle capture.

(3) The increase in the slot width will reduce the field strength above the trap and increase the field strength below, while increased slot spacing has little effect on the electric field near the trap.

References

[1] Li Q M, Wang J, Li B T, et al. 2016, Review on Metal Particle Contamination in GIS/GIL, High Voltage Engineering, chapter 42(03), P 849-860

[2] Wang J, Ni X R, Wang Z Y, et al. 2018, Trapping Probability and Designing Optimization of Particle Trap in DC GIL, High Voltage Engineering, chapter 44(12), P 4090-4097

[3] Ni X R, Wang J R, Wang J, et al. 2018, Polarity Effect of DC GIL Electrode Coating on Particle Inhibition, High Voltage Engineering, chapter 44(8), P 2695-2703

[4] Zhang Y T, Wu Z H, Xu J Z, et al. 2023, Optimization Design of Insulation Structure for Post Insulator in UHVAC GIS, Transactions of China Electrotechnical Society, chapter 38(01), P 258-269

[5] Li J, Yang L W, Yu J. 2022, Review on Metal Particle Contamination in GIS/GIL Equipment, Electrotechnics Electric, chapter (03), P 1-7

[6] Li X Y, Dan S H, You W R, et al. 2013, Review of Particle Trap in SF6 Gas Insulated Apparatus High Voltage Apparatus, chapter 49(9), P 150-154

[7] Hopkins M D, Bolin P C. 1981, Gas insulated transmission line with adhesive particle trap carrier, Westinghouse Electric Corp

[8] Wang J, Chang Y N, Wang J R, et al. 2020, Design and Optimization of Particle Traps in DC GIL Based on the Capture Effect Analysis, Proceedings of the CSEE, chapter 40(15), P 5050-5061

Prediction of the Spatial-Temporal Distribution of Charging Loads Considering Travel Behaviors of Electric Vehicles

Fuzhang Wu[1, 2 *], **Fan Zhang**[3, 4], **Jianlin Tang**[3, 4], **Fang Chi**[1, 2]

[1] Hubei Engineering and Technology Research Center for AC/DC Intelligent Distribution Network, School of Electrical Engineering and Automation, Wuhan University, Wuhan, Hubei Province, 430072, P. R. China

[2] School of Electrical Engineering and Automation, Wuhan University, Wuhan, Hubei Province, 430072, P. R. China

[3] Electric Power Research Institute, CSG, Guangzhou, Guangdong Province, 510663, P. R. China

[4] Guangdong Provincial Key Laboratory of Intelligent Measurement and Advanced Metering of Power Grid, Guangzhou, Guangdong Province, 510663, P. R. China

*Corresponding author's e-mail: wfzhang@whu.edu.cn

Abstract. The analysis of the spatial-temporal distribution of electric vehicle (EV) charging loads serves as the cornerstone for estimating charging resource supply and demand, as well as scheduling charging loads. The travel behavior of EVs is intrinsic to the charging load, and therefore, it is essential to consider this behavior in predicting charging loads. In this paper, we propose a novel charging load prediction method that takes into account the travel behavior of EVs. Firstly, we establish an EV travel decision model for users based on regret theory, which enables us to obtain the choice probabilities of EVs' travel paths and departure times for origin-destination (OD) pairs across different time periods. Secondly, we construct a traffic equilibrium model that incorporates traffic dynamics to analyze the transfer behavior of large-scale EVs within transportation networks. Thirdly, we calculate the distribution characteristics of EV charging loads based on time-space transfer behavior, power consumption rates, and charging conditions. Finally, we conduct case studies to demonstrate the effectiveness of our proposed method, thereby validating its practical application in the realm of EV charging load prediction.

1. Introduction

The connection of large-scale electric vehicles (EVs) brings not only great challenges but also opportunities to the operation of power systems. For example, the charging load of EVs will cause the peak-valley difference of power grids to increase [1], and the orderly charging and discharging scheduling can participate in the auxiliary services of power grids to support economic and safe operation [2]. Both the analysis of the impact of EV charging loads on power grids and the participation of EVs in the power grid interaction cannot be separated from the effective calculation of the distribution characteristics of charging loads. Therefore, the calculation of the spatial-temporal distribution of charging loads has very important theoretical and application value.

The existing methods on EVs charging load analysis or forecasting are mainly divided into two categories: data-based and model-based. Ref. [3, 4] describes the driving characteristics of EVs through statistics on different travel time and driving time, and then obtains the charging load distribution. Data-based forecasting methods often have low accuracy due to lack of large amounts of data and poor adaptability of results. The model-based charging load prediction starts from the description of the traffic behavior of EVs, and analyzes the charging load distribution law by mastering the travel process of EVs. The travel journey of EVs encompasses two primary aspects: user travel decision-making and vehicle driving transitions. In terms of user decision modeling, prevalent research primarily leans on the expected utility theory. Specifically, Ref. [5] analyzes EV charging consumption decisions based on the principle of utility maximization. However, these existing decision theories are grounded in the assumption of users' complete rationality, which often diverges from real-world user decision-making patterns. When it comes to describing the driving transitions of EVs, existing research primarily focuses on elucidating the transfer relationships between EVs at different locations based on the travel chain concept. These transitions are categorized into simple and complex chains, with the transition probabilities between various locations being described using Markov processes [6, 7]. Yet, this framework, while structured, overlooks the intricate interplay between EVs and the broader transportation network. In conclusion, while advancements have been made in understanding EV travel processes, there remains a need to refine these models by integrating realistic user decision-making and considering the broader context of the transportation network. This holistic approach will yield more accurate and insightful models, ultimately contributing to the development and optimization of EV systems.

In essence, the charging load of electric vehicles (EVs) is intrinsically linked to their travel patterns, which are shaped by users' travel decisions and the driving behaviors of EVs within the traffic network. EV users, as real economic agents, exhibit characteristics of bounded rationality, meaning their decisions are influenced by a combination of factors rather than pure economic rationality. Furthermore, the dynamic changes within the transportation network have a significant impact on EV travel. To address this complexity, this paper proposes a comprehensive prediction method for EV travel behaviors and charging load distribution. This method takes into account the bounded rational decision-making of users and the influence of dynamic traffic conditions. Firstly, a travel decision model for EV users with bounded rationality is constructed, based on regret theory, to capture the nuances of their decision-making processes. Secondly, by considering the probabilistic decisions of these bounded rational users regarding travel paths and departure times, a dynamic traffic equilibrium model is developed to simulate the driving process of EVs. Finally, by integrating power consumption data, the spatio-temporal distribution of EV charging load is derived.

2. Travel Decisions of EV Users with Bounded Rationality

2.1. Modeling of the User's Perceived Utility Based on the Regret Theory

The travel decisions of EV users encompass both departure time and path selection. Bounded rationality manifests in two ways: on the one hand, due to the inherent limitations of human cognitive abilities, decision-makers are unable to fully grasp all available information and instead rely on the partial information they have acquired to perceive potential outcomes. On the other hand, because of constraints in human processing capabilities, decision-makers often aim for "satisfactory" rather than "optimal" criteria in the decision-making process. This means that they seek solutions that are good enough based on their available resources and time, rather than striving for the absolute best outcome, which may be impractical or unattainable. In the context of EV travel decisions, bounded rationality suggests that users make choices based on limited information and their own personal satisfaction thresholds, rather than a comprehensive understanding of all potential outcomes and an exhaustive search for the best possible option. This has implications for the prediction of EV charging load distribution, as user decisions about when and which path to travel are influenced by these bounded rational factors.

In this paper, considering the bounded rationality of actual users, a travel decision model of EV users is constructed based on the regret theory [8]. Considering the decision maker's avoidance of regret and pursuit of the optimal solution, a monotonically increasing concave function is used to reflect the regret-joy degree of users when comparing two different decision schemes, as shown in (1).

$$G\left(\Delta T_{ij}\right)=1-\exp\left(-\delta\Delta T_{ij}\right) \tag{1}$$

where, the regret avoidance coefficient $0<\delta\leq 1$ reflects the strength of the decision maker's regret avoidance degree, and ΔT_{ij} represents the difference of the utility of the two decision schemes. The regret avoidance coefficient is related to the user's education level, economic conditions, population composition and other complex factors, and can be averaged through a large number of statistical data.

The utility of the decision scheme may depend on several evaluation indicators, such as the time required to travel, the cost of travel expenses, and the driving experience. For the o-th indicator, the utility difference between decision schemes i and j is ΔT_{ij}^{o}, expressed by (2).

$$\Delta T_{ij}^{o}=E_i^{o}-E_j^{o} \tag{2}$$

where, E_i^{o} and E_j^{o} are the expectation of the o-th evaluation indicator of the decision scheme i and j respectively.

According to the utility difference of all indicators, the utility difference of the decision scheme i and j can be obtained:

$$\Delta T_{ij}=\sum_{o=1}^{n}\omega_o\Delta T_{ij}^{o} \tag{3}$$

In (3), n is the total number of evaluation indicators for user decision making, and wo is the weight of the o-th evaluation indicator, which reflects the degree of importance users attach to this indicator.

Formula (3) is used to calculate the regret-joy value when the user chooses decision scheme i but does not choose other schemes, and the regret-joy value matrix $G=\left[G_{ij}\right]_{m\times m}$ is obtained, where $G_{ij}>0$ reflects the psychological activity of the decision maker's joy, and $G_{ij}<0$ reflects the psychological activity of the decision maker's regret. Finally, (4) is used to calculate the overall perceived utility of decision makers for decision scheme i as follows:

$$Q_i=\sum_{j=1}^{m}G_{ij} \tag{4}$$

where, m is the total number of decision schemes.

2.2. *Probabilistic Choice Model of EVs Travel*

In the process of travel decision-making, EVs will select travel options based on the perceived utility associated with various departure times and travel paths. To account for the fact that EV users often prioritize "satisfactory" criteria over "optimal" criteria, the probabilities of different travel choices for EVs are calculated using the Logit model. Let $\mu^{w}(t)$ and $\eta_p^{w}(t)$ represent the probabilities of EV users choosing a specific departure time t and a particular travel path p, respectively. Then, the selection probabilities of EVs for various decisions can be mathematically expressed as follows:

$$\begin{cases}\mu^{w}(t)=\dfrac{\exp\left(Q_{p,k}^{w}(t)\right)}{\sum\limits_{s\in T}\exp Q_{p,k}^{w}(s)}\\[2ex]\eta_k^{w}(t)=\dfrac{\exp\left(Q_{p,k}^{w}(t)\right)}{\sum\limits_{x\in R_w}\exp\left(Q_{x,k}^{w}(t)\right)}\end{cases} \tag{5}$$

where, T and R_w are the sets of travel time periods and travel paths respectively.

3. Spatiotemporal Distribution of EVs Charging Load Based on the Dynamic Traffic Equilibrium

Once the travel decisions of EV users have been obtained, a dynamic analysis of their travel process can be performed to ascertain the power consumption and spatiotemporal locations of the EVs. Subsequently, by integrating this information with the conditions of charging demand, it becomes possible to calculate the charging load distribution at charging stations.

3.1. Modeling of Vehicle Driving Transfer Behavior Considering Dynamic Traffic Effects

The following dynamic traffic equilibrium model is established to solve the spatiotemporal transfer dynamic process of electric vehicles.

$$f_p^w(t) = d_w \eta_k^w(t) \mu^w(t) \tag{6}$$

In (6), $f_p^w(t)$ is the traffic flow that the remaining travel demand of w-th origin-destination (OD) pair d_w allocates to path p at time t. It needs to satisfy the constraints of (7)~(9).

$$\begin{cases} \dfrac{dx_a(t)}{dt} = u_a(t) - v_a(t) = \sum_{w\in W}\sum_{a\in p} u_a^w(t) - \sum_{w\in W}\sum_{a\in p} v_a^w(t) \\ \int_0^t u_a^w(t)dt = \int_0^{t+t_a(t)} v_a^w(t)dt \end{cases} \tag{7}$$

$$d_w = \sum_t^T \sum_{p\in R_w} f_p^w(t) \tag{8}$$

$$x_a(t) \geq 0, f_p^w(t) \geq 0 \tag{9}$$

In (7)~(9), $u_a(t)$ and $v_a(t)$ are the traffic inflow and outflow of link a at time t respectively, $u_a^w(t)$ and $v_a^w(t)$ are the traffic inflow and outflow of link a on path p of OD pair w travel demand at time t respectively, W is the set of travel OD pairs.

3.2. Charging Load Distribution Calculation

Through dynamic traffic equilibrium analysis, it is possible to ascertain the precise travel path and location of each vehicle at every moment in time. Based on the well-established principles of classical traffic modeling, the time required for an EV, denoted as e, to traverse a specific link a at a given time t can be mathematically calculated as follows:

$$t_a(t) = t_a^0 \left[1 + 0.15 \left(\frac{x_a(t)}{C_a} \right)^4 \right] \tag{10}$$

where, t_a^0 is the free flow time of the EV e passing through link a at time t, and C_a is the traffic capacity of link a.

If the power consumption rate of EV e is S and its driving path is p based on the dynamic traffic equilibrium analysis, the power consumption of this trip is L_p^e, as shown in (11).

$$L_p^e = S \sum_{a\in p} L_a \tag{11}$$

where, L_p^e is the length of link a.

Usually, the charging scenario of EVs includes midway charging and destination charging, such as halfway charging when the State of Charge (SOC) is lower than the range anxiety value during the trip, and choosing to fully charge when arriving at the work place.

Therefore, according to the initial remaining SOC, travel power consumption, charging demand conditions and spatio-temporal transfer characteristics of EVs, its charging load p_t^e at any time and charging duration can be obtained. The charging load of charging station c at time t is equivalent to the aggregate charging load of all EVs that are being charged at the station during that specific time period.

$$P_c^t = \sum_{e\in E} p_e^t \tag{12}$$

where, E denotes the collection of EVs that are undergoing charging at charging station c during the specific time period t.

4. Case Studies

Case studies on the spatial-temporal distribution of charging loads considering travel behaviors of EVs are carried out based on a regional traffic network in Nanjing, China. According to the function of different areas, they can be divided into residential areas, commercial areas and work areas. The topology of the regional traffic network and the distribution of charging stations are shown in Figure 1.

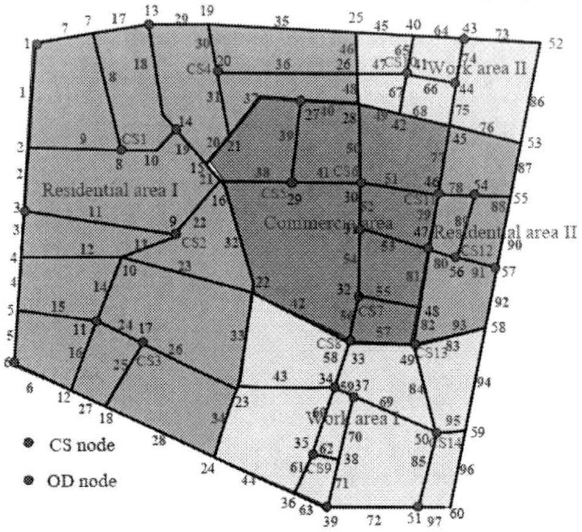

Figure 1. The topology of the regional traffic network and the distribution of charging stations

According to the travel rules of Nanjing residents, the distribution of travel demand at different time periods in this region is shown in TABLE I.

All EVs in the study are Nissan LEAF, and the charging power of the charging stations in residential and office areas is 7.3kW, and the charging power in commercial areas is 45kW.

Table 1. PT results and acceptance criteria

Categories	Periods	Travel Demand	Origin Points	Destination Points
Morning travel peak	7:00~9:00	8640 of each origin point	Evenly distributed in residential areas	Evenly distributed in work areas
Afternoon travel peak	12:00~14:00	3840 of each origin point	Evenly distributed in work areas	Evenly distributed in commercial areas
Evening travel peak	17:00~19:00	8640 of each origin point	Evenly distributed in work areas	60% of the destination points are evenly distributed in residential areas, 40% of the destination points are evenly distributed in commercial areas,

4.1. *Space-time Transfer Behavior of EVs*

Utilizing the advanced travel decision-making theory tailored for EVs, which takes into account the bounded rationality of users, we conduct a comprehensive analysis of the departure times and travel route selections for each OD pair's travel demand during the morning peak hours. In this analysis, a critical parameter, the regret avoidance coefficient (δ), is meticulously set at 0.3 to reflect the tendency of EV users to avoid making decisions that they might later regret.

Using the OD pair 10-51 as a representative example, Figure 2 displays the departure time distribution of electric vehicles (EVs). By thoroughly analyzing the distribution of departure times and the selection of travel routes for each OD pair, and subsequently integrating this data with the principles of dynamic traffic equilibrium, we can accurately ascertain the traffic flow distribution within the transportation network at any given point in time. To provide a concrete illustration, let's consider the specific instance of 8:00; Figure 3 depicts the traffic flow distribution within the network during this particular hour.

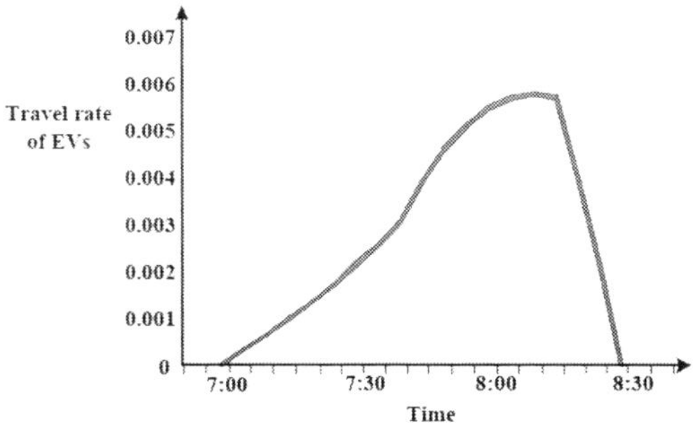

Figure 2. The departure time distribution of EVs OD pairs of 10-51

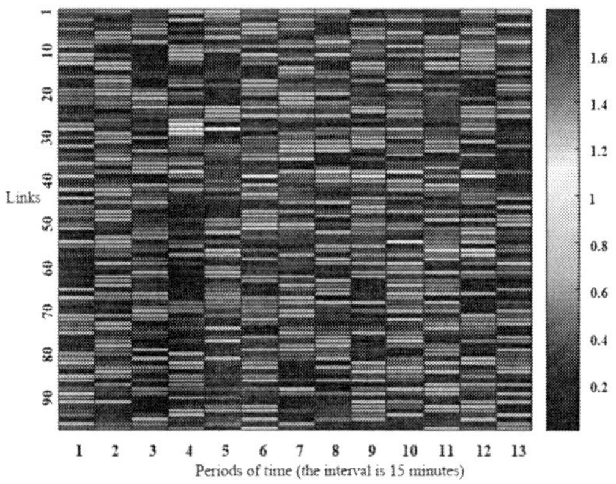

Figure 3. The traffic flow distribution in the traffic network at 8:00

By analyzing Figure 2 and Figure 3, it becomes apparent that the traffic flow distribution within the transportation network at any moment in time can be determined by integrating the travel decisions made by electric vehicle (EV) users with the dynamic nature of the traffic network. This integration offers a comprehensive portrayal of the EV travel transition process.

4.2. Distribution of EVs Charging Loads

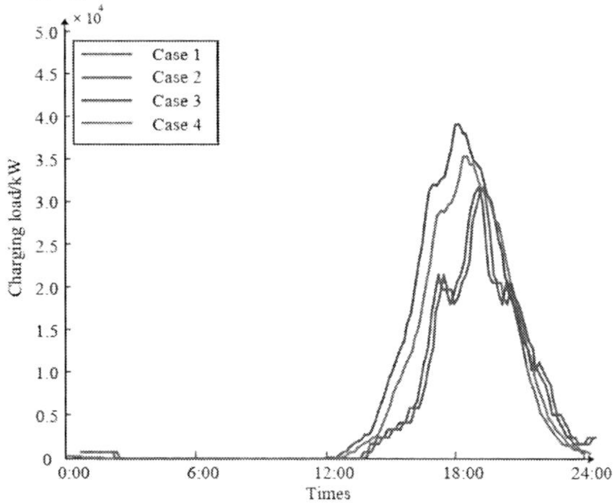

Figure 4. Daily charging load distribution of CS3 in residential areas

On the basis of obtaining the transfer behavior of EVs, the charging demand of a single EV can be obtained according to the initial charging state, driving range, power consumption and charging conditions of EVs. By deducting the daily driving process of EVs for each OD in the transportation network, the charging load of different charging stations at different times is obtained. The regret avoidance coefficient is set as 0.2, 0.3, 0.4 and 0.5 respectively, corresponding to cases 1 to 4, to study the influence of different users' bounded rationality degrees on the charging load distribution of charging stations. It is worth noting that the greater the regret avoidance coefficient, the higher the rational degree of EV users. Taking the charging station CS3 in residential areas as an example, its daily charging load distribution is shown in Figure 4.

As can be seen from Figure 4, with the increase in the rational degree of EV users, the daily charging loads of charging stations decrease as a whole. This is because the higher the degree of rationality of electric vehicle users, the stronger the regret avoidance of EV users, and they will pursue the optimal path to the maximum extent and reduce the travel power consumption.

5. Conclusions

In this paper, a spatial-temporal distribution prediction method of charging load considering EV driving behavior is proposed based on user decision making and dynamic traffic, and the bounded rationality of EV users decision making is emphasized. The case studies show that: 1) the charging load analysis method proposed in this paper is effective; 2) With the increase of user rationality, the charging load distribution of charging stations will decrease as a whole.

Acknowledgments

This work was supported by Guangdong Provincial Key Laboratory of Intelligent Measurement and Advanced Metering of Power Grid [grant number: GPKLIMAMPG-2023-KF-02].

References

[1] Qi T, Ye C, Zhao Y, Li L and Ding Y, 2023, Deep Reinforcement Learning Based Charging Scheduling for Household Electric Vehicles in Active Distribution Network, Journal of Modern Power Systems and Clean Energy, vol. 11, pp 1890-1901.

[2] Li K, Shao C, Zhang H and Wang X, 2023, Strategic Pricing of Electric Vehicle Charging Service Providers in Coupled Power-transportation Networks, IEEE Transactions on Smart Grid, vol. 14, pp. 2189-2201.

[3] Qian K, Zhou C, Allan M and Yuan Y, 2011, Modeling of Load Demand Due to EV Battery Charging in Distribution Systems, IEEE Transactions on Power Systems, vol. 26, pp. 802-810.

[4] Yang W, Xiang Y, Liu J and Gu C, 2018, Agent-based Modeling for Scale Evolution of Plug-in Electric Vehicles and Charging Demand, IEEE Transactions on Power Systems, vol. 33, pp. 1915-1925.

[5] Weckx S, D'Hulst R, Claessens B and Driesensam J, 2014, Multiagent charging of electric vehicles respecting distribution transformer loading and voltage limits, IEEE Transactions on Smart Grid, vol. 5, pp. 2857-2867.

[6] Yang Z, Li L, Yao R, Liu C, Liu Y and Zhou Z, 2024, Deep Reinforcement Learning-based Real-time Online Energy Management Strategy for Electric Vehicle Charging Stations, 2024 4th Power System and Green Energy Conference (PSGEC), Shanghai, China, pp. 490-494.

[7] Liu H, Shen H, Hu W, Ji L, Li J and Yu Y, 2023, Electric Vehicle Load Forecast Based on Higher Order Markov Chain, 2023 5th Asia Energy and Electrical Engineering Symposium (AEEES), Chengdu, China, pp. 1203-1207.

[8] Qian J, Lu Y, Yu Y, Zhou J and Miao D, 2024, Hierarchical Sequential Three-Way Multiattribute Decision-Making Method Based on Regret Theory in Multiscale Fuzzy Decision Systems, IEEE Transactions on Fuzzy Systems, vol. 32, pp. 4961-4975 .

Prediction of City Energy Consumption Based on Long-Range Energy Alternatives Planning System (LEAP) Model

Xuanyi Chen[1], Yutong Qian[1], Tingting Han[2*], Yue Pu[2]

[1] Business School, Hohai University, Nanjing, Jiangsu Province, 211100, China

[2] School of Electrical and Power Engineering, Hohai University, Nanjing, Jiangsu Province, 211100, China

*Corresponding author's e-mail: 241306020059@hhu.edu.cn

Abstract. Energy is the foundation of human economic and social development. Energy demand is closely related to energy security and national economic development, and fossil energy consumption is positively correlated with carbon emissions. This research takes a coastal city in eastern China as the example and takes the data of energy consumption in 2019 as the baseline to project the future energy consumption. The study was conducted using the Long-Range Energy Alternatives Planning System (LEAP) software and this paper develops a LEAP model with data of GDP, population, industrial structure, vehicle ownership, energy intensity, as well as power generation efficiency and other factors. The LEAP model is extrapolated until the year 2035 to acquire the predictions of the city's economic sectors: agriculture sector, industry sector, construction sector, transportation sector, residential sector, other sectors and local electricity production sector. The results show that the trend of energy demand in various sectors is differential. Among these sectors, the energy demand of agriculture and construction increase first and then decrease, the trend of energy demand in industry and transportation increase steadily, and the energy consumption of residential sector almost have no change.

1. Introduction

In response to environmental problems caused by global warming, low-carbon emission power generation and production have attracted wide attention around the world. With global temperatures and greenhouse gas emissions setting new records, there is an urgent need to advance the global low-carbon transition and achieve a 28% reduction in projected greenhouse gas emissions by 2030. A coastal city in eastern China has a high energy demand and energy consumption density. In the past 10 years, per capita consumption has ranked the top list in China and energy consumption has been growing. Environmental carrying capacity has also been rising year by year. Controlling carbon emissions associated with energy combustion is an essential part of achieving peak carbon neutrality.

In 2023, the CO2 emissions produced by energy consumption amounted to 12 billion tons, accounting for 40% of global greenhouse gas emissions. The energy consumption in China was 5.72 billion tons of standard coal equivalent (tce). Both of them ranked first in the world. China made commitment to implement dual-carbon actions and incorporated carbon neutrality into national policy. The coastal city of energy consumption ranked first in China and its total energy consumption was 109.51 million tce in 2023. The city holds an extremely important strategic position in China's economic landscape. Therefore, predicting the future energy demand of this coastal city accurately is of great significance for China to achieve its carbon neutrality target.

In terms of energy demand forecasting method, Liang [1] divides China into eight regions and utilized the input-output method to establish a multi-regional input-output model for China's energy demand. However, the prediction accuracy of the input-output method is low, and the compilation of its input-output tables is also complex. Li [2], based on the evolutionary trends in economic development and energy consumption over the past two decades, utilizes the energy consumption elasticity coefficient to forecast China's energy consumption in 2030. However, the prediction accuracy of elasticity coefficient method is low, and it can only provide trend analysis. Pradhan [3] uses a Computable General Equilibrium model to assess the coal consumption and technological progress in emerging energy technologies, but the mathematical calculation of this model is complex and requires high-quality data. Howells [4] focuses on assessing the projection of electrification in rural areas of developing countries and of regions of Africa using modeling techniques such as TIMES, which has limitation in long-term prediction. LEAP model has advantages in substitution forecasting, quantitative dynamics, determination of influencing factors and policy setting, and has been widely applied in different scales. Wu and Chen use LEAP model to conduct scenario analysis of carbon emission in a coastal province in China [5][6]. Maduekwe [7] uses the LEAP model to predict energy demand and greenhouse gas emissions of the road transport sector in Lagos, the capital of Nigeria. Nieves [8] utilizes LEAP model to analyze the energy demand and greenhouse gas emission produced in Colombia.

The shortcomings of former studies proceed as follows: first, these studies don't pay attention to the energy consumption in city level; Secondly, indicators they selected are single and not comprehensive. This paper takes the coastal city in eastern China as an example, takes the data of energy consumption in 2019 as the baseline and summarizes the latest policy documents. This paper builds a LEAP model corresponding to the city's actual condition to project the energy demand from 2023 to 2035 taking into consideration GDP, population, industrial structure, vehicle ownership, energy intensity, as well as power generation efficiency and other factors. Moreover, this article can help find out the path of low-carbon emission reduction in the city to complement China's carbon neutrality goal.

2. Principles and methods

The steps of constructing LEAP model are as follows: First, build model structure and database according to the characteristics of the research object. Second, set scenarios according to existing policies. Third, analyze energy consumption and greenhouse gas emissions under different scenarios. In this paper, the LEAP model is a comprehensive energy modeling tool, which can produce a scenario-based result and analyze the fuel demand, green gas emissions, cost-effectiveness of various regional sectors. Using LEAP model (Figure 1), policy analysts can synthesize and compare the indicators to be achieved in order to quantitatively create and evaluate different scenarios. The LEAP model requires data from the base year and evaluated data of short-term future years, and then extrapolates various parameters to estimate future trends of energy consumption.

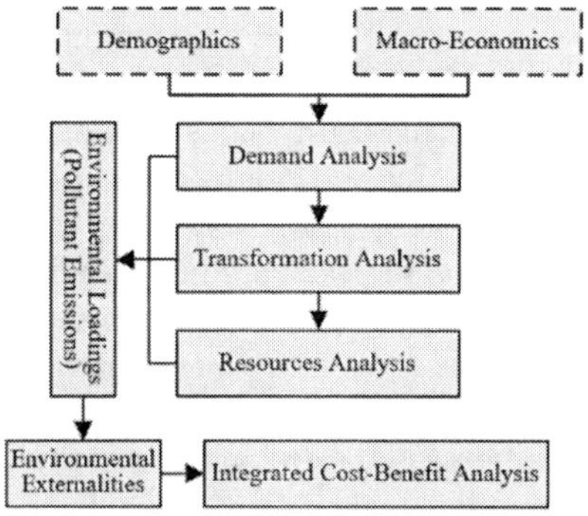

Figure 1. Process of LEAP model

2.1. *Energy consumption of terminal sectors*

The energy consumption of terminal sectors is the energy supplied to the inlet of end-use energy equipment. The energy consumption of terminal sectors is equivalent to the amount of energy consumption minus the loss in the energy transformation and the quantity used for energy processing and conversion to secondary energy. The energy loss is not considered in this paper because the loss is lower than the total energy consumption in the coastal city in eastern China. The calculation of energy consumption in terminal sectors as in (1):

$$TC_i = A_i \times I_i \tag{1}$$

where TC is the energy consumption for a specific terminal sector, in tce; i is the type of terminal sector. A is the level of activity as a parameter of the economic activity for which the energy is demanded. For agriculture, industry, construction, and other sectors, the variable is in ten thousand Yuan-GDP. For transportation sector, the variable is the number of vehicles. While for residential sector it is the number of people; I is the energy intensity for activity level. For agriculture, industry, construction, and other sectors, the unit is tce/10 thousand Yuan GDP; while for transportation sector the unit is in tce/vehicles and the unit is tce/people for residential sector.

For the transportation sector, the energy consumption is calculated in equation (2):

$$EV_k = s \times \frac{m}{fe} \tag{2}$$

where EV represents the energy consumption of transportation sector; s is the number of vehicles; m represents the distance; fe represents the specific fuel consumption and k is the type of fuel.

For the residents, the energy demands can be calculated as follows:

$$EP = N_u \times e_u + N_r \times e_r \tag{3}$$

where EP represents the residents' energy demand; N_u and N_r are the number of urban and rural populations respectively; e_u and e_r are urban and rural energy consumption per capita.

2.2. *Energy consumption of local electricity production sector*

The local electricity production sector is responsible for converting primary energy into secondary energy utilized by terminal sectors. The energy consumption signifies the energy required for electricity generation. The energy consumption is calculated by means of equation (4).

$$TC = G \times \partial_e \times E \tag{4}$$

where TC represents the energy consumption, in tce; G represents the power generation capacity of the local electricity production sector, and the unit is 100 million kilowatt-hours; E represents the conversion efficiency of the power plants, in percentage (%); ∂_e is the conversion coefficient of electric energy to standard coal, expressed in kilogram of standard coal/kWh.

2.3. *LEAP framework*

The LEAP model, as shown in Figure 2, is composed of three components: driving factors, energy transformation and energy demand. The energy demand sectors include terminal sector and transformation sector. The transformation contains local and external electricity production. Driving factors are those that significantly impact energy demand. The energy resources include primary energy (e.g. coal, oil, natural gas) and secondary energy (e.g. electricity).

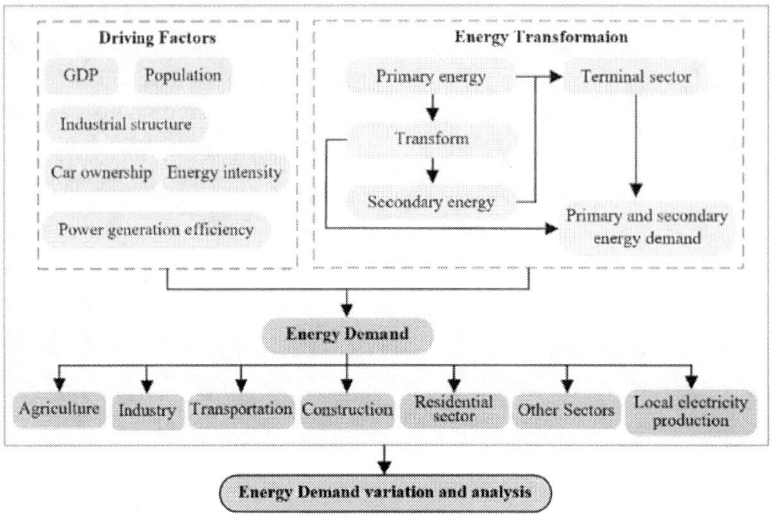

Figure 2. LEAP calculation flows

3. Case analysis

3.1. Parameters of LEAP model

Numerous factors can impact energy consumption. According to relevant planning files, the parameters of LEAP model are shown in Table 1. Given the accessibility of data, the base year for analysis is set at 2019, with prediction periods spanning from 2023 to 2035.

Table 1. Parameters of LEAP model

Key Parameters	2019 (base year)	2025	2030	2035
GDP (100 million yuan)	38155	54124	69077	84043
Population (10 thousand people)	2428	2457	2475	2488
Agricultural proportion (%)	0.27	0.22	0.19	0.15
Industry proportion (%)	25.35	23.28	20.51	17.7
Construction proportion (%)	1.65	1.5	1.3	1.15
Other sectors proportion (%)	72.7	75	78	81
Car ownership (10 thousand vehicles)	442	540	620	683
Local electricity production (GWh)	836	1117	1147	1200
Agriculture energy intensity (tce/10 thousand yuan)	0.60	0.54	0.50	0.46
Industry energy intensity (tce/10 thousand yuan)	0.59	0.50	0.45	0.41
Construction energy intensity (tce/10 thousand yuan)	0.40	0.34	0.30	0.28
Transportation energy intensity (tce/10 thousand vehicles)	5.92	5.32	4.99	4.75
Residential sector energy intensity (tce/10 thousand people)	0.53	0.47	0.44	0.42
Other sectors energy intensity (tce/10 thou-sand yuan)	0.07	0.06	0.05	0.05
Power generation efficiency (%)	35.97	41.5	47.2	52.5

3.2. Model validation

The data of actual energy consumption and predictions in 2020-2022 are shown in Table 2. The mean relative error of each industry in 2020-2022 can be calculated using equation (5), and the calculation result is 10% which indicates that the LEAP model proposed in this paper is reasonable in predicting the energy demand of each industry.

$$MRE = \frac{1}{3m} \sum_{i=1}^{m} \sum_{j=1}^{3} \left| \frac{y_{ij} - y'_{ij}}{y_{ij}} \right| \tag{5}$$

where MRE is mean relative error. m is the number of terminal sectors. i is the type of terminal sectors. y and y' represent the actual energy consumption and energy demand prediction.

Table 2. Actual energy consumption and Energy prediction in 2020-2022

	Sectors	2020	2021	2022
Actual energy consumption (10 thousand tce)	Agriculture	59.30	62.73	61.35
	Industry	5395.59	5573.98	5300.24
	Construction	263.23	295.38	284.78
	Transportation	2196.79	2307.76	1928.81
	Residential sector	1317.64	1388.09	1452.82
	Other sectors	1827.13	1936.01	1953.17
Energy demand prediction (10 thousand tce)	Agriculture	61.09	62.474	63.359
	Industry	5807.7	5962.9	6108.1
	Construction	258.05	264.12	269.59
	Transportation	2673.63	2726.53	2763.03
	Residential sector	1283.92	1284.64	1284.26
	Other sectors	2100.71	2198.8	2308.1

3.3. *Result and Analysis*
After the basic parameters are set, LEAP software is used to simulate and forecast the energy consumption of various sectors in 2020-2035. The following is an analysis of the future energy demand of different sectors in a coastal city in eastern China.

3.3.1. *Agriculture*
Agriculture activity level is the multiplication of GDP and agricultural proportion. The energy intensity of agriculture is agriculture energy consumption per unit activity level. The energy demand of agriculture in 2023-2035 is shown in Figure 3.

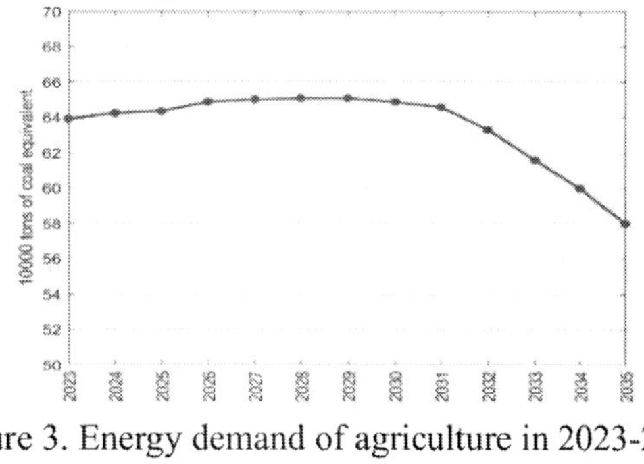

Figure 3. Energy demand of agriculture in 2023-2035

The agriculture energy demand from 2023 to 2029 shows an upward trend, from 639.1 thousand tce in 2023 to 650.8 thousand tce in 2029, and then decreases steadily after reaching a peak in 2029. The total energy consumption of agriculture in 2035 is 579.9 thousand tce.

3.3.2. *Industry*
The activity level of industry is the multiplication of GDP and industry proportion. The industry energy intensity is industry energy consumption per unit activity level. The energy demand of industry in 2023-2035 is shown in Figure 4.

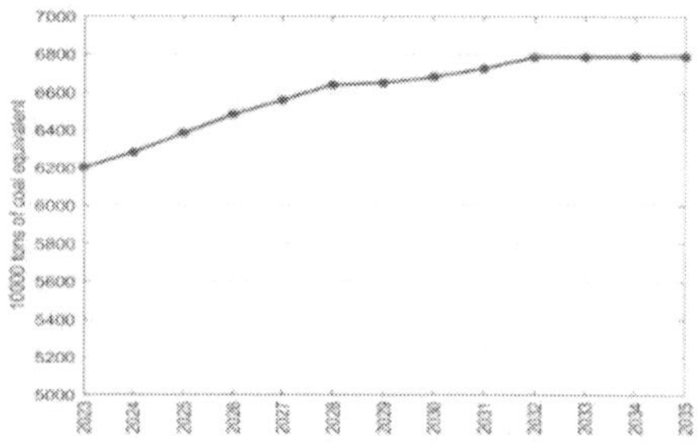

Figure 4. Energy demand of industry in 2023-2035

The future energy demand of industry increases steadily, from 62.035 million tce in 2023 to 67.907 million tce in 2035.

3.3.3. Construction

The activity level of the construction is the multiplication of GDP and construction proportion. The energy intensity of construction is construction energy consumption per unit activity level. The energy demand of construction in 2023-2035 is shown in Figure 5.

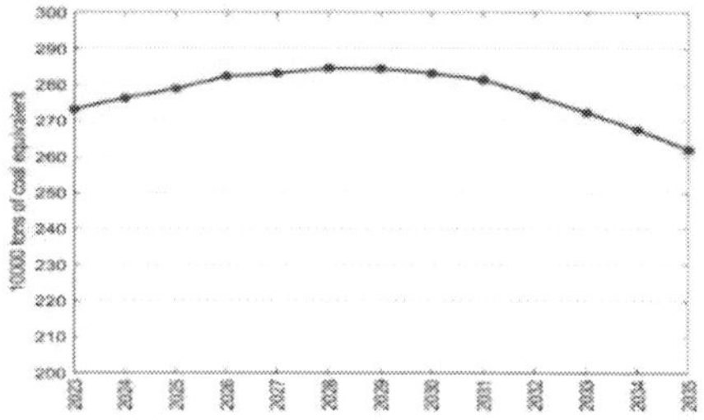

Figure 5. Energy demand of construction in 2023-2035

The trend of energy demand in the construction is consistent with that of agriculture, which increases first and then decreases. The energy demand of construction from 2023 to 2028 shows an upward trend, increasing from 2.7328 million tce in 2023 to 2.8457 million tce in 2028, reaching a peak in 2028. And then decreases steadily. The total energy consumption of construction in 2035 is 2.621 million tce.

3.3.4. Transformation

The activity level of transportation is vehicle ownership. The energy intensity of transportation is the amount of energy consumed per unit activity level. Figure 6 shows the energy demand of transformation from 2023 to 2035.

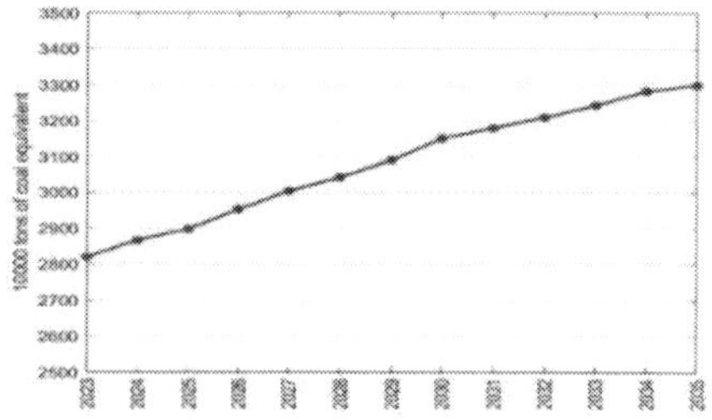

Figure 6. Energy demand of transformation in 2023-2035

The future energy demand of transportation increases steadily. The energy demand in 2023, 2025, 2030 and 2035 are 28.2003 million tce, 28.9843 million tce, 31.518 million tce and 33.012 million tce respectively.

3.3.5. Residential sector

The activity level of residential sector is the total population of the city. The energy intensity of residential sector refers to the amount of energy consumed per unit activity level residential sector. The

Journal of Physics: Conference Series **3012** (2025) 012051
doi:10.1088/1742-6596/3012/1/012051

energy demand of residential sector in 2023-2035 is shown in Figure 7.

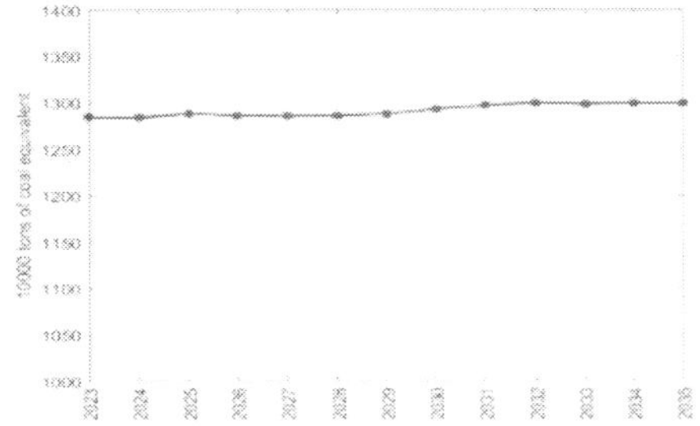

Figure 7. Energy demand of residential sector in 2023-2035

The energy demand of residential sector in the future remains almost steadily. The reason is that the population increases stably, but with the advance of technology, the per capita energy consumption decreases steadily. The energy demand of residents in 2023, 2025, 2030 and 2035 is 12.84443 million tce, 12.8804 million tce, 12.9285 million tce and 12.99 million tce, respectively.

3.3.6. Other sectors

The activity level of other sectors is the multiplication of GDP and sector proportion. The energy intensity of other sectors refers to the amount of energy consumed per unit activity level of other sectors. Figure 8 shows the energy demand of other sectors from 2023 to 2035.

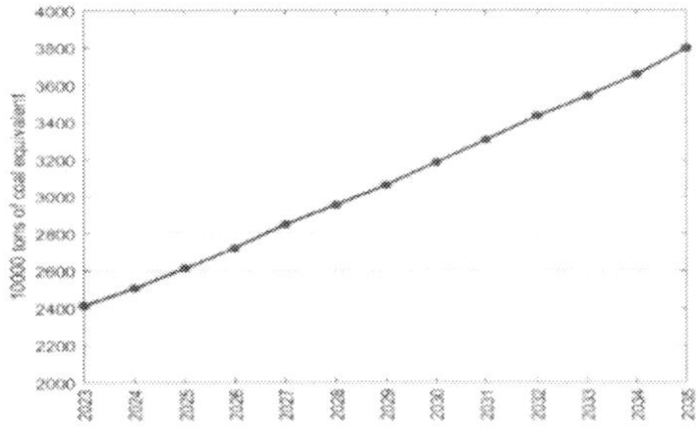

Figure 8. Energy demand of other sectors in 2023-2035

The energy demand trend of other sectors increases steadily in line with the transportation industry. The reason is that the increase of GDP and the expanding proportion of tertiary industry. The energy demand of other sectors in 2023, 2025, 2030 and 2035 is 24.127 million tce, 26.1419 million tce, 31.8606 million tce and 37.9951 million tce, respectively.

3.3.7. Local electricity production sector

The activity level of local electricity production sector refers to its generated power. The amount of energy consumed per unit activity level of local electricity production sector is inversely proportional to power generation efficiency. Figure 9 shows the energy demand of the local electricity production sector from 2023 to 2035.

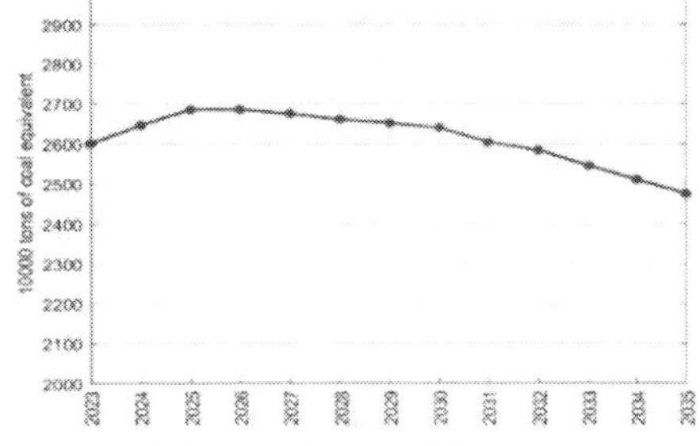

Figure 9. Energy demand of the local electricity production sector in 2023-2035

The energy consumption of local electricity production sector from 2023 to 2025 shows an upward trend, increasing from 26.0086 million tce in 2023 to 26.8605 million tce in 2025 and then decreasing steadily. In 2035, the energy consumption is 24.7585 million tce. Since the parameter settings of local generation in the local generation sector are not uniformly variable, the slope of energy demand in the local generation sector in 2025 suddenly changes.

4. Conclusion

This paper takes the energy consumption of a coastal city in eastern China in 2019 as example, comprehensively analyzes the factors influencing energy consumption, drawing on the latest policy documents and the energy consumption status. Then, this paper establishes a LEAP model to project the energy demand of various sectors. The conclusions can be list as follows:

First, an analysis of energy consumption in a coastal city in eastern China reveals a notable deceleration in the growth rate of energy consumption over recent years. In the future, the city needs to focus on the overall energy transformation.

Second, the future trend of energy consumption: The industry energy demand increases steadily. As the activity level of the transportation, car ownership increases due to the improvement of residents' income and policy support. Obviously, the energy demand of transportation increases. The trend of energy consumption in agriculture, construction and local electricity production sector decreases steadily after reaching the peak.

These results in this paper can provide support and guidance for decision-makers across various levels.

Acknowledgments

This work was supported in part by the 2023 Graduate Research and Innovation Program of Jiangsu, China (No. KYCX23_0680).

Reference

[1] Liang Q M and Fan Y 2007 Multi-regional input-output model for regional energy requirements and CO2 emissions in China Energy Policy vol 35 pp 1685-1700.

[2] Li H B and Zhang J J 2021 China's Energy Consumption Structure and Forecast on Natural Gas Demand vol 37 pp 71-78.

[3] Pradhan Basanta K and Ghosh Joydeep 2022 A computable general equilibrium (CGE) assessment of technological progress and carbon pricing in India's green energy transition via furthering its renewable capacity Energy Economics vol 106.

[4] Howells MI, Alfstad T, Victor DG, Goldstein G and Remme U 2005 A model of household energy services in a low-income rural African village Energy Policy vol33 pp 1833-1851.

[5] Wu W, Zhang T T and Xiao X M 2019 Research on Regional Low Carbon Development Path Based on LEAP Model: Taking Zhejiang Province as an Example Ecological Economy vol 12 pp 19-24.

[6] Xiao H J, Xie Z and Zhou J H 2018 Research on Evolutionary Game of International Transfer of Low-carbon Environmentally-sound Technology Journal of Huzhou Teachers College vol 5.

[7] Maduekwe M, Akpan U and Isihak S 2020 Road transport energy consumption and vehicular emissions in Lagos, Nigeria: An application of the LEAP model Transportation Research Interdisciplinary Perspectives vol 6.

[8] Nieves J A, Aristizabal A J and Dyner I 2019 Energy demand and greenhouse gas emissions analysis in Colombia: a LEAP model application Energy vol 169 pp 380-397.

Joint Scheduling Method for Offshore Wind-hydrogen and Coastal Energy System

Xiaoqing Cao[1], Zhenxi Fan[2*], Ke Zhu[2], Lei Zhang[2]

[1] Wuhan Huayuan Electric Power Design Institute Co., Ltd. Wuhan 430010, China

[2] China Three Gorges University Yichang 443000, China

[*]Corresponding author's e-mail: fanzx_0115@163.com

Abstract. Considering the rapid transformation of global energy structure leading by renewable energy, the joint scheduling of offshore wind power and hydrogen energy system is drawing more research attention. In this paper, a joint scheduling method of offshore wind hydrogen system and coastal power grid considering the fatigue load for wind farm clusters is proposed. First, a wind farm cluster generation transfer model taking into account the exotic increased use of wind energy is constructed by evaluating the wind energy coupling effect among offshore wind farms. Second, a wind farm cluster equilibrium load model is built with the convenience of maintenance in mind, and the link between wind farm fatigue load and fatigue cost is quantitatively examined. With the goal of reducing the cost of system operation, a joint scheduling method between the offshore wind-hydrogen system and the electric-hydrogen of the coastal grid is proposed. The results of the simulation show that the suggested scheduling strategy can successfully meet the needs of the coastal grid for power supply while also balancing the fatigue load of each wind farm group. The method realizes the efficient use of energy and optimizes the cost as well.

1. Introduction

With the purpose to promote energy transformation, many countries in the world are progressively promoting centralized renewable energy with abundant resources and efforts [1]. In terms of clean energy with higher density and lower environmental impact, offshore wind power is drawing much attention in the development planning. This is because the offshore wind energy can make full use of offshore resources, increase the proportion of wind power generation, add power sources to the energy supply of various countries, and improve the economic benefits of wind power [2].

In recent years, scholars from various countries have carried out studies on optimal scheduling of offshore wind power. However, there are still two issues that have not been well addressed:

•The research work does not fully consider the coupling effect of wind energy between wind farms, and the fatigue load coordination among different offshore wind farms has rarely been studied. Most of the research only focuses on the optimization of electric energy, ignoring the multi-energy complementarity, leading to the insufficient level of regulation of offshore wind farms [3-6].

•The existing model mainly relies on the AC grid to transmit electricity, while offshore wind power may use DC transmission. In terms of hydrogen energy transmission, trucks are mainly used to transport hydrogen, whereas offshore climate is different and the time of ship transmission will be affected accordingly [7-10].

Journal of Physics: Conference Series 3012 (2025) 012052 doi:10.1088/1742-6596/3012/1/012052

For these two reasons, in light of the fatigue load on wind farm clusters, this research suggests a combination electricity-hydrogen scheduling strategy for offshore wind hydrogen systems and coastal power grids. The following are this paper's primary innovations:

(1)Establish generation transfer modelling of wind farm clusters and equilibrium load modeling of wind farm clusters;

(2)Adopt the model of offshore wind power combining with hydrogen production by transportation mode and establish the model of electric-hydrogen hybrid energy export;

(3)Propose the joint scheduling method of offshore wind-hydrogen system and coastal grid electric-hydrogen considering the fatigue load of wind farm clusters.

2. Modeling of Offshore Wind-Hydrogen System

2.1. *The structure of offshore wind-hydrogen system*

An offshore wind farm, a hydrogen production device, a hydrogen storage unit, a converter station, a thermal power unit, a ship, and additional components (e.g., electrical load, hydrogen load) make up the offshore wind-hydrogen system. Fig.1 depicts the offshore wind-hydrogen system's structure and workflow in the following manner:

(1)Offshore wind farms convert wind energy into electricity, which is divided into two parts; one part of the electricity is used to produce hydrogen, and the other part of the electricity is transmitted to the coastal power grid through the converter station and flexible DC transmission technology.

(2)The hydrogen production unit converts the electricity into hydrogen, which is stored in an offshore hydrogen storage unit, awaiting transport by ship.

(3)The ship transports the hydrogen to the land-side hydrogen storage station to meet the hydrogen load demand of the coastal grid.

(4)Thermal power unit generate electricity from fossil fuels and, together with offshore wind farms, meet the energy demand of the coastal grid loads.

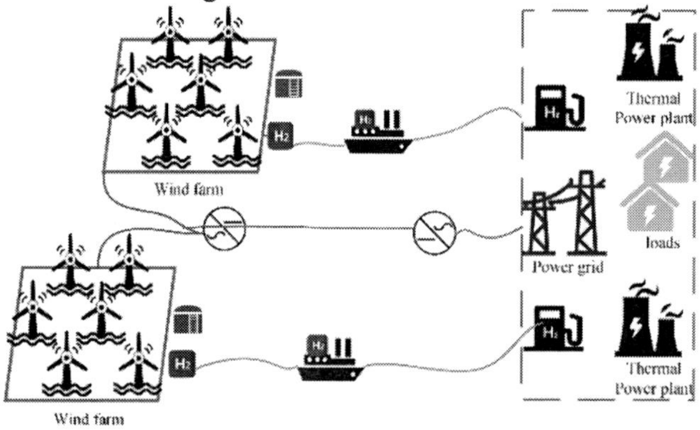

Figure 1. The structure of the offshore wind-hydrogen system

2.2. *Generation transfer modelling of wind farm clusters*

The correlation of wind speed between upstream and downstream wind farm is in the same dominant wind direction. The output of the wind farms has some degree of similarity and followability since there is a certain amount of space between the upstream and downstream wind farms. In the dynamic coordination of offshore wind farm cluster, the upstream wind farm reduces the absorption rate of wind energy through active wind abandonment behavior, which increases the amount of wind energy transferred to the wind farms downstream, hence increasing the power output of those wind farms. [11].

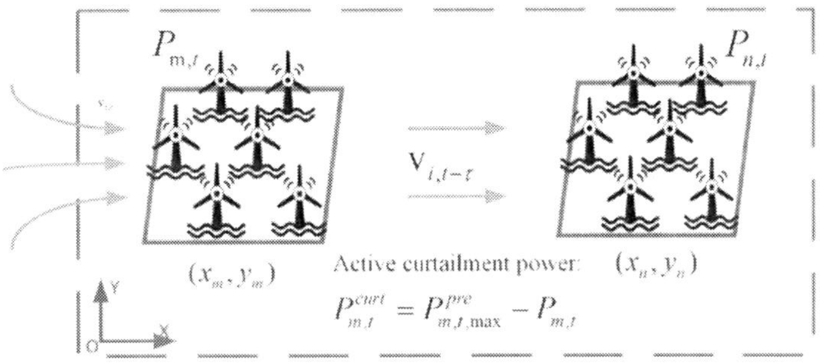

Figure 2. The schematic diagram of the wind farm cluster generation transfer

The schematic diagram of the wind farm cluster generation transfer is shown in Fig.2. $P_{m,t}^{\max}$ is the predicted maximum output of wind farm m at time t. $P_{m,t}$ and $P_{n,t}$ are the power outputs of wind farms m and n at time t. $P_{m,t}^{curt}$ is the active curtailment power of wind farm m at time t.

2.3. *Equilibrium load modelling of wind farm clusters*

Damage to the wind turbine generators is assessed using the fatigue load value in actual operation and thus to determine the repair time of the WTGs. The fatigue load model of a wind farm can be expressed as:

$$F_{m,t} = F_{m,t_0} + (1+\gamma)\frac{\int_{t_0}^{t} P_{m,t}dt}{P_m^{rate}T_{life}(1+R_{com})} \tag{1}$$

where $F_{m,t}$、F_{m,t_0} are the fatigue load values of the wind farm m at t、t_0 time, respectively; P_m^{rate} is the rated power of the wind farm m; T_{life} is the design service life, R_{com} is the compensation coefficient, with a value range of [0,1]; γ is the ratio of disturbance fatigue to work fatigue, which is an empirical parameter.

The relationship between power output and load cost in offshore wind farm is described in Fig.3. It is evident that the load cost rises as the load does, and thus there is a mutually exclusive link between the power output and the optimization goal.

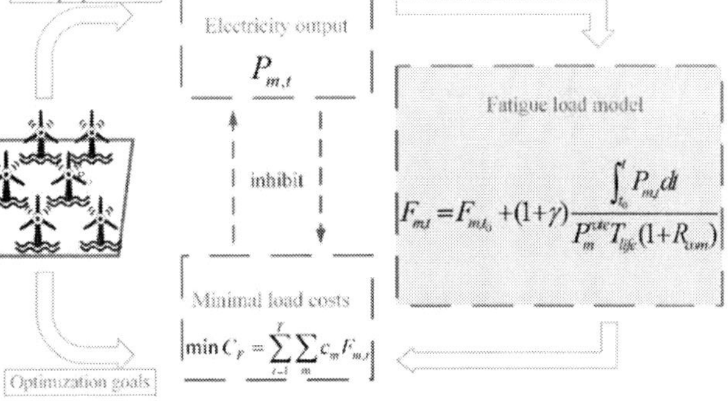

Figure 3. Relationship between offshore wind farm power output and load cost

2.4. *The modelling for other sections*

Other component models include the wind power pooling model considering active island mode, the modeling of continuous power output, and the modeling of discrete hydrogen energy delivery. The summary of their modeling situations is as follows:

•Wind power pooling model considering active island mode: the wind farm and the hydrogen production unit have a dynamic interaction that jointly defines the wind farm's external power output. The AC aggregation network is described by a branch-current model, where 0-1 variables are used to indicate the connection status of the branches;

•Modeling of continuous power output: the power from the AC convergence network is boosted and converted to DC for transmission through a voltage source converter (VSC), which functions within the limitations of both the AC and DC sides' active and reactive power;

•Modeling of discrete hydrogen energy delivery: the berthing and sailing status of ships is defined to ensure that the transfer of ships between berths meets time and space constraints. Based on these constraints, the ship's hydrogen charging and discharging operations and sailing power requirements are further managed to optimize the efficiency of hydrogen energy transportation.

3. MODELING OF ELECTRICITY-HYDROGEN SCHEDULING

The energy collaborative optimization scheduling mode of offshore wind-hydrogen system and coastal energy system is shown in Fig.4. The hybrid energy scheduling platform jointly dispatches wind farms, thermal power units and ships at the same time. By optimizing the spatial and temporal distribution of wind farms, thermal power units and ships, the system operation cost is minimized.

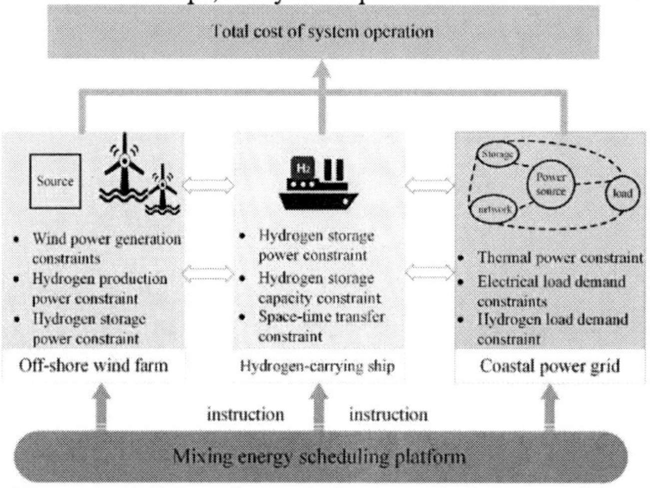

Figure 4. Collaborative optimization scheduling model

3.1. Objective function

The objective function is to achieve the economic optimization of the system under the condition of satisfying the demand of electricity-hydrogen load on land. Considering the benefit of hydrogen production from wind power, the objective function is expressed as:

$$C_{all} = \min(C_T + C_H + C_S + C_R + C_{Fa} - C_h) \tag{2}$$

$$C_T = \sum_{t=1}^{T} \sum_{g\in\Phi_{nc}} (a_{1,g}P_{g,t}^2 + a_{2,g}P_{g,t} + a_{3,g}) \tag{3}$$

$$C_H = \sum_{t=1}^{T} \sum_{h} c_h P_{h,t}, C_S = c_k \sum_{t=1}^{T} I_t S_k \tag{4}$$

$$C_R = \sum_{t=1}^{T} \sum_{p\in\Phi_w \cup \Phi_h} v_p (R_{p,t}^{up} + R_{p,t}^{down}) \tag{5}$$

$$C_{Fa} = \begin{cases} \sum_{t=1}^{T}\sum_{m} c_m F_{m,t}, 1-\kappa \le \frac{F_m}{F_n} \le 1+\kappa \\ \sum_{t=1}^{T}\sum_{m} c_m F_{m,t} + \sum_{t=1}^{T}\sum_{m} \delta_m F_{m,t}, \text{else} \end{cases} \tag{6}$$

$$C_h = \lambda_H (m_H - \overline{m}_H) \tag{7}$$

where C_{all} is the total cost of system operation; C_T is the operation cost of thermal power units; C_S is the cost of ship navigation; C_R is the reserve cost; C_{Fa} is the total cost of the fatigue load; C_h is hydrogen yield; Φ_{nc} is the set of thermal power units; c_h is the cost coefficient of hydrogen production device; $a_{1,g}$、$a_{2,g}$、$a_{3,g}$ are the cost coefficients of thermal power units; v_p is the reserve cost coefficient; c_k is the unit navigation cost coefficient of the ship; S_k is the distance between adjacent berths; $R_{p,t}^{up}$ and $R_{p,t}^{down}$ are the upper reserve and lower reserve of unit p at time t; Φ_w and Φ_h are wind farm collection and hydrogen production device collection respectively; λ_H is the unit hydrogen sales price; m_H is the total amount of hydrogen transported by ships to land during the scheduling period; $\overline{m}_H$ is the reference value of terrestrial hydrogen load; c_m is the load cost factor; δ_m is the load penalty factor; κ is a constant; F_m、F_n are the total fatigue loads for wind farm m and wind farm n, respectively. T is the scheduling time.

3.2. Constraints

The constraints in the mode are states as below:

(1)Steady power transmission constraints: To ensure that offshore wind farms can meet the power load demands of the land-side power grid at all times, the grid-connected power of offshore wind and the power fluctuations of offshore wind must satisfy the constraints that the deviation between the expected grid-connected power and the actual grid-connected power is within a predetermined range, and that the reserve power can balance the fluctuations in the power output of the wind farm.

(2)Spinning reserve constraints: Supplying reserves so that the difference between the anticipated and actual maximum output of wind farms and hydrogen production facilities is not exceeded by the spinning reserve up. And the spinning reserve down does not exceed their actual output.

(3)Hydrogen energy storage constraints: The amount of hydrogen produced by wind farms through the electrolysis of water needs to be limited by the capacity of the storage unit. Specifically, the amount of hydrogen produced by the hydrogen production device at a given moment must not exceed the storage capacity of the storage unit, while also taking into account the efficiency of the electrolysis process.

(4)Hydrogen device operating constraints: These constraints relate to the amount of hydrogen stored, the power of hydrogen production, the calorific value of hydrogen and the efficiency of hydrogen production, which restricteach other and together determine the efficiency and output of the hydrogen production process.

(5)Active power balance constraints: These constraints ensure that at any given moment, the power output of wind farms and thermal power units matches the power demand of the electrical load.

(6)Hydrogen load constraints: The total amount of hydrogen transported to land by ships should be higher than the baseline value of the land-based hydrogen load.

(7)Thermal unit operating constraints: Minimum and maximum output limits for the operation of thermal power units is set to ensure that they are operated within safe and efficient limits.

3.3. Solution algorithm

First, the power flow model must be convexified. Next, wind farm cluster equilibrium load modelling, thermal power unit operating costs, and hydrogen production device linearization must all be done.

The model developed in this paper is converted into a mixed integer second-order cone programming issue once the aforementioned procedure is finished. This problem is then programmed in the Matlab Yalmip environment and solved using the Gurobi solver.

4. RESULT AND DISCUSSION

4.1. Basic data

This paper takes two offshore wind farms and one land as the model to verify the algorithm. Because wind farm 2 is upstream, the production of wind farm 1, which is downstream, will be impacted by its wind abandonment. Assuming that the installed capacity of the two wind farms is 250MW, the maximum output curve of the two wind farms predicted for a day in winter is shown in Fig.5. The rated power of the hydrogen plant is 65MW, and the maximum amount of hydrogen loaded on the ship is 4 tons.

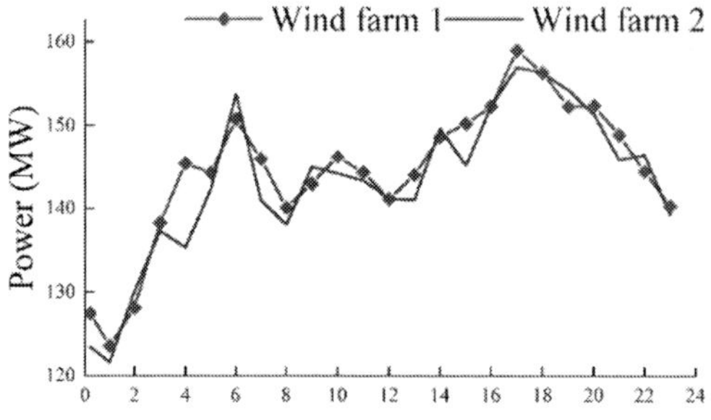

Figure 5. Wind Power Forecast Maximum Output Curve

4.2. Analysis of scheduling results

According to the power demand of coastal area and the current hydrogen price, the wind farm output is reasonably allocated, and the wind power grid-connected power and hydrogen production power in each scheduling period are coordinated to obtain the economic optimal scheduling scheme.

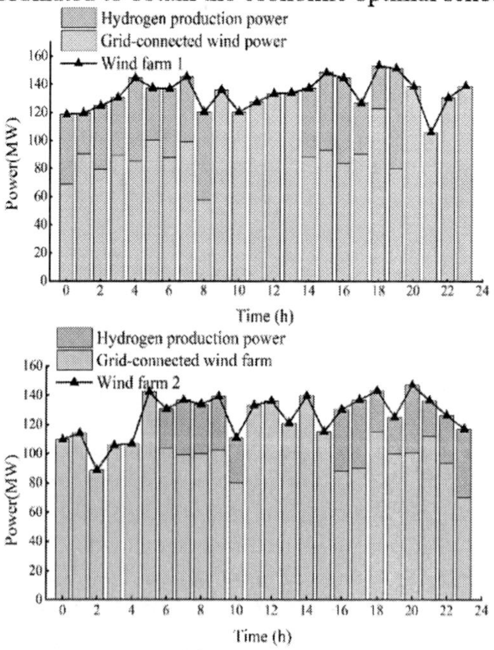

Figure 6. Wind farm power allocation

Fig.6 illustrates how the power distribution between the two wind farms, in the 0-8h, 14-19h within the hydrogen production device 1 power is higher, can be faster to produce hydrogen to meet the system hydrogen demand; subject to the time and space constraints of the hydrogen transportation ship, other time periods of the hydrogen production device is no longer running. Wind farms 1 and 2 have a combined electricity generating capacity of 3027.194MW and 3201.311MW, respectively.

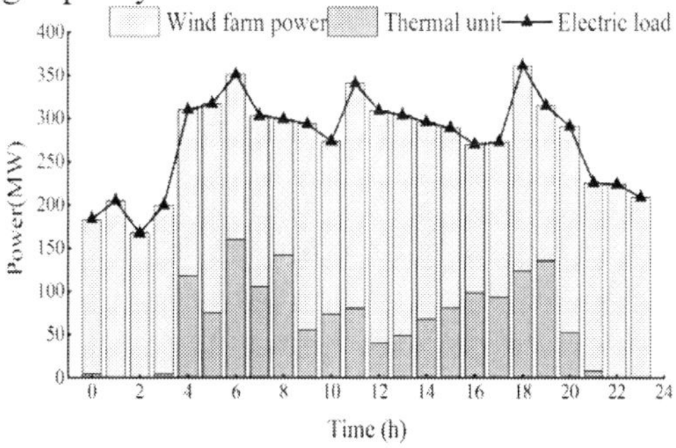

Figure 7. Electric power scheduling

As demonstrated by Fig. 7 power dispatch, wind farms and thermal power units work well together to satisfy the power load demand in the coastal area when the joint dispatch approach outlined in this study is used. During the whole dispatching cycle, the amount of wind farms connected to the grid was 5046.231MW, accounting for 76.32% of the total electric load. It enhances wind power consumption and reducing carbon emissions and at the same time secure power supply.

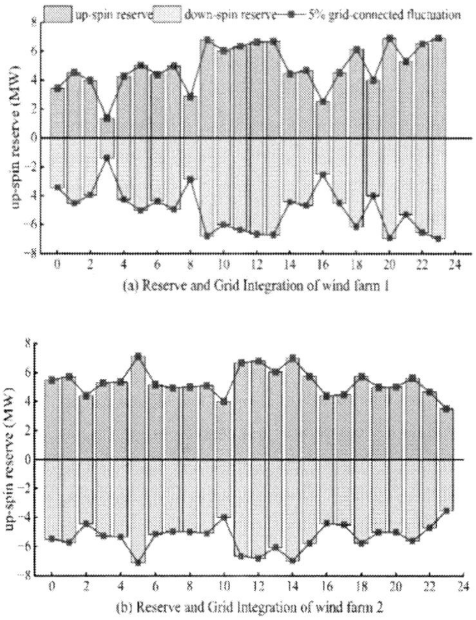

Figure 8. Grid connection and standby for wind farms

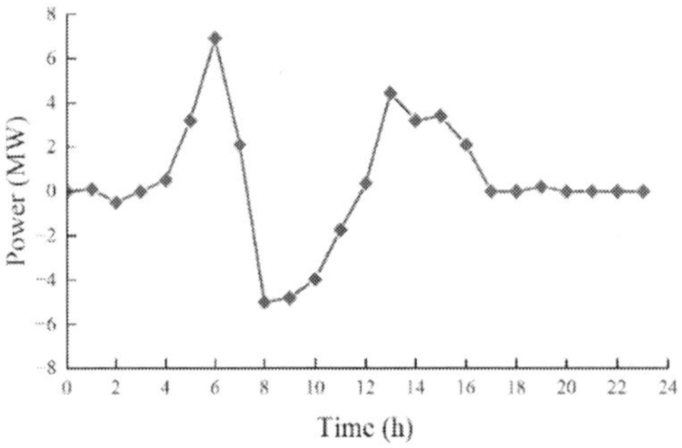

Figure 9. Deviation between expected and actual grid-connected power of wind farm

It is evident from Fig.8 and Fig.9 that the two wind farms' side standby can meet the demand for grid power fluctuation of ±5%, and that the difference between the actual and expected grid power of the wind power is always within the predefined range of ±7.5MW, ensuring the smooth transmission of offshore wind power.

From Table I we can observe the fatigue loads of wind farm daily operation. Compared with the traditional strategy with no deviation setting, the fatigue load variance for the two wind farms is 5.8%, falling within the 8% setting range. It can be seen that the wind farm cluster reaches a balance in load distribution, which verifies the equilibrium load model proposed is efficiency. And it also can reduce the maintenance cost caused by the number of frequent repairing.

Table 1. Daily operational fatigue loads for wind farms

Deviation setting		Fatigue load (M Nm)	Fatigue cost (yuan)
8%	Wind farm 1	4.403	16009.308
	Wind farm 2	4.162	15133.032
	Deviation	5.8%	5.8%
No deviation	Wind farm 1	6.697	224350.292
	Wind farm 2	4.434	16122.024
	Deviation	33.79%	33.79%

5. Conclusion

In this paper, a joint scheduling method for coastal energy and offshore wind-hydrogen system is proposed to realize the electric-hydrogen energy transfer by considering taking ships and DC transmission as carriers. It also considers the fatigue load of the wind farm group and the constraints of the system spinning reserve. Simulation results indicate that the proposed method in this paper realizes the smooth transmission of offshore wind power and can equalize the fatigue load value of wind farms, reduce the maintenance cost, and ensure that the wind farms can always meet the power demand of the land grid.

References

[1] J Zheng. Analysis on the development path of offshore wind power investment driven by national energy transformation. Chinese Market,2024, 20: 13-16.

[2] Brelik A, Nowaczyk P, Cheb K. The Economic Importance of Offshore Wind Energy Development in Poland [J]. Energies, 2023, 16, 23.

[3] P. Hou, W. Hu, M. Soltani, C. Chen, B. Zhang and Z. Chen, "Offshore Wind Farm Layout Design Considering Optimized Power Dispatch Strategy," in IEEE Transactions on Sustainable Energy, vol. 8, no. 2, pp. 638-647, April 2017.

[4] Xie P, Lu Q, Zhao Y, et al. Optimal Operation Scheduling of Multi-energy Complementary Systems Including Offshore Wind Power[J]. Recent Advances in Electrical & Electronic Engineering,2024,17: 295-303.

[5] J Zhang, Y Zhu, Z Zhao. Optimal Scheduling Strategy for Power Systems Containing Offshore Wind Farms Considering Wind Power Uncertainty[J]. Journal of Circuits, Systems and Computers, 2023, 32.
[6] H Liao, W Hu, X Wu, et al. Active power dispatch optimization for offshore wind farms considering fatigue distribution[J]. Renewable energy, 2020, 151: 1173-1185.
[7] X Liu, L Zu, Z Wei, et al. Two-layer optimal scheduling of integrated electric-hydrogen energy system with seasonal energy storage[J]. International Journal of Hydrogen Energy,2024, 82, 821131-1145.
[8] G Wu, T Li, W Xu, et al. Chance-constrained energy-reserve co-optimization scheduling of wind-photovoltaic-hydrogen integrated energy systems[J]. International Journal of Hydrogen Energy, 2023, 48: 6892-6905.
[9] L. Lin, X. Zheng and J. Gu, "Optimal Dispatching of Combined Heat and Power System Considering the Power Demand Elasticity of Hydrogen Storage Active Load," in IEEE Transactions on Industry Applications, vol. 58, no. 2, pp. 2760-2770, March-April 2022.
[10] Y Ma, Y Liu, X Bai, et al.: A multi-vessel optimization approach for cooperative operation and maintenance scheduling in offshore wind farm[J]. Renewable Energy, 2024,229: 120758-.
[11] C Zhu, X Bian, Y Zang, et al. Optimal scheduling method of offshore wind farm group with energy storage considering power time shift between adjacent offshore wind farms[J/OL]. Power System Technology, 2024, 48: 2394-2403.

Journal of Physics: Conference Series **3012** (2025) 012053 doi:10.1088/1742-6596/3012/1/012053

Electromagnetic Transient Modelling and Characteristic Analysis of Flexible Distribution Network with Series Parallel Transformers

Zheng Xiang[1], Bin Wang[1*], Ran Chen[2], Wei Bao[2], Qian Feng[2], Hua Huang[2]

[1]Department of Electrical Engineering, Tsinghua University, Beijing, China

[2]State Grid Shanghai Municipal Electric Power Company, ShangHai, China

[*]Corresponding author's e-mail: binw_ee@mail.tsinghua.edu.cn

Abstract. With the large-scale integration of distributed generation and diversified loads, traditional distribution networks can no longer meet the current operational demands. To ensure the reliability of load power supply and maintain power quality, the introduction of series parallel transformers has become a critical approach for efficient integration of distributed generation and ensuring load supply reliability in distribution networks. This paper aims to analyze the operational characteristics of distribution networks incorporating series parallel transformers by developing both component-based and instantaneous value-based electromagnetic transient models, thereby validating the effectiveness of series parallel transformers and the accuracy of the instantaneous value model. Firstly, the topology and working principles of series parallel transformers are elucidated. Subsequently, a distribution network model with series parallel transformers is constructed, and corresponding control strategies are proposed. Finally, an instantaneous value model is developed based on practical application scenarios, and a component-based simulation environment is established using the PSCAD platform to verify the accuracy and effectiveness of the model. The results demonstrate that series parallel transformers significantly enhance the reliability of power supply and the quality of electric energy in distribution networks, and the instantaneous value model accurately reflects the actual operating conditions of the system.

1. Introduction

To meet the national strategic objectives of "carbon peaking and carbon neutrality," it is imperative to accelerate the development of a clean, low-carbon, safe, and efficient energy system. Distributed Generators, represented by solar and wind power, are being widely integrated into distribution networks, while electric vehicles are rapidly proliferating and connecting to these networks on a large scale[1-2]. Consequently, traditional distribution networks operating in open-loop mode can no longer meet the urgent needs of current power users for reliable electricity supply and flexible distribution network operations. Specifically, issues such as bidirectional power flow, overloaded line capacity in some sections, voltage violations at nodes, frequent and severe load fluctuations, and the blurring of boundaries between sources and loads have emerged. The root cause of these problems is the low reliability of power supply in distribution networks under the traditional open-loop operation mode. Achieving interconnected operation of distribution networks is considered an effective solution to these challenges[3-4].

Flexible interconnection technologies not only enable interconnected power transfer within distribution networks but also offer capabilities for power flow regulation. As a result, key equipment and technologies tailored for flexible interconnection have garnered increasing attention from researchers worldwide, making flexible interconnected distribution networks a focal point in distribution network development[5-6]. The Phase-Shifting Transformer (PST) is a typical electromagnetic-based flexible interconnection device[7]. It can be directly connected to the grid and output compensating voltage, adjusting the phase angle of the output voltage within a certain range to control the active power distribution across distribution lines, thereby achieving a rational allocation of power flow. However, traditional PSTs employ slow and short-lifespan mechanical switches to adjust the compensating voltage magnitude, leading to drawbacks such as long response times and complex operation and maintenance. With advancements in power electronics, fully power-electronic-based flexible interconnection devices, exemplified by the Unified Power Flow Controller (UPFC), have been widely adopted[8]. Nevertheless, fully power-electronic solutions often suffer from high costs, large physical sizes, and operational challenges, making them unsuitable for widespread application in distribution networks.

To address practical needs, researchers have proposed various new topologies for flexible interconnection equipment based on PST configurations integrated with power electronic switching technologies. In 2002, K. K. Sen introduced the concept of the "Sen" Transformer (ST), which builds upon traditional three-winding transformers and On-Load Tap-Changing(OLTC) technology. It obtains the line voltage at the parallel point via a shunt transformer, combines and superimposes the three-phase voltages through the OLTC to generate three-phase compensating voltage, and then feeds it back into the line through a series transformer to adjust the magnitude and phase angle of the line voltage[9]. The advantage of the ST lies in achieving voltage compensation effects similar to that of a UPFC without relying on numerous power electronic devices or complex control strategies. However, its inability to achieve continuous regulation remains a limitation.

To overcome the shortcomings of the ST, a new type of series-parallel flexible interconnection equipment has been proposed, based on semi-controlled power electronic devices and the "Sen" Transformer. This equipment, known as the Thyristor Controlled Hybrid Transformer (TCHT), offers a wider compensation range and can rapidly achieve multi-level voltage regulation, making it more suitable for interconnected power transfer in distribution networks. Therefore, this paper will conduct an in-depth analysis of the topology and operating principles of TCHT, and establish both component-based and instantaneous-value-based electromagnetic transient models. The accuracy and effectiveness of these models are then verified. Ultimately, the findings confirm that TCHT can effectively regulate power flow, thereby enhancing the reliability of power supply and the quality of electrical energy in distribution networks. Furthermore, the proposed models are shown to accurately reflect the actual operating conditions of the system.

2. Methodology

2.1. Topology of the TCHT

The main circuit topology of the TCHT is illustrated in Figure 1. This circuit features a series-parallel configuration and primarily consists of three parts: the parallel excitation section, the thyristor-controlled valve-regulated voltage combination section, and the series voltage compensation section. The parallel excitation section employs three single-phase multi-winding transformers. The primary windings are connected in parallel to the distribution network system to obtain the system voltage. Each secondary winding corresponds to an array of three sub-windings. Specifically, the primary winding of phase A corresponds to the secondary sub-winding arrays Aa, Ab, and Ac; the primary winding of phase B corresponds to the secondary sub-winding arrays Ba, Bb, and Bc; and the primary winding of phase C corresponds to the secondary sub-winding arrays Ca, Cb, and Cc.

The thyristor-controlled valve-regulated voltage combination section comprises nine modules with identical structures, corresponding to the secondary sub-winding arrays of the parallel transformers. These modules are divided into three groups: Aa, Ba, and Ca are connected in series to form the phase

A compensation voltage; Ab, Bb, and Cb are connected in series to form the phase B compensation voltage; and Ac, Bc, and Cc are connected in series to form the phase C compensation voltage. All three compensation voltages are connected to the series voltage compensation section. Through series transformers, the three-phase compensation voltages generated by the thyristor-controlled valve-regulated voltage combination section are induced to the primary side and then connected in series to the distribution network system, achieving voltage compensation.

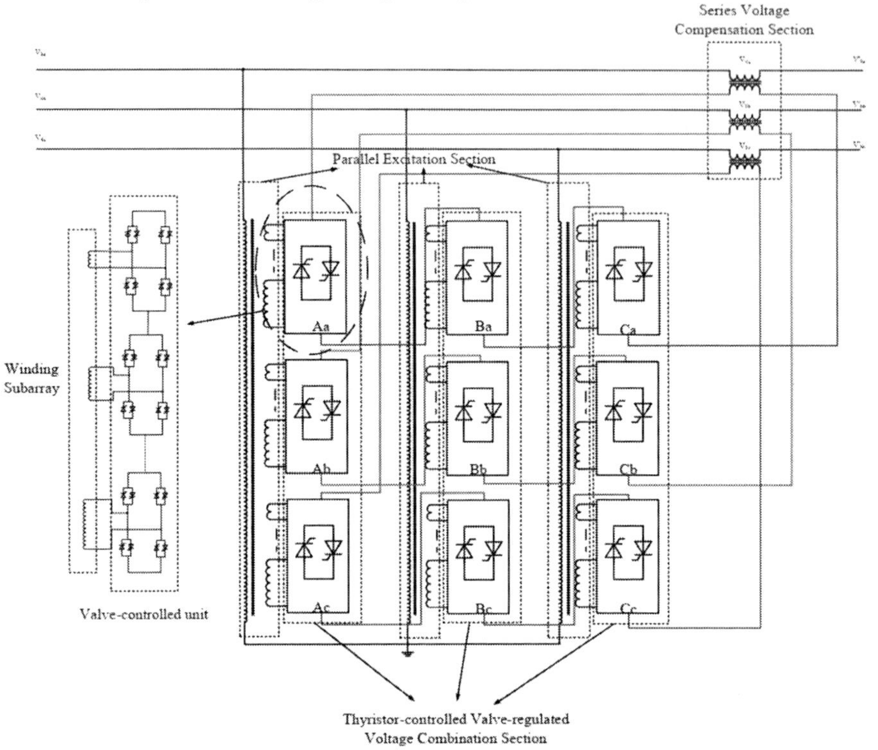

Figure 1. Main Circuit Topology of the TCHT

2.2. Operating Principle of the Thyristor-Controlled Valve-Regulated Voltage Combination

The sub-arrays of the secondary windings of the parallel transformers in the TCHT are composed of multiple thyristor bridges. Their structure is illustrated in Figure 2, where $U_1, U_2, \ldots, U_n$ represent the rated voltages of the n windings, respectively.

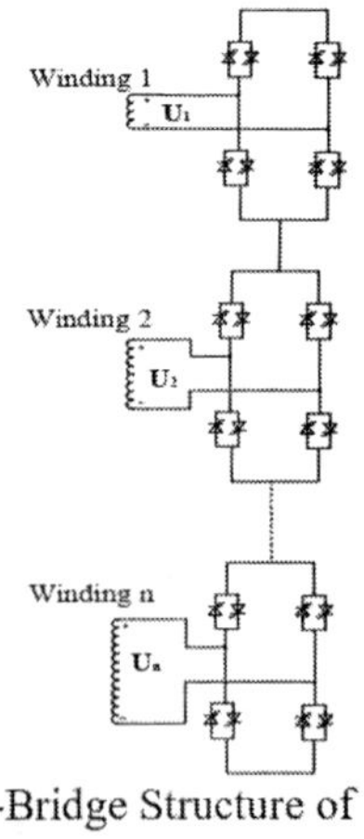

Figure 2. Cascaded H-Bridge Structure of the Winding Subarray

For a single thyristor bridge, when the upper thyristor of the left arm and the lower thyristor of the right arm conduct simultaneously, the switching function is assigned a value of +1, and its output voltage equals the winding voltage. When the lower thyristor of the left arm and the upper thyristor of the right arm conduct simultaneously, the switching function is -1, and its output voltage equals the negative of the winding voltage. If both the upper and lower thyristors of either the left or right arm conduct simultaneously, the switching function is 0, and the output voltage is zero. The total output voltage of the subarray is the sum of the output voltages from each thyristor bridge within it

When the ratios of the rated voltages of the windings in the subarray satisfy $U_1 : U_2 : \ldots : U_n = U : 3U : \ldots : 3^{n-1}U$, the total output voltage of the subarray changes incrementally by U, ranging from $-\left[\left(3^{n-1}\right)/2\right]U$ to $\left[\left(3^{n-1}\right)/2\right]U$. Therefore, the winding subarray has a total of 3n possible output voltage levels. For instance, when $n=2$, the ratio of rated voltages is 1:3. The mapping relationship between the total output voltage of the subarray and the switching functions of the thyristor bridges is shown in Table 1. In this paper, the ratio of the subarray's output voltage to U is termed the "winding subarray output ratio," denoted by ρ.

Table 1 Mapping Relationship Between Subarray Output Voltage and Switching Functions

Thyristor Bridge 1 Switching Function	Thyristor Bridge 2 Switching Function	Winding Subarray Output Voltage	Winding Subarray Output Ratio ρ
+1	+1	+4U	+4
0	+1	+3U	+3
-1	+1	+2U	+2
+1	0	+1U	+1
0	0	0	0
-1	0	-1U	-1
+1	-1	-2U	-2
0	-1	-3U	-3
-1	-1	-4U	-4

Each subarray has its own independent thyristor bridge and winding, allowing for independent regulation of the output voltage. For phase A, the output voltages of the subarrays are denoted as $\dot{U}_{Aa}$, $\dot{U}_{Ab}$, and $\dot{U}_{Ac}$, with corresponding output ratios ρ_{Aa}, ρ_{Ab}, and ρ_{Ac}. For phase B, the output voltages of the subarrays are denoted as $\dot{U}_{Ba}$, $\dot{U}_{Bb}$, and $\dot{U}_{Bc}$, with output ratios ρ_{Ba}, ρ_{Bb}, and ρ_{Bc} For phase C, the subarray output voltages are $\dot{U}_{Ca}$, $\dot{U}_{Cb}$, and $\dot{U}_{Cc}$, with output ratios ρ_{Ca}, ρ_{Cb}, and ρ_{Cc}. Therefore, the compensation voltage for each phase of the TCHT is:

$$\dot{U}_T = \begin{pmatrix} \dot{U}_{Ta} \\ \dot{U}_{Tb} \\ \dot{U}_{Tc} \end{pmatrix} = \begin{bmatrix} \dot{U}_{\mathrm{Aa}} + \dot{U}_{\mathrm{Ba}} + \dot{U}_{\mathrm{Ca}} \\ \dot{U}_{\mathrm{Ab}} + \dot{U}_{\mathrm{Bb}} + \dot{U}_{\mathrm{Cb}} \\ \dot{U}_{\mathrm{Ac}} + \dot{U}_{\mathrm{Bc}} + \dot{U}_{\mathrm{Cc}} \end{bmatrix} = \begin{bmatrix} \rho_{Aa} & \rho_{Ba} & \rho_{Ca} \\ \rho_{Ab} & \rho_{Bb} & \rho_{Cb} \\ \rho_{Ac} & \rho_{Bc} & \rho_{Cc} \end{bmatrix} \dot{U} \tag{1}$$

Based on Table 1 and Equation (1), it is easy to deduce that by series-connecting the output voltages of the three sub-windings in each phase to obtain the compensation voltage for each phase, a total of 93, that is, 729 voltage vector combinations are formed. Taking the compensation voltage of phase A as an example, its compensation range is illustrated in Figure 3.

Journal of Physics: Conference Series **3012** (2025) 012053 doi:10.1088/1742-6596/3012/1/012053

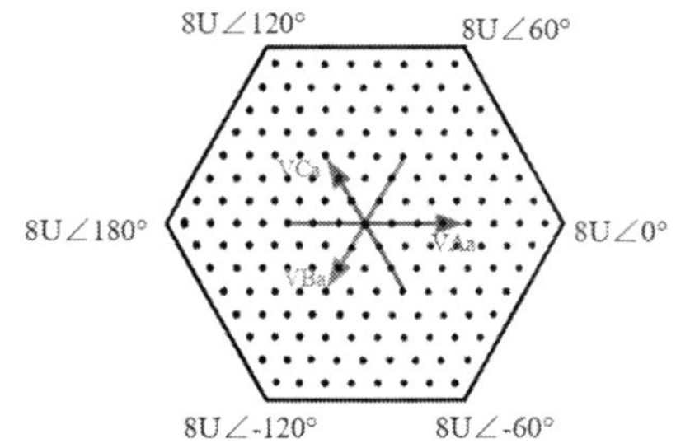

Figure 3. Voltage Compensation Range of Phase A

The compensation voltage ranges for phases B and C are identical to that of phase A. Furthermore, when n takes larger values, the boundaries of the lattice remain unchanged, but the density increases exponentially. This means that the adjustment step size of the compensation voltage $\dot{U}_T$ becomes increasingly finer.

2.3. Control Strategy

2.3.1. Soft Interconnection. The equivalent circuit of the distribution network interconnection system based on the TCHT is shown in Figure 4. To simplify the analysis, the equivalent circuits of the two distribution areas have been streamlined. In Figure 4, V_1 and V_2 represent the system bus voltages of Area 1 and Area 2, respectively; V_{L1} and V_{L2} denote the voltages at the ends of the feeders in Area 1 and Area 2, respectively. Circuit breakers BK1 and BK2 control the connection and disconnection of the TCHT.

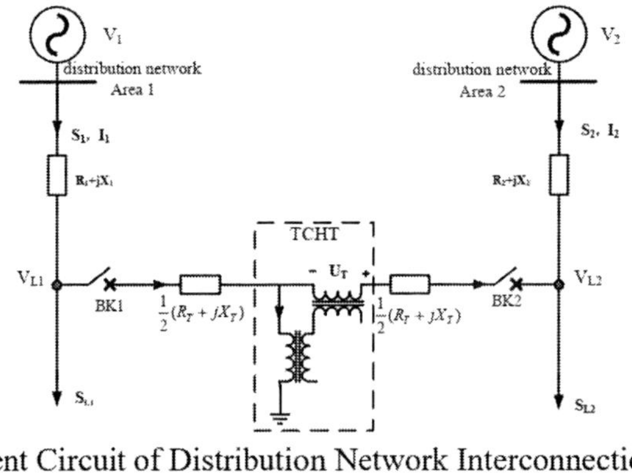

Figure 4. Equivalent Circuit of Distribution Network Interconnection Based on TCHT

When the TCHT is not in operation, distribution network Area 1 and Area 2 function independently. There exist certain deviations in both the magnitude and phase angle of the feeder end voltages V_{L1} and V_{L2} for the two areas, denoted as:

$$\Delta V = V_{L2} - V_{L1} \tag{2}$$

If the tie switch is used directly for interconnection, it will result in a significant inrush current. However, by using the TCHT and selecting the appropriate tap position of the secondary winding of the parallel transformer, the series compensating voltage U_T output by the TCHT can be made equal to ΔV. Consequently, when these two feeders are interconnected, no inrush current will be generated, demonstrating the characteristics of soft interconnection.

Therefore, the control strategy for soft interconnection involves real-time measurement of the voltages V_{L1} and V_{L2} at both ends of the feeder to calculate the voltage difference ΔV between them.

Based on this calculated voltage difference, the most appropriate tap position is selected so that the TCHT outputs the optimal compensation voltage.

2.3.2. Power Flow Regulation. Based on Table 1 and Equation (1), it can be concluded that by adjusting the output voltages of the winding subarrays, the compensation voltages of each phase can be modified, thereby regulating the power flow direction between the two distribution areas. Upon receiving a power flow regulation command, a rotational optimization method is employed to find the optimal compensation voltage switching point. Let the current switching point be denoted as K0, and the output ratios of the winding subarrays constituting the phase A compensation voltage $(\rho_{Aa}, \rho_{Bb}, \rho_{Cc})$ are represented by coordinates (k_a, k_b, k_c). Due to three-phase symmetry, we have $\rho_{Aa} = \rho_{Bb} = \rho_{Cc}$, $\rho_{Ac} = \rho_{Ba} = \rho_{Cb}$, and $\rho_{Ab} = \rho_{Bc} = \rho_{Ca}$. Therefore, by altering the coordinates (k_a, k_b, k_c), the magnitude of the compensation voltage can be adjusted. The method for finding the optimal switching point using rotational optimization is as follows:

Step 1: Take the power flow regulation command S_T^* as the target value for power flow adjustment in this optimization, that is, $P_{ref} = P_T^*$、$Q_{ref} = Q_T^*$. The allowable error in power flow is ΔS, and the deviation between the power flow change at each switching point and the target power flow value is: $\varepsilon_i = \sqrt{(P_{Ki} - P_{ref})^2 + (Q_{Ki} - Q_{ref})^2}$;

Step 2: Let the current switching point K_0 be the center point for this optimization. Real-time monitoring of the power flow transferred via the TCHT is performed to calculate the power flow regulation deviation at the current switching point. Determine whether this deviation is within the allowable error range ΔS for power flow regulation; if so, then K_0 is the optimal compensation voltage switching point.

Step 3: As illustrated in Figure 5, take K_0 as the center point and sequentially test the switching points $K1 = (k_a - 1, k_b, k_c)$, $K2 = (k_a, k_b - 1, k_c)$, $K3 = (k_a, k_b, k_c - 1)$, $K4 = (k_a + 1, k_b, k_c)$, $K5 = (k_a, k_b + 1, k_c)$, $K6 = (k_a, k_b, k_c + 1)$. Calculate and store the corresponding power deviation values ε_i.

Step 4: From the six power deviation values $\varepsilon_i (i = 1, 2, 3, 4, 5, 6)$ obtained in Step 3, select the one with the smallest deviation and compare it with the allowable power flow deviation ΔS. If $\min(\varepsilon_i) \le \Delta S$, the optimization process concludes, and the current coordinates are considered optimal. Otherwise, take the coordinate point corresponding to the smallest power deviation as the new center, and return to Step 2 to re-initiate the optimization.

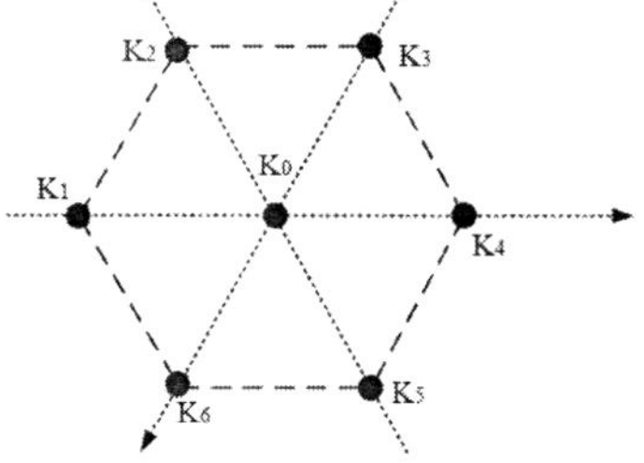

Figure 5. Principle of the Rotational Optimization Method

2.4. Instantaneous Value Model of the TCHT

The TCHT has a three-phase symmetrical structure. Taking phase A as a reference for analysis, the corresponding equations are written as follows.

$$\begin{cases} u_{L2A}\text{-}u_{L1A}\text{-}u_T = L_t \dfrac{di_1}{dt} \\ u_{eq} + L_{eq}\dfrac{di_2}{dt} + R_{eq}i_2 = \dfrac{u_T}{n_a} \end{cases} \tag{3}$$

In the equation, u_{L1A}, u_{L2A} are the phase A voltages at nodes V_{L1} and V_{L2}, respectively; u_T, i_1 represent the primary voltage and current of the series transformer; L_t is the equivalent leakage inductance of the series transformer; u_{eq}, L_{eq}, R_{eq} are the equivalent voltage, inductance, and resistance of the secondary circuit of the series transformer, respectively; i_2 is the current flowing through the secondary winding of the series transformer; and n_a is the turns ratio of the series transformer. From Equation (3), further derivation yields:

$$(n_a L_{eq} + \frac{L_t}{n_a})\frac{di_1}{dt} + n_a R_{eq} i_1 = \frac{u_{L2A} - u_{L1A}}{n_a} - u_{eq} \tag{4}$$

For Equation (4), the fourth-order Runge-Kutta method is applied, and MATLAB is used to calculate the current flowing through the TCHT.

3. Results

To validate the soft interconnection and power flow regulation functions of the TCHT proposed earlier, as well as the effectiveness of its instantaneous value model, this paper develops a simulation model based on the application scenario shown in Figure 4 using the PSCAD/EMTDC simulation platform. The main simulation parameters are listed in Table 2.

Table 2 Main Parameters Used in the Simulation

Parameter	Value
Voltage of Distribution Area 1(kV)	10∠0°
Voltage of Distribution Area 2(kV)	10∠-5°
Capacity of Parallel Transformer (MVA)	0.288
Capacity of Series Transformer (MVA)	0.288
Rated Primary Voltage of Parallel Transformer (kV)	5.7
Rated Primary Voltage of Series Transformer (kV)	1.44
Rated Secondary Voltage of Parallel Transformer (V)	90、270
Rated Secondary Voltage of Series Transformer (V)	720
Equivalent Load of Distribution Area 1 (MVA)	3+j1.5
Equivalent Load of Distribution Area 2 (MVA)	4.5+j2

3.1. Soft Interconnection

Based on the same medium-voltage distribution network equivalent model, the medium-voltage distribution network interconnection simulation is conducted using a normally open tie switch and a TCHT as interconnection devices. The current flowing through the interconnection devices at the moment of interconnection is monitored, as shown in Figure 6. A comparative analysis of the impact on both systems after interconnection is performed, verifying the soft interconnection function of the TCHT.

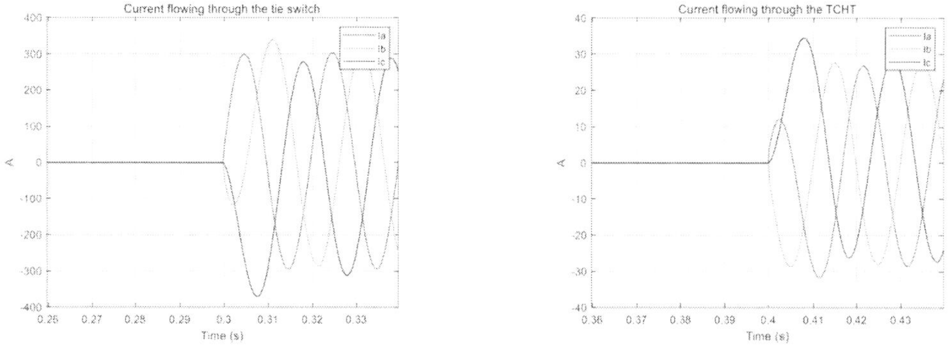

(a) Current flowing through the tie switch (b) Current flowing through the TCHT

Figure 6. Comparison of soft interconnection

As shown in Figure 6(a), the current waveform flowing through the tie switch exhibits an inrush current of 370A. After compensation by the TCHT, the current flowing through the TCHT, as shown in Figure 6(b), reduces the inrush current amplitude to 30A, which can be regarded as soft interconnection.

3.2. Power Flow Regulation

Compared to the traditional normally open tie switch, the greatest advantage of TCHT lies in its ability to regulate power flow in the line. The power flow regulation command is set to 2 MW and 1 Mvar, with the power error set to 0.5. Combined with the aforementioned optimization control strategy, the TCHT's ability to regulate power flow is verified.

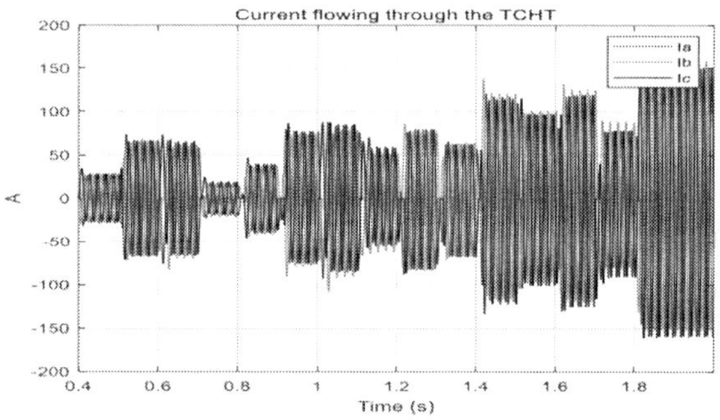

(a) Current flowing through the TCHT

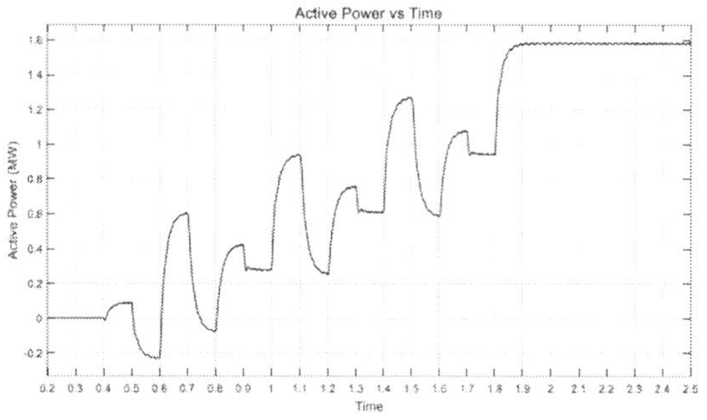

(b) Active power through the TCHT

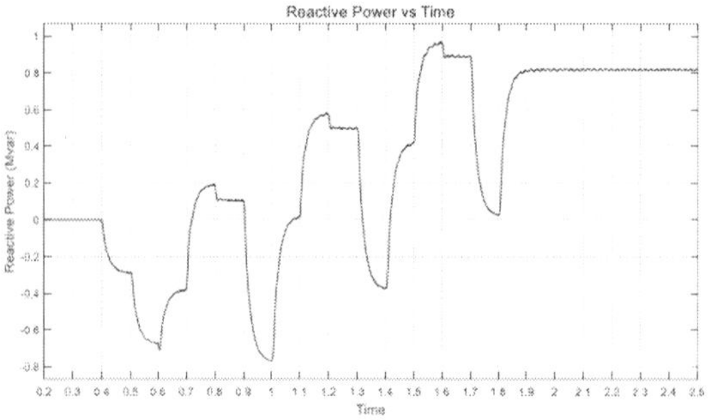

(c) Reactive power through the TCHT

Figure 7. Simulation results of TCHT power flow regulation

The simulation results indicate that the TCHT can follow the power reference values in real-time and achieve bidirectional control of power flow. The switching of the levels can be completed quickly within one power frequency cycle, without causing significant inrush currents. The final active power through the TCHT is 1.58 MW, and the reactive power is 0.81 Mvar, with power deviations $\varepsilon = 0.46 < 0.5$, meeting the requirements.

3.3. *Instantaneous value model*

With the help of simulation data, the A-phase current flowing through the TCHT is calculated using Equation (4) and compared with the simulated A-phase current, as shown in Figure 8.

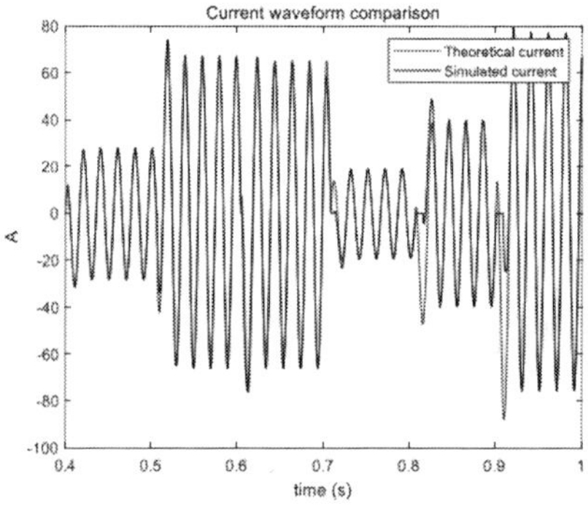

Figure 8. Current waveform comparison

As shown in Figure 8, the waveform based on the instantaneous value model closely matches the simulated waveform. During steady-state operation, the two simulation results overlap entirely, with only slight differences observed during the transient period. The discrepancy during the transient process arises because the thyristor turn-off requires current interruption and the maintenance of a certain reverse voltage, leading to a period where the current is zero. However, the current waveform based on the instantaneous value model neglects this period.

4. Conclusion

The TCHT can achieve soft interconnection between two distribution areas, enabling interconnected power transfer while following power reference values in real-time to realize bidirectional control of power flow. The proposed instantaneous value model, while ensuring accuracy, reduces computational complexity and improves efficiency. Therefore, TCHT can effectively enhance the reliability of power distribution and improve power quality. Due to the voltage regulation step size of 90V for the TCHT winding submodules in this study, there is some deviation between the power flow and the reference values during power transfer. In future engineering applications, increasing the number of H-bridge levels or reducing the voltage step size U could improve power regulation precision.

Acknowledgments
The authors would like to express their gratitude to the State Grid Corporation of China Headquarters for supporting the project "Research on Fault Ride Through Mechanism and Protection Coordination Method of Series Parallel Flexible Interconnection Equipment in Medium Voltage Distribution Network" (Project No. SGSHDK00PWJS2310566). All the support, information, and assistance provided in producing this paper are greatly appreciated.

References
[1] Liu H, Cao R Z, Han J, et al., 2024, 'Research Status and Future Prospect of Distributed Generation Hosting Capacity Assessment of Distribution Network', *Power Systems Automation*, pp. 1–13.
[2] Li X, Wang S Q, Li Q Y, et al., 2024, 'Joint Planning of Smart Distribution Grid Considering Distributed Generation and Electric Vehicle Charging Station', *Chinese Journal of Electrical Engineering*, pp. 1–9.
[3] Cao H, Jiang M, Rong F, et al., 2024, 'Impact of Distributed Generation on Static Stability of Radial Distribution Network', *Electric Power and Energy*, 45(04), pp. 509–512.
[4] Li J L, Di W F, Li Y X, et al., 2024, 'Optimized Allocation of Energy Storage Zones in Distribution Networks Containing High-permeability Photovoltaic Regions', *Journal of Electrical Engineering*, pp. 1–11.
[5] Hu P F, Zhu N X, Jiang D Z, et al., 2021, 'Research Progress and Prospects of Key Technologies of Flexible Interconnected Smart Distribution Network', *Power Systems Automation*, 45(08), pp. 2–12.
[6] Zhang Y J, Liu Z W, Deng F Q, 2020, 'Review on Research Status and Development of Flexible Interconnected Distribution Networks', *Guangdong Electric Power*, 33(12), pp. 3–13.
[7] Yu M Z, Li F, Li Z H, et al., 2023, 'Principle, Topology, Application and Development of Phase-shifting Transformer', *Journal of Wuhan University (Engineering Edition)*, 56(10), pp. 1224–1237.
[8] Liu J Q, 2020, 'Research on Power Flow Regulation Characteristics and Control Strategy of UPFC', PhD thesis, Zhejiang University.
[9] Sen K K, Sen M L, 2002, 'Introducing the Family of "Sen" Transformers: A Set of Power Flow Controlling Transformers', in *Proc. IEEE Power Engineering Society Summer Meeting*, Chicago, IL, USA, vol. 1, pp. 486.

Journal of Physics: Conference Series **3012** (2025) 012054 doi:10.1088/1742-6596/3012/1/012054

Frequency Domain Analysis and Suppression Measures of Ferroresonance in 35kV New Energy Access Station

Jing Wang*, **Yanhua Han, Hao Wu, Can Guo**

State Grid Shaanxi Electric Power Research Institute, Xi'an 710100, Shaanxi Province, China

*Corresponding author's e-mail: bjjdwj@163.com

Abstract. As the new power system has become the development trend of the power grid, the proportion of photovoltaic power stations in the power system has increased, and the characteristics of its output voltage wide band and high frequency have led to frequent fusing accidents of PT fuses induced by ferroresonance, which are significantly different from those of traditional ferroresonance. However, the existing research has not revealed its frequency domain characteristics, and the parameter configuration of traditional suppression measures is not fully applicable. In view of the above problems, this paper takes a 35kV new energy access station in China as the research object, builds the electromagnetic transient simulation model of the system, obtains the ferroresonance waveform, determines the harmonic component of photovoltaic power generation that has the most significant influence on the ferroresonance through fast Fourier decomposition, and comparatively analyzes the difference between the ferroresonance of the new energy access station and the traditional ferroresonance frequency domain. The parameter optimization suggestions of suppression measures are given.

1. Introductory

Due to the increasing growth of the total amount and density of urban power loads, the problems of environmental pollution and energy depletion of the traditional power system have gradually appeared, and the construction of a new type of clean and low-carbon power system has become the direction of development [1]. With the vigorous implementation of the dual-carbon strategy, the proportion of photovoltaic power plants in the power system has increased significantly in recent years [2]. Due to the difference in operation mechanism between PV power plants and traditional power plants, grid-connected PV power plants will bring new challenges to the safe and stable operation of the power system [3].

Power transformers and transformers and other transformer equipment is the most important equipment family in the power system, in which the potential transformer (PT) is 35kV and the following AC power system in the large-scale application of voltage sensing devices [4]. Therefore, PT equipment is also one of the non-linear equipment with iron core in the power system. In order to protect PT equipment, high voltage fuses are often installed on the high voltage side of the equipment to avoid PT burning due to system failure. In recent years, as the proportion of photovoltaic (PV) power generation continues to increase, the PV grid-connected system in the PT insurance meltdown accidents occur frequently, the cause of the accident is still unclear, the lack of effective inhibition measures [5].

Ferroresonance is one of the main causes of PT fuse blowing accidents [6,7]. Ferroresonance is when the nonlinear inductive characteristics of equipment and cables, overhead lines and other lines to the ground there is stray capacitance, constituting the basic circuit of ferroresonance, and PT nonlinear excitation inductance and system capacitance to the ground to reach a certain condition of the electromagnetic transient phenomenon, which results in the resonance of the over-voltage, overcurrent prone to cause the PT high-voltage side of the fuse breaking [8-12].

With the increase in the proportion of grid-connected photovoltaic power generation systems, due to the use of inverters as the grid-connected interface in photovoltaic power plants, the introduction of a large number of power electronic devices, whose output voltage is characterized by a wide bandwidth, high frequency, etc [13,14]. Since the ferroresonance in the resulting PV power plant is significantly different from the conventional ferroresonance, the parameter configuration of the existing suppression measures is not fully applicable.

To address the above issues, this paper takes a 35kV new energy access station as the research object, constructs the system electromagnetic transient simulation model based on the actual equipment parameters, and analyzes the harmonic components mainly generated by photovoltaic power generation according to the actual measurement data and the electromagnetic transient model of photovoltaic power generation, and researches the mechanism of ferroresonance under the injection of photovoltaic harmonics. Analyze the difference between the frequency domain characteristics of ferroresonance under the influence of photovoltaic power generation and the traditional ferroresonance, and combine with the simulation analysis to give the optimization suggestions for the parameters of the suppression measures.

2. Modeling Analysis of 35kV Photovoltaic Power Plant

2.1. Overall modeling of photovoltaic substations

The purpose of this paper is to investigate the impact of harmonics generated by PV power generation on the ferroresonance of the power system, as shown in Figure 1, based on the actual wiring diagram of a 35kV substation in China and the equipment parameters to build the electromagnetic transient simulation model of the PV power generation system into the station. Firstly, the electromagnetic transient model of PV power generation needs to be established to analyze the voltage harmonic characteristics of PV power generation in combination with the actual measurement data; then the substation system simulation model is established, including the line model, the nonlinear transformer model considering the core saturation, the nonlinear PT model considering the core saturation, the line model, and so on.

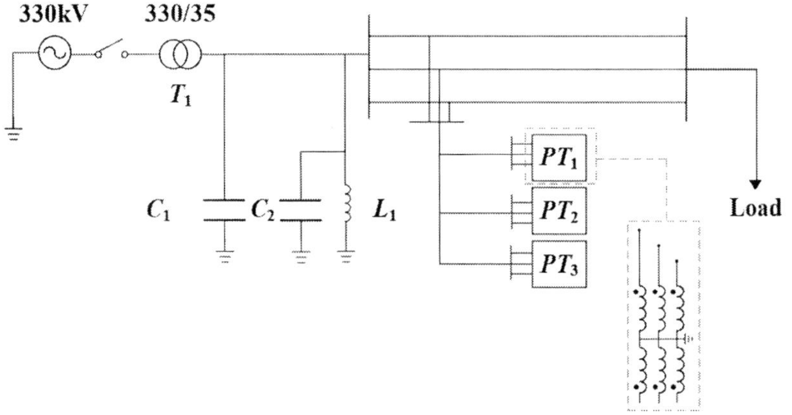

Figure 1. Schematic diagram of 35kV photovoltaic substation

Figure 1 shows the schematic diagram of the PV substation. In this case, the PV equivalent is a voltage source of 330kV with direct grounding. T_1 is a transformer of 330/35kV, C_1 and C_2 are capacitors in parallel in the line, L_1 is a reactor in parallel in the line, and PT_1, PT_2, and PT_3 are electromagnetic voltage transformers for the three-phase line, respectively. The load is a power load with parameters of 1512 kW and 1134 kVar.

2.2. Photovoltaic power equivalent modeling

PV power station is mainly composed of PV array, maximum power point tracking (MPPT), DC/AC inverter and other parts. Throughout the above process, the current harmonics of the PV panels are affected by the intensity of radiation, and the voltage passing through the non-ideal inverter also generates harmonics of the 2nd, 3rd, 5th, 7th, and 9th harmonics due to dead time, and harmonics distributed at the switching frequency due to modulation.

Literature [13] analyzed in detail the harmonic characteristics and main harmonic contents of PV power generation system, combined with the actual measurement data, a harmonic source is used to simulate the voltage harmonics generated by PV power generation, the expression of which is shown in equation (1):

$$
\begin{aligned}
u(t) = \\
& 466690\cos(2\times 50\pi t)+ \\
& 466690\times 0.045\cos(2\times 2\times 50\pi t)+ \\
& 466690\times 0.089\cos(2\times 3\times 50\pi t)+ \\
& 466690\times 0.051\cos(2\times 5\times 50\pi t)+ \\
& 466690\times 0.017\times\cos(2\times 7\times 50\pi t)+ \\
& 466690\times 0.006\times\cos(2\times 9\times 50\pi t)
\end{aligned}
\tag{1}
$$

2.3. Wiring modeling

The transmission line model equates the heat effect of current-carrying conductors to series resistance, the corona loss under high voltage to parallel conductance, and the magnetic field effect of AC current to inductance and the electric field effect of AC voltage to capacitance. As shown in equation (2), equation (3), equation (4) for the line parameter calculation equation.

$$L = (4.6\lg\frac{D}{r}+\frac{u_r}{2})\times 10^{-4} \tag{2}$$

$$C = \frac{0.0241}{\lg\frac{D}{r}}\times 10^{-6} \tag{3}$$

$$R = \rho\frac{L}{S} \tag{4}$$

Where D is the geometric mean distance, r is the wire radius, u_r is the relative permeability constant, and ρ is the wire material resistivity.

In this paper, we mainly study the effect of PV harmonic joining on ferroresonance overvoltage, ignoring the conductance, and use the medium-length line π-shaped equivalent circuit, the line length is 500m, the unit capacitance to ground is 0.03nF/m, and the unit-length equivalent inductance is 0.28μH/m. The parameters of inductance and capacitance are shown in Table 1.

Table. 1 Line Parameter

Component	Unit Length (nF/m)	Overall Length Value(nF)
Capacitors	0.03	1.5
Inductors	0.35	0.098

2.4. PT modeling

The T-equivalent circuit in which R_{s1}, R_{s2}, L_{s1}, L_{s2} are the resistance and leakage inductance of the winding of the primary and secondary sides of the PT, respectively, and L_m is the nonlinear excitation inductance, and R_m is the excitation resistance at the rated current. The nonlinear inductance of the PT used in this paper has the flux-current data shown in Table 2.

The first and second side winding resistance Rs_1 and R_{s2} values of the PT will be provided in the PT factory report, and the leakage inductance L_s can be calculated based on the short-circuit test data, as shown in equation (5).

$$\begin{cases} L_{s1} = \dfrac{\sqrt{|z_s|^2 - R_S^2}}{2\pi f} \\ Z_s = \dfrac{U_{rms}}{I_{rms}} \end{cases} \tag{5}$$

Where U_{rms} is the rms value of short-circuit test voltage, I_{rms} is the rms value of short-circuit test current, and f is the frequency of power supply, taken as 50Hz.

Table. 2 PT Excitation Curve Parameters

Number	Current (A)	Flux(Wb)	Number	Current (A)	Flux(Wb)
1	0	0	6	0.084	0.393
2	0.022	0.044	7	0.114	0.448
3	0.042	0.131	8	0.289	0.549
4	0.050	0.218	9	0.421	0.583
5	0.060	0.312	10	0.672	0.312

Table. 3 PT equivalent model parameters

Primary Winding Resistance R_{s1}/Ω	Core Loss Resistance R_m/Ω	Secondary Winding Resistance R_{s2}/Ω	Leakage Sensation L_s/mH
2000	4000	0.05	1.96

2.5. Transformer modeling

The main structure of the transformer has the core and winding, where the part connected to the power supply is the primary winding, and the part connected to the load is the secondary winding, and the two windings are connected by magnetic coupling, and the T-equivalent circuits are used with the same model as the PT model, as shown in Figure 2.

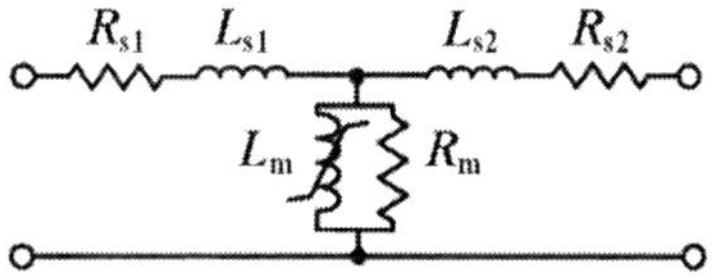

Figure 2. PT equivalent circuit model

The factory report of the transformer will provide the transformer's primary and secondary side winding resistance R_{s1}, R_{s2} values, leakage inductance L_s can be based on the short-circuit test data using the equation (5) calculated in Table 5; the transformer excitation curve parameters are shown in Table 4.

Tab. 4 Transformer Excitation Curve Parameters

Number	Current (A)	Flux(Wb)	Number	Current (A)	Flux(Wb)
1	0	0	6	1.599	112.531
2	0.282	2.250	7	1.608	126.044
3	0.338	9.003	8	3.165	157.555
4	1.053	81.028	9	11.974	175.561
5	1.207	94.533	10	53.005	180.063

Tab. 5 Transformer Equivalent Model Parameters

Primary Winding Resistance R_{s1}/Ω	Core Loss Resistance R_m/Ω	Secondary Winding Resistance R_{s2}/Ω	Leakage Sensation L_s/mH
455	16616	19.5	0.008

3. Simulation Studies

3.1. Construction of electromagnetic transient simulation model for 35kV photovoltaic substation

The electromagnetic transient simulation wiring model constructed based on the actual data of a 35kV substation is shown in Figure 1, which is a photovoltaic power generation substation, and in order to simulate the effect of harmonics generated by photovoltaic power generation on ferroresonance, combined with the actual measurement data, the photovoltaic power generation harmonics are added to the substation through a voltage source.

The components in the simulation model can be divided into 1) PT module, 2) line module, 3) transformer module, 4) shunt capacitor module, and 5) other modules including: power supply, switch and other auxiliary modules.

The simulation sets the parameters total time 3s, step size 100μs, t=0.1s, and the line switch is closed.

In order to comparatively analyze the effect of PV harmonics on ferromagnetic resonant overvoltage under different working conditions, in this paper, the control circuit parameters and simulation settings are unchanged, and only the harmonic joining composition is changed, which is divided into two parts before joining harmonics and after joining harmonics.

3.2. Frequency domain simulation waveform analysis

Conversion of the time domain to the frequency domain using the fast Fourier transform method. Comparing Figure 3(a) with Figure 3(b), it can be concluded from the frequency-amplitude plots before and after the addition of harmonics that, under the action of the fundamental voltage source, it contains only the industrial frequency voltage with an amplitude of 45 kV at 50 Hz. But with the addition of PV harmonics, in addition to the original 50Hz fundamental frequency component, a crossover resonant overvoltage of 48kV appears at 25Hz, and also contains a small amount of 2nd, 3rd, 5th, 7th and 9th harmonics.

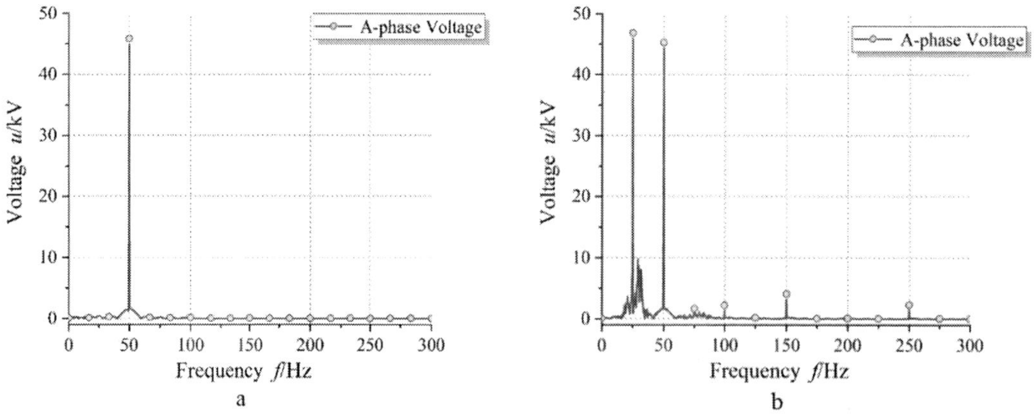

Figure 3. 35kV line voltage spectrum diagram (a) without harmonics;(b) with harmonics

Comparing Figure 4(a) with Figure 4(b), it can be concluded that under the action of the power supply containing only the fundamental wave, the frequency-dividing resonance did not occur, and the current on the primary side of the PT was small; whereas after the addition of the harmonics, a frequency-dividing resonance current with an amplitude of 42 mA was generated on the primary side of the PT.

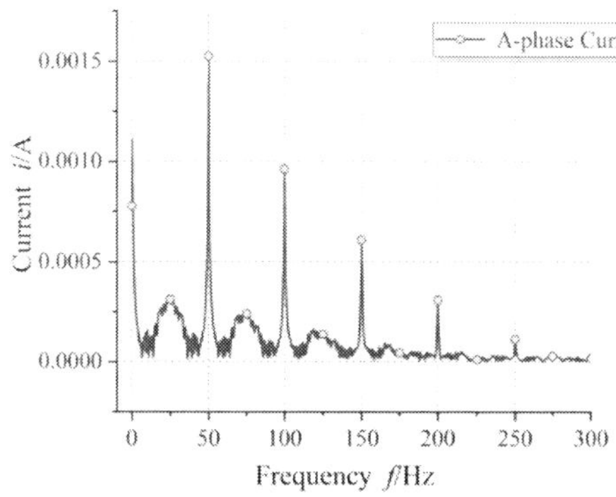

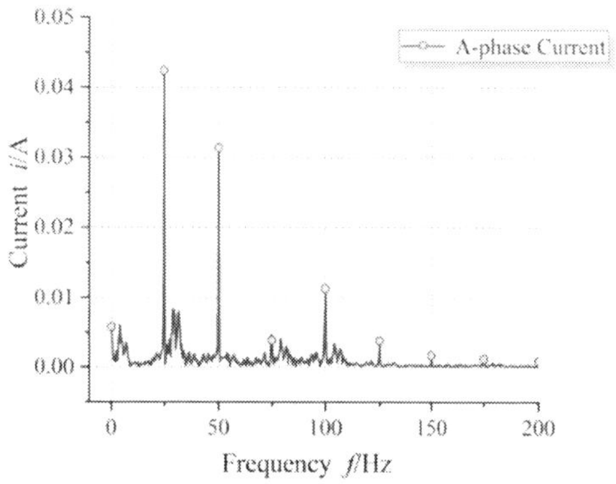

Figure 4. PT primary current spectrum (a)without harmonics; (b)with harmonics

4. Disincentives

Combined with the structure of the PT equipment, there are two main ideas for suppression measures: one is to use the damping effect to suppress the overcurrent amplitude; the other is to start from the nonlinear inductance of the PT iron core, to reduce the current flowing through the primary side of the PT, to avoid the PT from entering into deep saturation [15].

In this paper, the damping effect is used to suppress the overcurrent amplitude, as shown in Figure 5, where a damping resistor is connected in series in the PT grounding loop. As shown in Table 6 for the current peak value table of series connection of 10kΩ to 80kΩ resistor, its current peak value decreases with the increase of the resistance value of the series resistance, and the best current peak suppression effect is achieved when the 80kΩ resistor is connected in series with the peak size of 22.3mA. Therefore, series connection of resistor R_1 with the resistance value of 80kΩ in the PT grounding loop can effectively limit the overcurrent amplitude.

Table. 6 Resistance value and peak current

Number	Resistance R_1/kΩ	Peak Current /mA	Number	Resistance R_1/kΩ	Peak Current /mA
1	10	41.6	5	50	30.2
2	20	38.9	6	60	26.9
3	30	35.7	7	70	24.5
4	40	32.8	8	80	22.3

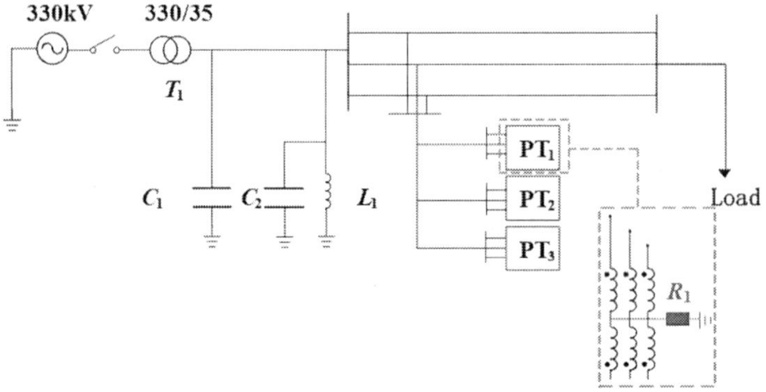

Figure 5. PT overcurrent suppression measures

As shown in Figure 6, with the addition of the 80kΩ linear resistor, the crossover resonant voltage amplitude at 25Hz is limited to about 1kV and the maximum current amplitude is limited to 1.25mA.

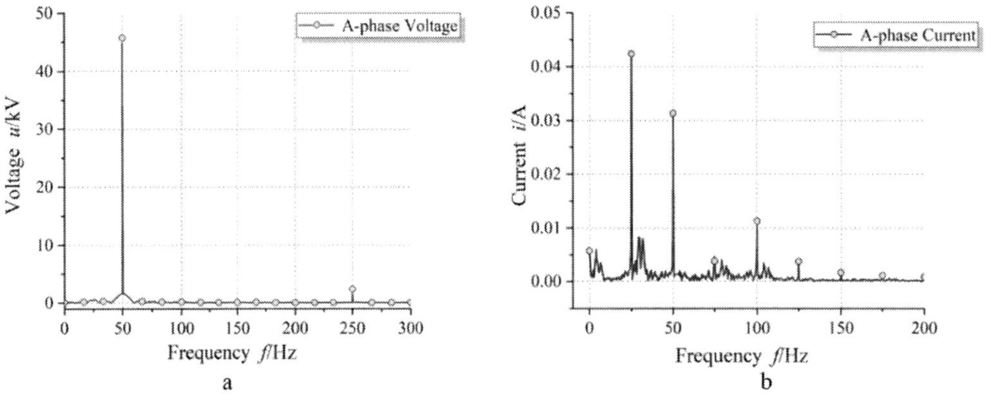

Figure 6. Three-phase voltage and current spectrum diagram after series resistor R_1

5. Conclusion

For the current new energy access station ferroresonance characteristics are unknown, difficult to inhibit the difficult problem, this paper to a 35kV new energy access station as a research object, extract the actual parameters required for equipment modeling, construct the system electromagnetic transient simulation model, according to the actual measurement data and electromagnetic transient simulation model, analyze the photovoltaic power generation harmonics on the ferroresonance amplitude, spectrum, and other characteristics of the time-frequency domain, based on the simulation test to put forward the Based on the simulation test, we propose the parameter configuration of the suppression measures, and obtain the following conclusions:

(1) The ferroresonance voltage amplitude is 50kV before adding harmonics, and the voltage amplitude is 120kV after adding harmonics, which is 2.4 times of that before adding harmonics. The maximum value of current is 40mA before adding harmonics, and the maximum value of current is 160mA after adding harmonics, which is 4 times of that before adding harmonics.

(2) Before adding harmonics, the voltage is only the fundamental voltage at 50Hz; after adding harmonics, the voltage generates a 48kV frequency-dividing resonance at 25Hz, and there are also higher harmonics with smaller amplitudes. Before adding harmonics, the current contains only integer harmonics; after adding harmonics, the current has a crossover frequency resonance at 25Hz with an amplitude of 40mA in addition to the high harmonics.

(3) In this paper, the current amplitude is limited by adding damping in the discharge circuit. 80kΩ linear resistor is added, and the current amplitude is reduced to 22.3mA, which is reduced by 0.16. The

voltage amplitude is reduced to 50kV, which is reduced by 0.58, and the waveform returns to normal after 0.5s.

Acknowledgments
The research work encompassed in this paper was funded by the Science and Technology Project of State Grid Shaanxi Electric Power Company Limited (5226KY220007). We would like to express our sincere gratitude to the Science and Technology Project of State Grid Shaanxi Electric Power Company Limited for providing the necessary resources and opportunities to conduct this study.

References

[1] Zhang Z G and Kang C Q 2022 Challenges and Prospects for Constructing the New-type Power System Towards a Carbon Neutrality Future on Proceedings of the CSEE vol 42, pp 2806-2819
[2] Zhang D H, et al 2023 Analysis of High-frequency Resonance and Suppression Strategy of Photo-voltaic Power Plant With Static Reactive Power Compensation Device on Proceedings of the CSEE vol 43, pp 9580-9594
[3] Wang B, et al 2023 Analysis on Transi-ent Voltage Stability of Power System with Photovoltaic Based on Monotone System Theory on Automation of Electric Power Systems vol 47, pp 19-29
[4] Yang M, et al 2021 Solution of the Inverse Problem of "Low-Frequency Overvoltage Excitation to Response" for Electromagnetic Potential Transformers on Transactions of China Electrotechnical Society vol 36, pp 3605-3613
[5] Chen R, et al 2022 Simulation and Analysis of Primary Side Current of Voltage Transformers in 20kV Distribution Networks on Guangdong Electric Power vol 35, pp 54-65
[6] Hong X H, et al 2020 Analysis and Test Research on Operation Fault of Electromagnetic Voltage Trans-former in Distribution Network on Electric Engineering vol 20, pp 118-119
[7] Yao Y H, et al 2020 Fault Analysis and Countermeasures for fuses blown of 10kV distribution network PT on Electrical Measurement & Instrumentation vol 57, pp 122-128
[8] Heidary A, et al 2020 Voltage Transformer Ferroresonance: An Inhibitor Device on IEEE Transactions on Power Delivery vol 35, pp 3731-2733
[9] Nogay H S, et al 2023 Diagnosis of Chaotic Ferroresonance Phenomena Using Deep Learning on IEEE Access vol 11, pp 58937-58946
[10] Chang Z, et al 2023 Phase Current Based Fault Section Location for Single-phase Grounding Fault in Non-effectively Grounded Distribution Network on 2023 IEEE/IAS Industrial and Commercial Power System Asia (I&CPS Asia), pp 01-06
[11] Wang B, et al 2022 Time domain analysis of single-phase series ferromagnetic resonance in a directly grounded neutral system on Proceedings of the CSEE vol 42, pp 2124-2133
[12] Zhu B J, et al 2023 Explosion Accident Analysis and Prevention of Resonance Overvoltage of 35 kV Potential Transformer on High Voltage Apparatus vol 59, pp 185-191
[13] Xie N, et al 2013 Dynamic Modeling and Characteristic Analysis on Harmonics of Photovoltaic Power Stations on Proceedings of the CSEE vol 33, pp 10-17
[14] Li T, et al 2023 Reactive Power Optimization for Time-Varying Tracking of Maximum Output Power of Grid-Connected Photovoltaic Power Station on Transactions of China Electrotechnical Society vol 38, pp 2921-2931
[15] Hao P 2018 the Ferroresonance Overvoltage and Suppression Measures in Distribution Network (Nanjing: School of Electrical Engineering Southeast University)

Journal of Physics: Conference Series **3012** (2025) 012055 doi:10.1088/1742-6596/3012/1/012055

Aggregated Dispatchable Region of Active and Reactive Power for Distributed Photovoltaic and Energy Storage

Rongbo Ye[1], Ruoying Yu[1*], Peng Yu[2]

[1] China Electric Power Research Institute, Nanjing, Jiangsu Province, 210019, PR China

[2] State Grid Shandong Electric Power Company, Jinan, Shandong Province, 250001, PR China

*Corresponding author's e-mail: 158074595@qq.com

Abstract. With the advancement of distributed photovoltaic (DPV) and energy storage (DES) technologies, small-scale photovoltaic and storage systems are becoming increasingly accessible, contributing to enhanced flexibility in power systems through aggregation. This paper proposes an aggregation method for DPV and DES resources, aimed at improving the aggregation efficiency of distributed resources and enhancing the grid's regulation capability. First, for DPV resources, the Minkowski summation method is employed to aggregate the dispatchable regions of active and reactive power. Second, for DES resources, this study calculates the charging and discharging power, as well as the capacity constraints of aggregated energy storage, using high-dimensional intra-polygon approximation combined with simple Minkowski summation. This approach effectively incorporates the time-series characteristics of energy storage devices. Additionally, the Minkowski summation method is applied to aggregate both photovoltaic and energy storage resources, resulting in the derivation of the active and reactive power dispatchable regions, thereby improving the regulation accuracy of the photovoltaic and storage system. The proposed aggregation method ensures both the precision of aggregation and the feasibility of subsequent disaggregation by accounting for the coupling of time and space.

1. Introduction

In recent years, as the proportion of DPV generation in the power system continues to rise, the inherent instability and unpredictability of this renewable energy source have posed significant challenges to the stable operation of the power grid. To enhance the efficiency and reliability of DPV power generation, DES technologies have emerged, with the integration of energy storage devices helping to mitigate power surpluses or deficits in the grid. However, due to the large number of distributed photovoltaic storage systems, each with relatively small individual capacities, directly involving these systems in grid regulation can lead to a substantial increase in the number of decision variables within the optimization problem. This increase complicates the solution process, making it difficult to achieve stable convergence, and introduces other operational challenges. Consequently, aggregating distributed resources and analyzing their response characteristics becomes crucial.

Extensive research has been conducted globally on the aggregation of distributed resources. In [1], heterogeneous loads were aggregated using the outer approximation of the Minkowski sum, which may result in infeasible outcomes. In [2], a quantitative flexibility description and pricing method were proposed to effectively aggregate system flexibility and decompose it in an economically optimal manner. In [3], the aggregation results were refined to the internal right-angled prismatic cone or the maximum internal hypercube of the actual flexible regulation capability region, yielding conservative results when the parameters of the aggregated units differ. In [4], homomorphic polytopes were used to approximate the flexible adjustment capacity of each unit to be aggregated, demonstrating the efficiency of homomorphic polyhedral aggregation. In [5], the internal structure of the aggregated objects was approximated as a simple Chinoy polyhedron, employing a straightforward aggregation principle.

Current research has given less attention to the aggregation and disaggregation of active and reactive power for each distributed energy resource. Further in-depth investigation into aggregation and disaggregation techniques is needed to provide the necessary technical support for flexible resources to effectively contribute flexibility to the power system. Therefore, the main contributions of this paper are as follows: (1) Considering the limitations of active power and battery capacity, the upper and lower limits of active power and capacity are derived after aggregation using Minkowski's aggregation method, based on high-dimensional polyhedral endohedral approximation; (2) Aggregating DPV and DES resources to define the dispatchable regions of active and reactive power for the combined optical-storage system using the Minkowski summation method, while ensuring the feasibility of subsequent disaggregation of the aggregated resources.

2. Aggregation methods for DPV and DES systems by voltage level and region

For the aggregation of DPV and DES systems, the spatial geographic location and voltage level are critical factors. Geographic proximity facilitates real-time data transmission and the issuance of scheduling commands, thereby simplifying the tasks of the power dispatch center in terms of unified information collection and task allocation. Voltage stratification within subregions involves grouping grid-connected photovoltaic and storage systems according to voltage levels, which enhances management and regulation. This hierarchical approach, applied across various voltage levels, refines the aggregation process of distributed resources. As shown in the figure below, the aggregation is classified based on distinct regions and voltage tiers.

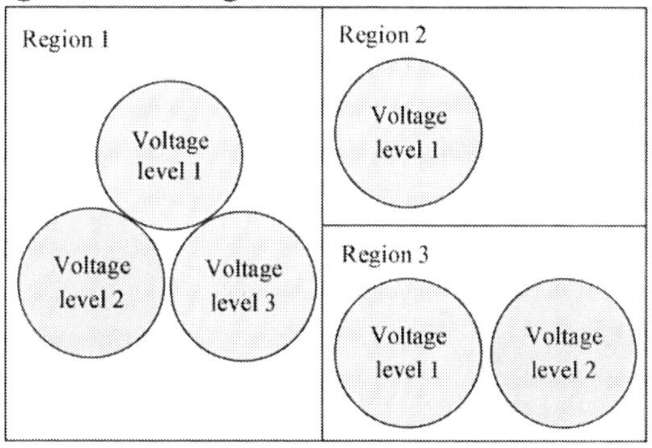

Figure 1. Schematic of sub-region and voltage level aggregation

After establishing clusters across various regions and voltage levels, we proceed to define the aggregation of active and reactive power for the photovoltaic system, energy storage system, and the photovoltaic-storage system within each cluster. This aggregation is performed using the method outlined in this paper.

3. DPV and DES active and reactive power dispatchable region aggregation

Both DPV and DES systems exhibit active-reactive power coupling, with energy storage systems also subject to active power and capacity limitations. Therefore, this paper takes into account the limitations of both photovoltaic and energy storage systems and analyzes the dispatchable regions of active and reactive power for their flexible aggregation. This analysis is carried out using the Minkowski summation method and high-dimensional polyhedral intra-approximation techniques.

3.1. *Photovoltaic aggregation*

Given the large number of DPV systems connected at each node and the relatively small dispatchable capacity of individual DPV resources, the aggregation of active and reactive power becomes essential. Thus, we aggregate multiple photovoltaic resources based on a single photovoltaic model using the Minkowski summation approach.

The active-reactive power dispatchable region of a photovoltaic system is semicircular in shape and cannot be directly summed using Minkowski summation. Therefore, we approximate the photovoltaic active and reactive power dispatchable region, which is semicircular, as a convex polygonal set. Subsequently, we perform Minkowski aggregation and summation of the active and reactive power of the photovoltaic systems, as illustrated in Figure 2.

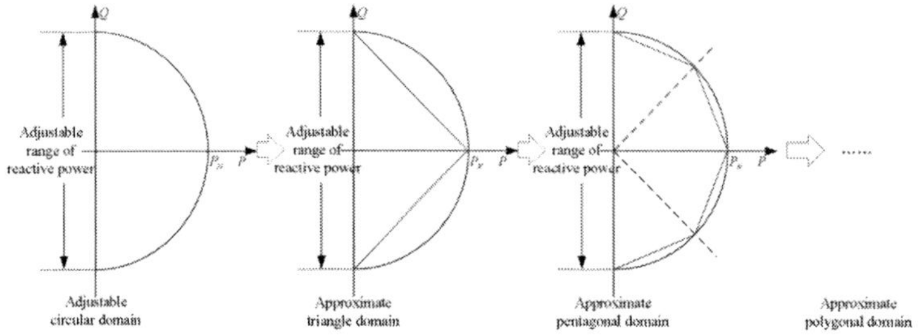

Figure 2. Schematic diagram of polygonal approximate aggregation

However, the polygonal approximation of the aggregation region results in a loss of some reactive power adjustability compared to the original dispatchable region. Specifically, there is a 36.31% reduction in the dispatchable region for triangular approximations, a 10% reduction for pentagonal approximations, and a 2.51% reduction for nonagonal approximations. To minimize this loss and make the polygonal approximation closer to the original dispatchable region, it is necessary to increase the number of sides of the polygon. This adjustment will reduce the discrepancy, ensuring that the loss of dispatchable region becomes negligible.

3.2. *Energy storage aggregation*

3.2.1. *Active power and battery capacity aggregation*

A conventional mathematical model of a single energy storage is given in the following equation:

$$\begin{cases} 0 \le P_{i,ch}^{t} \le P_{i,ch}^{\max} \\ 0 \le P_{i,dch}^{t} \le P_{i,dch}^{\max} \\ E_i^{t+1} = E_i^{t} + (\eta_{i,ch} P_{i,ch}^{t} - \dfrac{P_{i,dch}^{t}}{\eta_{i,dch}})\Delta t \\ E_i^{\min} \le E_i^{t} \le E_i^{\max} \end{cases} \tag{1}$$

where, $P_{i,ch}^{t}$ is the charging power of storage unit i at time t, $P_{i,dch}^{t}$ is the discharging power of storage unit i at time t, and E_i^{t} is the battery capacity of storage unit i at time t.

The active power feasible region for energy storage is defined as the set of all feasible active power curves, which can be represented in polyhedral form as follows:

$$\Omega_i = \left\{ P_i \in \mathbb{R}^{2m} \mid M_i P_i < N_i \right\} \tag{2}$$

where, $P_i = [P_{i,ch}, P_{i,dch}]^T$, $P_{i,ch} = [P_{i,ch}^{t}]_{1\times m}$ $P_{i,dch} = [P_{i,dch}^{t}]_{1\times m}$, M_i and N_i are the matrix of coefficients. M_i and N_i are expressed as respectively:

$$M_i = \begin{bmatrix} I_m & 0 \\ 0 & I_m \\ -I_m & 0 \\ 0 & -I_m \\ B_{i,ch} & -B_{i,dch} \\ -B_{i,ch} & B_{i,dch} \end{bmatrix} \tag{3}$$

where, $P_{i,ch}^{\max}$, $P_{i,dch}^{\max}$, $E_i^{\max}$, $E_i^{\min}$, E_i^0 are an m-dimensional column vector. $P_{i,ch}^{\max} = [P_{i,ch}^{\max}]_{m\times1}$, $P_{i,dch}^{\max} = [P_{i,dch}^{\max}]_{m\times1}$, $E_i^{\max} = [E_i^{\max}]_{m\times1}$, $E_i^{\min} = [E_i^{\min}]_{m\times1}$, $E_i^0 = [E_i^0]_{m\times1}$.

In this paper, an inner approximation aggregation method is used, under which the dispatch signals from the higher grid are feasible and disaggregated for realization.

Firstly, the basis aggregation matrix $\phi_k^0(M_k^0, N_k^0)$ is constructed, M_k^0 and N_k^0 are the average of all the energy storage parameters. Then, a series of same-type ϕ_k^0 approximations to each energy storage ϕ_i is generated by shrinking, expanding and translating the basis matrices, so that M of same-type polygons is conveniently computed as follows:[6]

$$\tilde{\phi}_i = \beta_i \phi_k^0 + \upsilon_i = \left\{ P_i \in \mathbb{R}^{2m} \mid M_k^0 (P_i - \upsilon_i) \le \beta_i N_k^0 \right\} \tag{4}$$

$$(\beta_i \phi_k^0 + \upsilon_i) \uplus = (\beta_j \phi_k^0 + \upsilon_j) = (\beta_i + \beta_j)\phi_k^0 + (\upsilon_i + \upsilon_j) \tag{5}$$

where, $\tilde{\phi}_i$ is ϕ_i the inner approximate feasible region, β_i and υ_i are the scaling and translation coefficients, respectively.

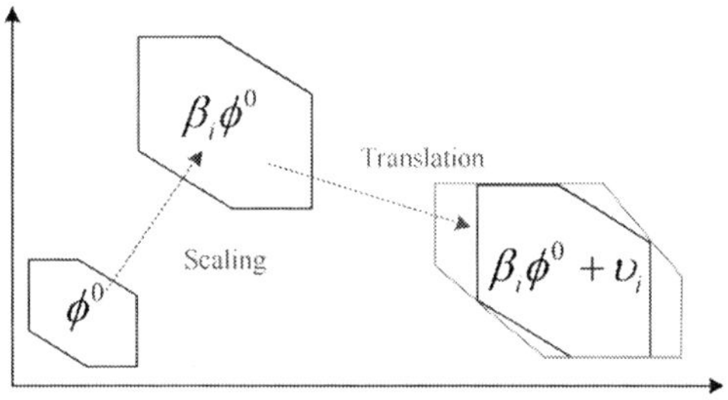

Figure 3. Schematic diagram of the intra-polygonal approximation

As shown in Figure 3, β_i for zooming the prototype and determining the size of the feasible region, and υ_i for translating the prototype and determining the position of the feasible region. Therefore, to obtain the maximum inner approximation region, the following problem is established:

$$\begin{aligned} & \max_{\beta_i, \upsilon_i} \beta_i \\ & \text{s.t. } \beta_i \phi_k^0 + \upsilon_i \subseteq \phi_i \, , \beta_i \ge 0 \end{aligned} \tag{6}$$

According to Farkas' derivation, the above set inclusion problem is equivalent to the following equation and inequality problem, the proof of which is referred to [7]:

$$\begin{aligned} & \max_{\beta_i, \upsilon_i, G} \beta_i \\ & \text{s.t.} G M_k^0 = M_i \, , G \ge 0 \\ & \beta_i G N_k^0 + M_i \upsilon_i \le N_i \, , \beta_i \ge 0 \end{aligned} \tag{7}$$

Due to its variable-solving problem, introducing the assistant variable $b_i = 1/\beta_i, r_i = -\upsilon_i / \beta_i$, the following equation can be further obtained:

$$\begin{aligned}&\min_{b_i,\upsilon_i,G} b_i\\&\text{s.t.} GM_k^0 = M_i, G \geq 0\\&GN_k^0 \leq b_i N_i + M_i r_i, b_i \geq 0\end{aligned} \tag{8}$$

The aggregation of the reactive power adjustable range is then performed using the Minkowski summation method, which ultimately results in the derivation of the energy storage system's active and reactive power adjustable regions, as well as the upper and lower limits of battery capacity.

3.2.2 Energy storage active and reactive power aggregation

Similarly to photovoltaic systems, energy storage systems must be approximated as a polygonal convex set within the circular region of the storage system's active-reactive power dispatchable region. The Minkowski summation method is then applied to aggregate the active and reactive power of the energy storage system. Finally, based on the derived active dispatchable region for each time period, the reactive power is constrained, allowing for the determination of the final interval for the active-reactive power dispatchable region.

3.3. *Photovoltaic and energy storage aggregation*

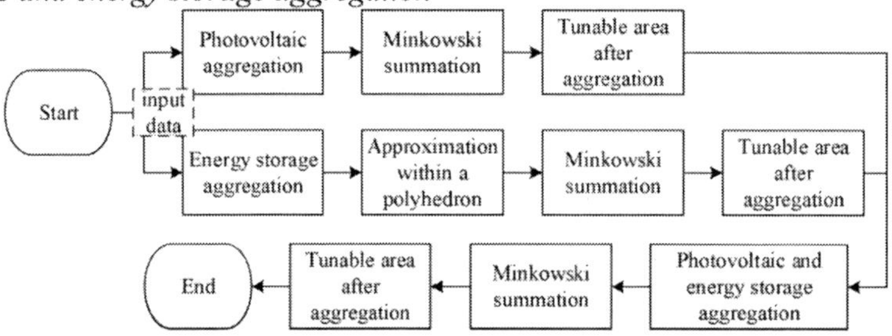

Figure 4. Aggregation flowchart for photovoltaics and energy storage

As illustrated in Figure 4 above, the aggregation of photovoltaic and energy storage systems is initially performed separately. After obtaining the aggregated dispatchable regions for both distributed photovoltaic and distributed storage systems, the flexible dispatchable region for the entire photovoltaic-storage system at this node is then derived using the Minkowski summation method.

4. Case studies

In this paper, the example data is as follows: a node within a specific voltage level and region is selected, and its connected resources include two distributed photovoltaic systems, with maximum power capacities of 0.1 MW and 0.25 MW, respectively. Additionally, three distributed energy storage systems are considered, with maximum power capacities of 0.1 MW, 0.2 MW, and 0.3 MW, and battery capacities of 0.5 MWh, 1 MWh, and 1.5 MWh, respectively.

4.1 *Photovoltaic aggregation*

In this paper, the example data are as follows: The photovoltaic day-ahead prediction curve is used to represent the maximum generation power in Maximum Power Point Tracking (MPPT) mode. The active-reactive power dispatchable region is then approximated as a polygon within the circular region. The aggregated active-reactive power dispatchable region is obtained through Minkowski summation, as illustrated in Figure 5 below.

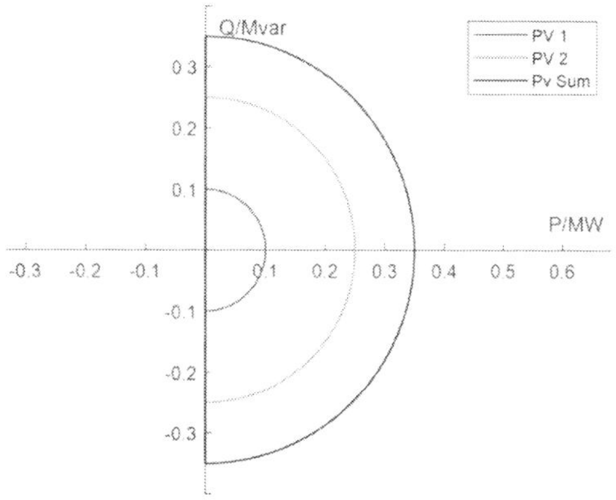

Figure 5. DPV aggregation dispatchable region

The example data in this paper are as follows: aggregation based on the PV rated power, the PV1 rated power of 0.1 MW and the PV2 rated power of 0.25 MW are selected so as to solve the aggregation of the power adjustable region by Minkowski summation.

4.2 *Energy storage aggregation*

As shown below, the active-reactive power of the energy storage is aggregated by the high-dimensional polyhedral intra-approximation and Minkowski summation methods:

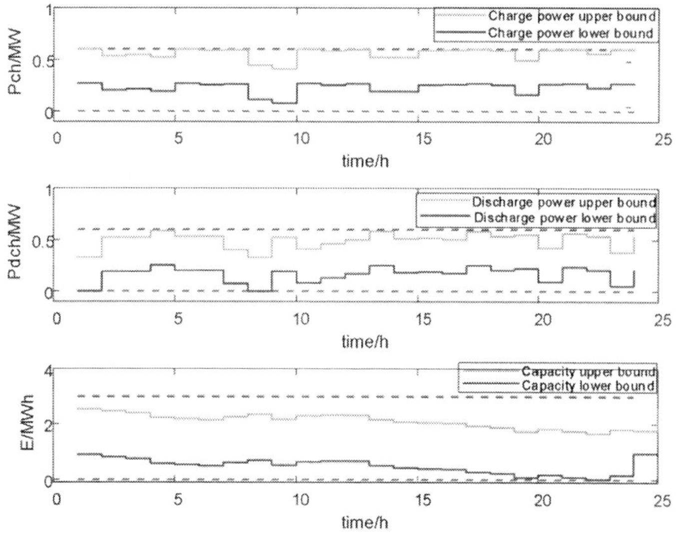

(a) Charge and discharge power upper and lower bound and capacity upper and lower bound for DES aggregation

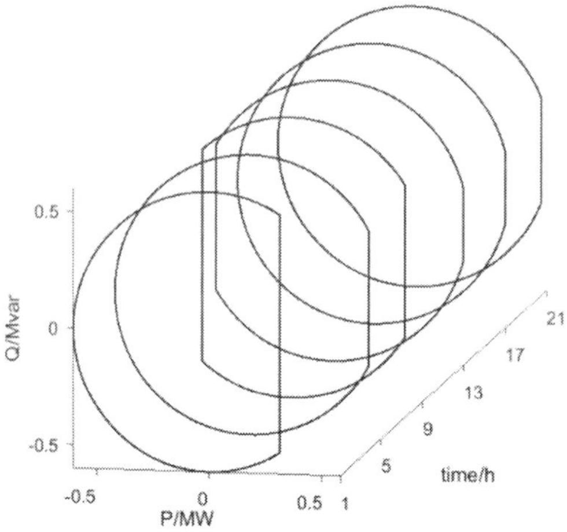

(b) Dispatchable region for DES aggregation considering constraints

Figure 6. Schematic diagram of the dispatchable region for storage aggregation

In this paper, the example data are as follows: As shown in the figure above, after accounting for the aggregation limitations of the storage's effective capacity, the dispatchable region, along with the battery capacity, charge, and discharge power at each time period across the 24 hours, becomes smaller, resulting in a reduction of the dispatchable region. However, the limitation imposed by the storage's effective capacity has been effectively considered, allowing for a more precise decomposition of the aggregation instructions.

The following three DES aggregate examples at some time are set up for comparative analysis:

Method 1 (M1): Original optimal scheduling without using the aggregation method;

Method 2 (M2): Polytope inner approximation for aggregation optimal scheduling of this paper;

Method 3 (M3): Box inner approximate method [8] for aggregation optimal scheduling.

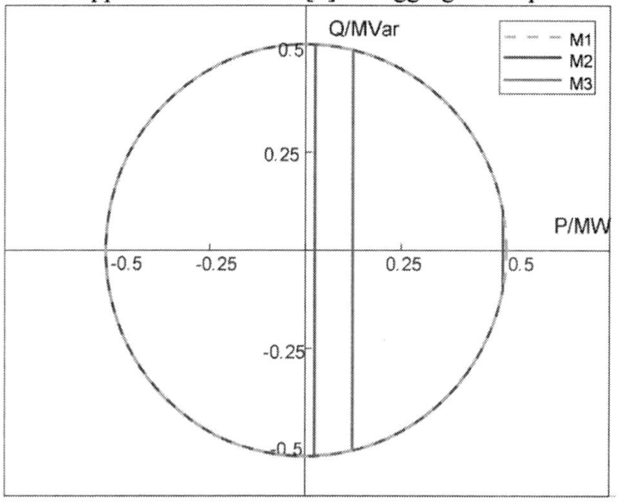

Figure 7. Comparison of the active and reactive power aggregation regions of the three methods at time 16

Figure 7 shows the effectiveness of the M2 method of aggregation, which is further verified by comparing the active and reactive power feasible regions of the different methods. It can be seen that at the time t1, the active and reactive power feasible region after M2 aggregation is approximated as the M1 feasible region, with only a small part of the feasible region lost, while the feasible region of

M3 is a long strip, with most of the active and reactive power feasible region lost.

4.3 *Photovoltaic and energy storage aggregation*

In this paper, the example data are as follows: After determining the photovoltaic and energy storage aggregation dispatchable regions, the flexible dispatchable region for the entire optical storage system at the node can be derived using the Minkowski summation method, as illustrated in the figure below.

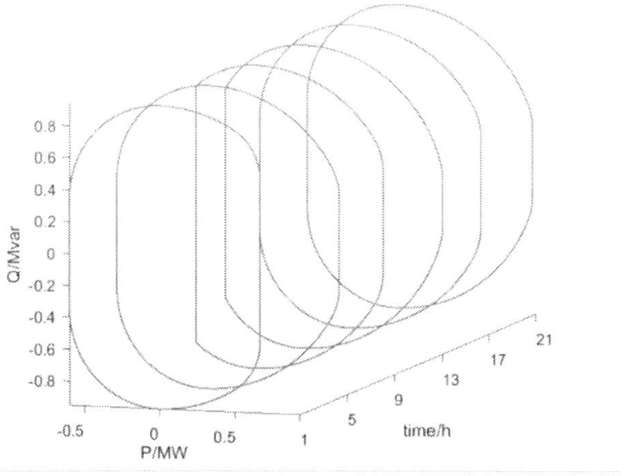

Figure 8. DPV and DES aggregation dispatchable region

In this paper, the example data are as follows: After aggregating the optical storage, it can be observed that the entire schedulable region is not circular; instead, it takes on an elliptical shape. This occurs because the adjustable domain of the photovoltaic system is represented as a semicircle, combined with the irregular circular adjustable domain of the energy storage system, resulting in the ellipse-like shape. Once the flexible schedulable region of the entire optical storage system is derived, scheduling and optimization can be performed within this region.

5. Conclusion

This paper proposes a method for aggregating the active and reactive power of DPV and DES.

First, for DPV, the Minkowski method is employed, resulting in a semicircular dispatchable region of varying sizes.

Second, for DES, the intra-polygon approximation method is utilized to derive the power and capacity limitation ranges for both charging and discharging. Subsequently, the Minkowski method is applied to determine the dispatchable region of active and reactive energy storage aggregation, which takes on an irregular, circular-like shape due to the power limitations imposed by the aggregation.

Finally, the Minkowski summation method is used to obtain the combined dispatchable region for both photovoltaic and storage systems' active and reactive power.

It is observed that the use of approximation methods during aggregation results in some loss of dispatchable region, and the resulting dispatchable region adopts an irregular shape due to the inherent characteristics of energy storage. However, through optimized scheduling, the feasibility of disaggregation can be ensured.

Acknowledgment

This work is supported by the Science and Technology Project of State Grid Corporation of China, Research and application of distributed photovoltaic and energy storage aggregation and regulation technology (Grant No. 5419-202316463A-3-2-ZN).

References

[1] Barot S, Taylor J A. A concise, approximate representation of a collection of loads described by polytopes[J]. International Journal of Electrical Power and Energy Systems, 2017, 84: 55-63.

[2] Müller F L, Szabó J, Sundström O, et al. Aggregation and disaggregation of energetic flexibility from distributed energy resources[J]. IEEE Transactions on Smart Grid, 2019, 10(2): 1205-1214.

[3] Hao H, Sanandaji B M, Poolla K, et al. A generalized battery model of a collection of thermostatically controlled loads for providing ancillary service[C]// 2013 51st Annual Allerton Conference on Communication, Control, and Computing (Allerton), October 3-4, 2013, Monticello, USA.

[4] X. Huang, P. Dong, X. He, M. Liu and W. Deng, "Research on Aggregation Flexibility Method of AC/DC Distribution Network Considering Flexibility Balance and Its Application," in IEEE Systems Journal, vol. 17, no. 3, pp. 3635-3645, Sept. 2023.

[5] A. Churkin, W. Kong, J. N. Melchor Gutierrez, E. A. Martínez Ceseña and P. Mancarella, "Tracing, Ranking and Valuation of Aggregated DER Flexibility in Active Distribution Networks," in IEEE Transactions on Smart Grid, vol. 15, no. 2, pp. 1694-1711, March 2024.

[6] M. Zhang, Y. Xu, X. Shi and Q. Guo, "A Fast Polytope-Based Approach for Aggregating Large-Scale Electric Vehicles in the Joint Market Under Uncertainty," in IEEE Transactions on Smart Grid, vol. 15, no. 1, pp. 701-713, Jan. 2024.

[7] L. Zhao, W. Zhang, H. Hao and K. Kalsi, "A Geometric Approach to Aggregate Flexibility Modeling of Thermostatically Controlled Loads," in IEEE Transactions on Power Systems, vol. 32, no. 6, pp. 4721-4731, Nov. 2017.

[8] M. Zhang, Y. Xu, X. Shi, and Q. Guo, "A Fast Polytope-Based Approach for Aggregating Large-Scale Electric Vehicles in the Joint Market Under Uncertainty," in IEEE Transactions on Smart Grid, vol. 15, no. 1, pp. 701-713, Jan. 2024.

Simulation Study on the Distribution Characteristic of Pyrolysis Gas in 10kV Switchgear

Zhaoguo Wu[1], Jing Zhao[1], Qian Wang[1] ,Ke Wu[2] ,Yingkai Long[1] ,Luobin Wang[3] ,Yuhuan Shao[4*]

[1] State Grid Chongqing Electric Power Research Institute, Chongqing, China

[2] State Grid Chongqing Electric Power Company Ultra-High Voltage Branch, Chongqing, China

[3] Sichuan Hantang Tenghui Technology Co., Ltd, Sichuan, China

[4] Chongqing University of Posts and Telecommunications, Chongqing, China

[*]Corresponding author's e-mail: cherryfa999@163.com

Abstract. With the rapid development of power distribution networks, a stable and reliable power supply system has become a key requirement. Switchgear plays an important role in the power system, but fires often occur inside the switchgear due to faults, and early fires are not easy to detect. Existing fire monitoring methods have deficiencies in sensitivity, weak anti-interference performance, etc., making it difficult to provide effective early warning. Based on the different stages of the fire, this paper explores the distribution pattern of pyrolysis gases in the cable compartment of the switchgear, focusing on the concentration changes of CO and CH_3Cl. The diffusion process of pyrolysis products under superheated defects was simulated by means of a finite element simulation model, and the mass share of the gas and its distribution uniformity at different time points were analyzed. The results show that the initial gas concentration is higher near the heating point, gradually diffuses to other areas over time, and eventually tends to be evenly distributed. To improve monitoring sensitivity, it is recommended to install the sensor in a non-corner position on the top of the switchgear cable compartment. The research in this paper provides theoretical basis and practical guidance for fire monitoring of 10kV switchgear.

1. Introduction

With the rapid development of power distribution networks, stable and reliable power supply systems have become the constant pursuit of distribution networks. Switchgear is mainly used to distribute power, control, protect and monitor power equipment in power systems. The 10kV medium-voltage switchgear has a large number of uses due to its reliable operation and easy operation. In switchgear, fires are often caused by various reasons (e.g. arc faults, short-circuit faults, overload operation, overloading, etc.) and are not easily detected in the early stages of the fire, which ultimately leads to significant economic losses[1,2].

At present, smoke detection, temperature detectors, flame detectors and gas detectors are commonly used to monitor fire hazards in switchgear. However, there are some shortcomings, such as smoke detection is easily interfered by dust, temperature detectors are not sensitive to early fires, flame detectors are easily interfered by other external light sources, and the monitoring range is limited. When such alarms sound, the defects often develop to an irreversible stage [3-5].

According to the stage of fire, it can be divided into pyrolysis period, smoldering period, combustion period and open flame period. During the pyrolysis period, the temperature is low, and the changes in molecules and particles are the main factors. The pyrolysis reaction begins, delaying the obvious flame or flame; During the smoldering period, the temperature gradually rises and some combustible gases are generated by decomposition, but the flame has not yet appeared. A small amount of pyrolysis products or smoke particles may appear at this stage; During the combustion phase, the temperature rises further, flames and smoke begin to appear, and there is an obvious combustion reaction. The visible smoke concentration increases, indicating that the fire has entered a dangerous stage; During the open flame period, the fire develops violently, with open flames and thick smoke appearing, a large amount of energy is released, the fire enters the out-of-control stage, the ambient temperature rises rapidly, posing a great threat to personnel and equipment. The overall change is from no open flame in the initial stage to a violent combustion reaction, and the fire hazard gradually increases over time. Before the defect produces smoke, the defect causes local overheating, and the surrounding material will begin to decompose at a certain temperature——pyrolysis[6,7].

Therefore, many researchers have paid attention to this stage and carried out very early monitoring of fire through the characteristics of this stage, such as pyrolysis particles, pyrolysis gases, etc. Mengzhu Hu et al. [8] studied the thermal aging behavior of FR-4 epoxy resin in high-voltage switchgear and explored the pyrolysis characteristic gases, providing a reference for the condition monitoring and early fault prediction of high-voltage switchgear; Jiamin Kong et al. [9] studied the distribution of gas products during the pyrolysis of cross-linked polyethylene (XLPE) and explored the effect of temperature on the distribution of gas products; Yalong Wang et al. [10] studied the pyrolysis characteristics of four PVC cable insulation materials with different flame retardant grades.

The effectiveness of very early fire monitoring through pyrolysis gases has been demonstrated. However, there has been no systematic research on the application scenarios of switchgear, the types of corresponding monitoring sensors, parameters, and alarm threshold settings. Therefore, this paper investigates the distribution characteristic of pyrolysis gas generated in the cable compartment of the switchgear through finite element simulation to provide theoretical basis for the installation method of the characteristic gas monitoring sensor of the 10kV switchgear as well as the method of threshold setting.

2. 10kV switchgear structure and finite element model

2.1. Typical 10kV switchgear structure

As shown in the Figure 1 for the 10kV switchgear structure schematic diagram.

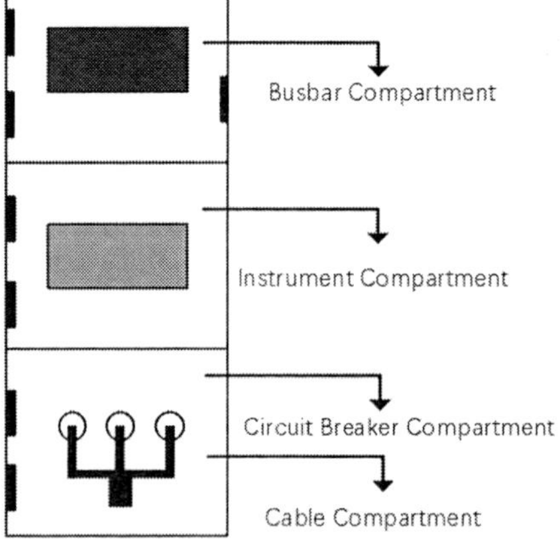

Figure 1. Schematic diagram of 10kV switchgear structure

The interior of the switchgear is mainly composed of circuit breaker compartment, busbar compartment, cable compartment, instrument compartment, and instrument compartment. The basic components of the original are: circuit breakers, fuses, contactors relays, current transformers, voltage transformers, energy meters and so on. We have analyzed these major charged materials and found that the heat generated in the early stages of a fire causes thermal decomposition of the skins of materials including XLPE (cross-linked polyethylene), PVC (polyvinyl chloride), silicone rubber, and PE (polyethylene), producing pyrolysis gases. In this paper, the pyrolysis of PVC material is taken as an example to study the pyrolysis gas distribution characteristic in the cable compartment.

2.2. *Finite element simulation model of switchgear cable compartment*

In this paper, finite element fluid computation software is used to model the switchgear cable compartment and a superheated defect is placed on the C-phase cable near the bottom surface of the switchgear, which generates the corresponding pyrolysis gas.

The finite element fluid calculation software is used to model the switchgear compartment. According to the actual situation of the switchgear cable compartment in the experiment, the box of the switchgear cable compartment is reasonably simplified. The size of the switchgear cable compartment is 0.8m×0.8m×1.1m. The obtained geometric figure is as Figure 2:

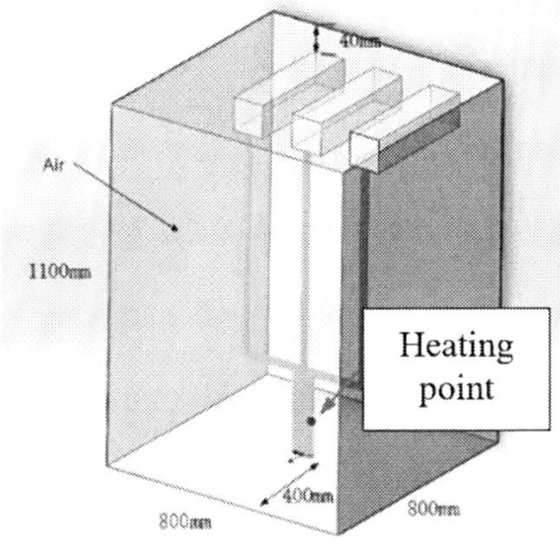

Figure 2. Switchgear cable compartment simulation model

3. Characteristic pyrolysis gas distribution pattern

According to Yalong Wang's research [10], the pyrolysis of PVC produces more than ten characteristic pyrolysis products, such as CH_3Cl, C_2H_4, CO, CO_2, etc., between 120°C and 140°C. In this paper, we choose CO and CH_3Cl as the characteristic pyrolysis gases. In this paper, CO and CH_3Cl were chosen as the characteristic pyrolysis gases.

3.1. *Distribution pattern of CO*

As can be seen from the Figure 3, at the beginning of the test (around 10 s), the maximum value of the mass share of CO distribution in the space of the switchgear cable chamber is 0.05 and the minimum value is 0.005, and the CO concentration at the bottom of the right side of the switchgear cable chamber (the heating point) is much higher than that at the top and the left side. With time, CO gradually spreads to other parts of the switchgear cable compartment space, and at 220 s, the maximum value of the mass share of CO distribution in the switchgear cable compartment space is 6×10^{-5} and the minimum value is 1.5×10^{-5}. In the steady state case (30 min in the simulation), the maximum value of the mass share of CO distribution in the cable compartment space of the switchgear is 3.11375×10^{-5} and the minimum value is 3.11315×10^{-5}.

Journal of Physics: Conference Series **3012** (2025) 012056 doi:10.1088/1742-6596/3012/1/012056

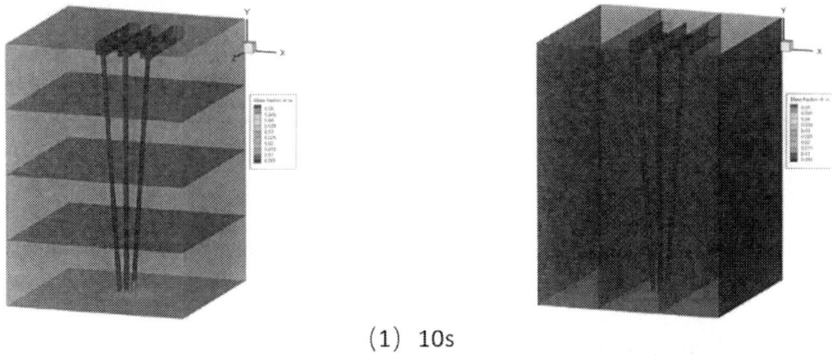

(1) 10s

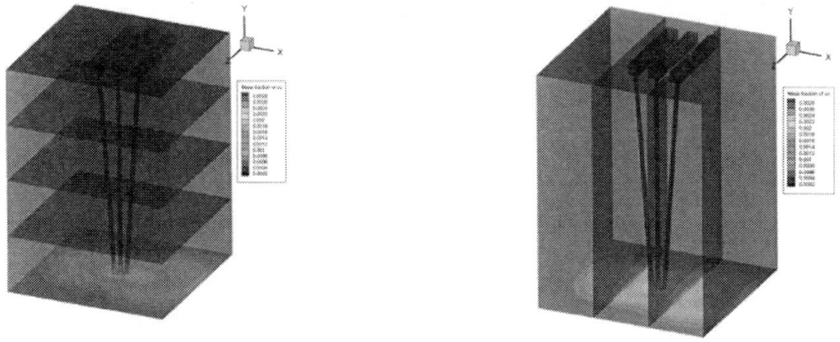

(2) 20s

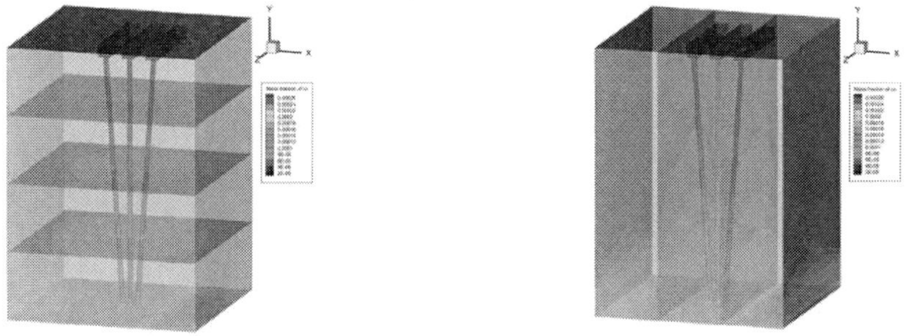

(3) 60s

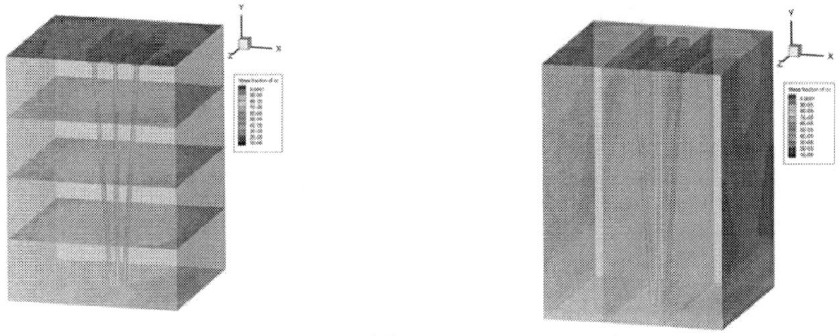

(4) 140s

(5) 220s

(6) 30min

Figure 3. CO distribution diffusion simulation diagram

3.2. The distribution pattern of CH_3Cl

As can be seen from the Figure 4, at the beginning of the test (around 10 s), the maximum value of the mass share of CH_3Cl distribution in the space of the switchgear cable chamber is 0.9 and the minimum value is 0.05, and the concentration of CH_3Cl at the bottom of the right side of the switchgear cable chamber (the heating point) is much higher than that at the top and the left side. With time, CH_3CL gradually spreads to other parts of the switchgear cable compartment space, and at 220 s, the maximum value of the mass share of CH_3Cl distribution in the switchgear cable compartment space is 0.0011, and the minimum value is 0.0003. In the steady state condition (30 min in the simulation, for example), the maximum value of the mass share of CH_3Cl distribution in the switchgear cable compartment space is 5.6125×10^{-4} and the minimum value is 5.6113×10^{-4}.

(1) 10s

Journal of Physics: Conference Series **3012** (2025) 012056 doi:10.1088/1742-6596/3012/1/012056

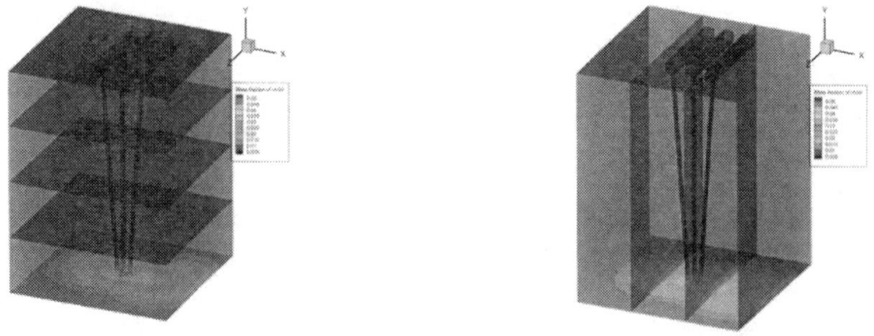

(2) 20s

(3) 60s

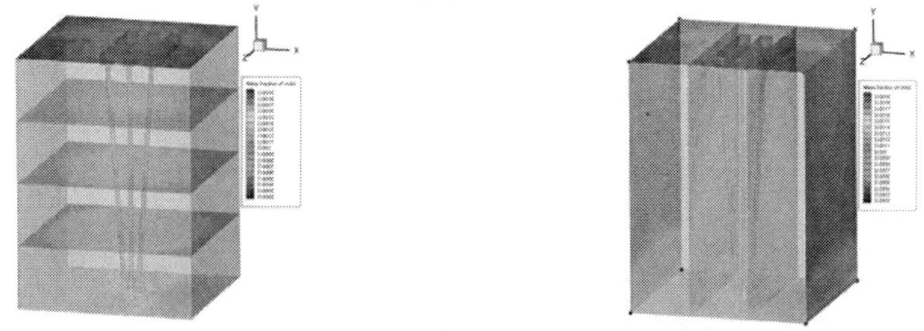

(4) 140s

(5) 220s

(6) 30min

Figure 4. CH_3Cl distribution diffusion simulation diagram

3.3. Overall rule

The maximum and minimum mass proportions of each component in the switchgear cable compartment at different times are shown in Table 1, and the difference multiple (difference multiple = maximum mass proportion/minimum mass proportion) is used to indicate the distribution uniformity of the component in the switchgear cable compartment space at that moment. Pyrolysis products of different compositions in the cable compartment of the switchgear were present at the beginning of the test only in the vicinity of the heating point, where the mass share was high, while the mass share of this composition in the other regions of the space was low, with a large multiplier of difference. As the pyrolysis products gradually diffuse outward from the point of generation, the maximum value of the compositional mass fraction share gradually decreases, and the difference multiplier shows an overall decreasing trend. To the steady state (30 min is used as the steady state value in this simulation), the difference multiplier tends to be close to 1, indicating that the distribution of this component in the cable compartment of the switchgear has become uniform.

Table 1.The maximum, minimum and difference multiples of the pyrolysis product composition in the switchgear cable compartment at different times

	Time	10s	20s	60s	140s	220s	30min
CO	Maximum	0.05	0.0028	2.6×10^{-4}	1×10^{-4}	6×10^{-5}	3.11375×10^{-5}
	Minimum	0.005	0.0002	2×10^{-5}	1×10^{-5}	1.5×10^{-5}	3.11315×10^{-5}
	Difference Multiplier	**10**	**14**	**13**	**10**	**4**	**1**
CH_3Cl	Maximum	0.9	0.05	0.0045	0.0019	0.0011	5.6125×10^{-4}
	Minimum	0.05	0.005	0.0005	0.0002	0.0003	5.6113×10^{-4}
	Difference Multiplier	**18**	**10**	**9**	**9.5**	**3.66**	**1**

4. Conclusion

In this paper, CO and CH_3Cl are first selected as the characteristic gases of PVC pyrolysis, and then a finite element simulation model of the switchgear cable compartment is built to simulate the generation and diffusion process of pyrolysis gases (CO and CH_3Cl) in the cable compartment of a 10kV switchgear. By analyzing the distribution characteristics of the gas concentration at different time points, the mass percentage of the gas and its changing characteristic were studied. The distributional properties of CO and CH_3Cl were studied accordingly and the following pattern was obtained: At the beginning of the test, the concentration at the bottom of the right side of the cable compartment of the switchgear (heating point) was slightly higher than that on the left side, and with the passage of time, the concentration of particles at the bottom gradually decreased, while the concentration of particles at the top gradually increased, and eventually tended to be more evenly distributed. According to the distribution cloud diagrams of each component at different moments, during the diffusion stage of pyroelectric particles, the concentration of the components in the four corner locations in the space of the switchgear cable compartment is relatively low, which is unfavorable for the detection of pyroelectric particles. In the case that the location of the heating fault is not clear, in order to obtain higher detection sensitivity and faster time response, it is recommended that the pyroelectric particle sensor be installed on the top of the switchgear cable compartment in the non-corner position

This article is conducting pyrolysis gas monitoring at different locations in the switchgear space to verify. This article only provides theoretical analysis and does not go into too much detail.

Acknowledgments

2024 Science Project of State Grid Chongqing Power Company (Yudian Science Project, No.11).

References

[1] M. R. Vallerotonda, S. Ansaldi, A. Pirone, Tommaso Bragatto, and Paolo Bragatto, "Accident Triggered by Electrical Failures in Seveso Sites," DOAJ (DOAJ: Directory of Open Access Journals), vol. 91, Jun. 2022.

[2] J. Jia, W. Wang, H. Liu, P. Qi, T. Guo, and G. Ma, "Fire Ignition Mechanism and Extinguishing Method of Cable in High Voltage Switchgear," in 2021 IEEE 2nd China International Youth Conference on Electrical Engineering (CIYCEE), IEEE, Dec. 2021, pp. 1–5.

[3] J. Fonollosa, A. Solórzano, and S. Marco, "Chemical Sensor Systems and Associated Algorithms for Fire Detection: A Review," Sensors, vol. 18, no. 2, p. 553, Feb. 2018.

[4] J. Zhang, P. Li, and T. You, "Infrared Thermography-based Temperature Detection of Key Components in Switchgear," May 2024, pp. 1–6.

[5] R. Bogue, "Sensors for fire detection," Sensor Review, vol. 33, no. 2, pp. 99–103, Mar. 2013.

[6] K. Wu et al., "Study on pyrolysis characteristics of polyethylene outer sheath of 110kV power cable under overheating defect," in 2021 IEEE International Conference on Emergency Science and Information Technology (ICESIT), Nov. 2021, pp. 307–310.

[7] I. Ahmad et al., "Pyrolysis Study of Polypropylene and Polyethylene Into Premium Oil Products," International Journal of Green Energy, vol. 12, no. 7, pp. 663–671, Mar. 2014.

[8] M. Hu et al., "Monitoring and prediction of thermal failure of high-voltage switchgear insulation materials by studying the thermodynamic properties of FR-4 epoxy resin and the characteristic gas of thermal decomposition," Journal of Physics Conference Series, vol. 2387, no. 1, pp. 012028–012028, Nov. 2022.

[9] J. Kong, K. Zhou, X. Ren, Y. Chen, Y. Li, and P. Meng, "Insight into gaseous product distribution of cross-linked polyethylene pyrolysis using ReaxFF MD simulation and TG-MS," Journal of Analytical and Applied Pyrolysis, vol. 169, p. 105847, Jan. 2023.

[10] Y. Wang, N. Kang, J. Lin, S. Lu, and K. M. Liew, "On the pyrolysis characteristic parameters of four flame-retardant classes of PVC sheathless cable insulation materials," Journal of Analytical and Applied Pyrolysis, vol. 170, pp. 105901–105901, Feb. 2023.

Analysis on Waveform Diversity of Arc High-impedance Fault in Distribution Network Overhead Line based on Measured Data

Yuhang Chen, Rong Zheng, Haitao Lin, Hong Xu*

State Grid Fuzhou Power Supply Company, Fuzhou, Fujian Province, 350003, P.R. China

*Corresponding author's e-mail: 71835522@qq.com

Abstract. In response to the strong diversity of waveforms associated with arc high-impedance faults in distribution networks and the weak generalization capability of existing fault identification models, this study undertook the following work: by reproducing faults with different causes, it obtained fault voltage and current data, including those from line-tree contact faults, ground medium-induced grounding faults, and polluted insulator flashover faults, while recording the fault process; this study elucidated the mechanisms behind fault arcs and flame formation, focusing on the analysis of fault phenomena and waveforms at different stages caused by line-tree contact; this study extracted the diversity of fault waveforms, highlighting static characteristics such as nonlinearity, thermal inertia, and zero-crossing offset, as well as dynamic characteristics like faintness, intermittency, and randomness. The analytical process and conclusions possess theoretical significance and can provide a basis and valuable insights for the identification of arc high-impedance faults in distribution networks, as well as for medium and type recognition.

1. Introduction

The signals of arc high-impedance grounding faults in distribution networks are weak and diverse, making them difficult to be captured by existing zero-sequence protection [1]. Most of the existing research focuses on their weakness and utilizes the arc characteristics of faults, such as the zero-crossing feature (i.e., the "shoulder" characteristic of fault current [2]), to propose detection and location methods, with less analysis on their diversity. Some studies [3] and [4] use a small amount of experimental data and simulation data to perform waveform augmentation, which greatly improves the generalization ability of the fault recognition model trained with the enhanced data. However, there is still a lack of research on the diverse characteristics of faults and the inherent laws of data.

Arcing high-impedance grounding faults in distribution networks have various causes [5]. The most common cause is trees toppled by typhoons, which then bridge with overhead lines. Long-term undetected single-phase tree-contact faults can easily lead to forest fires [6] or widespread power outages. They also include broken wire grounding faults that result in compound faults of broken wire-high-impedance grounding, and AC flashover of contaminated insulators [7]. The evolution and local characteristics of arcing high-impedance grounding fault waveforms vary significantly depending on the cause of the fault.

Reproducing the arc high-impedance grounding faults of different causes has the following significance: it fully explores the physical causes of waveform characteristics that are difficult to simulate, such as nonlinearity, zero-crossing offset, zero-crossing distortion [8] and [9]; it improves the understanding of arc high-impedance grounding arcs and provides a theoretical basis for medium identification of arc high-impedance grounding faults in distribution networks.

The existing experimental research on high-impedance arc grounding faults at home and abroad mainly focuses on tree-contact faults: Literature [10] first conducted research on tree-contact faults of distribution lines in 2014, and concluded that the tree-contact fault current is continuous and only at the level of 0.5 A; in recent years, domestic scholars have also carried out relevant research, literature [11] expounds the formation mechanism of tree-contact faults, and divides the faults into four stages through carbonization effect and cooking effect, literature [12] deduces the evolution model of tree-contact faults, and proposes a method for detecting the time limit of tree-contact faults.

Research on ground media, such as in reference [8], has proposed the DIST-C model for different types of ground faults, but it does not fully explore the fault evolution process and mechanism. Reference [13] uses three machine learning methods to simply distinguish between dry and wet ground fault waveforms.

There are many studies on flashover of polluted insulators at home and abroad [14], but most of them focus on DC, discussing its mechanism and model. Literature [15] gives some leakage current waveforms of AC pollution flashover of insulators.

Some experiments use a ball gap arc in series with a fixed resistance to simulate the high-resistance grounding of arc light, such as literature [16] and literature [17], which is different from the actual arc light high-resistance grounding fault arc scene. In view of the above problems of insufficient guiding theory and experimental data support for arc light high-impedance grounding fault diagnosis, this paper has done the following work:

(1) Through experiments to reproduce arc light high-impedance grounding faults under various causes, collect waveforms, and observe phenomena, point out the causes of arc light high-impedance grounding faults.

(2) Explain the physical reasons for the randomness of arc current, the intermittency of faults, thermal inertia, and the unevenness of nonlinearity, and conduct a quantitative analysis.

(3) Analyze the single-cycle characteristics of tree-contact faults and test the effectiveness of existing methods.

2. Fault Test

2.1. *Fault Test Platform*

In order to improve the understanding of arc light high-impedance grounding faults and fully collect fault data samples, this study has carried out reproduction tests of arc light high-impedance grounding faults under different causes in a 10 kV distribution network real-type test system. The topology of the test system is shown in Figure 1.

The test system mainly includes: the primary side equipment composed of a boosting transformer, four feeders, an arc suppression coil that can be accessed, and an arc high-impedance fault generator; the secondary side measurement/control equipment composed of 10 kV:100 V voltage transformer, 75 A: 5A current transformer, fault transient recorder, camera, terminal cabinet, remote control switch cabinet, communication cabinet, main control console, etc. The feeder parameters are shown in Table 1.

Table 1. Analog line parameters

Type	Phase sequence	Resistance (Ω/km)	Capacitance (μF/km)	Inductance (mH/km)
Overhead Line	Positive sequence	0.1700	0.0097	1.2000
	Zero sequence	0.2300	0.0060	5.4800
Cable Line	Positive sequence	0.2700	0.3390	0.2550
	Zero sequence	2.7000	0.2800	1.0190

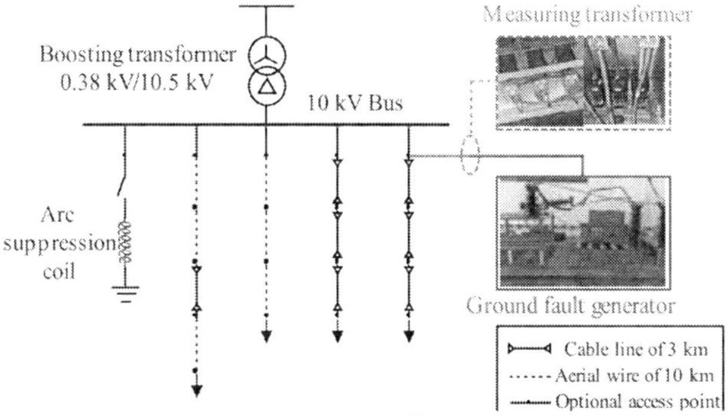

Figure 1. Example of a figure caption

The arc high-impedance grounding fault generator can be controlled remotely through the remote control switch cabinet to modify the access node and fault phase of the circuit breaker. The principle of the arc high-impedance grounding fault generator is shown in Figure 2: the cable is drawn out from the feeder node and connected to the steel-core aluminum stranded wire of the manipulator. The contact between the steel-core aluminum stranded wire of the manipulator and the high-impedance medium in the heat-resistant insulating box is controlled remotely by the servo motor. The fault current is introduced into the ground through the grounding body on the inner surface of the box to simulate the arc high-impedance grounding fault caused by the ground medium (hereinafter referred to as the ground-contact fault). The arc high-impedance grounding fault caused by tree contact (hereinafter referred to as the tree-contact fault) and the AC flashover of the insulator (hereinafter referred to as the pollution flashover fault) can also be simulated by clamping one end of the tree branch or the dirty insulator with the grounding wire and alligator clip.

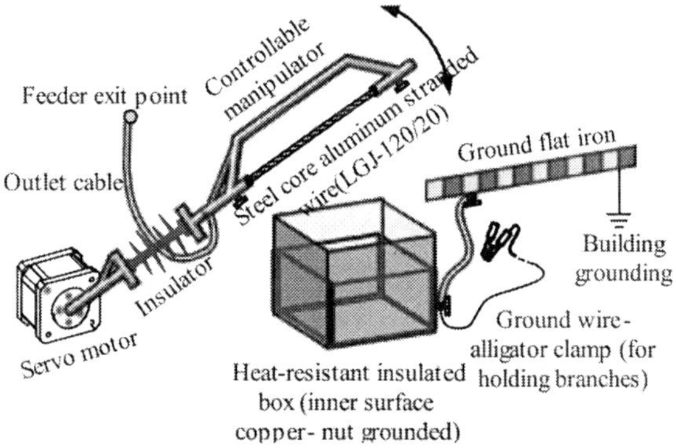

Figure 2. Schematic diagram of arc high-impedance fault generator

To eliminate the randomness of experimental phenomena, multiple experiments were conducted under random different conditions (temperature, humidity, dryness/wetness of test materials), including 20 sets of tree fault tests, 30 sets of ground fault tests, and 10 sets of pollution flashover fault tests.

2.2. Fault Test Results

2.2.1. Tree-Contact Fault Test. The problem of conflicts between trees and power lines is relatively serious in the operation and maintenance of rural distribution networks [18]. There are two main reasons for the frequent occurrence of tree-contact faults. The terrain in rural areas of southeast China is complex, with high forest coverage and severe wind disasters; the distribution lines in rural towns are mainly overhead bare wires, and the wind resistance of 10 kV concrete poles is insufficient [19].

In this study, mango tree branches (woody plants) and bamboo segments (gramineous plants) were used as test materials to continuously or intermittently touch LGJ-120/20 steel-core aluminum stranded wires to obtain fault voltage and current, and to observe the test phenomena.

The test results are shown in Figure 3. Similar to the literature [11], the fault process can be divided into four main stages based on these phenomena/waveforms. Stage I- Contact arcing occurs, and carbonization takes place; Stage II- The contact point ignites, large-area carbonization occurs, and a low-resistance Lichtenberg figure is formed; Stage III- The carbonization pattern almost runs through, and a large amount of moisture is released; Stage IV- Breakdown flashover occurs.

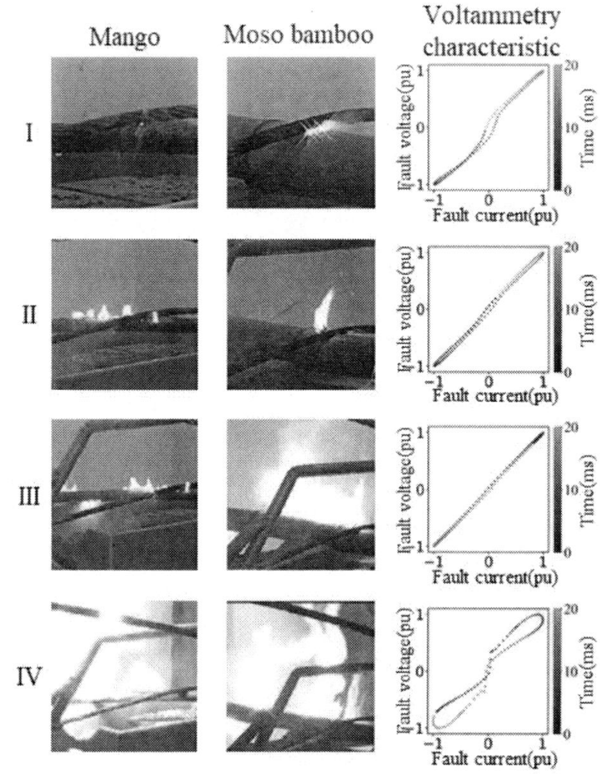

Figure 3. Characteristics of each stage of tree-contact fault

The fault volt-ampere characteristics are shown in Figure 3: in stage I, due to the nonlinearity of the contact point arc and the possible capacitive characteristics of the medium [20], the fault volt-ampere characteristic presents a counterclockwise closed loop, shaped like a distorted "s"; in stage II, because the contact point is wrapped in flames, the flames have conductive properties [21], replacing the arc to form a path, and the carbonization pattern destroys the insulating strength of the medium; in stage III, the pattern runs through, and a large area of fire causes the volt-ampere characteristic to present a straight line; in stage IV, the flashover arc clamps the fault point to ground voltage, showing the classic AC arc volt-ampere characteristic [22], that is, nonlinearity and thermal inertia.

Because the tree-contact fault has current continuity [10], the process of intermittent tree-contact faults is similar to that of continuous tree-contact faults, but because the power-on time is shorter, the heat accumulation is less obvious, and there is an arc phenomenon caused by tree-line contact-separation, as shown in Figure 4.

Figure 4. Arcing caused by wire tree separation

Phenomena of bamboo faults differ from those of wooden branches in several aspects. Due to the higher moisture content in bamboo, the fault current is larger, the fault progression is faster, the flame color [21] is whiter, and the flame brightness is higher; the internodes of bamboo culms are relatively closed and humid, and when heated to high temperatures, the internal steam splashes significantly after the bamboo wall is burned; in the initial stage, wooden branches mainly conduct electricity through internal vessels, while in the later stage of the fault, the surface carbonization pattern dominates; After bamboo conducts electricity, the current flow path is on the surface, and in intermittent tests, the arc is more pronounced when the tree line is detached.

2.2.2. Ground-contact Fault. In ground medium fault tests, faults can be divided into two main stages based on the significance of arc randomness, as shown in Figure 5. The stable arcing stage when the ground medium is moist, the fault phenomenon is stable arcing, and the medium becomes hot; the random arcing stage of the ground medium, the fault phenomenon is misty water, and the arcing distance becomes longer. The V-I characteristic is manifested as an increase in the V-I characteristic area, but it always maintains an "s" shape and has a certain thermal inertia.

2.2.3. Flashover Fault Test. Two types of ceramic insulators from the ZSW series were used in the experiment. The insulators were artificially polluted by spraying saturated salt water and applying diatomaceous earth, and a Nebulizer was used to continuously spray mist to simulate a humid climate. The test results are shown in Figure 6.

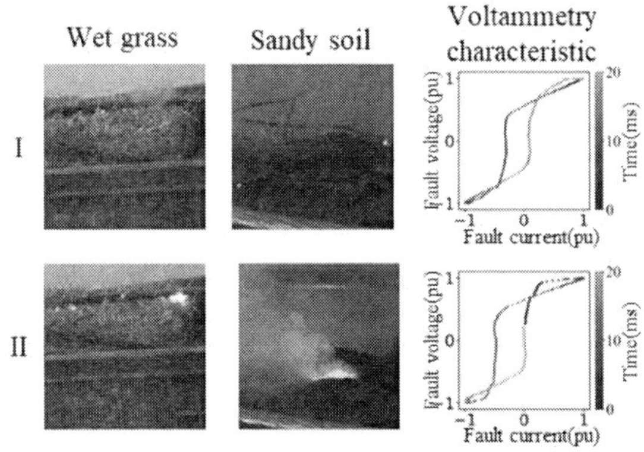

Figure 5. Characteristics of each stage of ground-contact fault

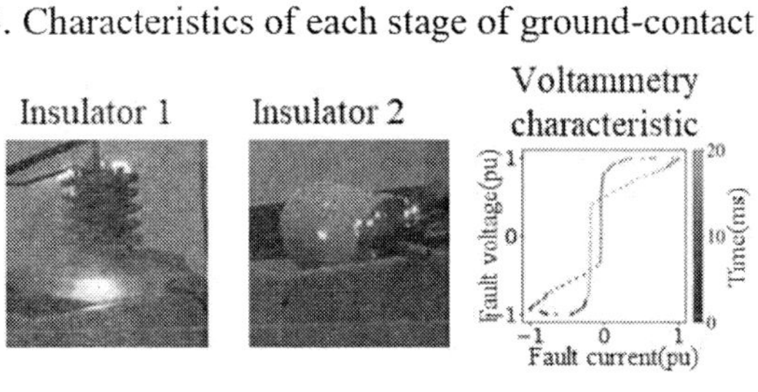

Figure 6. Insulator pollution flashover test characteristicslt

2.2.4. Fault Test Trends. The envelope of fault recording current waveform is shown in Figure 7. In fault tests, the fault current has a clear trend of increasing or decreasing, accompanied by flashovers or random enhancement of faults.

During the fault test, the fault current exhibits a distinct trend of strengthening or weakening, along with the development of the fault, flashovers may occur, or the randomness of the fault may increase.

Current research mostly focuses on the "zero-crossing" characteristics of fault currents, such as the zero-crossing duration, distortion offset, and distortion degree proposed in reference [8]. From the research in this section, it can be seen that the diverse changes of these indicators are essentially caused by the change of V-A characteristics in the fault process due to the dynamic change of physical characteristics, which further leads to the abnormal change of zero zero-cross of fault current, and has weak correlation with phase voltage drop of system side and neutral grounding mode.

The typical fault currents in each stage are shown in Figure 8. Among them, the zero-crossing phenomenon is not obvious in stage I and stage III of tree-contact faults, while the insulator flashover fault current has a relatively obvious asymmetry between the positive and negative half-cycles.

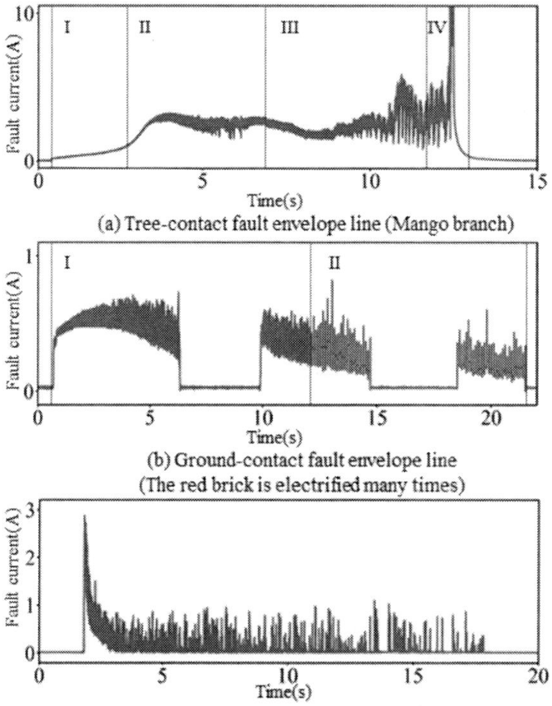

(a) Tree-contact fault envelope line (Mango branch)

(b) Ground-contact fault envelope line
(The red brick is electrified many times)

(c) Pollution flashover fault envelope line (Insulator)

Figure 7. Fault current envelope

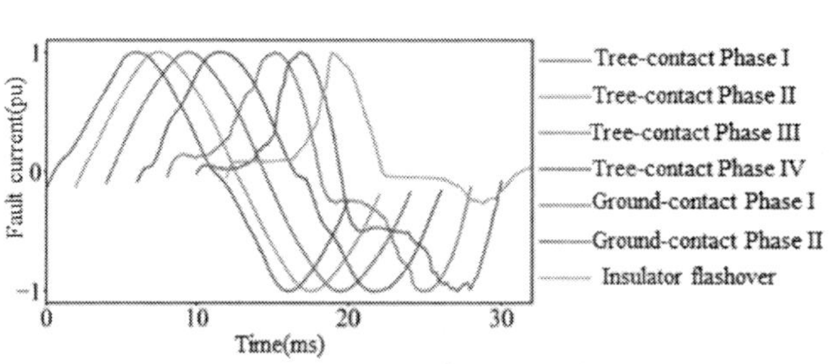

Figure 8. Diversity of Fault Currents

3. Discussion on Test Arc Phenomenon

3.1. Mechanism of High-Impedance Grounding Arc

Based on the average breakdown voltage of air at 30 kV/cm, the limiting distance for single-phase voltage arcing of a 10 kV line without contact is approximately.

$$d_{\max} = \frac{10\text{kV}}{\sqrt{3}} \cdot \frac{\sqrt{2}}{30\text{kV/cm}} \approx 0.27\text{cm} \tag{1}$$

This implies that it is difficult to generate flying arc grounding in the case of good ground insulation and no contact between overhead lines and grounding media. In the test phenomena obtained in the previous section, the arcs in tree-contact and ground-contact faults mainly originate from the contact sites.

According to the theory of electrical contact, the contact resistance R_s consists of the constriction resistance R_j and the film resistance R_b [11], and the specific expression is:

$$R_s = R_j + R_b = \frac{\rho}{2na_p} + \frac{\sigma}{\pi n a_p{}^2} \tag{2}$$

Where n is the number of conductive spots, a_p is the average radius of conductive spots, ρ is the material resistivity, and σ is the film tunnel resistivity. According to the calculation in literature [12], the contact resistance is on the order of $10^5 - 10^6\ \Omega$. In the previous test, the test current was in the range of $0.5 - 40\ \text{A}$, that is, the tower base resistance was only on the order of $10^4 - 10^5\ \Omega$, which is much smaller than the contact resistance.

In the contact between overhead lines and high-impedance media, after the AC voltage passes through zero, the loop resistance is almost equal to the contact resistance, and the phase voltage is fully applied across the contact gap. Then, as the AC voltage continues to rise, the gap is broken down, the fault current flows through the arc channel, and a high-impedance fault with arc light is formed. The principle is shown in Figure 9.

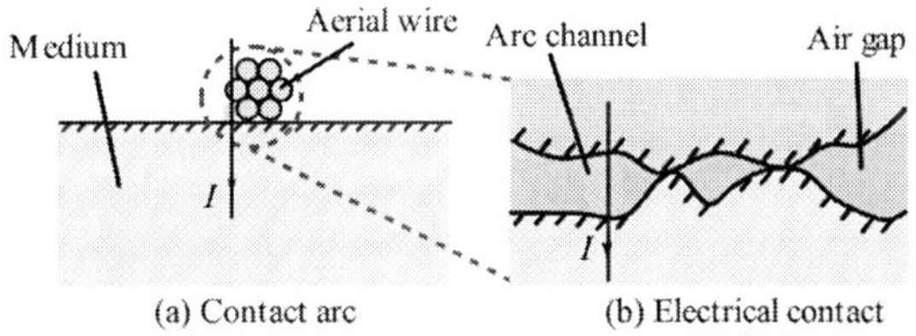

Figure 9. Contact arc generation principle diagram

The principle of pollution flashover arc is quite different from that of contact arc, which has been extensively studied in the academic community [14]. Most towers are effectively grounded at the cross arms in addition to the installation of lightning arresters to meet the lightning protection requirements. This ensures that when non-penetrating flashover occurs in insulators, the following two conditions are satisfied: arc-to-ground discharge; the fault equivalent impedance is relatively high.

As shown in Figure 10, after a mild pollution flashover occurs, the arc area will rotate with the disappearance of the wet area on the umbrella skirt. As the dry area expands, the arc gradually lengthens (while the probability of breakdown decreases), until the dry area becomes too large to break down. Mild pollution flashover faults of insulators are difficult to sustain and usually disappear with the evaporation of moisture, making them transient faults.

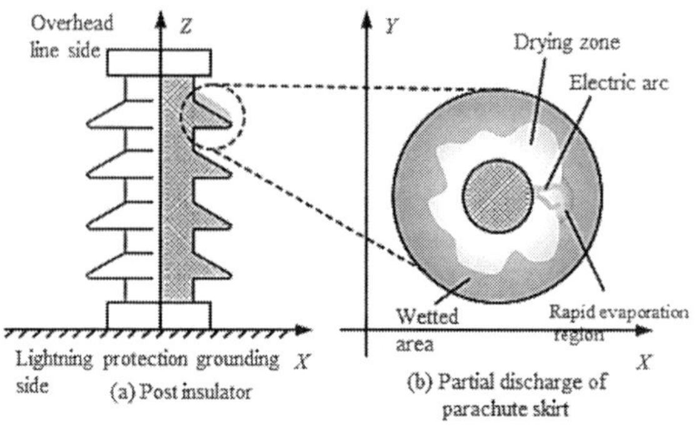

Figure 10. Schematic diagram of mild pollution flashover of insulator

3.2. *Generation conditions of high-impedance grounding arc*

A necessary condition for the generation of fault arc is the concentrated distribution of voltage drop, forming a large electric field intensity in a small space.

For insulators, the following conditions must be satisfied to generate AC flashover: the surface pollution greatly reduces the insulation resistance, so that the Joule heat generated by the leakage current cannot be ignored; the insulator surface is unevenly wet and dry, forming a large non-uniform electric field. In the case of complete wetting, the surface salt is washed away, the voltage drop of the insulator is evenly distributed, and flashover is not easy to occur. After flashover occurs under uneven wet and dry conditions, the moisture evaporates, making the insulator return to a dry state and no longer meet the necessary conditions for the formation of AC arc.

For non-insulating media, the large contact air gap resistance causes the voltage drop to concentrate, allowing the contact arc to form. In tree-contact faults, the appearance of flames short-circuits the contact resistance, causing the arc to disappear; while in ground-contact faults, the medium is not flammable, and the arc always exists. If the medium has a very high resistivity, the medium resistance can reach the same order of magnitude as the contact resistance, so that the voltage drop no longer concentrates around the contact point, and the arc is not easy to form. At this time, the leakage current is only in the range of 0.1-10 mA, such as asphalt.

4. Analysis of fault nature based on experimental data

Based on the volt-ampere characteristics of a single cycle test, it is not difficult to summarize the following local characteristics of high-impedance fault arc: nonlinearity: the instantaneous transition resistance change caused by the sudden increase in air conductivity after air breakdown, which is manifested in the segmented change of slope in the volt-ampere characteristics, and the nonlinearity is related to the air breakdown voltage and the degree of change in air breakdown conductivity; thermal inertia: during the arcing period, the higher the arc column temperature at the same voltage, the higher the conductivity, and for AC arcs, the temperature change lags behind, so the current magnitude is asymmetric during the voltage rise and fall process of a single positive/negative half cycle, which is manifested as clockwise in the volt-ampere characteristics; zero rest offset: during the zero rest period (when the voltage does not reach the breakdown voltage), there is a weak current with the same polarity as the voltage, and the local zero crossing of the volt-ampere characteristics shows a counterclockwise trend.

Different types of faults have different local waveform characteristics in different processes, which are summarized in Table 2. It can be seen that in tree-contact faults, the characteristics of arc high-impedance grounding faults vary greatly. Fourier transform is used to calculate the relative amplitude changes of harmonics in each stage, and the typical data calculation results are shown in Figure 11. The calculation formula of relative amplitude is:

$$r_i = \frac{V_i}{V_1} \tag{3}$$

where V_i is the amplitude of the i harmonic, and V_1 is the amplitude of the fundamental wave.

The total harmonic distortion (THD_R) of voltage and current in tree-contact faults is also calculated, and the statistical results are shown in Table 3. The calculation formula is:

$$THD_R = \frac{\sqrt{V_2{}^2+V_3{}^2+V_4{}^2+\cdots}}{\sqrt{V_1{}^2+V_2{}^2+V_3{}^2+\cdots}} \tag{4}$$

In addition, referring to the calculation method of the relative slope of volt-ampere characteristics at voltage zero-crossing in reference [9], a nonlinearity strength index (NLSI) is set, and the calculation formula is as follows:

$$\text{NLSI} = \frac{k_+ + k_-}{2} - 1 \tag{5}$$

where k_+ and k_- represent the slopes of the volt-ampere characteristic curve when the voltage changes from negative to positive and from positive to negative, respectively. The calculation method is as follows: for the discretized single-frequency cycle voltage data $[u_1, u_2, \cdots u_N]$ and current $[i_1, i_2, \cdots i_N]$ after min-max normalization, where N is the number of sampling points in a single cycle, if the voltage crosses zero at the M point, then:

$$k = \frac{u_{M+1} - u_M}{i_{M+1} - i_M} \tag{6}$$

The calculation results are still listed in Table 3.

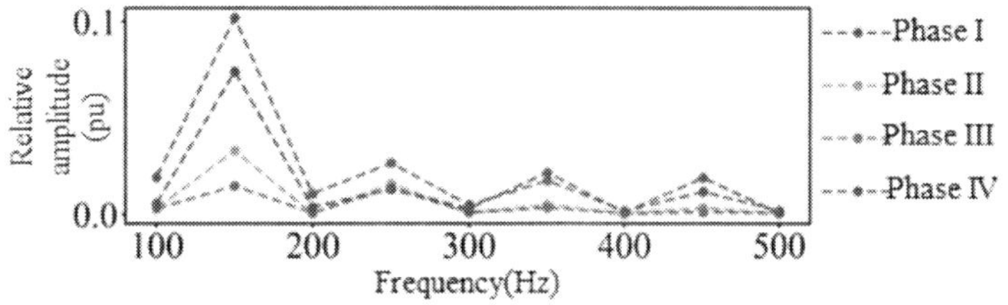

Figure 11. Harmonic components at each stage of tree-contact fault current

Table 2. Local characteristics of fault waveform

Fault Type		Nonlinearity	Thermal Inertia	Zero-Rest Offset
tree-contact fault	The Stage of I	possess	do not possess	possess
	The Stage of II	do not possess	do not possess	possess
	The Stage of III	do not possess	do not possess	do not possess
	The Stage of IV	possess	possess	possess
ground-contact fault	The Stage of I	possess	possess	possess
	The Stage of II	possess	possess	possess
Flashover Fault	Minor fault	possess	possess	possess
	Severe fault	possess	possess	possess

Table 3. Indicators of each stage of tree-contact fault

data type	The Stage of I	The Stage of II	The Stage of III	The Stage of IV
THDR	0.0203	0.0196	0.0175	0.1172
THDR	0.1075	0.0371	0.0205	0.0806
NLSI	1.7346	0.3944	0.0995	13.6927

It can be seen that, except for a small amount of harmonic distortion caused by the voltage source, during the tree-contact fault process: in stages I-III, the proportion of harmonics continues to decrease, corresponding to the disappearance of the arc zero-crossing characteristics caused by the ignition in the fault development stage, and the nonlinearity continues to weaken; while in stage IV, after complete flashover occurs, the phase voltage is clamped by the arc voltage, causing large voltage distortion, and the zero-crossing characteristics reappear. Except for stage IV, the nonlinearity indicators are relatively small, which brings difficulties to the diagnosis method of arc high-impedance grounding faults based on nonlinearity [2] and [9].

In addition to the above local characteristics, there are also some overall characteristics for long-term fault voltage and current waveforms. Intermittent: wind causes tree-contact faults to occur repeatedly; Random: the random occurrence of arcs in ground-contact faults leads to unstable fault currents, sometimes large and sometimes small; Weak: the fault current is small and difficult to be captured by traditional zero-sequence protection. It is worth mentioning that in some test processes, there will be large fault currents, such as stage IV of tree-contact faults, severe pollution flashover, and some ground-contact faults in literature [23], etc. These large current faults will cause serious fault phase voltage drop, large changes in zero-sequence voltage, and do not have the signal weakness of arc high-impedance grounding faults [24], belonging to arc low-impedance grounding faults.

5. Conclusion

In this study, various fault scenarios such as tree-contact faults, ground-contact faults, and pollution flashover faults were designed using an arc high-impedance grounding fault generation device. Some guiding conclusions were drawn through the analysis of data and phenomena:

(1) Under different causes of arc high-impedance grounding faults, the fault trends are different. The arc in the later stage may disappear on its own, or a more serious flashover event may occur, which is reflected in the change of the long-term fault current envelope. Arcs can be classified into contact arcs and dry-zone arcs of insulators based on their generation principles.

(2) The single-cycle characteristics of faults vary depending on the fault phenomena, and the fault scenario and evolution process determine the single-cycle V-A characteristics, thus causing diverse changes in fault currents. Faults exhibit one or more characteristics such as nonlinearity, thermal inertia, zero-crossing offset, weakness, intermittency, and randomness.

(3) Tree-contact faults pose difficulties for fault diagnosis algorithms based on nonlinearity, especially in the middle and late stages of tree-contact faults where the fault current has almost no zero-crossing feature. By constructing simple indicators, the harmonic components, total harmonic distortion, and nonlinearity indicators in typical tree-contact faults at various stages were calculated, and it was pointed out that the existing harmonic method [25] and nonlinearity indicator method [26] face failure problems.

Acknowledgments

This work is supported by the Science and Technology Project of Fuzhou Power Supply Company, State Grid Fujian Electric Power Co., Ltd, under Award 521310240005. Furthermore, the authors express their gratitude to the Referees and the Associate Editor for their valuable advice and comments.

References

[1] B. Wang, X. Cui, X.Z Dong, "Overview of Arc High Impedance Grounding Fault Detection Technologies in Distribution System," Proceedings of the Chinese Society of Electrical Engineering, vol. 40, no. 1, pp. 96-107, 2020.

[2] J.Z. Geng, "Analysis and Detection of High Impedance Grounding Fault in Neutral Point Effectively Grounding Distribution Network," Dian li xi tong zi dong hua, vol. 37, no. 16, pp. 85–91, 2013.
[3] W. Gao, W.X. He, M.F. Guo, H. Bai, "Detection Method of High-impedance Fault in Distribution Network Based on Uneven Small Samples from Actual Measurements," High Voltage Engineering, pp. 1-11, 2024.
[4] M.F. Guo, W.L. Liu, J.H. Gao, D.Y. Chen, "A Data-Enhanced High Impedance Fault Detection Method Under Imbalanced Sample Scenarios in Distribution Networks," IEEE Transactions on Industry Applications, vol. 59, no. 4, pp. 1-14, 2023.
[5] Y. Hu, K. Liu, T. Wu, Y. Liu, Z.M. Su, "Analysis of Influential Factors on Operation Safety of Transmission Line and Countermeasures," High Voltage Engineering, vol. 40, no. 11, pp. 3491-3499, 2014-11.
[6] D.C. Huang, P. Li, J.J. Ruan, Y.F Zhang., T. Wu, "Review on Discharge Mechanism and Breakdown Characteristics of Transmission Line Gap Under Forest Fire Condition," High Voltage Engineering, vol. 41, no. 2, pp. 622-632, 2015.
[7] Y.G. Sun, J. Ni, B. Wang, M. Xie, H.G. Wang, "Transmission Line High-impedance Fault Modeling Analysis for Single-end Fault Location," Automation of Electric Power Systems, vol. 40, no. 22, pp. 86-92, 2016.
[8] M. Wei, W. Liu, F. Shi, H. Zhang, Z. Jin, W. Chen, "Distortion-Controllable Arc Modeling for High Impedance Arc Fault in the Distribution Network," IEEE Transactions on Power Delivery, vol. 36, no. 1, pp. 52-63, 2021.
[9] F. Deng, Y.L. Chen, Z. Zeng, H.F. Shi, "Combined Data-knowledge Driven Detection Method for High Impedance Faults in Distribution Networks," Proceedings of the CSEE, pp. 1-14, 2024-10-30.
[10] Y. Liu, H. Yu, Y. Sheng, B. Wang, L. Li and L. Zhang, "Modelling of Tree Contact Faults in Distribution Networks," 2023 IEEE International Conference on Advanced Power System Automation and Protection (APAP), Xuchang, China, pp. 75-79, 2023.
[11] J.K. Qin, X. Ning, S.H. Fan, Z.D. Jia, "Discharge Ignition Mechanism of Tree-contacting Single-phase-to-ground Faults of Overhead Lines," Power System Technology, vol. 47, no. 3, pp. 1289-1298, 2023.
[12] D. Liang, Xu, Y. Tang, P.W. Wang, W. Wang, Z.Y. Sun, "10kV Model and Detection Method for Tree-contact Single-phase-to-ground Faults on 10kV Overhead Lines," Proceedings of the CSEE, vol. 41, no. 15, pp. 11, 2021.
[13] H. Lala, S. Karmakar. "Detection and Experimental Validation of High Impedance Arc Fault in Distribution System Using Empirical Mode Decomposition," IEEE Systems Journal, vol. 14, no. 3, pp. 3494-3505, 2020.
[14] Z.J. Zhang, X.J. Jiang, C.X. Sun, "Summary of Research on Flashover Mechanism of Polluted Insulators," Power System Technology, vol. 32, no. 16, pp. 37-42, 2008.
[15] T. Suda. "Frequency characteristics of leakage current waveforms of an artificially polluted suspension insulator," IEEE Transactions on Dielectrics and Electrical Insulation, vol. 8, no. 4, pp. 705-709, 2001.
[16] J.Q. Ran, Q. Yang, L.W. He, H.W. Liu, T. Zhou, S.Y. Wei, "Improved mathematical model for single-ground arcing fault in medium voltage distribution networks," CSEE Journal of Power and Energy Systems, pp. 1-10, 2022.
[17] T. Zhou, Q. Yang, T. Yuan, H.X. He, H.W. Liu, "A Novel Mathematical–Physical Arc Model and its Application to the Simulation of High-Impedance Arc Faults in Distribution Networks," IEEE Transactions on Power Delivery, vol. 39, no. 3, pp. 1794-1806, 2024.
[18] F.X Yao., S.H. Miao, Tu Q.Y., Sun Q., F.J. Li, "Risk Warning and Optimization Processing for Tree-Line Contradiction in Rural Distribution Network Considering Severe Convective Weather," Transactions of China Electrotechnical Society, vol. 38, no. 22, pp. 6188-6203, 2023.

[19] L.M. Zheng, S.W. Shu, B. Chen, H. Wu, J.Y. Huang, J. Qian, "Prediction Method for Amount of 10 kV Damaged Poles Under Severe Typhoon Environment Based on Meshing and Support Vector Machine," High Voltage Engineering, vol. 46, no. 1, pp. 42-51, 2020.

[20] J. Liang, Z.W. Qu, X.W. Jia, X.Y. Zhang, H.W. Liu, "Studies of Tree Capacitance on the Basis of Physiology," Scientia Silvae Sinicae, vol. 42, no. 1, pp. 90-95, 2006.

[21] T. Wu, J.J. Ruan, Y. Hu, "Study on Forest Fire Induced Breakdown of 500kV Transmission Line in Terms of Characteristics and Mechanism," Proceedings of the CSEE, vol. 31, no. 34, pp. 163-170, 2011.

[22] A.E. Emanuel, D. Cyganski, "High impedance fault arcing on sandy soil in 15 kV distribution feeders: contributions to the evaluation of the low frequency spectrum," Power Delivery IEEE Transactions on, vol. 5, no. 2, pp. 676-686, 1990.

[23] M.J. Wei, "Technology and Application of High Impedance Fault Diagnosis at Distribution Networks Based on Wideband Synchronous Measurements," Shandong University, 2022.

[24] Amin Ghaderi, Herbert L. Ginn, Hossein Ali Mohammadpour, "High impedance fault detection: A review," Electric Power Systems Research, vol. 143, no. 2, pp. 376-388, 2017.

[25] A.F. Sultan, G.W. Swift, D.J. Fedirchuk, "Detecting arcing downed-wires using fault current flicker and half-cycle asymmetry," IEEE Transactions on Power Delivery, vol. 9, no. 1, pp. 461-470, 1994.

[26] S. Gautam and S. M. Brahma. "Detection of High Impedance Fault in Power Distribution Systems Using Mathematical Morphology," IEEE Transactions on Power Systems, vol. 28, no. 2, pp. 1226-1234, 2013.

Identification of Inductance Parameters for BDDC in Rural Integrated Energy Systems Based on PIML

Xiaoxuan Guo[1[0000-0003-4754-2467]] and Haibo Bao[2*[0000-0001-8307-2672]]

[1] Electric Power Research Institute, Guangxi Power Grid Corporation, Guangxi, Nanning, 530023, China

[2] Nanning Power Supply Bureau, Guangxi Power Grid Corporation, Guangxi, Nanning, 530021, China

Abstract: In the integrated energy storage scenario, bidirectional DC converters play a key role in the energy conversion and management process. As one of the core elements affecting the performance of bidirectional DC converters, identifying the inductance parameters holds great significance for the stable and efficient operation of the system. The high-performance identification of parameters for bidirectional DC converters can improve system control and operation efficiency. In view of this, this article puts forward an approach for identifying inductance parameters of bidirectional DC converters based on physical information machine learning (PIML). The parameter identification approach grounded in physical information machine learning surmounts the hurdles of substantial training data requirements, accuracy limitations, and robustness issues that are typically encountered by data-driven methods. This article demonstrates through simulation and experimentation that the method has high accuracy and robustness, providing strong support for the optimization control and stable operation of bidirectional DC converters in rural integrated energy systems.
Keywords: rural integrated energy system; Bidirectional DC converter; Physical information machine learning; Identification of inductance parameters

1. Introduction

Amid the escalating global energy crisis, the advancement of renewable energy has gained ever-increasing significance. However, rural renewable energy has the characteristic of unstable output power. The integrated energy system, as an advanced energy supply mode, can effectively solve the above problems. In rural integrated energy systems, energy storage batteries are linked to high-voltage DC buses via bidirectional DC-DC converters (BDDC). When renewable energy power is insufficient, BDDC is needed to supplement the power deficit. When renewable energy power is excessive, BDDC needs to absorb the excess power. Therefore, BDDC in energy storage units plays an important role in maintaining power balance [1]. Inductance is one of the key components of bidirectional DC converters, and its parameters directly affect multiple performance indicators of the converter. During the operation of energy storage systems, the inductance components in BDDC undergo parameter changes. Therefore, it is urgent to find a scientific and effective method for identifying inductance parameters to achieve accurate and effective parameter identification of bidirectional DC-DC converters [2].

Currently, the parameter identification approaches for DC/DC converters are primarily classified into two types: model-based methods and knowledge-based methods. Reference [3] uses wavelet transform to perform time-frequency analysis on the output voltage of a boost converter, and derives an equivalent series resistance calculation model for aluminum electrolytic capacitors to achieve parameter identification. Reference [4] uses the Kalman filter algorithm to perform real-time parameter estimation on switching power converters, and proposes an adaptive tuning technique based on this, which has the merit of updating parameter vectors at various rates. The principle of knowledge-based parameter identification method is to use historical datasets, analyze and learn them, reveal hidden rules between data, and finally use intelligent algorithms to establish the relationship between output features and the parameters to be identified. Reference [5] regards the DC/DC converter as a black box and trains a wavelet convolutional neural network which is capable of precisely replicating the behavior of the DC/DC converter using a vast amount of experimental data. Reference [6] used a dual scale convolutional autoencoder based on capacitor voltage and a limit learning machine to identify the parameters of aluminum electrolytic capacitors and equivalent series resistance.

Therefore, existing research has the following limitations: 1) In most cases, frequency domain analysis is required, like s-domain or z-domain transformation. These analytical techniques pose challenges to the scalability of large and complex systems; 2) Due to inherent nonlinearity, model discretization techniques play a crucial role in streamlining the model. However, the imprecision resulting from the discrete form becomes more noticeable in intricate systems and might cause a reduction in estimation accuracy. Based on these, this research proposes an approach for identifying the inductance parameters of bidirectional DC converters in rural integrated energy systems using physical information machine learning. The approach has the following contributions and advantages: 1) The PIML implemented by bidirectional DC converters requires relatively small training data, which is an easily obtainable feedback signal for control. 2) The proposed method does not require complex s-domain or z-domain transformations and discretization techniques, and has a wide range of applications.

2. Construction of system dynamic model for BDDC

By studying the operating principle of the non isolated bidirectional half bridge DC/DC converter, the topology diagram adopted in this paper for the composite energy storage system is shown in Figure 1.

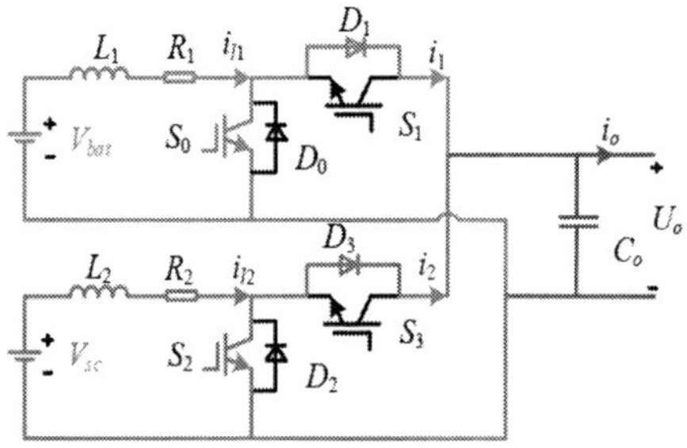

Figure 1. Topology diagram of composite energy storage system.

Due to the use of two BDDCs, the entire composite energy storage system has four operating modes. If a mathematical model is constructed for each working mode, it will greatly increase the computational cost. In order to simplify the model and reduce computational resources, this article has completed the establishment of a global mathematical model that can cover four modes. According to the working principle of bidirectional DC converters, the global state space model of the composite energy storage system is obtained as [7]:

$$\begin{bmatrix} \frac{di_{l1}}{dt} \\ \frac{di_{l2}}{dt} \\ \frac{dU_o}{dt} \end{bmatrix} = \begin{pmatrix} -\frac{R_1}{L_1} & 0 & -\frac{\mu_{01}}{C_o} \\ 0 & -\frac{R_2}{L_2} & -\frac{\mu_{23}}{C_o} \\ \frac{\mu_{01}}{C_o} & \frac{\mu_{23}}{C_o} & 0 \end{pmatrix} \begin{bmatrix} i_{l1} \\ i_{l2} \\ U_o \end{bmatrix} + \begin{pmatrix} \frac{1}{L_1} & 0 & 0 \\ 0 & \frac{1}{L_2} & 0 \\ 0 & 0 & -\frac{1}{C_o} \end{pmatrix} \begin{bmatrix} V_{bat} \\ V_{sc} \\ i_o \end{bmatrix} \tag{1}$$

3. Research on parameter identification method of BDDC based on PIML

3.1. System Dynamics Model of BDDC

For the parameter identification of transformers, this paper utilizes modern numerical differential equation solvers and considers a general system dynamics model:

$$u_t + N[u;\lambda] = 0, x \in \Omega, \quad t \in [0,T] \tag{2}$$

$$f = u_t + N[u;\lambda] \tag{3}$$

Among them, $u=u(x,t)$ is the solution of the differential equation, x is the spatial coordinate, and t is the time coordinate.

Two general solutions for this model are to solve $u(x,t)$ when λ is known and to solve the parameter λ value generating these data under the condition of given measurement data $u(x,t)$ when λ is unknown. Due to the need for parameter discrimination of bidirectional DC/DC converters in this article, λ is an unknown variable. Therefore, λ is selected as inductance L_1, L_2, resistance R_1, R_2, and capacitance C_o. u is selected as the output current i_{l1} of the battery pack, the output current i_{l2} of the supercapacitor pack, and the output voltage U_o.

Therefore, the dynamic model of BDDC can be described as:

$$\frac{di_{l1}}{dt} + N[i_{l1};\lambda] = 0 \tag{4}$$

$$\frac{di_{l2}}{dt} + N[i_{l2};\lambda] = 0 \tag{5}$$

$$\frac{dU_o}{dt} + N[U_o;\lambda] = 0 \tag{6}$$

By substituting the DC bus terminal voltage into it, we can obtain:

$$N[i_{l1};\lambda] = \frac{[R_1 i_{l1} + \mu_{01} U_o L_1 / C_o - V_{bat}]}{L_1} \tag{7}$$

$$N[i_{l2};\lambda] = \frac{[R_2 i_{l2} + \mu_{23} U_o L_2 / C_o - V_{sc}]}{L_2} \tag{8}$$

$$N[U_o;\lambda] = \frac{[i_o - \mu_{01} i_{l1} - \mu_{23} i_{l2}]}{C_o} \tag{9}$$

Among them, $\lambda=\{L_1, L_2, R_1, R_2, C_o\}$ are the parameters of the components to be identified, and μ_{01}, μ_{23} are the global control signals.

3.2. Implicit Runge Kutta time progression scheme

In the above system dynamics model, u represents the time-varying data in BDDC, therefore its spatial coordinate x is ignored. Assuming the system's state at time t_n is $u(t_n)$ and at time t_{n+1} is $u(t_{n+1})$, $t_n=t_{n+1}+T_s$, where T_s is the system sampling frequency. Assuming there are q intermediate states $u(t_{n+ki})$ in T_s, where $t_{n+ki}=t_n+ki$, $ki\in[0,1]$, $i=1, \ldots, q$. These intermediate states can be coupled with detectable states $u(t_n)$ and $u(t_{n+1})$ within the method of the implicit Runge Kutta approach [8]. This functional relationship is expressed as:

$$u_i(t_n)=u(t_{n+\mathrm{k}i})+T_s\sum_{j=1}^{q}a_{ij}N[u(t_{n+kj});\lambda] \tag{10}$$

$$u_i(t_{n+1})=u(t_n)-T_s\sum_{j=1}^{q}b_jN[u(t_{n+\mathrm{k}j});\lambda] \tag{11}$$

Among them, $i, j=1, 2, \ldots, q$, $u_i(t_n)=u(t_n)$, $u_i(t_{n+1})=u(t_{n+1})$, parameters $\{h_{ij}, l_j, k_j\}$ are provided by a Buchner table given the order q [9], expressed as follows:

$$\begin{array}{c|c} k & A \\ \hline & l^{\mathrm{T}} \end{array} = \begin{array}{c|cccc} k_1 & h_{11} & h_{12} & \cdots & h_{1q} \\ k_2 & h_{21} & h_{22} & \cdots & h_{2q} \\ \vdots & \vdots & \vdots & \ddots & \vdots \\ k_q & h_{q1} & h_{q2} & \cdots & h_{qq} \\ \hline & l_1 & l_2 & \cdots & l_q \end{array} \tag{12}$$

By substituting formula (10) into formula (11), the end state $u(t_{n+1})$ can be represented by the intermediate state $u(t_{n+ki})$ as follows:

$$u_i(t_{n+1})=u(t_{n+ki})+T_s\sum_{j=1}^{q}(h_{ij}-l_j)N[u(t_{n+kj});\lambda] \tag{13}$$

In summary, the expressions for the intermediate state $u(t_{n+ki})$, the initial state $u(t_n)$, the final state $u(t_{n+1})$ are represented by formulas (10) and (13), respectively. It is worth noting that the parameter λ involves the functional relationship of this construction.

From the perspective of a dynamic system, taking the battery pack current i_{l1} as an example, given the initial state $i_{l1}(t_n)$, BDDC will pass through each intermediate state $i_{l1}(t_{n+ki})$ and reach the end state $i_{l1}(t_{n+1})$ during the T_S period. The current i_{l2} and output voltage U_o of the supercapacitor group are also applicable to the above analysis.

Within the method, the relationship between the initial and final states of the supercapacitor and battery pack currents, as well as the system output voltage, has been clearly established, which holds great importance for the combination of data and physical models.

3.3. *PINN construction of DC/DC converter*

Deep learning can use its approximation ability to input and output data to obtain an approximate model with nonlinear relationships. Therefore, based on the initial and final states of the supercapacitor and battery pack currents, as well as the system output voltage, deep neural networks can be used to obtain their approximate intermediate states, which can then be used as prior information to input into formulas (4)~(6).

Figure 2 shows the Physics-informed Neural Network (PINN) structure illustration of BDDC. The inputs of data-driven section are battery pack current $i_{l1}(t_n)$, supercapacitor pack current $i_{l2}(t_n)$, output voltage $U_o(t_n)$, and switch signals μ_{01} and μ_{23} of two BDDCs. By adjusting the quantity of hidden layers and neurons in deep neural networks, the network's performance can be effectively changed. Therefore, configuring it reasonably can achieve the maximization of resource utilization efficiency. The output is the intermediate state of the input signal $[i_{l1}(t_{n+k1}), \ldots, i_{l1}(t_{n+kq}), i_{l2}(t_{n+k1}), \ldots, i_{l2}(t_{n+kq}), U_o(t_{n+k1}), \ldots, U_o(t_{n+kq})]$. The second part is the physical model part, which takes the intermediate state output by the data-driven part as input. The data is substituted into the forward equation formula (13) and backward equation formula (10) to obtain the predicted values of PINN, including $i'_{m1}(t_n)$, $i'_{m1}(t_{n+1})$, $i'_{m2}(t_n)$, $i'_{m2}(t_{n+1})$, $U'_o(t_n)$, $U'_o(t_{n+1})$.

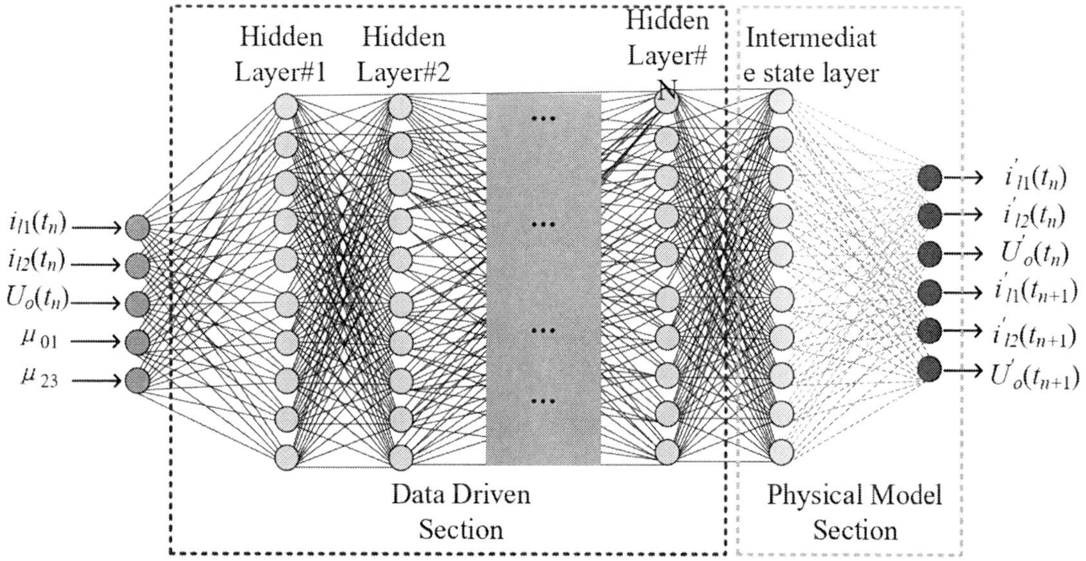

Figure 2. PINN structure illustration of DC/DC converter.

Thus, the error function can be obtained:

$$\begin{aligned} E(\theta) = &\sum_n \left[(i_{m1}(t_n) - (i_{m1}'(t_n))^2 + (i_{m1}(t_{n+1}) - (i_{m1}'(t_{n+1}))^2 \right] \\ &+ \sum_n \left[(i_{m2}(t_n) - (i_{m2}'(t_n))^2 + (i_{m2}(t_{n+1}) - (i_{m2}'(t_{n+1}))^2 \right] \\ &+ \sum_n \left[(U_o(t_n) - (U_o'(t_n))^2 + (U_o(t_{n+1}) - (U_o'(t_{n+1}))^2 \right] \end{aligned} \tag{14}$$

The parameter group θ is $\{w,b,\lambda\}$, where $\{w,l\}$ represents the weights and biases in the driving part of the neural network.

The chain rules of a common backpropagation training approach can be easily extended to the physical portion, facilitating the concurrent training of all parameters in PINN by means of standard backpropagation procedures.

4. Simulation and experimental verification

4.1 Simulation verification

The purpose of this article is to perform online identification of inductance component parameters in BDDC, using MATLAB/Simulink to continuously collect a set of 360 data points. The parameter identification method based on PINN is implemented by combining Python and TensorFlow backend. Train PINN using Adam optimizer with default parameters, learning rate η=0.001, exponential decay rate β_1=0.9, β_2=0.999.

In theory, data-driven neural networks should have sufficient expressive power to model intricate intermediate states. Therefore, it's essential to evaluate the network architecture depending on the varying quantity of hidden layers and neurons. Figure 3 shows the average error percentage of BDDC parameters under various quantities of hidden layers and neurons. Considering the computational resources and accuracy, the PINN data-driven part of this paper selects a deep neural network with 5 hidden layers and 50 neurons per layer for parameter identification of BDDC.

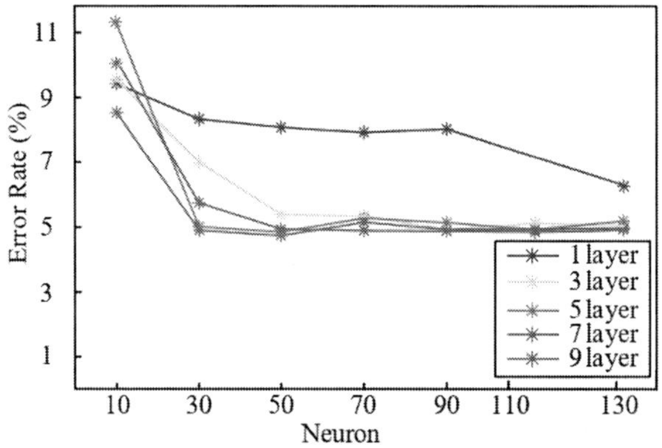

Figure 3. Error under different PINN configurations.

Import the trained PINN into MATLAB/Simulink, and transmit the current and voltage signals from the simulation of the composite energy storage system in real-time to the PINN for parameter identification. The parameter identification results are shown in Figure 4. The parameters of inductance elements L_1 and L_2 have been fluctuating around the actual values. The proposed PINN has a good learning effect on inductance elements, while inductance resistors R_1 and R_2, as well as filtering capacitor C_o, have some offset compared to the actual reference values.

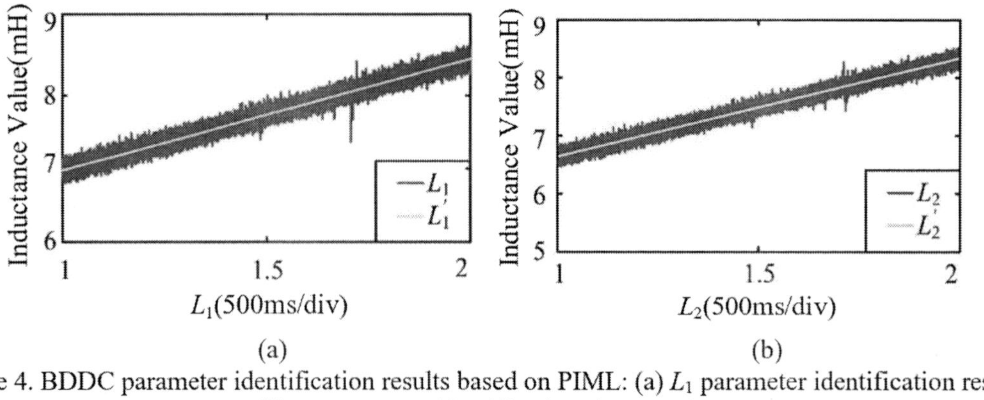

Figure 4. BDDC parameter identification results based on PIML: (a) L_1 parameter identification result. (b) L_2 parameter identification chart result.

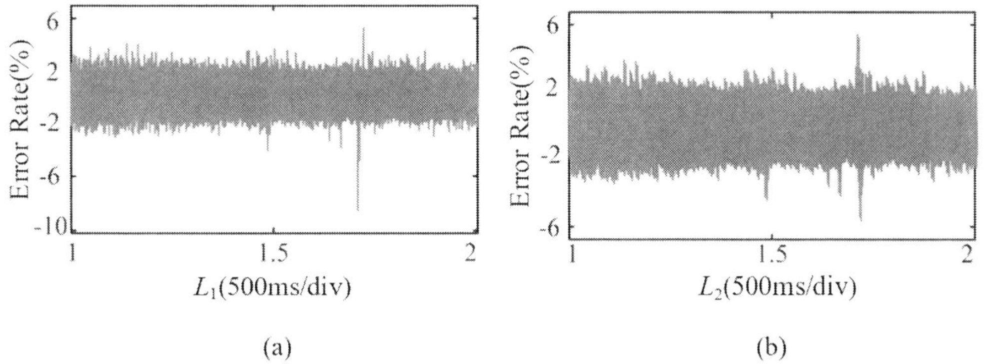

Figure 5. Percentage of identification error for inductive component parameters: (a) L_1 parameter identification error percentage. (b) L_2 parameter identification error percentage.

As shown in Figure 5, the error percentage of parameter identification remained stable within 3% from 1 to 1.7 seconds. At t=1.7s, the operating conditions changed, causing a sudden change in the current of both the battery pack and supercapacitor pack, resulting in an increase in the percentage of identification error, reaching up to 8%.

4.2 Experimental results of BDDC parameter identification

To further verify the advantages of the raised approach, an experimental platform for a composite energy storage system was built. Substitute the experimental data during normal operation into PINN as the training set to practice the neural network. Put the trained neural network into perform online parameter identification of the inductance components of BDDC on the experimental platform of the composite energy storage system. Figures 6 shows the experimental results of L_1 and L_2 parameter identification. The inductance parameter used in the experiment is 5mH. From the figure, it can be seen that the inductance components are all in a healthy state and have not undergone significant parameter changes. The deviation between the identification results and the actual values of the inductance parameters is within 10%. The experiment has demonstrated the effectiveness of the approach based on PILM.

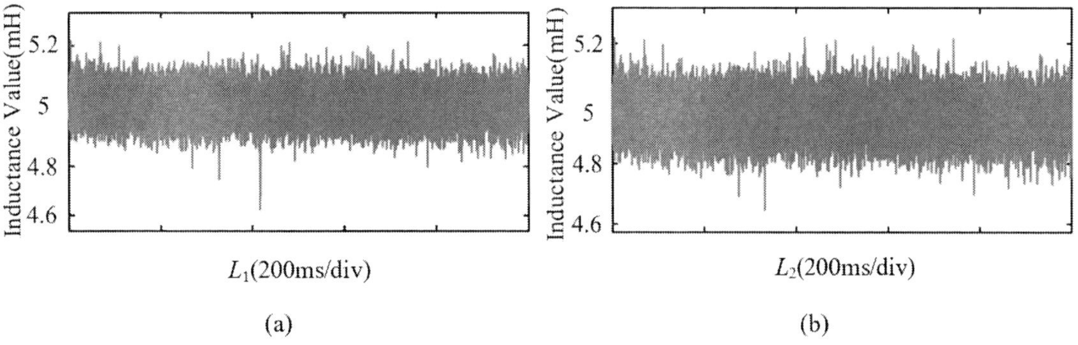

Figure 6. Experimental results: (a)L_1 parameter identification. (b)L_2 parameter identification.

5. Conclusion

This article proposes a method for BDDC parameter identification based on PIML to address the situation where the actual model and observer model are highly mismatched when the electronic component parameters in BDDC vary greatly. The method first establishes a system dynamics model for bidirectional DC/DC conversion, and then uses the implicit Runge Kutta approach to construct the relationship between the intermediate state, initial state, and final state of the current and output voltage of the battery pack and supercapacitor pack. Then use deep neural networks to obtain intermediate states and combine them as inputs with the physical model to obtain PINN. The efficacy of the raised approach has been validated via simulation and experimentation.

Acknowledgments

This work was supported by the Bagui Youth Top notch Talent Project, the Science and Technology Project of China Southern Power Grid Corporation (GXKJXM20230145) and the Science and Technology Project of Guangxi Power Grid (CG0400022001601303).

References

[1] Dang Chaoliang, Wang Fei, Mu Xiaoyu, et al. Vienna rectifier dual vector predictive constant frequency control with introduction of inductance parameter identification. Proceeding of the CSEE, 2022,42(S1):246-255 (in Chinese).

[2] Liu Yu, Liu Wensheng, Zhang Haomiao. Research on Identification of Inductance Parameters Based on Neural Network Algorithm in Weak Magnetic Control. Computer and Digital Engineering, 2020,48(10):2505-2511 (in Chinese).

[3] Lu W, Lu X, Han J, et al. Online estimation of ESR for DC-link capacitor of boost PFC converter using wavelet transform based time–frequency analysis method. IEEE Transactions on Power Electronics, 2019, 35(8): 7755-7764.

[4] Ahmeid M, Armstrong M, Gadoue S, et al. Real-time parameter estimation of DC–DC converters using a self-tuned Kalman filter. IEEE Transactions on Power Electronics, 2016, 32(7): 5666-5674.

[5] Rojas-Dueñas G, Riba J R, Moreno-Eguilaz M. Black-box modeling of DC–DC converters based on wavelet convolutional neural networks. IEEE Transactions on Instrumentation and Measurement, 2021, 70: 1-9.

[6] Li X, He L, Zhu Z, et al. Double-scale convolutional autoencoder and extreme learning machine for parameter identification of DC bus capacitor in power electronic Transformer. IEEE Transactions on Industrial Informatics, 2022, 19(8): 9102-9112.

[7] S. Ding, D. Tang, J. Hang, J. Zhao and S. Gui, "Robust Open-Switch Fault Diagnosis of Bidirectional DC/DC Converters Based on Extended Kalman Filter With Multiple Corrections," IEEE Transactions on Circuits and Systems I: Regular Papers, vol. 71, no. 9, pp. 4363-4374, Sept. 2024.

[8] R. Chen and X. Li, "Implicit Runge-Kutta Methods for Accelerated Unconstrained Convex Optimization," IEEE Access, vol. 8, pp. 28624-28634, 2020.

[9] A. Dély, F. P. Andriulli and K. Cools, "Large Time Step and DC Stable TD-EFIE Discretized With Implicit Runge–Kutta Methods," IEEE Transactions on Antennas and Propagation, vol. 68, no. 2, pp. 976-985, Feb. 2020.

Journal of Physics: Conference Series 3012 (2025) 012059 doi:10.1088/1742-6596/3012/1/012059

Study on Energy Allocation Strategy of Multiple Electrolyzers for Renewable Energy-Based Hydrogen Production System

Jianqiang Xu[1,2], Jiangong Zhu[1,2], Hao Yuan[2,3], Haoyu Zhang[1,2], Xuezhe Wei[1,2], Haifeng Dai[1,2], Chao Wang [1,2*]

[1] School of Automotive Studies, Tongji University, Shanghai 201804, China

[2] Clean Energy Automotive Engineering Center, Tongji University, Shanghai 201804, China

[3] Postdoctoral Station of Mechanical Engineering, Tongji University, Shanghai 201804, China

*Corresponding author's e-mail: wangchao_4@tongji.edu.cn

Abstract. In the context of the Carbon Peaking and Carbon Neutrality Goals, the utilization of renewable energy to produce hydrogen by water electrolysis is of great significance as it can solve the problems of renewable energy consumption and large-scale green hydrogen production. However, the renewable energy-based hydrogen production system faces many problems such as large fluctuation of renewable energy and high cost of hydrogen production. Therefore, this study proposes an average power strategy to enhance the system economy without significant improvement in the performance and cost of the existing power system and electrolyzers. Firstly, the wind-hydrogen capacity ratio is caried out, and according to the real data of a wind fam in Qinghai, a reasonable number of 1,000 standard cubic meters of alkaline electrolyzers is configured. Subsequently, the objective function is developed with economy as the main goal and considering the lifetime of the electrolyzer. Finally, the average power strategy and the common rated power strategy are introduced in detail, and they are compared through example analysis. The results show that, compared with the rated power strategy, the average power strategy improves the economic efficiency, renewable energy consumption rate, and the consistency of the electrolyzer lifetime.

1. Introduction

In recent years, renewable energy power generation has flourished[1]. By the end of 2023, China's installed renewable energy capacity had reached 1.45 billion kilowatts. However, the fluctuating, intermittent, and random characteristics of wind and solar power generation pose increasing challenges to the safe and stable operation of the power system[2]. It is necessary to accept and accommodate the rapidly increasing wind and solar power generation[3,4]. Water electrolysis based on renewable energy for hydrogen production has broad application prospects[5,6], which can solve the problem of renewable energy power consumption and make the hydrogen industry green and carbon-free.

Currently, the maximum power of a single electrolyzer is in the megawatt range. Large-scale renewable energy-based hydrogen production system often requires the use of multiple electrolyzers. It is necessary to improve the economics of hydrogen production by managing the energy distribution of multiple electrolyzers. Reference [7] proposed two control rules: Equal force and Doctor triage. These

Journal of Physics: Conference Series **3012** (2025) 012059 doi:10.1088/1742-6596/3012/1/012059

rules made a certain number of electrolyzers run at rated power according to the wind power output, and balance the operating time of each electrolyzer by cycling the order of electrolyzer operation. Reference [8] proposed a strategy to improve the efficiency of electrolyzer array based on subsection fuzzy control method. The artificial bee colony algorithm was used to calculate the optimal hydrogen production power. Finally, a modular control strategy was introduced to enhance and extend the lifetime of the entire electrolyzer array.

In order to balance the complexity of the strategy and the optimization effect of the system, this study proposes an average power-based energy control strategy for multiple electrolyzers system. Firstly, wind power consumption capacities of different numbers of the electrolyzers are investigated to solve the capacity matching issue between electrolyzers and wind farm. This ensures that 99.53% of wind power is utilized by the hydrogen production system, satisfying the power consumption demand. Subsequently, with economy as the primary objective and considering the lifetime of electrolyzers, a multiple electrolyzers operation control strategy is developed to ensure efficient operation of each electrolyzer. Finally, a detailed comparative analysis between the average power strategy and the rated power strategy is conducted through example analysis.

2. Wind-Hydrogen Capacity Ratio

2.1. Wind power data analysis

Figures 1 presents actual wind turbine data from a wind farm in Qinghai, China, spanning a period of one month. In this study, the wind power data undergoes normalization to extract wind power characteristics. Based on these characteristics, a virtual wind farm is established, with a unit rated power of 50MW, which is commonly seen in the industry. In subsequent sections, the real wind power data from this one-month period will be used as a representative sample to simulate wind power output for all 12 months of a year.

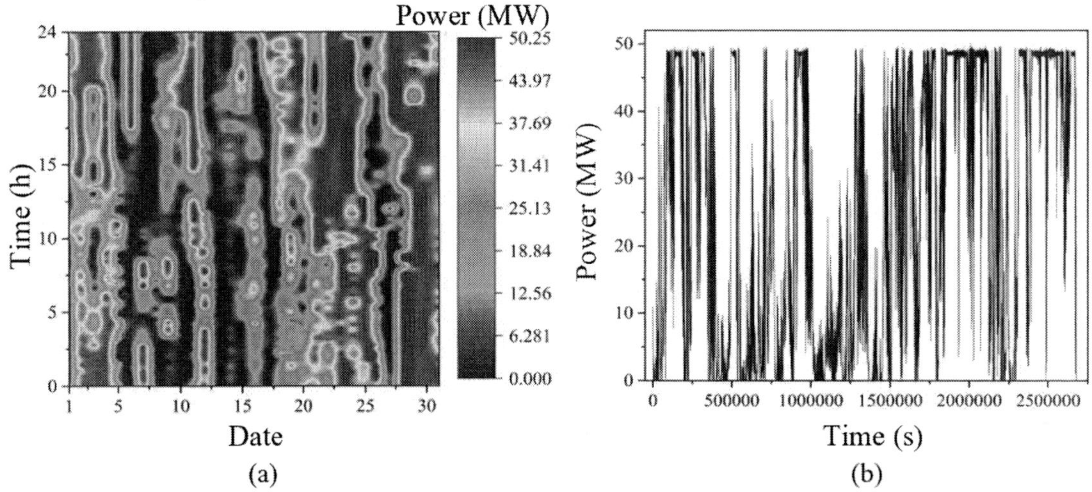

Figure 1. Wind power data of a wind farm in Qinghai: (a) Wind power data Heatmap. (b) Wind power output curve

2.2. Electrolyzer capacity configuration

In this study, 1000 Nm^3 alkaline electrolyzers (with a rated power of 5 MW) are used for hydrogen production. There exists a balance issue between efficiency and safety in electrolyzers. As the load increases, the efficiency of the electrolyzer gradually decreases, while the trend in hydrogen content within the oxygen shows an opposite pattern. The efficiency of the electrolyzer can be expressed by equation (1) of the National Standard 32311-2015, where W represents the unit energy consumption of the electrolyzer. To enhance the hydrogen production efficiency in this study, the electrolyzers operate predominantly at low loads. To ensure operational safety, it is necessary to appropriately raise the lower limit of the electrolyzers working load, setting the load range of the alkaline electrolyzer at 50% to 100%.

$$\eta = \frac{1.48 \times 2390}{1000W}\% \tag{1}$$

When considering the number of electrolyzers, the increase in the number of electrolyzers will increase the cost of equipment, but at the same time increase the renewable energy consumption rate and the amount of hydrogen production, so the balance problem needs to be considered. With reference to Figure 2, if 9 electrolyzers are used, the maximum wind power utilized by the electrolyzers is 90%, and the proportion of abandoned wind energy is 4.055%. From the perspective of renewable energy consumption, the wind power utilization rate should not be less than 95%. Therefore, at least 9 electrolyzers are required. The market price for 1000 Nm^3 alkaline electrolyzers is 7.55 million yuan per unit, with a lifetime of about 20 years. Assuming a wind power price of 0.332 yuan per kWh, if only 9 electrolyzers are used, the total value of wind energy abandoned due to excessive wind power output over the 20-year period would amount to 59.94813 million yuan, which far exceeds the cost of an electrolyzer. Therefore, 10 electrolyzers are employed to cover the maximum wind power output.

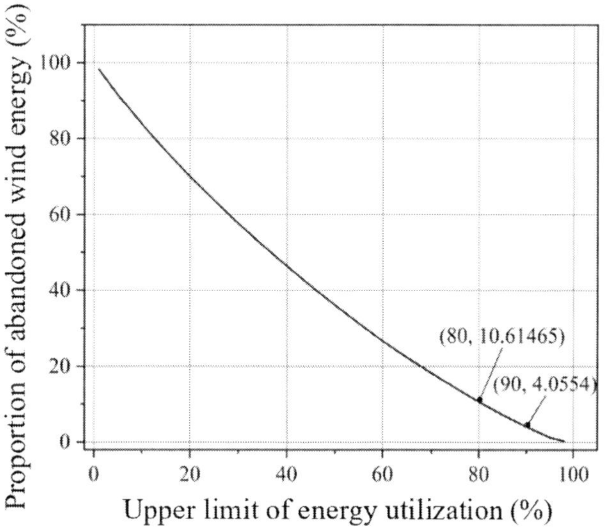

Figure 2. Curve of wind abandonment rate with upper limit of energy utilization

3. Objective function selection

The core components of the renewable energy-based hydrogen production system are the renewable energy power system and the hydrogen production system. In this study, the impact of the startup and shutdown processes of the electrolyzers is simplified, and it is assumed that the lifetime decay rate of the electrolyzers is basically the same across different power levels. With the economic benefit of the electrolysis hydrogen production system as the primary objective, and taking into comprehensive consideration the lifetime of the electrolyzers, a total revenue function is formulated, which can be expressed by equation (2).

$$E = \sum_{t=1}^{T} \left(E_{earn,t} - E_{cost,t} \right) \tag{2}$$

Where E represents the total revenue, T is the number of periods, $E_{earn,t}$ is the revenue of hydrogen production at time t, and $E_{cost,t}$ is the operating cost of electrolyzers at time t, both of them can be expressed by equations (3) and (4).

$$E_{earn,t} = \sum_{i=1}^{n_t} \Delta t \cdot \frac{P_i}{W_{di}} \cdot c \tag{3}$$

$$E_{cost,t} = \sum_{i=1}^{n_t} \Delta t \cdot (1+\alpha) \cdot C \tag{4}$$

Where i is the electrolyzer number, nt is the number of electrolyzers operating at time t, P_i is the power of the i electrolyzer, W_{di} is the unit hydrogen production energy consumption of the i electrolyzer, c is the selling price of hydrogen, C is the unit depreciation cost of the electrolyzers, α is the operational and maintenance coefficient of electrolyzers, and C can be derived from equation (5).

$$C = N \times \frac{Pr \times 10000}{365 \times 24 \times Li} \tag{5}$$

Where N is the total number of electrolyzers, Pr is the unit price of electrolyzers, and Li is the lifetime of the electrolyzers, and C is 43.09 yuan/hour after calculating.

In the standard state, the molar mass of hydrogen is 2g/mol, the charge required to electrolyze water to obtain 1mol of hydrogen is $2F$, and the charge required to make 1kg of hydrogen is Q_1. The calculate process is shown in equation (6).

$$Q_1 = \frac{2 \times F \times 1000}{3600 \times 2} \tag{6}$$

Where F is faraday constant, taken as 96485.33289C/mol, and Q_1 is in Ah/kg.

The unit energy consumption W_d of the electrolyzers is shown in equation (7).

$$W_d = \frac{Q_1 \times E}{1000 \times \eta} \tag{7}$$

Where E is the cell voltage of the electrolyzer, η is the efficiency of hydrogen production, and W_d is in kWh/kg.

The specific values of the symbols mentioned above are presented in Table 1.

Table 1. The value of each symbol

Symbol	Value	Unit
c	20.22	yuan/kg
N	10	/
Pr	7.55	million yuan
Li	20	year
C	43.09	yuan/hour
α	85	%
F	96485.33289	C/mol

4. Rule-based control strategies

4.1. Rated power strategy

The common control strategy for multiple electrolyzers is the rated power strategy, where wind power is prioritized to supply a portion of the electrolyzers, allowing them to operate at their rated power. If the remaining wind power is not sufficient to support another electrolyzer operating at its rated power, the decision to activate an additional electrolyzer to absorb the residual power depends on whether the remaining power meets the minimum power requirement of an electrolyzer. The rated power strategy has the advantages of low hydrogen content in the product oxygen, safe and reliable systems, and simple control logic. However, its disadvantages include lower electrolyzer efficiency at rated power and significant inconsistencies in electrolyzer lifetime.

4.2. Average power strategy

The electrolytic efficiency of an electrolyzer gradually decreases as its power increases. To enhance the economic viability of renewable energy-based hydrogen production system, it is useful to operate the electrolyzers at medium and low loads, while ensuring the consumption rate of renewable energy and the safe operation of the electrolyzers. Therefore, this study proposes a control strategy for electrolyzers based on average power.

The core idea of this strategy is to activate as many electrolyzers as possible and distribute the wind power evenly among them, allowing them to operate at lower loads. This approach aims to improve the efficiency of hydrogen production, increase hydrogen output, and enhance the overall system economy. Additionally, it helps to balance the working hours of each electrolyzer, preventing a few from being prematurely scrapped due to excessive use. The fundamental concept of this strategy can be represented by Equation (8), where P_l denotes the minimum power limit of an electrolyzer and n represents the number of electrolyzers that reach the minimum startup power. The specific strategy is illustrated in Figures 3.

$$\frac{P_w}{P_l} = n \cdots\cdots P_s \tag{8}$$

It should be noted that when employing the average power control strategy for electrolyzer operation, the working hours of each electrolyzer increase compared to the rated power strategy. Therefore, the average power strategy actually sacrifices a small amount of lifetime performance in exchange for higher hydrogen production efficiency and improved economy. However, alkaline electrolyzers possess certain advantages in terms of lifetime and durability, making the trade-off of a minor reduction in lifetime acceptable.

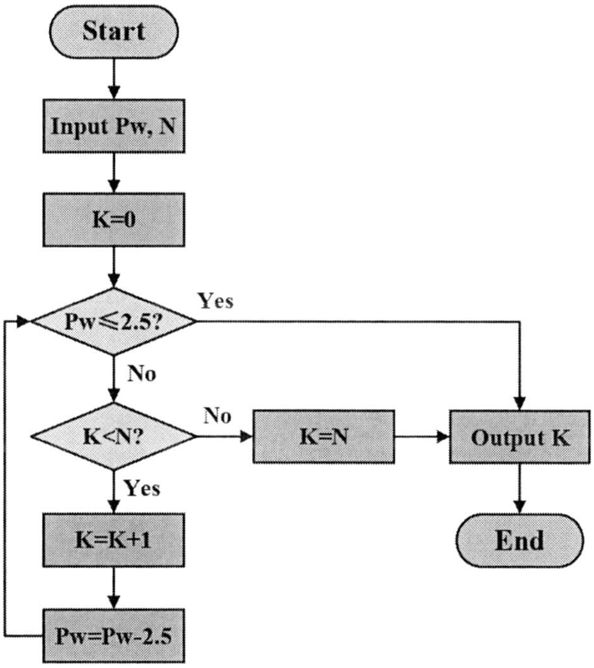

Figure 3. Flowchart of average power strategy

5. Example analysis

5.1. *Economic benefits and renewable energy consumption*

Using the wind power data from Figure 1 mentioned earlier as the power input, and taking the wind power of that month as a representative for all 12 months of the year, this study set the operation duration of the renewable energy-based hydrogen production system to 20 years. Simulations of the two control strategies are conducted using MATLAB software, and the results are presented in Table 2.

Table 2. Operating results of the two strategies

	Unit	Rated power strategy	Average power strategy	Improvement
Renewable energy consumption rate	%	98.28	99.53	1.25
Hydrogen production	ton	86783.13	89572.36	2789.23
Hydrogen revenue	million yuan	1754.7549	1811.1532	56.3983
Operating cost	million yuan	71.7457	87.6663	15.9206
Total revenue	million yuan	1683.0092	1723.4869	40.4777

As can be seen from the table, when the average power strategy is adopted, the renewable energy consumption rate increases from 98.28% under the rated power strategy to 99.53%, representing an improvement of 1.25%. The hydrogen production increases from 86,783.13 tons to 89,572.36 tons, with an additional 2,789.23 tons of hydrogen produced, marking an increase of 3.21%. The total revenue rises from 1,683.0092 million yuan to 1,723.4869 million yuan, an increase of 40.4777 million yuan, or 2.41%. Based on the above analysis, the average power strategy outperforms the rated power strategy in terms of both economic benefits and renewable energy consumption rate.

5.2. *Electrolyzer lifetime consistency*

Apart from economic benefits, the electrolyzer lifetime consistency is also an important factor that needs to be considered in renewable energy-based hydrogen production system. The lifetime of each electrolyzer is calculated using MATLAB software and normalized. Table 3 and Figure 4 present the working hours of electrolyzers under two different control strategies. As can be seen from Figure 4, when the average power strategy is adopted, the difference in electrolyzer working time is smaller, and the average working time is longer.

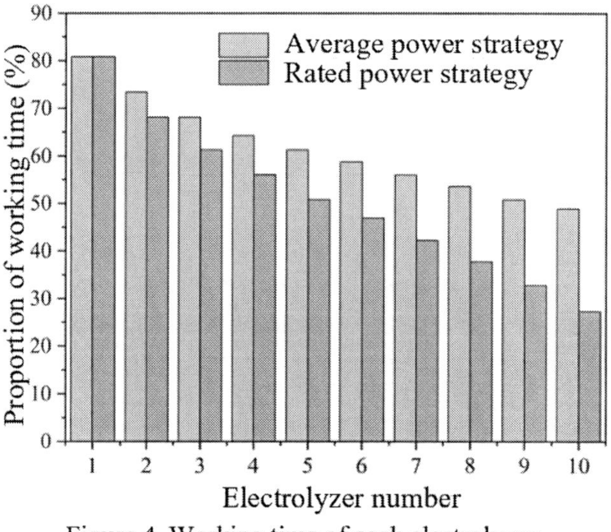

Figure 4. Working time of each electrolyzer

Table 3. Comparison of electrolyzer working time under two strategies

	Average power strategy	Rated power strategy
Average lifetime/%	61.58853	50.40379
Maximum difference/%	31.96286	53.52637
Standard deviation/%	9.704675	15.72952

Detailed data are presented in Table 3. When the rated power strategy is adopted, the maximum difference in working time between the longest-running electrolyzer and the shortest-running electrolyzer is 53.52637%. The standard deviation of the working times of all electrolyzers is 15.72952. In contrast, when the average power strategy is used, the maximum difference in working time between the longest-running electrolyzer and the shortest-running electrolyzer decreases to 31.96286%. The standard deviation of the working times of all electrolyzers also decreases to 9.704675, which is only 55.88% of that under the rated power strategy. It can be seen that the average power strategy is also superior to the rated power strategy in terms of the consistency of electrolyzer lifetime.

6. Conclusion

For the multiple electrolyzers hydrogen production subsystem in renewable energy-based hydrogen production system, this study proposes an energy control strategy based on average power. Initially, actual wind turbine data from a wind farm in Qinghai is analysed to establish a 50MW virtual wind farm. Wind-hydrogen capacity matching is conducted, and 10 electrolyzers are chosen to cover the maximum output power of the wind farm. Subsequently, an analysis of hydrogen production costs is performed, establishing a total revenue function that includes hydrogen production revenue and equipment usage costs. Finally, both the average power strategy and the rated power strategy are introduced, and MATLAB software is utilized to model and simulate the two strategies. The results indicate that the average power strategy outperforms the rated power strategy in terms of economic benefits, renewable energy consumption rate, and consistency of electrolyzer lifetime. Therefore, it is a promising energy control strategy for electrolyzers.

Acknowledgments

This study is supported by National Key R&D Program of China (2022YFB4002303), Qinghai Provincial Major Science and Technology Special Project (2023-GX-A2) and Shanghai Science and Technology Innovation and Science Popularization Project Funding Program (22120240468).

References

[1] Balu V, Krishnaveni K, Malla P, et al. Improving the power quality and hydrogen production from renewable energy sources based microgrid. Engineering Research Express, 2023, 5(3): 035037.

[2] Song C, Zhang B, Cai W, et al. Techno-economic analysis of hydrogen production systems based on solar/wind hybrid renewable energy. IOP Conference Series: The 6th International Symposium on Hydrogen Energy and Energy Technologies (HEET 2023), 2024, 2723(1): 012002.

[3] Guo X, Kang P, Zhang J, et al. Coordinated Control Strategy for System of Hydrogen Production from Renewable Energy Considering Regulating Characteristics of Electrolyzers. IOP Conference Series: 2023 2nd International Conference on Power System and Energy Technology, 2023, 2659: 012029.

[4] Zhou J, Meng X and Chen Y. Research on DC Power Supply for Electrolytic Water to Hydrogen Based on Renewable Energy. IOP Conference Series: 2022 2nd International Conference on Intelligent Power and Systems (ICIPS 2022), 2023, 2465: 012007.

[5] Deng W, Pei W, Zhuang Y, et al. Integration Control of Renewable Energy/Hydrogen Energy System Based on Flexible DC Interconnection. IOP Conference Series: 2021 International Conference on Advanced Electrical and Energy Systems, 2021, 2022(1).

[6] Nasser M, Megahed TF, Ookawara S, et al. A review of water electrolysis-based systems for hydrogen production using hybrid/solar/wind energy systems. Environmental Science and Pollution Research, 2022, 29(58): 86994-87018.

[7] Li Y, Deng X, Zhang T, et al. Exploration of the configuration and operation rule of the multi-electrolyzers hybrid system of large-scale alkaline water hydrogen production system. Applied Energy, 2023, 331: 120413.

[8] Niu M, Hong Z, li B, et al. Optimal control strategy of wind power to hydrogen system considering electrolyzer efficiency improvement. Acta Energiae Solaris Sinica, 2023, 44(09): 366-376(in Chinese).

Two-layer Capacity Planning Model for Integrated Energy Systems Based on RSOC

Yuan Gao[1], Wei Zhou[2], Wei Yang[1*], Mengxiao Wang[2], Weizhi Sun[2]

[1] Southwest Petroleum University, School of Electrical Information, Chengdu, Sichuan Province, 610000, PR China

[2] CNOOC Ltd Research Institute, Beijing, 100028, PR China

*Corresponding author's e-mail: 2356754496@qq.com

Abstract. In order to solve the problem of energy absorption in wind and solar power generation, a comprehensive energy system capacity planning method based on reversible solid oxide fuel cell (RSOC) is proposed in this paper. Firstly, the integrated energy station architecture with RSOC as the core equipment is proposed, and the RSOC output model and degradation model are established. Secondly, a two-layer capacity optimization configuration model is established. The upper layer takes the maximum total profit in the life cycle as the objective function, and the lower layer takes the minimum daily operating expenditure as the objective function, which is solved by particle swarm optimization algorithm, and the upper and lower layers are iterated to meet the convergence condition. Finally, the effectiveness and rationality of the proposed capacity optimization allocation method are verified by an example analysis, and the economic analysis of the two allocation methods is carried out.

1. Introduction

In recent years, China has vigorously developed clean energy power generation such as photovoltaic and wind power, with power generation ranking first in world[1]. However, the uncertainty on the load side has exacerbated the mismatch between supply and demand[2], which may lead to the abandonment of light and other phenomena[3]. In to the long-term demand for clean energy consumption, China has included hydrogen energy storage in the 14th Five-Year Plan[4], and the method of producing green with clean energy has great potential in hydrogen supply[5]. By reasonably configuring the scale of equipment and the capacity of energy storage devices[6], the safety, reliability, and stability the system can be maximized. Meanwhile, the factor of power output decay of RSOC equipment over long-term operation often leads to overly ideal planning schemes[7]. Therefore it is necessary to consider equipment degradation in capacity configuration research.

2. System architecture and RSOC model

2.1. System architecture

The RSOC operates in two modes, solid oxide electrolysis cell (SOEC) and solid oxide fuel cell (SOFC), these two modes can be flexibly switched during operation. When there is excess power generation, the RSOC can operate in SOFC mode to enhance the system's consumption capability by converting excess power to hydrogen through water electrolysis; when there is insufficient power generation[8], the RSOC can operate in SOEC mode, using clean such as hydrogen as fuel to produce power to meet the power load demand. Given the strong coupling capability of the RSOC, it can be combined with cog units to meet the system's demands for power, heat, and hydrogen, thereby improving the energy consumption capability. Considering the above operating characteristics of the RSOC this paper constructs an IES-RSOC architecture as shown in Figure 1. In this system, the energy sources are wind and solar power, hydrogen, natural, and purchased power. The system has three types of loads: power, heat, and hydrogen. The main equipment in the system includes RSOC, wind generator(WG), photovoltaic (PV), cogeneration unit (CHP), gas boiler (GB), methanation device (MET), and hydrogen storage device (HES).

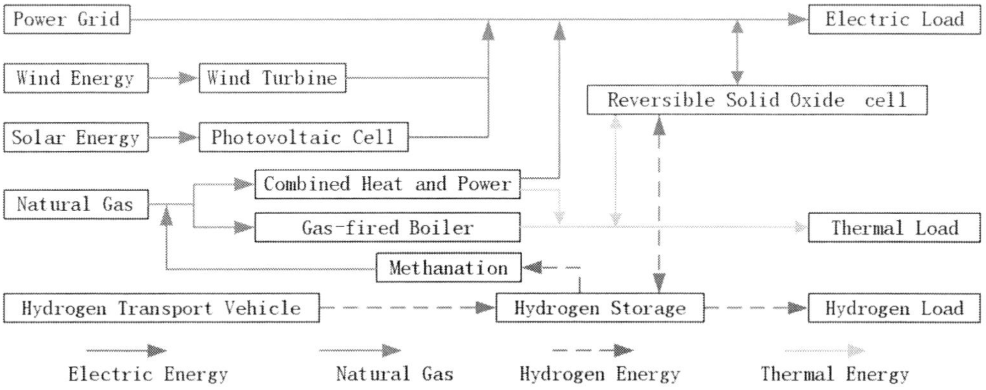

Figure 1. IES-RSOC structural diagram

2.2. *RSOC equipment model*

2.2.1. RSOC output model. RSOC in SOEC mode consumes electrical power and thermal power, as shown in the following formula[9].

$$\begin{cases} P_{\text{SOEC,th}}^{t} = P_{\text{SOEC,pe}}^{t} \eta_{\text{SOEC,th}}^{t} \\ P_{\text{SOEC,pe}}^{t} + P_{\text{SOEC,th}}^{t} = \text{L}_{\text{hv}} m_{\text{SOEC,H}_2}^{t} / \eta_{\text{SOEC,el}}^{t} \end{cases} \tag{1}$$

In the formula, $P_{\text{SOEC,th}}^{t}$, $P_{\text{SOEC,pe}}^{t}$, $m_{\text{SOEC,H}_2}^{t}$ are the heat consumption power, power consumption, and hydrogen production rate of the SOEC at time *t*, respectively. $\eta_{\text{SOEC,th}}^{t}$, $\eta_{\text{SOEC,el}}^{t}$ are the heat consumption coefficient of SOEC and electrolysis efficiency, respectively. L_{hv} is the higher heating value of hydrogen. is the higher heating value of hydrogen.

RSOC operating in SOFC mode will produce electrical power and thermal power, as shown in the following equation.

$$\begin{cases} P_{\text{SOFC,th}}^{t} = \text{L}_{\text{hv}} m_{\text{SOFC,H}_2}^{t} \eta_{\text{SOFC,th}}^{t} \\ P_{\text{SOFC,pe}}^{t} = \text{L}_{\text{hv}} m_{\text{SOFC,H}_2}^{t} \eta_{\text{SOFC,pe}}^{t} \end{cases} \tag{2}$$

In the formula, $P_{\text{SOFC,th}}^{t}$, $P_{\text{SOFC,pe}}^{t}$, $m_{\text{SOFC,H}_2}^{t}$ are the thermal power output, electrical power output, and hydrogen consumption rate of the SOFC at time *t*, respectively. $\eta_{\text{SOFC,th}}^{t}$, $\eta_{\text{SOFC,pe}}^{t}$ are the thermal efficiency and power generation efficiency of the SOFC, respectively.

2.2.2. RSOC Degradation Model. The operation efficiency of RSOC is highly related to the working hours. The IES planning cycle is long, and it does not consider the output factor of the equipment running for a long time, which often leads to overly ideal planning schemes. The main factors affecting the lifespan of RSOC are battery degradation caused oxidation and reduction, microstructure damage, and thermodynamic gradients.

$$\begin{cases} n_{\text{RSOC}} = (1-\lambda_1-\lambda_2-\lambda_3)n_{\text{RSOC,th}} \\ \lambda_1 = \mu_1 \\ \lambda_2 = \mu_2 T_{\text{RSOC}} \\ \lambda_3 = \begin{cases} \sigma_1 T_{\text{SOEC}}, T_{\text{SOEC}} \geq T_0 \\ \sigma_1 T_{\text{SOEC}} - \sigma_2 T_{\text{SFOC}}, T_{\text{SEOC}} < T_0 \end{cases} \end{cases} \tag{3}$$

In the formula, n_{RSOC}, $n_{\text{RSOC,th}}$ are the actual lifetime and ideal lifetime of the RSOC, respectively. λ_1, λ_2, λ_3 are the degradation factors caused by thermodynamics, redox, and microstructure of the RSOC, respectively. μ_1, μ_2 is a constant. T_{RSOC}, T_0 are the working hours of the RSOC and the upper limit of the reversible time of microstructure degradation, respectively. σ_1, σ_2 are the attenuation constants, respectively.

3. Two-layer capacity planning model

A bilevel capacity optimization planning model for the integrated energy system of RSOC is established, where the upper-level model aims to maximize the total profit the planning period of the system, and the lower-level model aims to minimize the daily operational expenses.

3.1. Upper layer model

The upper level model takes the maximum total profit as the objective function in the planning cycle, in which the total profit of the system is composed of three parts: operating income, operating expenditure and investment cost, and the relationship is shown as follows.

$$F_{\text{up}}^{\text{price}} = \max[\sum_{y=1}^{Y}\sum_{d=1}^{365}\frac{(F_{\text{in}}^{\text{price}} - F_{\text{out}}^{\text{price}})}{(1+v)^Y} - F_{\text{inv}}^{\text{price}}] \tag{4}$$

In the formula, $F_{\text{up}}^{\text{price}}$ is the total profit for the planning period. $F_{\text{in}}^{\text{price}}$, $F_{\text{out}}^{\text{price}}$, $F_{\text{inv}}^{\text{price}}$ are the system's daily operating income, daily operating expenses, and investment costs, respectively. Y is the system planning horizon, v is the discount rate.

The daily operational income of the system mainly comes from the revenue of electricity sales, heat sales, hydrogen sales, and oxygen sales. The sources of sales are the electricity generated by PV, WG, CHP, and SOFC. The sources of heat sales are the heat produced by CHP, GB, SOFC. The sources of hydrogen and oxygen sales are the hydrogen and oxygen produced by SOEC electrolysis. The relationship of the daily operational income is specifically shown the following equation.

$$F_{\text{in}}^{\text{price}} = \sum_{t=1}^{24}(\xi_{\text{hy,sell}}^{\text{price}}P_{\text{1,hy}}^t + \xi_{\text{ox,sell}}^{\text{price}}P_{\text{1,ox}}^t + \xi_{\text{el,sell}}^{\text{price}}P_{\text{1,el}}^t + \xi_{\text{hot,sell}}^{\text{price}}P_{\text{1,hot}}^t) \tag{5}$$

In the formula, $\xi_{\text{hy,sell}}^{\text{price}}$, $\xi_{\text{ox,sell}}^{\text{price}}$, $\xi_{\text{el,sell}}^{\text{price}}$, $\xi_{\text{hot,sell}}^{\text{price}}$ are the hydrogen selling price, oxygen selling price, electricity selling price, and heat selling price of the system at time *t*, respectively. $P_{\text{1,hy}}^t$, $P_{\text{1,ox}}^t$, $P_{\text{1,el}}^t$, $P_{\text{1,hot}}^t$ are the hydrogen load, oxygen sales, electrical load, and thermal load at time t, respectively.

The system investment cost mainly includes the initial purchase of equipment, equipment replacement, and equipment residual value. Due to the different service life of each, some equipment needs to be replaced within the life cycle, which requires certain equipment replacement costs. After the life cycle ends, there is still a certain residual value equipment recovery. The relationship between investment costs is shown in the following equation.

$$\begin{cases} F_{\text{inv}}^{\text{price}} = (1-\eta_{\text{eq,re}})F_{\text{inv}} \\ F_{\text{inv}} = \sum_{h}\sum_{r=1}^{r_h}\xi_{h,\text{inv}}P_h \\ r_h = \text{roundup}(Y/Y_h) \end{cases} \tag{6}$$

In the formula, $h \in \{\mathrm{RSOC, PV, WG, MET, CHP, GB, HES}\}$. $\eta_{\mathrm{eq,re}}$ is the residual value of the equipment. r_h is the number of times device h is replaced in the planning cycle. $\xi_{h,\mathrm{buy}}$ is the replacement unit price for device h. P_h is the design capacity of device h. $\mathrm{roundup}(\cdot)$ is the ceiling.

The upper constraint includes the operation constraint of wind power, photovoltaic, RSOC, gas boiler, cogeneration and other equipment.

3.2. *Lower layer model*

The lower-level model takes the minimization of the system's daily operating costs as the objective function, where the system's daily operating costs are composed energy purchase costs, operation and maintenance costs, penalty costs, and environmental costs. The relationship is specifically shown in the following equation.

$$F_{\mathrm{out}}^{\mathrm{price}} = \min(F_{\mathrm{buy}}^{\mathrm{price}} + F_{\mathrm{wel}}^{\mathrm{price}} + F_{\mathrm{pun}}^{\mathrm{price}} + F_{\mathrm{env}}^{\mathrm{price}}) \tag{7}$$

In the formula, $F_{\mathrm{buy}}^{\mathrm{price}}$, $F_{\mathrm{wel}}^{\mathrm{price}}$, $F_{\mathrm{pun}}^{\mathrm{price}}$, $F_{\mathrm{env}}^{\mathrm{price}}$ are the energy purchase cost, operation and maintenance cost, penalty cost, and environmental cost of the system, respectively.

The energy purchase cost of the system is the sum of the cost of hydrogen purchase, natural gas purchase, electricity purchase from the grid, and water for electrolysis in SOEC mode, as shown in the following equation.

$$\begin{aligned} F_{\mathrm{buy}}^{\mathrm{price}} = \sum_{t=1}^{24} (&\xi_{\mathrm{el,buy}}^{\mathrm{price}} P_{\mathrm{el,buy}}^{t} + \xi_{\mathrm{water,buy}}^{\mathrm{price}} \sigma_{\mathrm{water}} P_{\mathrm{SOEC,pe}}^{t} \\ &+ \xi_{\mathrm{hy,buy}}^{\mathrm{price}} P_{\mathrm{hy,buy}}^{t} + \xi_{\mathrm{gas,buy}}^{\mathrm{price}} P_{\mathrm{gas,buy}}^{t} + \xi_{\mathrm{co_2,buy}}^{\mathrm{price}} \sigma_{\mathrm{co_2}} P_{\mathrm{MET}}^{t}) \end{aligned} \tag{8}$$

In the formula, $\xi_{\mathrm{el,buy}}^{\mathrm{price}}$, $\xi_{\mathrm{water,buy}}^{\mathrm{price}}$, $\xi_{\mathrm{hy,buy}}^{\mathrm{price}}$, $\xi_{\mathrm{gas,buy}}^{\mathrm{price}}$ are the electricity purchase price, water purchase price, hydrogen purchase price, and natural gas purchase price of the system, respectively. σ_{water} is the water consumption coefficient for electrolysis of water in SOEC mode.

The cost of the system is the maintenance cost of the RSOC, PV, CHP, GB, WG, MET, and HES devices, which is specifically expressed as follows.

$$\begin{aligned} F_{\mathrm{wel}}^{\mathrm{price}} = \sum_{t=1}^{24} (&\xi_{\mathrm{RSOC,wel}}^{\mathrm{price}} P_{\mathrm{RSOC}}^{t} + \xi_{\mathrm{PV,wel}}^{\mathrm{price}} P_{\mathrm{PV}}^{t} + \xi_{\mathrm{WG,wel}}^{\mathrm{price}} P_{\mathrm{WG}}^{t} + \xi_{\mathrm{MET,wel}}^{\mathrm{price}} P_{\mathrm{MET}}^{t} \\ &+ \xi_{\mathrm{GB,wel}}^{\mathrm{price}} P_{\mathrm{GB}}^{t} + \xi_{\mathrm{CHP,wel}}^{\mathrm{price}} P_{\mathrm{CHP}}^{t}) + \sum_{t=1}^{24} \max[\xi_{\mathrm{HES,wel}}^{\mathrm{price}} (S_{\mathrm{HES}}^{t} - S_{\mathrm{HES}}^{t-1})] \end{aligned} \tag{9}$$

In the formula, $\xi_{\mathrm{RSOC,wel}}^{\mathrm{price}}$, $\xi_{\mathrm{PV,wel}}^{\mathrm{price}}$, $\xi_{\mathrm{WG,wel}}^{\mathrm{price}}$, $\xi_{\mathrm{GB,wel}}^{\mathrm{price}}$, $\xi_{\mathrm{MET,wel}}^{\mathrm{price}}$, $\xi_{\mathrm{CHP,wel}}^{\mathrm{price}}$, $\xi_{z,\mathrm{wel}}^{\mathrm{price}}$ are the operation and maintenance costs of the RSOC, PV, WG, GB, MET, CHP, and HES energy storage devices.

The penalty cost of the system is composed of the penalty cost of cutting off power, hydrogen, and heat loads, and the penalty cost of abandoning and light. The relationship is specifically shown in the following equation.

$$F_{\mathrm{cut}}^{\mathrm{price}} = \sum_{t=1}^{24} (\xi_{\mathrm{PV,loss}}^{\mathrm{price}} P_{\mathrm{PV,loss}}^{t} + \xi_{\mathrm{WG,loss}}^{\mathrm{price}} P_{\mathrm{WG,loss}}^{t} + \xi_{\mathrm{el,qe}}^{\mathrm{price}} P_{\mathrm{el,qe}}^{t} + \xi_{\mathrm{hy,qe}}^{\mathrm{price}} P_{\mathrm{hy,qe}}^{t} + \xi_{\mathrm{hot,qe}}^{\mathrm{price}} P_{\mathrm{hot,qe}}^{t}) \tag{10}$$

In the formula, $\xi_{\mathrm{PV,loss}}^{\mathrm{price}}$, $\xi_{\mathrm{WG,loss}}^{\mathrm{price}}$, $\xi_{\mathrm{el,qe}}^{\mathrm{price}}$, $\xi_{\mathrm{hy,qe}}^{\mathrm{price}}$, $\xi_{\mathrm{hot,qe}}^{\mathrm{price}}$ are the penalty cost coefficients for the system's curtailed solar power, curtailed wind power, electricity load shedding, hydrogen load shedding and heat load shedding, respectively. $P_{\mathrm{PV,loss}}^{t}$, $P_{\mathrm{WG,loss}}^{t}$, $P_{\mathrm{el,qe}}^{t}$, $P_{\mathrm{hy,qe}}^{t}$, $P_{\mathrm{hot,qe}}^{t}$ are the curtailed solar power, curtailed wind power, shed electric power, shed hydrogen power, and shed thermal power of the at time *t*, respectively.

The lower level constraints are mainly the electrical power balance constraints, thermal power balance constraints, hydrogen power balance constraints, natural gas power balance constraints, and system energy purchase constraints.

4. Model solving

The Particle Swarm Optimization (PSO) algorithm was proposed in 1955. Its fundamental principle is similar to the foraging behavior of bird flocks in nature. The PSO algorithm can gradually evolve the solution to a problem by fully utilizing group information. During the solution process, particles within the group can use collaboration and information sharing to find the optimal solution to the problem. Compared with the genetic algorithm, the PSO algorithm requires fewer parameters to be set, which can save some potential costs for problem-solving. In the PSO algorithm, the position update of individual particles only relies on the particle at the optimal position in the population, and each particle searches independently without sharing information with others. This makes the algorithm search under the guidance of the optimal solution of the current iteration throughout the process, which easily leads to the loss of diversity in the population and getting stuck in local optima. To obtain the best capacity planning scheme, this paper constructs a two-layer capacity planning model of IES-RSOC and uses the PSO algorithm to solve it. The upper-level model transmits the capacity planning scheme to the lower-level model, and the lower-level model solves the operating profit based on the capacity provided by the upper-level model and returns the profit value to the upper-level model.

5. Example setting and analysis

5.1. Example setting

The meteorological data was extracted based on the commercial software for the new energy industry and was divided into four typical days using the-means clustering algorithm. The reference power of the photovoltaic power generation is 3500kW, the reference power of the wind power is2500kW, the reference power of the electric load is 4000kW, the reference power of the thermal load is 000kW, and the reference power of the hydrogen load is 1500kW. The curves of clean energy and load are shown Figure 2. Two schemes were set up in this paper for comparison to verify the feasibility of the proposed model and method, as detailed below.

Scheme 1: Configure the integrated energy system of RSOC

Scheme 2: Integrated energy system with AWE and AFC configuration

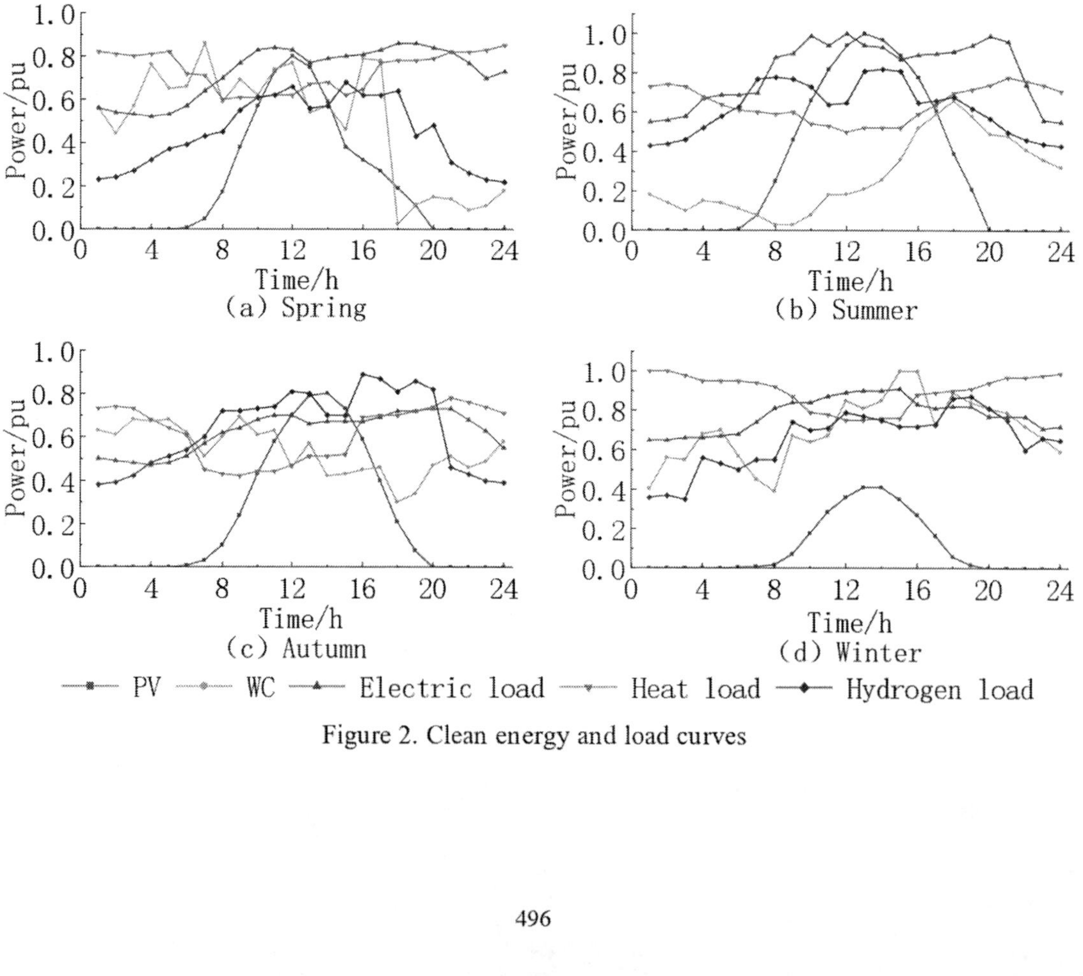

Figure 2. Clean energy and load curves

5.2. Example analysis

The economic benefits of the two schemes are shown in Table 1. It can be seen from the table that compared to Scheme , Scheme 1 has an increase of 3.2% in the average annual net present value, an increase of 0.8% in investment costs an increase of 6.4% in operating costs, a reduction of 5.0% in energy purchase costs, similar operating income, a reduction of 57.8% in penalty costs, and a reduction of 18.4% in wind and light abandonment penalties. Compared to Scheme , except for a slight increase in operating costs and investment costs, all other costs are reduced in Scheme 1. This is because the operation of Scheme 1 equipment is more complex, with a larger capacity configuration of high-maintenance energy storage equipment, leading to higher system operating costs. Additionally, the higher price of RS in Scheme 1 leads to a slight increase in investment costs. However, the RSOC efficiency in Scheme 1 is higher, and all other costs of system are lower than Scheme 2. Moreover, the average annual net present value of Scheme 1 is higher, indicating that Scheme 1 has a stronger ability generate economic benefits. Therefore, the integrated energy system based on RSOC can efficiently and economically accommodate clean energy. With the development of reversible solid oxide fuel cell hydrogen storage technology, the investment cost of hydrogen storage systems will continue to decrease, and their operating life will be longer, bringing more significant benefits to integrated energy systems.

Table 1. Comparison of economic benefits of two schemes

Scheme	Revenue /Yuan	Investment Cost/Yuan	Run Cost/Yuan	Purchase Cost/Yuan	Operation Cost/Yuan	Punishment Cost/Yuan	Abandoned Cost/Yuan
1	1.234×10^7	6.918×10^6	6.640×10^5	1.910×10^6	2.194×10^7	8.258×10^4	3.120×10^4
2	1.196×10^7	6.874×10^6	6.248×10^5	2.057×10^6	2.191×10^7	1.957×10^5	3.844×10^4

6. Conclusion

This paper establishes a two-layer capacity optimization configuration model of the integrated energy system based on RSOC, and proposes a particle swarm algorithm to optimize capacity configuration of each device in the system. Meanwhile, in order to verify the feasibility and rationality of the proposed capacity optimization configuration method, two schemes of traditional-hydrogen integrated energy system and RSOC integrated energy system are set up for capacity optimization configuration respectively. The following conclusions are obtained:

(1)In the case of capacity allocation with the goal of annual net present value, the RSOC integrated energy system has a stronger ability to accommodate clean energy is more suitable for systems with high clean energy penetration compared to the traditional electricity-hydrogen integrated energy system.

(2)The RSOC integrated energy system is more economically effective than the mainstream electric hydrogen integrated energy system, and with the development of fuel cell technology the cost of the RSOC system will be further reduced, and the economic benefits brought by RSOC will be more prominent.

References

[1] Gao Jie, Song Jie, Wang Jianxiao, et al. The architecture and key technologies of power-hydro coupling system for supporting China's energy security[J]. Automation of Electric Power Systems, 2023, 47(19): 1-7.

[2] Zhang Xian, Wang Caixia, Xie Kai, et al. Key Issues in Building China's Green Power Market Under "Dual Carbon" Goals [J]. Automation of Electric Power Systems, 2024, Vol.48(4): 25-33.

[3] Wang Tianyu, Li Gensheng, Song Xianzhi, et al. Research on the Development of Smart and Gas Fields with "Wind, Light, Heat, and Storage" Multi-Energy Synergy [J]. Chinese Journal of Engineering Science, 2024 Vol. 26 (4): 259-270.

[4] Wang Yongli, Xiang Hao, Guo Lu, et al. Research on Planning and Optimization of Distributed PV and Electroydrogen Hybrid Energy Storage Facing Multi-energy Complementarity [J]. Power Grid Technology, 2024, Vol. 48 (2): 64-576.

[5] Du Yida, Wang Er, Tan Zhongfu, et al. Research on the Operation Optimization of Power-Hydrogen Cou System under the Coupling of Electricity-Carbon-Gas-Green Certificate Market [J]. Power Grid Technology, 2023, Vol. 478): 3121-3135.

[6] Molina, Pablo, Montero-Chacon, Francisco, et al. Technological evolution of electrolizers for hydrogen production[J].DYNA,2024,Vol.99(1): 10-11.

[7] Mehdi Talaie, Farkhondeh Jabari, et al. Modeling and economic dispatch of a novel fresh water, methane and electricity generation system considering green hydrogenation and carbon capture[J].Energy Conversion and Management,2024,Vol.302: 118087

[8] J.M. Vidueira, A. Contreras; T.N, et al. PV autonomous installation to produce hydrogen via electrolysis, and its use in FC buses[J].International journal of hydrogen energy,2003,Vol.28(9): 927-937.

[9] Hassan I A, Ramadan H S, Saleh M A, et al. Hydrogen storage technologies for stationary and mobile applications: Review, analysis and perspectives[J]. Renewable and Sustainable Energy Reviews, 2021, 149.

A Backbone Network Searching Method for the Receiving-End Power Grid under Large-Scale Wind Power Integration and High Proportion of External Power Feed-In

Zheng Li*, Yuqiao Jia, Dajiang Wang, Xinyao Zhu, Shengjun Wu, Wenbo Li

State Grid Jiangsu Electric Power Co.Ltd Research Institute, Nanjing, China

*Corresponding author's e-mail: 403168450@qq.com

Abstract. As the proportion of external power imports and local renewable energy continues to increase, the receiving-end power grid faces growing challenges in ensuring the reliable supply of critical loads under extreme weather conditions such as typhoons. External power imports are substantial, and renewable energy outputs exhibit high uncertainty. Combined with transmission line damage caused by extreme weather events, these factors lead to transmission congestion. Consequently, receiving-end power grids with high proportions of renewable energy and external power imports face not only power supply challenges on the generation side but also transmission constraints due to damaged lines. Based on coal-fired power and other guaranteed power sources within the receiving-end grid, and considering the distribution characteristics of critical loads, identifying the backbone network of the receiving-end grid is of great significance for ensuring reliable power supply under extreme weather conditions. This paper proposes a backbone network searching method for receiving-end grids. The method aims to minimize the network scale while considering load loss constraints at nodes, prioritizing the avoidance of high-failure-probability transmission channels, and incorporating cable requirements. The effectiveness of the proposed method is validated by comparing the differences in backbone networks and critical load reliability under various scenarios in the IEEE benchmark system.

1. Introduction

In recent years, the frequent occurrence of extreme disasters, such as typhoons, has posed severe challenges to the secure and reliable power supply of electricity systems[1]. China has an extensive coastline, and many economically developed cities are located in the southeastern coastal regions. These cities are highly vulnerable to typhoon damage, and power supply interruptions can result in significant economic losses. Typhoon-induced disasters, including strong winds and heavy rainfall, can cause transmission tower collapses and wind turbine shutdowns, impacting not only the power generation side but also leading to transmission congestion, which obstructs power delivery[2]-[3]. Additionally, China's energy resources are unevenly distributed, with large energy bases located in the western regions, while electricity demand is concentrated in coastal cities. Numerous ultra-high-voltage (UHV) AC and DC transmission lines extend from the west to the southeastern coastal load centers, forming multiple large-scale receiving-end power grids. These receiving-end grids are characterized by a hollowed-out power source structure and weak support capabilities. When combined with the transmission congestion caused by typhoon disasters and insufficient output from wind and solar renewable energy sources, the pressure on receiving-end grids to maintain reliable power supply during typhoon events becomes immense[4].

In the fields of backbone network identification and reinforcement, numerous researchers have conducted relevant studies. Reference [5] proposed a disaster-resistant backbone network searching algorithm that integrates economic efficiency, system recoverability, and network resilience as optimization objectives. Reference [6] introduced a backbone network searching method that focuses on identifying critical components and facilities. Reference [7] constructed a disaster-resistant backbone network optimization model using network topology connectivity and system operational security as primary constraints, solved using an improved crisscross optimization algorithm. Reference [8] developed a multi-objective optimization method for backbone networks, balancing economic efficiency with multi-stage disaster-resistance enhancement. The objectives included economic efficiency, survivability, resilience, and recoverability, with constraints on connectivity, retention of key nodes, and critical branches. Reference [9] proposed evaluation metrics for backbone network identification based on improved comprehensive electrical betweenness and other indices. Reference [10] presented a two-stage optimization strategy for disaster-resistant backbone networks, employing differential reinforcement to mitigate losses from extreme disasters such as typhoons. These studies indicate that current research has yet to address the specific challenges of backbone network searching methods for receiving-end power grids. Given the substantial proportion of external power feed-in, receiving-end grids exhibit significantly different system characteristics compared to self-balancing grids. When extreme disasters cause widespread failures, the consequences are particularly severe. Therefore, studying backbone network searching strategies for receiving-end grids and identifying their backbone network is critical for enhancing grid resilience and mitigating the impact of extreme disasters.

Building on existing research on backbone network searching, this paper considers the impact of large-scale external power feed-in and renewable energy on the backbone network of receiving-end power grids. It proposes a backbone network searching method for receiving-end grids under the context of large-scale wind power integration and high proportions of external power feed-in. The method comprehensively accounts for the characteristics of external power feed-in, as well as constraints such as backbone network connectivity and critical load reliability. The proposed approach is of significant importance for enhancing disaster prevention and mitigation capabilities of receiving-end power grids.

2. Backbone Network Searching Model

In order to enhance the power supply reliability of receiving-end power grids under extreme typhoon disasters, this paper combines the distribution of guaranteed power sources and critical loads in the receiving-end grid to identify the backbone network. It proposes differentiated protection standards for the backbone network to ensure uninterrupted power supply to critical loads after severe damage to the receiving-end grid. The proposed method effectively improves the critical load reliability during typhoon disasters while maintaining economic efficiency. It is of significant importance for achieving carbon peak and carbon neutrality goals in receiving-end power grids with high proportions of renewable energy and external power feed-in in the future.

The backbone grid is the minimum scale grid structure that ensures the power supply requirements for critical power users. The prerequisites that the backbone grid structure needs to meet: i) Ensure continuous power supply to critical users; ii) Meet the network topology connectivity constraints; iii) Minimize the number of branches used. The searching model for the backbone grid structure of the receiving-end grid is established with the objective of minimizing the number of branches in the backbone grid, subject to constraints such as secure operation, topology connectivity, and ensuring power supply to critical users.

Objective function

$$min\ N_B \tag{1}$$

Constraints are as follows:

a) Node power balance constraint:

$$P_{i,t} + P_{i,t}^{R} - \sum_{l_i} f_{l_i,t} + P_{Tie\text{-}line,t} = \sum_{\xi=1}^{3} D_{i,t}^{\xi} \quad \forall i,\ \forall t \tag{2}$$

b) DC power flow constraint on the line:

$$f_{l,t}=b_l(\theta_{i,t}-\theta_{j,t}) \quad \forall l,\ \forall t \tag{3}$$

c) Line capacity constraint:

$$-z_l\overline{f_l}\le f_{l,t}\le z_l\overline{f_l} \quad \forall l,\ \forall t \tag{4}$$

d) Power generation output constraint:

$$u_{i,t}\underline{P_i}\le P_{i,t}\le u_{i,t}\overline{P_i} \quad \forall i,\ \forall t \tag{5}$$

e) Load supply reliability constraint:

$$\lambda\sum_t\overline{D_{i,t}^{\xi}}\le\sum_t D_{i,t}^{\xi}\le\sum_t\overline{D_{i,t}^{\xi}} \quad \forall i,\forall\xi \tag{6}$$

f) Power ramping constraint:

$$-\overline{R_i}\le P_{i,t}-P_{i,t-1}\le\overline{R_i} \quad \forall i,\ \forall t \tag{7}$$

g) Power generation minimum startup and shutdown time constraint:

$$u_{i,t}-u_{i,t-1}-u_{i,\varsigma}\le 0 \quad \forall\varsigma\in\{t,\ldots,\overline{T}+t-1\}\ \forall i,\ \forall t \tag{8}$$

$$u_{i,t-1}-u_{i,t}+u_{i,\varsigma}\le 1 \quad \forall\varsigma\in\{t,\ldots,\underline{T}+t-1\}\ \forall i,\ \forall t \tag{9}$$

h) Sequential power received from other grid:

$$P_{Tie-line,t}=\begin{cases}P_1 & t\in T_1\\ \vdots & \\ P_n & t\in T_n\end{cases} \tag{10}$$

i) Renewable energy generation output range constraint:

$$0\le P_{i,t}^{R}\le\overline{P_{i,t}^{R}} \quad \forall i,\ \forall t \tag{11}$$

j) Line operation status statistical constraint:

$$\sum_l z_l\le N_B \tag{12}$$

Where N_B represents the number of backbone grid lines; $P_{i,t}$ represents the power generation output at node i at time t; $P_{i,t}^{R}$ refers to the output of renewable energy power stations; $f_{l,t}$ represents the line power flow; $P_{Tie-line,t}$ represents the power receiving from other grid, specifically the tie line referring to the high voltage direct current (HVDC). In this article, we consider the temporal variation of the power input from outside the area, that is, the power magnitude varies in different time periods; $D_{i,t}^{\xi}$ represents the load sizes for the first, second, and third levels of load after load shedding, where ξ=1, 2, 3; $\overline{D_{i,t}^{\xi}}$ represents the load sizes before shedding for the first three levels of load; λ is the load guarantee rate; B_l is the line reactance; $\theta_{i,t}$ and $\theta_{j,t}$ are the voltage phase angles at both ends of the line; z_l is the line status as a 0-1 variable, where 1 indicates the line is operating and 0 indicates the line is disconnected; $\overline{f_l}$ represents the line capacity; $\overline{P_i}$ and $\underline{P_i}$ represent the maximum and minimum power outputs of the power sources; $\overline{R_i}$ is the maximum ramping power of the generator; $u_{i,t}$ is the generator status as a 0-1 variable, where 1 indicates the generator is operating and 0 indicates the generator is shut down; $\overline{T}$ and $\underline{T}$ are the minimum startup and shutdown times of the generator, respectively; $\overline{P_{i,t}^{R}}$ is the predicted output of renewable energy power stations; Δt is the time step size, set at 1 h.

3. Case Study Analysis

The load in the original system is increased by 5 times, and the power generation capacity is increased by 4 times; the line capacities are as shown in Table. 1. A DC line with a total capacity of 400 MW is connected at busbars 2 and 6. A 75 MW wind farm is connected at busbar 5. The improved IEEE 14-bus system schematic is shown in Figure 1.

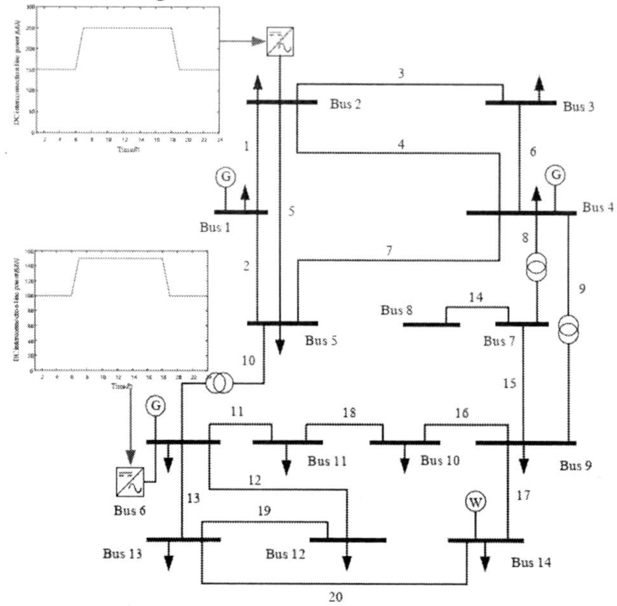

Figure. 1. Schematic of the Improved IEEE 14-Bus System

Table. 1. Line capacity parameters

LINE ID	CAPACITY (MW)	LINE ID	CAPACITY (MW)	LINE ID	CAPACITY (MW)
1	300	2/3/4/7	250	others	150

Under the parameters of the above case study, the backbone network is represented by the red lines in Figure 2. The resulting backbone network includes 14 lines, such as lines 1-5, 2-3, and others. This backbone network can meet the power supply requirements for all primary loads, with supply reliability rates for secondary and tertiary loads reaching 64.03% and 45.90%, respectively.

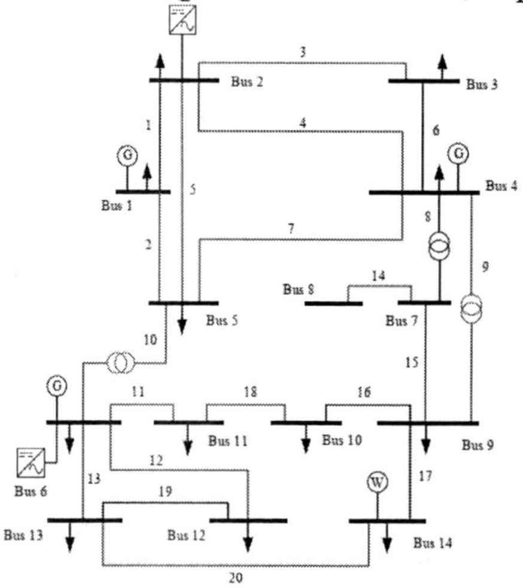

Figure. 2. Backbone Network of the Improved IEEE 14-Bus System

A. Backbone Network under Different External Power Feed-in Scales

External power feed-in is a special type of power supply for receiving-end grids, and its scale has a significant impact on the configuration of the backbone network. In this section, we quantitatively analyze the influence of external power feed-in scale on the backbone network of receiving-end power grids. The results of the backbone network lines under different scenarios are shown in Table. 2, and the supply reliability rates for secondary and tertiary loads are shown in Figure 3 and 4. It should be noted that, due to the constraint of a 100% reliability requirement for primary load supply, the primary load supply reliability in all scenarios meets the requirement and will not be listed separately.

Scenario 1: External Power Feed-in 280 MW
Scenario 2: External Power Feed-in 320 MW
Scenario 3: External Power Feed-in 360 MW
Scenario 4: Base Scenario, External Power Feed-in 400 MW
Scenario 5: External Power Feed-in 440 MW
Scenario 6: External Power Feed-in 480 MW

Table. 2. Backbone network of the system under different external power feed-in scales

Scenario 1		Scenario 2		Scenario 3		Scenario 5		Scenario 6	
Line ID	Line	Line ID	Line	Line ID	Line	Line ID	Line	Line ID	Line
2	1-5	2	1-5	2	1-5	2	1-5	**1**	**1-2**
3	2-3	3	2-3	3	2-3	3	2-3	2	1-5
4	2-4	4	2-4	4	2-4	4	2-4	3	2-3
5	2-5	5	2-5	5	2-5	5	2-5	4	2-4
7	4-5	7	4-5	7	4-5	**6**	**3-4**	5	2-5
8	**4-7**	**8**	**4-7**	**9**	**4-9**	7	4-5	**6**	**3-4**
10	5-6	10	5-6	10	5-6	**9**	**4-9**	7	4-5
11	6-11	11	6-11	11	6-11	10	5-6	**9**	**4-9**
12	6-12	12	6-12	12	6-12	11	6-11	10	5-6
13	6-13	13	6-13	13	6-13	12	6-12	11	6-11
14	7-8	14	7-8	14	7-8	13	6-13	12	6-12
15	7-9	15	7-9	15	7-9	14	7-8	13	6-13
18	10-11	18	10-11	18	10-11	15	7-9	14	7-8
20	13-14	20	13-14	20	13-14	18	10-11	15	7-9
-						20	13-14	18	10-11
-								20	13-14

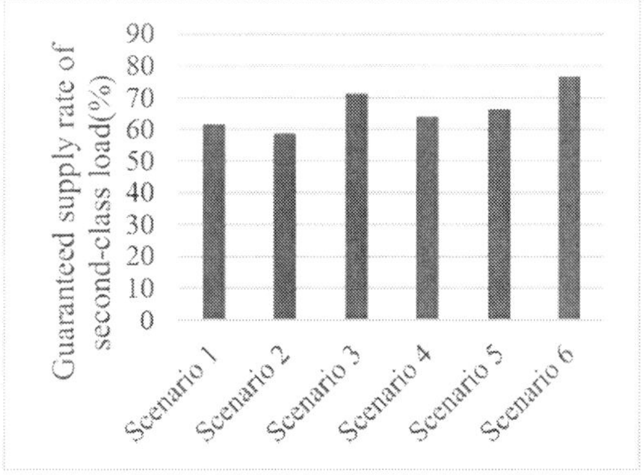

Figure. 3 Secodary Load Supply Reliability Rate

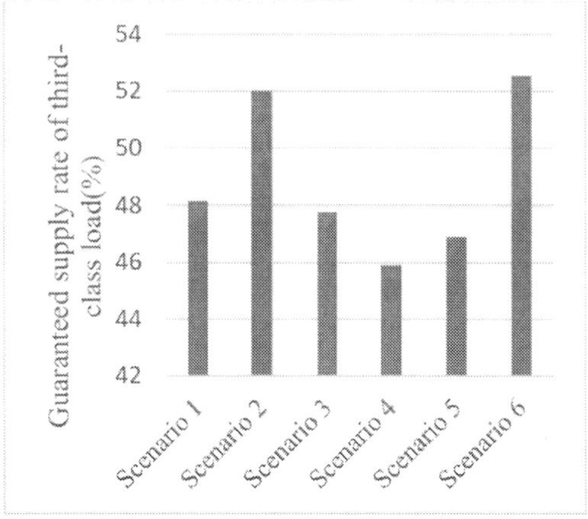

Figure. 4 Tertiary Load Supply Reliability Rate

From the above results, it can be seen that as the scale of external power feed-in increases, the number of transmission lines required for the backbone network gradually increases due to the need to allocate external power. When external power feed-in increases from 280 MW to 480 MW, the number of transmission lines in the backbone network increases from 14 to 16. Although the number of transmission lines does not change when the scale of external power feed-in increases slightly, remaining at 14 lines, the specific lines change. For example, the initial line 4-7 is replaced by line 4-9. From the secondary and tertiary load supply reliability rates, it can be observed that as the external power feed-in scale increases, the supply reliability does not consistently show an increasing trend but fluctuates. Due to the poor adjustability of external power, although it can serve as a power source for critical loads, when the external power feed-in scale is large, transmission congestion occurs in the backbone network, meaning that some of the power cannot be transmitted to the load side.

B. Impact of External Power Feed-in Access Points on Backbone Network

To further explore the impact of external power feed-in access points on the backbone network, this section quantitatively analyzes how different access point locations for external power feed-in affect the number of transmission lines in the backbone network and the reliability of power supply to various load levels. The following four scenarios are considered, with the number of transmission lines and load supply reliability for each scenario shown in Table. 3 and 4.

Scenario 7: External power feed-in at buses 3 and 6

Scenario 8: External power feed-in at buses 3 and 9

Scenario 9: External power feed-in at buses 3 and 4

Scenario 10: External power feed-in at buses 2 and 6

Table. 3. Backbone network under different external power feed-in access points

Scenario 7		Scenario 8		Scenario 9		Scenario 10	
Line ID	Line	Line ID	Line	Line ID	Line	Line ID	Line
1	1-2	1	1-2	1	1-2	2	1-5
2	1-5	2	1-5	2	1-5	3	2-3
3	2-3	3	2-3	3	2-3	4	2-4
5	2-5	6	3-4	9	4-9	5	2-5
7	4-5	9	4-9	10	5-6	7	4-5
8	4-7	11	6-11	11	6-11	9	4-9
10	5-6	12	6-12	12	6-12	10	5-6
11	6-11	13	6-13	13	6-13	11	6-11
12	6-12	14	7-8	14	7-8	12	6-12
13	6-13	15	7-9	15	7-9	13	6-13

14	7-8	16	9-10	18	10-11	14	7-8
15	7-9	18	10-11	20	13-14	15	7-9
18	10-11	-				18	10-11
20	13-14	-				20	13-14

Table. 4. Backbone network under different external power feed-in access points

Scenario	7	8	9	10
Secodary Load Supply Reliability Rate□%□	54.94	56.33	62.24	64.03
Tertiary Load Supply Reliability Rate□%□	57.10	45.51	53.70	45.90

From the above results, it can be seen that with different locations of external power feed-in access points, the backbone network and load supply reliability rates show variations. Specifically, in grid connection points with heavier loads, such as in Scenarios 8 and 9, the number of transmission lines in the backbone network is smaller. This is because part of the power is absorbed by the load at the grid connection point, reducing the demand for power transmission from the external feed-in. Therefore, fewer transmission lines are required in the backbone network. Regarding the supply reliability rates, although the load is the smallest at both access points in Scenario 10, the supply reliability rates for secondary and tertiary loads are not the highest in this scenario. Specifically, the secondary load supply reliability is the highest among the four scenarios, reaching 64.03%, while the tertiary load supply reliability is 45.90%.

C. Backbone Network Under Different Proportions of Renewable Energy

Considering that the proportion of renewable energy is expected to gradually increase in the future, this section further investigates the trend of changes in the backbone network of the receiving-end power grid under different wind power installed capacities. The following four scenarios are set. The load supply reliability rates for each scenario are shown in Table. 5.

Scenario 11: Wind farm installed capacity 75 MW

Scenario 12: Wind farm installed capacity 150 MW

Scenario 13: Wind farm installed capacity 225 MW

Scenario 14: Wind farm installed capacity 300 MW

Table. 5. Secondary and tertiary load supply reliability under different proportions of renewable energy

Scenario	11	12	13	14
Secodary Load Supply Reliability Rate□%□	64.03	60.00	72.37	72.37
Tertiary Load Supply Reliability Rate（%）	45.90	55.88	55.76	55.76

From the results, it can be seen that the wind power installed capacity does not affect the backbone network, as the backbone network under different wind power installed capacities remains the same as shown in Fig 2 (the backbone network in the base scenario). However, from the perspective of load supply reliability, when the wind power installed capacity increases to 225 MW, the supply reliability for secondary and tertiary loads stabilizes at 72.37% and 55.76%, showing no further change. On one hand, the randomness, intermittency, and fluctuation of wind power do not provide a stable electricity supply for the loads that require guaranteed supply. On the other hand, as the wind power capacity increases, although there are periods when output increases, the impact of transmission congestion results in some wind power being curtailed, leading to a loss of available electricity.

4. Conclusion

This paper explores the search method for the backbone network of receiving-end grids under extreme disasters. It proposes a backbone network search method for receiving-end grids with high external power feed-in under the background of large-scale wind power integration and analyzes scenarios from multiple perspectives, including external power feed-in scale, external power access points, and wind farm installed capacity. The effectiveness of the proposed method is validated through case studies. The findings provide guidance for disaster resilience planning and operation of receiving-end grids. Future work could further combine receiving-end grid planning and differentiated reinforcement to explore strategies for improving resilience based on the backbone network. Besides, considering the continuous increase of new energy in the future, the uncertainty of its output, weak support for grid frequency, voltage, etc., will have an increasingly significant impact on the backbone network. Subsequent research can combine the uncertainty of new energy output, adopt robust optimization methods, consider safety and stability constraints such as frequency and voltage, and establish a backbone grid search model and solution method.

References

[1] HOU Hui, GAO Fu, WEI Ruizeng, et al. Analysis and Evaluation on Damage to Transmission and Distribution Equipment in Guangdong Power Grid Caused by Typhoon Talim [J]. Journal of Global Energy Interconnection,2024,7(05):499-509.

[2] LIU Hanchen, WANG Chong, JU Ping, et al. A sequentially preventive model enhancing power system resilience against extreme-weather-triggered failures[J]. Renewable and Sustainable Energy Reviews, 2022, 156: 111945.

[3] CADINI F, AGLIARDI G L, ZIO E. A modeling and simulation framework for the reliability/availability assessment of a power transmission grid subject to cascading failures under extreme weather conditions[J]. Applied Energy, 2017, 185: 267-279.

[4] Zhang Yijing;Li Zhi;Shi Yanqiang ,et al. Optimal Frequency Control Method of Receiving Power Grid Based on Energy Storage Inertia Support [J]. Transactions of China Electrotechnical Society,2024,39(11):3556-3568.

[5] HAN Chang, LIN Zhenzhi, YANG Li, et al. Multi-objective Planning for Anti-disaster Backbone Grid Considering Economics and Network Frame Performance[J]. Automation of Electric Power Systems, 2019, 43(2):34-41.

[6] LIN Zhenzhi,WEN Fushuan,WANG Huifang,et al.CRITIC-based node importance evaluation for skeleton-network reconfiguration of power grids[J].IEEE Transactionson Circuits and SystemsII-Express Briefs,2018,65(2):206-210.

[7] Chunhui LUO, Tonghua ZOU, QU Gangju, Tao TANG. A Planning Method for Anti-disaster Backbone Grid Considering Economy and Reliability[J]. Power Generation Technology, 2023, 44(3): 425-430.

[8] LOU Yuanyuan,CHEN Changming,LIU Xinmiao,GONG Xianfu,CHEN Honglin,LIN Zhenzhi.Multi-Objective Planning Method for Backbone Grid Considering Economy and Multi-Stage Anti-Disaster Performance[J].Southern Power System Technology,2022,16(3):32-39.

[9] WU D,MA F,JAVADI M,et al.A study of the impacts offlow direction and electrical constraints on vulnerability assess-ment of power grid using electrical betweenness measures[J].Physica A:Statistical Mechanics&Its Applications,2017(466):295-309.

[10] CHEN Changming, GONG Xianfu, LOU Yuanyuan, CHEN Honglin, LIU Shengyuan, LIN Zhenzhi. Two-stage Optimal Planning of All-stage Anti-disaster Backbone Network Considering Differential Reinforcement[J]. Power System Technology, 2023, 47(2): 763-772.G. Eason, B. Noble, and I. N. Sneddon, “On certain integrals of Lipschitz-Hankel type involving products of Bessel functions,” Phil. Trans. Roy. Soc. London, vol. A247, pp. 529–551, April 1955.

Power Coordination Control of Series Brushless Doubly-Fed Wind Power System Integrated to DC Distribution Network

Wei Luo[1][0009-0006-9097-2971], Yan Le[2*][0009-0008-9974-569X], Yi Liu[2][0000-0003-1806-3045], Minglei Xie[3], Jinfeng Wang[4], Yuanzhe Zhu[5], Dayi Li[2][0000-0001-6717-1072], Wanlin Du[4], Fang Chen[5]

[1] Guangdong Power Grid Co., Ltd. Meizhou Power Supply Bureau, Meizhou 514000, Guangdong, China

[2] School of Electrical and Electronic Engineering, Huazhong University of Science and Technology, Wuhan 430074, Hubei, China

[3] China Southern Power Grid Electric Power Research Institute Co. LTD, Guangzhou 510000, Guangdong, China

[4] Electric Power Research Institute of Guangdong Power Grid Co. LTD, Guangzhou 510000, Guangdong, China

[5] China Southern Power Grid Energy Development Research Institute Co. LTD, Guangzhou 510000, Guangdong, China

*Corresponding author's e-mail: leyan@hust.edu.cn

Abstract: With the advancement of the construction of new power systems and new energy systems, the large-scale access of new energy is the future development trend of distribution networks. The brushless doubly-fed induction generator (BDFIG) is with characteristics of small inverter capacity, high reliability and low maintenance cost, and has broad application prospects in wind power systems integrated to the dc distribution network. The series BDFIG-DC wind power system can increase the output voltage and save the cost of building booster stations. Wind energy is constantly changing, and the wind energy captured by each BDFIG in series will be different and change continuously, so the output power of the series BDFIGs needs to be coordinated. In the coordination process, these series BDFIGs usually suffer voltage fluctuations, which affects the stability of the dc distribution network. In order to solve this problem, the characteristics of the series BDFIG-DC wind power system are analyzed at first. And then, a power coordination control method based on voltage feedforward is proposed, which introduces output voltage feedforward to control winding current closed loop to reduce the voltage fluctuation of each BDFIG in the series wind power system. Finally, the feasibility of the proposed method is verified by simulation.

1. Introduction

The distribution network serves as a critical component of the power system. With the ongoing development of modern power systems and renewable energy infrastructures, the large-scale integration of renewable energy sources and the rapid increase in new types of electrical loads represent key trends in the future evolution of distribution networks[1]. Concurrently, advancements in wind power generation technology have positioned wind energy as a cost-effective, widely available, zero-emission, and environmentally friendly renewable energy source, making it one of the fastest-growing sectors in the energy industry [2]. Among wind power technologies, the brushless doubly-fed induction generator (BDFIG) stands out due to its advantages, including reduced converter capacity, high operational reliability, and low maintenance costs, rendering it a promising solution for wind energy applications [3][4].

In the context of the series BDFIG-DC wind power system integrated into DC distribution networks, the variability of wind energy necessitates power coordination control among the series-connected BDFIGs. During the coordination of output power across generators, significant voltage fluctuations occur, which can compromise the stability of the distribution network. While this specific issue has not yet been addressed in the literature, researchers have explored similar power coordination challenges in related systems. For instance, reference [5] introduced an enhanced master-slave control strategy to achieve power distribution and voltage balance in a photovoltaic-based modular system, ensuring stability during voltage regulation. Reference [6] investigated a 750V photovoltaic-battery DC microgrid configuration and proposed a control strategy grounded in the distributed V-I droop control principle. Additionally, reference [7] presented a coordinated inertia control approach for a doubly-fed induction generator paired with a battery energy storage system in a wind power plant. Although these studies offer valuable insights into power coordination control, their solutions are not directly applicable to brushless doubly-fed induction generator (BDFIG) systems.

In order to improve the defects of the existing control methods, this paper proposes a power coordination control method based on voltage feedforward, it is more flexible than master-slave and equalization control and more accurate than droop control. With voltage feedforward in the control winding, this control method will suppress the output voltage fluctuation of each BDFIG, thereby improving the stability of the system. This paper analyzes the application of the series BDFIG-DC wind power system, clarifies the main factors affecting power coordination control, theoretically derives the feasibility of the control method, proposes a suppression method for output voltage and power fluctuations by introducing voltage feedforward to the CW current control loop to enhance the stability of the DC distribution network, and verifies the proposed method's feasibility in Simulink.

2. Basic Principle of BDFIG

BDFIG eliminates the need for brushes and slip rings by incorporating two distinct sets of stator windings with differing pole pairs: the power winding (PW) and the control winding (CW). These windings are electrically isolated but magnetically coupled through the rotor winding. Depending on the configuration of the windings, the BDFIG can operate in various modes, such as induction mode, cascade mode, and doubly-fed mode. In this study, the brushless doubly-fed power generation system applied to the distribution network operates in the doubly-fed mode, which offers the most significant advantages for this application.

The power coordination control method used in this paper needs to be derived under the static abc frame. The voltage and flux linkage equations of the BDFIG dynamic mathematical models under the static abc frame are as follows [8]:

$$\boldsymbol{u}_p = R_p\boldsymbol{i}_p + \frac{d\boldsymbol{\psi}_p}{dt} \tag{1}$$

$$\boldsymbol{\psi}_p = \left(L_{mp} + L_{\sigma p}\right)\boldsymbol{i}_p + L_{mp}\boldsymbol{i}_r \tag{2}$$

$$\boldsymbol{u}_c = R_c\boldsymbol{i}_c + \frac{d\boldsymbol{\psi}_c}{dt} - j\left(p_p + p_c\right)\omega_{rm}\boldsymbol{\psi}_c \tag{3}$$

$$\boldsymbol{\psi}_c = \left(L_{mc} + L_{\sigma c}\right)\boldsymbol{i}_c + L_{mc}\boldsymbol{i}_r \tag{4}$$

$$\boldsymbol{u}_r = R_r\boldsymbol{i}_r + \frac{d\boldsymbol{\psi}_r}{dt} - jp_p\omega_{rm}\boldsymbol{\psi}_r \tag{5}$$

$$\boldsymbol{\psi}_r = \left(L_{mp} + L_{mc} + L_{\sigma r}\right)\boldsymbol{i}_r + L_{mp}\boldsymbol{i}_p + L_{mc}\boldsymbol{i}_c \tag{6}$$

where u , i and ψ represent the effective value of phase voltage, phase current and phase flux linkage, respectively; the subscripts p c r indicate PW, CW and rotor winding, respectively; L_{mp} and L_{mc} are the phase excitation inductance of PW and CW sides, respectively; $L_{\sigma p}$, $L_{\sigma c}$ and $L_{\sigma r}$ represent the phase leakage inductance of PW, CW and rotor winding, respectively; R_p, R_c and R_r are the phase resistance of PW, CW and rotor winding, respectively.

3. Control Strategy Design

3.1. System analysis

In the series BDFIG-DC wind power system used in the distribution network, each BDFIG is connected in series with the dc output end. In the application of wind power generation, since each wind turbine captures different amounts of wind energy, the amount of power delivered to the dc grid by the BDFIGs participating in the series power generation is also different.

In order to make full use of wind energy and enable the generator set that captures more wind energy to bear more power, it is necessary to flexibly adjust the output power of each BDFIG. During the adjustment process, because only the output power is closed-loop controlled and the voltage is not controlled, the output voltage of each BDFIG will fluctuate significantly, which is not conducive to the stable operation of the distribution network. Therefore, both the output power and the output voltage should be controlled to reduce the fluctuation.

3.2. Control method derivation

To address the simultaneous reduction of output power and voltage fluctuations in each BDFIG during the adjustment process, this study introduces a power coordination control strategy for a series BDFIG-DC wind power system, incorporating a voltage feedforward mechanism. In the proposed system, the power winding (PW) of the BDFIG is linked to the DC bus capacitor via a three-phase uncontrolled rectifier.

Based on the principle of power conservation, the relationship between the DC bus voltage and the PW voltage can be derived as

$$U_{dc} = \left(\frac{1}{2} + \frac{3\sqrt{3}}{4}\right)\sqrt{3}U_p \approx 1.5822U_p \approx 2.2376u_p \tag{7}$$

where U_{dc} represents the dc bus voltage; U_p represents the PW phase voltage amplitude; u_p represents the PW phase voltage effective value.

Under actual working conditions, the rotor phase resistance of BDFIG is much smaller than the excitation reactance and rotor leakage reactance, and can be ignored in the mathematical model. In this case, the rotor flux is equivalent to 0. From the rotor flux equation (6) of the dynamic mathematical model of BDFIG under the stationary ABC shaft system, it can be deduced that

$$\boldsymbol{i}_r = -\frac{L_{mp}\boldsymbol{i}_p + L_{mc}\boldsymbol{i}_c}{L_{mp} + L_{mc} + L_{\sigma r}}. \tag{8}$$

If the entire BDFIG-dc system is equivalent to a two-port network, with PW as its port, it can be obtained that

$$\boldsymbol{u}_p = \boldsymbol{i}_p \boldsymbol{Z}_{load} \tag{9}$$

where u_p is the phase voltage on the PW side; i_p is the phase current on the PW side; and Z_{load} represents the impedance of the entire system.

Substituting (8) into the PW side flux equation (2), PW side flux can be derived as

$$\boldsymbol{\psi}_p = \left[\frac{L_{mp}\left(L_{mc} + L_{\sigma r}\right)}{L_{mp} + L_{mc} + L_{\sigma r}} + L_{\sigma p}\right]\boldsymbol{i}_p - \frac{L_{mp}L_{mc}}{L_{mp} + L_{mc} + L_{\sigma r}}\boldsymbol{i}_c. \tag{10}$$

Similarly, the PW voltage equation (1) under steady-state conditions can be obtained as

$$\boldsymbol{u}_p = R_p\boldsymbol{i}_p + j\omega_p\boldsymbol{\psi}_p. \tag{11}$$

Substituting (9) and (10) into (11), PW voltage derived as

$$\boldsymbol{u}_p = \frac{j\omega_p L_{mp} L_{mc} \boldsymbol{Z}_{load}}{\left[j\omega_p L_{mp}\left(L_{mc}+L_{\sigma r}\right)+\left(L_{mp}+L_{mc}+L_{\sigma r}\right)\left(j\omega_p L_{\sigma p}+R_p-\boldsymbol{Z}_{load}\right)\right]}\boldsymbol{i}_c . \tag{12}$$

Substituting the modulus value of (12) into (7), the dc bus voltage can be expressed as

$$U_{dc} \approx \left| \frac{j2.2376\omega_p L_{mp} L_{mc} \boldsymbol{Z}_{load}}{\left[j\omega_p L_{mp}\left(L_{mc}+L_{\sigma r}\right)+\left(j\omega_p L_{\sigma p}+R_p-\boldsymbol{Z}_{load}\right)\left(L_{mp}+L_{mc}+L_{\sigma r}\right)\right]}\boldsymbol{i}_c \right| . \tag{13}$$

In the dc grid-connected power generation system, the dc output current and dc output power and can be described as

$$P_{dc} = U_{dc}\frac{U_{dc}-U}{Z_{dc}} \tag{14}$$

where U represents the dc bus voltage; I_{dc} represents the dc bus current; Z_{dc} represents the dc bus resistance; P_{dc} represents the power output by the system to the dc bus.

Substituting (13) into (14), the dc output power can be derived as

$$P_{dc} \approx \left| \frac{\dfrac{j2.2376\omega_p L_{mp} L_{mc} \boldsymbol{Z}_{load}}{\left(L_{mp}+L_{mc}+L_{\sigma r}\right)}}{\dfrac{j\omega_p L_{mp}\left(L_{mc}+L_{\sigma r}\right)}{\left(L_{mp}+L_{mc}+L_{\sigma r}\right)}+\left(j\omega_p L_{\sigma p}+R_p-\boldsymbol{Z}_{load}\right)} \right| \times \left(\left| \frac{\dfrac{j2.2376\omega_p L_{mp} L_{mc} \boldsymbol{Z}_{load}}{\left(L_{mp}+L_{mc}+L_{\sigma r}\right)}}{\left(\dfrac{j\omega_p L_{mp}\left(L_{mc}+L_{\sigma r}\right)}{\left(L_{mp}+L_{mc}+L_{\sigma r}\right)}+\left(j\omega_p L_{\sigma p}+R_p-\boldsymbol{Z}_{load}\right)\right)} \right| i_c - U \right) \frac{i_c}{Z_{dc}} . \tag{15}$$

It can be seen from the above formula that the dc output power can be controlled by the CW current effective value. At this point, the connection between the dc output power and the entire system has been preliminarily established.

On the basis of the power outer loop, in order to improve the output dc voltage stability of the system, the output dc voltage is introduced as feedforward.

In steady state, the d-axis components of the CW current can be set as

$$i_{cd} = \left|\boldsymbol{i}_c\right| . \tag{16}$$

Substituting (13) into (16), d-axis components of the CW current can be derived as

$$i_{cd} = U_{dc}\left| \frac{\left(j\omega_p L_{mp}\left(L_{mc}+L_{\sigma r}\right)+\left(L_{mp}+L_{mc}+L_{\sigma r}\right)\left(j\omega_p L_{\sigma p}+R_p-\boldsymbol{Z}_{load}\right)\right)}{j2.2376\omega_p L_{mp} L_{mc} \boldsymbol{Z}_{load}} \right| . \tag{17}$$

Therefore, the feedforward coefficient can be derived as

$$k = \left| \frac{\left(j\omega_p L_{mp}\left(L_{mc}+L_{\sigma r}\right)+\left(L_{mp}+L_{mc}+L_{\sigma r}\right)\left(j\omega_p L_{\sigma p}+R_p-\boldsymbol{Z}_{load}\right)\right)}{j2.2376\omega_p L_{mp} L_{mc} \boldsymbol{Z}_{load}} \right| . \tag{18}$$

Since (18) is too complicated and not a constant, it should be simplified in practical engineering applications. In actual working conditions, PW phase excitation inductance is much greater than the sum of CW phase excitation inductance and the rotor leakage inductance, it can be rewritten as

$$L_{mp} \gg L_{mc} + L_{\sigma r} . \tag{19}$$

From the BDFIG equivalent circuit shown in Figure 1, PW side equivalent impedance can be obtained as [9]

$$\boldsymbol{Z}_{load} = \frac{\boldsymbol{U}_p}{\boldsymbol{i}_p} = j\omega_p L_{\sigma p} + R_p + j\omega_p L_{mp}\frac{\boldsymbol{i}_{mp}}{\boldsymbol{i}_p} . \tag{20}$$

Because of (19) and in actual working conditions, $|i_{mp}|$ and $|i_p|$ are in the same order of magnitude, it can be obtained that

$$\left| j\omega_p L_{mp} \frac{\boldsymbol{i}_{mp}}{\boldsymbol{i}_p} \right| \gg \left| j\omega_p \left(L_{mc} + L_{\sigma r} \right) \right| \tag{21}$$

$$\left| \boldsymbol{Z}_{load} \right| \gg \left| j\omega_p \left(L_{mc} + L_{\sigma r} + L_{\sigma p} \right) + R_p \right|. \tag{22}$$

Substituting (19), (21) and (22) into (18), the feedforward coefficient can be simplified as

$$k \approx \frac{1}{2.2376\omega_p L_{mc}}. \tag{23}$$

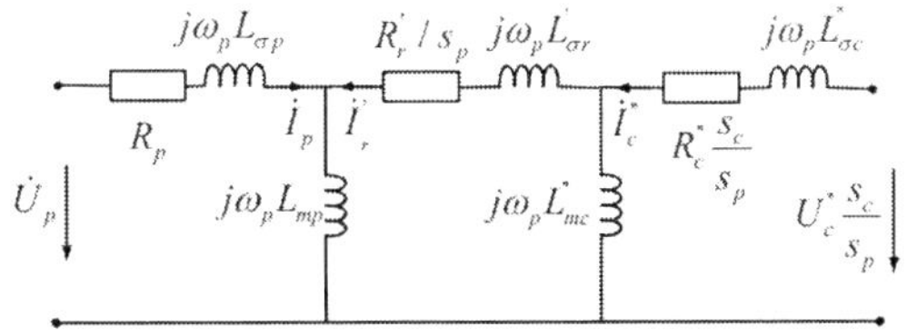

Figure 1. Equivalent circuit diagram of BDFIG

3.3. Control method design

The brushless double-fed motor series power coordination control system is connected in series to the common dc bus, including a series power coordination module, an output power control module, a CW current control module and an SVPWM generator connected to the CW side of the BDFIG. The PW side of BDFIG series power coordination control system is connected to the three-phase uncontrolled rectifier bridge, and the CW side is connected to the machine-side converter. Voltage feedforward is added before the CW current PI control to limit system dc bus voltage by feedforward compensation. The control principle block diagram is shown in Figure 2.

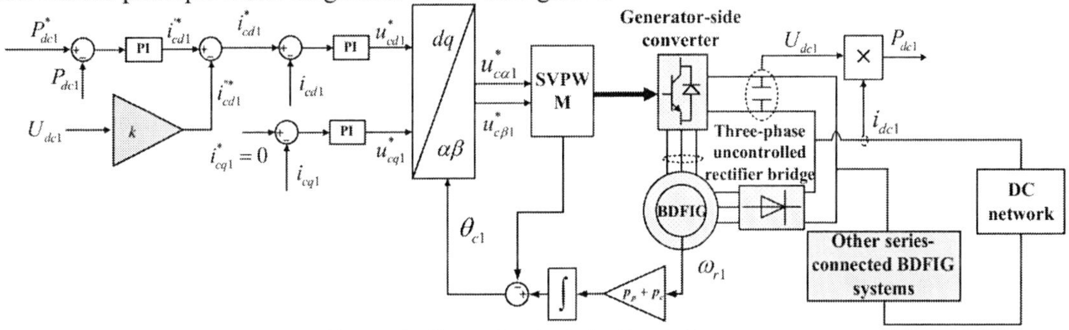

Figure 2. Control principle block diagram

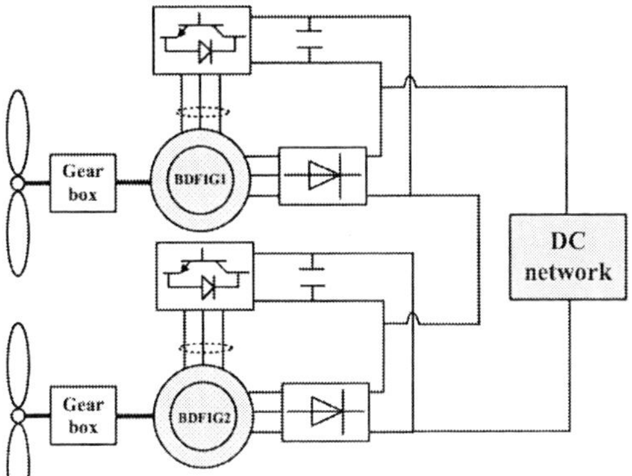

Figure 3. Basic structure of simulation system

4. Simulation Results

To validate the effectiveness of the proposed voltage feedforward-based power coordination control method for the series BDFIG-DC wind power system, a simulation model was developed by Simulink. The relevant machine parameters utilized in the simulation are provided in Table 1.

Table 1. BDFIG parameters applied in the simulation

Parameter	Value	Parameter	Value	Parameter	Value
PW pole pairs	1	PW rated frequency	50 Hz	CW resistance	0.571 Ω
CW pole pairs	3	PW rated power factor	0.8(lag)	CW self-inductance	0.0472 H
Natural synchronous speed	750 r/min	CW current range	0~25 A	CW mutual inductance	0.0281 H
Speed range	600~1200 r/min	PW resistance	2.430 Ω	Rotor resistance	0.238 Ω
PW rated voltage	380 V	PW self-inductance	0.8010 H	Rotor self-inductance	0.0769 H
PW rated current	8 A	PW mutual inductance	0.2730 H	Rotor type	Wound rotor

In the simulation system, the dc ports of BDFIG1 and BDFIG2 are connected in series, and the connection relationship is shown in the Figure 3. The duration of the simulation is 20 s., and the BDFIG speed is 900 r/min. At 10 s, the BDFIG1 output power reference value is changed from 400 to 500 W. At the same time, the BDFIG2 output power reference value remains unchanged. The simulation results are shown in Figures 4 and 5.

From Figure 4(a) and 4(b), it can be seen that the output voltage is greatly improved after applying the power coordination control method based on voltage feedforward proposed in this paper. The voltage fluctuation of BDFIG1 caused by power change drops by 63.2%. The voltage fluctuation of BDFIG2 caused by power change decreases by 68.4%. As shown in Figure 5(a) and 5(b), the output power waveform of the system is also significantly improved, and the stabilization time reduces by 55.2%. It can be seen that voltage feedforward significantly improves the power coordination control of the dc series brushless double-fed wind power generation system, limiting the output voltage and output power fluctuations, and is conducive to improving the stability of the system, which is consistent with the previous analysis.

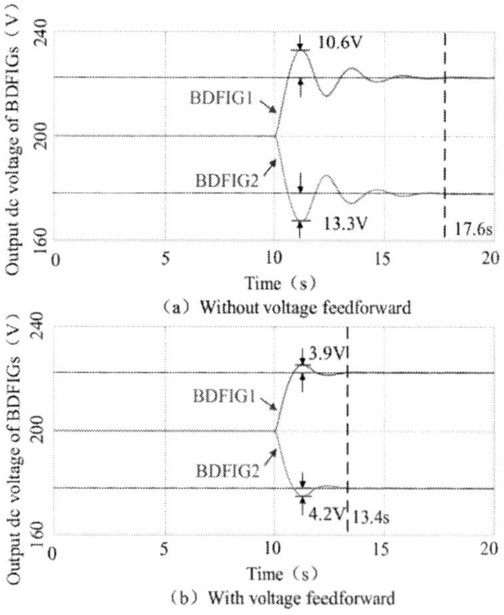

Figure 4. Waveform of output dc voltage

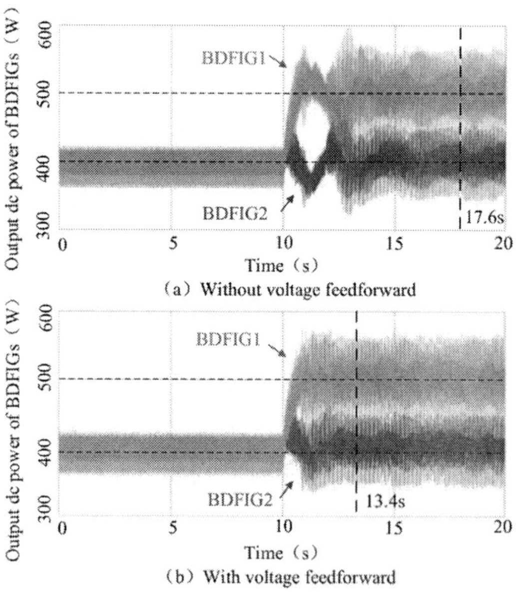

Figure 5. Waveform of output power

5. Conclusion

Firstly, this paper analyzes the application of series BDFIG-DC wind power system, clarifies the main targets that affect power coordination control, and theoretically derives the feasibility of the control method. And then, by introducing voltage feedforward to the CW current control loop, the suppression method of output voltage and output power fluctuations is proposed, which effectively suppresses fluctuations in both output voltage and output power, thereby enhancing the overall stability and reliability of the dc distribution network. Finally, to validate the theoretical findings and practical applicability of the proposed power coordination control method, a comprehensive simulation model is developed in Simulink. The simulation results confirm the feasibility and effectiveness of the proposed approach, confirm its potential for implementation in real-world wind power systems integrated into dc distribution networks. This study contributes to the advancement of control strategies for renewable energy systems, offering a robust solution for improving the stability and efficiency of wind power integration.

Acknowledgments

This work was supported by the Key Science and Technology Project of China Southern Power Grid Co., Ltd. under Grant 031400KC23120011.

Reference

[1] Peng F Z Liu C C Li Y Jain A K and Vinnikov D. 2024, Envisioning the future renewable and resilient energy grids—a power grid revolution enabled by renewables energy storage and energy electronics. IEEE Journal of Emerging and Selected Topics in Industrial Electronics (vol 5 no 1), pp. 8-26.

[2] Sun C Chen J and Tang Z. 2021, New energy wind power development status and future trends. 2021 International Conference on Advanced Electrical Equipment and Reliable Operation, pp. 1-5

[3] Zhu L Zheng H Ji H Li G Li H and Yao S 2021 Development trends and suggestions on offshore wind power of china 2021 IEEE Sustainable Power and Energy Conference pp 2043-2048.

[4] Zhang D Ma J Wang S Ji Y Jiang K and Qu C. 2022 Research on control for grid-connected brushless doubly-fed power generation system under different quantities, 2022 International Power Electronics Conference, pp. 1202-1206.

[5] Jiang S Sun K Wu H Shi H Dong X. 2018, An improved master-slave control for three-port converter based distributed dc grid-connected PV system. 2018 International Power Electronics Conference.

[6] Liao W D Liu M R Shang H L and Li S J. 2022, Analysis of 750V PV-battery dc microgrid based on distributed droop control strategy. 2022 Power System and Green Energy Conference.

[7] Chang F Li Y Peng Y Wang S Zhang X and Cao Y. 2023, Coordinated inertial control method of doubly fed induction generators and battery energy storage system for short-term frequency support. 2023 6th International Conference on Energy Electrical and Power Engineering, pp. 1192-1197.

[8] Poza J Oyarbide E Roye D and Rodriguez M. 2006, Unified reference frame dq model of the brushless doubly fed machine. IEE Proceedings-Electric Power Applications (vol 153 no 5), pp. 726-734.

[9] Ge J Xu W Liu Y and Xiong F. 2022, Novel equivalent circuit model applicable to all operation modes for brushless doubly fed induction machines. IEEE Transactions on Industrial Electronics (vol 69 no 12), pp. 12540-12550.

Review of the Power Quality Comprehensive Improvement Device and the Key Technology for Grid-Connected Microgrids

Wei Luo[1] [0009-0006-9097-2971], Yizheng Zhang[2*] [0009-0004-3340-0104], Yi Liu[2] [0000-0003-1806-3045], Minglei Xie[3], Jinfeng Wang[4], Yuanzhe Zhu[5], Dayi Li[2] [0000-0001-6717-1072], Wanlin Du[4], Fang Chen[5]

[1] Guangdong Power Grid Co., Ltd. Meizhou Power Supply Bureau, Meizhou 514000, Guangdong, China

[2] School of Electrical and Electronic Engineering, Huazhong University of Science and Technology, Wuhan 430074, Hubei, China

[3] China Southern Power Grid Electric Power Research Institute Co. LTD, Guangzhou 510000, Guangdong, China

[4] Electric Power Research Institute of Guangdong Power Grid Co. LTD, Guangzhou 510000, Guangdong, China

[5] China Southern Power Grid Energy Development Research Institute Co. LTD, Guangzhou 510000, Guangdong, China

*Corresponding author's e-mail: yzzhang18@hust.edu.cn

Abstract. As the development of high-end manufacturing, some high-tech industries, production automation and intelligent industries have high requirements for quality of power supply. Due to the high proportion of renewable energy accessed to the grid-connected microgrid, the power quality (PQ) problem of the microgrid should be concerned. To solve the power quality problem of microgrid, some devices and techniques have been proposed by some scholars. Aiming at the power quality problem of grid-connected microgrid, the existing power quality management devices and technologies are summarized and analysed in this paper, so as to provide reference for other scholars studying power quality management. Initially, the typical power quality problems of grid-connected microgrid are analysed, such as three-phase imbalance, harmonics distortion, etc. Secondly, the overview of PQ devices is introduced. Finally, the power quality improvement techniques for grid-connected microgrid are summarized.

1. Introduction

As the application of distributed generations, grid-connected inverters as the AC/DC conversion interfaces are connected to the grid at the load side, which has become an important factor affecting the PQ of the grid [1]. Load in the grid-connected microgrid are often with the nonlinear and unbalanced characteristics, which further reduces the PQ of the microgrid [2]. The deterioration of PQ not only affects the normal operation of the electrical equipment, but also increases the loss of microgrid. For the traditional power grid, the microgrid is a controllable unit, and for the user side, the microgrid can meet the specific needs of the user side. The capacity of the microgrid is smaller than that of large grid, so microgrid PQ problem will cause a greater impact. Additionally, there are many new energy interfaces in the microgrid, these new energy sources have strong randomness, volatility and intermittency, which also makes the output of new energy distributed generators have corresponding characteristics [3]. Due to the use of power electronic devices in distributed generators, switching on and off of IGBT and other devices in the process of power electronic transformation can cause switching subharmonics, further reducing the PQ of the grid [4]. The inertia of the traditional power system is gradually reduced, affecting the power supply quality of the existing power system [5].

This paper will summarize the main PQ problems faced by the existing grid-connected microgrid, and review the current typical PQ equipment and its key control technologies. The rest of this paper is organized as follows. In Chapter 2, the PQ problems in grid-connected microgrid are summarized. The overview of PQ management devices is induced in Chapter 3. The PQ improvement techniques for grid-connected microgrid is developed in Chapter 4. Finally, Chapter 5 concludes this article.

2. Power quality problems in grid-connected microgrid

PQ problems are defined as deviations in voltage, current and frequency that cause electrical equipment to malfunction or fail to work properly [6]. The PQ problems in grid-connected microgrids are often different from those of traditional power systems due to their diverse types and variable operation modes, reflected in the following aspects:

2.1. The imbalance of the three phases

Microgrids are usually connected with a number of single-phase loads and distributed power sources, which causes a certain degree of three-phase imbalance. When the grid is in grid-connected operation, the unbalanced voltage quality of the distribution network can seriously affect the operation of the microgrid [7]. Negative sequence currents can mis-trigger negative sequence protection in the power grid and threaten the stability operation of the grid. In addition, the unbalance makes the current flow in the neutral line, which increases loss.

2.2. Harmonic distortion

Harmonic distortion of current waveforms is a common category of PQ problems at the user side of microgrids, which can lead to vibration and heating of machines, transformers and other equipment [8]. There are many nonlinear electrical equipments and loads in the microgrid, such as rectifiers, frequency converters, etc., which lead to high harmonic components in the current injected into the microgrid [9]. Micro sources devices in the microgrid are usually connected to the microgrid through power electronic inverters based on PWM modulation technology. Harmonics usually lead to additional losses of components in the microgrid, which increase the cost of the operation. Additionally, many harmonics flow through the neutral line, causing the wire to overheat.

2.3. Deviation of frequency

In the traditional grid, the system frequency can quickly return to its rated value with appropriate adjustment when the load fluctuation occurs. Compared with the large capacity grid, the output power and load change of the micro source account for the big proportion of the total system capacity, which makes the microgrid show the characteristics of the weak grid, and the frequencyin microgrid is easily affected by the disturbance such as load mutation of the micro source [10]. Frequent disturbance can cause the imbalance of the power of the microgrid and deviation of the system frequency.

3. Overview of power quality management devices

To solve the PQ problem of grid-connected microgrid, some related PQ management devices have been proposed. The current power management devices are mainly divided into four categories: static var compensator (SVC), static synchronous compensator (STATCOM), active power filter (APF) and unified PQ conditioner (UPQC). The four types of electric PQ management devices and their subcategories are introduced as follow.

3.1. Static var compensator

SVC is usually used to compensate reactive power with advanced technology and reliable performance. SVC adopts thyristor as the control core, compared with other reactive power compensation equipment in the early stage, SVC has excellent dynamic reactive power adjustment ability, short response time, fast response speed and other characteristics, which can well meet the needs of the system for fast reactive power compensation. Although SVC also has problems such as harmonic generation, it has been widely used in microgrids due to its large adjustable range, reliable technology, small capacity limitation and high cost performance.

The main types of SVC conclude thyristor controller reactor (TCR), thyristor switched capacitor (TSC) and hybrid type. Figure 1 shows several typical SVC structures. As shown in Figure 1(a), The TCR can smoothly change the magnitude of its compensated reactive power by continuously adjusting the thyristor trigger Angle. However, a large number of odd harmonics can be generated by the reactor branch current when the trigger angle is greater than 90 degrees. So, multiple sets of filter compensators (FC) are usually equipped to suppress the harmonics generated by TCR. Figure 1 (b) shows the structure of TSC. Compared with the earlier similar devices, TCR uses thyristor instead of mechanical switch, which greatly improves the durability and effectively reduces the failure rate. And the time point of capacitor switching can be accurately determined, so as to effectively suppress the impact on the system when the capacitor switching. However, TSC is limited by its physical characteristics and can be only of graded compensation, so as shown in Figure. 1(c), it is more compatible with TCR to realize the continuous control of reactive power.

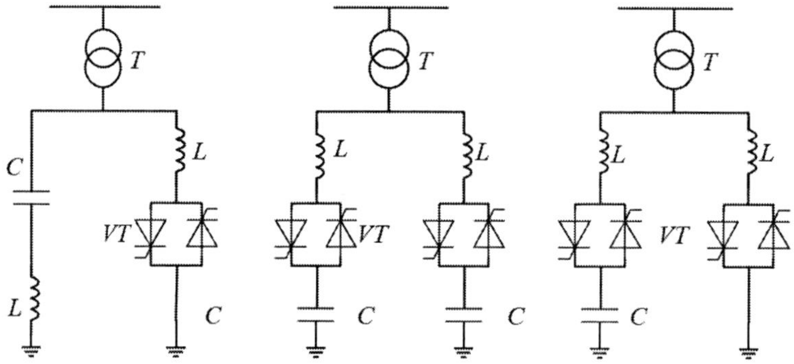

(a) TCR (b) TSC (c) TSC+TCR

Figure 1. Typical SVC structures

3.2. Static synchronous compensator

STATCOM is a kind of equipment widely used in microgrid to compensate reactive power. The load data can be collected by he designed controller in real time from STATCOM, so that the circuit absorbs or emits reactive power to achieve reactive power compensation. STATCOM is widely used in microgrid because of its fast response speed, low pollution to the grid, small loss, strong dynamic performance, and not affected by the system voltage. It can not only compensate reactive power, but also compensate unbalanced load and suppress voltage flicker. Compared with SVC, STATCOM adopts GTO, IGBT, IGCT and other fully controlled switches, which can better increase the response speed, improve the compensation efficiency, and make the operation performance more excellent. However, the cost of STATCOM is too high, and with the increase of compensation capacity, its operation and maintenance costs increase significantly, so STATCOM is not more competitive than SVC in the market in microgrid system. Figure. 2 illustrates the typical structure of STATCOM, it monitors the load reactive current change and adjusts its reactive power output to achieve dynamic reactive power compensation.

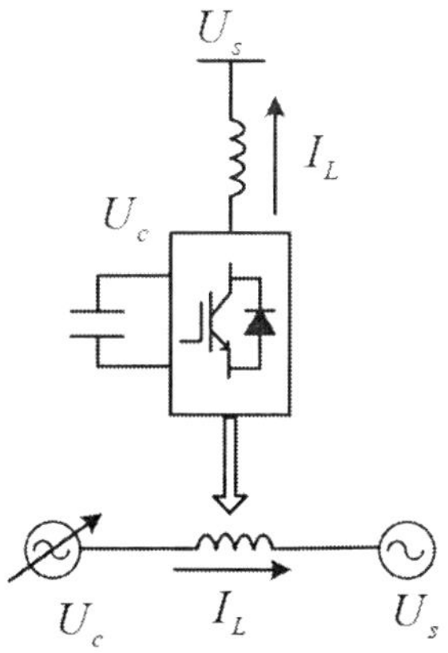

Figure 2. Typical STATCOM structures

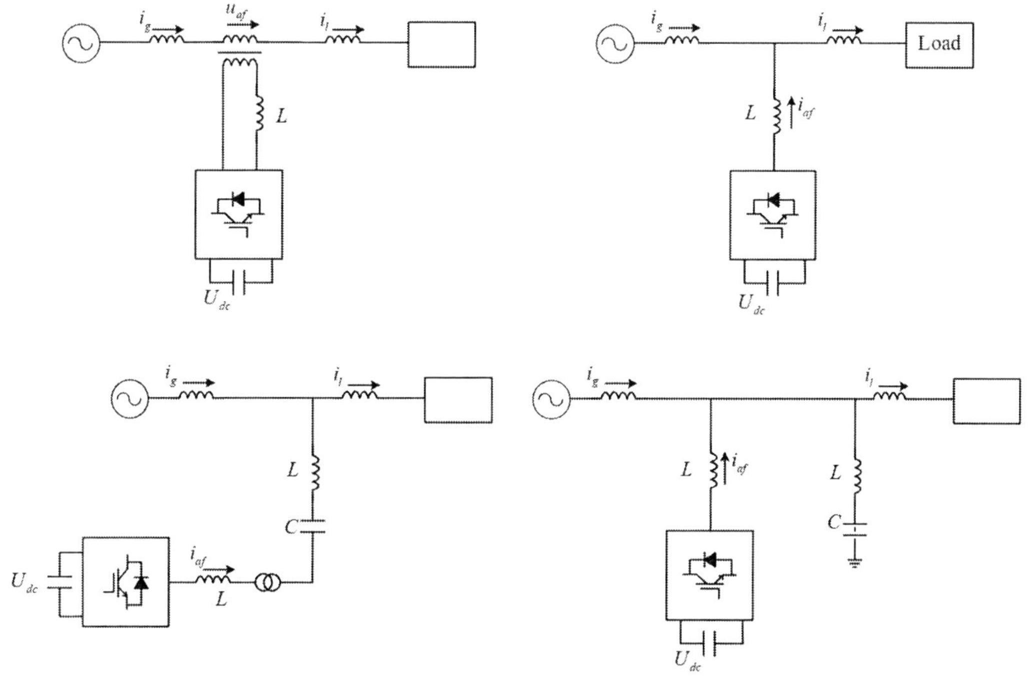

Figure 3. Typical APF structures

3.3. *Active power filter*

APF is mainly used to control the microgrid harmonics, which can dynamically suppress harmonics and compensate reactive power. APF can compensate harmonics with varying magnitude and frequency and reactive power, which is widely used in medium and low voltage distribution network. APF can be divided into three types according to its topology: series APF, parallel APF and hybrid APF. Figure 3 shows the topology of the four types of APF. As shown in Figure 3(a), the series APF is mainly aimed at voltage type harmonic sources. Figure 3 (b) demonstrates the structure of the parallel APF, the parallel APF is accessed to the grid by a filter inductor or LCL filter and generates a harmonic current to supress the harmonic components in the load, so as to ensure that the grid only provides the fundamental positive sequence current. Parallel APF is mainly aimed at current-type harmonic sources, which is equivalent to controlled current sources. The hybrid APF usually consists of active and passive power filters, Figure 3(c) shows a series APF in series with a passive power filter (PPF). The APF in the figure 3(c) is to prevent the harmonic current generated by the harmonic voltage of the supply from flowing into the parallel PPF. Figure 3(d) shows a parallel APF in parallel with a PPF. The parallel APF in the Figure 3(d) can decompose harmonics, while the PPF can compensate the high-frequency current.

3.4. *Unified power quality conditioner*

UPQC is composed of a series APF and a parallel APF through a DC capacitor, which combines the functions of a dynamic voltage restorer and APF, so as to compensate for a variety of PQ issues simultaneously. The typical topology of the UPQC is shown in Figure 4, the series converter is able to output a sine-wave current controlled by the series converter in the grid voltage, while bearing the voltage difference between the load current and a given reference value. UPQC can significantly reduce the number of required input equipment.

Overall, the performance of the four types of devices is summarized in Table I. In fact, considering the large investment and high loss of power grid facilities. Therefore, those classic devices are often used in the grid market.

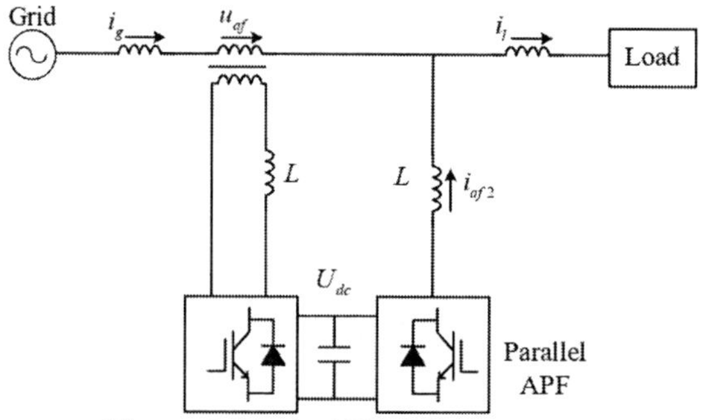

Figure 4. Typical UPQC structures

Table 1. The performance of the four types of devices

Devices	Main functions	Response time	Loss and noise	Cost
SVC	Compensating the reactive power	< 20 ms	High	Low
STATCOM	Compensating the reactive power	< 10 ms	High	Medium
APF	Compensating the reactive power and suppressing the harmonics	< 10 ms	Medium	Medium
UPQC	Compensating the reactive power and suppressing the harmonics	< 10 ms	Low	Low

4. Power quality improvement techniques for grid-connected microgrid

The improvement strategy of PQ in grid-connected microgrid focuses on the control of grid-connected inverters the control of power management devices in microgrid. The following can be introduced from those two parts.

4.1. *Control strategy of grid-connected inverters*

There are different types of micro power supply in the microgrid, corresponding to different forms of power output interface, the vast majority of the micro power supply using DC/AC converter interface, the control of the micro power mainly refers to the control of the converter interface, is the key to the microgrid control. Due to the characteristics of micro-power and the operation control requirements of micro-power in microgrid.

4.1.1. Droop control. Droop control is mainly used to control the amplitude and frequency when there are multiple parallel power sources in the microgrid. This method mainly controls the amplitude and frequency according to the two linear relationships. When the increase of load causes the frequency to decrease, the frequency will produce a negative change, and according to the proportional relationship, the active power needs to produce a positive change to balance the load change. Similarly, the relationship between voltage is the same. Then the active and reactive power output of the distributed power supply is adjusted by increasing or decreasing the corresponding amount of frequency, and finally the purpose of stable operation of the microgrid is achieved.

4.1.2. V/F control. Since distributed power supply inside the microgrid is the dominant power supply unit when it is off the grid, and the inverter must control the voltage and frequency inside whole system to maintain stable operation of microgrid. The control method is mainly through the control of frequency and voltage, to deal with the dynamic operation of the microgrid load conversion, so that the voltage and frequency of microgrid are constant. As displayed in Figure 5, Firstly, the required command current of the current loop is obtained through voltage outer loop, the error signal is obtained by comparing with the actual value and adjusting with PI. Finally, the switching signal can be obtained by the corresponding modulation method.

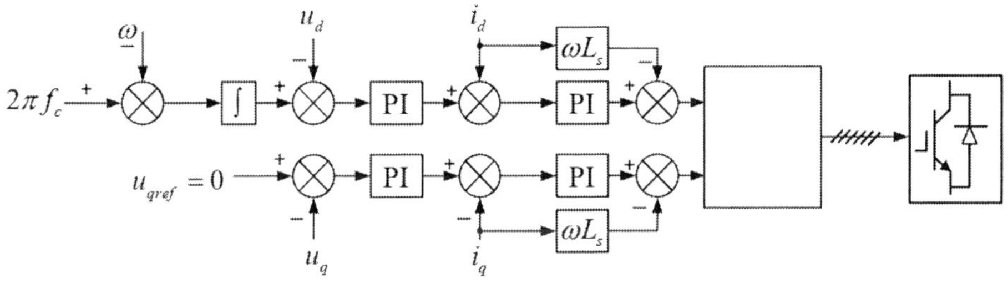

Figure 5. V/F control strategy

4.1.3. PQ control. Constant active power/reactive power control is also called PQ control, because the large grid is the main power supply of the microgrid during grid connection, the microgrid only needs to pay attention to the exchange of energy and power, so to control PQ, so PQ control is mainly used in grid-connected operation mode. In the grid-connected mode, the voltage frequency and phase are obtained from the large grid by using the PLL, and then the actual r P, Q of the inverter are obtained by decoupling through the PQ controller. By adding the reference values of P, Q, the desired reference value of the command current can be obtained by Park transformation.

4.2. Control strategy of power management devices

The control of power management device is mainly the control of active management equipment, such as STATCOM, APF, UPQC. Typical control algorithms for those devices can be introduced as follow.

4.2.1. Control strategy of STATCOM. When STATCOM is connected to the microgrid, there are high requirements for the current control methods of STATCOM system. The current control methods can be divided into direct current control (DCC) and indirect current control (ICC) according to the way the controller generates the PWM control signal of the switch tube. DCC directly converts the deviation value of the reference compensation current and the feedback compensation current into the duty cycle, and the common ones are hysteresis control, model predictive control, etc. ICC controls the current by controlling the power of the converter.

4.2.2. Control strategy of APF. The strategy of APF mainly includes DC side capacitor voltage control and output power tracking control. In structure of voltage inverter, in order to ensure that the system has good compensation current following performance, the dc side capacitor voltage is controlled to avoid over-compensation or under-compensation caused by large fluctuations. The PI control method is mainly used to maintain the DC side voltage. APF takes the inverse of the harmonic current obtained by the harmonic detection algorithm as the command current to obtain the PWM signal of the control power device, and controls the switching device on and off through the drive circuit to compensating the harmonic current. Due to the complexity of the microgrid, the harmonic environment is often accompanied by uncontrollable factors such as fluctuation and flicker, so it is required that the compensation current should follow the command current in real time, and the real-time and dynamic performance requirements are high. There are many kinds of compensating current tracking control methods. The commonly used control methods are space voltage vector control, single cycle control, deadbeat control, repeat control and predictive control.

4.2.3. Control strategy of UPQC. The control methods of UPQC can be divided into two classes: direct control strategy and indirect control strategy. UPQC that based on indirect control is also called traditional UPQC. Its principle is: taking the difference between the value of the load voltage as the control target, the voltage of the series converter of UPQC acts on the isolation transformer to ensure the stability of the load voltage. Taking the harmonic current of the grid current as the control target, the output current of the parallel converter of the control UPQC is cancel the harmonic component of the grid current. Under the indirect control strategy, the UPQC needs to extract the deviation of the load, which are used as signals for the serial-parallel converter, respectively. Therefore, the detection of the reference is in the UPQC performance of the indirect control.

5. Conclusion

To provide reference for other scholars studying PQ management, the typical PQ problems of grid-connected microgrid, as well as the PQ management equipment and technology, is summarized and reviewed in this paper. Firstly, the typical PQ problems of grid-connected microgrid are analysed. To solve those PQ problems, the corresponding PQ management devices are summarized and compared. This paper mainly introduces four categories PQ management devices are introduced in this paper, and their working principles, advantages and disadvantages are summarized, which can cope with different PQ issues in microgrids Furthermore, the PQ improvement techniques are also reviewed.

Acknowledgments
This work was supported by the Key Science and Technology Project of China Southern Power Grid Co., Ltd. under Grant 031400KC23120011.

References

[1] Lasseter R, Akhil A, Marnay C. 2002, Integration of Distributed Energy Resources. The CERTS Microgrid Concept (United States: Berkeley). chapter 4, PP. 123-124.

[2] T. Morstyn, B. Hredzak, and V. G. Agelidis. 2018, Control Strategies for Microgrids With Distributed Energy Storage Systems: An Overview. IEEE Transactions on Smart Grid (vol 9), pp. 3652–3666.

[3] A. A. Alkahtani, et, al. 2020, Power Quality in Microgrids Including Supraharmonics: Issues. Standards, and Mitigations (vol 8), pp. 127104-127122.

[4] J. M. Carrasco, et al. 2006, Power-Electronic Systems for the Grid Integration of Renewable Energy Sources: A Survey. IEEE Transactions on Industrial Electronics (vol 53), pp. 1002-1016.

[5] V. Jain and B. Singh. 2021, Power Quality Improvement in PV System Tied to Weak Grid. IEEE Transactions on Industry Applications (vol 57), pp. 1265–1273.

[6] P. Ray, P. K. Ray, and S. K. Dash. 2022, Power Quality Enhancement and Power Flow Analysis

of a PV Integrated UPQC System in a Distribution Network. IEEE Transactions on Industry Applications (vol 58), pp. 201–211.

[7] E. Espina, M. Espinoza-B, and D. Sáez. 2020, Cooperative Regulation of Imbalances in Three-Phase Four-Wire Microgrids Using Single-Phase Droop Control and Secondary Control Algorithms. IEEE Transactions on Power Electronics (vol 35), pp. 1978-1992.

[8] X. Liang and C. Andalib -Bin- Karim. 2018, Harmonics and Mitigation Techniques Through Advanced Control in Grid-Connected Renewable Energy Sources: A Review. IEEE Transactions on Industry Applications (vol 54), pp. 3100–3111.

[9] X. Wang and F. Blaabjerg. 2019, Harmonic Stability in Power Electronic-Based Power Systems: Concept, Modeling, and Analysis. IEEE Transactions on Smart Grid (vol 10), pp. 2858–2870.

[10] K. Liu, T. Liu, Z. Tang, and D. J. Hill, 2019, Distributed MPC-Based Frequency Control in Networked Microgrids With Voltage Constraints. IEEE Transactions on Smart Grid (vol 10), pp. 6343–6354.

Research of Carbon Ceramic Materials for Closing Resistors by Hot Pressing Sintering

Nianping Yan[1], Guilai Yin[1], Tian Chen[1], Jinyan He[2], Zhixiang Zhu[3], Jingkai Nie[3], Rongxia Huang[4], Zhuo Liu[2], Xilin Wang[2*] [0000-0002-5285-365X]

[1] Electric Power Science Research Institute, State Grid Jiangxi Electric Power Co. Ltd. Nanchang, 330096, P.R. China

[2] Tsinghua Shenzhen International Graduate School, Tsinghua University, Shenzhen, Guangdong, 518055, P.R. China

[3] China Electric Power Research Institute, Beijing, China, 100192, P.R. China

[4] School of Electro-mechanical Engineering, Guangdong University of Technology Guangzhou, P.R.China

*Corresponding author's e-mail:wang.xilin@sz.tsinghua.edu.cn

Abstract. The closing resistor is an important component of high-voltage circuit breakers. Its function is to withstand the L-C oscillation in the power grid during line closing, convert electrical energy into thermal energy, and release it to the external environment to achieve the purpose of suppressing closing overvoltage. The ceramic resistors currently used in circuit breakers are mainly resistance sheet products from a British company. The sintering method for preparing resistance sheets is pressureless sintering, but it cannot be ignored that this sintering method has problems such as high sintering temperature, long insulation time, and abnormal grain growth. Considering that hot pressing sintering can produce ceramic products with higher density and better performance, this paper uses hot pressing sintering to prepare carbon ceramic closing resistors, and compares their performance with resistance sheets of the same specification from a British company. The experimental results show that the density of resistance sheets prepared by hot pressing sintering is much higher than that of British company products, and their conductivity is better. Compared to products from British companies, the thermal conductivity has doubled and the bending strength has increased by nearly twice.

1. Introduction

In ultra-high voltage and ultra-high voltage systems, operating overvoltage directly affects the insulation level of power equipment. In voltage level lines of 550kV and above, when no-load long lines are switched to the power grid system through closing, significant overvoltage will be generated due to sudden changes in power grid parameters and energy oscillations of inductance and capacitance on the lines[1]. In order to suppress the operating overvoltage caused by line closing, it is necessary to install a closing resistor on the circuit breaker. When the line is closed, the closing resistor is turned on 8-11 milliseconds in advance. At this time, the closing resistor will withstand the L-C oscillation energy in the power grid and convert it into heat energy, which will be released into the external environment [2].

The ceramic resistors currently used in circuit breakers are mainly resistance sheet products from a British company. The raw materials for resistance sheets are clay, alumina, and carbon. After mixing, the materials are pressed into the desired shape, and then the resistors are sintered at high temperature in an inert atmosphere tunnel kiln. The sintered resistor is coated with electrodes on the upper and lower surfaces to provide electrical contact, and insulated with insulation paint on the side to improve insulation performance. The resistance value of the closing resistor in this system can be adjusted by changing the proportion of each component, especially the proportion of carbon powder. Due to the presence of carbon, the closing resistor needs to be sintered in an inert atmosphere.

In the current research on the sintering process of carbon ceramic resistors, research institutions represented by Xi'an Jiaotong University [3], Shanghai University [4], University of Pretoria in South Africa [5], Indian Institute of Technology [6], etc., mainly focus on pressureless sintering for the preparation of resistance sheets. Pressure free sintering is widely used due to its low cost and simple production method, but it cannot be ignored that this sintering method has problems such as high sintering temperature, long insulation time, abnormal grain growth, and easy cracking and deformation of the ceramic body.

Hot press sintering is a widely used and mature sintering method. The sintering device is shown in Figure 1. During the sintering process, uniaxial pressure is applied while heating the powder in the graphite mold. The pressure is usually between 10-30 Mpa. Under the pressure assisted sintering method, the sintering temperature can be reduced, the sintering time can be reduced, and ceramic products with high density and better performance can be obtained. Numerous studies have shown that hot pressed sintered products can improve the mechanical and thermal properties of ceramics, such as flexural strength and fracture toughness, compared to pressureless sintering [8].

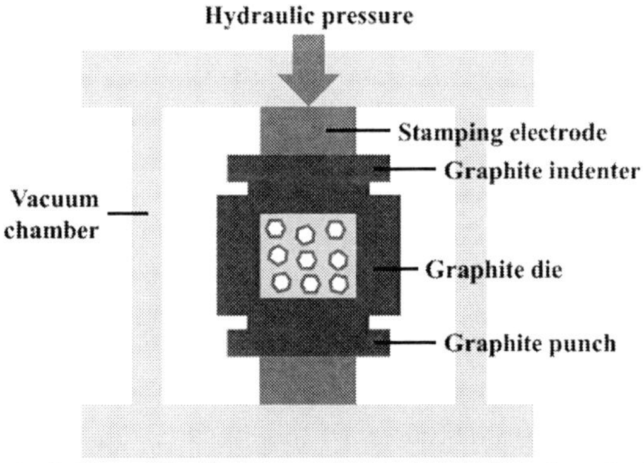

Figure 1. Schematic diagram of hot press sintering device [7]

2. Experimental

2.1. Sample Preparation and Test Description:

This experiment uses chemically pure raw materials, with a composition of 60 parts by weight of alpha alumina, 45 parts by weight of kaolin, and 15 parts by weight of graphite. Among them, alpha alumina comes from Guangzhou Nano Chemical Co., Ltd., with a purity of 99.9% and a particle size of 1μm; Kaolin comes from Hunan Shanlin Shiyu Mineral Products Co., Ltd., with a purity of 95% and a particle size of 1μm; Graphite comes from Henan Liugong Graphite Co., Ltd., mainly in the form of flake graphite, with a purity of 99.9% and a particle size of 1μm.

Weigh the raw material powder and put it into a ball milling tank. The grinding ball is zirconium ball, and the grinding medium is anhydrous ethanol. The material, medium, and ball are placed in a 1:4:20 weight ratio, and then placed on a planetary ball mill with a speed set at 400 r/min for 12 hours of ball milling. Pour the ball milled slurry into the tray, separate the grinding ball to obtain the slurry, and obtain the mixed raw material powder through the spray granulation method.

In the hot pressing sintering method, the powder is loaded into a graphite mold with a diameter of 100mm. The hot pressing sintering parameters are: argon atmosphere, sintering temperature of 1400 °C, holding time of 1 hour, and sintering pressure set to 20Mpa. The sintered resistor has a diameter of 100mm and a thickness of 25mm. It is sprayed with silver electrodes, aluminum electrodes, or copper aluminum alloy electrodes on the upper and lower surfaces through plasma spraying. Insulation high resistance coating is sprayed on the side and inner hole side walls, and dried in an oven to obtain a complete closing resistor sheet product. A performance comparison study was conducted on products of the same specifications from the British company, and the physical images of the two are shown in Figure 2.

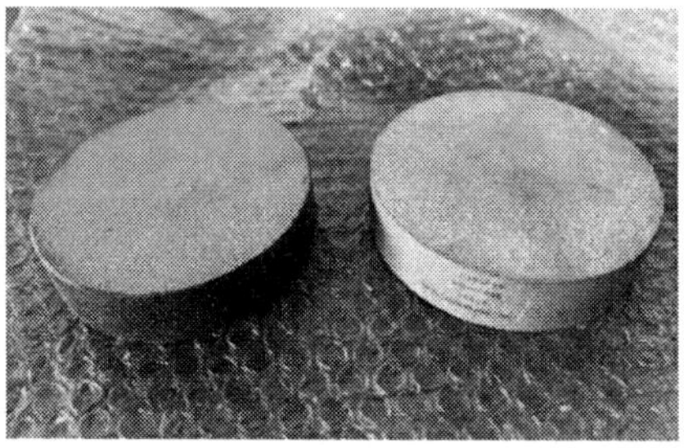

Figure 2. Physical images of hot pressed sintered resistor (left) and British company resistor (right)

2.2. Microscopic Characterization and Performance Testing:

In terms of microscopic characterization: 1) Observing and analyzing the microstructure of ceramic resistance using HITACHI SU8010 field emission scanning electron microscope (SEM), and labeling the distribution of conductive carbon elements in the resistance using X-ray energy dispersive spectroscopy (EDS). 2) Use Bruker D8 Advance in-situ X-ray powder diffraction (XRD) to analyze the phase composition of ceramic resistance. In terms of performance testing: 1) Resistivity and temperature coefficient of resistance of the sample are measured using both a heating platform and a KEITHLEY 2470 digital source meter, according to GB/T 6148-2005. 2) Bending strength at 25°C, according to GB/T 6569-2006, is measured by the three-point bending method using a CMT6104 microcomputer controlled electronic universal testing machine to measure the bending strength of resistance. 3) Ceramic density and porosity at 25°C were measured using the METTLER TOLEDO ME104 balance density component using the Archimedes drainage method according to GB/T 25995-2010. 4) Thermal performance, according to GB/T 22588-2008, the thermal conductivity of ceramic resistance at 200°C is measured using the LFA457 Micro Flash laser thermal conductivity analyzer and laser pulse method. For all performance testing samples, 5-20 samples are prepared for each type of resistance to be tested, and the average of all sample test results is taken as the final result.

3. Results and Discussion

3.1. Microscopic Characterization:

The microstructure and C-element labeling of hot pressed sintered resistance sheets (sample number HPS) and British company resistance sheets (sample number BRS) are shown in Figure 3, where the red part represents the labeled carbon element. The hot pressed resistance sheets have few pores, graphite is distributed in the system, and the direct gap with the aggregate is very small. The British resistor can observe that carbon black is relatively evenly distributed in the system. Due to the small particle size of carbon black, agglomeration occurs, and there are gaps between the aggregates, which affects the density of ceramics.

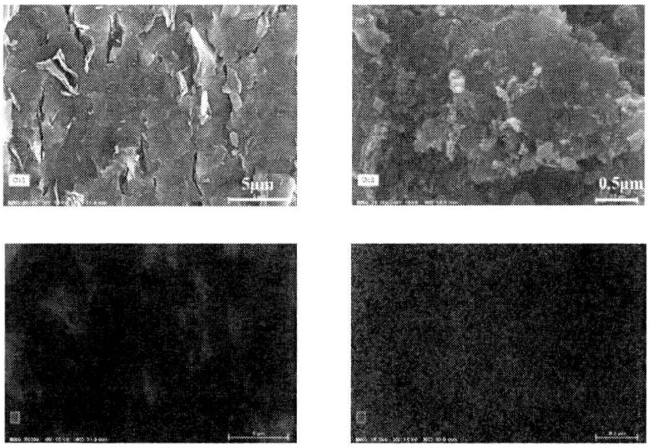

Figure 3. Microscopic morphology and C-element labeling diagram of hot pressed sintered resistor (left) and British resistor (right)

After cutting the resistor into small pieces and polishing them flat, phase analysis was performed using an in-situ X-ray powder diffractometer (XRD), and the results are shown in Figure 4.

The main substances in British company resistors include corundum (α-Al_2O_3), cristobalite, and mullite, with unclear carbon black peaks; The main substances in the hot pressed sintered samples include corundum (α-Al_2O_3) and mullite, with a more prominent graphite peak. The main difference between the two lies in the difference in conductive carbon materials. The hot pressed sample formula uses flake graphite, which is different from the carbon black added to British company products. The finished products of the two are mainly composed of corundum and mullite. The difference is that the British resistance sheet contains cristobalite, which is due to the different raw materials and the use of chemically pure raw materials, reducing intermediate products.

Density and Porosity: In the experiment, the sample is first dried and placed on a density component balance to record its mass in the air. Then, it is placed on a leak screen that has been submerged in water. After the water reading stabilizes, the mass in water is recorded. Take out the sample from the water, wipe the surface of the sample with gauze, and quickly measure the wet weight in the air after absorbing water. The testing environment is maintained at a temperature of 25°C. Finally, calculate the volume density and porosity of the ceramic resistance using the following formula:

$$D = \frac{m_1}{m_2 - m_1}\rho \tag{1}$$

$$P = \frac{m_3 - m_1}{m_3 - m_2} \times 100\% \tag{2}$$

Among them, D is the volume density, P is the porosity, ρ is density of the medium liquid, m_1 is the weight of the sample after drying, m_2 is the weight of the sample immersed in water, and m_3 is the weight of the sample in air after absorbing water. Calculate the volume density and porosity of the sample according to the formula, as shown in Figure 5.

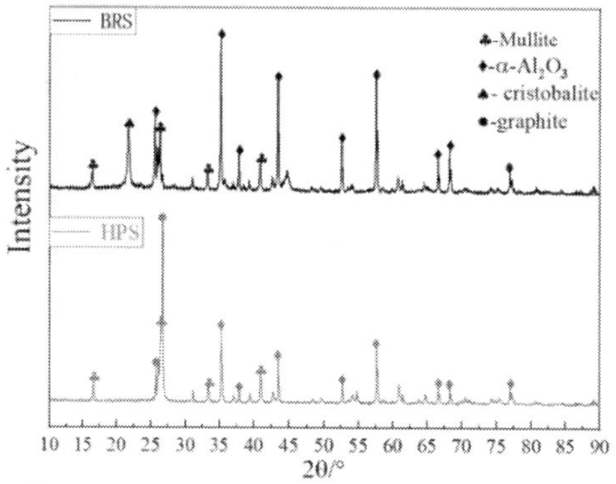

Figure 4. XRD patterns of hot pressed samples (HPS) and British company samples (BRS)

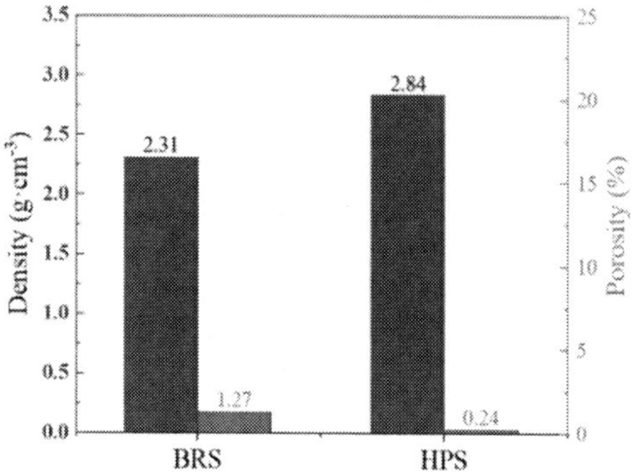

Figure 5. Density and porosity of hot pressed samples (HPS) and British company samples

According to the measurement results, the volume density values of the two are relatively close, and the porosity of the hot pressed sintered sample is only 0.24%, which is much higher than the density of the British company sample. This reflects that the hot pressed sintering method has a significant effect on improving the density of ceramics, and the improvement of microstructure also contributes to the improvement of macroscopic properties.

3.2. Electrical Performance:

Using the Keithley 2470 digital source meter, place the test electrode on the electrode coating of the resistance, and record the resistance value as R. Measure the resistance values at 25°C and 200°C. Finally, calculate the volume resistivity and temperature coefficient of resistance according to the formula, which is as follows:

$$\rho = RS / L \tag{3}$$

$$\alpha = \frac{R_2 - R_1}{R_1 \times (T_2 - T_1)} \tag{4}$$

Among them, ρ is the resistivity, R is the resistance value, S is the cross-sectional area, L is the thickness, α is the temperature coefficient of resistance, R_1 is the resistance value at T_1=25°C and R_2 is the resistance value at T_2=200°C. Calculate the volume resistivity and temperature coefficient of resistance of the sample according to the formulas, as shown in Tab. 1.

According to the measurement results, the electrical resistivity of the hot pressed sintered sample is much lower than that of the British company sample. On the one hand, this is because the conductivity of graphite is better than that of carbon black. On the other hand, compared to pressureless sintering, the hot pressed sintering method improves density, shortens the conductive path of carbon materials, and significantly improves conductivity. The temperature coefficient of resistance of the hot pressed sintered sample is higher than that of the British company sample.

3.3. Mechanical Property:

Cut two types of resistance sheet samples into test specimens with a width of 40mm and a thickness of about 3mm. Use the CMT6104 microcomputer to control the bending strength testing component of the electronic universal testing machine, set the span to 30mm, and place the test specimen on the workbench. Approximately using a three-point bending loading method, continuously apply bending load in the direction perpendicular to the thickness of the specimen at a loading rate of 1mm/min until the specimen is damaged and automatically stops. The testing environment is maintained at a temperature of 25°C. The calculation formula for bending strength is as follows:

$$\sigma = \frac{3P_{\max}L_0}{2bh^2} \tag{5}$$

Among them, σ is the bending strength, $P_{\max}$ is the maximum load when the sample is damaged, L_0 is the span of the bent sample, b is the average width of the sample, and h is the average thickness of the sample. Obtain the bending strength values of the samples according to the formula as shown in Table 1.

Table 1. Resistivity, temperature coefficient of resistance, and bending strength of hot pressed samples (HPS) and British company samples (BRS)

Sample	Resistivity at 25°C/ (Ω·cm)	Resistivity at 200°C/ (Ω·cm)	Temperature coefficient of resistance/ (10^{-3}/°C)	Bending strength/ MPa
BRS	402.9	347.8	-0.78	44.3
HPS	9.89	5.96	-2.27	120

The bending strength of the hot pressed sintered samples has increased by nearly twice compared to the samples from the UK company, reflecting the advantages of hot pressed sintering. While sintering is heating up, pressure is applied to directly sinter the powder into shape, resulting in a high degree of densification of the sintered body and good mechanical properties.

3.4. Thermal Properties

The thermal diffusion coefficient was measured using the MDTC-EQ-M04-01 laser thermal conductivity analyzer, and the specific heat capacity was measured using the MDTC-EQ-M06-01 differential scanning calorimeter. The thermal conductivity coefficient was calculated based on the following formula:

$$\lambda = \alpha \cdot \rho \cdot C_P \tag{6}$$

Among them, λ is the thermal conductivity coefficient (W/(m·K)), α is the thermal diffusion coefficient (mm^2/s), ρ is the volume density (g/cm^3), C_P is the specific heat capacity (J/(g·K)), and the thermal parameters are shown in Figure 6.

The thermal diffusion coefficient and bulk density of the hot pressed sintered sample are both higher than those of the British company sample, and the specific heat capacity of the two is basically the same. Therefore, the thermal conductivity of the hot pressed sintered sample is calculated to be 4.96 W/(m·K), which is twice that of the British company sample, and the thermal parameters are better. This also reflects that the improvement of the microstructure by the hot pressed sintering method helps to enhance the thermal performance of the resistance.

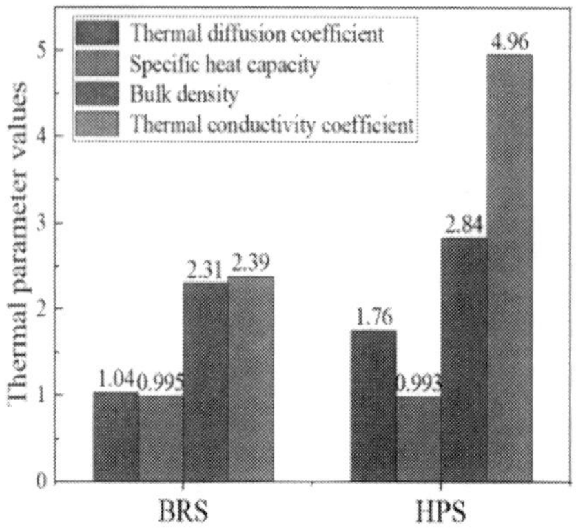

Figure 6. Thermal parameters of hot pressed samples (HPS) and British company samples (BRS)

4. Conclusion

This article uses hot pressing sintering technology to prepare carbon ceramic material samples for closing resistors, and compares and analyzes the electrical, mechanical, and thermal properties with the closing resistor products mainly used by British companies in circuit breakers today. The conclusions are as follows:

(1) The electrical resistivity at 25°C of the British company's resistor products is 402.9 Ω· cm, resistivity at 200°C is 347.8 Ω· cm, temperature coefficient of resistance is -0.78×10^{-3}/°C, porosity is 1.27%, density is 2.31 g/cm^3, thermal conductivity is 2.39W/(m·K), and bending strength is 44.3 MPa; The resistivity of the sample prepared by hot pressing sintering is 9.89 Ω· cm, resistivity at 200°C is 5.96 Ω· cm, temperature coefficient of resistance is -2.27×10^{-3}/°C, porosity is 0.24%, density is 2.84 g/cm^3, thermal conductivity is 4.96 W/(m·K), and bending strength is 120 MPa.

(2) The density of resistance sheets prepared by hot pressing sintering method is much higher than that of products from British companies, and their conductivity is better. Compared to products from British companies, the thermal conductivity has doubled and the bending strength has increased by nearly twice.

(3) Compared with traditional sintering, hot pressing sintering improves the microstructure of ceramics under pressure, and can produce better performance closing resistors without the need for additional raw material units. This sintering method has great application prospects.

Acknowledgments

This work was supported by the Jiangxi Key Research and Development Program (20243BBG71028), State Grid Corporation of China Science and Technology Project (5500-202455373A-3-2-HW) and the National Natural Science Foundation of China (52077118).

References

[1] J. He, J. Wu, P. Meng and J. Hu, "Feasibility analysis of operating overvoltage in deep restricted ultra-high voltage systems," High Voltage Engineering, vol. 44(1), pp. 210-217, January 2018.

[2] Y. Bao, K. Liu, D. Wen, Y. Li, H. Wang and H. Zhang, "Electrical-thermal coupling characteristics of pre-insertion resistors in AC filter circuit breaker for UHV grid," Mathematical Biosciences and Engineering: MBE, vol. 20(7), pp. 12056-12075, May 2023.

[3] R. Chen, J. Song, L. Hao, Z. Xu, K. Li and W. Chen, "Research on the mixing process of carbon ceramic resistors for high voltage switchgear," High voltage electrical appliances, vol. 57(11), pp. 196-201, November 2021.

[4] W. You, X. Ren, W. Yu, X. Xu, Q. Cheng and X. Wang, "The influence of sintering temperature and aggregate particle size on the linear resistance performance of ceramics," Electric porcelain lightning arrester, vol. 4, pp. 215-220, April 2021.

[5] F.O. Aramide and A.P. Popoola, "Effects of Ifon clay and sintering temperature on the phase

evolution and physico-mechanical properties of mullite-carbon ceramic composites," Journal of Ceramic Processing Research, vol. 19, pp. 87-97, September 2018.
[6] R. Kumar, P. Bhargava, "Fabrication of low specific resistance ceramic carbon composites by slip casting," Journal of Asian Ceramic Societies, vol. 3(3), pp. 262-265, January 2018.
[7] K. Lu, C. Wang, C. Wang, H. He, X. Fan, F. Chen and F. Qi, "Study of the hot-pressing sintering process of diamond/copper composites and their thermal conductivity," Journal of Alloys and Compounds, vol. 960, pp. 170608, January 2023.
[8] J. Lin, X. Zhang, W. Han, "Comparison of ZrB2–ZrO2f ceramics prepared by hot pressing and pressureless sintering," International Journal of Refractory Metals and Hard Materials, vol. 35, pp. 102-107, November 2023.

Journal of Physics: Conference Series **3012** (2025) 012065 doi:10.1088/1742-6596/3012/1/012065

Research on Multi-timescale Optimized Scheduling Strategy for Peak Load Regulation with Participation of Distributed Energy Storage Aggregators

Yinchao Wang[1], Wen Gu[1], Dong Han[1], Junxia Yu[1], Qizhen Guo[2*], Haijun Xing[2]

[1] State Grid Shanghai Electric Power Company Economic and Technical Research Institute, Shanghai, 200120, P.R. China

[2] Shanghai University of Electric Power, Shanghai, 200090, P.R. China

*Corresponding author's e-mail: 19811716052@163.com

Abstract. As environmental issues intensify and fossil fuel resources become increasingly scarce, the construction of a new type of power system has become an important strategy for China to address the energy crisis. In response to the system peak load regulation pressure brought about by the rapid development of renewable energies such as wind and photovoltaic power, the article proposes a collaborative scheduling strategy that integrates wind and photovoltaic power, thermal power units, pumped storage units, and distributed energy storage aggregators. Under the aggregator model, the strategy considers a multi-timescale optimization approach for distributed energy storage to participate in peak load regulation, using ultra-short-term and real-time forecasting data, aiming to balance the system's economic efficiency and the rate of new energy consumption. Case study validation shows that the strategy can effectively improve the rate of new energy consumption while ensuring the system's economic and operational stability. The research in the article provides a new perspective on the scheduling of distributed energy storage aggregators in the power system, which is of great significance for improving the system's peak load regulation capabilities and new energy consumption.

1. Introduction

In the face of growing environmental concerns and energy shortages, there is a global push for the widespread use of renewable energy sources such as solar and wind. However, the instability and large-scale grid integration of these energy sources have led to grid peaking and energy abandonment problems. By the end of 2023, China's installed power generation capacity reached 2.92 billion kilowatts (kW), with significant growth in wind and photovoltaic (PV) capacity, which is projected to exceed 1.2 billion kW in 2030[1]. In the meantime the proportion of thermal power is decreasing, increasing the pressure on peaking [2]. Distributed energy storage systems (DESS) are widely used due to their flexibility and versatility [3], but it is difficult to directly participate in grid regulation due to decentralization and small capacity [4], so aggregators are needed as an intermediate bridge for distributed energy storage to participate in system operation.

Distributed energy storage has numerous application scenarios. Reference [5] summarizes the planning method based on the energy storage access method, which helps to alleviate the pressure of grid peak regulation. Electric vehicles, a unique form of distributed energy storage, can also participate in regulation. Reference [6] proposes a multi-mode frequency regulation strategy for electric vehicle clusters based on model predictive control to realize fast response. On the user and microgrid side, distributed energy storage improves voltage quality through demand response. Reference [7] explored a demand response model for energy storage batteries based on time-of-use tariffs.The participation of distributed energy storage in system operation effectively helps alleviate peak pressure and balance supply and demand. Reference [8] developed a two-tier optimization model for energy storage participation in system peaking, taking into account both system economics and the entire lifecycle economics of storage.Reference [9] proposed a two-layer optimization model for energy storage that considers system carbon emissions and operational risks, improving the system's peak flexibility.

Compared with existing methods, if only focusing on the time scale before the day, it will not be able to take advantage of the wind power, photovoltaic, load power prediction accuracy with the shortening of the time scale step by step to improve the characteristics of intermittent energy access to the grid, the need for coordinated optimization of scheduling through a number of time scales, step by step to eliminate the impact on the operation of the system. In addition, taking the thermal power unit as the basic power source, and configuring a certain capacity of pumped storage units and distributed energy storage to cooperate with the thermal power unit, it can provide the system with more space for new energy consumption. Therefore, this paper innovatively proposes a coordinated scheduling strategy that considers wind power PV, thermal power units as well as pumped storage units with flexible regulating characteristics and distributed energy storage aggregators regulating distributed energy storage to jointly participate in system operation.And based on the ultra-short-term forecast data and real-time forecast data, a short-term three-phase coordinated scheduling model is constructed for the period of 1 hour before the day-1 hour during the day-15 min in real time. The multi-timescale optimized scheduling strategy maximizes the fast regulation capabilities of pumped storage units and distributed energy storage. Through example validation, the proposed multi-timescale coordinated scheduling model effectively reduces the peaking and standby pressure on thermal power units, minimizes wind and solar abandonment, and enhances renewable energy consumption.

2. Distributed Energy Storage Aggregators and Their Operating Models

Distributed Energy Storage Aggregators monitor and optimize by integrating various distributed energy storage resources using smart technologies in order to provide services including energy storage, dispatch, sales and trading. The primary objective of an energy storage aggregator is to optimize the energy storage system's benefits, ensure efficient energy use and flexible dispatch, and support energy market balance and demand response. Energy storage aggregators, by combining decentralized storage resources, can offer more flexible and reliable energy solutions, supporting the growth and adoption of clean energy.

Distributed energy storage transmits information such as equipment status and output expectations to the aggregator, which performs charging and discharging based on the storage's own status in response to commands from the grid control center. Under the aggregator's unified management, energy storage systems can store power during low-demand periods and discharge during peak times, balancing power shortages and surpluses. This helps reduce system load fluctuations, stabilize renewable energy sources like wind power, and address the system's peaking issues. Effective management of distributed energy storage usually adopts a centralized-distributed control strategy, with unified regulation and control by the aggregator. The control structure is divided into three layers: the grid dispatch center issues commands, the aggregator layer issues charging and discharging commands and monitors the equipment, and the user side performs operations. In the framework proposed in this paper, the aggregators independently manage the energy storage in their respective regions and participate in grid scheduling, which improves scheduling efficiency and resource utilization.

3. Multi Time Scale Scheduling Architecture

Based on the participation of distributed energy storage, the participation of pumped storage units is considered to analyze its impact on system stability and new energy consumption. Considering the large fluctuation of new energy output in different time scales, in order to solve the multi-time scale supply and demand balance problem brought by large-scale new energy grid connection, it is necessary to carry out multi-time scale optimal scheduling for the system. Based on the short-term, ultra-short-term and real-time forecast data of wind power, photovoltaic, and load, the intra-day and day-ahead scheduling schemes are corrected by the real-time scheduling plan to achieve the requirements of reducing the system operation cost and improving the system stability. The basic architecture of multi-time scale scheduling is shown in Figure 1:

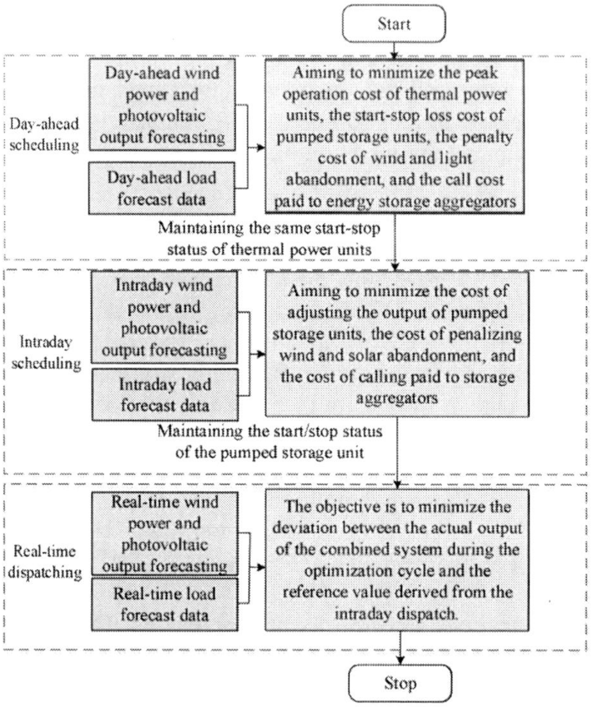

Figure 1. Multi-timescale scheduling flowchart

4. A multi-timescale scheduling model for distributed energy storage aggregators to participate in system peaking

4.1. Day-ahead optimization scheduling model

The day-ahead scheduling model developed in this paper aims to minimize system operating costs, which encompass the peak operation cost of thermal power units, the operating cost of pumped-storage units, penalties for wind and solar abandonment, and the fees paid to the energy storage aggregator,as follows:

$$F_1 = \min(C_{G,s} + C_{ps,s} + C_{loss,s} + C_{2,s}) \tag{1}$$

$$C_{G,s} = C_{gen,s} - I_{gen,s} \tag{2}$$

$$C_{ps,s} = \sum_{j=1}^{J}\sum_{t=1}^{T}\left\{\begin{aligned}&S_{on}\left[I^g_{j,t}(1-I^g_{j,t-1}) + I^p_{j,t}(1-I^p_{j,t-1})\right]\\&+S_{off}\left[I^g_{j,t-1}(1-I^g_{j,t}) + I^p_{j,t-1}(1-I^p_{j,t})\right]\end{aligned}\right\} \tag{3}$$

$$C_{loss,s} = C_{wloss,s} + C_{ploss,s} = \sum_{t=1}^{T}(c^{wp}P_{wp,t} + c^{pv}P_{pv,t}) \tag{4}$$

Where: $C_{G,s}$ is the cost function of thermal power units; $C_{\mathrm{ps},s}$ is the operating cost of pumped storage units; $C_{\mathrm{loss},s}$ is the cost of wind and light abandonment for day-ahead dispatch; $C_{2,s}$ is the peaking cost paid by the system to the distributed storage aggregator; $C_{\mathrm{gen},s}$ is the peaking operating cost of thermal power units on a typical day s; $I_{\mathrm{gen},s}$ is the peaking compensation benefit of thermal power units on a typical day s; S_{on} , S_{off} are the startup cost and shutdown cost of the pumped storage unit, respectively; $I^{g}_{j,t}$, $I^{p}_{j,t}$ are the 0-1 state variables of whether the pumped storage unit j is in power generation or pumped storage condition in time period t, with 1 being yes and 0 being no; $C_{\mathrm{wloss},s}$, $C_{\mathrm{ploss},s}$ are the cost of wind and light abandonment under a typical day s, respectively; c^{wp} is the unit cost of wind abandonment; c^{pv} is the unit cost of photovoltaic abandonment; $P_{\mathrm{wp},t}$ denotes the standard deviation of the prediction error of wind power output at time t; $P_{\mathrm{pv},t}$ is the standard deviation of the prediction error of photovoltaic output at time t; and T is the scheduling period.

In day-ahead scheduling, the constraints include thermal power unit output limitations, climbing capacity, spinning reserve and start/stop limitations, distributed energy storage charging and discharging power and energy state limitations, wind power output range limitations, pumped storage plant output and start/stop limitations as well as water quantity constraints, and power balance constraints for the total output of the system to meet the load demand.

4.2. *Intraday optimized scheduling model*

The hourly conventional unit startup and shutdown schedules for the next 24h are obtained through the day-ahead optimal dispatch model, and the pumped storage unit and thermal unit output startup and shutdown schedules are corrected in the intraday dispatch phase using the ultra-short-term wind-scenery and load forecast data. By adjusting the unit output size, the deviation of the wind and solar load forecast in the new time scale is absorbed. In order to give full play to the flexible adjustment capability of distributed energy storage and pumped storage units at 1h scale, to realize greater wind and solar energy consumption and further reduce system operation costs, the intra-day scheduling model aims to minimize the cost of adjusting the output of pumped storage units, the cost of penalizing wind and solar energy abandonment, and the cost of calling paid to the energy storage aggregator, as follows:

$$F_2 = \min(C_{\Delta\mathrm{ps},s} + C_{\mathrm{loss,s}} + C_{2,s}) \tag{5}$$

$$C_{\Delta\mathrm{ps},s} = \sum_{t=t_0}^{t_0+N_T-1} \sum_{j=1}^{J} c_{\Delta\mathrm{ps}} P^{\mathrm{hour}}_{\mathrm{ps},j,t} \tag{6}$$

where F_2 is the total operating cost of the intraday dispatch model; $C_{\Delta\mathrm{ps},s}$ is the pumped storage unit output adjustment cost; t_0 is the initial time period of the current dispatch cycle; N_T is the total number of time periods in the intraday dispatch cycle; $c_{\Delta\mathrm{ps}}$ is the unit output unit adjustment cost coefficient; $P^{\mathrm{hour}}_{\mathrm{ps},j,t}$ is the output adjustment of the pumping unit j in time period t compared to the previous day.

In the intraday scheduling phase, constraints include wind power output, thermal power unit operation, distributed energy storage, pumped storage unit operations, and overall system operation. The wind power output, distributed energy storage, and pumped storage unit constraints remain the same as in the previous day's model. The intraday scheduling uses the start-stop plan for thermal power units set by the previous day's schedule, with the system power balance constraints for the intraday operation as follows:

$$P^{\mathrm{h}}_{\mathrm{wp},t} + P^{\mathrm{h}}_{\mathrm{pv},t} + P^{\mathrm{h}}_{\mathrm{g},t} + \sum_{j=1}^{J} (P^{g,\mathrm{h}}_{j,t} - P^{p,\mathrm{h}}_{j,t}) + P^{\mathrm{h}}_{\mathrm{DASP},t} = P^{\mathrm{h}}_{l,t} \tag{7}$$

4.3. Real-time correction of scheduling models

In the real-time correction phase, the main focus is to quickly adjust the amount of prediction error between the intraday and real-time phases using pumped storage and distributed energy storage. The real-time dispatch model aims to minimize the deviation between the actual system output and the reference value from the intraday dispatch during the optimization cycle. The objective function is expressed as follows:

$$F_3 = \min(P_t^{\mathrm{real}} - P_t^{\mathrm{roll}})^2 \tag{8}$$

Where: P_t^{real} is the total system output during the real-time correction phase in time period t; P_t^{roll} is the total planned output of generation determined by the intra-day dispatch model. In the real-time scheduling phase, the constraints are the same as in the day-ahead and intraday constraint models, and will not be repeated here.

5. Case Study

The example system in the paper consists of an IEEE30 node transmission grid and four IEEE33 node distribution grids. Among them, D1-D4 are four distribution grids, which are connected to nodes 11, 13, 25 and 27 of the transmission grid, and a pumped storage unit is connected to node 11. A distributed energy storage aggregator is set to access node 27 of the system. The errors of day-ahead, intraday, and real-time forecasts of wind and PV output are 20%, 5%, and 2%, respectively, and the errors of day-ahead, intraday, and real-time forecasts of load are 3%, 1%, and 0.5%, respectively [10].

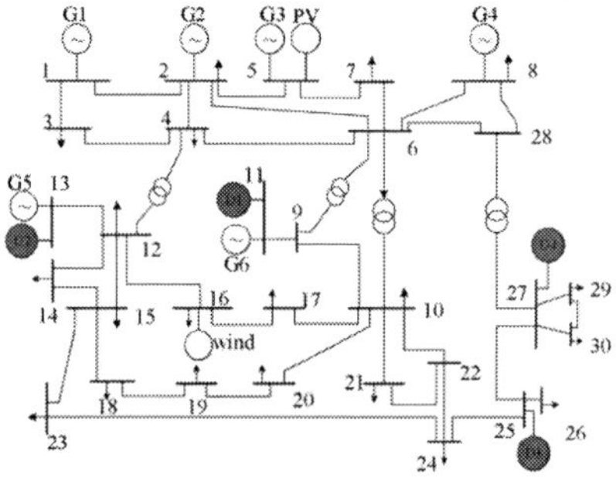

Figure 2. IEEE 30-node system

This example defines the following scenarios for comparison:

Scenario 1: Pumped storage units with distributed storage aggregators are not considered for participation, and peaking is done only through thermal units;

Scenario 2: Considering the participation of pumped storage units with distributed energy storage aggregators, only day-ahead optimized dispatch is considered;

Scenario 3: Considering the participation of pumped storage units and distributed storage aggregators, using the coordinated scheduling model with multiple time scales proposed in this chapter.

Table 1. Results of system operation indicators under different scheduling modes (ten thousand yuan)

Scenario	Total cost	Running costs of thermal power units	Wind abandonment penalty costs	Pumped storage unit	Aggregator cost
Scenario 1	128.13	122.56	5.57	0	0
Scenario 2	112.69	105.58	0.23	6.13	0.75
Scenario 3	102.7	96.36	0	5.69	0.65

As Table 1 shows the results of the system operation indexes under the three scenarios, it can be seen that Scenario 1 considers only thermal power peaking, and its thermal power's operation cost reaches 122.56 ten thousand yuan, and the wind abandonment penalty cost also reaches 5.57 ten thousand yuan. Scenario 2 includes the addition of pumped storage units and energy storage aggregators, resulting in a 16.98 ten thousand yuan reduction in thermal power unit operating costs and a significant decrease in wind and solar abandonment penalty costs. Compared with only thermal power for peaking, the addition of pumped storage units and energy storage aggregators not only reduces the total system operating cost but also realizes the significant consumption of wind and light, thus reducing the peak-to-valley difference of the system. Scenario 3 is the premise of considering pumped storage units and distributed energy storage aggregators, using the proposed multi-timescale scheduling model, compared with Scenario 2, the total operating cost of the system is reduced by 9.99 ten thousand yuan, and realizes all the wind and light to be fed into the Internet, and the operating cost of the thermal power unit is also reduced by 9.22 ten thousand yuan, which can be seen that the multi-timescale scheduling model proposed in this paper can significantly improve the economy of the system and reduce the peak-to-valley difference.

The two graphs in Fig. 3 show the distributed energy storage aggregator regulating storage charging and discharging in scenarios 2 and 3, respectively. As can be seen from the figure, the energy storage is charged at the load valley and discharged at the load peak to reduce the peak-to-valley difference of the grid load, so that the power generation and the load demand tend to be balanced. Although the aggregator regulation energy storage capacity in the article is not large, in the real-time correction phase of Scenario 3, the distributed energy storage utilizes its own fast regulation characteristics to continuously correct the prediction error between the day before and the day within the day, which ensures the safe and economic operation of the system and alleviates the system's peak regulation pressure.

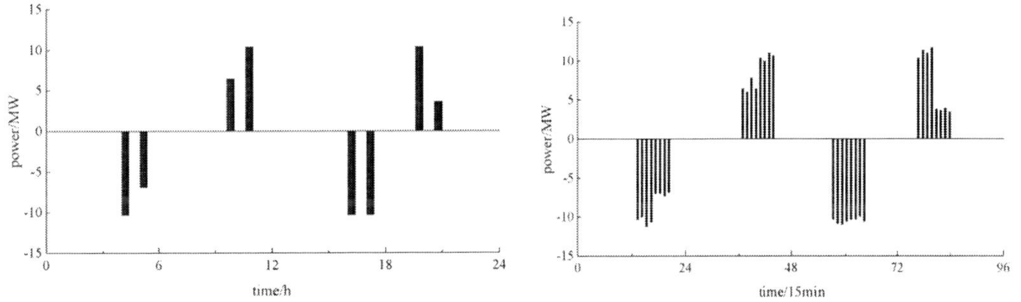

Figure 3. Distributed energy storage system operation in Scenario 2 and Scenario 3

Figure 4 shows the operation of pumped storage power plant for scenario 2 and the real-time 15-minute scale for scenario 3. From the figure, it can be seen that in the real-time scheduling phase, the pumped storage unit can give full play to its fast regulation capability to correct the day-ahead and intra-day output plans, effectively alleviating the peak pressure of thermal power units. It shows that the multi-timescale system peak dispatch model involving pumped storage units and aggregators proposed in this paper fully exploits the regulation characteristics of the pumped storage units, reduces the peak-to-valley difference of the system, and realizes its safe and economic operation.

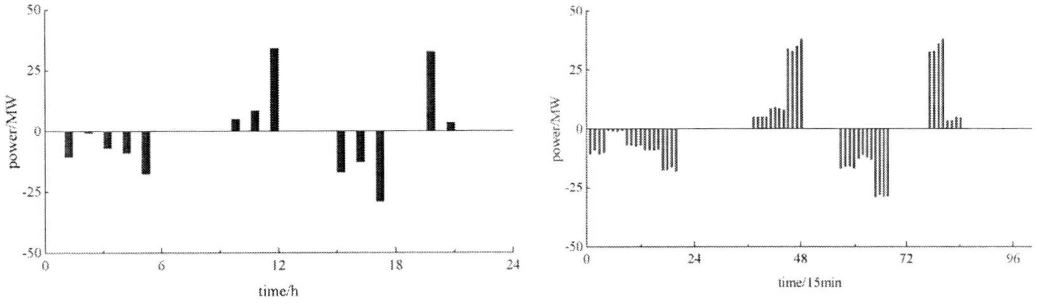

Figure 4. Pumped storage plant operation in scenarios 2 and 3

As shown in Figure 5, the distributed energy storage output at each node of the distribution network (D1) under Scenario 3 aligns with the trend of output regulated uniformly by the aggregator. The aggregator monitors the charging and discharging status of distributed energy storage at each node under its centralized management, executing charging and discharging commands based on the status of each unit in response to grid scheduling directives. In Scenario 3, the reduced time scale improves the accuracy of distributed energy storage output, which not only refines the day-ahead and intraday output plans but also minimizes resource waste, enhancing both the economic efficiency and stability of system operation.

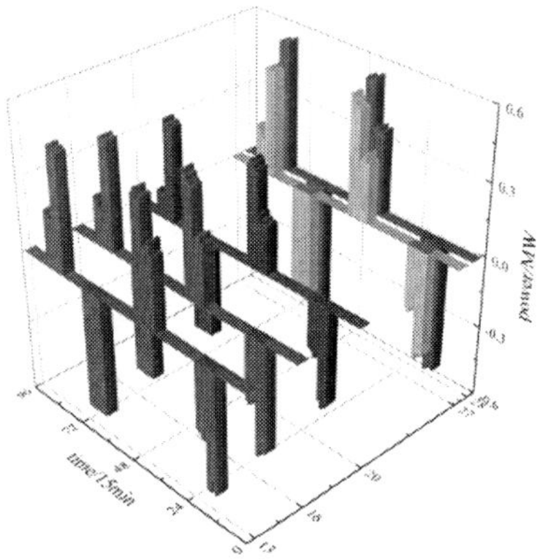

Figure 5. Operation status of distributed energy storage at each node in distribution network D1

The two graphs in Figure 6 show scenario 2 day-ahead scheduling results and scenario 3 real-time phase scheduling results, respectively.The figures show that the real-time phase leverages the fast regulation capabilities of energy storage and pumped storage units to maximize the consumption of new energy and minimize wind and solar abandonment. Additionally, real-time scheduling effectively tracks fluctuations in wind and solar output, making timely adjustments to the scheduling plan. This helps reduce system scheduling costs and ensures both the economic efficiency and stability of system operation.

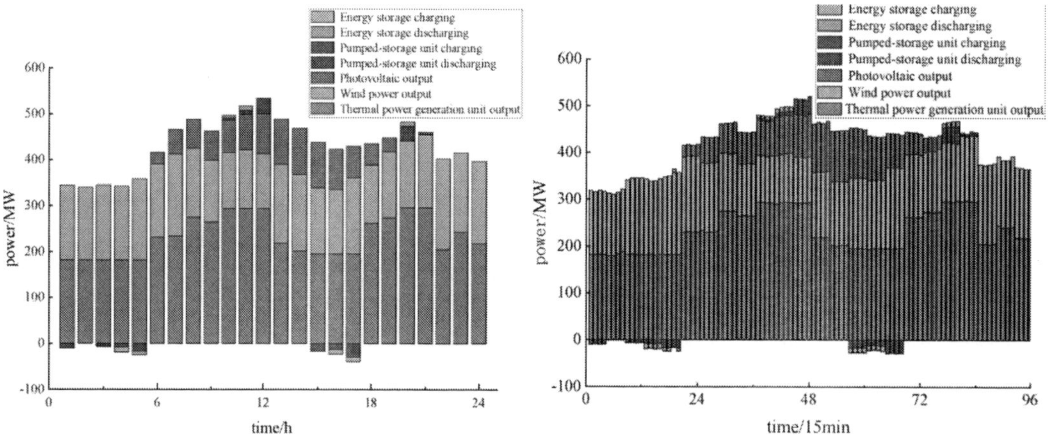

Figure 6. System optimization operation results in Scenario 2 and Scenario 3.

6. Conclusion

This paper proposes a coordinated scheduling strategy that integrates wind and solar power, thermal power units, pumped storage units with flexible regulation capabilities, and distributed energy storage aggregators to optimize system operation. The strategy incorporates a multi-timescale scheduling model for distributed energy storage participating in peak regulation under the aggregator mode, covering a 24-hour day-ahead period, 1-hour intraday intervals, and 15-minute real-time adjustments.

The analysis of a case study shows that the inclusion of distributed energy storage aggregators and pumped storage units significantly reduces system dispatch costs and minimizes wind power abandonment. Compared to a day-ahead-only scheduling approach, the proposed multi-timescale model allows for real-time adjustments to the day-ahead and intraday schedules, resulting in a reduction of total system operation costs by 99.9 million RMB. Additionally, it ensures full integration of wind and solar power and decreases thermal power unit operating costs by 92.2 million RMB. This approach not only enhances the system's economic performance but also ensures its safe and stable operation. Despite the model's excellent performance in terms of flexibility and new energy consumption, there are still limitations in its dependence on prediction accuracy and its adaptability to extreme conditions, and future work will improve the stability of the model.

Acknowledgments
This work is supported by Science and Technology Project of State Grid Shanghai Electric Power Co., Ltd (No. B3090R240008)

References

[1] Zheyuan Liu, Haijun Xing, Yangfan Luo, et al. Low-carbon economic dispatch of integrated energy system based on carbon emission flow theory[J].Journal of Electrical Engineering & Technology, 2023,18(3):1613–1624.

[2] Yang Cui, Shipeng Yu, Xuebin Wang, et al. Optimal configuration of thermal energy storage capacity considering the balance between system peak load demand and the revenue of solar thermal power stations [J]. Proceedings of the CSEE, 2023, 43(22): 8745-8757.

[3] Dongmei Zhao, Chenyu Xu, Ran Tao, et al. A review of flexible control of multi-energy distributed energy storage on the distribution side of new power systems [J]. Proceedings of the CSEE, 2023, 43(05): 1776-1799.

[4] Jianlin Li, Huimeng Ma, Xiaodong Yuan, et al. A review of key application technologies for large-scale distributed energy storage [J]. Power System Technology, 2017, 41(10): 3365-3375.

[5] Guodong Xu, Haozhong Cheng, Zifeng Ma, et al. A review of energy storage system planning methods for alleviating power grid peak load pressure [J]. Electric Power Automation Equipment, 2017, 37(8): 9-17.

[6] Kai Dong, Xinlei Cai, Yanlin Cui, et al. Aggregation modeling of electric vehicles based on Markov chains and multi-mode frequency regulation control strategy [J]. Power System Technology, 2022, 46(2): 633-634.

[7] Wenqi Li, Zhaohui Li, Haijing Yang, et al. Research on coordinated operation strategy of distributed energy storage under demand response [J]. Power Demand Side Management, 2018, 20(3): 11-14.

[8] Junhui Li, Jiahui Zhang, Cuiping Li, et al. Configuration Scheme and Economic Analysis of Energy Storage Systems Participating in Peak Shaving [J]. Proceedings of the CSEE, 2021, 36(19): 4148-4160.

[9] Zhiyao Zhang, Jingjie Huang, Nian Guang Zhou, et al. Bi-level Optimization Model for Energy Storage Participation in Low-Carbon Flexible Peak Shaving [J]. Southern Power System Technology, 2023, 17(04): 49-57.

[10] Yuqing Bao, Beibei Wang, Yang Li, et al. Rolling scheduling model considering large-scale wind power integration and coordinated optimization of multi-timescale demand response resources [J]. Proceedings of the CSEE, 2016, 36(17): 4589-4600.

Journal of Physics: Conference Series 3012 (2025) 012066 doi:10.1088/1742-6596/3012/1/012066

Research on Carbon Emission Prediction in China Based on Statistical Models

Junyi Liu[1], Aihua Chen[1*], Ziyuan Pan[1], Qinghua Li[1], Yuhang Chen[1], Xiaohai Lu[1], Minshan Zhu[1], Weijie Li[2], Yingqi Liang[3]

[1] Zhongshan Power Supply Bureau of Guangdong Power Grid Co., Ltd, Zhongshan, Guangdong Province, 528400, P.R. China

[2] Southern Power Grid Peak shaving and Frequency Regulation (Guangdong) Energy Storage Technology Co., Ltd, Guangzhou, Guangdong Province, 510000, P.R. China

[3] Electric Power Dispatching and Control Center of Guangdong Power Grid Co., Ltd, Guangzhou, Guangdong Province, 510000, P.R. China

[*]Corresponding author's e-mail: chenaihua188@163.com

Abstract. To address the requirements of China's dual-carbon target and provide scientific support for the development of the nation's electric power sector, this study introduces a carbon emissions prediction method based on statistical modeling. This approach utilizes a composite framework combining the STIRPAT model and ARIMA model, integrating factors such as installed grid power generation capacity, GDP, and population size to establish an analytical prediction model for China's level of carbon emissions. Initially, the study employs the STIRPAT model to examine the effects of population size on carbon emissions, revealing that a 1% increase in population size leads to a 2.498% rise in carbon emissions. This finding highlights population size as the most influential factor of carbon emissions. Subsequently, the ARIMA model is applied for carbon emissions forecasting. Model evaluation metrics indicate a mean absolute error (MAE) of 0.318, mean square error (MSE) of 0.001, root mean square error (RMSE) of 0.386, normalized mean absolute error (NMAE) of 0.007, and a relative error of 4.44%. These results demonstrate the model's strong fitting performance and its ability to accurately predict China's carbon emissions in the coming years. This predictive capability holds significant value for guiding the construction and optimization of the power grid.

1. Introduction

Over the past few years, carbon emissions have gained significant attention as a key driver of global climate change concerns. As a major contributor to carbon emissions, primarily from its electricity generation, heat production, industrial activities, and transportation sectors, China has taken on the responsibility to address this challenge. In 2020, China announced its important policies: to achieve carbon peaking by 2030 and carbon neutrality by 2060[1-2]. In order to monitor carbon emissions to promote environmentally friendly and green development of industries in China, the sustainably prediction of national overall carbon emissions should not be overlooked. The provision of accurate predictions could help identify trends, set achievable targets, and implement effective strategies for the purpose of reducing carbon emissions to reach China's Double-Carbon Policy[3]. Also, the result supports the transition to a low-carbon economy, fosters innovation in clean energy technologies, and enhances energy efficiency[4-6].

There were a large number of previous papers that concentrated on the carbon emission level on a particular province or specific industry such as transportation industry, agriculture or construction in China, with identifying different factors to influence carbon emission level. By using a STIRPAT model, some researchers associated positive effects of the China's transportation-related CO2 emissions with the rate of urbanization, population, economic scale, transportation development, carbon intensity of traffic, and road network length[7]. Using a combined Isomap–ACO–ET model, it is projected that Guangdong Province's agricultural carbon emissions will fall between 11,142,200 tons and 11,386,000 tons by 2030, taking into account key factors such as industrial structure, rural population and productivity efficiency[8]. Carbon emissions in China's construction industry are anticipated to increase at a slower pace in the future, driven by advancements in energy efficiency, greater use of clean energy, and internal structural reforms[9]. China's national carbon emission figures were predicted in past papers by concentrating mostly on influencing factors of urbanization, energy consumption and economic growth. Despite China's recent slowdown in economic growth, carbon emissions are projected to rise in the short term over the next five years. This increase is attributed to factors such as economic expansion, population size, energy consumption, and urbanization as analyzed using a new multivariable grey prediction model[10]. If China focuses on optimizing the energy consumption mix and moderating GDP (Gross Domestic Product) growth in the future, carbon emissions could peak at 9,575.94 million tons by 2029, and the main factors affecting carbon emissions were fossil energy use, population size, urbanization rate, and GDP [11].

Despite extensive research on carbon emissions, there has been a lack of in-depth analysis of the impacts of installed power generation capacity, population, and GDP on carbon emission levels. Moreover, many previous studies have overlooked a national perspective in carbon emission predictions. This research addresses these gaps by employing the ARIMA and STIRPAT statistical models to predict carbon emissions, offering a scientific basis and methodology for China's sustainable and low-carbon development. The study's innovative contribution lies in its focus on predicting China's total carbon emissions, rather than limiting the analysis to regional predictions, incorporating three key influencing factors: installed power generation capacity, population, and GDP. Additionally, it explores the relationships between these factors and carbon emission levels, providing valuable scientific support for China's efforts to reduce overall carbon emissions.

2. Modeling Description

2.1. The Modeling of the STIRPAT

The STIRPAT (Stochastic Impacts by Regression on Population, Affluence, and Technology) model is a statistical tool used to analyse the relationship between environmental impacts (such as carbon emissions and energy consumption) and factors like population, affluence, and technology. It extends the classical IPAT equation by incorporating stochastic elements, allowing the model to better accommodate variability in real-world data and providing more accurate quantification of environmental impacts. The STIRPAT model is a multivariable nonlinear regression model. By taking the logarithm of both sides, it can be transformed into a linear form, expressed as:

$$ln\ (I) = ln\alpha + \beta ln(P) + \gamma ln(A) + \delta ln(T) + ln(\epsilon) \quad (1)$$

In the STIRPAT model, I represents environmental impact (e.g., carbon emissions), P, A, and T stand for population, affluence, and technology, respectively. α is a constant term, β, γ, and δ are regression coefficients indicating the influence of each factor on carbon emissions, and ϵ is the error term. The STIRPAT model is commonly applied to analyze the effects of population, economic development, and energy consumption on carbon emissions and climate change. The STIRPAT model explores the connection between economic development and energy consumption, helping to understand how economic growth impacts energy demand and offering quantitative insights for evaluating the effectiveness of various energy policies.

2.2. *The Modeling of the ARIMA*

The ARIMA (Autoregressive Integrated Moving Average) model is a classic time series analysis method widely used in forecasting studies within the energy and environmental fields. By integrating Autoregression (AR), Differencing (I), and Moving Average (MA) processes, the ARIMA model effectively captures trends, periodic fluctuations, and random variations in time series data, making it suitable for forecasting scenarios without clearly defined exogenous driving variables. Given the time-dependent nature of carbon emission data, ARIMA has become a key tool for carbon emission forecasting. At its core, the ARIMA model constructs time series representations through the Autoregressive (AR) component, the Differencing (I) component, and the Moving Average (MA) component, capturing both long-term trends and short-term fluctuations in the data. The standard ARIMA model is expressed as:

$$Y_t = c + \sum_{i=1}^{p} \phi_i Y_{t-i} + \sum_{j=1}^{q} \theta_j \epsilon_{t-j} + \epsilon_t \tag{2}$$

Y_t denotes the observed value of the time series; c represents the constant term; ϕ_i is the coefficient of the autoregressive (AR) term, reflecting the influence of historical observations on the current value; θ_j is the coefficient of the moving average (MA) term, indicating the correction of past errors on the current value; and ϵ_t represents the error term, capturing white noise.

In this study, the ARIMA model is employed to forecast future carbon emissions based on historical data. Through model fitting, it captures the trends and seasonal variations in carbon emission data, providing policymakers with short-term predictions and supporting the formulation of carbon emission reduction strategies.

2.3. *Ridge Regression*

Ridge Regression, also known as Tikhonov Regularization, is an enhanced linear regression algorithm specifically designed to address multicollinearity issues. In the field of carbon emissions prediction, Ridge Regression is widely employed to tackle the challenges posed by high correlations among features, particularly when multiple economic and social indicators are involved. By introducing an L2 regularization term into the loss function, Ridge Regression effectively reduces the variance of parameter estimates, mitigating multicollinearity issues and improving model stability. This characteristic is especially critical in carbon emissions forecasting, where accurate predictions under varying scenarios are essential. The basic form of Ridge Regression can be expressed as:

$$\text{Minimize} \sum_{i=1}^{n} \left(y_i - \beta_0 - \sum_{j=1}^{k} \beta_j x_{ij}\right)^2 + \lambda \sum_{j=1}^{k} \beta_j^2 \tag{3}$$

Where y_i is the dependent variable of the ith observation; x_{ij} represents the jth independent variable of the ith observation; β_0 is the intercept; β_j beta is the coefficient of the jth independent variable; λ is the regularization parameter (also known as the ridge parameter), controlling the strength of the regularization term; and k is the number of independent variables.

2.4. *Granger Causality Test*

The Granger Causality Test, introduced by economist Granger in 1969 [12], is a statistical method for examining causal relationships between time series. This approach investigates whether one time series can provide predictive information for the future values of another time series. Specifically, if the past values of variable X can assist in forecasting the future values of variable Y, it is said that X Granger-causes Y (i.e., there is a Granger causal relationship from X to Y). The core idea of Granger causality is that if the historical information of variable X has predictive power over variable Y, and this predictive power surpasses that of using only the past information of Y, then X can be considered to Granger-cause Y. The Granger causality test is derived based on the following regression models:

$$Y_t = \alpha_0 + \sum_{i=1}^{p} \beta_i Y_{t-i} + \sum_{i=1}^{q} \gamma_i X_{t-i} + \epsilon_i \tag{4}$$

α_0, β_i are the parameters of the autoregressive part of Y_t, γ_i is the parameter of the effect of X_{t-i} on Y_t, p is the lagged order of Y_{t-i}, q is the lagged order of X_{t-i}, and ϵ_i is the error term.

2.5. Evaluation Criteria

Assessing a model's predictive performance typically involves comparing the predicted values of the target variable with the actual observed values. Based on this comparison, conclusions about the model's effectiveness can be drawn. In this study, the evaluation metrics employed include Mean Absolute Error (MAE), Mean Squared Error (MSE), Root Mean Squared Error (RMSE), Normalized Mean Squared Error (NMSE), and Normalized Mean Absolute Error (NMAE) as standards for measuring the model's predictive performance. Specifically: (1) MAE measures the average absolute deviation between predicted and actual values, reflecting the average size of the prediction errors; (2) MSE quantifies the average of the squared deviations between predicted and actual values, representing a measure of squared prediction errors; (3) RMSE as the square root of MSE provides an error measure in the same unit as the original data; (4) NMSE is the ratio of the model's predictive performance to that of a simple baseline model (e.g., a mean model), used to evaluate the model's relative performance; (5) The value of NMSE ranges between 0 and 1; if the model outperforms the baseline, its value should be significantly less than 1, with smaller values indicating better model performance. The flowchart of research is shown in the following Figure 1.

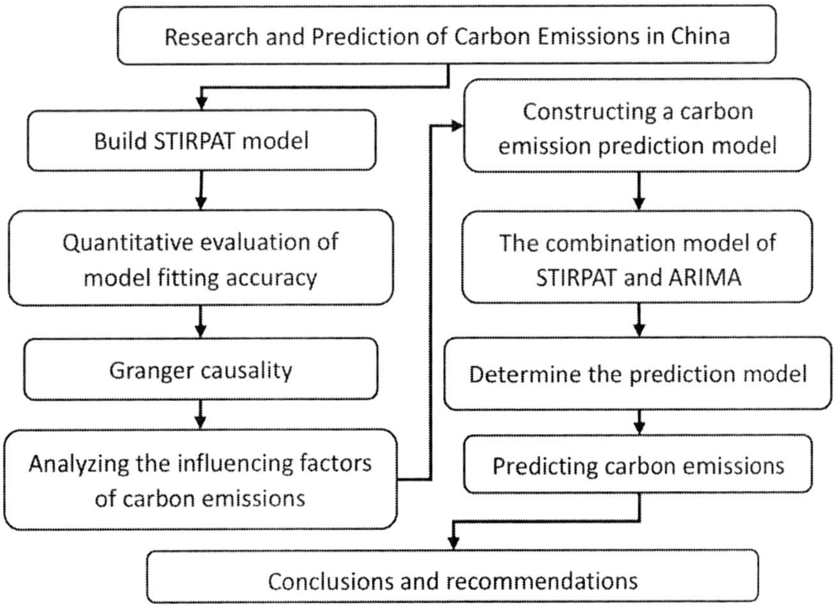

Figure 1. The Flowchart of Research

2.6. Model Building

2.6.1. Variable Selection. To thoroughly figure out the driving factors of carbon emissions in China, this paper draws upon the key variables developed in the STIRPAT model by Song and Wang [13].

Considering China's current socioeconomic development status, indicators closely related to the national carbon emission levels were selected. Details of the specific variables are provided in the table below:

Table 1. Driving Factors of Carbon Emissions in China

	Variables	Units
Population Scale	Total Population at Year-End	100 million people
Affluence	GDP	100 million yuan
Technological Level	Installed Power Generation Capacity	100 million kilowatts

In this study, within the theoretical framework of the STIRPAT model, we selected population scale (P), affluence (A), and technological level (T) as the key factors influencing China's carbon emissions. Specifically, population scale is quantified by the total population at year-end, which directly reflects the potential impact of population on environmental pressure. Affluence is measured by GDP, a comprehensive indicator of economic activity that reveals the association between economic development and carbon emissions. As for technological level, installed power generation capacity is chosen as a proxy variable, as it reflects not only the extent of energy usage but also the influence of energy structure and technological advancements on carbon emissions. By thoroughly analyzing these indicators, this study seeks to identify the main drivers of China's carbon emissions and offer scientific support for the development of environmental policies and emission reduction strategies.

2.6.2. Data Sources. The data of China's carbon emissions used in this study is sourced from the China Emission Accounts and Datasets (CEDS), covering the sample period from 2004 to 2023. GDP and population are obtained from the National Bureau of Statistics to ensure official and accurate data. As for the installed power generation capacity, the data is extracted from the China Energy Statistical Yearbook 2022 [14-15], a recognized authoritative source for statistical information in the power industry.

3. Results and Discussion

3.1. STIRPAT Model

This study seeks to examine the impact of three factors—population size (P), affluence (A), and technology level (T)—on China's carbon emissions and the degree of their contribution, using the STIRPAT model to analyze the data. Before model fitting, the raw data were log-transformed to stabilize the variance and tested for smoothness. The results showed that the raw data series were non-stationary, so first order differencing was performed to ensure the stationarity of the data and provide a prerequisite for the applicability of the STIRPAT model. The differenced data passed the smoothness test and was subsequently used to construct and fit the STIRPAT model. The methodology of this study adheres strictly to the requirements of time series analysis to ensure the reliability and validity of the model outcomes, offering a fresh perspective on understanding the factors driving carbon emissions of China. The data for P (population size), A (affluence), and T (technological level) from 2004 to 2023 were incorporated into the model to calculate their respective predicted values. The prediction results are detailed in the Figure 2 below:

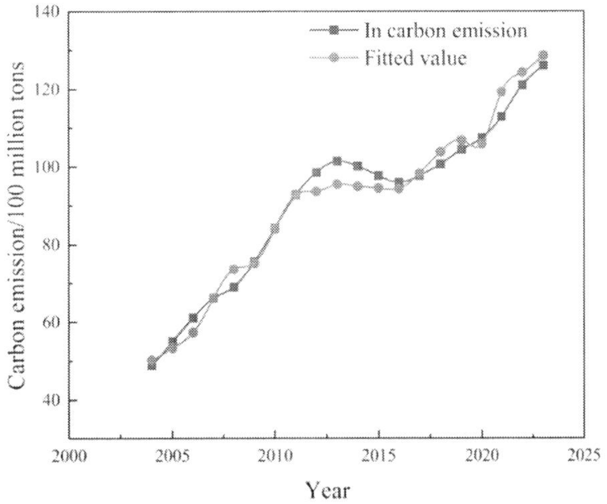

Figure 2. The Result of STIRPAT Model Fitting

This indicates a high level of consistency between the observed and predicted values, demonstrating that the model exhibits excellent fitting performance and can accurately capture the trends in carbon emissions.

Subsequent calculations of the model's MAE, MSE, RMSE, and NMAE were conducted, with the final results listed in the table as presented:

Table 2. STIRPAT Model Evaluation Indicators

MAE	MSE	RMSE	NMAE	Average relative error rate（%）
0.318	0.001	0.386	0.007	4.444

Based on the data provided in the table, the absolute error between the observed and predicted values was calculated to be 0.318, with a mean relative error rate of 4.444%. These results indicate that the model demonstrates excellent performance in fitting carbon emissions, accurately capturing and explaining the relationships between carbon emissions and the associated variables. The low values of absolute error and mean relative error rate further validate the reliability and effectiveness of the model, establishing it as a powerful tool for analyzing the dynamics and factors influencing carbon emissions.

3.1.1. Granger Causality Test. The Granger causality test is a widely used method in economics and time series analysis for exploring causal relationships between variables. In this study, Granger causality test is used to determine whether population size (P), affluence (A) and technology level (T) are significant influences on carbon emissions in China. The original hypothesis of the Granger causality test is: “This independent variable is not a Granger cause of the explained variable”. After the test, the p-values of all variables are below 0.05, indicating that there is enough evidence to reject the original hypothesis that all three variables are Granger causes of carbon emissions of China. This means that population, affluence and technology level all have a significant predictive effect on carbon emissions, and their historical values can significantly improve the predictive ability of future values of carbon emissions. This result highlights the critical impact of these three factors on the changes in carbon emissions of China.

After conducting the Granger causality test on the sample data, this study reached a consistent conclusion: within the considered lag periods, the test results for all variables rejected the null hypothesis, which states that "the variable is not a Granger cause of carbon emissions variations." This outcome indicates that population size, affluence, and technological progress are all identified as

Journal of Physics: Conference Series **3012** (2025) 012066 doi:10.1088/1742-6596/3012/1/012066

Granger causes of carbon emissions variations. Specifically, the historical values of these variables significantly enhance the ability to predict future carbon emissions, confirming their statistically significant predictive effect on carbon emissions changes. This finding highlights the critical role of population size, economic prosperity, and technological advancement in driving China's carbon emissions, providing essential insights for further policy development and emission reduction strategies.

Table 3. The Conclusion of Granger Causality Test

Categories of Variables	Null Hypothesis	Chi2	P-value	Conclusion
Population Scale	Not the Granger cause of carbon emissions	11.31	0.0456	Null hypothesis is rejected
Affluence	Not the Granger cause of carbon emissions	6.59	0.0253	Null hypothesis is rejected
Technological Level	Not the Granger cause of carbon emissions	13.28	0.0209	Null hypothesis is rejected

3.1.2. Discussion. The signs of the model coefficients reveal the directional impact of different factors on China's carbon emissions. In this model, the coefficients for power generation capacity, population size, and GDP are all positive, indicating that these factors exert a positive influence on China's carbon emissions. Specifically, as China's power generation capacity expands, population size increases, and GDP grows, the country's carbon emissions are expected to rise. These positive coefficients emphasize the direct influence of population and economic growth on energy consumption and carbon emissions, highlighting the environmental pressures that China encounters in its pursuit of economic growth. Examining the absolute values of the model coefficients allows for quantification of the relative influence of different factors on China's carbon emissions. Among these factors, population size has the most significant effect. Specifically, a 1% increase in population size (expressed as the natural logarithm, ln(population)) brings to a 2.498% rise of carbon emissions. This can be explained by the direct impact of population on energy consumption. Given that China is one of the most populous countries globally, its sustained the growth of population over these years suggests a further increase in energy consumption and, consequently, carbon emissions in the future. GDP (expressed as the natural logarithm, ln(gdp)) is a key indicator of the activity level of a nation's economy. A 1% increase in GDP results in a 0.372% rise in carbon emissions. This finding underscores the positive relationship between economic growth and energy consumption, indicating that as economic activities grow, energy consumption and related carbon emissions rise as well. Power generation capacity is another significant factor affecting China's carbon emissions. A 1% increase in power generation capacity results in a 0.317% increase of carbon emissions. Although China's renewable energy generation capacity has been growing in recent years, thermal power generation remains the dominant source of electricity. As a result, the increase in power generation capacity—particularly from thermal power—continues to contribute to higher carbon emissions.

3.2. ARIMA Model

Although the STIRPAT model effectively reveals the quantitative relationships between carbon emissions and their influencing factors, its prediction of carbon emissions relies on assumptions about the future values of independent variables. Some scholars have employed scenario analysis to set these future data points and subsequently predict China's carbon emissions. However, such methods may be influenced by subjective judgment, potentially affecting the objectivity of the predictions.

To enhance the objectivity of China's carbon emissions forecasts, this study developed a predictive model based on the evolving trends of carbon emissions in China. The model primarily employs time series analysis methods, specifically the ARIMA model. This approach is advantageous for its simplicity and flexibility, allowing it to accommodate various data characteristics and providing a more objective and scientific framework for predicting carbon emissions.

In this study, firstly this paper constructs an ARIMA model and determines its optimal lag order and parameters through a systematic approach. The time series data has been previously differenced to ensure its smoothness, which is a prerequisite for the application of the ARIMA model. Through the unit root test, it was determined that first-order differencing could bring the data to a smooth state, and therefore the differencing order (d) of the ARIMA model was set to be 1. Next, this paper uses the Akaike Information Criterion (AIC) and the Bayesian Information Criterion (BIC) to accurately determine the orders of the moving average term (q) and the autoregressive term (p) in the ARIMA model. By comparing the AIC and BIC values under different model configurations, we find that the model exhibits an optimal fit when both q and p are set to 1. This result implies that the first-order model is the optimal choice in terms of autoregression and moving average. The AIC and BIC values are minimized with the first-order model, aligning with the model selection criterion, thereby confirming the appropriateness of selecting the first-order autoregressive and moving average models.

In constructing the carbon emissions forecast model, this research used the carbon emissions sequence predicted by the STIRPAT model alongside the actual observed values as the base data. The training phase of the model utilized China's carbon emissions data from 2004 to 2013, while the testing phase incorporated data from 2014 to 2023. This division allowed the study to assess the model's predictive performance on a separate time series, ensuring the evaluation results' objectivity and reliability.

This approach not only verified the model's fitting performance on historical data but also assessed its effectiveness in forecasting future carbon emissions. The selection of the training sample ensured that the model could learn the underlying patterns of carbon emissions changes, while the use of the testing sample provided a direct evaluation of the model's generalization ability. This methodology, rooted in time series best practices, offers profound insights into the trends of China's carbon emissions.

In this paper, a preliminary analysis of carbon emissions trends in China from 2004 to 2023 was conducted, and a time series plot visually illustrated the trajectory of carbon emissions over this period. As shown in the graph, China's carbon emissions exhibited a clear upward trend from 2004 to 2023.

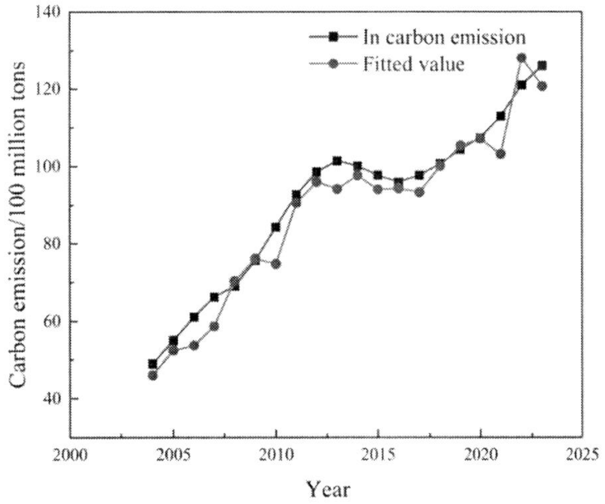

Figure 3. The Result of ARIMA Model Fitting

Building on this trend analysis, the study further applied the ARIMA model to statistically fit the historical carbon emissions data from 2003 to 2023. Based on the model's fitting results, predictions for carbon emissions from 2024 to 2026 were generated. This analysis not only deepens the understanding of historical data but also provides a scientific basis for forecasting future carbon emissions trends.

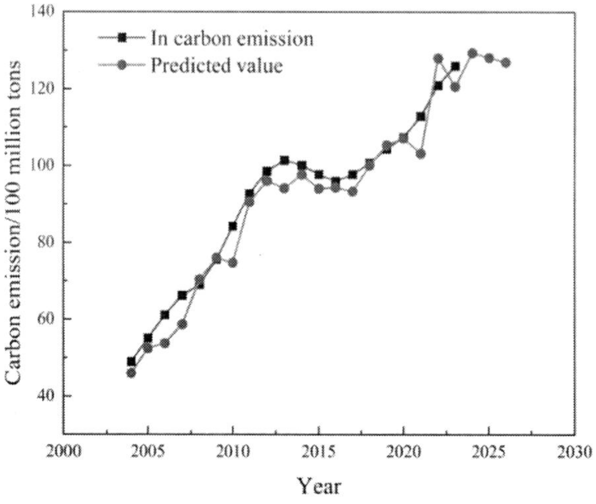

Figure 4. The Prediction of Carbon Emissions in China

Drawing on historical carbon emission data from 2003 to 2023 and the latest predictive models, this study offers a thorough assessment of China's carbon emission trends. During this 20-year period, China's carbon emissions are projected to increase from 4.2497 billion tons to 12.6 billion tons. This sustained upward trend is closely linked to factors such as national economic growth, industrialization, technological progress, and population expansion. According to the model results, it can be seen that with the active efforts of the Chinese government in policy guidance, energy structure optimization, carbon market development, and international cooperation, the upward trend of carbon emissions has been vigorously curbed. Overall carbon emissions in the next few years will fluctuate between 12.6 and 13 billion tons, which also indicates that China's carbon peak target may be achieved ahead of schedule.

4. Conclusion

This research first utilise the STIRPAT model to help identify the key drivers of carbon emissions in China: population, GDP and installed Power Generation Capacity. Among these, population size has the most substantial impact, with a 1% increase leading to a 2.498% rise in carbon emissions. This underscores the significant pressure that a growing population exerts on energy demand and environmental sustainability. Similarly, GDP growth demonstrates a positive correlation with carbon emissions, reflecting the increased energy consumption associated with economic expansion. Power generation capacity, while critical for meeting rising energy demands, also contributes to higher emissions, particularly due to the reliance on thermal power generation despite advances in renewable energy. These findings highlight the dual challenge China faces in balancing economic development with environmental sustainability.

In this study, the ARIMA model is applied to predict and analyze China's carbon emissions over the coming years, and the results show that the MAE, MSE,RMSE, and NMAE of the model are 0.318, 0.001, 0.386, 0.007, and 4.44%, respectively. These error indicators reflect the accuracy of the model in capturing the trend of carbon emission changes, and it is able to maintain high stability and reliability in long-term forecasting.

The carbon emission prediction model proposed in this dissertation provides an important reference for China's power grid construction, especially in the areas of energy demand forecasting and carbon emission reduction strategy formulation. By combining the accurate prediction of future carbon emission trends of installed power capacity, power grids can better plan the construction and upgrading of power infrastructure, optimize the power structure, promote the application of clean energy, and effectively reduce carbon emissions. This study not only supplies essential data to help China achieve its carbon neutrality target, but also presents important insights and strategies for combating global climate change.

References

[1] Zhang QQ, Li JX 2023 Building carbon peak scenario prediction in China using system dynamics model Environmental Science and Pollution Research vol 42 p 96019-96039.
[2] Wang R, Zhang Y, Xu H, Li Y, Li X, and Huang Q 2024 Port carbon emission peaking simulation and emission reduction pathway in China E3S Web of Conferences vol 536 p 7.
[3] Zhao ZY, Yang K, and Xu Y 2023 Low-carbon transformation of power structure under the "double carbon" goal: power planning and policy implications Environmental Science and Pollution Research vol 30 p 66961-66977.
[4] Li GM, Wang JC, and Qi ZH 2023 Research on renewable energy prediction technology: empirical analysis for Argentina and China Environmental Science and Pollution Research vol 30 p 21225-21237.
[5] Lian WW, Sun XY, and Wang YX 2024 Identification of clean energy development routes under carbon emission constraints: A path towards structural adjustment of the power system Journal of Cleaner Production Environmental Science and Pollution Research vol 434 p 140169.
[6] Wang SB, Yuan XL, and Jin ZG 2024 Prediction of Energy-Related Carbon Emissions in East China Using a Spatial Reverse-Accumulation Discrete Grey Model Sustainability vol 16 p 9428.
[7] Li C, Zhang ZC, and Wang LP 2023 Carbon peak forecast and low carbon policy choice of transportation industry in China: scenario prediction based on STIRPAT model Environmental Science and Pollution Research vol 30 p 63250-63271.
[8] Qi YW, Liu HL, and Zhao JB 2023 Prediction model and demonstration of regional agricultural carbon emissions based on Isomap-ACO-ET: a case study of Guangdong Province, China Scientific Reports vol 13 p 12688.
[9] Liu JB, Yuan XY, and Lee CC 2024 Prediction of carbon emissions in China's construction industry using an improved grey prediction model Science of Total environment vol 938 p 173351.
[10] Yin FF, Bo Z, and Yu L 2023 Prediction of carbon dioxide emissions in China using a novel grey model with multi-parameter combination optimization Journal of Cleaner Production vol 404 p 136889.
[11] Yue H, Bu LT 2023 Prediction of CO2 emissions in China by generalized regression neural network optimized with fruit fly optimization algorithm Environmental Science and Pollution Research vol 30 p 80676-80692.
[12] Danish, Ozcan B, and Ulucak, R 2024 An empirical investigation of nuclear energy consumption and carbon dioxide (CO2) emission in India: Bridging IPAT and EKC hypotheses Unclear Engineering and Technology vol 53 p 2056-2065.
[13] Guo FP, Zhang LJ, and Wang ZF 2022 Research on Determining the Critical Influencing Factors of Carbon Emission Integrating GRA with an Improved STIRPAT Model: Taking the Yangtze River Delta as an Example, International Journal of Environment Research and Public Health vol 19 p 8791.
[14] Gutknecht AJ, Barnett L 2023 Sampling distribution for single-regression Granger causality estimators Biometrika vol 110 p 933-952.
[15] Zhang XS, Guo ZX , Pan F, Yang YY, and Li CS 2022 Dynamic carbon emission factor based interactive control of distribution network by a generalized regression neural network assisted optimization Energy vol 283 p 129132.

Journal of Physics: Conference Series **3012** (2025) 012067 doi:10.1088/1742-6596/3012/1/012067

Study on Forecasting Electricity Consumption Based on Statistical Modeling

Aihua Chen[1*], Ziyuan Pan[1], Junyi Liu[1], Qinghua Li[1], Ganyu Wang[2], Shuxian Li[1], Chang Liu[1], Yingqi Liang[3]

[1] Zhongshan Power Supply Bureau of Guangdong Power Grid Co., Ltd, Zhongshan, Guangdong Province, 528400, P.R. China

[2] Electric Power Research Institute of Guangdong Power Grid Co., Ltd, Guangzhou Guangdong Province, 510080, P.R. China

[3] Electric Power Dispatching and Control Center of Guangdong Power Grid Co., Ltd, Guangzhou, Guangdong Province, 510000, P.R. China

[*]Corresponding author's e-mail: chenaihua188@163.com

Abstract. With the advancement of smart grid technology, accurate forecast of electricity consumption has become crucial for grid load management and power generation planning. To this end, this paper innovatively constructs a composite model that combines the theoretical advantages of the ARIMA model and the STIRPAT model with a view to more comprehensively forecasting future electricity consumption in Guangdong and analyzing its influencing factors for optimal grid management. First, this study logarithmized population size (P), wealth (A) and technology level (T) and explored the impact of these factors on electricity consumption using the STIRPAT model. The model has a goodness of fit of 97.47%, showing its effectiveness in capturing the major influences on electricity consumption. The significant effects of population, wealth and technology level on electricity consumption were confirmed through Granger causality test and precision test, where each 1% increase in population size, wealth and technology level leads to 3.31%, 0.402% and 0.057% increase in electricity consumption respectively. Further, this study constructed a composite model by constructing an ARIMA model and combining it with the STIRPAT model to predict electricity consumption under different technology level growth scenarios, including three scenarios of low growth (8%), steady growth (14%), and high growth (18%). By comparing the forecast results under these scenarios, this study provides the grid with insights on the future changes in electricity consumption, which is beneficial for the grid to deploy resources in advance to meet potential grid load challenges. This study not only provides a scientific basis for load forecasting in smart grids, but also provides scientific guidance for formulating effective energy policies, which has important practical application value.

1. Introduction

With the gradual advancement of smart grid construction, the distribution network is facing increasingly high requirements in ensuring power supply stability and enhancing carrying capacity. Traditional power systems often struggle to cope with increasingly complex and variable load demands when faced with uncertainties brought about by rapidly developing power infrastructure construction and population growth[1]. At this time, the importance of accurate electricity consumption forecasting in power system planning and scheduling becomes increasingly prominent. This not only ensures the safe operation of the power grid under different load conditions, but also provides sufficient margin for power supply to cope with potential load fluctuations and sudden demand in the future. Accurate electricity consumption forecasting models provide strong data support for power grid operation, ensuring that power supply can respond to changes in electricity demand in a timely and flexible manner [2-3].

At the same time, China's "dual-carbon" policy has set more stringent carbon reduction targets. As the primary sector for energy consumption and carbon emissions, the power industry plays a crucial role in the nation's efforts to reduce carbon emissions. According to statistics, the power industry accounts for approximately one-third of the country's total carbon emissions, highlighting its crucial role in achieving carbon peak and carbon neutrality goals. To implement this policy, accurate electricity consumption forecasts can provide power companies with a solid basis for decision-making, helping to minimize energy waste while enabling more efficient energy dispatch and production. With the growing share of renewable energy and the high volatility of electricity demand, accurately forecasting demand and flexibly adjusting supply—particularly through clean energy generation—has become crucial to reducing reliance on fossil fuels, advancing carbon emission reduction, and achieving a green transition. In recent years, many scholars have conducted research on electricity consumption prediction. Xiong et al used a random forest model to model and predict the monthly electricity consumption of five representative industries in a province in China[4]. The best-performing month's prediction had a relative error within 2%, providing valuable guidance for power grid peak shaving and valley filling. Shwetha et al addressed the challenge of accurately predicting individual household electricity consumption by constructing prediction models using ARIMA, CNN, and LSTM models based on the past electricity consumption of a single household, and achieved good results [5]. Wang et al proposed a hybrid model for predicting electricity consumption, which consists of three parts: Empirical Mode Decomposition (EEMD), Gated Recurrent Unit (GRU), and Gradient Boosting Regression Tree (GBRT) [6]. The model also considered the impact factors of weather temperature changes on residential electricity consumption. In order to accurately predict the electricity consumption of industrial enterprises in Zhejiang Province, Chen et al combined buffer operators and DGGM models to build a new seasonal grey prediction model (AWBO-DGGM) [7]. During testing, the model had a percentage error of about 5%, providing scientific guidance for the development of industrial enterprises in Zhejiang Province. Sun et al proposed a new grey power Markov prediction model, which effectively overcomes the disadvantages of fixed structure and poor adaptability to changes in raw data compared to previous models [8]. The model was used to predict the total electricity consumption in China from 2018 to 2023. Zhou et al proposed a long short-term memory model for predicting the electricity consumption of a single building, and introduced an RL agent learning strategy to dynamically adjust the LSTM model parameters based on prediction errors, providing a prediction method for the electricity consumption of buildings [9]. Nofal et al used the Jordan region as a case study and constructed a small-scale short-term electricity consumption prediction model using machine learning algorithms such as neural networks and decision trees [10]. This provides a scientific reference model for planning electricity consumption in special areas during war and natural crises. Although various scholars have conducted extensive research on electricity consumption prediction, there are currently few scholars studying the relationship between installed electricity capacity and electricity consumption, as there is a mutual influence between installed electricity capacity and electricity consumption. The installed capacity of power facilities determines the power generation capacity of the power system. When the installed capacity increases, the system can provide more power supply to meet a wider range of electricity demands. In the case of sufficient power supply, users' electricity consumption behavior often changes. Enterprises and households may expand their electricity consumption as power supply is no longer restricted, especially in industrial and high energy consuming sectors, where demand growth is more significant. In addition, when the power system has sufficient installed capacity, electricity prices may tend to stabilize or decrease, further stimulating the electricity demand of various users. It can be seen that the installed capacity not only affects the availability of electricity, but also promotes the increase of electricity consumption by ensuring supply stability and reducing electricity costs. Therefore, this article will combine GDP, population size, and explore the relationship between installed capacity and electricity consumption.

Accurate electricity consumption prediction is not only crucial for the safe and stable operation of the power grid, but also provides strong support for the government to formulate scientific and accurate carbon reduction policies. The electricity consumption prediction model enables the government to monitor real-time electricity demand across regions and industries, implement differentiated pricing policies, encourage demand response and flexible electricity usage, and promote the efficient operation of the electricity market along with the optimal allocation of resources. As a major electricity consumer in China and a pioneer in smart grid construction, Guangdong Province has a huge demand for industrial, commercial, and residential electricity. Accurate electricity consumption forecasting helps to plan the construction and layout of power facilities reasonably, ensures sufficient and stable power supply, and avoids power shortages or system overload caused by excessive demand. Through scientific electricity consumption forecasting, the government and power grid can better regulate the balance of electricity supply and demand, ensure the sustainable use of energy, and help Guangdong Province achieve its "dual carbon" goals as soon as possible, moving towards a low-carbon and green future.

2. Theoretical Modeling of Electricity Consumption Forecasting

With the continuous growth of the global economy and social development, the volatility of electricity demand is becoming increasingly significant. Accurately predicting electricity consumption has become one of the core issues in power system management. Traditional forecasting methods relied on simple historical data deduction, making it difficult to cope with complex demand changes and changing external factors. Statistical models can reveal the regularity and potential influencing factors of electricity consumption changes through in-depth mining and analysis of a large amount of historical electricity consumption data, and analyze the impact of different factors on electricity consumption. Therefore, this article will explore the feasibility of using statistical models to predict electricity consumption, focusing on analyzing how different factors affect fluctuations in electricity consumption, and further studying how to optimize electricity demand forecasting through these models, improve the efficiency of power system scheduling and management, and provide scientific basis for future energy planning.In this study, the number of population at the end of the year is used as a measure of population size (billion people); GDP per capita is used to measure the level of wealth; installed power generation capacity (billion kilowatts) is used as a measure of the technical level of the variable; and the explanatory variable is the total electricity consumption of the whole society of Guangdong (billion kilowatt-hours). The data in this paper are from the National Bureau of Statistics and the China Electricity Statistical Yearbook.

2.1. STIRPAT Model

The STIRPAT model (Stochastic Impacts by Regression on Population, Affluence, and Technology) is a statistical model based on regression analysis for exploring the relationship between environmental impacts (e.g., electricity consumption, carbon emissions, etc.) and population, wealth, and technology. The model extends the classic IPAT constant equation by introducing random effects, which allows it to more accurately reflect the volatility and uncertainty in the actual data and provide a more flexible and precise assessment of environmental impacts.

The STIRPAT model is derived from the IPAT Constant Equation proposed by Paul Ehrlich and John Holdren in the 1970's. The IPAT Constant Equation succinctly expresses the relationship between environmental impacts (I) and the three main drivers: population (P), wealth (A) and technology (T). It takes the form:

$$I = P \times A \times T \tag{1}$$

where I represents environmental impacts (e.g., energy consumption or carbon emissions), P represents population size, A represents the level of wealth (often expressed using GDP), and T represents the level of technology (e.g., technical indicators such as energy intensity).

However, the IPAT model only presents a simple multiplicative relationship between these factors and does not take into account the stochasticity and variability in the data. To overcome this limitation, Dietz and Rosa proposed the STIRPAT model in 1994. They added random effects to IPAT, allowing the model to quantitatively estimate the individual drivers of environmental impacts through regression analysis. The STIRPAT model is usually represented as a multivariate regression model that can be transformed into a linear form after a logarithmic transformation:

$$ln\ (I) = \alpha + \beta ln(P) + \gamma ln(A) + \delta ln(T) + ln(\epsilon) \quad (2)$$

Where I represents the environmental impact (e.g., electricity consumption), P, A, and T represent population, wealth, and technology level, respectively, α is a constant term, $\beta 1$, $\beta 2$, and $\beta 3$ are regression coefficients indicating the extent to which each factor affects electricity consumption, and ϵ is an error term reflecting the stochastic nature of the model and unexplained variations.

Through regression analysis, the STIRPAT model is able to quantify the contribution of different factors to environmental impacts such as electricity consumption. This enables policymakers to understand more precisely the impact of population growth, economic development and technological progress on energy demand, and to analyze and forecast scenarios accordingly.

2.2. ARIMA Model

The ARIMA (Autoregressive Integrated Moving Average) model is a classic method widely used in time series analysis, particularly suitable for processing data with temporal dependencies. This model combines autoregressive (AR), differencing (I), and moving average (MA) components to effectively capture long-term trends, short-term fluctuations, and random disturbances in time series data. The core idea of the ARIMA model is to model historical data, reveal the inherent time structure of the data, and predict future values based on this. The standard ARIMA model can be represented as:

$$Y_t = c + \sum_{i=1}^{p} \phi_i Y_{t-i} + \sum_{j=1}^{q} \theta_j \epsilon_{t-j} + \epsilon_t \quad (3)$$

Where Y_t represents the observed values of the time series; c is a constant term; ϕ_i is the coefficient of the autoregressive term, which represents the effect of the historical observations on the current values; θ_j is the coefficient of the moving average term, which represents the correction of the past error to the current values; and ϵ_t is the error term, which represents white noise.

2.3. Granger Causality Tests

The Granger Causality Test (GCT) is an important statistical method proposed by statistician Clive Granger in 1969 to test for causal relationships between time series. Its central goal is to analyze whether one time series (e.g., variable X) can provide useful predictive information about the future values of another time series (e.g., variable Y). Unlike the traditional concept of causality, Granger causality does not require a direct causal mechanism in the traditional sense, but rather infers predictive power through lagged effects in time series data. The core idea of Granger's causality test is that if the historical data of variable X has a significant predictive effect on the future value of variable Y, and this effect exceeds the predictive ability of Y's own past data alone, then it can be assumed that there is a causal relationship between X and Y. In other words, the past value of X contains unique information about the future change of Y, which can improve the prediction accuracy to a certain extent, thus showing causality. When the lag term of X is included in the regression model, if it can significantly improve the prediction accuracy of Y, then it can be concluded that X has Granger causality on Y.

The Granger causality test is derived based on the following regression model:

$$Y_t = \alpha_0 + \sum_{i=1}^{p} \beta_i Y_{t-i} + \sum_{i=1}^{q} \gamma_i X_{t-i} + \epsilon_i \tag{4}$$

α_0, β_i are the parameters of the autoregressive part of Y_t, γ_i is the parameter of the effect of X_{t-i} on Y_t, p is the lagged order of Y_{t-i}, q is the lagged order of X_{t-i}, and ϵ_i is the error term.

3. Results and Analysis

3.1. Model Parameter Estimation

This study aims to examine electricity consumption in Guangdong Province and investigate the effects of population size, wealth level, and technological advancement on it. Specifically, the dependent variable is electricity consumption in Guangdong Province, while the independent variables include population size (P), wealth level (A), and technology level (T). Population size is represented by the year-end population of Guangdong Province, wealth level is measured by GDP per capita, and technology level is indicated by the installed power generation capacity within the province. Data on electricity consumption and installed power generation capacity were sourced from the authoritative *China Electricity Yearbook*, while year-end population and GDP per capita figures were obtained from the National Bureau of Statistics of China.

Table 1. Driving Factors of Electricity Consumption in Guangdong

	Variables	Units
Population Scale	Total Population at Year-End	100 million people
Affluence	GDP	yuan
Technological Level	Installed Power Generation Capacity	100 million kilowatts

In this study, the model parameters were estimated using Ordinary Least Squares (OLS) and the estimation results are detailed in the table below.

Table 2. Estimates of Model Parameters

	lnα	β	γ	δ	mathematical equation
Estimated value	19.882	3.055	0.402	0.498	
P value	0.000	0.097	0.063	0.037	0.000

According to the statistical results in the table, the p-values of all independent variables are below the 0.1 significance threshold, indicating their statistical significance. Specifically, this means that the effect of the independent variables in the model on the dependent variable was statistically significant. To further assess potential multicollinearity in the model, the Variance Inflation Factor (VIF) was calculated for the independent variables to determine whether their correlations could affect the regression analysis results.

Table 3. VIF Value

	lnP	lnA	lnT
VIF	2.66	122.21	118.52

Based on the data in the table, this paper observed that two of the independent variables in the model have their Variance Inflation Factor (VIF) exceeding 10. This result suggested that although these independent variables were statistically significant, there was a serious problem of multicollinearity among them. Multicollinearity results in unstable regression coefficient estimates, increased sensitivity to data variations, larger standard errors, reduced statistical test efficacy, and greater complexity in model interpretation, potentially weakening predictive power. To address these problems, this paper used ridge regression, which reduced parameter variance through L2 regularization and enhances model stability and predictive power.

Table 4. Estimates of Model Parameters

	lnα	β	γ	δ
Estimated value	21.866	3.310	0.402	0.057

The estimates obtained after using ridge regression analysis revealed a high degree of fit between the model and the data. The Adjusted R-squared (Adjusted Goodness of Fit) value is 0.9747, the closer the Goodness of Fit is to 1 the better the fit. The significance of this statistical indicator suggested that the model was able to fit the analyzed data set extremely effectively. In addition, the effect of the independent variables on the dependent variable was statistically extremely significant, which further validated the rationality of the variable selection in the model and the reliability of the analyzed results.

3.2. *STIRPAT Model*

Prior to constructing the STIRPAT model, this study first preprocessed the key explanatory variables. Specifically, population size (P), affluence (A), and technology level (T) were natural log transformed to mitigate potential heteroskedasticity and nonlinear relationships. Subsequently, unit root tests were conducted on these variables to assess the stability of their time series. Stability is a prerequisite for performing STIRPAT model fitting because non-stationary data may lead to pseudo-regression phenomena, which may affect the accuracy of the model. After unit root test, the original data series was found to be non-stationary. Therefore, first order differencing was applied to the data to remove the non-stationarity in the time series. The differenced data was again passed through unit root test to confirm its smoothness. Based on ensuring the smoothness of the data, a STIRPAT model was built using these processed data that satisfied the smoothness condition. This step is a critical part of ensuring the validity and reliability of the model estimation results. Data on P (population size), A (affluence), and T (technology level) for the period 2003 to 2023 were each included in the model to calculate their corresponding projections. The results of the projections were presented in detail in the figure below. As can be seen from the above figure, there was a high degree of consistency between the predicted values and the actual observed values. Therefore, it could be concluded that the model exhibits a good fitting effect and was able to capture the trend of electricity consumption more accurately.

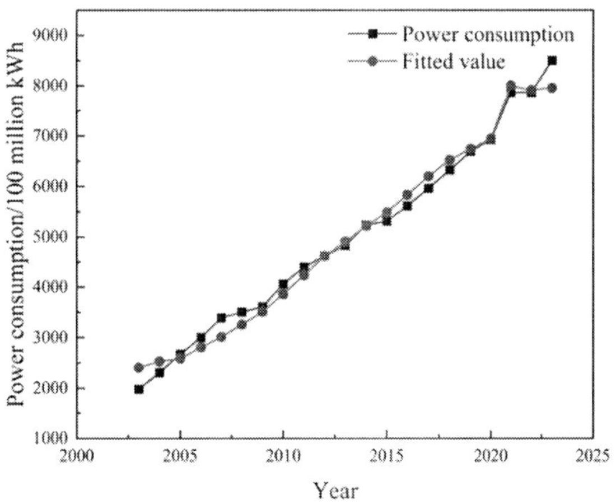

Figure 1. Data of STIRPAT Model Fitting

3.3. Causality Analysis of the STIRPAT Model

After conducting the Granger causality test on the sample data, the p-values of the three variables of this test are below 0.05, indicating that there was sufficient evidence to reject the original hypothesis, so the paper comes to the consistent conclusion that the original hypothesis was rejected for all the variables in the number of lags considered: "The variables are not the Granger causes of the changes in electricity consumption ". This result indicated that population size, affluence and technology level were all identified as Granger causes of electricity consumption changes. Specifically, the historical values of these variables significantly improved the ability to predict future values of electricity consumption, thus confirming that they were statistically significant predictors of changes in electricity consumption. This finding emphasized the key roles of population size, economic prosperity and technological advancement in changes in electricity consumption in Guangdong.

Table 5. Model Causality Analysis

	Original hypothesis	Chi2	p-value	Conclusion
Population size	Not a Granger Cause of Electricity Consumption	12.29	0.006	Rejection of the original hypothesis
Affluence	Granger reasons that are not electricity consumption	9.51	0.009	Reject the original hypothesis
Technology level	Granger reasons that are not electricity consumption	19.95	0.00	Reject the original hypothesis

3.4. Evaluation Indicators Model

This study further calculated the mean absolute error (MAE), mean square error (MSE), root mean square error (RMSE), and normalized mean absolute error (NMAE) of the model. As the table shown, this study calculated an absolute error of 0.045 between predicted and observed values and an average relative error rate of 0.045%. These values indicated that the model used performs well in fitting the electricity consumption and was able to accurately capture and explain the interrelationships between electricity consumption and various related variables. The low absolute error and average relative error rate values further confirmed the reliability and validity of the model, making it a powerful tool for analyzing the dynamics of electricity consumption and the factors affecting it.

Table 6. Model Evaluation Indicators

MAE	MSE	RMSE	NMAE	Average relative error rate (%)
0.045	0.004	0.064	0.003	0.045

3.5. Electricity Consumption Prediction Model

The STIRPAT model is widely recognized for its excellent fitting capabilities, but it lacks standalone predictive functionality. To address this limitation, this study innovatively integrated the STIRPAT model with the ARIMA model to forecast electricity consumption in Guangdong Province. Specifically, the ARIMA model was first employed to predict the population size (P) and wealth level (A) over a time series, generating future estimates for 2024 to 2028 based on historical trends. These projections provided reasonable approximations for population and wealth levels in the coming years. Subsequently, the predicted values of population size (P) and wealth level (A) were incorporated into the STIRPAT model. Building on this foundation, the study explored three scenarios for the growth rate of technology level (T): installed power generation capacity low growth (8%), installed power generation capacity stable growth (14%), and installed power generation capacity high growth (18%). These scenarios were used to simulate the impact of varying technological advancements on future electricity consumption in Guangdong Province. By employing this hybrid model, the study provided a comprehensive forecast of electricity consumption under different technology growth scenarios while accounting for the effects of population size (P) and wealth level (A) on electricity demand.

In this study, the Akaike Information Criterion (AIC) and Bayesian Information Criterion (BIC) were used to accurately determine the order of the moving average term (q) and autoregressive term (p) in the ARIMA model. By comparing the AIC and BIC values under different model configurations, we found that when q and p are both set to 1, the model exhibits the best fitting effect. This result implied that in terms of autoregression and moving average, the first-order model was the optimal choice.

The change in installed power generation capacity had a direct impact on the amount of electricity generated. Although its direct impact on electricity consumption is limited, installed power generation capacity was usually regarded as a reflection of economic conditions, and new capacity was planned based on the speed of economic development and changes in electricity consumption. This study examines three scenarios for the growth rate of installed power generation capacity:

1. **Low Growth Scenario** – an annual growth rate of 8%.
2. **Stable Growth Scenario** – an annual growth rate of 14%.
3. **High Growth Scenario** – an annual growth rate of 18%.

The forecast period spans from 2024 to 2028.

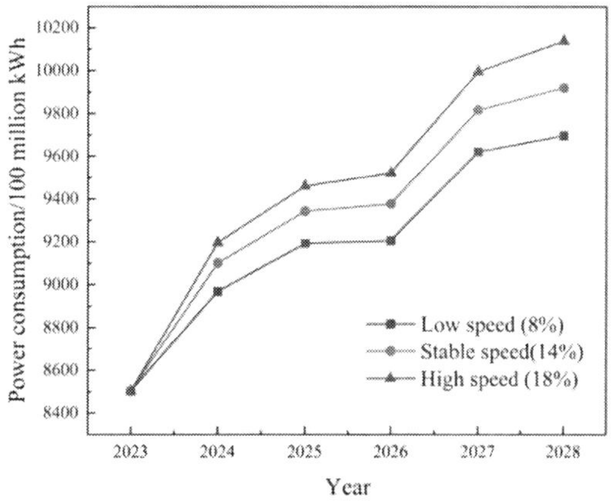

Figure 2. The Prediction of Electricity Consumption

The model prediction results showed that different growth rates of installed capacity also had a great impact on electricity consumption, which also provided a certain reference basis for the subsequent construction of the power grid.

4. Conclusion

4.1. Electricity Consumption Analysis of Factors Affecting Electricity Consumption

By analyzing the model coefficients, this study revealed the impact of key factors on Guangdong's electricity consumption. The coefficients of installed generating capacity, population size, and GDP per capita were all significantly positive, indicating that these factors positively contribute to electricity consumption. Specifically, a 1% increase in population size led to a 3.31% increase in electricity consumption, while a 1% increase in GDP per capita resulted in a 0.402% increase. Installed generating capacity also had a significant effect, with a 1% increase leading to a 0.057% rise in electricity consumption. These findings highlighted the direct contributions of economic development and population growth to electricity consumption in Guangdong.

4.2. Analysis of Electricity Consumption Forecasts
In forecasting electricity consumption, this study categorizes the growth rate of installed generating capacity into three modes: low growth (8% annually), stable growth (14% annually), and high growth (18% annually).Low Growth (8% annually): From 2024 to 2028, the cumulative growth of installed capacity is 727.118 GW, with an average annual increase of 181.7795 GW. The growth rate is relatively stable, showing a gradual acceleration.Stable Growth (14% annually): During the same period, the cumulative growth is 817.975 GW, with an average annual increase of 204.494 GW. This indicates a faster growth rate and a more significant acceleration trend compared to the low growth scenario.High Growth (18% annually): The cumulative growth reaches 9,406.94 GW, with an average annual increase of 2,351.73 GW. This represents the fastest growth rate and the most significant acceleration trend among the three scenarios.These results demonstrate that higher growth rates in installed generating capacity lead to significantly increased growth potential and faster growth rates, which are crucial for meeting future electricity demand and promoting renewable energy development.

References
[1] Chen HB, Pei LL and Zhao YF 2021 Forecasting seasonal variations in electricity consumption and electricity usage efficiency of industrial sectors using a grey modeling approach Energy vol 222 p 15.
[2] Nie Y, Zhang GX, Zhong LH, Su B and Xi X 2024 Urban-rural disparities in household energy and electricity consumption under the influence of electricity price reform policies Energy Policy vol 184 p 13.
[3] Nofal S 2023 Forecasting next-hour electricity demand in small-scale territories: Evidence from Jordan Heliyon vol 9 p 23.
[4] Shwetha BN, Sapna R and Raja SP 2022 Analysis of Individual Household Electricity Consumption Forecasting Using ARIMA Model, CNN and LSTM Model Cognition and Recognition: 8th International Conference, ICCR 2021, Revised Selected Papers. Communications in Computer and Information Science (1697) p 53-66.
[5] Song M, Zhang LY and Li MX 2022 The influence path and dynamic relationship between economic development, industrial structure upgrading, urbanization, urban-rural income gap, and electricity consumption in China Energy Science & Engineering vol10 p 4366-4381.
[6] Sun LQ, Yang YL, Ning T and Zhu JD 2022 A novel grey power-Markov model for the prediction of China's electricity consumption Environmental Science and Pollution Research vol 29 p 21717-21738.
[7] Wang Z, Yao DG, Shi Y, Fan ZX, Liang Y, Wang YY and Li H 2024 A two-stage electricity consumption forecasting method integrated hybrid algorithms and multiple factors Electric Power Systems Research vol 234 p 11.
[8] Xiong J, Niu Y and Liu W 2023 Application of Improved Random Forest Algorithm in Electricity Consumption Forecasting Process Automation Instrumentation vol 44 p 95-100.
[9] Yuan JH and Zhao DF 2022 A panel empirical modeling on the driving factors of provincial electricity consumption in China Environmental Science and Pollution Research vol 29 p 10345-10356.
[10] Zhou XL, Lin WY, Kumar R, Cui P and Ma ZJ 2022 A data-driven strategy using long short term memory models and reinforcement learning to predict building electricity consumption Applied Energy vol 306 p 15.

Power System Security Research based on a Novel Reward-Guided Diffusion Model for Extreme Weather Forecasting

Yicheng Zhang*, Dexi Xia, Fei Yuan

CHN Energy United Power Technology, CHN Energy, Beijing 100036, China

*Corresponding author's e-mail: 20049955@chnenergy.com.cn

Abstract. Recently, the frequency of extreme weather events impacting power grids has increased, presenting significant challenges to the security, quality, and stability of electrical systems. Effectively predicting the impact of extreme weather events on distribution networks has become crucial for maintaining the safe operation of power systems. Current ensemble forecasting methods utilize numerical weather prediction to forecast weather by simulating atmospheric laws with parameterized models. These methodologies require the simultaneous operation of numerous model instances, leading to substantial computational resources and thereby increasing operational costs for power systems. Simultaneously, despite ongoing enhancements in meteorological forecasting technology, ensemble forecasting tends to produce blurry predictions, often underestimating the severity of extreme weather events. Moreover, there is a lack of extreme weather data. To address these issues, a novel Reward-Guided Diffusion (RGD) model is proposed to augment data of the extreme weather for further forecasting. By integrating the diffusion model with the reward regression model which can capture finer-grained atmospheric features of the extreme weather, the diffusion model can be finetuned by the feedback from the reward regression model in an online manner. The extreme weather data can be augmented by the proposed RGD model and further enhance the prediction for occurrence of extreme weather events more accurately at the later stage. As a result, by providing more reliable early warning information to power systems, this method strengthens the grid's resilience to risks and ensures the secure and stable operation of electrical systems.

1. Introduction

With the intensification of global climate change, the occurrence of extreme weather events has significantly increased, posing unprecedented challenges to guaranteeing the secure and stable functioning of power distribution networks.The inherent randomness, volatility, and difficulty in predicting extreme weather make it challenging to accurately anticipate when and how severe these events will be. The Earth's climate system, characterized by its high variability and complexity, implies that even minor alterations in initial conditions can result in significantly divergent weather patterns, posing a substantial risk to the stability of the power grid. When the devastating force of severe weather strikes the power grid, it can trigger widespread power outages, severely disrupting daily life and socio-economic development. For instance, from January to February 2008, severe snowstorms struck southern China, causing widespread power outages across multiple provinces. Therefore, to strengthen the power grid's resilience against harsh weather circumstances, there is an immediate necessity to develop advanced extreme weather prediction technologies. This technologies will improve the accuracy of predicting extreme weather events, thereby facilitating the structured development and implementation of safety protocols to guarantee the grid's stable operation.

1.1. Ensemble forecasts
Ensemble forecasting utilizes sophisticated atmospheric physical models and high-performance computing to simulate atmospheric dynamics and weather evolution. Notably, numerical weather prediction (NWP) models integrate both recent and historical atmospheric data, employing iterative processes to run multiple forecasting models, thereby enabling the prediction of future weather conditions. To accommodate the processing of various initial conditions and the simultaneous execution of multiple models, this forecasting technique requires exceptionally high computational efficiency[6]. At the same time, the range of potential forecast outcomes produced by ensemble forecasting is typically evaluated by minimizing mean squared error (MSE). However, this approach often results in forecasts that are overly smooth, particularly for medium to long-term predictions, and it struggles to forecast extreme weather events, which present a threat to the stability of power systems.

1.2. Reward-Guided Diffusion (RGD) model for extreme weather forecasting
Diffusion models[4][11] are capable of generating high-resolution, high-quality images and have demonstrated strong performance in computer vision[3]. Existing research has applied diffusion models to extreme weather forecasting[2][5][12]. However, current diffusion models encounter challenges due to limitations such as imbalanced training data distribution, which results in generated images may not match the guiding conditional variables (such as T2, U10, V10, MSL, SST, PV, GH). Moreover, the mechanism of the tranditional diffusion model can not guarantee that synthetic high-resolution meteorological maps can capture the fine-grained characteristics only by gradually adding and removing noise in a Markov Chain. To address this issue, a Reward-Guided Diffusion (RGD) model is proposed to generate high-resolution meteorological maps, which can capture the fine-grained featrues with respect to the extreme weather characteristics, such as typhoons.

Our contribution can be outlined as follows:

- We proposed a novel Reward-Guided Diffusion (RGD) model for extreme weather forecasting. It achieves superior performance than NWP ensemble forecasts in medium to long-term forecasts ranging from 3 to 5 days.
- There are three stages for training the proposed RGD model . The first stage involves pre-training the latent diffusion model guided by the conditional embeddings encoded from variables (T2, U10, V10, MSL, SST, PV, and GH) and low-resolution meteorological maps obtained from NWP model. The second stage is the pre-training of the reward regression model, which will give each ynthetic high-resolution meteorological map a reward. The last stage involves updating the diffusion model weights through reward backpropagation from the pre-trained reward regression model for each generated image in an online learning manner. Thus the generated images are more consistent with the conditions, and the fine-grained the extreme weather characteristics. This process not only improves the quality of the generated images but also enhances their diversity. It ensures that the generated predictions align better with domain-specific prior knowledge, thereby improving the reliability of the forecasts.
- Experimental results on the ERA5 dataset show that our RGD model can fine-tune regression models through an online reward mechanism, effectively improving the prediction accuracy for extreme weather events such as typhoons, thereby enhancing the safety of power facilities and energy supply.

2. RGD model
As shown in Figure 1, the RGD framework includes the NWP model and conditional diffusion model, which is pretrained in the first stage and will be further refined using reward feedback from the reward regression model in the subsequent stage. The input to the NWP model is the previous time step $y_1, ..., y_k$. The output is the predicted low-resolution image y_{k+1} for the subsequent time step. y_{k+1}

and the atmospheric variables, including T2, U10, V10, MSL, SST, PV, and GH, are encoded and concatenated as conditional hidden embeddings $\boldsymbol{C}$. These embeddings $\boldsymbol{C}$ guide the diffusion model through cross-attention[10] to generate the high-resolution image x_{k+1}, which aligns with the input conditions.

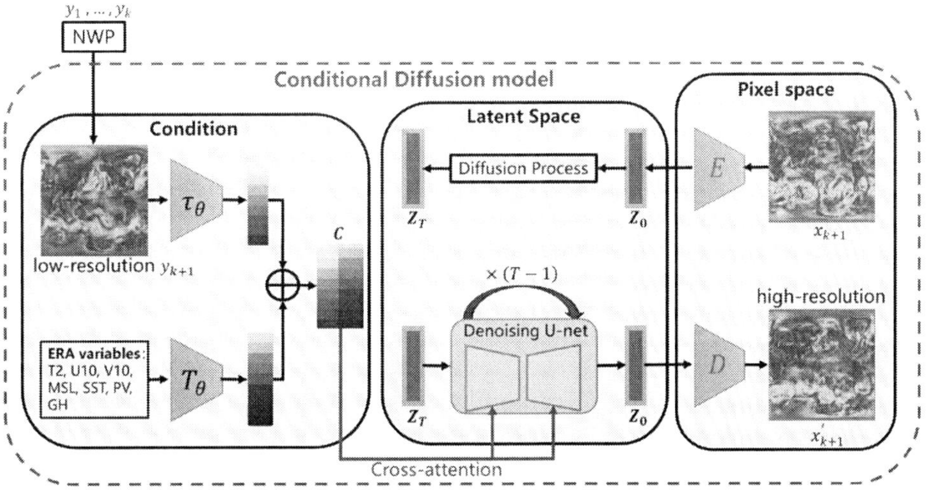

Figure 1. At the first training stage, the proposed model comprises two components: (a) NWP, which forecasts the low-resolution image y_{k+1} from the previous time steps $\{y_1, \ldots, y_k\}$; (b) The diffusion framework, which guided by y_{k+1} and ERA5 atmospheric variables to generate the high-resolution image x'_{k+1}.

2.1. *Conditional Diffusion model*

The probabilistic diffusion framework excels at reconstructing fine-grained, subtle features from conditions such as low-resolution images, which may be coarse or blurred. In this paper, we fine-tuned the conditional diffusion model using reward feedback to improve the outcomes generated by NWP, placing a greater emphasis on recovering the finer-grained features of extreme weather. The Diffusion Framework [8] is a stochastic model that captures a data distribution through a forward process of noise addition and a reverse denoising operation in a fixed Markov chain.At the first stage, the pretrained diffusion model learns the conditional distribution $P(z|C)$ using the conditional denoising Unet $\varepsilon_\theta(Z_t, t, C)$, setting the foundation for controlling the denoising process. This process is guided by condition embeddings C, which include the low-resolution image y_{k+1} generated from NWP and other representative atmospheric variables such as T2, U10, V10, MSL, SST, PV, and GH, using the cross-attention mechanism. The objective of the diffusion model can be expressed as:

$$\mathrm{L_{pre}} = \mathrm{E}_{\mathrm{E(x)},\mathrm{C},\varepsilon\sim\mathrm{N}(0,1),\mathrm{t}}[\|\epsilon - \epsilon_\theta(\mathrm{Z_t}, \mathrm{t}, \mathrm{C})\|_2^2] \tag{1}$$

Where:

- t is uniformly sampled from the set $\{1, \ldots, T\}$, representing the timestep in the forward diffusion process.
- x_t represents the noisy image decoded from noisy latent embedding Z_t at timestep t, which is progressively degraded by adding noise starting from the embedding Z_0, which is encoded from high-resolution image x_0.
- $E(x)$ is the pretrained image encoder E that maps the high-resolution image (ground-truth) x_{k+1}into latent space as hidden embeddings for the forward adding noise process.
- $\epsilon_\theta(Z_t, t, C)$ is the predicted noise by U-net[9] at timestep t given the noisy hidden embedding Z_t and the conditions C.
- ϵ is the actual noise added throughout the forward diffusion procedure.
- C represents the guided conditions (including the low-resolution image and ERA5 atmospheric variables) provided to control the denoising process.

- The final denoised latent representation Z_0, which has been transformed through the UNet, will be decoded into pixel space to recover the high-resolution image x'_{k+1}.

2.2. Reward regression model

The images generated by the first-stage diffusion fail to meet the required standards due to data imbalance, as extreme weather data is scarce. These high-resolution images exhibit inconsistencies with the guiding conditions and lack accurate representations of the fine-grained anomalous features typical of extreme weather events.Therefore, the proposed method pre-trains a reward regression model that provides real-time, precise reward feedback on whether the diffusion-generated high-resolution images meet the fine-grained features of extreme weather conditions, If these conditions are met, the feedback reward is a high positive value; otherwise, it is a low negative value. The data used to train the reward regression model is pre-annotated according to these conditional rules.

As shown in Figure 2, the reward regression model is designed to identify a hyperplane that distinguishes whether the generated images capture the specified fine-grained features of extreme weather, as described above. This is accomplished using shared CLIP hidden embeddings extracted from the pretrained CLIP encoder, which is trained through image-text contrastive learning[7] and can extract highly generalized features. A linear support vector machine (SVM) [1] is then employed for this regression task. The SVM establishes a decision boundary between samples that satisfy the extreme weather characteristics and those that do not.

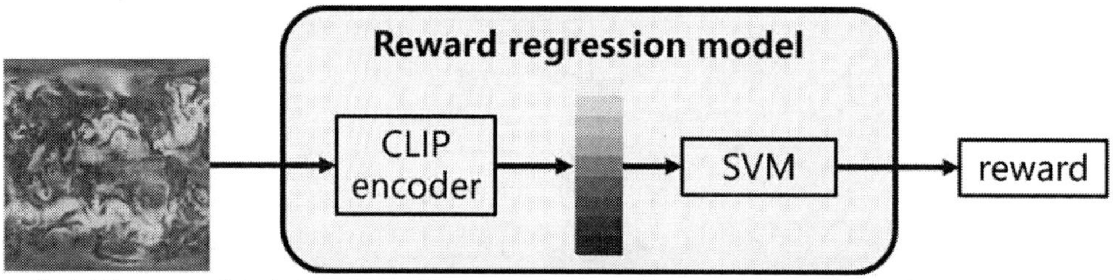

Figure 2. At the second training stage, the reward regression model consists of two components: (a) the CLIP encoder, which compresses the image at the pixel level into hidden embeddings; and (b) the SVM machine learning model, which determines whether the generated high-resolution image contains the fine-grained features of extreme weather based on the shared CLIP embeddings.

The pretrained reward regression model provides real-time reward feedback for the diffusion model's reverse denoising fine-tuning process. The reward feedback indicates whether the generated image meets the fine-grained features of extreme weather. If the features are captured, the reward is high; otherwise, it is low. In the next stage, the diffusion model adjusts the U-Net through backpropagated gradient updates based on the feedback from the reward regression model. This process helps generate high-resolution images that more accurately capture extreme weather characteristics by maximizing the reward.

2.3. RGD model : Reward Feedback online Learning Improves Diffusion

As shown in Figure 3, the reward obtained from the reward regression model is used as reliable feedback to improve the diffusion process. We fine-tune the pretrained conditional diffusion model by backpropagating gradients from the reward provided by the reward regression model to a randomly selected later time step t (in our case, $t \in [400,500]$) during the backward noise removal process. The reason for randomly selecting the time step t instead of using the last time step is that fine-tuning only the gradient of the last denoising step is unstable, which, in our initial experiments, resulted in extremely poor quality of the generated high-resolution images. Therefore, to prevent rapid overfitting and stabilize the fine-tuning process, the loss for fine-tuning the diffusion model is formulated as follows:

$$L_{reward} = -\lambda E_{C_i \sim C}(r(g_\theta(C_i))) \quad (2)$$

where θ denotes the U-net weights of the diffusion framework, $g_\theta(C_i)$ represents the images generated by θ under the conditional embeddings C_i. $r(*)$ represents the reward predicted by the reward regression model for each generated image, and λ represents the weight scale of the reward loss.

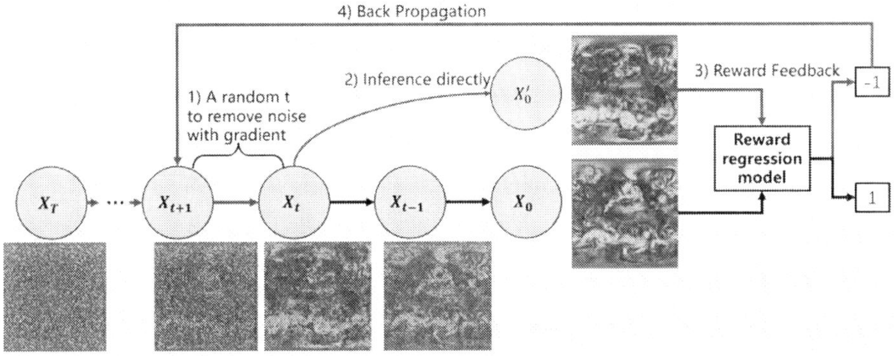

Figure 3. The visualization of the denoising process at the diffusion fine-tuning stage. The diffusion model in Figure 1 will be finetuned by the feedback from the Reward regression model in Figure 2.

3. Simulation example

3.1. Data Source

To verify the model's effectiveness, we used the ECMWF's ERA5 reanalysis data for typhoon prediction validation. The time span of the model data used is from 2014 to 2024, a period of 10 years. The dataset spanning from 2014 to 2022, a period of eight years, served as the model training set. The remaining data, comprising a two-year timeframe, was utilized as the prediction set to assess the efficacy of the model's outcomes. The model's input can be summarized as shown in the table below.

Table 1. ERA5 variables for input into DMs

Variable name	Short name	Period (year)	Predicted (days)	Level
Temperature (2m)	T2	2014-2024	5	Surface
Sea Surface Temperature	SST	2014-2024	5	Surface
Mean sea level pressure	MSL	2014-2024	5	Surface
Wind Speed U component (10m)	U10	2014-2024	5	Surface
Wind Speed V component (10m)	V10	2014-2024	5	Surface
Potential Vorticity	PV	2014-2024	5	850hPa
Geopotential Height	GH	2014-2024	5	200hPa/850hPa

3.2. Evaluation method

3.2.1. Implementation Details. Throughout the training procedure, NWP is fixed to generate low-resolution data, which will be used to train the diffusion framework. The batch size is set to 11 for training over 600 epochs during the first-stage pretraining and third-stage fine-tuning. In the second stage, the reward regression model is pretrained using a batch size of 32 and ten-fold cross-validation. At the third stage, the weights of the reward regression model and the autoencoder model at the pixel level are frozen, and only the weights of the conditional diffusion model are updated. The conditional diffusion model was trained on both low-resolution data as the condition and high-resolution data as the ground truth. Meanwhile, the reward regression model was only trained on high-resolution data because it needs to provide reward feedback for the generated high-resolution data from the diffusion model during each training epoch in an online manner.

3.2.2. Baseline. In the evaluation of Tropical Cyclone (TC) track accuracy, the selection of appropriate criteria is of paramount importance. As highlighted by Knapp et al. (2010) in the IBTrACS dataset, the choice of detection criteria can significantly influence the analysis of cyclone paths. To address this challenge, this study adopts a comprehensive approach, drawing upon a variety of cyclone tracking algorithms and threshold schemes. Specifically, we reference the Tempest Extremes cyclone tracker developed by Ilan Price, the Tropical Cyclone tracker algorithm of the ECMWF proposed by van der Grijn, and the TC threshold scheme of the FuXi-Extreme forecast. The application of these methods ensures physical consistency in the behavior of cyclones within the original forecasts. To minimize dependence on and sensitivity to specific models, we employ a uniform standard, applying the cyclone tracker to all model and ground truth analysis datasets.When predicting the next position of a TC, our tracker evaluates all vaules within a radius of 278 km, as detailed in Table 2. In this process, we screen for local minima that meet specific criteria and identify the candidate minimum value closest to the current estimated position.

Table 2. TC tracker's criteria

Variable name	Level	Radius	Threshold	Formula
Wind Speed (10m)	Surface	278km	>8 m/s	$\mathrm{WS10} = \sqrt{\mathrm{U10}^2 + \mathrm{V10}^2}$
Vorticity	850hPa	278km	$>5\times10^{-5}s^{-1}$	

3.2.3. Evaluation Metrics. The accuracy of various models are evaluated in predicting typhoon tracks and intensity using the MAE, RMSE and CSI metrics. The definitions of MAE and RMSE are the same as the standard explanations in mathematical formulas. The Critical Success Index (CSI) is used to quantify the success rate of the predictions, $CSI = \frac{TN}{TN+FN+MN}$ .In this case study, TN stands for the accurate prediction of typhoons, FN for false predictions, and MN for missed predictions. The specific criteria for typhoon prediction are determined based on the thresholds outlined in Table 2 for the tropical cyclone tracker.

4. Result

TC prediction evaluation mainly relies on two aspects: trajectory and intensity. Track accuracy is determined by MAE, which measures the average hourly distance difference between observed and forecasted typhoon center positions. The intensity evaluation focuses on the WS10 near the TC center, also with one-hour intervals, by calculating the RMSE between the observed and predicted WS10 values. Additionally, we use the CSI metric to statistically analyze the success rate of each model in predicting typhoons.

Track. The RGD model, along with NWP ensemble forecasts and ERA5, were assessed for their predictive accuracy regarding TC over a five-day forecast period. As shown in Figure 4, this evaluation reflects the outcomes corresponding to 17 significant typhoon landfall events that occurred along the Chinese coast from 2022 to 2024. The two methods exhibited comparable accuracy in predictions for the first 48 hours. However, beginning on the third day, the RGD model demonstrated superior predictive performance over the NWP ensemble forecast, with its accuracy advantage increasing progressively over time.

Intensity. In tropical cyclone intensity forecasting, Figure 5 reveals that the RGD model outperforms the NWP ensemble forecast in WS10 predictions. A comparison of RMSE values shows that the RGD model maintains the lowest RMSE throughout the entire forecast period. Statistical analysis of 17 major typhoon landfall events shows that the RGD model has an average RMSE of less than 15% in the first two days and less than 19% in the 3-to-5-day forecast, outperforming the NWP model.

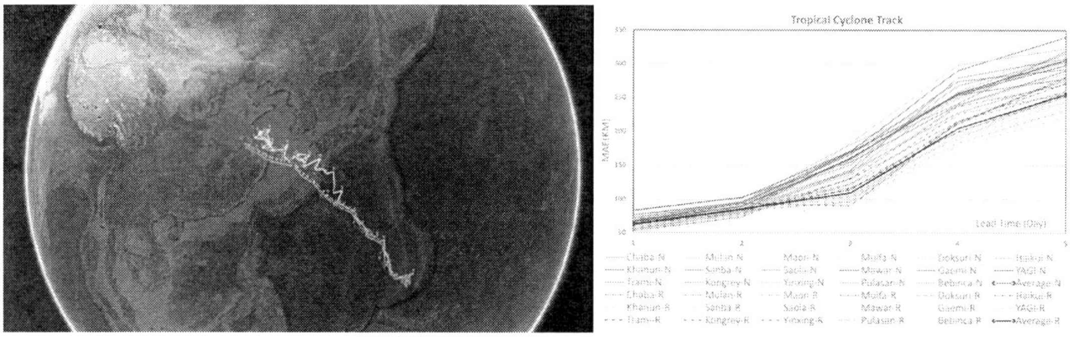

Figure 4 shows the path of Typhoon Bebinka from September 11 to 16, 2024. The NWP ensemble forecast, RGD model, and RGD model are represented by yellow, pink, and cyan lines, respectively. The chart indicates that the two methods perform similarly in predictions from 0 to 2 days. However, starting from the third day, the RGD model's predictive accuracy surpasses that of the NWP ensemble forecast, and this advantage becomes more pronounced as the prediction time increases.

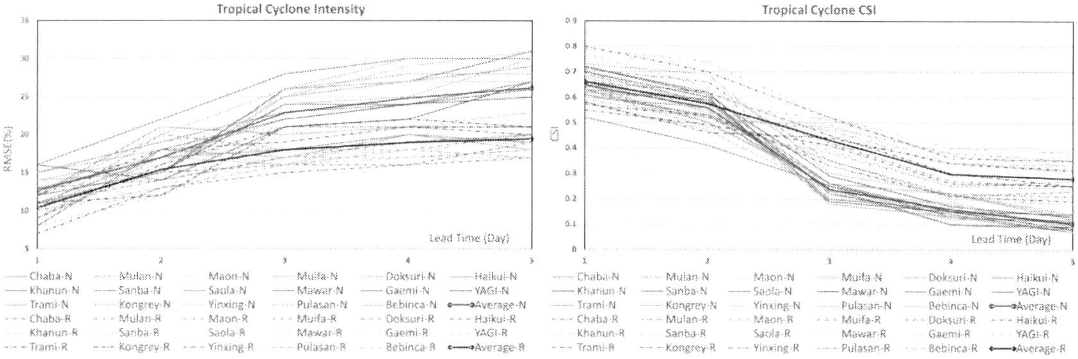

Figure 5: Intensity and CSI Analysis. A comparative analysis of the 17 aforementioned typhoon cases over a five-day forecast period.

CSI. As shown in Figure 5, during the first two forecast days, the performance of the RGD model is comparable to that of the NWP forecasts. However, from the third to the fifth day, the RGD model demonstrates superior performance.

Table 3. Experimental Results of Typhoon Invasions in Coastal Areas of China from 2023 to 2024

TC name	Track MAE(KM)		Track MAE(KM) (0-2day)		Track MAE(KM) (3-5day)		Intensity RMSE(%)		Intensity RMSE(%) (0-2day)		Intensity RMSE(%) (3-5day)	
	N*	R*	N*	R*	N*	R*	N*	R*	N*	R*	N*	R*
Chaba	160	136	77	70	243	202	19	15	17	13	21	17
Mulan	161	134	84	66	238	201	18	15	13	13	23	17
Ma-on	167	136	82	84	252	189	21	14	15	12	27	17
Muifa	166	132	86	69	246	196	18	17	10	15	25	19
Doksuri	161	133	76	72	245	193	19	16	15	14	24	19
Haikui	141	138	67	63	215	213	17	13	16	10	18	16
Khanun	180	127	93	73	267	181	19	15	13	14	24	17
Sanba	165	139	85	80	244	199	16	16	12	14	21	18
Saola	158	119	73	60	243	178	20	17	16	13	24	21
Mawar	160	127	93	71	228	184	18	15	16	13	20	17
Gaemi	158	124	81	70	234	177	17	16	17	11	17	21

	N*	R*	N*	R*	N*	R*	N*	R*	N*	R*	N*	R*
Yagi	157	135	75	81	238	190	20	19	15	17	24	22
Trami	162	138	83	74	241	202	15	16	13	12	17	21
Kongrey	155	136	73	80	238	192	19	18	15	15	23	20
Yinxing	149	130	68	71	230	189	18	17	15	13	21	21
Pulasan	147	128	74	83	220	173	14	16	13	14	16	19
Bebinca	166	124	78	75	268	165	19	14	18	10	20	19
Average	159	131	79	73	241	190	18	16	14	13	22	19

*N represents the NWP ensemble forecast; *R represents the RGD model.

5. Conclusion

We present RGD, a high-resolution data synthesis model that enhances conventional diffusion models by incorporating reward feedback from a pretrained reward regression model. This novel approach demonstrates superior performance in medium to long term forecasting and has been empirically validated through 17 typhoon landings between 2022 and 2024. The method provides a reliable foundation for ensuring the safe and stable operation of power systems.

6. Acknowledgement

This work is supported by the 18 MW Platform Offshore Wind Turbine Unit R&D and Design CHN Energy Technology Project GJNY-24-121.

References

[1] Awad, M., Khanna, R., 2015. Support Vector Machines for Classification, in: Efficient Learning Machines. Apress, Berkeley, CA, pp. 39–66.

[2] Chen, L., Du, F., Hu, Y., Wang, F., Wang, Z., 2023. SwinRDM: Integrate SwinRNN with Diffusion Model towards High-Resolution and High-Quality Weather Forecasting. https://doi.org/10.48448/zn7f-fc64

[3] Hatanaka, Y., Glaser, Y., Galgon, G., Torri, G., Sadowski, P., 2023. Diffusion Models for High-Resolution Solar Forecasts. https://doi.org/10.48550/arXiv.2302.00170

[4] Ho, J., Jain, A., Abbeel, P., 2020. Denoising Diffusion Probabilistic Models. https://doi.org/10.48550/ARXIV.2006.11239

[5] Hua, Z., He, Y., Ma, C., Anderson-Frey, A., 2024. Weather Prediction with Diffusion Guided by Realistic Forecast Processes. https://doi.org/10.48550/arXiv.2402.06666

[6] Merizzi, F., Asperti, A., Colamonaco, S., 2024. Wind speed super-resolution and validation: from ERA5 to CERRA via diffusion models. Neural Comput & Applic 36, 21899–21921.

[7] Radford, A., Kim, J.W., Hallacy, C., Ramesh, A., Goh, G., Agarwal, S., Sastry, G., Askell, A., Mishkin, P., Clark, J., Krueger, G., Sutskever, I., n.d. Learning Transferable Visual Models From Natural Language Supervision.

[8] Rombach, R., Blattmann, A., Lorenz, D., Esser, P., Ommer, B., 2022. High-Resolution Image Synthesis with Latent Diffusion Models, Presented at the 2022 IEEE/CVF Conference on Computer Vision and Pattern Recognition (CVPR), IEEE, New Orleans, LA, USA, pp. 10674–10685. https://doi.org/10.1109/CVPR52688.2022.01042

[9] Ronneberger, O., Fischer, P., Brox, T., 2015. U-Net: Convolutional Networks for Biomedical Image Segmentation, MICCAI 2015, Lecture Notes in Computer Science. pp. 234–241

[10] Vaswani, A., Shazeer, N., Parmar, N., Uszkoreit, J., Jones, L., Gomez, A.N., Kaiser, Ł., Polosukhin, I., n.d. Attention is All you Need.

[11] Watt, R.A., Mansfield, L.A., 2024. Generative Diffusion-based Downscaling for Climate. https://doi.org/10.48550/arXiv.2404.17752

[12] Zhong, X., Chen, L., Liu, J., Lin, C., Qi, Y., Li, H., 2024. FuXi-Extreme: Improving extreme rainfall and wind forecasts with diffusion model. Sci. China Earth Sci. 67, 3696–3708.

Journal of Physics: Conference Series **3012** (2025) 012069 doi:10.1088/1742-6596/3012/1/012069

Small-Signal Stability Analysis of the Grid-Forming MMC-HVDC System

Yifei Meng[1], Shanshan Wang[2*], Yuanyuan Sun[1], Jianhua Li[3], Jianyu Lu[3]

[1] School of Electrical Engineering, Shandong University, Jinan, Shandong Province, 250061, China

[2] China Electric Power Research Institute, Haidian District, Beijing, 100192, China

[3] East China Branch of State Grid Corporation of China, Pudong New Area, Shanghai, 200120, China

*Corresponding author's e-mail: ssyj1003@163.com

Abstract. Unlike grid-following control, grid-forming (GFM) control provides voltage and frequency support to modular multilevel converter-based high voltage direct current (MMC-HVDC) systems, thereby enhancing system stability. However, the impact of the internal harmonic dynamics of the modular multilevel converter (MMC) and the critical control loops of GFM control on the stability of grid-connected systems has not been fully clarified. Therefore, a small-signal model of the MMC-HVDC system is developed to investigate the influence of grid impedance characteristics and key control loops on system stability.

1. Introduction

Modular multilevel converter (MMC), with its many advantages, such as high output waveform quality, easy scalability, and low switching losses, has been widely applied in the field of voltage source converter-based high voltage direct current (VSC-HVDC) transmission[1]. However, the system strength decreases as the capacity of DC and renewable energy integration increases. Integrating modular multilevel converter-based high voltage direct current (MMC-HVDC) systems into the grid has induced several oscillation incidents, such as the 1270 Hz high-frequency oscillation in the Luxi HVDC project[3].

MMC's control methods can be classified into grid-following(GFL) control and grid-forming(GFM) control[4]. Currently, GFL control is widely used in practical engineering applications. However, as the system strength decreases, the GFL MMC-HVDC system is prone to instability. Existing studies show that the interaction and coupling between the PLL and weak AC systems are the key reasons for system instability. Ref. [5] analyzes the coupling between the PLL and grid impedance, pointing out that increasing the PLL control bandwidth will enhance the coupling between the PLL and grid impedance, reducing system stability. Ref. [6] highlights that under weak grid conditions, the coupling between the PLL and current control becomes more pronounced, and the PLL's dynamic characteristics significantly weaken the current control's stability margin.

To improve the stability of converter systems when connected to weak grids, researchers have proposed GFM control methods such as virtual synchronous generator (VSG) control. However, when GFM MMC is connected to a strong grid, it introduces low-frequency oscillations similar to those of synchronous generators[7]. Currently, a considerable amount of papers have studied the impact of control loop parameters and grid impedance characteristics on the stability of GFM MMC systems. Ref. [8] analyzes the impact of different parameters on stability and points out that increasing the droop control gain or reducing the voltage loop control gain is detrimental to system stability. Ref. [9] examines the impact of GFM control on power transmission capability under weak AC grid conditions, stating that higher frequency regulation gain will reduce system stability. However, most papers only establish simplified small-signal models of GFM MMC for analysis, which do not consider the impact of MMC arm circulating currents and circulating current suppressors on the system's stability. Therefore, this paper establishes a detailed modeling of GFM MMC and investigates the impact of power synchronization loop(PSL) and short circuit ratio(SCR) on the system's small-signal stability.

The paper is organized as follows: Section 2 establishes the small-signal model of the MMC-HVDC system. Section 3 investigates the influence of PSL and SCR on system stability. The concluding remarks are made in section 4.

2. Small-signal Model of GFM MMC-HVDC System

2.1. Main circuit of GFM MMC-HVDC system

This subsection mainly introduces the system main circuit and the topology of the MMC, and provides the state-space equations.

The system's main circuit can be shown in Figure 1. In Figure 1, P_s and Q_s are the active and reactive power transmitted from the point of common coupling (PCC) to the AC grid, respectively. The AC grid is modeled using a Thevenin equivalent, with an infinite voltage source in series and an equivalent impedance of Z_g. $u_s/u_g/u_v$ is the grid voltage/the voltage at the PCC/the MMC's AC output voltage. i_s is the phase current. On the DC side, a DC source replaces the MMC with constant DC voltage control at the other end.

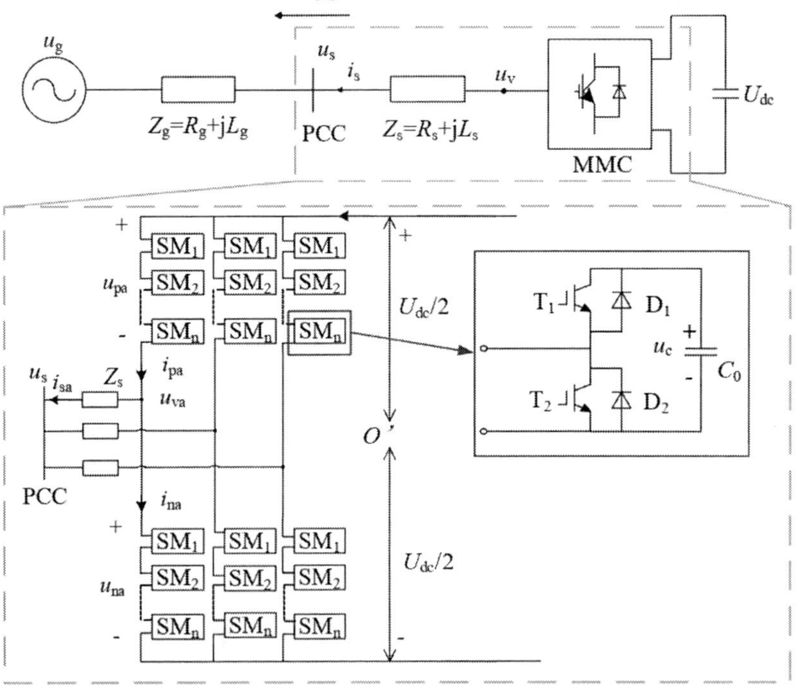

Figure 1 The main circuit of the MMC-HVDC system and topology diagram of MMC

The internal topology of the MMC can be shown in Figure 1. The number of submodules (SM) in the upper and lower arms of each phase is N. R_{arm}, L_{arm}, and C_0 represent the arm resistance, inductance, and MMC SM capacitance, respectively. The transformer impedance Z_T and the MMC arm impedance Z_{arm} can be merged and equivalently represented as Z_s=R_s+jX_s.

To summarize, the small signal model of the system's main circuit consists of the following three parts: 1) The Thevenin equivalent model of the AC grid; 2) The MMC model that reflects the internal dynamic characteristics; 3) The equivalent model of the DC side. The modeling process of parts 2) and 3) can be referenced in Ref. [10]. The state-space equation of part 1) can be given as

$$\begin{cases} \dfrac{di_s}{dt} = \dfrac{1}{L_s}(u_v - u_s - R_s i_s) \\ \dfrac{di_s}{dt} = \dfrac{1}{L_g}(u_s - u_g - R_g i_s) \end{cases} \tag{1}$$

2.2. *Control system of GFM MMC-HVDC system*

This subsection mainly introduces the control system and provides the state-space equations.

The Park's transformation matrix of the control system can be given as

$$T_1 = \frac{2}{3}\begin{bmatrix} \cos(\omega t) & \cos(\omega t - \frac{2}{3}\pi) & \cos(\omega t + \frac{2}{3}\pi) \\ -\sin(\omega t) & -\sin(\omega t - \frac{2}{3}\pi) & -\sin(\omega t + \frac{2}{3}\pi) \end{bmatrix} \tag{2}$$

Taking the fundamental frequency as an example, since the PSL's output may lag behind the actual voltage phase angle of PCC after a disturbance, there are two rotating dq frames: the grid dq frame and the GFM control dq frame. The transformation relationship between the two frames can be shown in Figure 2 and (3). In Figure 2, f is the electrical quantity, and the superscripts 'gf' and 'cf' denote the grid dq frame and the GFM control dq frame, respectively.

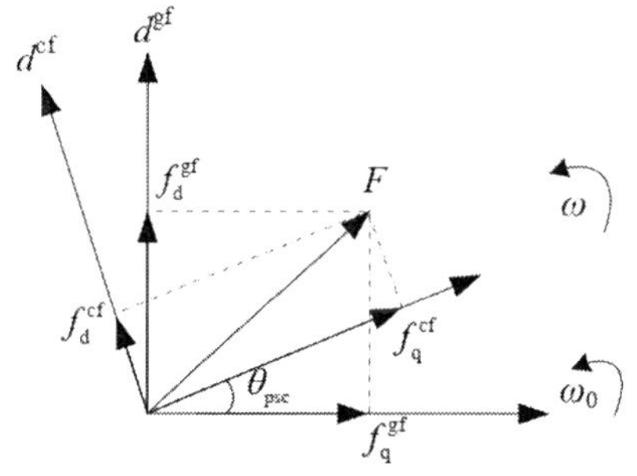

Figure 2 Relationship between the dq frames

$$\begin{bmatrix} f_d^{gf} \\ f_q^{gf} \end{bmatrix} = \begin{bmatrix} \cos\theta_{psc} & \sin\theta_{psc} \\ -\sin\theta_{psc} & \cos\theta_{psc} \end{bmatrix} \begin{bmatrix} f_d^{cf} \\ f_q^{cf} \end{bmatrix}, \begin{bmatrix} f_d^{cf} \\ f_q^{cf} \end{bmatrix} = \begin{bmatrix} \cos\theta_{psc} & -\sin\theta_{psc} \\ \sin\theta_{psc} & \cos\theta_{psc} \end{bmatrix} \begin{bmatrix} f_d^{gf} \\ f_q^{gf} \end{bmatrix} \tag{3}$$

Table 1 Explanations of the various variables

Names	Explanations	Names	Explanations
u_{cird}, u_{cirq}	dq components of the circulating voltage of the MMC	I_{cird}, I_{cirq}	dq components of the circulating current of the MMC
$u_{vdq}/u_{gdq}/u_{sdq}$	dq components of $u_v/u_g/u_s$	i_{sdref}, i_{sqref}	Reference dq components of i_s
ω	Output frequency of the PSL	u_{vdref}, u_{vqref}	Reference dq components of u_v
P_{ref}	Reference active power	ω_0	Grid nominal frequency
i_{sd}, i_{sq}	dq components of i_s	Q_{ref}	Reference reactive power
$I_{cirdref}$, $I_{cirqref}$	Reference dq components of I_{cridq}	U_{s0}	The base voltage of the system
θ_{psc}	The output phase angle of PSL	$u_{vabcref}$	Reference of u_{vabc}

The control system of the GFM MMC converter station can be shown in Figure 3, which consists of three parts: PSL, the circulating current suppression control(CCSC), and the outer loop-inner loop control. The explanations of the variables can be shown in Table 1.

In Figure 3, the state-space equation of the PSL, the CCSC, and the outer loop-inner loop control can be given by (4)-(7).

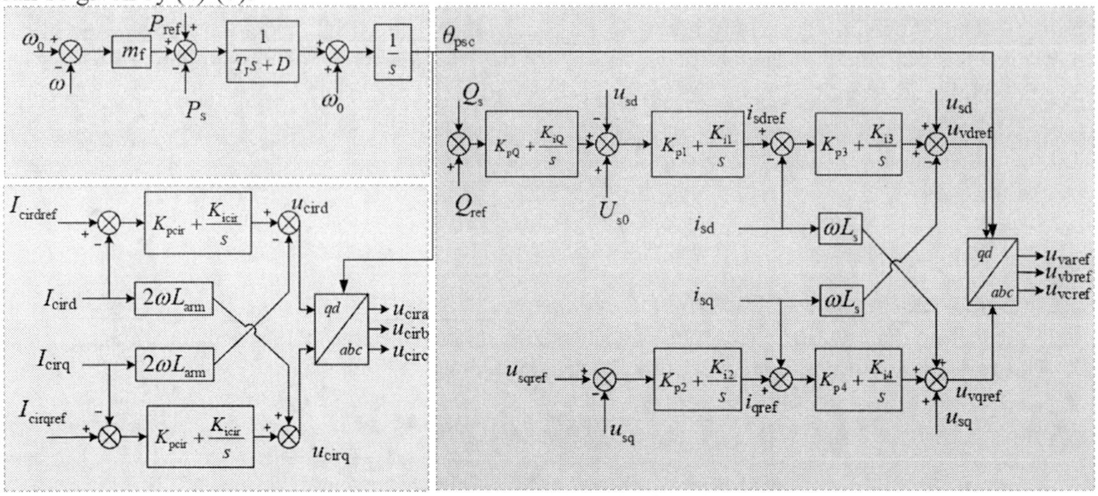

Figure 3 The control system of the grid-forming MMC station

$$\begin{cases} \dfrac{dx_{\mathrm{psc}}}{dt} = \dfrac{(P_{\mathrm{ref}} - m_{\mathrm{f}} x_{\mathrm{psc}}) - P}{T_{\mathrm{J}}} - \dfrac{D}{T_{\mathrm{J}}} x_{\mathrm{psc}} \\ \dfrac{d\theta_{\mathrm{psc}}}{dt} = x_{\mathrm{psc}} \end{cases} \tag{4}$$

$$\begin{cases} \dfrac{dx_1}{dt} = Q_{\mathrm{ref}} - Q_{\mathrm{s}} \\ \dfrac{dx_2}{dt} = K_{\mathrm{pQ}}(Q_{\mathrm{ref}} - Q_{\mathrm{s}}) + K_{\mathrm{iQ}} x_1 + U_{\mathrm{s0}} - u_{\mathrm{sd}} \\ \dfrac{dx_3}{dt} = -u_{\mathrm{sq}} \end{cases} \tag{5}$$

$$\begin{cases} \dfrac{dx_4}{dt} = K_{\mathrm{p1}}[K_{\mathrm{pQ}}(Q_{\mathrm{ref}} - Q_{\mathrm{s}}) + K_{\mathrm{iQ}} x_1 + U_{\mathrm{s0}} - u_{\mathrm{sd}}] + K_{\mathrm{i1}} x_2 - i_{\mathrm{sd}} \\ \dfrac{dx_5}{dt} = -K_{\mathrm{p2}} u_{\mathrm{sq}} + K_{\mathrm{i2}} x_3 - i_{\mathrm{sq}} \end{cases} \tag{6}$$

$$\begin{cases} \dfrac{dx_{\mathrm{cird}}}{dt} = I_{\mathrm{cirdref}} - I_{\mathrm{cird}} \\ \dfrac{dx_{\mathrm{cirq}}}{dt} = I_{\mathrm{cirqref}} - I_{\mathrm{cirq}} \end{cases} \tag{7}$$

x_1~x_5, x_{cridq}, and x_{psc} are intermediate variables within the PI regulator in the control system. The output equations for each loop can be listed according to the structure in Figure 3.

2.3. Small-signal Model of GFM MMC-HVDC system

The steady-state operating point of the system can be solved by coupling (1)-(7) and other output equations. (1)-(7) and other output equations can be linearised to obtain the small-signal model as

$$\frac{d\Delta \boldsymbol{x}}{dt} = \boldsymbol{A}\Delta \boldsymbol{x} + \boldsymbol{B}\Delta \boldsymbol{u} \tag{8}$$

$\Delta\boldsymbol{x}$=[Δu_{cdc}, Δu_{c1d}, Δu_{c1q}, Δu_{c2d}, Δu_{c2q}, ΔI_{dc}, ΔI_{sd}, ΔI_{sq}, ΔI_{cird}, ΔI_{cirq}, Δx_1, Δx_2, Δx_3, Δx_4, Δx_5, Δx_{cird}, Δx_{cirq}, Δx_{psc}, $\Delta\theta_{\mathrm{psc}}$]^{T}, $\Delta\boldsymbol{u}$=[ΔP_{ref}, ΔQ_{ref}, $\Delta I_{\mathrm{cirdref}}$, $\Delta I_{\mathrm{cirqref}}$]^{T}. u_{cdc} is the DC component of the submodule capacitor voltage. u_{c1d} and u_{c1q} are the fundamental frequency dq components of the SM capacitor voltage. u_{c2d} and u_{c2q} are the second harmonic dq components of the SM capacitor voltage.

Table 2 Parameters of the GFM MMC-HVDC system

Parameter	Value	Parameter	Value
SCR	2	Number of MMC's SM / N	12
AC nominal voltage / kV	10	MMC arm inductance / mH	10.6
DC nominal voltage / kV	±10	Transformer's reactance Z_{T} / p.u.	0.18
Impedance angle / °	50	Parameters of PSL: m_{f} / T_{j} / D	0.4, 0.01, 0.1428
Parameters of outer loop: K_{pQ} / K_{iQ} / K_{p1} / K_{i1}	0.02, 1.8, 0.08, 2	Parameters of outer loop: K_{p2} / K_{i2}	0.08, 2
Parameters of inner loop: K_{p3} / K_{i3} / K_{p4} / K_{i4}	15, 150, 15, 150	Parameters of CCSC: K_{pcir} / K_{icir}	50, 100

Then, the small-signal model will be validated. The parameters are shown in Table 2. The initial condition, i.e., at 0s, P is 5.0MW (1 p.u.), Q is 0Var. P_{ref} is made to step to 4.75MW (0.95p.u.) at 1s. The dynamic response of the relevant electrical quantities can be shown in Figure 4.

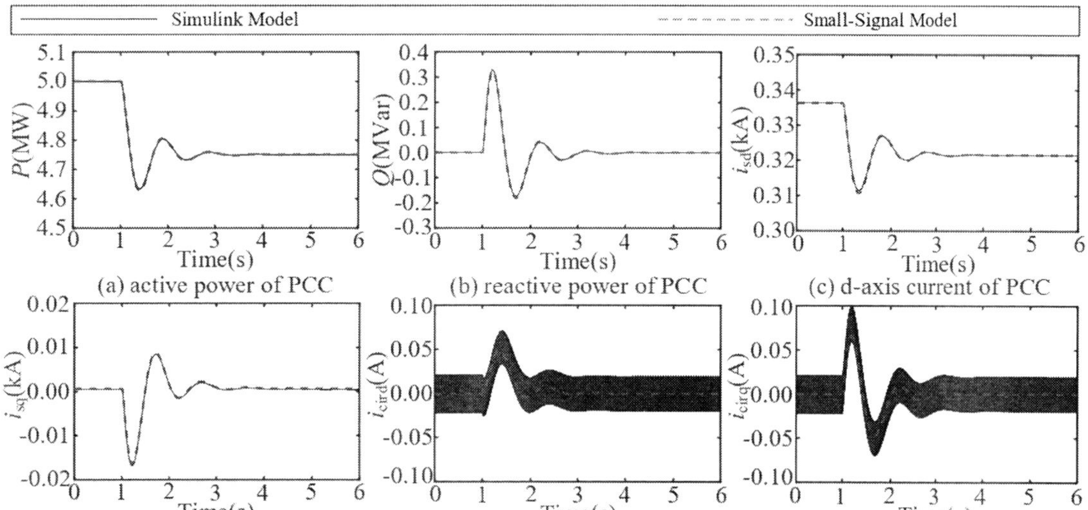

(a) active power of PCC (b) reactive power of PCC (c) d-axis current of PCC
(d) q-axis current of PCC (e) d-axis circulating current (f) q-axis circulating current

Figure 4 The dynamic response of P, Q, i_{sd}, i_{sq}, i_{cird}, i_{cirq} as P_{ref} step changes

In Figure 4, the small-signal model result is consistent with the simulation based on Simulink so that the model can be validated.

3. Analysis of the Small-signal Stability of the System

PSL is the key loop in the GFM control and requires investigating how its parameters affect system stability. Therefore, this section primarily studies the PSL and SCR, analyzing the impact of changes in key parameters on system stability, and uses factor participation analysis to identify the factors influencing system stability.

3.1. Research on the power synchronization loop

This subsection analyzes the impact of the damping coefficient D and inertia time constant T_j in the PSL on system stability.

3.1.1. Research on Damping coefficient D. Parameters are shown in Table 2. Let T_j=0.10, D decreases from 2 to 0.01, the eigenvalues can be shown in Figure 5.

In Figure 5, the eigenvalue corresponding to the system's unstable mode is $\lambda_{1,2}$. As D decreases, $\lambda_{1,2}$ progressively shift toward the right-half plane, leading to a decline in system stability. When D=0.4, $\lambda_{1,2}$ become 0.154±5.150i, leading to system instability. To validate the result, Set D=0.5 at 0s. At 1s, D changes to 0.4. Other parameters are shown in Table 2. The simulation is conducted in Simulink. And the participation factors in mode 1,2 at D=0.4 are obtained. The verification results of the eigenvalue analysis and the modal participation factors in mode 1,2 can be shown in Figure 6.

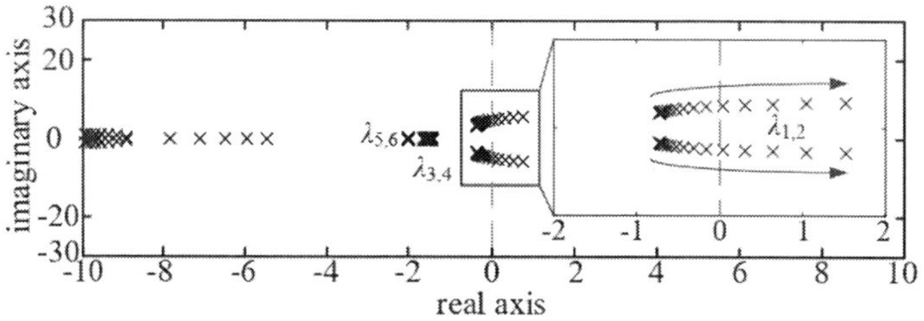

Figure 5 The eigenvalues of A with the decrease of D

In Figure 6(a), the system eventually becomes unstable with the change of D. The oscillation frequency $f_{m1,2}$(0.8241Hz) obtained from the simulation is nearly equal to the oscillation frequency $f_{1,2}$(0.8196Hz), so the calculation results can be verified. At this point, the system experiences low-frequency oscillations. In Figure 6(b), the control loop that has the greatest impact on mode 1,2 is PSL, while the outer loop also significantly affects mode 1,2. At this point, the impact of internal dynamics on stability is relatively small.

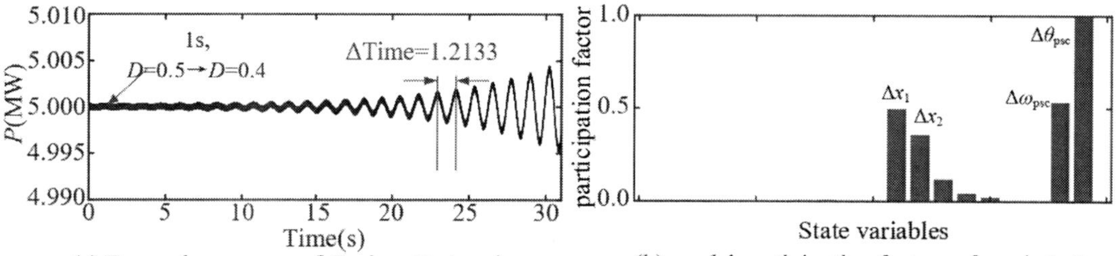

(a) Dynamic response of P when D step changes (b) modal participation factors of mode 1, 2

Figure 6 Verification results and the modal participation factors in mode 1,2 when D=0.4

3.1.2. Research on inertia time constant T_j. Parameters are shown in Table 2. Let D=0.10, T_j increases from 0.01 to 0.2, the eigenvalues can be shown in Figure 7.

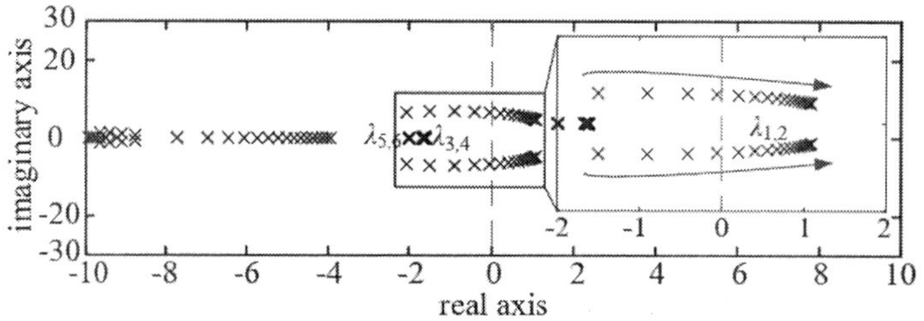

Figure 7 The eigenvalues of A with the increase of T_j

In Figure 7, the eigenvalue corresponding to the system's unstable mode is $\lambda_{1,2}$. As T_j increases, $\lambda_{1,2}$ progressively shift toward the right-half plane, leading to a decline in system stability. When T_j=0.06, $\lambda_{1,2}$ become 0.207±6.477i, leading to system instability. To validate the result, Set T_j=0.05 at 0s. At 1s,

T_j changes to 0.06. Other parameters are shown in Table 2. The simulation is conducted in Simulink. And the participation factors in mode 1,2 at T_j=0.06 are obtained. The verification results of the eigenvalue analysis and the modal participation factors in mode 1,2 can be shown in Figure 8.

In Figure 8(a), the system eventually becomes unstable with the change of T_j. The oscillation frequency $f_{m1,2}$(1.0346Hz) obtained from the simulation is nearly equal to the oscillation frequency $f_{1,2}$(1.0308Hz) from $\lambda_{1,2}$, so the calculation results can be verified. At this point, the system experiences low-frequency oscillations. In Figure 8(b), the control loop that has the greatest impact on mode 1,2 is PSL, while the outer loop also significantly affects mode 1,2. At this point, the impact of internal dynamics on stability is relatively small.

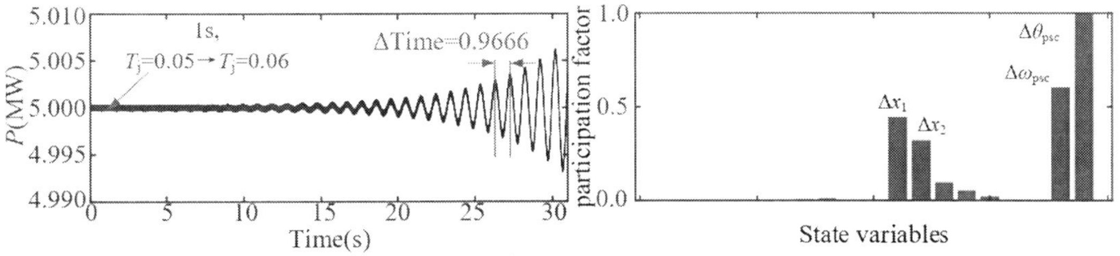

(a) Dynamic response of P when T_j step changes (b) modal participation factors of mode 1, 2

Figure 8 Verification results and the modal participation factors in mode 1,2 when T_j=0.06

3.2. Research on SCR

The effect of SCR on the stability of the system is analyzed in this subsection. Parameters are shown in Table 2. Let SCR increases from 1.5 to 10. The eigenvalues can be shown in Figure 9.

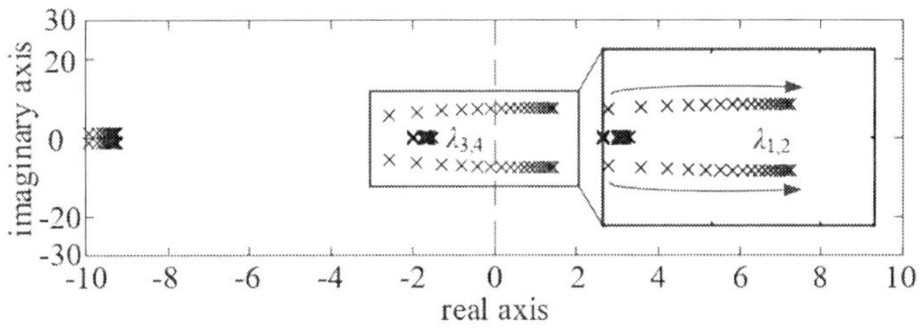

Figure 9 The eigenvalues of A with the increase of SCR

In Figure 9, the eigenvalue corresponding to the system's unstable mode is $\lambda_{1,2}$. As SCR increases, $\lambda_{1,2}$ progressively shift toward the right-half plane, leading to a decline in system stability. When SCR=4.5, $\lambda_{1,2}$ become 0.158±7.456i, leading to system instability. To validate the result, Set SCR=4 at 0s. At 1s, SCR changes to 4.5. Other parameters are shown in Table 2. The simulation is conducted in Simulink. And the participation factors in mode 1,2 at SCR=4.5 are obtained. The verification results of the eigenvalue analysis and the modal participation factors in mode 1,2 can be shown in Figure 10.

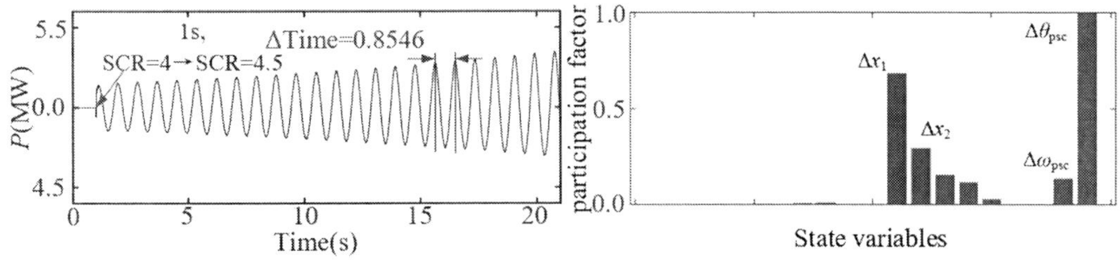

(a) Dynamic response of P when SCR step changes (b) modal participation factors of mode 1, 2

Figure 10 Verification results and the modal participation factors in mode 1,2 when SCR=4.5

In Figure 10(a), the system eventually becomes unstable with the change of SCR. The oscillation frequency $f_{m1,2}$(1.1701Hz) obtained from the simulation is nearly equal to the oscillation frequency $f_{1,2}$(1.1866Hz) from $\lambda_{1,2}$, so the calculation results can be verified. At this point, the system experiences low-frequency oscillations. In Figure 10(b), PSL is the control loop with the greatest impact on mode

1,2. Moreover, under higher SCR conditions, the outer loop control significantly impacts system stability.

4. Conclusion

In this thesis, a small-signal model of the GFM MMC-HVDC system considering internal harmonic dynamics has been developed. The discussion on the influence of PSL and SCR on system stability is approached. The conclusions are as follows:

(1) A too-small damping coefficient and too-large inertia time constant can both reduce system stability, even leading to system instability resulting in low-frequency oscillations. The dominant mode is primarily influenced by the power synchronization loop and is less affected by the internal dynamics of the MMC.

(2) Under constant control parameters, the GFM MMC-HVDC system can operate stably when the SCR is low but becomes unstable when the SCR is high. The system's dominant mode is primarily influenced by the power synchronization loop. At the same time, the outer loop control also has a significant effect on this mode.

Acknowledgments

This work is supported by the Research Project of East China Branch of State Grid Corporation of China (Research and Application of Key Technologies for Grid-Forming Operation and Control of New Energy Generators in New Power Systems at the Receiving End. No.52992424001R).

References

[1] Jiang Z, et al. 2019, Application and Design of Resistive SFCL in ±160 kV MMC-HVdc System, IEEE Transactions on Applied Superconductivity, vol 29, pp 1-5.
[2] Shan P, et al. 2025, Adaptive Parameter Tuning and Virtual Impedance Injection Control for Coupled Harmonic Mitigation of Photovoltaic Converter, IEEE Transactions on Power Electronics, vol. 40, pp 162-175.
[3] Changyue Z, et al. 2018, Analysis of Resonance Between a VSC-HVDC Converter and the AC Grid, IEEE Transactions on Power Electronics, vol 33, pp 10157-10168.
[4] Zheng X. 2020, Physical mechanism and research approach of generalized synchronous stability for power systems, Electric Power Automation Equipment, vol 40, pp 3-9(in Chinese).
[5] Xueguang Z, et al. 2018, An Improved Feedforward Control Method Considering PLL Dynamics to Improve Weak Grid Stability of Grid-Connected Inverters. IEEE Transactions on Industry Applications, vol 54, pp 5143-5151.
[6] Donghai Z, et al. 2020, Improved Design of PLL Controller for LCL-Type Grid-Connected Converter in Weak Grid. IEEE Transactions on Power Electronics, vol 35, pp 4715-4727.
[7] Yuting T, et al. 2022, Review on grid-forming converter control methods in high-proportion renewable energy power systems. Global Energy Interconnection, vol 05, pp 328-342.
[8] Qiyu L, et al. 2024, State Space Modeling and Stability Analysis of Grid-forming Inverter, Journal of North China Electric Power University, vol 51, pp 83-93(in Chinese).
[9] Arash A, et al. 2019, Impedance Analysis of Virtual Synchronous Generator-Based Vector Controlled Converters for Weak AC Grid Integration, IEEE Transactions on Sustainable Energy, vol 10, pp 1481-1490.
[10] Tan L, et al. 2016, Small-Signal Model of the Modular Multilevel Converter Considering the Internal Dynamics, Proceedings of the CSEE, vol 36, pp 2890-2899(in Chinese).

Simulation of Uncertainty Scenarios in Renewable Energy Output and Calculation Method for Their Probability of Occurrence

Naifeng Han, Tingyue Guo, Panlong Men, Shijin Liu, Jialin Lu, Zonghang Li, Anyu Zhou, Baoying Zhang, Huanna Niu*

College of Information and Electrical Engineering, China Agricultural University, Beijing, 100083, China

*Corresponding author's e-mail: 1642890250@QQ.com

Abstract. This article proposes a simulation of renewable energy output uncertainty scenarios and a method for calculating their probability of occurrence. Firstly, improve the traditional generative adversarial network to increase model training stability. Then, based on historical load data of wind power, photovoltaics, and the environment, a large number of scenarios with source load correlation are generated and reduced into future probability scenarios through fuzzy C-means clustering method. Finally, the example demonstrates the accuracy of the proposed method through comparison with other scenario methods.

1. Introduction

As the impact of large amounts of fossil energy on the global greenhouse effect intensifies, a new type of power system characterized by large-scale grid integration of new energy sources such as wind power and photovoltaics has become a development trend [1-2]. The output power of wind power and photovoltaic has randomness and volatility, but there is a correlation between the output power of both in a certain regional range. Renewable energy output uncertainty scenario simulation is of great significance for power system scheduling to maximize the benefits and ensure the safety of power grid operation [3].

There are two main approaches to the study of renewable energy correlation: the Copula function-based mathematical statistics method and the neural network-based machine learning method. Literature [4] used Copula function to calculate the joint probability distribution of the actual and predicted values of wind power output and photovoltaic output respectively, and used the clustering method to analyze the historical data according to the weather type and quarter respectively. Literature [5] proposed four simulation methods for generating wind and solar joint output power time series based on Markov chain and Copula theory. Literature [6] trained the subsequence and weather data as prediction inputs to an improved long and short-term memory network (LSTM) model optimized by Gray Wolf Algorithm (GWO) based on Refractive Learning Strategy (RL), predicted the subsequence of the day to be predicted separately and superimposed it to obtain the short-term regional wind power generation power. Literature [7] proposes a joint wind-scenery prediction model based on the improved attention mechanism and the long- and short-term memory method, which is able to effectively extract the coupling relationship between wind-scenery. For the mathematical statistical method of historical statistical data and mathematical model is difficult to effectively match and neural network training overfitting and other issues, literature[8] proposed generative adversarial networks (GAN) by the generator (Generator, G) and discriminator (Discriminator, D) composed of mutual game over the model can be a good solution to the above problems, but there is a training is not stable to generate the effect of the bad problem.

This paper proposes renewable energy output uncertainty scenario simulation and its occurrence probability calculation method. First, the generative adversarial network is improved, and then a large number of scenarios with source-load correlation are generated based on the historical load data of wind power and photovoltaic, which are then cut down into future probability scenarios based on the fuzzy C-mean (FCM) clustering method. Finally, the calculation example is compared by other scenario methods to prove the accuracy of the method proposed in this paper.

2. GAN-based joint scene generation method for renewable energy

2.1. GAN model improvement

Convolutional neural network (CNN) is a kind of deep feedforward neural network with local connection and weight sharing characteristics, including input layer, convolutional layer, pooling layer, fully connected layer and output layer. After inputting the data into the input layer, the convolutional layer extracts the characteristics of the data sequence by continuously iterating the model parameters and shares the weights, the pooling layer then reduces the dimensionality through a sliding window, and finally the fully connected layer outputs the results, the structure is shown in Figure 1.

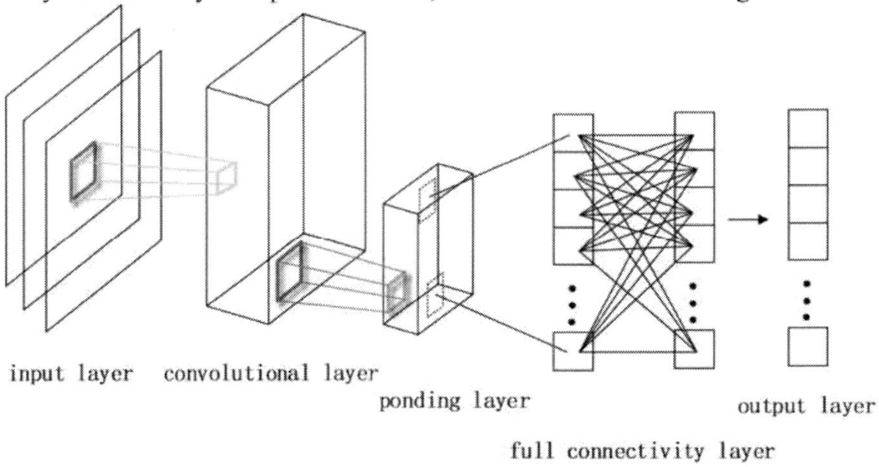

Figure. 1 CNN structure diagram

The improved GAN uses a convolutional layer instead of the CNN pooling structure in generator G. A regularization layer is introduced after the convolutional layer to avoid entering the saturation zone of the activation function and to enhance the training capability of the network. The discriminator D still uses a convolutional network, considering the final need to discriminate the input values, and finally the data is spread to output a discriminant value through a fully connected layer. Due to the lack of the original GAN to generate the scene of other relevant factors conditions, the generated results often do not meet the actual requirements, this paper on this basis to add wind speed, temperature, irradiation intensity, humidity conditions as input, together with the training of the generator G and discriminator D, to improve the GAN modeling framework is shown in Figure 2.

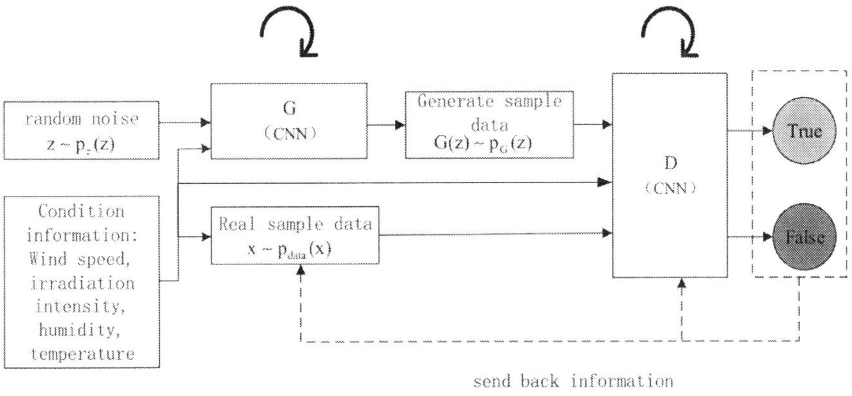

Figure. 2 Improved GAN model framework diagram

2.2. *Scenario Evaluation Indicators*

In order to verify the effectiveness of the improved modeling method, the coverage rate (CR) and the power error coefficient are used to measure the generated scenarios, and the formulas are shown below:

$$C_R = \frac{R_x}{R} \times 100\% \tag{1}$$

$$\varepsilon = \frac{1}{R}\sum_{r=1}^{R}\frac{\left|P_r^0 - P_r\right|}{P_r} \tag{2}$$

Where: R_x is the number of moments in the fluctuation range of the real scene power in the improved GAN-generated scene; R is the number of scene moments, take 24; P_r^0 is the power value at the moment of the improved GAN-generated scene; P_r is the real value of the power at the moment.

3. Renewable energy uncertainty scenario generation and occurrence probability calculation based on FCM clustering

Compared to the strong clustering of clustering algorithms such as K-means, FCM uses the concepts of fuzzy theory to represent the affiliation of each element to the clustering centers with a weight matrix rather than forcing all clustered elements to each clustering center.The objective function of FCM is shown in Eq. (3), and the steps of clustering the scene day are as follows:

$$\min J(W,C) = \sum_{i}^{N}\sum_{j}^{M} w_{ji}^{m}\left\|c_i - s_j\right\|^2 \tag{3}$$

where w_{ji} is the affiliation degree of the improved GAN generated scene s_j to the clustering center c_i; m is the fuzzification index.

1) Specify the final number of clusters N, select N samples from the data samples as the center of each initial clustering, and set the iteration accuracy ε and the maximum number of iterations $L_{\max}$.

2) Calculate the clustering center

$$c_i = \frac{\sum_{i}^{N} w_{ji}^{m} s_j}{\sum_{i}^{N} w_{ji}^{m}} \tag{4}$$

3) Update affiliation

$$w_{ji}^{(L)} = \frac{\sum_{i}^{N} w_{ji}^{m} s_j}{\sum_{i}^{N}\left(\frac{\left\|s_j - c_i\right\|}{\left\|s_j - c_L\right\|}\right)^{\frac{2}{m-1}}} \tag{5}$$

4) Determine whether the iteration meets the convergence accuracy, if it meets the iteration is stopped, if not, continue to iterate until the iteration to the maximum number of times L_{max} .

5) Output the final clustering centers and the probability of occurrence for each scenario day P_i , $\sum_{i}^{N} P_i = 1$.

4. Calculus analysis

In this paper, wind power output, photovoltaic output, wind speed, temperature, irradiation intensity, and humidity conditions in a park in Beijing from November 2020 to February 2022 are selected as the input data, and the data interval is 1 h. The improved GAN model is constructed by the deep learning framework Pytorch, and the GPUs are called to carry out the CUDA parallel computation to accelerate the training of the model. The experimental computer is equipped with CPU Intel Core i7 i7-13700 2.10 GHz, RAM 512GB, GPU NVIDIA RTX 4060, and 16GB of video memory. Renewable energy output uncertainty scenario clustering is modeled by MALAB2024A software platform YALMIP tool and solved by calling CPLEX solver.

The test set is fed into the trained generator G to generate the scenery power curve and random one day test curve data comparison, using and indicators to measure the effect of generator G, as shown in Figure 3, the test set is compared with the CNN, RNN and LSTM models trained based on the historical data and the indicators are shown in Table 1. At the same coverage rate of 95%, the improved GAN model has a lower power error coefficient than other models.

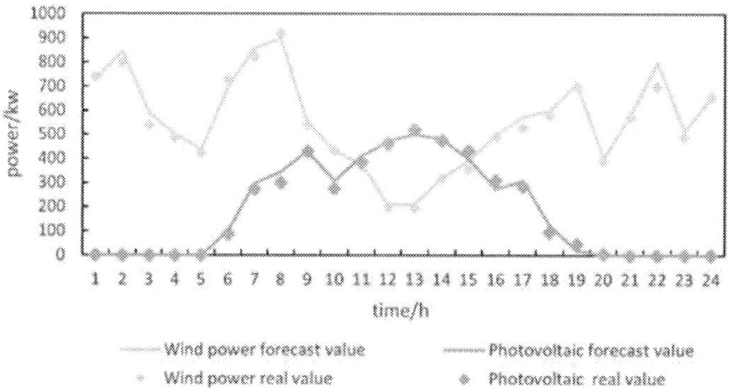

Figure. 3 Error Chart of Renewable Energy Curve Prediction

Table 1. Generate scene evaluation indicators

	CR (%)	Model	ε (%)
wind power	95	CNN	12.7
		RNN	11.9
		LSTM	10.0
		Improving the GAN	9.1
photovoltaic	95	CNN	13.2
		RNN	11.4
		LSTM	11.3
		Improving the GAN	10.5

Aiming at the problem of generating scenarios related to wind power and photovoltaic renewable energy, an improved GAN model is proposed to learn historical data of environmental conditions. Compared with traditional machine learning models, it can more accurately predict the source load curve. Based on this, the generated precise simulation scenes are aggregated into the typical scene days shown

Journal of Physics: Conference Series **3012** (2025) 012070
doi:10.1088/1742-6596/3012/1/012070

in Figure 4, providing scenario examples for wind and solar optimization and reducing the problem of inaccurate and random computing power scenarios.

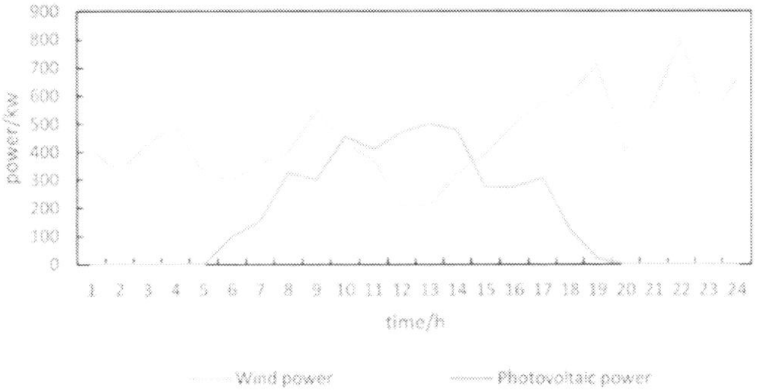

Probability of occurrence 0.254

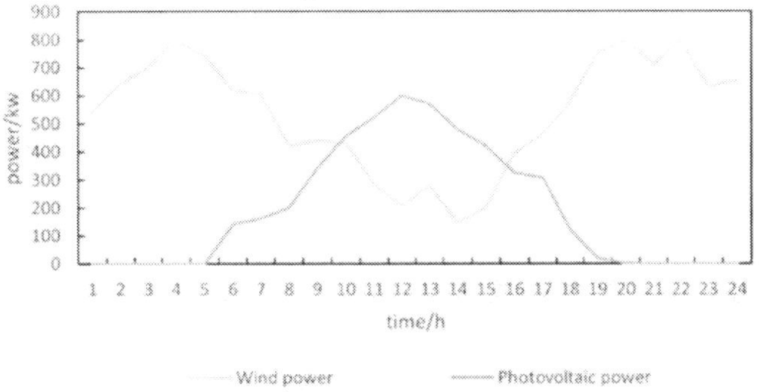

Probability of occurrence 0.468

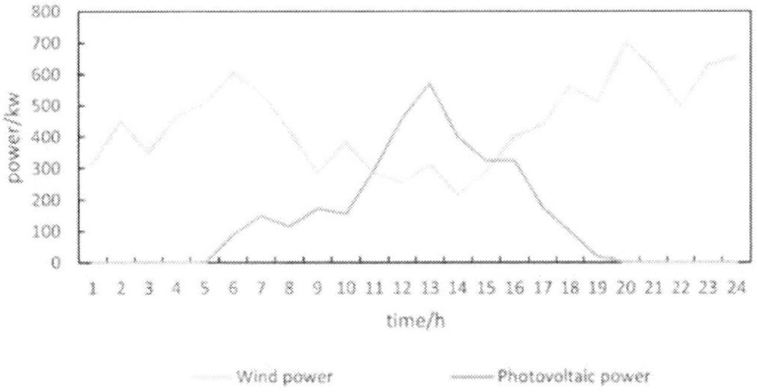

Probability of occurrence 0.278

Figure 4 Renewable energy scenario aggregation map

5. Conclusion

In this paper, we propose the simulation of renewable energy output uncertainty scenarios and the calculation method of their occurrence probability, and draw the following conclusions through case studies and analysis.

For wind power, photovoltaic renewable energy correlation scenario generation problems, proposed the use of improved GAN model for environmental conditions of historical data data learning, compared with the traditional machine learning model can be more accurate prediction of the source-load curve.

In this paper, the study of wind and solar correlation mainly analyzes wind and environmental historical data, and has not yet involved other renewable energy sources and environmental factors. Analyzing wind and other environmental factors will be the next step in the research.

References

[1] ZHANG Z and KANG C J 2022 Challenges and prospects for constructing the new-type power system towards a carbon neutrality future(Proceedings of the CSEE) pp2806-2819.
[2] JIANG T and TAO J 2024 Review of Load Response Management for Power Balance in the New Power System(Automation of Elec tric Power Systems) pp1-14.
[3] An Y and Jia C J 2024 Consider the complementary system of wind solar correlation and flexible carbon capture for low-carbon economic dispatch(Acta energiae solaris sinica) pp40-49.
[4] ZHONG J and LI M J 2017 Method of wind/ solar output forecast error analysis based on Copula theory(Advanced technology of electrical engineering and energy) pp39-46.
[5] XIAO B and LYU D J 2020 Simulation methods to generate time series for joint wind and photovoltaic power output basede on Markov chain and Copula theory(Modern electric power) pp245-255
[6] Zhang Y and Shi X J 2024 Short term prediction of wind and solar power in EEMDRLGWOLSTM region using reverse cloud grey correlation with similar days(Acta energiae solaris sinica) pp144-152.
[7] WANG J and REN Z J 2023 A distributed wind-solar joint prediction method considering data coupling(Distribution & Utilization) pp74-82.
[8] GOODFELLOW I J and POUGET-ABADIE J C 2014 A distributed wind-solar joint prediction method considering data coupling(Proceedings of the27th International Conference on Neural Information Processing Systems.Montreal) pp2672-2680.

An Artificial Intelligence Wind Power Prediction Model Based on Meteorological Parameters

Yue Zhou[1[0000-0002-6235-2787]], Wei Wang[1], Yucheng Su[2*], Yaying Pan[1], Tingting Gu[1], Luxuan Zhang[1], Jiayi Zhang[1], and Chendi Wang[1]

[1] Zhejiang Meteorological Service Center, Hangzhou, Zhejiang Province, 310000, PR China

[2] Zhuhai Public Meteorological Service Center, Zhuhai, Guangdong Province, 519000, PR China

[*]Corresponding author's e-mail: yuchengsu@yeah.net

Abstract. In the light of the problems of large fluctuation of wind power, complex influencing factors, and low accuracy of traditional single forecasting model. This paper investigates the correlation of the meteorological parameters and wind power, and combines the different characteristic of BiGRU, CNN, and attention mechanism, the hybrid CNN-BiGRU-Attention based wind power prediction model has been proposed. The model can provide more accurate wind power forecasts for wind farms. In addition, the experimental conclusions of this work illustrate that the forecast accuracy of the model which uses wind speed and wind power as inputs can superior to other models, with higher prediction accuracy and stability.

1. Introduction

Due to the increasingly serious issue of global warming and the rising demand for energy, renewable energy, as clean and sustainable energy, has garnered widespread attention. Wind energy, as an essential clean energy resource, has grown rapidly in recent years. However, due to the influence of environment and geographical conditions, wind speed is random and intermittent, resulting in low-quality wind power generation and low stability of the electricity system. Consequently, more accurate wind power forecasts have been a key factor in building a new power system that relies primarily on integrating new energy sources.

Plentiful studies are exploring various approaches to provide the wind power prediction accuracy. Generally, wind power and wind speed forecasts are categorized into four classes based on the prediction time scale: very short-term forecast (several seconds to 4h), short-term forecast (4h to 24h), medium-term forecast (1d to 7d), and long-term forecast (one week or more ahead) models. According to calculation theory and forecasting methods, wind speed and wind power forecast models have Bennett generally grouped into four categories: physics models, statistics models, the artificial intelligence method-based models, ensemble models.

In contrast, statistical methods are more suitable for very short-term wind speed and wind power forecasting using historical data [1]. Statistical models [2-4] which only use previous data to estimate the possible correlation of forecast wind power with past observations, which can achieve higher forecasting accuracy compared to physical models. However, statistical models are with high demand of data, which must be linearly correlated, smooth, and stable. Therefore, in the case of rapid changes in wind speed or wind power, the forecast accuracy will be lower and the stability will be poor. Such inherent linear assumptions of statistical models make them lack the ability to fit nonlinearities [5] and are not suitable for processing wind speed data with strong nonlinear features [6].

Artificial intelligence has advanced rapidly these years, and a variety of deep learning approaches have been employed in wind speed prediction [7]. In general, artificial intelligence methods have good self-learning and self-organizing capabilities, which can approximate any nonlinear function and perform well in very short-term prediction. Hao et al. [8] develop a prediction model using a two-stage hybrid decomposition method, and indicating that the model improves the forecast accuracy. Yan et al. [9] propose a hybrid EWT-SMA-RKELM-SGRU model using non-stationary weakening technique and demonstrate that this model can have good performance in wind power forecast. Almutairi and Alrumayh [10] propose a new deep learning method using lower-upper-bound-estimation and LSTM models, this model can alleviate the issue of get stuck in a local optimum caused by changes in LSTM parameters in traditional wind power prediction and promote the precision of wind power prediction. Aiming to resolve the problem of the effect of non-stationary wind power sequences in wind power forecasting, Khorramdel et al. [11] propose a kernel density estimation to obtain a high-quality characteristic signal, and employ a three-level adaptive function to accurately predict wind power generation in different seasons. Shahid et al. [12] improved the traditional LSTM model using a wavelet kernel algorithm, and strengthening the ability of deep learning to capture the inherent laws of wind power prediction.

Although there have been a number of studies related to the wind power forecast, the existing forecasting accuracy still cannot meet the needs of wind farms. Currently, there have been studies that show that meteorological factors have impacts on wind power, however, there are less studies that add meteorological factors into wind power modeling. This study, a wind power forecast model has been built and trained using the data from a wind farm ranging from January 2022 to December 2022. This study also investigates the correlation between wind power and meteorological feature and combines meteorological features with wind power prediction models. The results of this work illustrate that a new model that uses meteorological characteristics as input can significantly enhance the accuracy of wind power. The model build in this work can provide a basis for wind power prediction of wind farms.

2. Data and Method

2.1. Data preprocessing

The wind farm of this work is situated in Shengsi, Jiaxing City, Zhejiang. The wind turbines in this facility are located at an elevation of 95 m above sea level and with 4000KW rate power. In this study, we used the wind power data of Shengsi wind farm from January 2022 to January 2023 with the 15 min time resolution. In this work, meteorological data are from the automatic weather station located near the wind farm. The meteorological data used in this work is with 1 hour time resolution. This work, we choose 6 variables, including wind speed, precipitation, max daily temperature, min daily temperature, vapor pressure, and humidity. The wind power has been selected into 1h time resolution corresponding to the meteorological data and the abnormal data has also been removed. The data is divided into training set and testing set in a 7:3 ratio. In order to eliminate dimensional differences between different features and learning rate, the wind power and meteorological data are normalized.

2.2. Model introduction

2.2.1. Convolutional Neural Network (CNN). The CNN is an architecture designed for deep learning, which is mainly used to extract characteristics from spatially related data. The model employs convolutional layers to perform local receptive field operations, which enable the identification of spatial dependencies and hidden correlations in the input features. By applying convolutional filters, CNN can automatically extract meaningful representations while reducing the dimensionality of the input data, leading to more efficient and informative feature maps for downstream processing.

2.2.2. Bidirectional Long Short-Term Memory (BiLSTM). LSTM network is a kind of RNN which were developed to model lengthy dependencies in data. Different from RNN, LSTMs address the vanishing gradient problem through a unique gating mechanism that regulates the flow of information across time steps. In this work, we employ BiLSTM to further enhance the temporal modeling capability of the network. BiLSTM can work with input data in both forward and backward order, which allows the model to identify time-dependent relationships in the past and the future at the same time.

2.2.3. Bidirectional Gated Recurrent Unit (BiGRU). It is an alternative form of GRU, and GRU can get long-term dependencies among data series by introducing a gating mechanism, and simplify the calculation through their gating mechanisms. The BiGRU model can successfully catch bidirectional time dependencies by processing the input data in both forward and backward directions. Compared to BiLSTM, BiGRU requires fewer parameters, which makes it computationally more efficient while maintaining comparable performance.

2.2.4. Attention Mechanism. While CNN, BiLSTM, and BiGRU components provide strong capabilities for feature extraction and temporal modelling, the attention further strengthens the model by enabling it to concentrating the most relevant features and time steps. Attention can allocate weights dynamically to various input characteristics, allowing the model to prioritize significant information while disregarding irrelevant or less important data.

2.2.5. Entire Structure of the Proposed Model. In light of further enhancing the prediction ability of wind power model, this work combines the advantages of CNN, BiLSTM and BiGRU. The model takes wind power data and meteorological factors as inputs, CNN is used to preliminarily extract high-dimensional feature vectors from the input data. Then BiGRU or BilSTM performs bidirectional temporal training on the feature vectors to uncover deep-level features of the load data. Attention is then introduced to distribute different weights to the output states of BiGRU or BilSTM, highlighting the key features related to wind power. Finally, the model outputs the prediction results.

2.3. Evaluation indicators

For quantitatively compare the forecast accuracy, Mean Absolute Error (MAE), Root Mean Square Error (RMSE), and Mean Absolute Percentage Error (MAPE) were taken as evaluation indexes. The formulae are listed below:

$$RMSE = \sqrt{\frac{1}{N}\sum_{t=I}^{N} (o_i - m_i)^2} \tag{1}$$

$$MAE = \frac{1}{N}\sum_{t=1}^{N} |o_i - m_i| \tag{2}$$

$$MAPE = \frac{100\%}{N}\sum_{t=1}^{N} \left|\frac{o_i - m_i}{o_i}\right| \tag{3}$$

Here, o_i is the observation of wind power, m_i is the prediction of the model.

3. Results and Analysis

For comparing the prediction performances among each models, two wind power prediction models, CNN-BiGRU and CNN-BiLSTM, have been build. In the model, multi-step recursive prediction is used to predict the power of the 37th hour using wind power data of the previous 36h hours (3 days). Figure 1 gives the prediction results of different models at a time of large fluctuations in wind power, in which the blue line denotes the prediction results obtained by CNN-BiGRU, the orange line gives the prediction results by CNN-BiLSTM, and the violet line shows the observation of wind power. From Figure 1, it can be seen that both models can predict the trend of wind power, except at extreme values of wind power. In addition, both models have a forecast lag, with the forecast lagging about one hour behind the observations.

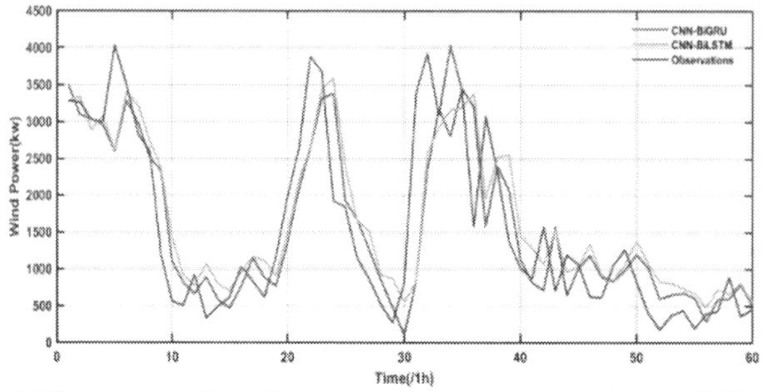

Figure 1 The comparison between observation and two different models

Table 1 shows evaluation indexes of two models for test sets. It can be seen from Table 1 that the prediction effect of CNN-BiGRU is better than that of CNN-BiLSTM, and the CNN-BiGRU model has been used in future studies in this paper.

Table 1 Forecasting results of different model

Model	RMSE	MAE	MAPE
CNN-BiGRU	730.80	534.23	13.07
CNN-BiLSTM	736.90	553.05	15.80

To further optimize the prediction effect of the model, the attention mechanism is introduced into the CNN-BiGRU model. The prediction results of CNN-BiGRU and CNN-BiGRU-attention are shown in Figure2. The prediction of wind power has been improved with the introduction of attention, while the lag between the model predicted and the observation has not been solved.

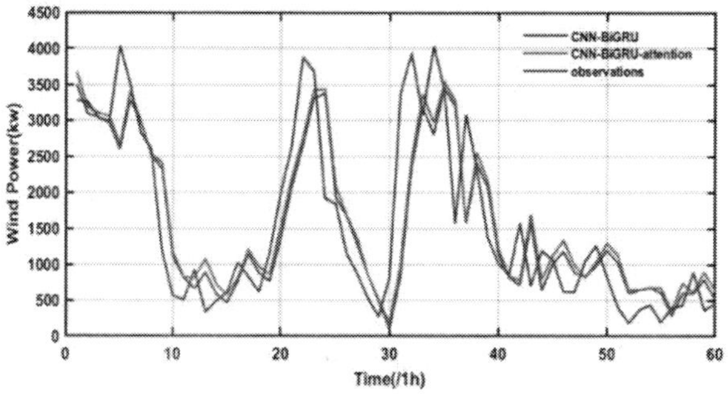

Figure 2 The comparison between observation and two different models

In order to further solve the issue of prediction lag, meteorological data has been introduced in this work. The selected meteorological data include wind speed, precipitation, max daily temperature, min daily temperature, vapor pressure, and humidity. In this study, the correlation coefficient method is employed to conduct feature selection on the meteorological data, and the related degree between six meteorological parameters and wind power are shown in the Figure 2. Among the six meteorological features, correlation between wind speed and wind force is the highest, with a correlation coefficient of 0.35. It is followed by max daily temperature and then followed by min daily temperature, with a correlation coefficient of 0.19, and then vapor pressure and humidity, with correlation coefficients of 0.15 and 0.1. The correlation coefficient between precipitation and wind power is the lowest, at 0.05. Based on the correlation coefficients, four input combinations were designed: 1) all meteorological parameters and wind power (C1); 2) wind speed, max daily temperature, min daily temperature, humidity, and wind power (C2); 3) wind speed and wind power (C3); 4) only wind power (C4).

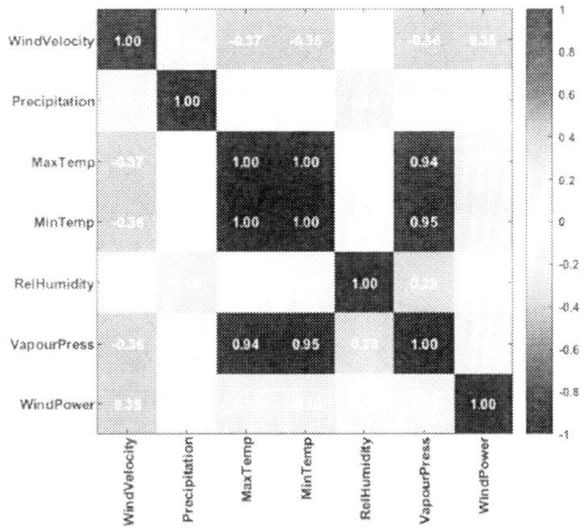

Figure 3 The correlation between meteorological features and wind power

Figure 4 shows the prediction results for different input combinations. The prediction accuracy is lower for the combination of wind power input only and the combination of all meteorological parameters and wind power input, and there is still room for further optimization. The prediction trends of the remaining input combination are more closely to the observation, and the C3 has the best prediction accuracy.

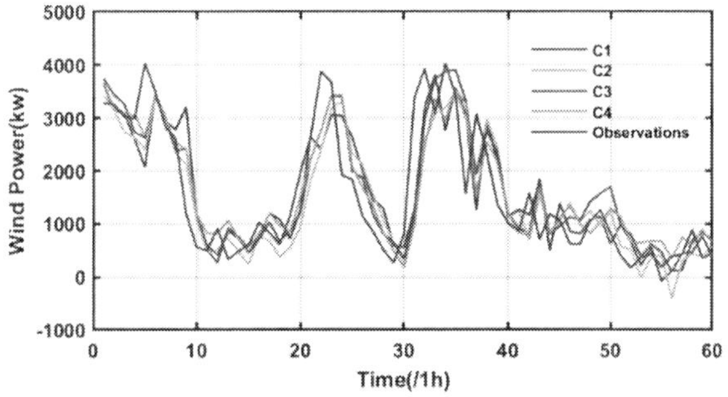

Figure 4 Comparison of observation and different combinations

With the aim of further confirming the forecast accuracy of model, all prediction models were compared, as shown in Table 2. It can be seen that C3, which considers both wind power and wind speed, provides best forecast with the lowest RMSE, MAE, and MAPE. Compared to C4, which only considers wind power, C1 has 6.9% higher RMSE, 6.6% higher MAE, and 28.9% higher MAPE. C1, with all meteorological factors considered, has the worst results, indicating that the relevant between meteorological parameters and wind power should fully considered when modelling.

Table 2 Forecasting results of different combinations

Model	RMSE	MAE	MAPE (%)
C1	781.73	588.10	11.7
C2	720.18	539.20	11.3
C3	676.79	496.72	9.6
C4	727.25	531.73	13.5

4. Conclusion

The accurate forecasting of wind power contributes significantly in power grid scheduling and the safe and stable operation of power systems. In this study, a wind power prediction model has been established by utilizing the CNN-BiGRU-attention model. The conclusions are shown below:

1)The wind power forecast accuracy of CNN-BiGRU and CNN-BiLSTM models can be improved by adding the attention mechanism.

2) Wind speed and wind power have a strong correlation with each other, while other meteorological factors, precipitation, max daily temperature, min daily temperature, vapor pressure, and humidity, have very little correlation with the wind power.
3) Using wind power and wind speed simultaneously as inputs of the model, more accurate wind power predictions can be obtained.

Acknowledgments
This study is assisted by Joint Funds of Natural Science Foundation of Zhejiang, China (Grant No. LZJMY23D050001), Science and Technology Research Project of Guangdong Meteorological Bureau (Grant No. GRMC2022Z06), and key program of Zhejiang Provincial Meteorological Bureau (Grant No.2020ZD15, 2023ZD11).

References
[1] Naik J Satapathy P and Dash P 2018 Short-term wind speed and wind power prediction using hybrid empirical mode decomposition and kernel ridge regression Applied Soft Computing 70 pp 1167-1188
[2] Poggi P Muselli M Notton G et al 2003 Forecasting and simulating wind speed in corsica byusing an autoregressive model Energy conversion and management 44(20) pp 3177-3196
[3] Wang J Zhou Q and Zhang X 2018 Wind power forecasting based on time series arma model Earth and environmental science 199 IOP Publishing
[4] Fattah J Ezzine L Aman Z et al 2018 Forecasting of demand using arima model International Journal of Engineering Business Management, 10
[5] Deng Y Wang B Lu Z 2020 A hybrid model based on data preprocessing strategy and errorcorrection system for wind speed forecasting Energy Conversion and Management 212(8) p 112779.
[6] Gao S He Y and Chen H 2009 Wind speed forecast for wind farms based on arma-arch model International Conference on Sustainable Power Generation and Supply IEEE
[7] Khosravi A Machado L and Nunes R 2018 Time-series prediction of wind speed using machine learning algorithms: A case study osorio wind farm, brazil Applied Energy 224 pp 550-566
[8] Hao Y and Tian C 2020 A hybrid framework for carbon trading price forecasting: the role of multiple influence factor Journal of Cleaner Production 262
[9] Yang R Liu H Nikitas N et al 2022 Short-term wind speed forecasting using deep reinforcement learning with improved multiple error correction approach Energy 239
[10] Almutairi A and Alrumayh O 2022 An intelligent deep learning based prediction model for windpower generation Computers and Electrical Engineering 101 p 108000
[11] Khorramdel B Chung C Y and Safari N 2018 A Fuzzy Adaptive Probabilistic Wind Power Prediction Framework Using Diffusion Kernel Density Estimators IEEE Transactions on Power Systems 33(6) pp 7109-7121
[12] Shahid F Zameer A Mehmood A et al 2020 A novel wavenets long short term memory paradigm for wind power prediction Applied Energy 269 p 115098

Study on Atmospheric Neutron Single Event Effect of 650 V Silicon Carbide Diode

Hong Zhang[1,2] [0000-0003-0096-4152], **Teng Ma**[1*] [0009-0005-4453-5669]

[1] Reliability Physics and Application Technology of Electronic Component Key Laboratory, Guangzhou 510610, China

[2] South China University of Technology, Guangzhou 510610, China

*Corresponding author's e-mail: mateng_ceprei@163.com

Abstract. Based on the Atmospheric Neutron Irradiation Spectrometer, the atmospheric neutron single event effect under five bias voltages for silicon carbide diodes with a rated voltage of 650 V was studied. As the bias voltage decreases, the single event burnout cross section of silicon carbide diodes gradually decreases. The electrical testing results show that the turn-on characteristics of silicon carbide diodes with single event burnout degrade and the breakdown characteristics are completely lost. The results of damage localization and characterization show that there are obvious cavity on the surface of silicon carbide diodes induced by single event burnout, and the current leakage path formed at the cavity is the main reason for the degradation of the electrical characteristics.

1. Introduction

The significant advantages in bandgap width, breakdown electric field, thermal conductivity, power density, etc. have attracted people to continuously develop and apply silicon carbide power devices[1-3]. With the continuous maturity of device technology, the performance of silicon carbide power devices continues to improve while the cost gradually decreases. Silicon carbide power devices are increasingly being applied by companies in electronic power systems such as inverters and transformers. The use of silicon carbide power devices can effectively promote the miniaturization and lightweight development of equipment, and therefore has great potential for application in the aerospace field.

Atmospheric neutron have the characteristics of wide energy spectrum, uncharged, strong penetrability, and wide distribution, which makes power devices inevitably affected by atmospheric neutron in various application fields, and this influence will last throughout their entire life cycle. Several studies have been conducted both domestically and internationally on the neutron single event effect in silicon carbide power devices[4-6]. The nuclear reaction products and cross sections of atmospheric neutron are affected by neutron energy, which in turn affects the single event effect cross section of silicon carbide power devices. Meanwhile, the improvement of device technology and optimization of structure may also have an impact on the sensitivity of neutron single event effect. The single event effect and its mechanism induced by atmospheric neutron in silicon carbide power devices need further research.

The Atmospheric Neutron Irradiation Spectrometer (ANIS) is the only neutron beam from the Chinese Spallation Neutron Source that has not been slowed down, and has an approximate differential energy spectrum with real atmospheric neutron[7]. Based on the ANIS, this paper conducted a study on the atmospheric neutron single event effect of silicon carbide diodes. The characteristics of the single event effect were analyzed through online testing, offline testing, and microscopic characterization.

2. Experimental setup

The silicon carbide diode produced by CREE company was selected for the atmospheric neutron irradiation test, with the device model C6D10065A and a rated voltage of 650 V. In the experiment, the neutron flux rate (>1MeV) is about 6.15×10^5 n/(cm^2·s^1). Due to the large neutron beam spot, multiple devices can be irradiated simultaneously. In the experiment, atmospheric neutron irradiation tests were conducted with bias voltages of 100%, 94.4%, 88.9%, 77%, and 50% of the rated voltage, using the ratio of device voltage to its rated voltage as a reference, as shown in Table 1. When conducting irradiation tests on silicon carbide diodes, a bias voltage is applied to the cathode.

Table 1. Neutron irradiation bias conditions

Type\\V_{Bias}	100%	94.4%	88.9%	77%	50%
Bias voltage	650 V	614 V	578 V	500 V	325 V

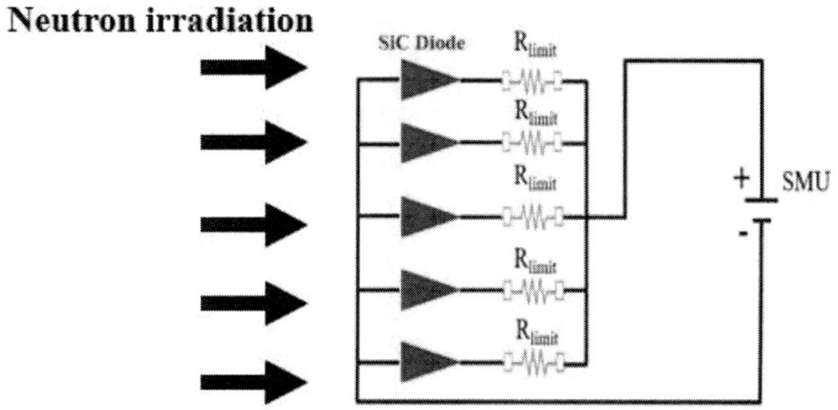

Figure 1. Schematic diagram of atmospheric neutron single event effect test for silicon carbide diode.

In the experiment, apply reverse bias voltage to 5 devices of the same model at the same time. When the cumulative flux is less than 1×10^9 n/cm^2, if all devices occur single event burnout, stop the experiment. When the cumulative flux exceeds 1×10^9 n/cm^2, if there are devices that have not been burnout, the irradiation experiment should be stopped when the devices number that occur burnout exceeds 2. The real-time current of silicon carbide diodes in the experiment is monitored to count the number of single event burnout events. Figure 1 is a schematic diagram of atmospheric neutron single event effect testing for silicon carbide diodes. During the experiment, real-time current monitoring was conducted on 5 devices simultaneously. The resistance value of the protective resistor R_{limit} in the test circuit was 1.5 MΩ. The sourcemeter (SMU) applied reverse bias voltage to the devices while also recording and storing the total current in real time. In theory, when a silicon carbide diode experiences single event burnout, its reverse resistance approaches zero, and the current in this branch is mainly limited by R_{limit}, there will be a significant increase in the total current of the circuit.

3. Experimental results and analysis

3.1. Online testing

The atmospheric neutron irradiation test of silicon carbide diode was conducted, and the real-time variation of the total current in the test circuit is shown in Figure 2. The total current increases step by step, and each step represents a single event burnout of a device. When the single event burnout occurs, the current of this branch rapidly increases to about 0.385 mA due to the influence of the protective resistor. Within 1200 seconds, all 5 devices with a bias voltage of 578 V experienced single event burnout.

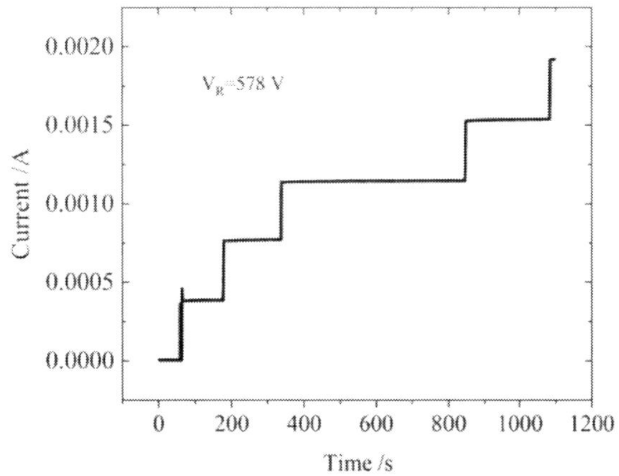

Figure 2. Real time variation of total current in the test circuit during atmospheric neutron irradiation test

By combining the accumulated neutron flux, the number of devices experiencing single event burnout, and the area of the active region of silicon carbide diodes, the single event burnout cross section of the device can be calculated. The atmospheric neutron single event burnout cross section of silicon carbide diodes under different bias voltages is shown in Figure 3. When the bias voltage is the rated voltage, the neutron single event burnout cross section is about 5.44×10^{-9} cm^2. As the bias voltage gradually decreases, the single event burnout cross section of silicon carbide diodes shows a decreasing trend. When the bias voltage is 325 V, for the gray dots in Figure 3, there was actually no single event burnout during the experiment. We considered increasing the atmospheric neutron flux as much as possible, but no single event burnout occurred. Therefore, we assume that under cumulative neutron flux, one device with this bias voltage occurred single event burnout. By calculating the maximum possible single event burnout cross section, which is the gray point data in Figure 3, it is approximately 1.22×10^{-10} cm^2.

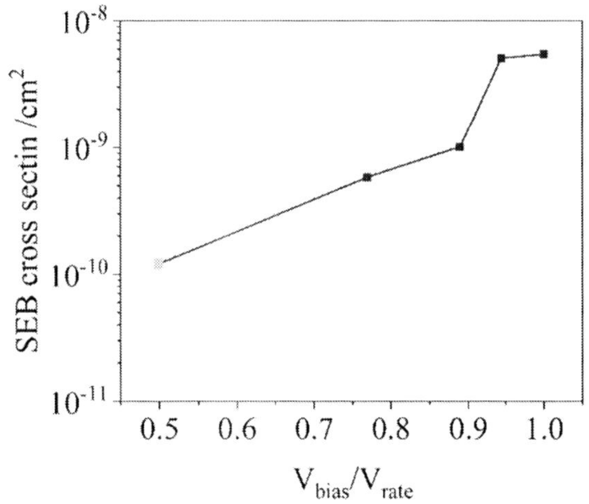

Figure 3. Single event burnout cross section of silicon carbide diode

3.2. *Offline testing*

Figure 4 shows the changes in forward and breakdown characteristics of a silicon carbide diode before and after neutron single event burnout. After the occurrence of single event burnout, the current before the forward turn-on of silicon carbide diode increases significantly. As shown in Figure 5b, the reverse breakdown characteristics of silicon carbide diode have significantly degraded, and the reverse current of the device has reached 50 mA when the reverse voltage is below 10 V. After single event burnout occurred, silicon carbide diodes still have turn-on characteristics, but the breakdown characteristics have been completely lost, making it unable to have the normal switching capability required in electronic power systems. It is worth noting that single event burnout induced by atmospheric neutron and high-energy protons has similar effects on the electrical properties[8]. At the same time, there was no single event leakage current in silicon carbide power devices under both types of particle irradiation, which may be related to the secondary ion distribution induced by the single event effect and the mechanism of single event leakage current.

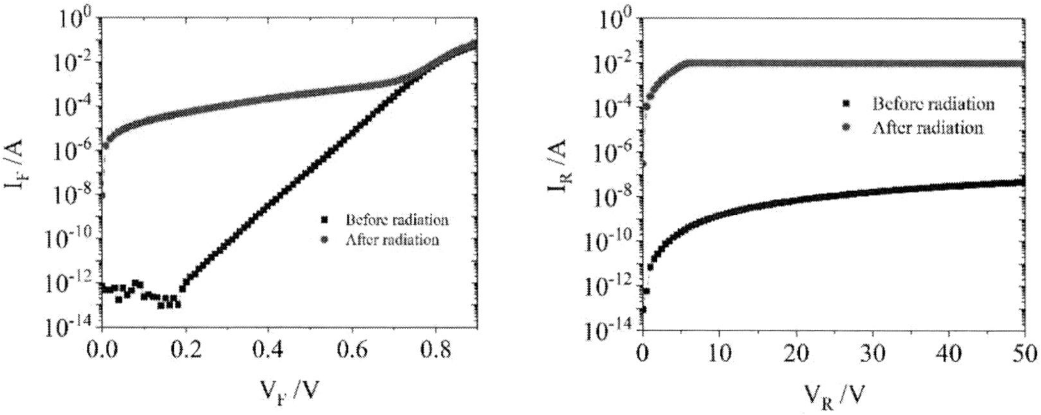

(a) forward characteristics (b) breakdown characteristics

Figure 4. Electrical characteristics of silicon carbide diode before and after single event burnout

Microscopic characterization of silicon carbide diodes with single event burnout. Using Emission Microscope (EMMI) to test devices that have experienced single event burnout, a obvious leakage points appeared on the surface of the device, as shown in Figure 5a. The red marked positions in the figure are significant leakage points detected in EMMI analysis. Furthermore, using Scanning Electron Microscopy (SEM) to characterize the area, it was found that there was a cavity formed by the single event burnout. As shown in Figure 5b, the size of the cavity is greater than 10 μm. As shown in Figure 5c, based on the Focused Ion Beam (FIB) test, further analysis was conducted on the single event burnout area, and the damage depth was about 9 μm, which theoretically exceeds the epitaxial layer thickness of commercial 650 V silicon carbide diodes. The damage area induced by single event burnout to the anode, epitaxial layer, and part of the substrate, thereby affecting the electrical characteristics of silicon carbide diodes. There are no obvious abnormalities in current leakage and morphology in other areas of silicon carbide diode.

Journal of Physics: Conference Series **3012** (2025) 012072 doi:10.1088/1742-6596/3012/1/012072

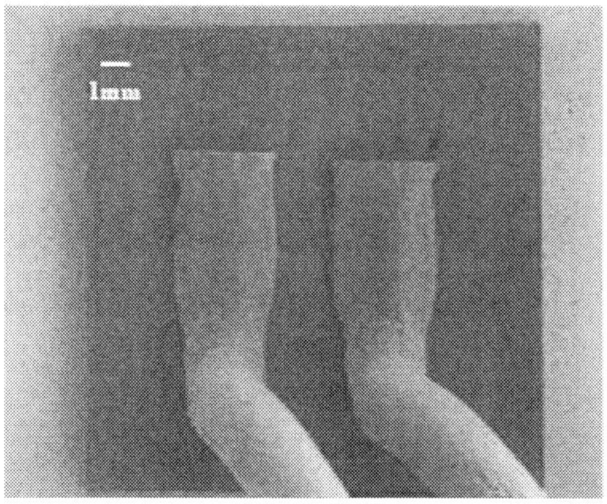

(a) EMMI

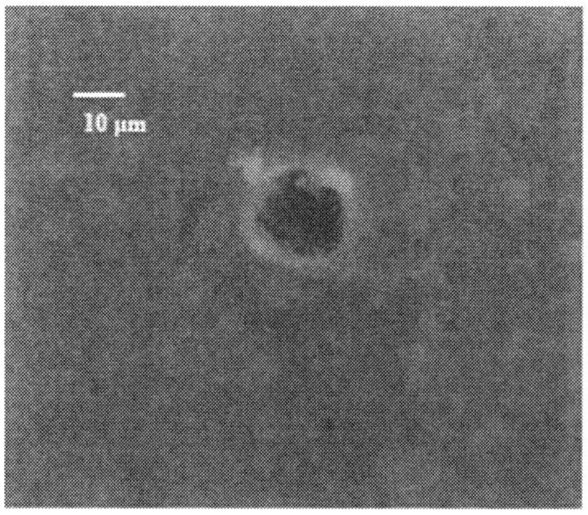

(b) SEM

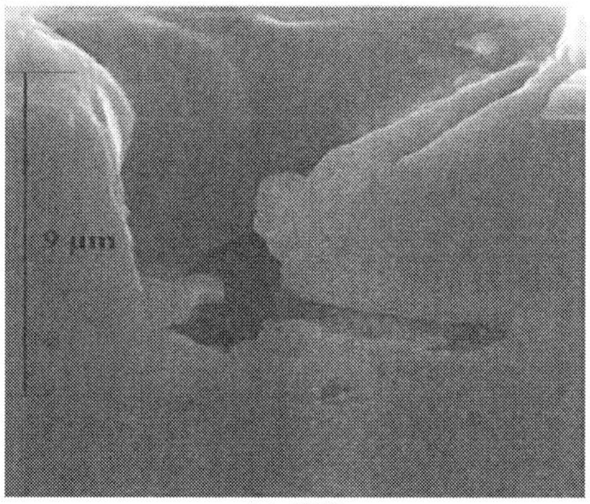

(c) FIB

Figure 5. Microscopic analysis of silicon carbide diode

The schematic diagram of silicon carbide diode is shown in Figure 6. Silicon carbide diodes have a vertical structure, and the abnormal electrical characteristics that occur during online and offline testing mainly come from the current leakage caused by the leakage point. In online testing, reverse voltage forms a current flowing from the cathode to the anode. Due to the single event burnout, a low resistance channel is formed in the epitaxial layer. In the forward characteristic testing, the low resistance channel formed by single event burnout before the schottky junction turn-on is the main path for forward current. As the forward voltage increases, the current channel formed after the schottky junction turn-on becomes the main current path, indicating that strong damage in local areas does not affect the overall turn-on characteristics of silicon carbide diode. The breakdown characteristic test is similar to online testing, where low resistance channel cause the breakdown characteristics of silicon carbide diodes to be lost.

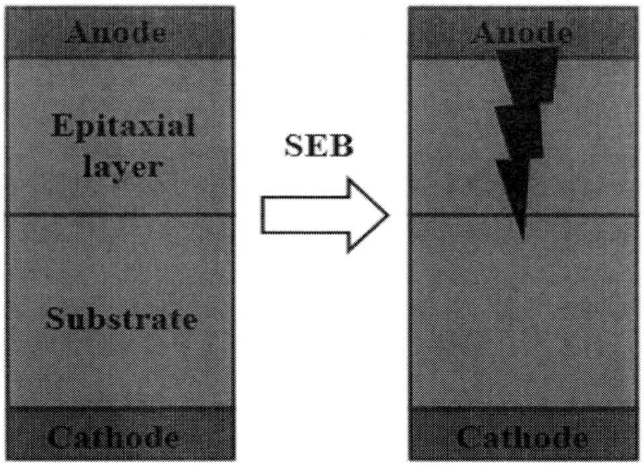

Figure 6. Schematic diagram of single event burnout

Silicon carbide power devices, as a new type of power device, have certain application potential in electric vehicle power electronics systems and grid power transportation. The changes in application scenarios, especially changes in altitude, pose different atmospheric neutron radiation risks. Based on the EXPACS model[9], the atmospheric neutron radiation flux rates in three typical areas of Guangzhou, Beijing, and Lhasa were calculated, and the failure in time (FIT) of silicon carbide diodes under reverse bias voltage was obtained, the reverse operating voltage of silicon carbide diodes is considered to be 50% of the rated voltage, as shown in Table 2. The FIT (10^9 h) of silicon carbide diodes is about 10 times higher than in Beijing and about 15 times higher than in Guangzhou. In high-altitude areas, atmospheric neutron radiation may be one of the important factors affecting the reliability and stability of silicon carbide power devices.

Table 2. Atmospheric neutron flux rate and FIT in typical areas

	Average altitude/km	Neutron flux rate (>1MeV)/(n/cm2/h)	FIT (10^9 h) under bias voltage of 325 V
Guangzhou	0.0011	3.96	0.97
Beijing	0.0435	156.6	1.46
Lhasa	3.65	13140	15.43

4. Conclusion

Based on the ANIS, a study was conducted on the atmospheric neutron single event effect of 650-V silicon carbide diodes. In online testing, the current of a silicon carbide diode that experiences single event burnout instantly increases to the limiting current. The electrical characteristic test results show that after single event burnout occurred, the breakdown characteristics of silicon carbide diode are completely lost, the turn-on characteristics are degraded to a certain extent, and the forward current before the diode turn-on increases. Combining the EMMI and SEM for single event burnout damage localization and characterization, it was found that there was obvious current leakage point on the surface of the device. Further characterization revealed the presence of obvious cavity, which led to abnormal electrical characteristics of silicon carbide diode.

Acknowledgments

This work was supported in part by the Guang Dong Basic and Applied Basic Research Foundation under Grant (No.2023A1515110958).

References

[1] Elasser A 2002 Silicon carbide benefits and advantages for power electronics circuits and Systems (Proceedings of the IEEE vol 90) (New York: IEEE) chapter 6 pp 969-986
[2] Hudgins J L 2003 An assessment of wide bandgap semiconductors for power devices (IEEE Transactions on power electronics vol 18) (New York: IEEE) chapter 3 pp 907-914
[3] Dimitrijev S 2003 Advances in SiC power MOSFET technology (Microelectronics reliability vol 43) (Amsterdam: Elsevier) chapter 2 pp 225-233
[4] Shoji T 2014 Experimental and simulation studies of neutron-induced single-event burnout in SiC power diodes (Japanese Journal of Applied Physics vol 53) (Tokyo: JSAP) chapter 4S pp 04EP03
[5] Akturk A 2016 Single event effects in Si and SiC power MOSFETs due to terrestrial neutrons (IEEE Transactions on Nuclear Science vol 64) (New York: IEEE) chapter 1 pp 529-535
[6] Niskanen K 2020 Impact of electrical stress and neutron irradiation on reliability of silicon carbide power MOSFET(IEEE Transactions on Nuclear Science vol 67) (New York: IEEE) chapter 7 pp 1365-1373
[7] Yu Q 2022 Physical design of an atmospheric neutron irradiation spectrometer at China spallation neutron source (Nuclear Engineering and Design vol 386) (Amsterdam: Elsevier) pp 111579
[8] Zhang H 2021 Advances in SiC power MOSFET technology (Microelectronics reliability vol 124) (Amsterdam: Elsevier) pp 114329
[9] Sato T 2016 Analytical model for estimating the zenith angle dependence of terrestrial cosmic ray fluxes (PLoS ONE vol 11) (San Francisco: Public Library of Science) chapter 8 pp e0160390

Research on the Method for Extracting Three-Dimensional Vibration Signals of Oil-Immersed Core Reactors Based on Visual Enhancement

Kejun Han[1], Zhenjie Zhao[2], Xinbing Wang[1], Xuecheng Li[1], Hui Xu[1], Fan Zhang[2*], Shengchang Ji[2]

[1] Shandong Power Equipment Co., Ltd., Jinan 250011, China

[2] State Key Laboratory of Electrical Insulation and Power Equipment , Xi'an Jiaotong University, Xi'an 710049, China

[*]Corresponding author's e-mail: zhangfanee@xjtu.edu.cn

Abstract. The oil-immersed core reactor is widely used in power systems, and vibration, along with the internal mechanical defects it causes, is one of the key issues faced in equipment operation and maintenance. By analyzing the vibration signals of the reactor, it is possible to assess the equipment's condition and diagnose defects. Traditional methods of vibration measurement have certain limitations. Therefore, this paper proposes and validates a method for extracting three-dimensional vibration signals of oil-immersed core reactors based on visual enhancement. A vibration measurement platform for the reactor has been established. The effectiveness of the proposed method in extracting the vibration modes of reactors operating at 95%, 100%, and 105% of rated voltage has been investigated. The results show that the measured vibration modes have a structural similarity of 0.9139 with those obtained by sensors, and the vibration mode extraction results do not fluctuate with changes in operating voltage, verifying the reliability of this method for extracting vibration modes of oil-immersed core reactors.

1. Introduction

With the continuous development of the power grid, oil-immersed core reactors have been widely applied in power systems due to their ability to balance capacitive reactive power and limit line overvoltage [1,2]. During normal operation, the coils and core inside the reactor's enclosure are subjected to continuous alternating electromagnetic forces. Under prolonged vibration, the internal structure of the reactor gradually experiences degradation of insulation performance and mechanical instability, which, in extreme cases, can lead to severe failures [3,4]. According to on-site statistics, mechanical defects caused by long-term vibrations in oil-immersed core reactors have become the leading threat to their safe operation [5-7].

The vibration of a reactor can be classified into winding axial and radial vibrations, as well as core vibrations. The winding vibrates due to the Lorentz force [8], while the core vibrates because of the Maxwell force [9,10] and the Magnetostriction [11]. Vibration waves propagate through the fasteners and transformer oil to the enclosure wall.

Traditional vibration signal measurement methods mainly fall into two categories: one is contact measurement using piezoelectric accelerometers, and the other is non-contact measurement using laser vibrometers [12]. The contact measurement method allows flexible sensor placement and offers high measurement accuracy, but the sensors add extra mass to the equipment shell, which can alter the structural characteristics of the device to some extent, thus affecting the measurement results. Laser vibrometers, on the other hand, can measure the vibration of an object's surface directly through a laser beam, avoiding the issue of added sensor mass. However, laser vibrometers are expensive. For complex electrical equipment like oil-immersed core reactors, there is an urgent need to develop a vibration mode extraction method with strong synchronization and high spatial resolution to better meet the demands of modern condition monitoring and defect diagnosis.

This paper proposes a method for extracting three-dimensional vibration signals of oil-immersed core reactors based on visual enhancement. The method first calibrates a stereo imaging system consisting of binocular cameras and establishes a three-dimensional coordinate system. Through the calibration algorithm, the two-dimensional coordinates in the images are converted into three-dimensional coordinates in the real world, laying the foundation for subsequent image processing and data analysis. Then, by analyzing the phase information in the images, small movements are magnified, making the vibration displacement information more prominent. Next, the three-dimensional digital image correlation method is applied, where sub-region matching is performed on the processed images to track the displacement changes of corresponding pixel points, thus calculating the vibration information on the surface of the measured object. Finally, post-processing is carried out on the data to obtain the vibration extraction results.

2. Fundamental theory

2.1. Binocular stereo vision

As shown in Figure 1, a binocular stereo vision system consists of two cameras that observe the same object from different viewpoints, capturing two images. These cameras are typically mounted at a certain baseline distance to generate enough disparity for depth calculation. To obtain the three-dimensional shape information of an object's surface, binocular stereo vision technology mainly involves the following steps: (1) Two cameras capture the same scene simultaneously, obtaining the left view and the right view. (2) Perform geometric and color calibration on the two cameras. (3) Extract salient feature points from the left and right views. Then, find corresponding feature point pairs in the two images through feature matching algorithms. (4) Calculate the positional differences of feature point pairs between the left and right images. Derive the distance from the object points to the cameras through geometric relationships. (5) Integrate the depth information of all feature points to generate the three-dimensional shape of the object's surface.

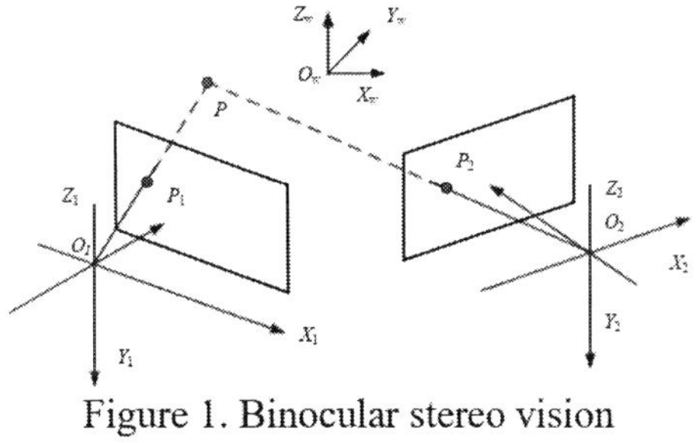

Figure 1. Binocular stereo vision

2.2. Phase-based motion magnification method

Gabor transform is introduced to analyze the amplitude and phase variations of local images. The Gabor transform is generated based on the simulation of the human visual system. The basis of the Gabor transform is the design of the Gabor filter. A two-dimensional Gabor function can be expressed as:

$$g(x,y;\lambda,\theta,\psi,\sigma,\gamma)=\exp\left(-\frac{x_\theta^2+\gamma^2 y_\theta^2}{2\sigma^2}\right)\times\exp\left(i\left(2\pi\frac{x_\theta}{\lambda}\right)+\psi\right) \tag{1}$$

λ is the wavelength of the sine wave, and ψ is the phase shift of the tuning function. γ is the spatial aspect ratio, which determines the shape of the Gabor function. σ represents the standard deviation of the Gaussian function, which determines the size of the acceptable region of the Gabor filter kernel. Here, x and y represent spatial positions. θ represents the direction of the parallel stripes in the Gabor filter kernel. The two-dimensional Gabor function can also be expressed in simplified forms as the real part G_θ and the imaginary part H_θ :

$$g(x,y;\lambda,\theta,\psi,\sigma,\gamma)=G_\theta+iH_\theta \quad (2)$$

By utilizing the directional and frequency-selective characteristics of the Gabor filter, local amplitude and phase variations in different directions can be extracted. Then, the displacement signal is obtained based on the relationship between the local phase and displacement.The phase difference between the two frames is calculated as:

$$\phi_{\theta(x,y,t_0+\Delta t)}-\phi_{\theta(x,y,t_0)}=2\pi\frac{\Delta x}{\lambda} \quad (3)$$

This gives the relationship between the pixel displacement and phase difference. The displacement magnification is then achieved by multiplying the phase difference at times t_0 and $t_0+\Delta t$ by the magnification factor.

2.3. Digital image correlation

Digital Image Correlation (DIC) can analyze the surface deformation of structures and materials when subjected to forces or motion. In this paper, three-dimensional digital image correlation (3D-DIC) is primarily used to extract vibration data from the oil tank of the oil-immersed core reactor.

The 3D-DIC measurement starts with calibration, and after establishing a unified world coordinate system, the photographs are converted into speckle digital images (pixel coordinates and grayscale values). As shown in Figure 2, (a) represents the result captured by the binocular camera, which is then converted into a grayscale image as shown in (b) for processing. The numbers represent the grayscale level, which indicates the light intensity of each pixel (ranging from 0 to 255).

Next, the image is divided into multiple individual subregions. Once the matching is successful, the displacement of each point in the target area is calculated based on the coordinate differences of each pixel in the unified coordinate system. As shown in Figure 2(c), $f(x, y)$ represents the grayscale value of a pixel in the reference subregion, and $g(x, y)$ represents the grayscale value of a pixel in the deformed subregion. μ_f and μ_g represent the mean grayscale values of the reference and deformed subregions, respectively, while σ_f and σ_g represent the corresponding grayscale value variances.

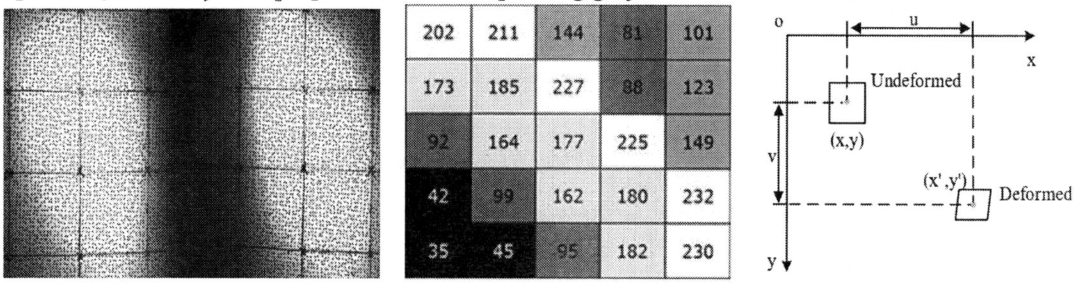

(a) Stereo camera captured photos (b) Grayscale image (c) Deformed and undeformed subregions

Figure 2. DIC

3. Vibration measurement system

The measurement platform mainly consists of two high-speed cameras, a light source, and a computer, as shown in Figure 3. The camera parameters used in the measurement are listed in Table 1. Based on the characteristics of the vibration data from the oil-immersed core reactor tank wall, it can be observed that the vibration data is mainly distributed at 100 Hz and its harmonics, with the vibration amplitude being on the micrometer scale. The normal component of the vibration mode is of greater research value. The resolution of the camera images is constrained by the shooting speed. It is considered that the vibration data collection requirements can be met if the shooting rate is no less than 1000 fps, allowing for precise capture of the vibration information on the oil tank.

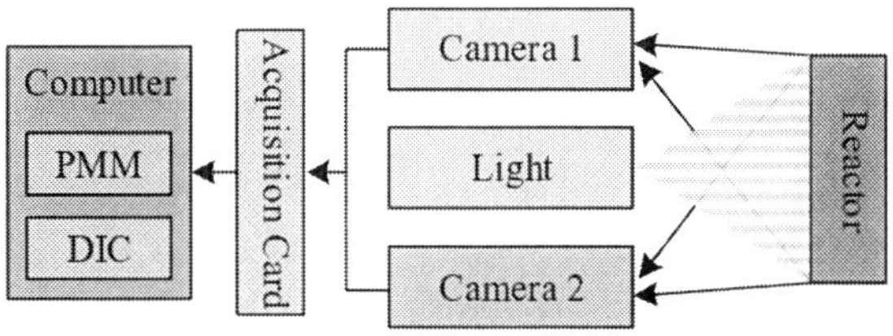

Figure 3. Measurement system

Table 1. Binocular camera parameters

Parameters	Values
Model	SH6-504
Maximum resolution	2560×2016
Maximum shooting speed /fps	50000
Exposure time /μs	1

4. Experimental process and result analysis

4.1. Experiment process

The main parameters of the parallel reactor used in the experiment are shown in Table 2. The experimental platform consists of two parts: the reactor vibration platform and the vibration measurement system. Figure 4(a) shows the equivalent circuit of the reactor vibration platform. In the experiment, a symmetric three-phase circuit is used. The specific configuration is as follows: the voltage is gradually increased in steps using a voltage regulator. After being stepped up by a transformer, the voltage is supplied to the reactor. The parallel capacitor is mainly used to provide compensating current to ensure the system's stability and voltage quality.

Table 2. Reactor parameters

Parameters	Values
Model	BKS-76/10
Type	Three-phase oil-immersed
Cooling method	ONAN
Rated capacity /kvar	76
Rated voltage /kV	10
Rated current /A	4.39
Actual reactance /Ω	1301.22

Figure 4(b) shows the specific reactor under test. To enhance the measurement accuracy, a speckle pattern was sprayed in the testing area. This method improves the precision of Digital Image Correlation (DIC) technology, making the vibration measurement more accurate. Additionally, vibration data from the same position are collected using a sensor array for comparison.

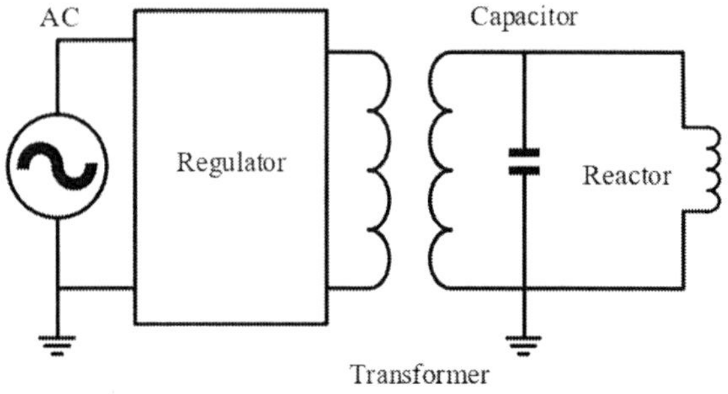

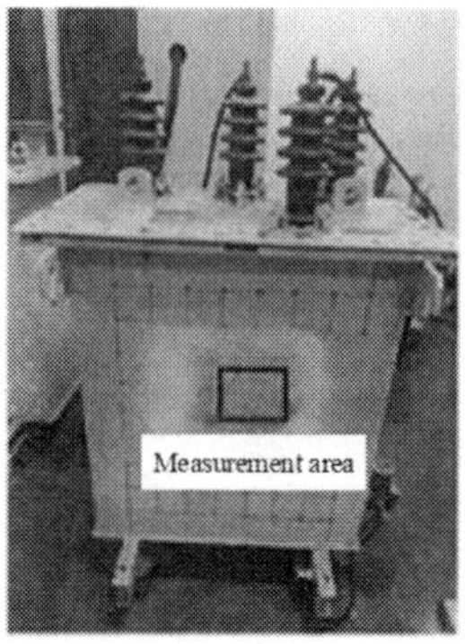

(a) Equivalent circuit of the reactor vibration platform (b) reactor

Figure 4. Reactor vibration measurement experiment

4.2. *Results and analysis*

In the experiment, the camera captures images every 0.5 seconds, containing 50 cycles of the main frequency component. Longer duration data are analyzed to reduce errors. As shown in Figure 5(a), the vibration displacement waveform obtained without amplification is non-sinusoidal and contains high noise. Figure 5(b) shows the vibration extraction results after 20x amplification. In the time-domain waveform, it closely resembles a sine wave; in the frequency domain, the main frequency component accounts for a large proportion, which aligns with the characteristics of reactor vibration signals. Figure 5(c) presents comparison images from other positions, where the blue solid line represents the amplified result and the red dashed line represents the unamplified result. The results indicate that the phase-based motion amplification technique is effective in amplifying small displacement information from the reactor's box wall, significantly improving the extraction results of DIC measurements.

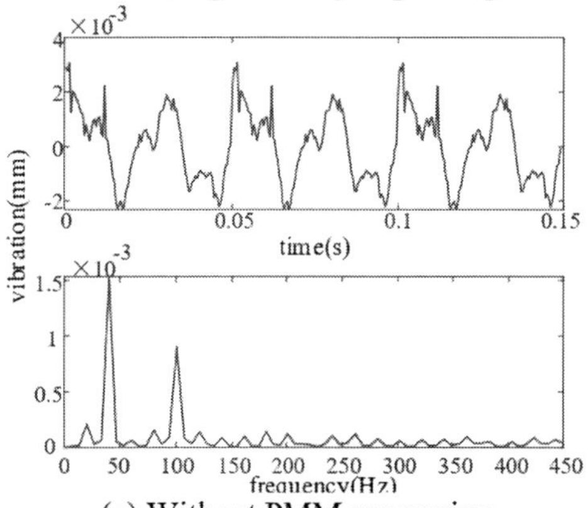

(a) Without PMM processing

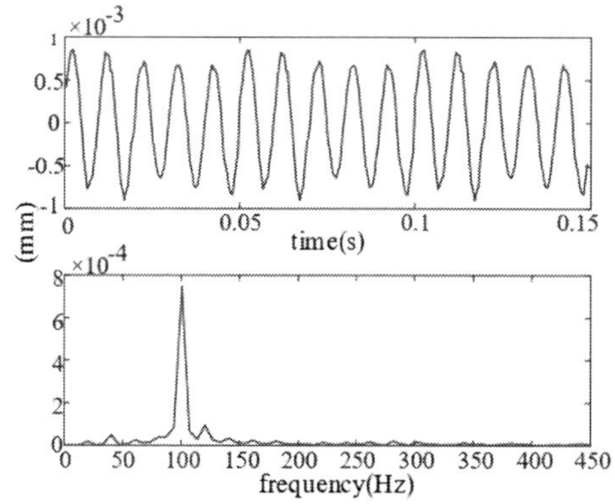

(b) After PMM processing

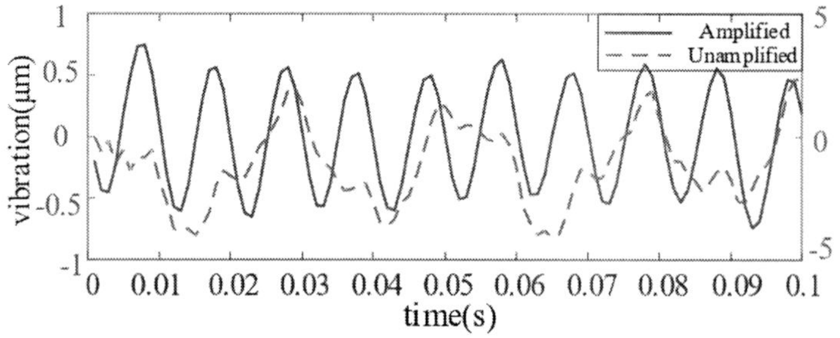

(c) Results from different positions

Figure 5. Comparison of measurement effects for reactor vibration signals

Firstly, the ODS diagrams obtained without PMM amplification and those obtained with amplification are analyzed. As shown in Figure 6(a) is the box wall vibration mode extracted using only

Journal of Physics: Conference Series **3012** (2025) 012073 doi:10.1088/1742-6596/3012/1/012073

DIC without the PMM technique. The maximum vibration displacement is 2.240 micrometers. From the amplitude distribution, the region with high vibration intensity is concentrated in the center of the capture area, occupying a relatively large spatial area, and is close to a square shape. Figure 6(b) shows the vibration mode obtained after amplification using DIC technology, where the maximum vibration displacement is 1.305 micrometers. In terms of amplitude distribution, the high-intensity vibration area is located on the right side and occupies a smaller space, more closely resembling a rectangular shape. Comparing the two results, it is evident that for small amplitude vibration signals that exceed the DIC measurement accuracy, the proposed measurement method can effectively extract the vibration information.

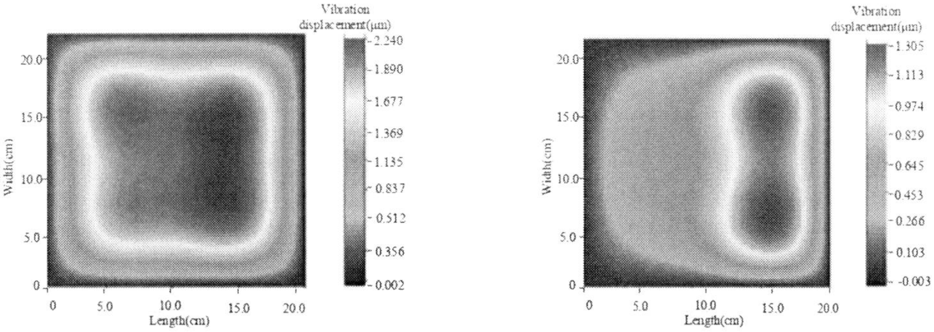

(a) Without PMM processing (b) After PMM processing

Figure 6. ODS of reactor

Next, a comparative analysis is conducted between the box wall vibration modes measured by the sensor array and those obtained using the visual method. As shown in Figure 7(a), the ODS diagram obtained from the sensor shows that the maximum vibration displacement is 1.080 micrometers, with high-intensity vibration concentrated on the right side, forming a nearly rectangular shape. In Figure 7(b), the ODS diagram obtained using the visual method shows a maximum vibration displacement of 1.045 micrometers, and the vibration distribution closely resembles the result obtained from the sensor. The Structural Similarity Index Measure (SSIM) is used to assess the similarity between the ODS diagrams, and by calculating the similarity between the ODS diagrams extracted using both methods, the following results are obtained:

$$\mathrm{SSIM} = 0.9139$$

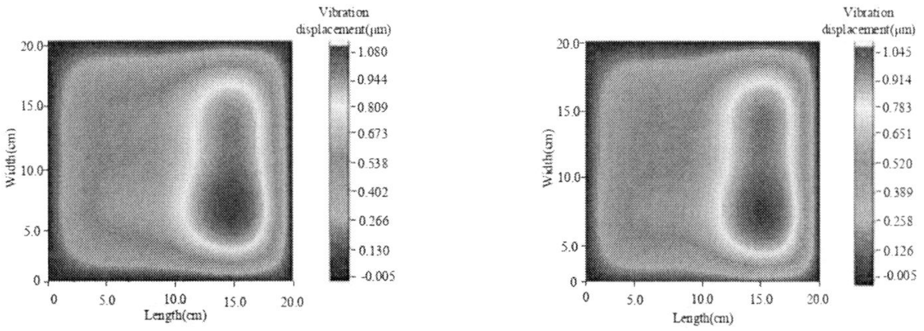

(a) Sensor measurement method (b) Visual measurement method

Figure 7. Comparison of the effects of two measurement methods

Considering that the parallel reactor may experience voltage fluctuations during actual operation, it is necessary to verify the stability of the vibration mode results measured by the method proposed in this paper. Therefore, the reactor's vibration modes are tested by varying the voltage applied across its terminals. In the experiment, voltages of 95%, 100%, and 105% of the rated voltage are applied, and the vibration modes are extracted using the visual method to generate the ODS diagrams. The results obtained are shown in Figure 8.

Journal of Physics: Conference Series **3012** (2025) 012073 doi:10.1088/1742-6596/3012/1/012073

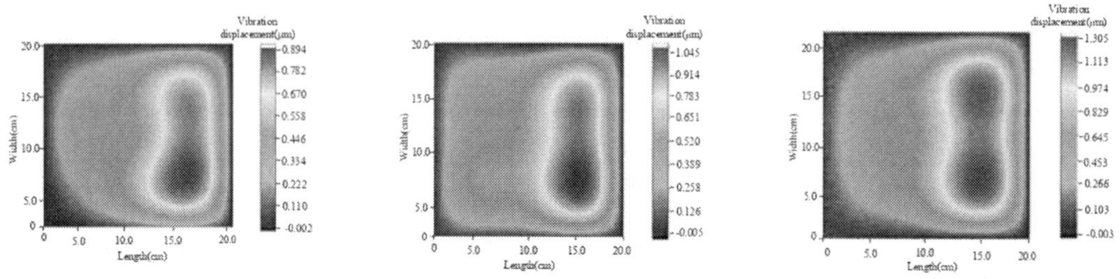

(a) 95% U_N (b) 100% U_N (c) 105% U_N

Figure 8. ODS of reactor under different voltages

A comparative analysis is performed on the ODS diagrams of the reactor's box wall operating at 95% U_N, 100% U_N, and 105% U_N. The maximum vibration displacement increases as the operating voltage rises: 0.8940 micrometers at 95% U_N, 1.045 micrometers at 100% U_N, and 1.305 micrometers at 105% U_N. From the vibration distribution, in all three diagrams, the high-intensity vibration regions are mainly concentrated on the right side. Although there are slight differences, the overall shape resembles a rectangular prism. Visual observation alone makes it difficult to analyze the changes in the ODS diagrams. Therefore, the results at 100% rated voltage are chosen as the reference, and the similarity between the ODS diagrams at 95% and 105% rated voltage and those at 100% rated voltage is calculated. The results are shown in Table 3.

Table 3. ODS similarity under different voltages

Voltage	95%U_N	100% U_N	105% U_N
SSIM	0.9382	1	0.9585

Based on the SSIM index, a quantitative analysis further demonstrates that the similarity of the ODS diagrams under voltage fluctuations is high, proving that the vibration mode extracted using the visual method remains stable even during voltage fluctuations in the parallel reactor. This validates the stability and reliability of the method for extracting 3D vibration signals from oil-immersed core reactors based on visual enhancement.

5. Conclusion

With the increasing capacity of high-voltage parallel reactors in power systems and the adoption of multi-gap core structures, the vibration issues of reactors are becoming increasingly severe. The operation of reactors in the electromagnetic field generates vibrations, which may lead to loosening of internal components and an increase in operational resistance. As a result, the failure rate of equipment increases and its lifespan is reduced.

This paper proposes a method for extracting 3D vibration signals from oil-immersed core reactors based on machine vision enhancement, combining the characteristics of reactor vibration modes, phase amplification, and digital image correlation principles. A comparison between the visual method and the sensor method for extracting the reactor's box wall vibration modes shows that the similarity of the ODS diagrams obtained by both methods is high, confirming the high accuracy of the proposed visual measurement method, which meets the measurement standards. Comparing the vibration modes extracted under voltage fluctuations using the visual method, the vibration modes obtained at 95% U_N, and 105% U_N, are nearly identical to those at the rated voltage (100% U_N,), demonstrating the stability of the method in vibration mode extraction.

References

[1] Han X, Sun X, Chen H, et al. The Overview of Development of UHV AC Transmission Technology in China[J]. Proceedings of the Chinese Society of Electrical Engineering, 2020, 40(14): 4371-4386-4386.

[2] Zhang J, Li X, Liu X, et al. Ultra High Voltage AC Grid Reactive Voltage Control [J]. Automation of Electric Power Systems, 2017, 41(5): 159-167.

[3] Liu Y, Lai T, Liu J, et al. Vibration Acoustic Signature Characteristics and Looseness Degree Acoustic Detection Method of Saturated Reactor in UHVDC Converter Valve [J] Transactions of China Electrotechnical Society 2023, 38(5): 1375-1389.

[4] Wu J, Ma H, Wu S, et al. Development of Vibration Signal-Based Fault Diagnosis Method and Monitoring System for High-Voltage Shunt Reactors [J] Electrical Measurement & Instrumentation 2020, 57(1): 113-120.

[5] Zhang F, Zhang Y, Xu L, et al. Analysis of Partial Discharge Fault in a 750kV High-Voltage Reactor [J]. High Voltage Apparatus, 2015, 51(03): 135-139.

[6] Magdaleno-Adame S, Escarela-Perez R, Olivares-Galvan J C, et al. Temperature Reduction in the Clamping Bolt Zone of Shunt Reactors: Design Enhancements[J]. IEEE Transactions on Power Delivery, Power Delivery, IEEE Transactions on, IEEE Trans. Power Delivery, 2014, 29(6): 2648-2655.

[7] Zhang F, Ji S, Ma H, et al. Operational Modal Analysis of Transformer Windings[J]. IEEE Transactions on Power Delivery, Power Delivery, IEEE Transactions on, IEEE Trans. Power Delivery, 2020, 35(3): 1285-1298.

[8] Wang F, Duan R, Geng C, et al. Research on Vibration Characteristics of Power Transformer Windings Based on the Magnetic-Mechanical Coupled Field Theory [J]. Proceedings of the CSEE, 2016, 36(09): 2555-2562.

[9] Minteer T M. Electrostatic stress: insightful analysis and manipulation of Maxwell's stress equation for electrostatics[J]. IEEE Transactions on Dielectrics and Electrical Insulation, Dielectrics and Electrical Insulation, IEEE Transactions on, IEEE Trans. Dielect. Electr. Insul., 2014, 21(1): 64-73.

[10] Minteer T M. Magnetostatic Stress: Insightful Analysis and Manipulation of Maxwell's Stress Equation for Magnetostatics[J]. IEEE Transactions on Magnetics, Magnetics, IEEE Transactions on, IEEE Trans. Magn., 2013, 49(11): 5387-5398.

[11] He Z, Sha Y, Zhao Z, et al. The calculation of magnetic domain and magnetostriction in stressed grain-oriented silicon steel[J]. Journal of Applied Physics, 2020, 127(3): 1-6.

[12] Gao L, Ji S, Zhu L, et al. Study on Vibration Mechanism and Impact Factors of Dry-type Air-core Reactor in HVDC Converter Stations[J]. Proceedings of the Chinese Society of Electrical Engineering, 2022, 42(16): 5960-5970-5970.

Modular Modeling of Electromechanical Transients for Grid-Forming Energy Storage Systems

Shumin Sun[1], Yuejiao Wang[1], Zhijie Zheng[2], Jingqi Yang[3*], Yan Cheng[1], Bo Chen[2], Chunyi Wang[2]

[1] China State Grid Shandong Electric Power Research Institute, Jinan, China

[2] State Grid Shandong Electric Power Company, Jinan, China

[3] Electric Power Research Institute, Beijing, China

*Corresponding author's e-mail: yangjingqi@epri.sgcc.com.cn

Abstract. As power systems increasingly incorporate renewable energy sources and power electronic converters, the overall system inertia declines, making grid-forming energy storage systems essential for enhancing system stability and flexibility. This paper presents a modular electromechanical transient modelling approach for grid-forming energy storage systems. The proposed model comprises initial value calculation, power loop, injected current calculation, current limit compensation, and active/reactive auxiliary control modules, enabling precise simulation of the electromechanical transient behavior under various operating conditions and disturbances. Through simulations on a single-machine infinite bus system, the impact of frequency and voltage regulation parameters on both dynamic performance and steady-state levels is investigated. The results demonstrate that appropriate parameter selection can significantly improve the dynamic and steady-state responses of grid frequency and voltage under disturbances, providing valuable insights for the design and optimization of grid-forming energy storage systems.

1. Introduction

As renewable energy sources and power electronic converters become deeply integrated into modern power systems, the overall system inertia declines significantly, posing substantial challenges to maintaining both voltage and frequency stability. Traditional synchronous generators inherently provide rotational inertia, which naturally buffers against rapid changes in frequency. However, as these conventional power plants are replaced or supplemented by inverter-based resources, the grid's ability to resist sudden fluctuations diminishes. In this context, grid-forming (GFM) energy storage systems, which emulate the dynamic behavior of synchronous machines, have emerged as a key technological means to ensure stable operation. By providing virtual inertia, autonomous frequency regulation, and voltage support, these GFM systems help maintain system reliability and security, even as conventional synchronous generation recedes.

Over the past few years, research efforts have intensified in developing accurate, flexible, and scalable electromechanical transient models that capture the intricate dynamics of GFM energy storage systems under a broad spectrum of operating conditions[1]. Reference [1] derived a comprehensive frequency-domain model for a GFM-based battery energy storage system (BESS) and integrated it with the impedance model of an offshore wind power plant (WPP). The combined model was validated through electromagnetic transient simulations, reinforcing confidence in its predictive capabilities. Similarly, reference [2] explored the potential of both grid-following and grid-forming BESSs to enhance voltage stability under short-circuit conditions, shedding light on their efficacy during severe fault scenarios. Further, references [3]-[5] investigated how BESSs, capable of operating flexibly in either grid-following or grid-forming modes, contribute to frequency regulation in low-inertia power systems, highlighting their adaptability to various network conditions. Meanwhile, reference [6] introduced a hierarchical control strategy for GFM energy storage systems at both the power plant and converter levels, providing a structured framework to manage their complex, multi-timescale dynamics. Reference [7] proposes an energy storage control strategy that embeds a phase-locked loop (PLL) into the inverter control, which provides improved synchronization and fault current-limiting capabilities. Reference [8] presents a method for enhancing the transient synchronization stability of grid-forming energy storage systems during sudden grid voltage drops and rises, thereby supporting their voltage ride-through capability.

Building on the aforementioned foundational works, this study proposes a modular electromechanical transient modelling approach specifically tailored for grid-forming energy storage systems. The proposed method decomposes the overall model into a series of functionally distinct submodules—initial value calculation, equivalent voltage source dynamic equations, injected current calculation, dynamic internal potential vector compensation, and active/reactive auxiliary controls. This modelling approach overcomes the challenge of accurately representing low-inertia systems under high-penetration conditions and addresses the complexity arising from multi-loop control coupling. By systematically adjusting the control parameters within these submodules, the influence of various control settings on frequency and voltage regulation performance can be thoroughly investigated, thereby providing a more flexible and precise theoretical and technical foundation for the design and optimization of GFM energy storage systems.

The remainder of this paper is organized as follows:

1) Section 2 presents the proposed modular electromechanical transient model, detailing the functionality of its submodules for initial value calculation, power loops, injected current computation, current limit compensation, and active/reactive auxiliary controls.

2) Section 3 focuses on analyzing the system's dynamic response to frequency and voltage disturbances, examining the impact of various control parameters on steady-state and transient performance.

3) Finally, Section 4 summarizes the main contributions and provides an outlook on future research directions.

2. Electromechanical Transients for Grid-Forming Energy Storage Systems

The structured model of a grid-forming flexible converter primarily consists of the following components: initial state calculation, power loop, injected current calculation, power calculation, current limit compensation control, and active/reactive power auxiliary controls for system stability participation. The information interaction between each component is shown in Figure 1.

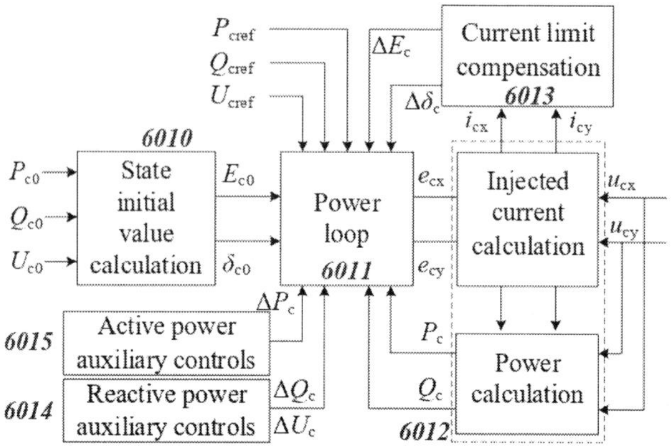

Figure 1. Structural framework of the electromechanical transient model for grid-forming energy storage systems

In the figure 1, P_{c0}, Q_{c0}, and U_{c0} represent the steady-state active power, reactive power, and node voltage at the grid connection point of the grid-forming energy storage derived from power flow calculations. Ec0 and δc0 denote the initial values of the internal potential magnitude and phase angle, respectively. e_{cx}, e_{cy}, u_{cx}, u_{cy}, i_{cx}, and icy represent the x- and y-axis components of the internal potential, grid connection point voltage, and injected current in the synchronous rotating xy-coordinate system. P_{cref}, Q_{cref}, and U_{cref} are the power and voltage reference values corresponding to equations (1) and (2). P_c and Q_c are the active and reactive power injected into the grid connection point. ΔP_c, ΔQ_c, and ΔU_c are the additional power regulation quantities for system stability control contributed by the grid-forming energy storage.

The figure above shows the electromechanical transient simulation model for grid-forming energy storage systems, with the implementation of each component detailed as follows:

Module 6010: state initial value calculation. For the grid-forming converter described in equations (1) and (2), the state variables include the internal potential magnitude E_c, the power angle δ_c, and the angular frequency ω_c. The initial values of these variables must be calculated before performing the electromechanical transient simulation.

$$\begin{cases} \dfrac{d\delta_c}{dt} = \omega_0\omega_c \\ J\dfrac{d\omega_c}{dt} = \dfrac{P_{\mathrm{cref}} - P_c}{\omega_c} - D_\delta(\omega_c - \omega_0) \end{cases} \tag{1}$$

$$K_e\frac{dE_c}{dt} = Q_{\mathrm{cref}} - Q_c - D_u(U_{\mathrm{cref}} - U_c) \tag{2}$$

During steady-state operation, the initial value of ω_c, denoted as ω_{c0}, is set to 1.0 pu. The initial values E_{c0} and δ_{c0} are calculated based on the steady-state injected power P_{c0}, Q_{c0}, and the x- and y-axis components of the node voltage, u_{cx0} and u_{cy0}, as shown in equations (3) and (4). Here, i_{cx0} and i_{cy0} represent the x- and y-axis components of the steady-state injected current.

$$\hat{I}_{c0} = i_{cx0} - ji_{cy0} = \frac{\dot{S}_{c0}}{\dot{U}_{c0}} = \frac{P_{c0} + jQ_{c0}}{u_{cx0} + ju_{cy0}} \tag{3}$$

$$\begin{cases} i_{cx0} = (P_{c0}u_{cx0} + Q_{c0}u_{cy0}) / U_c^2 \\ i_{cy0} = (Q_{c0}u_{cx0} - P_{c0}u_{cy0}) / U_c^2 \end{cases} \tag{4}$$

$$\dot{E}_{c0} = \dot{U}_{c0} + (r_c + jx_c)\dot{I}_{c0} \tag{5}$$

$$\begin{cases} e_{cx0} = u_{cx0} + r_c i_{cx0} - x_c i_{cy0} \\ e_{cy0} = u_{cy0} + x_c i_{cx0} + r_c i_{cy0} \end{cases} \tag{6}$$

$$E_{c0} = \sqrt{e_{cx0}^2 + e_{cy0}^2}, \delta_{c0} = \arctan(e_{cy0} / e_{cx0}) \tag{7}$$

Module 6011: dynamic simulation of voltage source external characteristics. This can be implemented using the dynamic equations shown in equations (1) and (2), or by adopting the equations corresponding to control methods such as droop control or synchronous power control.

Module 6012: calculation of injected current for grid-forming energy storage. Based on the internal potential e_{cx}, e_{cy} and the grid-connected node voltage u_{cx}, u_{cy} updated at each time step, the injected current i_{cx}, i_{cy}, and the power P_c, Q_c for the current time step are calculated using equations (8) and (9).

$$\begin{cases} i_{cx} = \left[r_c(e_{cx} - u_{cx}) + x_c(e_{cy} - x_{cy}) \right] / Z_c^2 \\ i_{cy} = \left[r_c(e_{cy} - u_{cy}) - x_c(e_{cx} - x_{cx}) \right] / Z_c^2 \end{cases} \tag{8}$$

$$\begin{cases} P_c = e_{cx} i_{cx} + e_{cy} i_{cy} \\ Q_c = e_{cy} i_{cx} - e_{cx} i_{cy} \end{cases} \tag{9}$$

Module 6013: dynamic compensation control of internal potential vector for limiting converter current. When the converter current magnitude I_c reaches the limit I_{clim}, appropriate magnitude and phase angle compensation values ΔE_c, $\Delta\delta_c$ are added. These values can be calculated based on the differentiated requirements of the system's transient performance.

Module 6014: active power quxiliary control of grid-forming energy storage for system stability. This includes auxiliary low-frequency oscillation damping control, which uses the active power P_{line} of the AC branch as the input signal, and emergency power boost or reduction control in response to stability control system commands. The module outputs ΔP_c, which is added to the reference value P_{cref} in Module 6011.

Module 6015: reactive power auxiliary control of grid-forming energy storage for system stability. This includes emergency reactive power boost or reduction control in response to stability control system commands. The module outputs ΔU_c or ΔQ_c, which are added to the reference values U_{cref} or Q_{cref} in Module 6011.

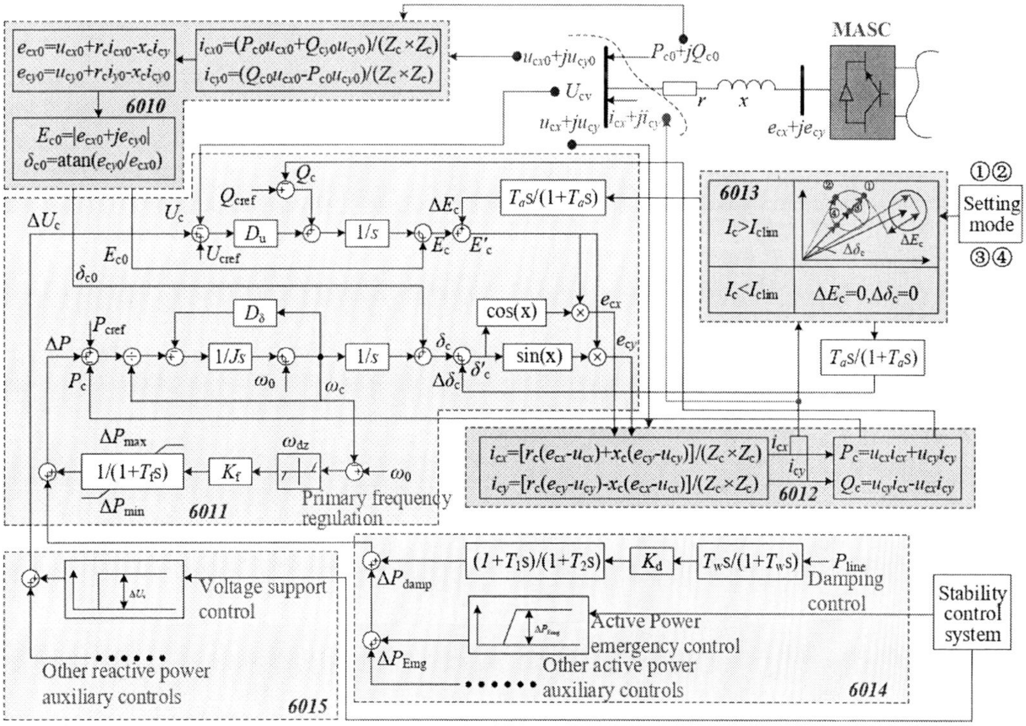

Figure 2. Grid-forming energy storage electromechanical transient simulation model

3. Case Studies

Based on the electromechanical transient model of the grid-forming energy storage system, a single-machine infinite bus system was constructed for simulation. The system structure is shown in Figure 3.

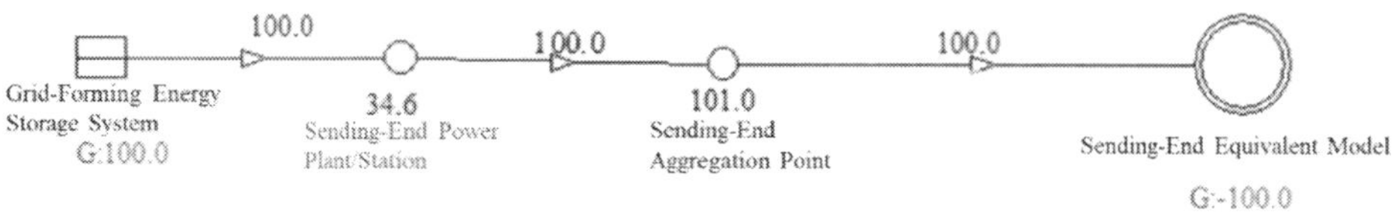

Figure 3. Structure of the single-machine infinite bus system

In this case study, the grid-forming energy storage system is linked through transmission lines—each with specified lengths and parameters—first to a sending-end power plant or station, then to a sending-end aggregation point, and finally to an ideal infinite bus. The grid-forming energy storage system operates at a nominal voltage of 220 kV with a capacity of 100MW/200MWh. The transmission lines between the nodes have length of 100 km, respectively, with impedance values of 0.005 + j0.05 p.u. on a 100 MVA, 220 kV base. The sending-end power plant/station is also connected at 220 kV and has a rating of 100 MW. The sending-end aggregation point aggregates power from the sending-end plant and connects to the ideal infinite bus, which is fixed at 220 kV and 50 Hz.

3.1. *Impact of Different Parameters on Frequency Regulation Performance*

In the equivalent system illustrated in Figure 3, a governor dynamic model is considered. When the grid load suddenly decreases by 10%, the system's equilibrium point shifts accordingly. Without primary frequency control capabilities in the grid-forming energy storage converter, its internal control strategy will not actively adjust active power output, causing the system frequency to rise from the nominal value (50 Hz) to approximately 50.7 Hz. However, when primary frequency control is considered, as soon as the system frequency begins to deviate from its nominal value, the converter's frequency control loop can quickly detect this change and adjust its active power output (either increasing or decreasing the amount injected into the grid) to counteract the power imbalance caused by the load change, thereby reducing the amplitude of frequency fluctuations.

The frequency deviation amplification factor K in the virtual governor significantly influences the converter's primary frequency control capability. Considering two cases where K equals 5 and 25, respectively, it can be observed that the larger the frequency deviation amplification factor, the greater the modulation of active power by the converter, and consequently, the smaller the frequency fluctuations in the system (Stable to 50.35Hz and 50.14Hz respectively).

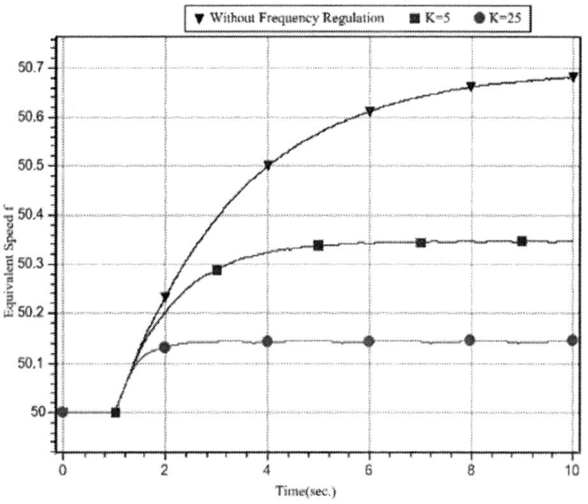

Figure 4. Impact of governor parameters on steady-state frequency regulation

3.2. Impact of Different Parameters on Voltage Regulation Performance

To study how the excitation controller regulator gain affects the steady-state voltage regulation capabilities of a grid-forming energy storage system, we consider a scenario in which the grid voltage experiences a sudden drop to 0.9 pu. Under this disturbance, the system's ability to swiftly and effectively restore the voltage to its nominal level becomes a critical measure of performance. To capture the influence of different control intensities, three distinct regulator gain coefficients—0.1, 0.5, and 2.5—are chosen.

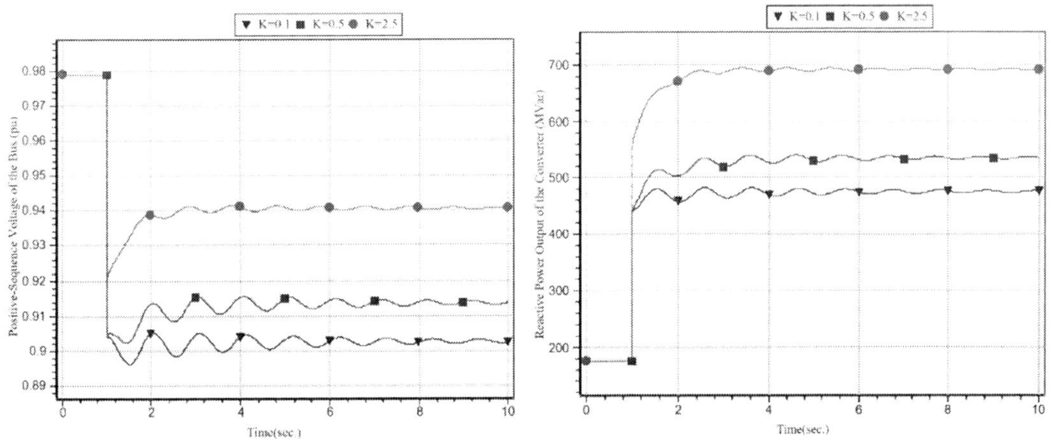

(a) Bus voltage (b) Reactive Power Output of the Converter

Figure 5. Impact of excitation regulator gain on steady-state voltage regulation

As illustrated in Figure 5, when the grid voltage drops, the grid-forming energy storage converter dynamically adjusts its reactive power output to maintain the grid bus voltage level close to its rated value. When the excitation controller gain K equals 0.1, the converter provides reactive power support of 474 MVar; When the excitation controller gain K equals 0.25, the converter provides reactive power support of 535 MVar; When the excitation controller gain K equals 0.5, the converter provides 692MVar of reactive power support. A larger gain coefficient means the excitation controller will more strongly amplify and respond to the voltage deviation, thereby enabling the converter to provide greater reactive power support and facilitating a more effective steady-state voltage recovery.

To study the influence of the excitation controller regulator gain on the transient voltage regulation performance of a grid-forming energy storage system, dynamic simulations were conducted under different transient open-circuit time constant ($T_{d'0}$) settings. A 0.2-second three-phase instantaneous short-circuit fault was applied at the converter's output.

As shown in Figure 6, after clearing the fault, the recovery time for $T_{d'0}$ equals 1s is 0.32 seconds, for $T_{d'0}$ equals 10s it is 0.34 seconds, and for $T_{d'0}$ equals 50s it is about seconds. The greater the $T_{d'0}$, the slower the voltage recovery following a short-circuit fault on the converter's output. After the fault clears, the excitation voltage undergoes an instantaneous step change, and due to the effect of $T_{d'0}$, the virtual internal potential increases only gradually. The larger the $T_{d'0}$, the more pronounced the delay in the rise of the internal potential. This extended delay curtails the converter's ability to provide the necessary reactive power for voltage support in the immediate aftermath of a fault, thereby weakening the unit's transient reactive support capability and deteriorating the overall AC voltage recovery performance.

A smaller $T_{d'0}$ corresponds to a more responsive excitation control, which is able to ramp up the virtual internal potential in a shorter time frame. As a result, the converter can quickly inject or absorb reactive power to stabilize the voltage, minimizing the magnitude and duration of any post-fault voltage dip. In contrast, as $T_{d'0}$ increases, the excitation controller's response becomes more sluggish. The internal potential takes longer to build up to its required level, and the converter's capacity to provide immediate reactive power support diminishes.

Journal of Physics: Conference Series **3012** (2025) 012074 doi:10.1088/1742-6596/3012/1/012074

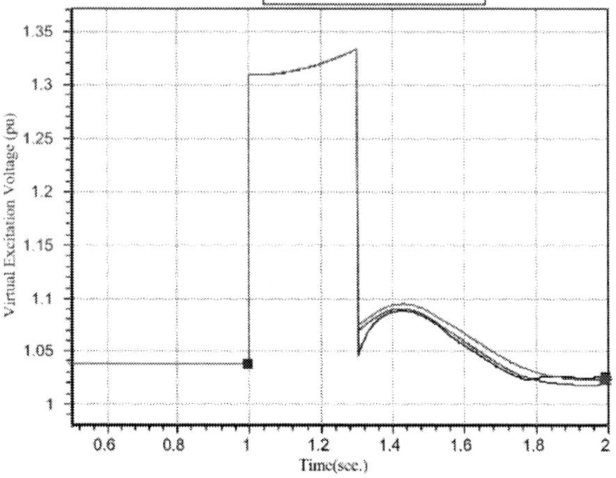

(a) Excitation voltage

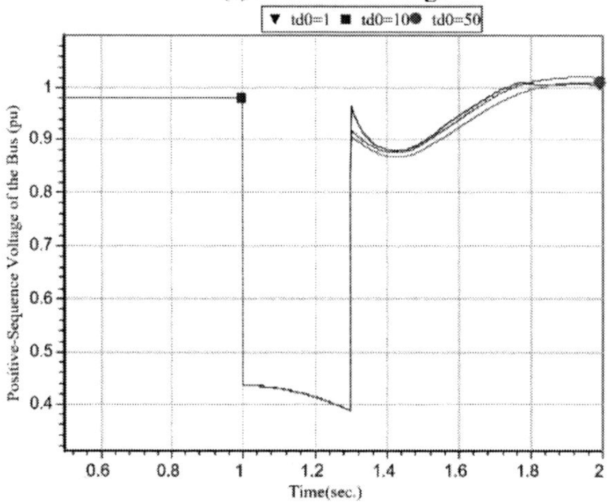

(b) Bus voltage

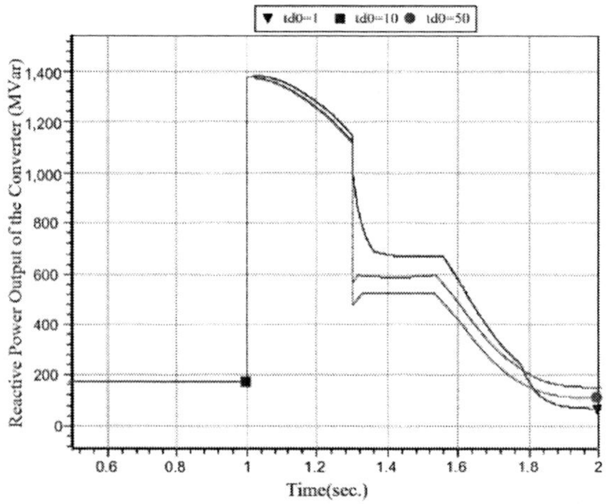

(c) Reactive power

Figure 6. Impact of excitation regulator gain on transient voltage regulation

4. Conclusions

This paper has presented a modular electromechanical transient modelling approach for grid-forming energy storage systems, in which the complex system is decomposed into modules for initial value calculation, power loop, injected current calculation, current limit compensation, and active/reactive auxiliary controls. The proposed framework enables accurate and efficient simulation of system dynamics, as demonstrated by the significant improvements in frequency response and voltage recovery observed when key parameters are properly tuned. By addressing the complexity arising from multi-loop control coupling and providing a clear pathway for parameter sensitivity analysis, this modelling approach effectively overcomes the challenge of accurately representing low-inertia systems under high-penetration conditions. Future research can build upon these results by extending the modular framework to multi-machine scenarios, and examining the performance of alternative converter topologies or hierarchical control structures, thereby further enhancing the stability of power systems with high penetrations of inverter-based resources.

References

[1] F. Zhao, X. Wang, Z. Zhou, et al, "Control Interaction Modeling and Analysis of Grid-Forming Battery Energy Storage System for Offshore Wind Power Plant," IEEE Transactions on Power Systems, vol. 37, no. 1, pp. 497-507, Jan. 2022.

[2] Al Kez, Dlzar, Aoife Foley, and Faraedoon Ahmed. "Exploring Transient Voltage and Frequency Responses in Low Inertia Power Systems: A Comparative Study of Grid Following and Grid Forming Battery Energy Storage Systems." Available at SSRN 4512292 (2023).

[3] Y. Zuo, Z. Yuan, F. Sossan, et al, "Performance assessment of grid-forming and grid-following converter-interfaced battery energy storage systems on frequency regulation in low-inertia power grids," Sustainable Energy, Grids and Networks, Volume 27, 2021.

[4] S. Chen, Q. Meng, P. Lin, et al, "Grid Forming Hybrid Energy Storage System for Flexible Frequency Regulation in Islanded AC Microgrids," 2024 4th Power System and Green Energy Conference (PSGEC), Shanghai, China, 2024, pp. 1337-1341.

[5] E . Rakhshani, J. M. Ruiz, X. Benavides, et al, "Enhancing Low-Inertia Power Systems with Grid Forming Based Hybrid Energy Storage Technology," 2024 12th International Conference on Smart Grid (icSmartGrid), Setubal, Portugal, 2024, pp. 464-468.

[6] T. Zheng, S. Qing, H. Zhao, et al, "Transient Angle Stability Improvement Strategy of Power Grid Based on Grid-Forming Energy Storage System," 2022 4th International Conference on Power and Energy Technology (ICPET), Beijing, China, 2022, pp. 1017-1022.

[7] Cai Y, Yu L, Wu M, et al, "Grid-Forming Control for Solar Generation System with Battery Energy Storage," Energies, 2024, 17(15):3642.

[8] X. Wu, S Wu, F Wang, et al., "Grid-Forming Control and Experimental Validation for High Voltage Transformerless Battery Energy Storage System," in IEEE Transactions on Power Electronics, doi: 10.1109/TPEL.2024.3510361.

Diode Clamped Shared Redundancy Topology for Middle Arm Reuse MMC

Wenjun Liu [1,2*], **Yan Chen**[1,2], **Genxu Chen**[1,2], **Bicheng Xu**[1,2], **Yaoqiang Wang**[1,2]

[1] School of Electrical and Information Engineering, Zhengzhou University, Zhengzhou, China

[2] Henan Engineering Research Center of Power Electronics and Energy Systems, Zhengzhou, China

*Corresponding author's e-mail: lwenjun1990@zzu.edu.cn

Abstract: Given the significant quantity of submodules (SMs) employed in modular multilevel converters (MMCs), a lightweight MMC named CM-MMC (Capacitor multiplexing MMC) is proposed in this paper that can cut down SM count by 25% by arm reuse, significantly reducing system costs while ensuring efficient operation and DC fault current interruption capability. However, its increased bridge arm number may lead to increased redundancy cost since enough redundant SMs must be equipped for each bridge arm. Enhancing the CM-MMC's reliability while reducing the redundancy cost, an innovative redundancy sharing scheme is proposed, enabling the upper, middle, and lower arms to utilize shared redundant capacitors. Additionally, a simulation model in Matlab/Simulink is created, and the proposed topology's feasibility is validated through simulation results.

1. Introduction

The superior output waveform quality and scalability of Modular Multilevel Converters (MMCs) make them widely used in HVDC systems [1]. However, traditional MMCs are large in size and costly. To address these issues, ref [2] proposed a bridge-arm reuse MMC topology, which enhances the utilization of submodules (SMs). This paper introduces a lightweight MMC, referred to as CM-MMC (Capacitor Multiplexing MMC), which achieves a 25% reduction in the number of SMs by reusing arms.

Nevertheless, owing to the substantial quantity of SMs in an MMC, the malfunction of a single SM may disrupt the entire system, possibly resulting in a shutdown[3]. Therefore, redundancy protection is essential, as it ensures the converter can continue to operate safely and stably, even in the event of an SM failure. Redundancy protection is typically classified into two types: cold standby and hot standby. Ref [4] considers factors such as the number of faulty SMs, voltage/current stress, and power loss, and designs a preprocessing unit to implement a hierarchical redundancy strategy. This approach ensures that the system can continue to operate normally during SM failures by employing a classification-based selection strategy. Ref [5] introduces an analytical method for evaluating the reliability of MMCs under various redundancy schemes. Ref [6] integrates fault-tolerant operation strategies with voltage balancing, sorting, and modulation techniques, taking into account modulation boundaries and redundancy structures. It presents two fault-tolerant control approaches designed to achieve either optimal dynamic performance or a balanced stress distribution. Ref [7] introduces a shared redundancy protection technique aimed at minimizing the number of redundant SMs and the costs associated with them. While this method allows redundancy sharing between the upper and lower arms, it demands that thyristors

Journal of Physics: Conference Series **3012** (2025) 012075 doi:10.1088/1742-6596/3012/1/012075

endure higher voltage stresses, as the number of arms increases, the method of adding redundant SMs between two arms increases the system's complexity. Hence, the CM-MMC has three arms per phase, this redundancy topology is not suitable for multi-arm MMCs. This paper proposes a redundancy protection scheme based on a quasi-DNPC[8][9](Dual Active Neutral Point Clamped) topology, which facilitates redundancy sharing across the upper, middle, and lower arms in a CM-MMC, while also reducing the number of redundancy capacitors. This approach offers valuable insights for the redundancy design of future multi-arm MMCs.

The paper is organized as follows: Section 2 introduces the topology of the CM-MMC and the conventional cold standby redundancy protection. Section 3 presents the redundancy sharing strategy across the upper, middle, and lower arms of the CM-MMC. Section 4 validates the feasibility of the proposed topology through simulations. Section 5 provides a conclusion and summary of the paper.

2. The Topology of CM-MMC and Current Status of Redundancy Protection

2.1. Topology and Operating Principle of CM-MMC

Fig. 1 shows the CM-MMC configuration. Each phase unit has an upper, middle, and lower arm, with SMs distributed in a 1:1:1 ratio, each arm containing N SMs. The upper and lower arms employ half-bridge SMs (HBSMs), and the middle arm uses full-bridge SMs (FBSMs) for DC fault interruption.

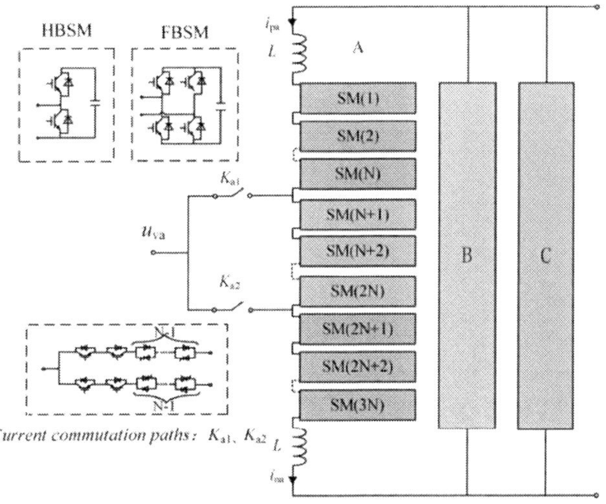

Figure 1. CM-MMC Topology Structure.

A selector switch is employed that consists of two current commutation paths. Each of the current commutation paths is equipped with two anti-series connected IGBIs for bidirectional current interruption and (N-1) pairs of diodes and thyristors connected in reverse parallel to withstand the voltage during its turn-off mode. By alternately switching these two commutation paths, the middle arm is shared between the upper and lower arms, making each equivalent arm contain (2N) SMs, thus reducing the total SM count by 25%. The upper arm reuse state is defined as "Mode I", and the lower arm reuse state is defined as "Mode II".

Consider phase A , with the current in the upper arm denoted as i_{pa} , denote the lower arm current as i_{na} , denote the upper arm voltage as u_{ppa} , denote the middle arm voltage as u_{ma}, denote the the lower arm voltage as u_{nna} , the AC side voltage is u_{va}, the AC side current is i_{va}, and the DC side voltage is u_{dc}. Hence, the equivalent upper arm voltage (u_{pa}) and lower arm (u_{na}) can be represented as:

$$\begin{cases} u_{pa} = u_{ppa} + u_{ma}, u_{na} = u_{nna}, \text{Mode I} \\ u_{pa} = u_{ppa}, u_{na} = u_{nna} + u_{ma}, \text{Mode II} \end{cases} \tag{1}$$

In an ideal state, the equivalent voltages of the upper (u_{pa}) and lower (u_{na}) arm in the CM-MMC are as follows:

$$\begin{cases} u_{pa} = \frac{1}{2}u_{dc} - L\frac{di_{pa}}{dt} - u_{va} \\ u_{na} = \frac{1}{2}u_{dc} - L\frac{di_{na}}{dt} + u_{va} \\ i_{va} = i_{pa} - i_{na} \end{cases} \tag{2}$$

In an ideal state, phase A AC voltage (u_{pa}) is expressed as:

$$u_{va} = \frac{u_{na} - u_{pa}}{2} - \frac{L}{2}\frac{di_{va}}{dt} \tag{3}$$

2.2. Traditional Cold Standby Protection Strategy

Fig. 2(a) illustrates the conventional MMC cold standby (Cold-RSM) configuration. Each arm consists of *N* SMs and *M* cold standby redundant SMs (RSM). Each SM has a mechanical bypass switch, which isolates the faulty SM and activates the redundant module. Taking phase A as an example, under normal operating conditions, the bypass switches T^a_1 to T^a_{3N} are in the off state, and all the series-connected SMs are involved in the operation of the circuit. Meanwhile, T^a_{3N+1} to T^a_{3N+3M} are in the closed state, so the RSMs are bypassed and do not participate in the operation. As shown in Fig. 2(b), when a fault occurs in an SM, it will be bypassed, and an RSM from the faulty arm will be activated. The RSM will restore normal operation through a fast charging system.

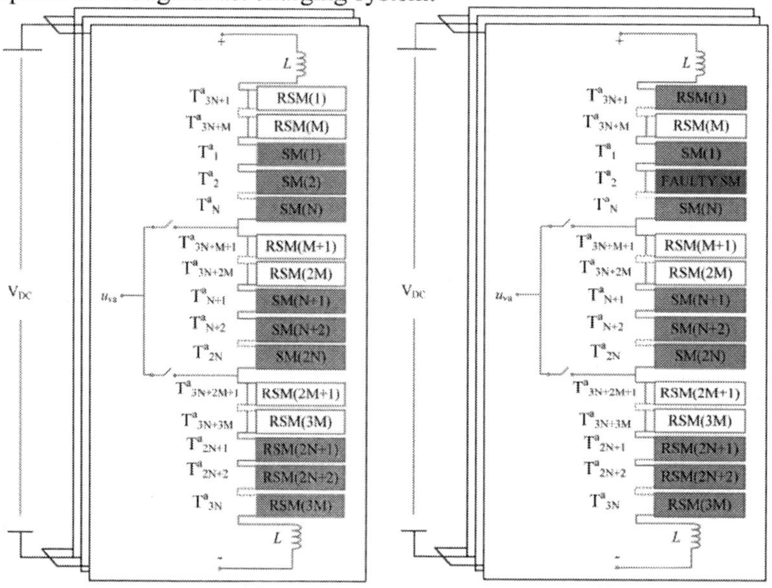

Figure 2. (a) Traditional cold standby redundancy. (b) An RSM replaces the faulty SM.

A key drawback of incorporating RSMs in a traditional MMC is the high cost associated with them. Moreover, the RSMs in the traditional design are generally limited to use within the designated arm and cannot be shared across arms for redundancy purposes. For example, in a CM-MMC, it is assumed that each arm contains four SMs and is equipped with one RSM. An upper arm SM fault in phase A can be replaced by the arm's RSM. However, If a second SM in the upper arm fails, the MMC cannot handle the fault due to the absence of an extra RSM. Even though the RSMs in the middle and lower arms remain unoperational, they are unable to replace the faulty SM in the upper arm due to structural constraints.

3. CM-MMC Redundancy Sharing Topology

3.1. Redundancy Sharing Topology Structure

Fig. 3 illustrates the redundancy sharing topology of the CM-MMC, where M redundant capacitors are positioned between the upper and middle arms. As shown in Fig. 4(a), the redundant capacitors are connected through a clamp-type topology, functioning like a double-pole, multi-position switch, as shown in Fig. 4(b), to enable redundancy sharing between the upper and middle arms. Fig. 4(c) shows that the M SMs near the lower arm in the middle arm are connected using the clamp-type topology, functioning like a single-pole, multi-position switch, as shown in Fig. 4(d), to restore the normal operation of the lower arm.

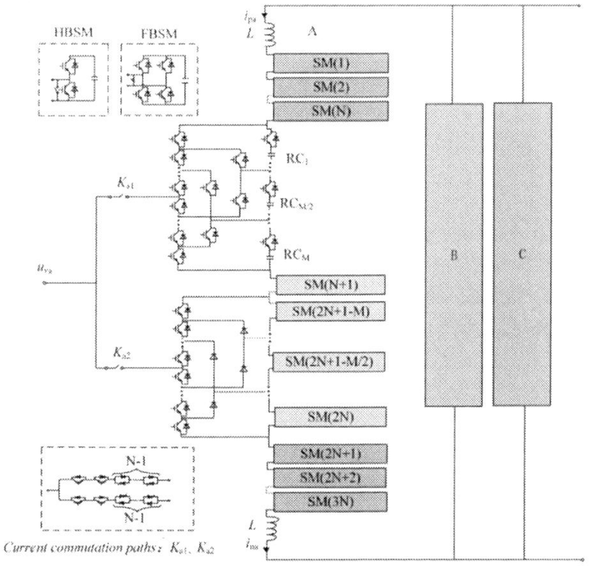

Figure 3. CM-MMC Redundancy Sharing Topology.

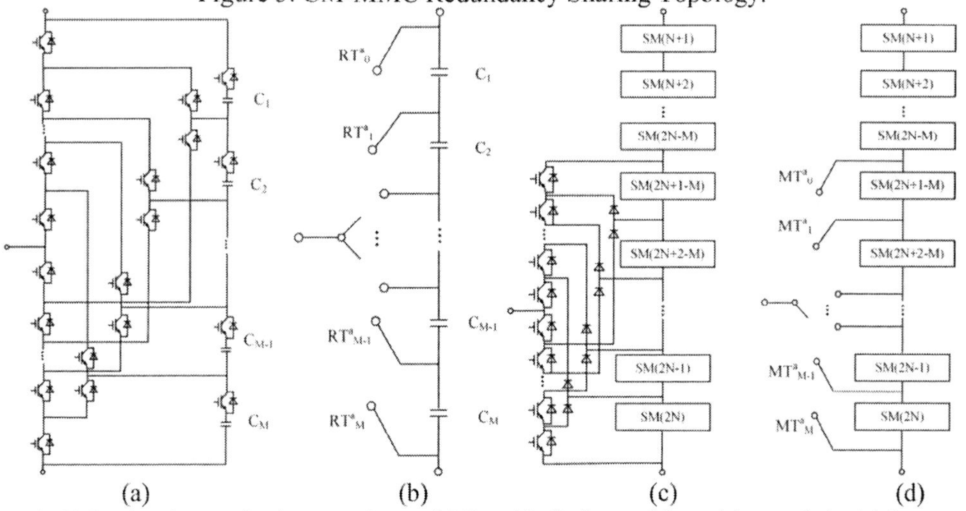

Figure 4. (a) Redundancy sharing topology. (b) Double-Pole, multi-position switch. (c) Lower arm fault transfer topology. (d) Single-Pole, multi-position switch.

As shown in Fig. 4(a), in terms of control strategy, to implement PWM modulation control for the redundant capacitors, this paper replaces the traditional clamping diodes with IGBTs, thereby enhancing the flexibility and accuracy of the system's control. To provide the redundant capacitors with similar functions of insertion, removal, and locking as the half-bridge modules, each redundant capacitor unit is connected in series with an IGBT, enabling the redundant capacitors to smoothly switch between different operating states.

As shown in Fig. 4(c), in the redundancy topology designed in this paper, the M SMs near the lower arm in the middle arm are configured using a clamping topology and voltage clamping is achieved through the generalized DNPC topology. Since these SMs are FBSMs and do not require additional PWM modulation control, the design simply replaces the capacitor portion in the DNPC topology with SMs from the middle arm.

In the fault handling mechanism, when SMs in the upper arm fail, the redundant capacitors are switched to the upper arm. When SMs in the middle arm fail, the redundant capacitors are switched to the middle arm. When SMs in the lower arm fail, the system switches to transfer normal SMs from the middle arm to the lower arm to continue operation. At this point, the middle arm loses one SM, effectively transferring the faulty SM from the lower arm to the middle arm. To stabilize the system's voltage, the redundant capacitors are switched to the middle arm, and these redundant capacitors are quickly charged to restore normal operation, thereby replacing the faulty SM in the lower arm. This design significantly improves the flexibility and fault response capability of the redundancy topology.

3.2. *Case Study*

Taking three cold standby redundant capacitors as an example, the operating principle of the redundancy topology is described. Under normal operation, the control signals for IGBT S_{R1}~S_{R6} are set to 1, while the control signals for the other IGBT devices are set to 0. As shown in Figure 5(a), when an SM in the upper arm fails, the control signal for IGBT S_{R7} changes from 0 to 1, transmitting the PWM modulation signal to S_{R1} and S_{C1}, causing capacitor C_1 to charge rapidly and restoring the system to normal operation.

As shown in Figure 5(b), when an SM in the middle arm fails, the control signal for IGBT SR_{12} changes from 0 to 1, transmitting the PWM modulation signal to S_{R6} and S_{C3}, causing capacitor C_3 to charge rapidly and restoring the system to normal operation. Other fault handling processes are similar, with fault isolation and system recovery achieved by activating the corresponding redundant capacitors.

Taking the situation where the last three SMs of the middle arm are clamped as an example, the working principle of fault transfer in the lower arm is explained. In the absence of faults in the lower arm, the control signals for IGBT S_{M1} to S_{M3} are set to 0 (off), while the control signals for IGBT S_{M4} to S_{M6} are set to 1 (on). As shown in Figure 5(c), when an SM in the lower arm fails, the control signal for S_{M3} changes from 0 to 1, and the control signal for S_{M6} changes from 1 to 0. At this point, the SM ($2N$) module is switched to the lower arm, replacing the faulty module. At the same time, a redundant capacitor is transferred to the middle arm to supplement the SM that was shifted from the middle arm to the lower arm. Through quick charging of the redundant capacitors, the system restores normal operation.

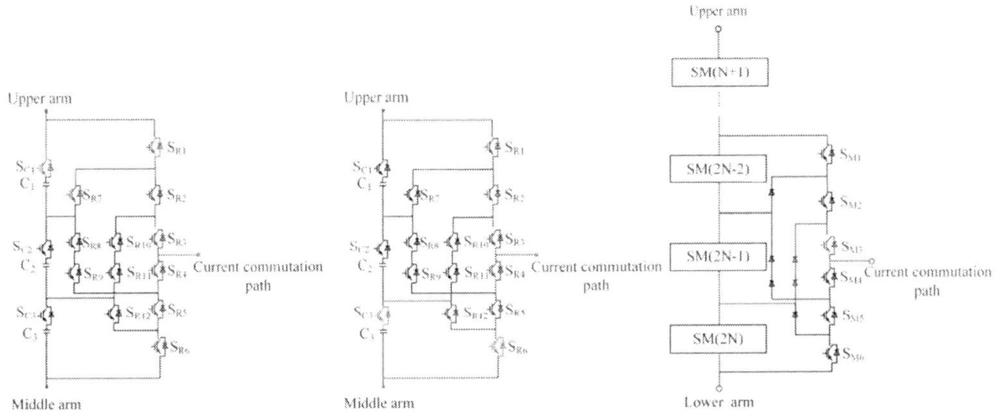

(a) (b) (c)

Figure 5. (a) A SM fault in upper arm. (b) A SM fault in middle arm. (c) A SM fault in lower arm.

3.3. CM-MMC Redundancy Sharing Control Strategy

Fig.6 illustrates the protection system's working principle. By continuously monitoring the status of each SM in real-time, the system is able to detect any faulty SMs. The redundancy protection mechanism is triggered when the number of faulty SMs is equal to or less than *M*. If the upper arm fails, the required of redundant capacitors will be switched to it. If the middle arm fails, the required number of redundant capacitors will be switched to the middle arm to enable redundancy sharing. If the lower arm fails, the non-faulted SMs will shift from the middle arm to the lower arm, and redundant capacitors will be switched to the middle arm, thereby completing the fault transfer. The redundant capacitors are rapidly charged to bring the system back to normal operation.

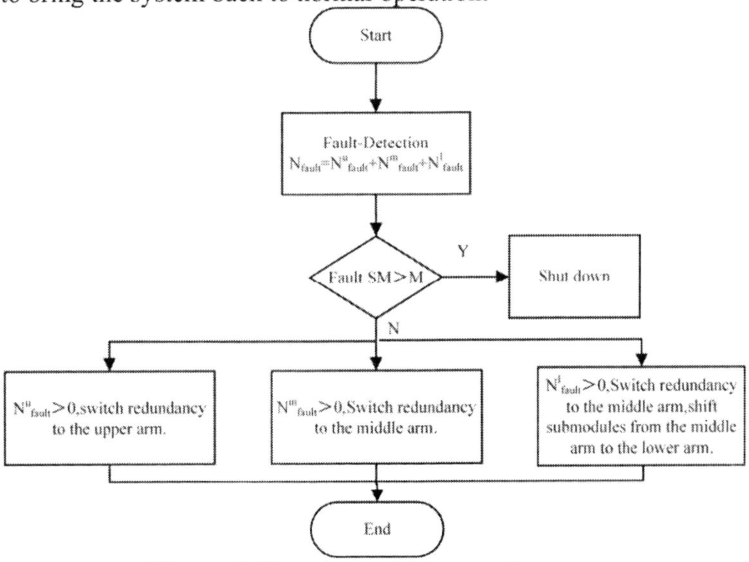

Figure 6. Flowchart of the control system.

4. Simulation Results

To evaluate the proposed topology's feasibility, a simulation model is developed and validated in MATLAB, with the simulation parameters provided in Table 1. Consider phase A as an example, where each arm in the CM-MMC has four SMs. Three redundant capacitors are placed between the upper and middle arms in phase A, and the last three middle-arm SMs use clamping topology. In this configuration, *M*=3 can handle any three SM failures in phase A.

To assess the feasibility of the redundancy sharing strategy, one upper-arm SM in phase A is bypassed at 0.4 s, 0.6 s, and 0.8 s, with a total of three SMs bypassed, as shown in Fig. 7(a) and (b). To verify the fault transfer capability of the lower arm, one SM in phase A's lower arm is bypassed at 0.4 s, 0.6 s, and 0.8 s, totaling three SMs bypassed, as shown in Fig. 7(c) and (d). The simulation shows that regardless of the faulty SM's position in phase A, the proposed topology is capable of handling up to three faulty SMs, thus confirming the feasibility of the proposed topology.

Table 1. Simulation parameters

Parameters	Values
DC voltage (V)	800
SM capacitor voltage rating (V)	100
AC line voltage rms(V)	380
SM count per arm	4
Redundant capacitor count in phase A	3
SM capacitor value (mF)	2

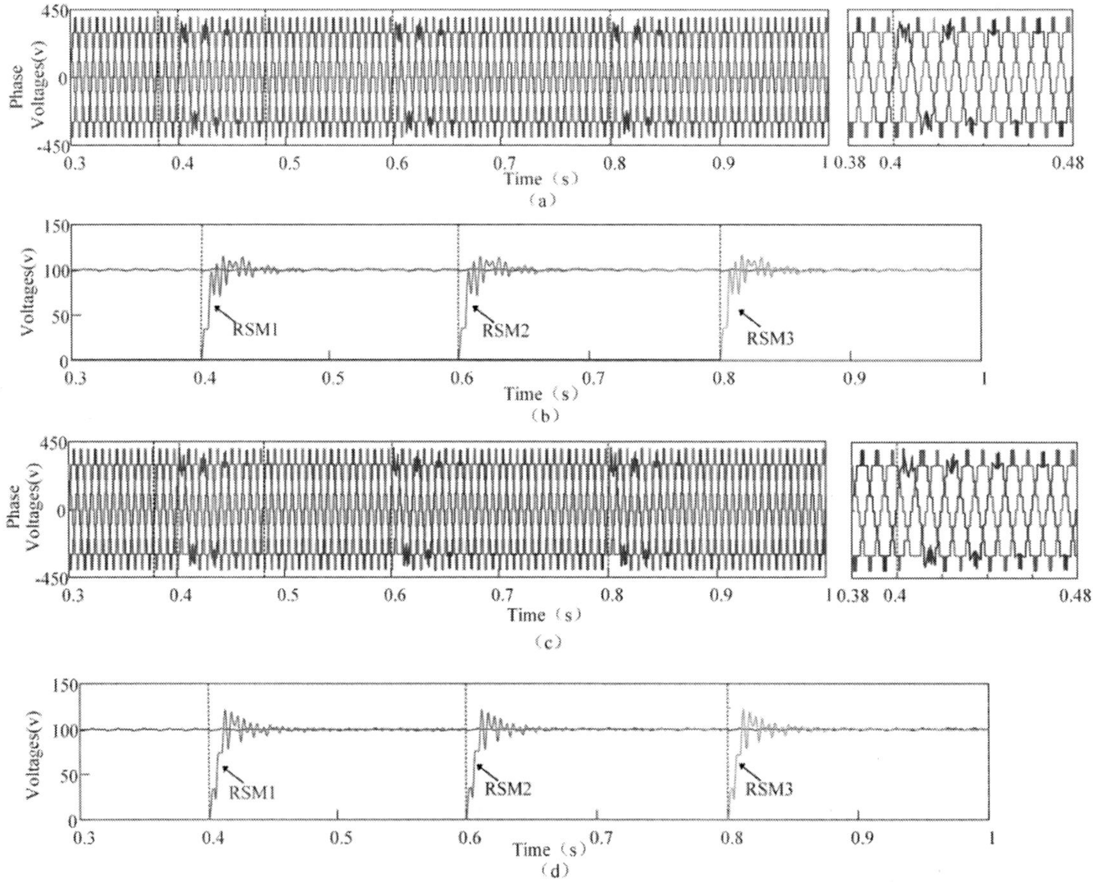

Figure 7. (a) Upper arm fault, phase voltage. (b) Upper arm fault, RSM voltage. (c) Lower arm fault, phase voltage. (d) Lower arm fault, RSM voltage.

5. Conclusion

This paper presents a novel redundancy topology that enhances the utilization of cold standby SMs. In contrast to conventional cold standby approaches, it enables redundancy sharing among the upper, middle, and lower arms, offering valuable insights for redundancy configurations in future multi-arm MMCs. The proposed topology's effectiveness is validated by simulation results.

Acknowledgments

This work is supported in part by the Henan Province Key Scientific Research Project for University and Colleges (25A470001) and the National Natural Science Foundation of China (52007171).

References

[1] Farias J V M, Cupertino A F, Pereira H A, et al. On converter fault tolerance in MMC-HVDC systems: A comprehensive survey[J]. IEEE Journal of Emerging and Selected Topics in Power Electronics, 2020, 9(6): 7459-7470.

[2] Wang C, Tao J, Wang Y. Research on topology and control of arm multiplexing modular multilevel converter[C]//Proceedings of the CSEE. 2022, 42(9): 3373-3385.

[3] He J B, Yang Q, Wang Z. On-line fault diagnosis and fault-tolerant operation of modular multilevel converters—A comprehensive review[J]. CES Transactions on Electrical Machines and Systems, 2020, 4(4): 360-372.

[4] Chen Y, Li Z, Zhao S, et al. Design and implementation of a modular multilevel converter with hierarchical redundancy ability for electric ship MVDC system[J]. IEEE Journal of Emerging and Selected Topics in Power Electronics, 2016, 5(1): 189-202.

[5] Guo J, Wang X, Liang J, et al. Reliability modeling and evaluation of MMCs under different redundancy schemes[J]. IEEE Transactions on Power Delivery, 2017, 33(5): 2087-2096.

[6] Wang J, Tang Y. A fault-tolerant operation method for medium voltage modular multilevel converters with phase-shifted carrier modulation[J]. IEEE Transactions on Power Electronics, 2019, 34(10): 9459-9470.

[7] Farzamkia S, Rizi H S, Huang A Q, et al. Shared redundancy strategy to improve the reliability and fault-tolerant capability of modular multilevel converter[J]. IEEE Transactions on Industrial Electronics, 2022, 70(4): 3326-3336.

[8] Wang K, Wu M, Zheng Z, et al. A Generalized Derivation Method of Tree-Type Active Neutral-Point-Clamped Multilevel Topologies[J]. IEEE Transactions on Power Electronics, 2024.

[9] Wang K, Zheng Z, Xu L, et al. A generalized carrier-overlapped PWM method for neutral-point-clamped multilevel converters[J]. IEEE Transactions on Power Electronics, 2020, 35(9): 9095-9106.

Journal of Physics: Conference Series **3012** (2025) 012076 doi:10.1088/1742-6596/3012/1/012076

Dynamic Characteristics of Wind Turbine in Transition Region Based on Reference Bias Control

Yanqin Huang[1], Ruibo Liu[1], Xiaokun Li[1], Haoran Wang[1], Fenglei Zhu[2], Xiangtian Deng[2*]

[1] Three Gorges intelligent control technology Co Ltd., Wuhan 430070, China

[2] School of Automation Wuhan University of Technology, Wuhan 430070, China

[*]Corresponding author's e-mail: dengxt@whut.edu.cn

Abstract. Wind turbines usually operate in corresponding intervals based on different wind speeds. When wind turbines transition from the current operating range to the next pre-operating range, the switching of control strategies has strong nonlinearity. The reference bias control (RBC) strategy is proposed to achieve decoupling control of the system. However, the internal logic and mechanism of this RBC strategy lack in-depth discussion. This paper takes a 5MW wind turbine based on the RBC strategy as an example to analyze its dynamic characteristics in transition regions. Firstly, a linearized wind turbine model in the transition region is established. Secondly, under the RBC strategy, the transition region is re-divided into three sub-intervals based on the torque and pitch controllers' features. Closed-loop transfer functions are set up for each sub-interval to analyze their dynamic trajectories. Finally, a simulation model combing MATLAB and the FAST platform is built to verify the decoupling control effect of this RBC strategy. The simulation results show that the RBC strategy can enhance the dynamic response characteristics of the wind turbine system to wind speed by adding a feedforward path in the transition region.

1. Introduction

The penetration rate of new energy units represented by wind turbines in the electric system is gradually increasing. Moreover, the capacity of individual wind turbines is gradually increasing. At present, wind turbines with variable speed and variable pitch are widely used in renewable energy stations. Conventionally, the control strategies employed by wind turbines vary at different wind speeds [1]. The transition region exists when the wind turbine unit switches from the current operating range to the next operating range. In this region, there is a coupling between the pitch and torque controller [2]. Therefore, designing corresponding decoupling control strategies can improve the dynamic characteristics of wind turbines.

Usually, the wind turbine's pitch angle is zeroed when wind speed falls below the rated reference. At this time, the speed of the wind turbine is mainly adjusted through the torque controller to achieve maximum power tracking. In another case, the torque value reaches up to rated level under the wind speed is higher than the rated reference. The operational state regulation is changed through the pitch controller. Therefore, in the transition region, these two controllers may act simultaneously [3]. The tower of a wind turbine bears the maximum aerodynamic thrust, and there is a risk of overspeed for the turbine blades. Therefore, it is necessary to design a controller to minimize tower load and control impeller speed within a threshold while meeting maximum power output.

To avoid mutual coupling between these two controllers, a switch logic has been proposed to switch the controller, allowing only one control loop to be activated during the operation of the wind turbine [4]. On this basis, an SR flip-flop is designed to achieve switching control in the transition region [5]. However, under this switch logic, the two controllers will frequently switch near the switching point. This will lead to a decrease in control performance. A more effective decoupling method, called the reference bias control (RBC) strategy, is proposed in [6]. This control strategy introduces the deviation between the torque command and the rated value into the pitch controller. It presents the deviation between the pitch angle command and the rated value into the torque controller, thereby achieving the decoupling of the two controllers.

The other researches use advanced control methods in transition regions to achieve better control performance. A generalized model predictive control strategy is proposed to achieve segmented optimal control across the entire wind speed domain [7]. An LQR control based on feedforward instructions with estimated wind speed is adopted to improve system response time [8]. Another nonlinear method using an integral sliding mode is proposed to achieve speed control in the transition region [9]. Besides, LiDAR-assisted systems are employed to provide the controller with advanced wind speed, which allows the torque controller or pitch controller to act in advance [10-11]. These control algorithms have certain optimization effects on the control performance in the transition region. However, the controller design process is cumbersome and heavily dependent on the model, and the compatibility of their algorithms needs to be improved. Therefore, based on the coupling characteristics of wind turbines in transition region control, this paper studies the dynamic characteristics of wind turbines under the RBC control strategy. Besides, the characteristics of the torque controller and pitch controller are analyzed in detail. According to its time constant, the region area is re-divided into three sub-intervals. By constructing the corresponding linearized model, the transfer function of each subinterval is derived to analyze the dynamic characteristics of the system. A simulation model of the NREL 5MW wind turbine is built to verify the dynamic characteristics of the wind turbines based on RBC control strategy.

2. Analysis and linear modelling of wind turbine

2.1. The Problem Description in the Transition Region

The transition region refers to the working area of a wind turbine between the endpoint state of the maximum power tracking control strategy and the rated operating point. In this area, wind turbines need to provide appropriate electromagnetic torque to achieve the transition of the unit state.

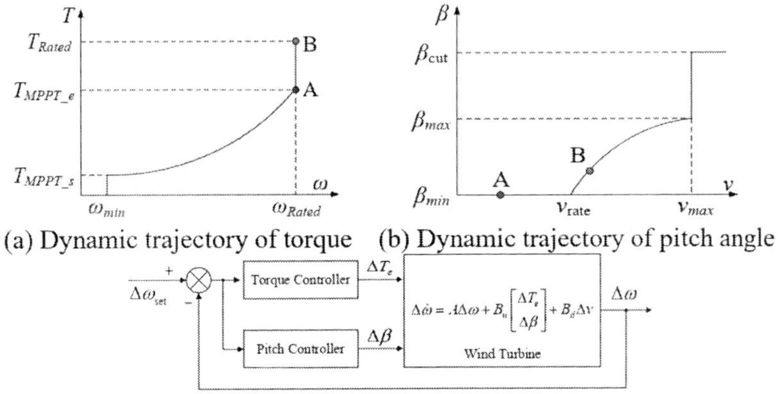

(a) Dynamic trajectory of torque (b) Dynamic trajectory of pitch angle

(c) Linearization model

Figure 1. The switching state of wind turbines in the transition region

According to the Fig. 1, the transition process can be simply described as the process of the wind turbine switching from state A to state B. The wind power generation system is a relatively complex electromechanical coupling system. This means that this system contains multiple degrees of freedom.

Besides, each subsystem will generate unique motion modes during the power generation process. They will constrain and interfere with each other, and there are many uncertain factors between them.

2.2. *The Modeling in Transition Region*

Traditionally, wind turbines provide active power to the grid by capturing wind energy. According to the operating characteristics based on the aerodynamics theory, its expression can be described as:

$$P_a = 0.5 \cdot C_{\mathrm{p}}(\lambda,\beta) \cdot \rho \cdot S \cdot v^3 \tag{1}$$

$$\lambda = \omega_r \cdot R/v \tag{2}$$

where P_a is the active power captured by wind turbines. $C_p(\lambda, \beta)$ is the utilization rate of wind energy. ρ, ω_r and v are air density, the rotation speed of the impeller, and wind speed, respectively. S is a swept area. R is the radius of the impeller.

In the transmission system, the generator side is the high-speed shaft and the impeller side is the low-speed shaft. Generally, the speed ratio between the high-speed shaft and the low-speed shaft is the gear ratio of the gearbox. Due to the flexibility of the shaft system in actual operation, its motion state can be represented as:

$$J_{\mathrm{r}}\dot{\omega}_r = T_{\mathrm{a}} - C_{\mathrm{d}}\dot{\gamma} - K\gamma \tag{3}$$

$$J_{\mathrm{g}}\dot{\omega}_{\mathrm{g}} = \left(\left(C_{\mathrm{d}}\dot{\gamma} + K\gamma\right)/N\right) - T_{\mathrm{e}} \tag{4}$$

$$\dot{\gamma} = \omega_r - \omega_g/N \tag{5}$$

where ω_g is the rotation speed of the generator. T_a and T_e are the pneumatic torque and electromagnetic torque, respectively. γ is the. C_d and K are the angle difference, damping coefficient, and stiffness coefficient, respectively.

Obviously, the variable pitch actuator is a key subsystem for adjusting wind energy utilization efficiency. Its sensitivity and accuracy have a significant impact on the output power. However, this article does not delve into the implementation principle of variable pitch actuators and instead represents them using a first-order dynamic model, as shown in (6).

$$T_{\mathrm{ac}} \cdot \dot{\beta} + \beta = \beta^* \tag{6}$$

where T_{ac} is the time constant of the pitch dynamics, β^* represents the pitch command.

Wind turbines installed at the top of the tower will generate rolling and tangential forces during operation. Then, the tower needs to withstand various loads. To accurately evaluate the load of wind turbines, this paper models a second-order damping amplitude system based on the front and rear vibration characteristics of the tower, as expressed in (7).

$$M_{\mathrm{t}}\ddot{x}_{\mathrm{t}} + C_{\mathrm{t}}\dot{x}_{\mathrm{t}} + K_{\mathrm{t}}x_{\mathrm{t}} = F_{\mathrm{areo}} + \Delta F_{\beta} \tag{7}$$

where x_t is the forward and backward displacement of the tower. F_{areo} is the thrust of wind speed. M_t is the equivalent mass between impeller and generator. C_t and K_t are damping ratio and stiffness coefficient, respectively. The calculation methods for these parameters are described in [5]. Usually, C_t is relatively small. This aspect is often overlooked in some studies. However, this article focuses on the refined modeling of wind turbines. Therefore, when designing control parameters, consider the impact of the parameter on system dynamics. Through this modeling and parameter calculation method, the dynamic response of the tower under wind force can be more accurately predicted and analyzed. The determination of system characteristics provides a design reference for the control parameters selection of the different controller.

2.3. *The Linear Modelling in Transition Region*

To facilitate the analysis of control problems in the transition region of wind turbines, a linearization tool in the FAST is adopted to linearize the system model of the wind turbine in the transition region. Combining (1) to (7), the nonlinear system model of the wind turbine is given in (8).

$$M(x_t,u,t)\ddot{x}_t + f\left(x_t,\dot{x}_t,u,u_{\mathrm{d}},t\right) = 0 \tag{8}$$

where $M(x_t,u,t)$ is the quality matrix. f is a nonlinear force function. The motion state of wind turbines in the transition region can be described as the process in which aerodynamic torque and electromagnetic torque jointly affect the motion state of the transmission chain. Because the operating range of the transition region is small, linearizing the system at the rated operating point can reflect the current working state of the system. Assuming the wind speed is rated, the generator reaches rated speed and rated torque, and pitch angle keep to zero. Besides, the damping coefficient of the rotor is ignored due to its low friction coefficient. By local linearization, the state space equation can be obtained as:

$$\Delta\dot{\omega} = -5.8e^{-2}\cdot\Delta\omega + \begin{bmatrix}-2.1e^{-5} & -0.9829\end{bmatrix}\cdot\begin{bmatrix}\Delta T_e & \Delta\beta\end{bmatrix}^T + 0.211\cdot\Delta v \tag{9}$$

$$\Delta y = \Delta\omega \tag{10}$$

where $\Delta\omega$ is the speed deviation in the low-speed shaft. ΔT_e is the torque deviation in a high-speed shaft. $\Delta\beta$ and Δv are the deviations in pitch angle and wind speed, respectively. Under the aforementioned linearization model of the wind turbine, the control structure of the transition region is shown in Fig. 1(c). The rotational speed of the blade is affected by pitch angle and electromagnetic torque. Its speed control system is an overdrive system. To track the power target in different wind speed regions, the traditional strategy is to have the torque controller dominate the speed control in the low wind speed region and the pitch controller dominate the speed control in the high wind speed region. However, in the transition region, the two controllers will be coupled together due to the speed state of the impeller. In this coupled interval, analyzing it as an overdrive system is more conducive to optimizing control performance.

3. Analysis of reference bias control strategy

When a MIMO system is decomposed into multiple SISO systems, there is a decoupling method called feedforward compensation. The basic principle is to treat the control signal of the coupling loop as a disturbance signal and add it to the feedback error of the main signal loop. By using feedforward compensation, the coupling impact of the main loop is minimized. The basic control logic of the RBC algorithm is similar to it. Fig. 2(a) gives its structure and its signal flow is given in Fig. 2(b).

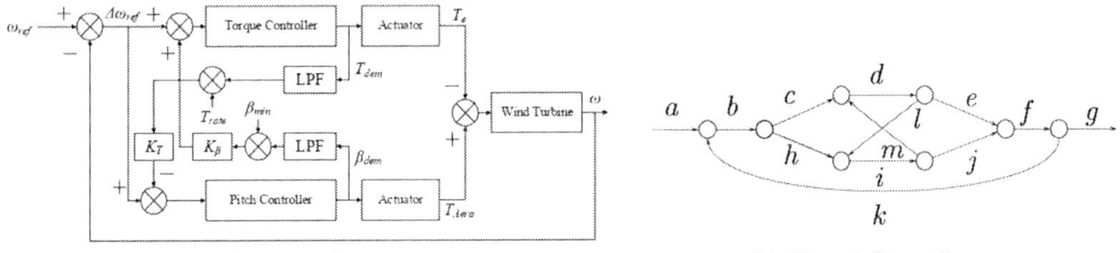

(a) Topology (b) Signal flow diagram

Figure 2. The topology and Signal flow diagram of this RBC control strategy

3.1. The Structure

This RBC control strategy is a decoupling method between controllers by introducing additional bias signals into the reference speed of the controllers. This method subtracts the torque command from the rated value, and multiplies it by a gain to generate a bias signal, and adds it to the pitch controller. Similarly, the difference between the pitch command and the rated value is multiplied by a gain to generate a bias signal, which is added to the torque controller. The bias signal of the speed will generate an additional value of the command after passing through the controller. Due to physical constraints, the instruction will maintain saturation at the limiting position. To achieve the goals of stable speed and rated power generation, the transition region can be divided into three sub-regions according to the changes in the state variables, as given in Fig. 3.

During the dominant area of torque control in the transition region, the torque controller needs to ensure that the torque command can reach the rated value. Then, the speed error caused by the command of pitch angle belongs to the disturbance term of the torque circuit. The additional

Journal of Physics: Conference Series **3012** (2025) 012076 doi:10.1088/1742-6596/3012/1/012076

compensation values should be generated by the torque controller to offset the speed error caused by the pitch angle in advance. Fig. 3(a) gives its structure. In this paper, this state is defined as Region I. At this time, the bias signal gain of torque controller is equal to the static gain from the pitch command to the impeller speed. To avoid system oscillation, K_β is set to 5.65.

Similarly, during the operation area in the pitch dominant area, the speed error caused by the torque command belongs to the disturbance term of the pitch circuit, and the pitch controller should generate additional compensation values to offset the speed error caused by the torque in advance, as given in Fig. 3(b). This state is defined as Region III. The gain of the speed bias signal of the pitch controller is equal to the static gain of the torque command to the impeller speed. K_T can be set to 1.2×10^{-4}.

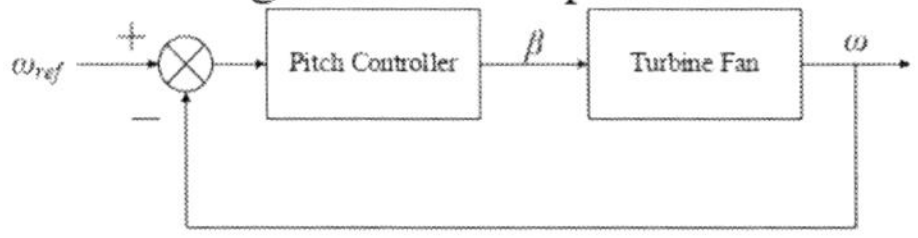

(a) Dominant by torque controller

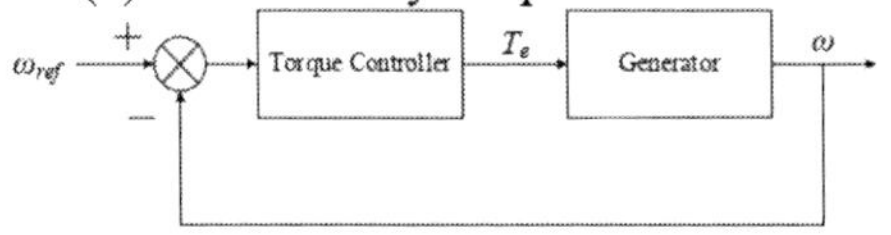

(b) Dominant by pitch controller

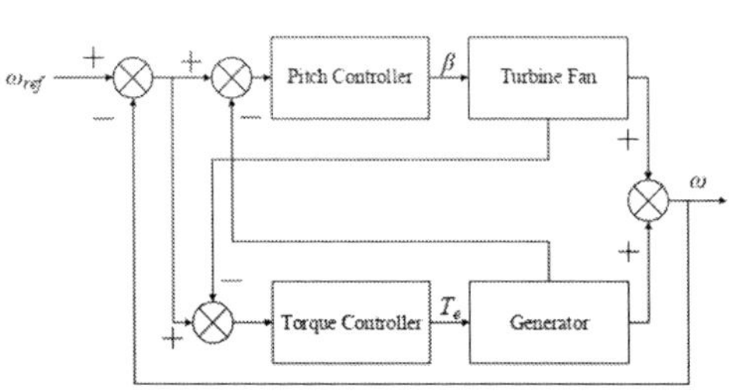

(c) Simultaneous control

Figure 3. The structures of three operating states in transition region

In the remaining areas, the pitch controller and torque controller need to work simultaneously, as given in Fig. 3(c). This state is defined as Region II. In this subinterval, the control strategy design for these two controllers is the most complex. Because both controllers are not in a saturated state. The speed feedback controller is the result of the joint action of these two controllers. The speed error of both controllers contains information about control commands.

3.2. Mechanism analysis

According to Fig. 2(b), the proposed control system contains 4 forward paths and 13 branches. The Table I gives the gains of each branch of the proposed wind turbine system.

Table 1. The gain value of each branch

Branch Number	Gain	Branch Number	Gain	Branch Number	Gain
a	1	e	G_1	i	C_2
b	1	f	1	j	G_2
c	1	g	1	k	-1
d	C_1	h	1	l	$-K_2$
m	$-K_1$	-	-	-	-

Based on the Mason gain formula, the transfer function of the wind turbine system in each subinterval can be obtained by (11) ~ (13).

$$\Phi_I(s) = C(s)/R(s) = (C_1G_1)/(1+C_1G_1) \tag{11}$$

$$\Phi_{II}(s) = C(s)/R(s) = \sum_{k=1}^{m} P_k\Delta_k \Big/ \Delta = \sum_{k=1}^{4} P_k\Delta_k \Big/ 1-\sum_{i=1}^{4} L_i \tag{12}$$

$$\Phi_{III}(s) = (C(s)/R(s)) = C_2G_2/(1+C_2G_2) \tag{13}$$

where Φ_I, Φ_{II}, and Φ_{III} are the closed-loop gain in each subinterval. Then, according to the actual parameters of the NREL 5MW wind turbine, the dynamic response curve of the speed closed-loop can be obtained in Fig. 4 when considering the dynamics of the variable pitch actuator.

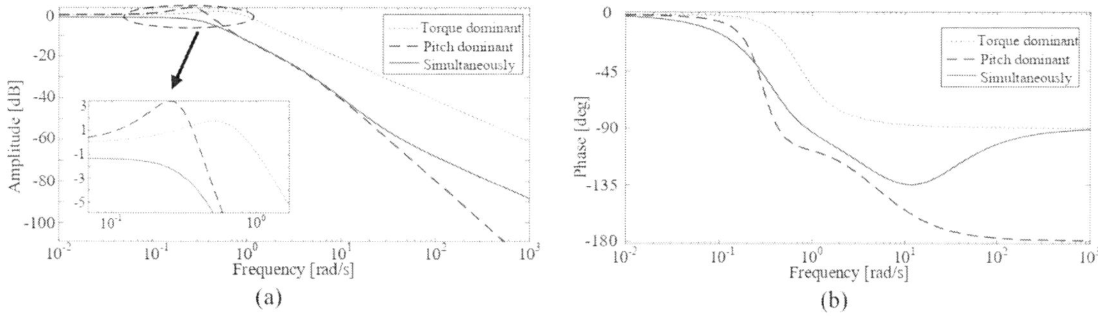

Figure 4. The topology of the typical direct-driver wind turbine and its aerodynamic characteristics. (a) Topology. (b) aerodynamic characteristics.

According to the Bode curve in Fig. 4(b), when the wind turbine enters the transition operating region from low wind speed, it first passes through the area dominated by the torque controller. At this time, if the speed is too high, it will generate a pitch angle signal and enter the area where the two controllers work together. Then, the wind turbine completely enters the area dominated by the pitch controller when the torque approaches saturation. This process can complete the entire decoupling control. It should be noted that when two control instructions act simultaneously, the closed-loop bandwidth of the proposed wind turbine system is 0.336 rad/s. This value is close to the closed-loop bandwidth of the pitch control loop. Therefore, although the RBC algorithm can achieve decoupling control for different controller, the system dynamics are still slow and need to be improved. This feature will be validated in the next section.

4. Simulation verification

An NREL-5MW model of the wind turbine is built to analyze the effectiveness of the partition control method based on the RBC control strategy. Besides, a simulation model is constructed by MATLAB/Simulink and FAST platforms. Table II shows the main parameters of the selected wind turbine.

Table 2. The parameters of the proposed wind turbine

Variable	Description	Parameters
S_N	Capacity Reference	5 MVA
ω_r	Rated Rotor Speed	12.1 rpm
v	Rated Wind Speed	11.4 m/s
N	Ratio of Gearbox	97
R	Blade Radius	63 m
H	Tower Height	87.6 m

4.1. The process of increasing wind speed

Fig. 5(a) shows the dynamic trajectory of the wind speed in the case. Before reaching the rated wind speed, the torque controller dominates the operating state of the wind turbine. Under this control mode, when the change rate in blade speed exceeds the threshold, the pitch angle will open before the electromagnetic torque reaches the rated value, as shown in Fig. 5(c) and Fig. 5(d). At the same time, the decoupling loop of the RBC algorithm will make the electromagnetic torque reach the rated value faster to suppress wind turbine overspeed and reduce the force on the tower frame. On the other hand, due to the large bandwidth of the torque control loop, the torque command value rises to a position close to the rated state in the early stage. At this time, the pitch loop can be opened in a short time and the duration of Region II is relatively short, about 0.2 seconds. For the pitch control loop, the change rate of pitch speed is slow and the rotational speed of generator cannot be effectively suppressed. Considering the bandwidth limitation of pitch control loop, in the early stage of overspeed, the speed error signal is small and the equivalent gain of the pitch angle command decreases, resulting in a smaller pitch angle command. Moreover, the wind turbine system is greatly affected by parameters near the rated operating point. The above two reasons lead to a significant rotor overspeed problem.

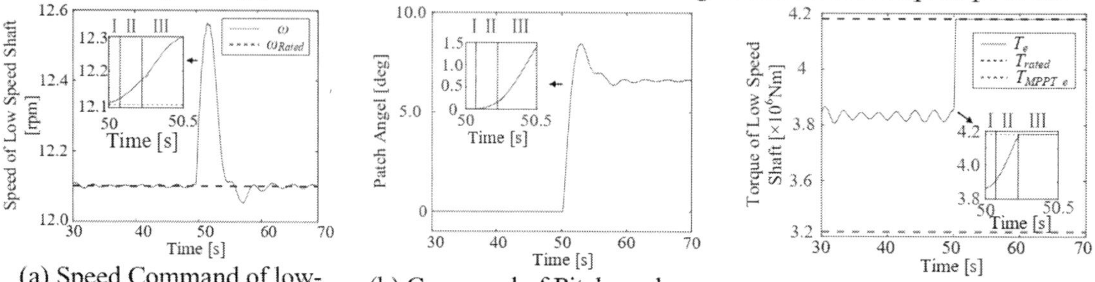

(a) Speed Command of low-speed shaft (b) Command of Pitch angle (c) Torque Command of low-speed shaft

Figure 5. The dynamic trajectories of torque command and propeller angle command during the process of wind speed increase.

4.2. *The process of decreasing wind speed*

The other set of wind conditions is used to verify the dynamic process of the wind turbine system. Fig. 6(a) gives the wind speed. According to the simulation results given in Fig. 6(c) and Fig. 6(d), the electromagnetic torque will act before the pitch angle reaches its minimum value during this process. Besides, the decoupling circuit will make the pitch angle reach its minimum value faster.

As previously analyzed, the bandwidth of pitch control loop is small, which causes the adjustment effect on the speed is slow and results in a faster drop in speed. The torque loop participates in decoupling control earlier, and at this time, the pitch angle command changes slowly, and the speed drop speed begins to slow down. The duration of region II is relatively long, about 2.3 seconds. When the propeller angle drops to 0°, the torque loop recovers the control gain and the decoupling algorithm proceeds to region I. Due to the slow variation of pitch commands, the time for torque control to fully dominate speed control is delayed, resulting in a more significant decrease in speed. In summary, according to the simulation results, when using the RBC algorithm in the transition region, the wind turbine switches between the three working states I, II, and III mentioned above. The control effect of impeller speed varies in three working states, and the controllers that take effect in different states are different, so the control effect of speed is different. The RBC algorithm effectively solves the coupling problem between controllers during the process of increasing and decreasing wind speed. When the speed fluctuation does not exceed the speed setting offset range, the impeller speed is dominated by one of the controllers. When the speed fluctuation exceeds the deviation range of the command, both controllers will participate in the speed regulation simultaneously.

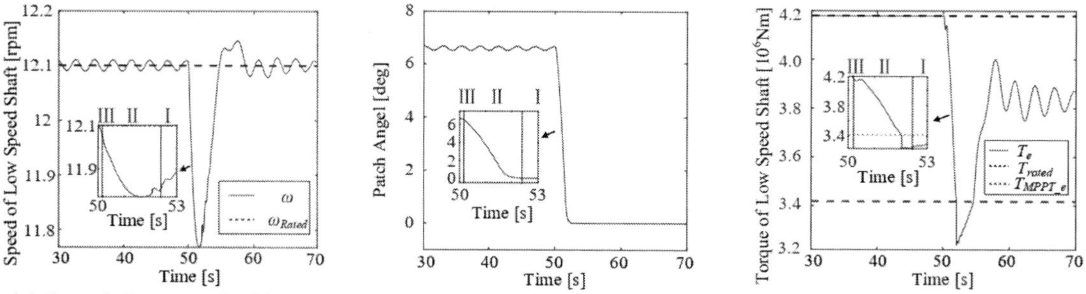

(a) Speed Command of low-speed shaft (b) Command of Pitch angle (c) Torque Command of low-speed shaft

Figure 6. The dynamic trajectories of torque command and propeller angle command during decreasing wind speed.

5. Conclusion

In response to the coupling problem between different controllers in the transition region of wind turbines, this paper establishes a linearized model for wind turbines in the transition region using the RBC control strategy. According to the operating characteristics of the torque controller and pitch controller, the region is divided into three parts, and their respective closed-loop transfer functions are derived. The amplitude and phase characteristic curves indicate that the effects of each controller are different in different regions. Finally, based on the simulation results under different wind speed step conditions, the conclusion drawn in this article can provide the parameter selection range for each control loop and analyze the rationality of setting each subinterval.

Acknowledgments

This work is supported by Hubei Province Key Research and Development Plan Project (No. 2023DJC201) and Three Gorges intelligent control technology Co., LTD. Research Foundation (Grant No. TGIC-KY-2024-04-00).

References

[1] Wu. B, Lang. Y, Zargari. N. Power conversion and control of wind energy systems. 1st ed. Wiley-IEEE Presss, 2011.
[2] S. Xiao, G. Yang and H. Geng, "Nonlinear pitch control design for load reduction on wind turbines," 2014 International Power Electronics Conference (IPEC-Hiroshima 2014 - ECCE ASIA), Hiroshima, Japan, 2014, pp. 543-547.
[3] A. González-González, A. J. Cortadi, D. Galar, "Condition monitoring of wind turbine pitch controller: A maintenance approach", Measurement, Jul. 2018, vol. 123, pp. 80-93.
[4] Bossanyi, Ervin A. "The Design of closed loop controllers for wind turbines." Wind Energy, 2000, vol. 3, pp. 149-163.
[5] Nam. Y, Kim. J, Paek. I, Moon. Y, Kim. S, and Kim. D. "Feedforward Pitch Control Using Wind Speed Estimation," Journal of Power Electronics, 2011, vol. 11, pp. 211-217.
[6] E. C. Navarrete, M. Trejo Perea, J. C. Jáuregui Correa, R. V. Carrillo Serrano and G. J. R. Moreno, "Expert Control Systems Implemented in a Pitch Control of Wind Turbine: A Review," IEEE Access, 2019, vol. 7, pp. 13241-13259.
[7] Jeon. T, Kim. et, al. I. "Improvements to and Experimental Validation of PI Controllers Using a Reference Bias Control Algorithm for Wind Turbines", Energies, 2022, vol.15, pp. 8298.
[8] Kim K, Kim H-G, Song Y, Paek I, "Design and Simulation of an LQR-PI Control Algorithm for Medium Wind Turbine," Energies. 2019, vol. 12, No.12, pp. 2248.
[9] Saravanakumar, R. and Debashisha Jena. "Validation of an integral sliding mode control for optimal control of a three blade variable speed variable pitch wind turbine," International Journal of Electrical Power & Energy Systems, 2015, vol. 69, pp. 421-429.
[10] Y. Huang, G. Zhu, X. Deng, J. Lu, C. Jia and H. Geng, "A LiDAR-Assisted Reference Feedforward Bias Control Algorithm for Wind Turbines in the Transition Region," 2023 IEEE 2nd International Power Electronics and Application Symposium (PEAS), Guangzhou, China, 2023, pp. 2132-2137.
[11] R. Cai et al., "A Feedforward Pitch Control Strategy Using LiDAR Measurement Considering Wind Evolution Under Extreme Gust Conditions," 2023 IEEE 2nd International Power Electronics and Application Symposium (PEAS), Guangzhou, China, 2023, pp. 2128-2131.
[12] Y. Huang, L. Wang, S. Zhang, D. Wang, X. Deng and G. Zhu, "Impacts of Phase-Locked Loop Dynamic on the Stability of DC-Link Voltage Control in Voltage Source Converter Integrated to Weak Grid," IEEE Journal on Emerging and Selected Topics in Circuits and Systems, vol. 12, no. 1, pp. 48-58, March 2022.
[13] Y. Huang, X. Yuan, J. Hu and P. Zhou, "Modeling of VSC Connected to Weak Grid for Stability Analysis of DC-Link Voltage Control," IEEE Journal of Emerging and Selected Topics in Power Electronics, vol. 3, no. 4, pp. 1193-1204, Dec. 2015.

Journal of Physics: Conference Series **3012** (2025) 012077 doi:10.1088/1742-6596/3012/1/012077

A Series/Parallel Full-bridge MMC with Submodule Capacitor Voltage Self-balancing Capability

Wenjun Liu[1,2*], Genxu Chen[1,2], Yimiao Wang[3], Bicheng Xu[1,2], Yan Chen[1,2]

[1] School of Electrical and Information Engineering, Zhengzhou University, Zhengzhou, China

[2] Henan Engineering Research Center of Power Electronics and Energy Systems, Zhengzhou, China

[3] Guangdong power grid limited liability company Zhaoqing power supply bureau, Zhaoqing, China

[*]Corresponding author's e-mail: lwenjun1990@zzu.edu.cn

Abstract. Modular Multilevel Converters (MMC) are the key equipment in High Voltage Direct Current (HVDC) systems, but they still face challenges such as DC fault clearance and excessive computation for submodule capacitor voltage sequencing. This paper presents an improved full-bridge MMC (MF-MMC) with submodule capacitor voltage self-balancing and DC fault clearance capability. By using two reverse-series IGBTs, adjacent full-bridge submodules are reconfigured from series to parallel, establishing a parallel path between adjacent capacitors. The MF-MMC enables voltage self-balancing between capacitors of different submodules within the same arm without the need for monitoring the capacitor voltage, significantly reducing the complexity and computational load of the controller. At the same time, the parallel connection of capacitors reduces the output of individual capacitors, thereby decreasing voltage fluctuations. A dynamic distribution voltage balancing control strategy is proposed to optimize the switching of power devices, reducing overall switching losses. Finally, MATLAB experimental results validate the effectiveness of this topology in terms of submodule parallel capabilities and DC fault clearance.

1. Introduction

The Modular Multilevel Converter (MMC) is widely used in High-voltage DC (HVDC) transmission systems due to its low harmonic distortion , modular configuration and flexible control.[1] Since clearing DC faults is a significant challenge for HVDC systems, and DC circuit breaker solutions are costly and bulky, using MMCs for fault interruption has become an attractive option.[2] The MMC based on Full Bridge Submodules (FBSMs) has garnered considerable attention due to its ability to clear DC faults.

One of the greatest challenges for MMC is submodule capacitor balance.[3] The current methods for balancing submodule voltages can be classified into two types. One is to improve the control method of submodule capacitor voltage. One is to improve the sorting algorithm of submodule capacitor voltage. In [4], an improved cyclic modulation method is proposed, where the generalized cyclic matrix is introduced to derive the optimal switching matrix for cyclic modulation. This approach minimizes the voltage fluctuations of the submodules to the greatest extent. In [5], a low-capacitance operation and its layered control strategy are proposed. The strategy features a model predictive control with a single objective, along with an enhanced algorithm for sorting the capacitor voltages.

In [6], the integration of an LC resonant branch into the ISOS DAB DC-DC converter is employed to achieve voltage balance on the input side, while the control strategy adjusts the phase shift ratio to eliminate voltage imbalance on the output side, resulting in voltage balance at both the input and output of the converter. In [7], a novel MMC with three active switches (SHB) is proposed. By paralleling capacitors between the submodules, the hardware and control complexity, as well as system cost, are reduced, achieving balanced capacitor voltages. In [8], by connecting the midpoints of the input and output of the three-level converter, common-mode (CM) noise is reduced, enabling the converter to achieve ideal automatic balancing of the DC link capacitor voltages.

In this paper, a modified full bridge submodule (MF-MMC) is proposed by improving the FBSM, and an improved dynamic distribution voltage equalizing strategy is proposed accordingly. It does not need to monitor the capacitor voltage of submodules to achieve self-balancing capacitor voltage, and has the ability of DC fault clearing. Compared with the voltage equalization strategy proposed in previous literatures, the switching of the power switches is reduced, resulting in less switching loss.

2. MF-MMC Topology and operation modes

The improved three-phase full-bridge MMC (MF-MMC) consists of two arms per phase, with each arm composed of N submodules in series with the arm inductance. In MF-MMC, a branch formed by two IGBTs connected in reverse series is inserted to provide a parallel path for directly connecting the adjacent submodule capacitors. The improved three-phase MMC topology are shown in Figure 1. In the figure, U_{dc} represents the DC voltage, and U_{va}, U_{vb}, and U_{vc} are the three-phase AC voltages. L_0 is the arm inductance, i_{pa} and i_{na} represent the currents of the upper and lower arms of phase a, respectively.

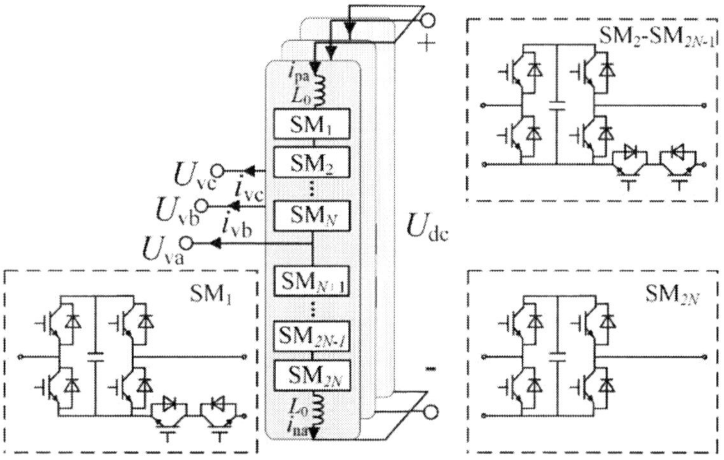

Figure 1. MF-MMC topology.

When adjacent submodules are connected in parallel, the parallel branch is activated, and the parallel submodules are defined as a submodule group. The capacitors within the same group are connected in parallel to collectively provide the required output voltage level. As a result, the current through each capacitor is reduced, significantly decreasing voltage fluctuations compared to simply connecting submodule capacitors in series. Moreover, by ensuring that no submodule capacitors remain idle, the utilization of the capacitors is improved. After the capacitors are paralleled, voltage self-balancing is achieved. The submodule group, as a whole, can operate in parallel, positive input, or reverse input states, as shown in Figures 2 (a), (b), and (c). The overall state depends on the first and last submodules in the group, with the corresponding switching state shown in Table 1. Compared to the traditional MMC, the MF-MMC introduces a unique parallel state. Since this parallel state requires a pair of reverse-series IGBTs for control, it can be considered that the control degree of freedom shifts from individual submodules to the switching set of the two reverse-series IGBTs between adjacent submodules.

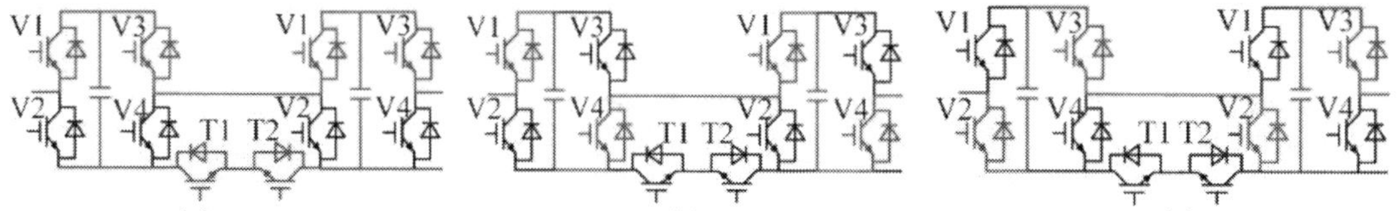

Figure 2. Current flowing path of a submodule group in different states.

Table 1. Switch state sets of MF-MMC

Mode	V1	V2	V3	V4	T1	T2
Parallel	1	0	1	0	1	1
Positive input	1	0	0	1	0	0
Reverse input	0	1	1	0	0	0

3. Modulation method of MF-MMC

3.1. Switching function of MF-MMC bridge arm

The switching function of the submodules are set to be S, which can be expressed as:

$$S = \begin{cases} 0, \text{Parallel} \\ 1, \text{Positive input} \\ -1, \text{Reverse input} \end{cases} \tag{1}$$

As can be seen, the modification of the converter topology does not introduce additional switching actions to the submodule. Assume that the number of submodules in a submodule group is M. Regardless of the state of the overall group, the submodules of the group are all in parallel state, excluding the first and last one. Then depending on the state of the submodule group, the states of the first and last submodules are different. When the submodule group is in parallel state, positive input state, and reverse input state, the switching functions of the submodules in the submodule group is as follows:

$$\begin{cases} S_1 = 0,\ S_2 = S_{M-1} = \cdots = 0,\ S_M = 0, \text{Parallel} \\ S_1 = 0,\ S_2 = S_{M-1} = \cdots = 0,\ S_M = 1, \text{Positive input} \\ S_1 = -1,\ S_2 = S_{M-1} = \cdots = 0,\ S_M = 0, \text{Reverse input} \end{cases} \tag{2}$$

Taking one bridge arm of a certain phase for example, the current of the bridge arm is evenly distributed in each capacitor of the same submodule group, assuming that the number of submodules in group i is M_i (i=1, ..., n), the relationship between the arm current and the capacitor voltage in the submodule can be represented as:

$$i_{\text{arm}} = M_i C \frac{du_{\text{Cap},i}}{dt}, i = 1, \ldots, n \tag{3}$$

where i_{arm} is the current of the bridge arm, C is the capacitance value, and $u_{\text{Cap},i}$ is the capacitor voltage.

As can be seen from the above equation, the current flowing through the capacitor of each input submodule is $1/M_i$ times the arm current, which reduces the capacitor voltage fluctuation and improves the voltage balancing performance.

3.2. Modulation strategy and voltage equalization control

Compared with the traditional MMC, MF-MMC only make changes in the submodule level topology, and the modulation strategy of the traditional MMC is still suitable for MF-MMC. In order to reduce switching losses, nearest level approximation modulation (NLM) is generally used in the project for high voltage high power flexible HVDC transmission. In this paper, NLM is used as the modulation strategy to design the corresponding voltage equalization control and submodule conduction method.

Set the level number instruction as n. When $0<n<N$, in order to avoid excessive charge/discharge caused by single or several submodules in parallel, submodules should be grouped so that each group should have as many submodules as possible. Also, in order to improve the voltage equalization effect, the number of submodules contained in each group should be as equal as possible.

Based on the idea of averaging, this paper designs a dynamic distribution voltage equalization control strategy, and the flow chart is shown in Figure 3.

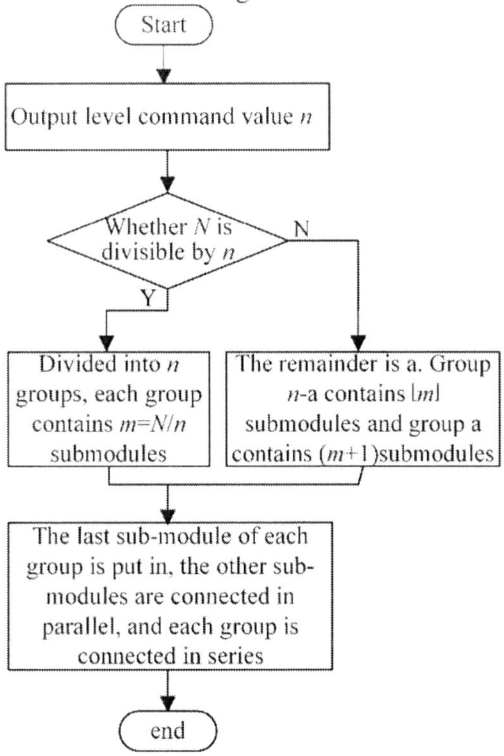

Figure 3. Flow chart of dynamic distribution-voltage balance control.

4. DC fault clearance

MF-MMC has the same external characteristics as FBSM-MMC except that the current circuit branches when the submodules are in parallel. Therefore, the topological characteristics at the submodule level determine that MF-MMC has DC fault traversal capability.

In the closed state, all the switching tubes are turned off, regardless of the current direction, the submodule capacitors are in the charging state, and the fault current rapidly decreases with the increase of the capacitor voltage in the loop, and eventually decays to zero.

When the DC bipolar short-circuit fault occurs, the submodule is blocked, and the AC system feeds energy to the DC side through the diode. Taking the highest A phase voltage and the lowest C phase voltage after locking as an example, the fault current flows through the upper A phase bridge arm and the lower C phase bridge arm, the process is shown in Figure 4, and the loop voltage equation can be expressed as:

$$u_{AC} = 2N(U_d + U_c) + 2u_L \tag{4}$$

where u_{AC} is the instantaneous value of MMC AC two-phase line voltage measurement; U_d diode voltage sum; u_L is the impedance voltage of the bridge arm; U_C indicates the average capacitance voltage of a submodule.

During normal operation, the adjustment system is set to 1, and the relationship between capacitor voltage and AC/DC voltage of the submodule can be expressed as:

$$\begin{cases} U_{dc} = 2NU_c \\ U_{ac} = \dfrac{1}{2}U_{dc} \end{cases} \tag{5}$$

where U_{dc} indicates the MMC DC voltage, and U_{ac} indicates the peak AC voltage.

From equations (4) and (5), it can be obtained:

$$U_d = \frac{1}{2N}(u_{AC} - U_{dc} - 2u_L) \tag{6}$$

The analysis of equation (6) shows that $U_d<0$, the diode in the submodule withstands the reverse voltage and shuts off, the fault current gradually decreases to zero, and energy from the AC side is no longer transferred to the DC side; the DC fault current is cleared.

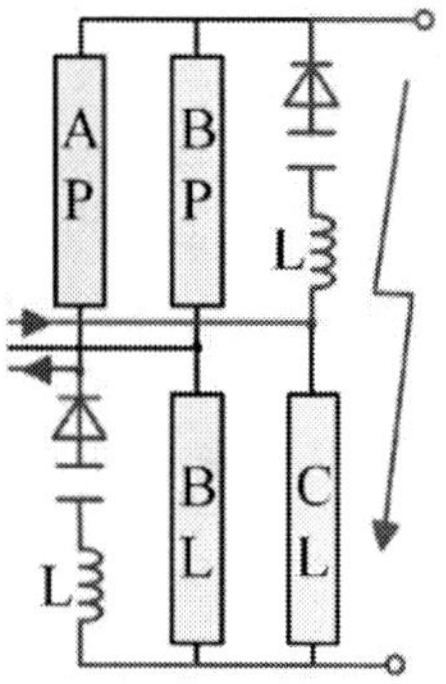

Figure 4. Current circuit of MF-MMC in fault current clearing stage.

5. Verifications

To verify the self-voltage balancing and DC fault self-clearing capabilities of the proposed MF-MMC, an 11-level MF-MMC is built using MATLAB. The system parameters are shown in Table 2.

Table 2. Parameters of MF-MMC system

Parameter	Quantitative value
Dc side voltage /kV	40
Submodule capacitance rated voltage/kV	4
Submodule capacitance /mF	6
Converter bridge arm inductance /mH	1
Number of submodules of the bridge arm	10

In normal operation, the output voltage waveform of the AC side is an 11-level step wave, as shown in Figure 5 (a). The current waveform on the AC side, which demonstrates good waveform quality in Figure 5 (b). Figure 5 (c) shows the capacitor voltage diagram of submodules in the MF-MMC one-phase upper bridge arm. As can be seen from Figure 5 (c), the capacitor voltage of the submodule fluctuates up and down at 4 kV, and the capacitor is effectively balanced to its rated value, with a ripple of about 0.375%. The self-voltage balancing characteristic of MF-MMC is verified, there is no need for submodule capacitor voltage monitoring under dynamic distribution voltage balancing control, which effectively reduces the computational complexity of the controller and the real-time requirements of the sensor, and can achieve good voltage balancing effect and save the secondary cost caused by voltage balancing.

A bipolar short circuit on the DC side was introduced at t = 0.5s, After 2 milliseconds, all IGBTs were deactivated. The simulation results of the MF-MMC system during the fault are shown in Figure 6. As seen in Figures 6 (a) and 6 (b), after a short circuit fault occurs on the DC side, the current direction between the fault point and the AC side changes rapidly, the submodule capacitors discharge, and energy from the AC side is fed back to the DC side. The DC side current increases sharply, and the submodule capacitor voltage drops rapidly. As shown in Figure 6 (c), when t = 0.502s, all IGBTs are turned off, and the three-phase AC current drops to zero. The submodule capacitors are put into reversely through the diodes, and the energy stored in the inductors change the capacitors. Due to the clamping effect of the diodes and the equal voltage characteristic of the MF-MMC, the submodule capacitor voltage remains constant, and the fault current rapidly drops to zero. This validates that the submodule has the capability to interrupt DC fault currents.

After the DC fault is eliminated, the converter is turned back on at t= 0.7s. From Figure 6 (b), (c), during fault recovery, the current on the AC side quickly recovers after a short oscillation. The capacitive voltage of the submodule returns to normal operation after 0.15s, and the capacitive voltage of the MF-MMC remains balanced throughout the recovery process, which verified the applicability of the proposed dynamic distribution voltage equalization method in the transient process.

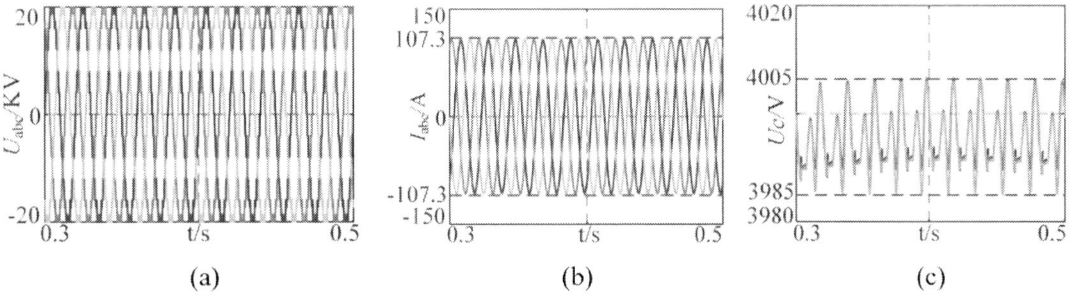

Figure 5. Waveform of MF-MMC during normal operation.

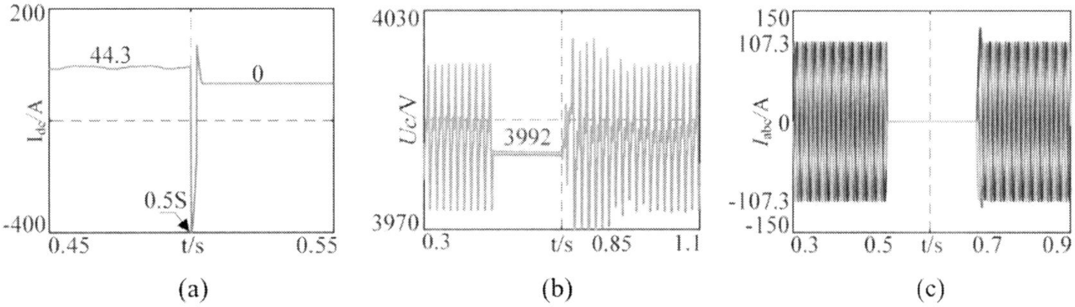

Figure 6. Waveform of MF-MMC during DC fault.

6. Conclusion

Building on the traditional MMC (FBSM), this paper presents an MF-MMC topology that incorporates self-clearing of DC faults and self-balancing of capacitor voltages. This solution effectively resolves the problems of DC faults and voltage imbalance in capacitors within high-voltage DC systems. By adding a small number of switching devices, the parallel branches are added compared with FBSM, and the dynamic distribution voltage equalization strategy is designed according to its parallel operation mode. Compared with MMC topology which adopts sequential voltage balancing control or modified topological circuit, it can ensure good voltage balancing effect of submodule capacitance and improve system reliability on the basis of no need to monitor and sort capacitor voltage information and simple topological modification without causing complex circuits.

Acknowledgments

This work is supported in part by Henan Province Key Scientific Research Project for University and Colleges (25A470001) and National Natural Science Foundation of China (52007171).

References

[1] Zhu Y, Jiang S, Pou J, et al. Virtual arm impedance emulation and stability improvement in modular multilevel converters[J]. IEEE Transactions on Power Electronics, 2024.

[2] Fang X, Li G, Wang C, et al. A reverse-biased voltage controlling method for mitigating arm overcurrent and submodule overvoltage in hybrid MMCs during DC faults[J]. IEEE Transactions on Power Electronics, 2023.

[3] Zheng T, Wang K, Zheng Z, et al. A Novel Sensorless Capacitor Voltage Balancing Method for the Modular Multilevel DC Transformer with Diode-Clamped Circuits[J]. IEEE Transactions on Power Electronics, 2024.

[4] Fan S, Xiang X, Gu Y, et al. Optimal Circulant Modulation for Submodule Voltage Ripple Minimization With Inherent Balancing Capability in Modular Multilevel DC-DC Converters[J]. IEEE Transactions on Power Electronics, 2023.

[5] Liu S, Niu D, Zhang L, et al. Low-Capacitance Operation and Its Hierarchical Control for STATCOM Based on T-type Converter with Modular Multilevel DC-Link[J]. IEEE Transactions on Power Electronics, 2024.

[6] Liu Y, Hu H, Wang X, et al. Voltage Balance Scheme for Input-Series–Output-Series DAB DC–DC Converter With Bidirectional Power Flow[J]. IEEE Transactions on Power Electronics, 2024.

[7] Liu C, Goetz S M, Fang J. Modular Multilevel Converters With Three-Active-Switch Symmetrical-Half-Bridge Submodules and Parallel Connectivity[J]. IEEE Transactions on Power Electronics, 2024.

[8] Cao Y, Fan B, Bai Y, et al. Common-mode noise reduction and capacitor voltage auto-balance using bridged midpoints and coupled inductor in a 3-L buck-boost converter[J]. IEEE Transactions on Power Electronics, 2023.

Two-stage Synergistic Autonomous Optimization of Interconnected Multi-energy Systems based on Interaction Mechanisms

Jin Yu[0009-0000-8588-8587], **Hanhua Cao**[0000-0003-1554-0993], **Yilan Luo***[0009-0003-6211-3481]

School of Information and Intelligent Engineering, Guangzhou Xinhua University, Suzhou, Guangdong Province, 523142, P.R.China

*Corresponding author's e-mail: 1010093796@qq.com

Abstract. The rapid development of the energy industry has made the complexity of managing Multi-Energy Systems (MES) a key challenge. Optimizing MES operations is central to enhancing energy efficiency and sustainability. This paper proposes a framework based on interactive mechanisms to achieve collaborative autonomous optimization of MES. The framework allows MES to autonomously optimize while exchanging signals with a coordinator. Through numerical simulations, we minimize operational costs and effectively manage information privacy and operational authority. The results show that the framework significantly improves the coordination of MES operations, promotes the consumption of renewable energy, and reduces dependence on the main power grid. The CAFIL model encourages local consumption of renewable energy, alleviates grid congestion, complements the power demand of the main grid, and reduces load fluctuations. As the load increases, energy efficiency decreases. This framework is applicable for assessing the evolution of MES operations.

1. Introduction

In the transformation expansion of the energy industry, the integration of multiple energy systems (MES) has become the key to sustainable and efficient energy management.MES improves energy efficiency and reduces environmental impacts through the synergistic operation of multiple energy carriers. However, their complexity and inter-dependencies pose challenges, especially in system response prediction and optimization. Design theory and engineering experience are lagging behind in the management of operating costs, information privacy, and operating privileges of interconnected MES.

There have been numerous studies on the management and optimization of MES. For instance, HU Bo et al. [1] proposed an optimal dispatch method for integrated energy systems based on the flexibility of thermal load, which provided valuable insights but did not fully account for the complex interactions among MESs. QIU Di et al.[2]developed a multi-scenario modeling and day-ahead optimization approach for park multiple energy systems based on discrete hybrid automaton, yet their work was less focused on the co-optimization of multiple energy devices. BAHRAMARA S et al.[3]modeled the microgrid operator's participation in day-ahead energy and reserve markets, considering stochastic decisions in the real-time market, Zhang et al.[4] achieved low-carbon scheduling through Stackelberg game theory. Wang et al.[5] proposed a leader-follower game-driven multi-time-scale optimization of energy hubs. Li et al. [6]enhanced the collaborative efficiency of microgrids based on virtual energy storage. Furthermore, Zhang et al.[7] integrated carbon quotas with multi-agent game theory, thereby expanding the dimensions of system optimization.Li et al. [7]proposed a Stackelberg game-based optimal scheduling framework for multi-micro grid systems, explicitly incorporating photovoltaic generation accommodation and integrated demand response to achieve enhanced operational coordination and energy efficiency,but their research did not adequately address the issues of MES information privacy and operation privileges. While these studies have contributed to the development of MES, they have not yet achieved the MES co-autonomous optimization that is essential for the efficient operation of modern energy systems.

A hierarchical two-tier optimization framework is established for interconnected multi-energy communities, wherein experimental investigations of system dynamics are synergistic-ally combined with market incentive analyses to elucidate the inter dependencies among energy conversion assets and real-time pricing signals.

Numerical models are developed to consider real-time price signals, renewable energy variability and operational constraints effects. Numerical simulations based on the model are conducted to investigate the efficiency and coordination of MES operations under different market conditions. The two-phase framework is found to significantly improve MES operational coordination, promote renewable energy consumption, and reduce dependence on the main grid.The CAFIL model relieves grid congestion, supplements main grid power demand, and mitigates load volatility. The results emphasis the importance of a synergistic approach in optimal MES operations, showing the potential of the framework to cope with the complexity of modern energy systems.

2. Multi-energy system frameworks and models

This paper discusses two key energy management roles: the multi-energy system operator and the system coordinator. The operator is responsible for centralizing the management of multiple energy systems for optimal operation. However, IMES information segregation may lead to individual MES optimization solutions that do not meet security standards. The system coordinator, as a higher-level unit, coordinates IMES operations, manages interconnected systems, and connects IMES to the grid. The paper proposes an autonomous optimization framework for IMES operational co-operation based on an interaction mechanism, which allows MESs to optimize autonomously while minimizing operating costs and exchanging excitation/response signals with the system coordinator to achieve co-optimization, effectively dealing with the issues of information privacy and operating privileges.

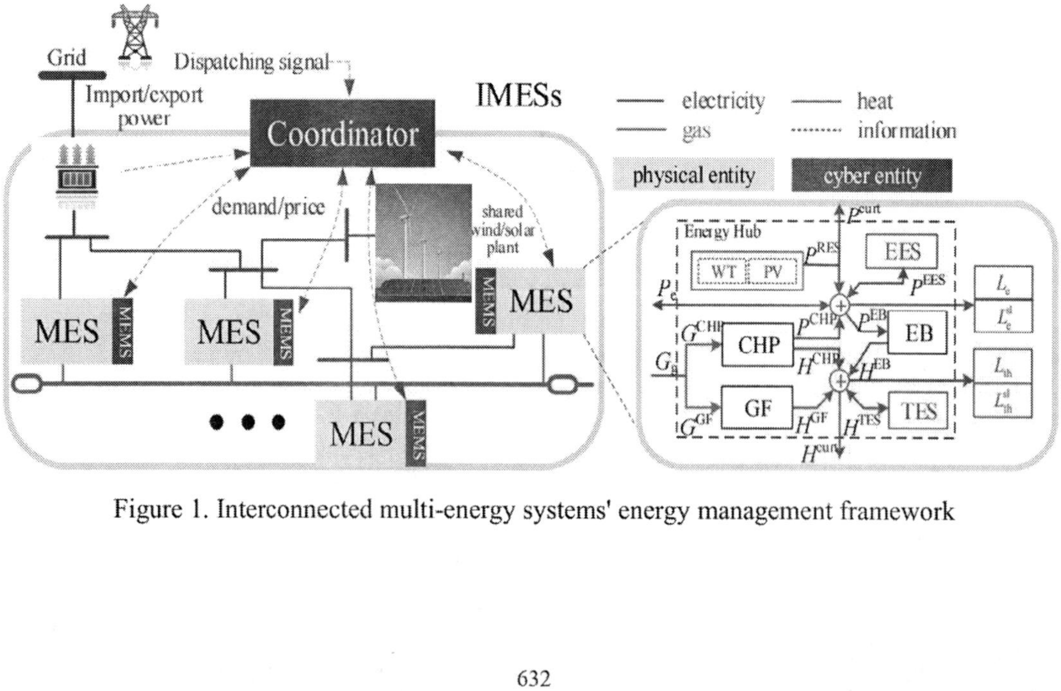

Figure 1. Interconnected multi-energy systems' energy management framework

2.1. MES Autonomous Optimisation Framework

Figure 1's lower visualization presents a synergistic MES architecture where the CHP unit operates in concert with hybrid heating systems (GF/EB combination) and is augmented by dual-energy storage platforms (EES for electricity, TES for heat retention). The system dynamics are mathematically characterized by:

$$\begin{pmatrix} Pe_t+P_t^{res} \\ 0 \end{pmatrix} + \begin{pmatrix} \eta_{ge}^{CHP} & 0 \\ \eta_{gth}^{CHP} & \eta_{gth}^{GF} \end{pmatrix}\begin{pmatrix} G_t^{CHP} \\ G_t^{GF} \end{pmatrix} + \begin{pmatrix} -1 \\ \eta_{eth}^{EB} \end{pmatrix} Pe_t^{EB} + \begin{pmatrix} P_{dch,t}^{EES}-P_{ch,t}^{EES} \\ H_{dch,t}^{TES}-H_{ch,t}^{TES} \end{pmatrix} - \begin{pmatrix} P_t^{curt} \\ H_t^{curt} \end{pmatrix} = \begin{pmatrix} L_{e,t}^{sl} \\ L_{th,t}^{sl} \end{pmatrix} + \begin{pmatrix} L_{e,t} \\ L_{th,t} \end{pmatrix} \tag{1}$$

Pe_t denotes the bidirectional power flow through the MES-grid interface at discrete time index t, where positive values indicate export to the grid.When$Pe_t < 0$, the MES sells excess power to the grid; conversely, it buys power from the grid. Meanwhile,P_t^{res} is introduced as the renewable energy power output at time point t. Define G_t^{CHP} and G_t^{GF} as the natural gas consumption of the CHP unit and the natural gas furnace, respectively. Introduce η_{ge}^{CHP} and η_{gth}^{CHP} as the efficiency of the CHP in terms of heat and power conversion. Define Pe_t^{EB} as the power consumption of electric boilers. Introduce η_{gFh}^{GF} and η_{elh}^{EB} as the GF gas-to-heat conversion efficiency and EB electron-thermal conversion efficiency, respectively. Define $Pe_{ch,t}^{EES}$ and $Pe_{dch,t}^{EES}$ as the charging and discharging power of EES, $H_{ch,T}^{TES}$、$H_{dch,t}^{TES}$ as the absorbed and released thermal power of TES, Pe_t^{curt} and H_t^{curt} as the curtailed renewable power and thermal power, $L_{e,,}^{s}$ and$L_{th,t}^{sl}$ as the transferable electrical and thermal loads, $L_{e,t}$ and $L_{th,t}$ as the fixed electrical and thermal loads.

At each dispatch time slot, each Manufacturing Execution System (MES) automatically seeks to optimize by minimizing the expected running costs during the dispatch time slot, after updating the local Renewable Energy Sources (RES) output and load demand forecasts. During time period t, operating costs consist of two distinct elements. The first involves expenses related to electricity procurement, while the second pertains to gas purchasing.

$$Fe_t = \rho_{e,t}Pe_t + \rho_{g,t}(G_{g,t}^{CHP} + G_{g,t}^{GF}) \tag{2}$$

At a specific time point t, $\rho_{e,t}$ and $\rho_{g,t}$ represent the market prices of electricity and natural gas, respectively. The dispatch process not only has to satisfy the constraints specified in equation (1), but also has to follow other additional restrictive conditions. The electric power constraint

$$-\bar{P}e^{out} \leqslant f(Pe_t) \leqslant \bar{P}e^{in},\ \forall t \tag{3}$$

where $\bar{P}e^{out}$ and $\bar{P}e^{in}$ are the maximum output electric power and maximum input electric power, respectively.

CHP、GF、EB Power Constraints

$$\begin{aligned} &Pe^{CHP} \leqslant Pe_t^{CHP} = \eta_{ge}^{CHP}G_t^{CHP} \leqslant \bar{P}e^{CHP},\ \forall t \\ &\underline{Pe}^{GF} \leqslant Pe_t^{GF} = \eta_{gth}^{GF}G_t^{GF} \leqslant \bar{P}e^{GF},\ \forall t \\ &Pe^{EB} \leqslant Pe_t^{EB} \leqslant \bar{P}e^{EB},\ \forall t \end{aligned} \tag{4}$$

In Eqs. $\bar{P}e^{CHP}$、$\bar{P}e^{GF}$ and $\bar{P}e^{EB}$ denote the upper limit of the output force of CHP, GF and EB, respectively; Pe^{CHP}、$\underline{Pe}^{GF}$、$\underline{Pe}^{EB}$ is the lower limit of the corresponding output.

Climbing constraints for CHP and EB

$$|Pe_{t+1}^{CHP} - Pe_t^{CHP}| \leqslant \Delta Pe^{CHP},\ \forall t \tag{5}$$

$$|Pe_{t+1}^{EB} - P_t^{EB}| \leqslant \Delta Pe^{EB},\ \forall t \tag{6}$$

Journal of Physics: Conference Series **3012** (2025) 012078 doi:10.1088/1742-6596/3012/1/012078

The equations ΔPe^{CHP} and ΔPe^{EB} are the creep rates of CHP and EB, respectively.
Maximum charging and discharging power constraint

$$\begin{cases} 0 \leqslant Pe_{ch,t}^{EES} \leqslant \bar{Pe}_{max}^{ch,EES}, \ \forall t \\ 0 \leqslant Pe_{dch,t}^{EES} \leqslant \bar{Pe}_{max}^{dch,EES}, \ \forall t \end{cases} \tag{7}$$

$$\begin{cases} 0 \leqslant H_{ch,t}^{TES} \leqslant \bar{H}_{max}^{ch,TES}, \ \forall t \\ 0 \leqslant H_{dch,t}^{TES} \leqslant \bar{H}_{max}^{dch,TES}, \ \forall t \end{cases} \tag{8}$$

Constraints related to transferable loads

$$0 \leqslant L_{e,t}^{sl} \leqslant \bar{L}_{e}^{sl}, \ \forall t \in \Omega_e \tag{9}$$

$$\sum\nolimits_{t \in \Omega_{th}} L_{th,t}^{sl} = L_{th}^{sl} \tag{10}$$

$$0 \leqslant L_{lh,t}^{sl} \leqslant \bar{L}_{lh}^{sl}, \ \forall t \in \Omega_{th} \tag{11}$$

$$0 \leqslant P_t^{curt} \leqslant P_t^{res}, \forall t \tag{12}$$

$$0 \leqslant H_t^{curt}, \forall t \tag{13}$$

2.2. *Non-collaborative autonomous (NCA) model*

Manufacturing Execution Systems (MES) operate under decentralized control architecture, where autonomous agents dynamically respond to Real-Time Pricing (RTP) signals through self-optimized scheduling protocols (as defined in Eq. 14), bypassing traditional centralized coordination mechanisms.

$$\begin{cases} \text{P1:} \quad \min \sum\nolimits_{t=t_c}^{t_c} Fe_t \\ \text{s.t.} \quad (1),(3)-(13) \end{cases} \tag{14}$$

2.3. *Collaborative autonomous (CA) model*

The collaborative autonomous (CA) model involves decomposing the optimization problem into a two-tier structure and solving interactively using a two-tier framework. A hierarchically structured control paradigm is formulated in this work, prioritizing iterative efficiency through bidirectional interaction layers for intranet rolling optimization scenarios.The IMES system is not only equipped with power trading functionality, but also capable of profit maximization. In this assumption, the feed-in tariff is aligned with the real-time tariff (RTP). The operational flow in the co-autonomous mode is shown in Figure 2, which is clearly divided into two phases:

Stage A: The current stage uses the sub-gradient method to interactively solve the inter-period optimization problem. In the short-term price prediction phase, where the target is to generate hourly locational marginal prices (LMPs), a rolling optimization framework is omitted due to its incompatibility with deterministic forecasting models trained on high-resolution temporal data.

Stage B:At the second operational echelon, power exchange decisions are dynamically optimized through:RES/load intro-hour variability modeling,Sliding window price adaptation,Constrained metaheuristic search ,Hourly decision epoch synchronization.At time period t_c, it is assumed that the tariff forecast for the future time period $t_{c+1}, t_{c+2}, \cdots, t_e$ is accurate and only the current tariff needs to be updated. Since there is only one decision variable, the optimization can be performed using the particle swarm algorithm. The search range of the particle swarm algorithm is$\lambda_e, \bar{\lambda}_e$, whereλ_e、$\bar{\lambda}_e$ represents the minimum and maximum values of the search space, respectively, and its search procedure is described below.

$$\begin{cases} SP_n: \min \sum_{t=t_c}^{t_e} \left[\rho_{e,t} Pe_{t,n} + \rho_{g,t}\left(G_{g,t,n}^{CHP} + G_{g,t,n}^{GF}\right)\right] \\ \text{s.t. } (1),(3)\text{-}(14) \end{cases} \tag{15}$$

SP_n is the individual autonomous optimization problem P1 for MES, just $\mu_{e,t}$ replacing $\rho_{e,t}$.

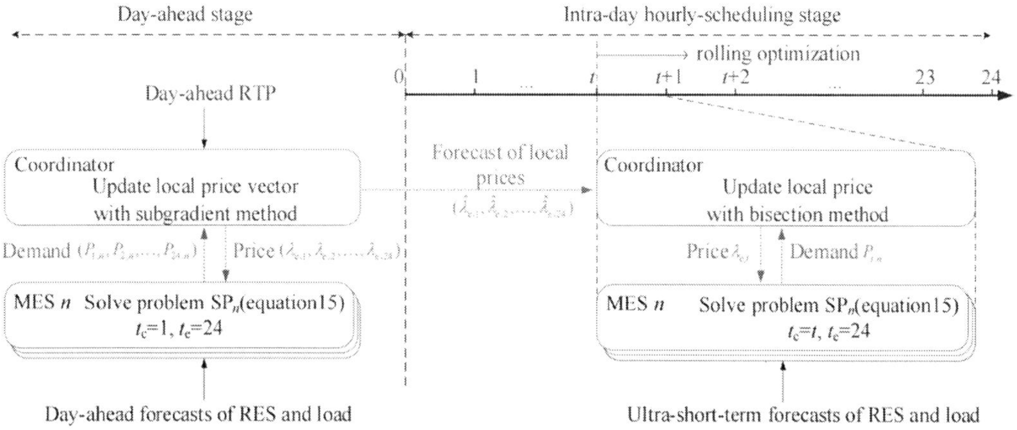

Figure 2. Operational processes under the collaborative autonomy model

2.4. Collaborative autonomous mode with feed-in limitation (CAFIL)

The introduction of feed-in tariff limitation (CAFIL) in the collaborative autonomous mode (CA), as opposed to the traditional collaborative autonomous (CA) model, has the core purpose of incentivising local consumption of electricity generated from renewable energy sources (RES) in order to reduce the dependence and impact on the main grid. One of the strategies to achieve this goal is by setting a low feed-in tariff, which in this paper is set at zero.

3. Analysis of examples

3.1. Parameterize

This paper assumes that natural gas prices remain constant over the course of a day, while electricity prices can be released or accurately predicted a day in advance. The renewable energy system (RES) and load demand forecast errors follow a normal distribution. The control period is 1 hour, and the upper limit of the market electricity price is \$1.0/kW-h and the lower limit is \$0.2/kW-h.
Real-time electricity pricing metrics are sourced from PJM Interconnection's public database, with natural gas costs stabilized at \$3.3 per cubic meter under current market conditions. Renewable energy generation forecasts exhibit progressively reduced error margins of ±30%, ±10%, and ±5% across day-ahead, intra-day, and real-time prediction intervals respectively, reflecting enhanced temporal resolution in operational planning; and the load prediction errors are ±20%, ±8%, and ±3%, respectively. The study constructed a small interconnected multi-energy system including two industrial and one residential multi-energy systems. The main transformer power limit is 4.4 MW.

3.2. Effectiveness analysis of co-optimization

The system implements three control architectures:
(1) Decentralized Autonomy (NCA) with self-contained decision-making,
(2) Coordinated Autonomy (CA) employing federated learning protocols,
(3) The constrained collaborative autonomy (CAFIL) integrates feed-in limitations through distributed consensus protocols, where autonomous agents coordinate generation schedules while respecting grid capacity thresholds.Three modes of transformer power simulation results are shown in Figure 3.

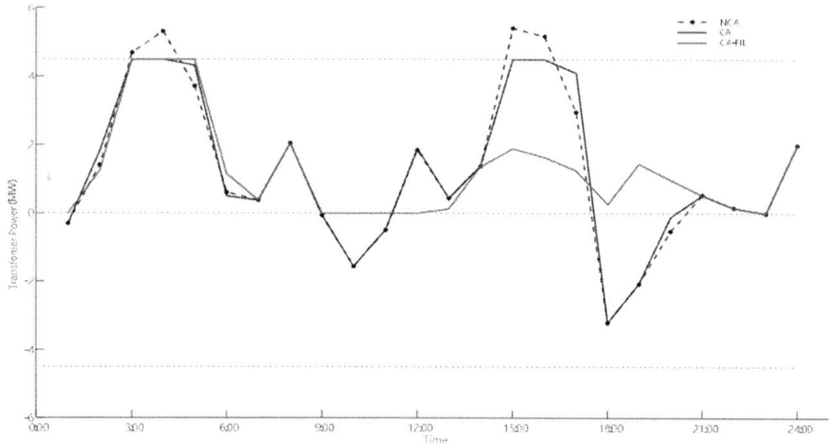

Figure 3. Energy consumption of each MES under three modes

In NCA mode, each MES is optimised with its own optimum, and the global optimum cannot be achieved. As a result, the main transformer operates beyond its nominal capacity during early morning (03:00-04:00 am) and afternoon (03:00-04:00 pm) periods, the integrated microgrid energy system (IMES) strategically dispatches excess renewable generation to the upstream network in mid-morning (10:00-11:00 am) and evening (6:00-8:00 pm) intervals. The IMES system sells the redundant RES power to the main grid during 10:00-11:00 and 18:00-20:00. After coordinating IMES in CA mode, the main transformer overload protection is successful.

In CA mode, the clearing tariff is increased above the RTP price at 02:00-04:00 in order to avoid congestion of the main transformer caused by power purchases from the grid. This will limit the consumption of MES2 and MES3 and reduce the charging power of EES if possible. MES1, on the other hand, will increase its CHP output during this period, selling power to other MESs at a profit.

Compared with static tariff models, the CA-FIL scheme demonstrates asymmetric price modulation, with a +19% tariff during grid import to prevent transformer overload and a -32% tariff during [01:00, 09-12:00, 23:00] to suppress IMES exports. It also has differentiated MES control, including CHP de-rating and TES charge limitation for MES1/3, and load elevation for RES absorption for MES2. Besides, there are temporal market constraints, such as the arbitrage prohibition from 15:00 to 21:00, which guarantees 100% RES utilization. The strategy intentionally engineers inverse load curves (IMES valley demand vs. grid peak demand), achieving a 17.2% reduction in the grid load fluctuation index (LFIg).

3.3. Analysis of the effect of MES autonomy optimisation in the CA-FIL model

This section focuses on the MES autonomous optimisation in CA-FIL mode.

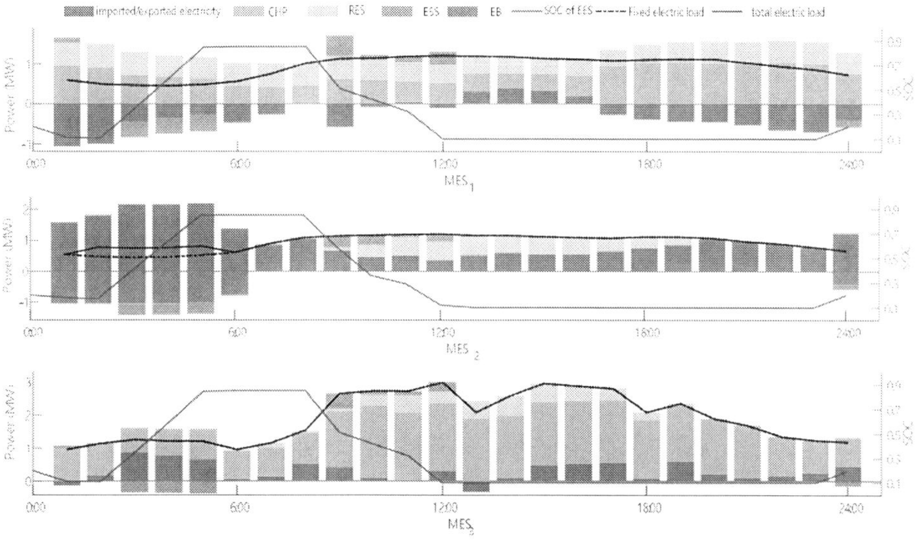

(a) Electric power

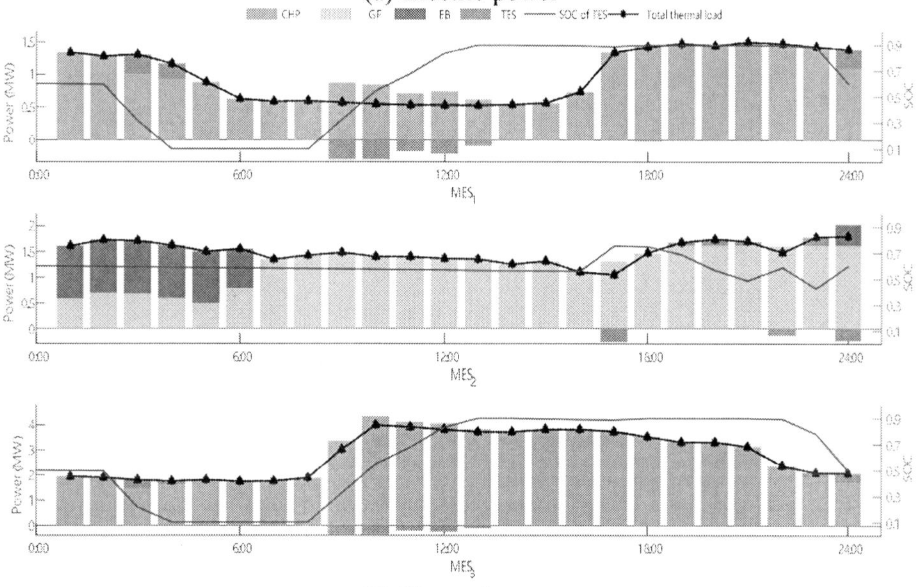

(b) Thermal power

Figure 4. Result of each MES's (a) electric and (b) thermal power

Figure 4 illustrates the operation of the energy system inside the MES for one day. the CHP heat output of MES1 and MES3 is affected by the heat load throughout the day, and MES1 and MES3 generate excess power at midnight and noon respectively, which has to be sold to other MESs. MES2 does not have a CHP unit installed, and has to purchase power from the main grid, which results in a higher power consumption for the whole day.

In summary, energy complimentary helps IMES to adapt to price fluctuations, utilities the optimization capabilities of each MES, reduces total energy costs, protects the main transformer from overloads and increases renewable energy consumption.

4. Conclusion

This paper explores the problem of collaborative and IMES, with a focus on enhancing the efficiency of multi-energy systems (MES) through mechanism design, while addressing the issues of information privacy and operational privileges to enhance the overall system performance. The main conclusions drawn from the study include:

(1)The proposed framework significantly improves the operational synergy of MES, effectively regulating energy flows and costs while protecting information privacy and operational autonomy.

(2)The Collaborative Autonomous (CA) model not only enhances the operational efficiency of individual MES but also optimizes the overall performance of the interconnected system. It achieves a balance between the local generation and consumption of energy, leading to a more resilient and sustainable energy supply.

(3)The Co-Autonomous Model (CAFIL) promotes local renewable energy consumption under feed-in tariff constraints, reduces dependence on the main grid and eases grid congestion.

(4)The CAFIL model complements the main grid power demand, helping to mitigate load fluctuations and improve grid stability.

Acknowledgments
This work was sponsored in part by Young Innovative Talents Program of Guangdong General Colleges and Universities (No. 2023KQNCX125), University-level Research Project of Guangzhou Xinhua University in 2024 (Project No. 2024KYYBZK02).

References

[1] HU Bo, CHENG Xin, SHAO Changzheng, et al. Optimal dispatch of integrated energy system based on flexibility of thermal load[J]. Journal of Shanghai Jiao Tong University, 2023, 57(7): 803-813.

[2] QIU Di, LIU Dong, GAO Fei, et al. Multi-scenario modeling and day-ahead optimization of park multiple energy system based on discrete hybrid automaton[J]. Proceedings of the CSEE, 2023, 43(1): 135-147.

[3] BAHRAMARA S, SHEIKHAHMADI P, CHICCO G, et al. Modeling the microgrid operator participation in day-ahead energy and reserve markets considering stochastic decisions in the real-time market[J].

[4] ZHANG Hong, ZHANG Ruifang, ZHOU Jiancheng, et al. Low-carbon economic dispatch of integrated energy system in campus based on stackelberg game and hybrid carbon policy[J]. Acta Energiae Solaris Sinica, 2023, 44(9): 9-17

[5] WANG Rui, CHENG Shan, XU Jianyu, et al. Multi-time scale optimal scheduling strategy of energy hub based on master-slave game and hybrid demand response[J]. Electric Power Automation Equipment, 2023, 43(1): 32-40.

[6] LI Dongdong, CHEN Tianxu, SHEN Yunwei, et al. Collaborative optimization of island microgrid based on joint virtual energy storage system strategy[J]. Proceedings of the CSEE, 2023, 43(20): 7842-7856.

[7] ZHANG Yan, LI Ruifang, ZHAO Jian, et al. Multi agent game collaborative optimization of integrated energy system considering carbon quota and integrated demand response[J]. Renewable Energy Resources, 2024, 42(3): 388-397.

[8] Li, Jie, Shengyuan Ji, Xiuli Wang, Hengyuan Zhang,et al.A Stackelberg Game-Based Optimal Scheduling Model for Multi-Microgrid Systems Considering Photovoltaic Consumption and Integrated Demand Response[J]. Energies, 2024,17(23),6002

Research on Grounding Grid Electrical Impedance Imaging Algorithm Based on Sparse Optimization Model

Liang Zhang[0000-0002-6211-878X], Wenbo Ma*, Chaohai Ji, Qilun Su, Qinchuan Zhang, Siming Lv

School of Electrical Information, Southwest Petroleum University, Chengdu, Sichuan 610500, China

*Corresponding author's e-mail: 202322000133@stu.swpu.edu.cn

Abstract. The resistivity of the grounding grid conductor and the soil differ greatly in magnitude, which makes it difficult to converge the grounding grid impedance imaging. In order to expand the convergence range, improve the inverse problem pathology of grounding grid impedance imaging, and the noise interference in the measurement process, this paper proposes a grounding grid impedance imaging method based on a sparse optimization model, aiming to improve the image reconstruction accuracy and enhance the anti-noise ability. First, based on the impedance imaging theory, the simulation model of the forward and inverse problems is established; then the grounding grid model is established and the voltage data is obtained; finally, the piecewise quadratic approximation threshold algorithm (PQA) is used to reconstruct the image in MATLAB, and compared with the existing traditional methods. The experimental results show that the proposed algorithm is superior to other algorithms in terms of image quality and anti-interference ability. When the noise signal-to-noise ratio is higher than 20dB, the sparse representation method can effectively reduce the influence of noise on image reconstruction.

1. Introduction

As a crucial infrastructure to ensure the safe and reliable operation of substations, the proper functioning of grounding grids is directly tied to the overall stability and safety of the power system. Grounding grids are essential for safely dissipating fault currents, lightning strikes, and other electrical surges into the earth, preventing damage to electrical equipment and ensuring the safety of personnel. However, over time, grounding grids are prone to deterioration due to prolonged usage and adverse environmental factors, such as soil erosion and corrosion. These issues have become increasingly severe, raising concerns about the integrity of grounding grids and posing a significant risk to the safety of power systems [1]. Corrosion reduces the effectiveness of grounding grids, compromising their ability to conduct fault currents and potentially destabilizing the entire electrical network.

Traditional methods for detecting corrosion in grounding grids, such as large-scale excavation and maintenance after power outages, are inefficient, costly, and disruptive to substation operations. These methods cause extended downtime, affecting both industrial and domestic electricity consumption. Additionally, the need for invasive procedures causes further damage to the infrastructure, increases repair costs, and prolongs the restoration of normal operations. As a result, there is an urgent need for a more effective, non-invasive, and accurate diagnostic method for assessing grounding grid corrosion.

Such methods would help ensure the safety of substations and the continuous operation of electrical equipment [2-3].

Various non-destructive testing techniques have been proposed for assessing the condition of grounding grids, including the transient electromagnetic method, electrochemical method, and electrical network method. The electrical network method models the grounding grid as a resistor network and calculates the resistance value of each branch by rotating the excitation source. These resistance values are then compared with initial nominal resistance values to assess the corrosion state. However, accurately measuring port impedance remains a challenge in this method [4]. The electromagnetic method is also susceptible to interference from the substation ' s electromagnetic environment, which can distort the results of the detection. Moreover, the electrical network method requires detailed knowledge of the grounding grid's topological structure, which is often unavailable due to the loss of grid drawings or incomplete records. This lack of information can lead to misdiagnoses, especially when dealing with complex or modified grid structures.

Electrical impedance tomography (EIT) has shown significant promise for grounding grid imaging. EIT offers non-invasiveness, simplicity, cost-effectiveness, and high resolution for capturing the electrical properties of grids [5-6]. However, challenges such as the limited number of accessible measurement nodes, the "soft field effect," and the ill-posed nature of the inverse problem complicate its application. These issues often result in reconstructed images that fail to meet the desired accuracy, making it difficult to evaluate corrosion damage [7]. To address these challenges, this study proposes a novel impedance imaging approach that combines sparse optimization models with EIT, utilizing sparse representation of the reconstructed image as a constraint in the objective function. This method introduces the Piecewise Quadratic Approximation Threshold Algorithm (PQA) to effectively solve the inverse problem.

The proposed algorithm enhances the accuracy of corrosion detection and allows for precise localization of corrosion within the grounding grid. To validate its effectiveness, the study used professional simulation software (COMSOL and MATLAB) to construct an inverse problem imaging model for the grounding grid. By comparing the performance of different algorithms, the results demonstrated that the proposed method significantly improves impedance imaging quality, providing an effective solution for corrosion diagnosis and ensuring the safe operation of the power system.

2. EIT Corrosion Diagnosis Method for Grounding Grid

Grounding grid electrical impedance imaging technology uses the difference in material conductivity to reconstruct the electrical impedance distribution image of the underground or inside an object. This technology injects current into the grounding grid lead and measures its voltage response, thereby revealing the structural characteristics inside the grounding grid, such as corrosion or fracture.

The schematic diagram of diagnosing the corrosion of the grounding grid using EIT technology is shown in Figure 1.

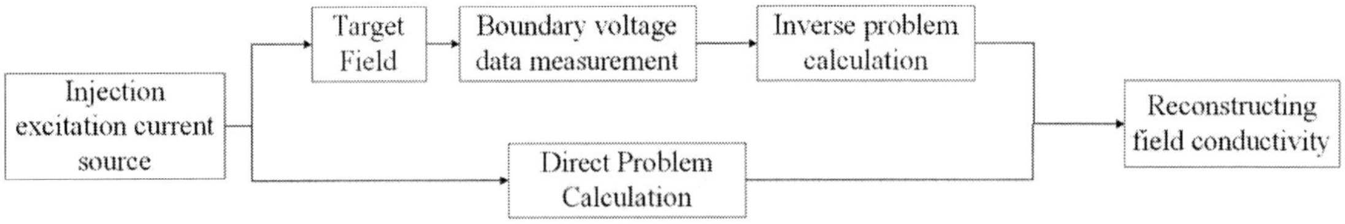

Figure 1. Schematic diagram of the corrosion diagnosis by EIT technology.

the EIT direct problem and the EIT inverse problem. The EIT direct problem involves solving the "effect" with a known "cause." In EIT technology, the direct problem is typically described as follows: The potential distribution is determined by injecting a current and measuring the boundary voltage, where the conductivity distribution is known. This process is well-posed, meaning that the solution to the direct problem is unique and stable [8].

The solution process for the inverse problem is also referred to as image reconstruction. The EIT inverse problem entails inferring the "cause" from a known "effect." In EIT technology, the inverse problem is usually described as follows: when the boundary voltage and injected current are known, the conductivity distribution in the field is determined through inversion. This process is ill-posed, meaning that the solution to the inverse problem is typically not unique or stable. A small disturbance can lead to significant fluctuations in the solution, causing the inferred result to lose its practical significance.

3. EIT reconstruction algorithm under sparse optimization model

There are several key points to pay attention to when solving the EIT sparse model problem: first, an appropriate sparse dictionary should be selected to efficiently capture feature information and enhance sparsity; then, a reasonable EIT sparse imaging model should be constructed. In addition, solving sparse models requires selecting appropriate algorithms and balancing multiple factors such as the convergence of the algorithm, computational complexity, and imaging effect. Common algorithm choices include greedy strategies such as orthogonal matching pursuit (OMP), the Lasso algorithm in the convex optimization category, and non-convex optimization algorithms [11-12].

3.1. Reconstructing sparse representation of images

3.1.1. Sparse representation mathematical model

In order to reconstruct the impedance image of the grounding grid more accurately, this paper introduces a sparse representation method. In this technique, we first select a soil area of the same size as the measurement area that is not affected by the grounding grid as an empty field. This empty field represents the natural impedance distribution when there is no interference from the grounding grid. Next, we designate the area with the grounding grid as the imaging area, measure the electrical impedance data of this area, and compare it with the impedance distribution of the empty field to obtain the impedance difference. Sparse representation theory allows us to decompose complex impedance difference data under a specific sparse basis to obtain a set of sparse coefficient vectors. This set of vectors concisely expresses the main features of the impedance difference under the sparse basis, thereby greatly simplifying the complexity of the problem. With this method, the corrosion locations of the grounding grid can be identified, and a higher quality impedance image can be generated.

Therefore, the model based on the L_1-L_2 norm is constructed as follows:

$$\min_{D} \|\Delta\sigma - D\alpha\|_2^2 + \lambda_1 \|\alpha\|_1 + \lambda_2 \|\alpha\|_2 \tag{1}$$

Where $\Delta\sigma$ is the reconstructed conductivity distribution, defined as the difference between the impedance under empty soil conditions and the impedance under grounding grid conditions. D represents the sparse dictionary, α is the sparse coefficient, and λ_1 and λ_2 are the regularization coefficients.

3.1.2. Solution of sparse coefficients

The key step in solving the sparse representation of impedance images is to solve for the sparse coefficients, which requires selecting an appropriate solving algorithm and completing it under a given dictionary.

The piecewise quadratic approximation threshold algorithm (PQA) efficiently solves the sparse coefficients through piecewise quadratic approximation of the objective function. Compared to traditional threshold algorithms, PQA incorporates higher-order information, resulting in more accurate updates during each iteration and a faster convergence speed. In the iterative process of PQA, the sparse coefficients are updated by gradually adjusting the threshold based on the current gradient information and historical data, ultimately yielding coefficients that meet the sparsity requirements.

(1) Initialization: initial sparse coefficient $\alpha(0)$; and set the maximum number of iterations T and threshold ε to judge convergence.

(2) Construct quadratic approximation: In each iteration, we need to make a quadratic approximation of the objective function. For a given current estimate of u(t), we approximate the objective function through Taylor expansion, focusing on the piecewise quadratic approximation of the L1 norm. For the L_1 norm term $\|\alpha\|_1$, we use piecewise quadratic approximation to calculate its impact on u. Specifically, for each u_i (i.e., each element of u), define a quadratic approximation: Update the sparsity factor:

$$\|\alpha_i\|_1 \approx \begin{cases} \frac{1}{2v}(\alpha_i - v)^2 & if \quad \alpha_i > v \\ 0 & if \quad \alpha_i = 0 \\ \frac{1}{2v}(\alpha_i + v)^2 & if \quad \alpha_i < -v \end{cases} \tag{2}$$

Where v is a parameter used to control the accuracy of the quadratic approximation and can usually be set to a small value.

(3) Piecewise quadratic approximation of the objective function: By performing a quadratic approximation on the objective function, we obtain:

$$f_{approx}(\alpha) = \|D\alpha - \Delta\sigma\|_2^2 + \lambda_1 \sum_i \frac{1}{2v}(\alpha_i - v)^2 + \lambda_2 \|\alpha\|_2^2 \tag{3}$$

(4) Optimization solution: For each u_i, the new $u^{(t+1)}$ is obtained by solving the quadratic optimization problem:

$$\alpha_i^{(t+1)} = soft - threshold\left(\alpha_i^{(t)} - \eta \frac{\partial f_{approx}}{\partial u_i}\right) \tag{4}$$

Where η is the iteration step size and $soft - threshold$ is defined as:

$$soft - threshold(x - \alpha) = sign(\mathrm{x}) \cdot \max(|x| - \alpha, 0) \tag{5}$$

(5) Convergence check: After each iteration, check whether the objective function value has converged. The convergence criterion is typically that the change in the objective function value is less than a preset threshold ε, or that the number of iterations exceeds the maximum limit.

$$\left\|\alpha^{(t+1)} - \alpha^{(t)}\right\|_2 < \varepsilon \tag{6}$$

(6) Update iteration: Repeat steps 3, 4, and 5 until convergence conditions are reached and finally output the results.

3.2. *EIT Sparse Imaging Algorithm*

Based on the calculated voltage, the EIT inverse problem and the measured voltage, this paper constructs a mathematical model of EIT imaging, and incorporates the sparse expression theory of impedance maps into this model. It not only significantly improves the clarity of imaging contours, but also effectively suppresses common artifacts in EIT reconstructed images, thus greatly improving the overall quality of imaging. The specific model is as shown in Equation 11:

$$\min \frac{1}{2}\|U(\Delta\sigma) - V_{measured}\|_2^2 + \|\Delta\sigma - D\alpha\|_2^2 + \lambda_1 \|\alpha\|_1 + \lambda_2 \|\alpha\|_2 \tag{7}$$

where D is the discrete cosine transform (DCT) dictionary.

(1) First, the field is segmented based on the EIT positive problem, and then the dictionary D is generated according to this segmentation.
(2) The conductivity $\Delta\sigma = D\alpha$ is updated using the dictionary D and the iteratively obtained sparse coefficient α. The calculated voltage $U(\Delta\sigma)$ is derived from the solution of the EIT direct problem.
(3) Substitute the calculated voltage and the measured voltage into equation (11) to determine if the set threshold has been reached.
(4) The coefficient α is solved iteratively according to the PQA algorithm.

4. Imaging results and analysis

4.1. Simulation model establishment

In order to compare the imaging effects of the algorithm proposed in this paper with those of other algorithms, a 2×2 model is established based on COMSOL to obtain the boundary voltage data of the grounding grid in a 16-electrode adjacent excitation mode. The specific model as shown in Figure 2

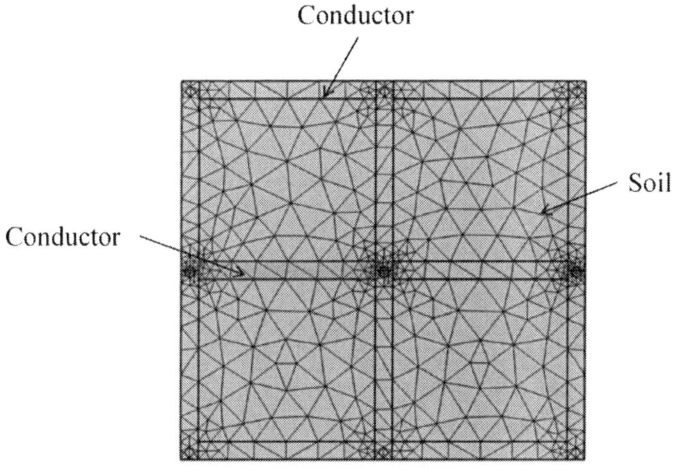

Figure 2. Grounding grid model.

The cross-section of the flat steel measures 80 mm × 80 mm, with a side length of 80 cm. The position of corrosion is indicated in the figure 2. The conductivity of the flat steel in the corroded area is designated as 1.6×10^6 S/m , while the conductivity in the uncorroded area is designated as 1.6×10^7 S/m, based on the consulted data. The soil conductivity is designated as 2×10^{-4} S/m .

The voltage data obtained by injecting 1A current through the grounding grid down conductor was imported into MATLAB for image reconstruction. The inverse problem model was divided into 3200 small triangles with 1681 nodes as shown in Figure 3.

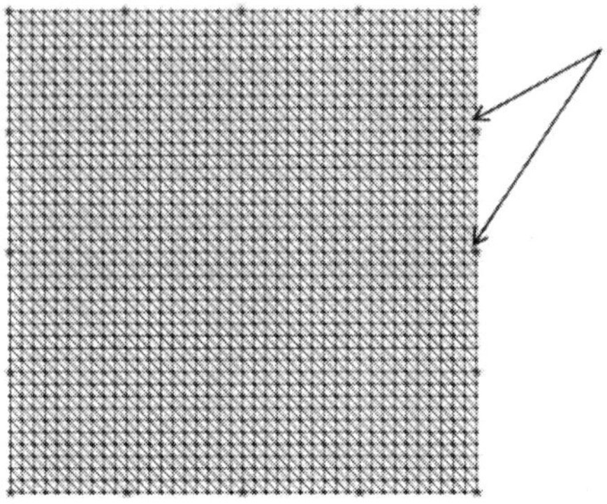

Figure 3. Meshing grid of inverse problem.

4.2. *Algorithm Comparison*

The optimal regularization parameters of each algorithm are obtained by the L curve method and the generalized cross-validation method. Then the data obtained by COMSOL is imported into MATLAB for imaging analysis. The results are shown in Figure 4:

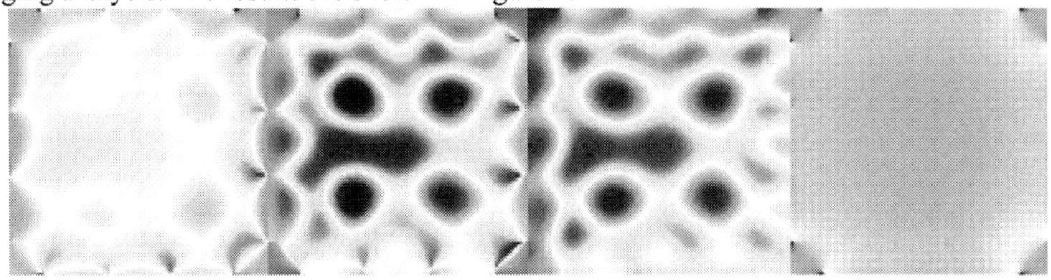

Figure 4. Grounding grid reconstruction image.

It can be seen from Figure 4 that the Tikhonov algorithm and the NOSER algorithm are unable to image the field impedance of the grounding grid. The image reconstruction effects of the TV algorithm and sparse representation algorithm can clearly present the topological structure of the grounding grid, with fewer misdiagnoses and higher imaging accuracy compared to the other three algorithms. However, the TV algorithm has more misdiagnosis than the sparse representation algorithm.

4.3. *Noise Impact*

In order to evaluate the robustness of the proposed electrical impedance tomography method based on sparse representation in practical applications, a noise experiment was designed in this paper. The error of measurement data in a real environment was simulated by adding noise. The experiment proved that the algorithm can still accurately identify the grounding grid topology and determine the corroded area under the influence of noise.

Assuming that the measured voltage data contains Gaussian noise, the four algorithms used above are used to reconstruct the grounding grid impedance image under the conditions of signal-to-noise ratio of 10dB and signal-to-noise ratio of 20dB. Figures 5 and 6 are the reconstructed images with noise signal-to-noise ratio of 10dB and 20bB respectively.

Journal of Physics: Conference Series **3012** (2025) 012079 doi:10.1088/1742-6596/3012/1/012079

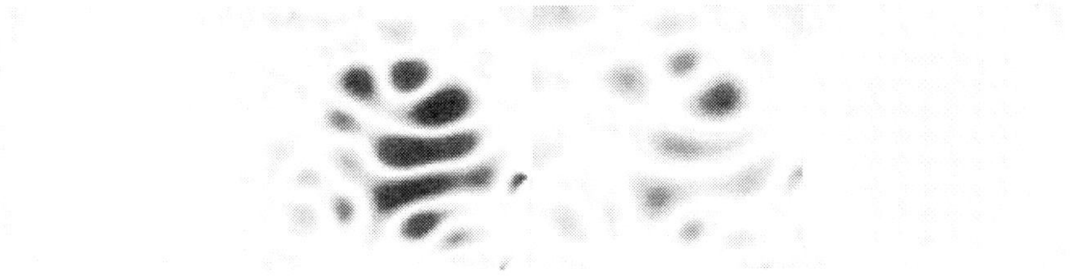

Tikhonov TV Sparse representation NOSER

Figure 5. Imaging with noise (SNR = 10 dB).

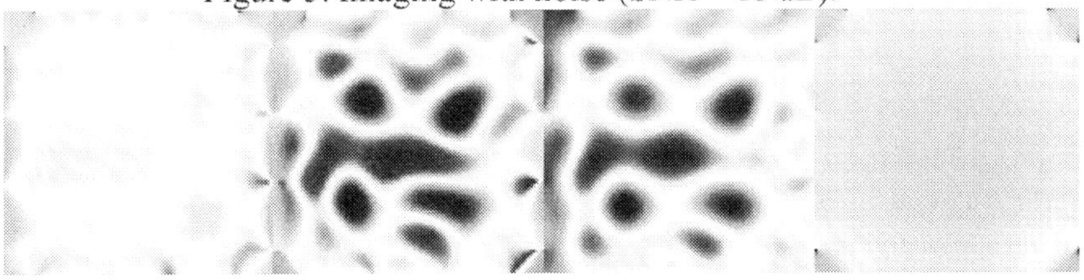

Tikhonov TV Sparse representation NOSER

Figure 6. Imaging with noise (SNR = 20 dB).

When the signal-to-noise ratio is 20 dB, the results of the reconstructed image show that the imaging method based on sparse representation still shows the best effect. In order to more comprehensively evaluate the quality of the reconstructed grounding grid image, this paper further conducts quantitative analysis of the reconstructed image. Specifically, this paper introduces the voltage relative error as an evaluation index to measure the difference between the reconstructed image and the real voltage data.The expression is as shown in Equation 12:

$$EV = \sqrt{\frac{\left(U(\sigma)-V\right)^T \left(U(\sigma)-V\right)}{V^T V}} \tag{8}$$

The comparison of the voltage relative errors of the four algorithms when all algorithms have the optimal regularization parameters is shown in Figure 7. From the figure 7 that the sparse representation reconstruction algorithm needs multiple iterations to converge and the voltage relative error is the smallest among the four algorithms after convergence.

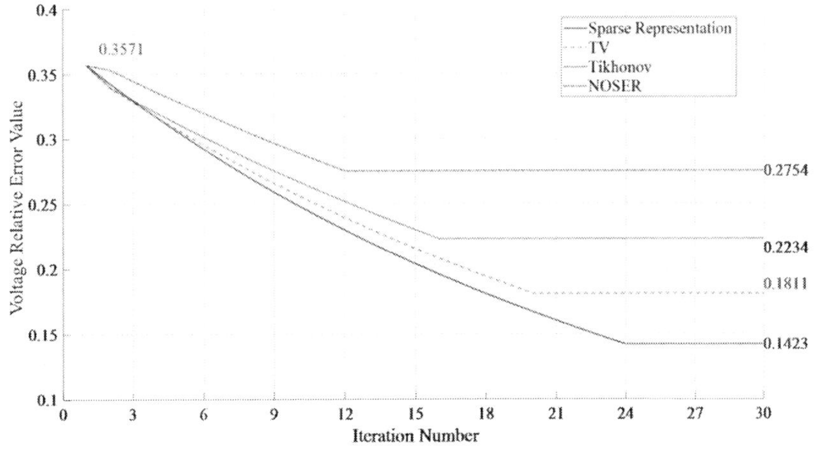

Figure 7. Voltage relative error diagram

This experiment verifies the robustness of the sparse representation-based impedance tomography method under noisy conditions. The algorithm can still maintain good sparsity under high noise conditions, proving the potential of the sparse representation-based impedance tomography method in the practical application of grounding grid corrosion diagnosis.

5. Conclusions

This paper uses sparse representation theory to concisely represent the impedance difference characteristics, and adopts the piecewise quadratic approximation threshold (PQA) algorithm to solve it, achieving high-precision conductivity distribution reconstruction. Experimental and simulation validations demonstrate that this method outperforms traditional algorithms in terms of image quality and noise immunity, particularly significantly improving imaging accuracy when the signal-to-noise ratio exceeds 20dB. This provides a novel approach for non-invasive detection of grounding grid corrosion, which is crucial for the safe and stable operation of power systems.

References

[1] Fu Zhihong, Song Siyu, Wang Xiujuan, Li Junqiang, Tai Heng-Ming. Imaging the Topology of Grounding Grids Based on Wavelet Edge Detection [J]. IEEE Transactions on Magnetics, 2018, 54(4): 6200808.

[2] Zhanlong Zhang, Huarui Ye, Yihua Dan, Ziang Duanmu, Yiqiao Li, Jun Deng. Novel Method for Comprehensive Corrosion Evaluation of Grounding Device [J]. IEEE Access, 2020, 8: 72102-72111.

[3] Xiujuan Wang, Zhihong Fu, Yao Wang, Renkuan Liu, Lin Chen. A Non-Destructive Testing Method for Fault Detection of Substation Grounding Grids [J]. Sensors (Basel), 2019, 19(9): 2046.

[4] Liang Zhang; Wei Qiu; He Yin; Kaiqi Sun; Lawrence C Markel; DaHan Liao; Zhi Li; Ben W Mcconnell; Yilu Liu. Immunity Study: Port Impedance Measurement of PMU and PCI testing under EMP[J]. IEEE Transactions on Industry Applications, 2023, 59(6): 1-9.

[5] Yang Li, Nan Wang, Li-Feng Fan, Yong-Qian Wang, Peng-Fei Zhao, Lan Huang, Zhong-Yi Wang. A fast approach to determine excitation eigenfrequencies for TD-EIT and FD-EIT [J]. Measurement Science and Technology, 2023, 34(10): 105501.

[6] Tushar Kanti Bera. Applications of Electrical Impedance Tomography (EIT): A Short Review[J]. IOP Conference Series: Materials Science and Engineering, 2018, 331: 012004.

[7] Xiaoheng Yan, Shangfei Huang, Waldemar Tomasz Smolik, Weihua Chen, Songhan Yang. A Detection Method for Fast Electrical Impedance Imaging of Grounding Grid Based on Optimized Differential-Multigrid-Homotopy Algorithm [J]. IEEE Transactions on Instrumentation and Measurement, 2023, 72: 1.

[8] Duc Minh Nguyen, Tomas Andersen, Pierre C. Qian, Tony Barry, Alistair L. McEwan. Electrical Impedance Tomography for monitoring cardiac radiofrequency ablation: a scoping review of an emerging technology [J]. Medical Engineering & Physics, 2020, 84: 36-50.

[9] Hamilton SJ.Beltrami-net: domain-independent deep D-bar learning for absolute imaging with electrical impedance tomography (a-EIT).[J].Physiological Measurement,2019,40(7): 074002

[10] Xuechao Liu, Tao Zhang, Jian'an Ye, Xiang Tian, Weirui Zhang, Bin Yang, Meng Dai, Canhua Xu, Feng Fu. Fast Iterative Shrinkage-Thresholding Algorithm with Continuation for Brain Injury Monitoring Imaging Based on Electrical Impedance Tomography [J]. Sensors (Basel, Switzerland), 2022, 22(24): 9934.

[11] Nan Wang, Yang Li, Peng-Fei Zhao, Lan Huang, Zhong-Yi Wang. Fast electrical impedance tomography based on sparse Bayesian learning [J]. Applied Soft Computing, 2023, 143: 110384.

[12] Vincent Liu;Raksha Kumaraswamy;Lei Le;Martha White.The Utility of Sparse Representations for Control in Reinforcement Learning[J].Statistics, 2019, 33:4384-4391.

Day-ahead Optimal Scheduling Strategy for Offshore Wind to Hydrogen System

Shuangrui Yin*, Xianzhi Chu

Shanghai Electric Wind Power Group Co.,Ltd., Shanghai, 200233, PR China

*Corresponding author's e-mail: yinshr@shanghai-electric.com

Abstract. Hydrogen production using offshore wind power provides a new solution for renewable energy utilization and green hydrogen development. Firstly, a hydrogen production system regulation model is established in this paper on the basis of analysing the efficiency of the proton exchange membrane (PEM) based electrolyzer and the switching of operating states. Secondly, considering the high-pressure gaseous hydrogen storage mode, the operation constraints and power consumption of the hydrogen storage system are linearized and characterized. Thirdly, the day-ahead optimal scheduling strategy for offshore wind to hydrogen system is proposed by comprehensively considering the system's trading revenue by participation in the electricity and hydrogen market, the system's operation cost and the aging state of the system's key equipment. Finally, an offshore wind to hydrogen pilot project in China is taken as an example to verify the effectiveness of the strategy.

1. Introduction

With the continuous development of renewable energy and hydrogen energy industry, green hydrogen has gradually become an important carrier of global energy green low-carbon transition, and offshore wind power hydrogen production is one of the main ways to support the development of green hydrogen energy upstream industry chain [1]. On the one hand, offshore hydrogen production by wind power can effectively solve the problem of high construction cost and insufficient utilization rate of power transmission lines for offshore wind farms, especially for far-sea wind farms [2]. By utilizing partial or complete wind power for localized offshore hydrogen production, it is possible to ensure efficient utilization of wind energy while reducing the planning capacity and construction costs of power transmission grids for wind power export. On the other hand, the comprehensive application of multi-energy coupling technologies such as wind power hydrogen production and hydrogen storage can enhance the ability of offshore wind power to support the power system, alleviate power supply-demand imbalance, and improve the flexibility and operational stability of the energy system [3].

Energy management in offshore wind-to-hydrogen production systems is critical for ensuring economic operation and energy efficiency. An energy management control strategy with 12 operational modes is proposed in [4] for a wind-hydrogen coupling system composed of wind turbines, electrolyzers, fuel cells, and supercapacitors. In [5], a bi-level optimization scheduling model incorporating lithium batteries and hydrogen storage tanks is developed to address energy management issues in multiple hybrid energy storage systems. Moreover, experts and scholars have expanded the energy conversion processes and forms to the more complicated integrated energy system from the basic wind-hydrogen coupling system, exploring more complex control strategies. For instance, in the energy production phase, various power generation methods such as wind power, photovoltaics, and wave energy are comprehensively considered, along with the integration of seawater-source heat pumps and cold or thermal storage devices [6]. Alternatively, the system's dispatchable range is analyzed after the system being further expanded to incorporate hydrogen utilization processes such as ammonia synthesis and methanol production [7].

This work focuses on the offshore wind to hydrogen system and addresses the energy management issues under grid-connected operation. First, the operational and control characteristics of proton exchange membrane (PEM) based hydrogen production system and high-pressure gaseous hydrogen storage system are analyzed, and corresponding mathematical models are established. Second, a day-ahead optimal scheduling model is developed, with the objective of minimizing the system's daily operational net cost and the aging state of key devices. Finally, a case study is conducted to analyze the scheduling results of the proposed strategy.

2. Subsystem regulation model

2.1. PEM hydrogen production system

The regulation characteristics of PEM-based hydrogen production system mainly consider four parts: electric-hydrogen efficiency conversion, hydrogen production power limitation, working state transition, and system electrical load composition.

Electrolyzer efficiency decreases as the operating power of electrolyzer increases [8], whereas the rectifier efficiency of hydrogen production power source increases with the growth of input power. For the EMS system scheduling, we only need to focus on the relationship between the input power of the hydrogen production power source $P_{\mathrm{el}}^{\mathrm{act}}$ and the final hydrogen production $W_{\mathrm{elH_2}}$ (unit kg/h). From the simulation data, it can be observed that within the preset operating load range [10%, 100%], the relationship between the two can be represented by a linear fit, i.e.,

$$W_{\mathrm{elH_2}}(t) = \eta_{\mathrm{el}}^{\mathrm{P2H}} P_{\mathrm{el}}^{\mathrm{act}}(t) \tag{1}$$

where $\eta_{\mathrm{el}}^{\mathrm{P2H}}$ denotes the electricity-hydrogen conversion efficiency of the PEM electrolyzer. The input power $P_{\mathrm{el}}^{\mathrm{act}}(t)$ in the hydrogen production state satisfies the following constraint,

$$L_{\mathrm{el}}(t) P_{\mathrm{el}}^{\mathrm{min}}(t) \le P_{\mathrm{el}}^{\mathrm{act}}(t) \le L_{\mathrm{el}}(t) P_{\mathrm{el}}^{\mathrm{max}}(t) \tag{2}$$

where the binary variable L_{el} indicates the operating state of the PEM, which is taken as 1 for operation and 0 for vice versa.

Compared with alkaline electrolyzer, the dynamic response of PEM-based electrolyzer is faster, and there is no need to consider its climbing constraints. Moreover, the start-stop time is greatly shortened. However, compared to electrical equipment such as wind turbines and electrical energy storage systems, there is still a certain response delay. And at the control level, the start-up and shutdown costs need to be considered. Therefore, its conventional operating states can be classified into idle state, cold standby state, production state, and hot standby state. [9]. The hot standby state generally has a short duration and is not considered in the scheduling plan. It can be unified with the production state as the operating state. The general day before the scheduling time particle is 1 hour, then the working state L_{el} and idle state I_{el}, cold standby state S_{el} in the unit time scale can be completed conversion, without considering the start-stop time constraints. In general, the time granularity of day-ahead scheduling is 1 hour. Then the transition of the operating state L_{el} to the idle state I_{el} or cold standby state S_{el} can be completed within a single time unit, without the need to consider start-up and shutdown duration constraints. The mathematical model for the working state transitions of the hydrogen production system is as follows,

$$L_{\mathrm{el}}(t) + S_{\mathrm{el}}(t) + I_{\mathrm{el}}(t) = 1 \quad t = t_0, t_1, \cdots, t_{\mathrm{DA}} \tag{3}$$

$$\begin{cases} U_{\mathrm{el}}^{\mathrm{start}}(t) \le I_{\mathrm{el}}(t) \quad t = t_0, t_1, \cdots, t_{\mathrm{DA}} \\ U_{\mathrm{el}}^{\mathrm{start}}(t-1) \le S_{\mathrm{el}}(t) + L_{\mathrm{el}}(t) \quad t = t_1, \cdots, t_{\mathrm{DA}} \\ U_{\mathrm{el}}^{\mathrm{start}}(t-1) \ge I_{\mathrm{el}}(t-1) + S_{\mathrm{el}}(t) + L_{\mathrm{el}}(t) - 1 \ t = t_1, \cdots, t_{\mathrm{DA}} \end{cases} \tag{4}$$

$$\begin{cases} U_{\text{el}}^{\text{cold}}(t) \leq S_{\text{el}}(t) \quad t = t_0, t_1, \cdots, t_{\text{DA}} \\ U_{\text{el}}^{\text{cold}}(t-1) \leq L_{\text{el}}(t) \quad t = t_1, \cdots, t_{\text{DA}} \\ U_{\text{el}}^{\text{cold}}(t-1) \geq S_{\text{el}}(t-1) + L_{\text{el}}(t) - 1 \ \ t = t_1, \cdots, t_{\text{DA}} \end{cases} \tag{5}$$

Considering that the state constraints of full start-up and cold start-up have time-coupling characteristics, the initial values of the operating states of the hydrogen production system, $S_{\text{el}}(t_0)$ and $L_{\text{el}}(t_0)$, need to be given in the day-ahead scheduling based on the day-ahead optimization results of the previous day.

Based on the above analysis, the electrical load composition of the PEM-based hydrogen production system is as follows,

$$P_{\text{el}}^{\text{total}}(t) = P_{\text{el}}(t) + P_{\text{el}}^{\text{aux}}(t) \tag{6}$$

$$P_{\text{el}}(t) = P_{\text{el}}^{\text{act}}(t) + L_{\text{el}}(t)P_{\text{el}}^{\text{wsb}} + S_{\text{el}}(t)P_{\text{el}}^{\text{csb}} + U_{\text{el}}^{\text{start}}(t)P_{\text{el}}^{\text{start}} + U_{\text{el}}^{\text{cold}}(t)P_{\text{el}}^{\text{cold}} \tag{7}$$

$$P_{\text{el}}^{\text{aux}}(t) = \big(L_{\text{el}}(t) + S_{\text{el}}(t)\big)P_{\text{el\_0}}^{\text{aux}} + \alpha_{\text{el}}^{\text{aux}} W_{\text{elH}_2}(t) \tag{8}$$

Equation (6) indicates that the power consumption $P_{\text{el}}^{\text{total}}$ of the whole hydrogen production system consists of the power consumption P_{el} of the core electrolysis water equipment and the power consumption $P_{\text{el}}^{\text{aux}}$ of the auxiliary machines, such as seawater desalination equipment. And the expressions of P_{el} and $P_{\text{el}}^{\text{aux}}$ are given by equations (7) and (8), where $P_{\text{el}}^{\text{start}}$ denotes the average power consumption of the process for complete start-up ($I_{\text{el}}(t) \rightarrow S_{\text{el}}(t)$ or $I_{\text{el}}(t) \rightarrow L_{\text{el}}(t)$). $P_{\text{el}}^{\text{cold}}$ denotes the average additional power consumption of the process for cold start-up ($S_{\text{el}}(t) \rightarrow L_{\text{el}}(t)$), with a value lower than that of $P_{\text{el}}^{\text{start}}$. And $P_{\text{el\_0}}^{\text{aux}}$ is the fixed power consumption of the auxiliary equipment. $\alpha_{\text{el}}^{\text{aux}}$ is the power consumption coefficient of the auxiliary equipment related to hydrogen production, which depends on the system performance parameters, e.g., desalination and purification efficiencies.

2.2. *Hydrogen storage system*

Considering the high-pressure gaseous hydrogen storage method, the hydrogen generated by the PEM-based electrolyzer needs to be pressurized by a compressor into a hydrogen storage unit or buffer tank (uniformly called hydrogen storage tank in this paper). The hydrogen storage system can be thought of as having the intake rate controlled by the compressor, while the exhaust rate is limited to the flow controlled by the inlet of the hydrogen pipeline or the hydrogen carrying vessel. The compressor is also the main power consumption equipment of the hydrogen storage system. Therefore, the control of the hydrogen storage system is actually the control of the inlet compressor and the outlet valve.

Considering the special scenarios such as experimental tests in the actual engineering operation, there may be a situation where hydrogen is discharged directly into the atmosphere, i.e., for the compressor inlet rate $W_{\text{hs}}^{\text{in}}(t)$ (in kg/h),

$$W_{\text{hs}}^{\text{in}}(t) + W_{\text{H}_2}^{\text{aban}}(t) = W_{\text{elH}_2}(t) \tag{9}$$

where $W_{\text{H}_2}^{\text{aban}}(t)$ refer to the rate of hydrogen discharged directly to the atmosphere from the hydrogen production system, which is affected by the compressor inlet rate. The hydrogen storage system inlet and exhaust rates $W_{\text{hs}}^{\text{in}}(t)$ and $W_{\text{hs}}^{\text{out}}(t)$ need to satisfy the upper bound constraints,

$$\begin{cases} 0 \leq W_{\text{hs}}^{\text{in}}(t) \leq W_{\text{hs}}^{\text{in}_{\max}}(t) \\ 0 \leq W_{\text{hs}}^{\text{out}}(t) \leq W_{\text{hs}}^{\text{out}_{\max}}(t) \end{cases} \tag{10}$$

In addition, the storage capacity of the hydrogen storage tank, $M_{\text{hs}}(t)$ (in kg), should match the time-sequential change,

$$M_{\text{hs}}(t) = M_{\text{hs}}(t-1) + \left(\eta_{\text{hs}}^{\text{in}} W_{\text{hs}}^{\text{in}}(t) - \frac{W_{\text{hs}}^{\text{out}}(t)}{\eta_{\text{hs}}^{\text{out}}}\right)\Delta t \tag{11}$$

where $\eta_{\mathrm{hs}}^{\mathrm{in}}$, $\eta_{\mathrm{hs}}^{\mathrm{out}}$ are the conversion efficiencies for the inlet and outlet of the hydrogen storage system.

The number of moles of hydrogen corresponding to $M_{\mathrm{hs}}(t)$ is, $n_{\mathrm{hs}}(t) = M_{\mathrm{hs}}(t)/M_{\mathrm{H_2}}$. $M_{\mathrm{H_2}}$ denotes the molar mass of hydrogen in kg/mol.

High-pressure gas hydrogen storage is to pressurize hydrogen into a high-pressure vessel, and the compression factor z needs to be introduced to correct the ideal gas equation of state [5],

$$p_{\mathrm{hs}}(t)V_{\mathrm{hs}} = n_{\mathrm{hs}}(t)RT_{\mathrm{hs}} \times z \tag{12}$$

where: $p_{\mathrm{hs}}(t)$ is the gas pressure in the hydrogen storage tank. V_{hs} denotes the hydrogen storage tank volume in m^3; and T_{hs} denotes the degree Kelvin in the hydrogen storage tank. The compression factor z satisfies,

$$\left(z + \frac{27p_{\mathrm{hs}}(t)}{64T_{\mathrm{hs}}^2 \times z}\right)\left(1 - \frac{p_{\mathrm{hs}}(t)}{8T_{\mathrm{hs}} \times z}\right) = 1 \tag{13}$$

Linearly simplifying the above expression, the compression factor can be approximated as,

$$z = z_0 + z_1 T_{\mathrm{hs}} + z_2 p_{\mathrm{hs}}(t) \tag{14}$$

Then, the gas pressure in the hydrogen storage tank can be expressed as:

$$p_{\mathrm{hs}}(t) = \frac{z_0 + z_1 T_{\mathrm{hs}}}{V_{\mathrm{hs}} - z_2 n_{\mathrm{hs}}(t)RT_{\mathrm{hs}}} n_{\mathrm{hs}}(t)RT_{\mathrm{hs}} \tag{15}$$

The gas pressure in the hydrogen storage tank at any time meets the following constraints,

$$p_{\mathrm{hs}}^{\mathrm{min}} \le p_{\mathrm{hs}}(t) \le p_{\mathrm{hs}}^{\mathrm{max}} \tag{16}$$

where: $p_{\mathrm{hs}}^{\mathrm{max}}$, $p_{\mathrm{hs}}^{\mathrm{min}}$ denote the upper and lower operating pressure limits of the hydrogen storage tank. From the above operating constraints, it can be seen that the pressure inside a hydrogen storage tank with the fixed volume is related to the amount of hydrogen inside the tank and the temperature. Assuming that the temperature inside essentially remains constant, the operating pressure constraints can be converted to constraints for hydrogen volume inside the tank:

$$M_{\mathrm{hs}}^{\mathrm{min}} \le M_{\mathrm{hs}}(t) \le M_{\mathrm{hs}}^{\mathrm{min}} \tag{17}$$

High-pressure gaseous hydrogen storage in the storage process can be considered that the tank does not consume power, only the compressor energy consumption is considered. And the compressor power consumption is related to the hydrogen flow rate and the pressure before and after compression, which can be expressed as [10],

$$P_{\mathrm{H_2}}^{\mathrm{cpr}}(t) = W_{\mathrm{hs}}^{\mathrm{in}}(t) P_{\mathrm{cpr}}^{\mathrm{ref}} \frac{\ln(p_{\mathrm{c}}/p_0)}{\ln(p_{\mathrm{c}}^{\mathrm{ref}}/p_0)} \tag{18}$$

where $P_{\mathrm{cpr}}^{\mathrm{ref}}$ is the baseline power consumption of the compressor at the reference pressure $p_{\mathrm{c}}^{\mathrm{ref}}$ in kWh/kg. p_{c} is the regular operating pressure of the compressor. And p_0 is the standard atmospheric pressure. Although the gas pressure in the hydrogen storage tank rises continuously with the increase of hydrogen storage, it can be assumed that the compressor can maintain a constant exhaust pressure during the working process. Thus the power consumption of the compressor is proportional to the amount of hydrogen compression, which can be simplified and expressed as follows:

$$P_{\mathrm{H_2}}^{\mathrm{cpr}}(t) = a_{\mathrm{cpr}} \times W_{\mathrm{hs}}^{\mathrm{in}}(t) \tag{19}$$

Total power consumption of the hydrogen storage system is,

$$P_{\mathrm{hs}}(t) = U_{\mathrm{hs}}^{\mathrm{in}}(t) P_{\mathrm{hs,0}} + P_{\mathrm{H_2}}^{\mathrm{cpr}}(t) \tag{20}$$

where $P_{\mathrm{hs,0}}$ denotes the remaining fixed auxiliary power consumption except for the compression load. $U_{\mathrm{hs}}^{\mathrm{in}}(t)$ is the 0-1 working state variable of high pressure gaseous hydrogen storage system.

Journal of Physics: Conference Series **3012** (2025) 012080 doi:10.1088/1742-6596/3012/1/012080

3. Day-ahead scheduling model

The day-ahead scheduling strategy for offshore wind-hydrogen coupled system in grid-connected mode is studied in this work. On the basis of the above PEM-based hydrogen generation and storage subsystems, the optimization model is established and the grid-connected interactive power, the power generation and consumption plan of the controllable equipment are obtained during the full operation day. Set time resolution Δt_{da} to 1h, that is, a total of 24 scheduling periods for the operation day.

3.1. Objective

The offshore wind to hydrogen production, storage and transportation system can obtain energy sales revenues by participating in the electricity and hydrogen markets. The equipment aging during operation and the cost of using resources other than electricity consumption are considered as operation and maintenance (O&M) cost in the optimization objective, and the specific mathematical expressions are as follows,

$$\min F_{\mathrm{con}}^{\mathrm{DA}} = \sum_{t=t_0}^{t_{\mathrm{DA}}}\left(-c_{EM}^{s}(t)P_{EM}^{s}(t) - c_{HM}W_{HM}(t)\right)\Delta t_{\mathrm{da}} + \sum_{t=t_0}^{t_{\mathrm{DA}}}\left[c_{\mathrm{wd}}^{\mathrm{om}}P_{\mathrm{wd}}(t) + \right.$$
$$\left. + c_{\mathrm{el}}^{\mathrm{om}}W_{\mathrm{elH_2}}(t)\right]\Delta t_{\mathrm{da}} + \sum_{t=t_0}^{t_{\mathrm{DA}}}\left(c_{\mathrm{el}}^{\mathrm{start}}U_{\mathrm{el}}^{\mathrm{start}}(t) + c_{\mathrm{el}}^{\mathrm{cold}}U_{\mathrm{el}}^{\mathrm{cold}}(t)\right)$$
$$+ \sum_{t=t_0}^{t_{\mathrm{DA}}}\left[c_{\mathrm{cpr}}^{\mathrm{om}}P_{\mathrm{hs}}(t) + c_{\mathrm{hs}}^{\mathrm{om}}\left(W_{\mathrm{hs,g}}^{\mathrm{in}}(t) + W_{\mathrm{hs,g}}^{\mathrm{out}}(t)\right)\right]\Delta t_{\mathrm{da}} \tag{21}$$

The first term in the above equation represents the system's revenue from electricity and hydrogen sales on the scheduling day, where $c_{EM}^{s}(t)$ is the wind power feed-in tariff in time period t. c_{HM} is the hydrogen sales price, which generally includes the green hydrogen subsidy according to the local policy. $P_{EM}^{s}(t)$ and $W_{HM}(t)$ are the optimized grid-connected generation power and hydrogen production rate in time period t, respectively. The second term represents the O&M cost of the wind power system and the PEM-based hydrogen production system, where $c_{\mathrm{wd}}^{\mathrm{om}}$ and $c_{\mathrm{el}}^{\mathrm{om}}$ are the O&M cost coefficients of the above two systems. The third term represents the full start-up and cold start-up costs of the hydrogen production system, where $c_{\mathrm{el}}^{\mathrm{start}}$ and $c_{\mathrm{el}}^{\mathrm{cold}}$ are the unit prices of the start-up costs corresponding to the above two start-up methods. The fourth item represents the O&M cost of the high-pressure gaseous hydrogen storage system (including compressor), where $c_{\mathrm{cpr}}^{\mathrm{om}}$ and $c_{\mathrm{hs}}^{\mathrm{om}}$ are the O&M cost coefficients of the compressor and the high-pressure gaseous storage system related to the hydrogen storage and exhaust process, respectively.

3.2. Constraints

3.2.1. Power balance

$$P_{\mathrm{wd}}(t) - P_{EM}^{s}(t) = P_{\mathrm{el}}^{\mathrm{total}}(t) + P_{\mathrm{hs}}(t) \tag{22}$$

The left side of the above equation is the remaining generation power after subtracting the feed-in sales power from the wind power output. The right side of the equation is the total power consumed by the hydrogen production and storage systems in the system.

3.2.2. Hydrogen flow balance

$$W_{HM}(t) = W_{\mathrm{hs}}^{\mathrm{out}}(t) \tag{23}$$

The above equation represents the flow equilibrium of hydrogen sales. Although the actual revenue from hydrogen sales will be delayed for a long time due to the shipping and land transportation time, it is assumed that the hydrogen delivered from the storage is counted as revenue since the market price of hydrogen is stable over a long period. Note that it is implied here that hydrogen transfer occurs at least once a day. If it does not, the exhaust volume need to be replaced by the product of the intake rate and conversion efficiency of the hydrogen storage system to ensure that the income from hydrogen production is included in the optimization objective.

3.2.3. Wind power system operational constraints

$$P_{wd}^{min}(t) \leq P_{wd}(t) \leq P_{wd}^{max}(t) \tag{24}$$

Where $P_{wd}^{max}(t)$ and $P_{wd}^{min}(t)$ are the upper and lower limits of controllable generation power reported by the wind power system. When the wind power system operates normally, $P_{wd}^{max}(t) = P_{wd}^{f}(t)$, $P_{wd}^{min}(t) = 0$, i.e., the wind power output can be flexibly adjusted within the predicted value.

3.2.4. Operational constraints on hydrogen production and storage systems

The operational constraints of the PEM-based hydrogen production system are defined by equations (1)-(8) in Section 2.1. The operational constraints of the hydrogen storage system are defined by equations (10)-(11), (17), and (19)-(20) in Section 2.2. Additionally, incorporating the hydrogen transport vessel's schedule, constraints on the hydrogen storage state during transportation periods can be added. For instance, if it is assumed that the transport vessel arrives during the final scheduling period of each day, it transfers the hydrogen from the high-pressure gaseous storage system, and the tank is emptied within one hour (to the minimum storage state, as complete evacuation to a vacuum is not possible). Then $\mathrm{M_{hs}}(t_{\mathrm{DA}}) = \mathrm{M_{hs}^{min}}$ can be set.

4. Case study

Taking an offshore wind-to-hydrogen pilot project in China as an example, the wind power installed capacity is 9 MW, and the agreed electricity selling price is 0.453 CNY/kWh. The predicted daily output curve is shown in Figure 1. The project is equipped with a PEM-based electrolyzer with a hydrogen production rate of 1000 Nm³/h (89 kg/h), corresponding to a rated hydrogen production input power of 4250 kW. A high-pressure gaseous hydrogen storage system is configured with a storage capacity range of [50–845] kg. The maximum inlet flow rate matches the PEM-based electrolyzer at 89 kg/h. The maximum outlet flow rate is 845 kg/h during the designated transfer period for the hydrogen transport vessel (23:00–24:00). During other periods, the maximum outlet flow rate is 5 kg/h, primarily determined by the hydrogen consumption rate for by-products such as methanol.

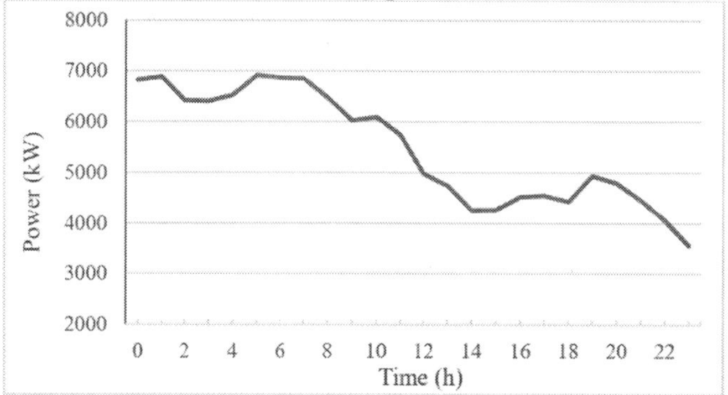

Figure 1. Forecast output curve of wind power

Based on the above data, the day-ahead optimization scheduling model established in Section 3 was solved using a mixed-integer linear programming solver. The optimization results indicate that the hydrogen production system operates during the 0:00–10:00 period and switches to idle mode after 10:00. Accordingly, the hydrogen storage system stores hydrogen during the production period, whereas the compressor remains inactive during other periods. The optimization results show the hourly variations for the storage system's average inlet and outlet flow rates and storage states of hydrogen throughout the day, as illustrated in Figure 2. Figure 3 depicts the utilization of wind power for different stages on the scheduling day.

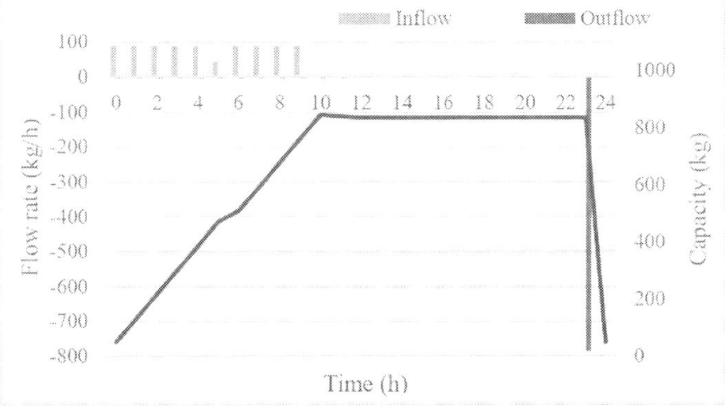

Figure 2. Day-ahead optimal scheduling results of hydrogen storage system

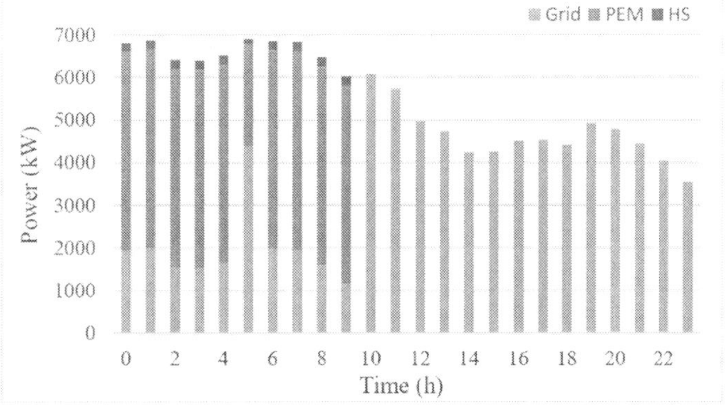

Figure 3. Day-ahead optimal scheduling results of wind power usage

From Figures 2 and 3, it is easy to observe that, given the case study's assumption of sufficient external transmission capacity, wind power grid integration is unrestricted, and no wind curtailment occurs. However, due to the relatively small capacity of the hydrogen storage tanks and the low subsequent hydrogen utilization rate (5 kg/h during 0:00–12:00), as well as constraints imposed by the hydrogen transport vessel schedule (in this case, the vessel arrives at the offshore hydrogen production platform during the last scheduling period of the operational day to replace and transfer all hydrogen storage tanks), the hydrogen storage system reaches its maximum capacity within half a day under the given wind power output conditions. This, in turn, causes the hydrogen production system to switch from the working state to the idle state, with all wind power being fed into the grid during this period.

Through deductive analysis, it is clear that if the grid integration capacity of wind power is limited, wind curtailment will inevitably occur under the current system configuration. This indicates that for offshore wind to hydrogen projects including hydrogen storage, transportation, and comprehensive utilization, ensuring long-term economic viability and renewable energy utilization requires careful consideration during the planning phase. Specifically, the capacity of offshore wind farm transmission channels, offshore hydrogen production and storage systems, and hydrogen utilization processes such as ammonia synthesis and methanol production must be optimized and balanced.

Additionally, it is worth noting that during 5:00–6:00, the hydrogen production power and corresponding inlet flow rate of the hydrogen storage system show a noticeable drop compared to other periods in the optimization results. This is caused by the multiple solutions inherent in the mathematical programming model. When the hydrogen storage system capacity is not an integer multiple of the maximum hourly inlet flow rate, there will always be one scheduling period where the inlet flow rate does not reach its maximum. In actual operations, when wind power is sufficient, priority is given to maintaining rated power hydrogen production and maximizing the hydrogen storage rate until the storage tanks are full. This is equivalent to swapping the optimization results for 5:00-6:00 with those for 9:00-10:00, and this adjustment would be implemented during intra-day scheduling.

5. Conclusion

This paper takes the offshore wind to hydrogen system as the research object, a day-ahead optimal scheduling model is established based on the analysis of regulation models for two key subsystems, i.e., PEM-based production and hydrogen storage subsystems. The goal is to achieve optimal economic operation of the system while comprehensively considering the depreciation of equipment lifespan. The case study results validate the effectiveness of the day-ahead scheduling model and the subsystem regulation models, while highlighting the importance of equipment capacity planning in offshore wind power hydrogen production, storage, and subsequent utilization processes.

Acknowledgments

This work is supported by the project "Key Technology Development and Engineering Demonstration of Offshore Wind Power Hydrogen Production and Comprehensive Utilization" (CBZ04N23-04).

References

[1] IEA. Global Hydrogen Review 2024. [2024-12-20].
[2] North Sea Wind Power Hub, https://northseawindpowerhub.eu/. [2024-12-01].
[3] Du T, Li J, Ge L, and Zhang B, Optimal operation model of seaport integrated energy system with seawater desalination for hydrogen production and thermal inertia, Electric Power Automation Equipment, vol. 43(12), pp. 143-150, Dec. 2023.
[4] Deng H, Chen J, Teng Y, Zhang B, and Fu J, Energy management strategy of wind power coupled with hydrogen system, ACTA ENERGIAE SOLARIS SINICA, vol. 42(1), pp. 256-263, Jan. 2021.
[5] Li B, Miao H, and Li J, Multiple hydrogen-based hybrid storage systems operation for microgrids: A combined TOPSIS and model predictive control methodology, Applied Energy, vol. 283, 2021: 116303.
[6] Wang R, Research on optimal scheduling of electricity-cold-heat-hydrogen-fresh water island integrated energy system based on power prediction, Shandong University, 2023.
[7] Zhou B, Zhu W, Zhu J, Qiu Y, Zang T, He G, and Chen G, Multi-stage dispatchable region analysis of wind and solar power-based hydrogen production and ammonia synthesis system, Proceedings of the CSEE, vol. 44(01), pp.160-174, 2024.
[8] Lu Y, Research on optimal scheduling of multiple hydrogen generating units under wind/photovoltaic/energy-storage hydrogen production system, Hebei University of Technology, 2021.
[9] Wan Z, Research on operation control strategy of new energy hydrogen production system, Dalian University of Technology, 2021.
[10] Tian T, Ma Z, Cui Q, Shu J, Tan L, and Wang H, Multi-objective Optimization of a Hydrogen-Battery Hybrid Storage System for Offshore Wind Farm Using MOPSO, Journal of Electrical Engineering & Technology, vol. 18, pp. 4091–4103, 2023.

Research and Design of Orderly Charging Strategy for Electric Vehicle Charging Stations

Yong Yu[1*[0009-0008-7692-9883]], **Gaofeng Hua**[1[0009-0002-8953-9741]], **Daodong Hu**[1[0009-0000-7086-5673]], **Shanshan Ma**[2[0009-0003-8322-8122]], **Haihao Wei**[2[0009-0001-5476-6314]]

[1] XJ Software Technology Co., Ltd, Xuchang, Henan Province, 461000, P.R. China

[2] XJ Electric Co., Ltd. Xuchang, Henan Province, 461000, P.R. China

*Corresponding author's e-mail: yu1183688986@163.com

Abstract. As the development of new energy electric vehicles progresses, the contradictions between the increasing charging demands and issues such as long charging queue times, high charging costs, and charging safety are becoming increasingly prominent. In order to solve these issues and enhance the efficiency and safety of charging stations, this paper investigates and designs an orderly charging strategy. By examining the relationships between voltage, current, State of Charge (SOC), and charging time during the EV charging process, empirical values for the reasonable termination of charging at different electricity pricing periods and under conditions of limited charging infrastructure are established. Furthermore, orderly charging methods for different charging piles are designed, tailored to the service vehicle targets of various charging stations, thereby improving the benefits of these stations. The implementation of an orderly charging strategy can achieve a win-win situation for both charging users and charging stations, while also playing a positive role in peak shaving and valley filling for the power grid, reducing load fluctuation rates.

1. Introduction

In recent years, with the depletion of fossil energy and the rise of clean new energy technologies, electric vehicles and their corresponding charging stations, charging piles and charging technologies have developed rapidly[1-2]. The Internet of Things (IoT) technology enables bidirectional communication between electric vehicles , charging piles, and charging platforms, facilitating real-time data exchange regarding the status of charging stations, electricity consumption, and user behavior. This supports the continuous improvement and development of charging service infrastructure, but there are still many problems. On the one hand, it is difficult for users to charge, find charging piles, and have long queues. On the other hand, charging piles have low utilization, long-term idleness, and high operating costs[3-4]. In addition, the large-scale use of charging facilities has a certain impact on the load safety of power grid operation[5-6]. In particular, the disordered charging load of electric vehicles will bring a series of problems to the distribution network, such as voltage drop, line overload, power grid load peak, network loss, etc.[7-9]. Therefore, it is increasingly necessary to study orderly charging strategies to solve charging difficulties, improve the efficiency of charging piles and reduce the impact of charging on power grid load[10-12].

There is no strict definition of orderly charging of electric vehicles at present. The orderly charging strategy proposed in this paper refers to a charging control method designed under the conditions of many vehicles and few charging piles, power fluctuations at charging stations, and reasonable use of peak and valley electricity prices. By analyzing and studying the relationship curves between charging voltage, current, soc and charging time during vehicle charging[13-14], an optimal soc value for terminating charging is formulated to improve the charging queue efficiency; the charging time is arranged as much as possible during the valley period with low electricity prices to reduce the charging operation cost[15-16], and the peak of the power grid is reduced and valley filling is carried out by limiting the power of charging piles to improve the safety of power grid operation.

2. Design of an ordered charging scheme

This paper comprehensively defines orderly charging from three aspects. First, the orderly charging of electric vehicles in queues is designed to address the shortage of charging piles. The charging efficiency of the charging piles is provided and the charging order of the vehicles waiting in queue is planned by calculating the remaining charging time of the electric vehicles and the soc when charging is terminated. On the other hand, according to the difference in peak and valley electricity prices, the orderly charging time is reasonably arranged to reduce the charging cost. Third, an orderly charging strategy is designed to reduce the peak and valley of the grid load curve, reduce the variance of the load curve, and coordinate the interaction between the charging facilities and the grid load.

Discrete charging piles are difficult to manage due to their dispersion. Therefore, orderly charging is currently mainly used in centralized charging stations. The scale of charging piles in centralized charging stations is generally 20-50, and the main service objects are electric buses. According to the needs of operators, charging stations are mainly divided into dedicated charging stations and public charging stations. Dedicated charging stations only provide charging services for special vehicles (electric buses, etc.) designated by operators and are not open to the public; public charging stations not only serve special vehicles, but also operate to the public and can charge all social electric vehicles. This article will conduct orderly charging research on these two types of charging stations.

The charging station service system is shown in Figure 1. The charging piles in the station are connected to the charging service platform through the Internet of Things or the Internet such as 4G/5G. The charging service platform is provided by the operator to the operation monitoring client. Charging users can use charging cards and mobile phone apps to charge. In order to coordinate and interact with the charging station and the power grid load, the charging service platform is connected to the charging station distribution equipment, which sends the available power information of the charging station to the charging service platform in real time.

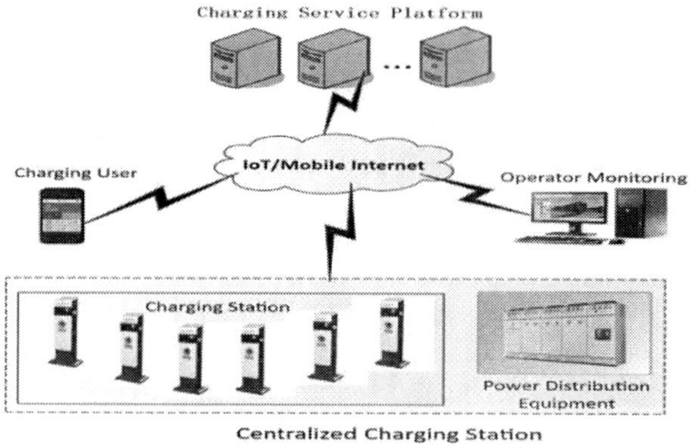

Figure 1. Charging station service system

The orderly charging strategy designed in this paper for centralized charging stations has the following functions.

A. Reduce vehicle queuing time by calculating the remaining charging time of the charging vehicle and selecting the optimal soc value for terminating charging to improve charging station service efficiency and reduce the queuing time of charging users.

B. Reduce charging costs by designing an orderly charging strategy to allow vehicles to charge during off-peak hours when electricity prices are low. Currently, most charging stations divide charging time into three periods: normal, off-peak, and peak, according to the power grid price. These periods correspond to normal, preferential, and high-price billing, respectively. The charging prices and tiered periods are shown in Table 1.

Table 1. Charging tariff and ladder period

Time Slot	Electricity Price	Service Charge
8:00 AM to 11:00 AM And 6:00 PM to 11:00 PM	1.09 CNY/kWh	0.8 CNY/kWh
7:00 AM to 8:00 AM And 11:00 AM to 6:00 PM	0.68 CNY/kWh	0.8 CNY/kWh
11:00 PM to 7:00 AM	0.36 CNY/kWh	0.8 CNY/kWh

C. Optimize the grid load. During peak periods of electricity consumption, the increased load pressure on the grid will cause network losses and reduce the quality of electricity. This article designs a charging pile power adjustment strategy to enable the charging station power to be dynamically adjusted according to the grid demand.

3. Structured charging strategy for dedicated charging stations

At present, dedicated centralized charging stations are mostly used for charging electric buses. This article takes a charging station with 30 60kw DC piles serving 100 electric buses as the design object. In order to improve the efficiency of charging piles and reduce the queuing time of vehicles, this article analyzes more than 30 charging data. The conclusion is that the charging process data curves of similar charging piles are basically the same. This article extracts a vehicle charging process data from 33% to 100% of SOC as a sample for graphic processing and analysis. The charging process data is shown in Table 2 (charging voltage unit is V, power unit is A, time unit is minute).

Table 2. Charging tariff and ladder period

Time	0	1	2	3	4	5	6	7	8	9	10
Soc(%)	33	34	35	36	37	38	39	40	41	42	43
Charging Voltage	568. 1	573	574. 4	574. 4	574. 4	574. 9	574. 4	574. 4	574. 4	574. 4	574. 4
Charge Current	98	97.8	96.9	97	97.5	97.1	97.5	96.9	97.5	97.8	97.1
Time	11	12	13	14	15	16	17	18	19	20	21
Soc(%)	44	45	46	47	48	49	50	51	52	53	54
Charging Voltage	574. 4	574. 4	574. 4	574. 4	574. 4	574. 4	574. 4	574. 4	574. 4	574. 4	574. 4
Charge Current	97.1	98	98	96.9	97.1	97.1	96.9	97	97	97	97.1
Time	22	23	24	25	26	27	28	29	30	31	32
Soc(%)	55	56	56	57	58	59	60	61	62	63	64
Charging Voltage	574. 9	574. 9	574. 9	574. 9	574. 9	574. 9	574. 9	574. 9	574. 9	574. 9	574. 9
Charge Current	97.7	97.7	97.6	98	97.8	97.5	97.6	97	97	97.4	97.4
Time	33	34	35	36	37	38	39	40	41	42	43
Soc(%)	65	66	66	67	68	69	70	71	72	73	74
Charging Voltage	574. 9	574. 9	574. 9	574. 9	574. 9	574. 9	574. 9	574. 9	574. 9	574. 4	575. 4
Charge	97	97.4	97.1	97.8	96.8	96.9	97.1	96.8	97.6	98	97

Current											
Time	44	45	46	47	48	49	50	51	52	53	54
Soc(%)	75	76	76	77	78	79	80	81	82	83	83
Charging Voltage	575.4	575.4	575.4	575.4	575.4	575.4	575.4	575.9	575.4	575.9	575.9
Charge Current	97	98	98	97.8	96.8	97.7	97.3	97.8	98	97.1	98
Time	55	56	57	58	59	60	61	62	63	64	65
Soc(%)	84	85	86	87	87	89	90	91	92	93	93
Charging Voltage	575.9	575.9	575.9	575.9	584.6	586.1	580.7	579.8	579.8	579.8	580.3
Charge Current	96.8	96.8	97	97.7	101.3	101	100.2	99.1	98.6	98.4	98.4
Time	66	67	68	69	70	71	72	73	74	75	76
Soc(%)	94	95	95	96	97	97	98	98	98	99	99
Charging Voltage	580.3	574.4	573.5	573.5	573.9	573.9	573.9	573.9	573.9	74.4	74.4
Charge Current	99.5	73.4	73.2	73	72.5	72.4	71	70.9	71.7	71.5	71.7
Time	77	78	79	80	81	82	83	84	85	86	87
Soc(%)	99	99	99	99	99	99	99	99	99	99	99
Charging Voltage	74.9	74.9	74.4	575.4	575.9	576.4	576.9	576.9	577.3	578.3	578.8
Charge Current	71.4	71.5	71	71.1	71.4	71.3	70.8	71.5	71.6	70.9	70.9
Time	88	89	90	91	92	93	94	95	96	97	98
Soc(%)	99	99	99	99	99	99	99	99	99	99	99
Charging Voltage	579.8	580.7	582.2	584.2	581.2	581.2	583.2	586.6	580.3	580.7	579.8
Charge Current	70.8	70.9	47.7	47.8	47.8	47.7	24.7	24.2	24.2	24.2	10.2
Time	99	100	101	102	103	104					
Soc(%)	99	99	99	99	99	99					
Charging Voltage	575.9	575.4	576.4	577.8	580.3	581.7					
Charge Current	10.1	10	10.2	10.2	10.1	1.3					

As shown in Figure 2, the relationship curve between the charging voltage, current, SOC and charging time of one of the charging piles is shown in Figure 3. The relationship curve between the charging current, SOC and charging time is shown in Fig 3. It can be seen from the figure that the voltage remains basically stable during the entire charging process. In the range where SOC is less than 95%, the charging current fluctuates slightly. After SOC reaches 94%-95%, the output current of the charging pile begins to drop significantly, especially between 99%-100%, the current changes many times, and the charging time is relatively long. The analysis of the charging process data changes is shown in Table 3.

3.1. Design of peak-off-peak periods and queue-orderly charging strategy

The vehicle operation mode in the charging station is generally to use it during the day and charge it at night, and charging is concentrated in the valley electricity price period. However, due to special circumstances, when charging is required during the peak or flat electricity price period during the day, in order to reduce the charging cost and not affect the operation of the vehicle, a charging strategy is adopted to set the upper limit of the soc value of the vehicle to stop charging. The charging is terminated when the soc of the vehicle reaches 80% and 90% during the peak and flat electricity price periods, respectively, and the soc upper limit is not set during the valley period. On the other hand, in the case of many cars and few charging piles, in order to improve the efficiency of charging piles and increase the operating efficiency of charging stations, the upper limit of the soc charging of vehicles is also set. The definition of charging queue in this article is queuing for charging due to urgent need for the car. For the charging in the valley period at night without demand, queuing in turn does not fall into the scope of queuing charging defined in this article.

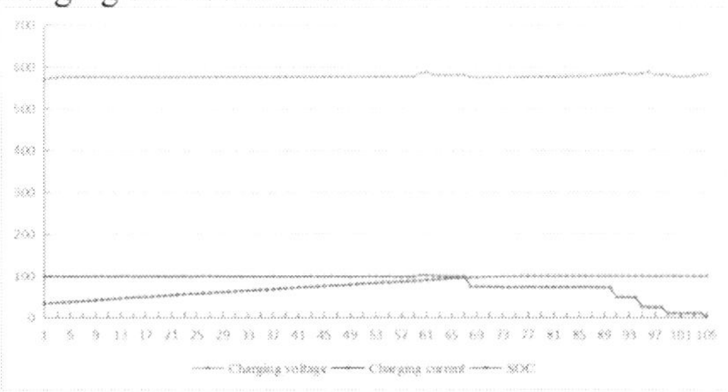

Figure 2. Charging process curve

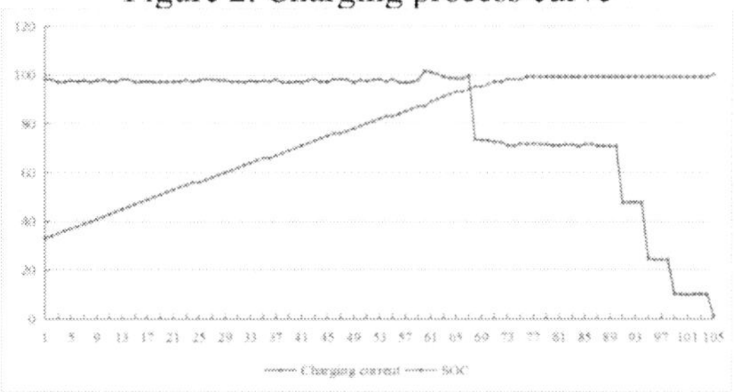

Figure 3. Charging process curve

Table 3. Data analysis of charging process

Soc	Charging Voltage	Charge Current	Charging Duration
30%-95%	574V-586V	97A-101A	66min
95%-99%	574V-580V	71A-73A	23min
99%-100%	574V-580V	47A-48A,24A-25A,10A-11A	15min

Assuming that the number of charging piles in the station is N and the number of vehicles waiting in line for charging is M, there is a relationship between the proportion of vehicles in the line (rate):

$$\text{rate} = N/M \tag{1}$$

The relationship between charging period, rate and charging termination SOC is shown in Table 4. The specific setting method is:

When the charging period is the peak period and the rate is greater than 1.5, the upper limit of the SOC reached by the vehicle charging is 80%;

When the charging period is the normal period and the rate is 0.75-1.5, the upper limit of the SOC reached by the vehicle charging is 90%;

During the valley period, when the rate is in the range of 0.4-0.75, the upper limit of the SOC reached by the vehicle charging is 95%;

During the valley period, when 0.4≥rate>0, the upper limit of the SOC reached by the vehicle charging is 99%.

Table 4. SOC partition of charging termination

Charging Termination Soc	Charging Period	Queued Vehicle Proportion
80%	Peak Period	rate >1.5
90%	Off-Peak Period	1.5≥ rate >0.75
95%	Valley Period	0.75≥ rate >0.4
99%	Valley Period	0.4≥ rate >0
100%	Valley Period	0

3.2. Design of power-limited and structured charging strategy for electric vehicle charging stations

There are no strict requirements for the charging priority of vehicles at dedicated charging stations. When the output power of the charging station is limited by the grid load during charging, an orderly charging strategy is designed to adjust the output power of the charging piles while ensuring that the vehicles have equal charging privileges. The charging pile power adjustment is divided into two types: downward adjustment and upward adjustment. When the total output power of the charging station is greater than the power limit value, it needs to be adjusted downward, otherwise it needs to be adjusted upward. Assuming that the power limit value and the total output power in the station are P_{limit} and P_{charge} respectively, the downward adjustment strategy is designed as follows:

A. Assuming that the number of charging piles with a vehicle soc greater than 90% and a charging current greater than 10A is a, the charging voltage and current are U_{charge} and I_{charge}, and the maximum charging current is I_{max}, then there is a relationship.

$$P_1 = \sum_1^a U_{charge} * I_{charge} \quad (2)$$

$$P_2 = \sum_1^a U_{charge} * 10 \quad (3)$$

$$P_3 = \sum_1^b U_{charge} * 1 \quad (4)$$

$$P_4 = \sum_1^c U_{charge} * 1 \quad (5)$$

In the above formula, P_1 is the charging power of a charging pile, P_2 is the charging power after the current of a charging pile is reduced to 10A, and P_3 is the power value after the current of the charging pile with a current of I_{max} is reduced by 1A. If P_{charge}-P_{limit}<=P_1-P_2, it means that there is no need to reduce the current of the charging piles with soc greater than 90% to 10A. At this time, the current of the b charging piles with the largest current is reduced by 1A. If P_{charge}-P_{limit} <=P_3, the charging service platform will issue a power reduction instruction to the charging pile with a current of I_{max}, and the power regulation requirement will be met if the current is reduced by 1A.

B. If P_{charge}-P_{limit} > P_3, continue to calculate the current reduction value. In the above formula, P_4 is the power value after the current of the charging pile with a current of I_{max} is reduced by 1A and then the current of the charging pile with a current of I_{max}-1 is reduced by 1A. P_{charge}-P_{limit} <= P_3 + P_4, which means that the current value of the charging pile has been calculated, and then the charging service platform sends a power reduction instruction to the charging pile. P_{charge}-P_{limit} > P_3 + P_4, then the current value of the charging pile is calculated by analogy in this way;

C. If P_{charge}-P_{limit} >P_1-P_2, it means that after limiting the charging current of charging piles with soc greater than 90% to 10A, the output power of the charging station is still greater than the limited power, and the power of the charging piles with soc of 80%-89% of the charging vehicle is further reduced. The lower limit of the charging current is 20A. The charging service platform calculates the lower current value of the charging pile in the same way as steps A and B;

D. If the output power of the charging station is still greater than the limited power of the charging station after the above power limit adjustment is performed, then by analogy, the charging piles are continued to be limited in power until the charging output power of all charging stations is equal to the limited power. The lower limit value of the charging current is shown in Table 5. The soc is divided into 10 intervals, and a current limit lower limit value is set for each interval.

Table 5. Lower limit of charging output current limit

Charging Vehicles Soc	Minimum Current Limit Value
90%-99%	10A
80%-89	20A
70%-79%	30A
60%-69%	40A
50%-59%	50A
40%-49%	60A
30%-39%	65A
20%-29%	70A
10%-19%	75A
0-9%	80A

When the charging pile power is increased, the charging service platform calculates the current reduction value in the opposite way to the downward calculation strategy. Assuming that the power limit value and the total output power in the station are P_{limit} and P_{charge} respectively, the upward strategy is designed as follows:

A. The charging service platform calculates the total required power value P_{need} and the current total power usage P_{use} of the charging pile when the vehicle soc is between 0% and 9%, and compares the value of P_{need} - P_{use} with P_{limit} - P_{charge};

B. If P_{need} - P_{use} < P_{limit} - P_{charge}, it means that the power of all charging piles with SOC between 0-9% cannot be increased to the charging demand power. Assuming that the minimum charging current in the charging pile is I_{min}, the calculation strategy is to first increase the current of the charging pile with a current of I_{min} by 1A. If the power increase condition is not met at this time, then increase the current of the charging pile with the adjusted current of I_{min}+1 by 1A. Then, the lower current value of the charging pile is calculated by analogy in this way.

C. If P_{need}-P_{use}>P_{limit}-P_{charge}, it means that the charging current of the charging pile at 0-9% will be increased to the demand value, and the output power of the charging station is still less than the limited power. The power of the charging pile with the soc of the charging vehicle at 10%-19% will continue to be increased. The charging service platform calculates the increased current value of the charging pile in the same way as steps A and B.

3.3. Implementation of structured charging strategy for dedicated charging stations

The dedicated charging station has designed an orderly charging strategy from three aspects: charging cost at peak and valley electricity prices, charging queue efficiency with few piles and many vehicles, and interaction with grid load. The strategy is implemented as follows:

A. During peak and flat electricity price periods, all charging piles are occupied. Later, vehicles that need to charge urgently for some reason can apply for queuing charging through a universal mobile app;

B. The charging service platform sends the charging termination soc strategy for different electricity price periods and ordered charging and the current number of queued vehicles to each charging pile. The charging pile calculates the charging termination soc value based on the actual situation and stops charging when the vehicle soc reaches the specified value;

C. The charging service platform collects the power limit value from the power distribution equipment of the charging station in real time, and selects the power regulation strategy based on the total output power and total demand power of the charging piles in the station. Assuming that the power limit value, total output power and total demand power in the station are P_{limit}, P_{charge}, and P_{need} respectively, there is a relationship:

$$P_{charge} = \sum_1^n U_{charge} * I_{charge} \tag{6}$$

$$P_{need} = \sum_1^n U_{need} * I_{need} \tag{7}$$

In the above formula, U_{charge} and I_{charge} are the charging output voltage and current of the charging pile, U_{need} and I_{need} are the required voltage and current of the charging vehicle, and n is the number of charging piles being charged.

D. When $P_{limit} \geq P_{need}$, the charging piles in the station are set to unlimited power state;

E. When $P_{charge} \geq P_{limit}$, the output power of the charging piles needs to be lowered. First, the power of the charging piles with high soc of the charging vehicle is lowered, that is, the power of the charging piles with soc greater than 90% is lowered. The charging service platform calculates the power reduction value of this part of the charging piles. The calculation method is to cyclically reduce the charging output current by 1% and then compare the output power of the charging station with the power limit value until the output power of the charging station is equal to the power limit value. In the cyclic calculation, the current cannot be less than the lower limit value of the current limit after the current is reduced; if the output current of the charging pile with soc greater than 90% is reduced to the lower limit value and still does not meet the power regulation requirements, the power of the charging piles with soc of 80%-89% will continue to be lowered according to the above 1% cyclic calculation method, and so on until the output power of the charging station is equal to the power limit value;

F. When $P_{need} > P_{limit} > P_{charge}$, the charging pile power increase strategy is started. First, the charging pile power with soc in the range of 0-9% is increased, and the charging current of the charging pile is increased in sequence according to the soc setting interval until the output power of the charging station is equal to the power limit value. The charging service platform calculates the charging pile power increase value by cyclically increasing the charging output current by 1% and then comparing the charging station output power with the power limit value. In the cyclic calculation, the output power of the charging pile after the increase cannot be greater than the vehicle charging demand power.

4. Structured charging strategies for public charging stations

Compared with dedicated charging stations, public charging piles are open to the public in addition to serving the dedicated vehicles of the station, allowing all social vehicles to charge, which not only increases the utilization rate of charging piles, but also brings additional benefits to operators. For dedicated vehicles and social charging vehicles in the station, the article designs an orderly charging strategy in which the charging priority of dedicated vehicles is higher than that of social vehicles, and the principle of first-come-first-served priority is followed among social vehicles.

4.1. Design of structured charging strategies for public charging stations

Public charging stations provide charging services for both special-purpose vehicles and social vehicles, and their orderly charging strategy is designed as follows:

A. In order to reduce operating costs, special vehicles in the station are generally charged during the night off-peak period. Therefore, social vehicles are prohibited from charging half an hour before the off-peak period, and all social vehicles are suspended for ten minutes before the off-peak period;

B. The soc value of social vehicle charging suspension is not limited during the peak-off-peak electricity price period, and the soc value of special vehicle charging suspension is limited according to the special charging station strategy;

C. When $P_{need} > P_{limit}$, the standby charging pile is set to the prohibited charging state (only prohibited for social piles), and when $P_{limit} \geq P_{need}$, the charging pile in the station is set to the unlimited power state;

D. When $P_{charge} > P_{limit}$, the power reduction strategy is started. First, the standby charging pile is set to the prohibited charging state (only prohibited for social piles), and then the social vehicle charging pile with a later charging start time is first reduced in power according to the charging start sequence. The charging current can be reduced to 10A at the lowest. If the output current of the charging piles for all social vehicles is reduced to 10A and still does not meet the power limit requirement, the charging piles for special vehicles are reduced in power according to the orderly charging strategy of the special station;

E. If $P_{need} > P_{limit} > P_{charge}$, the power needs to be increased, and the power is allocated to special vehicles first, and then the excess power is allocated to social vehicles. According to the charging start order, the social vehicles with the earliest charging start will be increased first, and the maximum power can be increased to the vehicle charging demand until the power is fully allocated;

F. When there are few piles and many vehicles, they need to queue up for orderly charging. At this time, it is set that the social charging vehicle will stop charging when the charging soc reaches 95%, and the charging pile and charging power will be provided to the vehicles in the queue, so as to improve the utilization rate of the charging pile and reduce the user's queuing time.

4.2. Implementation of queue-orderly charging strategy for public charging stations

When all the charging piles in the station are occupied, or the charging piles in standby are prohibited from charging, users can use the mobile app to apply for queuing charging. After receiving the application, the charging platform generates a verification code and calculates the charging pile with the shortest queuing time, and pushes the verification code and pile number to the user's mobile app. The verification code is also sent to the queued charging pile for identity verification when the queuing user charges.

When implementing the queued orderly charging strategy, the charging service platform needs to calculate the waiting time for the current queuing user. Assuming that the charging vehicle starts charging and the current soc is Soc_1 and Soc_2 respectively, the charging time is T_1, the charging time required to increase the soc by 1% is T_2, and the remaining charging time is T. Based on the fact that the time required for the soc to increase by 1% when the soc of the charging vehicle is less than 95% is basically the same, there is a relationship:

$$T_2 = (Soc_2 - Soc_1)/T_1 \tag{8}$$

$$T_{left} = (95\% - Soc_2)/T_2 \tag{9}$$

When the number of vehicles in the queue is greater than the number of charging vehicles or $P_{need} > P_{limit}$, the charging service platform will not be able to accurately calculate the waiting time in the queue. The platform will only push the verification code to the mobile phone APP. When the number of vehicles in front is less than the number of charging vehicles and $P_{limit} \geq P_{need}$, the platform will push the queue time and charging pile number to the user's mobile phone app. The orderly charging process of queuing is shown in Figure 4, and the specific implementation is as follows:

A. Charging users make a reservation and queue up through the mobile app. The charging platform sends the generated queue verification code and pile number to the mobile app, and the verification code to the charging pile;

B. If the charging pile reserved by the queuing user is in idle state, the charging pile can be plugged into the charging pile. After the charging pile detects the vehicle connection status, it prompts the user to enter the verification code on the human-machine interface;

C. Charging pile authentication verification code. If the verification code is incorrect, it prompts the charging pile to be booked and prohibited from use. If the authentication is passed, the user is prompted to swipe the card or scan to start the charging process;

D. The platform authenticates the type of charging user. If the charging vehicle is a dedicated vehicle user of this station, the charging pile will be started immediately to start charging. If the charging power required by the vehicle is greater than the limited power of the charging pile, the charging service platform will adjust the power according to the orderly charging strategy of the public charging station;

E. If the charging vehicle is not a dedicated vehicle of this station, if the charging pile is not in a prohibited charging state, it will directly start charging, and the charging power is the power limit value;

F. If the charging pile is in a state where charging is prohibited, the electronic lock of the charging gun will be closed and the charging pile will enter a charging waiting state. The charging service platform will calculate the charging waiting time and give it to the mobile app and the charging pile until the charging pile is allocated power and charging is started. The charging power is the power limit value. Users do not need to wait on site while waiting for charging, as the charging pile can execute automatically.

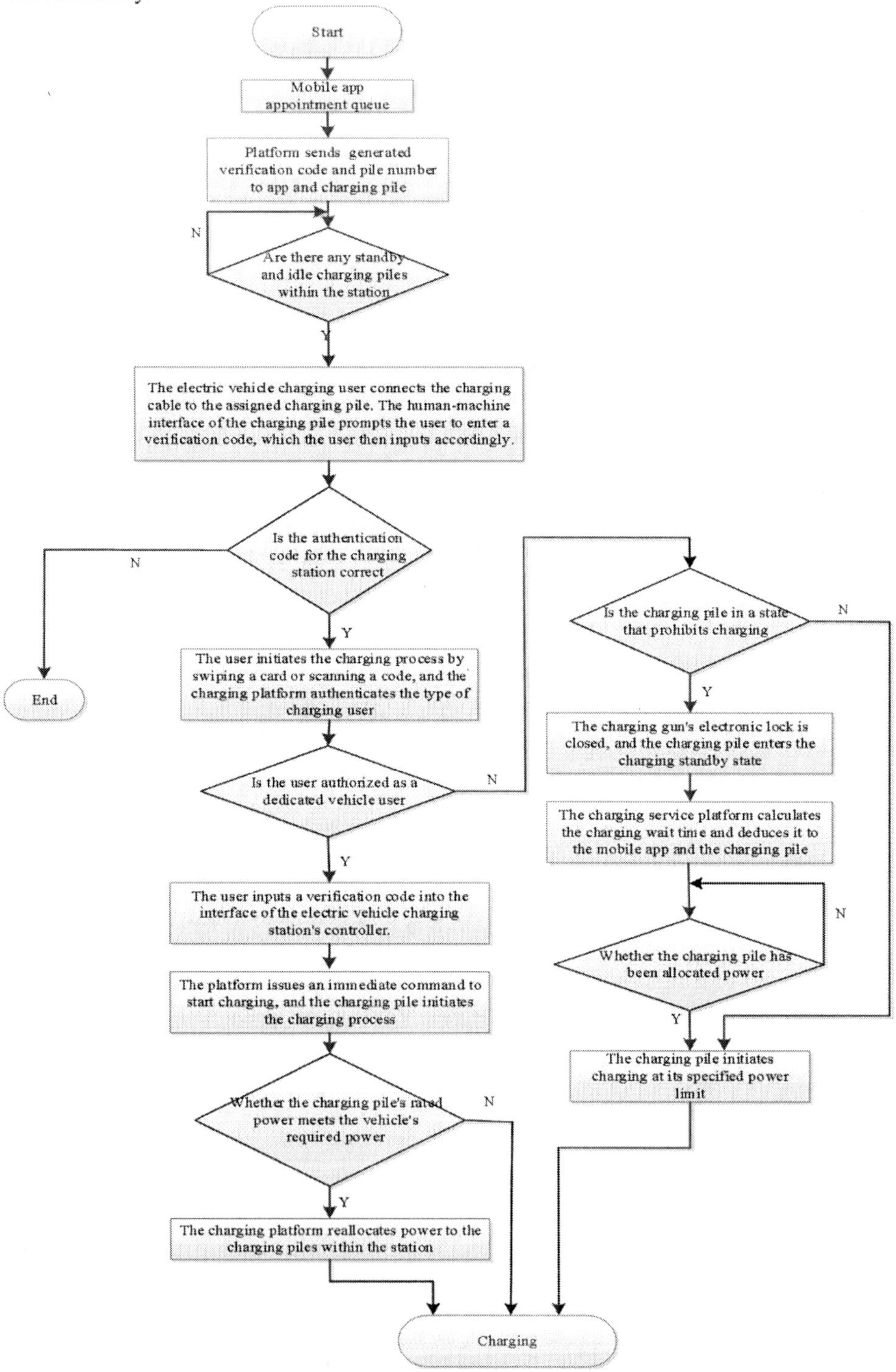

Figure 4. Orderly charging process of public charging station

5. Conclusion

The article designs structured charging strategies for dedicated and public centralized charging stations from three aspects: grid load, time-of-use electricity pricing, and queuing charging, to ensure the safe operation of the charging station's grid, improve the efficiency of charging pile usage, and increase the operational benefits of the charging station.

Structured charging of charging piles is still in its infancy in research. With the development and application of V2G (Vehicle-to-Grid) and charging/discharging technologies, structured charging will gradually become the development direction for charging piles and stations.

References

[1] Chen Z and Xiao XN and Lu XY 2013 Multi-objective optimization for capacity configuration of PV-based electric vehicle charging stations (Transactions of China Electrotechnical Society vol 28) pp 238-248

[2] Hu DD and Yang XH and Ke X 2018 Design and implementation of charging pile group operation management unit (Power System Protection and Control vol 46) pp 135-141

[3] Li HF and Chen ZG and Guo W 2014 Discussion on the technical scheme of community electric vehicle charging service platform (Power System Protection and Control vol 46) pp 138-144

[4] Xiao XN and Wen JF and Tao S 2014 Study and recommendations of the key issues in planning of electric vehicles charging facilities (Transactions of China Electrotechnical Society vol 29) pp 1-10

[5] Zhang CQ and Liu JH and Yi D 2016 Research on the effect of the distributed power grid-connected load property on electric energy metering (Electrical Measurement & Instrumentation vol 53) pp 75-80

[6] Wen CK and Chen JC and Teng JH 2012 Decentralized plug-in electric vehicle charging selection algorithm in power systems (IEEE Transactions on Smart Grid vol 3) pp 1779-1789

[7] Xiao JK and Qiu XY and Pan YJ 2019 Interactive scheduling strategy between electric vehicles and power grid based on group optimization (Electrical Measurement & Instrumentation vol 56) pp 1-7, 26

[8] Chen Z and Liu Y and Zhou T 2020 Optimal time-of-use charging pricing strategy of EVs considering mobile characteristics (Electric Power Automation Equipment vol 40) pp 96-102

[9] Li LN and Liu HJ and Zhang ZY 2018 Research on coordinated control of the interaction of electric vehicles in dc microgrid (Electrical Measurement & Instrumentation vol 55) pp 117-122

[10] Li B and Liu C and Chen HM 2016 Coordinated Charging of Plug-in Electric Buses in Fast Charging Stations Based on Mixed-Integer Programming (Power System Technology vol 40) pp 2623-2629

[11] An Y and Li HQ and Kan LF 2019 Charging and discharging optimization strategy of electric vehicles in residential areas considering users' overall satisfaction (Electrical Measurement & Instrumentation vol 56) pp 7-13

[12] Huang XQ and Chen W and Xie QB 2018 The influence of users' charging selection on charging schedule power grid (Transactions of China Electrotechnical Society vol 33) pp 3002-3011

[13] Guo BF and Zhang P and Wang WX 2019 Research on SOC estimation of LiFePO battery based on OCV-SOC curve cluster (Chinese Journal of Power Sources vol 43) pp 1125-1128, 1139

[14] Wang Q and Sun YK and NI FY 2016 A new method of battery state of charge prediction in the hybrid electric vehicle (Transactions of China Electrotechnical Society vol 31) pp 189-196

[15] Li M and Deng YJ and Yang HM 2016 Electric Vehicle Route Selection and Charging Navigation Strategy Considering Time-of-Use Price (Southern Power System Technology

vol 10) pp 61-66
[16] Liu ZZ and Yang Y and Qu DM 2017 Coordinated charging strategy for electric vehicle aggregator based on time-of-use price (Electric Machines and Control vol 21) pp 1-7

An Incipient Fault Identification Method for Medium-Voltage Cables Based on Shield Current

Hui Qin[1], Chizhi Huang[1], Chi Zhang[1], Jinfeng Hu[2], Jian Dai[2], Yunxiang Xu[1], Chuanyi Liu[1], Xuanhao Liu[1], Wenhai Zhang[3], Zhiyao Li[3*]

[1] China Electric Power Research Institute

[2] State Grid Wum Power Supply Company

[3] Sichuan University

[*]Corresponding author's e-mail: li-zhiyao@qq.com

Abstract. Practical experience has shown that cables can experience transient, repetitive arc breakdowns before permanent failures. Timely detection and identification of these faults have significant value for fault early warning. Incipient faults cause noticeable disturbances in shielding layer grounding currents due to instantaneous breakdowns from the cable core to the shielding layer. Therefore, this paper proposes a trigger detection algorithm based on periodic RMS value and mutation of single-ended shielding layer grounding currents in low-resistance grounded systems. For triggered data, a method for incipient fault identification is introduced, using RMS value, cross-correlation, and fault duration. The algorithm identifies whether the disturbance is an incipient fault and locates the fault section, corresponding to fault section localization in medium-voltage distribution networks.

1. Introduction

Medium-voltage cables are widely used due to their large capacity, small footprint, and low maintenance. In China, the total length of 10kV cables exceeds one million kilometers [1]. A large amount of measured data indicates that incipient faults in medium-voltage cables are manifested as repetitive instantaneous power disturbances (kHz level) before permanent faults. Although these disturbances have small amplitudes and short durations and do not trigger protective actions [2]-[3], they can be recorded by devices with waveform monitoring capabilities in the distribution network.

Efficient identification of incipient faults can be achieved through power disturbance monitoring and analysis. Power disturbances propagate long distances and contain information about fault conditions. With the development of distribution network informatization, many disturbance waveform monitoring devices can monitor kHz-level power disturbances, supporting medium-voltage cable condition sensing. [4]-[5]. Compared to busbar data, shielding layer grounding currents offer better anti-interference, avoiding traditional high-cost sensors.

Cable monitoring mainly relies on limited characteristic quantities like dielectric loss, insulation resistance, partial discharge, and temperature. However, research has not covered the full cable life cycle, and characteristic quantities don't encompass the entire abnormal state process. Medium-voltage cables, typically three-core, have directly grounded shielding layers at both ends, complicating monitoring due to issues like complex current distribution and interference. Given their number and distribution, economy is key. Exploring efficient online monitoring for medium-voltage cables is a recognized challenge [6]-[10].

2. Fault detection

2.1. Principle

The waveform recording device first collects the grounding current of the shielding layer. Under normal operating conditions, the shielding layer's current is almost zero. When an incipient fault occurs in the cable, there is an instantaneous and recoverable breakdown between the cable core and the steel armor layer of the shielding layer, causing an instantaneous increase in the shielding layer's current. Secondly, since the incipient fault is a "self-clearing" fault, the voltage and current operating state will return to normal after the fault is eliminated. Based on these two characteristics, this article will detect incipient faults through the following two trigger algorithms: RMS value and instantaneous value.

2.2. RMS Value Trigger Algorithm

The calculation formula for the RMS value of the grounding current in the end - side shielding layer is presented as follows:

$$I_{rmsi,e}(m) = \sqrt{\frac{1}{N_b}\sum_{n=m-N_b+1}^{m} i_{i,e}^2(n)} \tag{1}$$

In this context, N_b represents the total sampling points of a half-wave signal, $i_{i,e}$ denotes the sampling signal, n and m indicate the sampling sequence and current sampling point respectively, with the RMS value calculated on a point-by-point sliding basis. The rationale for selecting the end current over the beginning is elaborated in the fault identification algorithm in section IV.

Trigger detection criterion 1:

The RMS value amplitude of the grounding current in the terminal shielding layer exceeds the limit:

$$I_{rmsi,e}(m) > I_{index1} \tag{2}$$

Regarding the setting of thresholds, this paper provides a reference threshold for low-resistance grounding systems, defined as $I_{index1}=\sigma_1 I_L$, where I_L represents the RMS value of the rated current. I_{index1} is the trigger threshold. σ_1 is the coefficient. The setting of this threshold is related to system parameters, such as the distance to the fault location, the fault transition resistance, and the system impedance. When there is no fault in the cable, the shielding layer current only consists of equivalent capacitive current, with a very small amplitude. However, when a fault occurs, the shielding layer current may include the fault current itself or the induced current from the fault current.

Trigger detection criterion 2:

When the mutation of the RMS value exceeds the limit:

$$I_{rmsi,e}(m) - I_{rmsi,e}(m-1) > I_{index2} \tag{3}$$

Where $I_{index2}=\sigma_2 I_L$ represents the trigger threshold. σ_2 is the coefficient. The setting of this threshold, as mentioned above, needs to balance accuracy and reliability. Instantaneous Value Trigger Algorithm. The instantaneous mutation is calculated by subtracting the sampling points at corresponding positions of adjacent cycles of the shielding layer's current. The formula for calculating the instantaneous mutation of the single-ended the shielding layer's current is as follows:

$$i_{insi,e}(m) = i_{i,e}(m) - i_{i,e}(m - N_a) \tag{4}$$

Where N_a represents the number of sampling points in one cycle of the signal, $i_{i,e}$ denotes the sampling signal of the shielding layer's current, and m indicates the sequence number of the current sampling point.

Trigger detection criterion:

When the instantaneous mutation of the single-ended grounding current in the shielding layer exceeds the limit:

$$i_{insi,e}(m) > i_{index3} \tag{5}$$

Where i_{index3} is the trigger threshold for the instantaneous mutation of the shielding layer's current, and $i_{index3}=\sigma_3 I_L$. σ_3 is the coefficient.

3. Fault identification

3.1. Principle

Since fault section identification is achieved solely through the grounding current of the shielding layer at a single end, selecting whether the grounding current at the beginning or the end of the cable should be used as the identification target becomes an issue that needs to be addressed.

Figure 1(a) shows grounding current directional characteristics at cable section ends. Positive direction is current flow from cable to ground in shielding layer. Faults cause unbalanced three-phase currents, with fault current flowing same as zero-sequence current.

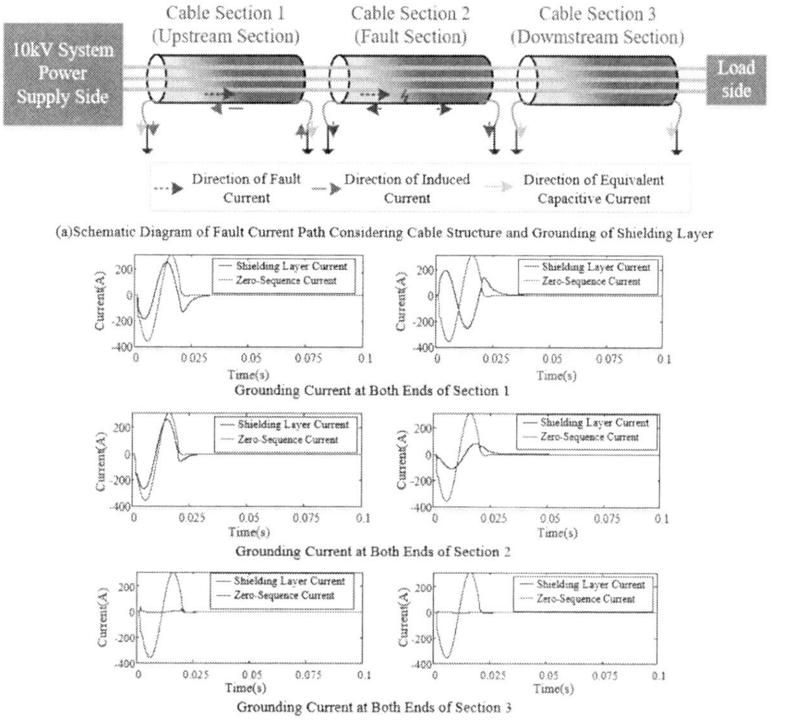

(a)Schematic Diagram of Fault Current Path Considering Cable Structure and Grounding of Shielding Layer

(b)Diagram of Shielding Layer Grounding Current and Zero-Sequence Current Signals at the Beginning and End of Different Sections

Figure 1. Diagram of Characteristic Directions of Sheath Grounding Currents at the Beginning and End of Different Sections

In the upstream section, shielding layer current is mainly induced by fault current, larger than equivalent capacitive current. Combined current direction matches induced current. Grounding currents at both ends are opposite, with beginning positive and end negative. In the fault section, fault breakdown current flows through shielding layer from cable to ground. Both ends have positive direction. Comparative analysis of shielding layer grounding currents shows directional characteristics at the end. Beginning grounding current is always positive. End grounding current is negative only in upstream fault section, positive otherwise. Thus, the proposed method selects end grounding current as identification signal input.

Journal of Physics: Conference Series **3012** (2025) 012082 doi:10.1088/1742-6596/3012/1/012082

From the diagrams of the grounding current direction characteristics in the sheaths at the beginning and end of different sections, particularly focusing on the end sections in Figure 1(b), the following two points can be clearly observed:

1) The grounding current amplitude in the shielding layer at the downstream part of the fault section shows significantly smaller value compared to those in the fault and upstream sections, allowing distinction.

2) The direction of the grounding current at the termination of the fault section coincides with the zero - sequence current, while upstream, it is in opposition. This difference helps with the identification of the fault section.

The incipient fault identification algorithm relies on: fault duration (1/4 to 4 cycles), grounding current amplitude and direction. Extraction algorithms, threshold setting, and judgment will be elaborated for overall fault identification.

3.2. *Characteristic of fault duration*

When it comes to incipient faults, there may be a "self-clearing" situation, which means that after the fault is eliminated, the current operating state will return to normal. Therefore, besides the starting time, a fault will definitely have an ending time.

Criterion for Determining the Starting Sampling Point of a fault:

$$I_{rmsi,e}(n_{start}-1) < 1.05I_L \&\& I_{rmsi,e}(n_{start}) > 1.05I_L \tag{6}$$

Where, n_{start} represents the starting sampling point of the fault, $I_{rmsi,e}$ denotes the RMS value of the grounding current, and the moment corresponding to the starting sampling point n_{start} is recorded as the fault initiation time t_{start}. Criterion for Determining the Ending Sampling Point of a fault:

$$I_{rmsi,e}(n_{end}-1) > 1.05I_L \&\& I_{rmsi,e}(n_{end}) < 1.05I_L \tag{7}$$

Where n_{end} represents the sampling point at which the fault ends, $I_{rms,e}$ denotes RMS value of the grounding current, and the moment corresponding to the fault sampling point n_{end} is recorded as the fault end time t_{end}.

Calculate the fault duration Δt_{break} and the count of sampling points throughout the fault duration N_{break}.

$$\Delta t_{break} = t_{end} - t_{start} \tag{8}$$

$$N_{break} = N_{end} - N_{strart} \tag{9}$$

Identification Criterion:

A reasonable threshold value N_{index} is set to distinguish between incipient faults and permanent breakdown faults. When N_{break} is less than this specific threshold, it indicates that the fault is an incipient fault.

$$N_{break} < N_{index} \tag{10}$$

The duration of an incipient fault ranges from 1/4 to 4 cycles, thus N_{index} can be established as the number of sampling points within these 4 cycles, specifically,

$$N_{index} = 4\frac{f_s}{f_0} \tag{11}$$

Where, f_s represents the sampling frequency of the monitoring device, and f_0 represents the fundamental frequency.

3.3. *Characteristic of the RMS value*

Observations show the grounding current amplitude in downstream and non-main sections is smaller during incipient faults. A threshold can be set to distinguish downstream faults. When the RMS value of shielding layer grounding current is below this threshold, it indicates a downstream or non-main fault section.

$$I_{rmsi,e}(m) < I_{index4} \tag{12}$$

Where I_{index4} represents the threshold for distinguishing sections based on the RMS value. The setting of this threshold is related to system parameters, and there are significant differences in the values for low-current grounding systems and low-resistance grounding systems. Therefore, the setting of the threshold requires separate system analysis. $I_{rmsi,e}(m)$ denotes the RMS value of the half-wave of the fault grounding current within window m, and the calculation method is shown in (1), which will not be repeated here.

In this paper, a reference threshold for low-resistance grounding systems is provided, where $I_{index4}=\sigma_4 I_{rmsi0}$, with I_{rmsi0} being the RMS value of the half-wave of the zero-sequence current in section i. Based on statistical simulation analysis results and considering the reliability of the identification algorithm, σ_4 is set to 8%.

3.4. Characteristic of the Cross-Correlation

The grounding currents at shielding layer ends in upstream and fault sections are nearly opposite in direction, reflected in fault grounding current waveform phase. Since cross-section communication is impossible and fault identification relies on local devices, a common reference is needed. This paper uses zero-sequence current as the reference. In fault sections, grounding current and shielding layer current are nearly in-phase (phase difference = 0); in upstream sections, they are nearly out-of-phase (phase difference = π).

By analyzing cross-correlation of shielding layer and zero-sequence currents, their temporal relation and phase difference can be quantified to infer fault location and direction. Significant peaks and short delays in cross-correlation indicate simultaneous changes, helping determine fault direction. The Cross-Correlation Function (CCF) of the two signals is calculated as:

$$R_{I_0 i_{i,e}}(\tau) = \sum_{n=0}^{N-1} I_0(n) \cdot i_{i,e}(n+\tau) \tag{13}$$

Where R represents the cross-correlation value between the zero-sequence current $I_0(n)$ and the grounding current at the shielding layer's end $i_{i,e}(n)$ in section i at a delay τ. $i_{i,e}(n+\tau)$ is the value of $i_{i,e}(n)$ after a delay of τ. The optimal delay τ between these two signals is determined by finding the delay corresponding to the maximum value in the CCF result.

Identification Criterion:

A suitable threshold is set in order to differentiate the fault section from the section located upstream of the fault site. Once the RMS value of the grounding current within the shielding layer is lower than a specific threshold, it shows that the section in focus is where the fault lies:

$$\tau\% \frac{4f_0}{f_s} < \tau_{index} \tag{14}$$

Where τ_{index} is the threshold for distinguishing sections based on the cross-correlation between the current at the shielding layer's end and the zero-sequence current, f_s is the sampling frequency of the monitoring device, and f_0 is the fundamental frequency. In this paper, the recommended setting for the threshold τ_{index} is:

$$\tau_{index} = \frac{5 \cdot f_s}{8 \cdot f_0} \tag{15}$$

4. Detection and identification process and steps

This method comprises fault detection and identification modules. The detection module monitors real-time data, calculates RMS grounding current per cycle, triggers warnings based on criteria, and records waveforms if triggered. The identification module analyses recorded waveforms to distinguish incipient vs. permanent faults and locate the fault segment. Steps are detailed below.

4.1. Fault Detection

Step 1: Detect grounding current at shielding layer end using high-frequency CTs, then calculate per-cycle RMS value of grounding current according to (1).

Step 2: Apply detection criteria (2)-(5) to trigger fault warning. If triggered, start waveform recording and store within the device.

4.2. Fault Identification

Step 1: Calculate fault duration Δt_{break} and sampling points N_{break} using (6)-(9). Determine fault type (permanent or incipient) using (10) and a predefined threshold.

Step 2: Calculate $I_{rms,e}$, using (1). Determine if fault segment is downstream or in non-trunk segments using predefined threshold and (12).

Step 3: Use (13) to determine optimal delay τ between shielding layer current and zero-sequence current based on cross-correlation.

Step 4: Use (14) and (15) with predefined thresholds to determine if fault waveform is in fault segment or upstream.

5. Simulation Verification

5.1. Simulation Model Construction

To simulate and verify the effectiveness and reliability of the proposed incipient fault identification method, a simulation model for incipient faults in 10kV three-core armored cables in a distribution network was established in PSCAD/EMTDC software, as shown in Figure 2. A 110kV power supply with a frequency of 50Hz was selected, and the transformer employed a Y/Y connection with a transformation ratio of 110kV/10.5kV and a rated capacity of 45MV·A. The neutral point was grounded through a small resistance with a resistance value of 6Ω. The 10kV busbar had five feeder lines, among which feeder line 1 was relatively complex with four sections. The other four feeder lines had consistent parameters except for differences in length and load size. The fault simulation module employs the Kizilcay arc model that shown as (16), utilizing resistors and switches to create controllable-duration faults between the cable conductor and the shielding layer.

$$\frac{dg}{dt} = \frac{1}{\tau_k}\left(\frac{|i_f|}{u_0 + r_0|i_f|} - g\right) \tag{16}$$

Here, g represents conductance, t represents time, τ_k represents the time constant, r_0 is the characteristic arc resistance, and u_0 is the characteristic arc voltage.

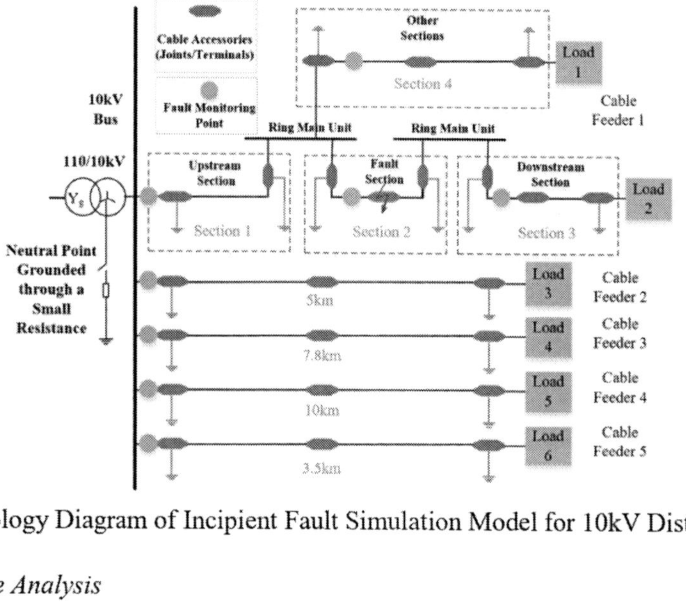

Figure 2. Topology Diagram of Incipient Fault Simulation Model for 10kV Distribution Network

5.2. Typical Case Analysis

Simulation data validation results show that under favorable noise conditions, this algorithm's fault identification accuracy is 95.7%, with clearly distinguishable feature distributions for different fault types and sections. Below are typical case analyses. Each case in detail records the specific fault section, fault distance, and fault duration, providing basic data for further analysis. The fault simulation schemes for each typical case are shown in Table 1.

Journal of Physics: Conference Series **3012** (2025) 012082 doi:10.1088/1742-6596/3012/1/012082

Case Numbers 1 and 2 document two fault events in Section 2 of Cable Feeder 1, both at 6 km but with durations of 0.02 and 0.10 seconds, respectively, demonstrating the method's ability to distinguish incipient from permanent faults. Case 3 records a 0.01-second incipient fault in Section 3 of the same feeder at 5 km, showing identification capability for different distances and durations. Case 4, a 0.04-second fault in Cable Feeder 4 at 7 km, demonstrates the method's ability to identify incipient faults on different feeders. Figure 3 showcase the distribution of characteristics extracted from the shielding layer grounding currents of various sections for Cases 1 to 4 using the algorithm proposed in this chapter. These include the distribution of fault duration characteristics, periodic RMS value characteristics, and cross-correlation characteristics. The fault sections are marked with yellow boxes.

Table 1 Fault Simulation Schemes for Typical Cases

Case Number	Fault Section	Fault Distance/km	Fault Duration/s
1	Cable Feeder 1 - Section 2	6	0.02
2	Cable Feeder 1 - Section 2	6	0.10
3	Cable Feeder 1 - Section 3	5	0.01
4	Cable Feeder 4	7	0.04

In Figure 3, there are three bar graphs, each being representative of one of the three fault characteristics introduced in Chapter 3. Specifically, the red is for fault duration, the blue is for the RMS value of fault current, and the green is for the cross-correlation coefficient. Each color includes four subgraphs, corresponding to Cases 1 through 4 in Table 1. The numbers on the horizontal axis represent the cable feeders shown in Figure 2, such as "2" for Feeder 2 and "1-1" for the first part of Feeder 1. The sections highlighted by yellow boxes indicate where faults are set, like "1-2" in the first subgraph of the red bar graph. The vertical axis indicates the characteristic values of the fault features, for example, the vertical axis of the red bar graph represents the characteristic values of fault duration.

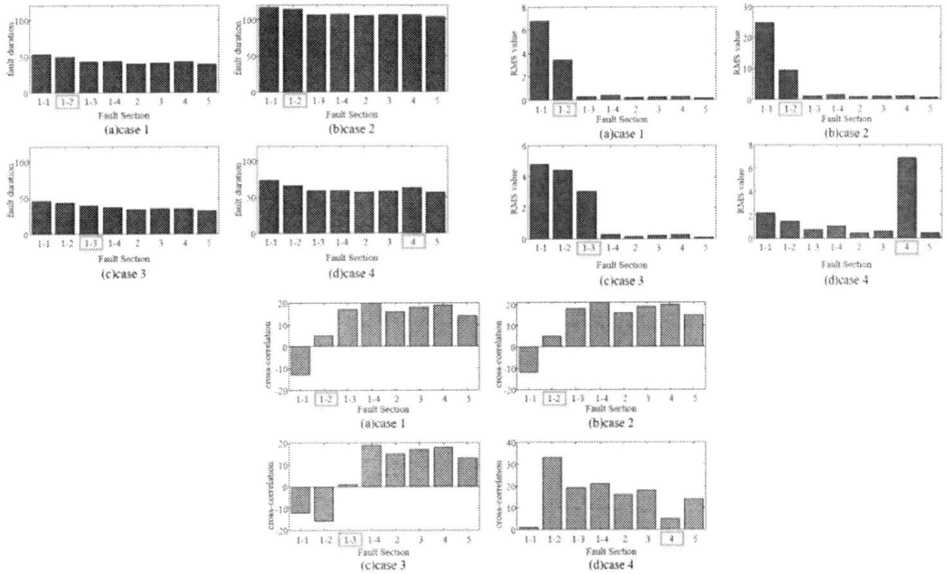

Figure 3. Characteristic Distribution of Fault Duration, RMS Values, Cross-Correlation for Shield Layer Grounding Current in Various Sections of Cases 1-4

Analyzing the red bar graphs, it is evident that Case 2, as the only permanent fault, has a fault duration significantly longer than those of Cases 1, 3, and 4, thereby verifying the feasibility of the fault duration algorithm.

Upon analyzing the blue bar graphs, it can be observed that the RMS values of the fault section and its upstream section are significantly higher than those of the downstream section, validating the effectiveness of the fault RMS value algorithm.

By analyzing the green bar graphs, it is apparent that only the upstream section has a negative cross-correlation coefficient, while the fault section and the downstream section have positive values, confirming the accuracy of the fault correlation algorithm.

6. Conclusion

During incipient faults, shielding layer grounding currents at both ends of different sections exhibit distinct characteristics. In the upstream section, induced currents cause opposite directions; at the fault section, currents are fault currents with the same direction; in downstream or other sections, currents are mostly equivalent capacitive currents with the same direction. A trigger detection algorithm based on periodic RMS value and mutation of single-end grounding current is proposed. By analyzing fault duration, RMS value, and cross-correlation, precise identification of disturbance data is achieved. This algorithm discriminates between incipient faults and identifies the fault section. Simulation results show a 95.7% fault identification accuracy under good noise levels.

In future work, multi - source information beyond shielding layer grounding currents will be integrated to enhance diagnostic reliability, and a shift from sectional to precise fault location will be made for more efficient system maintenance.

Acknowledgments

This paper was supported by the Science and Technology Project of State Grid Corporation of China (Project No._5400-202318550A-3-2-ZN_).

References

[1] National Energy Administration, China Electricity Council. "2022 National Annual Report on Power Reliability" [R]. June 2022: 151 - 152.

[2] Sidhu T S, Xu Z. Detection of Incipient Faults in Distribution Underground Cables[J]. IEEE Transactions on Power Delivery, 2010, 25(3): 1363-1371.

[3] Kulkarni S, Santoso S, Short T A. Incipient Fault Location Algorithm for Underground Cables[J]. IEEE Transactions on Smart Grid, 2014, 5(3): 1165-1174.

[4] Xu W, Huang Z, Xie X, Li C. Synchronized Waveforms – A Frontier of Data-Based Power System and Apparatus Monitoring, Protection, and Control[J]. IEEE Transactions on Power Delivery, 2022, 37(1): 3-17.

[5] State Grid Corporation of China. "Technical Specification for Transient Wave-recording Fault Indicators" [S]. Q/GDW 11814 - 2018, 2018.

[6] Xiong S, Liu Y, Fang J, Cong Z, Yan Y, Jiang X. Detection Method of Incipient Faults of Power Distribution Lines [J]. High Voltage Engineering, 2020, 46(11): 3970-3976.

[7] Qu K, Zhang W, Xiao X, Chu T. Condition Monitoring for Incipient Faults of Cable Joint Insulation Based on Waveform Attenuation Factor [J/OL]. High Voltage Engineering, 2022, 48(01): 318-327.

[8] Niu Y, Sheng L, Gao M, Wei Y, Zhou D. Incipient sensor fault detection for time-varying stochastic systems under Kalman filtering framework[C]. in 2023 CAA Symposium on Fault Detection, Supervision and Safety for Technical Processes (SAFEPROCESS), Sep. 2023, 1–6.

[9] Wang M, Cheng F, Chen K, Mi J, Xu Z, Qin G. Incipient Fault Detection with Feature Ensemble Based on One-Class Machine Learning Methods[C]. in 2023 62nd IEEE Conference on Decision and Control (CDC), Dec. 2023. 4867–4872.

[10] Zhang Y, Jia K, Li Y, Bi T. Incipient Fault Detection based Active Protection of Photovoltaic DC Integration System[J]. IEEE Transactions on Power Delivery, 2024, 39(2): 1198-1208.

Transient Overvoltage Characteristics Analysis of Renewable Energy Power Stations with SVG

Zhe Cheng[1*], Leilei Guan[1], Guangyao Qiao[1], Hongjin Pan[2], Yunfei Xu[1], Qiuyu Shi[1]

[1] State Key Laboratory of Advanced Power Transmission Technology (China Electric Power Research Institute Co., Ltd), Beijing, 102209, P.R. China

[2] State Grid Jiangsu Electric Power Co., Ltd. Electric Power Research Institute, Nanjing, 211103, P.R. China

*Corresponding author's e-mail: cz960918@163.com

Abstract. Thanks to their abundant resources, easy accessibility and pollution-free nature, wind and solar energy have been widely developed globally. The interaction among wind farms, photovoltaic (PV) power stations, and the power grid is becoming increasingly close. Power grid failures disrupt the stable operation of wind and photovoltaic power system, and even lead to the disconnection of wind and PV power stations. The stable operation of the power grid has set more stringent requirements for the grid connection of wind and photovoltaic system. The ability of wind and photovoltaic stations to pass through low voltage ride-through (LVRT) and high voltage ride-through (HVRT) is one of the basic conditions for renewable energy grid connection. This paper investigates the transient overvoltage characteristics with a system consisting of doubly fed induction generator (DFIG) units, photovoltaic systems, and static var generator (SVG). The analysis results reveal that the SVG can remarkably improve the response characteristics of wind turbines and photovoltaics in the event of faults. Furthermore, by providing dynamic reactive power compensation, it can successfully mitigate the adverse impact that grid voltage dips have on wind turbines and photovoltaics.

1. Introduction

The rapid growth of wind power and solar energy industries is an important direction for the development and utilization of clean energy, and it is a key support for achieving China's strategic goals of "carbon peak and carbon neutrality." According to the data from the National Energy Administration, the cumulative grid-connected capacity of wind power nationwide has reached 480 million kW with a year-on-year increase of 19.8%. The installed capacity of PV power generation nationwide reached 770 million KW with a year-on-year increase of 48.4% [1]. The transient overvoltage issues in power systems with large scale renewable plant resulted from the inherent attributes of such systems [2]: 1) The capacity of clean energy changes in accordance with the environment, which exacerbates the reactive power imbalance, causing voltage fluctuations to become prominent; 2) The changes in active power during the operation of large-scale renewable energy power stations, such as power regulation and system faults, thereby causing serious transient voltage problems, especially overvoltage issues; 3) The grid strength of high-proportion renewable energy systems is weak, making the system more prone to temporary overvoltage. These voltage transient problems might result in severe power cuts, causing the collapse of the entire power grid, and present a severe menace to the steady operation of the system. Therefore, renewable energy systems with LVRT and HVRT capabilities have attracted widespread attention.

To achieve the LVRT and HVRT capabilities of wind power and photovoltaic power systems, there are mainly two ways: adding auxiliary devices and improving system control strategies [3]. Reference [4] suggests that a Crowbar protection circuit can be added on the wind turbine rotor side to stabilize the DC bus voltage, but this makes the doubly fed induction generator (DFIG) absorb a large amount of reactive power. Reference [5] proposes a variable pitch angle control strategy for this issue, and the wind turbine grid-side converter switches to the STATCOM working state, which is favorable for the recovery and rebuilding of the powers system. Reference [6] proposes a reactive current distribution algorithm for the DFIG during the LVRT period and an improved power control strategy, achieving the LVRT of DFIG. References [7] propose the control strategies based on virtual impedance and variable damping for the grid voltage surges on generator rotors, thus enhancing the HVRT performance of the units. Literature [8] implemented the HVRT of wind turbines with the DC discharge circuit technology.

To enhance the LVRT capability of photovoltaic (PV) power stations, literature [9] puts forward an enhanced maximum power point tracking (MPPT) method to achieve LVRT for photovoltaic power stations. However, this algorithm results in an error where photovoltaic power generation deviates from the maximum power point during normal operation. Literature [10] proposes a nonlinear control algorithm for the LVRT control strategy of photovoltaic power stations, but the nonlinearity of the algorithm makes the control structure more complex. To enhance the HVRT capability of photovoltaic power stations, literature [11] proposes a power-voltage hybrid control strategy. This strategy improves the voltage stability of the DC system by switching the control strategy of the Boost circuit. Literature [12] proposes a mixed energy storage system consisting of super-capacitors and batteries, thus enhancing the HVRT capability of renewable energy.

The above literatures focus on the study of LVRT and HVRT capabilities, considering only wind power or photovoltaic power stations, with few studies analyzing the voltage ride-through strategies and transient overvoltage characteristics of renewable energy power stations simultaneously composed of wind power, PV, and static var generator (SVG). In light of this, this paper investigates the transient overvoltage characteristics of wind power – PV renewable energy power stations with SVG, while considering the dynamic characteristics of LVRT and HVRT.

2. Modeling of wind power generation system and PV power station

In this paper, the wind power generation system containing DFIG generator is used, and the structure is displayed in Figure 1. DFIG generator has the characteristics of high temperature and corrosion resistance compared with PMSG generator, and there is a gear box to realize the speed increase of the paddle. In this paper, we take the DFIG motor as an example, the wind power generation system containing DFIG generator, wind turbine, gear box, converter and so on. The wind turbine is used to capture the wind energy and transfer the mechanical power to the rotor by means of the gear box, and the rotor side and grid side are connected through the bidirectional converter to realize the bidirectional transfer of energy.

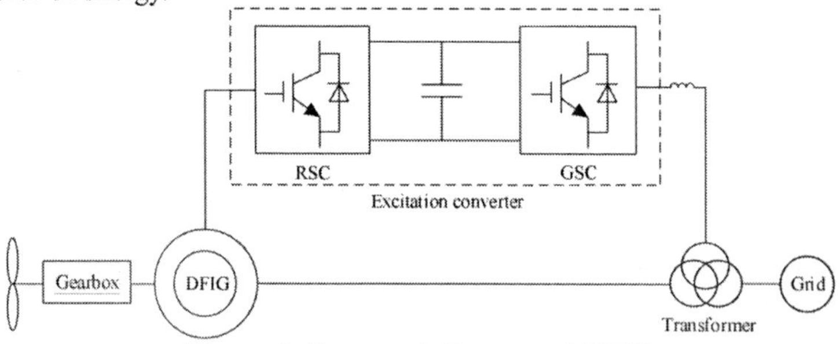

Figure 1. Structural diagram of DFIG

2.1. *Mathematical model of wind power generation system*

To build the DFIG mathematical model in Figure 1, it is assumed that the transformer operates in an ideal state and the environmental effects on the system are not considered. The mathematical model of DFIG motor can be represented with the following:

$$\begin{cases} U_{gd} = -R_g i_{gd} - p\psi_{gd} + \omega_1 \psi_{gq} \\ U_{gq} = -R_g i_{gq} - p\psi_{gq} - \omega_1 \psi_{gd} \\ U_{zd} = R_z i_{zd} + p\psi_{zd} - \omega_2 \psi_{zq} \\ U_{zq} = R_z i_{zq} + p\psi_{zq} + \omega_2 \psi_{zd} \end{cases} \tag{1}$$

$$\begin{cases} P_g = U_{sd} i_{gd} + U_{gq} i_{gq} \\ Q_g = U_{gq} i_{gd} - U_{gd} i_{gq} \\ P_z = U_{zd} i_{zd} + U_{zq} i_{zq} \\ Q_z = U_{zq} i_{zd} - U_{zd} i_{zq} \end{cases} \tag{2}$$

In equations (1) and (2), the indices g and z represent the stator and rotor of the motor; the subscripts d and q represent the d-axis and q-axis; L_g and L_z are the self-inductance of the stator and rotor; ω_1 and ω_2 represent the angular velocity of the dq reference frame, the angular velocity of the dq reference frame relative to the rotor, and the rotor angular velocity, respectively; p represents the differential operator.

2.2. *Mathematical model of PV power station*

The schematic diagram of the PV power station model is shown in Figure 2, which includes a photovoltaic array, MPPT algorithm, and a boost converter. The voltage output from the photovoltaic array is first stepped up by the boost converter, and then it is converted from DC to AC through a three-phase bridge inverter circuit. The output of the MPPT algorithm is used to control the on-off of the three-phase bridge inverter circuit, ensuring that the output power reaches its maximum.

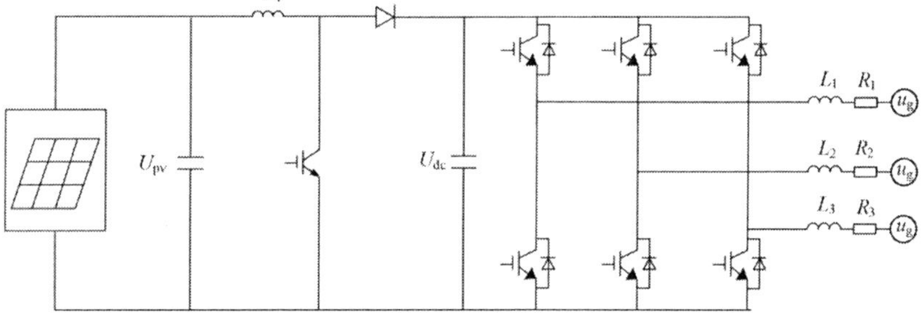

Figure 2. PV grid-connected topology

The mathematical model of PV power station is:

$$\begin{cases} u_{pd} - u_{gd} = ri_{pd} + L\dfrac{\mathrm{d}i_{pd}}{\mathrm{d}t} - \omega L i_{pq} \\ u_{pq} - u_{gq} = ri_{pq} + L\dfrac{\mathrm{d}i_q}{\mathrm{d}t} + \omega L i_{pd} \end{cases} \tag{3}$$

In the equation: u_{pd}, u_{pq}, i_{pd}, i_{pq} represent the d-q axis components voltage and current at the grid side; u_{gd} and u_{gq} are the d-q axis components of the inverter-side voltage; L and r represent the filter inductance and resistance, respectively. Based on the instantaneous power theory, align the d-axis component of the inverter-side voltage with the grid-side voltage u_{pd}, that is, $u_{gd} = u_g$, $u_{gq} = 0$. The active power as well as the reactive power fed into the grid by the inverter are:

$$\begin{cases} P = \frac{3}{2} u_g i_{pd} \\ Q = -\frac{3}{2} u_g i_{pq} \end{cases} \tag{4}$$

When grid side voltage dip takes place due to a fault, the active power delivered by the inverter to the grid side will decrease. If the front-stage circuit is still operating in the MPPT mode, there will be a power imbalance between the DC and AC sides. The power surplus will accumulate on the DC side, causing the DC voltage to become too high and overvoltage protection, and even lead to the disconnection of the photovoltaic system from the grid. If the losses of the inverter and the AC-side filter are ignored, the balance equation for the DC bus voltage is:

$$P_{pv} - P = \frac{1}{2} C \frac{du_{dc}^2}{dt} \tag{5}$$

Where P_{pv} stands for the active power of the photovoltaic side; P represents the power provided by the inverter to the grid side; C represents the capacitance of the DC bus; u_{dc} represents the DC voltage .

3. HVRT and LVRT Control Strategies for wind power and PV systems

3.1. HVRT and LVRT Control Strategies for Wind Power Generation Systems

In renewable energy power stations, the grid-connected inverter of DFIG wind turbines uses a voltage source converter (VSC), and the grid-side converter adopts a grid-following control mode. When high/low voltage faults occur in the system, the grid-connected inverter switches control modes and adopts a reactive power priority control strategy. China requires that renewable energy power stations must ensure they do not disconnect from the grid and supply a definite amount of reactive power support. The requirements for reactive current are as follows:

$$\Delta I_t = K_1 \times (0.9 - U_t) \times I_N,\ (0.2 \le U_t \le 0.9) \tag{6}$$

where K_1 is the reactive current ratio coefficient of wind farm, and the value range is 1.5~3.

When a high voltage fault occurs in the system, the grid-connected inverter also needs to switch control modes to remain connected to the grid for a certain period of time. However, contrary to low voltage faults, high voltage ride-through faults require the grid-connected inverter to absorb reactive power. Typically, this is achieved by increasing the reference figure for the DC-side voltage to stabilize the fluctuations in the DC-side voltage during the HVRT period. The requirements for reactive current are as follows:

$$\Delta I_t = K_2 \times (U_t - 1.1) \times I_N,\ (1.1 \le U_t \le 1.3) \tag{7}$$

where K_2 is the wind farm reactive current ratio coefficient, and the value range should be greater than 1.5.

3.2. HVRT and LVRT Control Strategies for Photovoltaic Power Generation Systems

During the LVRT period, photovoltaic power generation systems should have the capability to provide dynamic reactive power support, with the increment of reactive current as:

$$\Delta I_t = K_3 \times (0.85 - U_t) \times I_N,\ (0 \le U_t \le 0.85) \tag{8}$$

where K_3 is the reactive current ratio coefficient of the PV power generation system, and the value range is 1.5~3.

During the HVRT period, photovoltaic power generation systems should have the capability to provide dynamic reactive power support, with the increment of reactive current as:

$$\Delta I_t = K_4 \times (U_t - 1.1) \times I_N,\ (1.1 \le U_t \le 1.3) \tag{9}$$

where K_4 is the reactive current ratio coefficient of PV power generation system, with its value not less than 1.5.

3.3. SVG-based overvoltage suppression

The constant voltage/constant reactive power outer loop control module is the core part of SVG voltage control, which mainly adjusts the target value of reactive current based on the target values of voltage or reactive power. The DC-side capacitive voltage balance control module, based on the current values of the DC capacitive voltages in the SVG phase units, combined with the control target values, forms a closed-loop feedback control. It generates the active current required for the balance control of the DC capacitive voltage, serving as the reference value for active loop in the main control. The current inner loop control module integrates the current reference value and, after PI adjustment, forms a reference voltage signal. This signal, combined with the feedback signal from the voltage detected at the SVG point of grid connection, forms a closed-loop control, thereby achieving dynamic adjustment of the compensation current output by the SVG.

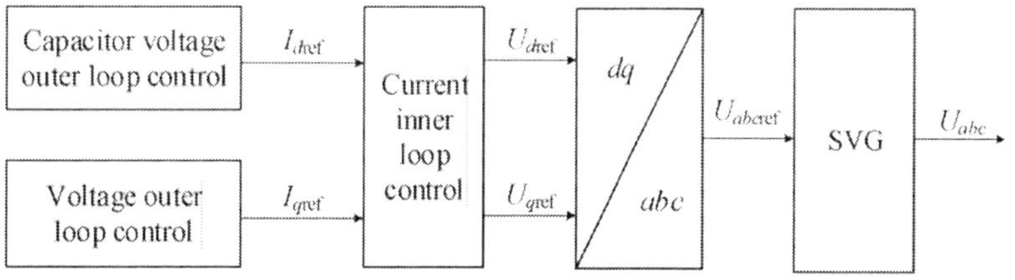

Figure 3. SVG control diagram

4. Simulation Analysis

To confirm the efficacy of the tactics put forward in this paper, PV grid-connected system models and wind power generation system models are established on the PSCAD/EMTDC simulation platform. A single PV module with a capacity of 3MW operates at a rated state with the temperature of 28°C and the irradiance of 1000W/m²; a single DFIG module with a capacity of 5MW operates at a rated state with the wind speed of 11.1m/s. The total capacity of the renewable energy power station is 250MW, with a total capacity of 150MW for wind power systems and 100MW for the photovoltaic power system. In engineering applications, the capacity of SVG is generally one-third of the aggregate capacity of power system, with the rated reactive power of the SVG being 75 Mvar. Build a system model of the renewable energy power station in PSCAD/EMTDC, as presented in the below.

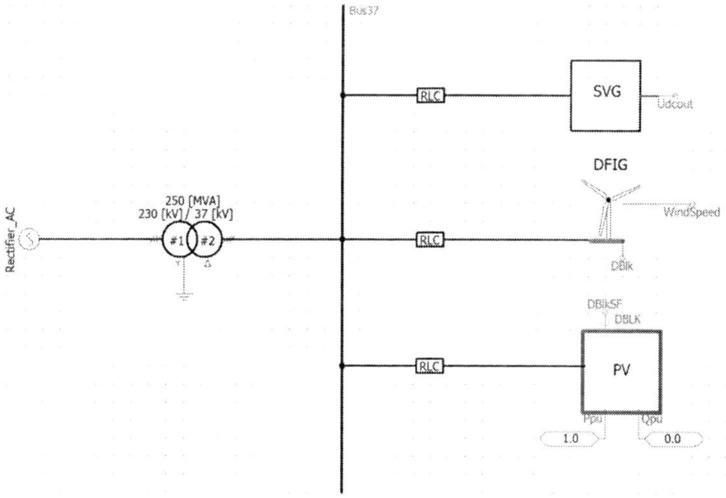

Figure 4. System structure diagram of renewable energy plant

Set a 3-phase fault under different short-circuit ratios at the PCC bus at the 2s moment. The voltage and reactive power waveform diagrams at the point of grid connection of the DFIG, PV and SVG showing in Figure 5-7.

The value of equipments in hybrid renewable energy power station are listed in Table 1.

Table 1 Parameters of model circuits for photovoltaic and wind turbine power generation units

	parameters	value
Photovoltaic power generation unit	DC side capacitance/mF	10
	Multiplication module inductance/mH	3.63
	DC voltage reference value/V	1500
	Boost outer loop PI: K_{pb}; T_{pb}	0.1; 1
	VSC outer loop PI: K_{pv}; T_{pv}	0.25; 0.1
Wind turbine power generation unit	DC side capacitance/mF	12
	Multiplication module inductance/mH	17.43
	DC voltage reference value/V	1500
	Rotor resistance/p.u	0.0054
	DC voltage control PI: K_{pD}; T_{pD}	5; 0.1
	Reactive power control PI: K_{pQ}; T_{pQ}	1; 0.1
SVG	Bridge arm inductance/mH	50
	DC voltage reference value/V	98
	PLL: K_{pll}; T_{pll}	100; 555.56
	Rated capacity/Mvar	75

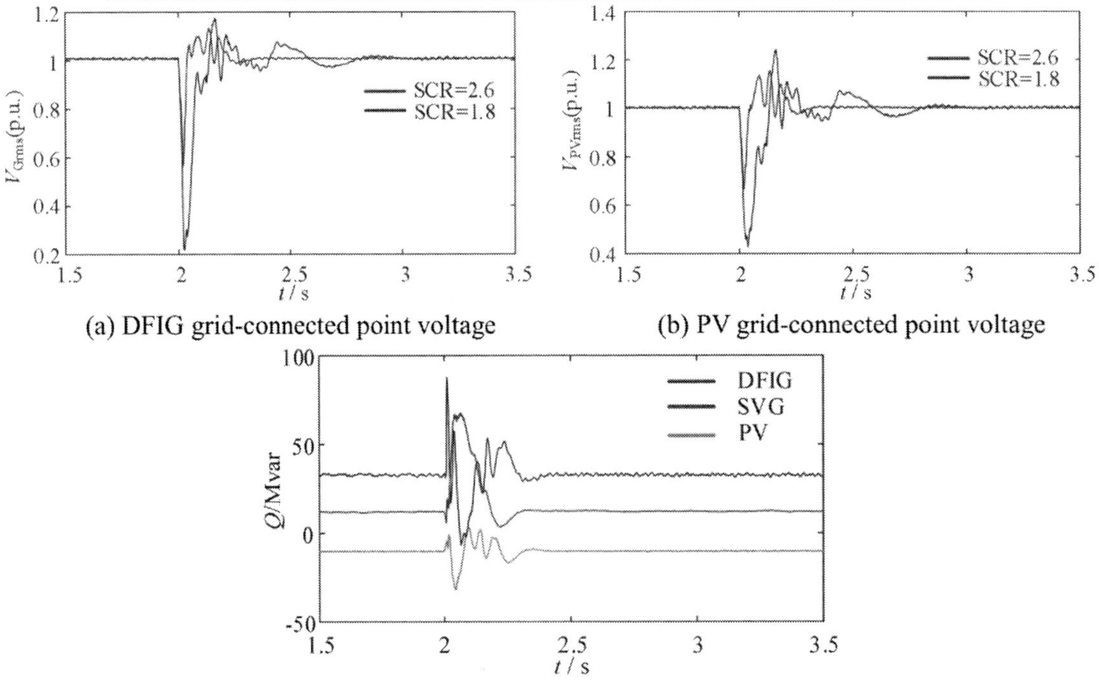

(a) DFIG grid-connected point voltage (b) PV grid-connected point voltage

(c) Reactive power waveform diagram of the DFIG, PV and SVG

Figure 5. Waveform diagrams of the DFIG, PV and SVG connected to the power system.

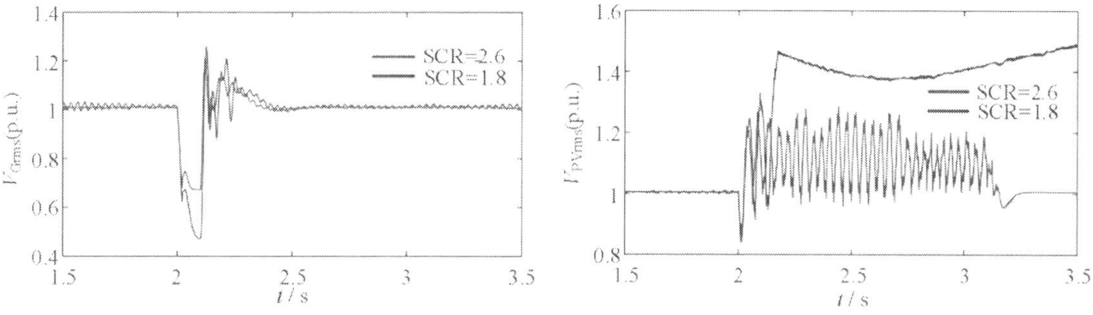

(a) DFIG +SVG connected (b) PV+SVG connected

Figure 6. Waveform diagrams of the DFIG and SVG connected to the power system.

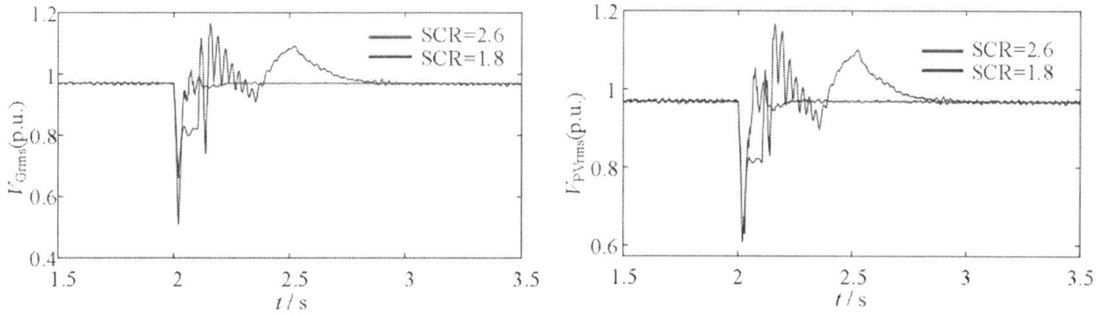

(a) Voltage waveform at the of DFIG (b) Voltage waveform of PV

Figure 7. Waveform diagrams of the DFIG PV without the connection of the SVG

It can be clearly found through the analysis of Figure 5 that the larger the value of SCR of the system is, the higher the stability level of the system will be. Correspondingly, the less impact a fault will have on the system. As can be seen from Figure 5(a), when SCR is 1.8, the transient overvoltage of DFIG reaches 1.18p.u. Meanwhile, its recovery time is approximately 0.8 seconds, and the voltage drop during the fault is significantly higher compared to that when the SCR is 2.6. Figure 5(c) reveals that in the steady state, the SVG supplies 10 Mvar of reactive power to the system. This not only helps balance the system's reactive power but also contributes to the improvement of the power factor.

When a fault takes place, the reactive power of the DFIG and the PV power generation system experiences a significant drop. At this moment, the SVG responds promptly by increasing its reactive power output. By doing so, it effectively alleviates the voltage drop of the bus and enhances the fault ride-through ability of the renewable energy power stations. From Figure 6(a), it can be observed that when only the DFIG and the SVG are connected to the system, the overvoltage characteristics of the system are more pronounced compared to those shown in Figure 5. Moreover, the dynamic reactive power compensation provided by the PV power station is of great significance in reducing the transient overvoltage at the point of grid connection of the DFIG. As depicted in Figure 6(b), when a malfunction takes place where the photovoltaic power generation system has the SCR of 1.8, the system becomes unstable. This fact clearly indicates that a rational allocation of clean energy power system has the potential to strengthen the system resilience. The continuous oscillation of the voltage at the point of grid connection when SCR = 2.6 in Figure 7(b) results from the continuous switching of high-voltage ride-through strategies. Figure 7 displays the voltage oscillogram at the point of grid connection without the SVG connected. It is evident that in the steady state, the voltage at the point of grid connection of the renewable stations does not reach 1.0 p.u. and remains at around 0.95 p.u. As the renewable energy power plants are connected through a 10km transmission line and lack the reactive power compensation of the SVG, resulting in the voltage of the collecting bus less than 1.0 p.u.

5. Conclusion

The higher the SCR of the system is, the better its stability will be. When a fault occurs, not only will the transient overvoltage be lower, but also will the recovery time be shorter.

Apart from the SCR, the transient overvoltage in renewable energy power stations is affected by the system configuration. Compared with individual wind energy systems or photovoltaic energy systems, hybrid renewable energy power stations have better stability. The connection of SVG can remarkably improve the fault response characteristics and voltage stability of the renewable plant. In the absence of SVG connection, the voltage at the grid-connection point of the renewable energy station will drop, which will affect the power supply quality.

Acknowledgments

This work is funded by the Technology Research Service Project of State Grid Corporation "Research on Quality Promotion of Ultra-High-Voltage Key Material and Equipment" (Project Number: SGZB0000TGJS2400863).

References

[1] ZHU Fahua, WANG Yushan, XU Zhen, et al. Research on the development path of carbon peak and carbon neutrality in China's power industry.[J] Electric Power Environmental Protection, 2021, 37 (03): 9-16.

[2] HAN Minxiao, ZHAO Zhengkui, ZHENG Jinghong, et al. Development of dynamic voltage support for power grid with large-scale renewable energy generation[J]. Power System Technology, 2023, 47 (04): 1309-1327.

[3] Yang Bohua. Research on low voltage ride through technology for wind power and photovoltaic power generation system based on STATCOM[D]. Jilin University, 2023.

[4] Liu Yindi. Simulation research on low voltage ride through comprehensive protection of DFIG[J]. Value Engineering, 2022, 41 (23): 93-95.

[5] ZHANG Long, YANG Junhua, CHEN Sizhe, et al. Research on reactive power control of doubly-fed induction generator during the grid voltage sag[J]. Electrical Measurement & Instrumentation, 2015, 52 (14): 66-70+109.

[6] LI Hui, FU Bo, YANG Chao, et al. Reactive current allocation and control strategies improvement of low voltage ride though for doubly fed induction wind turbine generation system [J]. Proceedings of the CSEE, 2012, 32 (22): 24-31+22.

[7] Xie Zhen, Zhang Xing, Yang Shuying, et al. High voltage ride-through control strategy of doubly fed induction wind generators based on virtual impedance [J]. Proceedings of the CSEE, 2012, 32 (27): 16-23+177.

[8] REN Yongfeng, HU Hongbin, XUE Yu, et al. Low voltage ride-through capability improvement of PMSG based on chopper circuit and reactive priority control [J]. High Voltage Engineering, 2016, 42 (01): 11-18.

[9] FU Wenwen, ZENG Chengbi, MIAO Hong, et al. Low-voltage ride through of photovoltaic system based on improved maximum power point tracking control[J]. Chinese Journal of Power Sources, 2016, 40 (05): 1052-1055.

[10] MA Liang, LI Xuyang, LU Yuanhong, et al. LVRT of photovoltaic grid-connected inverter adopting non-linear control[J]. Electronic Design Engineering, 2016, 24 (13): 82-84.

[11] YAO Chuan, RUAN Xinbo, CAO Weijie, et al. An input voltage feedforward control strategy for two-switch buck-boost DC-DC converters [J]. Proceedings of the CSEE, 2013, 33 (21): 36-44+191.

[12] Bao Jian, Yang Peihao, He Zhaohui, et al. Research on improving high voltage ride through capability of new energy hybrid energy storage system [J]. Thermal Power Generation, 2021, 50 (08): 79-86.

Journal of Physics: Conference Series 3012 (2025) 012084 doi:10.1088/1742-6596/3012/1/012084

Research on Optimal Dispatching of Wind-Solar-Hydrogen-Storage Microgrid Based on Fusion of Flamingo Algorithm and Genetic Algorithm

Qi Long*, **Chengqian Zhang**, **Wei Yang**

School of Electrical Engineering and Information, Southwest Petroleum University, Chengdu, Sichuan Province,610500, PR China

*Corresponding author's e-mail: 1597598378@qq.com

Abstract. Wind, solar, hydrogen and storage microgrids are of great significance in promoting energy transformation and improving energy efficiency. To this end, this paper constructs a wind-solar-hydrogen-storage microgrid system model and proposes an optimization strategy that integrates the Flamingo algorithm and genetic algorithm to improve the energy scheduling performance of the wind-solar-hydrogen-storage microgrid. This method dynamically adjusts the search strategies of the two algorithms to give full play to the global search advantages of the Flamingo algorithm and the local optimization capabilities of the genetic algorithm, thereby significantly improving the efficiency and stability of system scheduling. Experimental results show that compared with the traditional Flamingo algorithm, the fusion algorithm can find better solutions more quickly, show stronger global search ability and good convergence, and provide a practical optimization solution for the efficient operation of microgrids.

1. Introduction

Urbanization and globalization have accelerated, leading to a sharp rise in energy demand. Excessive reliance on traditional fossil energy has not only aggravated environmental pollution, but also led to resource shortages[1].Therefore, developing clean and efficient renewable energy has become the key to addressing climate change and the energy crisis[2]. Although wind and solar power are widely used, their volatility still poses challenges to grid access[3]. Microgrids effectively alleviate this problem by integrating renewable energy and energy storage equipment, but how to optimize their scheduling strategies to ensure system stability and reliability remains the focus of current research[4].

The integrated application of system optimization, energy dispatch, intelligent control, and energy storage technology is the primary focus of both domestic and international research on wind, solar, and hydrogen storage microgrids. The goal is to increase the system's stability and the efficiency of renewable energy utilization, particularly in addressing issues related to wind and solar energy[5]. Significant progress has been made in the coordinated dispatch of instability, load fluctuations, grid interconnections, and hydrogen storage.

This paper constructs a wind-solar-hydrogen-storage microgrid operation model, which successfully addresses the issues of energy volatility and supply and demand mismatch while offering a theoretical foundation for the coordinated optimization of renewable energy and energy storage systems[6]. Simultaneously, a multi-objective optimization dispatching approach was created that combined the Flamingo algorithm and the genetic algorithm was proposed, which significantly improved the dispatching efficiency and stability of the wind, photovoltaic and hydrogen storage microgrid[7]. Coordination problems offer creative solutions and have significant theoretical and practical utility in addressing the instability of energy storage and renewable energy.

2. Wind, solar, hydrogen storage microgrid system

2.1. Wind, solar, hydrogen storage microgrid structure

Wind turbines, solar arrays, hydrogen energy storage devices (such as electrolyzers, hydrogen storage tanks, and fuel cells), loads, converters, and inverters are the primary components of the microgrid system under study in this work. Figure 1 depicts the structure of the microgrid.

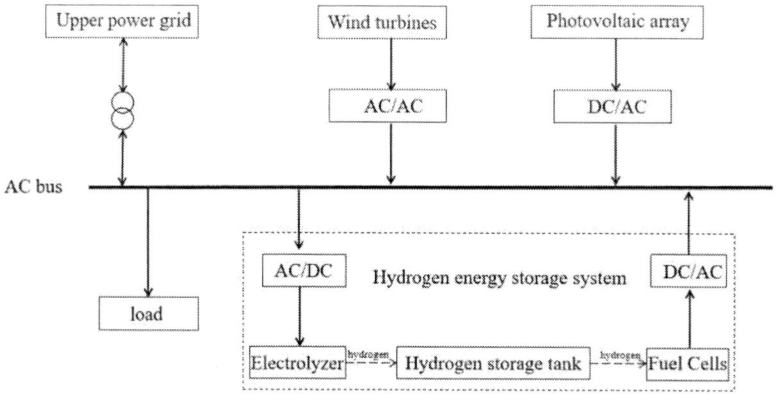

Figure 1. Microgrid construction with wind, sun, and hydrogen storage

2.2. Modeling wind power generating mathematically

$$P_W = \begin{cases} 0, v < v_C \\ P_R \cdot \dfrac{v - v_C}{v_R - v_C}, v_C \le v \le v_R \\ P_R, v_R \le v \le v_F \\ 0, v > v_F \end{cases} \tag{1}$$

P_W is the output of wind power; P_W is the wind power's rated power; v_C is the wind speed cut-off; v_R is the wind speed rating; v_F is the wind speed cut-off.

2.3. Mathematical Modeling of Photovoltaic Power Generation

$$P_{PV} = P_{STC} \cdot G_{AC} \cdot \frac{1 + k \cdot (T_C - T_R)}{G_{STC}} \tag{2}$$

P_{PV} is the output of the photovoltaic; G_{AC} is the intensity of the light; P_{STC} is the highest test power under standard test conditions (sunlight incident intensity 1000W/m2, ambient temperature 25°C); G_{STC} is the light intensity under standard test conditions, which is 1000W/m2; k is the coefficient of power temperature; T_C is the solar panel's operating temperature; T_R is the temperature that is referred to[8].

2.4. Mathematical modeling of technique for storing hydrogen energy

In a fuel cell powered by hydrogen, hydrogen and oxygen undergo an electrochemical reaction, generating a high and intense current density, which can output a large power in a small device. It is usually composed of a hydrogen storage tank, a hydrogen supply system, a fuel cell stack, etc. Its mathematical model expression is:

1) Hydrogen supply system model formula:

$$L_H(t) = P_q(t)N_e/(VF) \tag{3}$$

2) Fuel cell stack model formula:

$$P_Q(t) = L_H(t)C\eta \tag{4}$$

3) Hydrogen storage tank model formula:

$$Q_H(t) = \int_{t_0}^{t_0+\Delta t} L_H(t_0)dt + Q_H(t_0) \tag{5}$$

In the formula, $P_q(t)$ and $P_Q(t)$ are the hydrogen fuel cell's rated capacity and output power at that moment, respectively; N_e is the quantity of series-connected hydrogen fuel cells; V and F are the hydrogen fuel cell's rated voltage and Faraday constant, respectively; $L_H(t)$ and $L_H(t_0)$ are the rates of hydrogen consumption at beginning time t_0 and time t, respectively; $Q_H(t)$ and $Q_H(t_0)$ are the amounts of hydrogen stored in the tank at beginning time t_0 and time t, respectively; C and η are the electrolyte concentration and efficiency of the hydrogen fuel cell's power generation, respectively. Δt is the variance in time.

3. Research on optimization scheduling strategy and algorithm

3.1. *Optimizing scheduling strategies*

3.1.1. *Microgrid Multi-Objective Optimization Model*

1) Microgrid operating costs

$$f_a = \sum_{t=1}^{T} C_{grid}(t) \tag{6}$$

$$\begin{cases} C_{buy}(t) = P_{buy}(t)c_{buy}(t) \\ C_{sell}(t) = P_{sell}(t)c_{sell}(t) \\ C_{grid}(t) = C_{sell}(t) + C_{buy}(t) \end{cases} \tag{7}$$

At time t, the cost of connection between the main grid and the microgrid is denoted by $C_{grid}(t)$; $P_{sell}(t)$ and $P_{buy}(t)$ represents the power bought and sold by the main grid and the microgrid at time t; The microgrid's and the main grid's electricity purchase and sales prices at time t are denoted by $c_{sell}(t)$ and $c_{buy}(t)$, respectively.

2) Microgrid maintenance costs

$$f_b = C_w \cdot P_w(t) + C_p \cdot P_p(t) + C_h \cdot P_h(t) \tag{8}$$

In the formula, $P_w(t)$, $P_p(t)$, and $P_h(t)$ are, respectively, the instantaneous power of solar, wind, and hydrogen energy storage; C_w , C_p , and C_h are the unit power maintenance expenses for hydrogen energy storage, solar power, and wind power, respectively.

3) Depreciation lifespan attenuation cost of microgrids

$$f_c = \sum_{i=1}^{N} \frac{C_i}{L_i \cdot T} \tag{9}$$

C_i is the entire amount spent on category i equipment ; L_i is the category i equipment's service life, T is the entire period of time in a year, N is the total number of different kinds of equipment.

4) Microgrids' expenditures for protecting the environment

$$\begin{cases} f_d = \sum_{t=1}^{T} C_{GRID}(t) \\ C_{GRID}(t) = \sum_{k=1}^{n} P_{buy}(t) C_k \gamma_{grid.k} \end{cases} \tag{10}$$

$C_{GRID}(t)$ is the price of cleaning up pollution produced by the main electrical grid; $\gamma_{grid.k}$ is the release of type k pollutants brought on by the main power grid's functioning; C_k is the type k pollutant treatment cost coefficient.

3.1.2. *Objective function of microgrid optimal dispatch*

Minimizing the overall cost, including the cost of environmental protection and microgrid operation and maintenance, is the goal of optimal dispatch. Consequently, the goal function is:

$$Z = f_a + f_b + f_c + f_d \tag{11}$$

Z represents the microgrid's total cost in the calculation, which includes expenses for upkeep, operation, and environmental preservation.

3.1.3. *Constraints*

1)Limitations on power balancing

$$P_L(t) = P_{PV}(t) + P_W(t) + P_{bess}(t) + P_{grid}(t) \tag{12}$$

In the formula, $P_{bess}(t)$ and $P_L(t)$ are, respectively, the load and energy storage power at time t.

2) Transmission constraints of tie lines

$$P_{grid}^{min}(t) \le P_{grid}(t) \le P_{grid}^{max}(t) \tag{13}$$

3) Hydrogen storage equipment capacity constraints

$$Q_H(t) = Q_H(t-1) + \eta_i P_i(t) - P_0(t) / \eta_0 \tag{14}$$

$$Q_{min,\ H} \le Q_H(t) \le Q_{max,H} \tag{15}$$

In the formula, $Q_{max,H}$ and $Q_{min,\ H}$ are, respectively, the maximum and minimum capacities of the hydrogen storage apparatus; η_i and η_0 represent the equipment's relative hydrogen storage and release efficiency; $P_i(t)$ and $P_0(t)$ are, respectively, the powers that store and release hydrogen at time t.

4) Hill climbing constraint

$$|P(t) - P(t-1)| \le \Delta P_{max} \tag{16}$$

ΔP_{max} is the maximum allowable rate of change of system power within a certain time step.

3.2. *Fusion algorithm based on Flamingo and genetics*

The fusion of Flamingo algorithm and genetic algorithm is to combine the benefits of both to increase the effectiveness and precision of solving optimization challenges. A swarm intelligence algorithm called the Flamingo algorithm mimics the flamingos' foraging habits. It has strong global search capabilities and can find better solutions in complex search spaces. Genetic algorithms simulate the principles of natural selection and genetics, perform optimization through operations such as crossover and mutation, and are good at quickly exploring potential solutions in a wide search space. By combining these two algorithms, the Flamingo algorithm's global search capability and the genetic algorithm's local optimization features can be fully utilized, increasing the algorithm's convergence speed and solution quality while preventing it from reaching the local optimal solution. Combining the selection, crossover, and mutation stages with the foraging behavior of flamingos makes the optimization process more stable and efficient.

Combining genetic and Flamingo algorithms involves incorporating the genetic algorithm's crossover and mutation operators into the Flamingo algorithm. After the Flamingo algorithm has been optimized, the specific implementation will carry out crossover and mutation operations of the genetic algorithm. The specific process of the fusion algorithm based on Flamingo and genetics is displayed in the figure below.

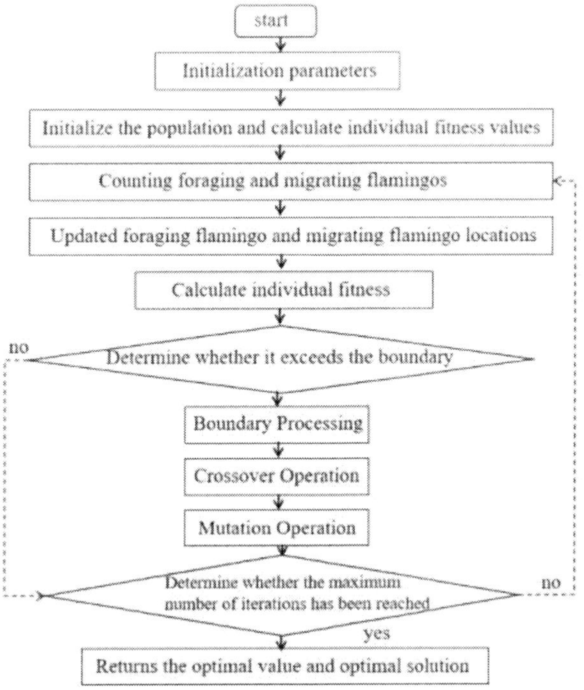

Figure 2. Flowchart of the fusion algorithm based on Flamingo and genetics

3.3. Simulation outcomes of microgrid optimization dispatch for wind, solar, and energy storage

3.3.1. Study Parameters

This study describes a wind-solar-storage microgrid system that incorporates energy storage, photovoltaics, and wind turbines. Table 1 displays the microgrid's operating characteristics; Table 2 displays the energy storage parameters; and Table 3 displays the costs and operational specifications of each pollutant.

Table 1 Unit parameters

Parameter Type	Wind power	Photovoltaic	Mainnet
Maximum power/kW	102	52	32
Minimum power/kW	0	0	-32
Reaching the power limit/(MW/min)	3	2	5
Maintenance and operation unit price(yuan/ kW·h)	0.03	0.015	0.05

Journal of Physics: Conference Series 3012 (2025) 012084 doi:10.1088/1742-6596/3012/1/012084

Table 2 Parameters of energy storage

type	parameter	Numeric	parameter	Numeric
Storage of Hydrogen Energy	*Maximum capacity/kW·h*	152	*initial capability for storing energy(kW·h)*	52
	Minimum capacity/kW·h	6	*Maximum power output/kW*	32
	Maximum power input/kW	32	*Rate of storage and discharge*	0.92

Table 3 Pollutant emission coefficients and costs

Pollutant Type	Governance costs/(yuan/kg)	Coefficient of pollution emission(g/kW·h) PV	WT	Grid
CO_2	0.023	0	0	886
SO_2	6	0	0	1.8
NOx	8	0	0	1.6

3.3.2. *Simulation Results*

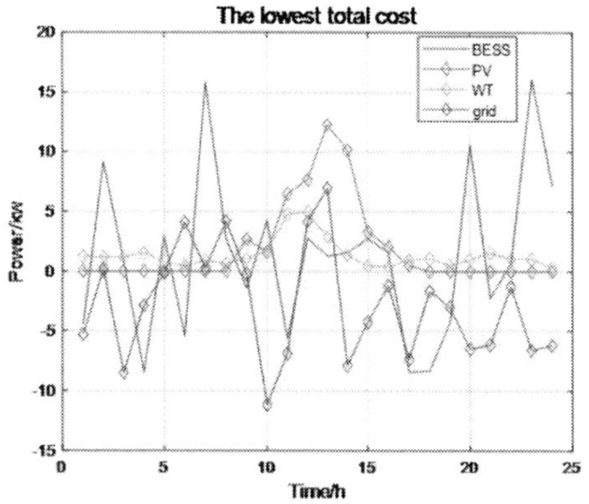

Figure 3. Output of each part when the cost of Flamingo algorithm is the lowest

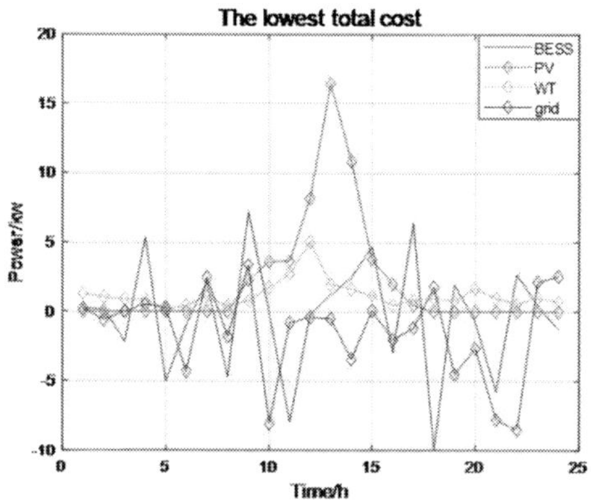

Figure 4. The output of each part when the fusion algorithm cost is the lowest

Figures 3 and 4 demonstrate that, in the event that the lowest cost is attained, the fusion algorithm of Flamingo and genetic algorithm shows higher optimization efficiency in the power output performance of each part compared with the traditional Flamingo algorithm. Specifically, the maximum utilization powers of photovoltaic and wind power reached 17kW and 5kW respectively, particularly at midday, when the photovoltaic system's electricity output dramatically rose. The fusion algorithm greatly boosts the system's optimization performance and efficiently increases the use efficiency of clean energy.

Journal of Physics: Conference Series 3012 (2025) 012084 doi:10.1088/1742-6596/3012/1/012084

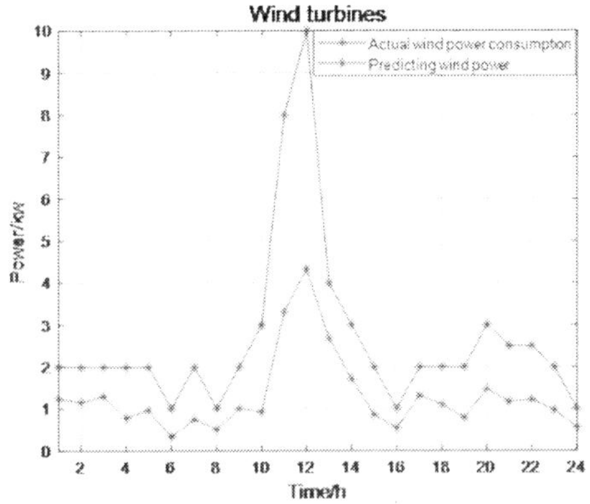

Figure 5. The prediction and actual consumption of wind power when the Flamingo algorithm has the lowest cost

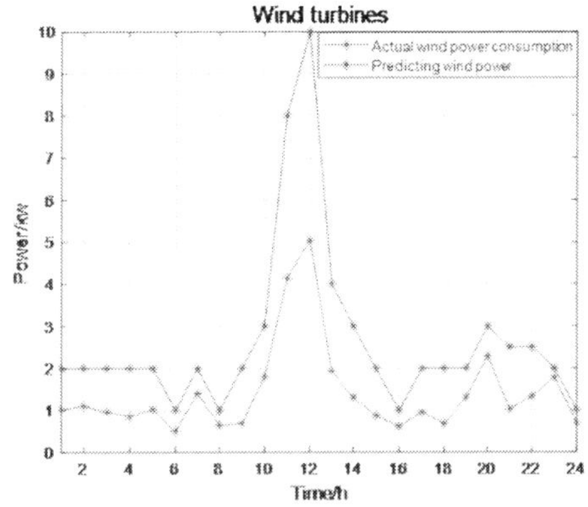

Figure 6. The prediction and actual consumption of wind power when the fusion algorithm has the lowest cost

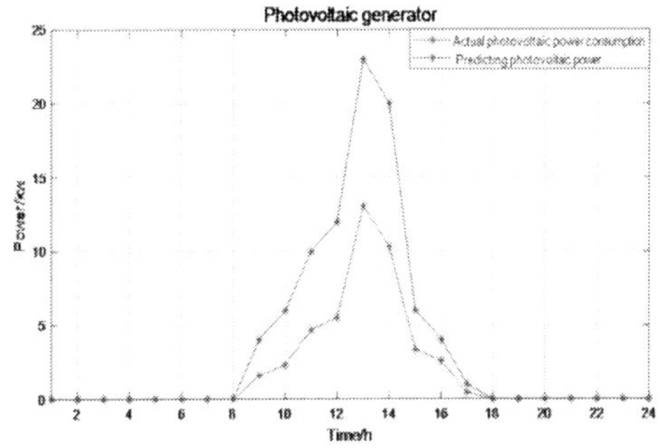

Figure 7. The prediction and actual consumption ,of photovoltaic power when the cost of Flamingo algorithm is the lowest

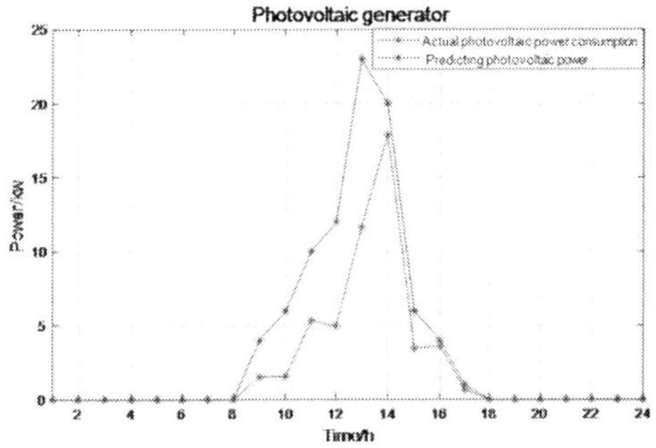

Figure 8. The prediction and actual consumption of photovoltaic power when the fusion algorithm has the lowest cost

Figures 5, 6, 7, and 8 demonstrate that, in contrast to the conventional Flamingo algorithm, the Flamingo and genetics fusion method considerably lowers the discrepancy between the actual consumption value and the anticipated value of wind and photovoltaic power. Improved the correlation between the two.

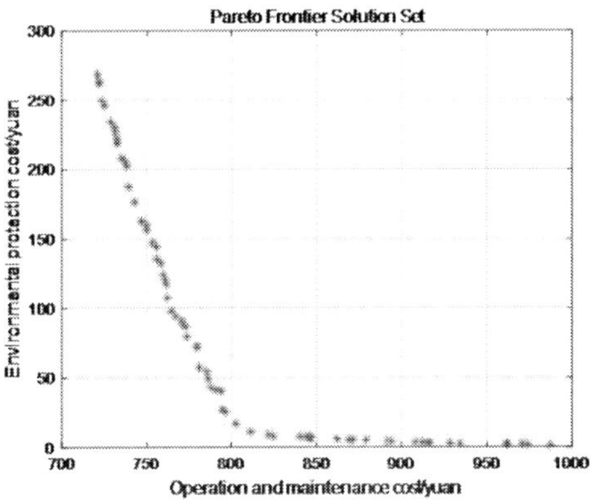

Figure 9. Solution set of Flamingo algorithm

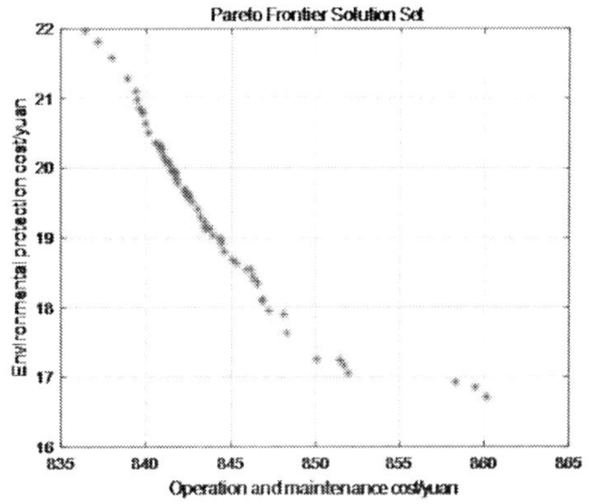

Figure 10. Solution set of fusion algorithm

Figures 9 and 10 demonstrate how the fusion algorithm of Flamingo and genetics dramatically lowers environmental costs as well as operation and maintenance costs as compared to the conventional Flamingo algorithm. Figure 9 shows that the cost of operation and maintenance might reach 990 yuan, and the environmental cost is up to 270 yuan; In Figure 10, the operation and maintenance cost is significantly reduced to 860 yuan, and the environmental cost is significantly reduced to a maximum of 22 yuan, and some points are even lower than 17 yuan. The fusion algorithm shows excellent optimization performance in terms of economy and environmental protection, providing a more balanced solution for microgrids' effective functioning.

3.4. Analysis and comparison
When compared to the classic Flamingo algorithm, the Flamingo and genetics fusion method is noticeably better. The algorithm successfully raises the clean energy use efficiency, narrows the deviation between the actual consumption values and predicted values of photovoltaic and wind power, and reduces environmental and operation and maintenance costs. With its advantages in global search, the fusion algorithm can more efficiently find solutions close to the global optimal solution, and therefore shows more economical cost-effectiveness in practical applications.

4. Conclusion
This paper proposes a multi-objective optimal scheduling scheme based on Flamingo and genetic fusion algorithms, demonstrating its significant application potential in wind, photovoltaic and hydrogen storage microgrids. This system can successfully strike a compromise between the expenses of environmental conservation, economic rewards, and grid operation reliability. The fusion algorithm of Flamingo and genetic algorithm shows excellent convergence and global search ability by dynamically adjusting the position and number of Flamingo, and can find the optimal solution in a shorter time. This method has strong adaptability and can cope with diverse application scenarios and user needs, thus providing an effective optimization strategy for power system scheduling problems. Furthermore, this algorithm is expected to promote the process of intelligent and sustainable development of power systems, and establish a strong theoretical and technical basis for future power systems' intelligent evolution and optimization. Future research can focus on further improving the computational efficiency of the algorithm, exploring its application in large-scale power systems, and optimizing its robustness and adaptability in uncertain environments.

References
[1] Liang, F. (2024, May). Planning of wind-photovoltaic-storage-hydrogen-water for a zero-carbon microgrid on an independent island. In Journal of Physics: Conference Series (Vol. 2771, No. 1, p. 012004). IOP Publishing.
[2] Bibih, M., Choukri, K., & Chakir, H. E. (2024). Assessing the potential of green hydrogen production from wind power: case of microgrid. Procedia Computer Science, 236, 25-32.
[3] Hao, P., Huang, C., Tian, P., Wang, Q., & Wang, C. Three-Stage Optimal Scheduling of Distribution Network with Multi-Microgrid Incorporating Wind-Solar-Hydro-Hydrogen-Storage. Available at SSRN 4753766.
[4] Zhu, N., Ma, X., Guo, Z., Shen, C., & Liu, J. (2024). Research on the optimal capacity configuration of green storage microgrid based on the improved sparrow search algorithm. Frontiers in Energy Research, 12, 1383332.
[5] Wei, W., Ji, L., Xu, X., Liu, J., Zhang, L., & Xiao, S. Optimal Configuration of Stand-Alone Microgrids with Hybrid Electric-Hydrogen Energy Storage on Multi-Time Scales. Available at SSRN 4724694.
[6] Durgam, R., & Devarakonda, N. (2023). A quasi-oppositional based flamingo search algorithm integrated with generalized ring crossover for effective feature selection. IETE Journal of Research, 69(12), 8623-8639.
[7] Kousalya, C. G., kumar, V. S., Diaz, P. M., & Jiju, M. J. E. (2024). Feature selection on real-time application using hybrid of flamingo search algorithm and improved non-dominated sorting genetic algorithm III. Journal of Experimental & Theoretical Artificial Intelligence, 1-19.
[8] Cui, Y., Wang, Z., Yang, L., Li, J., Chang, G., Song, L., & Yue, G. (2024). Multi-objective optimization design of the solid oxide fuel cells using response surface methodology and genetic algorithm. Applied Thermal Engineering, 242, 122503.

A Chebyshev's Inequality-based Distributionally Robust Chance-Constrained Model for Distribution Network Reactive Power Optimization

Kun Yang[1], Dan Yang[1], Zhi Zhang[1], Shengjian Zeng[1], Conghao Wang[1], Jian Li[1], Man Huang[1], Bin Li[1], Xijun Gou[1], Xin Wang[1], Qian Liu[2*], Yan Zou[2]

[1] Guangyuan Power Supply Company, State Grid Sichuan Electric Power Company, Guangyuan, Sichuan Province, 628000, P.R. China

[2] Southwest Petroleum University, Chengdu, Sichuan Province, 610500m P.R. China

*Corresponding author's e-mail: 18064578879@163.com

Abstract. This paper primarily examines the effects of integrating distributed generation (DG) on voltage and reactive power control within distribution networks. It introduces a distributionally robust chance-constrained (DRCC) model that employs Chebyshev's inequality to optimize reactive power. To account for the variability inherent in renewable energy resources, a moment-based fuzzy set approach is utilized. Furthermore, the DRCC model is reformulated using a method grounded in Chebyshev's inequality. The numerical results demonstrate that the proposed approach significantly enhances the voltage profile and minimizes energy losses.

1. Introduction

As active distribution network (DN) technology advances, DN have acquired a range of voltage and reactive power regulation technologies, including Group Switched Capacitors (referred to as Capacitor Banks, CB) and Distributed Generation (DG) [1]. The coordination of these multiple regulation methods is essential for maintaining voltage quality, which is critical to the safe and economic operation of the distribution network [2][3].

Many researchers have explored voltage and reactive power control in active DN, proposing diverse control strategies. These encompass enhanced traditional optimization algorithms, intelligent optimization algorithms, model predictive control (MPC), and collaborative control schemes. For instance, reference [4] addresses reactive power and voltage optimization problems by refining algorithms. Reference [5] employs intelligent optimization algorithms to optimize reactive power distribution, adjust compensation devices, and control voltage and reactive power. It adopts multi-objective optimization strategies to reduce power system losses, enhance voltage quality, and improve system stability. Literature [6-7] develops mathematical models using MPC and multi-agent system collaborative control to achieve optimal reactive power control, considering voltage quality and reliability, while reducing system losses and enhancing voltage stability and power quality.

Voltage and reactive power control in DN can be categorized into three primary approaches: demand response-based collaborative optimization, equipment coordination control for real-time optimization, and intelligent algorithm-based multi-objective optimization. Reference [8] presents a demand response-based collaborative optimization method, which involves constructing a multi-temporal-spatial framework to integrate various loads into optimization scheduling, thereby reducing compensation costs and enhancing system economic efficiency. Reference [9] suggests employing linear programming for real-time optimization of equipment coordination control, aiming to minimize voltage deviations and ensure the stability of the distribution network. Reference [10] introduces an attention-based multi-objective particle swarm optimization (AMOPSO) algorithm, which focuses on pertinent information, constricts the search space, effectively evades local optima, reconciles conflicts among different objectives, and identifies the optimal solution.

Despite advancements in the research of voltage and reactive power control in DN, several challenges persist in practical applications. These challenges encompass: the prevalent use of simplified models that do not sufficiently account for the uncertainty and dynamics of distribution network operations; the limitations of traditional optimization algorithms in handling high-dimensional, nonlinear, and multi-constraint problems; and the inability of current strategies to effectively manage the randomness and characteristics of distributed power sources.

To address the issue of unnecessary movements in equipment, this paper establishes a distributionally robust chance-constrained (DRCC) model to optimize reactive power within distribution network. To account for the variability inherent in renewable energy resources, a moment-based fuzzy set approach is utilized. Furthermore, the DRCC model is reformulated using a method grounded in Chebyshev's inequality. The numerical results demonstrate that the proposed approach significantly enhances the voltage profile and minimizes energy losses.

2. Problem formulation

In this section, a mathematical model for distribution network reactive power optimization based on branch power flow equation is constructed, and a voltage reactive power control model is created.

2.1. Basic model for distribution network reactive power optimization

Distribution networks (DNs) incorporate a range of flexible control devices, including capacity banks (CBs) and distributed generation (DG) systems. In DNs, CBs and DGs collaborate to maintain the safety, stability, and efficiency of the power grid. DG systems primarily provide clean, reliable power close to load centers, reducing transmission losses and enhancing grid flexibility. They also support voltage and frequency stability, optimize power resource allocation, and contribute to improved energy efficiency and environmental sustainability.

The DN model is delineated by Equations (1) through (6). Equations (1) and (2) represent the power balance equations at the distribution network nodes.

$$p_{ij} - x_{ij} i_{ij}^2 = \sum_{m \in n(i)} p_{jm} + p_{\mathrm{L},j} - p_{\mathrm{DG},j} \tag{1}$$

$$q_{ij} - x_{ij} i_{ij}^2 = \sum_{m \in n(j)} q_{jm} + q_{\mathrm{L},j} - q_{\mathrm{DG},j} - q_{\mathrm{CB},j} \tag{2}$$

where p_{ij} represents the active power of branch (i, j); x_{ij} represents the current size of branch (i, j); $n(j)$ represents the set of sub-nodes of node j; $p_{\mathrm{L},j}$ represents the active load of node j; and $p_{\mathrm{DG},j}$ represents the active output of distributed power at node j. q_{ij} is the reactive power of branch (i, j); x_{ij} is the reactive impedance of branch (i, j); $q_{\mathrm{L},j}$ is the reactive load of node j; $q_{\mathrm{DG},j}$ is the reactive output of distributed power supply of node j; $q_{\mathrm{CB},j}$ is the reactive power of capacitor group of node j. The branch voltage balance equation is denoted as equation (3).

$$U_j = U_i - 2\left(r_{ij} p_{ij} + x_{ij} q_{ij}\right) + \left(r_{ij}^2 + x_{ij}^2\right) i_{ij}^2 \tag{3}$$

where U_j is the amplitude of the voltage at node j. The line apparent power equation is shown in Equation (4).

$$u_i^2 i_{ij}^2 = p_{ij}^2 + q_{ij}^2 \tag{4}$$

The security constraints of the power distribution network are shown in equation (5) and equation (6).

$$U_{j,\min} \le U_j \le U_{j,\max} \tag{5}$$

$$i_{ij}^2 \le i_{ij,\max}^2 \tag{6}$$

where $U_{j,\min}$ and $U_{j,\max}$ are the voltage safety boundary of the node. $i_{ij,\max}$ is the maximum current of the branch (i, j).

2.2. Model reformulation

The model discussed previously is non-convex, posing challenges in terms of solvability. Consequently, a second-order cone programming (SOCP) based approach is utilized to render the proposed model convex, thereby enabling efficient solution via commercial solvers. The following sections provide detailed descriptions.

Based on the SOCP method, the original model can be converted by equation (7) to equation (13).

$$\begin{cases} U_{j,t} = u_{j,t}^2 \\ I_{ij,t} = i_{ij,t}^2 \end{cases} \tag{7}$$

$$p_{ij,t} - r_{ij} I_{ij,t} = \sum_{m \in n(j)} p_{jm,t} + p_{\mathrm{L},j,t} - p_{\mathrm{DG},j,t} \tag{8}$$

$$q_{ij,t} - x_{ij} I_{ij,t} = \sum_{m \in n(j)} q_{jm,t} + q_{\mathrm{L},j,t} - q_{\mathrm{DG},j,t} - q_{\mathrm{CB},j,t} \tag{9}$$

$$U_{j,t} = U_{i,t} - 2\left(r_{ij} p_{ij,t} + x_{ij} q_{ij,t}\right) + \left(r_{ij}^2 + x_{ij}^2\right) I_{ij,t} \tag{10}$$

$$U_{i,t} I_{ij,t} = p_{ij,t}^2 + q_{ij,t}^2 \tag{11}$$

$$U_{j,\min} \le U_{j,t} \le U_{j,\max} \tag{12}$$

$$0 \le I_{ij,t} \le I_{ij,\max} \tag{13}$$

where $U_{j,t}$ represents the square of the amplitude of the voltage of node j at t; $I_{ij,t}$ represents the square of the amplitude of the current of branch (i, j) at t. $p_{ij,t}$ is the active power of branch (i, j) at time t; $p_{\mathrm{L},j,t}$ is the active load of node j at time t; $p_{\mathrm{DG},j,t}$ is the active output of distributed power supply of node j at time t. $q_{\mathrm{L},j,t}$ represents the reactive power of branch (i, j) at time t; $q_{\mathrm{DG},j,t}$ represents the reactive load of node j at time t; $q_{ij,t}$ represents the reactive output of distributed power supply of node j at time t; and $q_{\mathrm{CB},j,t}$ represents the reactive power of capacitor group of node j at time t. The constraints for the controllable devices are shown in

$$q_{\mathrm{CB},j,t} = n_{\mathrm{CB},j,t} q_{\mathrm{CB0},j} \tag{14}$$

$$p_{DG,j,t}^2 + q_{DG,j,t}^2 \le S_{DG,j}^2 \tag{15}$$

$$q_{SVC,\min} \le q_{SVC,j,t} \le q_{SVC,\max} \tag{16}$$

$$0 \le P_S^{\mathrm{dis}}(t) \le U_S(t) P_S^{\max} \tag{17}$$

$$0 \le P_S^{ch}(t) \le \left[1 - U_S(t)\right] P_S^{\max} \tag{18}$$

Journal of Physics: Conference Series 3012 (2025) 012085 doi:10.1088/1742-6596/3012/1/012085

$$\eta\sum_{t=1}^{N_T}\left[P_S^{ch}(t)\Delta t\right]-\frac{1}{\eta}\sum_{t=1}^{N_T}\left[P_S^{\text{dis}}(t)\Delta t\right]=0 \tag{19}$$

$$E_S^{\min}\leq E_S(0)+\eta\sum_{t'=1}^{t}\left[P_S^{ch}\left(t'\right)\Delta t\right]-\frac{1}{\eta}\sum_{t'=1}^{t}\left[P_S^{\text{dis}}\left(t'\right)\Delta t\right]\leq E_S^{\max} \tag{20}$$

where, $S_{DG,j}$ represents the distributed power capacity of node j. $q_{SVC,j,t}$ is the reactive power of SVC. $P_S^{ch}(t)$ and $P_S^{\text{dis}}(t)$ represent the charging and discharging power of the energy storage, respectively. $P_S^{\max}$ denotes the maximum charging and discharging power of the energy storage (i.e., the rated power). $U_S(t)$ indicates the charging and discharging status of the energy storage, where a value of 1 signifies discharging and a value of 0 signifies charging. $E_S^{\max}$ and $E_S^{\min}$ are the maximum and minimum state of charge (SoC) allowed for the energy storage during the scheduling process, and N_T represents the scheduling period. The above model formulates a mixed integer nonlinear programming problem, and the equation (11) is non-convex. To convexify the equation (11), the second-order cone relaxation is adopted as shown in equation (21).

$$\left\|\begin{matrix}2p_{ij,t}\\2q_{ij,t}\\U_{i,t}-I_{ij,t}\end{matrix}\right\|_2\leq(U_{i,t}+I_{ij,t}) \tag{21}$$

where $\|\cdot\|_2$ denotes the Euclidean norm.

2.3. *Chance constraints of the renewable energy resources*

Considering the inherent uncertainty in photovoltaic (PV) power generation, this study utilizes a chance-constrained approach to achieve a balance between the economic benefits and the risks associated with the unpredictability of PV power generation, as illustrated in equation (22).

$$\|_t^{Pv}\left\{p_t^{Pv}\leq\xi_t^{Pv}\right\}\geq1-\beta^{Pv} \tag{22}$$

where $\|_t^{Pv}$ represents the probability distribution of uncertain PV power generation around ξ_t^{Pv} ; β^{Pv} is the pre-determined risk tolerance that will limit the opportunities for PV power generation.

Using moment-based ambiguity set, the uncertainties of the PV generation can be captured by equation (23).

$$\mathrm{P}_t^{PV}=\left\{\|_t^{PV}:[\xi_t^{PV}]=\mu_t^{PV},\left[\left(\xi_t^{PV}-\mu_t^{PV}\right)^2\right]=\sigma_t^{PV}\right\} \tag{23}$$

where μ_t^{PV} and ξ_t^{PV} respectively denote the mean and variance of the PV generation, which can refer to a specific random variable. The expectation operator is used to calculate the expected value under a given probability distribution. In equation (23), the mean μ_t^{PV} and variance σ_t^{PV} are calculated based on empirical data samples ξ_t^{PV} . In order to develop a reactive power optimization model based on DRO, the chance constraints are represented as DRCC as shown below:

$$\inf_{\|_t^{PV}\in\mathrm{P}}\|_t^{PV}\{A_t^{PV}(z)\xi_t^{PV}\leq b_t^{PV}(z)\}\geq1-\beta^{PV} \tag{24}$$

where z collects the decision variables of the reactive power optimization problem of DN. $A_t^{PV}(z)$ and $b_t^{PV}(z)$ are affine functions on the decision variable z. $\mathcal{P}_t^{PV}$ is the uncertainty set of the PV power generation forecast error ξ_t^{PV}.

2.4. Reformulation of the chance constraints

Equation (24) is non-convex, posing a challenge for commercial solvers. To address this, we employ Chebyshev's inequality to reformulate Equation (24) into a set of more manageable constraints, as elaborated upon subsequently.

$$\sqrt{A_t^{PV}(z)\sigma_t^{PV}A_t^{PV}(z)} \le \sqrt{\frac{\alpha^{PV}}{1-\alpha^{PV}}}[b_t^{PV}(z)-\mu_t^{PV}A_t^{PV}(z)] \tag{25}$$

2.5. Objective function

The objective function, as delineated in Equation (26), aims to minimize both transmission losses and the aggregate operational costs.

$$\begin{aligned}\min \pi_{\text{loss}} \sum_{t\in N_T(i,j)\in N_L} r_{ij} I_{ij,t}\Delta t + \pi_{CB}\sum_{t\in T_l}\sum_{j\in N_{CB}} n_{CB,j,t} + \pi_{SVC}\sum_{t\in T_l}\sum_{j\in N_{SVC}} q_{SVC,j,t} + \\ \pi_{ESS}\sum_{t\in T_l}\sum_{j\in N_{ESS}}\left(P_{S,j}^{ch}(t)+P_{S,j}^{\text{dis}}(t)\right)\end{aligned} \tag{26}$$

where π_{loss} denotes the cost for transmission loss, π_{SVC} signifies the cost for adjusting SVCs, π_{CB} corresponds to the cost for adjusting CBs, and π_{ESS} indicates the cost for adjusting ESSs.

3. Case study

3.1. Simulation setup

In this paper, the effectiveness of the proposed strategy is verified by an improved IEEE-33 node test system as shown in Figure 1. CBs, SVCs, ESSs, and DGs are added as voltage reactive power control equipment. Their configurations are shown in Table 1 and Table 2. The daily load curve and PV generation are shown in Figure 2, and the node voltage is limited to 0.93~1.07p.u. The simulation time of is 24h.

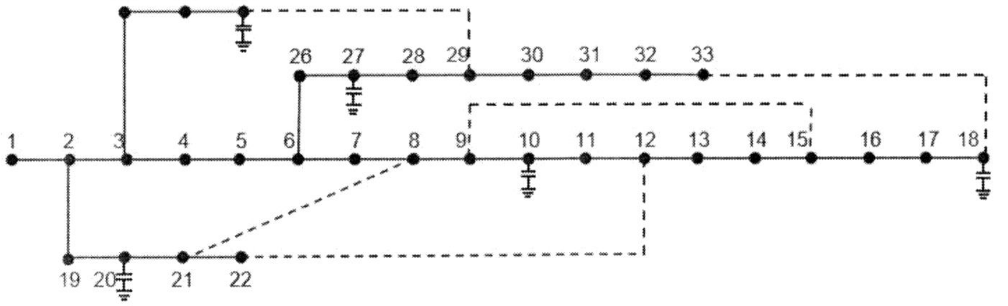

Figure1. Topology structure of IEEE-33 node system

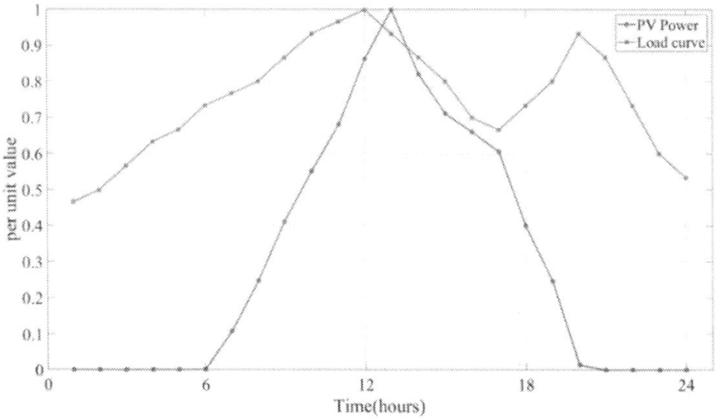

Figure 2. Daily load curve and PV generation

Table 1. Parameters of Reactive Power Compensation

Access node	Maximum active power / KW
10	20
18	20
20	100
25	100
27	125

Table 2 Parameters of DG allocation

Access node	Maximum active power / KW
13	15
31	15
24	15

3.2. *Optimization results*

The initial voltage profiles without optimization are shown in Figure 3.

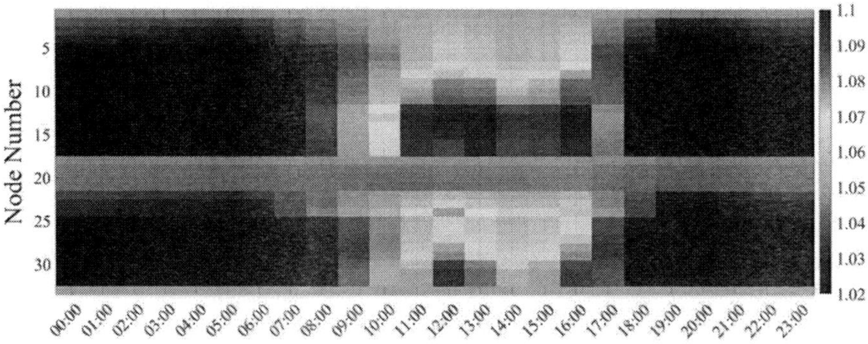

Figure 3. Initial voltage profile before optimization

Figure 3 illustrates the voltage levels across various nodes over time. The horizontal axis represents time, marked in hours, while the vertical axis denotes the node numbers. The color gradient, ranging from blue to red, indicates the voltage levels, with blue signifying lower voltages and red indicating higher voltages. From Figure 3, one can observe that there are distinct daily patterns where voltage levels peak during certain hours. For instance, nodes 5, 15, and 25 show a noticeable increase in voltage, indicated by the red and orange hues, around midday (approximately 12:00 to 14:00), suggesting higher PV generation during these times. The maximum voltage reaches 1.1p.u. After the optimization, the results are shown in Figure 4.

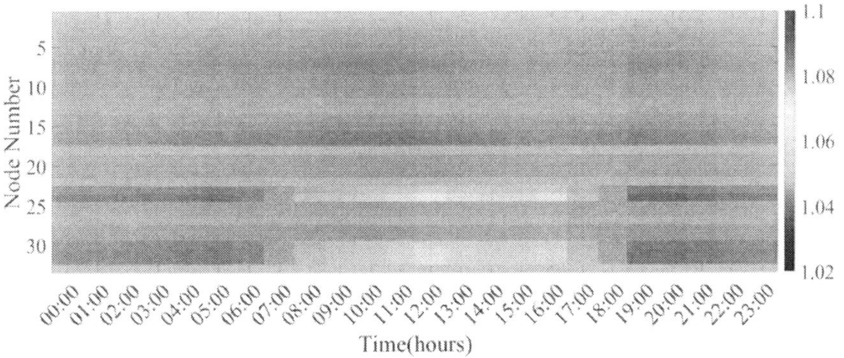

Figure 4. Voltage distribution of each node after optimization

Figure 4 demonstrates improved voltage stability, with voltage values converging towards the ideal value of 1.0 p.u. The fluctuation range has narrowed from 1.02 to 1.1 to 1.02 to 1.04, signifying a reduction in voltage deviation, particularly during peak hours. Notably, early morning (e.g., 00:00 to 05:00) and evening (19:00 to 23:00) are characterized by predominantly low values (blue area), suggesting potential low load and low voltage in the grid during these times. Voltage values at nodes near the top (e.g., nodes 1-10) and bottom (e.g., nodes 25-33) fluctuate markedly, as indicated by color changes. In contrast, nodes in the middle section (e.g., nodes 15-25) display a relatively stable voltage. The comparison of the network loss before and after optimization is depicted in Figure 5.

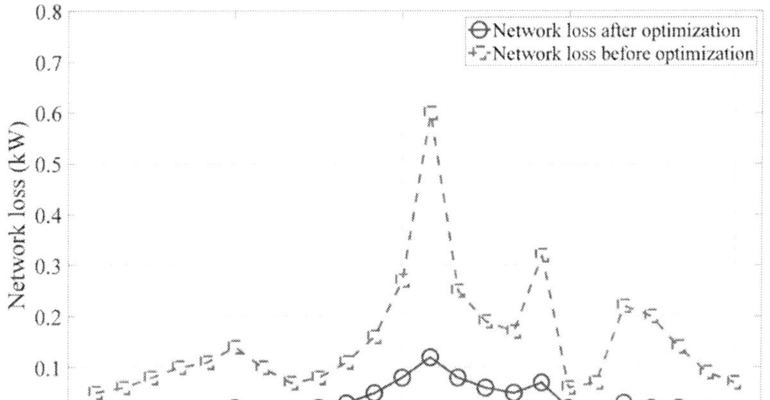

Figure 5. Comparison of network loss

As shown in Figure 5, the network loss before optimization, represented by the red squares and dashed line, shows a significant fluctuation throughout the day, with a notable peak around midday (around 12:00 to 14:00 hours). In contrast, the network loss after optimization, indicated by the blue circles and solid line, remains consistently low across the entire day. The proposed method has a substantial impact on reducing network loss. The peak loss before optimization is approximately 0.6 kW, while after optimization, the loss does not exceed 0.1 kW at any time, indicating a significant improvement in network efficiency.

4. Conclusion

This paper proposes an active distribution network voltage reactive power control model that considers the collaborative optimization of DG, CB, ESS, and SVC. It takes into account the operational costs of equipment and the total network loss. To account for the variability inherent in renewable energy resources, a moment-based fuzzy set approach is utilized. Furthermore, the DRCC model is reformulated using a method grounded in Chebyshev's inequality. The numerical results demonstrate that the proposed approach significantly enhances the voltage profile and minimizes energy losses.

References

[1] Yu Hengkang. 2024. Research on the optimization method of reactive voltage control for UHV power grid [J]. Modern industrial economy and informatization, 2024,14(11):261-262+267.DOI:10.16525/j.cnki.14-1362/n. 11.085.

[2] Xu Jie, Yu Bin, Bai Long, et al. 2024, "Power Factor Optimization of DG Distribution Networks Based on Forward and Backward Methods" [J]. Automation and Instrumentation, 39(11):13-17+110.DOI:10.19557/j.cnki.1001-9944.2024.11.004.

[3] Gao Jinlan, Diao Nan, Hou Xuecai, et al. 2024, "Innovative Particle Swarm Optimization for Reactive Power Optimization in Active Distribution Networks" [J/OL]. Automation Technology and Application 2024,(11):19-23[2024-12-27]. https://doi.org/10.20033/j.1003-7241. 11-0019-05.

[4] Wen Xing, Lin Hong. 2024, Consideration of the Coordination and Optimization of DRs Distribution Network Voltage Active and Reactive Power [J]. Control Engineering, 31(11):1992-1999. DOI:10.14107/j.cnki.kzgc. 20230003.

[5] Li Bin and Luo Xiaoyi, 2024, Research on the Impact of Distributed Power Sources on the Voltage Reactive Optimization of Power System [J], Electrical Technology, Issue 25,55-61+78.

[6] Xu Tao, Li Hugang, Yang Longyu, et al. 2024, Reactive Power Optimization Based on Attention Mechanism and Multi-objective Particle Swarm Hybrid Driven in Distribution Networks [J]. Power Grid and Clean Energy, 40(07):146-154.

[7] Xia Zhenglong, Chen Yu, Lu Liangshuai, et al. 2025, Research on Dynamic Reactive Power Optimization of Distribution Network Based on Multi-objective Whale Optimization Algorithm [J]. Journal of Henan Normal University (Natural Science Edition), 53(01):116-126.DOI:10.16366/j.cnki.1000-2367.2023.12.04.0002.

[8] Chen Meifu, Xia Mingchao, Chen Qifang, et al. 2018, "A Review on the Coordination and Scheduling of Sources, Grids, Loads, and Storage in Active Distribution Networks" [J]. "Power Construction", 39(11):109-118.

[9] Chen Qi, Zhao Xu and Gan De. 2017, Active-reactive scheduling of interactive load and battery energy storage for active distribution system [J]. Modern Power System Protection and Control, 2(1):1-11.

[10] Wang Xiuqiang, Liu Guangyi, Zeng Yuan, et al. 2014, Analysis of the Effectiveness of Reactive Power Control in Distribution Network Voltage under Distributed Access to Power Network [J]. Power System Protection and Control,42(01):47-53.

Design of High-speed Signal Acquisition Device Based on AVR-PSS Control Strategy

Lin Tang[1], Jing Zhang[1], Jian Cheng[1], Hongyi Hua[2], Yuting Wang[2], Peng Chen[2], Chuan Qin[2], Xuemin He[3*]

[1] Baihetan Power Plant, Three Gorges Jinshajiang Yunchuan Hydropower Development Co., Ltd, Liangshan 615400, Sichuan Province, China;

[2] China Three Gorges Corporation, Wuhan 430014, Hubei, China;

[3] Three Gorges Intelligent Industrial Control Technology Co., Ltd, Wuhan 430070, China

[*]Corresponding author's e-mail: he_xuemin@126.com

Abstract. In the field of hydropower, the peer distributed excitation control system, as the core control equipment of the generator set and the power grid, plays a decisive role in the stable operation of the generator set and the power system. In the actual operation process, most of the imported equipment is used as the control system, so the independent controllability is very important. In order to meet this challenge, first of all, this paper designs a domestic high-speed signal acquisition device. Then, the automatic voltage regulator(AVR) and power system stabilizer (PSS) control is proposed. Secondly, the signal collected by the acquisition device is used to realize AVR-PSS control, and finally improve the stability of the generator and power system.

1. Introduction

As a super-large and complex industrial system, the energy system's core controller - the programmable logic controller (PLC) - plays a crucial role in ensuring the reliability and safety of energy conversion and utilization [1]. The application scope of PLC technology covers almost all industrial sectors. However, in China's PLC application market, over 95% is occupied by foreign products, leaving domestic enterprises at a disadvantage [2].

In the hydropower field, the excitation system, as the core control equipment of the generator set and the power grid, its software and hardware's autonomy and controllability are crucial for the safety of the power system [3]. For the control of the excitation system, high-speed signal acquisition devices achieve precise data collection and processing, and are used in AVR and PSS control to optimize the overall performance of the power system. Literature [4] achieved a perfect match between domestic excitation regulators and imported Siemens power rectifier cabinets, optimizing and improving the PSS function. All excitation regulators are built-in PSS2B, which has a good damping effect on low-frequency oscillations of 0.1 to 2Hz. Literature [5] designed a compound excitation system combining fuzzy PID control and power system stabilizer PSS (FPID+PSS), which not only improves the stability and robustness of the generator terminal voltage but also leverages the PSS's suppression characteristics for low-frequency oscillations of generator speed, electrical power, and power angle, thereby enhancing the dynamic performance of the entire power system. Literature [6] established a parallel PSS-AVR model, proving that the parallel model can effectively suppress low-frequency oscillations of the system, and the phase angle required for compensation is significantly smaller than that of the PSS-AVR series model. Literature [7] proposed a data-driven model predictive control (MPC) method and applied it to the AVR+PSS control structure, enhancing the system's oscillation damping performance and voltage stabilization capability. Literature [8] proposed a power system stabilizer architecture, whose control strategy utilizes real-time estimation of the inertia center speed from wide-area measurements, enabling synchronous generators to better adapt to changes in system dynamics.

This paper designs a fully domestic high-speed signal acquisition device and applies it to the excitation control system to sample stator voltage and stator current, enabling both the excitation regulator cabinet controller and the intelligent power cabinet controller to implement the proposed AVR+PSS control. The strategy was tested and verified at the Hydropower Plant, demonstrating the effectiveness of the AVR+PSS control strategy based on the high-speed signal acquisition device for the stability of the excitation control system.

2. Autonomous and controllable high-speed signal acquisition device

The high-speed signal acquisition device is used for real-time collection of key parameters in the generator excitation system, such as anode voltage and current. Its hardware consists of an MCU and FPGA; the MCU handles logic control and data processing, while the FPGA, with its parallel processing capability, performs rapid signal acquisition and preliminary processing to meet high-speed and high-precision requirements.

2.1. Hardware architecture design

In this paper, E-SA4000A, a national production device control module developed based on Zhaoyi innovative MCU and Gaoyun FPGA, aims to build a hardware platform for high-speed signal acquisition device, focusing on analog data acquisition and high-speed communication functions.

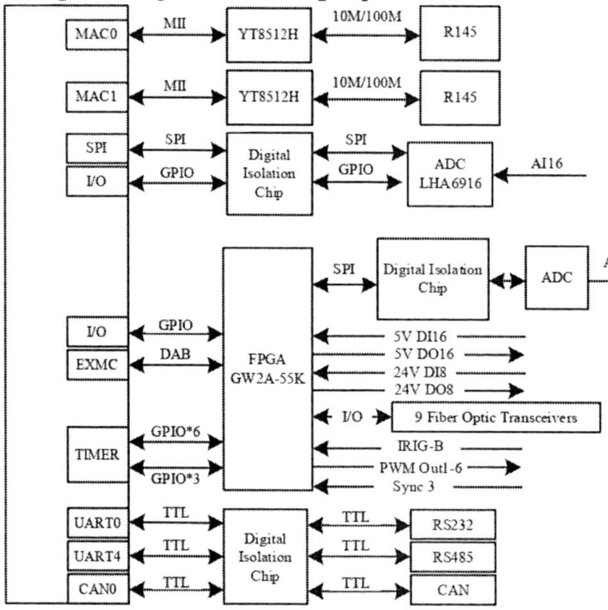

Figure 1. High speed signal acquisition module hardware block diagram

After completion of manufacturing, the E-SA4000A will undergo localization certification and electromagnetic compatibility tests to ensure its performance and reliability.

The hardware design block diagram of high-speed signal acquisition module is shown in Figure 1. The acquisition module includes analog signal acquisition module and switching signal acquisition module.

2.2. Data processing and conversion

The analog data acquisition of high-speed signal acquisition device includes key signals such as three-phase stator voltage (*Va, Vb, Vc*), three-phase stator current (*Ia, Ib, Ic*), system voltage Vs and excitation current *Ifd*. These raw sampling values are obtained by the MCU from the FPGA, which is responsible for parallel control of each ADC chip for sampling. Based on Fourier algorithm, active power, reactive power and angular velocity are calculated by three-phase stator voltage and current. The sample diagram shows that the sampling frequency of frequency conversion is 12 times of the signal frequency, the sampling point corresponds to the signal waveform, and the sampling period is related to the signal period, which is 1/12 of the signal period.

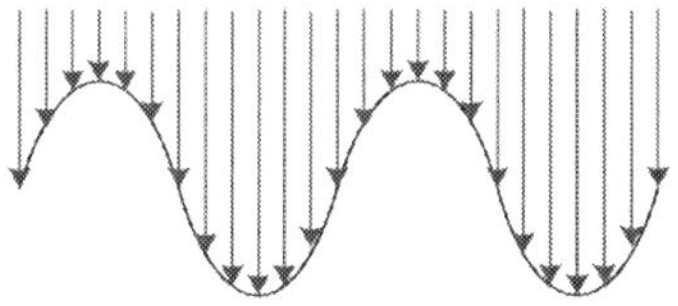

Figure 2. Variable frequency sampling pattern

$$V_i = k \cdot \frac{1}{6} \sum_{n=0}^{11} (V_n \cdot \sin \frac{n\pi}{6}) \quad (1)$$

$$V_r = k \cdot \frac{1}{6} \sum_{n=0}^{11} (V_n \cdot \cos \frac{n\pi}{6}) \quad (2)$$

$$\dot{V} = V_r + jV_i \quad (3)$$

Where is the signal vector, Vi is the virtual part of the signal, Vr is the effective value of the real part of the signal, k is the conversion coefficient of the signal from voltage to unit value, Vn is the result of the n_{th} frequency conversion sampling.

3. Peer distributed excitation control strategy

In the peer-to-peer distributed excitation control system, the excitation regulator is configured with 2 control channels, and each excitation power cabinet is configured with 1 control channel. The power cabinet control channel not only completes the conventional trigger pulse generation function, but also has the excitation control function completely consistent with the two control channels of the excitation regulator. The operation mode of each control channel mainly includes: constant voltage operation mode (AVR), constant excitation current operation mode (FCR), constant reactive power operation mode, constant power factor operation mode and constant control Angle operation mode. In order to enhance the stability of the power system, the excitation regulator in this paper also adopts a more complex control strategy, power system stabilizer (PSS).

3.1. Data processing and conversion

AVR mode is a closed-loop automatic regulation mode, its core task is to maintain the generator voltage constant. In this operation mode, the excitation regulator usually takes the terminal voltage as the feedback quantity, and realizes the voltage stability control through PID regulation. Its control block diagram is shown in Figure 4.

α _max and α_min: are the limits of control Angle. V0_max: indicates the upper limit of voltage in no-load state. V0_min: Lower limit of the voltage set in no-load condition. V1_max: indicates the upper limit of the voltage set in the grid-connected state. V1_min: Lower limit of the voltage set in the grid-connected state. Ifd0_max: indicates the upper limit of excitation current in no-load state. Ifd0_min: lower limit of excitation current under no-load condition. Ifd1_max: indicates the upper limit of the set excitation current in grid-connected state. Ifd1_min: lower limit of the set excitation current value in grid-connected state. Q_max: indicates the upper limit of the specified reactive power. Q_min: Lower limit of the specified reactive power. cos_max: indicates the upper limit of the power factor. cos_min:

Lower limit of the power factor. 1.35Van/0.9Vf0: The control scalar scaling coefficient of the excitation control algorithm.

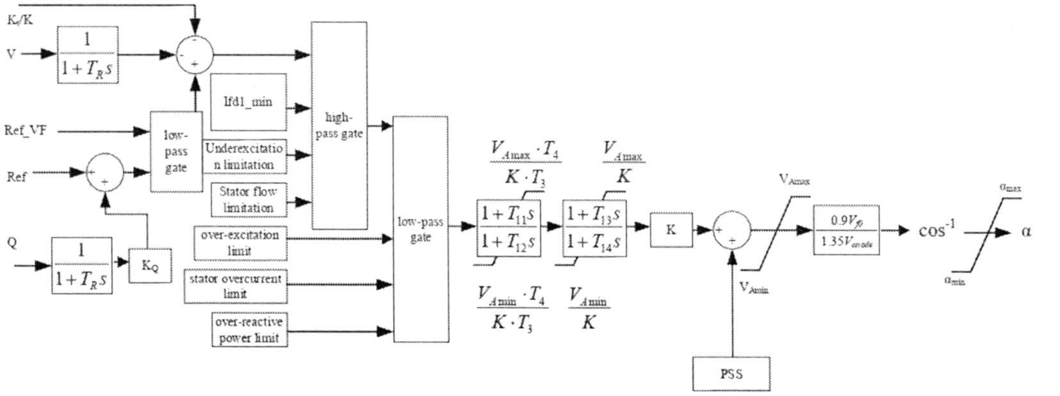

Figure 3. Excitation control system AVR control block diagram

Where *T11*, *T12*, *T13*, *T14*, and *K* are PID parameters.

3.2. System stabilizer

The electromagnetic torque of synchronous generators can be decomposed into two components: the synchronous torque component and the damping torque component. Insufficient synchronous torque leads to slipping out of step, while inadequate damping torque causes oscillatory loss of step. The development of the electric power industry has led to the expansion of power system scales and the increase in unit capacity. The excitation regulators of large generating units are typically composed of thyristors and integrated circuits, and rapid excitation regulation is widely applied. This reduces the time constant of the excitation system, resulting in decreased system damping and a tendency towards weakly or underdamped states, thereby inducing low-frequency oscillations. To suppress low-frequency oscillations, power system stabilizers have been introduced, with PSS2B being commonly used.

3.2.1. Principle of low frequency oscillation suppression by PSS. PSS input signals are generally composed of speed deviation $\Delta\omega$, frequency deviation Δf, mechanical power deviation ΔPm, electromagnetic power deviation ΔPe, and the specific structure is as follows:

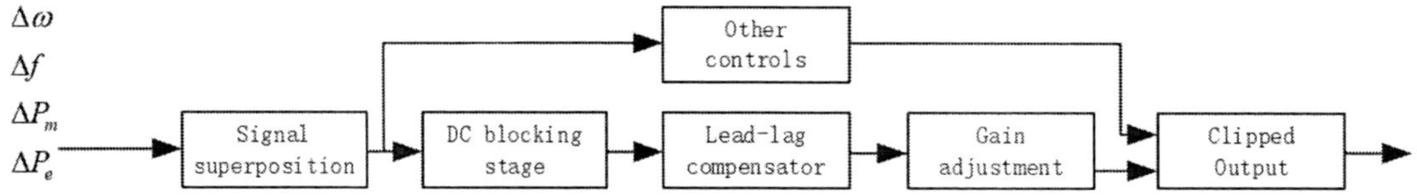

Figure 4. PSS structure block diagram

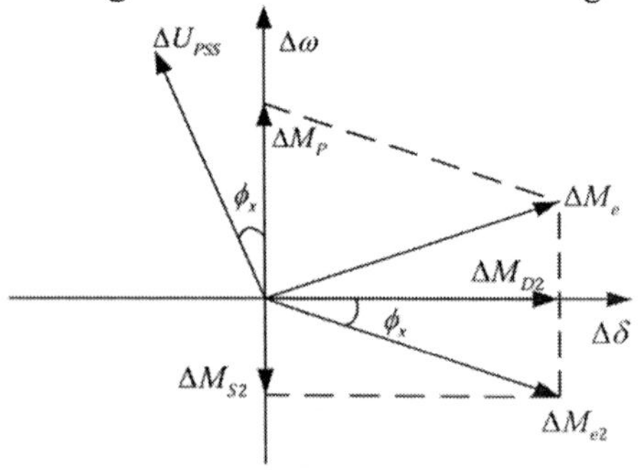

Figure 5. Damping torque phasor

By inference, the focus is on how to achieve the desired positive damped torque through the input signal UPSS. Assuming the PSS transfer function is *Gpss (s)* and the output signal is *Δω*, the expression for the additional torque introduced by PSS is:

$$\Delta M_p = \frac{K_2 K_3 G_e G_{PSS}}{1 + K_3 T'_{d0} s + K_6 K_3 G_e} \Delta\omega = G_X G_{PSS} \Delta\omega \tag{4}$$

$$G_X = \frac{K_2 K_3 G_e}{1 + K_3 T'_{d0} s + K_6 K_3 G_e} \tag{5}$$

When the system transmits heavy loads over long distances, the excitation regulator generates an additional torque (ΔMe2) that lags behind the power Angle deviation (Δδ). Let the lag phase Angle be Øx, and the corresponding damping torque vector diagram can be drawn accordingly. The additional Upss signal at the voltage superposition point of the excitation system regulator is generated by the PSS controller. Therefore, a leading phase Angle Øx can be generated by properly configuring the PSS parameter. This phase Angle compensates for the excitation lag phase and adjusts the PSS gain to provide a larger pure damping torque (ΔMp) for the system. ΔMe2 and ΔMp can form a synthetic torque (ΔMe) in the first quadrant, where both components of the synthetic torque are positive, which is beneficial to the stable operation of the system.

Therefore, only the phase provided by *Gpss* can compensate the lag phase of *Gx*, and the additional damping *ΔMp* provided by PSS can be in phase with the speed deviation *Δω,* at which time PSS will inject positive damping into the system.

3.2.2. Principle of PSS control. In the study of power system stability, the application of traditional PSS1A and PSS1B power system stabilizers (PSS) may lead to reactive power inverse phenomenon, that is, a significant or unacceptable decrease in reactive power occurs when the input power of prime mover is increased. On the contrary, when the prime mover input power is reduced, there is a significant or unacceptable increase in reactive power

To solve this problem, PSS2B model is adopted in this paper, which can effectively avoid the phenomenon of reactive power reversal by introducing synthetic acceleration power signal as input.

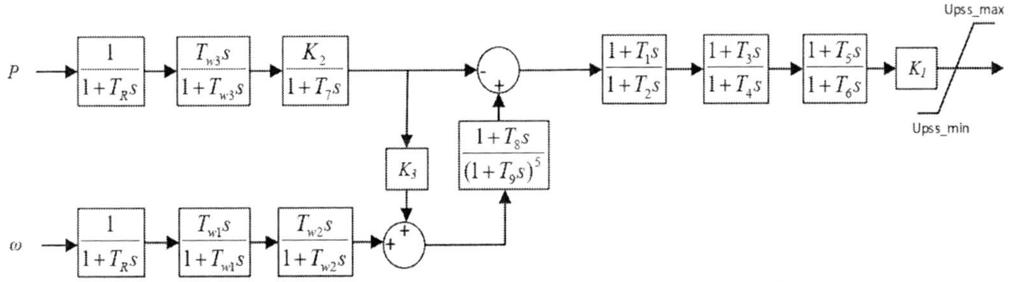

Figure 6. Excitation control system PSS2B control block diagram

According to the control block diagram of the type PSS2B power system stabilizer, the model consists of two forward channels for synthesizing the mechanical power change *ΔPm* and the electrical power change *ΔPe* of the generator. The difference between these two signals is the generator's acceleration power deviation *ΔPa*. In the case of unidirectional load change of the unit, *ΔPa* approaches to zero, and the PSS2B model has no influence, so as to avoid the inverse modulation of reactive power. When the prime mover of the generator performs a fast switching operation, the value of the frequency deviation *Δω* is relatively small, while the value of the electrical power change *ΔPe* is larger. After the two signals are synthesized, *ΔPm* is not zero, but after processing through a low-pass filter and subtracting *ΔPe*, the resulting *ΔPa* value is still small, resulting in a negligible role for the PSS2B model. Therefore, no matter in the case of unidirectional load change or the prime mover fast switching operation, the PSS2B model will not cause the reactive power reversal phenomenon of the unit. PSS2B model only plays a role in the active power oscillation of the unit to suppress the low-frequency oscillation of the power system.

Journal of Physics: Conference Series **3012** (2025) 012086 doi:10.1088/1742-6596/3012/1/012086

4. Experimental verification

A peer-to-peer distributed excitation control system has been tested in hydropower plant. The specific parameters are shown as follows:

The rated capacity SN of the test unit is 194.3MVA, the rated terminal voltage is 13.8kV, the rated terminal current is 8129A, the rated excitation current is 1990A and the rated excitation voltage is 465V. AC auxiliary power supply:380V, DC power supply:220V.

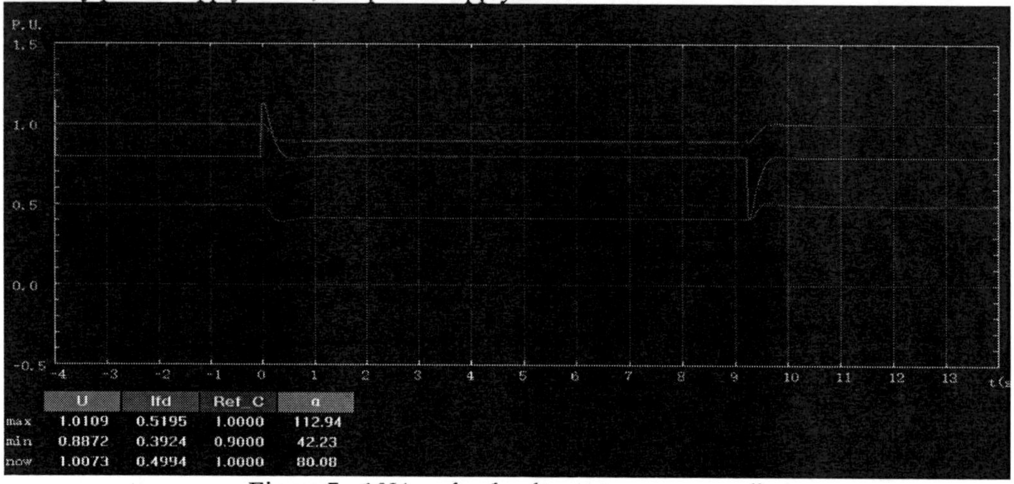

Figure 7. -10% no-load voltage step response diagram

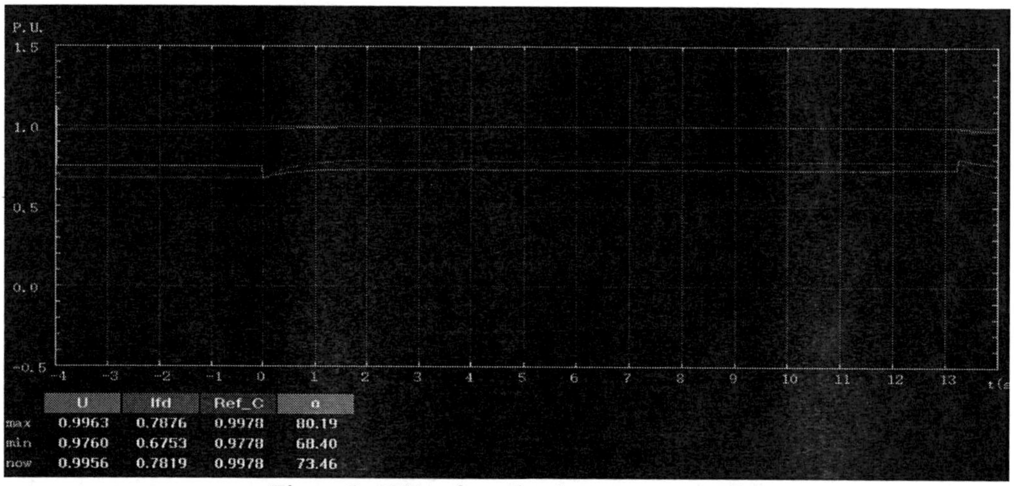

Figure 8. +2% voltage load step response diagram

The no-load step test verifies the working state of the power system when it oscillates under the constant terminal voltage operation mode. The terminal voltage is raised to the rated level, and the operation mode is set as the terminal voltage operation mode. The -10% voltage step response test is carried out. Load step test unit is connected to the grid, the terminal voltage is increased to the rated, the operation mode is set to the terminal voltage operation mode, the given value is modified, and the +2% voltage load step response test is carried out.

In Figure 7, the given voltage *Ref_C* decreases from 1pu to 0.9pu at 0s and recovers to 1pu at 9.2s

The voltage *U* changes rapidly with the given value of voltage after a response of about 0.5s, and the excitation current also drops at 0s and returns to the original state at 9.6s. The control Angle α quickly returns to the original state after oscillation when the voltage *U* changes, and generally remains unchanged.In Figure 8, when the voltage set *Ref_C* rises from 0.98pu to 1pu at 0s and recovers to 0.98pu at 13.2s, the terminal voltage U changes rapidly with the voltage set after a response of about 0.5s, and the excitation current also rises at 0s and recovers to the original state at 13.8s. The control Angle α decreases at 0s and returns to the original state at 13.2s.

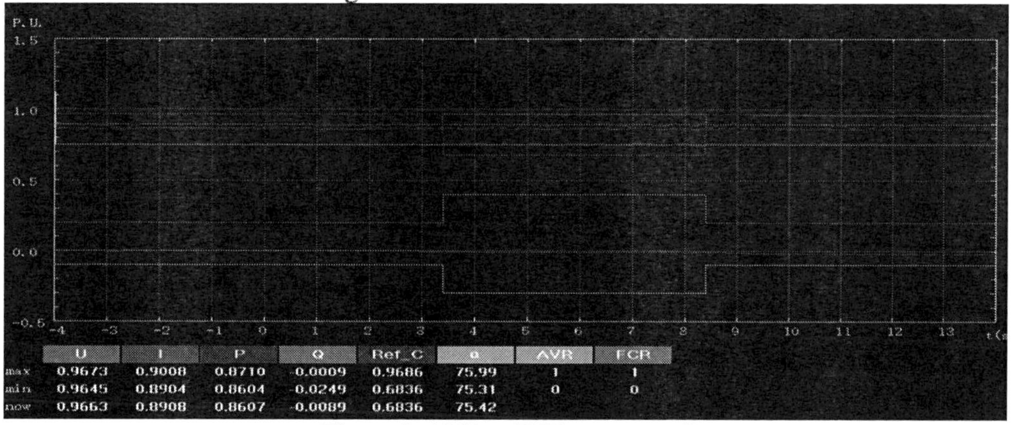

Figure 9. AVR to FCR response diagram

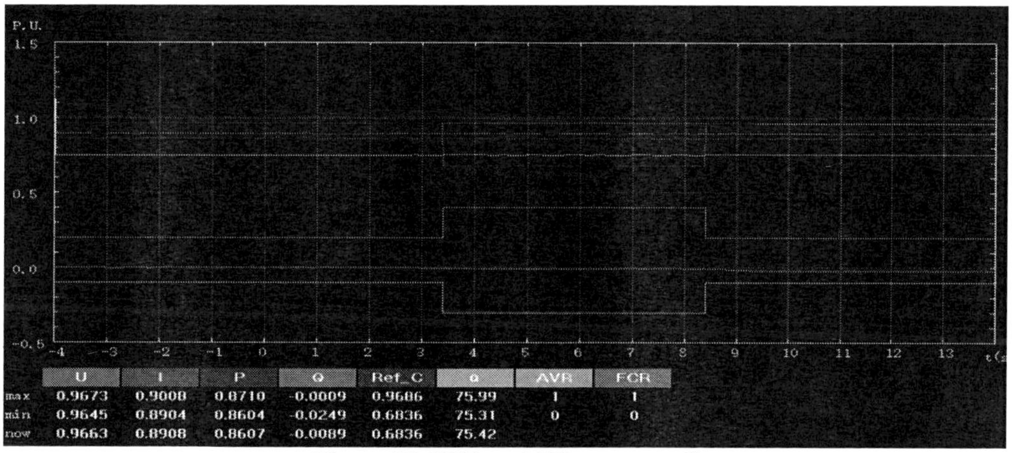

Figure 10. FCR to AVR response diagram

After the experimental unit is boosted and connected to the grid, the terminal voltage is first raised to the rated voltage and the operation mode is set to the excitation current mode to ensure stable operation. Then switch the operation mode to the terminal voltage mode and continue to maintain stable operation.

In Figure 9, when the operation mode is switched from AVR to FCR at 3.3s, the given voltage reference value Ref_C is reduced from 0.96pu to 0.68pu, and the terminal voltage U, current I, active power P and reactive power Q can all remain in the original state. In Figure 10, the operation mode is switched from FCR to AVR at 3.3s, at which time the given voltage reference value Ref_C is reduced from 0.97pu to 0.68pu, and the terminal voltage U, current I, active power P and reactive power Q can all remain in the original state.

5. Conclusion

A high-speed signal acquisition device has been designed and tested at the Gezhouba Power plant. Through optimized design and cost-effective components, it provides high-precision data critical to generator stability, fault diagnosis, real-time monitoring, analysis, prediction and optimization, while reducing costs and ensuring reliability. The AVR+PSS control strategy proposed in this paper shows obvious advantages in no-load step test and load step test, which can respond quickly and recover to the initial state when the system voltage fluctuates. Compared with the traditional control strategy, the operation efficiency, safety and reliability of the excitation control system are greatly improved. The traditional control strategy is slow in response, and it is difficult to ensure the stable operation of the system under complex conditions. Future research can further focus on enhancing the performance of high-speed signal acquisition equipment, especially improving the sampling frequency and accuracy in complex power environments. By combining artificial intelligence and big data technology, upgrading control strategies, such as using artificial intelligence algorithms to achieve accurate fault prediction and diagnosis, and using big data to optimize control strategies, significantly improve the stability and intelligence level of power systems.

Acknowledgments
Funded by the scientific research project of the Ningnan Baihetan Power Plant of the Three Gorges Jinshajiang Yunchuan Hydropower Development Co. Ltd. (5323020068)

References

[1] Fan Dapeng, Chen Lingyu, Wang Xiaoyu, et al. Architecture and key technology of autonomous controllable programmable control system [J]. Optics and Precision Engineering, 2023, 31 (01): 1-13.

[2] Rao Zhibo. The development of domestic PLC has a long way to go [J]. Automation Expo, 2019,37(04):22-23.

[3] Xiao Quan, Zhou Xiaoshan, Wang Ruiqing, et al. A Hardware Implementation Scheme of Autonomous Controllable Excitation Control Module [J]. Electric Drive,2023,53(08):3-8.

[4] Wang Benhong, Zhang Yuandong, Li Qingmao, et al. Localization of speed regulation and excitation System of Three Gorges Left Bank Hydro-turbine [J]. Mechanical and Electrical Technology of Hydropower Station,2022,45(07):85-88.

[5] Liu Wenfeng, Meng Wei. Research on Compound excitation System with fuzzy PID combined with PSS [J]. China Electrochina,2024,(09):27-32

[6] Deng Ruomei, Li Xiaocong, Pu Tingjie. Theory and Practice of phase compensation for excitation System using PSS-AVR parallel Model [J]. Electrical Switches,2019,57(02):52-57.

[7] W. Han and A. M. Stanković, "Model-Predictive Control Design for Power System Oscillation Damping via Excitation – A Data-Driven Approach," in IEEE Transactions on Power Systems, vol. 38, no. 2, pp. 1176-1188, March 2023.

[8] R. T. Elliott, P. Arabshahi and D. S. Kirschen, "A Generalized PSS Architecture for Balancing Transient and Small-Signal Response," in IEEE Transactions on Power Systems, vol. 35, no. 2, pp. 1446-1456, March 2020.

Design of a Fully Decentralized Distributed Architecture-Based Excitation System for Large Hydropower Units

Hongyi Hua[1], Jing Zhang[2], Yuting Wang[1], Peng Chen[1], Jian Cheng[2], Chuan Qin[1], Lin Tang[2], Xuemin He[3*]

[1] China Three Gorges Corporation, Wuhan 430014, Hubei, China;

[2] Baihetan Power Plant, Three Gorges Jinshajiang Yunchuan Hydropower Development Co., Ltd, Liangshan 615400, Sichuan Province, China;

[3] Three Gorges Intelligent Industrial Control Technology Co. , Ltd. , Wuhan 430070, China

*Corresponding author's e-mail: he_xuemin@126.com

Abstract. This paper proposes a new type of peer - to - peer distributed excitation control system to address the limitations of the traditional centralized architecture and enhance the redundancy and reliability of the system. This system adopts the fiber - optic point - to - point communication technology, which improves the communication speed and reduces the influence of electromagnetic interference. Verified by on - site engineering in hydropower stations, this architecture can continue to operate when a fault occurs at any node, significantly improving the stability of the entire system.

1. Introduction

The traditional excitation system's power cabinets and the de - excitation cabinets lack independent CPU modules and belong to the centralized control architecture. Under the control of the regulator, the excitation power cabinets output excitation current to maintain the stability of the generator terminal voltage and rationally distribute reactive power[1], while the de - excitation cabinets provide de - excitation and over - voltage protection functions. Due to the lack of independent CPU modules, when the regulator fails, the entire system will be out of operation, resulting in relatively low system reliability.

The distributed control architecture is widely used because of its advantage in improving reliability. As the complexity of power systems increases, the coordination of distributed energy resources and intelligent devices becomes more challenging. To address the limitations of the centralized architecture, Literature [2] introduces a distributed application architecture in the Grid APPS - D platform. Literature [3] presents a decentralized control system for grid - connected photovoltaic inverters, and Literature [4] proposes a communication - free distributed control architecture suitable for islanded micro - grids with renewable energy generation and energy storage functions.

With the increasing requirements for power grid stability, Literature [5] established a small - signal model of voltage - source converter (VSC) to study the stability of DC - link voltage control. This model is helpful for representing the dynamic characteristics of the DC - link voltage of the VSC integrated into a weak grid. The influences of grid strength, operating point and the interactions among control loops on the performance of VSC are all taken into account. Literature [6] studied the impact

Journal of Physics: Conference Series **3012** (2025) 012087 doi:10.1088/1742-6596/3012/1/012087

of the dynamic characteristics of phase - locked loop (PLL) on the stability of DC - link voltage control (DVC) in voltage - source converter (VSC) connected to a weak grid. At the same time, the traditional excitation system has been gradually optimized[7]. Independent CPU modules are introduced into the excitation power cabinets and the de - excitation cabinets, and data transmission with the regulation cabinets is carried out to form a distributed excitation control system[8]. The distributed architecture distributes the control functions among multiple nodes, reducing the impact of single - point failures and improving the reliability of the system. Literature [9] introduces a highly redundant excitation system based on distributed control. After both of the two control channels in the excitation regulation cabinet are out of operation, the one with the highest preset priority in the intelligent power cabinets can be selected as the online control module to coordinate all intelligent power cabinets to operate in the constant anode voltage mode, achieving 2 + N redundancy, where 2 represents the two control channels in the excitation regulation cabinet and N represents the number of power cabinets. Literature [10] adopts a dual - CAN - bus redundant communication design scheme to realize the design of a distributed excitation system.

However, in the current distributed excitation systems, the controllers are not of equal rank. The control channels of the regulation cabinets have a higher priority, while the power cabinets have a lower priority and can only compete for control rights after the high - priority controllers are out of operation. Therefore, designing an excitation system architecture with equal - priority is of great significance for improving system reliability. This paper proposes a new type of peer - to - peer distributed excitation control system to enhance the redundancy and reliability of the system.

2. Key Technologies of Decentralized Distributed Architecture-Based Excitation System

2.1. Overall Design Architecture

The peer - to - peer distributed excitation control system adopts the distributed control technology of fiber - optic point - to - point communication, which can avoid electromagnetic interference from affecting signal transmission and increase the communication rate. The regulating cabinet has two AVR + FCR automatic control channels, and each power cabinet has an independent AVR + FCR automatic control channel. The control cores of all channels are the same, and the controller hardware and structure are interchangeable and can be shared. The high - speed signal acquisition device broadcasts the collected analog signals to the control module, so as to make the software and hardware configurations of 2 + N controllers consistent, and improve the system redundancy, reliability and the utilization rate of spare parts. The system block diagram is shown in Figure 1.

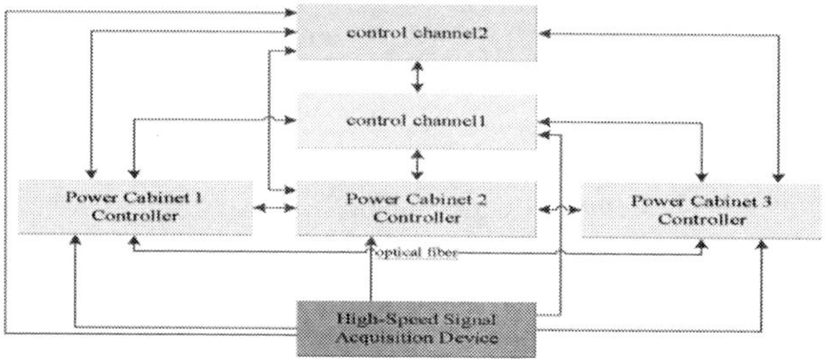

Figure 1. Block Diagram of Decentralized Distributed Architecture-Based Excitation System

2.2. Excitation Regulation Cabinet Architecture

The excitation regulator is a programmable intelligent controller. This regulator as a whole adopts the distributed control technology of fiber - optic point - point communication. Its regulation cabinet is equipped with two AVR + FCR control channels.

2.2.1. CPU Module of Excitation Regulator. When considering the real - time and reliability requirements of excitation control, a strategy of decoupling communication from control is adopted, and a hardware architecture of CPU, MCU and FPGA is selected. Specifically, the CPU is mainly responsible for external communication processing; the ARM mainly undertakes real - time control tasks; and the FPGA is responsible for I/O expansion and optical fiber communication. The overall hardware block diagram of the main controller module is shown below.

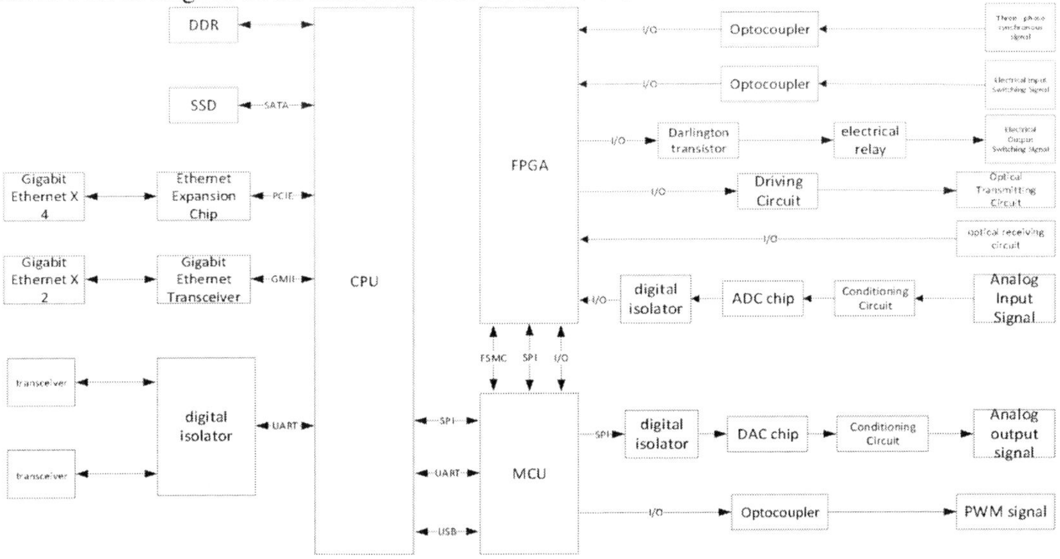

Figure 2. Hardware Block Diagram of the Controller Module

2.2.2. Analog Signal Interface Module of Excitation Regulator. The analog signal interface module is applied to the excitation regulation cabinet. Its main functions are to collect the generator terminal voltage, stator current, synchronization signal, excitation current signal and system voltage signal, and to perform electrical isolation and conditioning on these signals. The processed signals are transmitted to the control channel for control calculation.

The circuit design of this module includes a power supply circuit, an analog quantity acquisition circuit and a signal conditioning circuit. The power supply circuit provides power for other circuits and an isolated power supply for the analog quantity sampling channels. The analog quantity acquisition circuit is responsible for the preliminary collection and arrangement of analog signals. The signal conditioning circuit mainly realizes functions such as current - to - voltage conversion, filtering, amplification, shaping and comparison.

2.2.3. Digital Signal Input/Output Module of Excitation Regulator. The 16-channel digital signal input module in the excitation regulation cabinet collects and isolates on-off quantity signals. Its circuit includes power, photoelectric isolation, and signal conditioning circuits. The 16-channel digital signal output module is used in the cabinet with similar circuits and 2 input channels, each with 16-channel on-off quantity input for isolation output.

2.3. Excitation Power Cabinet Architecture

The power cabinet uses a thyristor rectifier with cooling by a gravity heat pipe and an auxiliary fan. The number of parallel branches is 3 with (n - 1) redundancy. Secondary equipment is installed based on the existing primary equipment. After the transformation, each power cabinet is equipped with an independent controller, a signal interface module, a high-speed signal acquisition device, and a thyristor gate trigger module. The control operation channels use the same controller module. The high-speed signal acquisition device broadcasts analog quantity signals to achieve the same control functions. When a channel in the regulating cabinet fails, the program switches to the independent operation of the control operation channel in the power cabinet automatically and seamlessly, with strong interference and vibration resistance.

2.3.1. Excitation Power Cabinet Controller. The excitation power cabinet controller uses the same controller as the regulating cabinet. It converts the control signal into the control pulse and supplies it to the interface board, and has the measurement and calculation functions of the anode synchronous voltage, excitation voltage, and excitation current.

2.3.2. Independently - Controllable High - speed Signal Acquisition Device. The high - speed signal acquisition device is placed in the power cabinet and is used to collect data in the analog signal interface module of the power cabinet, such as the anode voltage, anode current, synchronization signal, excitation voltage, excitation current signal, etc. of the generator excitation system. Its hardware structure is composed of MCU (Microcontroller Unit) and FPGA (Field - Programmable Gate Array).

2.3.3. Power Cabinet Analog Signal Interface Module. The power cabinet signal interface module is applied to the excitation power cabinet. It is mainly used to collect, electrically isolate and condition the anode voltage (AV), anode current (AC), synchronization signal (SS), excitation voltage (EV), and excitation current signal (ECS) of the generator excitation system. After processing, the signals are transmitted to the power cabinet controller for control operations.

2.3.4. Controllable Cabinet Gate Trigger Module. The thyristor gate trigger module is applied in the excitation power cabinet and is mainly used to trigger the three-phase full-bridge thyristor. When cooperating with the pulse amplifier module, it can realize the triggering of the thyristor with a rated current of less than 4000A.

3. Implementation Methods of the Control Architecture for Decentralized Distributed - Architecture - Based Excitation System

In the peer - to - peer distributed excitation system, the intelligent control modules (ICMs) of intelligent power cabinets are interconnected point - to - point by optical fibers, and the human - machine interface (HMI) is connected to each ICM via Ethernet. The ICM performs fixed - frequency sampling on 3 - phase anode V&I, 3 - phase stator V&I and output V&I of power cabinets. Through RMS algorithm and Fourier algorithm, the ICM interpolates and filters 3 - phase stator V or 3 - phase anode V signals to get signal period.

When using RMS algorithm, sample three - phase voltage signals u_a,u_b,u_c, Let sampling sequence be $u_a(n), u_b(n), \cdot u_c(n)_{(n=0,1,\cdots,N-1)}$ Then the formula for calculating the RMS value of three - phase voltage is:

$$U_{rms} = \sqrt{\frac{1}{N}\sum_{n=0}^{N-1}\left[u_a^2(n) + u_b^2(n) + u_c^2(n)\right]} \tag{1}$$

After interpolation and filtering, a relatively smooth root mean square (RMS) value curve $U_{rms}(m)$ is obtained, where m is the serial number of the processed sampling point. For a periodic signal, suppose the signal period is T, the sampling period is T_s, and the number of sampling points is N, then

$$T = NT_s \tag{2}$$

Define the autocorrelation function $R_{U_{rms}}(k)$ of the root - mean - square value curve as:

$$R_{U_{rms}}(k) = \frac{1}{M-k}\sum_{m=0}^{M-k-1} U_{rms}(m)U_{rms}(m+k) \tag{3}$$

When $k = nT/T_s \quad (n = 1, 2, \cdots)$, the autocorrelation function will have a peak. The value corresponding to the first peak (except 0) can be used to estimate the signal period.

Journal of Physics: Conference Series **3012** (2025) 012087 doi:10.1088/1742-6596/3012/1/012087

If the signal period is obtained by using the Fourier algorithm, then for a discrete sequence, its DFT is defined as

$$X(k)=\sum_{n=0}^{N-1}x(n)e^{-j\frac{2\pi}{N}kn},(k=0,1,\cdots,N-1) \tag{4}$$

Perform discrete Fourier transform on the three - phase voltage signals respectively. In the frequency domain of the signals after interpolation and filtering, there will be obvious spectral components at the discrete point frequency k_1 corresponding to the fundamental frequency f_1 . If the sampling frequency f_s and the discrete frequency point k_1 corresponding to the fundamental frequency are known, the formula for calculating the period is

$$T=\frac{N}{k_1 f_s} \tag{5}$$

In the excitation system, the intelligent control module (ICM) can simplify the online module design, remove the analog quantity acquisition function, and reduce the computational burden by borrowing external data processing to achieve high-speed sampling, processing, and communication to meet the real-time control requirements. Each ICM is equipped with the same excitation control program, and the control priority is pre-set. The one with the highest priority obtains the online control right and independently executes the program (such as intelligent current sharing, PSS2B addition, constant terminal voltage, etc. control).

The modules communicate with each other point-to-point via optical fibers, and exchange status information (stator voltage, PID intermediate quantity, control angle, etc.) every power frequency cycle. The modules are connected to the human-machine interface (HMI) via Ethernet, and the PID intermediate quantity is used for undisturbed switching control output, which is calculated by the standby module.

The modules determine the priority according to the ascending order of the power cabinet serial number, and the one with the lower serial number has priority. If the PT of the online module is disconnected, the one that is not faulty and has the highest priority obtains the control right. Count the number of normal communications between each module and other modules and sort them. If the number of normal communications of the online module is not the largest, then the one with the largest number of normal communications and the highest priority obtains the control right. If it also has a PT disconnection, then the one with the largest number of normal communications and the highest priority obtains the control right.

4. AVR and FCR Control

The power cabinet and the excitation cabinet adopt the AVR + FCR operation mode. Among them, the AVR adopts the series PID control algorithm, and its main - loop model is shown in the following figure:

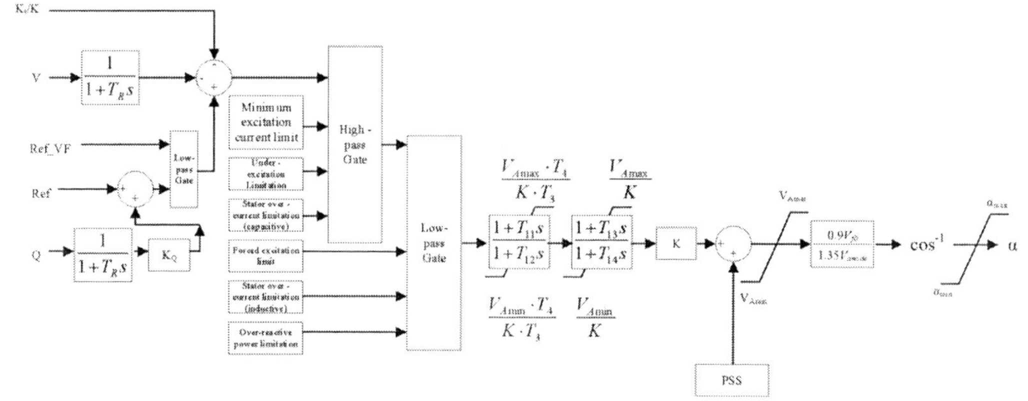

Figure 3. The main - loop model of AVR control

FCR is a backup mode for the widely - used AVR function of generators. In the field current regulation mode, when the regulator receives an “increase” or “decrease” signal, it will adjust the FCR set value I_{fref}, thus being able to adjust the excitation current of the generator and indirectly achieving the purpose of adjusting the generator voltage and reactive power.

5. Experimental Verification

The peer - to - peer distributed excitation system architecture designed in this paper has been subjected to on - site engineering verification in a hydropower plant. The following figure shows the channel switching experiment waveform diagram:

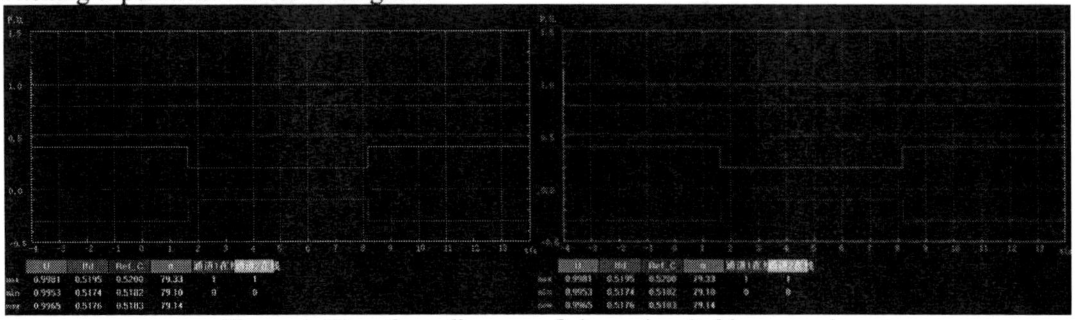

Figure 4. Waveform diagram of channel switching experiment

(1). Raise the generator - terminal voltage to the rated value, set the "operation mode" to "generator - terminal voltage" mode on the display screen, and it should be able to operate stably after channel switching.

(2). Set the "operation mode" to "excitation current" mode on the display screen, and it should be able to operate stably after channel switching.

(3). In case of a working channel failure, it should switch to the standby channel and maintain normal functions. In case of a standby channel failure, it should not affect the normal functions of the system and manual shutdown. In case of excitation system communication failure or interference, or standby channel tracking failure, it should not affect the normal functions.

Meanwhile, the peer-to-peer distributed excitation system architecture designed in this paper underwent load step tests. The load step test is an experiment that assesses the system's response capability when there is a sudden change in load. This experiment evaluates the system's adaptability and stability to load variations by simulating sudden increases or decreases in load. During the load step test, one can observe the fluctuations in the system's output voltage or current at the moment of load change, as well as the time required for the system to return to a stable state. The following is a description of the waveform from the load step test shown in Figure 5 and Figure 6.

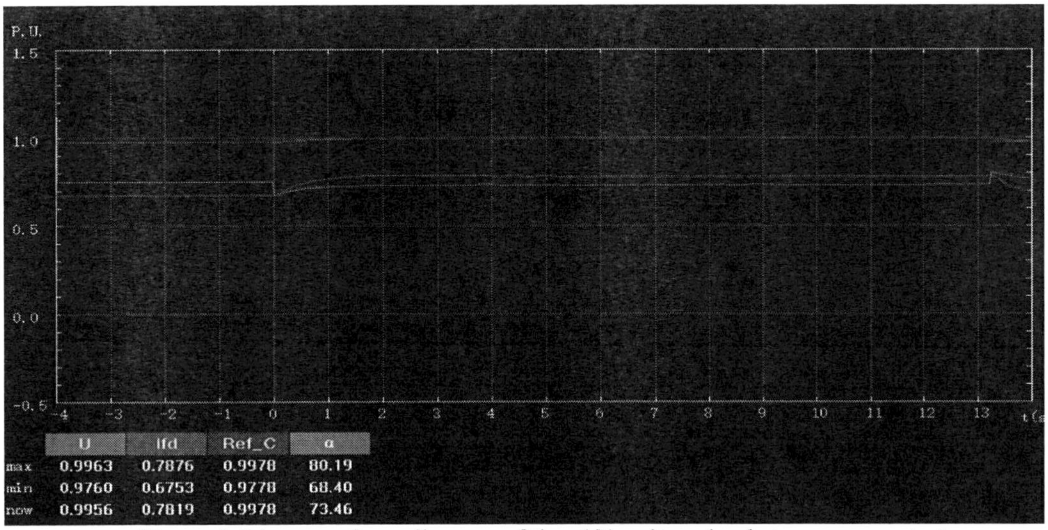

Figure 5. The waveform diagram of the -1% voltage load step response test

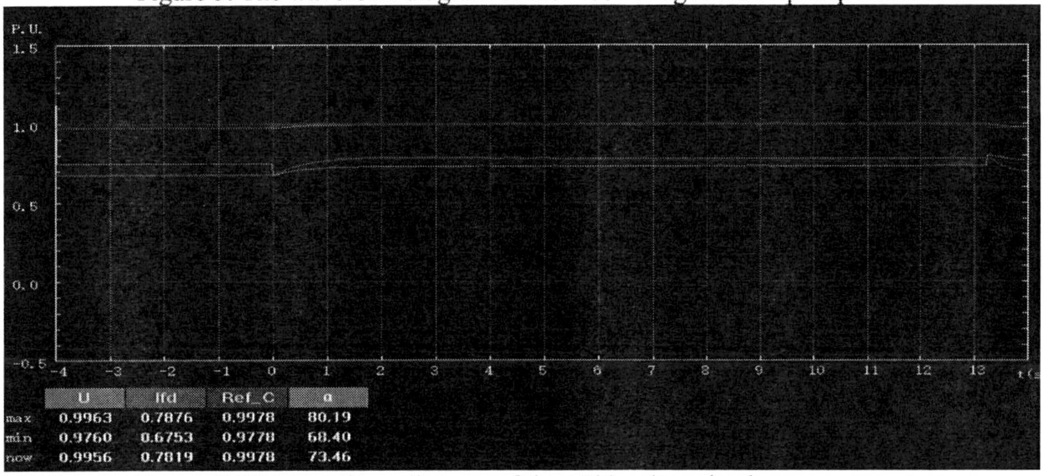

Figure 6. The waveform diagram of the +1% voltage load step response test

(1). Boost the unit voltage and connect it to the grid.
(2). Raise the terminal voltage to the rated value, and set the "Operating Mode" on the display screen to "Terminal Voltage" mode.
(3). In the test window, modify the set value to conduct a -1% rated terminal voltage load step test, and observe the step process.
(4). In the test window, modify the set value to conduct a +1% rated terminal voltage load step test, and observe the step process.

It can be seen from the experimental results that whether it is the load reduction experiment or the load increase experiment, the peer-to-peer distributed system can respond quickly and maintain the stability of the system.

Finally, the operating performance of the excitation system designed in this paper is compared with that of the traditional excitation system. The following Figure 7 is the test chart of the operating temperature of the power cabinet, in which the left side is the power cabinet of the peer-to-peer distributed excitation system and the right side is the traditional power cabinet. It is obvious that the operating temperature of the peer-to-peer distributed excitation system designed in this paper is lower, and it has higher reliability in terms of operation safety and other aspects.

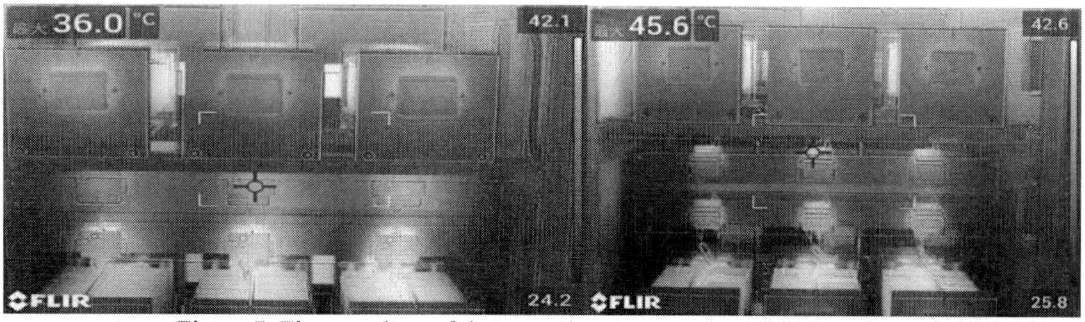

Figure 7. The test chart of the operating temperature of the power cabinet

6. Conclusion

This paper proposes a new type of peer-to-peer distributed excitation control system to address the limitations of the traditional centralized architecture and enhance the redundancy and reliability of the system. This system adopts the fiber-optic point-to-point communication technology, which improves the communication speed and reduces the influence of electromagnetic interference. Verified by experiments, this architecture can ensure that the system can still continue to operate when a fault occurs at any node, significantly improving the stability of the overall system. Future work will focus on further optimizing the response time of the system and exploring more applications of this technology in the smart grid environment. Moreover, peer-to-peer distributed excitation technology can be integrated with blockchain technology. Blockchain technology has the potential to enhance the security and transparency of peer-to-peer distributed excitation systems. By utilizing blockchain, decentralized energy transactions and power grid management can be achieved, which strengthens the system's resistance to attacks and ensures the integrity of data. In addition, edge computing technology can be applied to peer-to-peer distributed excitation systems to reduce data processing latency, thereby improving the real-time performance and efficiency of the system.

Acknowledgments

Funded by the scientific research project of the Ningnan Baihetan Power Plant of the Three Gorges Jinshajiang Yunchuan Hydropower Development Co. Ltd. (5323020068)

References

[1] Yu Minghao. Analysis on the Operating Characteristics of Flexible Excitation System [D]. Huazhong University of Science and Technology, 2020. DOI: 10.27157/d.cnki.ghzku.2020.002262.

[2] P. Sharma *et al.*, "GridAPPS-D Distributed App Architecture and API for Modular and Distributed Grid Operations," in *IEEE Access*, vol. 12, pp. 39862-39875, 2024, doi: 10.1109/ACCESS.2024.3374331.

[3] S. Farzamkia, H. S. K. Kedlaya and A. Q. Huang, "A Decentralized Control System for Series-Connected Grid-Integrated Photovoltaic Inverters," 2024 IEEE 15th International Symposium on Power Electronics for Distributed Generation Systems (PEDG), Luxembourg, Luxembourg, 2024, pp. 1-6, doi: 10.1109/PEDG61800.2024.10667454.

[4] N. L. Díaz, J. C. Vasquez and J. M. Guerrero, "A Communication-Less Distributed Control Architecture for Islanded Microgrids With Renewable Generation and Storage," in IEEE Transactions on Power Electronics, vol. 33, no. 3, pp. 1922-1939, March 2018, doi: 10.1109/TPEL.2017.2698023.

[5] Y. Huang, X. Yuan, J. Hu and P. Zhou, "Modeling of VSC Connected to Weak Grid for Stability Analysis of DC-Link Voltage Control," in *IEEE Journal of Emerging and Selected Topics in Power Electronics*, vol. 3, no. 4, pp. 1193-1204, Dec. 2015, doi: 10.1109/JESTPE.2015.2423494.

[6] Y. Huang, L. Wang, S. Zhang, D. Wang, X. Deng and G. Zhu, "Impacts of Phase-Locked Loop Dynamic on the Stability of DC-Link Voltage Control in Voltage Source Converter Integrated to Weak Grid," in *IEEE Journal on Emerging and Selected Topics in Circuits and Systems*, vol. 12, no. 1, pp. 48-58, March 2022, doi: 10.1109/JETCAS.2022.3149805.

[7] Duan Xiaoning. A Brief Analysis of Modern Synchronous Generator Excitation System [J]. Heilongjiang Science and Technology Information, 2009, (13): 32.

[8] Zhang Xiaoyi. Research and Implementation of High - speed Communication Network for Distributed Excitation System [D]. Hebei University of Technology, 2019. DOI: 10.27105/d.cnki.ghbgu.2019.000162.

[9] Yu Xiang, Zeng Tao, Gao Jinsong, et al. Distributed Excitation Control System Based on Fiber - optic Network [J]. Large Electric Machine and Hydraulic Turbine, 2017, (04): 71 - 75.

[10] Han Xuejing, Guo Jia, Li Zhijun. Design of a Distributed Excitation System [J]. Large Electric Machine and Hydraulic Turbine, 2012, (01): 56 - 58.

Journal of Physics: Conference Series **3012** (2025) 012088 doi:10.1088/1742-6596/3012/1/012088

Multi-source Fusion Considering Response Priority Flexibly Participate In Optimization Strategy of Reserve

Yue GUO*, Xuyang WANG, Boyu ZHOU, Yu SONG

State Grid Economic and Technological Research Institute Co., Ltd, Beijing 102209, China

*Corresponding author's e-mail: 1942819809@qq.com

Abstract: In order to solve problems of single source of reserve resources and possibility of low configuration or redundant configuration in the process of spare capacity allocation, a low-carbon economic dispatching method with multi-source fusion and flexible participation in reserve power considering response priority is proposed. Firstly, an optimization model with multi-source integration and flexible participation considering economy, carbon emissions and response time is established, the response priority of multiple reserve powers is sorted by analytic hierarchy process. Then, optimize distribution of reserve capacity and a multi-objective reserve optimal scheduling strategy is formulated. Finally, the improved IEEE-30 node system is used as a simulation example, and the experimental results show that the proposed method can effectively reduce response time and carbon emissions.

1. Introduction

As the penetration rates of new energy such as wind power and photovoltaic power, as well as new - type loads like flexible loads and electric vehicles in the distribution network gradually increase [1], how to configure a reasonable reserve capacity to ensure safe and reliable operation of power system has attracted extensive research [2]. Coordination and optimization are carried out considering multiple reserve resources [3], the coupling connection among various resources is strengthened [4], and complementary advantages are achieved by using the operating characteristics of different reserve sources [5].

In terms of the optimization goal of reserve capacity allocation, most of the existing studies consider it from the perspective of reducing the reserve cost of the system. Jiang et al. [6] proposed an optimization model for the water-solar-storage multi-energy complementary power generation system to optimize total reserve cost of system. Liu et al. [7] proposed a reserve capacity cost-sharing method for uncertainties such as load and new energy, and optimized upward and downward reserve costs respectively. Du et al. [8] proposed a reserve scheduling model considering uncertainty of wind power output and output characteristics of AGC units, which incorporated reserve cost into objective function to obtain optimal allocation coefficient of reserve capacity. Lu et al. [9] proposed to rank the priority of invocation for each type of reserve resource, dividing different types of reserve resources into three levels, and dividing the bearing coefficients of different reserves under condition of meeting reserve requirements.

In summary, based on existing research, this paper comprehensively considers the influence of differences in reserve costs, response times, and carbon emission intensities of different reserve resources on reserve allocation results. It introduces a reserve response sequence factor to quantitatively represent the importance of different reserve resources in the reserve allocation process. The numerical value of the reserve response sequence factor is solved based on the analytic hierarchy process and is used as the importance weight of each reserve resource participating in the reserve. Finally, the effectiveness of the proposed method is verified through a numerical example analysis.

2. Optimization Model

2.1. Objective function

The reserve capacity optimization model considering multi-source coordination takes the minimization of reserve cost and carbon emissions as the optimization objectives. And convert the carbon emissions into additional costs. Its objective function is:

$$\min F = C^{\text{reserve}} + C^{\text{carbon}} \tag{1}$$

Where C^{reserve} represents the reserve cost and C^{carbon} represents the additional cost of carbon emissions. The specific expressions for each part are as follows:

1) The reserve costs of various reserve resources for source-grid-load-storage are as follows:

$$C^{\text{reserve}} = \sum_{i\in\Omega_X} C_i + \sum_{j\in\Omega_Y} C_j \tag{2}$$

$$C_i = \sum_{t=1}^{n} R_{i,t} \times \Delta T_i \times \eta_i \tag{3}$$

$$C_j = \sum_{t=1}^{n} \left(R_{j,t}^{C} + R_{j,t}^{D}\right) \times \Delta T_j \times \eta_j \tag{4}$$

Where C_i represents the reserve cost of the *ith* reserve resource, $i \in \Omega_X$; $\Omega_X = [\text{G}, \text{GAS}, \text{H}, \text{WD}, \text{PV}, \text{Grid}, \text{TL}]$, G represents thermal power, GAS represents gas turbines, H represents hydropower, WD represents wind power, PV represents photovoltaic power, Grid represents power purchase from the power grid, and TL represents demand response; C_j represents the reserve cost of the *jth* reserve resource, $j \in \Omega_Y$; $\Omega_Y = [\text{PHES}, \text{SP}, \text{Li}]$, PHES represents pumped-storage hydroelectric energy reserve; SP represents supercapacitor reserve; Li represents lithium-battery reserve; $R_{i,t}$ represents the reserve amount provided by the *ith* reserve resource at *t*; η_i represents the cost coefficient of the *ith* reserve resource; ΔT_i represents the duration of *ith* reserve resource; $R_{j,t}^{C}$ represents the charging amount of the *jth* reserve resource at *t*; $R_{j,t}^{D}$ represents the discharging amount of the *jth* reserve resource at *t*; η_j represents the cost coefficient of the *jth* reserve resource; ΔT_j represents the duration of *jth* reserve resource; and *n* represents total scheduling duration.

2) Thermal power units, gas turbines and power grid electricity purchases will generate carbon emissions during the operation process. The expressions for converting carbon emissions into additional costs are as follows:

$$C^{\text{carbon}} = \sum_{t=1}^{n} \begin{pmatrix} R_{\text{G},t} \times \varepsilon_{\text{G}} \times \Delta T_{\text{G}} + \\ R_{\text{GAS},t} \times \varepsilon_{\text{GAS}} \times \Delta T_{\text{GAS}} + \\ R_{\text{Grid},t} \times \varepsilon_{\text{Grid}} \times \Delta T_{\text{Grid}} \end{pmatrix} \times \eta_{CO_2} \tag{5}$$

Where ε_{G} represents the carbon emission coefficient of the thermal power unit; ε_{GAS} represents the carbon emission coefficient of the gas turbine; $\varepsilon_{\text{Grid}}$ represents the carbon emission coefficient of CO_2 indirectly generated by power grid electricity purchase; ΔT_{G} represents the duration of thermal power unit; ΔT_{GAS} represents the duration of gas turbine; ΔT_{Grid} represents the duration of power purchase from the grid; η_{CO_2} represents the additional cost coefficient of carbon emission.

2.2. Constraints

$$P_{\text{load},t} = P^{e}_{\text{load},t} + P^{\text{up}}_{\text{TL},t} - P^{\text{dn}}_{\text{TL},t} \tag{6}$$

$$R_{\text{G},t} + R_{\text{GAS},t} + R_{\text{H},t} + R_{\text{WD},t} + R_{\text{PV},t} + R_{\text{Grid},t} + R^{D}_{\text{PHES},t} + R^{D}_{\text{SP},t} + R^{D}_{\text{Li},t} = P_{\text{load},t} + R^{C}_{\text{PHES},t} + R^{C}_{\text{SP},t} + R^{C}_{\text{Li},t} \tag{7}$$

Where $P^{e}_{\text{load},t}$ is the load forecast value; $P_{\text{load},t}$ is the load demand after peak-load shifting and valley-filling through demand response; $P^{\text{up}}_{\text{TL},t}$ is the increased load amount of the transferable load at t; $P^{\text{dn}}_{\text{TL},t}$ is the decreased load amount of the transferable load at t.

3. Model solving method

3.1. Framework

For reserve capacity optimization model with multi-source collaboration, setting priority can improve the efficiency of the model. In this paper, the Analytic Hierarchy Process (AHP) is used to obtain the priority order of the response order of each standby resource. Based on AHP principles, the hierarchical staged model for determining the priority order of reserve responses is depicted in Figure 1.

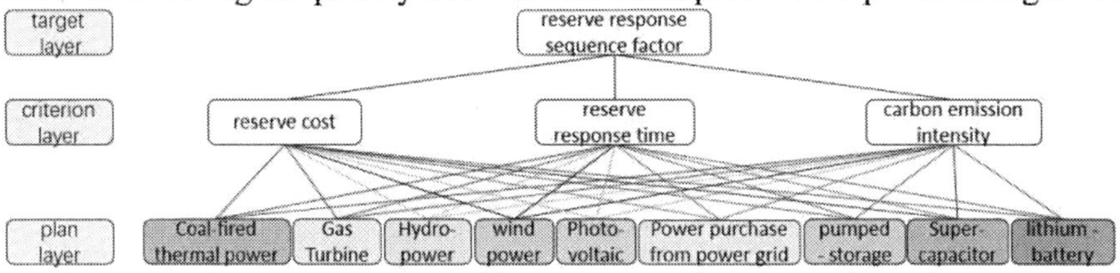

Figure1. Schematic diagram of the structure of analytic hierarchy process

According to the factors considered in the target problem, it is stratified according to different attributes: the first layer is target layer, the second layer is criterion layer, and the third layer is scheme layer. Among them, the demand response is to transfer part of the transferable load demand from peak hours to trough hours and participate in reserve by peak-load shifting and valley-filling of load demand. Therefore, the demand response does not participate in calculation of priority order of the subsequent scheme layer.

Through the analytic hierarchy process, the weight coefficients of various reserve resources can be determined, and the reserve levels in different application scenarios can be obtained.

3.2. Calculation Method of Response Priority Based on AHP

Suppose that to determine the proportion of n factors in the target Z, each factor should be compared with all factors, including itself, n times in pairs, to form a square matrix, the judgment matrix $\boldsymbol{A}$. The method used in this paper to divide the judgment matrix scale is the 1~9 scale method proposed by Saaty [10].

$$\boldsymbol{A} = \left(a_{ij}\right)_{n\times n} \tag{8}$$

Where $a_{ij} > 0, a_{ij} = 1/a_{ij}\,(i \neq j), a_{ii} = 1(i, j = 1, 2, ..., n)$.

The judgment matrix at the scheme layer is formed by comparing each scheme in pairs according to a certain criterion, that is, the reserve cost, response time, and carbon emission intensity of each scheme are compared in pairs, and the resulting judgment matrix is recorded as $\boldsymbol{A}_1$, $\boldsymbol{A}_2$, and $\boldsymbol{A}_3$. Since there are three different electricity purchase prices for peak, flat, and valley periods in power grid power purchase, the judgment matrix of reserve cost will distinguish these three periods. The judgment matrix for the peak period is denoted as $\boldsymbol{A}_{11}$, the judgment matrix for the flat period is denoted as $\boldsymbol{A}_{12}$, and the judgment matrix for the valley period is denoted as $\boldsymbol{A}_{13}$.

The indicators of the criterion layer include reserve cost, response time, and carbon emission intensity. The judgment matrix of the criterion layer is used to measure the proportion among these three factors, which has a certain degree of subjectivity. Considering that cost is a basic measurement indicator during the system operation process, this paper divides the proportion among reserve cost, response time, and carbon emission intensity into four situations. In the first situation, the three indicators are equally important, that is, 1:1:1. In the second situation, the proportion of reserve cost is higher than the other two, that is, 2:1:1. In the third situation, the proportion of reserve cost and response time is higher than that of carbon emission, that is, 2:2:1. In the fourth situation, the proportion of reserve cost and carbon emission is higher than that of response time, that is, 2:1:2.

The calculation of the ranking weights of the relative importance of the nine schemes under the three indicators of reserve cost, response time, and carbon emission intensity is called single-level ranking. In this paper, the arithmetic mean method is used to solve the weights, and the calculation formula is shown below [11].

$$\omega_i = \frac{1}{n}\sum_{j=1}^{n}\frac{a_{ij}}{\sum_{i=1}^{n} a_{ij}}, i = 1,2,\ldots,n \tag{9}$$

Where ω_i denotes the weight of the *ith* scheme under the *jth* criterion calculated by the arithmetic mean method.

The calculation of the comprehensive weights of each scheme layer under the target layer is called hierarchical overall ranking [12]. The calculated comprehensive weight is the reserve response order factor matrix of each type of reserve resource. Since the criterion layer sets four different weight proportions, and the judgment matrix of standby cost is divided into three different time periods, 12 reserve response order factor matrices will be obtained.

$$\omega_z = \left[\omega_1^{(2)}, \omega_2^{(2)}, \ldots, \omega_9^{(2)}\right], \quad z = 1,2,\ldots,12 \tag{10}$$

$$\omega_i^{(2)} = \sum_{j=1}^{3}\sum_{i=1}^{9}\omega_{ij}^{(1)}\omega_j^{(1)} \tag{11}$$

Where ω_z represents the *zth* reserve response priority order; $\omega_i^{(2)}$ represents the weight of the *ith* scheme to the target layer; $\omega_j^{(1)}$ represents the weight of the *jth* criterion of the criterion layer relative to the target layer; $\omega_{ij}^{(1)}$ represents the scheme weight of the *ith* scheme relative to the *jth* criterion.

The reserve response sequence factors of each reserve resource are sorted from small to large, and the priority order of various types of resources participating in reserve is obtained.

4. Solution method

The multi-source fusion flexible participation reserve optimization model built in this paper considering response priority is a mixed integer linear programming model, so it can be directly calculated based on the yalmip platform in Matlab environment and by calling the cplex solver. The specific solution process is shown in Figure 2.

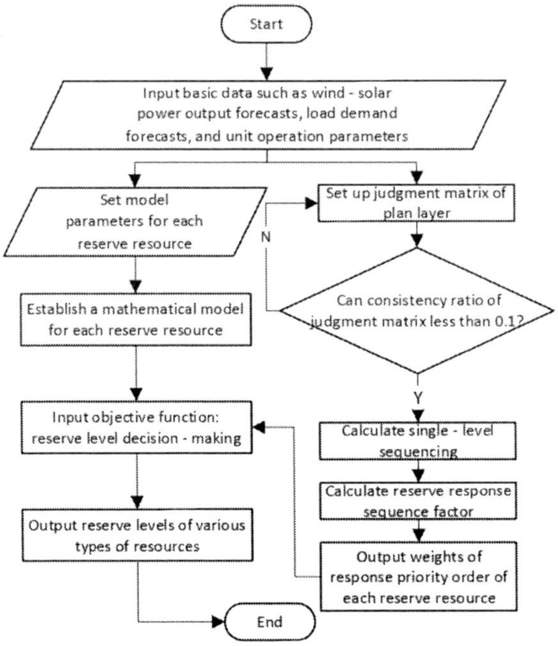

Figure 2. Algorithm flowchart

5. Results and discussion

In order to verify the improvement effect of the multi-source fusion participation reserve optimization model considering response priority proposed in this paper, an improved IEEE-30 node system is used for case analysis. The load demand forecast curve and the reserved reserves for wind power and photovoltaic power in each period are shown in Figure 3.

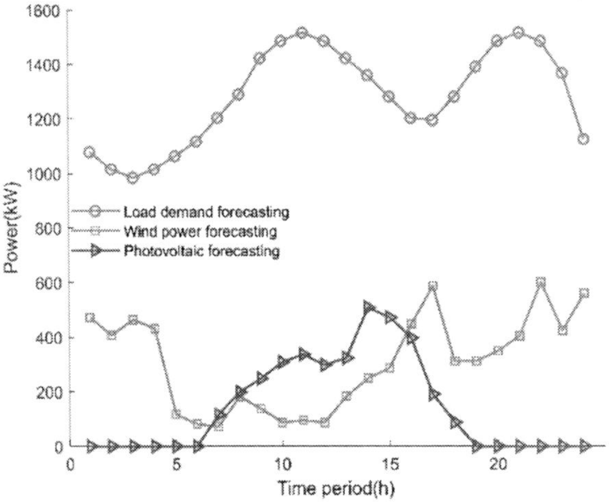

Figure 3. Wind power, photovoltaic and load forecasting curves

5.1. Calculation of reserve response factors under different criterion layer weights

Based on the data presented in Table 1, the nine alternative schemes under the three criteria of reserve cost, response time and carbon emission intensity are compared in pairs to obtain the judgment matrix A_{11} 、A_{12} 、A_{13} 、A_2 、A_3 . Taking the criterion layer weight of 1:1:1 as an example, the alternate response order factor is obtained according to the response priority calculation method introduced in Chapter 3.2, as shown in Table 2.

Table 1. Criteria layer comparison parameters for each type of reserve resource

Alternate sources	Reserve costs/ (Yuan/kWh)	Response time	Carbon emission intensity KgCO2/kWh
thermal power units	0.73	<5min	0.97
gas turbine	0.7	<5min	0.5
hydropower	0.35	<1min	-
Wind power	0.5	<50ms	-
photovoltaic	0.4	<1.5s	-
Grid power purchases	1.4/0.84/0.36	<2min	0.85
Pumped hydro storage	0.16	120s	-
Supercapacitors	0.07	10ms	-
Lithium battery energy storage	0.1	1s	-

Table 2. Calculation results of alternate response sequence

Criterion layer weight	1: 1: 1		
	Peak	Flat	Valley
Thermal	0.2789	0.2883	0.3079
gas turbine	0.1733	0.1828	0.2023
hydropower	0.0529	0.0550	0.0593
Wind power	0.0553	0.0608	0.0740
photovoltaic	0.0529	0.0562	0.0645
Grid power purchases	0.2564	0.2233	0.1539
Pumped hydro storage	0.0745	0.0758	0.0780
Supercapacitors	0.0251	0.0260	0.0272
lithium battery	0.0309	0.0318	0.0330

5.2. 5Analysis of the Optimization Effect of Multi-Source Reserve Considering Response Priority

Comparing the data in Table 3, it is not difficult to find that when the weight ratio of the criteria layer is 2:1:1, the reserve cost is the lowest, which is only 21,922.17 yuan. This is because when the ratio is 2:1:1, the proportion of reserve cost is the highest, so the reserve cost at this time is the lowest among the four criteria layers weights. When the weight of the criteria layer is 1:1:1 and 2:2:1, the carbon emissions are significantly lower compared to the other two. The comparison of response times is shown in Figure 4. It can be seen that the response time required for the weights of 1:1:1 and 2:2:1 is less than that of the other two. Through comprehensive comparison of reserve cost, response time, and carbon emissions, when the weight of the criteria layer is 1:1:1, the optimization effect is the best among the four weights. The carbon emissions and response time are greatly improved at the expense of a little reserve cost.

Table 3. Comparison of the calculated results under different criterion layer weights in Model 3

Model	1:1:1	2:1:1	2:2:1	2:1:2
Reserve cost/Yuan	23279.19	21922.17	23300.47	22225.48
Carbon emissions/kg	9333.423	9633.396	9354.4	9518.809

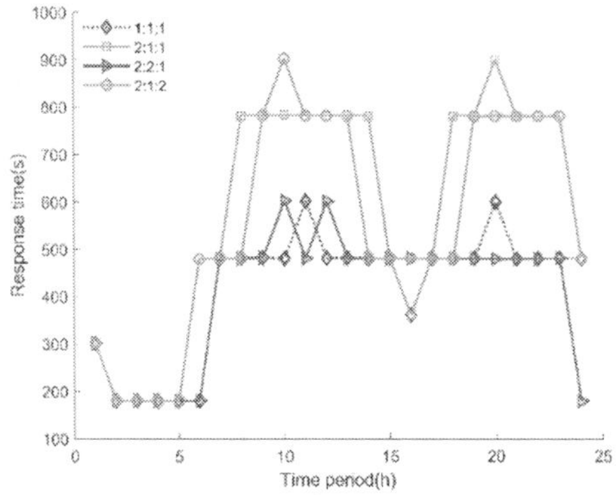

Figure 4. Comparison of response time under different criterion layer weights in model 3

The distribution results of the reserve capacity with a ratio of 1:1:1 in the criterion layer are analyzed. As can be seen from the negative semi-axis of Figures 5 and 6, the change in load demand between each period is small after the peak-to-valley difference reduced by demand response peak shaving and valley filling and the secondary peak shaving and valley filling of energy storage energy time-shift characteristics, which alleviates the backup pressure from the root.

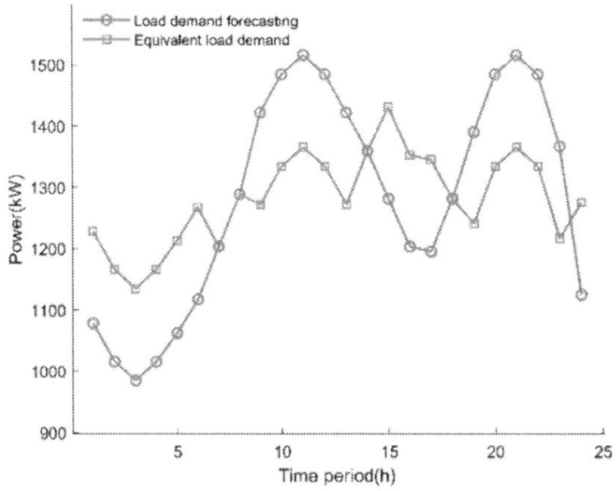

Figure 5. Effect of peak shaving and valley filling in demand response

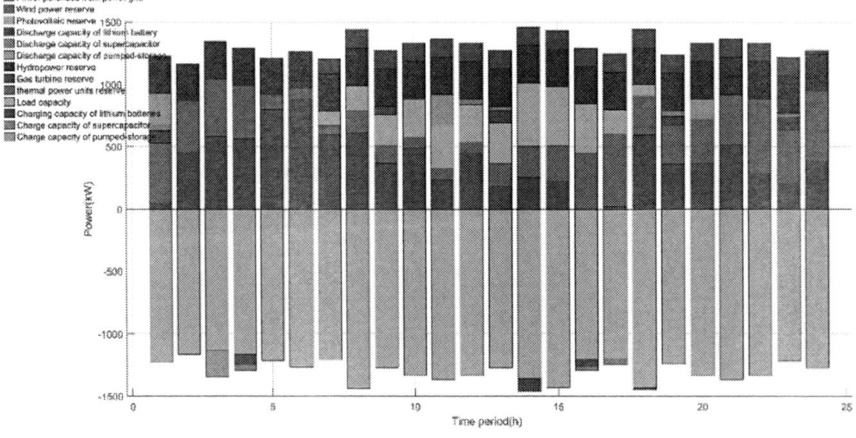

Figure 6 Spare capacity allocation

5.3. Comparative Analysis of Different Scheduling Models

In order to verify the optimization effect of the model proposed in this paper, the following will compare the three reserve capacity allocation methods.

Model 1: The standby sources are conventional units and power grid - purchased electricity.

Model 2: The standby sources are multiple types of standby resources for power generation, grid, load and storage.

Model 3: The standby sources are multiple types of standby resources for power generation, grid, load and storage, and the response priorities of each standby resource are divided.

Taking the situation where the weights of the criterion layer are 1:1:1 as the representative of Model 3, the simulation results are compared with those of Model 1 and Model 2. The comparison results are shown below.

A comprehensive comparative analysis of Table 6 and Figure 7 shows that model 1 is the highest among the three models in terms of reserve cost, response time and carbon emissions, because the reserve source of model 1 is relatively single. Multi-source synergy and complementarity are used to divide the response priority of each reserve resource, so as to obtain model 3. Comparing model 2 and model 3, it is not difficult to find that the reserve cost required by model 3 is 1731.64 yuan higher than that of model 2, but its carbon emission and response time are lower than those of model 2. In summary, the use of multi-source collaboration and response prioritization can optimize the reserve effect.

Table 6 Comparison of spare capacity allocation effects of different models

Model	1	2	3
Reserve cost/Yuan	24457.64	21568.83	23279.19
Carbon emissions/kg	11967.34	9835.6	9333.423

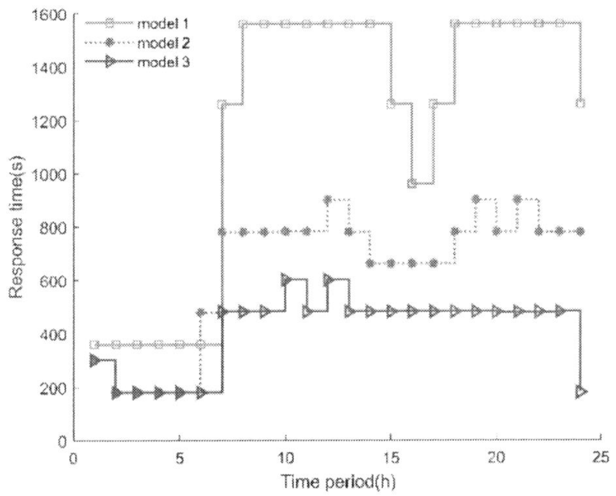

Figure 7 Comparison of response time of different models

6. Conclusions

In order to alleviate the problem of reserve shortage in the new power system and simultaneously reduce reserve costs and carbon emissions, this paper proposes an optimization model for multi-source integration flexibly participating in reserve considering response priority. Prioritize responses based on the reserve cost, response time, and carbon emissions of each standby resource to make the allocation process of standby capacity more efficient. According to the simulation results, in terms of reserve cost, the reserve cost of the model proposed in this paper is slightly higher than that of the model only considering multi-source cooperation, but the optimization effects of the model in this paper in terms of response time and carbon emissions are better than those of the traditional model and the model only considering multi - source cooperation.

The core work of this paper is to explore the multi-objective optimization strategy of source-grid-load-storage coordination and complementarity in the reserve market, which can be applied to the operation of the power system, but further research needs to be done on the reserve reservation ratio of each entity and other objectives that need to be optimized.

Acknowledgments

This work was supported by Headquarters Science and Technology Project of State Grid Co., Ltd, China. (5100-202356016A-1-1-ZN).

References

[1] QIU W and WANG M 2021 Comprehensive evaluation of shared energy storage towards new energy accommodation scenario under targets of carbon emission peak and carbon neutrality (Electric Power Automation Equipment vol 41) chapter 10 pp 244–255

[2] LI H and LIU D 2021 Analysis and reflection on the development of power system towards the goal of carbon emission peak and carbon neutrality (Proceedings of the CSEE vol 41) chapter 18 pp 6245–6255

[3] CHEN Z and WANG L 2020 Robust optimization of multiple reserve resource considering reserve capacity of wind power and demand response (Automation of Electric Power Systems vol 44) chapter 10 pp 50-58

[4] Wan Q and Yuan R 2024 Research on power grid prevention-emergency control strategy with wind and solar storage participation (Acta Energy Solaris Sinica vol 45) chapter 4 pp 404-415

[5] Wang Y and Wen X 2023 Real-time dispatch of hydro-photovoltaic (PV) hybrid system based on dynamic load reserve capacity (Energy Vol 285) chapter 129420 pp 0360-5442

[6] JIANG W and LIU J 2020 Reserve optimization for offline multi-energy complementary generation system in short time scale (Power System Technology vol 44) chapter 7 pp 2492-2502

[7] LIU S and YANG Y 2023 Reserve capacity determination and its cost allocation considering stochastic characteristics of renewable energy (Automation of Electric Power Systems vol 47) chapter 4 pp 10-18

[8] DU G and ZHAO D 2023 Stochastic reserve scheduling of high proportion wind power system (Power System Technology vol 47) chapter 4 pp 1378-1387

[9] LU Y and KUANG L 2024 Operating reserve capacity allocation strategy and optimization model with coordinated participation of source-network-load-storage (Electric Power Automation Equipment vol 44) chapter 7 pp 156-164

[10] HE J and CHENG H 2012 Comprehensive power quality assessment on distribution network planning containing Micro-Grid (Power System Technology vol 36) chapter 8 pp 209-214

[11] ZHAO X and ZHAO C 2005 Fuzzy synthetic evaluation of power quality based on changeable weight (Power System Technology vol 29) chapter 6 pp 11-16

[12] HE J and BAI D 2013 An optimal algorithm of comprehensive cost for under-frequency load-shedding (Power System Technology vol 37) chapter 12 pp 3461-3466

Reinforcement Learning Optimization Scheduling of Micro-integrated Energy Systems Under Uncertain Factors

Fulin Wang[1],Bozhao Wang[2],Wentao Tang[1,*]
[1]Jingchu University of Technology, Jingmen, China.
[2]China Energy Research Society, Beijing, China.

E-mail: twt@whu.edu.cn

Abstract. This study focuses on the optimal scheduling of integrated energy systems under uncertain factors. In view of the impact of uncertainty in wind power and photovoltaic output on system scheduling, an optimization strategy based on the improved Soft Actor-Critic algorithm is proposed. This challenge is met by simulating uncertainty data and improving the Soft Actor-Critic objective function. A mathematical model of a micro-integrated energy system is constructed in this paper, covering equipment and uncertainty models, and an optimal scheduling model is constructed based on the Soft Actor-Critic algorithm, including design states, actions, reward space, and improved objective functions. The experimental verification part compares the performance of the Soft Actor-Critic algorithm before and after improvement. The results show that the improved method has stronger adaptability in dealing with system uncertainty.

1. Introduction

In recent years, the integrated energy system has become an important way to improve energy efficiency due to its potential to achieve synergistic interconnection and complementary advantages of multiple heterogeneous energy sources.

There have been many studies on the scheduling of integrated energy systems, such as IMSA [1], which was proposed to solve the complex dynamic scheduling problems of spatiotemporally coupled thermal power and hydrothermal systems. At present, the research on reinforcement learning in integrated energy systems is also becoming increasingly active, providing new concepts and methods for solving complex energy scheduling problems. Reference [2] attempts to apply Q learning to the operation and management of community battery energy storage systems under the framework of microgrids. Reference [3] attempts to apply Q learning to the energy management of hybrid buses, thereby affecting the stability and accuracy of scheduling results.

In order to cope with the challenges in the scheduling of integrated energy systems, the Deep Deterministic Policy Gradient (DDPG) algorithm was introduced for research. Research[4] showed that DDPG has poor policy stability when facing fluctuations in renewable energy generation, is prone to policy oscillation, and is difficult to obtain a stable optimal solution.

In addition, other deep reinforcement learning algorithms such as proximal policy optimization (PPO) and asynchronous dominant behavior criticism algorithm (A3C) have also been applied to the scheduling of integrated energy systems [5-6]. In reference [7], researchers used the PPO algorithm to optimize the scheduling of microgrid systems containing various types of energy equipment, improving the stability of the scheduling strategy. In reference [8], researchers studied the use of the A3C algorithm to adaptively handle the uncertainty of sources and loads in integrated energy systems.

Wind power and photovoltaic power are important renewable energy sources that are greatly affected by weather and show obvious uncertainty fluctuations. In the scheduling of integrated energy systems, this uncertainty makes it difficult to maintain the balance between energy supply and demand. Reference [9] analyzed the power system in a certain region and found that the wind power output fluctuated greatly, which put the power grid at risk of insufficient power supply during peak hours and also led to a large amount of waste of renewable energy. Reference [10] considered the uncertainty of intermittent energy and load in the integrated energy system, constructed a PPO model in a data-driven way, and realized the dynamic scheduling decision of the integrated energy system. However, A3C and PPO still have shortcomings in terms of high computational complexity and handling continuous actions.

In view of this, this paper takes the optimization scheduling problem of the integrated energy system considering the uncertainty of wind power and photovoltaic power generation as the research object, and uses the improved Soft Actor-Critic(SAC) algorithm to solve it. The main contributions of this study are two-fold: first, using probability models such as Gaussian and Beta distribution to simulate uncertain data to provide more realistic samples for algorithm training; second, improving the objective function of the SAC algorithm, and fully incorporating uncertainty factors into the reward design, rewarding the robustness of the strategy to uncertainty, thereby improving the overall performance and stability of the system.

2. Description of Micro Integrated Energy System

An integrated energy system[11] is an energy system that couples, converts and coordinates multiple energy forms. It realizes the cascade utilization and mutual complementation of energy, improves energy utilization efficiency, and meets the diverse energy needs of users.

This paper mainly discusses the micro-integrated energy system (MIES), which consists of three parts: production capacity equipment, energy storage equipment and load, as shown in Figure 1.

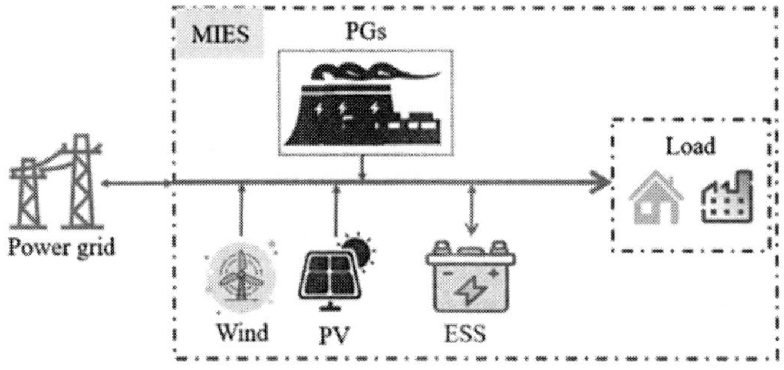

Figure 1. MIES schematic diagram.

2.1. Mathematical model of equipment components

MIES mainly consists of three parts: production capacity, energy storage and load.

Journal of Physics: Conference Series **3012** (2025) 012089 doi:10.1088/1742-6596/3012/1/012089

2.1.1. Generator Sets It mainly generates electricity by consuming fuel, and its output power is related to the fuel input and power generation efficiency, that is:

$$P_{PG}(t) = \eta_{PG} m(t) H_f \tag{1}$$

In the formula,$H_f = 40\,\mathrm{MJ/kg}$ is the calorific value of fuel,$\eta_{PG} = 0.45$ is the power generation efficiency,and $m(t)$ is the fuel input.

2.1.2. Energy storage device The energy storage device in this article focuses on the ability to store electricity. The change in its state of charge is related to the charging and discharging power. The calculation formula is:

$$\mathrm{SOC}(t) = \mathrm{SOC}(t-1) - \frac{\zeta_{\mathrm{ess}} \cdot P_{\mathrm{ESS}}(t) \cdot \Delta t}{E_{\mathrm{ESS}}} \tag{2}$$

Where E_{ESS} is the capacity of the energy storage device, Δt is the scheduling time interval, and $P_{ESS}(t)$ is the charge and discharge power. There is efficiency loss in the charge and discharge process. When charging, the actual charged energy is $\varsigma_{ch} P_{ESS}(t)\Delta t$; when discharging, the actual released energy is $\frac{P_{ESS}(t)\Delta t}{\varsigma_{dis}}$.$\varsigma_{ch} = 0.9$ and $\varsigma_{dis} = 0.93$ are the charge and discharge efficiencies, respectively.

2.2. Causes and hazards of uncertainty factors

The sources of uncertainty in MIES are extensive and complex. This study considers the uncertainty of wind power and photovoltaic output caused by the impact of weather variability and seasonal climate change.

2.3. Mathematical description of uncertainty

In order to effectively simulate the uncertainty of wind power and photovoltaic power output, a data-driven approach can be used to collect historical wind power and photovoltaic power output data, and a probability model can be used to establish a relationship model between wind power and photovoltaic power output and meteorological factors.

2.3.1. Wind power uncertainty model For wind power, the Gaussian model is used to describe the uncertainty of wind power:

$$p(P_w|v) \sim \varepsilon(u_w(v), \sigma_w^2(v)) \tag{3}$$

The expected active power P_w can be calculated based on the probability distribution of wind speed and the wind power curve. $\sigma_w^2(v)$ is the standard deviation, which reflects the degree of power fluctuation and can be determined through statistical analysis of historical data.

2.3.2. Photovoltaic power uncertainty model For photovoltaic power output, Beta distribution is used to describe the uncertainty of photovoltaic power, and its probability density function is:

$$f(P_{PV}(t)) = \frac{\Gamma(\alpha+\beta)}{\Gamma(\alpha)\Gamma(\beta)} \left(\frac{P_{PV}(t) - P_{PV,\min}}{P_{PV,\max} - P_{PV,\min}}\right)^{\alpha-1} \left(1 - \frac{P_{PV}(t) - P_{PV,\min}}{P_{PV,\max} - P_{PV,\min}}\right)^{\beta} \tag{4}$$

In this,β and α are distribution parameters, which can be obtained by fitting historical data, $P_{PV,\min}$ and $P_{PV,\max}$ are the minimum and maximum values of photovoltaic power, respectively. The Monte Carlo method is used to simulate and generate uncertain power data.

Journal of Physics: Conference Series **3012** (2025) 012089 doi:10.1088/1742-6596/3012/1/012089

2.4. System Constraints

The constraints of the IES dynamic scheduling problem include power balance constraints, energy storage device charge and discharge capacity constraints, generator power ramp constraints, and equipment operation constraints.

$$\sum_{i \in N} P_{i,PG}(t) + \sum P_{PV}(t) + \sum P_{wind}(t) + P_{GD}(t) + \sum P_{ESS}(t) = \sum P_L(t) \tag{5}$$

$$P_{i,PG}(t) - P_{i,PG}(t-1) \leq P_{PG,U} \tag{6}$$

$$P_{i,PG}(t) - P_{i,PG}(t+1) \leq P_{PG,D} \tag{7}$$

$$P_{ESS,\min} \leq P_{ESS}(t) \leq P_{ESS,\max} \tag{8}$$

$$SOC_{i,\min}(t) \leq SOC_i(t) \leq SOC_{i,\max}(t) \tag{9}$$

$$P_{GD,\min} \leq P_{i,GD}(t) \leq P_{GD,\max} \tag{10}$$

Expression (5) defines the power balance constraint. Expressions (6) and (7) represent the ramp-up and ramp-down constraints of the generator sets. The operating cost of the ESS is not considered, and it can schedule its discharge and charging power in advance, where (8) and (9) define the corresponding power limits. Expression (10) describes the export/import power limits of the power grid.

2.5. MIES Scheduling Objective Function

During the operation of the system, it is hoped that the operating cost will be as low as possible while improving energy efficiency. The scheduling objective function is as follows:

$$F = \min\left(C_{\text{total}} + \frac{1}{\varepsilon_{\text{total}}}\right) \tag{11}$$

C_{total} is the total cost of system operation, including energy procurement cost, electricity purchase cost, equipment maintenance cost, fuel cost, etc., and $\varepsilon_{\text{total}}$ is the comprehensive energy efficiency of the system.

$$C_{total} = \sum(\zeta_{GD}(t)P_{GD}(t) + \rho_{fuel}m(t) + C_{cons}(t))\Delta t \tag{12}$$

$$\varepsilon_{total} = \frac{\sum P_L(t)}{\sum(P_{PV}(t) + P_{wind}(t) + P_{PG}(t) + P_{GD}(t))} \tag{13}$$

where $\zeta_{GD}(t)$ is the grid electricity price, ρ_{fuel} is the fuel price, and $C_{cons}(t)$ is a constant for the equipment maintenance cost.

3. Micro-integrated energy system optimization scheduling model based on reinforcement learning

3.1. SAC algorithm principle

SAC is a reinforcement learning algorithm [12] with a policy network and a dual soft Q network, and uses entropy regularization to enhance policy exploration and balance immediate and long-term rewards.

In MIES, SAC uses system states (such as load demand, renewable energy generation, and equipment status) to generate scheduling actions. The environment provides feedback through reward signals based on economic costs, efficiency, and emissions. Through this process, SAC learns the optimal dynamic scheduling policy to ensure efficient and cost-effective MIES operation.

3.2. State, action and reward design

MIES scheduling problem can be modeled as a finite MDP.In this study, the current active power of photovoltaic, wind power, and generator, the load demand power, and the uncertainty index (standard deviation) of photovoltaic and wind power. We define the state space at time t as $s_t = \{P_{\mathrm{PV}}(t), P_{\mathrm{wind}}(t), P_L(t), SOC(t), \sigma_w^2, \sigma_{\mathrm{pv}}^2\}$.The intelligent agent can judge the energy supply and demand situation and uncertainty level of the system based on these state information, and then make reasonable decisions. At the same time, the action at time t is defined as $a_t = \{P_{\mathrm{PG}}(t), P_{\mathrm{ESS}}(t)\}$,that is, by controlling the power generation power of the generator and the charge and discharge power of the energy storage device to adjust the energy supply and demand balance of the system, and achieve the optimization scheduling goal. The reward function consists of two parts: the operating cost reward and the reward introduced by uncertainty. The basic reward is related to the operating cost of the system.

$$R_t = -(\kappa_1 F - \kappa_2 \sigma_w^2 - \kappa_3 \sigma_v^2) \tag{14}$$

κ_1,κ_2, and κ_3 are 0.7, 0.15, and 0.15 respectively. The reward introduced by uncertainty is used to motivate the agent to reduce the impact of uncertainty on the system, where σ_w and σ_v are the fluctuations of photovoltaic and wind power respectively.

3.3. SAC Improvement Method

In the basic SAC algorithm, the strategy objective function already contains an entropy regularization term to encourage exploration. However, in order to further improve and adapt to the uncertainty of energy system scheduling, the objective function is improved as follows: the entropy regularization weight is dynamically adjusted to make it related to the degree of uncertainty of the system. The entropy weight dynamic adjustment strategy is as follows:

$$\alpha_t = \alpha_{max} \cdot e^{-\beta t} + \alpha_{min} \tag{15}$$

$\alpha_{min} = 0.01$ and $\alpha_{max} = 0.2$ are the maximum and minimum entropy weights, respectively,$\beta = 0.001$ is the decay rate, and t is the current number of training steps. The purpose is to improve the utilization efficiency of the strategy and avoid over-exploration. At the same time, in high uncertainty areas, the entropy constraint is further relaxed to encourage the strategy to sample more possibilities. For high-volatility scheduling scenarios, the entropy target is dynamically adjusted:

$$H_{\mathrm{target}} = H_{\mathrm{base}} \cdot (1 + \mu \cdot \sigma_t) \tag{16}$$

where σ_t is the uncertainty of system fluctuations,calculated from $\sigma_t \to \sqrt{\sigma_w^2 + \sigma_v^2}$, $\mu = 0.1$. To sum up, the overall interaction process of the system is shown in Figure 2.

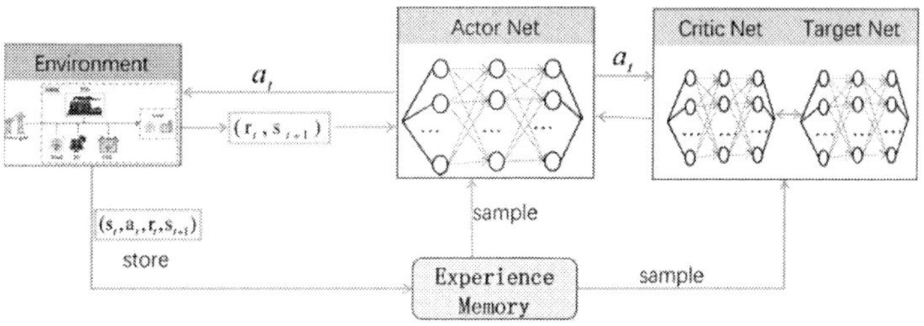

Figure 2. SAC scheduling diagram.

Journal of Physics: Conference Series **3012** (2025) 012089 doi:10.1088/1742-6596/3012/1/012089

4. Experiment and simulation

4.1. Example setting

Consider a micro-integrated energy system, including two photovoltaic array devices with the same parameters, two wind power generation devices with the same parameters, three generators with different capacities, an energy storage device and several power loads, as shown in Figure 1. Part of the power load and photovoltaic and wind turbine output data in MIES comes from a micro-project park, and the other part is simulation data, which can fully reflect the uncertainty in MIES. The equipment parameters in the system are shown in Table 1.

Table 1. Main parameters of equipment power.

Device	Power lower limit (KW)	Power Cap(KW)
PG1	0	140
PG2	0	300
PG3	0	500
ESS	0	300
PV	0	50
Wind	0	60
$P_{PG,D}$,$P_{PG,U}$	-100	100

The uncertainty description of photovoltaic and wind power output is shown in Section 2.3. The Monte Carlo method is used to generate wind power and photovoltaic uncertainty power data, simulating 365 days a year, 24 hours a day, and generating data for several scenarios every hour. Through analysis, the average wind speed is determined to be 5m/s, the standard deviation is 1.5m/s, and the Beta distribution parameters of light intensity are $\alpha = 2$ and $\beta = 3$.

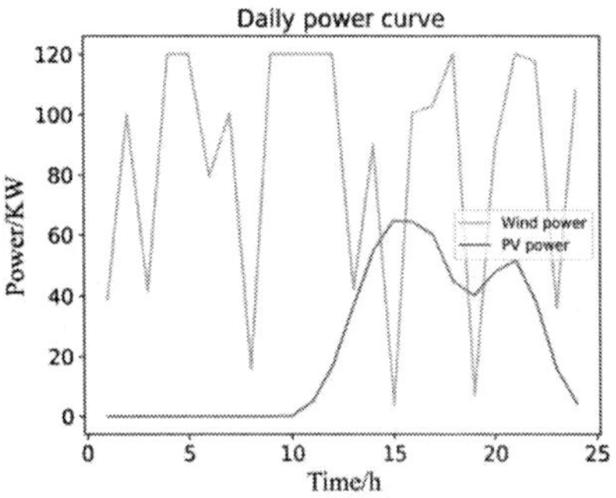

Figure 3. Daily distribution curve of photovoltaic and wind power.

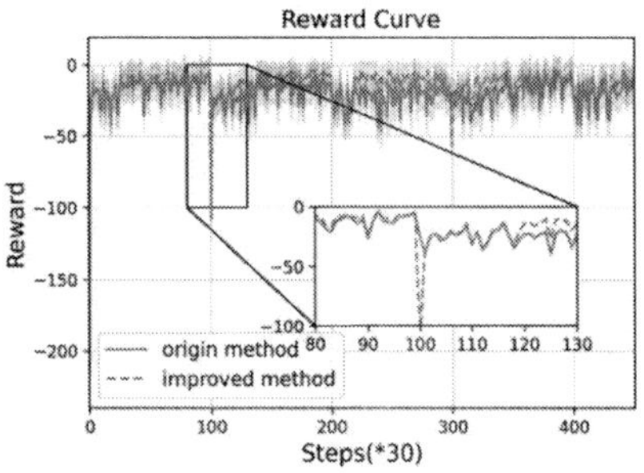

Figure 4. Reward value characteristic curve during training process.

4.2. Result Analysis

In the experiment, the training process randomly samples photovoltaic and wind power data in hours. The optimization scheduling results of the basic version and the improved version of the SAC reinforcement learning algorithm are compared. Figure 4 shows the exploration trend of the agent. Through comparison, it is found that the reward value of the proposed method fluctuates greatly, indicating that its exploration range is large and the final reward value is high.

Figure 5 shows the daily operating cost generated by the improved algorithm scheduling. The system has low load demand from 6:00 to 8:00, and outputs the remaining power to the grid to obtain part of the revenue. When there is surplus energy production, the energy storage device can be controlled more effectively to store energy and avoid energy waste. By reasonably arranging the power generation time and power of the generator, the energy efficiency of the system can be further improved.

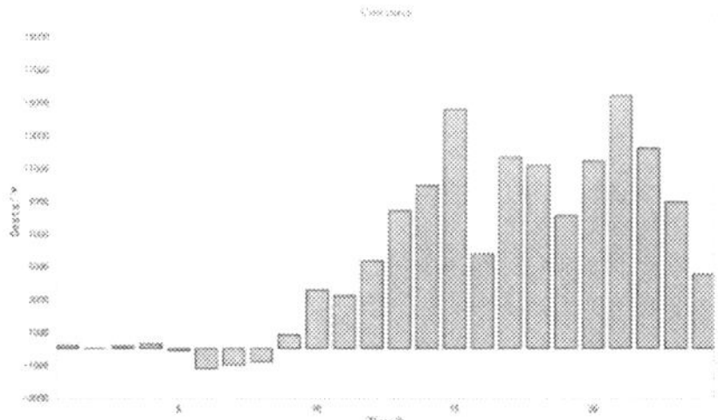

Figure 5. System operation cost curve.

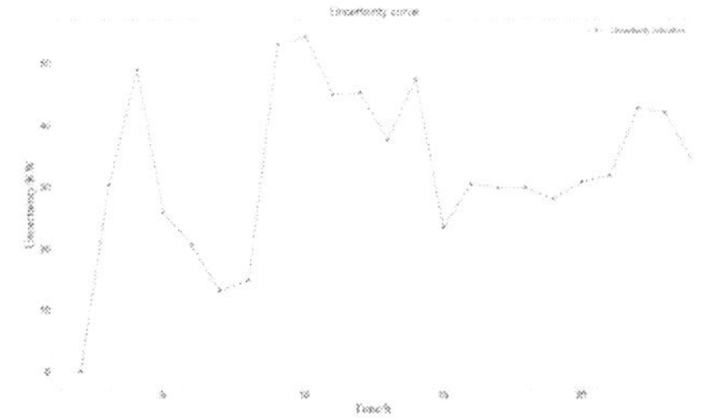

Figure 6. Input uncertainty curve curve.

Figure 6 shows the uncertainty power fluctuations of a random sample of a day during the test, and the specific explanation is shown in Section 3.3.

Figure 7 shows the dynamic dispatch result of the system power when the time scale is 1 h. As can be seen from Figure 7, during the period of 0:00-6:00, the system load demand is very low, and the energy storage device and wind turbine generator power generation can meet the system energy demand, especially during the period of 4:00-6:00, the excess power generated by the system is output to the main grid, generating energy sales. During the period of 9:00-18:00, the MIES system itself adopts the reinforcement learning algorithm scheduling strategy. The production capacity of each device is large, and the load needs to come from part of the system's own production capacity, and the rest is provided by the power grid. During the period of 19:00-23:00, the system load needs to reach the peak during this period, so more external energy is required. It should be noted that during the period of 13:00-16:00, the energy storage device stores energy to ensure that part of the energy can come from the energy storage device during the period of low load demand.

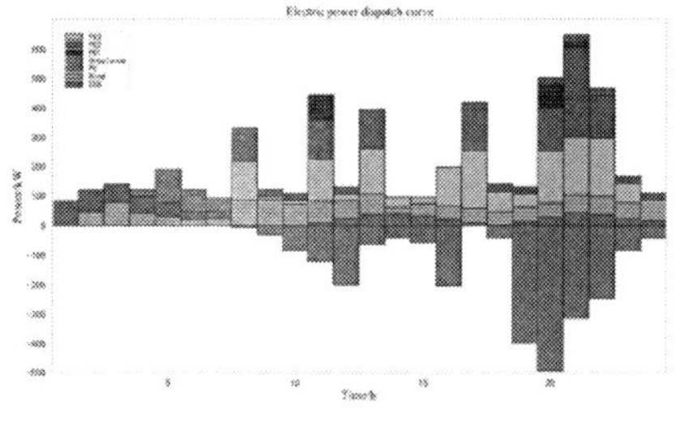

Figure 7. Input uncertainty curve curve.

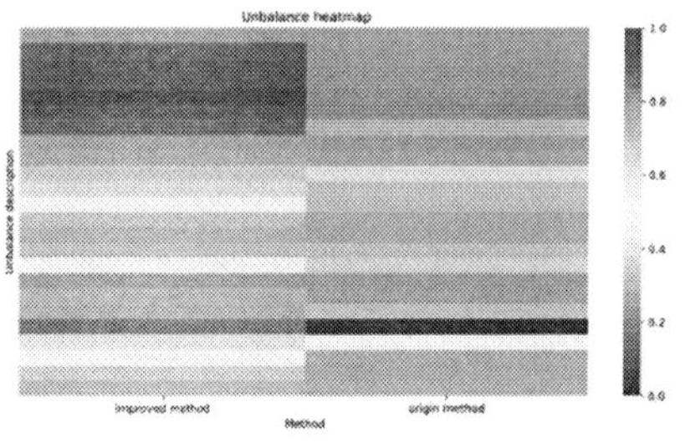

Figure 8. Power interaction heat map.

In Figure 8, the fluctuation of the power interaction between MIES and the power grid under the two methods can be intuitively analyzed from the perspective of color distribution. Red indicates that the power grid inputs power to the system. Blue is the opposite. The red area in the original method is darker, reflecting the strong volatility of the power interaction, resulting in lower efficiency. On the contrary, the blue area of the proposed method is darker and has a wider distribution range, and the color distribution is uniform, indicating that the power interaction is more stable and can handle uncertain situations more effectively.

5. Conclusion

This article constructs an improved SAC algorithm optimization scheduling model driven by curiosity mechanism. The experiment shows that this method can effectively handle the uncertainty of wind and photovoltaic output, providing effective theoretical methods and practical references for the optimization and scheduling of actual energy systems. In the future, we will consider studying the impact of demand response strategies on load uncertainty.

Acknowledgements

Funding: This work was supported by the National Natural Science Foundation of China(Grant Nos. 42174189, 41974176, 62203339); the Hubei Province Higher Education Institutions Outstanding Young and Middle-aged Science and Technology Innovation Team Project (Grant No.T2021028) and the Major Science and Technology Plan Project of Jingmen City(Grant Nos. 2022ZDYF017, 2023YFZD024, 2024YFZD076); Science and Technology Research Project of Hubei Provincial Department of Education(Grant No.B2023225).

Author Declarations

The authors declare no conflict of interest.

References

[1] Lai W, Song Q, Zheng X, et al. A new version of membrane search algorithm for hybrid renewable energy systems dynamic scheduling[J]. Renewable Energy, 2023, 209: 262-276.

[2] Bui V H, Hussain A, Kim H M. Double deep Q-learning-based distributed operation of battery energy storage system considering uncertainties[J]. IEEE Transactions on Smart Grid, 2019, 11(1): 457-469.

[3] Wu J, He H, Peng J, et al. Continuous reinforcement learning of energy management with deep Q network for a power split hybrid electric bus[J]. Applied energy, 2018, 222: 799-811.

[4] Wang, X.; Chen, S.; Yan, D.; Wei, J.; Yang, Z. Multi-agent deep reinforcement learning–based approach for optimization in microgrid clusters with renewable energy. In Proceedings of the 2021 International Conference on Power System Technology (POWERCON), Haikou, China, 8–9 December 2021; Volume 4, pp. 413–419.

[5] Shengren H, Salazar E M, Vergara P P, et al. Performance comparison of deep RL algorithms for energy systems optimal scheduling[C]//2022 IEEE PES Innovative Smart Grid Technologies Conference Europe (ISGT-Europe). IEEE, 2022: 1-6.

[6] Zhou X, Wang J, Wang X, et al. Optimal dispatch of integrated energy system based on deep reinforcement learning[J]. Energy Reports, 2023, 9: 373-378.

[7] Sun Y, Wu Y, Wu Y, et al. Optimizing intelligent startup strategy of power system using PPO algorithm[J]. Intelligent Decision Technologies, 2024 (Preprint): 1-14.

[8] Bijing L I U. Optimal Dispatch of Integrated Energy System Based on Deep Reinforcement Learning[J]. Modern Electric Power, 2024, 41(4): 710-717.

[9] Basit A, Hansen A D, Giannopoulos G. Real-time impact of power balancing on power system operation with large scale integration of wind power[J]. Journal of Modern Power Systems and Clean Energy, 2017, 5(2): 202-210.

[10] Weishan L,Xiaojun W,Qingkai S, et al. Dynamic dispatch of integrated energy system based on deep reinforcement learning under uncertain environment[J]. Power System Protection and Control, 2022, 50(18): 50-60.

[11] Wu J, Yan J, Jia H, et al. Integrated energy systems[J]. Applied Energy, 2016, 167: 155-157.

[12] Haarnoja T, Zhou A, Abbeel P, et al. Soft actor-critic: Off-policy maximum entropy deep reinforcement learning with a stochastic actor[C]. International conference on machine learning. PMLR, 2018: 1861-1870.

A Fault Diagnosis Model of Pumped Storage Units Based on Nonlinear Dynamics Methods

Lei Yang[1][0009-0009-4652-7178], Fei Chen[2*][0009-0004-5637-7082], Li Li[1][0009-0003-2211-5036], Qiang Peng[3], Jun Xie[3] and Zhigao Zhao[3][0000-0003-3911-8076]

[1] Power China Zhongnan Engineering Corporation Limited., Changsha 410000, Hunan, P.R. China

[2] Wuhan University, School of Water Resources and Hydropower Engineering, Wuhan 430072, Hubei, P.R. China

[3] Hubei Energy Group Luotian Pingtanyuan Pumped Storage Co., Ltd, Huanggang 438000, Hubei, China

*Corresponding author's e-mail: 3128179451@qq.com

Abstract. This paper proposes a fault feature extraction method of pumped storage units (PSUs) named tensor-weighted fuzzy dispersion entropy (TWFDE) based on tensor learning and nonlinear dynamics indexes. Intelligent fault diagnosis model for pumped storage units built around TWFDE and random forests (RF). Firstly, a weighted fuzzy dispersion entropy (WFDE) index is introduced, quantifying signal complexity and amplitude dynamics. Secondly, tensor learning extends WFDE to TWFDE, enabling time-frequency multi-dimensional extraction of signal anomaly information. Finally, integrated with the RF, this approach achieves diagnostic accuracies of 98.428% for hydraulic anomalies, demonstrating its superiority over other methods. This method enhances the safety and energy efficiency of pumped storage systems.

1. Introduction

The energy revolution is a precursor to Industry 5.0, with the high-quality development of renewable energy being a key trend [1]. By the end of 2023, global installed renewable capacity reached 3870 GW, expected to hit 11,000 GW by 2030 [2]. However, the randomness and volatility of renewable energy have put pressure on grid systems, making integration a major challenge [3]. Pumped storage offers large capacity, high efficiency, and rapid response, supporting peak shaving, frequency regulation, and black start. However, due to excessive working head and equipment degradation, have raised concerns about the safety and stability of pumped storage units (PSUs), making early failure detection in complex environments a key focus for intelligent operation and maintenance [4].

Most PSU research focuses on modeling individual components, neglecting the interactions between parts, leading to models that capture only general trends. With advances in sensors and computing, PSPSs now have comprehensive monitoring systems, enabling full-scale PSU monitoring. Data-driven models based on these signals are crucial for intelligent fault diagnosis. Fault diagnosis methods for PSUs include signal processing, artificial feature extraction, and deep learning. Signal processing involves extracting characteristic frequencies using techniques like FFT and Hilbert transform. Artificial feature extraction uses time and frequency metrics analyzed with statistical or machine learning methods

[5]. Deep learning methods, like the multiscale convolutional neural network (MCNN), leverage neural networks for end-to-end fault detection, such as identifying turbine cavitation [6].

Nonlinear dynamics methods, such as entropy, fractal dimension, and Lempel-Ziv complexity (LZC), have become popular in rotating machinery fault diagnosis [7]. Dispersion entropy (DE) is key in extracting fault features from machinery. As a variant of the DE, fuzzy dispersion entropy (FDE) reduces information loss and improves noise resistance, enhancing fault detection in nonlinear dynamics [8]. Multi-scale FDE methods, like RTSMFFDE, have been applied successfully to diagnose bearing faults in noisy environments [9].

FDE shows great potential for faulty information extraction in PSUs, but some limitations remain. 1) The symbolization process loses signal amplitude information, reducing sensitivity to amplitude changes [10];2) Multi-scale FDE variants focus on vector feature extraction but lack the ability to analyze signal complexity across both time and frequency domains. To address these, a weighted strategy can enhance amplitude sensitivity, and tensor learning can enable multidimensional feature extraction from both time and frequency domains.

This paper develops a fault diagnosis method for PSUs using nonlinear dynamics. Key contributions include:

(1) Introducing root mean square (RMS) into FDE to propose weighted fuzzy dispersion entropy (WFDE), capturing dynamic behavior and signal amplitude variations.

(2) Extending WFDE to tensor WFDE for extracting tensor structural information and fault features in both time and frequency domains.

(3) Creating an intelligent fault diagnosis model by combining TWFDE and random forests (RF), validated through case studies on hydraulic anomalies in a micro PSPS.

2. Methodology

2.1. Weighted Fuzzy dispersion entropy

WFDE combines the strengths of the FDE and RMS. By incorporating RMS into FDE, WFDE effectively captures the dynamic variations of signals while simultaneously reflecting abrupt changes in signal amplitude and energy. **It** can be summarized in the following steps:

Step 1. The time series signal X is transformed into a standard normal distribution signal V using the normal cumulative distribution function (NCDF).

Step 2. Each u_j^{nc} is assigned a fuzzy membership degree through a fuzzy membership function. For categories $c=1$ or $c=nc$, the trapezoidal membership function is used, while the triangular membership function is applied to other categories.

where $\mu_{\chi_c}\left(v_j^{nc}\right)$ denotes the degree of membership when u_j^{nc} is a category c.

Step 3. Through phase space reconstruction, U is transformed into $N-(m-1)\times\lambda$ vectors, expressed in the following form:

$$u_i^{m,nc}=\left\{u_i^{nc},u_{i+\lambda}^{nc},u_{i+2\lambda}^{nc},\cdots,u_{i+(m-1)\lambda}^{nc}\right\}\ i=1,2,\cdots,N-(m-1)\lambda \tag{1}$$

where $u_i^{m,nc}$ is the *i*-th vector. m and λ denote embedding dimension and delay time.

Step 4. Each vector $u_i^{m,nc}$ is assigned to the dispersion mode $\pi_{\omega_0\omega_1\cdots\omega_{m-1}}$, which ω_l mean the *l*-th data point of $u_i^{m,nc}$. Meanwhile, the membership degree of each dispersion mode is calculated according to **Eq. 2**.

$$\mu_{\pi_{\omega_1\omega_2\cdots\omega_{m-1}}}=\prod_{l=0}^{l=m}\mu_{M_{\omega_l}}\left(u_{j+(l-1)\lambda}^{nc}\right) \tag{2}$$

where $\mu_{\pi_{\omega_1\omega_2\cdots\omega_{m-1}}}$ means the membership degree of the vector $u_i^{m,nc}$.

Step 5. The probability of different dispersion patterns is obtained by **Eq. 3**:

$$p\left(\pi_{\omega_1\omega_2\cdots\omega_{m-1}}\right)=\frac{\sum_{i=1}^{N-(m-1)\times\lambda}\mu_{\pi_{\omega_1\omega_2\cdots\omega_{m-1}}}\left(u_i^{m,nc}\right)}{N-(m-1)\times\lambda} \tag{3}$$

Step 6. The initial entropy is calculated based on Shannon entropy, and RMS is introduced as a weighting factor to represent the amplitude and energy characteristics of the signal, resulting in the WFDE.

$$WFDE(X,m,nc,\lambda)=\frac{-\sum_{\pi=1}^{nc^m}p\left(\pi_{\omega_1\omega_2\cdots\omega_{m-1}}\right)\cdot\ln p\left(\pi_{\omega_1\omega_2\cdots\omega_{m-1}}\right)}{\ln\left(nc^m\right)}\times\left(\frac{1}{N}\sum_{j=1}^{N}x_j^2\right) \tag{4}$$

2.2. Tensor weighted fuzzy dispersion entropy

We introduce tensor learning theory to extend WFDE into TWFDE, enabling the extraction of signal features in both time and frequency domains. It is divided into the following key steps:

Step 1. For any signal X, it can be decomposed into the low-frequency operator $\zeta_0(x)$ and the high-frequency operator $\zeta_1(x)$.

Subsequently, the components obtained from the signal decomposition can be further separated into the high-frequency and low-frequency components, following a process of successive division: one becomes two, two becomes four, four becomes eight, and so on. Each component represents a distinct frequency component of the signal X, with the total number of components being 2^k (k is the number of hierarchical layers). Different component $X^{k,e}$ can be obtain by **Eq.5**:

$$X^{k,e}=\zeta_{f_k}^{k}\cdot\zeta_{f_{k-1}}^{k-1}\cdot\zeta_{f_{k-2}}^{k-2}\cdots\zeta_{f_1}^{1}\cdot X \tag{5}$$

where e denotes the indices of the components, and $e=\sum_{\theta=1}^{k}2^{k-\theta}f_\theta$. f_θ is the discriminative measure for the high-frequency ($f_\theta=0$) and low-frequency components ($f_\theta=1$) at the $\theta-th$ layer. The value of e ranges from 0 to 2^k-1.

Step 2. Based on the theory of coarse graining, the coarse-grained time series of $X^{k,e}$ at different time scales are computed, which can represent the temporal information of the signal. The calculation of the coarse-grained time series is as follows:

$$\begin{aligned} X_{\tau,\alpha}^{k,e}&=\frac{1}{\tau}\sum_{q=(\alpha-1)\tau+1}^{\alpha\tau}x_q \qquad 1\le\alpha\le\left\lfloor\frac{N}{\tau}\right\rfloor \\ X_{\tau}^{k,e}&=\left\{X_{\tau,\alpha}^{k,e}\right\} \qquad 1\le\tau\le\tau_{\max} \end{aligned} \tag{6}$$

where $X_{\tau,\alpha}^{k,e}$ mean the $\alpha-th$ data point of the $X_{\tau}^{k,e}$. $X_{\tau}^{k,e}$ is the coarse-grained time series of $X^{k,e}$ when scale τ, and $\tau_{\max}$ is the maximum scale factor. Iterating over all $X^{k,e}$, we get $2^k\times\tau_{\max}$ subcomponents.

Step 3. The WFDE value for different $X_{\tau}^{k,e}$ is calculated based on **Section 2.1**, and their collection used as the TWFDE of the signal.

$$\begin{aligned} TWFDE_{\tau}^{k,e}&=WFDE\left(X_{\tau}^{k,e},m,nc,\lambda\right) \\ TWFDE&=\left\{TWFDE_{\tau}^{k,e}\right\} \end{aligned} \qquad 1\le\tau\le\tau_{\max},\ 0\le e\le 2^k-1 \tag{7}$$

3. Process of the proposed method

This paper proposes an intelligent fault diagnosis method for PSUs, centered on the core feature extraction tool TWFDE. The intelligent diagnostic model contains the following steps:

Step 1 Data acquisition of pumped storage units

Sensors are strategically positioned in critical areas of the PSU. The signals they gather are routed to a data acquisition system. Calibration settings are applied within the main software to transform the electrical signals into measurable physical quantities. The resulting data is then saved in a database for further analysis.

Journal of Physics: Conference Series **3012** (2025) 012090 doi:10.1088/1742-6596/3012/1/012090

Step 2 Extracting features of signals using the TWFDE

The signal is then decomposed into components of varying frequencies using a multi-level decomposition technique. These components are subsequently converted into coarse-grained time series over different time intervals, applying coarse-graining principles. The WFDE values of these coarse-grained time series are extracted and used as the signal's distinguishing features.

Step 3 Intelligent classification using Random forests

The features extracted by TWFDE are divided into training and testing sets in a specified ratio, and the training set is used to fully train the RF model. Finally, the accuracy of the model is validated using the testing set.

4. Case study

4.1. Tensor weighted fuzzy dispersion entropy

This section highlights case studies from China's inaugural micro pumped storage power plant (PSPP), exploring its dynamic experimental features, including hydraulic anomaly detection and mechanical fault diagnosis. The micro PSPP, designed for scalability from its prototype, integrates a comprehensive hydraulic, mechanical, and electrical system. As shown in **Figure 1**, the plant is equipped with four units: Units 1# and 2# operate at a fixed speed of 1000 r/min, Unit 3# has a variable speed range (920-1080 r/min), and Unit 4# is currently being installed as another fixed-speed unit.

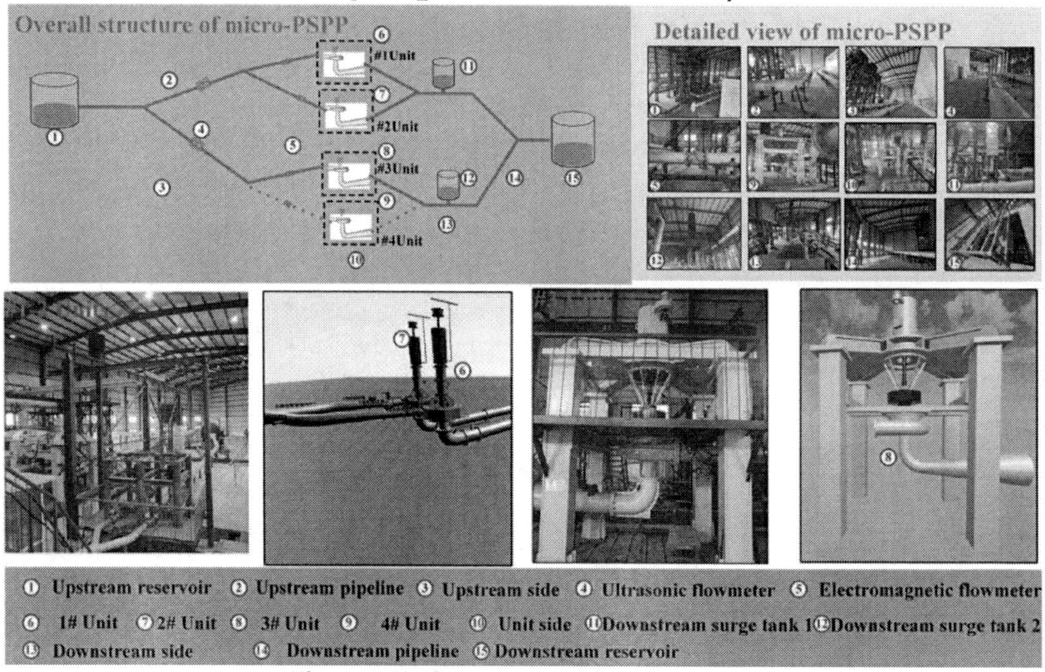

Figure 1. Basic structure of the micro PSPP

4.2. Hydraulic anomaly identification experiment

Hydraulic anomalies like vortex ropes are primary contributors to abnormal vibrations in PSUs. Vortex ropes pose a risk when present in under-loaded conditions during tasks, potentially striking the draft tube wall and threatening the safety of the PSU. This study targets Unit #2 of the micro-PSPP, where pressure pulsation is used to monitor vortex rope formation, and the TWFDE method is employed to extract fault-related data. Coupled with random forests (RF), this method enables quick detection of unusual flow behaviors. The research also details the acquisition and storage of pressure pulsation data throughout different vortex rope stages. Using a pressure transducer and high-speed camera, pressure data is collected and stored for further analysis. The flow patterns are divided into five phases: normal, vortex rope initiation, development, progression, and cylindrical vortex rope. Spiral vortex ropes, which occur in the initial stages, are the most hazardous, while cylindrical vortex ropes gradually evolve into spiral forms.

We extract pressure pulsation signals corresponding to five distinct stages, which we label as Stages 1 through 5. This results in a total of 250 data sets, with 50 sets for each stage. The data acquisition system operates at a sampling frequency of 1200 Hz, with each data set containing 4096 samples. As shown in **Figure 2**, there is a significant rise in the pressure pulsation signal amplitude from Stage 1 to Stage 2, suggesting that the formation of vortex ropes greatly amplifies the pulsations. From Stage 2 to Stage 4, the amplitude continues to increase, indicating the gradual development and intensification of vortex ropes, reaching their peak effect on the PSU in Stage 4. Finally, from Stage 4 to Stage 5, the amplitude decreases.

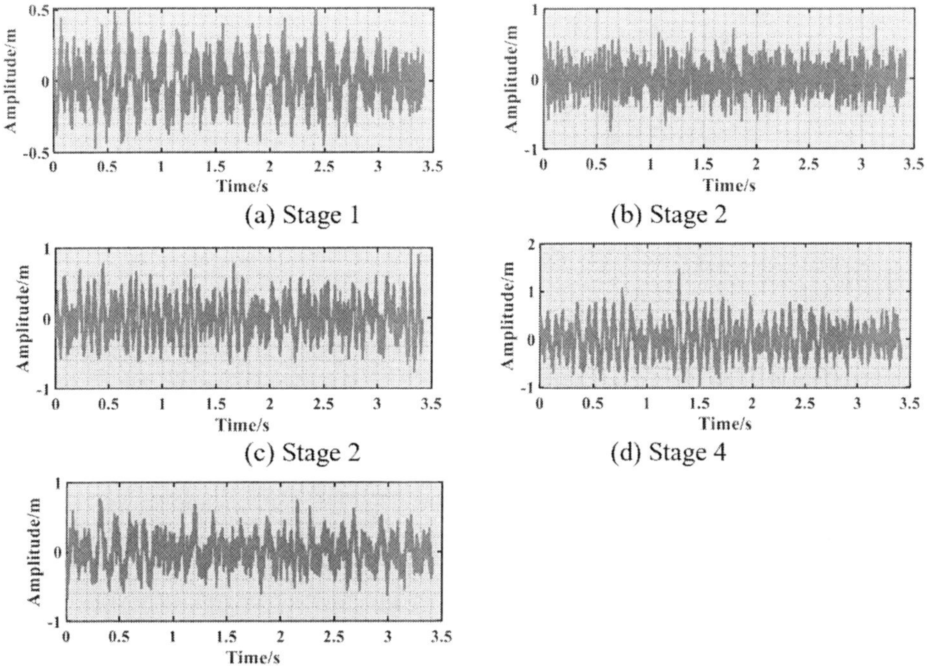

(a) Stage 1 (b) Stage 2

(c) Stage 2 (d) Stage 4

(e) Stage 5

Figure 2. Time-domain waveforms of pressure pulsation signals corresponding to different stages

We utilize TWFDE to extract features from the pressure pulsation signals, which are then used to train an RF model for anomaly detection. The results are presented in Figure 3. In Figure 3(a), the average feature values show clear distinctions across the five stages. The feature values rise from Stages 1 to 4, particularly at e=0, demonstrating that TWFDE successfully captures the increasing complexity of the flow. The differences between Stages 3 and 5 are minimal, likely due to Stage 5's instability. Figure 3(b) shows the feature distributions, with Stages 3 and 5 displaying similar patterns. In Figure 3(c), TSNE downscaling reveals an overlap between Stages 3 and 5, while Stage 1 remains easily distinguishable. These results validate the method's efficiency in distinguishing between normal and abnormal flow conditions. The model achieves an impressive 98.4% accuracy rate, with only two instances of misclassification where Stage 3 was incorrectly identified as Stage 5, and it successfully detected all abnormal stages (2-5), demonstrating its high effectiveness in recognizing abnormal flow behaviors in the draft tube.

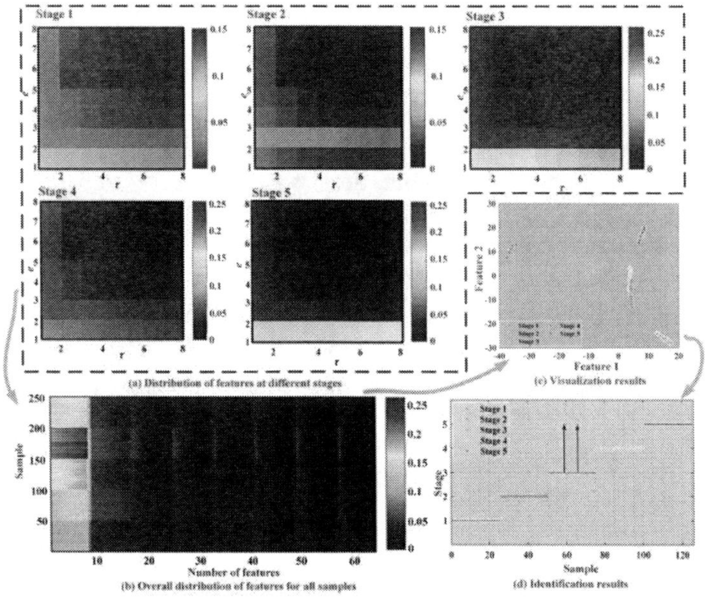

Figure 3 Identification process and results of the proposed method

5. Conclusions

This study presents TWFDE as a technique for detecting abnormal signals in PSUs and introduces an intelligent fault diagnosis model tailored to hydraulic systems. Key contributions include: (1) A novel nonlinear dynamics metric, WFDE, merges RMS and FDE to analyze both signal complexity and amplitude. It exhibits outstanding performance in detecting dynamic variations, resisting noise interference, and adapting to parameter changes; (2) WFDE is extended to TWFDE by incorporating tensor learning, enabling the extraction of features from both the temporal and spatial dimensions; (3) The intelligent fault diagnosis model, which combines TWFDE and random forests (RF), achieves a recognition accuracy of 98.4% for hydraulic anomalies. This proposed method effectively detects potential faults, contributing to the safe and efficient operation of PSUs. Future research will focus on expanding its use in fault detection, lifespan evaluation, and multi-sensor integrated fault diagnosis.

Acknowledgments

The work was supported by National Natural Science Foundation of China (Numbers 52209114), the scientific research foundation of China Postdoctoral Science Foundation (2021M702527), Powerchina Zhongnan Engineering Corporation Limited Research Project (PTTY-SS-01), and the China Postdoctoral Innovation Talents Support Program (BX20220243).

References

[1] Sharma R, Gupta H. Harmonizing sustainability in industry 5.0 era: Transformative strategies for cleaner production and sustainable competitive advantage. J Clean Prod 2024; 445: 141118.

[2] Yao L X, Guan Z W, Wang Y, Hui H X, Luo S Y, Jia C Y, You X X, Xiao X Y. Evaluating the feasibility of concentrated solar power as a replacement for coal-fired power in China: A comprehensive comparative analysis. Appl Energ 2025; 377: 124396.

[3] ALAhmad A K, Verayiah R, Ramasamy A, Marsadek M, Shareef H. Optimal planning of energy storage system for hybrid power system considering multi correlated input stochastic variables. J. Energy Storage 2024, 82: 110615.

[4] Chen P, Li C S, Zhang X Y. Degradation trend prediction of pumped storage unit based on a novel performance degradation index and GRU-attention model. Sustainable Energy Technol 2022; 54: 102807.

[5] Feng J J, Liu B X, Luo X Q, Zhu G J, Li K, Wu G K. Experimental investigation on characteristics of cavitation-induced vibration on the runner of a bulb turbine. Mech Syst Signal Pr 2023; 189: 110097.
[6] Li F, Wang C G, Liu Z L, Huang Y Y, Wang T Z. Sparrow search algorithm-optimized variational mode decomposition-based multiscale convolutional network for cavitation diagnosis of hydro turbines. Ocean Eng 2024; 312: 119055.
[7] Wang T, Khoo S Y, Ong Z C, Siow P Y, Wang T. Distance similarity entropy: A sensitive nonlinear feature extraction method for rolling bearing fault diagnosis. Reliab Eng Syst Safe 2025; 255: 110643.
[8] Rostaghi M, Khatibi M M, Ashory M R. Azami H. Fuzzy dispersion entropy: a nonlinear measure for signal analysis. IEEE Trans. Fuzzy Syst. 2021; 30(9): 3785-3796.
[9] Rostaghi M, Rostaghi R, Humeau-Heurtier A, Azami H. Refined composite multivariate multiscale fuzzy dispersion entropy: Theoretical analysis and applications. Chaos Soliton Fract 2024; 185: 115128.
[10] Wang Z, Shang P J. Generalized entropy plane based on multiscale weighted multivariate dispersion entropy for financial time series. Chaos Soliton Fract 2021; 142: 110473.

AUTHOR INDEX